ALTITUDE CORRECTION TABLES 10°-90°—

SUN

OCT.–MAR. App. Alt.	Lower Limb	Upper Limb	APR.–SEPT. App. Alt.	Lower Limb	Upper Limb
° ′	′	′	° ′	′	′
9 34	+10·8	−22·7	9 39	+10·6	−22·4
9 45	+10·9	−22·6	9 51	+10·7	−22·3
9 56	+11·0	−22·5	10 03	+10·8	−22·2
10 08	+11·1	−22·4	10 15	+10·9	−22·1
10 21	+11·2	−22·3	10 27	+11·0	−22·0
10 34	+11·3	−22·2	10 40	+11·1	−21·9
10 47	+11·4	−22·1	10 54	+11·2	−21·8
11 01	+11·5	−22·0	11 08	+11·3	−21·7
11 15	+11·6	−21·9	11 23	+11·4	−21·6
11 30	+11·7	−21·8	11 38	+11·5	−21·5
11 46	+11·8	−21·7	11 54	+11·6	−21·4
12 02	+11·9	−21·6	12 10	+11·7	−21·3
12 19	+12·0	−21·5	12 28	+11·8	−21·2
12 37	+12·1	−21·4	12 46	+11·9	−21·1
12 55	+12·2	−21·3	13 05	+12·0	−21·0
13 14	+12·3	−21·2	13 24	+12·1	−20·9
13 35	+12·4	−21·1	13 45	+12·2	−20·8
13 56	+12·5	−21·0	14 07	+12·3	−20·7
14 18	+12·6	−20·9	14 30	+12·4	−20·6
14 42	+12·7	−20·8	14 54	+12·5	−20·5
15 06	+12·8	−20·7	15 19	+12·6	−20·4
15 32	+12·9	−20·6	15 46	+12·7	−20·3
15 59	+13·0	−20·5	16 14	+12·8	−20·2
16 28	+13·1	−20·4	16 44	+12·9	−20·1
16 59	+13·2	−20·3	17 15	+13·0	−20·0
17 32	+13·3	−20·2	17 48	+13·1	−19·9
18 06	+13·4	−20·1	18 24	+13·2	−19·8
18 42	+13·5	−20·0	19 01	+13·3	−19·7
19 21	+13·6	−19·9	19 42	+13·4	−19·6
20 03	+13·7	−19·8	20 25	+13·5	−19·5
20 48	+13·8	−19·7	21 11	+13·6	−19·4
21 35	+13·9	−19·6	22 00	+13·7	−19·3
22 26	+14·0	−19·5	22 54	+13·8	−19·2
23 22	+14·1	−19·4	23 51	+13·9	−19·1
24 21	+14·2	−19·3	24 53	+14·0	−19·0
25 26	+14·3	−19·2	26 00	+14·1	−18·9
26 36	+14·4	−19·1	27 13	+14·2	−18·8
27 52	+14·5	−19·0	28 33	+14·3	−18·7
29 15	+14·6	−18·9	30 00	+14·4	−18·6
30 46	+14·7	−18·8	31 35	+14·5	−18·5
32 26	+14·8	−18·7	33 20	+14·6	−18·4
34 17	+14·9	−18·6	35 17	+14·7	−18·3
36 20	+15·0	−18·5	37 26	+14·8	−18·2
38 36	+15·1	−18·4	39 50	+14·9	−18·1
41 08	+15·2	−18·3	42 31	+15·0	−18·0
43 59	+15·3	−18·2	45 31	+15·1	−17·9
47 10	+15·4	−18·1	48 55	+15·2	−17·8
50 46	+15·5	−18·0	52 44	+15·3	−17·7
54 49	+15·6	−17·9	57 02	+15·4	−17·6
59 23	+15·7	−17·8	61 51	+15·5	−17·5
64 30	+15·8	−17·7	67 17	+15·6	−17·4
70 12	+15·9	−17·6	73 16	+15·7	−17·3
76 26	+16·0	−17·5	79 43	+15·8	−17·2
83 05	+16·1	−17·4	86 32	+15·9	−17·1
90 00			90 00		

STARS AND PLANET

App. Alt.	Corrⁿ
° ′	′
9 56	−5·3
10 08	−5·2
10 20	−5·1
10 33	−5·0
10 46	−4·9
11 00	−4·8
11 14	−4·7
11 29	−4·6
11 45	−4·5
12 01	−4·4
12 18	−4·3
12 35	−4·2
12 54	−4·1
13 13	−4·0
13 33	−3·9
13 54	−3·8
14 16	−3·7
14 40	−3·6
15 04	−3·5
15 30	−3·4
15 57	−3·3
16 26	−3·2
16 56	−3·1
17 28	−3·0
18 02	−2·9
18 38	−2·8
19 17	−2·7
19 58	−2·6
20 42	−2·5
21 28	−2·4
22 19	−2·3
23 13	−2·2
24 11	−2·1
25 14	−2·0
26 22	−1·9
27 36	−1·8
28 56	−1·7
30 24	−1·6
32 00	−1·5
33 45	−1·4
35 40	−1·3
37 48	−1·2
40 08	−1·1
42 44	−1·0
45 36	−0·9
48 47	−0·8
52 18	−0·7
56 11	−0·6
60 28	−0·5
65 08	−0·4
70 11	−0·3
75 34	−0·2
81 13	−0·1
87 03	0·0
90 00	

App. Alt.	Additi Corr
VENUS	
Jan. 1–Jan. 10	
°	′
0	+0·5
6	+0·6
20	+0·7
31	
Jan. 11–Feb. 14	
0	+0·6
4	+0·7
12	+0·8
22	
Feb. 15–Feb. 21	
0	+0·5
6	+0·6
20	+0·7
31	
Feb. 22–Mar. 9	
0	+0·4
11	+0·5
41	
Mar. 10–Apr. 4	
0	+0·3
46	
Apr. 5–May 19	
0	+0·2
47	
May 20–Dec. 31	
0	+0·1
42	
MARS	
Jan. 1–Sept. 3	
0	+0·1
60	
Sept. 4–Dec. 31	
0	+0·3
34	+0·2
60	+0·1
80	

[illegible]	−2·8	[illegible]
2·6	−2·9	8·6
2·8	−3·0	9·2
3·0	−3·1	9·8
3·2	−3·2	10·5
3·4	−3·3	11·2
3·6	−3·4	11·9
3·8	−3·5	12·6
4·0	−3·6	13·3
4·3	−3·7	14·1
4·5	−3·8	14·9
4·7	−3·9	15·7
5·0	−4·0	16·5
5·2	−4·1	17·4
5·5	−4·2	18·3
5·8	−4·3	19·1
6·1	−4·4	20·1
6·3	−4·5	21·0
6·6	−4·6	22·0
6·9	−4·7	22·9
7·2	−4·8	23·9
7·5	−4·9	24·9
7·9	−5·0	26·0
8·2	−5·1	27·1
8·5	−5·2	28·1
8·8	−5·3	29·2
9·2	−5·4	30·4
9·5	−5·5	31·5
9·9	−5·6	32·7
10·3	−5·7	33·9
10·6	−5·8	35·1
11·0	−5·9	36·3
11·4	−6·0	37·6
11·8	−6·1	38·9
12·2	−6·2	40·1
12·6	−6·3	41·5
13·0	−6·4	42·8
13·4	−6·5	44·2
13·8	−6·6	45·5
14·2	−6·7	46·9
14·7	−6·8	48·4
15·1	−6·9	49·8
15·5	−7·0	51·3
16·0	−7·1	52·8
16·5	−7·2	54·3
16·9	−7·3	55·8
17·4	−7·4	57·4
17·9	−7·5	58·9
18·4	−7·6	60·5
18·8	−7·7	62·1
19·3	−7·8	63·8
19·8	−7·9	65·4
20·4	−8·0	67·1
20·9	−8·1	68·8
21·4		70·5

1·5	− 2·2
2·0	− 2·5
2·5	− 2·8
3·0	− 3·0
See table ←	
m	′
20	− 7·9
22	− 8·3
24	− 8·6
26	− 9·0
28	− 9·3
30	− 9·6
32	− 10·0
34	− 10·3
36	− 10·6
38	− 10·8
40	− 11·1
42	− 11·4
44	− 11·7
46	− 11·9
48	− 12·2
ft.	′
2	− 1·4
4	− 1·9
6	− 2·4
8	− 2·7
10	− 3·1
See table ←	
ft.	′
70	− 8·1
75	− 8·4
80	− 8·7
85	− 8·9
90	− 9·2
95	− 9·5
100	− 9·7
105	− 9·9
110	− 10·2
115	− 10·4
120	− 10·6
125	− 10·8
130	− 11·1
135	− 11·3
140	− 11·5
145	− 11·7
150	− 11·9
155	− 12·1

ALTITUDE CORRECTION TABLES 0°–35°—MOON

App. Alt.	0°–4° Corrⁿ	5°–9° Corrⁿ	10°–14° Corrⁿ	15°–19° Corrⁿ	20°–24° Corrⁿ	25°–29° Corrⁿ	30°–34° Corrⁿ	App. Alt.
′	° ′	° ′	° ′	° ′	° ′	° ′	° ′	′
00	0 33·8	5 58·2	10 62·1	15 62·8	20 62·2	25 60·8	30 58·9	00
10	35·9	58·5	62·2	62·8	62·1	60·8	58·8	10
20	37·8	58·7	62·2	62·8	62·1	60·7	58·8	20
30	39·6	58·9	62·3	62·8	62·1	60·7	58·7	30
40	41·2	59·1	62·3	62·8	62·0	60·6	58·6	40
50	42·6	59·3	62·4	62·7	62·0	60·6	58·5	50
00	1 44·0	6 59·5	11 62·4	16 62·7	21 62·0	26 60·5	31 58·5	00
10	45·2	59·7	62·4	62·7	61·9	60·4	58·4	10
20	46·3	59·9	62·5	62·7	61·9	60·4	58·3	20
30	47·3	60·0	62·5	62·7	61·9	60·3	58·2	30
40	48·3	60·2	62·5	62·7	61·8	60·3	58·2	40
50	49·2	60·3	62·6	62·7	61·8	60·2	58·1	50
00	2 50·0	7 60·5	12 62·6	17 62·7	22 61·7	27 60·1	32 58·0	00
10	50·8	60·6	62·6	62·6	61·7	60·1	57·9	10
20	51·4	60·7	62·6	62·6	61·6	60·0	57·8	20
30	52·1	60·9	62·7	62·6	61·6	59·9	57·8	30
40	52·7	61·0	62·7	62·6	61·5	59·9	57·7	40
50	53·3	61·1	62·7	62·6	61·5	59·8	57·6	50
00	3 53·8	8 61·2	13 62·7	18 62·5	23 61·5	28 59·7	33 57·5	00
10	54·3	61·3	62·7	62·5	61·4	59·7	57·4	10
20	54·8	61·4	62·7	62·5	61·4	59·6	57·4	20
30	55·2	61·5	62·8	62·5	61·3	59·6	57·3	30
40	55·6	61·6	62·8	62·4	61·3	59·5	57·2	40
50	56·0	61·6	62·8	62·4	61·2	59·4	57·1	50
00	4 56·4	9 61·7	14 62·8	19 62·4	24 61·2	29 59·3	34 57·0	00
10	56·7	61·8	62·8	62·3	61·1	59·3	56·9	10
20	57·1	61·9	62·8	62·3	61·1	59·2	56·9	20
30	57·4	61·9	62·8	62·3	61·0	59·1	56·8	30
40	57·7	62·0	62·8	62·2	60·9	59·1	56·7	40
50	57·9	62·1	62·8	62·2	60·9	59·0	56·6	50

H.P.	L	U	L	U	L	U	L	U	L	U	L	U	L	U	H.P.
′	′	′	′	′	′	′	′	′	′	′	′	′	′	′	′
54·0	0·3	0·9	0·3	0·9	0·4	1·0	0·5	1·1	0·6	1·2	0·7	1·3	0·9	1·5	54·0
54·3	0·7	1·1	0·7	1·2	0·7	1·2	0·8	1·3	0·9	1·4	1·1	1·5	1·2	1·7	54·3
54·6	1·1	1·4	1·1	1·4	1·1	1·4	1·2	1·5	1·3	1·6	1·4	1·7	1·5	1·8	54·6
54·9	1·4	1·6	1·5	1·6	1·5	1·6	1·6	1·7	1·6	1·8	1·8	1·9	1·9	2·0	54·9
55·2	1·8	1·8	1·8	1·8	1·9	1·9	1·9	1·9	2·0	2·0	2·1	2·1	2·2	2·2	55·2
55·5	2·2	2·0	2·2	2·0	2·3	2·1	2·3	2·1	2·4	2·2	2·4	2·3	2·5	2·4	55·5
55·8	2·6	2·2	2·6	2·2	2·6	2·3	2·7	2·3	2·7	2·4	2·8	2·4	2·9	2·5	55·8
56·1	3·0	2·4	3·0	2·5	3·0	2·5	3·0	2·5	3·1	2·6	3·1	2·6	3·2	2·7	56·1
56·4	3·4	2·7	3·4	2·7	3·4	2·7	3·4	2·7	3·4	2·8	3·5	2·8	3·5	2·9	56·4
56·7	3·7	2·9	3·7	2·9	3·8	2·9	3·8	2·9	3·8	3·0	3·8	3·0	3·9	3·0	56·7
57·0	4·1	3·1	4·1	3·1	4·1	3·1	4·1	3·1	4·2	3·1	4·2	3·2	4·2	3·2	57·0
57·3	4·5	3·3	4·5	3·3	4·5	3·3	4·5	3·3	4·5	3·3	4·5	3·4	4·6	3·4	57·3
57·6	4·9	3·5	4·9	3·5	4·9	3·5	4·9	3·5	4·9	3·5	4·9	3·5	4·9	3·6	57·6
57·9	5·3	3·8	5·3	3·8	5·2	3·8	5·2	3·7	5·2	3·7	5·2	3·7	5·2	3·7	57·9
58·2	5·6	4·0	5·6	4·0	5·6	4·0	5·6	4·0	5·6	3·9	5·6	3·9	5·6	3·9	58·2
58·5	6·0	4·2	6·0	4·2	6·0	4·2	6·0	4·2	6·0	4·1	5·9	4·1	5·9	4·1	58·5
58·8	6·4	4·4	6·4	4·4	6·4	4·4	6·3	4·4	6·3	4·3	6·3	4·3	6·2	4·2	58·8
59·1	6·8	4·6	6·8	4·6	6·7	4·6	6·7	4·6	6·7	4·5	6·6	4·5	6·6	4·4	59·1
59·4	7·2	4·8	7·1	4·8	7·1	4·8	7·1	4·8	7·0	4·7	7·0	4·7	6·9	4·6	59·4
59·7	7·5	5·1	7·5	5·0	7·5	5·0	7·5	5·0	7·4	4·9	7·3	4·8	7·2	4·7	59·7
60·0	7·9	5·3	7·9	5·3	7·9	5·2	7·8	5·2	7·8	5·1	7·7	5·0	7·6	4·9	60·0
60·3	8·3	5·5	8·3	5·5	8·2	5·4	8·2	5·4	8·1	5·3	8·0	5·2	7·9	5·1	60·3
60·6	8·7	5·7	8·7	5·7	8·6	5·7	8·6	5·6	8·5	5·5	8·4	5·4	8·2	5·3	60·6
60·9	9·1	5·9	9·0	5·9	9·0	5·9	8·9	5·8	8·8	5·7	8·7	5·6	8·6	5·4	60·9
61·2	9·5	6·2	9·4	6·1	9·4	6·1	9·3	6·0	9·2	5·9	9·1	5·8	8·9	5·6	61·2
61·5	9·8	6·4	9·8	6·3	9·7	6·3	9·7	6·2	9·5	6·1	9·4	5·9	9·2	5·8	61·5

DIP

Ht. of Eye	Corrⁿ	Ht. of Eye	Corrⁿ	Ht. of Eye	Corrⁿ
ft.	′	ft.	′	ft.	′
4·0		24		63	
	−2·0		−4·9		−7·8
4·4		26		65	
	−2·1		−5·0		−7·9
4·9		27		67	
	−2·2		−5·1		−8·0
5·3		28		68	
	−2·3		−5·2		−8·1
5·8		29		70	
	−2·4		−5·3		−8·2
6·3		30		72	
	−2·5		−5·4		−8·3
6·9		31		74	
	−2·6		−5·5		−8·4
7·4		32		75	
	−2·7		−5·6		−8·5
8·0		33		77	
	−2·8		−5·7		−8·6
8·6		35		79	
	−2·9		−5·8		−8·7
9·2		36		81	
	−3·0		−5·9		−8·8
9·8		37		83	
	−3·1		−6·0		−8·9
10·5		38		85	
	−3·2		−6·1		−9·0
11·2		40		87	
	−3·3		−6·2		−9·1
11·9		41		88	
	−3·4		−6·3		−9·2
12·6		42		90	
	−3·5		−6·4		−9·3
13·3		44		92	
	−3·6		−6·5		−9·4
14·1		45		94	
	−3·7		−6·6		−9·5
14·9		47		96	
	−3·8		−6·7		−9·6
15·7		48		98	
	−3·9		−6·8		−9·7
16·5		49		101	
	−4·0		−6·9		−9·8
17·4		51		103	
	−4·1		−7·0		−9·9
18·3		52		105	
	−4·2		−7·1		−10·0
19·1		54		107	
	−4·3		−7·2		−10·1
20·1		55		109	
	−4·4		−7·3		−10·2
21·0		57		111	
	−4·5		−7·4		−10·3
22·0		58		113	
	−4·6		−7·5		−10·4
22·9		60		116	
	−4·7		−7·6		−10·5
23·9		62		118	
	−4·8		−7·7		−10·6
24·9		63		120	

MOON CORRECTION TABLE

The correction is in two parts; the first correction is taken from the upper part of the table with argument apparent altitude, and the second from the lower part, with argument H.P., in the same column as that from which the first correction was taken. Separate corrections are given in the lower part for lower (L) and upper (U) limbs. All corrections are to be **added** to apparent altitude, *but 30′ is to be subtracted from the altitude of the upper limb.*

For corrections for pressure and temperature see page A4.

For bubble sextant observations ignore dip, take the mean of upper and lower limb corrections and subtract 15′ from the altitude.

App. Alt. = Apparent altitude = Sextant altitude corrected for index error and dip.

ALTITUDE CORRECTION TABLES 35°-90°—MOON

App. Alt.	35°–39° Corrⁿ	40°–44° Corrⁿ	45°–49° Corrⁿ	50°–54° Corrⁿ	55°–59° Corrⁿ	60°–64° Corrⁿ	65°–69° Corrⁿ	70°–74° Corrⁿ	75°–79° Corrⁿ	80°–84° Corrⁿ	85°–89° Corrⁿ	App. Alt.
′	° ′	° ′	° ′	° ′	° ′	° ′	° ′	° ′	° ′	° ′	° ′	′
00	35 56·5	40 53·7	45 50·5	50 46·9	55 43·1	60 38·9	65 34·6	70 30·1	75 25·3	80 20·5	85 15·6	00
10	56·4	53·6	50·4	46·8	42·9	38·8	34·4	29·9	25·2	20·4	15·5	10
20	56·3	53·5	50·2	46·7	42·8	38·7	34·3	29·7	25·0	20·2	15·3	20
30	56·2	53·4	50·1	46·5	42·7	38·5	34·1	29·6	24·9	20·0	15·1	30
40	56·2	53·3	50·0	46·4	42·5	38·4	34·0	29·4	24·7	19·9	15·0	40
50	56·1	53·2	49·9	46·3	42·4	38·2	33·8	29·3	24·5	19·7	14·8	50
00	36 56·0	41 53·1	46 49·8	51 46·2	56 42·3	61 38·1	66 33·7	71 29·1	76 24·4	81 19·6	86 14·6	00
10	55·9	53·0	49·7	46·0	42·1	37·9	33·5	29·0	24·2	19·4	14·5	10
20	55·8	52·8	49·5	45·9	42·0	37·8	33·4	28·8	24·1	19·2	14·3	20
30	55·7	52·7	49·4	45·8	41·8	37·7	33·2	28·7	23·9	19·1	14·1	30
40	55·6	52·6	49·3	45·7	41·7	37·5	33·1	28·5	23·8	18·9	14·0	40
50	55·5	52·5	49·2	45·5	41·6	37·4	32·9	28·3	23·6	18·7	13·8	50
00	37 55·4	42 52·4	47 49·1	52 45·4	57 41·4	62 37·2	67 32·8	72 28·2	77 23·4	82 18·6	87 13·7	00
10	55·3	52·3	49·0	45·3	41·3	37·1	32·6	28·0	23·3	18·4	13·5	10
20	55·2	52·2	48·8	45·2	41·2	36·9	32·5	27·9	23·1	18·2	13·3	20
30	55·1	52·1	48·7	45·0	41·0	36·8	32·3	27·7	22·9	18·1	13·2	30
40	55·0	52·0	48·6	44·9	40·9	36·6	32·2	27·6	22·8	17·9	13·0	40
50	55·0	51·9	48·5	44·8	40·8	36·5	32·0	27·4	22·6	17·8	12·8	50
00	38 54·9	43 51·8	48 48·4	53 44·6	58 40·6	63 36·4	68 31·9	73 27·2	78 22·5	83 17·6	88 12·7	00
10	54·8	51·7	48·2	44·5	40·5	36·2	31·7	27·1	22·3	17·4	12·5	10
20	54·7	51·6	48·1	44·4	40·3	36·1	31·6	26·9	22·1	17·3	12·3	20
30	54·6	51·5	48·0	44·2	40·2	35·9	31·4	26·8	22·0	17·1	12·2	30
40	54·5	51·4	47·9	44·1	40·1	35·8	31·3	26·6	21·8	16·9	12·0	40
50	54·4	51·2	47·8	44·0	39·9	35·6	31·1	26·5	21·7	16·8	11·8	50
00	39 54·3	44 51·1	49 47·6	54 43·9	59 39·8	64 35·5	69 31·0	74 26·3	79 21·5	84 16·6	89 11·7	00
10	54·2	51·0	47·5	43·7	39·6	35·3	30·8	26·1	21·3	16·5	11·5	10
20	54·1	50·9	47·4	43·6	39·5	35·2	30·7	26·0	21·2	16·3	11·4	20
30	54·0	50·8	47·3	43·5	39·4	35·0	30·5	25·8	21·0	16·1	11·2	30
40	53·9	50·7	47·2	43·3	39·2	34·9	30·4	25·7	20·9	16·0	11·0	40
50	53·8	50·6	47·0	43·2	39·1	34·7	30·2	25·5	20·7	15·8	10·9	50

H.P.	L U	L U	L U	L U	L U	L U	L U	L U	L U	L U	L U	H.P.
′	′ ′	′ ′	′ ′	′ ′	′ ′	′ ′	′ ′	′ ′	′ ′	′ ′	′ ′	′
54·0	1·1 1·7	1·3 1·9	1·5 2·1	1·7 2·4	2·0 2·6	2·3 2·9	2·6 3·2	2·9 3·5	3·2 3·8	3·5 4·1	3·8 4·5	54·0
54·3	1·4 1·8	1·6 2·0	1·8 2·2	2·0 2·5	2·3 2·7	2·5 3·0	2·8 3·2	3·0 3·5	3·3 3·8	3·6 4·1	3·9 4·4	54·3
54·6	1·7 2·0	1·9 2·2	2·1 2·4	2·3 2·6	2·5 2·8	2·7 3·0	3·0 3·3	3·2 3·5	3·5 3·8	3·7 4·1	4·0 4·3	54·6
54·9	2·0 2·2	2·2 2·3	2·3 2·5	2·5 2·7	2·7 2·9	2·9 3·1	3·2 3·3	3·4 3·5	3·6 3·8	3·9 4·0	4·1 4·3	54·9
55·2	2·3 2·3	2·5 2·4	2·6 2·6	2·8 2·8	3·0 2·9	3·2 3·1	3·4 3·3	3·6 3·5	3·8 3·7	4·0 4·0	4·2 4·2	55·2
55·5	2·7 2·5	2·8 2·6	2·9 2·7	3·1 2·9	3·2 3·0	3·4 3·2	3·6 3·4	3·7 3·5	3·9 3·7	4·1 3·9	4·3 4·1	55·5
55·8	3·0 2·6	3·1 2·7	3·2 2·8	3·3 3·0	3·5 3·1	3·6 3·3	3·8 3·4	3·9 3·6	4·1 3·7	4·2 3·9	4·4 4·0	55·8
56·1	3·3 2·8	3·4 2·9	3·5 3·0	3·6 3·1	3·7 3·2	3·8 3·3	4·0 3·4	4·1 3·6	4·2 3·7	4·4 3·8	4·5 4·0	56·1
56·4	3·6 2·9	3·7 3·0	3·8 3·1	3·9 3·2	3·9 3·3	4·0 3·4	4·1 3·5	4·3 3·6	4·4 3·7	4·5 3·8	4·6 3·9	56·4
56·7	3·9 3·1	4·0 3·1	4·1 3·2	4·1 3·3	4·2 3·3	4·3 3·4	4·3 3·5	4·4 3·6	4·5 3·7	4·6 3·8	4·7 3·8	56·7
57·0	4·3 3·2	4·3 3·3	4·3 3·3	4·4 3·4	4·4 3·4	4·5 3·5	4·5 3·5	4·6 3·6	4·7 3·6	4·7 3·7	4·8 3·8	57·0
57·3	4·6 3·4	4·6 3·4	4·6 3·4	4·6 3·5	4·7 3·5	4·7 3·5	4·7 3·6	4·8 3·6	4·8 3·6	4·8 3·7	4·9 3·7	57·3
57·6	4·9 3·6	4·9 3·6	4·9 3·6	4·9 3·6	4·9 3·6	4·9 3·6	4·9 3·6	4·9 3·6	5·0 3·6	5·0 3·6	5·0 3·6	57·6
57·9	5·2 3·7	5·2 3·7	5·2 3·7	5·2 3·7	5·2 3·7	5·1 3·6	5·1 3·6	5·1 3·6	5·1 3·6	5·1 3·6	5·1 3·6	57·9
58·2	5·5 3·9	5·5 3·8	5·5 3·8	5·4 3·8	5·4 3·7	5·4 3·7	5·3 3·7	5·3 3·6	5·2 3·6	5·2 3·5	5·2 3·5	58·2
58·5	5·9 4·0	5·8 4·0	5·8 3·9	5·7 3·9	5·6 3·8	5·6 3·8	5·5 3·7	5·5 3·6	5·4 3·6	5·3 3·5	5·3 3·4	58·5
58·8	6·2 4·2	6·1 4·1	6·0 4·1	6·0 4·0	5·9 3·9	5·8 3·8	5·7 3·7	5·6 3·6	5·5 3·5	5·4 3·5	5·3 3·4	58·8
59·1	6·5 4·3	6·4 4·3	6·3 4·2	6·2 4·1	6·1 4·0	6·0 3·9	5·9 3·8	5·8 3·6	5·7 3·5	5·6 3·4	5·4 3·3	59·1
59·4	6·8 4·5	6·7 4·4	6·6 4·3	6·5 4·2	6·4 4·1	6·2 3·9	6·1 3·8	6·0 3·7	5·8 3·5	5·7 3·4	5·5 3·2	59·4
59·7	7·1 4·6	7·0 4·5	6·9 4·4	6·8 4·3	6·6 4·1	6·5 4·0	6·3 3·8	6·2 3·7	6·0 3·5	5·8 3·3	5·6 3·2	59·7
60·0	7·5 4·8	7·3 4·7	7·2 4·5	7·0 4·4	6·9 4·2	6·7 4·0	6·5 3·9	6·3 3·7	6·1 3·5	5·9 3·3	5·7 3·1	60·0
60·3	7·8 5·0	7·6 4·8	7·5 4·7	7·3 4·5	7·1 4·3	6·9 4·1	6·7 3·9	6·5 3·7	6·3 3·5	6·0 3·2	5·8 3·0	60·3
60·6	8·1 5·1	7·9 5·0	7·7 4·8	7·6 4·6	7·3 4·4	7·1 4·2	6·9 3·9	6·7 3·7	6·4 3·4	6·2 3·2	5·9 2·9	60·6
60·9	8·4 5·3	8·2 5·1	8·0 4·9	7·8 4·7	7·6 4·5	7·3 4·2	7·1 4·0	6·8 3·7	6·6 3·4	6·3 3·2	6·0 2·9	60·9
61·2	8·7 5·4	8·5 5·2	8·3 5·0	8·1 4·8	7·8 4·5	7·6 4·3	7·3 4·0	7·0 3·7	6·7 3·4	6·4 3·1	6·1 2·8	61·2
61·5	9·1 5·6	8·8 5·4	8·6 5·1	8·3 4·9	8·1 4·6	7·8 4·3	7·5 4·0	7·2 3·7	6·9 3·4	6·5 3·1	6·2 2·7	61·5

NICHOLLS'S CONCISE GUIDE

VOLUME 2.

NICHOLLS'S CONCISE GUIDE

TO THE

NAVIGATION EXAMINATIONS

VOLUME 2.

FIRST REVISION BY
H. H. BROWN, D.S.C.*, R.D., F.R.A.S., EXTRA MASTER.

REVISED BY
EDWARD J. COOLEN, F.N.I., D.M.S., Grad.I.P.M., F.I.Mgt.
Liveryman of the Honourable Company of Master Mariners,
World Underwater Federation (C.M.A.S.) International Diver,
Director, Marine Services Division,
Coolen Consultancy & Management Services, London W2 4ER.
Formerly:
Senior Lecturer, School of Navigation, Tower Hill, London,
British Technical Co-operation Officer, Overseas Development Administration,
13th Commandant of the Bangladesh Marine Academy, Chittagong.

GLASGOW
BROWN, SON & FERGUSON, LTD., NAUTICAL PUBLISHERS
4-10 DARNLEY STREET, G41 2SD.

9th Edition	-	-	-	*1953*
10th Edition	-	-	-	*1963*
Revised -	-	-	-	*1968*
Reprinted -	-	-	-	*1974*
Reprinted -	-	-	-	*1976*
Reprinted -	-	-	-	*1978*
11th Edition	-	-	-	*1984*
12th Edition	-	-	-	*1995*

ISBN 0 85174 635 7
ISBN 0 85174 480 X (11th Edition)

Printed and Made in Great Britain

PREFACE

CONTINUING demand makes this the twelfth edition of *Nicholls's Concise Guide, Volume 2.* It is complimentary to, and a continuation of, the work contained in *Volume 1.*

Numerous changes, additions and new features were included in the previous edition so that these, coupled with the greatly enlarged section on tide calculations, which has now been added, make for a more balanced and greatly improved volume.

The total of these changes, additions and new features are summarised in the following paragraphs.

In the enlarged section on tide calculations examples are now worked on the latest *Admiralty Tidal Prediction Forms*, NP 204, and graphical methods of interpolation are demonstrated. A check-list system of instructions for Tide Calculations, additional diagrams and the latest tide curves are included. A comprehensive and enlarged set of *Admiralty Tide Table* extracts, with separate index, has also been added. Highly detailed referencing to these changes may now be found in the main index in which the section on tides has been greatly enlarged in this edition.

Sections on the Sailings, Magnetic Compass, Ocean Passages have been revised. Those relating to Radar Plotting, Tide Calculations and Meridian Passage Star have been rewritten. Marine Surveying and information relating to charts has been updated.

To take account of modern practice, a list of symbols and abbreviations has been added, reference to quadrantal notation replaced by 360 degree notation and terminology relating to celestial navigation updated.

Numerous figures have been redrawn and new ones added.

A check-list system of instructions has also been included for Radar Plotting, Passage Planning and for finding suitable stars for Meridian Altitudes. Where appropriate, the various steps involved in calculations and procedure have been included in the worked examples and directly linked to the check-lists and proformas provided. It is hoped that these inclusions will reduce learning times and be of assistance to students working privately at sea and also to those in the developing world where there may be a shortage of technical teachers.

A careful selection of extracts from official and other publications is also included. Here the aim has been to make the volume self-contained, so far as possible. For those with limited budgets, this feature could be important. Examples include extracts from *Collision Avoidance Rules*, the *Admiralty Method of Tidal Prediction, Ocean Passages for the World, Symbols and Abbreviations Chart 5011*, and the more comprehensive *Admiralty Tide Table* extracts.

Summary lists of the more advanced formulae used in navigation and astronomy have also been included.

As in previous editions, track and practice charts are included in the rear wallet. The metric practice chart BA 5053 can be used for the exercises on the English Channel.

I again record my thanks to all those who assisted with the previous, eleventh edition: Captain P. R. Jackson, Captain A. G. Blance, Captain D. B. Fantham, Captain T. C. Haile and particularly to my close and very able friend the late Captain P. R. Lewis, and also to Alison Stenhouse who executed much of the artwork. I am most grateful to them all and to the publishers, Brown, Son & Ferguson, Ltd. of Glasgow, for their care and attention in the production of this new edition.

For kind permission to reprint from:

Admiralty Chart BA 5053, *Ocean Passages for the World* NP 136, *Admiralty Tide Tables*, Volume 1-3, the *Admiralty Method of Tidal Prediction* NP 159, and the *International Collision Regulations*, 1972.

Grateful acknowledgement is made to:

The Hydrographic Dept. Ministry of Defence, Taunton.
The Controller of her Majesty's Stationery Office, and the
International Maritime Organisation (IMO formerly IMCO).

I hope that this new edition will be of assistance to students and officers in the Merchant Navy wherever they may be and particularly to those in the developing world, some of whom I have been privileged to serve.

Edward Coolen,
March 1995

CONTENTS

DEPARTMENT OF TRANSPORT

SYLLABUSES FOR THE CLASS 5 - CLASS 2 (AND COMMAND ENDORSEMENT) CERTIFICATES OF COMPETENCY WRITTEN EXAMINATIONS IN NAVIGATION AND NAVIGATIONAL AIDS

Reproduced by kind permission of the Department of Transport

Index

CLASS 5 CERTIFICATE OF COMPETENCY
Chartwork and Practical Navigation

Chartwork

(*a*) Given variation and the deviation of the magnetic compass, or gyro error, to convert true courses into compass courses and vice versa.

Given a sample table of deviations to extract the deviation thence to convert true course into magnetic and compass courses.

To find the compass course between two positions.

The effect of current on speed.

Allowance for leeway.

Given compass course steered, the speed of the ship and direction and rate of the current, to find the true course made good.

To find the course to steer allowing for a current.

Given the course steered and distance run, to determine the set and rate of the current experienced between two positions.

(*b*) To fix a position on a chart by simultaneous cross bearings, by bearing and range, by positional information from radio aids to navigation or by any combination, applying the necessary corrections. The use of lattice charts.

(*c*) To fix the position by bearings of one or more objects with the run between, allowing for a current, and to find the distance at which the ship will pass off a given point.
The use of position lines obtained by any method, including terrestrial position lines and position circles. The use of transit bearings.

(*d*) Elementary knowledge of passage planning and execution. Landfalls in thick and clear weather. The selection of suitable anchorages, approaching anchorage and entering narrow waters. The use of clearing marks and horizontal and vertical danger angles. Distance of sighting lights.

(*e*) To find the time and height of high and low water at Standard ports and at Secondary ports by tidal differences using Admiralty Tide Tables, Vol. 1; the use of tide curves as published in the Admiralty Tide Tables.

(*f*) The interpretation of a chart or plan; particularly the information given about Lights, Buoys, Radio Beacons and other navigational aids; Depth and height contours; Tidal streams; Traffic lanes and separation zones.
Recognition of the coast and radar responsive targets.
Chart correction.
Depths and nature of bottom. Use of soundings.

Note: Questions may be set in this paper requiring descriptive answers as well as work on the chart provided. Oral questions, which may relate to any part of the syllabus, may also be asked.

Practical Navigation

(*a*) Practical questions on plane and Mercator sailing.

(*b*) The use of the traverse tables to obtain the position of the ship at any time, given compass courses, variation, deviation and the run recorded by log or calculated by time and estimated speed, allowing for the effects of wind and current.

(*c*) To find compass error by observation of heavenly bodies and, for a magnetic compass, the deviation for the direction of the ship's head.

CLASS 4 CERTIFICATE OF COMPETENCY

Coastal Navigation

(*a*) Given variation and the deviation of the magnetic compass or gyro error, to convert true courses into compass courses and vice versa.
Given a sample of deviations, to extract the deviation thence to convert true courses into magnetic and compass courses.
To find the compass course between two positions.
The effect of current on speed.
Allowance for leeway.
Given compass course steered, the speed of the ship and direction and rate of the current, to find the true course made good.
To find the course to steer allowing for a current.
Given the course steered and distance run, to determine the set and rate of the current experienced between two positions.

(*b*) To fix a position on a chart by simultaneous cross bearings, by bearing and range, by positional information from radio aids to navigation or by any combination, applying the necessary corrections. The use of lattice charts.

(*c*) To fix the position by bearings of one or more objects with the run between, allowing for a current and to find the distance at which the ship will pass off a given point.
The use of position lines obtained by any method, including terrestrial position lines and position circles. The use of transit bearings.

(*d*) Elementary knowledge of passage planning and execution. Landfalls in thick and clear weather. The selection of suitable anchorages, approaching anchorage and entering narrow waters. The use of clearing marks and horizontal and vertical danger angles. Distance of sighting lights.

(*e*) To find the time and height of high and low water at Standard ports and at Secondary ports by tidal differences, using Admiralty Tide Tables, Volume 1.

The use of tables and tide curves to find the time at which the tide reaches a specified height or the height of the tide at a given time and thence the approximate correction to be applied to soundings or to charted heights of shore objects.

(*f*) The interpretation of a chart or plan; particularly the information given about Buoys, Lights, Radio beacons and other navigational aids; Depth and height contours; Tidal streams; Traffic lanes and separation zones.

Recognition of the coast and radar responsive targets.

Chart correction.

Depths and nature of bottom. Use of soundings.

The use of gnomonic charts.

CLASS 4 CERTIFICATE OF COMPETENCY

Ocean and Offshore Navigation

(*a*) Practical problems on plane and Mercator sailing.

(*b*) The use of the traverse tables to obtain the position of the ship at any time, given compass courses, variation, deviation and the run recorded by log or calculated by time and estimated speed, allowing for the effects of wind and currents, if any.

(*c*) To find the latitude by meridian altitude of a heavenly body above or below the Pole. Latitude by observation of Polaris.

(*d*) From an observation of any heavenly body near or out of the meridian, to find the direction of the position line and a position through which it passes.

(*e*) To obtain a position by the use of position lines obtained from two or more observations with or without run.

(*f*) To find the true bearing of a heavenly body, the compass error and thence the deviation of the magnetic compass for the direction of the ship's head.

(*g*) To calculate the approximate time (to the nearest minute) of the meridian passage of a heavenly body; to calculate an approximate altitude for setting on the sextant to obtain the meridian altitude of a heavenly body.

(*h*) Great circle and composite sailing.

(*i*) Position fixing; fixed and variable errors, area of probability. Obtaining a position from two or more observations of any type, with or without run.

(*j*) (i) To use an azimuth mirror, pelorus (bearing plate) or other instrument for taking bearings.

(ii) To use a sextant for taking vertical and horizontal angles; to read a sextant both on and off the arc.

(iii) To correct a sextant into which has been introduced one or more of perpendicularity, side or index errors.

(iv) To find the index error of a sextant.

(v) The care, winding, rating and comparing of chronometers.

(vi) Use, care and limitations of the magnetic and gyro compass and associated equipment, including automatic pilot.

CLASS 3 CERTIFICATE OF COMPETENCY

Bridging Examination

Principles of Navigation

(*a*) The shape of the earth. Poles, equator, meridians, parallels of latitude. Position by latitude and longitude. Direction, bearing, distance, units of measurement. Difference of latitude, difference of longitude, departure, mean latitude, difference of meridional parts and the relationship between them.

Great circles, great circle course and distance, small circles on a sphere.

(*b*) The celestial sphere; definitions on the celestial sphere, apparent motion on the celestial sphere. Declination. Azimuth. Sidereal hour angle. The position of a body on the celestial sphere; azimuth with altitude of declination with sidereal or local hour angle. The rising, culmination and setting of heavenly bodies.

(*c*) Solar system, earth-moon system, planetary motion. Earth's rotation and movement in orbit, mean sun, ecliptic, first point of Aries. Equinox and solstice, sunrise, sunset, twilight.

(*d*) Time; Greenwich and other standard time, zone time, mean time, apparent time, sidereal time, equation of time, relationship between longitude and time.

(*e*) Local hour angle of a heavenly body in time and arc.

Greenwich hour angle of Sun, Moon planets and Aries.

(*f*) Correction of sextant altitudes; dip, refraction, horizontal parallax, parallax in altitude, semi-diameter and augmentation.

(*g*) Geographical position of heavenly body. A circle of position and its practical application, i.e. position line. Intercept.

(*h*) Simple properties of Mercator and gnomonic charts. Latitude and longitude scales; measurement of distance. Rhumb lines. Great circles and composite tracks.

(*i*) The relationship between the tides and the phases of the moon.

(*j*) Principles of position fixing by measurement of difference of distance from two or more fixed points. The hyperbolic lattice on navigational charts.

(*k*) Properties of the free gyroscope. Relationship between applied force and precession. Drift and tilt. Outline principle of developing a North-seeking instrument by gravity control.

(*l*) Basic principles of position fixing by satellite.

CLASS 2 CERTIFICATE OF COMPETENCY
Navigation

Candidates will be required to show a full understanding of the techniques involved and must be able fully to relate the various aspects one to another, and show their ability to make full use of all the navigational and meteorological information which is available to a ship's Master.

1. (*a*) Voyage planning. The selection of ocean routes. Shore-based weather routeing and self-routeing. Use of prognostic surface weather and wave charts. Use of current atlases and other charts and navigational publications relevant to ocean navigation.
   (*b*) Planning and executing a coastal passage. Approaching the coast. Approaching and entering harbour.
   (*c*) Navigation in pilotage waters, whether with or without a pilot on board.
   (*d*) Approaching and passing through traffic separation schemes and adjacent areas.

2. (*a*) The optimum course and speed for two ships wishing to rendezvous for any purpose.
   (*b*) Search and rescue procedures.
   (*c*) Navigational procedures when approaching off-shore installations, and when working with helicopters or small craft.
   (*d*) Navigation in extreme weather conditions.
   (*e*) Navigation in the vicinity of and the rules for avoiding tropical storms. Reports to be made under international conventions.
   (*f*) Navigation in the vicinity of ice. Reports to be made under international conventions.

3. Bridge procedures at sea, in harbour, and whilst berthing or anchoring.

Navigation Instrumentation

(*a*) The use and understanding of aids to navigation, including all those radio and electronic aids which are installed in a high proportion of British Merchant Ships. Questions will be directed towards ensuring that candidates have a proper understanding of the essential principles and operation of these aids and can make full use of their capabilities whilst appreciating their limitations.
(*b*) The interpretation and use of navigation and meteorological information.
(*c*) The interpretation and use of information from navigational aids, including the use of radar in collision avoidance.
(*d*) The effects of systematic and random errors in position fixing by any method.
(*e*) Advantages and disadvantages of various navigation systems and methods; considerations underlying the choice of navigational systems for differing trades and geographical regions.
(*f*) The construction, siting, care and maintenance of the magnetic compass and associated equipment. Causes of deviation, the production of a table of deviations. Co-efficients, A, B, C, D and E. The evaluation of the approximate co-efficients from given data and their relationship with the deviation. An appreciation, without calculations, of the effects of permanent magnetism and induced magnetism. Principles of compass adjustment and methods of adjustment. Heeling Error, Gaussin error and retentive error.

(*g*) Principle of the gyro compass. Errors associated with the gyro compass, including latitude, course and speed error and correction; ballistic deflection and its relation to change of speed error; rolling error and how it is minimized. The principal parts of a gyro compass; fundamental differences in the construction and operation of the better known gyro compasses.
(*h*) Principles of operation and use of gyro and transmitting magnetic compasses, repeater systems automatic pilot, projector compasses, rate of turn indicators.
(*i*) Principle and practical application of echo sounding devices logs and speed indicators.
(*j*) An outline knowledge of recent developments in navigation aids.

MASTER (LIMITED EUROPEAN) ENDORSEMENT
Magnetic Compass and Navigational Aids

(*a*) The earth's magnetic field, poles and equator. The earth's total magnetic force, angle of dip, horizontal and vertical components. Variation.
(*b*) Deviation. Elementary ideas of the effect of hard iron, vertical and horizontal soft iron on the compass. The means used to compensate for these effects with special reference to Limited European ships. Siting of magnetic compasses with reference to proximity of magnetic material and electrical appliances. Precautions to be taken with electric wiring in vicinity of the compass.
(*c*) To find the magnetic bearing of a distant object from compass bearings taken on equidistant headings and construct a table of deviations.
(*d*) The practical application of electronic and radio aids to navigation. Appreciation and interpretation of data. Position fixing; fixed and variable errors, area of probability. The use of Radar as an aid to collision avoidance.
(*e*) The use of navigational publications and information commonly supplied or available to ships.

MASTER (EXTENDED EUROPEAN) ENDORSEMENT
Magnetic and Gyro Compasses and Navigational Aids

(*a*) The earth's magnetic field, poles and equator. The earth's total magnetic force, angle of dip, horizontal and vertical components. Variation.
(*b*) Deviation, its cause and effect. An understanding of the effects of semi-permanent and induced magnetic fields on the deviation of the compass. The means used to compensate for these effects.
(*c*) Siting of magnetic compasses with reference to proximity of magnetic material and electrical appliances. Precautions to be taken with electric wiring in vicinity of the compass.
(*d*) To find the magnetic bearing of a distant object from compass bearings taken on equidistant headings and construct a table of deviations.
(*e*) The principle of the free gyroscope.
Tilt and drift.
Precession. Control and damping. Correction for latitude, course and speed error.
Care and maintenance of different types of compasses.

(*f*) The practical application of electronic and radio aids to navigation.
Appreciation and interpretation of data.
Position fixing; fixed and variable errors, area of probability.
The use of Radar as an aid to collision avoidance.

(*g*) The use of navigational publications and information commonly supplied or available to ships.

SYMBOLS and ABBREVIATIONS

True	T
Magnetic	M
Compass	C
Gyro	G
Course	Co
Azimuth	az
Chronometer	chron
Sextant Altitude	Sext alt
Index Error	IE
Height of eye	HE
Lower limb	LL
Upper limb	UL
Course steered	
,, made good	
Tidal stream vector	
Range	
Position line	P/L
Transferred position line	
Clearing line	
In transit	
Left hand edge	
Right hand edge	
Abeam to starboard	
,, ,, port	
Observed position or fix	Obs Pos
Dead reckoning position	DR
Estimated position	EP
Chosen position	CP

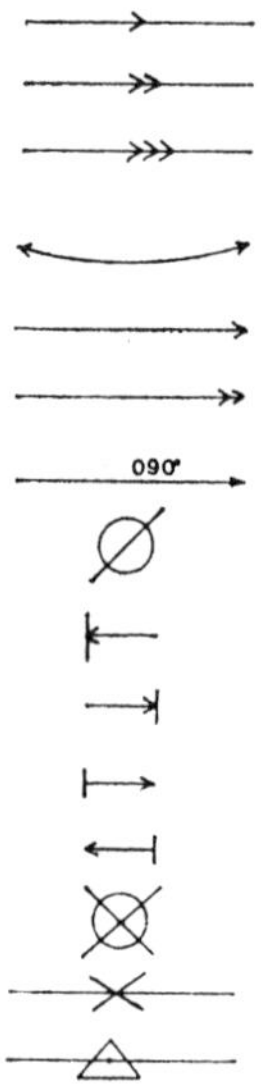

DEFINITIONS

CP	The chosen position is that position nearest to the observer chosen so that the latitude is an integral degree and the longitude is such that the local hour angle of the body at the time of observation is also an integral degree.
DR	The dead reckoning position is that obtained by allowing for courses steered and distances run only.
EP	The estimated position is that obtained by allowing for courses and distances and for leeway and current if any.
Sext alt	The sextant altitude is that read off a sextant.

NICHOLLS'S CONCISE GUIDE

CHAPTER I.

PLANE, PARALLEL AND MERCATOR SAILINGS.

CHAPTER VII of Volume I dealt with Plane, Parallel and Mercator Sailings. In this chapter it is proposed to refer to some further applications of these sailings and to comment on their use.

Let us first consider the formulae used in problems which involve use of one or other of the three sailings:

Parallel Sailing — Dep. = D. Long. × cos Lat.

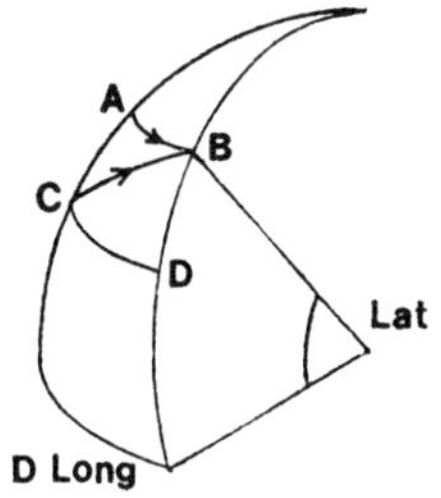

Plane or Mean Latitude Sailing — Dep. = D. Long. × cos Mn. Lat.

$$\frac{\text{Dep.}}{\text{D. Lat.}} = \tan \text{Co}$$

Dist. = D. Lat. × sec Co.

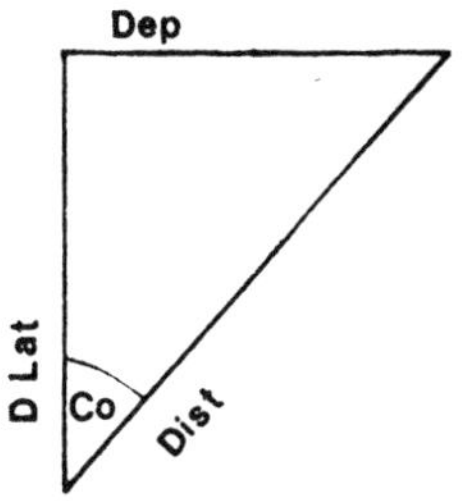

Mercator Sailing

$$\frac{\text{D. Long.}}{\text{D.M.P.}} = \tan \text{Co.}$$

Dist. = D. Lat. × sec Co.

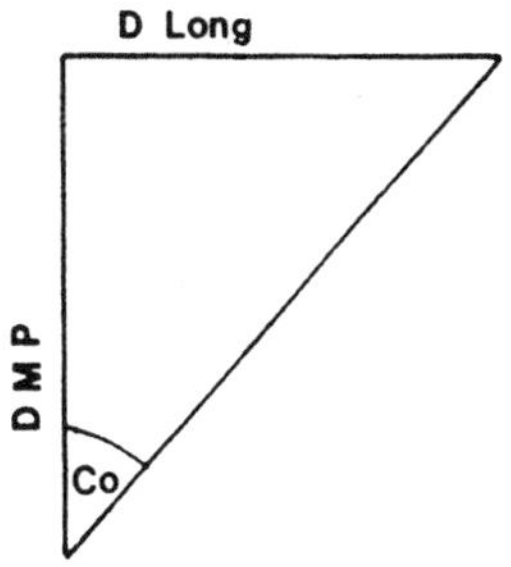

When sailing along a parallel (A to B) the latitude does not change. When sailing obliquely (C to B) the latitude does change and to convert the D. Long. to Departure it becomes necessary to use the arithmetic mean of the latitudes of the initial and final positions (C and B). Ideally, and for very accurate results, corrections should then be applied to make allowance for the compression of the earth and to convert the Mean Latitude applicable to a plane triangle to the latitude applicable to a sphere. Mean Latitude so corrected is referred to as the Middle Latitude.

Until recently many volumes of nautical tables included corrections for converting Mean to Middle Latitude. Unfortunately none of these correction tables has proved entirely satisfactory and recent editions of volumes, such as *Norie's Nautical Tables*, no longer include them. In addition, it is no longer required to apply the Mean to Middle Latitude correction in D.o.T. Examinations. It follows that you should always:

Use the Mean Latitude when working Plane Sailing problems.

Despite the lack of suitable Mean to Middle Latitude correction tables, the Plane Sailing (and Traverse Tables) gives acceptable results over short distances. Under certain conditions the Mean Latitude Sailing can give better results than the Mercator Sailing but, in the majority of cases, the Mercator Sailing is the more accurate one to use. In summary:

Use the Mercator Sailing where Accurate Results are required.

It must be added, however, that recent research indicates that, even when using the most suitable method, the calculated distance is liable to be in error by up to $\frac{1}{2}\%$.

Use of the Modified Distance Formula.

At large course angles, say over 70°, there is a danger of inaccurate interpolation when lifting the secant course from the tables. In cases where the interpolation becomes critical it is advisable to modify the standard formula as follows:

	Dist. = D. Lat. × sec Co.	(Standard Formula)
but	sec Co. = tan Co. × cosec Co.	
so	Dist. = D. Lat. × tan Co. × cosec Co.	(Modified Formula)

In the cases mentioned, the modified formula should be used for both Mean Latitude and Mercator Sailings. The student is advised to study the tables and verify that, at large angles, cosecant interpolation can be carried out quickly and with maximum accuracy. The modified formula can also be used to some advantage with course angles over 50°, at about which value the interval between successive cosecants becomes noticeably smaller than that between secants.

Another version of the modified formula can be used in conjunction with Mean Latitude Sailing. It will be recalled that:

$$\frac{\text{Dep.}}{\text{D. Lat.}} = \tan \text{Co}$$

so	Dep. = D. Lat. × tan Co.	
but	Dist. = D. Lat. × tan Co. cosec Co.	(Modified Formula-1)
so	Dist. = Dep. × cosec Co.	(Modified Formula-2)

As the departure will have been already found in a Mean Latitude Sailing this modification can be used to further simplify the calculation of distance.

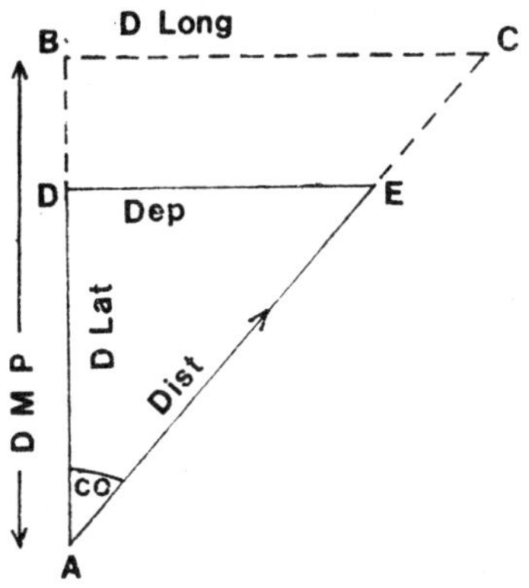

The adjoining figure is a general one for these types of problems, it shows the plane triangle *A D E* expanded into the mercatorial sailing triangle *A B C*. Angle *A*, the course, is common to both triangles, *A E* is the distance. The difference of latitude *A D* is expanded to *A B*, the difference of meridional parts on the Mercatorial projection, and departure *D E* is expanded into difference of longitude *B C*.

From fig:

$$\text{Tan Co.} = \frac{\text{D. Long.}}{\text{D.M.P.}} = \frac{\text{Dep.}}{\text{D. Lat.}} \text{ which is a case of similar triangles as } \frac{BC}{AB} = \frac{DE}{AD}$$

and this relationship may be used in finding the exact difference of longitude corresponding to a departure when sailing on an oblique course, viz. ·

$$\text{D. Long.} = \frac{\text{D.M.P.} \times \text{Dep.}}{\text{D. Lat.}}$$

Example.—From Lat. 45° N., Long. 40° W., a ship steered 030° T., for 800 miles. Required the latitude and longitude arrived at.

Given N. 30° E., 800 miles → D. Lat. 692·8′ N., Dep. 400′ E.

Lat.	45° 00·0′ N.	M.P.	3013·4
D. Lat.	11 32·8 N.		
Lat.	56 32·8 N.	M.P.	4113·4
		D.M.P.	1100·0

As stated above:

$$\text{D. Long.} = \frac{\text{D.M.P.} \times \text{Dep.}}{\text{D. Lat.}} = \frac{1100 \times 400}{692 \cdot 8} = 635 \cdot 3' = 10° \ 35 \cdot 3'$$

Initial Long.	40° 00′ W.
D. Long.	10 35·3 E.
Final Long.	29 24·7 W.

Position reached Lat. 56° 32·8′ N., Long. 29° 24·7′ W.

The above Dep. could also be converted into the correct D. Long. as follows:—

Dep. = D. Long. × cos Mn. Lat.

so D. Long. = Dep. × sec Mn. Lat.

Lat. - -	45° 00′ N.				
½ D. Lat. -	5 46·4 N.				
Mean Lat. -	50 46·4 N.	log sec	0·19901	Initial Long.	40° 00′ W.
Dep. - -	400′	log	2·60206	D. Long.	10 32·5 E.
D. Long. -	632·5′ ←	log	2·80107	Final Long.	29 27·5 W.

If the problem is worked by Mean Latitude Sailing the D. Long. so calculated will be 632·5′ instead of 635·3′ as obtained by Mercator Sailing—a difference of 2·8′.

Example.—Given Lat. 35° 15′ S., Course 140° T., D. Long. 12°. Find the latitude reached and the distance sailed.

The above information is indicated on the figure. It is required to find the difference of meridional parts *A B*, to apply to the meridional parts for latitude *A* to get the meridional parts for C, thence the latitude of *C* from the table of meridional parts; also *A E* the distance sailed.

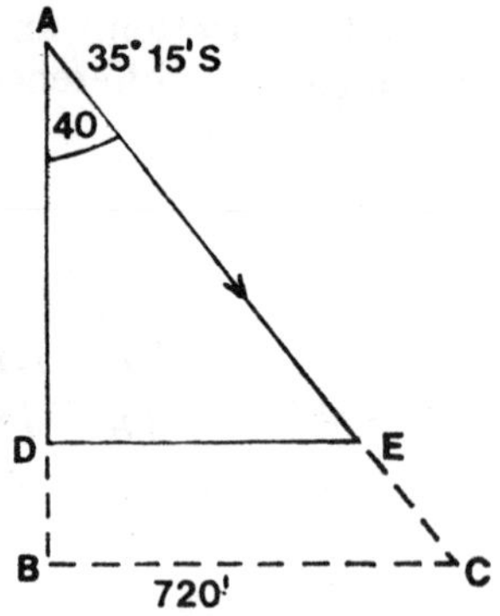

$$\frac{\text{D. Long.}}{\text{D.M.P.}} = \tan \text{Co.}$$

So D.M.P. = D. Long. cot Co. = 720 cot 40°

	(*B C*)	D. Long.	720	log	2·85733
	(∠*A*)	Course	40°	log cot	0·07619
	(*A B*)	D.M.P.	858·1	log	2·93352
From Tables Lat. 35° 15′ →		M.P.	2249·1		
From Tables Lat. 46 06 ←		M.P.	3107·0		
D. Lat. 10 51	(*A D*)				

Dist. = D. Lat. sec Co.

(*A*)	Course 40°		log sec	0·11575
(*A D*)	D. Lat.	651	log	2·85318
(*A E*)	Dist.	849·8	log	2·92933

Answer.—Lat. 46° 06′ S., Distance 849·5 miles.

Example.—Calculate the course and distance from Lat. 50° 00′ N., Long. 45° 00′ W. to Lat. 50° 31′ N., Long. 13° 49′ W., (i) Using the Mercator Sailing, (ii) using the Mean Latitude Sailing, and (iii) compare your results.

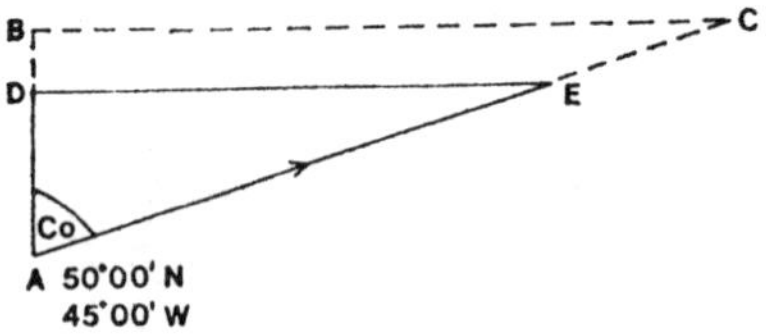

From	50° 00′ N.	M.P.	3456·53		45° 00′ W.
To	50 31 N.	M.P.	3504·89		13 49 W.
D. Lat.	31 N.	D.M.P.	48·36	D. Long.	31 11 E.
				,,	1871′ E.

$$\frac{\text{D. Long.}}{\text{D.M.P.}} = \tan \text{Co.}$$

D. Long.	1871′		log	3·27207
D.M.P.	48·36		log	1·68449
Co.	N. 88° 31′ E.	←	log tan	1·58758
,,	,,		log cosec	0·00015
D. Lat.	31		log	1·49136
Dist.	1199·75	←	log	3·07909

Dist. = D. Lat. × tan Co. cosec Co.

Answer (i) Course 088° 31′ T., Distance 1199·75 miles (by Mercator Sailing).

(ii) From	50° 00′ N.		45° 00′ W.	Lat.	50° 00′ N.
To	50 31 N.		13 49 W.	½ D. Lat.	15·5 N.
D. Lat.	31 N.	D. Long.	31 11 E.	Mn. Lat.	50° 15·5′ N.
			1871′ E.		

Dep. = D. Long. × cos. Mn. Lat.	D. Long.	1871′		log	3·27207	
	Mn Lat.	50° 15·5′		log cos.	9·80572	
$\frac{\text{Dep.}}{\text{D. Lat.}}$ = tan Co.	Dep.			log	3·07779	3·07779
	D. Lat.	31′ N.		log	1·49136	
Dist. = D. Lat. × tan Co. × cosec Co.	Co.	N. 88° 31′ E.	←	log tan	1·58643	
or				log cosec	0·00015	0·00015
Dist. = Dep. × cosec Co.	D. Lat.	31′ N.		log	1·49136	
	Dist.	1196·58′	←	log	3·07794	3·07794

Answer (ii) Course 088° 31′ T., Distance 1196·58 miles (by Mean Latitude Sailing).

By Mercator Sailing	Co.	088° 31′ T.	Dist.	1199·75 miles
„ Mean Latitude Sailing	„	088° 31′ T.	„	1196·58 „
Answer (iii) Difference.		Nil		3·17 miles

N.B.—Using advanced techniques the actual distance between the two positions has been calculated as being 1200·2 International Nautical Miles (6076·12 feet). In this case the Mercator Sailing gives very satisfactory results.

A LONG DISTANCE TRAVERSE.

Example.—A vessel set out from Lat. 20° N., Long. 30° W., and steered 315° for 600 miles, then 045° for 1200 miles, and again 315° for 600 miles. Required the position arrived at.

This type of question is intended to draw attention to the limitations of plane sailing for long stretches of distance in computing the longitude of the position arrived at.

If this example were worked out as a traverse we would get the latitude of arrival quite correctly but not the longitude, because the several East and West departures would cancel each other and indicate that the vessel had made no easting or westing which is not the case. For example—

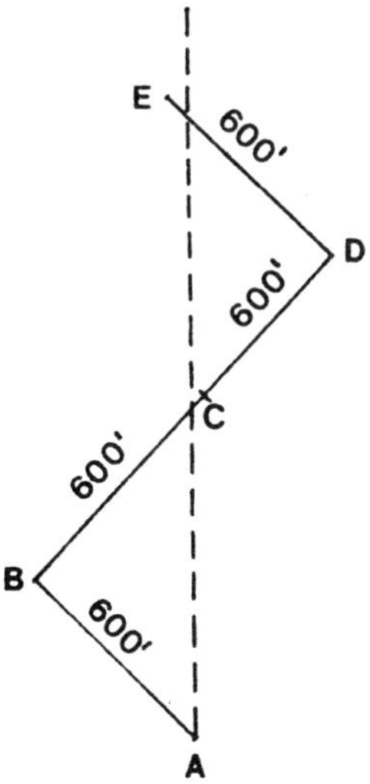

Courses	Dist.	D. Lat.	Dep.
315° T.	600	424·3 N.	424·3 W.
045° T.	1200	848·6 N.	848·6 E.
315° T.	600	424·3 N.	424·3 W.
	D. Lat.	1697·2 N.	Dep 0·0

Lat. left	20° 00′ N.	Long. left	30° 00′ W.
D. Lat.	28 17·2 N.	Dep. 0, D. Long.	00 W.
Lat. arrd.	48 17·2 N.	Long. arrd.	30 00 W. which is wrong.

The departure in this case being zero we might assume the D. Long. to be zero which would be wrong.

We shall now work the question by inspection using the mean latitudes to convert the several departures into their corresponding differences of longitude.

			Co.	Dist.	D. Lat.	Dep.
Refer to figure,	1st Course	*A* to *B*	315° T.	600	424·3 N.	424·3 W.
	2nd ,,	*B* to *C*	045° T.	600	424·3 N.	424·3 E.
	3rd ,,	*C* to *D*	045° T.	600	424·3 N.	424·3 E.
	4th ,,	*D* to *E*	315° T.	600	424·3 N.	424·3 W.

	Lat.	Mean Lat.	Dep.	D. Long. by Inspection	Long.
A	20° 00·0′ N.				30° 00·0′ W.
D. Lat.	7 04·3 N.	23° 32·2′	424·3	462·8	= 7 42·8 W.
B	27 04·3 N.				37 42·8 W.
D. Lat.	7 04·3 N.	30 36·6	424·3	492·5	= 8 12·5 E.
C	34 8·6 N.				29 30·3 W.
D. Lat.	7 04·3 N.	37 40·8	424·3	536·1	= 8 56·1 E.
D	41 12·9 N.				20 34·2 W.
D. Lat.	7 04·3 N.	44 45·1	424·3	597·4	= 9 57·4 W.
E	48 17·2 N.				30 31·6 W.

Position reached, E. Lat. 48° 17·2′ N., Long. 30° 31·6′ W., not 30° 00′ W., as above.

Note.—The departures are the same but the corresponding differences of longitude increase as the latitude increases.

Example.—At 1200 L.M.T., Lat. 35° 10′ S., Long. 164° 30′ E., ship steaming 268° C. (deviation 8° E., variation 14° E.) at 10 knots until 2000 hrs. when a distress call signal was received from a ship in Lat. 37° 15′ S., Long. 159° 30′ E., urgently asking for assistance. Find the true course to steer and the G.M.T. on arrival at the ship if speed is increased to 12 knots.

	Lat.		Long.
1200 position	35° 10′ S.		164° 30′ E.
(290° T. 80 miles) D. Lat.	27·4 N.	D. Long.	1 32 W.
My position at 2000	34 42·6 S.		162 58 E.
Her position	37 15·0 S.		159 30 E.

Course by Mercator 228° T., dist. to go 227·7 miles at 12 knots = steaming time 18h 54m.

L.M.T. - - - - - - - - - -	0d 12h 00m
Long. 164° 30′ E. - - - - - - -	— 10 58
G.M.T. - - - - - - - - - -	0 01 02
Steaming time (8h + 18h 54m) = - - - -	1 2 54
G.M.T. on arrival at vessel - - - - -	1 03 56

TERRESTRIAL AND CELESTIAL POSITION LINES COMBINED.

Ushant 33 M light (Lat. 48° 27·7′ N., Long. 5° 07·7′ W.) bearing 184° T., approx. distance 20 miles. Ship then steered 052° C. (deviation 5° E., variation 17° W.) for 30 miles by log; estimated set and drift of current 210° T., 4 miles. Sights of a star were then worked with this E.P. and gave azimuth 330° T., intercept 6′ away. Required the ship's Observed Position at sights.

(1) Work a three course traverse from Ushant and find the E.P. for working the star sights.

The Traverse. See Figure.		Distance	D. Lat.	Dep.
Reversed Bearing	004° T.	20 miles	20·0′ N.	1·4′ E.
Course	040° T.	30 „	23·0 N.	19·3 E.
Current	210° T.	4 „	3·5 S.	2·0 W.
A to *C*, Co. and Dist.	025° T.	43·6 „	39·5 N.	18·7 E.

	Lat.		Long.
Ushant Position *A*	48° 27·7′ N.		5° 07·7′ W.
025° T., 43·6 miles D. Lat.	39·5 N.	Dep. 18·7 = D. Long.	28·5 E.
Estimated Position *C*	49 07·2 N.		4 39·2 W.
C to F			
184° T., 7·2 miles D. Lat.	7·2 S.	D. Long.	0·8 W.
Ship's Observed Position *F*	49 00·0 N.		4 40·0 W.

In Figure—

A B is the reversed bearing and approx. distance of ship from Ushant.

B C is the course and distance made good.

C is the E.P.

P′ L′ the transferred position line of Ushant 004°—184°.

C D the intercept 6′ away.

D the Intercept Terminal Position.

P L the star position line.

F the ship's position.

(2) Now obtain the ship's position by solving triangle *CDF* and finding distance *CF* by inspection of the Traverse Tables. The intercept is Away so reverse the azimuth (330° T.) to 150° T.

In triangle *CDF*, *C*=34° (184°—150°), use as Co; *CD*=6′ use as D. Lat.; *CF*=7·2′ found under Dist.

(3) Now *C* to *F* is 184° T. distance 7·2′ giving D. Lat. 7·2′ S. and Dep. 0·5′ W.

(4) Using Mean Lat. 49° and Dep. 0·5′ gives D. Long. 0·76′ W. (0·8′ W.).

(5) Apply D. Lat. 7·2′ S. and D. Long. 0·8′ W. (found at (3) and (4) above) to the E.P. found from the three course traverse. This is the ship's Observed Position (*F*) at the star sight.

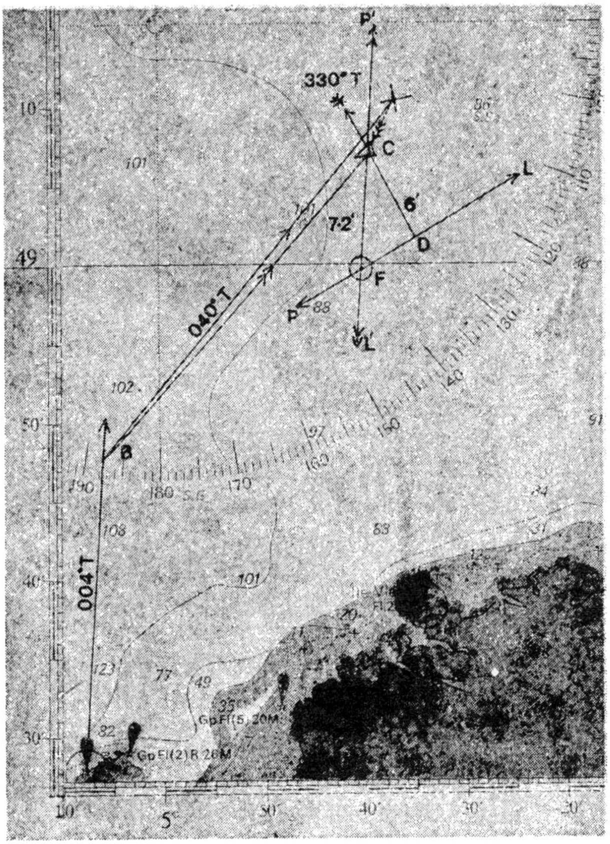

Reproduced by kind permission of the Hydrographer of the Navy.

Though, in this case, all plotting could have been carried out on the chart, most navigators use a combination of plotting and Traverse Table inspection to work out star sights.

On small scale charts it is seldom possible to plot sights accurately.

Examples for Exercises.

SAILINGS—PLANE, PARALLEL AND MERCATOR.

1. Two ships are 28 miles apart in Lat. 44° S., both steam due North keeping due East and West of each other until they are 37 miles apart. How far have they sailed?
2. Given course 280° T., dep. 198 miles, D. Long. 240′. Required the latitudes and the distance sailed.
3. Given course 310° T., D. Long. 10°, left Lat. 57° 05′ N. Find the latitude reached.
4. Given course 210° T., D. Long. 11° 30′, left Lat. 37° 10′ S. Find the latitude reached and distance sailed.
5. From the equator sailed a 210° T. course and altered Longitude 12° 10′. Find the latitude reached and distance sailed.
6. A vessel sailed from Lat. 17° 11′ N. to Lat. 16° 10′ N., and changed her Longitude 7° 13′ W. Find the distance sailed.
7. A vessel left Lat. 40° 30′ N., Long. 48° 30′ W. and arrived in Lat. 48° 30′ N. after making 600 miles of westing. Find the longitude reached.
8. Find the true course and distance from Lat. 37° 14′ N., Long. 30° 00′ W., to Lat. 31° 26′ N., Long. 34° 00′ W.
9. Given Lat. 42° 20′ N., course 270° T. for 3 days 2 hours, D. Long. 17° 10′. Find the vessel's speed.
10. Ship *A* in Lat. 17° S., Long. 20° W., steers due West at 17 knots. A ship *B* in Lat. 18° S., Long. 21° W., wishes to intercept *A* in 4 hours. Find her course and speed to be able to do so.
11. From Lat. 52° 30′ N., Long. 168° 20′ W., a ship steered 215° T., 500 miles, then 135° T., 500 miles. Find the position arrived at.
12. From Lat. 50° S., Long. 70° E., a ship steered 050° T., 1000 miles, required the position reached.
13. On 8th May, at noon, apparent time at ship, in Lat. 39° 16′ N., Long. 72° 51½′ W., find the mercatorial course and distance to a position 10 miles, 180° T. from Fastnet (Lat. 51° 23½′ N., Long. 9° 36½′ W.), also the apparent time at ship on arrival allowing a speed of 10 knots.
14. On 11th January, at noon, apparent time at ship, in Lat. 49° 17′ N., Long. 45° 50′ W., steaming 260° at 15½ knots. At 7h 30m P.M., apparent time at ship, an S.O.S. was received from a disabled ship in Lat. 46° 00′ N., Long. 47° 20′ W. Find the true course and distance to steam and the G.M.T. of arrival at your maximum speed of 17½ knots.

BOW AND BEAM BEARINGS WITH CURRENT.

At 10 P.M. a point in Lat. 43° 10′ N., Long. 9° 13′ W., bore 4 points on the port bow, and at 11 P.M. it was abeam, ship steaming 15 knots. Compass course 231°, wind N.W., leeway 4°, variation 19° W., deviation 6° E., current 110° T. at 3 knots Required the true bearing, distance off when abeam, and the ship's position.

1. Correct the compass course for deviation and variation to get the true direction of ship's head from which to get the directions of the bow and beam bearings.
2. Resolve a two course traverse to find the course and distance made good during the interval between the bearings, viz., the compass course corrected for deviation, variation and leeway, if any, and the set and drift of current.
3. The question now resolves itself into solving a running fix either by calculation, projection on a chart or on squared paper.

Refer to figure and note that—

A is the point of land.
B the position of ship at first bearing.
B A the bow bearing and *A C* the beam bearing.
B C the true direction ship is heading.
B D the true course adjusted for leeway.
B F the course and distance made good between the bearings.
F is the ship's position relatively to *A*.

To Find the Bow and Beam Bearings.

Course 231° C. (deviation 6° E., variation 19° W.) gives ship's head 218° T.

Apply 45° and 90° to the left of 218° T. to get the bearings *A B* and *A C*, viz., 173° and 128°, to be laid off on the chart.

To Find the Course and Distance made Good between the Bearings.

Ship's head 218° T., leeway 4°, 214° T.

		Distance	D. Lat.	Dep.
B D course	214° T.	15 miles	12·4′ S.	8·4′ W.
D F current	110° T.	3 ,,	1·0 S.	2·8 E.
Course made good	203° T.	14·5 miles	13·4 S.	5·6 W.

The above work need not be shown either on the chart or on squared paper unless asked for.

The Question now Reads.—A point of land bore 173° T. and after making good 203° T., 14·5 miles, the point bore 128° T. Find the ship's distance from the point and her latitude and longitude.

The heavy lines on the figure represent the actual charting to be laid off as a running fix. *A F* is the distance from the point, 10·2 miles by measurement. It is now required to transfer position *A* to *F*.

	Lat.		Long.	
Position *A*	43° 10′ N.		9° 13′ W.	
308° T. 10·2′ mls. = D. Lat.	6·3 N.	D. Long.	11 W.	Dep. 8·0
Position *F*	43 16·3 N.		9 24 W.	

When a chart is not available the triangle *A B F* being oblique may be solved as follows:—

Given *B F*=14·5 miles, ∠*A*=45°, ∠*F B A*=30°, find side *A F*.

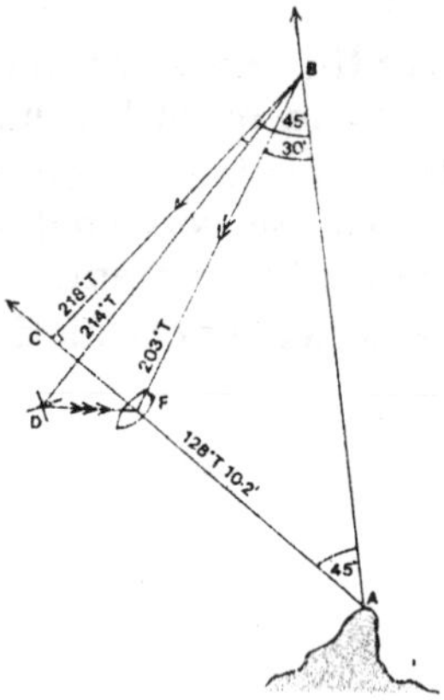

$$\frac{A\ F}{\sin B} = \frac{B\ F}{\sin A} \quad \therefore A\ F = B\ F \sin B \operatorname{cosec} A$$
$$= 14{\cdot}5 \sin 30° \operatorname{cosec} 45°$$

14·5	log	1·161368
30°	sin	9·698970
45°	cosec	10·150515
A F 10·2 =	log	1·010853

The ship is therefore 10·2 miles, 308° T. from *A*.

		Lat.	Long.
308° T., 10·2′ *A*		43° 10′ N.	9° 13′ W.
D. Lat. 6·3′ N.		6·3 N.	11 W.
Dep. 8·0′ W.			
D. Long. 11′ W.	*F*	43 16·3 N.	9 24 W

Some students may prefer to find the distance *A F* by inspection from the Traverse Tables by dropping *F H* perpendicular to *A B*.

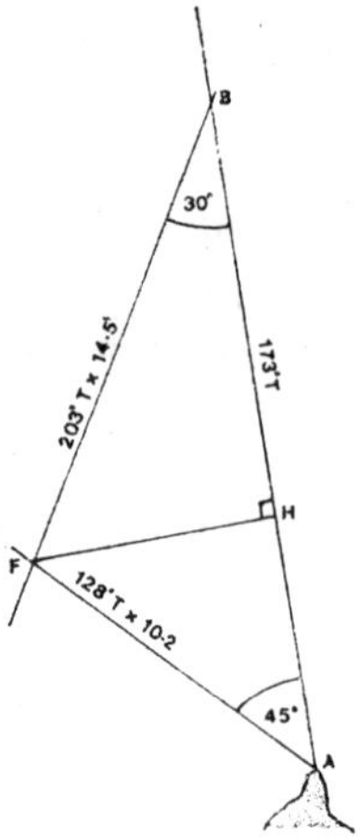

In △ *F B H*, *B F*=14·5 (dist) ∠*B*=30° (course), *F H*=dep (7·25).

In △ *A H F*, ∠*A*=45° (course), *H F*=7·25 (dep), *A F*=dist (10·2).

Note.—The problem may be set giving the leeway in knots as in Examples 5 and 6, in which case it will be applied as if it were a course dead before the wind. There would then be a three course traverse to solve to get the course and distance made good, viz., true course steered and speed, set and drift due to leeway and the set and drift due to current. When leeway is left out the case is simplified to that extent.

Examples for Exercise.

1. At 1625 hrs. Dungeness lighthouse (Lat. 50° 54·9′ N., Long. 00° 58·4′ E.), bore 4 points on the starboard bow and at 1713 hrs. it was abeam. Course 230° C. (deviation 4° E., variation 14° W.), speed 10 knots, wind West, leeway 8°, tidal stream 250° T., 3 knots. Find the distance off when lighthouse was abeam and the ship's position on the beam bearing.

2. Dungeness lighthouse (Lat. 50° 54·9′ N., Long. 00° 58·4′ E.) bore 45° on the port bow at 0820 hrs., log 37, and at 0920 hrs. it was abeam, log 48. The ship steamed East by compass (deviation 6° W., variation 14° W.), wind N.E., leeway 5°, tidal stream 222° T., 4 knots. Find the ship's position on the beam bearing.

3. Beachy Head lighthouse (Lat. 50° 44′ N., Long. 00° 15′ E.) bore 4 points on the port bow at 2220 hrs. and at 2327½ hrs. it was abeam. Course 086° C. (deviation 20° E., variation 14° W.), speed 8 knots, wind North, leeway 4°, tidal stream West (magnetic), 2½ knots. Find the distance off on the beam bearing and the ship's position.

4. Cape Gris Nez (Lat. 50° 52·2′ N., Long. 01° 35′ E.) bore 45° on the port bow and after steaming 7 miles it was abeam. Course 195° C. (deviation 14° E., variation 14° W.), wind N.W., leeway 5°, tide set 300° M., 2·5 miles during the interval. Find the beam distance and ship's position.

5. South Foreland light (Lat. 51° 08½′ N., Long. 01° 22½′ E.) bore 45° on the starboard bow and after steaming for one hour at 9 knots it was abeam. Course 207° C. (deviation 7° E., variation 14° W.), wind West, leeway 2 knots, current 050° T., 3 knots. Find the ship's position. (See Note on p. 9.)

6. North Foreland light (Lat. 51° 22½′ N., Long. 01° 26·8′ E.) bore 4 points on the port bow, log 17, and after 30 minutes steaming it was abeam, log 21. Course 324° C. (deviation 10° E., variation 14° W.), wind S.W., leeway 2 knots, tide North (magnetic), 5 knots. Find ship's position.

7. Flamborough Head (Lat. 54° 7′ N., Long. 0° 4·6′ W.) bore 310° T., and after steaming 340° T., at 9 knots for 40 minutes it bore 240° T. Find the ship's position allowing for a tidal stream setting 010° T., 3 knots.

8. Needles light (Lat. 50° 39·7′ N., Long. 1° 35·4′ W.) bore 328° C., and after steaming 275° C. (deviation 12° E., variation 17° W.) for 3 hours at 6 knots, it bore 043° C. by the same compass; current setting 107° Magnetic, 2 knots. Find the ship's position on the second bearing.

9. From an observation of the sun, the position line was 120°—300° T., and it was drawn through Lat. 46° 28′ N., Long. 179° 58′ W. The ship then steered 292° T. for 20 hours at 10 knots through a current setting 180° T. at 2 knots when a point of land in 47° 03′ N., 176° 13′ E., bore 050° T. Find the ship's position.

10. Ship *A*, in Lat. 17° S., Long. 0°; ship *B* in Lat. 18° S., Long. 1° W. *A* steers due west at 17 knots. Find *B*'s course and speed to meet *A* in 4 hours.

11. Ship in position *A* Lat. 31° 40′ N., Long. 31° 41′ E., at apparent noon arrives at *B*, Lat. 33° 20′ N., Long. 24° 12′ E., at apparent noon next day. Find true course and average speed noon to noon.

12. At noon A.T.S. Jan. 31, Lat. 30° 15′ N., Long. 162° 47′ E.
At noon A.T.S. Feb. 1, Lat. 30° 20′ N., Long. 168° 32′ E.
Find steaming distance and average speed from noon to noon.

13. In D.R. Lat. 51° 10′ N., obs. Long. was 10° 14′ W., the Sun bearing 220° T. The ship then ran 090° for 24 miles when a point in Lat. 51° 23′ N., Long. 9° 36′ W., bore 070° T. Find the ship's position.

CHAPTER II.

THE MAGNETIC COMPASS—VARIATION AND DEVIATION.

THE error which can affect a magnetic compass can be divided into two parts, (*a*) Variation and (*b*) Deviation.

Variation varies with location and time. The earth's magnetic field is not the simple pattern shown in experiments using a bar magnet. However, if this type of field is compared with the actual earth's field, it will be found that the differences have a definite geographical distribution which coincides with currents within the earth. To find variation, charts showing curves of equal variation (isogonals), or the nearest information on the navigational chart, should be used and up-dated for the current year. On standard charts details of variation are included either in the compass rose or on printed isogonals.

Deviation varies essentially with the ship's head but also to some extent with time and location. The earth's field induces magnetism within the ship which in turn creates a secondary field. The ship's aspect with respect to the earth's magnetic field determines the amount of deviation. That is, deviation varies with the course. Deviation is made up of two main components—permanent magnetism associated with "hard" iron and temporary magnetism associated with "soft" iron. In the case of permanent magnetism, some parts of the ship's structure retain their magnetic characteristics for a very long time. In the case of temporary magnetism, the deviation will change, at least in theory, every time the ship changes course.

Changes in the "permanent" magnetism do take place slowly but the process can be accelerated by such factors as large structural alterations or completely change if the ship is struck by lightning. Permanent deviation is corrected by permanent compensating magnets. However, the strength of the earth's field varies with latitude and this alters the directive force of the compass which, in turn, affects the compensation of the permanent magnets.

The temporary component of magnetism results in a time lag between altering course and the "soft" iron adjusting to the change in the field. The time to adjust may range from a few minutes to a week. In the latter case the ship has probably been on a long straight course presenting the same aspect to the earth's field for a considerable time, e.g. two weeks.

The compasses are normally swung approximately once a year by a professional adjuster. The deviations are recorded on a card usually for every 10° of compass course. The watchkeeping officers should, when possible, check the deviation at each course alteration and at least once every watch when on a steady course. Each observation together with the date, ship's head and deviation should be recorded in a book and compared with the card prepared by the adjuster. Whenever a significant difference is noted, the record of observations should be analysed and it may be considered necessary to swing the ship during the passage.

In coastal waters the best method is to select a transit of fixed objects about 7 to 8 miles away. At this distance the objects should remain in transit whilst the ship is being swung. The compass bearing of the transit must be taken at equidistant headings, preferably every 10 degrees. The magnetic bearing of the transit is then compared with compass bearing on each heading. In the open ocean azimuths of a heavenly body are substituted for the transits.

When in coastal waters, where a suitable transit is not available, the compass can be swung using a single fixed object, again 7 to 8 miles away. This method assumes that the magnetic field of the ship is symmetrical and that on reciprocal headings the deviations will be equal and opposite. In the majority of cases this is nearly true for merchant ships. The bearings are taken on equidistant headings as for the transits but in this case the compass bearings are summed up and then divided by the number of observations to obtain an average bearing. This average bearing is then taken as the magnetic bearing and used for determining the deviation on individual compass headings.

To Find the Magnetic Bearing of a Distant Object.

Worked Example.

Having taken the following bearings of a distant object by the standard compass, find the magnetic bearing of the object and thence the deviations.

Ship's head by Standard Compass.	Bearing of object by Standard Compass.	Deviation.	Ship's head by Standard Compass.	Bearing of object by Standard Compass.	Deviation.
000°	128°	10° E.	180°	146°	8° W.
010°	130°	8° E.	190°	144½°	6½° W.
020°	132°	6° E.	200°	143°	5° W.
030°	134½°	3½° E.	210°	141°	3° W.
040°	137½°	½° E.	220°	139°	1° W.
050°	140°	2° W.	230°	137°	1° E.
060°	142½°	4½° W.	240°	135°	3° E.
070°	145°	7° W.	250°	132½°	5½° E.
080°	147°	9° W.	260°	130½°	7½° E.
090°	148°	10° W.	270°	128½°	9½° E.
100°	149½°	11½° W.	280°	127°	11° E.
110°	151°	13° W.	290°	125°	13° E.
120°	152°	14° W.	300°	124°	14° E.
130°	153°	15° W.	310°	123°	15° E.
140°	152½°	14½° W.	320°	124°	14° E.
150°	151°	13° W.	330°	124½°	13½° E.
160°	149½°	11½° W.	340°	125½°	12½° E.
170°	148°	10° W.	350°	127°	11° E.

The sum of the bearings of the distance object = 4968.

The average bearing = 4968 ÷ 36 = 138°.

Therefore, the magnetic bearing is 138° and the deviations are then inserted in column three.

Note if the bearings of the distant object lie both sides of North add 360° for those in the North-East quadrant, e.g. 010° write as 370°. Then sum the bearings to obtain the average bearing and, if necessary, subtract 360°.

Use of the Deviation Card and Deviation Curve.

Deviations can be presented either on a Deviation Card, in tabular form or on a Deviation Curve, in graphical form. In both cases the ship's head by compass is normally shown at intervals of ten degrees.

When using a deviation card to obtain the deviation for an intermediate heading it becomes necessary to interpolate between the two values which lie either side of the required value. Interpolation is effected by linear ratios.

When using a deviation curve, the deviation for a particular compass heading can be picked off directly and no interpolation is required. Providing the curve has been carefully drawn and the deviation picked off with equal care, the deviation curve is likely to give the best results but, in practice and D.o.T. Examinations, deviation cards are more generally used.

When using deviation cards it is good practice to construct a third column of magnetic headings which are listed opposite their corresponding deviations. Thus, when reading from left to right, we have the compass heading, deviation and magnetic heading shown alongside one another. The card can now be used to quickly convert courses from either compass to magnetic or from magnetic to compass.

In all cases, very careful interpolation, preferably to the nearest $\frac{1}{4}°$ or 0·2°, is essential. The final result may then be rounded up or down to the nearest $\frac{1}{2}°$ which is the degree of accuracy required in D.o.T. Examination answers.

EXAMPLE 1. (Using deviation card No. 1)

Find the magnetic course for a course of 228° C.

From the card.

Ship's Head by Compass.	Diff.	Dev.	Diff.
220°	10°	8° E.	2°
228°	2°	y°	x°
230°		10° E.	

$\frac{x°}{2°} = \frac{2°}{10°}$ so $x = \frac{4}{10} = 0{\cdot}4° \simeq \frac{1}{2}°$

10° E. — $\frac{1}{2}°$ = $9\frac{1}{2}°$ E.

228° C. + $9\frac{1}{2}°$ E. = $237\frac{1}{2}°$ M. *Answer.*

EXAMPLE 2.

Find the compass course for a course of $237\frac{1}{2}°$ M.

Ship's Head by Compass.	Dev.	Diff.	Mag.	Diff.
220°	8° E.	2°	228°	12°
	y	x°	$237\frac{1}{2}°$	$2\frac{1}{2}°$
230°	10° E.		240°	

$\frac{2\frac{1}{2}°}{12°} = \frac{x°}{2°}$ so $x = \frac{5}{12} \simeq \frac{1}{2}°$

10° E. — $\frac{1}{2}°$ = $9\frac{1}{2}°$ E.

$237\frac{1}{2}°$ M. — $9\frac{1}{2}°$ E. = 228° C. *Answer.*

Using Sun Azimuths to Construct a Table of Deviations.

When a vessel is "swung" using sun azimuths to find the deviation, the procedure is much the same as when using a distant object. In this case, however, it will be necessary to note the G.M.T. of each azimuth which is then worked out using ABC tables.

The ABC tables when entered with latitude and hour angle and declination of the sun, give the true azimuth. The angle between the compass bearing and the true azimuth of the sun gives the error, and the difference between the error and the variation is the deviation for the direction of the ship's head at the time.

Worked Example.

On 8th May in Lat. 49° N., Long. 5° W., the variation being 16° W., the vessel's head was steadied on various compass headings in order to find the deviations. The compass bearings of the sun were taken at the times indicated.

Latitude 49° N., Sun Declination 17° N. Refer to ABC Tables.

Head by Compass	8th May G.M.T.	Bearing by Compass	True Az. from Tables	Error of Compass	Var.	Dev.
000° C.	0920 hrs.	128° C.	115·2° T.	12·8° W.	16° W.	3·2° E.
045°	0928	125°	117·3°	7·7° W.	,,	8·3° E.
090°	0936	122°	119·4°	2·6° W.	,,	13·4° E.
135°	0944	132°	121·5°	10·5° W.	,,	5·5° E.
180°	0952	141°	123·8°	17·2° W.	,,	1·2° W.
225°	0900	150°	126·0°	24·0° W.	,,	8·0° W.
270°	0908	159°	128·4°	30·6° W.	,,	14·6° W.
315°	0912	154°	130·9°	23·1° W.	,,	7·1° W.

Examples for Exercise.

(1) You have taken the following compass bearings of a distant object. Find its magnetic bearing, and thence the deviation on those points on which the bearings were observed.

Ship's head by Standard Compass.	Bearing of object by Standard Compass.	Deviation required.	Ship's head by Standard Compass.	Bearing of object by Standard Compass.	Deviation required.
000°	255°		180°	214°	
045°	244°		225°	211°	
090°	236°		270°	249°	
135°	230°		315°	251°	

Required: (*a*) The Magnetic Bearings, (*b*) the Table of Deviations, and

(*c*) the Compass Courses to steer in order to make the following magnetic courses: 298° M, 158° M, 292° M, 139° M, and

(*d*) Use the Table of Deviations to find the Magnetic Courses corresponding to the following Compass Courses:

022° C 067° C 203° C 247° C.

The following examples are not given in full, as the foregoing example is sufficient to show the method and order in which this problem may be set at the examinations.

(2)

Ship's head by Standard Compass.	Bearing of distant object by Standard Compass.	Deviation required.	Ship's head by Standard Compass.	Bearing of distant object by Standard Compass.	Deviation required.
000°	160°		180°	182°	
045°	154°		225°	186°	
090°	162°		270°	181°	
135°	178°		315°	175°	

Required: (*a*) The Magnetic bearing,

(*b*) the Compass Courses to make the following Magnetic Courses: 022° M and 292° M, and

(*c*) With the ship's head on 270° C. find the Magnetic bearings which correspond to the following Compass bearings: 045° C, 315° C.

(3)

Ship's head by Standard Compass.	Bearing of distant object by Standard Compass.	Deviation required.	Ship's head by Standard Compass.	Bearing of distant object by Standard Compass.	Deviation required.
000°	003°		180°	357°	
045°	343°		225°	009°	
090°	340°		270°	021°	
135°	345°		315°	022°	

Required: (*a*) The Magnetic bearing,

(*b*) the Table of Deviations, and

(*c*) the Magnetic Courses to make the following Compass Courses: 078° C and 275° C.

(4) On the 28th September in Lat. 49° N., Long. 5° W., the vessel's head was steadied on various Compass Headings in order to find the Deviations. Variation was 30° W. and Declination Sun was 2° S. Compass bearings of the Sun were taken as follows:

Head by Compass	G.M.T.	Bearing by Compass	Head by Compass	G.M.T.	Bearing by Compass
000° C.	0908 hrs.	132° C.	180° C.	0940 hrs.	132° C.
045°	0916	140°	225°	0948	127°
090°	0924	142°	270°	0956	128°
135°	0932	139°	315°	1004	137°

Required: The Table of Deviations.

Deviation Card No. 1

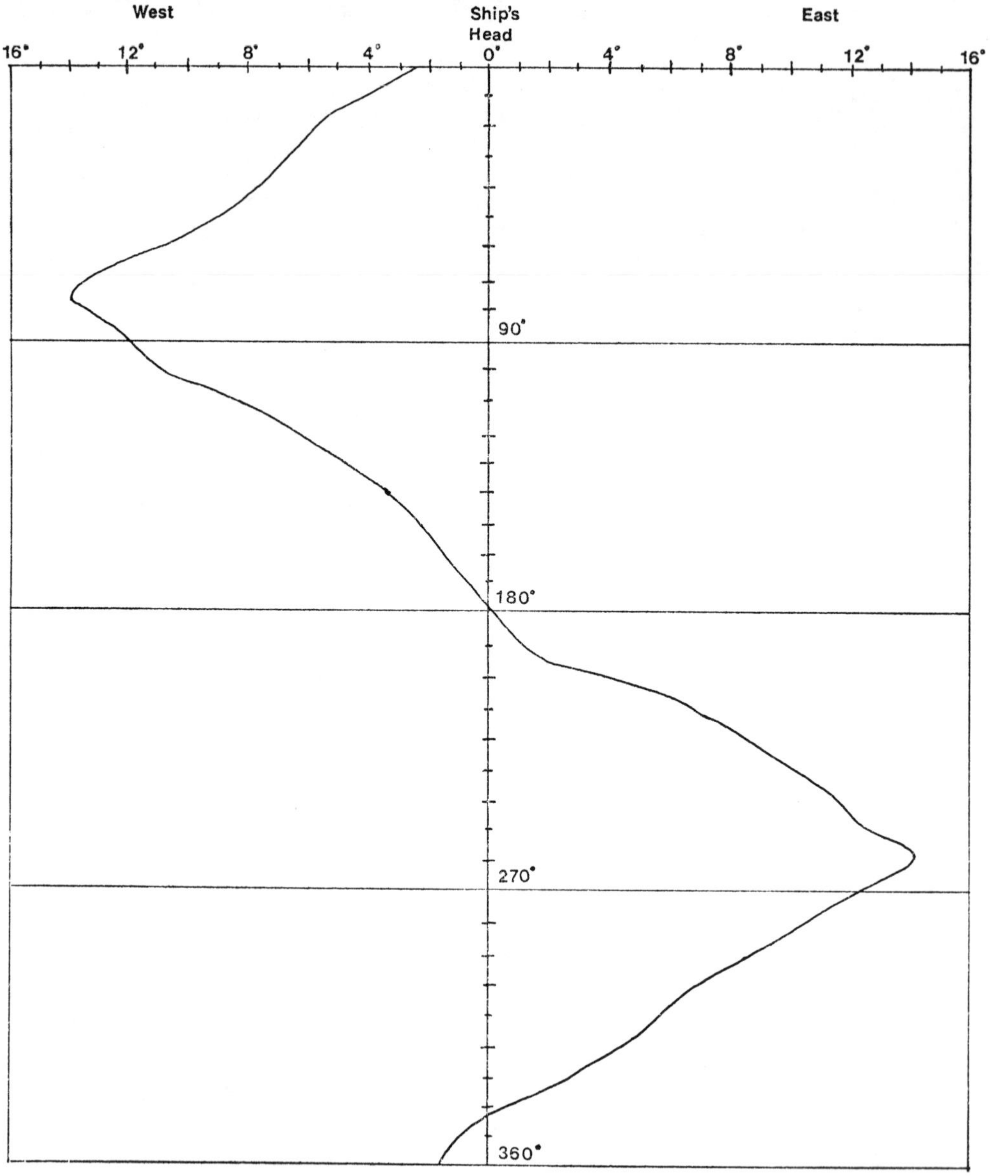

[To Face Page 17

DEVIATION CARD No. 1.

Ship's Head by Compass.	*Deviation.*
000°	2° W.
010°	3½° W.
020°	5½° W.
030°	6½° W.
040°	7½° W.
050°	8½° W.
060°	11° W.
070°	13½° W.
080°	13½° W.
090°	12° W.
100°	11° W.
110°	8½° W.
120°	6¼° W.
130°	5° W.
140°	3½° W.
150°	2½° W.
160°	1¾° W.
170°	1° W.
180°	0°
190°	1° E.
200°	4° E.
210°	6½° E.
220°	8° E.
230°	10° E.
240°	11½° E.
250°	12½° E.
260°	14° E.
270°	12° E.
280°	10¼° E.
290°	8½° E.
300°	6¾° E.
310°	5½° E.
320°	4° E.
330°	2½° E.
340°	½° E.
350°	1° W.
360°	2° W.

TO FIND THE COURSE TO STEER TO BRING A POINT A GIVEN DISTANCE OFF AT A GIVEN ANGLE ON THE BOW.

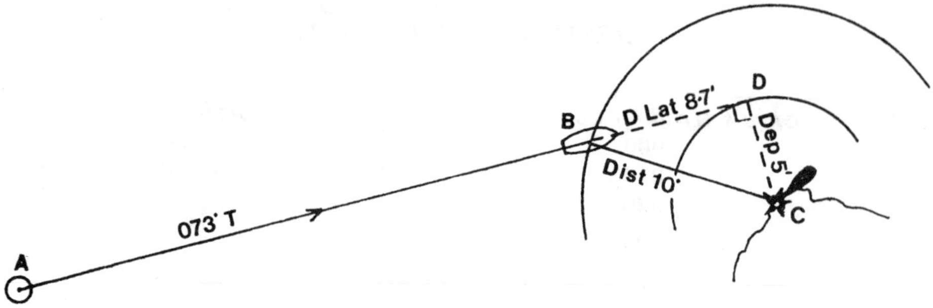

ENGLISH CHANNEL CHART.

Example.—Find the course to steer from (*A*) Lat. 49° 37′ N., Long. 3° 22′ W., to fetch the Casquets light (*C*) 30° on the starboard bow when 10 miles off. Find also the distance off when it comes abeam and the distance to go to bring it on the bow bearing.

1. With centre Casquets (*C*) and radius 10 miles (*C B*) describe a circle.
2. Enter Traverse Table with 30° as course, 10 miles as distance and we get 5 miles in Departure Column.
3. With centre Casquets and radius 5 miles (*C D*) describe a circle.
4. From *A* draw (*AD*) a tangent to the 5 mile circle cutting the 10 mile circle at *B*. Angle *C B D* will be 30°.

Answers.—*A B* is the course to steer, 073° T, *C D* the beam distance 5 miles, *A B* the distance to go 55½ miles. The Traverse Table gives *D B* in difference of latitude column, 8·7 miles, should this distance be required.

Example.—Bristol Channel. Find the course from Breaksea Lightvessel to bring Bull Point 7 miles off when it is 40° on the port bow. Give also the distance to go to bring Bull Point on this bearing and the distance off when light comes abeam.

Answer.—Course 263° T, Distance to go 31 miles, Beam distance 4·5 miles.

TO FIND THE COURSE TO STEER TO BRING A POINT ABEAM USING THE SHIP'S RADIUS OF ACTION.

Example.—A ship steaming at 14 knots is at position *A*, 250° T, 30 miles from Lizard Head Point. Find the course to steer to bring the Point on the port beam in 1½ hours.

1. Plot position *A*.
2. The ship's radius of action in 1½ hours is 21 miles. With centre *A* and radius 21 miles describe a circle.
3. Draw a tangent to the circle from the Lizard (*C B*).
4. Join *A B* which is the required course of 116° T.

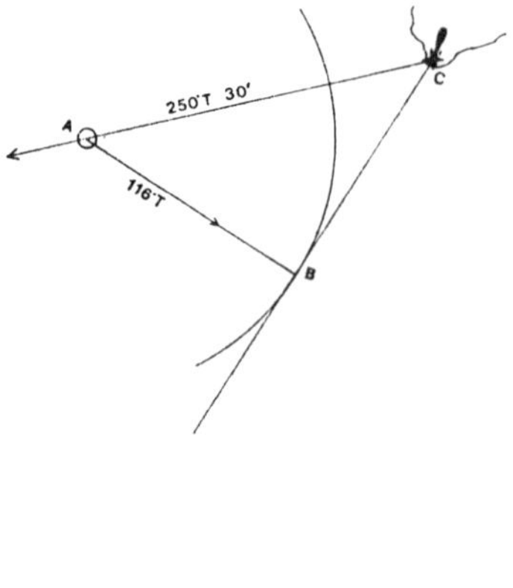

Given three true bearings of a point, the true course steered and the distance steamed between each bearing, and the ship's position on the first bearing.

To find the ship's position on the third bearing, the true course made good, and the set and drift since the first bearing was taken.

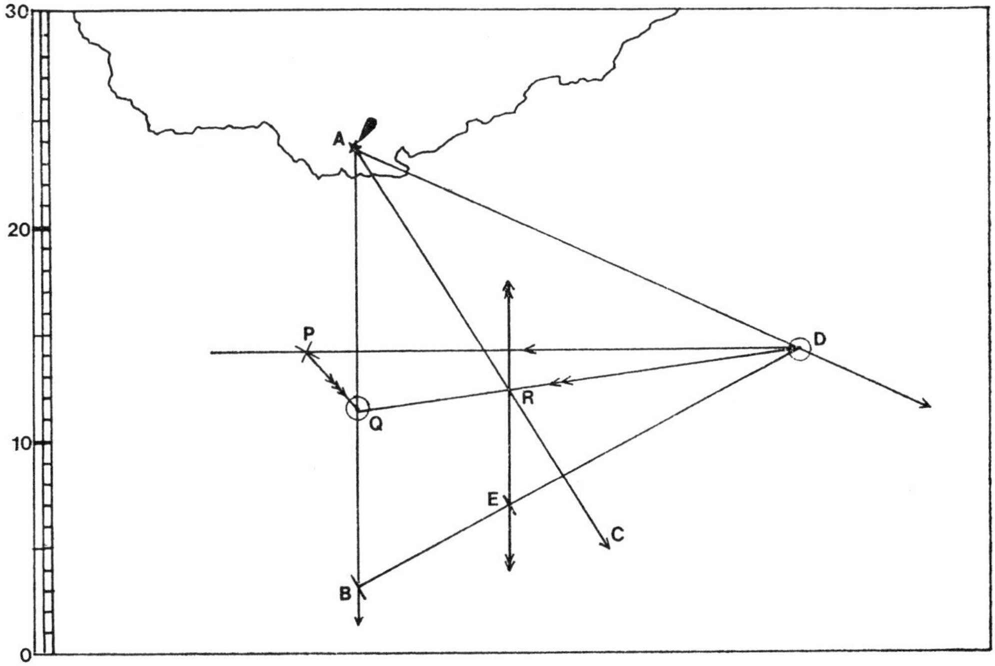

Example.—A vessel was steering 270° T. when she picked up a light *A* dipping on the horizon, distance off 23·8 miles, and bearing 295° T. After running 16 miles on her course it bore 328° T, and after running another 8 miles it bore 000° T.

Find the ship's position when abeam, the course and distance made good, and the set and drift since the light was first sighted.

Lay off all the true bearings from *A*, and on the first bearing make *AD*=23·8 miles, then *D* is the position of the ship on the first bearing, as she is 23·8 miles from the light. The ship steamed 24 miles between the first and third bearings, therefore with 24 miles on the compasses and the point of the compasses on *D* describe an arc cutting the third bearing at *B*. The distance steamed between second and third bearing was 8 miles, therefore from *B* make *B E*=8 miles. Draw a line parallel to the third bearing through *E*, and where this meets the second bearing at *R* is the ship's position on the second bearing. Join *D R* and extend it to *Q* on the third bearing. Then D *R Q* is the course the ship has made good, and *Q* her position on the third bearing, that is when abeam.

To find the set and drift of the current since the vessel was at *D*, lay off the true course and distance steamed from *D* by making *D P* 270° T. × 24 miles, then *P* is the ship's D.R. position and *Q* is where she actually is. *P Q* is, therefore, the set and drift since her position was ascertained at *D*.

Course made good 263° T.
Distance 21·7 miles.
Set 137° T.
Drift 3·5 miles.

Example.—A ship steering 275° T. observes point *A* to be bearing 307° T. After running 7 miles it bears 342° T. and after having done another 5 miles on the same course it bears 023° T. At the time of taking the third bearing point *J* was observed to bear 308° T. Required the ship's position when the third bearing was taken, the course and distance made good between the bearings, and the set and drift of the current.

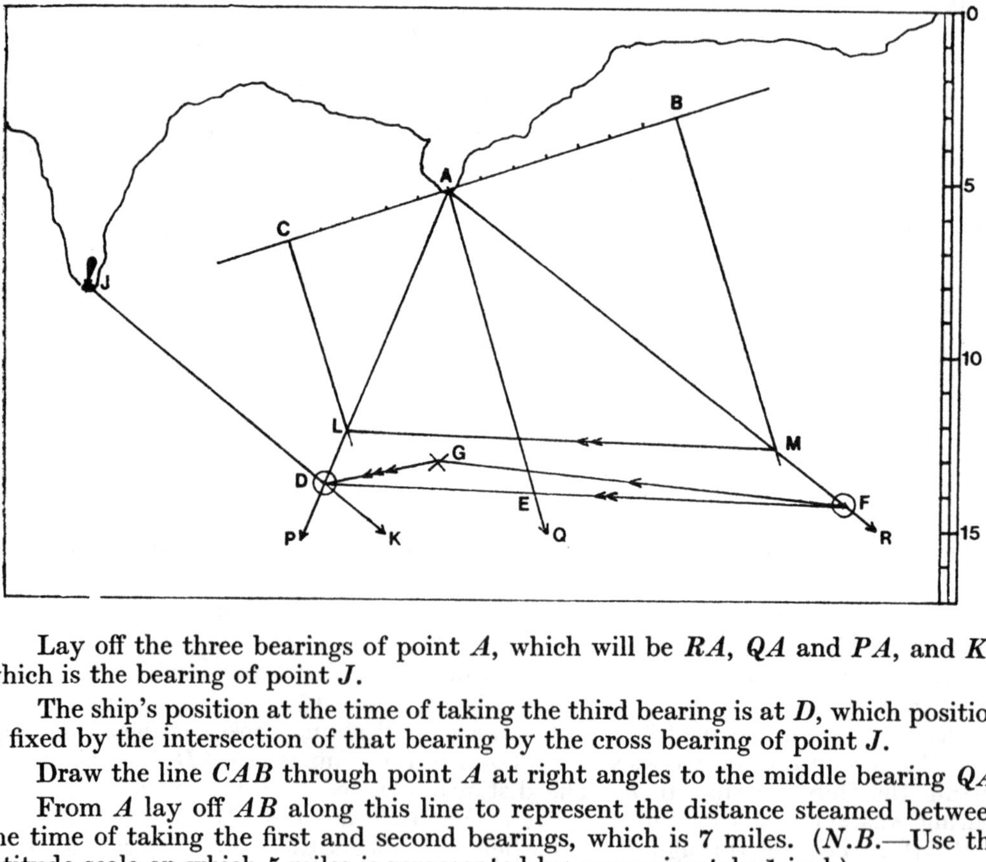

Lay off the three bearings of point *A*, which will be *RA*, *QA* and *PA*, and *KJ* which is the bearing of point *J*.

The ship's position at the time of taking the third bearing is at *D*, which position is fixed by the intersection of that bearing by the cross bearing of point *J*.

Draw the line *CAB* through point *A* at right angles to the middle bearing *QA*.

From *A* lay off *AB* along this line to represent the distance steamed between the time of taking the first and second bearings, which is 7 miles. (*N.B.*—Use the latitude scale on which 5 miles is represented by approximately 1 inch).

From *A* lay off *AC* along this line in the other direction. This represents the distance steamed between the time of taking the second and third bearings, which is 5 miles (1 inch).

Draw the lines *BM* and *CL* parallel to *AQ*.

These lines intersect the first and third bearings at *M* and *L* respectively.

Then *ML* represents the course which the ship is making good over the ground, though not necessarily the actual distance. In this case it certainly cannot be the distance, because the ship was somewhere on the first bearing *RA* at the start, and was at position *D* when the third bearing was taken at the finish.

Draw *FD* parallel to *ML*.

FD represents the actual track and distance made good over the ground between the times of taking the first and third bearings. This is about 272° T, 16 miles (3·2 inches).

Knowing that the ship was at *D* when the third bearing was taken she must have been at *E* when the second bearing was taken, and at *F* at the time of taking the first one.

Now, by laying off from *F* the course and distance **which she has steamed,** this will be 12 miles along the line *AB*, we should expect to find her at D.R. position *G*.

Having found that **her real position is at** *D*, the line *GD* must represent the set and drift of the current which is 259° T., 4 miles (0·8 inch).

Worked Example I.

(Worked on the St. George's Channel chart inset.)

Question 1.—**A vessel steering 187° C. observed Bailey Head light, height 81 feet, to be dipping on the starboard beam. Height of eye 22 feet. Required the true course, also the ship's position and her distance from Bailey Head light.** (Use Dev. Card No. 1).

Instructions for Working.—The extreme range (or dipping distance) at which a light is just visible is found by adding the distance of the sea horizon from the light to the distance of the sea horizon from the observer. In both cases it is either the height of light or height of eye which determines the distance of the respective sea horizons.

This method applies only to lights of adequate power and ignores meteorological conditions which affect visibility. Accuracy may also be affected by the fact that charted heights ignore fractions.

Distance of the sea horizon can be found using the formula $1{\cdot}17\sqrt{h}$ where h is the height of eye. In practice, extreme range is found by inspection of Distance of Sea Horizon Tables as follows:—

From tables:

Distance of Sea Horizon for Light 81 ft.	10·55	miles
,, ,, ,, ,, ,, Height of Eye 22 ft.	5·50	,,
Dipping Distance (off Bailey Head Lt.)	16·05	,,

With this distance as radius and Bailey Head Light as centre describe the circle *B P C*, the circumference of this circle will then be a **position line**—*i.e.*, the ship must be somewhere on this line. As Bailey Head Light was abeam on the starboard side we can get its true bearing by applying 90° to the right of the true course. Thus we have two position lines, the circle of equal distance and the true bearing, and as the vessel is on both of these lines, she must be at their point of intersection at *P*.

The ship's course being given by compass, it must be turned into a true one by applying the deviation (taken from the card), and the variation.

Compass course - -	187° C.		
Dev. from card - -	$+\frac{3}{4}$ E.		
Magnetic course - -	$187\frac{3}{4}$° M.	Ship's position	Lat., 53° 24′ N.
Corrected variation - -	—17 W.		Long., 5° $41\frac{1}{2}$′ W.
True course - - -	$170\frac{3}{4}$° T. → 171° T.		

From the ship's position at *P*, project the course line *P H* in a 171° T. direction.

question, a
rved to be
v Head Lt.
et and drift
l Lt.

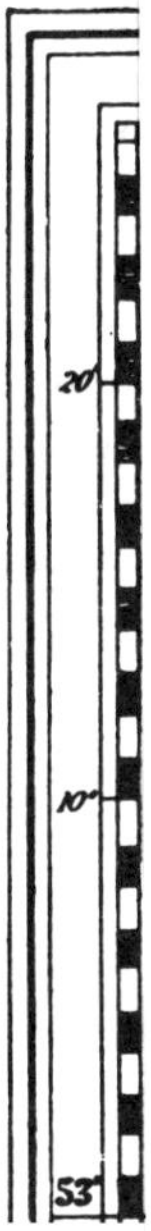

se line the
s gives the
ı be laid off
an and the
m the true

171° T.
85 (+)

256° T.

171° T.
115 (+)

286° T.

Position of
).R. to the

N.
W.

WORKED EXAMPLE I. (*Continued*).

Question 3.—**Whilst on the last-mentioned course Rosslare Point Lt. bore 60° on the starboard bow, and after running for 1h 15m at 8 knots the same light bore 20° abaft the starboard beam. Find the ship's position and her distance from Rosslare Point Lt. when the second bearing was taken, allowing for a current as ascertained in Question 2. Required also the distance from Tuskar Rock Lt. when abeam.**

Position of Rosslare Point Lt. : Lat. 52° 21′ N., Long. 6° 22·7′ W. Put it on the chart if it is not marked on your copy.

INSTRUCTIONS FOR WORKING.—Lay off the two bearings of Rosslare Point Lt. as explained when laying them off in Question 2. From any position (*K*) on the first bearing draw a line parallel to the ship's true course, and on it measure off from *K* to *L* the distance steamed in 1¼ hours, which is 10 miles.

From *L* lay off the current ascertained in the preceding question. This was 219° T. at 2·75 knots, which in 1¼ hours will be 3·45 miles. This will reach the mark *M*. Through *M* draw a line parallel to the first bearing. The point where this line intersects the second bearing at *N* is the ship's position at the time of taking that bearing. Now, *K M* is the course and distance **made good** between the time of taking the two bearings, and is the result of the course and distance steamed and the current experienced during that time. If we draw a line parallel to *K M* through *N* (the ship's position when the second bearing was taken) it will represent the vessel's **actual track** over the ground. This is the line *Q N O U*. The beam bearing of an object is always at right angles to the **direction of the ship's head,** which is represented by the line *K L*. We, therefore, draw a line at right angles to *K L*, and the point where this line (*O R* on chart) meets the actual track of the vessel at *O* will be the ship's position when abeam of Tuskar Light.

True course - - -	171° T.
Angle on bow - - -	60 (+)
First true bg. of Rosslare pt.	231° T.

True course - - -	171° T.
Angle on bow - -	110 (+)
	281° T.
Second true bg. Rosslare pt.	281° T.

Ship's position on second bearing Lat. 52° 19′ N.
Long. 6° 05′ W.

Distance from Rosslare Point, 12½ miles.

Distance from Tuskar Rock when abeam, 3 miles.

WORKED EXAMPLE I. (*Continued*).

Question 4.—**Find the course to steer by compass from the last position with Tuskar Rock Lt. abeam' to reach a position 000° T. from Strumble Head Lt. distant 5 miles. Make proper allowance to counteract the effect of a current setting 238° M. at $2\frac{1}{2}$ knots' and allow 5° leeway for a N.N.E. wind, the ship's speed being $8\frac{1}{2}$ knots. Give the distance steamed and the time taken to reach the destination.** (Deviation Card No. 1.)

INSTRUCTIONS FOR WORKING.—Mark off the position of the destination at a distance of 5 miles due North (true) from Strumble Head Lt. Call it *S*. Join *O S* with a straight line. If there was not any current, *O S* would be the true course.

To find the course to steer to counteract the effect of a current setting 238° (magnetic), lay off a line in a 238° magnetic direction from *O* to represent the set of the current.

No time being specified in the question, we must choose some convenient interval of time on which to base our calculations, let us say two hours. Now, in two hours the current will set 5 miles, so we measure off 5 miles along the current line from *O* and make the mark *T*. *O T*, then, represents the set of the current in two hours.

The ship's speed being $8\frac{1}{2}$ knots, in two hours she will do 17 miles. With this distance in the compasses, place one leg on *T* and mark the position where the other leg meets the line *O S*. Name this mark *W*. Join *T W*, then *T W* is the course to steer to counteract the current *O T*. As the vessel will have to steer this course until she reaches her destination, draw a line parallel to *T W* through *S*. This line meets *O T* (extended) at *X*, and *XS* represents the distance which the ship will have to steam to reach *S*. The time occupied is found by proportion, as shown below.

The correction for leeway is not projected on the chart, but is applied to the true course *after* it has been taken from the chart. Remember that it must be allowed **to windward**—in other words, **towards the wind**—in order to counteract the effect of the ship being set to leeward or away from the wind.

It is important that the leeway should be applied before the deviation is taken from the card, as the deviation changes with the direction of the ship's head.

True course - -	083°
Leeway (N.N.E. wind)	—5
	078°
Variation - - -	+17° W.
Magnetic course -	095°
Deviation - - -	+10° W.
Compass course - -	105°
Distance steamed -	51·75 miles
Time occupied - -	6 hrs. 05·4 mins.

TO FIND THE TIME TAKEN TO REACH DESTINATION

As $8\frac{1}{2}$: : $51\frac{3}{4}$: : 1 hr. to time.

8·50)51·75(6·09 hours.
51·00
7500
7650

·09 hours
60
5·40 minutes

Worked Example II.

(To be worked on the practice chart of Tasmania.)

Working inset.

Question 1.—**The ship's D.R. position was Latitude 40° 40′ S.′ Longitude 143° 40′ E. D.F. bearing of the ship from Hobart RC Station in Lat. 42° 54′ S., Long. 147° 15′ E. was 132·5° T. and D.F. bearing of the ship from Cape Grim RC Station in Lat. 40° 42′ S., Long. 144° 43′ E. was 092·5° T. Required the position of the ship by these D.F. bearings.**

Important Note:—In practice D.F. bearings should **never** be taken across land.

The D.F. bearings are great circle bearings and have to be converted into their equivalent mercatorial bearings before laying them off on the chart. This is done by applying the half convergency correction. Convergency can be calculated using the formula

$$\text{Convergency} = \text{D. Long.} \times \cos \text{Mn. Lat.}$$

but, in practice, Half Convergency Correction Tables are generally used.

The tables are entered with the Mean Latitude and Difference of Longitude between the ship and the RC station. Care must be exercised to apply the correction the right way; this is best decided by making a sketch.

Hobart R.C. station	42° 54′ S.	147° 15′ E.
D.R. ship	40 40 S.	143 40 E.
	2)83 34 S.	D. Long 3 35′ W. ≏ 3·6°
Mean Lat.	41 47 S.	
,, ,,	≏ 41·8°	

From Half Convergency Tables:

		D. Long. 2°	D. Long. 4°
Mean Lat.	39°	0·6°	1·3°
	42°	0·7°	1·4°

Use corrections for Mean Lat. 42° as follows:—

D. Long		Half Convergency	
2°	→	0·7°	
1·6°		x°	0·7°
2° 3·6		C	
4°	→	1·4°	

To Interpolate for D. Long 3·6°:

$3{\cdot}6° \sim 2° = 1{\cdot}6°$

$$\frac{1{\cdot}6}{2} = \frac{x}{0{\cdot}7}$$

$$\frac{1{\cdot}6 \times 0{\cdot}7}{2} = x \qquad = 0{\cdot}56°$$

Half Convergency for 2°	= 0·7
,, ,, ,, 3·6°	= 1·26°
D.F. Bearing	132·50°
Half Convergency	— 1·26
Mercatorial Bearing	131·24°
or	131° T.

How to apply the Half Convergency.

Make a thumbnail sketch showing the ship and D.F. station in their relative position. The straight line joining them represents the rhumb line, or mercatorial bearing. Draw a curve between the two points to represent the great circle, or D.F. bearing, being careful to draw the bulge of the curve on that side of the rhumb line next to the pole of the hemisphere. It is then obvious from the figure whether the correction should be added or subtracted. It is plus in this case because the mercatorial bearing is more clockwise than the D.F. bearing.

For the Cape Grim bearing we have—

	Lat.	Long.		
Station	40° 42′ S.	144° 43′ E.	D.F. bearing	092·5°
Ship	40 40 S.	143 40 E.	½ convergency	— ·3
	2)81 22	1 03	Mercatorial bearing	092·2
Mn Lat.	40 41	D. Long 63′		or 092° T.

Convergency may also be found from the Traverse Tables as follows:—

Convergency = D. Long. × sin Mn. Lat.

and Dep. = Dist. × sin Course

so by substitution we get

Lat. as Course - - - -	40° 41′
D. Long. in Dist. Col. - - -	63′
Convergency in Dep. Col. - -	41·3′

$\therefore \frac{1}{2}$ convergency = + 20·6′ = 0·3°

The position of the ship is fixed by the intersection of these two lines drawn from the D.F. stations (point *J* on the chart).

WORKED EXAMPLE II. (*Continued*).

Question 2.—**From this position obtained by the D.F. bearings, a course was set to pass 15 miles off Cape Sorell Lt. when abeam. Using Deviation Card No. 1, find the course to steer by compass and also the distance to steam to bring the light abeam.**

INSTRUCTIONS.—Take, from the latitude scale abreast of Cape Sorell Lt., 15 miles on your compasses, and describe an arc round the light. From the point of departure *J* draw a tangent to this arc. Now draw a line at right angles to this tangent on to the light, and obtain the beam bearing of the light (*K* on the chart). The tangent to the arc is the course, which must be corrected, and turned into a compass course. The distance to bring Cape Sorell Lt. abeam (*J K*) measured on the latitude scale is found to be 106·5 miles.

True Course - -	155° T.		140° C.
Variation - - -	—8 E.	Dev.	$-3\frac{1}{2}$ W.
Magnetic course -	147° M.		$136\frac{1}{2}$ M.
Deviation - - -	$+2\frac{1}{2}$ W.		
Compass Course -	$149\frac{1}{2}$° C.		150° C.
		Dev.	$-2\frac{1}{2}$ W.
			$147\frac{1}{2}$° M.

Distance to bring light abeam, $106\frac{1}{2}$ miles.

Question 3.—**While on this course Cape Sorell Lt. bore 105° C. and after continuing for $1\frac{1}{2}$ hours at 12 knots, the light was found to be exactly abeam. Required the ship·s position when abeam' and her distance from Cape Sorell Light.**

INSTRUCTIONS.—Correct the compass bearing of Cape Sorell Light, using the deviation found in the preceding question. There is no need to correct the second bearing of the light because it is abeam, or at right angles to the course. Plot these lines, and from where the first bearing intersects the course line, measure along the course line 18 miles (the distance steamed between the bearings) and through this point draw a line parallel to the first bearing of the light. Where it crosses the second or beam bearing is the position of the ship when the light is abeam (*L* on chart).

CAPE SORELL LT.		CAPE SORRELL LT.	
Compass Bearing -	105° C.	Abeam, requires no correction.	
Deviation - - -	—$2\frac{1}{2}$° W.		
Magnetic Bearing -	102° M.	Lat. - - -	42° $18\frac{1}{2}$′ S.
Variation - - -	+8 E.		
True Bearing - -	$110\frac{1}{2}$° T.	Long. - - -	144° 48′ E.

Distance from Cape Sorell Light, 18 miles.

WORKED EXAMPLE II. (*Continued*).

Question 4.—**Proceeding on this course for a further 3 hours at the same speed, the western extremity of Point Hibbs was observed to be bearing 355° C., and at the same instant the Signal Staff on Rocky Point bore 109° C. Find the position of the ship and the set and drift experienced since determining the position by the running fix off Cape Sorell Light.**

INSTRUCTIONS.—As we are still continuing on the same course as in the last question, we draw a line parallel to the original course line from *L*. From *L*, measure along this course line a distance of 36 miles (3 hours' steaming), this gives us the D.R. position *M*. After applying the same compass error to the given compass bearings of Point Hibbs and Rocky Point Signal Staff, to turn them into true bearings, we plot them and their point of intersection *N* gives us the vessel's Observed Position. Joining up the D.R. *M* with the Observed Position *N*, we obtain the set and drift of the current during the previous three hours.

WESTERN EXTREMITY OF POINT HIBBS.			ROCKY POINT SIGNAL STAFF.	
Compass Bearing	355° C.		Compass Bearing	109° C.
Deviation	—2½ W.		Deviation	—2½ W.
Magnetic Bearing	352½° M.		Magnetic Bearing	106½° M.
Variation	+8 E.		Variation	+8 E.
True Bearing	360°	000½° T.	True Bearing	114½° T.

Lat. 42° 54½′ S. Long. 145° 15′ E.
Set, 128·5° T. Drift, 6 miles.
Rate 2 knots.

Question 5.—**From the position thus found by cross-bearings, reset the course by compass (Deviation Card No. 1) to a position 9 miles, 264° T. from South-west Cape, to counteract the effect of the current ascertained in the previous questions, and also find what distance will be made good in 3 hours towards the destination, and the distance from South-west Cape at the end of that time.**

INSTRUCTIONS.—To reset the course to the position 9 miles, 264° from South-west Cape, we plot this position on the chart (*O*) and draw a line from the Observed Position, *N*, found by cross-bearings in the previous question, to *O*. This gives us the course to be made good *N O*. Now, from the position by cross-bearings (*N*) lay off the current similar to that found in the last question; this will be a continuation of the line *M N*.

From *N* measure in the direction *M N* (continued) the distance the ship will be set in 3 hours; this is 6 miles, and reaches the point *P*. Take the distance steamed in 3 hours—that is, 36 miles—in the compasses, and with one leg on *P* mark the point *Q* on the line *N O*. A line from *P* to *Q* indicates the true course to steer which must be converted to a compass course. *N Q* represents the distance made good in 3 hours.

True Course	151° T.	Compass Course	140° C.
Variation	—8 E.	Deviation	—3½ W.
Magnetic Course	143° M.	Magnetic Course	136½° M.
Deviation	+3 W.		
		Compass Course	150° C.
Compass Course	146° C.	Deviation	—2½° W.
		Magnetic Course	147½° M.

Distance made good in 3 hours, 42 miles.
Distance from South-west Cape, 13 miles.

WORKED EXAMPLE III.

Question 1.—**Approaching the south-west coast of Tasmania on a compass course of 112½° C. a Lt. Gp. Fl. (2) ev. 30 secs. was observed to be bearing 087° C. 1·2 hours later it bore 060° C.′ and 48 minutes later still it was abeam. Required the position of the ship when abeam, also the rate of the current, the set being estimated to be 203° T. and the speed of the ship 10 knots by revolutions.** (Use Deviation Card No. 1.)

INSTRUCTIONS.—As the course by compass is given (112½° C.), we take the deviation direct from the card for that point and apply it to same, together with the variation, and obtain the true course. This same error is also applied to the three bearings taken of the Lt. Gp. Fl. (2) ev. 30 secs. to turn them into true bearings. It is a coincidence that the variation and deviation are equal in amount but opposite in name, thus making no compass error, and the compass bearings are, therefore, true ones.

Draw these three bearings on to the light, and through the light draw a line at right angles to the middle of the three bearings. Now take the distance steamed between the first and second bearings; in 1 hour 12 minutes this will be 12 miles, and measure it along this line to the left of the light, reaching point *R*, also take the distance steamed between the second and third bearings; in 48 minutes this will be 8 miles, and measure it along this line to the right of the light, reaching point *S*. From *R* and *S* lay off two lines parallel to the middle bearing, and let them intersect the first and third bearings in points *T* and *U*. These two points joined together will give the actual direction in which the ship is moving over the ground.

Draw from the point *T* on the first bearing the true course being steamed 113° and measure along it the distance steamed in the two given times, 1h 12m + 0h 48m = 2 hours, which at 10 knots = 20 miles. Name this point *V*. From *V* draw the given set of the current 203° T. on to the course being made good, reaching the point *W*. You will thus obtain the amount the tide has set you during the two hours (*V W*) and also the actual distance made good (*T W*). We then draw through the point *W* a line parallel to the first bearing, and the point where this parallel line meets the third bearing is the actual position of the ship at the time of taking the third bearing or when abeam (*X*).

Compass Course	112·5° C.	
Deviation	—8·0 W.	Dev. and var. practically the same value, therefore the error is assumed to be nil.
Magnetic Course	104·5° M.	
Variation	+8·5 E.	The three bearings of the light are, therefore, both compass and true.
True Course	113° T.	

First bearing 087° T., second bearing 060° T., third bearing 023° T. or abeam.

Position of the ship, Lat. 43° 50½′ S., Long. 146° 11′ E.

Drift 2 miles. Rate, 1 knot.

WORKED EXAMPLE III. (*Continued*).

Question 2.—**When abeam of this light (Lt. Gp. Fl. (2) ev. 30 secs.) the course was reset to pass 6 miles off Whale Head when abeam, and after proceeding for 2·5 hours at 10 knots, cross bearings were taken to fix the ship's position as follows :—Piedra Blanca 115° C. and Whale Head 022° C., both by the same compass. Required the course to steer by compass, the position of the ship by the cross bearings, and the set and drift of the current affecting the ship since the course was reset.** (Deviation Card No. 1.)

INSTRUCTIONS.—From the position (*X*) the course is reset to pass 6 miles off Whale Head. Draw a circle of 6 miles radius round Whale Head, and draw a tangent to it; measure 25 miles (distance steamed in 2·5 hours) from *X* along it, and obtain the expected position (*Y*). Correct the true course and obtain the compass error, and apply same to the compass bearings of Piedra Blanca and Whale Head to turn them into true bearings. Plot them on the chart, and where they intersect at *Z* is the Observed Position. The line *Y Z* is the set and drift experienced in the 2·5 hours.

True Course	079° T.		080° C.
Variation	—8·5° E.	Dev.	−13·5° W.
Magnetic Course	070·5° M.		066·5° M.
Deviation	+13° W.		
			090° C.
Compass Course	083·5° C.	Dev.	−12° W.
			078° M.

	Piedra Blanca.		Whale Hd.
Compass Bearing	115° C.	Compass Bearing	022° C.
Error	−4·5° W.	Error	−4·5° W.
True Bearing	110·5° T.	True Bearing	017·5° T.

Lat. 43° 45½′ S. Long. 146° 50′ E.
Set 133° T. Drift 4 miles. Rate 1·6 knots.

WORKED EXAMPLE III. (*Continued*).

Question 3.—**Reset the course again to maintain a distance of 5·5 miles from Tasman Head to counteract the effect of a current similar to that determined in the previous question, and when Tasman Head bears 270° T. alter the course to an anchorage with Iron Pot Lt. dead ahead and distant 2 miles.**

Using Dev. Card No. 1, give the course to steer and the distance to steam to the position where Tasman Head bears 270° T., and also give the course to steer to the anchorage (both by compass). If on the final course the tide is dead astern and the anchorage is made in exactly 2 hours, required the rate of the current, the speed of the ship being 10 knots throughout.

INSTRUCTIONS.—Draw a circle round Tasman Head 5·5 miles in radius, and from the "fix" by cross bearings in the previous question (*Z*) a tangent is drawn to this circle. As the course requires to be carried on until Tasman Head bears 270° T., we produce it to intersect a line drawn 090° T. from Tasman Head (*A*). This will obtain for us the real course and distance to be made good (*ZA*) to arrive at that position. As the tide found in the previous question must be counteracted, the direction is laid off from the starting point (*Z*). It should be observed that no specified time is given in the question, therefore any interval of time may be assumed. To save the time taken in calculating by proportion and to reduce the possibility of a mistake being made, it is usual to assume the same time used in determining the amount of the current. Extend the line *Y Z*, measure along it from *Z*, 4 miles, the distance set in 2·5 hours (*Z B*), and from *B*, 25 miles is fitted on to the course line to be made good (*Z A*) at point *C*. *B C* connected together gives the true course to be steered, and a parallel line to it (shown dotted) from the extended current line to the position (*A*) off Tasman Head gives the distance to be steamed to reach that position.

From the position of the ship (*A*) when Tasman Head bears 270° T., draw a line to Iron Pot Lt., measure 2 miles from it, and mark the point *D*. This will be the anchorage, or destination. Correct this true course to obtain the compass course, and measure the distance along it; this will be found to be 26·5 miles. As the ship makes the anchorage in 2 hours, steaming at 10 knots, she must have experienced 6·5 miles of fair tide or at the rate of 3·25 knots.

True Course - - -	054° T.		050° C.
Variation - - - -	—9 E.	Dev.	−8·5° W.
Magnetic Course - - -	045° M.		041·5° M.
Deviation - - - -	+9·75° W.		
			060° C.
Compass Course - - -	054·75° C.	Dev.	—11° W.
or	055° C.		
			049° M.

Distance to steam to the position off Tasman Head, 37·5 miles.

True Course - - -	344° T.		330° C.
Variation - - - -	—9° E.	Dev.	+2·5° E.
Magnetic Course - - -	335° M.		332·5° M.
Deviation - - - -	—1·75° E.		
Compass Course - - -	333·25° C.		340° C.
or	333° C.	Dev.	+0·5° E.
			340·5° M.

Distance on last course, 26·5 miles.

Distance in 2 hours at 10 knots, 20 miles.

Favourable Set. Drift, 6·5 miles. Rate, 3·25 knots.

Example for Exercise I.

(To be worked on the Chart of Tasmania.)

Proceeding through Bass Strait to the Eastward in hazy weather, soundings were taken at intervals of one hour, the ship steaming at a reduced speed of 6 knots. At 1800 hrs. 44 fathoms, mud ; at 1900 hrs. 41 fathoms ; 2000 hrs. 38 fathoms ; at 2100 hrs. 27 fathoms, coarse sand and shells. (All corrected soundings.)

The course being made good was estimated to be 102° T.

Soon after obtaining the cast at 2100 hrs. the fog lifted, and Goose Island light was observed to be bearing 49° from ahead on the port bow. Speed was increased to 11 knots, and at 2200 hrs. the light was found to be abeam. From the position thus obtained set a course by compass (using Deviation Card No. 1) to pass Swan Is. light at a distance of 5 miles, and state how long you will take and how far you will have to steam to bring it abeam. The tide may be assumed to be ebbing at 2 knots in approximately the same direction as the course of the ship.

Still steering on the same course and at the same speed, some time later Swan Is. light was observed to be bearing 276° C., and at the same time Eddystone Point light bore 204° C. Required the position of the ship at that time, and her distance from Swan Is. light.

From the position obtained by these cross bearings, reset the course by compass to pass Cape Forestier light 10 miles distant when abeam on the approaching course. Proper allowance to be made to counteract the effect of a current estimated to be setting 135° T. at 1·5 knots. Required also the time you will take to reach the position abeam of Cape Forestier light, the ship steaming at 12 knots by revolutions.

Answers.

Course 114° T. (variation 9° E., deviation 8° W.) 113° C.

Distance over the ground to bring Swan Is. light abeam is 21·5 miles.

As the ship is steaming at 11 knots, but making 13 knots over the ground, the time she will take to bring the light abeam will be $\frac{21\cdot5}{13}$, which is 1·65 hours, or 1h. 39m.

As she is doing 11 knots through the water, the distance she will steam will be 11 × 1·65, which is 18·15 miles.

Lat. 40° 46′ S., Long. 148° 29½′ E., distance 16½ miles.

Course 181° T. (variation 9·5 E., deviation ¾° W.), compass course 172¼° C. or 172° C.

Example for Exercise II.

(To be worked on the Chart of Tasmania.)

Using D.R. latitude 40° 50′ S., the longitude obtained by an observation of a celestial object on the prime vertical was 143° 47′ E., and at the same time a cast of the deep-sea lead gave 70 fathoms (corrected), sand and shells. The course was set to pass 20 miles outside Cape Sorell light allowing for 5° leeway, wind S.W., and a current estimated to be setting 060° T. at 2 knots. Required the course to steer by compass and the approximate distance that will be made good in 8 hours, ship steaming at 10 knots. (Use Deviation Card No. 1.)

Cape Sorell light was observed dipping bearing 119° C. with height of eye 42 feet, at which time the course was reset to 162° C. Required the position of the ship and her distance from Point Hibbs when abeam. No leeway to be allowed, the wind having decreased considerably in strength. The current was considered to be setting 140° T. at rate 2 knots, and the ship to be making 10 knots, as in the previous question.

From this position reset the course to pass 3 miles off South West Cape to a position with Maatsuyker Is. light bearing 070° T. Give the compass course to steer and the distance. After running on this course for 3 hours at 10 knots, Rocky Point was found to be abeam, and at the same time Hill 2445 (De Witt Range) bore 100° C. Required the position of the ship and also the set and drift of the current since the course was reset off Point Hibbs.

From this position obtained by the cross bearings find the course to steer and also the distance to steam to reach the final point off Maatsuyker Is. light making due allowance for a similar set and drift to that just ascertained.

Answers.

True course to counteract the current, 165° T.

True course to counteract current and leeway, 170° T.

Course 170° T. (variation 8° E., deviation 1·5° W.), 163·5° C. Distance approximately 77 miles.

Course 162° C. (deviation 1·5° W., variation 8° E.), 168·5° T.

Lat. 42° 40·2′ S., Long. 144° 59·5′ E., beam distance from Point Hibbs 12 miles.

Course 142° T. (variation 8° E., deviation $3\frac{3}{4}$° W.), $137\frac{3}{4}$° C. or 138° C., distance 80 miles.

Lat. 43° 04·5′ S., Long. 145° 22·5′ E.

Set 248° T., Drift 2 miles, Rate 0·66 knots.

Course 136° T. (variation 8° E., deviation $4\frac{1}{2}$° W.), $132\frac{1}{2}$° C. or 133° C. Distance to steam 51·5 miles.

EXAMPLES FOR EXERCISE III.

(To be worked on the Chart of Tasmania.)

From a position in Lat. 41° 25′ S., Long. 149° 4′ E., set a course by compass (using Deviation Card No. 1) to bring Tasman Is. light abeam, distant 3 miles. Whilst on this course C. Forestier light was observed to be bearing 227° C., after proceeding for 9 miles it bore 254° C., and after doing another 6 miles it bore 280° C., whilst at the same moment the southern extremity of Schouten Is. bore 247° C., all these bearings being taken by the same compass.

Find the position of the ship at the time of taking the cross-bearings, and also find the set and drift of the current experienced since taking the first bearing of Cape Forestier light, the ship having been steaming for 1·25 hours at 12 knots until fixing her position by cross-bearings. Give also the distance made good from the initial point of departure.

Reset the course to pass 3 miles off Tasman Is. light as before, to counteract the effect of the current determined in the previous question.

Give the compass course to steer, the distance to steam, and the time occupied in bringing the light abeam, ship steaming at rate 12 knots.

When abeam of Tasman Is. light reset the course to pass 2 miles off C. Raoul, and when Wedge Is. opens just clear of Tasman Peninsula, alter course to reach an anchorage in Trumpeter Bay, N. Bruny Is.

Give the courses to steer by compass, making due allowance for a current setting 200° T. at 2 knots on the first course, and 160° T. at 2 knots on the second course.

Give, also, the time occupied in reaching the anchorage since resetting the course off Tasman Is., the ship maintaining a speed of 12 knots.

ANSWERS.

Course 201·5° T. (variation 9·5° E., deviation $1\frac{1}{4}$° E.), $190\frac{3}{4}$° C. or 191° C.

Lat. 42° $16\frac{3}{4}$′ S., Long. 148° 42′ E., set 169° T., drift $3\frac{1}{4}$ miles, distance made good 54 miles.

Course 212° T. (variation $9\frac{1}{2}$° E., deviation $3\frac{1}{2}$° E.), 199° C., distance 56 miles at 12 knots→4h 40m.

Course 275° T. (variation $9\frac{1}{2}$° E., deviation 13° E.), 252·5° C., distance 14 miles at 12 knots→1h 10m.

Course 302° T. (variation $9\frac{1}{2}$° E., deviation 10° E.), $282\frac{1}{2}$° C., distance 20 miles at 12 knots→1h 40m.

Steaming time on last two courses, 2h 50m.

Example for Exercise IV.

(To be worked on the English Channel Chart.)
(Deviation Card No. 1, page 23.)

Ques. 1.—D.R. Lat. 50° 23′ N., Long. by observation of a star was 6° 41′ W., bearing 240° T. At the same time, which was one hour after high water at Dover, a corrected sounding gave 54 fathoms, gravel and shells. Set a compass course from this position to bring Lundy Island South light bearing 000° T., distant 3 miles. Make allowance for the tide as indicated by the chart, and for 5° leeway, wind N.N.W. force 7.

Find, also, the distance to steam and the time to reach destination, ship steaming at 14 knots.

Ques. 2.—From a position with Morte Point bearing 150° T., distant 5 miles, set the course by compass to make Flatholm Island light dead ahead distant 3 miles. No leeway, but the ebb tide was estimated to be setting 280° T. at 3 knots.

Find time required to reach the destination when steaming at 14 knots.

Answers.

1. Tidal stream as estimated from chart 042° T., 3 knots. Course 060° T. (leeway 5°, var. 20° W., dev. 12½° W.). Course 087½° C., distance to steam 74 miles, time occupied 5h 17m.

2. Course 084° T. (var. 20° W., dev. 8° W.). Course 112° C., distance to steam 52 miles, time occupied 3h 43m.

Exercise V.

Ques. 1.—Making the Channel, D.R. position 48° 55′ N., 6° 17′ W. Sights of a star gave intercept 5′ towards, and azimuth 340° T. Steered 092° C. (var. 16° W., dev. 4° E.) for 3 hours at 10 knots, Ushant W/T D.F. gave bearing of ship 350° T.

Find position.

Ques. 2.—After steaming 8 miles on the same course Ushant gave a second bearing 000° T., and after steaming another 8 miles gave a third bearing 010° T. A Pole Star latitude then gave 49° 11′ N.

Find the longitude and the course made good.

Ques. 3.—Altered course to 071° C. (var. 15° W., dev. 3° E.) and after logging 120 miles on this course, got a glimpse of Portland Bill light dipping, bearing 330° T. (H.E. 15 ft.). Continued on same course for 7 miles ; Anvil Point light then bore 018° T. Find position.

Ques. 4.—Set course from above position to sight Beachy Head light 30° on port bow when on range at 16 miles. Give also the distance Beachy Head should be when it comes abeam, if ship maintains the same course.

Answers.

1. Lat. 49° 07·5′ N., Long. 5° 20′ W.
2. Long. 4° 56′ W.; Course 085° T.
3. Lat. 50° 21′ N., Long. 2° 05·5′ W.
4. 080° T.; beam distance 8 miles.

DAY'S WORK.

(Answers at end of Book.)

1.—At noon, a point in Lat. 47° 00′ N., Long. 178° 35′ W., bore 115° C., distant 12 miles, dev. 6° W. Ship then sailed the following compass courses:—

273° C.	70 miles.	Dev. 2° W.
317° C.	45 miles.	Dev. 5° E.
286° C.	20 miles.	Dev. 9° E.
335° C.	82 miles.	Dev. 4° E.

The var. throughout was 19° W. and a current set 025° T. at 1·5 knots during the day.

Find the position at the end of the run and the course and distance made good during the 24 hours.

2.—At noon in Lat. 32° 24′ N., Long. 53° 12′ W., course was set 260° C. Var. 18° W., Dev. 1° E., Speed 12 knots to counteract the effect of a current estimated to set 297° T. at 2 knots. After 18 hours steaming, stellar observations fixed the ship's position in Lat. 31° 22′ N., Long. 57° 32′ W.

Find the compass course to steer and the speed required in order to reach the position Lat. 31° 00′ N., Long. 59° 00′ W., in 6 hours, allowing for a similar set and drift to that experienced between noon and star sights. Var. 15° W., Dev. 0°.

3.—At 1330 hours a point of land in Lat. 49° 18′ N., Long. 4° 20′ W., bore 348° T., distant 10 miles, log 15, ship steering 210° C., var. 7° W., dev. 10° W. The following extracts were taken from the log:—

Log 45 altered course 230° C., dev. 12° W.

Log 46 a/c 240° C., dev. 7° W., log 79 a/c 245° C., dev. 7° W., this course being maintained until 2200 hours when log read 98. Current set 180° T., at 2 knots from 1330 until 1600 hours and 060° T., at 1·5 knots from 1800 until 2200 hours. Var. 7° W.

Find the ship's position at the end of the run.

4.—May 1st, at 0130 hours, log 142, in Lat. 49° 30′ N., Long. 4° 50′ W., course set 220° C., dev. 10° E., var. 15° W. At 0400 hours, log 162, a/c to 000° C., dev. 7° W. At 0730 hours, log 180, a/c to 220° C., dev. 10° E., this course being maintained till noon, log 216. Allow 5° leeway throughout for a westerly wind. Current from 0130 hrs. to 0600 hrs., 080° T., 2 knots, and thereafter till noon 260° T., 2·5 knots.

Find D.R. position at noon.

5.—Thursday at noon, point in Lat. 47° 30′ N., Long. 179° 25′ W., bearing 322° T., distant 12 miles. During the next 24 hours ship steered 210° C., dev. nil, at 11 knots. Var. during first 10h was 12° E. and for the remainder of the day 10° E. At 0800 hours on Friday the Observed Position was 44° 40′ N., 177° 00′ E.

Find the set, drift experienced to 0800 and, assuming the current remains unchanged, find the E.P. at noon on Friday.

6.—From Lat. 55° 43·3′ N., Long. 4° 58·9′ W., at 1600 hours a vessel steered 198° C. error 14° W., at 1650 hrs., with log 30, a point in Lat. 55° 15·2′ N., Long. 5° 6·2′ W., was 4 points on the starboard bow. At 1710 hrs. this point was abeam, log 33. Between the bearings a current set 223° T.; at 3 knots; var. 17° W.

Find the beam position and the bearing and distance of the ship from the point.

7.—At noon a point in Lat. 34° 50′ S., Long. 20° 01′ E., bore 060° T., 8 miles off, ship's head 114° C., dev. 6° E., var. 27° W., log zero. 1800 hrs. a/c to 249° C., dev. 6° E., log 42; 2400 hrs. a/c 105° C., dev. 5° E., log 82; 0600 hrs. a/c to 103° C., dev. 5° E., log 122. At noon, position by observation was 34° 59′ S., 22° 20′ E., log 164.

Find set and drift of current; and course and distance made good in 24 hours.

8.—In a D.R. position of Lat. 20° 10′ S., Long. 50° 37′ W., the Sun bore 070° T., intercept 5·7′ towards. Ship steered 310° T. for 44 miles when Lat. by ex-mer. alt. was 19° 45′ S. Ship continued on her course for 4 miles more till noon.

Find noon position.

9.—At an assumed position of Lat. 30° 3′ S., Long. 24° 17′ W., Sun bore 060° T., intercept 11·5 towards. Ship steamed 090° T., 41 miles, when the latitude by ex-meridian was found to be 29° 53′ S.

Find position at second observation and if ship continued on same course for another 3 miles till noon, find her position at noon.

10.—At noon D.R. Lat. 25° 00′ S., Long. 3° 00′ E., the ship then sailed as follows:

Time	Comp. Co.	Var.	Dev.	Wind	Leeway	Distance
1200–1600 hrs.	095° C.	25 W.	10° E.	East	0	43 miles
1600–2000	064°	,,	3° E.	,,	0	50 ,,
2000–2400	032°	,,	8° W.	,,	4	48 ,,
2400–0400	057°	,,	Nil	,,	2	44 ,,
0400–0800	005°	,,	15° W.	,,	5	53 ,,
0800–1200	082°	,,	5° E.	,,	2	40 ,,

At 2000 hrs. the obs. position was 24° 30′ S., Long. 4° 18′ E. Find the set and drift experienced till 2000 hrs. and allow a similar set and drift throughout the remainder of the day.

Find the E.P. the following noon, the course and distance made good and the average speed.

THE CAUSES AND EFFECTS OF THE TRADE WINDS, MONSOONS, ETC.

If all parts of the globe had the same temperature the atmospheric pressure would be practically uniform all over the world, and the result would be calm, fine weather. Such is not the case, however, and those parts that come more directly beneath the sun's rays—*i.e.* the tropics—have a much higher temperature than the temperate zones, where the sun shines more obliquely on the earth's surface. The atmosphere, being composed mainly of nitrogen, oxygen, and water vapour, is very elastic, and in common with other gases expands when heated and contracts when cooled. Cold air is thus comparatively heavier and so causes a greater pressure, whilst warm air is lighter and causes a lower pressure. Expansion and contraction which occurs is in the same ratio as the Absolute temperature.

From this we have two effects: the air at the equator expands, whilst the air in the colder regions contracts. This upsets the equilibrium of the atmosphere, with the result that between 20° N. and 20° S. we have a low pressure, between Latitudes 20° and 40° N. and S. we have a comparatively high pressure, whilst from Latitude 40° the pressure decreases towards the poles in both hemispheres. The air in the tropics rises and at the same time it expands, and in consequence of this the upper strata of air move from the equator towards the poles, whilst the lower strata at the earth's surface move from about Latitude 30°, where the pressure is greatest, towards the equator. A current of air in moving from Latitude 30° to the equator would be meeting the earth where it is rotating at greater speed, thus impressing these winds with North-easterly and South-easterly directions. They are indicated in Plates I. and II. by the blue and red colouring, and constitute the N.E. and S.E. trade winds. From these plates we see that they blow in all the oceans with the exception of the Indian Ocean, North of the equator, when they are substituted by the monsoons.

The general tendency of the atmosphere is to move from localities of high pressure to those of lower pressure, but in doing this it is influenced by the different speeds in rotation of the earth's surface. In the N. hemisphere currents of air are always deflected to the right, and in the S. hemisphere to the left. Given a high pressure at any particular place, the air moving outwards and turning to the right in the N. hemisphere sets up a circular clockwise motion. In the S. hemisphere, being deflected to the left, its circular motion would be anti-clockwise. This system of wind is called Anticyclonic. If now on Plates I. and II. we take Latitude 30° N. and S. to be the high pressure, we find that the wind is acting in conformity with the above. To the Eastward of these areas of high pressure we have the N.E. and S.E. trades, and to the Northward and Southward of Latitude 30° we have Westerly winds.

These winds are consistent in all the oceans with the exception of the Indian Ocean, where the great continent of Asia so closely approaches the equatorial region as to upset the system. When the sun is North the land directly under the sun's rays in Northern India gets much hotter than the sea at the equator. The mean barometric pressure in this locality falls to about 1000 millibars—*i.e.* about 11 millibars lower than the mean equatorial pressure. The wind, instead of blowing towards the equator, now blows from the equator, and with considerable violence; this is known as the S.W. monsoon, shown on Plate II.

When the sun is South there is a high pressure over Asia in about Latitude 30°, and well to the Eastward of that continent. This sets up an anticyclonic system similar to that in the Atlantic and Pacific, but the high pressure being exceptionally high and to the Eastward we get very strong N.E. winds in the China Sea and moderate winds in the Indian Ocean.

The sun being South at the same time creates a low-pressure area in about Latitude 10° to 15° S., which is also affected by the Northern part of Australia, where the low-pressure area extends still further South. The N.E. monsoon, as this wind is called, blows therefore right across the equator to about 10° S. On crossing the equator, however, it is deflected to the left, as all winds are, so that it blows from the N.W., and is now called the N.W. monsoon (see Plate I.).

The approximate limits of the trade winds are as follows:

North Atlantic	6° N. to 28° N.
South Atlantic	2½° N. to 27° S.
North Pacific	10° N. to 27° N.
South Pacific	6° N. to 27° S.
South Indian Ocean	January, 15° S. to 30° S.; July, 0° S. to 25° S.

In the Atlantic and Pacific these limits move slightly with the sun, the polar limits about 5° and the equatorial limits about 3°, the above latitudes being taken as the mean.

Between the N.E. and S.E. trades we have an area of light variable winds and calms, with heavy rain squalls. This is called the Inter Tropical Convergence Zone (ITCZ). They are very similar in the Atlantic and the Pacific, and fairly consistent in both oceans. In the Indian Ocean they are both irregular and inconsistent on account of the monsoons; there is generally, however, an area of variable winds between the S.E. trades and the N.W. monsoons in about Latitude 10° S., and a similar area between the S.E. trade and the S.W. monsoon in about Latitude 2° S.

The student is advised to study carefully these two plates, I. and II., which have been reproduced from the *Admiralty Manual of Navigation*, by kind permission of the Controller of His Majesty's Stationery Office.

The various meteorological instruments are described and illustrated in *Nicholls's Seamanship and Nautical Knowledge*, and in *Meteorology for Masters and Mates.*

SUMMARY OF TRADE WINDS AND MONSOONS.

Name or Direction of Wind.	Average Limits of Locality.	Season.	Remarks.
	Atlantic Ocean.		
N.E. Trade,	27° N. to 8° N.		Polar limits vary 3° or 4° N. or S., and equatorial limits 5° or 6°, according as the Sun is N. or S.
S.E. Trade,	6° N. to a line joining Cape of Good Hope and Martin Vaz and Trinidad Islands.		
Anti-Trades (W'ly Winds)	Between 35° and 60° N. and S.		
E'ly Winds,	English Channel.	Spring.	
	Pacific Ocean.		
N.E. Trade,	27° N. to 7° N., and as far W. as the Mariana Islands.		Limits vary, like those in the Atlantic.
S.E. Trade,	5° N. to 27° S.		
Anti-Trades,	Between 35° and 60° N. and S.		Similar to Atlantic.
	Indian Ocean and China Seas.		
S.E. Trade,	7° S. to 27° S. in Indian Ocean.		Limits vary, as in the other oceans.
N.W. Monsoon,	Equator to 10° S. in Indian Ocean, and extending in the Pacific as far as the New Hebrides.	Nov. to March.	Not a steady monsoon, but varies between N.W. and W.S.W.
N.E. Monsoon,	East Coast of Africa, Arabian Sea, Bay of Bengal, and China Sea.	Oct. to April.	Stronger in China Seas. Is steadier in Dec., Jan. and Feb.
S.W. Monsoon,	China Seas.	April to Oct.	Stronger in East Indies. Least liable to change in June, July and August, when it blows fiercely over the East Indian Seas.
Anti-Trades,	35° to 60° S.		Similar to Atlantic.

OCEAN CURRENTS.

Whilst inshore currents are mainly generated by tidal effects, those in the open ocean are caused mainly by the prevailing winds. In consequence, they follow the general circulation around the mid-ocean permanent high pressure area. In Northern Oceans the general current circulation is clockwise and in the Southern Oceans it is anti-clockwise.

Due to the earth's rotation the direction of a current will be subject to Coriolis effect. In the Northern hemisphere the current will take up a direction to the right of that of the wind. In the Southern hemisphere the current will take up a direction to the left of the wind.

A body of water impinging on a coast takes the path of least resistance, and, therefore, follows the contours of the coastline, flowing through openings between islands and continents, sometimes with great velocity when a channel is comparatively small and a large volume of water is being impelled through it, such, for example, as the Straits of Florida, where it attains a speed of 4 to 5 knots. Ocean currents, however, vary from ½ to 2 knots, although in some congested localities they may be accelerated to 3 or 4 knots by pressure of water in the rear, or by seasonal winds increasing the drift of the upper layers, but it is the surface current, of course, that directly affects the ship.

Refer to Map II. and trace out the general circulation in each of the oceans.

North Atlantic.—We have to describe a right-handed circulation. Beginning at the equator the current flows westward into the Gulf of Mexico, follows the coastline, flows rapidly through the Straits of Florida and up the East coast of the United States, then broadens out and bends to the Eastward, now known as the Gulf Stream. Midway across the Atlantic the waters broaden out still more, and part flows Northward towards the British Isles and Iceland, the other part bending to the South-East and Southward, down the African coast, when it is now called the N.E. Trade or African drift, and joins the equatorial set to the Westward again.

Cold-water currents flow Southward down the East coast of Greenland and the West side of Davis Straits, along the coast of Labrador and Newfoundland. The cold Arctic current meets the warmer waters of the Gulf Stream and gets sandwiched between it and the coast and eventually loses itself by sinking out of sight.

Westerly winds bank up the water in the Bay of Biscay, with the result that a variable N.W. offshoot flows across the chops of the English Channel and is known as the Rennel current.

It will be noticed that a South equatorial current flows obliquely across the equator from the South Atlantic and connects up with the West-going North equatorial current. They leave in their rear a triangular-shaped area off the Guinea Coast, where counter-currents prevail. The Guinea current flows South and Eastward into the Gulf, and a counter-current setting Eastward is sometimes experienced in the region of the doldrums, between 5° and 9° North of the equator.

South Atlantic.—The equatorial current setting Westerly bends to the S.W. and Southward down the coast of Brazil, then turns more and more to the Eastward when off the River Plate. This East-going water is joined by the cold Cape Horn current and together they flow towards the Cape ; part of the water continues to flow Eastward and part flows Northward up the African coast, forming a S.E. trade drift, and eventually forms the equatorial current again.

North Pacific.—The equatorial current flows to the West between the parallels of 10° and 20° N. and is deflected to the N.W. and Northward on reaching the Philippine Islands, then North and North-Eastward up the China coast towards Japan. It is then called the Kuro Siwo current, which is similar in character to the Gulf Stream of the North Atlantic. The water flows Eastward across the Pacific towards the coast of North America, then bends sharply to the Southward and is then called the California current; thereafter it runs S.W. and West and rejoins the equatorial current.

A counter-current flows Eastward between the equator and 10° N. until it is deflected by Central America, part of it bending to the N.E. and part to the S.E.

A cold-water current flows Southerly towards Japan and gets lost between the warmer waters of the Kuro Siwo and the coast.

South Pacific.—We note that the South equatorial current flows West and divides on approaching the Australian coast, part flowing to the N.W. past New Guinea and part flowing S.W. and South down the East coast of Australia, then bending sharply to the Eastward flowing part New Zealand and across the Pacific to South America, where it again divides into an East-going current round Cape Horn and a North-going set up the coast of Chile, when it is called the Humbolt or Peruvian current, before it eventually joins up again with the equatorial set to the West.

Indian Ocean.—The equatorial current flows West between the parallels of 5° and 20° S. and splits on Madagascar; part flows round the North end of the island and South through the Mozambique Channel and joins the other part, which, meantime, has flowed down the East coast towards the South end of the island. The current flows down the coast of Natal and West round the Cape, when it is called the Algulhas current. This current extends from the coast to about 100 miles off, and is deflected on meeting the East-going waters of the South Atlantic, part of the water slipping into the Atlantic and the remainder bending to the South and S.E. to join the general East-going set in the Southern Ocean.

The currents in the Northern area of the Indian Ocean and in the Bay of Bengal and Arabian Sea are greatly influenced by the monsoons. A portion of the West-going equatorial current which passed the North end of Madagascar flows North up the African coast and eddies to the Eastward, forming a counter-current just South of the equator. The strength and the existence of this current is controlled by the prevailing monsoon. During the S.W. monsoon it is more pronounced than during the N.E. monsoon, as, in the former case, it is assisted by the S.W. wind, and in the latter case it is retarded by the N.E. wind. Currents due to the wind are called drift currents and are confined tc the surface layers of the water. In the Bay of Bengal and Arabian Sea the currents set to the S.W. during the N.E. monsoon from October to March and to the N.E. during the S.W. monsoon from April to September.

OCEAN PASSAGES AND SHIPS' ROUTEING.

General Information, References and International Requirements.

Full information relating to ocean passages may be found in British Admiralty publication, NP 136, *Ocean Passages for the World,* Third Edition, together with the current Supplement. When approaching land the mariner should adopt the procedure appropriate to coastal passage planning with particular reference to Traffic Separation Schemes. Details of such schemes are shown on British Admiralty Charts, referred to in *Sailing Directions* and listed in Annual Notice No. 17 which is contained in the Annual Summary of Admiralty Notices to Mariners.

Details of traffic separation schemes adopted by IMCO are set out in the IMCO publication, *Ships' Routeing,* 4th Edition with Amendments, and in subsequent IMCO Resolutions.

Compliance with Rule 10 of the, *International Regulations for the Prevention of Collisions at Sea,* is mandatory for all ships when operating in or near schemes which have been adopted by IMCO. This rule is repeated at the end of this section.

Further information about ships routeing is contained at Chapter 4 of *The Mariner's Handbook* and appropriate reference should also be made to the *Loadline Rules—Zones, Areas and Seasonal Periods* BA Chart, D 6083, and to Routeing Charts.

Ocean Passages for the World—NP 136.

NP 136 specifies that the recommended routes are intended for vessels with sea-going speeds of 15 knots and moderate draughts. Such routes should also be considered by all ships, particularly those operating in high latitudes. Special requirements of vessels drawing more than 12 metres are not covered.

Ocean passages for the World bears much the same relationship to Ocean Charts as the *Sailing Directions* bear to Coastal Charts. The volume is divided into two parts of which Part 1 covers Power Vessel Routes and Part 2, Sailing Vessel Routes.

Part 1 contains:

Chapter 1. Planning a passage.
2. North Atlantic Ocean.
3. South Atlantic Ocean.
4. Gulf of Mexico and Caribbean Sea.
5. Mediterranean Sea and Black Sea.
6. Red Sea, Indian Ocean and Persian Gulf.
7. Pacific Ocean, China and Japan Seas and Eastern Archipelago.
8. Miscellaneous information for power vessels.

Climatic and Weather Routeing.

As a result of higher fuel costs and the consequent greater emphasis placed on Weather Routeing, ships frequently follow routes which differ from the climatic routes recommended in NP 136. The need to modify the climatic route for variations in weather and other factors such as owners' requirements, fuel consumption, risk of damage and sensitive cargoes, is brought out in the supplementary pages on Passage Planning and in the following extract on General Planning from NP 136 :—

General Planning.

The best track.—The art of passage planning has been practised from time immemorial. The selection of the best track for an individual voyage demands skilled evaluation of all the factors controlling the voyage and modification of the shortest route accordingly.

In the past, most passage planning has been done with the aid of statistics on weather, currents, and climate which, together with the experience of previous voyages, have enabled the publication of suggested routes for a wide variety of passages. These statistic-based or "climatic" routes, usually depending on factors which can vary seasonally, serve the mariner's purpose up to a point, but they do not take into account short-term variations in the statistical pattern, which can be detected and even forecast by modern methods, and can therefore be incorporated in the plan or transmitted to the vessel at sea with great benefit to the immediate conduct of the voyage.

Each chapter of routes for power vessels contains a review, based on all available statistics and experience, of the usual climatic and other conditions affecting the area concerned. Having made a first study of the projected passage with the aid of the routes recommended as a result, the required route should be adjusted to meet such factors as urgency, risk of damage, and fuel consumption. In addition, the growing availability of shore-based routeing advice, together with forecasts of weather, currents, swell, and ice movements should be taken into account. A great deal of information is thus available to the shipmaster in most parts of the world, for application in aid of the successful prosecution of the voyage.

Exercises.

The climatic routeing exercises which follow are intended to familiarise the student in the use of NP 136, *Ocean Passages of the World, Main Ocean Route Chart,* 5307, *World Climatic Charts,* 5301 and 5302, (or similar charts showing wind and current direction) and, if required, with the *Load Line Rules—Zones, Areas and Seasonal Periods,* BA Chart, D 6083.

Example.—Describe the route to be followed on a passage from New York to Fastnet. State the winds and currents likely to be experienced, and comment on any special considerations which should be taken into account. List your references.

(*a*) **Atlantic Ocean**—General.

Regulation 8(e) of Chapter V. of the *International Convention for on Safety of Life at Sea,* 1974 (IMCO) directed that all ships proceeding on voyages in the vicinity of the Grand Banks of Newfoundland avoid, as far as possible, the fishing banks of Newfoundland north of latitude 43° N. and pass outside regions known or believed to be endangered by ice.

Diagram 18 of NP 136 shows Standard Alter-Course Positions and Approach Routes for Transatlantic Voyages. There are three such positions: CR (Cape Race) May–Nov.; BN (Banks North) April–May and Nov.–Feb.; and BS (Banks South) Feb.–April. Westbound tracks are slightly to the North of their corresponding Eastbound tracks.

(*b*) **New York to Fastnet** (NP 136, 2.61—2.67)

Follow the rhumb line tracks to Alter-Course Positions BN or BS, according to season, then the great circle to the English Channel.

Current: SW—variable at first, North Atlantic current (NE'ly) remainder.
Wind: SW becoming W'ly.
Ref: NP 136, 2.61—2.67.

EXAMPLES FOR EXERCISE.

In each of the following questions refer to the relevant chartlets provided and to the charts and volumes previously mentioned.

For each passage give a brief general description of any special considerations, state the route to be followed, the currents and winds likely to be experienced and the relevant references with which to enter NP 136.

Summary answers are provided at the end of these exercises.

Atlantic Ocean.

1. New York to Fastnet.
2. Cape Race to Malin.
3. U.K. to the Gulf of Mexico.
4. U.K. to Valparaiso via the Panama Canal.
5. U.K. to the Cape of Good Hope.

Indian Ocean.

6. Suez to Bombay.
7. Bombay to Hong Kong (NE Monsoon).
8. Hong Kong to Aden (SW Monsoon).
9. Bombay to Aden (SW Monsoon).
10. Cape Town to Java.
11. Cape Town to Fremantle.
12. Chittagong to Colombo (SW Monsoon).
13. Fremantle to Sunda Straits.

Pacific Ocean.

14. Yokohama to San Francisco.
15. Yokohama to Honolulu.
16. Philippines to Honolulu.
17. Fiji to Wellington.
18. Honolulu to the Panama Canal.

China Sea.

19. Singapore to Shanghai.
20. Singapore to Yokohama (NE Monsoon).

ANSWERS.

Atlantic Ocean.

1. New York to Fastnet.
As per the example previously given.

2. Cape Race to Malin Head.
Current: N. Atlantic current (NE'ly).
Wind: SW–Westerly.
Ref.: NP 136 no reference. Great Circle track.

3. U.K. to the Gulf of Mexico.
Current: Gulf stream setting Easterly, North Equatorial setting NW, Mexican currents various.
Wind: Westerlies, Variables, NE Trades and Variables.
Ref.: NP 136 2.81, 2.86, 8.11.

4. U.K. to Valparaiso via the Panama Canal.
Current: As for (3) above. From Panama the Peruvian current will be against the ship.
Wind: As for (3) above thence winds light and southerly.
Ref.: NP 136, 8.02, 7.195.

5. U.K. to the Cape of Good Hope.
Current: Rennel current (NW); Portuguese current (S); Canary current (S); Guinea current (ESE); Equatorial current (W); and South Atlantic current (NW).
Wind: Westerlies; Variables, 35° N to 30° N; NE Trades; ITCZ; SE Trades, then Variables to the Cape.
Ref.: Chart 5307.

Indian Ocean.

6. Suez to Bombay.
Current: In the Red Sea currents are influenced by the prevailing monsoon. The general set is Southerly, but stronger during the SW monsoon than during the NE monsoon and in the Southern part of the Red Sea than in the Northern part, the drift varying from 20 to 30 miles per day. The current sets NNW in the Straits of Bab-el-Mandeb during the NE monsoon.
Wind: According to Monsoon.
Ref.: NP 136, 6.52, 6.76.

7. Bombay to Hong Kong (NE Monsoon).
Ref.: NP 136, 6.81, 6.156, 7.116.

8. Hong Kong to Aden (SW Monsoon).
Ref.: NP 136, 7.116, 6.154.

9. Bombay to Aden (SW Monsoon).
Ref.: NP 136, 6.77.

10. Cape Town to Java.
Current: Agulhas current, then, in crossing the Indian Ocean circulation, a Southerly current will be experienced, and afterwards, on reaching the parallel of about 20° S, a Westerly current.
Wind: Variable when rounding the Cape, then the SE Trades.
Ref.: NP 136, 6.150.

11. Cape Town to Fremantle.
Current: Easterly.
Wind: Westerly.
Ref.: NP 136, 6.161.

12. Chittagong to Colombo (SW Monsoon).
Current: NE'ly and variable.
Ref.: NP 136, 6.71.

13. Fremantle to Sunda Straits.
Current: Northerly up the coast and Westerly after clearing NW Cape.
Wind: Variable, then SE Trades.
Ref.: None.

Pacific Ocean.

14. Yokohama to San Francisco.
Current: Easterly.
Wind: Westerly.
Ref.: NP 136, 7.305.

15. Yokohama to Honolulu.
Current: Kuro Shio (NE), variable later.
Wind: NE Trades.
Ref.: NP 136, 7.213.

16. Philippines to Honolulu.
Current: NW, more Westerly later.
Wind: NE Trades.
Ref.: NP 136, 7.210.

17. Fiji to Wellington.
Current: South sub tropical (Westerly), variable then northerly.
Wind: Seasonal.
Ref.: NP 136, 7.92.

18. Honolulu to the Panama Canal.
Current: Westerly, changing southerly, then easterly on approaching the land.
Wind: NE Trades.
Ref.: NP 136, 7.218, 7.265.

China Sea.

19. Singapore to Shanghai.
Current: Kuro Shio (NNE).
Wind: Monsoon.
Ref.: NP 136, 7.119.

20. Singapore to Yokohama (NE Monsoon).
Current: Kuro Shio (NNE).
Wind: NE.
Ref.: NP 136, 7.122.

COLLISION AVOIDANCE RULE WHICH REFERS TO THE USE OF TRAFFIC SEPARATION SCHEMES

The following Rule from the *International Regulations for Preventing Collisions at Sea,* 1972 (Incorporating the 1981 Amendments), refers to the use of *Traffic Separation Schemes.*

Rule 10

Traffic Separation Schemes

(*a*) This Rule applies to traffic separation schemes adopted by the Organization.

(*b*) A vessel using a traffic separation scheme shall:

(i) proceed in the appropriate traffic lane in the general direction of traffic flow for that lane;

(ii) so far as practicable keep clear of a traffic separation line or separation zone;

(iii) normally join or leave a traffic lane at the termination of the lane, but when joining or leaving from either side shall do so at as small an angle to the general direction of traffic flow as practicable.

(*c*) A vessel shall so far as practicable avoid crossing traffic lanes, but if obliged to do so shall cross as nearly as practicable at right angles to the general direction of traffic flow.

(*d*) Inshore traffic zones shall not normally be used by through traffic which can safely use the appropriate traffic lane within the adjacent traffic separation scheme. However vessels of less than 20 metres in length and sailing vessels may in all circumstances use inshore traffic zones.

(*e*) A vessel, other than a crossing vessel or a vessel joining or leaving a lane shall not normally enter a separation zone or cross a separation line except:

(i) in cases of emergency to avoid immediate danger;

(ii) to engage in fishing within a separation zone.

(*f*) A vessel navigating in areas near the terminations of traffic separation schemes shall do so with particular caution.

(*g*) A vessel shall so far as practicable avoid anchoring in a traffic separation scheme or in areas near its terminations.

(*h*) A vessel not using a traffic separation scheme shall avoid it by as wide a margin as is practicable.

(*i*) A vessel engaged in fishing shall not impede the passage of any vessel following a traffic lane.

(*j*) A vessel of less than 20 metres in length or a sailing vessel shall not impede the safe passage of a power-driven vessel following a traffic lane.

(*k*) A vessel restricted in her ability to manoeuvre when engaged in an operation for the maintenance of safety of navigation in a traffic separation scheme is exempted from complying with this Rule to the extent necessary to carry out the operation.

(*l*) A vessel restricted in her ability to manoeuvre when engaged in an operation for the laying, servicing or picking up of a submarine cable, within a traffic separation scheme, is exempted from complying with this Rule to the extent necessary to carry out the operation.

PASSAGE PLANNING

Research into shipping casualties has shown that the most important contributing factor is that of human error; the incidence has been estimated at 85%.

Among the recommendations for improving this situation is one for PASSAGE PLANNING, by which is meant the preliminary assessment of all relevant information by the Master and Officers concerned. Reference to this subject is made in the following publications:—

"*A Guide to the Planning and Conduct of Sea Passages*"—Dept. of Trade.
M Notices Nos. 756 and 854—Dept. of Trade.
"*Bridge Procedures Guide*"—International Chamber of Shipping.

These publications refer mainly to coastal passages, where the safety aspect is paramount. It will be convenient, therefore, to consider Ocean Passages under a separate heading.

Check Lists.—It is recommended that suitable reminders should be prepared for any procedure that has to be followed. In the case of passage planning, the lists given below would be useful:—

1. Publications to be consulted.
2. Factors to be taken into account.
3. Information to be shown on the charts or in the bridge notebook.

The Organisation of Passage Planning.

The full procedure consists of four stages, but strictly speaking, the last stage is not part of the planning phase.

1. *Appraisal.*—The Master, in consultation with his Deck Officers, will consider all the relevant information and make a general decision on the track to be followed. It could be useful to lay off the tracks on a small scale chart so that the overall situation may be assessed.

2. *Planning.*—The Navigating Officer may now lay off the courses on charts of suitable scale and complete the details of the plan, which should be from "berth to berth". Those parts under pilotage should be prepared in such a manner that the Bridge Officer may check on the ship's safety independently of the Pilot.

3. *Execution.*—When the time of departure is known, tactics may be considered. Conditions of light or darkness, the state of the tide etc. must be assessed, and, if necessary, the Master consulted with a view to adjusting speed or modifying the track so as to obtain more favourable circumstances.

4. *Monitoring.*—At this stage the plan is in operation with the Officer of the Watch checking progress. He should, of course, call the Master in all cases of doubt, but also bear in mind the possible necessity for immediate action in cases of emergency.

PASSAGE PLANNING (COASTAL)—STAGE 1 (APPRAISAL).

List of Publications to which reference may need to be made.

(*N.B.*—Check that the publications are corrected to date, or that the latest supplement is consulted.)

1. **Passage Record Book.**
Advisable to retain past plans for future reference. Value is enhanced by comments on the success, or otherwise, of the plan used.
2. **Sailing Directions.**
For routes, approaches, pilotage, hazards, port facilities, local weather, etc., etc.
3. **Distance Tables.**
Admiralty, or other.
4. **Routeing Instructions.**
IMCO or local. Details of Traffic Schemes.
5. **Catalogue of Charts and Publications.**
To check on those available, especially for the largest scale charts for coastal navigation.
6. **Tide Tables.**
Depths of water. Port entry.
7. **Tidal Atlas.**
Tidal streams.
8. **List of Lights and Fog Signals.**
Information is more detailed, and may be more recent than on the charts.
9. **Mariners' Handbook (NP 100)**
General information on navigation and weather.
10. **Nautical Almanac.**
Sunrise/set, moonrise/set.
11. **Admiralty List of Radio Signals.**
Vol. I. Communications.
II. Radio beacons.
III. Weather reports.
IV. Meteorological stations.
V. Radio aids to navigation.
VI. Port communication.
12. **Ships' Manoeuvring Data.**
Turning circle. Allowance for "squat".
13. **Bridge Procedures Guide.**
Published by the International Chamber of Shipping. Gives check lists for various procedures while coasting and when under pilotage.
14. **Notices to Mariners.**
For corrections to charts and publications.
15. **Radio Warnings.**
To ensure that information is up to date.
16. **Weather Reports.**
For the latest weather information.
17. **Special Publications.**
As appropriate to some particular area,
e.g. English Channel—Chart 5500.
Gulf of Mexico—*Shipping Safety Fairways.*

PASSAGE PLANNING (COASTAL)—STAGE 2 (PLANNING).

List of Information to be noted on the charts and in the bridge note book.

1. **Tracks.**
   Courses to be shown alongside in 360° notation.
2. **Areas of Danger.**
   Also indications of hazards on the next chart. Take care not to obscure information on the chart.
3. **Position Fixing.**
   Suitable notes to include ranges of lights, radar conspicuous areas, etc. Indicate where special accuracy is required, giving alternative methods for independent checks.
4. **Transits.**
   Also clearing bearings and clearing ranges.
5. **Underkeel Clearance.**
   Especially where information is sparse or unreliable.
6. **Alterations of Course.**
   Indicate "wheel over" positions. Allow for the turning circle and any current.
7. **Speed.**
   Indicate where changes must be made to a "safe speed" or to make the ETA. Allow for squat.
8. **Tides.**
   Note times of HW and LW at reference ports and at destination, and if springs or neaps.
9. **Communication Points.**
   For reporting progress to navigation control centres.
10. **Nautical Almanac.**
    Times of sunrise/set, moonrise/set.
11. **Contingency Plans.**
    Alternative tracks around areas where there may be special hazards due to fog, possible engine failure etc.

An Example of Passage Planning.

In this particular case the Master has decided to lay off his projected passage on a small scale chart (referring as necessary to the large scale charts). As well as giving an overall picture, it enables him to check distances along the routes selected. He will also be able to make notes, at appropriate places, for the guidance of the Navigating Officer and the Officers of the Watch.

Plymouth towards Swansea—Appraisal.

(*Vessel*—General cargo carrier, 12,000 GT, speed 16 knots. Equipped with: magnetic and gyro compasses, D/F, radar and satellite navigation).

Publications—As per standard list, plus:—Chart 5500—*English Channel Passage Planning Guide.*

Factors to be considered—

Probable time of sailing—Evening of 17 April.
Distance (from tables)—199 miles at 16 knots, $12\frac{1}{2}$ hours approx.
Approximate ETA—Forenoon, 18 April.
Probable Draft—F 7·2 m, A 7·9 m.
Swansea—Entrance channel: dredged to 3·0 m.

LW	0351	0·8 m. on 18th
HW	1017	8·3

N.B.—Full speed will be needed to make the tide.

Distances along the tracks = 225 or 217 miles

Latest time at 16 knots—14 hours (allowing for slowing down)		
Latest time at Pilot	1000	on 18th
Clear Breakwater	2000	on 17th

N.B.—Try to be ready for sailing at 1900.
Reconsider speed if there is a later finish.

Memo:—Contact Swansea Agents re: berth, draft, and latest time at the Pilot.

An Example of Passage Planning (cont.)

Master's Guidance Notes on the Passage Plan.

Sunset 1955/17	Sunrise 0605/18.		
Devonport	HW 2129 on 17th	Springs.	
Dover	HW 0244	18th	
Swansea	LW 0351	HW 1017 on 18th.	

Numbers refer to locations on the chart.

1. Take the Westerly track only if weather, traffic and tidal streams are favourable. Parallel Index on Eddystone Lt Ho.
2. If radar indicates heavy traffic, continue on the 260° course, then alter course so as to enter the traffic stream at a fine angle.
3. Parallel Index on the Wolf Lt Ho.
4. Master to consider progress and weather reports. Westerly track if there is any chance of fog. Ensure that Satnav is in operation.
5. Keep further East in bad weather, to avoid rough water on the shoal.
6. Send ETA to the Pilot, Make allowance for cross tides on the next track.

The Plan and Notes will now be given to the Navigating Officer who will select appropriate charts, lay off and check the courses.

When the sailing time is known, tidal streams can be assessed, and the ETA and speed considered.

The Plan and Notes, amended if necessary, will be placed on the chart table for the guidance of the Officers of the Watch.

Also to be prepared are the Pilotage Plans for departure and arrival.

After completing the passage the Plan should be filed for future reference. Any "post mortem" comments that can be made will greatly enhance its value as a guide to planning on another occasion.

PASSAGE PLANNING (OCEAN)—NOTES.

1. Objectives.

Your aim could be :—

1. The least time
2. The least damage
3. Maintaining schedule
4. The most economical speed
5. A constant speed as per the charter party.

A compromise is often desirable—put the most important objective first.

2. Appraisal.

In making a general decision about the route, the following factors may have to be considered. Some may have an overriding importance so that the choice of an alternative is forced; on other occasions their relative merits must be assessed.

3. ETA and Owners Requirements.

The owners or operator should indicate any urgency regarding the ETA, and the required speed. They should also indicate if there is any cargo offering en route.

4. Canals.

The use of canals involves an economic decision regarding dues and delays.

5. Bunkers.

Consider the quantity of bunkers on board; requirements and reserves; prices and availability and the relative distances to bunkering ports; ship s stability in the worst condition should also be considered.

6. Delays.

Delays and stoppages may be caused by political or industrial action and natural disasters.

7. Limits.

The insurance policy and articles of agreement should be checked for high latitude limits. Take account of war risk zones.

8. Load Line Zones.

Select the most advantageous zone if likely to be fully laden.

9. Draft.

Check the depths of water en route.

10. Sensitive Cargoes.

Cargoes which include passengers, live-stock, deck cargo and noxious substances may be considered sensitive. In such cases it is even more important to avoid heavy weather areas.

11. Bulk Liquid Cargoes.

Warm sea temperatures are desirable for economy of tank heating.

12. Seasonal Influences.

Consider alternative routes which take account of: monsoons, ice, fog, gales or the high incidence of tropical storms.

13. Weather, Currents and Climatic Routeing.

Take advantage of favourable weather and currents and minimise what is unfavourable. Make reference to NP 136, *Ocean Passages for the World,* and the preceding pages (44-49) which deal with ocean passages and ships routeing.

14. Weather Routeing (Shore-Based).

After considering the factors above you can now decide on the broad outlines of your route, and at this point weigh up the possible economic advantages of shore-based routeing, providing that it is covered by the owner's rules and that it is suitable to the area. In general such schemes are more appropriate to temperate zones where the weather may change from day to day than to tropical latitudes where conditions are more predictable.

INFORMATION RELATING TO CHARTS.

The publications issued by the Hydrographic Department, Admiralty, give a tremendous amount of information in detail regarding the navigation of coasts and of pilotage waters, anchorages, inland waters and harbours, topographical features, depths of water, shoals and hidden dangers, clearing marks, tidal streams, the winds and currents to be expected, port facilities, trade and administration and indeed all the available information that might be of value or of interest to mariners navigating the areas described. The *Merchant Shipping* (*Carriage of Nautical Publications*) *Rules* 1975, which are based on Regulation 20 of the 1974 SOLAS Convention, require U.K. registered ships to carry suitably detailed charts, Sailing Directions and other publications appropriate to the voyage—see Annual Notice to Mariners No. 18 for further details. Navigation students should make a careful study of the remarks contained in these publications and which relate to practical navigation, pilotage and the use of charts.

Reliance on a chart must manifestly depend upon the date and accuracy of the survey on which it is based, and on whether or not it has been corrected up to date from the information given in the *Notices to Mariners* as issued by the Hydrographic Department, Taunton. The degree of dependence to be placed on a chart is also influenced by the scale and the amount of detail shown along the coast. Large scale charts should be used when coasting as corrections cannot always be inserted satisfactorily on charts drawn on a small scale, nor can positions be laid off so accurately on them.

The closeness of the soundings is an indication of the reliability of a chart and as depths of water have hitherto been obtained solely by the laborious use of the lead and that the lead only covers about 5 cms. (2 inches) of the bottom, it follows that inequalities situated between two lines of soundings may have escaped notice. Blank spaces among soundings mean that no soundings have been obtained at these spots. When the sounded depths are deep it may be assumed that in the blanks the water is also deep; but when the sounded depths are shallow such blanks should be regarded with suspicion, especially in coral regions and off rocky coasts. A wide berth should therefore be given to every rocky shore or patch and this rule should be followed, viz., that instead of considering a coast to be clear, unless shown to be foul, the contrary should be assumed.

The twenty metre contour line is a caution against unnecessarily approaching the shore or bank within that line on account of the possibility of the existence of undiscovered inequalities of the bottom which nothing but an elaborate detailed survey could reveal.

It has been pointed out by the Hydrographic Department that when transferring the ship's position from one chart to another it should be done whenever possible by taking the bearing and distance from a distinguishing feature common to both, such as a light or a point of land. Chart graduations may differ slightly due to one being constructed on later and more complete geographical data than the other.

RADAR AS AN AID TO NAVIGATION.

Radar, as the name implies, is a radio direction and range finder. The principle upon which it depends is similar to that of ranging by sound wave echoes, because radio waves are also reflected from solid objects. The equipment consists of the motor generator, the transmitter and receiver, the directional aerial with turning mechanism and the plan position indicator (P.P.I.). The set generates ultra short radio waves which are transmitted in short pulses and directed from the aerial in a narrow beam. When the beam strikes an object in its path some of the radio waves are reflected back to the set and an echo is received. The time interval from transmission to reception of a pulse is measured electronically and half of this interval multiplied by the speed of radio waves gives the range of the target. The range is indicated to scale as the distance from the centre of the P.P.I. to the bright spot of light produced by the echo. The echo displayed persists for a few seconds so that all targets appear continuously as the beam scans the horizon at a rate of 20 revolutions per minute.

A radar set should be capable of accurately indicating ranges from 50 yards for a small object to 20 miles for a coastline target where the ground rises to a height of 200 feet, and a cargo steamer of 5000 g.r.t. should be detected at 7 miles.

The bearing of a target can be read off to an accuracy of one degree from the azimuth scale at the rim of the P.P.I. and expressed from the fore and aft line of an unstabilised display as 0°-360°. If the display is stabilised then the bearings may be read off as 0°-360° True.

The ship's position can be plotted on the chart from the radar bearing and range of a prominent shore target which is also conspicuous on the radar picture, but it is preferable to use ranges only, from three suitable objects. Special reflectors are fitted on certain lighthouses and buoys to increase their efficiency as radar targets. As this is standard practice for modern buoys of GRP construction, the chart abbreviation, "Ra. Refl." is now obsolescent and is found only on the older style Fathom Charts. On Metric Charts only the symbol 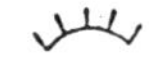 for a radar reflector is charted. Radar reflectors are not charted for IALA System A buoys. The chart abbreviation, "Ra. conspic", is also used to indicate radar conspicuous targets.

Further information regarding radar symbols and abbreviations may be found at sections M and L70 of the latest Book Edition of Admiralty Chart No. 5011, *Symbols and Abbreviations.*

RADAR PLOTTING.

On the Plan Position Indicator (PPI) radar display, the range and bearing of a ship's echo can be obtained from a single observation but correct action to avoid a close quarters situation cannot be determined from a single observation alone. To avoid a close quarters situation it is necessary to make a plot consisting of several observations. These observations of range and bearing should be taken at suitable intervals.

Two methods of plotting can be used. The first method is a Relative Motion Plot in which the observing ship remains at the centre of the PPI. The second method is the True Motion Plot in which the observing ship moves along her course line on the PPI.

In the Relative Motion Plot, the target's relative track is immediately obtained after plotting the first set of observations. When produced, the relative track gives the target's Closest Point of Approach (CPA) to the observing ship. The Relative Plot does not, however, give the target's true course and speed without further plotting—this is demonstrated in Example 1 (a).

In the True Motion Plot, the target's true course and speed are immediately obtained after plotting the first set of observations but the True Plot does not give the CPA without further plotting—this is demonstrated in Example 1 (b).

As will be appreciated, both methods have certain advantages.

Example 1(*a*).

TO DETERMINE CPA AND RISK OF COLLISION BY RELATIVE PLOT

(Refer to Radar Plot 1)

The following radar bearings and ranges were obtained when steering 080°T at 12 knots.

	Target X		Target Y		Target Z	
	Bearing	*Range*	*Bearing*	*Range*	*Bearing*	*Range*
0900	143°T	10 miles	023°T	10 miles	310°T	9 miles
0906	140°T	8 miles	023°T	8 miles	310°T	8 miles

Relative Motion Plot :

1. The observing ship remains at the centre of the plot.
2. Plot the target's positions as shown, the first position of each target being labelled '*O*' and the last '*A*'.
3. *OA* is the relative track of each target.
4. Produce each track (*OA*) towards the centre. The perpendicular from the centre of the plot to each track gives the target's CPA.
5. The time interval to CPA is found by:

$$\frac{\text{Distance to CPA}}{OA} \times \text{Plotting Interval in minutes}$$

Target '*X*' will have a CPA of 2·0 miles.
CPA will occur in :

$$\frac{7{\cdot}7}{2} \times 6 = 23{\cdot}1 \text{ minutes}$$

Target '*Y*' is on a steady bearing and CPA is a collision which will occur in :

$$\frac{8}{2} \times 6 = 24 \text{ minutes}$$

Target '*Z*' is on a steady bearing and CPA is a collision which will occur in :

$$\frac{8}{1} \times 6 = 48 \text{ minutes}$$

If the same information is used in a True Motion Plot the true course and speed of the target are obtained immediately but the CPA and collision threat priority cannot be obtained without further plotting. This is demonstrated in Radar Plot 2.

Radar Plot 1
(Relative Motion)

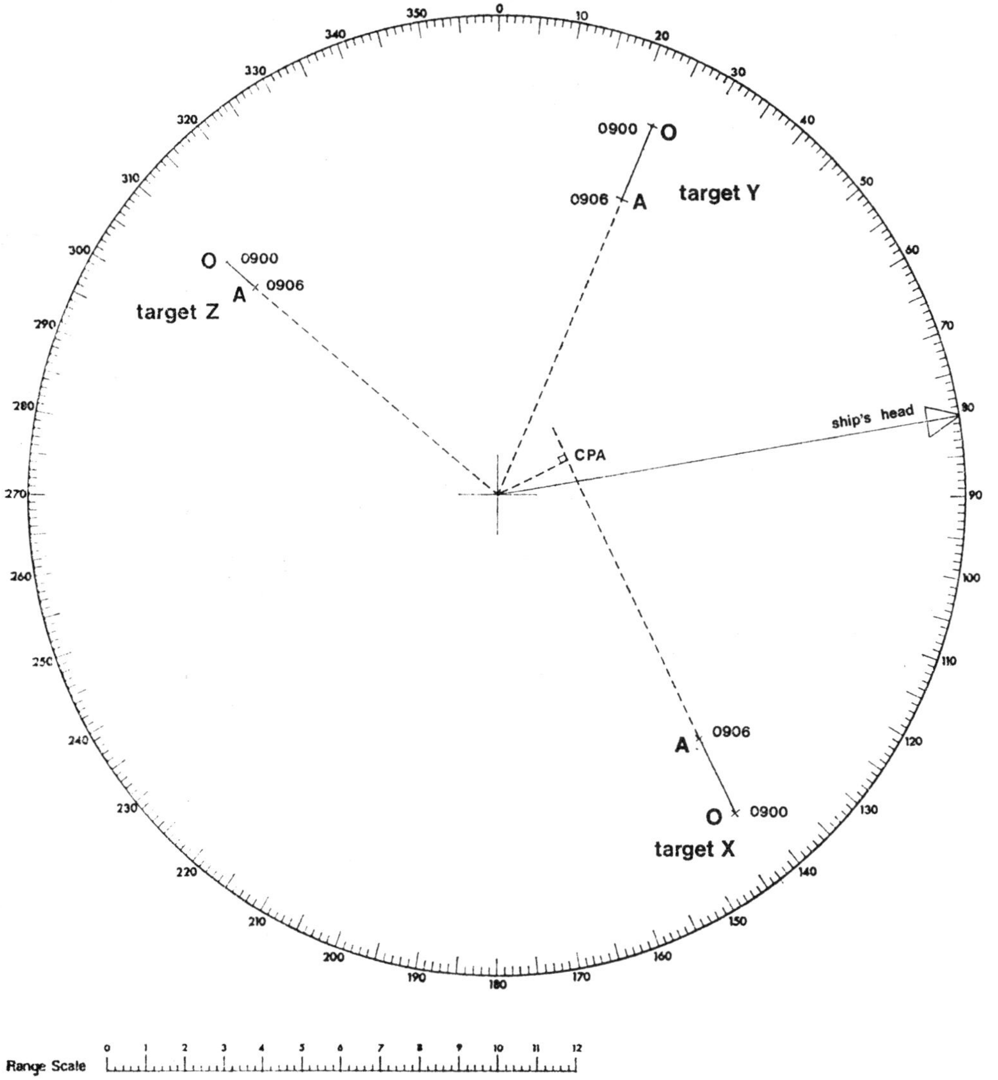

Example 1(*b*).

TO FIND THE TARGET'S TRUE COURSE AND SPEED BY TRUE MOTION PLOT

(Refer to Radar Plot 2)

True Motion Plot:

1. The observing ship moves along its course line of 080°T at a speed of 12 knots i.e. 1·2 miles in 6 minutes. The observing ship's position is '*B*' at 0900 hrs. and '*C*' at 0906 hrs.
2. Using the bearings and ranges of the targets at 0900, lay off the positions of the 3 targets from '*B*'. Label these positions '*W*'.
3. Using the bearings and ranges of the targets at 0906, lay off the positions of the 3 targets from '*C*'. Label these positions '*A*'.
4. *WA* is the true course and distance made good in 6 minutes.

Target '*X*'	is steering	010°T at 21·5 knots
Target '*Y*'	,,	168°T at 16·0 knots
Target '*Z*'	,,	101°T at 20·0 knots

RADAR AS AN ANTI-COLLISION AID—INFORMATION REQUIRED

A minimum of 3 radar bearings and ranges, with suitable intervals between them, are necessary before reliable information can be obtained from a radar plot.

When radar is being used as an anti-collision aid the following information will be required from the plot:

1. True Course and Speed of the Target,
2. Aspect (which is the target ship's relative bearing of the observing ship),
3. Closest Point of Approach of the Target (CPA),
4. Time of CPA, and
5. Observer's Relative Bearing of the Target at CPA.

How to obtain this information is explained in Example 2.

Radar Plot 2

(True Motion)

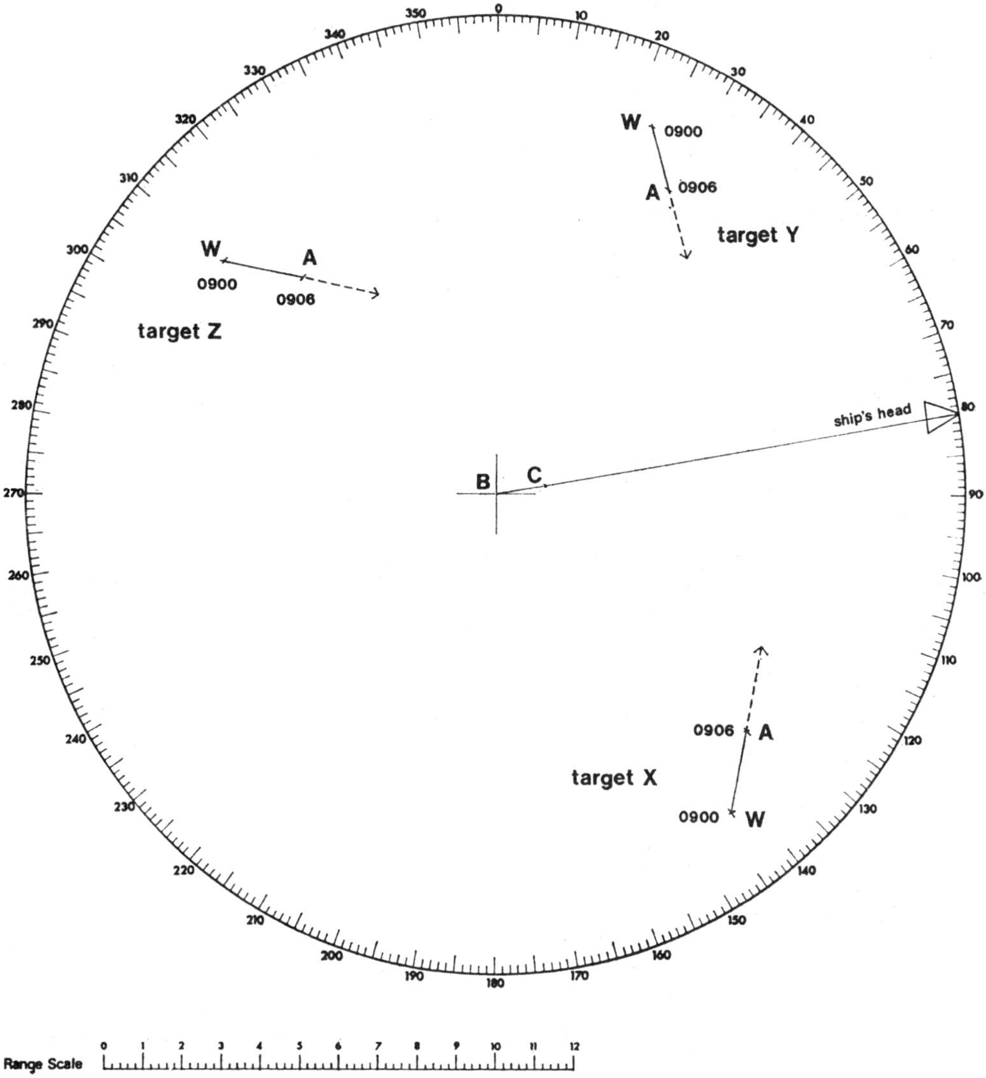

Example 2.

TO FIND THE TARGET'S TRUE COURSE, SPEED, ASPECT, CPA AND THE OBSERVER'S RELATIVE BEARING OF THE TARGET AT CPA

The following radar information was obtained when steering 080°T at 10 knots :

	Relative Bearing	Range
0900	050°	7·0 miles
0906	046°	5·6 miles
0912	038°	4·2 miles

Required**:** (*a*) The true course, speed and aspect of the target at 0912.
(*b*) The time and range of the target at CPA.
(*c*) The relative bearing of the target at CPA.

To obtain the required information the plot can be either :

Method (*a*) Relative Plot, Ship's Head Up.
,, (*b*) Relative Plot, North Up.
,, (*c*) True Plot, North Up.

Example 2, *Method* (*a*)

RELATIVE PLOT, SHIP'S HEAD UP

(Refer to Radar Plot 3)

1. The observing ship remains at the centre of the plot.
2. Plot the positions of the target as shown in Radar Plot 3.
3. The position of the target at 0900 is labelled '*O*' and at 0912 it is labelled '*A*'.
4. *OA*, the relative track of the target, is identified by an encircled arrow always from '*O*' to '*A*' as shown. If the target has not altered course or speed its 0906 position should lie on *OA*.
5. *CZ*, the perpendicular from the centre of the plot to *OA* produced, is the range of the target at CPA, i.e. 2·0 miles.
6. The time to CPA is :

$$\frac{AZ}{OA} \times 12 \text{ minutes}$$

$$\frac{3{\cdot}7}{3{\cdot}0} \times 12 = 15 \text{ minutes (approx)}$$

CPA will occur at 0927 hrs. (i.e. 0912 + 15).
7. The relative bearing at CPA is the angle between the ship's course line and *ZC*, i.e. 337° Relative.
8. From '*O*' lay off *OW* equal to the reciprocal of the observing ship's course (080°T) and the distance it steams in 12 minutes (2 miles), i.e. *WO*, arrowed as shown, is the observing ship's course and distance between 0900 and 0912.
9. *WA*, arrowed as shown, is the target's course and distance between 0900 and 0912. *WA* is at an angle of 72° to the observing ship's course so the target's course is 008°T, (i.e. 080°–72°). *WA* is 2·8 miles long so the target has steamed 2·8 miles in 12 minutes which gives a speed of 14·0 knots.
10. Angle *XAC*, the aspect, is the target's relative bearing of the observing ship, Red 68°.

Radar Plot 3
(Relative - Ship's head up)

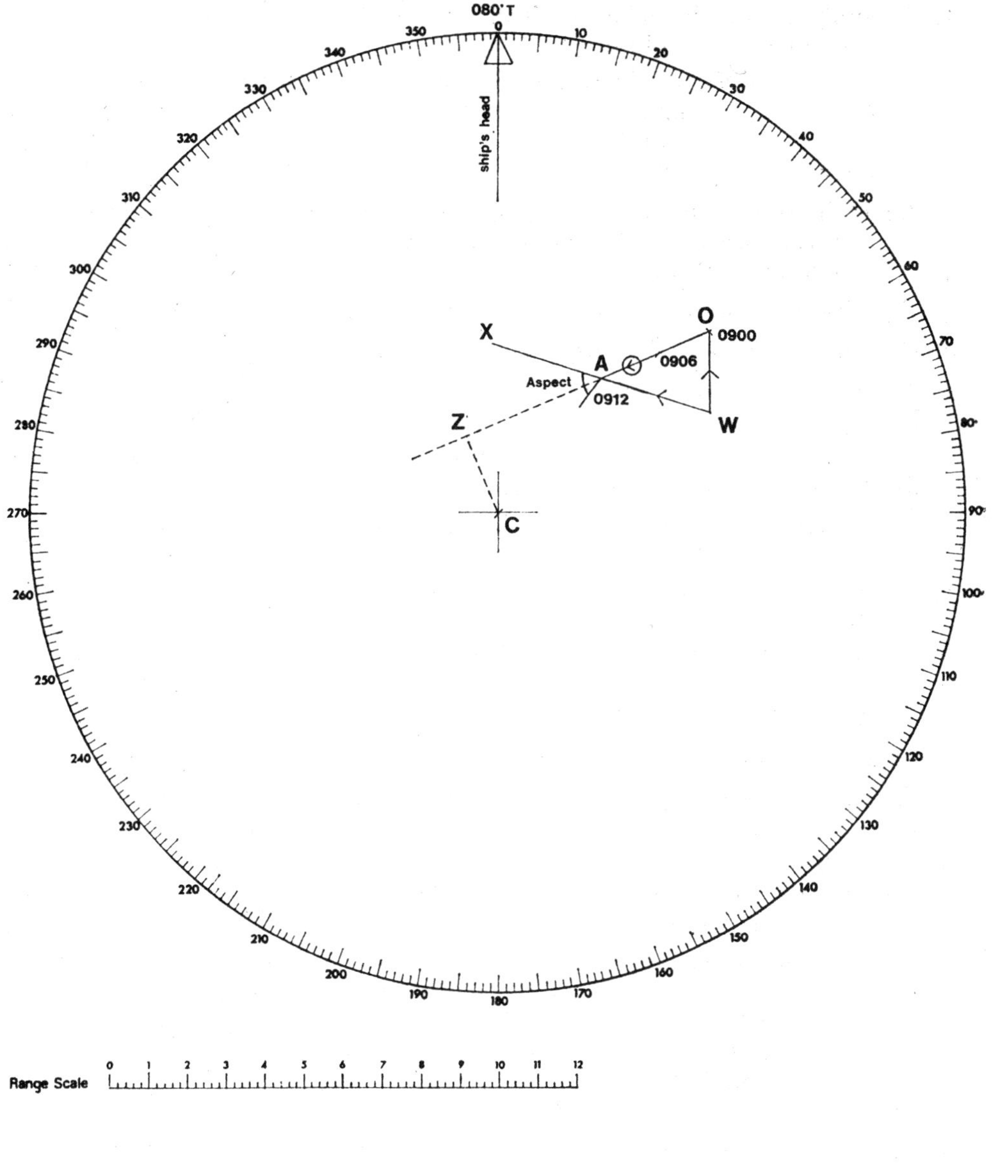

Example 2, *Method* (*b*)

RELATIVE PLOT, NORTH UP

(Refer to Radar Plot 4)

1. By applying the observing ship's course, 080°T, to the relative bearings, the true bearings can be obtained:

	Relative Bearing*	Course	Bearing	Range
0900	050°	080°T	130°T	7·0 miles
0906	046°	,,	126°T	5·6 miles
0912	038°	,,	118°T	4·2 miles

*N.B.: If the Radar Display is compass stabilised, any bearings obtained will be compass bearings. To obtain true bearings, compass error, if present, should be applied.

2. The observer's ship is at the centre of the plot, heading 080°T.
3. Plot the positions of the target as shown in Radar Plot 4.
4. The position of the target at 0900 is labelled '*O*' and at 0912 it is labelled '*A*'.
5. *OA*, the relative track of the target, is identified with an encircled arrow, always from '*O*' to '*A*' as shown. If the target has not altered course or speed its 0906 position should lie on *OA*.
6. *CZ*, the perpendicular from the centre of the plot to *OA* produced, is the range of the target at CPA, i.e. 2·0 miles.
7. The time to CPA is:

   $$\frac{AZ}{OA} \times 12 \text{ minutes}$$

   $$\frac{3{\cdot}7}{3{\cdot}0} \times 12 = 15 \text{ minutes (approx)}$$

   CPA will occur at 0927 hrs. (i.e. 0912 + 15).
8. The relative bearing at CPA is the angle between the ship's course line and *ZC*, i.e. 337° Relative.
9. From '*O*' lay off *OW* equal to the reciprocal of the observing ship's course (080°T) and the distance it steams in 12 minutes (2 miles) i.e. *WO*, arrowed as shown, is the observing ship's course and distance between 0900 and 0912.
10. *WA*, arrowed as shown, is the target's course and distance between 0900 and 0912. The course of 006°T is obtained directly from the azimuth scale on the outer edge of the plotting sheet, i.e. the direction *WA* does not have to be first applied to the observing ship's course as was the case for a ship's head up plot—see Example 2(*a*) 9. *WA* is 2·8 miles long so the target has steamed 2·8 miles in 12 minutes which gives a speed of 14·0 knots.
11. Angle *XAC*, the aspect, is the target's relative bearing of the observing ship, i.e. Red 68°.

Radar Plot 4
(Relative-North up)

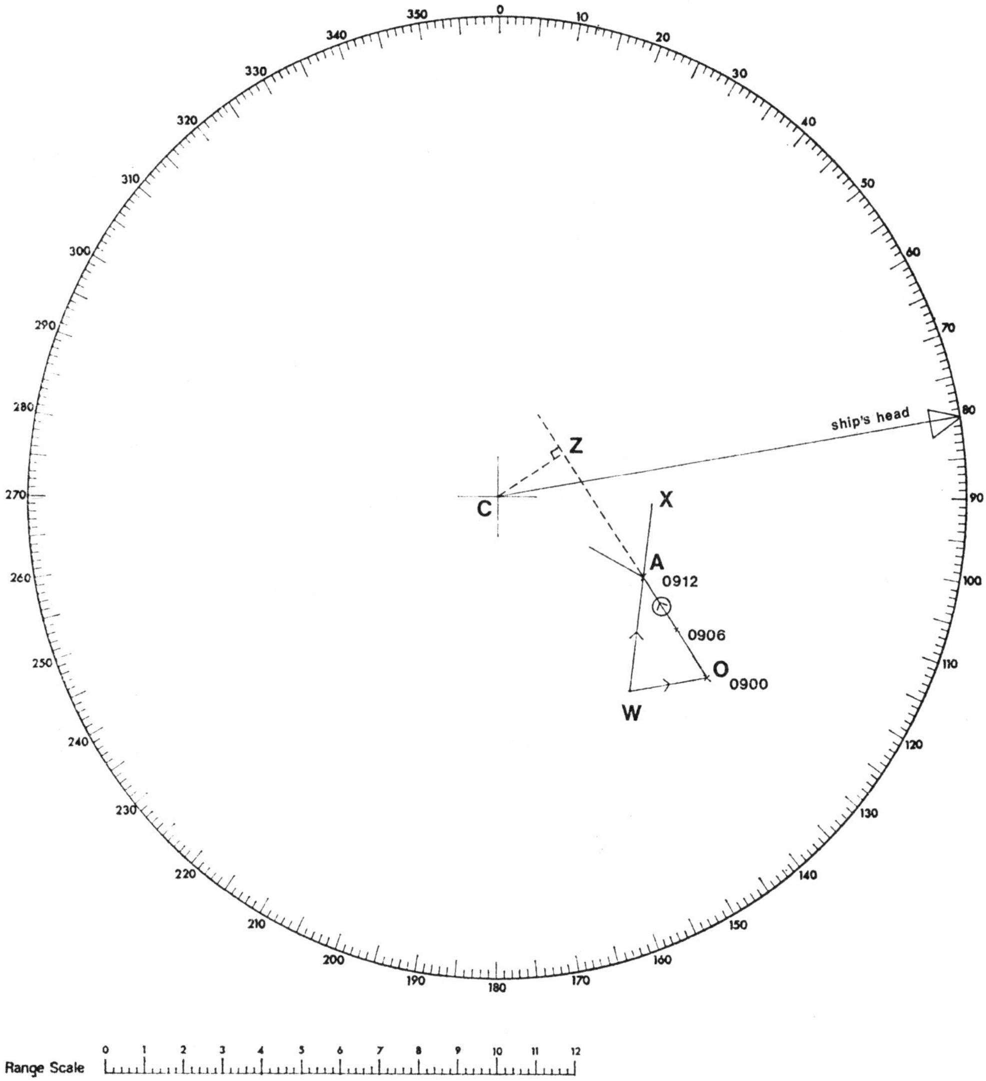

Example 2, *Method* (*c*).

TRUE MOTION PLOT, NORTH UP

(Refer to Radar Plot 5)

1. With speed 10 knots, plot the observing ship's position at 0900, 0906 and 0912 along her course line, 080°T. Label these positions '*E*', '*D*' and '*C*' respectively, as shown in Radar Plot 5.
2. From these points lay off the target's positions at 0900 and 0912.
3. The position of the target at 0900 is labelled '*W*' and at 0912 it is labelled '*A*'.
4. *WA*, arrowed as shown, is the true course (006°T) of the target and its length (2·8 miles) is the distance steamed in 12 minutes which gives a speed of 14 knots.
5. To obtain the CPA draw *WO* equal to *EC* in magnitude and direction. *OA* is the target's relative track and *CZ*, the perpendicular to *OA* produced, is the CPA which is 2 miles. CPA will occur in

   $$\frac{AZ}{AO} \times 12 \text{ minutes}$$

   $$\frac{3{\cdot}7}{3{\cdot}0} \times 12 = 15 \text{ minutes (approx)}$$

   CPA will occur at 0927 hrs.
6. The Aspect, which is the target's relative bearing of the observing ship, is the angle between *AC* and *WA* produced, Red 68°.

AVOIDING A CLOSE QUARTERS SITUATION—THE FORECAST PLOT

Having determined the course and speed of the target, action to avoid a close quarters situation can be decided by means of a forecast plot. Such action will consist of either an alteration of course or an alteration of speed. Any alteration of course or speed, which has been obtained from a forecast plot, is assumed to be instantaneous. This assumption will result in a difference between the observed and estimated CPA.

The following example illustrates the construction and use of a forecast plot.

Radar Plot 5

(True-North up)

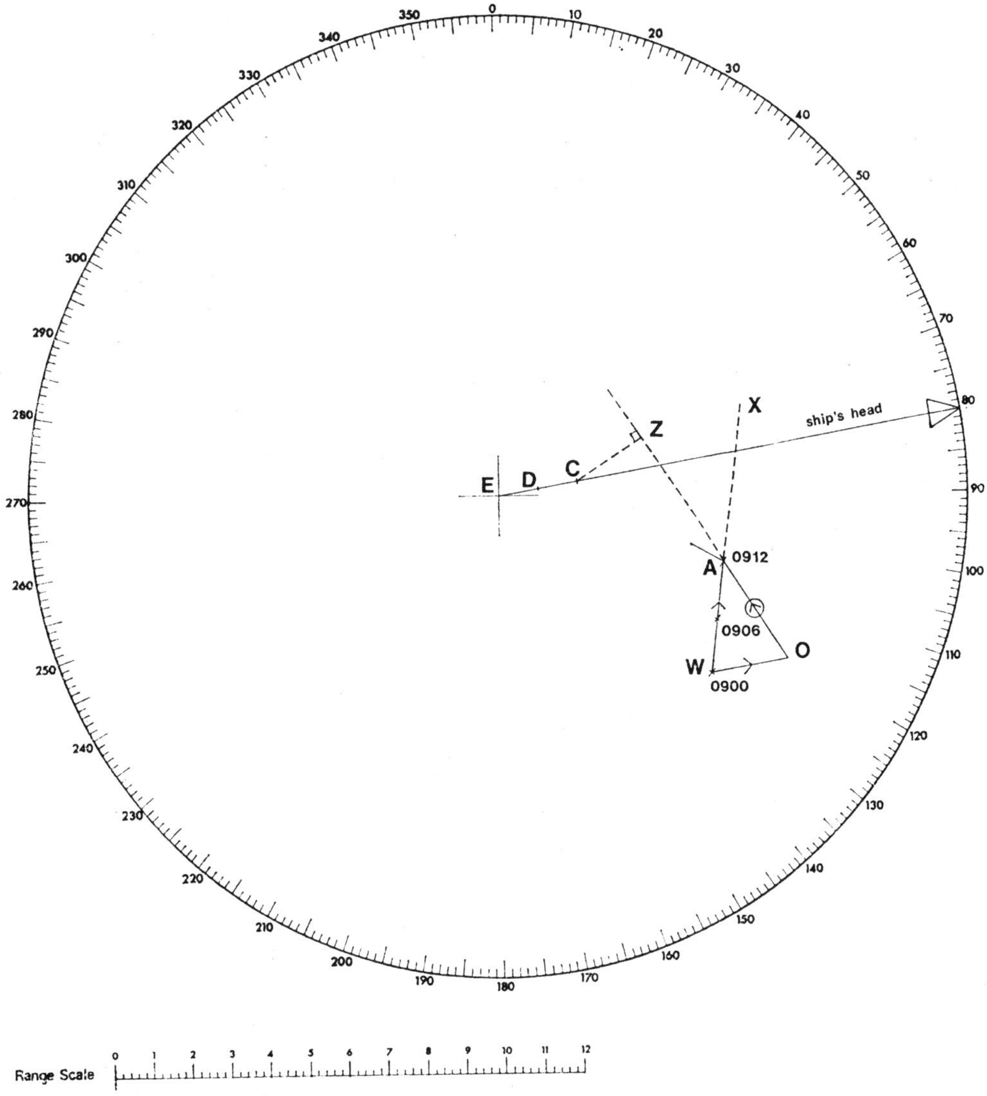

Example 3.

TO FIND THE COURSE OR SPEED ALTERATION NECESSARY TO INCREASE THE DISTANCE OFF AT CPA

(Refer to Radar Plot 6)

The following radar information was obtained when steering 340°T at 10 knots.

	Bearing	Range
0400	035°T	7·0 miles
0406	035°T	6·0 miles
0412	035°T	5·0 miles

Required: (*a*) A complete assessment of the situation at 0412.

(*b*) The single alteration of course or speed necessary at 0415 so that the target's CPA will be 2 miles to Port, assuming an instantaneous alteration.

ASSESSMENT

1. By Relative Motion, North Up Plot, as in Radar Plot 6, the assessment† of the situation is as follows:
   (*a*) The target is on a steady bearing, 55° on the starboard bow.
   (*b*) CPA will be a collision in

$$\frac{5}{2} \times 12 \text{ minutes} = 30 \text{ minutes}$$

   and will occur at 0442 hrs. (0412 + 30).
   (*c*) The target's course is 280°T at 9·5 knots.
   (*d*) Aspect is Red 65°.

†Students should verify this assessment by working through all stages of Radar Plot 6, as previously explained at Example 2 Method (*b*).

TO ACHIEVE A NEW CPA (2 MILES) BY A SINGLE ALTERATION OF COURSE

(Refer to Radar Plot 6)

2. Plot the target's 0415 position, on OA produced. Label this position 'A'. AA^1 is the distance made good along the relative track in 3 minutes.
3. From A^1 draw a tangent to the 2 mile circle and transfer this tangent back through 'A'.
4. With centre 'W' and radius WO draw an arc to cut the transferred tangent at O^1. WO^1 is the new course to steer, 027°T.
5. The new CPA will occur in

$$\frac{A^1Z}{O^1A} \times 12 \text{ minutes}$$

$$\frac{4}{3} \times 12 \text{ minutes} = 16 \text{ minutes}$$

   CPA will occur at 0431 hrs.

Radar Plot 6

(Relative-North up)

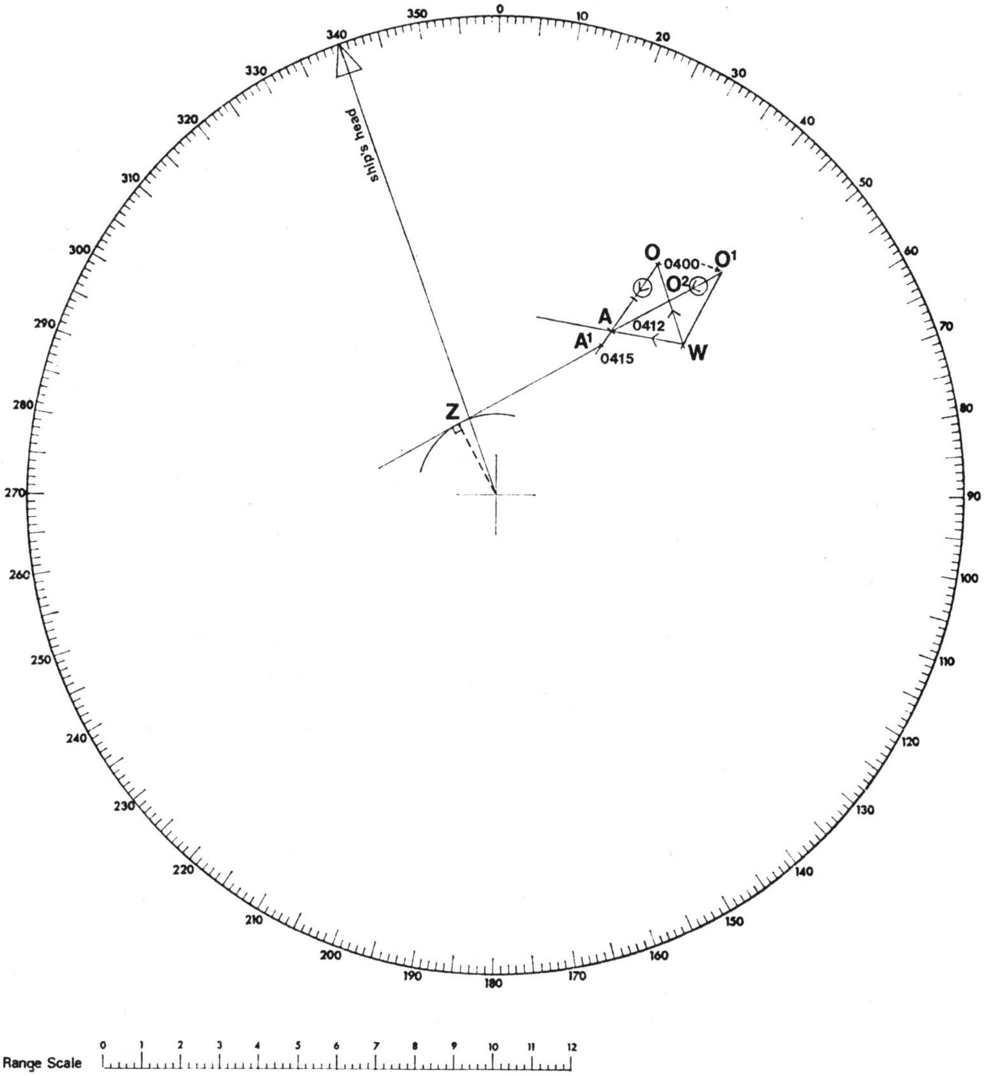

TO ACHIEVE A NEW CPA (2 MILES) BY SPEED REDUCTION ONLY
(Refer to Radar Plot 6)

6. The single alteration of speed is found from WO^2, where O^2 is the point of intersection of WO and O^1A. WO^2 is 1·2 miles long which, over the 12 minute period, represents a speed of 6 knots. The required speed reduction is 4 knots.
7. The new CPA will occur in

$$\frac{A^1Z}{O^2A} \times 12 \text{ minutes}$$

$$\frac{4\cdot1}{1\cdot6} \times 12 = 31 \text{ minutes (approx)}$$

CPA will occur at 0446 hrs.

Because course alterations of 47° and speed reductions of 4 knots cannot be carried out instantaneously, the actual CPA's in this example will be less than 2 miles.

It should be noted that, though a CPA of approximately 2 miles was achieved by both methods, the time of CPA was different in each case. At the reduced speed the time of CPA was later.

THE EFFECT OF TIDE

When using radar in the anti-collision mode, tidal effects are ignored and it is assumed that both the observing ship and the target are subjected to the same set. When a radar plot is made of a stationary target, such as a light vessel which is known to be anchored, the set and drift of the current can be obtained.

Example 4.

TO FIND THE SET AND RATE OF CURRENT FROM A STATIONARY TARGET AND THENCE THE COURSE TO STEER TO PASS A CHOSEN POINT AT A CHOSEN DISTANCE OFF
(Refer to Radar Plot 7)

Whilst steering 000°T at 12 knots the following information was obtained from radar observations of a light vessel:

	Bearing	Range
0600	030°T	4·0 miles
0615	070°T	2·8 miles

At 0615, a second light vessel was observed right ahead at a range of 5 miles.

Required: (*a*) The set and rate of the current at 0615.
(*b*) The course to steer so that the second lighthouse will be passed bearing 270°T distant 1 mile.

1. By Relative Motion, North Up Plot, as in Radar Plot 7, the set is in the direction *AW*, 233°T.

   Because the target is known to be a stationary light vessel, its apparent motion (*WA*) must be due to the current setting the observer in the opposite direction (*AW*). The drift is 0·8 miles in 15 minutes which gives a rate of 3·2 knots.

Radar Plot 7

(Relative-North up)

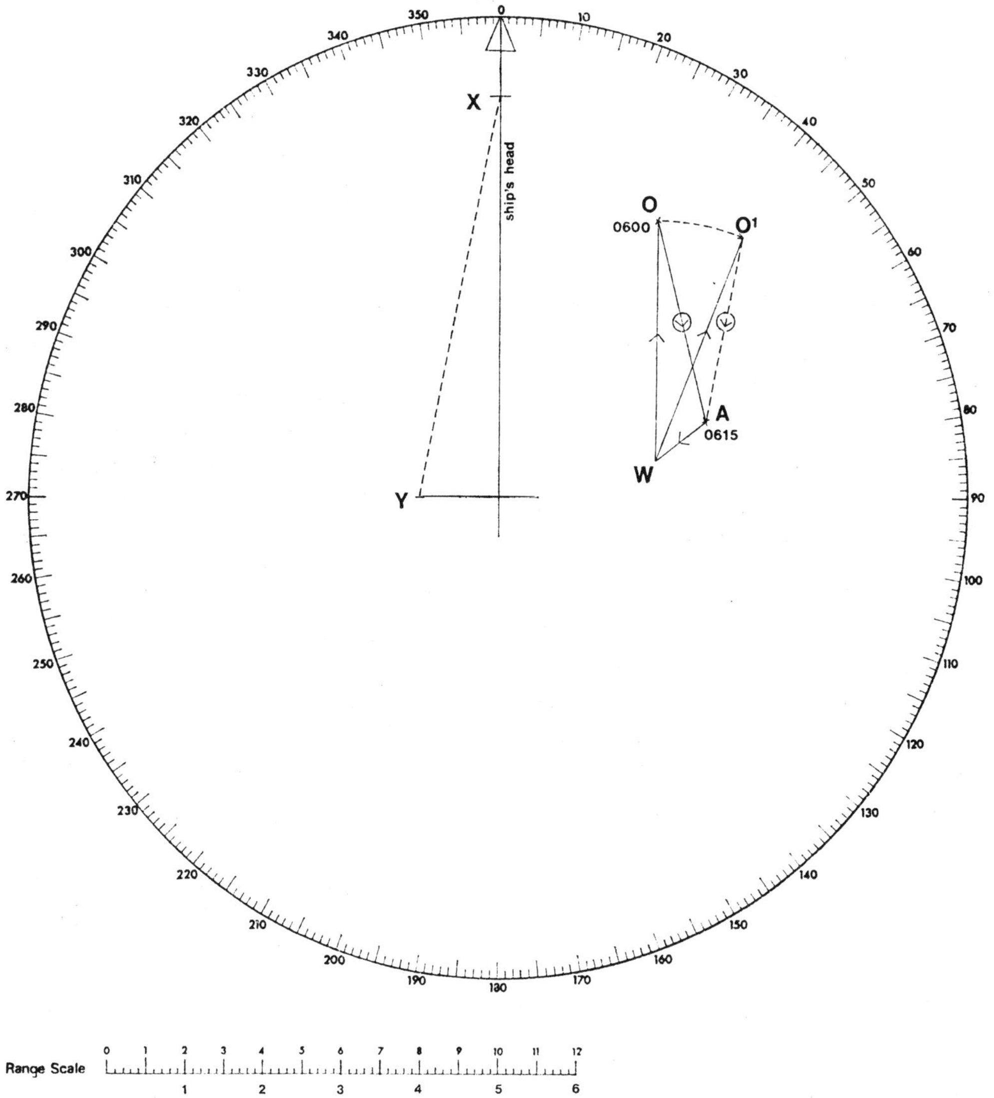

2. Plot the position of the 2nd light vessel at 0615 and its required position when bearing 270°T. Label these positions '*X*' and '*Y*', as shown in Radar Plot 7.
3. Draw a line through '*A*' parallel to *XY* and with centre '*W*' and radius *WO* draw an arc to cut this parallel line at O^1. WO^1 is the course to steer, 021°T, to bring the light vessel to the chosen beam position.
4. The time to CPA is

$$\frac{XY}{O^1A} \times 15 \text{ minutes}$$

$$\frac{5{\cdot}1}{2{\cdot}3} \times 15 = 33 \text{ minutes}$$

CPA will occur at 0648 hrs.

RELATIVE MOTION AND TRUE MOTION RADAR PLOTS—COMPARISON

Assuming the previous examples have been carefully studied, it should now be apparent that, having plotted the first set of radar observations, the Relative Motion Plot does not give the target's true course and speed without further plotting. The Relative Plot also requires the construction of a new triangle whenever the observing ship, or the target, alters course—this can be a disadvantage when several targets are involved. What the Relative Plot does give is an immediate indication of the Target's Closest Point of Approach (CPA) which is a vital piece of information.

In the True Motion Plot, further plotting is required to obtain the target's CPA. In the True Plot, however, both the observing ship's and the target's course lines are continuously produced with each plotted observation. Thus, as soon as a target ship alters course the alteration is immediately apparent by a change in direction of the plotted course line. Similarly, any change of speed is immediately apparent by a change in the target's rate of progress along the course line.

From the foregoing it will be seen that the True Plot tends to give a better overall understanding of the developing situation and is generally better suited to handling several targets at the same time. But it must be again stressed that the True Plot does not give the target ship's CPA without further plotting. If the True Plot is used the CPA should be obtained either by "on plotting" both own ship and target ship's positions along their course lines and, preferably, by making use of relative plotting to construct a course-speed triangle (*OAW*) as previously explained.

Many serving Officers and Masters prefer the True Plot which they consider gives them a better understanding of the developing situation than can be obtained from the Relative Plot. Others prefer the Relative Plot because it gives an immediate indication of risk of collision and the target's closest point of approach. A thorough knowledge of both methods is clearly essential and the student should become familiar with both before forming any personal opinion regarding their relative merits.

RADAR PLOTTING TERMS

True Bearing.

The direction of one object from another expressed in degrees clockwise from True North.

Sea Speed.

The speed of own ship through the water track expressed in knots.

Ground Speed.

The speed of own ship over the ground track expressed in knots.

Relative Track of Target.

The direction of motion of a target as observed on a relative motion display.

True Track of Target.

The direction of motion of a target as observed on a true motion display.

Aspect Symbol ° Red or ° Green.

The relative bearing of own ship from the target ship expressed in degrees 0 to 180 "RED" or "GREEN".

NOTE: When plotting a target ship on a radar the deduced aspect may not be the same as the visual aspect. Several factors can cause this variation. The effects of current and leeway which may be different on both own and target ship; the accuracy of speed input on a true motion display and the variation between the actual water or ground track of own ship and that displayed on the PPI.

Detection.

The recognition of the presence of a target.

Acquisition.

The selection of those targets requiring a tracking procedure and the initiation of their tracking.

Tracking.

The process of observing the sequential changes in the position of a target, to establish its motion.

PLOTTING DIAGRAMS

Radar data obtained from a stabilized or unstabilized relative motion display, or true motion sea stabilized display.

RELATIVE PLOT

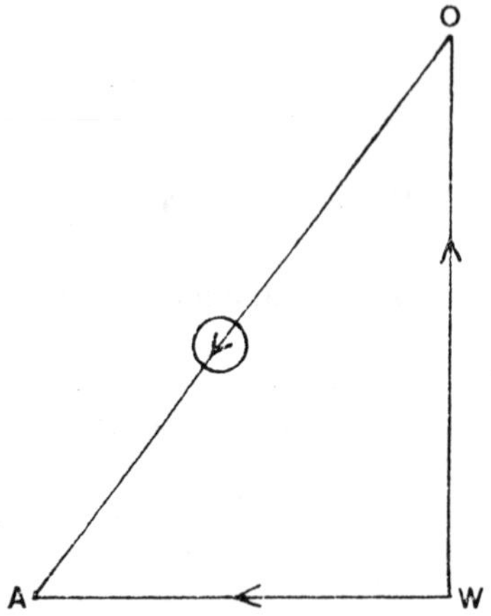

O = Initial position of target on plot at appropriate range and bearing from own ship.

O —(→)— A = target's relative track/speed vector.

W —→— O = own ship's course/speed vector over time increment *OA*.

W —→— A = target's true track/speed vector over time increment *OA*.

TRUE PLOT

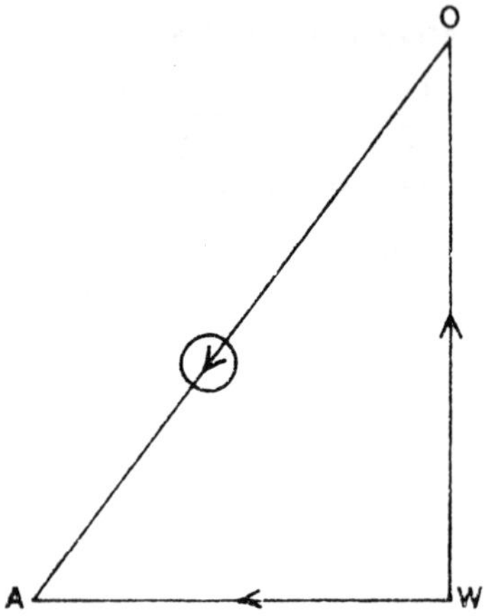

W = Initial position of target on plot at appropriate range and bearing from own ship.

W —→— A = target's true track/speed vector.

W —→— O = own ship's course/speed vector over time increment *WA*.

O —(→)— A = target's relative track/speed vector.

NOTE: When the target is a fixed terrestrial point A —→— W is the vector representing the set and drift of a current or tidal stream.

When the radar data is taken from a true motion ground stabilized display a vector A —→— W for a fixed terrestrial point represents an error of speed input.

COLLISION AVOIDANCE RULES WHICH REFER TO THE USE OF RADAR

The following Rules from the *International Regulations for Preventing Collisions at Sea*, 1972, refer to the use of Radar:—

RULE 6

Safe Speed

Every vessel shall at all times proceed at a safe speed so that she can take proper and effective action to avoid collision and be stopped within a distance appropriate to the prevailing circumstances and conditions.

In determining a safe speed the following factors shall be among those taken into account:

(*a*) By all vessels:
- (i) the state of visibility;
- (ii) the traffic density including concentrations of fishing vessels or any other vessels;
- (iii) the manoeuvrability of the vessel with special reference to stopping distance and turning ability in the prevailing conditions;
- (iv) at night the presence of background light such as from shore lights or from back scatter of her own lights;
- (v) the state of wind, sea and current, and the proximity of navigational hazards;
- (vi) the draught in relation to the available depth of water.

(*b*) Additionally, by vessels with operational radar:
- (i) the characteristics, efficiency and limitations of the radar equipment;
- (ii) any constraints imposed by the radar range scale in use;
- (iii) the effect on radar detection of the sea state, weather and other sources of interference;
- (iv) the possibility that small vessels, ice and other floating objects may not be detected by radar at an adequate range;
- (v) the number, location and movement of vessels detected by radar;
- (vi) the more exact assessment of the visibility that may be possible when radar is used to determine the range of vessels or other objects in the vicinity.

RULE 7

Risk of Collision

(*a*) Every vessel shall use all available means appropriate to the prevailing circumstances and conditions to determine if risk of collision exists. If there is any doubt such risk shall be deemed to exist.

(*b*) Proper use shall be made of radar equipment if fitted and operational, including long-range scanning to obtain early warning of risk of collision and radar plotting or equivalent systematic observation of detected objects.

(*c*) Assumptions shall not be made on the basis of scanty information, especially scanty radar information.

(*d*) In determining if risk of collision exists the following considerations shall be among those taken into account:

(i) such risk shall be deemed to exist if the compass bearing of an approaching vessel does not appreciably change;

(ii) such risk may sometimes exist even when an appreciable bearing change is evident, particularly when approaching a very large vessel or a tow or when approaching a vessel at close range.

Rule 8

Action to Avoid Collision

(*a*) Any action taken to avoid collision shall, if the circumstances of the case admit, be positive, made in ample time and with due regard to the observance of good seamanship.

(*b*) Any alteration of course and/or speed to avoid collision shall, if the circumstances of the case admit, be large enough to be readily apparent to another vessel observing visually or by radar; a succession of small alterations of course and/or speed should be avoided.

(*c*) If there is sufficient sea room, alteration of course alone may be the most effective action to avoid a close-quarters situation provided that it is made in good time, is substantial and does not result in another close-quarters situation.

(*d*) Action taken to avoid collision with another vessel shall be such as to result in passing at a safe distance. The effectiveness of the action shall be carefully checked until the other vessel is finally past and clear.

(*e*) If necessary to avoid collision or allow more time to assess the situation, a vessel shall slacken her speed or take all way off by stopping or reversing her means of propulsion.

Rule 19

Conduct of Vessels in Restricted Visibility

(*a*) This Rule applies to vessels not in sight of one another when navigating in or near an area of restricted visibility.

(*b*) Every vessel shall proceed at a safe speed adapted to the prevailing circumstances and conditions of restricted visibility. A power-driven vessel shall have her engines ready for immediate manoeuvre.

(*c*) Every vessel shall have due regard to the prevailing circumstances and conditions of restricted visibility when complying with the Rules of Section I of this Part.

(*d*) A vessel which detects by radar alone the presence of another vessel shall determine if a close-quarters situation is developing and/or risk of collision exists. If so, she shall take avoiding action in ample time, provided that when such action consists of an alteration of course, so far as possible the following shall be avoided;

(i) an alteration of course to port for a vessel forward of the beam, other than for a vessel being overtaken;

(ii) an alteration of course towards a vessel abeam or abaft the beam.

(*e*) Except where it has been determined that a risk of collision does not exist, every vessel which hears apparently forward of her beam the fog signal of another vessel, or which cannot avoid a close-quarters situation with another vessel forward of her beam, shall reduce her speed to the minimum at which she can be kept on her course. She shall if necessary take all her way off and in any event navigate with extreme caution until danger of collision is over.

RADAR PLOTTING

(Answers on page 378)

Examples for Exercise.

1. Own course 090° T., speed 12 knots. Target bore 315° Relative, range 5 miles and 10 minutes, later it bore 320° Relative, 4 miles. Find the course and speed of the target vessel and the nearest approach assuming no alteration of course and speed.

2. Heading 235° T., at 10 knots, target bearing 010° Relative, range 7 miles. 15 minutes later it bears 012° Relative, range 3 miles. What is the course and speed of the target vessel and what alteration of course would you make to allow a nearest approach of 2 miles.

3. Describe a True Plot and a Relative Plot. From the following information make a True Plot:—Own course 065° T., speed 6 knots. A target was observed bearing 080° Relative, range 5 miles, and after 30 minutes it bore 125° Relative, range 7 miles. Find the course and speed of the target vessel.

4. Heading 333° T., at 20 knots, a target which was known to be a light vessel bore 330° Relative, range 8 miles. After 18 minutes it bore 285° Relative, range 3 miles. Find the set and rate of the current in the vicinity.

5. Course 075° T., speed 10 knots, target bearing 045° Relative, range 8 miles. 3 minutes later it bears 043° Relative, range 7 miles, and after a further 3 minutes it bears 041° Relative, range 6 miles. It is desired to let this vessel cross ahead of you, 3 miles distant, without altering own course. What should the speed be reduced to? Find also the course, speed and aspect of target vessel at last plotted point.

6. Course 300° T., speed 15 knots, targets were observed as follows:—

	"*X*" *Relative Brg.*	*Range*	"*Y*" *Relative Brg.*	*Range*
0000	330°	5 miles	045°	7 miles
0004	331°	3 ,,	046°	6 ,,
0008	320°	2 ,,	047°	5 ,,
0012	293°	1·4 ,,	058°	3·7 ,,

Have the target vessels made any alteration of course or speed? If so, state what each has done.

What will be the nearest approach of each target vessel and would any avoiding action be necessary on your part?

7. Course 105° T., speed 6 knots the following plot was made:

	Relative	Range
1200 target bearing	020°	7 miles
1203	030°	6·5 ,,
1206	030°	6 ,,
1209	030°	5·5 ,,

At 1212 course was altered 90° to starboard (turning time 4 minutes). What report should be made at 1200 and at 1209?

What will be the expected bearing and range of target at 1220, and what will be the nearest approach after the alteration of course? Is a further alteration of course necessary?

8. Course 090° T., speed 20 knots, the following plot was made:—

Target 'X'	*Relative Brg.*	*Range*	*'Y'*	*Relative Brg.*	*Range*
1800	340°	8 miles		050°	7 miles
1803	331°	6 ,,		050°	6 ,,
1806	317°	4 ,,		049°	5 ,,

What report should be made to the O.O.W. at 1800 and at 1806? At 1809, what alteration of course will be necessary to allow "Y" to cross your course 2 miles ahead of you, and what will be his expected bearing and range at 1812? Make no allowance for turning time.

9. Course 180° T., speed 10 knots, the following observations of a target were made:—

	Relative Brg.	*Range*
2200	310°	7 miles
2203	310°	6·3 ,,
2206	310°	5·6 ,,
2209	303°	4·7 ,,
2212	294°	4 ,,

Is any avoiding action necessary on your part? What will be the CPA of the target and when? State what action has been taken by the target vessel.

10. Course 260° T., speed 12 knots, the following observations of a target were made:—

	Relative Brg.	*Range*
0000	290°	6 miles
0003	290°	5 ,,
0006	290°	4 ,,
0009	284°	3·4 ,,
0012	277°	2·9 ,,

Is avoiding action necessary? And if not why not? What is the course and speed of the target at 0006 hrs. and what is her speed at 0012 hrs? Has the target vessel made any alteration of course? What will be its CPA?

11. Course 333° T., speed 15 knots.

Target 'X'	*Relative Brg.*	*Range*	'Y'	*Relative Brg.*	*Range*
0100	315°	5 miles		045°	8 miles
0103	316°	4·4 ,,		045°	7·2 ,,
0106	317°	4 ,,		046°	6·6 ,,
0109	318°	3·4 ,,		046°	5·9 ,,

What report should be made to the O.O.W. at 0109? What are the courses and speeds of target vessels "*X*" and "*Y*"? At 0112 speed was reduced to 5 knots. How far ahead of you will "*X*" and "*Y*" be when they cross your course?

12. The following radar observations were made whilst steering 225° T. at 10 knots.

	Target Bearing	*Range*
1515	220° T	8 miles
1518	221° T	7 ,,
1521	221° T	6 ,,
1524	220° T	5 ,,

At 1524 the observing ship altered course to 285° T. and made the following observation:

1530	198° T	4 miles

At 1530 the observing ship altered course back to 225° T. and made a further observation:

1536	170° T	2·9 miles

Make no allowance for swinging time.

What will be the target's CPA?

Has target altered course since first observed, and if so, give the alteration, also her original course and speed and last course and speed.

13. During fog patches, ship steering 045° T., at 6 knots, targets were observed at 3 minute intervals as follows:—

Target 'X'		*Target 'Y'*		*Target 'Z'*	
Bearing	Range	Bearing	Range	Bearing	Range
020° T.	5 miles	085° T.	5·5 miles	145° T.	7 miles
022° T.	4·2	091° T.	4·3	145° T.	6·4
024° T.	3·4	101° T.	3·3	145° T.	5·8
023° T.	2·9	120° T.	2·4	141° T.	6·1
020° T.	2·6	152° T.	2·0	137° T.	6·4

Give the courses and speed of the targets and state what action you would take if any. What action has been taken by the targets?

14. While steering 045° T., at 10 knots a Ramark beacon is observed 035° T., range 9 miles. 12 minutes later the same beacon bore 034° T., range 7 miles and again 12 minutes later it bore 032° T., range 5 miles. Find the set and drift of the current and also the alteration of course necessary to pass the beacon 2 miles off on your port hand.

15. Steaming at 12 knots on a course of 225° T., targets "*X*" and "*Y*" were observed as follows:—

	Target 'X'	*Target 'Y'*
0000	205° T. 8·5 miles	270° T. 8 miles
0005	204° T. 7·3	270° T. 6·8
0010	204° T. 6·2	270° T. 5·6
0015	203° T. 5·1	270° T. 4·4

"*X*" is known to be a Light Vessel and you wish to pass 3 miles off leaving it to port.

Find (*a*) the alteration of course necessary at 0015 to pass 3 miles off. (*b*) the set and drift of the current. (*c*) the course and speed of "*Y*" and how far she will cross ahead of you, also will she cross your course before or after you pass the Lt. V.

CHAPTER III.

GREAT CIRCLE SAILING.

GREAT circle sailing is the method used for making long ocean passages between two positions which are widely separated on the earth's surface. More specifically, it is used to find the courses which a ship must steer and the track which must be followed when sailing along the lesser arc of a great circle between the two positions.

The lesser arc of the great circle contained between any two places is the shortest possible distance between them; it is, in fact, the track that would be followed if steering straight for a destination as if it were in sight. If a vessel were steering exactly on the great circle course she would be heading directly towards her destination for the whole of the time, and would be a mile nearer to her port for every mile that she made good.

It is, however, not practicable to keep a vessel *exactly* on the great circle track, because to do so would involve continually changing course. A vessel may, however, save considerable distance on a passage by keeping as nearly as practicable on the track of the great circle, and this is the usual procedure.

The position of a suitable number of points, separated by regular intervals of longitude along the great circle, are calculated by means of right-angled spherical trigonometry. The vessel is then steered direct from point to point, and thus very nearly attains her object of sailing or steaming the shortest possible distance. The great circle course from any one place to another is calculated by means of spherical trigonometry and may also be taken out by inspection from ABC Tables, Alt-Azimuth Tables, Weir's Azimuth Diagram or any Sight Reduction Tables such as NP 401/HD 605.

The advantages obtained from the use of great circle sailing as compared with Mercator's sailing are greatest in high latitudes, when there is a large difference of longitude between the points of departure and destination. The advantages are least, and the differences between the great circle and mercatorial courses are least, when the places of departure and destination are in low latitudes, or when the ship's course is near North or South in any latitude. The equator and all meridians are great circles, so that if the ship is steering true North or South it is already on a great circle and the same applies when steering true East or West along the equator.

Unless there is a definite reason for not following it, the great circle track is the proper one to use, and is the best, because it is the shortest. There are cases, however, when it is necessary or advisable to depart from it. It may pass through higher latitudes than are desirable for the ship to go to; it may pass over an island or other land, when, of course, it would be necessary to depart from it; also, it might pass through regions where ice may be expected or where head winds are prevalent. In the latter case the distance saved might be more than counterbalanced by the extra strain put on the ship and by the loss of time and speed caused by heading into strong winds and heavy seas.

The statement that a ship when on the great circle course is always heading direct for her destination, yet continually changing her course, sounds contrary, but is nevertheless quite correct.

A great circle drawn between any two places on a globe will show clearly that the great circle track cuts each successive meridian at a different angle; the ship is, therefore, steering on different true courses though heading in one and the same direction all the time.

Worked Example I.

Find the initial and final courses, also the distance on the great circle, from Honolulu in Lat. 21° 18′ N., Long. 157° 52′ W., to a point off Vancouver Island in Lat. 48° 34′ N., Long. 126° 4′ W. Required also the latitude and longitude of the vertex and the latitude of the points where the great circle cuts the meridian of 150° W., and thence every 10° of longitude to destination.

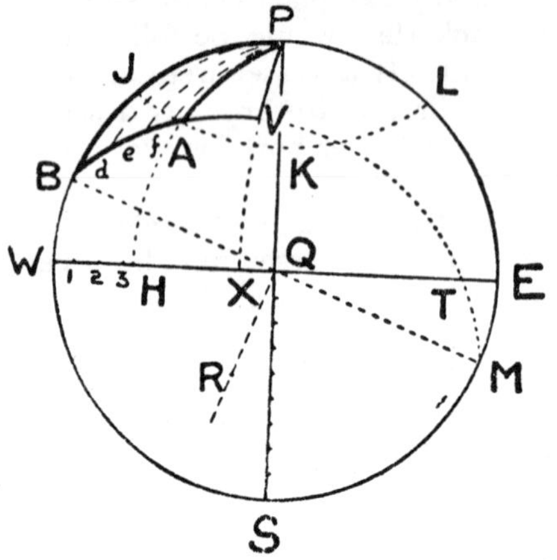

In figure:

P E S W represents the globe.
P and *S* the poles.
W Q E the equator.
B Honolulu, *A* Vancouver.
P B the co-latitude of *B*.
P A the co-latitude of *A*.
∠ *B P A* the D. Longitude *B* to *A*.
Arc *B A* the G.C. track and distance.
∠ *P B A* the initial course.
∠ *P A V* the final course.
d, e and *f* the successive points asked for
V the vertex of the great circle.

Construction of Figure.

It is desirable to draw a fairly accurate figure for great circle problems in order to see clearly the relative positions of the various points.

The simplest method of drawing an approximate figure on the equidistant projection is to take nine equal parts on the compasses from any convenient scale (we have used the equidistant marks on *Q S*) and describe a circle with this radius to represent the globe: then draw in *W Q E* the equator, *P Q S* a central meridian, and proceed as follows:—

1. Mark the spot *B*, the place of lesser latitude, on the circumference by placing the centre of a protractor at *Q* and making ∠ *W Q B*=Latitude *B*=21° 18′ N.
P B W then becomes the meridian of Honolulu, Long. 157° 52′ W.

2. Measure *W H*=D. Longitude=31° 48′ from the scale and draw in the meridian *P H*. The centre of all the meridian circles is on the equator, produced if necessary. *P H* is now the meridian of Longitude 126° 04′ W. passing through *A*, the destination.

3. Run in the parallel of Latitude 48° 34′ N., passing through *A* by making *Q K* =$48\frac{1}{2}$° from the scale of equal parts, and laying off ∠ *W Q J* and ∠ *E Q L*=$48\frac{1}{2}$° with the protractor. By trial, find on *Q P*, produced if required, the centre of the parallel *J K L* and draw it in. It cuts meridian *P H* at *A* and fixes the point of destination.

4. To draw in *B A* the great circle. Join *B Q* with the straight edge of the protractor and produce to *M*. Make ∠ *B Q R*=90°, then on *Q R* find, by trial, the centre of the great circle passing through *B* and *A*, and produce it to *M*. It will be found that *B Q M* is a diameter of the circle *P E S W*, as all great circles cut the circumference of the primitive at diametrically opposite points.

5. The meridian passing through the vertex of a great circle is always 90° from the longitude of the spot where it cuts the equator at *T*. Measure *T X*=D. Longitude

90°, and draw in the meridian *P X*; it cuts the great circle at *V*, the vertex, which is the turning point, and represents the maximum latitude reached on this track. The great circle course at the vertex is always due East or due West.

6. Make *W*1 equal to 8° to represent the D. Longitude between *B* 157° 52′ W. and 150° W.; then 1 to 2=D. Longitude 10°; 2 to 3=D. Longitude 10°. Draw in the meridians through those numbered spots 1, 2 and 3, cutting the great circle at *d*, *e* and *f* respectively. It is required to find the latitudes and longitudes of those points for laying off on a Mercator chart.

In triangle P B A:

1. Given *P B*, *P A* and ∠*B P A*, find side *B A*.

2. Given *P B*, *P A* and *B A*, find 1st ∠*B*, 2nd ∠*P A B*, to be subtracted from 180° to give ∠ *P A V*.

In triangle P A V:

3. Given ∠*V*=90°, side *P A* and ∠*P A V*, find 1st *P V* the co-latitude of vertex; 2nd ∠*A P V* the D. Longitude between *A* and vertex to be applied to Longitude *A*.

Lat. *A* 48° 34′ N.,	co-lat. 41° 26′ (*P A*)		Long. *A*,	126° 4′ W.
Lat. *B* 21 18 N.,	co-lat. 68 42 (*P B*)		Long. *B*,	157 52 W.
	(*P A*∼*P B*) 27° 16′		D. Long.	31° 48′ (*P*)

1. To find the Distance.

Hav *B A*=Hav *P* sin *P A* sin *P B*+hav (*P A*∼*P B*)

P	31° 48′	log hav	8·87537
P A	41 26	log sin	9·82069
P B	68 42	log sin	9·96927
		log hav	8·66533
		nat hav	0·04627
(*P A*∼*P B*)	27° 16′	nat hav	0·05556
B A	37° 13·2′	nat hav	0·10183
	60		
Distance	2233·2 miles		

2. 1st To find angle *B* (the initial course)

$$\text{Hav } B = \frac{\text{hav } PA - \text{hav } (BA \sim BP)}{\sin BA \sin BP}$$

P A	41° 26′	nat hav	0·12514
(*B A*∼*B P*)	31 28·8	nat hav	0·07359
		nat hav	0·05155
		log hav	8·71221
B A	37° 13·2′	log cosec	0·21833
B P	68 42	log cosec	0·03073
B	35° 12·5′	log hav	8·96127

Initial course, N. 35° 12·5′ E.
or 035° 12·5′ T.

2nd To find angle *A* (the final course)

$$\text{Hav } A = \frac{\text{hav } PB - \text{hav } (AB \sim AP)}{\sin AB \sin AP}$$

P B	68° 42′	nat hav	0·31837
(*A B*∼*A P*)	4 12·8	nat hav	0·00135
		nat hav	0·31702
		log hav	9·50109
A B	37° 13·2′	log cosec	0·21833
A P	41 26	log cosec	0·17931
A	125 44	log hav	9·89873

Final course, S.125° 44′ E.
or 054° 16′ T.

If angles *A* and *B* are both acute—that is, both less than 90°—the vertex of the great circle will be inside the triangle, but if either of them be over 90° it will be outside the triangle. In this case angle *A* is over 90°, and, therefore, *V* is on the great circle outside the triangle *P A B*, as shown in the diagram.

3. To find the Latitude and Longitude of the Vertex.

In triangle PAV. By Napier's Rules:—

1st Sin PV = cos *comp* PA cos *comp* A
Sin PV = sin PA sin A

PA	41° 26′	sin	9·82069
A	54 16	sin	9·90942
PV	32° 29·5′	sin	9·73011
	90 00		
Lat. vertex	57° 30·5′ N.		

2nd Sin *comp* PA = tan *comp* A tan *comp* APV
Cos PA = cot A cot APV
Cos PA × tan A = cot APV

PA	41° 26′	cos	9·87490
A	54 16	tan	0·14300
APV	43° 49′	cot	10·01790
Long. A	126 04 W.		
Long. vertex	82° 15′ W.		

4. To find the Latitude of Points 10° of Longitude apart for laying off on Mercator's Chart.

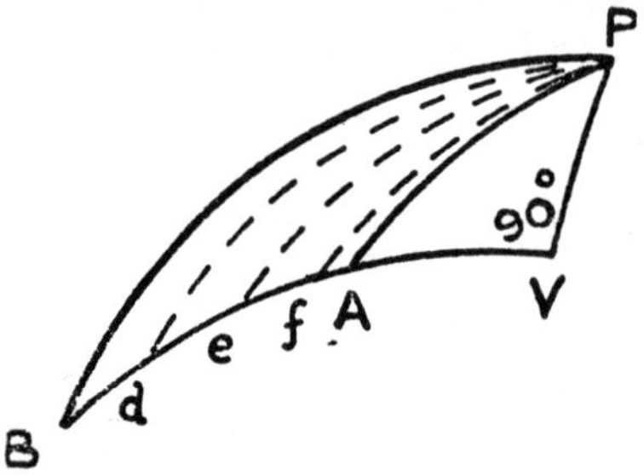

Let d be the first point in Long. 150° W.

Then in triangle PdV, given side PV and angle dPV.'

Required to find side Pd=co-lat. of d.

By Napier's Rules:—

Sin *comp* dPV = tan PV tan *comp* Pd
∴ cos dPV = tan PV cot Pd
i.e. cos dPV × cot PV = cot Pd
or cos dPV × tan lat. V = tan lat. d
similarly cos ePV × tan lat. V = tan lat. e
and cos fPV × tan lat. V = tan lat f

Long. of vertex	Long. of successive points	Polar angle of points	Latitudes as calculated
82° 15′ W.	150° W. (d)	67° 45′ (dPV)	30° 44′ N.
,,	140 W. (e)	57 45 (ePV)	39 57·5 N.
,,	130 W. (f)	47 45 (fPV)	46 33 N.

dPV	67° 45′	cos	9·57824	ePV	57° 45′	cos	9·72723	fPV	47° 45′	cos	9·82761
Lat. V	57 30·5	tan	0·19595	V	57 30·5	tan	0·19595	V	57 30·5	tan	0·19595
Lat d	30° 44′ N.	tan	9·77419	e	39° 57·5′ N.	tan	9·92318	f	46° 33′ N.	tan	10·02356

Worked Example II.

Find the initial and final courses and the distance on the great circle from Otago harbour, Tairoa Head, in Lat. 45° 47′ S., Long. 170° 45′ E., to Callao, in Lat 12° 4′ S., Long. 77° 14′ W. Required also the latitude and longitude of the vertex and the position of a succession of points along the great circle at a distance of 10° of longitude apart, beginning with a point in Long. 172° 54′ E.

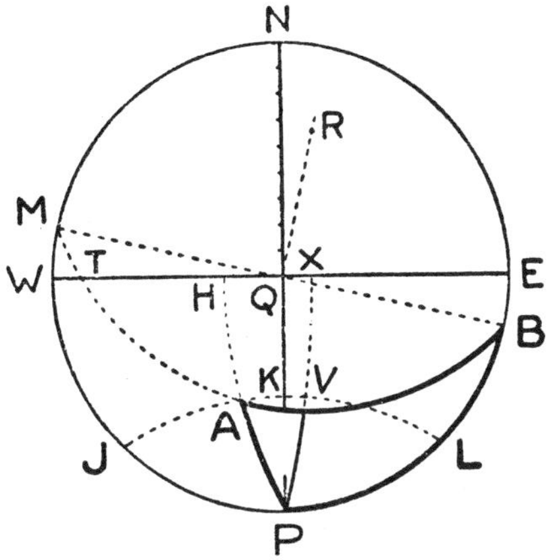

Construction of figure:

The geometrical construction of the figure is the same as in the previous example.

P B represents the meridian of Callao, the place of lesser latitude marked on the circumference of the circle.

Lay off *E H*, the D. Long. and draw in the meridian of Otago, *P H*.

J K L is the parallel of Lat. 45° 47′ S. and *A* represents the position of Otago.

X R is a locus perpendicular to the diameter *B X M*, and spot *R*, found by trial, is the centre of the great circle *M A B* which cuts the equator at *T*. *T X* is D. Long. 90°, giving the position of *P X* the meridian through vertex *V*.

The initial and terminal courses may be found by the sine formula with a considerable saving of labour, but care must be taken when the angles are near 90° as neither the figure nor the formula will indicate whether the angles fall below or above 90°.

In triangle P A B.

(i) Given *P A*, *P B* and included angle *P*, find *A B* by haversine formula.

(ii) Given ∠*P*, *A B* and *P B*, find ∠*A* by sine formula.

(iii) Given ∠*P*, *A B* and *P A*, find ∠*B* by sine formula.

In triangle P A V.

(iv) Given *P A*, ∠*A* and ∠*V*=90°, find first *P V* the co-lat of vertex; second, ∠*A P V* the D. long between *A* and vertex.

A Lat. 45° 47′ S.	co-lat. 44° 13′ (*PA*)	*A* Long.	170° 45′ E.
B Lat. 12 04 S.	co-lat. 77 56 (*P B*)	*B* Long.	77 14 W.
	(*PA*∼*P B*) 33° 43′		247 59 E.
			360 00
		∠*A P B*	112 01 W.

(i) Hav *A B* = hav *P* sin *P A* sin *P B* + hav (*P A*∼*P B*)
(ii) Sin *A* = sin *P* sin *B P* cosec *A B*
(iii) Sin *B* = sin *P* sin *A P* cosec *A B*

P	112° 01′	log hav	9·83723	sin	9·96711	sin	9·96711
P A	44 13	log sin	9·84347			sin	9·84347
P B	77 56	log sin	0·99030	sin	9·99030		
		log hav	9·67100				
		nat hav	0·46881				
(*P A*∼*P B*)	33° 43′	nat hav	0·08410				
A B	96 4·4	nat hav	0·55291	cosec	0·00244	cosec	0·00244
	60						
				sin	9·95985	sin	9·81302
Distance	5764·4 miles			*A*	65° 44·5′	*B*	40° 33·2′

Initial course S. 65° 44·5′ E. **Final course N. 40° 33·2′ E.** Distance 5764·4 miles.

The initial and final courses may be found simultaneously by means of the tangent formula from the two given sides and their included angle. Although this means introducing another formula, yet it has the advantage of using the quantities given in the question and not the calculated side *A B*.

$$\text{Tan } \tfrac{1}{2}(A+B) = \frac{\cos \tfrac{1}{2}(PA \sim PB)}{\cos \tfrac{1}{2}(PA - PB)} \cot \frac{P}{2}$$

$$\text{Tan } \tfrac{1}{2}(A \sim B) = \frac{\sin \tfrac{1}{2}(PA \sim PB)}{\sin \tfrac{1}{2}(PA + PB)} \cot \frac{P}{2}$$

then $\tfrac{1}{2}(A+B) \pm \tfrac{1}{2}(A \sim B) = A$ and B.

$\tfrac{1}{2}(PA \sim PB)$	16° 51·5′	cos	9·98092	sin	9·46241
$\tfrac{1}{2}(PA + PB)$	61 04·5	sec	0·31546	cosec	0·05787
$\frac{P}{2}$	56 00·5	cot	9·82885	cot	9·82885
$\tfrac{1}{2}(A+B)$	53 08·9	tan	10·12523	tan	9·34913
$\tfrac{1}{2}(A \sim B)$	12 35·6				

A	65 44·5	Initial course	S. 65° 44·5′ E.	or	114° 15·5′ T.
B	40 33·3	Final „	N. 40 33·3′ E.	or	040° 33·3′ T.

To find the Latitude and Longitude of the Vertex. By Napier's Rules:—

Sin *P V* = cos *comp P A* cos *comp A*
Sin *P V* = sin *P A* sin *A*

P A	44° 13′	sin	9·84347
A	65 45	sin	9·95988
P V	39 29	sin	9·80335
	90 00		
	50° 31′ S. Lat. of vertex.		

sin *comp P A* = tan *comp A* tan *comp A P V*
cos *P A* = cot *A* cot *A P V*
cos *P A* × tan *A* = cot *A P V*

P A	44° 13′	cos	9·85534
A	65 45	tan	0·34634
A P V	32 09	cot	0·20168
Long. *A*	170 45 E.		
	202° 54′ E.		
	360 00		
Long. of vertex	157° 6′ W.		

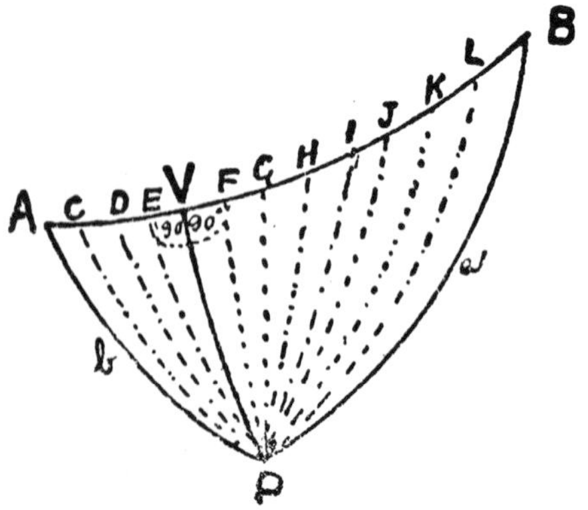

To find the Latitude of Points 10° of Longitude apart.

Let *C* be the first point in Long. 172° 54′ E.

Then in triangle *P C V*, given side *P V*, angle *V P C*, and ∠*V* = 90°.

Required to find side *P C* = co-lat. of *C*.

By Napier's Rules:—

Sin *comp P* = tan *P V* tan *comp P C*
∴ cos *P* = tan *P V* cot *P C*
i.e., cos *P* × cot *P V* = cot *P C*
or cos *P* × tan lat *V* = tan lat *C*

The same investigation applies to the other triangles to get the latitudes of the successive points.

Long. of vertex	Long. of successive points	Polar angle of points	Latitudes as calculated below
157° 06′ W.	172° 54′ E. (C)	30° *VPC*	40° 26′ S.
„	177 06 W. (*D*)	20 *VPD*	48 45¼ S.
„	167 06 W. (*E*)	10 *VPE*	50 05 S.
„	157 06 W. (*V*)	0 Vertex	50 31 S.
„	147 06 W. (*F*)	10 *VPF*	50 05 S.
„	137 06 W. (*G*)	20 *VPG*	48 45¼ S.
„	127 06 W. (*H*)	30 *VPH*	46 26 S.
„	117 06 W. (*I*)	40 *VPI*	42 55 S.
„	107 06 W. (*J*)	50 *VPJ*	37 58 S.
„	97 06 W. (*K*)	60 *VPK*	31 15 S.
„	87 06 W. (*L*)	70 *VPL*	22 33 S.

VPC	30° 00′	cos	9·937531	*VPD*	20° 00′	cos	9·972986	*VPE*	10° 00′	cos	9·99335 1
Lat. *V*	50 31	tan	10·084153	*V*	50 31	tan	10·084153	*V*	50 31	tan	10·08415 3
Lat. *C*	46° 26′	tan	10·021684	*D*	48° 45¼′	tan	10·057139	*E*	50° 5′ S.	tan	10·077504

Points *F*, *G*, and *H* have the same latitudes as *E*, *D* and *C* respectively, because their polar angles are the same but on the other side of the vertex. The latitudes of *I*, *J*, *K* and *L* are given above and the student may work them out in verification.

GNOMONIC CHART.

The adjoining illustration represents a section of a chart drawn on the gnomonic projection to facilitate laying off great circle tracks. The meridians are straight lines radiating from the pole, the parallels of latitude are conic sections, the great circle is a straight line.

Example.—Required to lay off the great circle track from Rathlin Island (55° 18′ N., 6° 10′ W.) to Belle Isle (51° 58′ N., 55° 30′ W.).

Join the two places on the chart by a straight line. It will be noted that the great circle track crosses successive meridians at a different angle, and that the latitudes and longitudes of points along the great circle taken from the gnomonic chart may be transferred to the Mercator chart on which the track will appear as a curve.

We have marked four points on the track and have read off their positions from the gnomonic chart as follows: *A*, 56° 18′ N., 15° W.; *B*, 56° 38′ N., 25° W.; *C*, 56° 00′ N., 35° W.; *D*, 54° 35′ N., 45° W., the respective Mercator courses being *A*, N. 87° W.; *B*, S. 84° W.; *C*, S. 76° W.; *D*, S. 67° W.

Another track is shown from Sable Island to Brest, the position of its vertex from the chart being Lat. 49° 56′ N., Long. 24° 20′ W.

GREAT CIRCLE SAILING.

In the following examples find the initial, or the final, course; the great circle distance; the latitude and longitude of vertex and of successive points as stated:

1. *Lizard to Barbados.*—(*A*) Lat. 49° 50′ N., Long. 5° 12′ W.; (*B*) Lat. 13° 06′ N., Long. 59° 20′ W. Points every 10° apart, commencing with a point in Long. 15° 12′ W.
2. From (*B*) Lat. 36° 55′ N., Long. 140° W. to (*A*) Lat. 51° 40′ N., Long. 170°. E. Points every 10° apart, commencing at the meridian of Long. 150° W.

3. *Algoa Bay to Cape Comorin.*—(*A*) Lat. 34° 02′ S., Long. 25° 30′ E.; (*B*) Lat. 8° 09′ N., Long. 78° 06′ E. Points every 10° apart, commencing at a point in Long. 35° 30′ E.

4. *Cape of Good Hope to Straits of Sunda.*—(*A*) Lat. 34° 22′ S., Long. 18° 29′ E; (*B*) Lat. 6° 40′ S., Long. 104° 29′ E. Points every 10° apart, beginning with a point 10° of Longitude eastward of vertex, and thence to destination.

5. *Bermuda to English Channel.*—(*A*) Lat. 32° 15′ N., Long. 64° 45′ W.; (*B*) Lat. 49° 45′ N., Long. 5° 12′ W. Points every 10° from Long. 55° W.

6. *Vancouver to Hawaii.*—(*A*) Lat. 48° 25′ N., Long. 123° 15′ W.; (*B*) Lat. 21° 20′ N., Long. 157° 48′ W. Points every 10° from Long. 135° W.

7. *Hawaiian Islands to Chile.*—(*A*) Lat. 20° N., Long. 160° W.; (*B*) Lat. 32° 30′ S., Long. 72° W. Points every 20° from 140° W.

8. *Sandy Hook to Straits of Gibraltar.*—(*A*) Lat. 40° 26′ N., Long. 73° 58′ W.; (*B*) Lat. 36° 00′ N., Long. 5° 33′ W. Points every 10° from 60° W.

9. *From A to B.*—(*A*) Lat. 36° 56′ N., Long. 76° 00′ W.; (*B*) Lat. 36° 56′ N., Long. 9° 00′ W. Points every 12½° from 67° 30′ W.

10. *From A to B.*—(*A*) Lat. 10° 37′ N., Long. 89° 56′ E.; (*B*) Lat. 40° 33′ S., Long. 55° 22′ E. Points every 5° from 85° E.

COMPOSITE GREAT CIRCLE SAILING.

Composite great circle sailing is the art of sailing a vessel on a composite track which is the shortest possible distance between any two places when a maximum latitude has been fixed upon, and it has been decided that the ship shall not go beyond that latitude.

Should a vessel, after considering the great circle track from one place to another, find that it would take her into too high a latitude, and therefore decide not to follow it, she should fix on what is to be her maximum latitude, and calculate and lay out a composite track.

This consists of the arcs of two separate great circles, and between them the arc of the parallel of latitude decided upon, which is the arc of a small circle, the three arcs together forming the composite track.

The great circle track from the point of departure to the maximum latitude is calculated, and the ship is sailed along it until she reaches the point where it touches the maximum latitude decided upon, which point is the vertex of the great circle on which she has been sailing. She then leaves the great circle, and sails along the parallel of maximum latitude until she reaches the vertex of another great circle (previously calculated), touching the maximum latitude and passing through the point of destination. She then leaves the parallel of latitude and sails along the arc of the second great circle to her destination.

Composite sailing is not a case of calculating a great circle direct from departure to destination, sailing along it until the maximum latitude is reached, then proceeding along that parallel of latitude, and picking up and following the arc of the same great circle at its other end. The arcs of the two great circles (one at each end of the track) are not in any way connected with one another, except by the arc of the small circle between them.

A composite track could be used with advantage by a vessel bound from, say, the River Plate to Melbourne, in which case the ordinary great circle track would pass through too high a latitude.

Worked Example I.

Required the composite route from Tristan da Cunha in Lat. 37° 3′ S., Long. 12° 18′ W., to Tasmania in Lat. 44° S., Long. 145° E., the maximum lat. to be 52° S. Find the longitudes at which the ship must meet and leave that latitude, also the initial and final great circle courses and the total distance along the composite track. Required also the longitude from vertex and the latitudes at which the track crosses the meridian of 20° E. and thence every 10° of longitude to destination.

Construction of figure:

A figure (in pencil) should be drawn in order to keep in view the relative positions of the points of departure and destination, the given parallel of maximum latitude and the two distinct great circles each having its vertex on the parallel of the latitude.

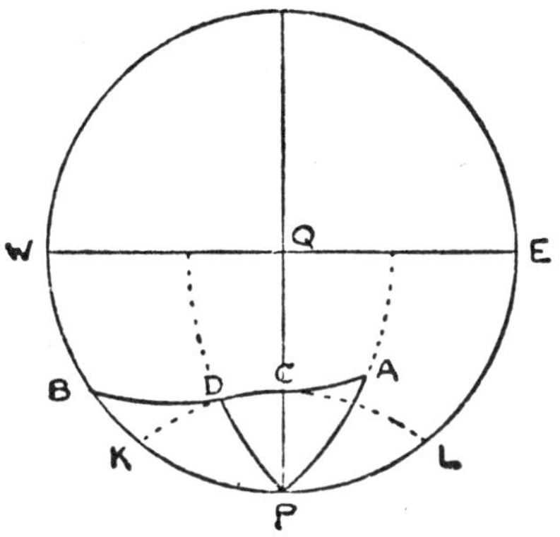

W P E	represents the South Hemisphere and
K D C L	the parallel of maximum latitude, 52° S.
P W	the meridian of Tristan da Cunha (*B*), 12° 18′ W.
P A	the meridian of Tasmania (*A*), 145° E.
D	the vertex of the 1st G.C., arc *B D*.
C	the vertex of the 2nd G.C., arc *C A*.
D C	an arc of the parallel *K L*.

These three arcs together form the composite route, viz., *B D*, *D C* and *C A*.

The calculation resolves itself into the solution of two right angled spherical triangles by Napier's Rules, and the conversion of a D. Long. into its corresponding departure by parallel sailing as follows.

In triangle P B D:

Given *P B* the co-latitude of *B*, *P D* the co-latitude of *D*, $\angle D = 90°$. Find (i) $\angle B$ the initial course, (ii) *B D* the great circle distance, (iii) $\angle BPD$ the D. Long. between *B* and *D*.

(i)

Sin *P D* = cos *comp P B* cos *comp B*
Sin *P D* = sin *P B* sin *B*
Sin *P D* × cosec *P B* = sin *B*
or, cos lat. *D* sec lat. *B* = sin *B*
cos 52° sec 37° 3′ = sin *B*

(ii)

Sin *comp P B* cos *P D* cos *B D*
Cos *P B* × sec *P D* = cos *B D*
or, sin lat. *B* cosec lat. *D* = cos *B D*
sin 37° 3′ cosec 52° = cos *B D*

(iii)

Sin *comp B P D* = tan *comp P B* tan *P D*
cos *B P D* = cot *P B* tan *P D*
or, cos *B P D* = tan lat. *B* cot lat *D*
cos *B P D* = tan 37° 3′ cot 52°

(i)

Lat. *D*	52° 00′		cos	9·789342
Lat. *B*	37 03		sec	0·097937
$\angle B$	50 28·8		sin	9·887279

Initial course S. 50° 28·8′ E. or 129° 31·2′ T.

Similarly in triangle *P C A* given *P C* the co-lat. of *C*, *P A* the co-lat of *A*, $\angle C = 90°$.

Find (i) $\angle A$ the final course, (ii) *C A* the great circle distance, (iii) $\angle CPA$ the D. Long. between *C* and *A*. The investigation is the same as in triangle *P B D* and we get

(i) Cos 52° sec 44° = sin *A*
(ii) Cosec 52° sin 44° = cos *C A*
(iii) Cot 52° tan 44° = cos *C P A*

(ii)		(iii)	
cosec	0·103468	cot	9·892810
sin	9·779966	tan	9·877903
cos	9·883434	cos	9·770713
B D	40° 7′ 43″	$\angle$ *B P D*	53° 51′ 23″
	60	Long. *B*	12 18 00 W.
Dist.	2407·7 miles	Long. *D*	41 33 23 E.

(i)				(ii)		(iii)	
Lat. *C*	52° 00′	cos	9·789342	cosec	0·103468	cot	9·892810
Lat. *A*	44 00	sec.	0·143066	sin	9·841771	tan	9·984837
∠ *A*	58 51·3	sin	9·932408	cos	9·945239	cos	9·877647
Final course			N. 58° 51·3′ E.	*C A*	28° 10′ 19″	∠ *A P C*	41° 1′ 12″
		or	058° 51·3′ T.		60		145 0 0 E.
				Dist.	1690·3 miles	Long *C*	103 58 48 E.

To find distance along parallel of maximum latitude.

Long. *D*	41° 33′ 23″ E.		*D C* = *D*. Long. × cos lat.		
Long. *C*	103 58 48 E.	Lat.	52° 0′	cos	9·789342
	62 25 25	*D*. Long	3745·4	log	3·573498
	60	*D C*	2305·9	log	3·362840
D. Long.	3745·4				

To find the total distance.

Total distance = *B D* + *D C* + *C A*

B D	2407·7 miles
D C	2305·9 ,,
C A	1690·3 ,,
	6403·9 miles

To find the latitudes of the points along the track.

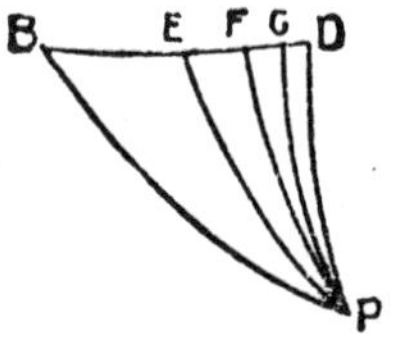

If *E* be the first point in Long. 20° E., then in triangle ***E P D***, given side ***P D***, ∠***E P D***, ∠***D***=90°, find side ***P E*** the co-lat. of ***E***.

Sin *comp* ***E P D***=tan ***P D*** tan *comp* ***P E***

cos ***E P D***=tan ***P D*** cot ***P E***

cos ***E P D*** cot ***P D***=cot ***P E***

or, cos D. Long. tan lat *D*=tan lat *E*.

The same formula applies to the other points by substituting the proper values of the successive polar angles.

Longitude of vertex	Longitudes of points	Polar angles of points	Latitudes as calculated
(*D*) 41° 33′ 23″ E.	20° E. (*E*)	21° 33′ 32″ (*E P D*)	49° 58′ 06″ S.
,,	30 E. (*F*)	11 33 23 (*F P D*)	51 25 46 S.
,,	40 E. (*G*)	1 33 23 (*G P D*)	51 59 23 S.
	The points in Long. 50°, 60°, 70°, 80°, 90° and 100° E. are all in 52° 00′ 00″ S.		
(*C*) 103 58 48 E.	110° E. (*H*)	6° 01′ 12″ (*H P C*)	51° 50′ 46″ S.
,,	120 E. (*J*)	16 01 12 (*J P C*)	50 53 38 S.
,,	130 E. (*K*)	26 01 12 (*K P C*)	48 59 46 S.
,,	140 E. (*L*)	36 01 12 (*L P C*)	45 59 30 S.

	° ′ ″				° ′ ″				° ′ ″		
∠ *E P D*	21 33 23	cos	9·968509	*F P D*	11 33 23	cos	9·991106	*G P D*	1 33 23	cos	9·999839
Lat. (*B*)	52 0 0	tan	0·107190		52 0 0	tan	0·107190		52 0 0	tan	0·107190
Lat. (*E*)	49 58 6	tan	0·075699	(*F*)	51 25 46	tan	10·098296	(*G*)	51 59 23	tan	10·107029

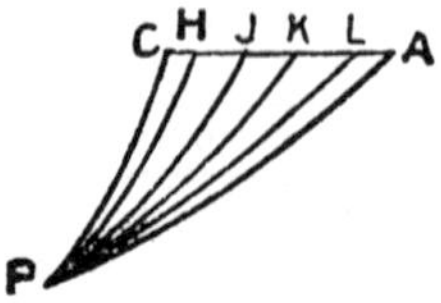

	° ′ ″				° ′ ″		
∠ *H P C*	6 1 12	cos	9·997598	*J P C*	16 1 12	cos	9·982798
Lat. (*C*)	52 0 0	tan	0·107190		52 0 0	tan	0·107190
Lat. (*H*)	51 50 46 S.	tan	10·104788	(*J*)	50 53 38 S.	tan	10·089988
∠ *K P C*	26 1 12	cos	9·953587	*L P C*	36 1 12	cos	9·907848
Lat. (*C*)	52 0 0	tan	0·107190		52 0 0	tan	0·107190
Lat. (*K*)	48 59 46	tan	10·060777	(*L*)	45 59 30	tan	10·015038

Construct a figure for the composite great circle route from a point in Japan (*A*) Lat. 35° 40′ N., Long. 141° E., to a point off San Francisco (*B*) Lat. 37° 48′ N., Long. 122° 40′ W., the maximum latitude to be 45° N. as given in exercise 2.

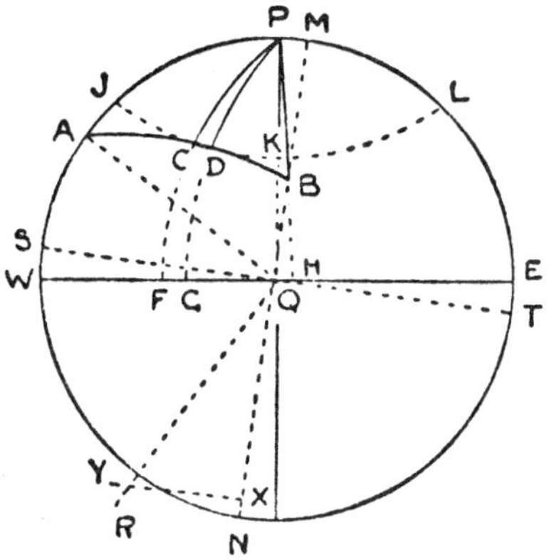

1. Describe a circle with nine equal parts as radius.

2. Draw in the parallel of maximum latitude *J K L*, 45° N.

3. Assume *P W* to be the meridian of *A*, 141° E., and make ∠*W Q A*=Lat. 35° 40′ N.

4. Lay off arc of the equator *W H*, the D. Long. between *A* and *B*=96° 20′, and draw in the meridian *P H*.

5. Lay off the parallel of latitude of *B*, 37° 48′ N., (not shown on figure) cutting meridian *P H* at *B*.

6. After having calculated the D. Long. of vertex *C* from *A*, viz., 44° 9′, and of vertex *D* from *B*, viz., 39° 8′, the meridians of *C* and *D* can be drawn by making arc *Wf*=44° 9′ and arc *H G*=39° 8′, and drawing in the meridians *P C F* and *P D G*, cutting the maximum parallel at *C* and *D* respectively.

7. Join *A Q* and draw *Q R* at right angles to *A Q*, then find, by trial, point *R* the centre of the great circle arc *A C*.

8. To draw in the great circle arc *B D* proceed as follows:—

Join *B Q* and produce it to meet the circumference of the circle at *M* and *N*. Draw in the diameter *S Q T* at right angles to *M Q*. Then, by trial, find on *B N* the point *X* which is the centre of a circle passing through *B*, *S* and *T*. A line drawn through *X* at right angles to *B N* provides a locus on which the centre *Y* of the great circle arc *B D* can be found by trial.

9. Harden in the arc *C D* of the small circle *J L*. The composite track is then *A C*+*C D*+*D B*.

COMPOSITE GREAT CIRCLE SAILING.

Examples for Exercise.

(1) Required the initial and final courses and the distance along the composite great circle route from Wellington Isle (Lat. 49° 7′ S., Long. 75° 50′ W.) to New Zealand (Lat. 45° 54′ S., Long. 170° 46′ E.), the maximum latitude to be 55° S. Find the longitude of the points where the track meets and leaves the maximum latitude, also the longitude from vertex and the latitudes of the points where the track crosses the meridian of 80° W. and thence every 10° of longitude to destination.

2. Required the initial and final courses and the distance along the composite route from the East Coast of Japan (Lat. 35° 40′ N., Long. 141° E.) to a point off San Francisco (Lat. 37° 48′ N., Long. 122° 40′ W.), the ship not to go higher than Lat. 45° N. Find the longitude of the points where the track meets and leaves the maximum latitude, also the longitude from vertex and the latitudes of the point where the track intersects the meridians passing through the 150th degree of East longitude and every 10° of longitude in succession until arriving off San Francisco.

3. *New Zealand to Cape Horn.*—(*A*) Lat. 40° 20′ S., Long. 180°; (*B*) Lat. 56° 10′ S., Long. 67° 10′ W.; maximum Lat. 56° 10′ S. Points every 10° from 170° W.

4. *New Zealand to Chile.*—(*A*) Lat. 45° 30′ S., Long. 170° 13′ E.; (*B*) Lat. 33° 01′ S., Long. 71° 40′ W.; maximum Lat. 50° S. Points every 10° from 180°.

GNOMONIC CHART.

Lay off on the gnomonic chart (see inset) the following exercises:—

1. *Composite Great Circle Track.*—Belle Isle (Lat. 52° N., Long. 55½° W.) to the Naze (Lat. 58° N., Long. 7° E.), 1st vertex, Lat. 60° N., Long. 25° W., 2nd vertex, Lat. 60° N., Long. 8° W. Plot the track on the chart and lift as closely as the smallness of the scale will permit the latitudes corresponding to the following longitudes to obtain points along the great circles for plotting on a Mercator chart. Long. 50° W., 45° W., 40° W., 35° W., 30° W., 5° W., 0°, 5° E.

2. *Great Circle.*—Lay off on the gnomonic chart the great circle from Sable Isle to Lisbon, and with a protractor find the respective courses where the great circle crosses the meridians of 50° W., 40° W., 30° W., 20° W., 10° W.

See page 82 for more examples in Great Circle Sailing.

ABC TABLES AND GREAT CIRCLE SAILING.

It has been remarked that great circle sailing depends on the solution of an oblique spherical triangle having two of its sides and their included angle given, which is analogous to the time azimuth problem in the *P Z X* triangle of nautical astronomy, so that ABC and other tables may be adapted to the solution by inspection of great circle questions as follows :

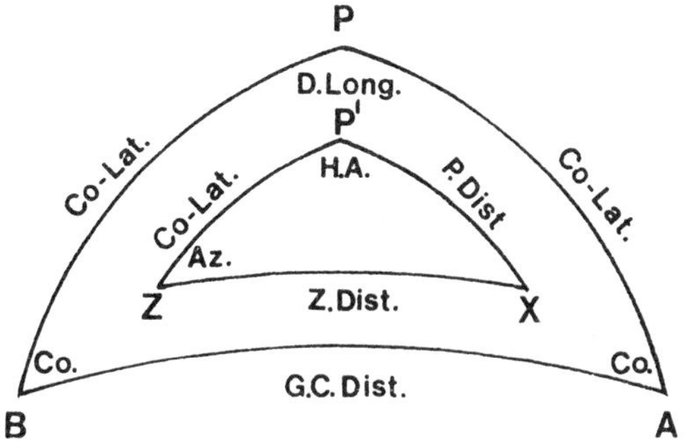

The inner triangle represents a *P Z X* triangle and the outer one a great circle triangle similar in all respects so that in $\triangle XP'Z$:

Z	Azimuth	$= \angle B$	the initial GC course	in $\triangle APB$
P'	Hour Angle	$= \angle P$	the D. Long.	,, ,, ,,
$P'Z$	Co. Lat. Z	$= PB$	the Co. Lat. of B	,, ,, ,,
$P'X$	Polar Dist. X	$= PA$	the Co. Lat. of A	,, ,, ,,
X	Angle of Posn.	$= \angle A$	the Final GC course	,, ,, ,,
ZX	Zenith Dist. X	$= AB$	the GC distance	,, ,, ,,

To Find Great Circle Courses by ABC Tables

The ABC Tables are entered with the arguments :

Lat.	and	Hour Angle	in	Table A
Dec.	,,	Hour Angle	,,	Table B
Lat.	,,	(A ± B)	,,	Table C

By substituting :

Initial GC Lat.	for	Lat.	in	Tables A & C
Final GC Lat.	,,	Dec.	,,	Table B, and
GC D. Long.	,,	Hour Angle		Table A & B

We can obtain :

The Initial GC Course in the Azimuth Column of Table C.

Conversely, to obtain the Final GC Course we substitute :

Final GC Lat.	for	Lat.	in	Tables A & C
Initial GC Lat.	,,	Dec.	,,	Table B, and
GC D. Long.	,,	Hour Angle		Table A & B

The following example is worked using the same data as for the great circle courses calculated in Example 1 on page 83.

Worked Example.

Find the initial and final great circle courses from
Initial position in Lat. 21° 18′ N. Long. 157° 52′ W to
Final ,, ,, ,, 48° 34′ N. ,, 126° 04′ W.

Enter the ABC Tables with:

Initial Lat.	21·3° N.	as	Lat.
Final Lat.	48·6° N.	as	Dec.
D. Long.	31·8° E.	as	Hour Angle.

Lat. 21·3° N & HA 31·8°	A	0·63 S	(change name)
Dec. 48·6° N & HA 31·8°	B	2·15 N	(same name)
A ± B =	C	1·52 N	

Lat. 21·3° N & C 1·52 N gives Azimuth N 35·3 E , or
Initial Course 035·3° T (035° 12·5′ T by calculation)

Now enter the ABC Tables with:

Final Lat.	48·6° N	as	Lat., and
Initial Lat.	21·3° N	as	Dec.

Lat. 48·6° N & HA 31·8°	A	1·83 S
Dec. 21·3° N & HA 31·8°	B	0·74 N
A ± B =	C	1·09 S

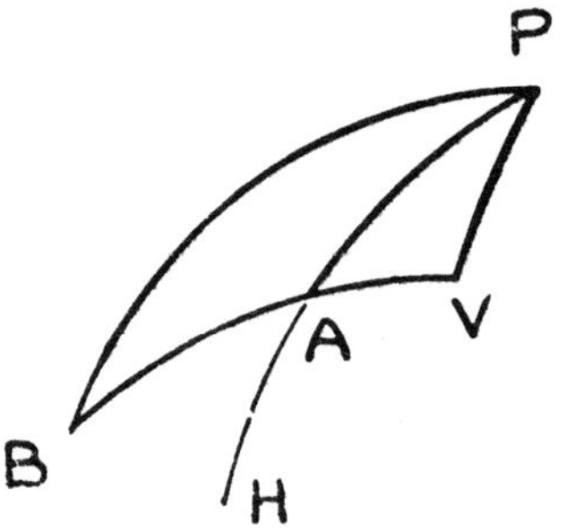

Lat. 48·6° N B C 1·09 S gives Azimuth S54·2° W
(∠HAB in the diagram), so
Final Course N 54·2° E or 054·2° T
(∠PAV in the diagram)
(054° 16′ T by calculation)

Great circle courses can also be found using Weir's Azimuth Diagram, Alt-Azimuth Tables or any Sight Reduction Tables such as HD 605/NP401.

Methods to Find Great Circle Distances by Tables

Alt-Azimuth and Sight Reduction Tables give both the azimuth and altitude. Since the great circle distance is equivalent to corresponding zenith distance (*ZX*), which is 90°-Altitude, these tables can also be used to find the great circle distance as well as the initial and final course.

It should be noted that Sight Reduction Tables are normally used in conjunction with an assumed or Chosen Position. The difference between *DR* and the assumed or Chosen Position (CP) is normally allowed for either in the plotting or by special corrections. In consequence, when obtaining a great circle distance by these methods a similar adjustment should be made for the difference in minutes between the initial GC position and the CP Lat. and the D. Long. between the initial position and the CP Longitude. Plotting is probably the most reliable method.

In practice, rhumb, great circle and composite great circle tracks are well charted for the ocean routes most commonly followed. Distances are also charted and included in a variety of distance tables. For further information the reader is referred to the section on Ocean Passages which is included at page 44 of this volume and which includes details of Admiralty publication, NP 136, "*Ocean Passages for the World*".

DOT Examination Requirements

D.o.T. Examination requirements are that great circle courses and distances are obtained by calculation but it is quite permissible, even desirable, to check the accuracy of results by other methods. The ABC Tables provide a quick and reliable method for checking courses.

Convergency.

The meridians on a globe converge on leaving the equator and intersect at the poles, so that the angles between successive meridians and a great circle crossing them are either progressively increasing or decreasing. The difference between the great circle courses at two points is called the angle of convergency. For example, in Figure, *BA* is a great circle and $\angle P\,d\,A - \angle P\,B\,A$ = convergency of meridians *P d* and *P B*.

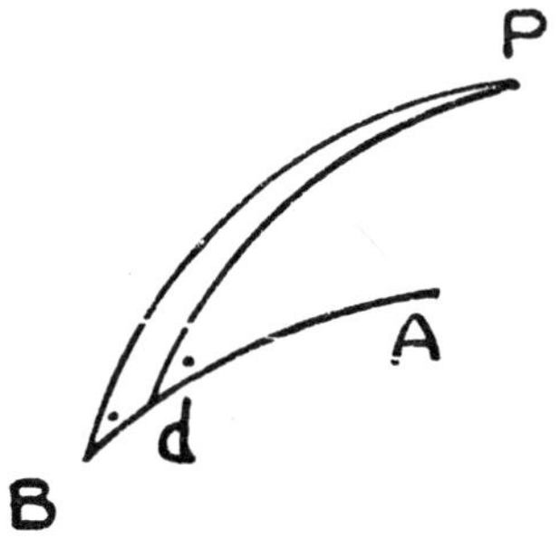

Reference is made to convergency in Volume I, regarding the conversion of radio DF bearings into mercatorial bearings suitable for laying off on a chart, viz., Convergency = D. Long. × cos Mn. Lat., half the correction (Half Convergency) being added to DF bearings under 180°, and subtracted from DF bearings over 180° in the North hemisphere; the rule being reversed in the South hemisphere.

The radio DF convergency formula is modified to meet the case of great circle courses by introducing

$$\text{D. Long} = \frac{\text{Dep.}}{\cos \text{Mn. Lat.}} \text{ as in parallel sailing}$$

and Dep. = Dist. × sin Co as in plane sailing,

so that convergency

$$= \text{D. Long.} \times \sin \text{Mn. Lat.}$$
$$= \frac{\text{Dep.}}{\cos \text{Mn. Lat.}} \times \sin \text{Mn. Lat.}$$
$$= \text{Dep.} \times \tan \text{Mn. Lat.}$$
$$= \text{Dist.} \times \sin \text{Co} \times \tan \text{Mn. Lat.}$$

Example.—Between Lat. 30° N. and 32° N. the distance along the great circle was 300 miles and the course 050° T, required the convergency.

Convergency = Dist. sin Co. tan Mn. Lat.

,, = 300 sin 50° tan 31°

,, = 138′ = 2° 18′

Example.—Calculate the convergency between the two following points on a great circle.

A Lat.	54° 20′ N.	Long.	50° W.
B ,,	56 00 N.	,,	45 W.
D. Lat.	1 40	D. Long.	300′

Convergency = D. long. nat sin Mn. lat. = 300′ sin 55° 10′.
= 300′ × ·8208 = 246·2′ = 4° 06·2′ *Ans.*

THE RELATIONSHIP OF VERTEX TO GREAT CIRCLE ARCS.

(1) The vertex is 90° removed from the point where the great circle crosses the equator, and (2) the course on crossing the equator is equal to the co-latitude of vertex.

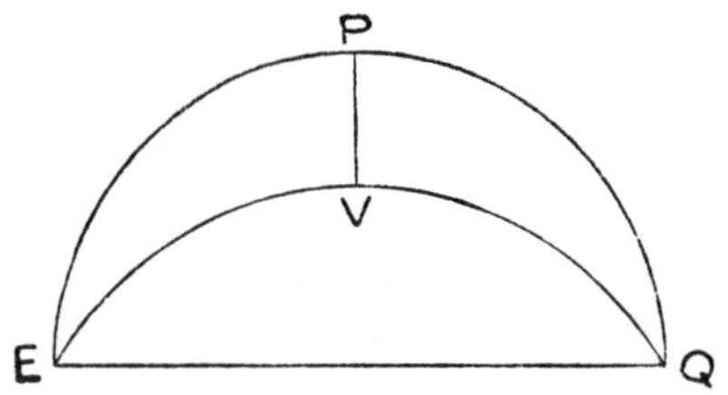

(1) In Figure, *E Q* is the equator, *P* the pole, *E V Q* is half a great circle cutting the equator at *E* and *Q*.

Let *V* be the vertex of the great circle, then

$\triangle PVE = \triangle PVQ$

because $PE = PQ = 90°$

$\angle PVE = \angle PVQ = 90°$

$PV = PV$ (common)

$\therefore$ $EV = VQ$

But the two great circles *E P Q* and *E V Q* intersect at *E* and *Q* and so bisect each other, and as $EVQ = 180° \therefore EV = VQ = 90°$, thus vertex is 90° removed from the great circle crossing point on the equator.

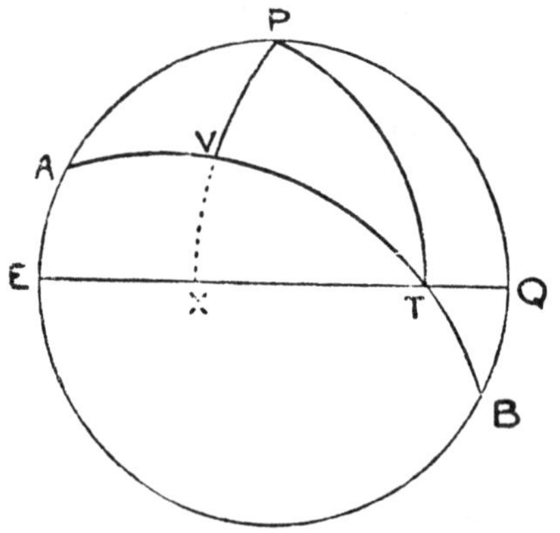

(2) In Figure, *A B* is a great circle cutting the equator at *T*

P T the meridian of *T*

$\angle PTV$ the course on crossing the equator

P V the meridian of vertex

It is required to prove that arc $PV = \angle PTV$.

Because $PT = 90°$, and $TV = 90°$, just proved, then *T* is the pole of the great circle *P V*, therefore $PV = \angle PTV$, that is to say, the co-lat. vertex = great circle course at the equator.

The pole of a great circle is a point in a sphere 90° removed from every part of the great circle. The student, for example, recognises that the earth's geographical pole is the pole of the equator and that it is 90° removed from every point on the equator, so that in $\triangle PTX$, $PT = 90°$, and *P X* 90°, therefore *P* is the pole of the great circle *E Q* and arc *T X* the D. Long. = $\angle TPX$ the angle at the pole which it subtends. So for similar reasons the arc *P V* subtends, and is equal to, its polar angle *P T V*.

QUESTIONS.

1. State the advantages of great circle sailing and the disadvantages if any.
2. Describe how a great circle track is plotted on a Mercator chart.
3. How may the difference between the great circle and the rhumb line joining two places in the same latitude be illustrated on a globe?
4. Why do vessels not always follow the great circle track in regions where it is theoretically more advantageous to do so?
5. What modification may be made when it is not advisable to follow strictly the great circle track?
6. Explain the procedure of plotting a composite great circle track.
7. What is a gnomonic chart and how is it applied to the purposes of practical navigation?
8. Show by a figure the similarity in principle of great circle sailing and time azimuth problems, and state how ABC Tables may be adapted to the solution of the former.
9. What is "convergency"? In what way is it associated with great circle courses?
10. What association exists between radio bearings and great circle courses?
11. A great circle course on crossing the equator in Long. 138° W. was S. 35° E. Draw a suitable figure and write down the latitude and longitude of both vertex.
12. Under what conditions will the course along the great circle (*a*) Change from a NE'ly to a SE'ly (or NW'ly to SW'ly) direction and (*b*) Maintain a NE'ly (or NW'ly) direction?

GREAT CIRCLE SAILING.

(Answers on page 380)

1. From Lat. 33° 50′ S., Long. 18° 10′ E., to Lat. 5° 10′ S., Long. 33° 50′ W., find initial co., dist., vertex and points every 10° from 10° E.

2. From Lat. 46° 30′ N., Long. 45° 00′ W., to Lat. 51° 23′ N., Long. 9° 36′ W., find initial co., dist., vertex and points every 5° from 40° W.

3. From Lat. 36° 06′ N., Long. 5° 21′ W., to Lat. 36° 56′ N., Long. 76° 00′ W., find initial co., final co. and dist.

4. From Lat. 36° 15′ S., Long. 56° 35′ W., to Lat. 15° 57′ S., Long. 5° 40′ W., find initial co., dist. and vertex.

5. From Lat. 48° 26′ N., Long. 161° 24′ E., to Lat. 37° 54′ N., Long. 142° 46′ W., find initial and final courses, dist., vertex and points every 10° from 170° E.

6. From Lat. 34° 10′ S., Long. 25° 47′ E., to Lat. 20° 00′ S., Long. 105° 10′ E., find initial co., dist. and vertex.

7. From Lat. 45° 47′ S., Long. 170° 45′ E., to Lat. 12° 04′ S., Long. 77° 15′ W., find initial co., dist., vertex and points every 20° from 180°.

8. From Lat. 56° 30′ S., Long. 67° 30′ W., to Lat. 37° 00′ S., Long. 20° 00′ E., find initial co., dist. and vertex.

9. From Lat. 25° 48′ S., Long. 45° 05′ E., to Lat. 32° 00′ S., Long. 115° 31′ E., find initial co., dist., vertex and points every 20° from 55° E.

10. From Lat. 49° 50′ N., Long. 6° 30′ W., to Lat. 51° 50′ N., Long. 55° 30′ W., given initial co. N. 67° 04′ W., final co. S. 74° 00′ W., dist. 1825½ mls, find vertex and points every 12° from 12° W.

11. From Lat. 6° 47′ S., Long. 105° 10′ E., to Lat. 34° 55′ S., Long. 20° E., find initial co., dist., vertex and lat. of point where meridian of 80° E. crosses the track.

12. From Lat. 35° 55′ N., Long. 6° 30′ W., to Lat. 40° 05′ N., Long. 70° 00′ W., find initial and final courses, dist., vertex and points every 10° from 10° W.

13. From Lat. 45° 30′ S., Long. 170° 13′ E., to Lat. 33° 01′ S., Long. 71° 40′ W., max. Lat. 50° S., find initial and final courses and total dist. and lat. of points every 10° from 180°.

14. From Lat. 40° 20′ S., Long. 180°, to Lat. 56° 10′ S., Long. 67° 10′ W., max. Lat. 56° 10′ S., find initial and final courses and the dist.

15. From Lat. 39° 10′ S., Long. 180°, to Lat. 57° 00′ S., Long. 70° 00′ W., max. Lat. 60° S., find initial and final courses, dist., and posn. of vertices.

16. From Lat. 35° 00′ S., Long. 180°, to Lat. 57° 10′ S., Long. 70° 00′ W., max. Lat. 60° S., find initial and final courses, dist. and long. of vertices.

CHAPTER IV.

TIDES

THE moon exerts a gravitational attraction on all the portions of the earth. This attractive force depends on the distance between the moon and the point of the earth's surface. In Fig 1 the attractive force at *C*, the point nearest to the Moon will be greater than that at point *A*, the point furthest from the Moon. If the attractive force at the centre of the earth (*P*) is taken as the average attractive force then the difference between this and the force at any point on the earth's surface is the Tide Generating Force due to the Moon.

This force can be resolved into vertical and horizontal components. The vertical component has little effect upon the sea level. However the horizontal component acting along the earth's surface is the Tractive Force which produces the tidal movements. There is a similar Tractive Force due to the sun but this is about 0·46 of the lunar force.

The tide raising forces at a point on the earth's surface change as the earth revolves on its axis, and as the relative position of the earth, moon and sun change. The extent to which the tides respond to the external forces vary in all parts of the world and frictional forces in water of varying depths moving around the land set up internal forces, which further complicate matters.

In a simple case the moon and the sun each produce a high water at the point on the earth nearest them and at the opposite point, with two low waters about midway between them. It follows that the angular distance separating high water from low water is about 90° of longitude.

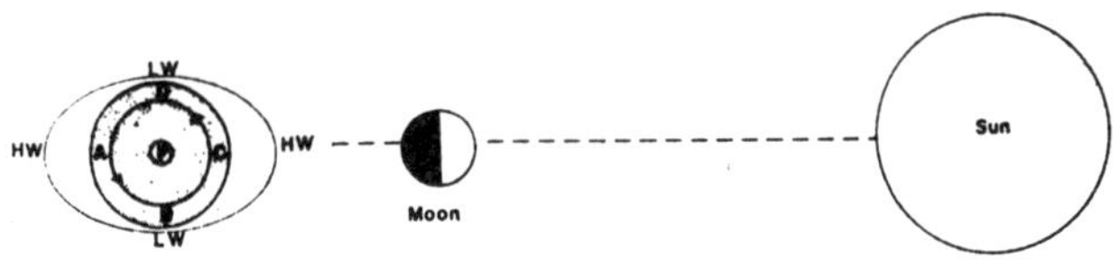

FIG 1 New Moon, Spring Tides

In Fig 1 the earth's centre is at *P* looking down on the pole, the sun and moon are supposed to be over the equator, in conjunction at new moon and the earth revolves in the direction indicated within the surrounding water.

An observer having HW at *A* will be carried round and experience in succession, LW at *B*, HW at *C*, LW at *D*, and HW again at *A*. This daily alternating rise and fall of tide is called a half-day or semidiurnal tide. The constituent due to the moon alone is called the Mean Lunar Semidiurnal Tide which is expressed as (M_2—M) for the moon. The Suffix 2 indicates two high waters per lunar day. The constituent due to the sun alone is called the Mean Solar Semidiurnal Tide, abbreviated to S_2 indicating two high waters per solar day.

SPRINGS AND NEAPS

The maximum time raising force occurs when the moon and sun are on the observer's meridian.

At new moon, Fig 1, the forces of the moon and sun are combined and give rise to a Spring Tide; the same occurs again about 14½ days later at the period of full moon, when moon and sun are in opposition and the forces are again in the same straight line as shown in Fig 2. The range at springs is expressed as (M_2+S_2).

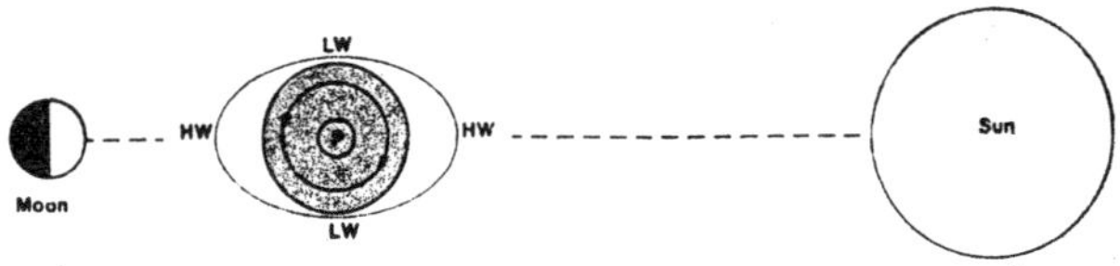

FIG 2 Full Moon, Spring Tides

Neap Tides occur about seven days after spring tides when the moon is in quadrature (Fig 3), at which time the HW due to M_2 coincides with the LW due to S_2 and the result is a tide having a smaller range. The range at neaps is expressed as (M_2-S_2).

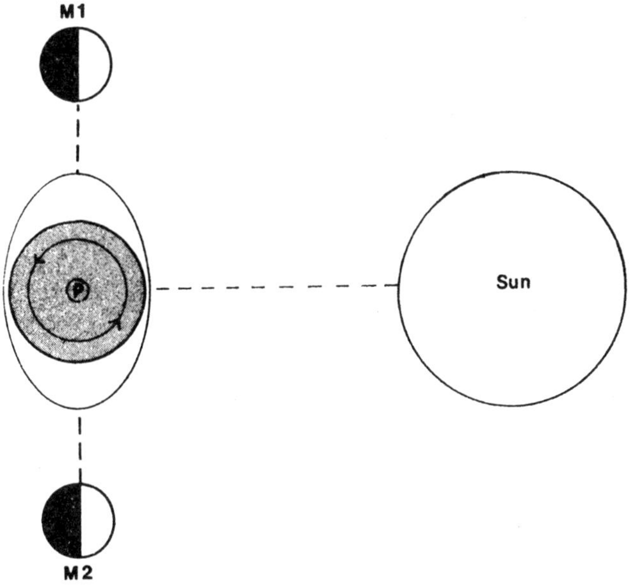

FIG 3 Quadrature, Neap Tides

The mean lunar day is about 50 minutes longer than the solar day, so that normally, an interval of about 24 hrs. 50 mins. separates the high water of one day from the corresponding high water of the next day at a given place, the average interval between high water and the next low water is, therefore, about 6 hrs. 12 mins.

PRIMING AND LAGGING

During the changes from springs to neaps when the interval between high waters mentioned above exceeds the average of 24 hrs 50 min the tide is said to LAG, and when the interval is less than 24 hrs 50 min the tide is said to PRIME.

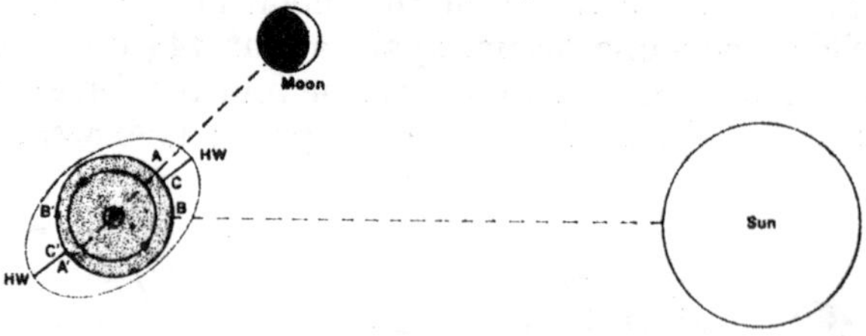

FIG 4 Priming

In Fig 4 the lunar HW during the first quarter would tend to be in the direction *AA'* and the solar HW along *BB'*, but the actual HW is likely to be in the direction *CC'*, which is a meridian nearer to that of the moon than to the sun's meridian. A place at *C* or *C'* will experience high water before the moon's meridian passage, and the tide is said to be priming. Generally the tide will also prime in the third quarter.

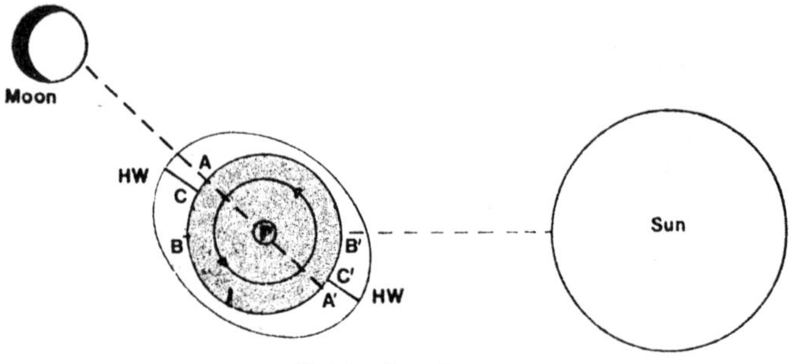

FIG 5 Lagging

When the moon is in the second quarter (Fig 5), the lunar HW would tend to appear at *A* and *A'* under the moon but due to the solar forces acting on the water the actual HW is likely to occur in the direction *CC'* so that a place will transit the moon's meridian before high water is experienced, and the tide is said to be lagging. This effect is also likely to occur in the fourth quarter.

The declination of the moon changes over a period of approximately one month and the sun's declination changes over a period of one year. These variable factors set up tide raising forces which sometimes act together and sometimes in opposition.

If the moon remained stationary over the equator there would tend to be a permanent low water at the poles and high waters on the upper and lower meridians

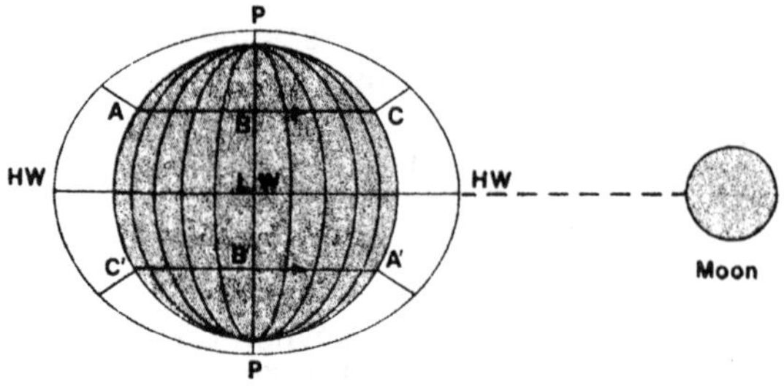

FIG 6 Moon on Equator

of the moon, with corresponding low waters 90° of longitude East and West of those meridians (Fig 6) where *HW* represents high water at the equator with the trough of low water *PP′* extending from pole to pole. *ABC* and *A′B′C′* are two parallels of latitude. The figure represents high water at *A*, low water at *B* and high water at *C*. Similarly for *C′B′* and *A′*, but these tide heights are obviously not so great as at places on the equator where the maximum range of tide from low water to high water is experienced.

Declination is mainly responsible for a difference in the heights of the two daily high waters (Fig 7), where latitude *C* equals moon's declination. The tide shown is a higher high water at *C*, where the moon is over the upper meridian, than at *A* on the lower meridian. At position *A* it is a lower high water, and this place, on being carried round by the earth's rotation, gets low water on reaching *B*, which is on the low water trough *BB′*, and, later gets the higher high water on reaching *C*. The same applies to the parallel *C′B′A′*. This gives a diurnal component of the tide due to the moon's declination. The declination of the sun produces a similar diurnal force effecting the tide and the average component of the combined forces is called constituent K_1.

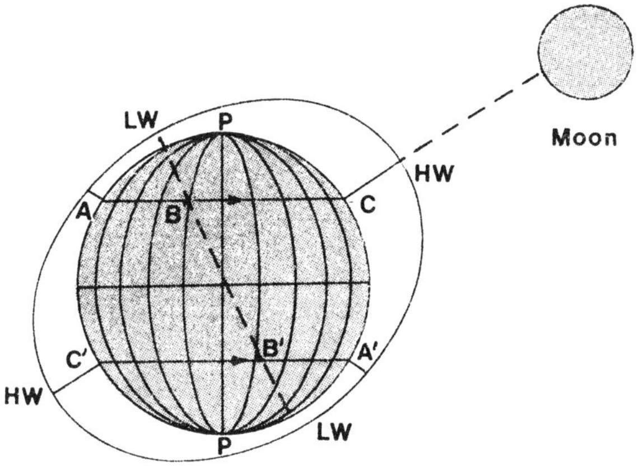

FIG 7 Moon in North Declination

The variable tide raising factors sometimes act together, and sometimes in opposition, and at every stage during their respective cycles, different lunar and solar components of a tide are formed. A composite tide may be built up and plotted on a graph, by computing the response of the water to each constituent, for a specified place and time. This is the Admiralty Method of Tidal Prediction.

Range of Tide and Duration of Tide

The Range of Tide is the height difference between a HW height and the preceding or following LW height at a particular place. The largest range occurs at Spring Tides and the smallest range at Neap Tides.

Duration of Tide is the time interval (time difference) between a HW time and the preceding or following LW time at a particular place.

DEPTH AND SOUNDINGS

It is frequently necessary to know the precise depth of water beneath the keel or over some charted danger.

Depth of water can be measured by hand using the Hand Lead Line, or by use of the Patent Sounding Machine, or by some type of Echo Sounding Device.

In the case of the hand lead line the depth so measured is from the surface water level whereas with an echo sounding device depth is measured from a transducer situated in the bottom of the vessel.

In both cases it is necessary to apply a correction to the measured depth in order that comparison can be made with the depth shown on the chart.

Charted Depth

The depth shown on a chart is measured from a level known as Chart Datum (CD). On a modern chart, Chart Datum is normally placed at the level of the Lowest Astronomical Tide (LAT) which is the lowest level which can be predicted to occur under average meteorological conditions and under any combination of astronomical conditions. It follows that the depth shown on a modern chart is the least depth which can be expected under normal conditions.

Abbreviations Used in Examples

The following abbreviations will be used in examples:

LW — Low Water.
HW — High Water.
ATT — Admiralty Tide Tables.

LW and HW Heights

LW and HW heights are always measured from Chart Datum. Unless preceded by a minus sign, LW heights are always additive to the charted depth. LW heights preceded by a minus sign indicate that LW falls below datum and that the charted depth is reduced at LW.

Example 1

(a)	Charted Depth	10·0 m	(b) Charted Depth	10·0 m
	LW from ATT	1·0 m	LW from ATT	−1·0
	Predicted Depth at LW	11·0 m	Predicted Depth at LW	9·0 m

Rise of Tide

(Refer to Figure 4)

Note that Rise of Tide is the vertical distance measured from LW to the actual water level.

Height of Tide

(Refer to Figure 4)

Note that Height of Tide is the vertical distance measured from Chart Datum to the actual water level.

Height of Tide = LW Height + Rise of Tide.

As already mentioned, if LW falls below chart datum it is expressed as a negative quantity. In such cases apply LW as a negative quantity in the equation.

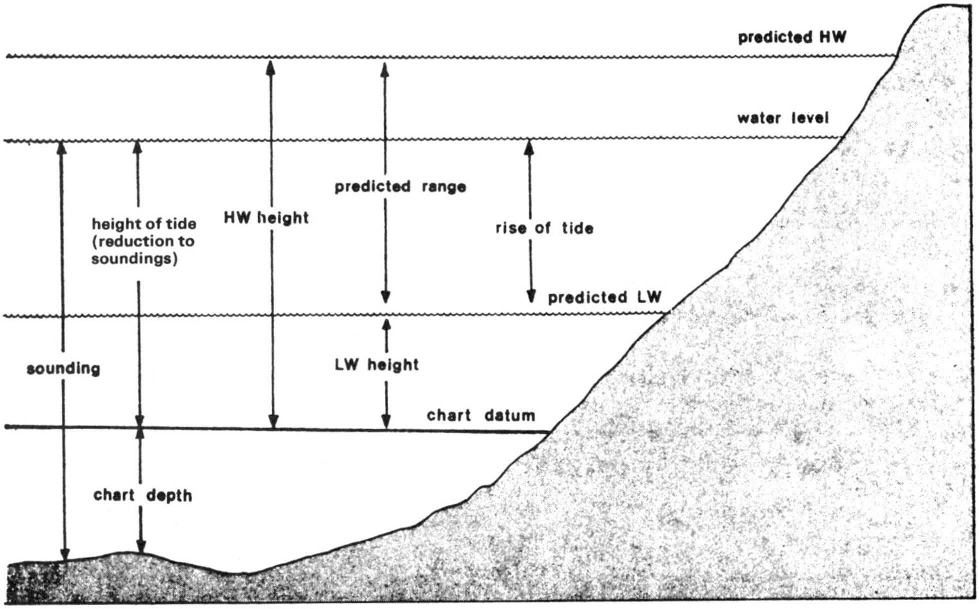

Figure 4. Rise of Tide and Height of Tide (Reduction to Soundings).

Example 2

(a) LW from ATT	1·0 m	(b) LW from ATT	−1·0 m
Rise of Tide	4·0 m	Rise of Tide	4·0 m
Height of Tide	5·0 m above CD.	Height of Tide	3·0 m above CD.

Height of Tide is also referred to as Correction to Leadline and Reduction to Soundings.

From the foregoing it should be clear that:

Predicted Depth = Charted Depth + LW Height + Rise of Tide

OR

Predicted Depth = Charted Depth + Height of Tide

Study Figure 4 and note that the actual depth of water has been increased by the Height of Tide above Chart Datum.

Reduction to Soundings

(Refer to Figure 4)

In order to compare a lead line sounding with that shown on the chart, the depth by sounding must be reduced by the Height of Tide.

Providing the water level is *above* Chart Datum:

Leadline Depth − Height of Tide = Charted Depth

Example 3

(a) Sounding by hand lead line 15 m
LW 2·0 m
Rise of Tide 3·0
Height of Tide 5

Depth for comparison with chart 10 metres

Example 3 a

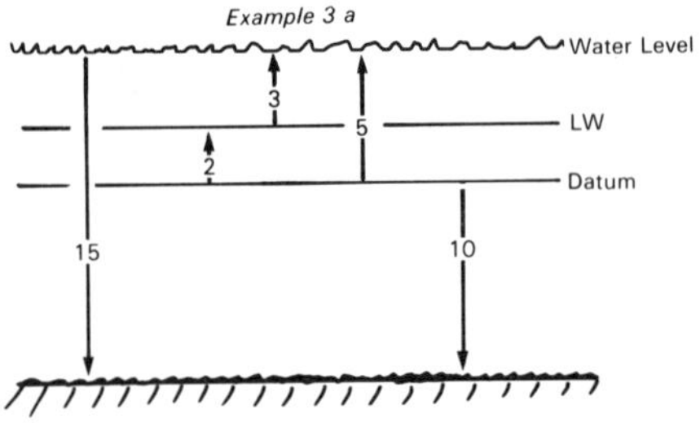

(b) Sounding by hand lead line 20 m
LW −1·0 m
Rise of Tide 4·0
Height of Tide 3

Depth for comparison with chart 17 metres

Example 3 b

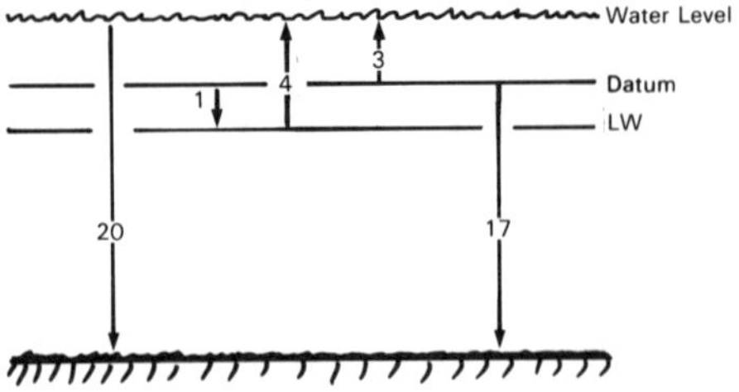

In the case of a sounding obtained by an Echo Sounding Device the recorded depth must first be corrected for the distance between the transducer (at the bottom of the vessel) and the surface water level. The corrected depth is then reduced by the Height of Tide to obtain the depth for comparison with that shown on the chart.

Providing the water level is *above* Chart Datum:

(Depth by Echo Sounder + Draft) − Height of Tide = Charted Depth

Example 4

Depth obtained by Echo Sounder		25 m
Draft		5
Depth from surface Water Level		30
LW	2·5 m	
Rise of Tide	4·5	
Height of Tide		7
Depth for comparison with Chart		23 metres

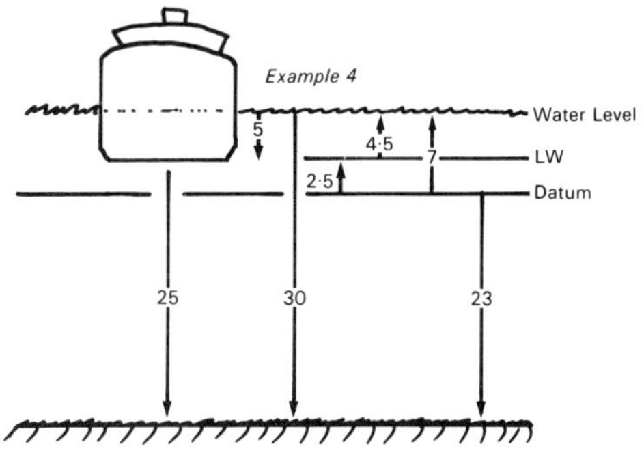

Example 5

Depth obtained by Echo Sounder		25 m
Draft		5
		30
LW	−0·5 m	
Rise of Tide	4·5	
Height of Tide		4
Depth for comparison with Chart		26 metres

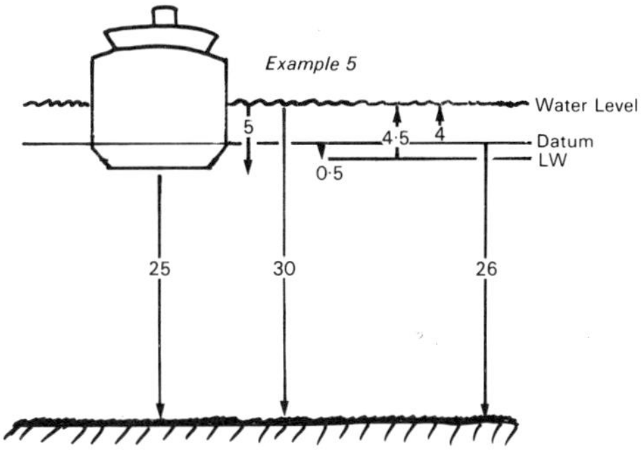

DRYING HEIGHTS AND CHARTED HEIGHTS

(Refer to Figure 5)

In addition to depths of water, the chart also shows the HEIGHT ABOVE CHART DATUM of rocks and sand banks which cover and uncover at different states of the tide. Such heights are referred to as DRYING HEIGHTS and are indicated by a figure with a line beneath, e.g. $\underline{3}$ indicates Drying Height 3 ft (Fathom Chart) and $\underline{1}_3$ would indicate a Drying Height of 1·3 metres (Metric Chart). In the latter case, the chart tells us that the rock projects 1·3 m vertically above Chart Datum and that a Height of Tide of 1·3 m will be required before the rock is just covered.

Example 6

Find the depth of water over a rock marked $\underline{6}$ near Avonmouth (p.m.) on September 29th. The calculated rise of tide was 5·5 m.

Drying Height (rock)		6 m
LW	3·5 m	
Rise of Tide	5·5	
Height of Tide		9
Depth of Water over rock		3 metres

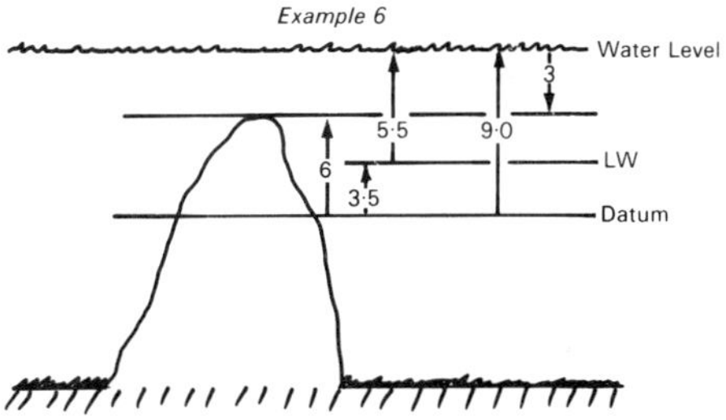

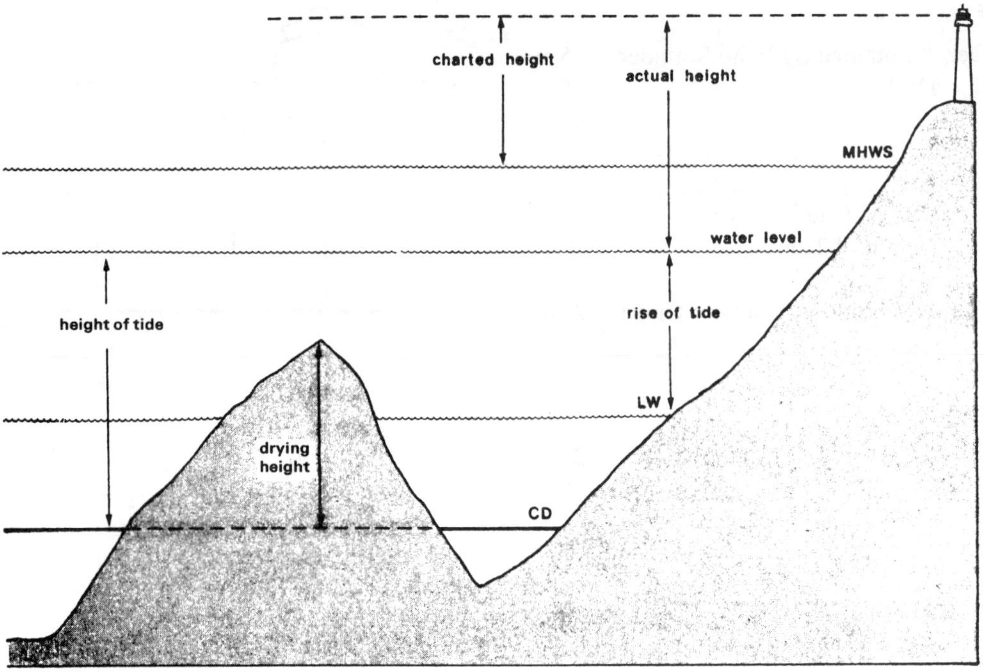

Figure 5. Charted Height and Drying Height.

Charted Heights

(Refer to Figure 5)

The charted height of lighthouses, hills and other high objects is given above the level of Mean HW Springs (MHWS).

In the case of a lighthouse the charted height is measured from MHWS to the centre of the focal plane of the light.

Example 7

A lighthouse near Portsmouth has a charted height of 19 m. Find the actual height of the lighthouse (p.m.) on July 29th. Rise of Tide for the required time was calculated as 2 m.

Charted Height of Lighthouse (above MHWS)			19 m
MHWS (from ATT Table V) Level above CD	4·7 m		
LW (from ATT)	1·7		
Rise of Tide	2·0		
Height of Tide		(−)3·7	
Water Level BELOW MHWS			1
Actual Height of Lighthouse (above water level)			20 m

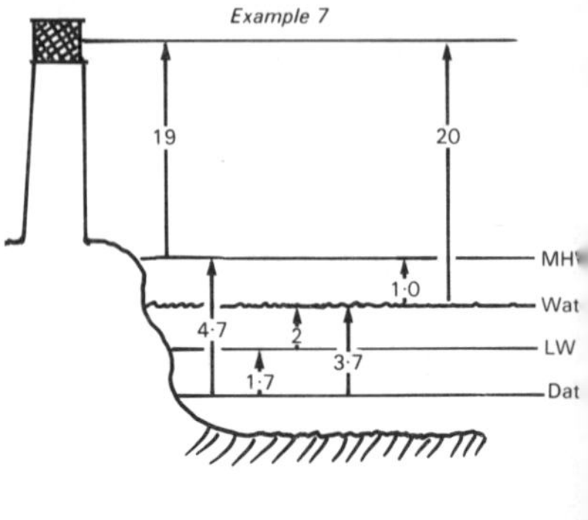

TIDAL LEVELS

(Refer to Figure 6)

Tidal levels for Standard Ports are listed in Table V of the *Admiralty Tide Tables* (see Appendix). They are given under abbreviated headings which are listed below and shown on the accompanying diagrams. As can be seen in the Appendix, these abbreviated headings are used to tabulate Secondary Port Time and Height Differences included in ATT Part II.

Study Figure 6 and note the various levels and ranges. Also note that the vertical distance between MHWS and MLWS is referred to as the Spring Range, between MHWN and MLWN as the Neap Range and between the Predicted HW and LW (for the day in question) as the Predicted Range. Refer to the ATT Extracts in the Appendix and note that the Mean Spring and Neap Ranges for a Standard Port are shown in the box included in the Tidal Curve for that port.

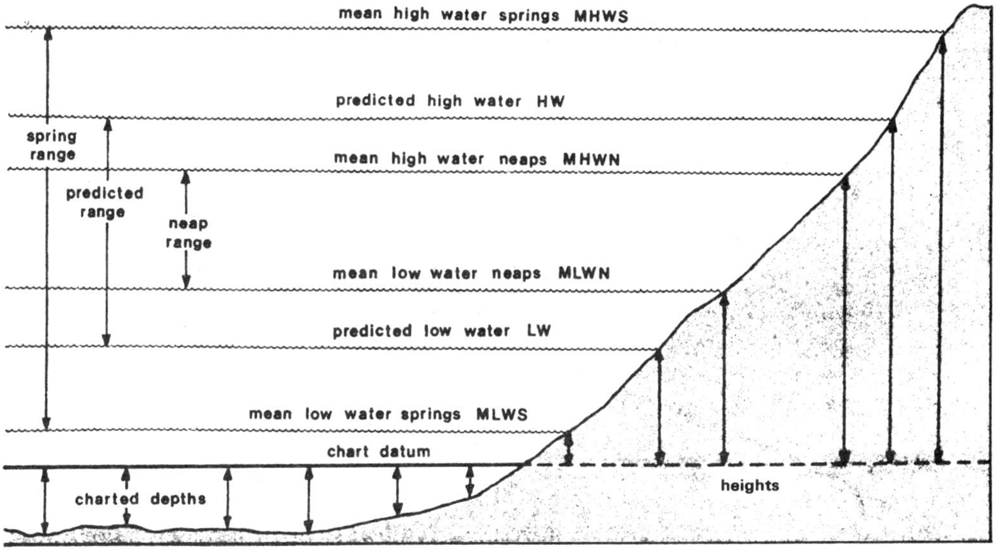

Figure 6. Tidal Levels.

MHWS (Mean High Water Springs) and **MLWS** (Mean Low Water Springs) are the average heights of two successive spring high and low waters in the same period when the range of tide is greatest.

MHWN (Mean High Water Neap) and **MLWN** (Mean Low Water Neap) are the average heights of two successive neap high and low waters in the same period when the range of tide is least.

MHHW (Mean Higher High Water) and **MLHW** (Mean Lower High Water) are the means of the higher and the lower of the two daily high waters taken over a long period of time.

MLLW (Mean Lower Low Water) and **MHLW** (Mean Higher Low Water) are the means of the lower and the higher of the two daily low waters.

HAT (Highest Astronomical Tide) and **LAT** (Lowest Astronomical Tide) are the highest and lowest levels which can be predicted under normal meteorological conditions. These levels will not occur every year.

MSL (Mean Sea Level) is the average level of the sea surface over a long period, preferably 18·6 years, or the average level which would exist in the absence of tides.

STANDARD PORTS AND SECONDARY PORTS

There are two types of port—Standard Ports and Secondary Ports. Secondary Ports are based on Standard Ports which have a similar tidal curve.

The Admiralty Tide Tables include the Tidal Curve and full details of daily predictions for Standard Ports. They also list the Time and Height Differences of Secondary Ports from the chosen Standard Port. To give full daily predictions for each Secondary Port would clearly require a very large volume indeed and the above method makes for a more convenient size.

ADMIRALTY TIDE TABLES

Admiralty Tide Tables (ATT) are published in three volumes:

Vol. 1. European Waters (including Mediterranean Sea).
Vol. 2 Atlantic and Indian Oceans (including tidal stream predictions).
Vol. 3. Pacific Ocean and Adjacent Seas (including tidal stream predictions).

Each of these volumes is divided into two parts:

Part I. Standard Ports.
Part II. Secondary Ports.

Standard Ports are listed in the Index to Standard Ports printed on the inside Front Cover of each volume.

Secondary Ports are listed in the Geographical Index at the rear of each volume and which also includes the Standard Ports.

Each port is assigned an Index Number.

TYPES OF TIDAL PROBLEM

Tidal problems fall into 4 main types in which it is necessary to find:

1. Time of HW and LW on a particular day.
2. Height of Tide at an Intermediate Time between HW and LW.

   OR

   Reduction to Soundings (Correction to Leadline) at an Intermediate Time.
3. Time at which a Required Height of Tide (or Depth of Water) is reached.
4. Correction to Apply to the Charted Height of a Lighthouse or other shore object.

Each type of problem is included in the following pages.

TIDE CALCULATIONS

Accuracy and Interpolation

It must be borne in mind that meteorological conditions which differ from average may cause differences between the predicted and actual tide. These effects are discussed in the Introduction to the tide tables. Despite these limitations, calculations should be worked as accurately as possible within the limits of the tables and, particularly in examinations, all interpolation shown.

Time Zones, Zone Time, Standard Time and Time Differences

The Time Zone for the port in question is clearly shown in the top, left hand corner of prediction pages in *Admiralty Tide Tables.* Daily predictions are given in the normal Standard Time of the port. Before attempting calculations, time zones, zone time and standard time should be understood.

The world is divided into 24 **Time Zones**. Each Time Zone is 15° of longitude in width. The "Zero Time Zone" extends from 7½°W to 7½°E—this zone keeps GMT. In each of the remaining 23 zones the time differs from GMT by a whole number of hours and is numbered, in sequence, 1–12: East of Greenwich with a Negative (−) prefix, West of Greenwich with a positive (+) prefix.

e.g. 60°E lies in the 52½°E—67½°E Zone (Zone −4)
10°W „ „ „ 7½°W—22½°W „ (Zone +1)

To obtain the **Zone Time** for a position or place, subtract the Zone number **algebraically** from GMT.

Example—Given GMT 1200, what is the correct Zone Time to keep in Longitude 60°E (Zone −4)?

GMT − (Zone) = Zone Time
1200 − (−4) = 1600 i.e. 4 hours ahead of GMT

conversely,

Zone Time + (Zone) = GMT
1600 + (−4) = 1200

For convenience on land, a **Standard Time** is adopted throughout a given country. In most cases Standard Time or Legal Time is that of the Zone in which the country mainly lies. Countries like USA and Australia, which extend over several Time Zones, generally adopt several Standard Times.

To prolong daylight hours, many countries also adopt (for part of the year) a form of Daylight Saving Time or **Summer Time**. For such periods, the time of the Eastward zone is usually adopted, e.g. BST (British Summer Time), which is kept in the UK from a date in March to a date in October, is the time for Zone −1, i.e. GMT −(−)1 = GMT +1 hour.

Time Differences for Secondary Ports, when applied to the printed times of HW and LW at Standard Ports will give the times of HW and LW at the Secondary Port in the Zone Time tabulated in the tables for the Secondary Port.

To Find the Times of HW and LW

Example 8

Find the Standard Times of HW and LW at SHEERNESS on January 7th.

1. Turn to "Index of Standard Ports" in ATT Volume 1 (See page 393). Find that Sheerness is a Standard Port for which daily predictions are given.
2. Turn to the daily predictions for Sheerness. (See page 403).

TIME ZONE GMT

JANUARY

7	0325	1·2	22	0433	1·3
	0952	5·3		1059	4·9
TH	1609	0·9	F	1648	1·1
	2237	5·5		2316	5·1

Four times are given with the height of the tide at each instant. The High Waters are the higher figures, 5·3 metres (m) and 5·5 m. The Low Waters are the lower figures, 1·2 m and 0·9 m.

	LW	HW
Times	0325	0952
	1609	2237

Note that the times are Standard Times for the Time Zone indicated. In this case the Time Zone is GMT.

Example 9

Find the BST (British Summer Time) of HW and LW at Greenock on August 18th and also the Duration and Range of the AM rising tide.

1. Check that Greenock is a Standard Port. (See page 393).
2. Turn to the daily predictions for Greenock. (See page 406).
3. Note and apply the time difference: GMT + 1 hr = BST.
4. Subtract the AM LW times and heights from the following HW times and heights to obtain the Duration and Range of Tide.

TIME ZONE GMT

18	0442	0·3
	1145	3·1
W	1704	0·4
	2345	3·4

High Waters are 3·1 m and 3·4 m and Low Waters are 0·3 m and 0·4 m.

	LW	HW	LW	HW
GMT	0442	1145	1704	2345 Aug. 18th
Time Diff.	+01	+01	+01	+01
BST	0542 (18th)	1245	1804	0045 (19th)

HW	12 45	3·1 m	
LW	05 42	0·3 m	
Duration	7 h 03 m	2·8 m	Range

Note the time difference has made the BST of the second High Water occur on the following day which is August 19th.

Admiralty Tidal Prediction Form—NP 204

In the following pages, tide calculations will be worked on *Admiralty Tidal Prediction Form,*

NP 204. Copies of this form, together with instructions, graph paper for interpolation and a selection of Standard Port Tidal Diagrams (with curves), are contained in the booklet edition of NP 204. The form is also included in the back pages of *Admiralty Tide Tables* (ATT).

Refer to the copies of NP 204 on page 420 (i) and in Example 10. Note that the form contains boxes numbered 1–16. Boxes 1–5 are completed for Standard Ports. Boxes 1–16 are completed for Secondary Ports.

To Find the Height of Tide at an Intermediate Time between HW and LW

(European Standard Port—**Checklist 1**)

1. Check that the port is in the ATT "Index to Standard Ports". Turn to the appropriate page. Complete heading in NP 204. In boxes 1–4, write down the Times and Heights of HW and LW which lie either side of the required time. Subtract LW height from HW height to obtain Predicted Range—enter in box 5.
2. Turn to the Tidal Diagram for the port (first page of predictions for that port). Plot the heights of HW (top LH scale) and LW (bottom LH scale) which occur either side of the required time. Join by a sloping line.
3. Write down the HW time in the HW box below the curve. In adjacent boxes, enter other times (differing by one hour intervals from HW time) to "embrace" the required time. It may be helpful to write down the required Time Interval from HW (i.e. HW Time ~ Required Time).
4. Plot the required time on the time scale (note, 10 minute divisions). Through this point draw a vertical line to intersect the appropriate curve—see (5).
5. Note Spring and Neap Ranges shown in the box (top right).
   Predicted Range equal to or greater than Spring Range—Use Spring Curve only.
   Predicted Range equal to or less than Neap Range—Use Neap Curve only.
   Other Ranges—Interpolate (and draw in the appropriate part of the curve as explained below.)
6. From vertical intersection with curve, proceed horizontally to sloping line (2) then vertically to the Height Scale.
7. Read off the Height of Tide (also known as Reduction to Soundings or Correction to Leadline).

Interpolating between Spring and Neap Curves

To interpolate between curves (i.e. where the Predicted Range lies between the Spring and Neap Ranges):

$$\text{Express: } \frac{\text{Spring Range} - \text{Predicted Range}}{\text{Spring Range} - \text{Neap Range}} \text{ as a fraction.}$$

Using this fraction, sketch in an intermediate curve, parallel to the Spring curve this fractional distance away from the Spring curve towards the Neap curve, e.g. Fraction $\frac{1}{4}$; draw curve $\frac{1}{4}$ of the distance away from the Spring curve towards the Neap curve.

Where there is an appreciable change in duration between Spring and Neap tides, interpolating between curves may result in a slight error—the error is greatest near LW. **Do not extrapolate.**

Example 10

(European Standard Port)

Find the Height of Tide at Londonderry at 1300 Standard Time on August 22nd. (Predictions on page 408).

Follow the instructions in **Checklist 1.** (Step 1)

TIME ZONE GMT

AUGUST

	TIME	M
22 SU	0424	0·4
	1012	2·6
	1614	0·3
	2228	2·8

STANDARD PORT LONDONDERRY TIME/[redacted] REQUIRED 1300

SECONDARY PORT — DATE AUG 22 TIME ZONE GMT

	TIME		HEIGHT		
STANDARD PORT	HW	LW	HW	LW	RANGE
	1 1012	2 1614	3 2·6	4 0·3	5 2·3
Seasonal change	Standard Port		6	6	
DIFFERENCES	7*	8*	9*	10*	
Seasonal change	Secondary port		11	11	
SECONDARY PORT	12	13	14	15	
Duration	16				

Interpolation (Step 5)

Predicted Range equals Spring Range—use Spring Curve.

NB At this port the Spring and Neap curves coincide throughout the tidal cycle—so interpolation "between curves" is, in any case, not possible.

Plotting (Steps 2, 3, 4, 6 and 7)

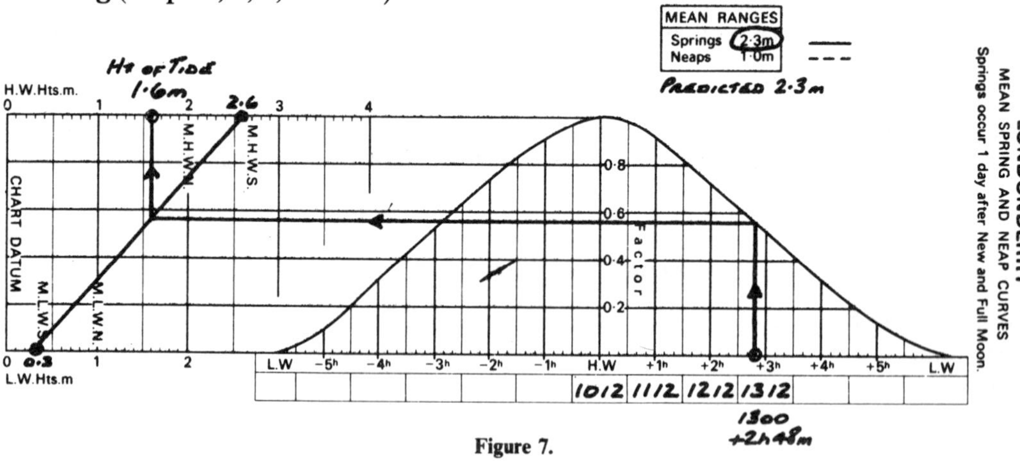

Figure 7.

Answer 1·6 metres is the Ht of Tide above Chart Datum.

Example 11

(European Standard Port)

Find the Height of Tide at Avonmouth at 1500 Standard Time on November 28th. (Predictions on page 407).

Follow the instructions in **Checklist 1.** (Step 1)

TIME ZONE GMT

NOVEMBER

28	0431	11·5	
SU	1104	2·7	
	1658	12·2	
	2339	2·1	

STANDARD PORT *Avonmouth* TIME/[redacted] REQUIRED *1500*

SECONDARY PORT *—* DATE *Nov. 28* TIME ZONE *GMT*

	TIME		HEIGHT		
	HW	LW	HW	LW	RANGE
STANDARD PORT	1 *1658*	2 *1104*	3 *12·2*	4 *2·7*	5 *9·5*
Seasonal change	Standard Port		6	6	
DIFFERENCES	7*	8*	9*	10*	
Seasonal change	Secondary port		11	11	
SECONDARY PORT	12	13	14	15	
Duration	16				

Interpolation (Step 5)

Spring Range	12·3 m	Spring Range	12·3 m
Neap Range	6·5 m	Predicted Range	9·5 m
S – N	5·8 m	S – P	2·8 m

$$\frac{S-P}{S-N} = \frac{2\cdot8}{5\cdot8} = 0\cdot48 \simeq \tfrac{1}{2} \text{ (Fractional Distance from Spring Curve).}$$

Plotting (Steps 2, 3, 4, 6 and 7)

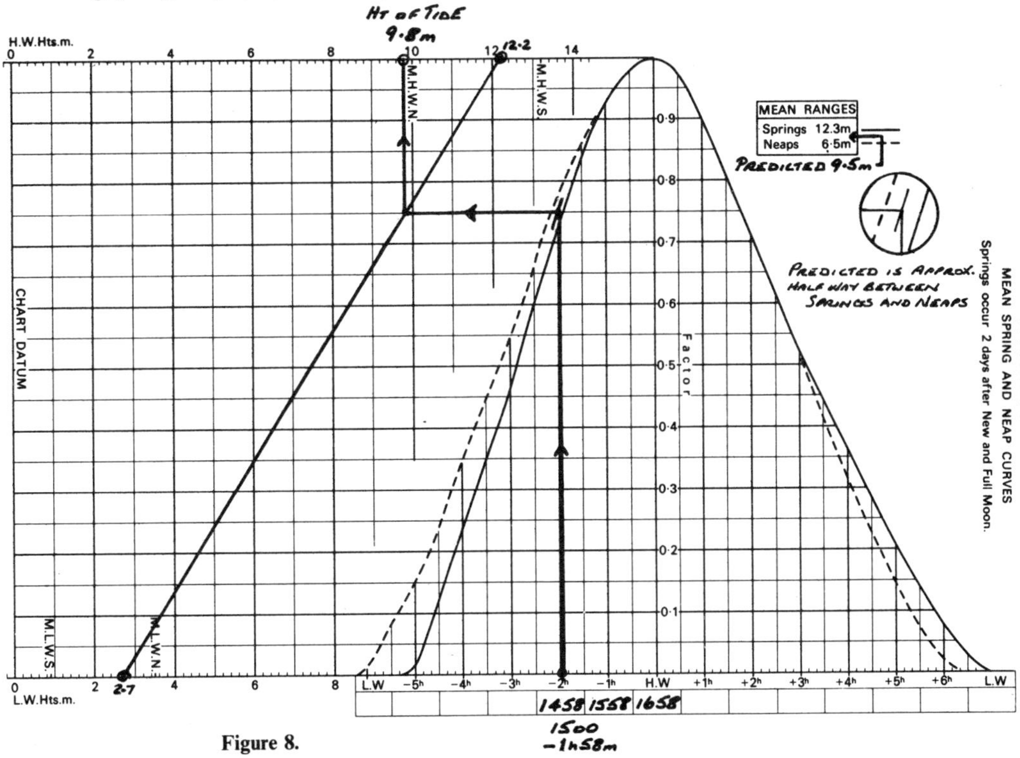

Figure 8.

Answer 9·8 metres is the Ht of Tide at 1500 GMT.

To Find the Time at which a Required Depth is Reached

(European Standard Port—**Checklist 2)**

1. Check that the port is in the ATT "Index to Standard Ports". Turn to the appropriate page. Complete heading in NP 204. In boxes 1–4, write down the Times and Heights of HW and LW which lie either side of the required time. Subtract LW height from HW height to obtain Predicted Range—enter in box 5.
2. Using a sketch to assist, find the Height of Tide necessary for the required depth.
3. Turn to the Tidal Diagram for the port. On the diagram, plot heights of HW (top LH scale) and LW (bottom LH scale) which occur either side of the required time. Join by a sloping line.
4. Write down the HW time in the HW box below the curve. In adjacent boxes, enter other times (differing by one hour intervals from HW time) to "embrace" the required time. If helpful, write down the required Time Interval from HW (i.e. HW time ~ Required Time).
5. From the required Height of Tide on the HW height scale, proceed vertically down to the sloping line, then horizontally to the appropriate curve—see (6).
6. Note Spring and Neap Ranges shown in the box, (top right).
   Predicted Range equal to or greater than Spring Range—Use Spring Curve only.
   Predicted Range equal to or less than Neap Range—Use Neap Curve only.
   Other Ranges—Interpolate—as previously explained, and draw in the appropriate part of the curve.
7. From the intersection at the appropriate curve, proceed vertically down to the time scale.
8. Read off the required time (or time interval) using the 10 minute divisions to assist.

Example 12

(European Standard Port)

Find the Standard Time when a ship of draft 5·5 m will have 2 m clearance under the keel over a 4·5 m shoal off Portsmouth on the a.m. ebb tide of June 29th.

Follow the instructions in **Checklist 2.** (Step 1)

TIME ZONE **GMT**

JUNE

	TIME	M
29	0530	4·0
	1058	1·5
TU	1820	4·2
	2337	1·7

STANDARD PORT PORTSMOUTH/HEIGHT REQUIRED 3·0 m

SECONDARY PORT — DATE June 29 TIME ZONE GMT

	TIME		HEIGHT		
STANDARD PORT	HW	LW	HW	LW	RANGE
	1 0530	2 1058	3 4·0	4 1·5	5 2·5
Seasonal change	Standard Port		6	6	
DIFFERENCES	7*	8*	9*	10*	
Seasonal change	Secondary port		11	11	
SECONDARY PORT	12	13	14	15	
Duration	16				

Sketch and Preliminary Calculations (Step 2)

Draft	5·5 m
Clearance	2·0
Required Depth	7·5
Charted Depth	4·5
Required Height of Tide	3·0 m

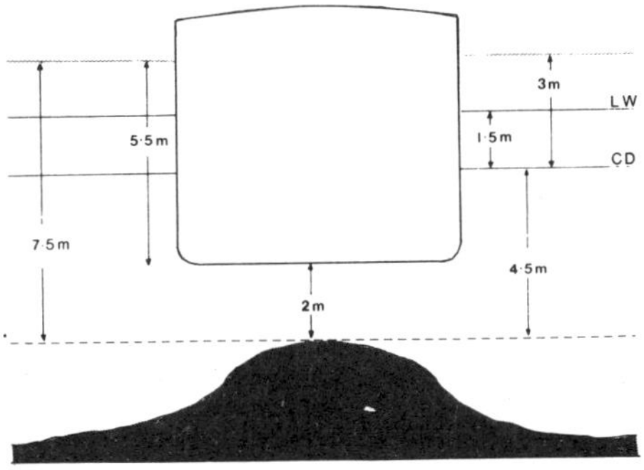

Figure 9.

Interpolation (Step 6)

Spring Range	4·1 m
Neap Range	2·0 m
S − N	2·1 m

Spring Range	4·1 m
Predicted Range	2·5 m
S − P	1·6 m

$$\frac{S-P}{S-N} = \frac{1{\cdot}6}{2{\cdot}1} = 0{\cdot}76 \simeq \tfrac{3}{4} \text{ (Fractional Distance from Spring Curve).}$$

Plotting (Steps 3, 4, 5, 7 and 8)

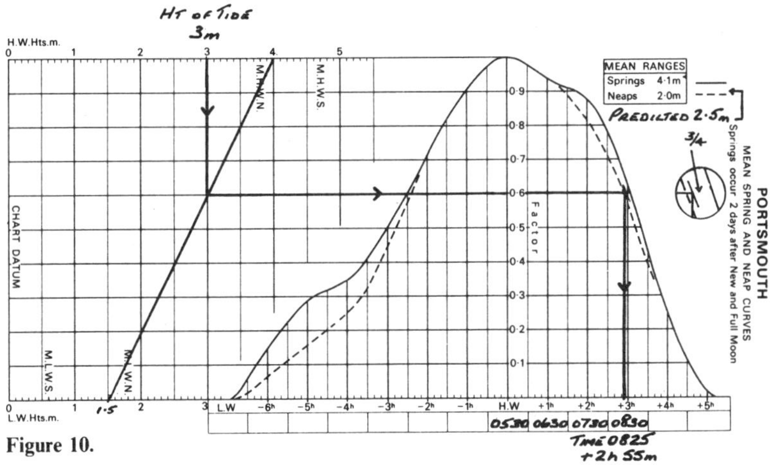

Figure 10.

Answer 0825 GMT (0925 BST).

To Find the Corrected Height of a Lighthouse

(European Standard Port)

Use earlier instructions (Checklist 1), "To Find the Height of Tide at an Intermediate Time between HW and LW". Note that the charted height of a lighthouse or other shore object, is given above MHWS. This level can be found from the Table of Tidal Levels on the chart, if included, or, for Standard Ports, from ATT, Table V (see page 395). For Secondary Ports, MHWS is obtained by applying the height difference in Part II to the Standard Port MHWS level given there.

Example 13

A lighthouse off Sheerness has a charted height of 15 m. Find its height above water level at 0930 GMT on April 22nd for use with a vertical sextant angle. (Predictions on page 402).

Follow the instructions in **Checklist 1.** (Step 1)

TIME ZONE GMT
APRIL

22 TH	0523	0·6
	1140	5·7
	1740	0·8
	2353	5·6

STANDARD PORT *SHEERNESS* TIME/[redacted] REQUIRED *0930*

SECONDARY PORT *—* DATE *APR. 22* TIME ZONE *GMT*

	TIME		HEIGHT		
STANDARD PORT	HW	LW	HW	LW	RANGE
	1 *1140*	2 *0523*	3 *5·7*	4 *0·6*	5 *5·1*
Seasonal change	Standard Port		6	6	
DIFFERENCES	7*	8*	9*	10*	
Seasonal change	Secondary port		11	11	
SECONDARY PORT	12	13	14	15	
Duration	16				

Interpolation (Step 5)

Predicted Range is equal to Spring Range—use Spring Curve only.

Plotting (Steps 2, 3, 4, 6 and 7)

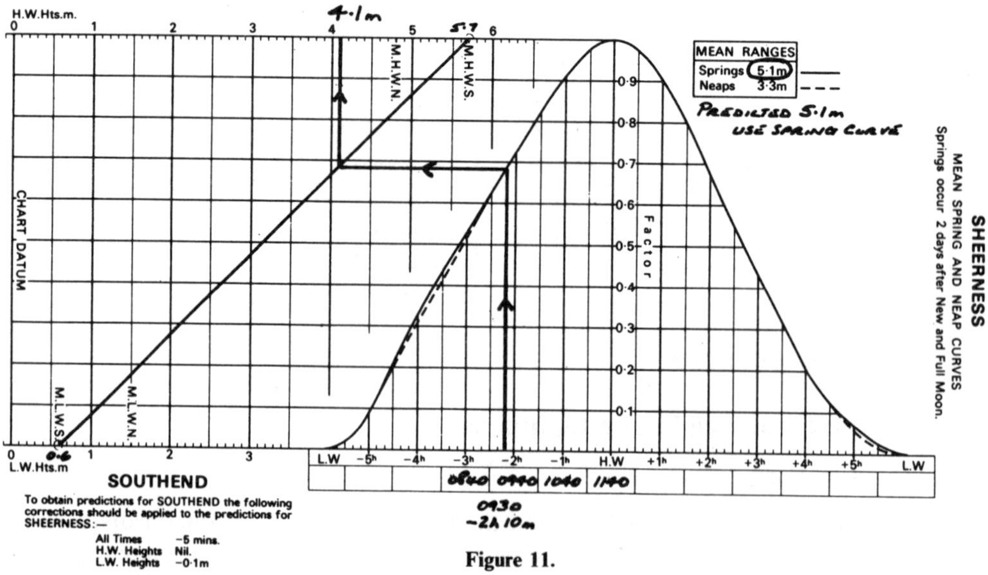

Figure 11.

Sketch and Final Calculations

Charted Height Lt.Ho.	15·0	
MHWS	5·7	(from Table V — page 395).
Height above CD	20·7	
Height of Tide	4·1	
Corrected Height Lt.Ho.	16·6 m	

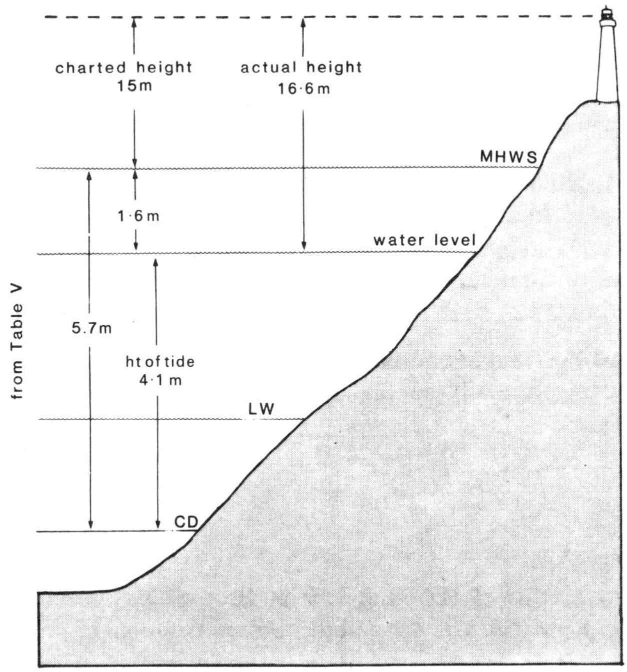

Figure 12.

Answer 16·6 metres is the corrected height.

Exercises on European Port tides are included at the end of the chapter.

SECONDARY PORT TIDE CALCULATIONS

European Secondary Ports—ATT Volume 1

The method of finding the time and heights of HW and LW at any Secondary Port given in the *Admiralty Tide Tables* is by means of simple differences in times and heights from predictions for the Standard Port indicated, which is the most suitable, not necessarily the closest, Standard Port.

To Find HW and LW Times and Heights

(European Secondary Ports—**Checklist 3**)

1. Find the Secondary Port in the Geographical Index at the back of ATT. Note its number.
2. Refer to ATT Part II with Secondary Port number and note the relevant Standard Port, its number and page number in Part I.
3. Refer to Part I. Find HW and LW times and heights at the Standard Port for the required day. Enter in NP 204, boxes 1–4, and complete box 5 with the predicted range.
4. Refer back to Part II, bottom RH page "Seasonal Change in Mean Level". Using its number to assist, locate the correct line and find the Standard Port's Seasonal Change (for the month). Enter this change in boxes 6, with its sign changed (see note).
5. Note the time and height differences for the Secondary Port and the related reference times and reference heights at the Standard Port.
6. Interpolate between Secondary Port time differences and height differences using graph paper (see Example 15) or by calculation. Suitable graph paper is included in ATT but any type can be used.
7. Enter the interpolated Secondary Port time and height differences in boxes 7–10.
8. Using its number to locate the right line, find and enter the Secondary Port's Seasonal Change for the month in boxes 11.
9. Apply differences to Standard Port times and heights to find times and heights at the Secondary Port (boxes 12–15).

Note The Standard Port height quoted in ATT already includes the Seasonal Change which must be subtracted algebraically so that subsequent interpolation is carried out on the unadjusted tidal heights.

Example 14

(European Secondary Port)

Find the times and heights of HW and LW at Brodick Bay for the AM rising tide on August 1st. (Refer to pages 415, 414, 406 and the table in Example 15.) 415, 414, 406

Follow the instructions in **Checklist 3**.

STANDARD PORT GREENOCK (404) TIME/HEIGHT REQUIRED AM RISING

SECONDARY PORT BRODICK BAY (408) DATE AUG 1ST TIME ZONE GMT

	TIME		HEIGHT		
STANDARD PORT	HW	LW	HW	LW	RANGE
	1 0939	2 0329	3 2·8	4 0·8	5 2·0
Seasonal change		Standard Port (404)	6 +0·1	6 +0·1	SIGN CHANGED
DIFFERENCES	7* 0000	8* +0005	9* −0·2	10* 0·0	
Seasonal change		Secondary port (408)	11 0·0	11 0·0	
SECONDARY PORT	12 0939	13 0334	14 2·7	15 0·9	
Duration	16 —				

Note that in this case (i.e. at Brodick Bay No. 408) the Secondary Port time and height differences at HW and LW remain constant for different Standard Port times and heights—compare these differences with those at Lamlash (409) where differences are not constant.

Answer HW 0939 GMT 2·7 m LW 0334 GMT 0·9 m at Brodick Bay.

Interpolating Secondary Port HW and LW Time and Height Differences using Graph Paper—Checklist 4.

1. Complete NP 204, heading and boxes 1–6, taking data from ATT Parts I and II.
2. In Part II, encircle and "arrow" the Standard Port HW and LW reference times and heights. Also encircle the Secondary Port time and height differences (as shown in Example 15).
3. Apply Standard Port Seasonal Change in Mean Level to HW and LW (boxes 3 and 4) to give the corrected Standard Port HW and LW—note the result required for later plotting—see (10).
4. Take a piece of graph paper. Mark up the RH side edge at hourly intervals with a scale of Standard Port HW reference times (previously encircled). Mark the top edge with a scale for the Secondary Port time differences (previously encircled)—see Example 15. The top scale may require negative (−), zero and positive (+) values.
5. Plot the Secondary Port HW time differences (using top scale) against their Standard Port HW reference times (RH side scale). Join plotted points by a sloping line.
6. Plot the predicted Standard Port HW time on the RH scale. From this point move horizontally to the sloping line—then vertically to the top scale and read off the corresponding Secondary Port HW time difference. Enter in box 7.
7. Repeat (5) and (6) but use LW reference times and LW time differences. Read off the Secondary Port LW time difference against the Standard Port LW time. Enter in box 8.
8. Mark up the LH side of the graph with a suitable scale for Standard Port heights from MLWS to MHWS (previously encircled values). Mark the bottom edge with a scale for Secondary Port height differences (previously encircled). It may be necessary to have negative (−), zero and positive (+) values.
9. Using the LH and bottom scales, plot Secondary Port HW height differences against the Standard Port MHWS and MHWN heights. Join by a sloping line—extend each way to allow possible extrapolation.
10. Plot the corrected Standard Port HW height on the LH scale. From this point move horizontally to the sloping line—then vertically down to the bottom scale and read off the corresponding Secondary Port height difference. Enter in box 9.
11. Repeat (9) and (10) but use Standard Port MLWS and MLWN heights and Secondary Port LW height differences. Read off the Secondary Port LW height difference corresponding to the Standard Port LW height. Enter in box 10.
12. Enter Secondary Port Seasonal Change in Mean Level (for the month) in boxes 11.
13. Add the columns and complete boxes 12–15 which give HW and LW times and heights at the Secondary Port.

Example 15

Find the predicted times and heights of HW and LW at Bowling (406) for the falling PM time on June 4th—predictions for Greenock were HW 1041 3·1 m, LW 1629 0·5 m. Time Zone GMT. The Predicted Seasonal Change in Mean Level for ports numbered 399–407, on June 1st, was −0·1 m. (Check this data by reference to the extracts.)

No.	PLACE	Lat. N.	Long. W.	TIME DIFFERENCES High Water (Zone G.M.T.)		Low Water		HEIGHT DIFFERENCES (IN METRES) MHWS	MHWN	MLWN	MLWS	M.L. Z_0 m.
404	GREENOCK	(see page 94)		0000 and 1200	0600 and 1800	0000 and 1200	0600 and 1800	3·4	2·9	1·0	0·4	2·00
	River Clyde											
405	Port Glasgow .	55 56	4 41	+0011	+0005	+0010	+0018	+0·2	+0·1	0·0	0·0	⊙
406	Bowling .	55 56	4 29	+0032	+0016	+0043	+0107	+0·6	+0·5	+0·3	+0·1	⊙
408	Brodick Bay	55 35	5 08	0000	0000	+0005	+0005	−0·2	−0·2	0·0	0·0	1·86
409	Lamlash .	55 32	5 07	−0016	−0036	−0024	−0004	−0·2	−0·2	⊙	⊙	⊙
410	Ardrossan .	55 38	4 49	−0020	−0020	−0030	−0010	−0·2	−0·2	−0·1	−0·1	⊙
411	Irvine .	55 36	4 41	−0020	−0020	−0030	−0010	−0·3	−0·3	−0·1	0·0	⊙

Interpolation

Follow the instructions in **Checklist 4.**

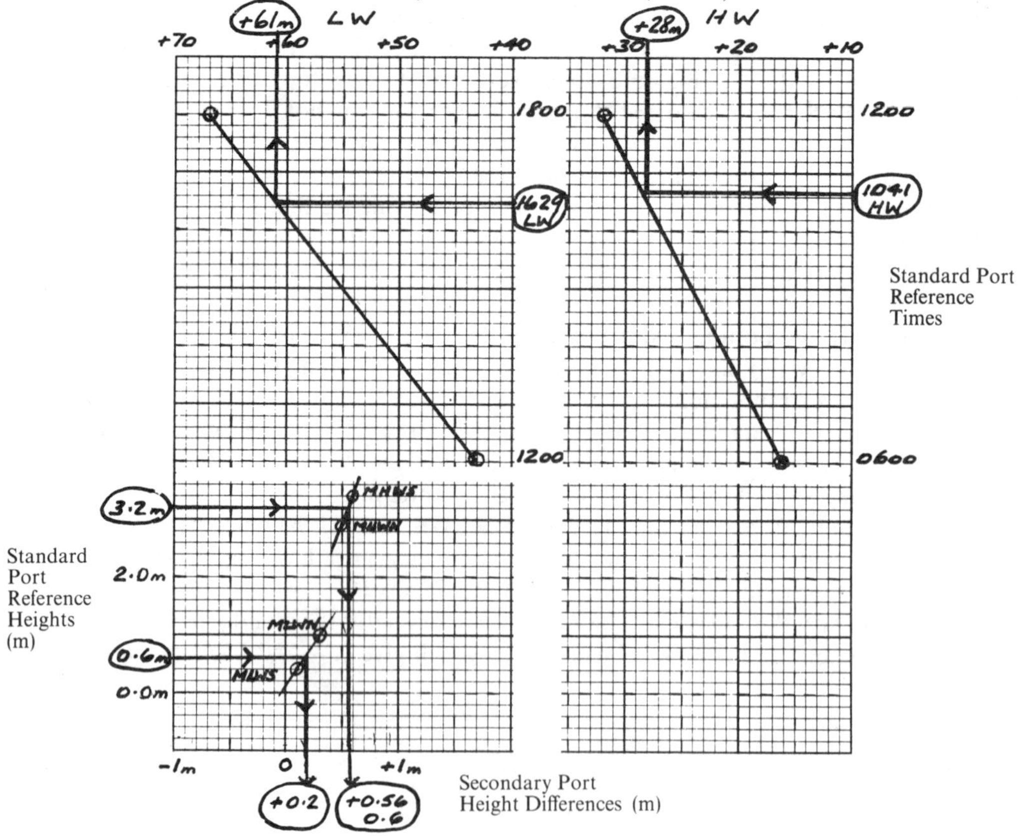

Figure 13.

STANDARD PORT GREENOCK (404) TIME/HEIGHTS REQUIRED PM FALLING

SECONDARY PORT BOWLING (406) DATE JUNE 4 TIME ZONE GMT

	TIME		HEIGHT		
STANDARD PORT	HW	LW	HW	LW	RANGE
	1 1041	2 1629	3 3·1	4 0·5	5 2·6
Seasonal change	Standard Port		6 +0·1	6 +0·1	SIGN CHANGED
DIFFERENCES	7* +0028	8* +0101	9* +0·6	10* +0·2	NB PLOT 3·2+0·6
Seasonal change	Secondary port		11 −0·1	11 −0·1	
SECONDARY PORT	12 1109	13 1730	14 3·7	15 0·7	
Duration	16 —				

Answer HW 1109 3·7 m LW 1730 0·7 m at Bowling.

Height of Tide at Intermediate Times—European Secondary Ports

At European Secondary Ports, the Height of Tide at an intermediate time (between HW and LW) is found by plotting Secondary Port data on the relevant Standard Port Tidal Diagram. Interpolating the curve for the day, if necessary, is carried out by comparing the Standard Port Range, for the time and date in question, with the Standard Port Spring and Neap Ranges, i.e. when interpolating, the procedure is identical for Standard and Secondary Ports. This gives acceptable accuracy because Secondary Ports are "paired" with Standard Ports because of the similarity of their tidal curves.

This method is only to be used when:

1. Adjacent Standard Port tidal curves are similar, and
2. Secondary Port HW and LW time differences are nearly the same.

In other cases, indicated in Part II by the letter "*c*", use the *Admiralty Method of Tidal Prediction*, NP159.

Swanage to Selsey Secondary Ports—Special Instructions

Special instructions apply to places between Swanage and Selsey. In this area there is a rapid change of tidal characteristics and distortion of the tidal curve. Tidal curves for these places have either a Spring and Neap curve or a Spring, Neap and Critical curve—see page 124 (h). On these curves times are related to LW not HW. Where three curves are shown, interpolation is carried out between either, the Critical and Spring curve, or the Critical and Neap curve. This type of calculation is demonstrated in ATT Volume 1 and Example 18.

Use of the Admiralty Method of Tidal Prediction (NP 159)

For Secondary Ports on stretches of coast where there is little change of shape between adjacent Standard Port curves and where the duration of rise or fall at the Secondary Port is not markedly different from the duration of rise or fall at the relevant Standard Port (i.e. where HW and LW time differences in Part II are nearly the same) intermediate times and heights

may be obtained by using the Spring and Neap Curves for the Standard Port. In other cases, where use of the *Admiralty Method of Tidal Prediction* (NP159) is recommended and where intermediate heights are important, these are indicated by the letter "*c*" in the final column of the Secondary Port entry in Part II. For those wishing to use a pocket calculator rather than the graphical method (shown later) a calculator "box diagram"(see page 420 (q)**and detailed instructions are included at the front of ATT. The "box diagram" is also helpful when programming a programmable calculator.**

Offshore Areas and Places Between Secondary Ports

Tidal predictions for certain offshore areas and stretches of coast between Secondary Ports should be obtained by use of Co-Tidal Charts. Full instructions for their use are included on the body of these charts which are listed in the *Catalogue of Admiralty Charts and Hydrographic Publications.*

To Find the Height of Tide at an Intermediate Time Between HW and LW

(European Secondary Port)

Refer to earlier comment on restrictions in the use of this method and use of NP 159.

Instructions for Secondary Ports are similar to those given earlier in **Checklist 1**, "To Find the Height of Tide at an Intermediate Time", for a Standard Port—see page 107. But, in step (2), use the Secondary Port HW and LW heights to plot the sloping line on the relevant Standard Port tidal diagram. And, in step (3), the Secondary Port HW time and time interval are entered below the curve.

Interpolating the predicted curve, step (5), is carried out using Standard Port Predicted, Spring and Neap Ranges—exactly as before. Steps (6) and (7) are also unchanged.

Example 16

(Refer to Figure 14)

Find the height of tide at 1400 GMT at Bowling (406) on the PM falling tide on June 4th. Use the times and heights given in Example 15.

Bowling Predictions:	HW 1109	3·7 m	LW 1730	0·7 m
Standard Port (Greenock):	Range 2·6 m			

Interpolating between Curves for Standard Port, Greenock

Spring Range	3·0 m	Spring Range	3·0 m
Neap Range	1·9 m	Predicted Range	2·6 m
S − N	1·1 m	S − P	0·4 m

$$\frac{S-P}{S-N} = \frac{0{\cdot}4}{1{\cdot}1} = 0{\cdot}36 \simeq \frac{1}{3} \text{ (Fractional Distance from Spring Curve)}$$

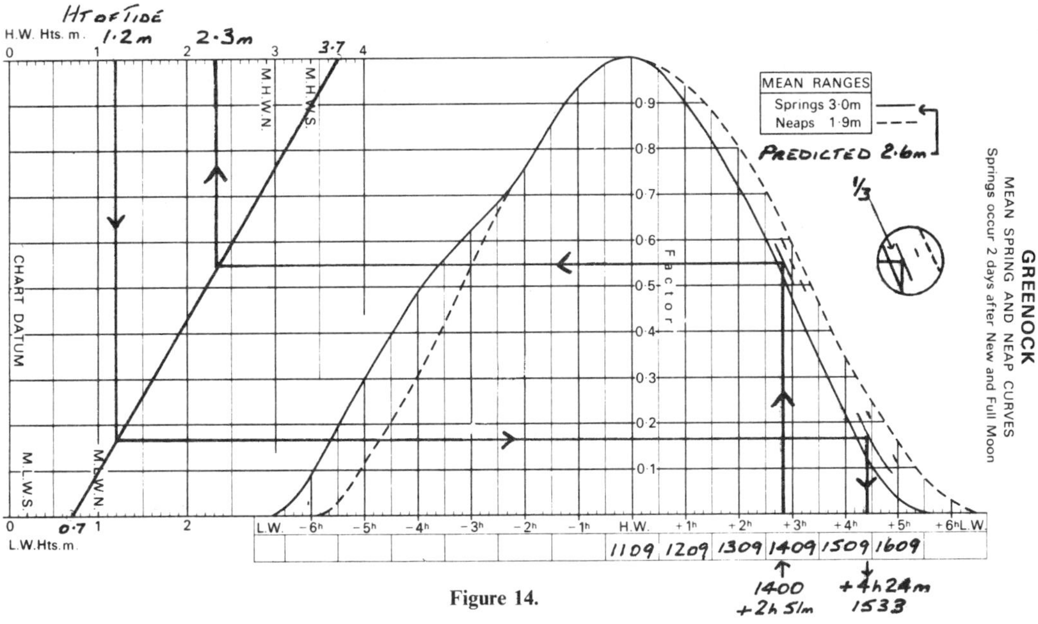

Figure 14.

Answer 2·3 metres Ht of Tide at 1400 GMT.

To Find the Time when the Tide Reaches a Required Height or Gives a Required Depth

(European Secondary Port)

Instructions for Secondary Ports are similar to those given previously in **Checklist 2**, "To Find the Time at which a Required Depth is Reached", for a Standard Port—see page **110**. But, in step (3), use the Secondary Port HW and LW heights to plot the sloping line on the relevant Standard Port tidal diagram. And, in step (4), the Secondary Port HW time and time interval are entered below the Standard Port curve.

Interpolating the predicted curve, step (6), is carried out using Standard Port Predicted, Spring and Neap Ranges—exactly as before. Steps (7) and (8) are also unchanged.

Example 17

(Refer to Figure 14)

Find the time when the PM tide will have fallen to 1·2 metres above Chart Datum at Bowling (406) on June 4th. Use the times and heights given in Example 15.

Interpolating between Curves for Standard Port, Greenock

Spring Range	3·0 m	Spring Range	3·0 m
Neap Range	1·9 m	Predicted Range	2·6 m
S − N	1·1 m	S − P	0·4 m

$$\frac{S-P}{S-N} = \frac{0{\cdot}4}{1{\cdot}1} = 0{\cdot}36 \simeq \frac{1}{3} \text{ (Fractional Distance from Spring Curve)}$$

Answer 1533 GMT Ht of Tide 1·2 metres.

Example 18

(Swanage to Selsey Port)

Given the following data, find height of tide at 1500 GMT at Swanage (35) and thence the depth of water over a wreck of charted depth 29 metres.

65 **PORTSMOUTH** . . . (see page 14)			0000 and 1200	0600 and 1800	0500 and 1700	1100 and 2300	4·7	3·8	1·8	0·6		
35 Swanage	50 37	1 57	−0250	+0105	−0105	−0105	−2·7	−2·2	−0·7	−0·3	1·49	*j

TIME ZONE GMT

0111	4·4
0637	1·1
1339	4·5
1905	1·0

STANDARD PORT *Portsmouth (65)* TIME/HEIGHT REQUIRED *1500*

SECONDARY PORT *Swanage (35)* DATE *Example* TIME ZONE *GMT*

	TIME		HEIGHT		
STANDARD PORT	HW	LW	HW	LW	RANGE
	1 —	2 *1905*	3 *4·5*	4 *1·0*	5 *3·5*
Seasonal change	Standard Port		6 *0·0*	6 *0·0*	
DIFFERENCES	7* —	8* *−0105*	9* *−2·6*	10* *−0·4*	
Seasonal change	Secondary port		11 *0·0*	11 *0·0*	
SECONDARY PORT	12 —	13 *1800*	14 *1·9*	15 *0·6*	
Duration	16 —				

Note that Swanage is a European Secondary Port for which there are special instructions and three tidal curves which are related to the time of LW (not HW as with other Secondary Ports)—refer to earlier comment.

Interpolating between Curves for Swanage

Interpolate between the Critical Range at Portsmouth and either the Spring or Neap Range at Portsmouth as appropriate—in this case Predicted Range (at Portsmouth) lies between the Critical and Spring Ranges.

Ranges at Portsmouth

Spring Range	4·1 m	Predicted Range	3·5 m
Critical Range	2·9 m	Critical Range	2·9 m
S − C	1·2 m	P − C	0·6 m

$$\frac{P-C}{S-C} = \frac{0{\cdot}6}{1{\cdot}2} = \frac{1}{2}$$ (Fractional Distance from Critical Curve towards Spring Curve)

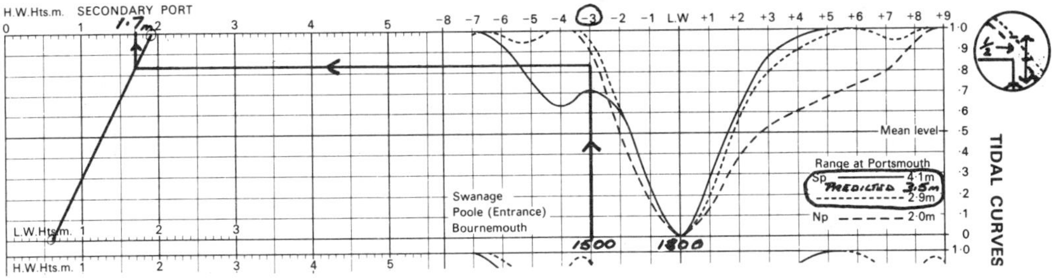

Figure 15.

Plotting

1. Plot the Secondary Port HW and LW heights and join by a sloping line.
2. Plot the Time Interval (Required Time – LW Time) on the scale beneath the curves, i.e. 1500—1800 = (–) 3 hrs (before LW).
3. From this point, draw a vertical line terminating at a point the correct fractional distance between the Critical and Spring Curves—as shown.
4. From the above point, draw a horizontal to intersect the sloping line and then a vertical to intersect the Height Scale.
5. Read off the height of tide above chart datum and apply to the charted depth of the wreck, i.e. 29 m + 1·7 = 30·7 metres.

Answer 30·7 metres is the actual depth over the wreck.

ATLANTIC and INDIAN OCEAN PORTS (ATT Volume 2), and PACIFIC OCEAN and ADJACENT SEAS (ATT Volume 3)

HW and LW Times and Heights at Standard Ports

Times and heights of HW and LW at Standard Ports are found directly from the daily predictions in ATT Volumes 2 and 3. The Zone Time used for the predicted times is usually the Standard Time for the area and is given at the top of each page. Care should be taken to check that this time is the actual Time Zone in use on the date in question—if not, correct as necessary.

HW and LW Times and Heights at Secondary Ports

Times and heights of HW and LW at Secondary Ports are found by applying simple differences in times and heights to the predictions at the most suitable (not necessarily the closest) Standard Port.

In many Pacific Ports, the heights of HW and LW show irregularities due to the predominantly diurnal effect. In extreme cases this can produce only one HW and one LW each day. Where the effect is less, it may cause two high waters at different levels and two low waters at different levels. These are usually termed Higher HW (HHW) and Lower HW (LHW); Higher LW (HLW) and Lower LW (LLW)—in calculations, these should be taken into account according to ATT footnotes for the ports affected.

For certain ports, indicated by the letter "*p*", no time differences are listed in ATT Part II. Predictions for these ports can only be made by using the *Admiralty Method of Tidal Prediction*, NP 159.

To Find HW and LW Times and Heights

(Secondary Ports—Atlantic, Indian and Pacific Oceans—**Checklist 5**)

1. Use NP 204 to record the date, required time or height and the Secondary Port name, i.e. complete NP 204 heading, as far as possible and then use the form to record, and total, data extracted from the tide tables.
2. Find the Secondary Port in the Geographical Index of ATT Volume 2 or 3. Note its number.
3. Refer to ATT Part II with Secondary Port number and note the relevant Standard Port, its number and page number in Part I.
4. Refer to Part I and find HW and LW times and heights at the Standard Port for the required day.
5. Refer back to Part II. Find the Standard Port Seasonal Change in Mean Level, change its sign and add it to the Standard Port heights. (NB. Standard Port heights quoted in ATT already include this Seasonal Correction which must be subtracted algebraically so that subsequent interpolation is carried out on the unadjusted tidal heights—this is achieved by changing the sign and adding.)
6. Note the time and height differences for the Secondary Port (interpolate using graph paper, if necessary).
7. To find the Secondary Port HW's and LW's, apply interpolated Secondary Port time and height differences to the Standard Port HW and LW times and heights and add, algebraically, the Secondary Port Seasonal Change in Mean Level.
8. Complete NP 204, box 16, by finding the Duration of Tide (time intervals between each HW and the preceding or following LW)—which will be required for any subsequent intermediate time or height calculations.

Example 19

(Indian Ocean Secondary Port)

Find the time and heights of HW and LW at Akyab (4520) on September 28th.

Form NP 204 is somewhat cramped for entering two HW's and two LW's—make up a new grid or use two forms, if desired. Predictions page 420 (c)–420 (d).

STANDARD PORT: Bassein River Entrance (4539) TIME/HEIGHT REQUIRED: HW + LW

SECONDARY PORT: Akyab (4520) Volume 2 DATE: Sept 28 TIME ZONE: -0630

	TIME		HEIGHT		
	HW	LW	HW	LW	RANGE
STANDARD PORT	1 0655 1928	2 0008 1337	3 2·0 1·8	4 1·3 1·2	5 0·7 0·6
Seasonal change	Standard Port		6 −0·1	6 −0·1	Sign Changed
DIFFERENCES	7* +0004	8* +0001	9* −0·1	10* −0·3	
Seasonal change	Secondary port		11 +0·1	11 +0·1	
SECONDARY PORT	12 0659 1932	13 0009 1338	14 1·9 1·7	15 1·0 0·9	
Duration	16 6h 50m 6h 39m 5h 54m				

N.B. In this particular example the Seasonal Change in Mean Level at the Standard and Secondary Ports are the same and hence cancel out—this is *not* always the case.

Standard Curves Diagram—ATT Volumes 2 and 3

(Refer to Figure 17 and page 420 (s)).

Curves for individual Standard Ports are not included in ATT Volumes 2 and 3. Instead, a single "Standard Curves Diagram", (identical in both volumes), is included. This diagram shows the height of tide attained at a given time or time interval from HW. Subject to certain restrictions, this diagram is used for all Atlantic, Indian and Pacific Ocean Ports.

The Diagram includes three Standard Curves which correspond to Durations of Tide of 5, 6 and 7 hours. It assumes that the tidal curve of the place under consideration closely resembles a cosine curve. Where the Duration of Tide has an intermediate value (say 6¼ hours), for which there is no curve, it becomes necessary to interpolate between adjacent curves (i.e. for 6¼ hrs, draw in part of a new curve ¼ of the way from the 6 hour curve towards the 7 hour curve).

Intermediate Times and Heights between HW and LW

The Standard Curves Diagram can be used to find either a height of tide at a particular place or the time at which a tide reaches a required height. It is used in much the same way as a European Standard Port Tidal Diagram. After any necessary interpolation, the Standard Curves give acceptable results for intermediate times and heights providing that:

1. Duration of Tide is between 5 and 7 hours, and
2. There is no Shallow Water Correction shown in Part III.

If either of these conditions is not met, intermediate times and heights at places listed in ATT Volumes 2 and 3 must be predicted using the *Admiralty Method of Tidal Prediction*, NP 159, which is explained later.

To Find the Height of Tide at an Intermediate Time between HW and LW using Standard Curves

(Secondary Ports—Atlantic, Indian and Pacific Oceans—**Checklist 6**)

1. Complete NP 204 to box 15 for the tide concerned—see Checklist 5.
2. Subtract earlier Secondary Port tide time from later time to obtain Duration of Tide (box 16).
3. On the Standard Curves Diagram, plot the heights of HW and LW and join by a sloping line.
4. Write down the HW time and sufficient others to "embrace" the required time in the boxes below the curves. If helpful, also write down the Time Interval from HW.
5. Plot the required time on the Time Scale and proceed vertically to the curve for the predicted Duration of Tide (interpolate, as necessary, between curves).
6. Proceed horizontally to the sloping line then vertically to the Height Scale. Read off the height of tide.

Example 20

(Pacific Ocean Secondary Port—refer to Figure 17)

Find the height of tide at Port Hardy (9020, Volume 3) at 1755 Standard Time on May 1st. (Predictions on page 420 (i)).

Interpolation

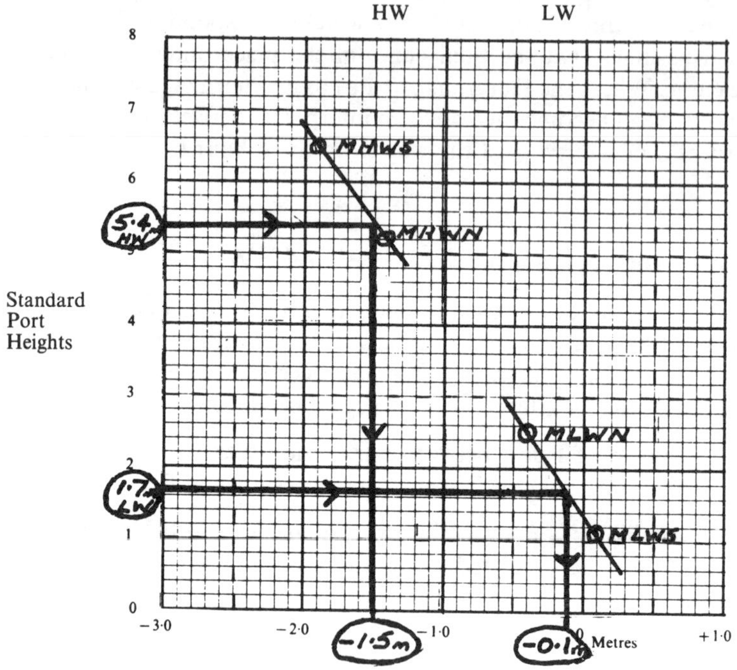

Figure 16. Secondary Port Height Differences.

NP 204

STANDARD PORT Prince Rupert (8850) TIME/[redacted] REQUIRED 1755

SECONDARY PORT Port Hardy (9020) Volume 3 DATE May 1st TIME ZONE +0800

	TIME		HEIGHT		
	HW	LW	HW	LW	RANGE
STANDARD PORT	1 2110	2 1430	3 5·4	4 1·7	5 3·7
Seasonal change	Standard Port		6 0·0	6 0·0	
DIFFERENCES	7* −0025	8* −0030	9* −1·5	10* −0·1	
Seasonal change	Secondary port		11 −0·1	11 −0·1	
SECONDARY PORT	12 2045	13 1400	14 3·8	15 1·5	
Duration	16 6h 45m				

NB Draw in 6¾ hr curve

Answer 2·9 metres Ht of Tide at 1755 Standard Time

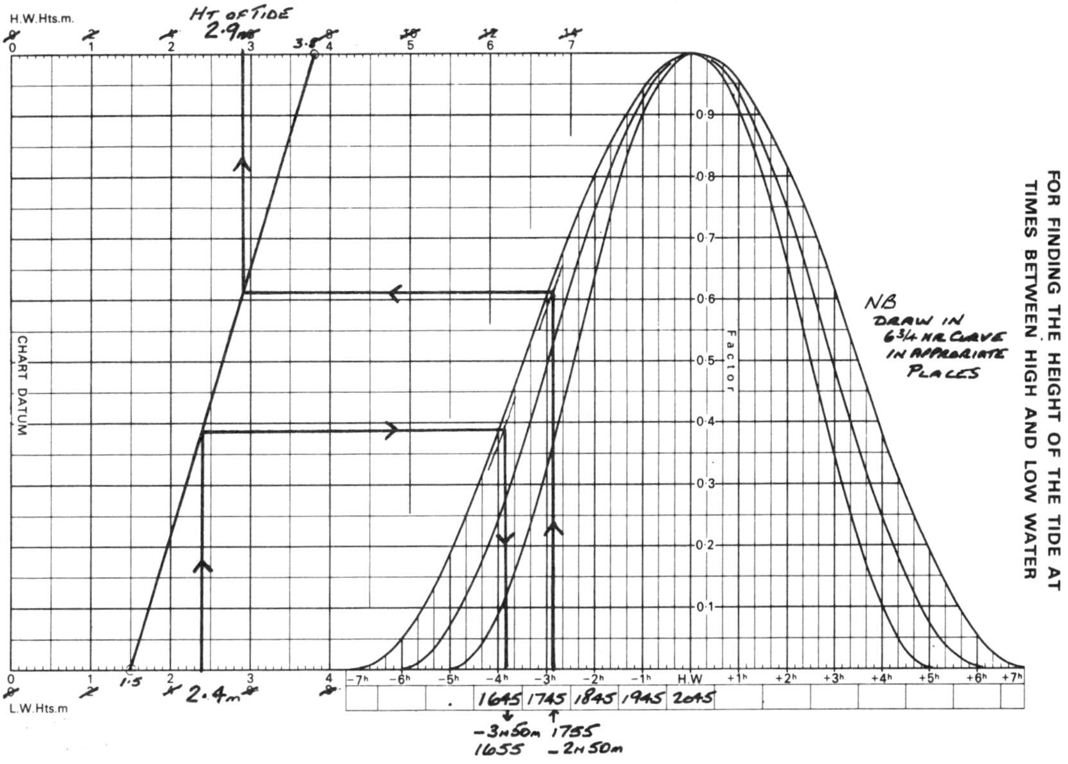

Figure 17.

To Find the Time at which a Tide Reaches a Required Intermediate Height using Standard Curves

(Secondary Ports—Atlantic, Indian and Pacific Oceans—**Checklist 7**)

1. Complete NP 204 to box 16. Plot HW and LW heights and join by a sloping line.
2. In the boxes beneath the curves, write in the HW time and sufficient others to "embrace" the likely time.
3. Plot the required Height of Tide, proceed vertically to the sloping line, then horizontally to the appropriate curve for the predicted Duration of Tide. From this point proceed vertically down to the Time Scale and read off the time.

Example 21

(Pacific Ocean Secondary Port—refer to Figure 17)

Find the time when the rising PM tide reaches a height of 2·4 metres at Port Hardy (9020, Volume 3) on May 1st. Use HW and LW predictions found in Example 20.

Secondary Port HW and LW predictions found in the previous example.

Plotting for this example is shown in Figure 17.

Answer 1655 Standard Time.

NP 159—Calculator Method. For those wishing to use a pocket calculator rather than the graphical method (demonstrated on subsequent pages) a calculator "box diagram" is included on page 420 (q).

Detailed instructions for the completion of the NP 159 "box diagram", and notes relating to the use of this method with calculators and programmable calculators, are included in: *Admiralty Tide Tables* Volume 1, pages xxvi–xxviii and in Volumes 2 and 3, on pages xviii–xx.

THE ADMIRALTY METHOD OF TIDAL PREDICTION—NP 159

A harmonic curve is a symmetrical curve which recurs periodically. The forces which raise the tides are periodic and they can be represented by harmonic curves.

If each force was represented by its harmonic curve then the predicted tidal curve would be the sum of these harmonic curves. Lord Kelvin proposed the harmonic system of tidal prediction. Previously this method required a complex mechanical tide predicting machine. The Admiralty Method has been developed so that the navigator can improve on the accuracy of the tidal curve. This is particularly necessary in those places where there is an appreciable Shallow Water effect.

Harmonic Constants for the port are given in Part II of the Tide Tables and in Table VII data is given to adjust them for the day and year. Detailed instructions are given in NP 159.

Four constituents are used, namely:

M_2 the lunar semidiurnal tractive force and the part of tide due to it.
S_2 the solar semidiurnal tractive force and the part of the tide due to it.
K_1 the combined lunar and solar diurnal force and tide due to it.
O_1 the principal variation in lunar diurnal force and tide due to it.

NP 159 consists of three forms:

Form A includes a Table of data for M_2, S_2, K_1, O_1 and the Mean Level and a plotting circle to produce the various parts of the tide height due to these Harmonic Constants.

Form B which is used to plot the predicted tide heights and on which a tidal curve may be drawn.

Form C which consists of a Table and Plotting Circle to produce the tide heights due to Shallow Water Effect.

A knowledge of the method is required in Class 1 examinations but not the full calculation and the plot of the tidal prediction. This method can also be used with a calculator—see NP 159 Box Diagram on page 420 (q).

Admiralty Method of Tidal Prediction

1. In those places where the Shallow Water effect is known and is appreciable, the predicted tidal curve can be improved by the addition of corrections based on the quarter and sixth diurnal constituents, using Form C. In these cases, the appropriate angles f_4 and f_6 and factors F_4 and F_6 are tabulated for the place concerned in Admiralty Tide Tables, Part II. If no data is tabulated for these angles and factors, Form C is not required.
2. On Forms A and B, all working and plotting should be done to 2 decimal places, to provide predictions correct to 1 decimal place. If the range of the tide to be predicted is either very large or very small, it may be desirable either to double or to halve the height scale of the Graph Sheet (Form B). If this is done, the same factor MUST be applied to the scale of the Plotting Circle on Form A.
3. On Form C, the working and plotting should be done to 3 decimal places, to provide Shallow Water Corrections correct to 2 decimal places for entering on Form B. The scale of the Plotting Circle on this form is considerably greater than that on Forms A and B; like the latter, it may be amended, if desired, for ease of plotting. When transferring the total Shallow Water Corrections at the foot of Form C to the Graph Sheet, the scale of Form B MUST be used.

4. The multiplications required on Forms A and C may be done long-hand, by slide-rule or by logarithms. A table of logarithms to a sufficient degree of accuracy is included inside the front and back covers of this book.
5. The data used in the Example does not refer to any particular port.
6. The Admiralty Method of Tidal Prediction is also suitable for use on an electronic calculator. Details are published in the Instructions section of each volume of Admiralty Tide Tables.

FORM A—Instructions (refer to example Form A across).

Enter PLACE, DATE and ZONE TIME (from Admiralty Tide Tables, Part II).

TABLE—Row 3 From ATT, Part II and Tables VI and VIa where appropriate, for the place enter g and H of M_2, S_2, K_1 and O_1 and Mean Level (Z_0).

Row 4 From ATT, Table VII, for the date, enter Tidal Angle (A) and Factor (F) for M_2, S_2, K_1 and O_1.
From ATT, Part II, enter Seasonal Correction to ML.

Row 5 Add $\alpha°$, g° and A° (subtracting 360° or 720° if necessary), to get m°, s°, k° and o°.
Multiply H and F, to get M, S, K and O.
Add Z_0 and Seasonal Correction, to get A_0.

Plotting Circle

1. From centre C, lay off length M in direction m°, to get point M_2.
From point M_2, lay off length S in direction s°, to get point H_2.
Draw line ②——② through centre C and perpendicular to CH_2.
With centre C and radius CH_2, describe a circle (Semi-Diurnal circle).
2. From centre C, lay off length K in direction k°, to get point K_1.
From point K_1, lay off length O in direction o°, to get point H_1.
Draw line ①——① through centre C and perpendicular to CH_1.
With centre C and radius CH_1, describe a semi-circle to cut all the short (pecked) radii from 06 to 18 (Diurnal semi-circle).
3. Take off the angle h_2 (= direction of H_2 from C) and length H_2 (= CH_2) and enter them in the appropriate spaces in the small panel to the right of the Circle. These are required later, for use with Shallow Water Corrections (Form C), if these are appreciable.

Place EXAMPLE PORT Date 10 JULY 1975 **FORM A**
[Example]

Lat 31° 24′ N Long 121° 30′ E Zone Time −0800

			M_2		S_2		K_1		O_1		M.L.	
TABLE	α°		012		000		180		192			
from ATT, Part II	g°	H	008	0·94	060	0·42	215	0·23	162	0·14	2·35	Mean level (Z_0)
from ATT, Table VII	A°	F	019	1·17	012	0·82	161	1·28	229	1·05	+0·2	Seasonal Correction
	α+g+A	H×F	039	1·100	072	0·344	196	0·294	223	0·147	2·55	ADD
			−m°	−M	−s°	−S	−k°	−K	−o°	−O	$-A_0$	

logs:	M_2	S_2	K_1	O_1
	$\bar{1}$·9731	$\bar{1}$·6232	$\bar{1}$·3617	$\bar{1}$·1461
	0·0682	$\bar{1}$·9138	0·1072	0·0212
	0·0413	$\bar{1}$·5370	$\bar{1}$·4689	$\bar{1}$·1673

h_2 =	047
H_2 =	1·40

PLOTTING CIRCLE

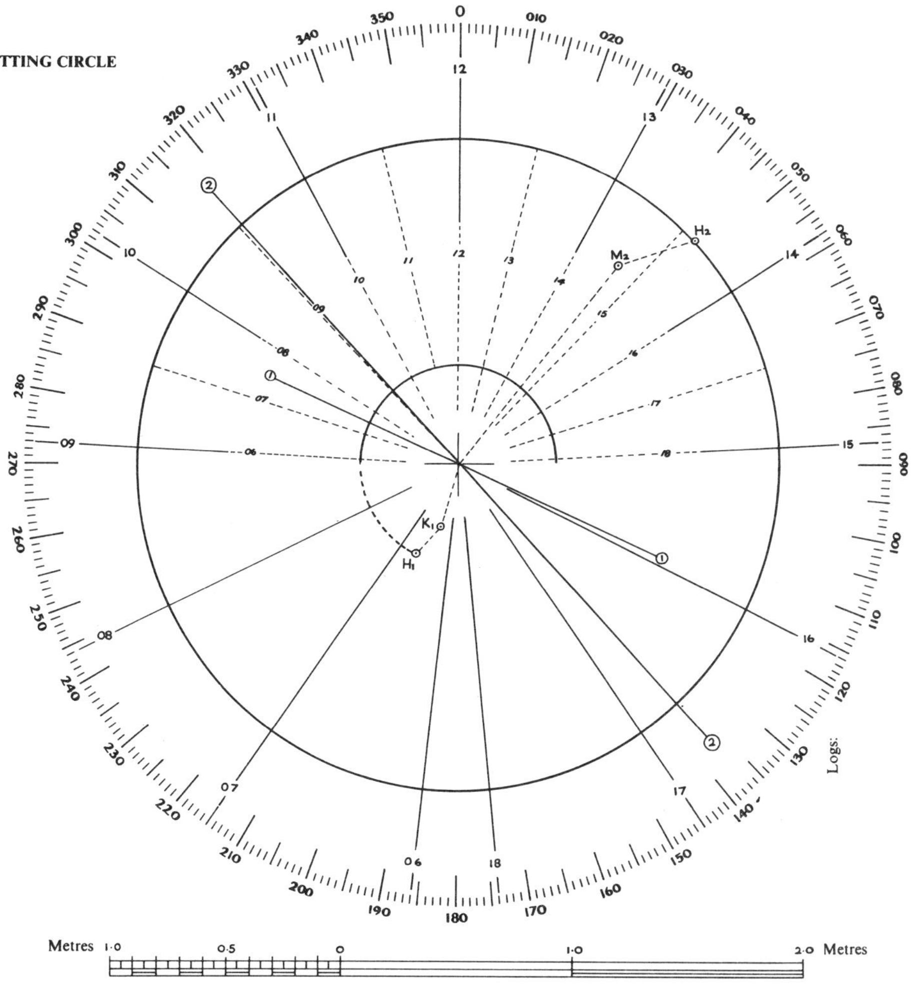

Admiralty Method of Tidal Prediction

FORM B—Instructions (refer to example Form B across).

1. Enter PLACE, DATE and ZONE TIME (from Admiralty Tide Tables, Part II).
2. Draw a line across the sheet at height A_0.
3. From Form A take off, with dividers, the shortest (perpendicular) distance from the intersection of the Semi-Diurnal circle with the long (firm line) radius 06 to the line ②——②. Transfer this length to Form B, laying it off from the line A_0 along the vertical marked 06 at the top, upwards ($+^{ve}$) if the intersection and H_2 are on the same side of ②——② and downwards ($-^{ve}$) if they are on opposite sides of ②——②.
4. Repeat this process of each of the intersections of the S.D. circle with the long (firm line) radii 07 to 18, laying off the distances, with appropriate signs, from the line A_0 along the verticals marked 07 to 18 at the top. Mark these points ⊡ if there are Shallow Water Corrections, otherwise ⊙.
5. At this stage, the Shallow Water Corrections (if any) should be applied. For the method of doing this, see the instructions for Form C on page 6.
6. After applying the Shallow Water Corrections, return to Form A. Take off, with dividers, the shortest (perpendicular) distance from the intersection of the Diurnal semi-circle with the short (pecked line) radius 06 to the line ①——①. Transfer this length to Form B, laying it off from the point ⊙ on the vertical marked 06 at the top, upwards ($+^{ve}$) if the intersection and H_1 are on the same side of ①——① and downwards ($-^{ve}$) if they are on opposite sides of ①——①.
7. Repeat this process for each of the intersections of the D. semi-circle with the short (pecked line) radii 07 to 18, laying off the distances, with appropriate signs, from the points ⊙ already plotted on the appropriate verticals 07 to 18 on Form B, marking these points +.
8. The result is a series of 13 points +. These points are joined by a smooth curve, which is the predicted tide curve from 0600 to 1800 on the date at the head of the form
9. To obtain predictions for further periods of 12 hours on either side of these times, see instructions for extending the predictions on page 124 (j).

FORM B
[Example]

Place EXAMPLE PORT Date 10 JULY 1975

Lat 31° 24' N Long 121° 30' E Zone Time − 0800

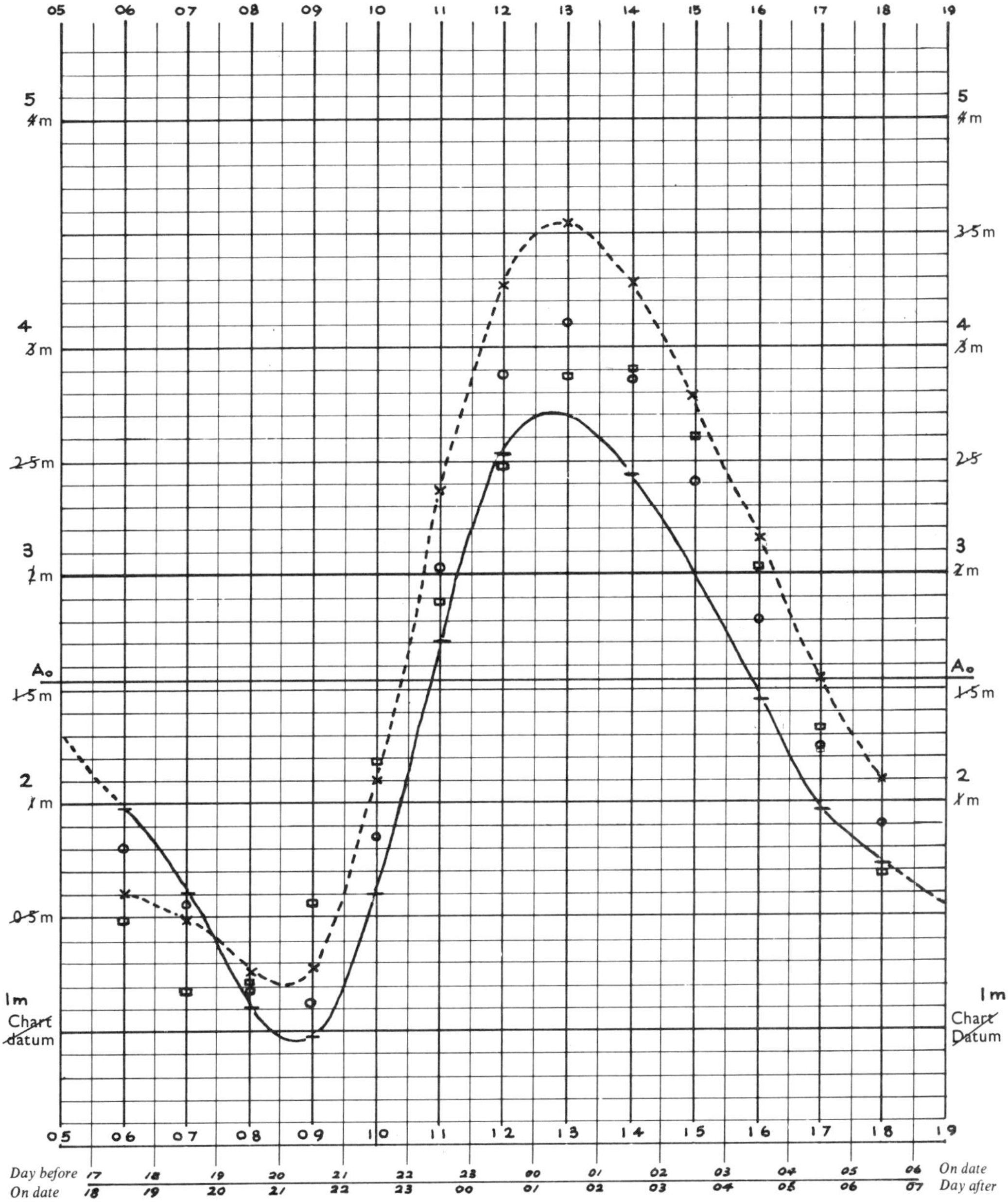

Times at top and bottom of main framework relate to continuous curve (drawn first)
Lower set of times (in italics) relate to pecked curve (drawn second).

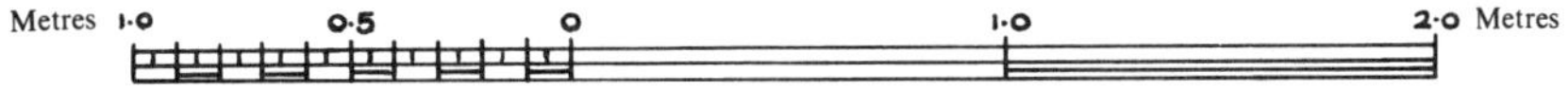

Admiralty Method of Tidal Prediction

FORM C—(Shallow Water Corrections) **instructions** (refer to example Form C across).

1. Enter PLACE, DATE and ZONE TIME (from Admiralty Tide Tables, Part II).
2. Transfer the values of h_2 and H_2 from the small panel on Form A to the appropriate spaces in the Table at the top of the form. Hence evaluate, and enter in the appropriate spaces, $2h_2$ and $3h_2$ (subtracting 360° or 720° if necessary), ${H_2}^2$ and ${H_2}^3$.
3. From ATT, Part II, for the place, obtain the angles f_4 and f_6 and factors F_4 and F_6 and enter them in the appropriate spaces in the Table.
4. Evaluate, and enter in the appropriate spaces in the Table, h_4 $(= f_4 + 2h_2)$, h_6 $(= f_6 + 3h_2)$ (subtracting 360° if necessary), H_4 $(= F_4 \times {H_2}^2)$ and H_6 $(= F_6 \times {H_2}^3)$. If no data is given for F_6, then H_6 is assumed to be zero.
5. On the Plotting Circle, lay off distance H_4 in direction h_4 from the centre C to get point H_4. Draw the line ④——④ through centre C and perpendicular to CH_4. With centre C and radius CH_4, describe the Quarter-Diurnal semi-circle to cut the continuous radii A and P to U.
6. Take off the shortest (perpendicular) distance from the intersection of the Q.D. semi-circle with the radius A to the line ④——④ and enter it in the Q.D. row of the 1200 column of the Table at the foot of the Form, calling it positive ($+^{ve}$) if the intersection and H_4 are on the same side of ④——④ and negative ($-^{ve}$) if they are on opposite sides.
7. Repeat this process for the intersections of the Q.D. semi-circle with the radii P, Q, R, S, T and U, entering the values, with appropriate signs, in the 06, 07, 08, 13, 14 and 1500 columns of the Table respectively, as indicated therein.
8. Complete the Q.D. row of the Table by entering P, Q, R, S, T and U, with OPPOSITE signs, in the 09, 10, 11, 16, 17 and 1800 columns respectively, again as indicated therein.
9. Repeat step 5, with H_6 and h_6 replacing H_4 and h_4, to get point H_6, the line ⑥——⑥ and the Sixth-Diurnal semi-circle cutting the continuous radius A and the pecked radii V to Z.
10. Take off the shortest (perpendicular) distance from the intersection of the 6-D. semi-circle with the continuous radius A to the line ⑥——⑥ and enter it, with appropriate sign, in the 6-D. row of the 1200 column of the Table.
11. Repeat this process for the intersections of the 6-D. semi-circle with the pecked radii V, W, X, Y and Z, entering the values, with appropriate signs, in the 06, 07, 11, 15 and 1600 columns of the Table respectively, as indicated therein.
12. Complete the 6-D. row of the Table by entering V, W, A, X, A, Y and Z, with OPPOSITE signs, in the 08, 09, 10, 13, 14, 17 and 1800 columns respectively, again as indicated therein.
13. Sum, algebraically, the Q.D. and 6-D. values in each of the 13 columns and enter the results in the bottom line of the Table, to obtain the total Shallow Water Correction for each hour.
14. Using the scale on the Graph Sheet (Form B), apply the total S.W.C. for each hour to the appropriate S.D. points ⊡ already plotted on the Graph Sheet, applying them upwards if the correction is positive ($+^{ve}$) and downwards if it is negative ($-^{ve}$). Mark these points ⊙.

FORM C
[Example]

Place EXAMPLE PORT Date 10 JULY 1975

Lat 31° 24′ N Long 121° 30′ E Zone Time −0800

TABLE

	h_2 − 047	$2h_2$ − 094	f_4 − 290	$h_4 - f_4 + 2h_2$ − 024	$3h_2$ − 141	f_4 − 168	$h_6 - f_6 + 3h_2$ − 309
	H_2 − 1·40	H_2^2 − 1·960	F_4 − 0·185	$H_4 - F_4 \times H_2^2$ − 0·363	H_2^3 − 2·744	F_4 − 0·038	$H_6 - F_6 \times H_2^3$ − 0·104
logs:	0·1461	0·2922	$\bar{1}$·2672	$\bar{1}$·5594	0·4383	$\bar{2}$·5798	$\bar{1}$·0181

PLOTTING CIRCLE

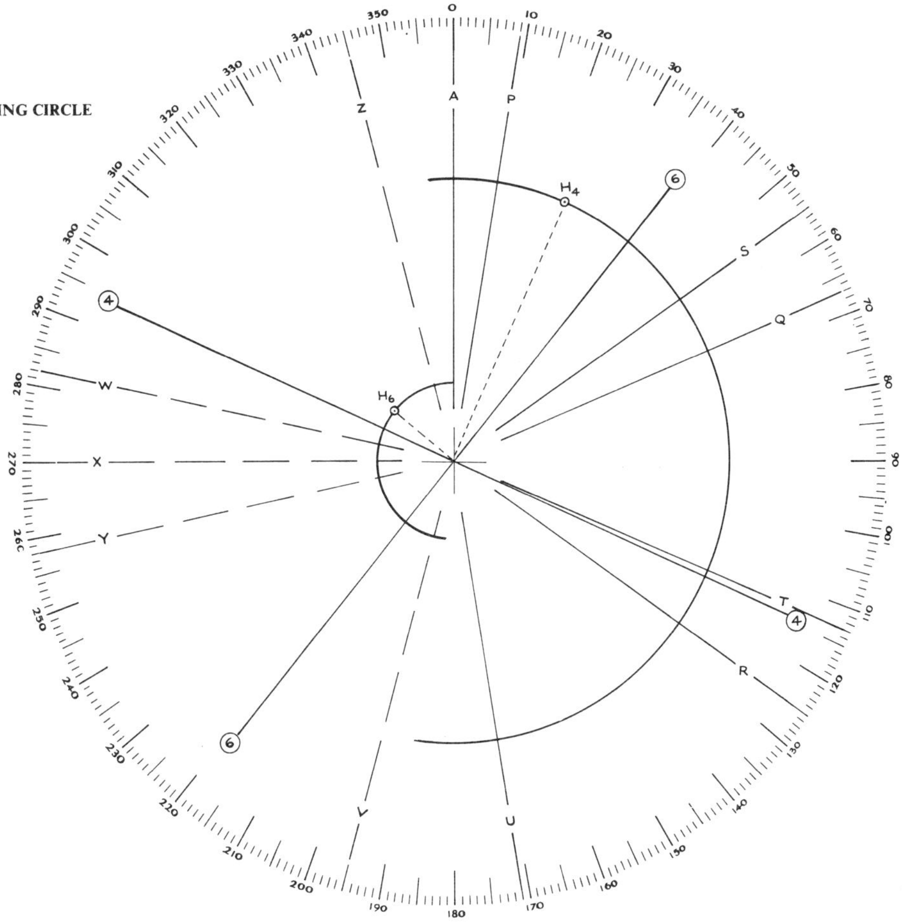

Metres ~~0·1~~ 0·2   ~~0·05~~ 0·1   0   ~~0·1~~ 0·2   ~~0·2~~ 0·4   ~~0·5~~ 0·6 Metres

TABLE

Time	0600	0700	0800	0900	1000	1100	1200	1300	1400	1500	1600	1700	1800
QD	P^{+} ·350	Q^{+} ·266	R^{-} ·068	$-P^{-}$ ·350	$-Q^{-}$ ·265	$-R^{+}$ ·068	A^{+} ·336	S^{+} ·314	T^{+} ·010	U^{-} ·304	$-S^{-}$ ·314	$-T^{-}$ ·010	$-U^{+}$ ·304
6-D	V^{-} ·042	W^{+} ·088	$-V^{+}$ ·042	$-W^{-}$ ·088	$-A^{-}$ ·062	X^{+} ·080	A^{+} ·062	$-X^{-}$ ·080	$-A^{-}$ ·062	Y^{+} ·062	Z^{+} ·082	$-Y^{-}$ ·062	$-Z^{-}$ ·082
Sum	+ ·308	+ ·354	− ·026	− ·438	− ·328	+ ·148	+ ·398	+ ·234	− ·052	− ·242	− ·232	− ·072	+ ·222

Admiralty Method of Tidal Prediction

Extending the predictions

The preceding instructions enable the predicted tide curve from 0600 to 1800 on the date to be drawn, using the time scale at the top and bottom of the main framework of Form B. This tide curve can be simply extended for a further 12 hours in both directions, giving predictions from about 1800 on the day before to about 0600 on the day after the date at the head of the form. These extended predictions are not as accurate as those for the central 12 hours, but are adequate for most purposes.

Previously, the diurnal tide at, say, hour 06 has been taken from the Plotting Circle on Form A with dividers and laid off in the appropriate direction along the 06 hour vertical from the already plotted point ⊙ (the combined S.D. and S.W.C. point, or, if there is no S.W.C., the S.D. point). This gives the height of the whole tide at 0600 on the date. Now lay off the same length from the same point ⊙ in the OPPOSITE direction along the 06 hour vertical, marking it ×. This point gives the approximate height of the whole tide $12\frac{1}{2}$ hours both before and after 0600 on the date—i.e. at 1730 on the day before and at 1830 on the date, as indicated on the lower time scales.

Repeat this process for the hours 07 to 18; the result will be 13 points ×. Join these points by a smooth curve—a pecked line to distinguish it from the continuous line curve already drawn. This pecked curve, read in conjunction with the appropriate lower time scale, extends the original continuous tide curve for a further 12 hours in both directions.

If H_1, the diurnal tide, is negligibly small, then the curve obtained by joining the points ⊙ is the predicted tide curve from 0600 to 1800 on the date at the head of the form, using the time scale at the top and bottom of the main framework of Form B. It is also the predicted tide curve for the 12 preceding and the 12 subsequent hours, using the appropriate one of the two time scales beneath the graph.

It may sometimes be desirable to extend the continuous curve to the edges of the Graph Sheet, at 0500 and 1900 on the date (e.g. when high or low waters occur near 0600 and 1800). The height of the tide at 0500 on the date appears on the pecked curve $1\frac{1}{2}$ hours from the right-hand edge of the Graph Sheet; this height is transferred to the 05 hour vertical on the left-hand edge, to give another point on the continuous curve. Similarly, a further point can be obtained at 1900 on the right-hand edge, enabling the continuous curve to be drawn from 0500 to 1900 on the date.

Additional uses

When the tidal curve for the day is drawn, the times and heights of HW and LW can be read off. Additional information is available from the curve, such as:

(a) When will the height of the tide be greater than 3 metres above datum?

(b) If the ship grounds at 0600 when will she refloat?

Answers:

HW 1250 3·7 m. LW 0845 0·95 m, 2200 1·2 m.

(a) Periods between 1118 to 1500 hours

(b) 1020.

Tides Around the United Kingdom

The main undulation from the Atlantic Ocean approaches the English Channel and Ireland from a SW'ly direction, the line of the crest of the progressing wave being NW and SE, so that it reaches the SW point of Ireland at almost the same time as it arrives at Ushant.

The tidal wave, on striking the SW point of Ireland, is split into two parts, one running northerly, causing high water successively on the West coast of Ireland and Scotland, the other running to the SE, causing high water successively along the South coast of Ireland and the North coast of France. When the tide wave reaches shallow water it generates tidal streams which have the same period as the wave, flowing in the direction of the rapidly moving crest and then ebbing in the opposite direction at the trough. The velocity of the stream is maximum at HW and LW.

In the Irish Sea, tides are raised by the tidal streams entering both North and South channels at the same time, and the resulting movement is similar to that observed when a basin of water is tilted. The level rises at one end of the basin and falls at the other while at a nodal point in between, there is no change in level. This is called a standing oscillation. The tidal streams set towards Liverpool when the water is rising at that port, and away from Liverpool as the water falls there.

In the North Sea the tide wave from the Atlantic sets up a standing oscillation type of tide having more than one high water area at the same time.

The high water crest proceeds S'ly from the vicinity of 60° N. to about 56° N. and a crest then continues from the next tidal system to meet a third system about latitude 53° N. The average rate of the tidal streams are moderate, not exceeding a knot and a half but that part of the stream that enters from the Pentland Firth acquires great rapidity, amounting in spring tides to 8 knots.

In the English Channel the tide waves are of the standing oscillation type and the crest of high water due to constituent M_2 runs transversely across the channel as shown by the co-tidal lines on Admiralty Chart No. 5058. Over a considerable length of the coast between Portland and Selsey Bill a double tide is experienced due to large tide waves from quarter diurnal and other constituents. In the Solent two distinct high waters are experienced with an interval of from one to two hours. At Portland a double low water is experienced from similar causes.

ADMIRALTY TIDAL STREAM ATLASES (NW EUROPE AND HONG KING)

Tidal Stream Atlases (ATSA) published by the Admiralty include some 13 volumes in respect of NW Europe and one volume in respect of Hong Kong

These Atlases give detailed tidal stream information for the areas covered and include full instructions for the prediction of tidal stream rates. A "Computation of Rates Diagram" is included to aid accurate interpolation. Refer to the diagram on page 420 (dd)).

Each atlas contains 13 charts (A4 size) which show, in pictorial form, the tidal streams operating at hourly intervals from High Water (HW) at a selected Standard Port.

The 13 charts included correspond to time intervals of 6 hrs–1 hr *Before* HW, HW, and 1 hr–6 hrs *After* HW. Specimen charts are shown on pages 420 (w)–420 (cc).

Direction of Tidal Stream

The Direction of Tidal Streams is indicated by arrows. No compass rose is included on the charts. If a precise indication of direction is required it is necessary to use the parallel ruler. (The side of the chart may be used as the meridian.) A "Douglas" protractor is a convenient alternative.

Rate

The Rate of a Tidal Stream is indicated in two ways—numerically and pictorially. In the numerical presentation the Neap and Spring rates are given in tenths of a knot. The two rates are separated by a comma which indicates the approximate position at which observations were obtained, e.g. 14,22 indicates 1·4 knots at Neaps and 2·2 knots at Springs at the position (,).

In the pictorial presentation the strength (rate) of the tidal stream is indicated by the length and thickness of the arrow, e.g. a short thin arrow represents a weak tidal stream.

Plotting Positions

If accurate data is required it is essential to accurately plot the position required. A latitude and longitude scale is provided.

Interpolation

Interpolation between Neap and Spring rates can be carried out in the usual way or by use of the "Computation of Rates" diagram which is printed on the inside front cover of each atlas. This diagram is a "purpose built" graph designed to be used for interpolating rates quoted in (and only in) the particular atlas in which it is contained, i.e. it should not be used for interpolating rates from areas covered by other atlases.

Use of the "Computation of Rates" Diagram

(Refer to page 420 (dd)).

1. Select the Tidal Stream Atlas for the required area. (The Area Coverage of Admiralty Tidal Stream Atlases is shown on the last page of each atlas.)
2. Select the appropriate volume of the Admiralty Tide Tables (ATT Vol. 1 for European Waters).
3. Write down the required Position, Date, Time and the Standard Port (as stated at the top of each page) on which the Tidal Stream Atlas predictions are based.
4. Refer to ATT and write down the HW and LW times and heights for the day.
5. Find the interval between the Required Time and the nearest HW. Label Before HW or After HW, as appropriate.
6. Find the ranges between successive HW and LW heights for the day and calculate the Mean Range for the day.
7. Compare the Mean Range (found above) with the Mean Spring and Neap Ranges at the Standard Port.

N.B.—Mean Spring and Neap ranges are shown in the box included on the tidal curve.

If equal to Springs use Spring Rate.
If equal to Neaps use Springs Rate.
In other cases use the Computation of Rates Diagram.

8. Find the tidal stream chart which corresponds to the Time Interval found at (5).
9. Plot the required position on the chart.
   N.B.—Having regard for the small scale, **accurate plotting** is **essential.**
10. Read off the Neap and Spring rates from the chart and write them down.
11. Using the parallel ruler (or a Douglas protractor) find the direction of tidal stream
12. Turn to the Computation of Rates Diagram on the inside front cover.
13. Plot the Neap Rate on the dotted line marked Neaps.
    N.B.—Each dot represents 0·1 knots.
14. Plot the Spring Rate on the dotted line marked Springs.
15. Joint the Neap and Spring rates by a straight line.
16. Draw a horizontal line from the point on the vertical scale which corresponds to the Mean Range for the day, found at (6), to intersect the Neap/Spring rates line at "×".
17. The predicted rate for the day will be found on the horizontal scale vertically below "×".

Admiralty Tidal Publications

Other publications include the Admiralty Manual of Tides, NP 120; Tidal Handbooks 1–3, NP 122; and the 3 volumes of Tide Tables already mentioned. The full list is included below and reference should be made to these publications for further information.

NP No.	
	ADMIRALTY TIDE TABLES (published annually in three volumes)
201	Volume **1:** European Waters (including Mediterranean Sea)
202	Volume **2:** Atlantic and Indian Oceans (including Tidal Stream Tables)
203	Volume **3:** Pacific Ocean and Adjacent Seas (including Tidal Stream Tables)
	ADMIRALTY TIDAL HANDBOOK SERIES
122(1)	No. 1 The Admiralty Semi-graphic Method of Harmonic Tidal Analysis
122(2)	No. 2 Datums for Hydrographic Surveys
122(3)	No. 3 Harmonic Tidal Analysis (short periods)
	TIDAL STREAM ATLASES
250	English and Bristol Channels
257	Approaches to Portland
337	The Solent and adjacent waters
219	Portsmouth Harbour and Approaches
233	Dover Strait
249	Thames Estuary (with co-tidal charts)
251	North Sea, Southern Part
252	North Sea, Northern Part
253	North Sea, Eastern Part
220	Rosyth Harbour and Approaches
209	Orkney and Shetland Islands
218	North Coast of Ireland and West Coast of Scotland
256	Irish Sea
264	The Channel Islands and adjacent coasts of France
265	France, West Coast
217	Hong Kong

MISCELLANEOUS PUBLICATIONS

120	Admiralty Manual of Tides
159	Admiralty Method of Tidal Prediction, using Harmonic Constants, book of forms with instructions and example
204	Admiralty Tidal Prediction Form, with instructions
112	Harmonic Analysis Form (H.206) for one month
171	Harmonic Analysis (H.224) for short period observations
164	Dover, Times of high water and mean ranges (published annually)
215	Tides in the South China Seas

MISCELLANEOUS ATLASES AND CHARTS

Chart 5057, Dungeness to Hoek van Holland, co-tidal and co-range chart
Chart 5058, British Isles and adjacent waters, co-tidal and co-range lines
Chart 5059, Southern North Sea, co-tidal and co-range chart
Chart 5084, Malacca Strait, co-tidal and co-range chart
Chart 5081, Persian Gulf, co-tidal atlas

The above publications are obtainable from the Agents for the sale of Admiralty Charts, a list of whom is published annually in Admiralty Notice to Mariners No. 2, or from the Hydrographic Department, Ministry of Defence, Taunton, Somerset.

Standard Port Tide Calculations

Exercise 1 (European Waters)

Find the times (as indicated) of HW and LW at:

1.	Portsmouth	GMT	on May 29th.
2.	Greenock	GMT	August 18th.
3.	Londonderry	BST	July 9th.
4.	Wilhelmshaven	ZT	January 17th.
5.	Avonmouth	GMT	September 11th.

Find the Height of Tide (Reduction to Sounding) for the given times:

6.	Sheerness	at 0830	GMT	on March 8th.
6.	Londonderry	1600		August 26th.
8.	Greenock	0600		May 5th
9.	Avonmouth	0900		December 28th.

10. Find the time when the height of tide above Chart Datum (CD) will be 3·0 m on the a.m. flood tide of January 18th at Wilhelmshaven.
11. Find the time when a ship drawing 4·7 m will have 1·5 m clearance under the keel over a charted depth of 2·5 m on the p.m. ebb tide at Portsmouth on June 14th.
12. A lighthouse off Greenock has a charted height of 21 m. Find its height above the water line at 0630 BST on May 18th. (From Table V—Greenock—MHWS +3·4 m.)
13. Between what Standard Times can a ship drawing 11·5 m cross a bar of charted depth 3 m with a clearance of one metre under the keel on the a.m. tide at Avonmouth on November 13th.
14. Off Wilhelmshaven, a sandbank is charted with a symbol $\underline{0}_5$. Find the distance that the sandbank is above the water at 1000 ZT on February 11th.

Secondary Port Tide Calculations

Exercise 2 (European Waters, Atlantic and Indian Oceans)

Find the times of high and lower water at:

1. Sandown on June 15th.
2. Moville July 24th.
3. Tilbury March 10th.
4. Bowling May 1st.
5. Barry November 27th
6. Akyab September 9th
7. Lubec (U.S.A.) February 8th

(Standard Tidal Curves are included in the Appendix.)

Exercise 3 (European Waters)

From the predictions found in the first five examples of Exercise 2 find the height of tide:

1. At Sandown on June 15th, at 2000 hours GMT
2. At Moville on July 24th, at 1400 hours GMT.
3. At Tilbury on March 10th, at 0600 hours GMT
4. When will the tide be 2·5 m above datum at Bowling on May 1st?
5. When will the tide be 6·0 m above datum at Barry on November 27th?

Exercise 4 (Pacific Ocean and Adjacent Seas)

1. Find the times and heights of HW and LW at Kildala (No. 8947) on May 1st, 1982. (Standard Port Prince Rupert, No. 8850.)
2. Find the time and heights of HW and LW at Namu (No. 8943) on May 1st, 1982. (Standard Port Prince Rupert, No. 8850.)

Exercise 5 (Atlantic, India Ocean and Pacific Ocean)

(Harmonic Constants)

NP 159, Forms A, B & C, are included in the Appendix

1. Find the times of HW and LW at Welshpool in February 8th. The ship grounded at 0815, when will she refloat?
2. Find the times and heights of HW and LW at Elephant Point on March 13th.
3. Find the times and heights of HW and LW at Quebec on May 16th.
4. Find the times and heights of HW and LW at Port Hedland on July 25th.
5. Find the times and heights of HW and LW and the period when the tide will be 3 metres and more above chart datum at Daly River on April 24th.

Answers on Page 124 (r).

Tidal Steam Rate Calculations

Exercise 1

Use extracts from Admiralty Tidal Stream Atlas, NP 249, Thames Estuary and ATT Vol. I, Predictions for Sheerness. All times are in GMT.

1. Predict the rate of Tidal Stream off Walton-on-the-Naze at approximately 0600 on February 8th.

2. Predict the rate of Tidal Stream east of Maplin Sands at approximately 0900 on March 18th
3. Predict the set and rate of Tidal Stream for a position off Margate (51° 20′ N. 1° 34′ E.) at 0952 on January 7th.
4. Predict the set and rate of Tidal Stream in position east of Gunfleet Sands (51° 47′ N. 1° 25′ E.) at 0748 on March 22nd.
5. Predict the set and rate of Tidal Stream in a position SE of Orfordness (52° 04′ N. 1° 37′ E.) at 1703 on April 12th.

Tide Questions

1. How many high waters and low waters are usually experienced each day? Where are Standard Port HW and LW times found?
2. What is meant by the term Standard Time? Why is a correction for Longitude required when LMT is required?
3. State a rule for the converting Standard Time into GMT
4. What is meant by the term Drying Height.
5. What effect has the solar tide on the lunar tide?
6. What influence has the rise and fall of the tide on: (a) The sighting of guns mounted ashore when firing at a floating target; (b) The range of visibility of a light vessel's lights from a ship and from the shore?
7. How is the Duration of Tide ascertained?
8. Define the terms (a) Chart Datum; (b) Mean Level; (c) Height of Tide; and (d) Rise of Tide.
9. How is the mean tide level obtained?
10. What is meant by the term "Range of Tide"?
11. Define the terms (a) Neap Rise; (b) Mean Rise; (c) Spring Rise. How often do spring tides and neap tides occur?
12. When is the greatest depth of water to be found, at Spring LW or at Neap LW?
13. Explain the terms (a) Conjunction; (b) Opposition; and (c) Quadrature. Make a sketch to aid explanation.
14. What information is required in the process of reducing the actual depth of water to the soundings given on the chart?
15. To what level are chart soundings referred?
16. What correction should be made to a depth obtained by Echo Sounder?
17. How is the factor found?
18. A lighthouse is given on the chart as 40 m. Suppose the spring rise to be 10 m and the height of tide at a given time to be 3 m, what will be the height of the lighthouse above sea level at that time? If the charted sounding at the place is 6 m, what will be the depth of water?
19. What is meant by the term Charted Height?
20. Explain the difference between a Standard Port and a Secondary Port.

ANSWERS

Tides

Exercise 1 (European Standard Port Tides)

			HW	HW	LW	LW
1. Portsmouth	May 29th	GMT	0345	1640	0910	2148
2. Greenock	August 18th	GMT	1145	2345	0442	1704
3. Londonderry	July 9th	BST	1050	2309	0500	1712
4. Wilhelmshaven	January 17th	ZT	0559	1830	1209	—
5. Avonmouth	September 11th	GMT	—	1210	0608	1834
6. Sheerness	March 8th	0830 GMT, 3·1 m				
7. Londonderry	August 26th	1600 GMT, 1·6 m				
8. Greenock	May 5th	0600 GMT, 1·6 m				
9. Avonmouth	December 28th	0900 GMT, 5·4 m				
10. Wilhelmshaven	January 18th	0405 ZT				
11. Portsmouth	June 14th	1850 GMT				
12. Greenock	May 18th	21·8 m				
13. Avonmouth	November 13th	0325 to 0723 GMT				
14. Wilhelmshaven	February 11th	0·3 m				

Secondary Port Tide Calculations

Exercise 2

	HW		LW	
	Time	Ht (m)	Time	Ht (m)
1. Sandown	0453	3·3	1037	1·3
	1744	3·5	2320	1·5
2. Moville	0944	2·2	0342	0·3
	2202	2·3	1532	0·3
3. Tilbury	0112	6·6	0751	0·1
	1336	6·8	1958	0·5
4. Bowling	0542	3·8	0029	1·2
	1912	3·3	1259	0·9
5. Barry	0309	9·0	0906	3·6
	1538	9·7	2138	2·8
6. Akyab	0057	2·6	0705	0·6
	1304	2·4	1921	0·5
7. Lubec (USA)	1140	6·9	0545·	0·7
	—	—	1815	0·3

Exercise 3

1. 3·1 m	2. 0·8 m	3. 0·6 m	
4. 0235	0833	1641	2200
5. 0633	1154	1857	

Exercise 4

1. Kildala	HW	0708	3·6 m	HW	2023	3·5 m
	LW	0109	2·0 m	LW	1349	1·2 m
2. Mamu Harbour	HW	0719	3·6 m	HW	2034	3·5 m
	LW	0114	1·9 m	LW	1354	1·1 m

Exercise 5 (Harmonic Constants)

1.	HW	1145	6·5 m	LW	0600	0·5 m
	HW	—	—	LW	1800	0·1 m
	Ship refloats about 1520					
2.	HW	0515	5·6 m	LW	0000	0·9 m
	HW	1730	5·6 m	LW	1230	0·9 m
3.	HW	1200	3·9 m	LW	0636	1·3 m
	HW	—	—	LW	2000	1·3 m
4.	HW	0130	6·6 m	LW	0730	1·7 m
	HW	1345	6·5 m	LW	2000	1·6 m
5.	HE	0600	6·8 m	LW	0000	0·4 m
	HW	1800	6·8 m	LW	1200	1·0 m
	Periods 0210 to 0915 and 1405 to 2105.					

Tidal Stream Rate Calculations

Exercise 1

1. Predicted Range 5·3 m (Springs) Rate 0·8 knots (6 hrs after HW).
2. Predicted Range 3·2/3·1 (Neaps) Rate 1·7 knots (3 hrs after HW).
3. Predicted Range 4·4, Set 010°T (approx), Rate 2·5 knots (HW).
4. Predicted Range 3·8, Set 220°T (approx), Rate 1·65 knots (3 hrs before HW).
5. Predicted Range 4·9, Set 030°T (approx), Rate 2·7 knots (2 hrs after HW).

CHAPTER V.

FORMULAE, TIME EQUATIONS AND SOME THEORY.

It is assumed that the reader is conversant with the earlier work of Volume I, but before proceeding with further navigation problems it might be as well to again record the accepted trigonometrical formulae. This also seems a convenient place to refer to some of the special Tables used in solving some of the problems, such as the ABC Tables and Norie's Ex-meridian Tables.

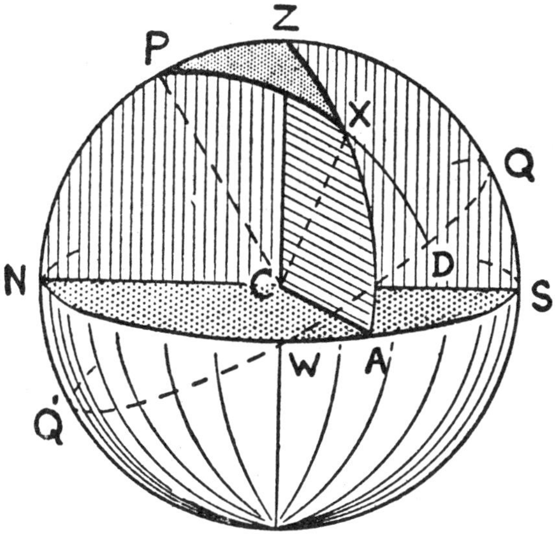

This figure is a diagram drawn on the plane of a meridian with the zenith on top.

$N\,W\,S$ is the horizon, Z the zenith, P the pole, X a star, $Q^1 W\,Q$ the equator.

Referring to the body X we have,

$A\,X$ its altitude $= \angle A\,C\,X$ $\qquad$ $Z\,X$ its zenith distance $= \angle Z\,C\,X$

$D\,X$ its declination $\qquad$ $P\,X$ its polar distance $= \angle P\,C\,X$

$P\,Z$ the co-latitude of observer $= \angle Z\,C\,P$

$\angle Z\,P\,X$ the hour angle of X = arc of equator $Q\,D$

$\angle P\,Z\,X$ the azimuth of X = arc of horizon $N\,A$

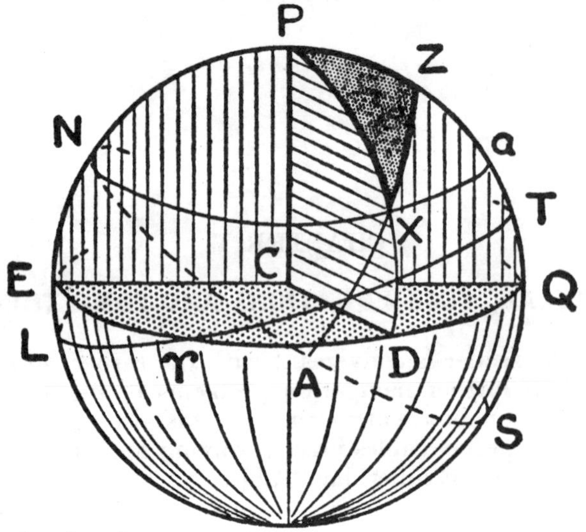

This diagram is also drawn on the plane of the meridian but turned round a little to bring the pole on top.

E D Q is the equator, *N A S* the horizon, *L* ♈ *T* the ecliptic, *P* the pole, *Z* the zenith, *X* a star moving along its diurnal arc *a X*

A X	= altitude of *X*	*Z X*	the zenith distance
D X	the declination	*P X*	the polar distance
P Z	the co-latitude of observer	*Q Z*	the latitude

∠*Z P X* the hour angle = arc of equator *Q D*

∠*P Z X* the azimuth = arc of horizon *N A*

♈ *D* is the right ascension of *X* and ♈ *D* (measured westwards from ♈) is the Sidereal Hour Angle of *X* (360°−*RA*=*SHA*).

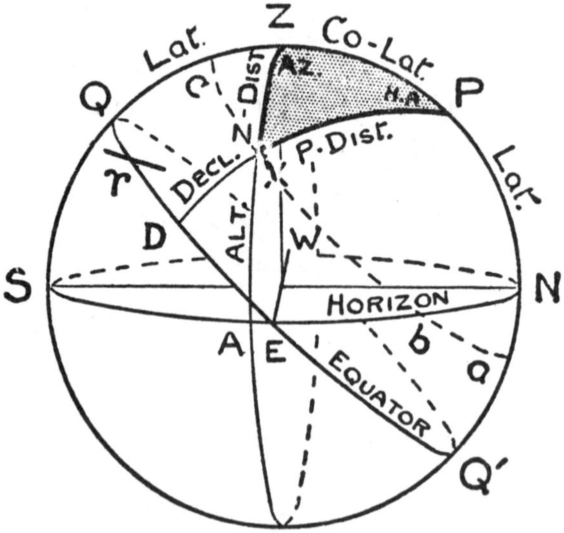

A diagram drawn on the plane of the meridian.
N E S W is the horizon.

N S the projection of the meridian on to the horizon, *W E* the projection of the prime vertical on to the horizon. *Q E Q′* the equator, ♈ First Point of Aries, *Z* the zenith, *P* the pole, *X* a star.

a b X c is the diurnal path of *X* parallel to the equator *Q E Q′*. The star crosses the lower meridian below the horizon at *a*, rises at *b*, and *Eb* is then its rising amplitude; the star is shown in altitude at *X* before culminating at *c*. When it is on the upper meridian at *c*, arc *Qc* is its declination, arc *Zc* its meridian zenith distance, so that $Qc+cZ=ZQ$ the latitude of observer.

The position of *X* relative to the equator is defined by means of its siderial hour angle (♈*D*, measured westward from ♈) and declination (*D X*); and relatively to the observer by (i) its zenith distance (*Z X*) and azimuth ($\angle PZX$); or, (ii) its hour angle ($\angle ZPX$) and zenith distance; or, (iii) its polar distance (*P X*) and hour angle; or (iv) its azimuth and polar distance.

SOLUTION OF RIGHT-ANGLED SPHERICAL TRIANGLES.
NAPIER'S RULES FOR CIRCULAR PARTS.

The rules are briefly as follows:

1. Let the triangle consist of five parts, viz., the three sides and two angles, the 90° being always **omitted.**
2. The **complements** of the hypotenuse and the two angles are used.
3. Adjacent part means the one next. Opposite part means the one past adjacent.
4. By selecting **any** one of three parts as a middle part the other two must form either two adjacent parts or two opposite parts when one of the following formulae will apply:—

sine of middle part = { product of **tangents** of **adjacent parts.** / product of **cosines** of **opposite parts.** }

which may be abbreviated to

I. sine mid part = tan adj × tan adj.

II. sine mid part = cos opp × cos opp.

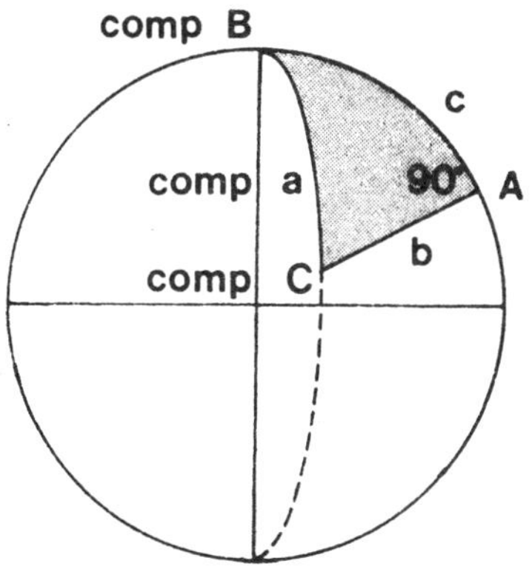

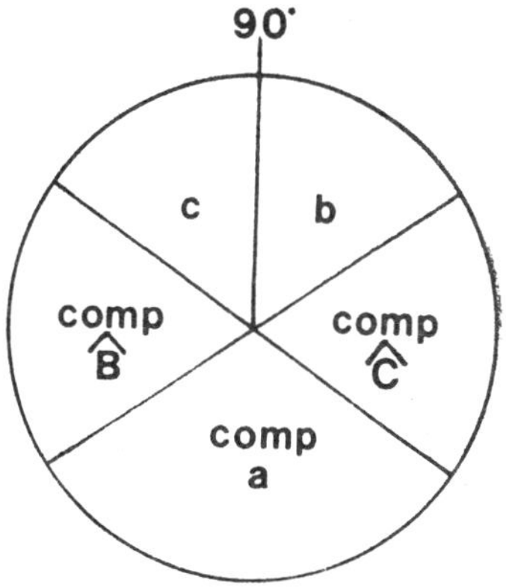

1. In figure, the five parts of the triangle *A B C* starting at *B* and named in rotation are angle *B*, side *c*, side *b*, angle *C* and side *a*, the 90° part may be indicated as shown.

Some beginners draw a circle and divide it off into five sectors as shown above, then by writing in the successive sectors the names of the parts in the order just given they are able to spot more readily the relative position of the adjacent, the opposite and the middle parts.

Napier's Rules are also applicable to quadrantal triangles, the only modification being that the 90° side is omitted, the complements of the remaining two sides are used and the complement of the angle opposite to the quadrant side. The investigation is similar to that of right-angled triangles.

The angles and their opposite sides are always of like affection, but the angle opposite to the 90° side is more than 90° when its two adjacent sides are of the same affection, but less than 90° when the sides are of unlike affection.

"The Extra Minus Rule": If, in the original formula (before any transposition) both adjacent parts or both opposite parts are both angles or both sides, an extra minus sign must be inserted. The algebraic result of all the signs in the formula will indicate whether the answer is in the 1st or 2nd quadrant.

Example.—In quadrantal triangle $A\ B\ C$, given $a=90°$, $A=60°$, $B=50°$, find sides c and b and angle C.

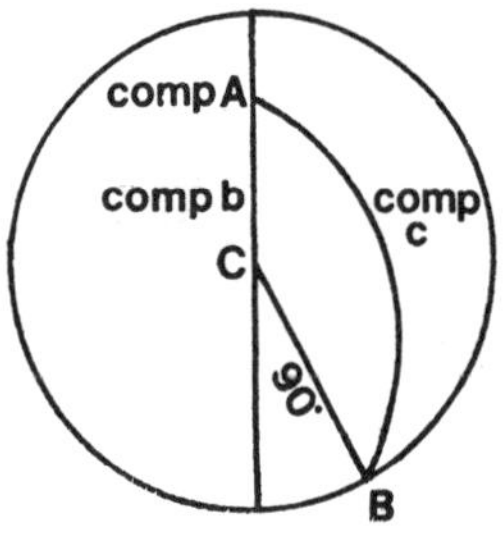

(i) *To find side c.*

c = mid part, A and B = adjacents

comp sin c = *comp* tan A tan B

$$\begin{array}{lcl} & + & \quad + \qquad + \\ (-)\cos c & = & \cot A \tan B \\ - & & \\ \cos c & = & \cot A \tan B \end{array}$$

NOTE:—The two adjacent parts are two angles, so put a (—) sign on the left side. The two + signs on the right side would make cos c a plus, but when combined with the additional (—) sign it comes out minus, that is, side c is greater than 90°.

(ii) *To find* $\angle C$

A=mid part, B and C=opposites.

comp sin A = cos B cos C

$$\begin{array}{lcl} + & & + \quad + \\ (-)\cos A & = & \cos B \cos C. \\ \cos C & = & \cos A \sec B. \end{array}$$

NOTE:—Angles B and C are under 90° and the product of their cosines makes cos $A+$, but the opposites are both angles so that the additional (—) added to the left side indicates that $\angle C$ is over 90°.

(iii) *To find side b.*

B = mid part, A and b = opposites

Sin B = *comp* cos A *comp* cos b

Sin B = Sin A sin b

Sin b = Sin B cosec A.

The rule of signs fails in this case as sine and cosec are + up to 180°. But $\angle B$ is less than 90°, therefore side b is less than 90°.

			(i)		(ii)		(iii)
A	60°	cot	9·76144	cos	9·69897	cosec	0·06247
B	50°	tan	0·07619	sec	0·19193	sin	9·88425
		cos	9·83763	cos	9·89090	sin	9·94672
			46° 31½′		38° 56′	side b	62° 12′
			180 00		180 00		
	side c		133 28½	$\angle C$	141 04		

OBLIQUE SPHERICAL TRIANGLES.

(i) Given three sides to find an angle.

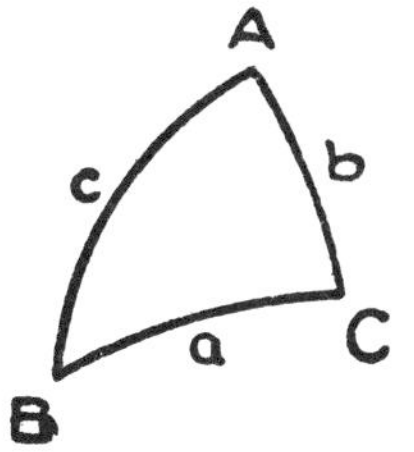

$$\text{hav } A = \frac{\text{hav } a - \text{hav } (b \sim c)}{\sin b \sin c}$$

When given two sides and included angle to find the third side, the above formula is modified to

hav a=hav A sin b sin c+hav $(b\sim c)$

(ii) Of two angles and two sides opposite to each other, given any three to find the fourth.

$$\frac{\sin A}{\sin a} = \frac{\sin B}{\sin b} = \frac{\sin C}{\sin c}$$

(iii) Given two sides and the included angle to find the other two angles.

$$\tan \tfrac{1}{2}(B+C) = \frac{\cos \frac{1}{2}(b \sim c)}{\cos \frac{1}{2}(b+c)} \cot \tfrac{1}{2} A$$

$$\tan \tfrac{1}{2}(B \sim C) = \frac{\sin \frac{1}{2}(b \sim c)}{\sin \frac{1}{2}(b+c)} \cot \tfrac{1}{2} A$$

(iv) The connection between any four adjacent parts is given by

$$\cot c \sin b - \cos A \cos b = \cot C \sin A$$

It is convenient to refer to these spherical formulae as the haversine formula, the sine formula, the tangent formula and the four adjacent parts formula respectively. See Chapter X for proofs.

(i) The haversine formula has been discussed in Volume I., Chapter VI., Trigonometry, page 120, and the method of its application in solving practical examples in navigation is presented throughout the text book. The formula satisfies nearly all the requirements of the PZX triangle of nautical astronomy.

(ii) Passing reference is made to the sine formula in Volume I., page 449. It will be noted that the shape and application of this formula is similar to the corresponding one for plane triangles; it is useful, sometimes, in finding the azimuth of a body in ex-meridian and position line questions when Tables are not available.

(iii) This compound formula is one of "Napier's Analogies". It applies to any spherical triangle in which two sides with the angle included between them is given and the remaining two angles are required, as in great circle sailing.

(iv) This formula connects any four adjacent parts of the spherical triangle and is introduced into the calculation of the A B C Azimuth Tables, which we shall describe directly in connection with the PZX triangle.

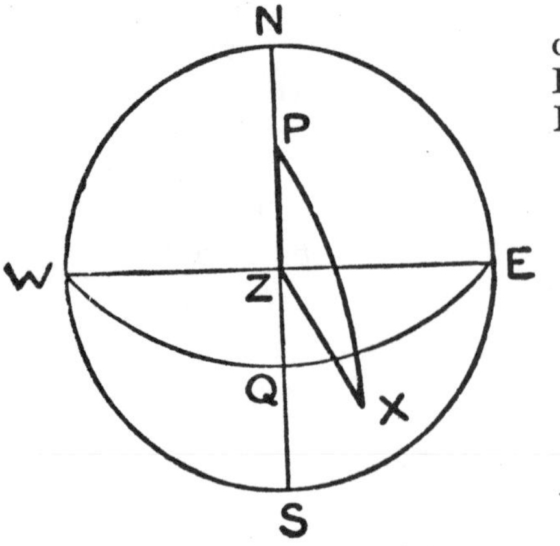

(ii) *Example.*—Given hour angle 30°, polar distance 110°, zenith distance 66° 12′, body East of meridian, calculate its azimuth. Latitude of ship 40° N.

$$\frac{\sin Z}{\sin PX} = \frac{\sin P}{\sin ZX}$$

P	30°	sin	9·69897	
PX	110°	sin	9·97299	
ZX	66° 12′	cosec	0·03860	
Z	30 54	sin	9·71056	Az. S. 30° 54′ E.

NOTE.—The formula gives us the smaller of the two angles at Z, viz., $\angle SZX$.

The examination syllabus states that candidates may use short method tables and, if so, they will be expected to know something of the principle on which they are constructed. We give here the formula from which the following have been compiled in case the question should arise at the examinations.

A B C Tables and *Norie's Ex-Meridian Tables.*

A B C TABLES.

(iv) Given hour angle 30°, decl. 20° S., Lat. 40° N., calculate the quantities given in the A B C Tables.

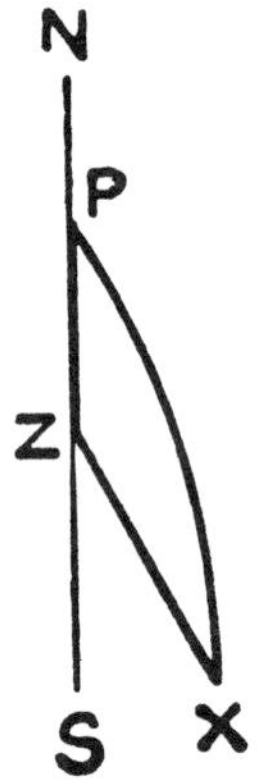

In the PZX triangle we have under consideration the adjacent parts—PX the polar distance, P the hour angle, PZ the co-latitude, and Z the azimuth—in their order round the triangle. Given the first three parts, find Z the fourth part.

$$\overset{B}{\cot PX \sin PZ} - \overset{A}{\cos P \cos PZ} = \overset{C}{\sin P \cot Z}$$

The Tables contain, respectively, the natural numbers corresponding to each part of the equation, but due regard must be paid to algebraic signs when combining the values. The above formula is camouflaged in *Norie's Nautical Tables* by dividing throughout by sin P sin PZ as follows: *see* also page 238.

$$\frac{\cot PX \sin PZ}{\sin P \quad \sin PZ} - \frac{\cos P \cos PZ}{\sin P \sin PZ} = \frac{\sin P \cot Z}{\sin P \sin PZ}$$

$$\cot PX \operatorname{cosec} P - \cot P \cot PZ = \cot Z \operatorname{cosec} PZ$$

or, tan decl. cosec hour angle—cot hour angle tan lat. = cot az. sec. lat.

± Table B — Table A = Table C

Table C = $(A \pm B)$ cos lat. = cot az.

Hour angle 30°	cosec	0·30103	cot	0·23856
Decl. 20° S.	tan	9·56107		
Lat. 40° N.		....	tan	9·92381
	log	9·86210	log	0·16237
Table	B —	·728		
Table	A —	1·453	A	1·453
Table C =	$(A+B)$ —	2·181	log	0·33865
	Lat.	40°	cos	9·88425
	Az. S.	30° 54′ E.	cot	10·22290

The quantity from Table B is give the sign + when the latitude and declination are of the same name, and — when of contrary name. In this example B is — ·728.

EX-MERIDIAN TABLES.

The altitude of a body changes most rapidly when it is on the prime vertical and, to an observer on the equator, a body whose declination is 0° rises, culminates and sets on his prime vertical and it changes its altitude 1′ in 4 seconds of time. In other latitudes and for other declinations, the rates at which the altitudes of heavenly bodies change depends on their position with regard to the observer. When near the meridian, for example, the change of altitude varies as the square of the hour angle. See Chapter XII for investigation of formula.

Table I. in Norie's Ex-Meridian Tables contains a factor (A) which is the rate of change of altitude in one minute of time. The quantity is computed from the formula

$$A = 1{\cdot}9635'' \times \cos \text{lat.} \times \cos \text{decl} \times \text{cosec (lat.} \pm \text{decl.)}$$

latitude + declination when they are contrary names and minus when they are of the same name.

Table II. gives the reduction to be subtracted from the true zenith distance of the body to get its meridian zenith distance. It is derived from the formula

$$\text{Reduction} = A \times \text{hour angle}^2$$

Example.—Calculate the reduction to the meridian as given in Norie's or in Burton's Tables for Lat. 33° 07′ N., Decl. 29° 52·3′ S. Hour angle 6° 28′.

$A = 1{\cdot}9635 \cos \text{lat.} \cos \text{dec cosec} (L+D)$

Constant	1·9635	log	0·29303
Lat.	33° 07′ N.	cos	9·92302
Decl.	29 52·3 S.	cos	9·93810
$(L+D)$	62 59·3	cosec	0·05017
A	1·601″	log	0·20432

Reduction $= A \times \text{Hour angle}^2$

H.A.	6° 28′	log	1·412460
			2
(H.A.)²	....		2·824920
A	1·601	log	0·204450
	6′0)107′0″	log	3·029370
Reduction	17·8′		

Table I.	Lat. 33° 07′ N., Decl. 29° 52′ S. gives A	1·5
Table II.	H.A. 6° 28′ and A 1·0 gives	11·2′
Table II.	H.A. 6° 28′ and A 0·6 gives	6·6
	Reduction	17·8
From worked example, Chap. VII Fomalhaut true Z.D.	63	19·5
	M.Z.D. 63	01·7

HOUR ANGLE EQUATIONS.

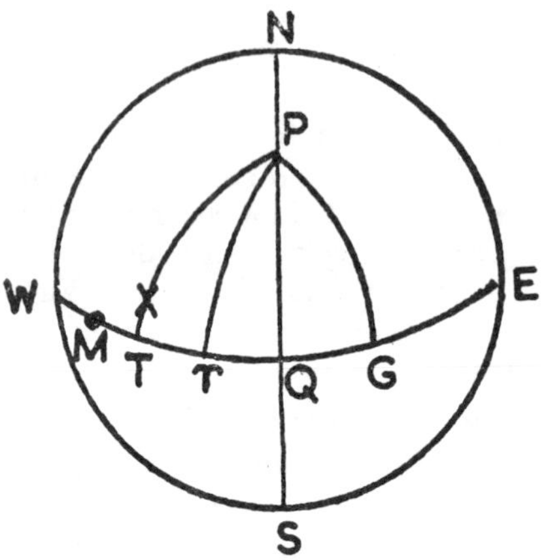

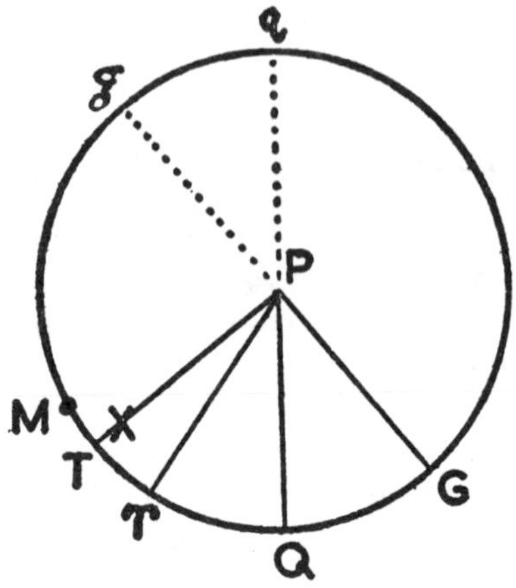

In this general figure, *NESW* represents the rational horizon, *WQE* the celestial equator, *P* the pole, *Z* the zenith. ♈ the First Point of Aries, *M* the mean sun.

PG the meridian of Greenwich
PQ ,, ,, ,, Observer
P♈ ,, ,, ,, Aries
PT ,, ,, ,, a celestial body, true sun, star, moon or planet.

Arc *GQ* = Longitude of observer West of Greenwich (Greenwich hour angle of observer).
,, *G*♈ = Greenwich hour angle of Aries (Greenwich sidereal time).
,, *GT* = Greenwich hour angle of body (geographical longitude of body).
,, *QT* = Local hour angle of body.
,, *Q*♈ = Local hour angle of Aries (local sidereal time).

If the body is a star, then

Arc ♈*T* = sidereal hour angle.

If the body is the true sun, then

Arc *GT* = G.H.A. Sun.
,, *GT* + 12 hours = Greenwich apparent time.
,, *GM* + 12 hours = Greenwich Mean time.
,, *QT* + 12 hours = Local apparent time.
,, *QM* + 12 hours = Local Mean time.
,, *MT* = Equation of time (plus to apparent time in fig.)

The student can verify the following equations by referring to Volume I., or to the respective arcs of the equator in the general figure keeping in mind always that time is 12 hours + the hour angle of the sun, and that 360° may be added to, or rejected from, hour angles as desired.

1. L.M.T. = G.M.T. − West Longitude
   $(QM+12h) = (GM+12h) - GQ$
   $QM = QM$

2. Eq. Time	=	L.M.T.	−	L.A.T.
MT	=	$(QM+12h)$	−	$(QT+12h)$
	=	MT		
3. G.H.A. Star	=	G.H.A. Aries	+	S.H.A. Star
GT	=	$G\Upsilon$	+	ΥT
	=	GT		
4. L.H.A. Body	=	G.H.A. Body	−	West Longitude
QT	=	GT	−	GQ
	=	QT		

The arc GT (G.H.A.) is tabulated at hourly intervals of G.M.T., for the Sun, Moon and planets, and the arc $G\Upsilon$ is also given for the same interval, in the daily pages of the *Nautical Almanac*. The arc ΥT (S.H.A. of Star) is given is the star list for the middle of each month and requires no correction for the day of the month. The arc MT (Equation of time) is given at intervals of 12 hours.

CHAPTER VI.

MERIDIAN PASSAGE

TIME OF MERIDIAN PASSAGE STAR

The times of the meridian passage of the sun, moon, planets and the First Point at Aries (♈) are listed in the daily pages of the *Nautical Almanac.* Times of meridian passage for individual stars are not included.

There are several methods for finding the time of a star's meridian passage:—

1. The Inspection Method using GHA Star at Meridian Passage.
2. Time of Meridian Passage Aries.
3. The Planet Diagram in the *Nautical Almanac.*

The first method, which involves inspection of the *Nautical Almanac* using the observer's west longitude as GHA star, has been used for example and the alternative approach using LHA star is also explained.

The second method, which involves use of the Time of Meridian Passage Aries, is less convenient than inspection methods but where a high degree of accuracy is not required, quick, approximate results can be obtained.

Use of the Planet Diagram contained in the *Nautical Almanac* is explained in the next section of this volume which deals with "Stars near the Meridian".

Method of Inspection Using the Observer's West Longitude to Obtain the Time of Meridian Passage Star.

The method of inspection relies on the following facts :—

1. GHA is measured Westwards from Greenwich.
2. At meridian passage :—
   (*a*) the star must lie on the Observer's meridian and
   (*b*) the star's GHA must equal the Observer's West longitude.
   (N.B.—An East longitude can be subtracted from 360° and expressed as a West longitude which exceeds 180°).
3. At Meridian Passage :—

GHA* = Observer's West Longitude
GHA* = GHA ♈ + SHA*
Observer's West longitude = GHA ♈ + SHA*
Observer's West longitude −SHA* = GHA ♈

In the *Nautical Almanac* it can be seen that a star's SHA remains constant over a period of several days. So, by subtracting the SHA star from the Observer's West Longitude we can obtain the GHA♈ at the time of meridian passage.

Now, at any particular moment in time, GHA Aries is made up of its value for a whole number of hours plus the increment for the remaining minutes and seconds. By subtracting the next lowest GHA value for hours we obtain the increment which can be converted into time using the incremental tables at the rear of the *Almanac.* The resulting GMT is then adjusted for longitude to give the LMT of meridian passage.

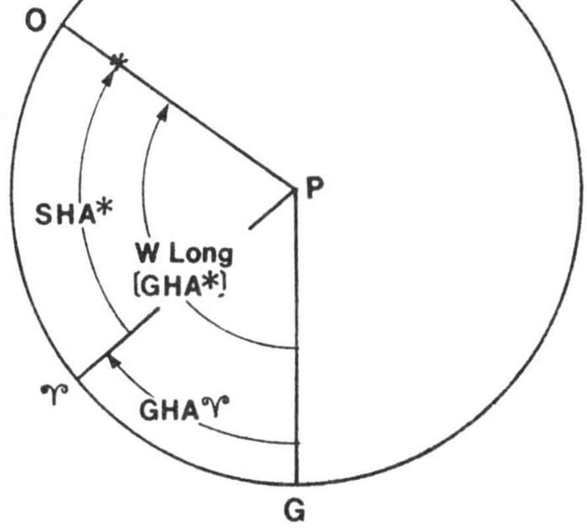

Fig.1 West Longitude

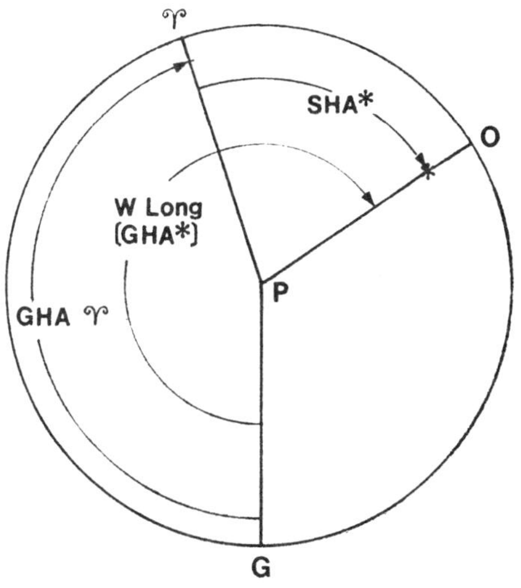

Fig.2 East Longitude

Alternative Method of Inspection Using the Star's LHA at Meridian Passage.

$$
\begin{array}{lrl}
\text{At Meridian Passage} & \text{LHA*} & = 360^\circ \\
 & \text{LHA*} & = \text{LHA}\Upsilon + \text{SHA*} \\
 & 360^\circ & = \text{LHA}\Upsilon + \text{SHA*} \\
 & 360^\circ - \text{SHA*} & = \text{LHA}\Upsilon \\
\text{Now} & \text{LHA}\Upsilon \pm \text{Long} & = \text{GHA}\Upsilon \\
\text{So} & (360^\circ - \text{SHA*}) \pm \text{Long} & = \text{GHA}\Upsilon
\end{array}
$$

From the above, it can be seen that by applying the Observer's Longitude to LHA Aries we obtain the GHA Aries for which the corresponding GMT can be extracted from the *Almanac* as previously explained.

This method gives precisely the same degree of accuracy as obtained by the longitude method and, for additional exercise, students may care to re-work the examples using this method.

TO FIND THE TIME OF MERIDIAN PASSAGE STAR—WEST LONGITUDE

Example 1 (W Long > SHA *)

Find the LMT of meridian passage for the star Vega on January 11th in DR 5° 00′ N, 90° 00′ W. Refer to Figure 3.

Observer's West Longitude − SHA * = GHA ♈ at Meridian Passage.

W Long	90°	00′					
SHA Vega	81	07·6					
GHA ♈	8	52·4					
From *Almanac* GHA ♈	5	41·7	gives	11d	17h		
			,,			12m	41s
,, ,, Increment	3°	10·7′					
			GMT	11	17	12	41
			W Long		6	—	—
Meridian Passage Vega			LMT	11d	11h	12m	41s
					11h	13m (approx.)	

In Example 1 the Observer's West longitude was greater than the SHA star. When the West longitude is less than the star's SHA it becomes necessary to add 360° to the Observer's longitude. Alternatively, Longitude ~ SHA can be subtracted from 360°. Both methods are demonstrated in the following example:—

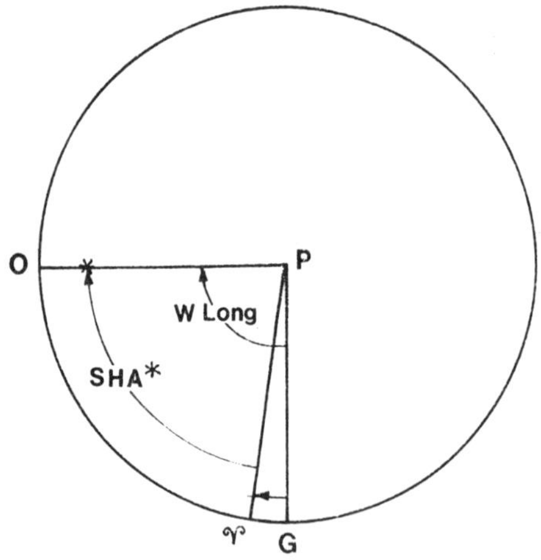

Fig.3 Example 1

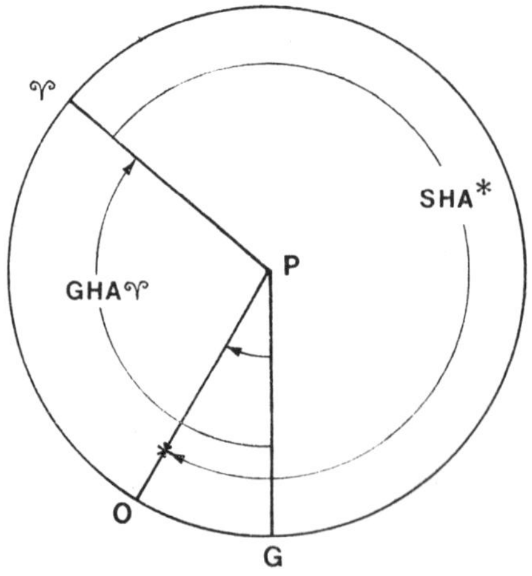

Fig.4 Example 2

Example 2 (W Long < SHA *)

Find the LMT of meridian passage for the star Sirius, am at ship, on October 11th in DR 40° 00′ N. 30° 00′ W. Refer to Figure 4.

Observer's West Long − SHA * = GHA ♈ at Meridian Passage.

	360°					
W Long	30°					
	390° 00′					
SHA Sirius	259° 09·9					
GHA ♈	130° 50·1					
From *Almanac* GHA ♈	124° 21·9	gives	11d	07h		
		,,			25m	48s
,, ,, Increment	6° 28·2′					
		GMT	11	07	25	48
		W Long		2	—	—
Meridian Passage Sirius		LMT	11d	05h	25m	48s
				05h	26m (approx.)	

Alternatively, GHA♈ can be found by subtracting longitude ∼SHA from 360° as follows:—

W Long	30° 00′
SHA Sirius	259° 09·9′
W Long ∼ SHA	229° 09·9′
	360°
GHA ♈	130° 50·1′

TO FIND THE TIME OF MERIDIAN PASSAGE STAR—EAST LONGITUDE

When the Observer is in East longitude it becomes necessary to express the East longitude as West longitude. To express East longitude as West longitude subtract East longitude from 360°—this is demonstrated in the following example:—

Example 3 (East Longitude)

Find the LMT of meridian passage for the star Acrux, pm at ship, on June 16th in DR 40° 00′ S. 155° 00′ E. Refer to Figure 5.

Observer's West Long − SHA * = GHA ♈ at meridian passage *.

	360°					
E Long	155° 00′					
W Long	205° 00′					
SHA Acrux	173° 55·3′					
GHA ♈	31 04·7′					
From *Almanac* GHA ♈	24 05·1′	gives	16d	08h		
		,,			27m	58s
,, ,, Increment	6 59·6					
		GMT	16	08	27	58
		E Long		10	20	—
Meridian Passage Acrux		LMT	16d	18h	47m	58s
				18h	48m (approx.)	

In this example the resulting West longitude was greater than the SHA. Had the SHA exceeded the West longitude it would have been necessary to add 360° as explained at Example 2.

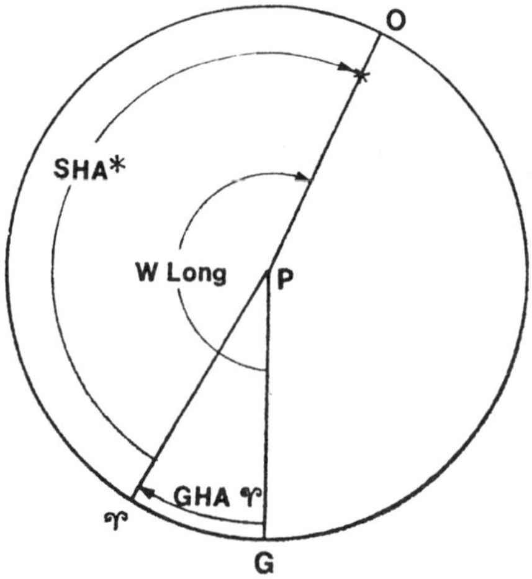

Fig.5 Example 3

FINDING A STAR WHOSE MERIDIAN PASSAGE OCCURS AT A SUITABLE TIME FOR SIGHTS

(Further information on this subject is included at page 148).

A star sight can only be taken when both the star and the horizon are clearly visible. This normally occurs only during twilight hours. In consequence, when deciding on a suitable star for a meridian observation it is generally necessary to limit the choice to that star whose meridian passage occurs during twilight hours. Nautical twilight occurs when the sun is between 6° and 12° below the horizon and during this period bright stars are visible and the horizon is sufficiently clear for star sights.

The times of twilight can be found in the daily pages of the *Nautical Almanac*. A study of these times will quickly show that the duration of twilight is dependent on the latitude of the observer. This, together with the fact that the clarity of the horizon depends heavily on atmospheric and sea conditions, results in situations where it is possible to obtain good star sights at times both earlier and later than those falling within nautical twilight, but in the following examples it has been assumed that nautical twilight is the correct period to be used in calculation.

TO FIND A STAR SUITABLE FOR MERIDIAN ALTITUDE AT TWILIGHT

1. Obtain GMT of nautical twilight.
2. Obtain DR longitude for this GMT.
3. Calculate LHA Aries for the GMT and longitude found above.
4. Subtract LHA Aries from 360° to obtain the SHA of an imaginary star whose meridian passage occurs precisely at this time.
   At meridian passage:

$$\begin{aligned} \text{LHA} * &= 360°, \\ \text{LHA} \,♈ + \text{SHA} * &= \text{LHA} *, \\ \text{LHA} \,♈ + \text{SHA} * &= 360°, \\ \text{SHA} * &= 360° - \text{LHA} \,♈ \end{aligned}$$

5. Inspect the list of stars in the *Nautical Almanac* and select a star (or stars) of similar SHA to that of the imaginary star found at 4. (Magnitude and altitude should be taken into account in making the final selection).
6. Find the LMT of meridian passage for the selected star and compare this time with the time and duration of twilight found at 2.
7. Decide if a meridian or ex-meridian sight is possible.
8. Prepare for sights by pre-calculating the meridian altitude of the selected star and note the time and altitude of meridian passage.

Example 4 (Refer also to Example 3 and Figure 5)

Find the LMT of meridian passage of a star suitable for a latitude sight during evening twilight on the 16th June.

DR 40° 00′ S. 155° 00′ E.

Nautical twilight	LMT	16d 17h 45m	
E Long		10h 20m	
	GMT	16d 07h 25m	(step 1)
	Long	155° 00′ E. (given)	(step 2)
GHA♈ 16d 07h		9° 02·6′	
Increment 25m		6° 16·0′	
GHA ♈		15° 18·6′	
E Long		155° 00·0′	
LHA ♈		170° 18·6′	(step 3)
		360°	
SHA (of a suitable star)		189° 41·4′	(step 4)

From the *Nautical Almanac*:

	SHA		*Declination*		*Magnitude*	
Acrux	173°	55·3′	S 62°	52·5′	1·1	
Denebola	193	15·8	N 14	48·3	2·2	(step 5)
Gagrux	172	46·8	S 56	53·2	1·6	
Gienah	176	34·8	S 17	18·8	2·8	

Denebola, SHA 183° 15·8′, is the most suitable star for a meridian altitude. Calculate the LMT at meridian passage Denebola, pm at ship, on 16th June.

Observer's West Longitude − SHA* = GHA ♈ at meridian passage.

	360°							
E Long	155°	00′ E						
W Long	205°							
SHA Denebola	183°	15·8′						
GHA ♈	21°	44·2′						
From *Almanac* GHA	9°	02·6′	gives	16d	07h	—	—	
			,,			50m	38s	
,, ,, Inc.	12°	41·6′						
			GMT	16	07	50	38	
			E long		10	20	—	
Meridian Passage Denebola			LMT	16d	18h	10m	38s	(step 6)

	18h	11m (approx.)
Nautical Twilight	17h	45m
		26m

Meridian passage occurs 26m after the start of Nautical Twilight.

Depending on the horizon available, either a meridian altitude or good ex-meridian altitude should be obtained. (step 7)

The meridian altitude should now be calculated or found by inspection from suitable tables. (step 8)

TIME OF MERIDIAN PASSAGE ARIES

The meridian passage of Aries is listed in the daily pages of the *Nautical Almanac* and the time quoted is for the middle day of the 3 days quoted. As each star is related to Aries by its sidereal hour angle the time of a star's meridian passage can be found by converting its SHA into time and applying this time to the time of meridian passage Aries:

Meridian passage * = Meridian passage ♈ + (360° − SHA *)

The rate at which the earth turns relative to the fixed stars is 23h 56·1m of Mean Solar Time. This means that Aries will reach the Greenwich meridian nearly 4 minutes earlier each day. In consequence, when converting SHA into time, the conversion should be made at the rate of 23h 56m to 360°.

Accurate results can be obtained by this method but the correction necessary for longitude and the 4 minute time difference is somewhat inconvenient.

Where speed and accuracy are required it is preferable to use inspection methods.

The approximate LMT of a star's meridian passage is frequently required in order to check the corresponding date at Greenwich. It is also used to check if an individual star will be on, or near, the meridian during twilight hours when a suitable horizon is available. For approximate purposes the 4 minute time correction can be safely ignored providing the limits of accuracy are appreciated.

TO FIND THE APPROXIMATE TIME AND DATE OF MERIDIAN PASSAGE STAR USING THE TIME OF MERIDIAN PASSAGE ARIES

In the previous examples it should be noted that no change has occurred between the GMT date and the LMT date, but when taking a sight it is always good practice to check that no such change has occurred. Failure to make this check can result in extracting *Almanac* data for the wrong day.

In question papers an indication of the LMT, or approximate time, of sights is often given, whilst in practice the approximate LMT of sights should be known. By carefully applying the longitude correction to the LMT and date at ship, the corresponding GMT and Greenwich date are easily found.

In solving problems where there is no such indication of LMT, or the approximate time of the observation, it is first necessary to obtain the corresponding time and date at Greenwich. In other cases a quick method for finding the approximate time of meridian passage star is sometimes useful.

As already mentioned, the *Nautical Almanac* lists the meridian passage Aries for the second of the three days listed on the daily pages.

If we neglect the small time difference between the Sidereal and Mean Solar Day:

Time of meridian passage * = Time of meridian passage ♈ + (360° − SHA *)

[The quantity (360° − SHA *) is known as Right Ascension (RA) and is expressed in hours and minutes.]

Providing this method is carefully used and providing careful corrections are made for longitude and the difference between the Sidereal and Mean Solar Day, this method can achieve results of similar accuracy to those achieved by the inspection method. In practice, such corrections are inconvenient and introduce unnecessary complications. But for approximate results, where such corrections are ignored, the method is quick and convenient.

Example 5

Find the approximate time of meridian passage Acrux on June 16th in DR 40° 00′ S. 155° 00′ E.

Meridian passage * = Meridian passage ♈ + (360° − SHA *)

Meridian passage ♈ from *Almanac*				16d	06h	24m	(approx.)
	360°						
SHA Acrux	173° 55·3						
(360° − SHA *)	186° 04·7	expressed as time			12h	24m	,,
			LMT	16d	18h	48m	,,
			E Long		10h	20m	,,
			GMT	16d	08h	28m	(approx.)
The result by inspection (see Example 3) was			GMT	16d	08h	27m	48s

To achieve a result comparable to that achieved by inspection would have required considerable more work and increased the chances of error.

Whenever consulting the Nautical Almanac always check the approximate Time and Date at Greenwich.

Circumpolar Stars—Upper and Lower Meridian Passage.

The apparent motion of a star (and all celestial bodies) is to describe a circle around the pole once in a sidereal day. During this period the star will cross (or transit) the observer's meridian and will also cross the observer's anti-meridian which is 180° removed from his meridian. The time interval between the two transits is equal to half a sidereal day.

It will be recalled that the latitude of the observer is equal to the altitude of the elevated pole so that when latitude is greater than polar distance (90° − Dec) a body will be continuously above the horizon. Under these conditions a star will cross the observer's meridian and anti-meridian and, in consequence, will have two meridian altitudes (though these will only be visible if occurring during the hours of twilight or darkness).

When the star is at its Higher Meridian Altitude it is said to be, "Above the Pole" or at, "Upper Meridian Passage".

When the star is at its Lower Meridian Altitude it is said to be, "Below the Pole" or at, "Lower Meridian Passage".

The terms "Upper (or Lower) Meridian Transit and, "On the Meridian Above (or Below) the Pole", are also used.

Stars which are continuously above the observer's horizon during a diurnal revolution (and which hence transit the observer's meridian "above" and "below" the pole) are referred to as "Circumpolar Stars".

It should be noted that circumpolar stars do not rise or set and have no amplitude.

Conditions Necessary for Circumpolarity.

Refer to Figures 9 and 10 and verify the following conditions which are necessary for circumpolarity:

Latitude of Observer must be greater than Polar Distance Star

$$\text{Lat} > (90^\circ - \text{Dec})$$

Time of Meridian Passage Across Any Specified Meridian.

The time of a star's upper and lower meridian passage across any specified meridian will be the same for all latitudes. This fact is quite independent of whether, or not, the star is visible at the time of lower meridian passage. It should be noted that when a star is visible at lower meridian passage it will cross the observer's meridian from West to East between the horizon and the elevated pole.

To Find the Latitude at Lower Meridian Passage.

Refer to Figures 9 and 10 and verify that:

Latitude at Lower Meridian Passage equals True Altitude plus Polar Distance Star.

$$\text{Lat} = \text{True Alt} + (90° - \text{Dec})$$

To Find the Time of Lower Meridian Passage.

The interval of time between the upper and lower meridian passage of a star is half a Sidereal Day or 11h 58m 02s of Mean Solar Time. It follows that the lower Meridian passage of a star (whether visible or not at the time of its passage below the pole) will be found as follows:—

LMT Lower Mer Passage = LMT Upper Mer Passage ± 11h 58m 02s.

An alternative method is as follows:—

LMT Lower Mer Passage = LMT when LHA * is 180°.

As previously explained, at meridian passage LHA star is equal to 360°. Similarly when a star is on the observer's anti-meridian LHA star is equal to 180°.

At Lower Meridian Passage	LHA *	= 180°
	LHA *	= LHA ♈ + SHA *
	180°	= LHA ♈ + SHA *
	180° − SHA *	= LHA ♈
Now	LHA ♈ ± Long	= GHA ♈
So	(180° − SHA *) ± Long	= GHA ♈

From the above it can be seen that by applying the longitude to LHA Aries we obtain the GHA Aries for which the corresponding GMT can be extracted from the *Almanac* as previously demonstrated.

Example 6

Find (*a*) The LMT of the Upper and Lower Meridian passage of the star Capella on December 24th in DR 56° 30′ N. 38° 18′ W., and

(*b*) The observer's latitude if the Sextant Meridian Altitude of Capella (below the pole) was 12° 30′ N. IE 2′ off the arc. HE 9m.

(*a*) *Check Greenwich Date:*

Meridian Passage * = Meridian Passage ♈ + (360° − SHA *)

Mer Pass ♈ from *Almanac*		LMT	24d	17h	49m	(approx.)
	360°					
SHA Capella	281° 34·5′					
(360° − SHA *)	78° 25·0′ in time			5h	14m	(approx.)
	Meridian Passage *	LMT	24d	23h	03m	
		W Long		2h	33m	
		GMT	25d	01h	36m	(approx.)

N.B.—Use Data for 25th and refer to Figures 6 & 7.

December 25th at Greenwich

Observer's West Longitude − SHA * = GHA ♈ at Meridian Passage

W Long	38°	18′					
	360°						
	398	18					
SHA Capella	281	34·5					
GHA ♈	116	43.5					
from *Almanac*	108	02·5	gives	25d	01h		
Increment	8	41	,,			34m	38s
Meridian Passage Capella			GMT	25	01	34	38
			W Long		2	33	12
Upper Meridian Passage			LMT	24	23	01	26
½ Sidereal Day (−)					11	58	02
Lower Meridian Passage			LMT	24d	11h	03m	24s

In this case the half sidereal day was subtracted because had it been added it would have given the LMT of Lower Meridian Passage on the 25th instead of the 24th which was the date required.

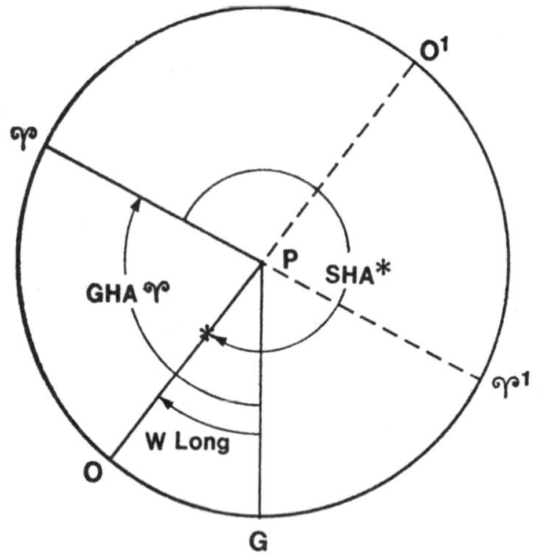

Fig.6 Upper Meridian Passage

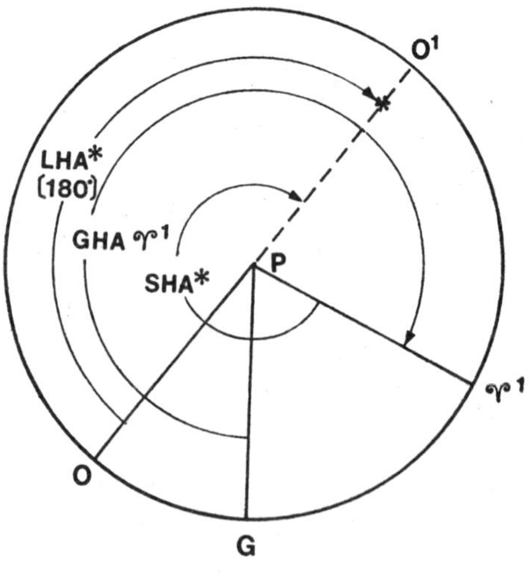

Fig.7 Lower Meridian Passage

(*b*) Latitude at Lower Meridian Passage = True Alt + (90° − Dec)
= $X^1N + X^1P$
= NP (the altitude of the elevated pole)
= QZ (the Observer's Latitude)

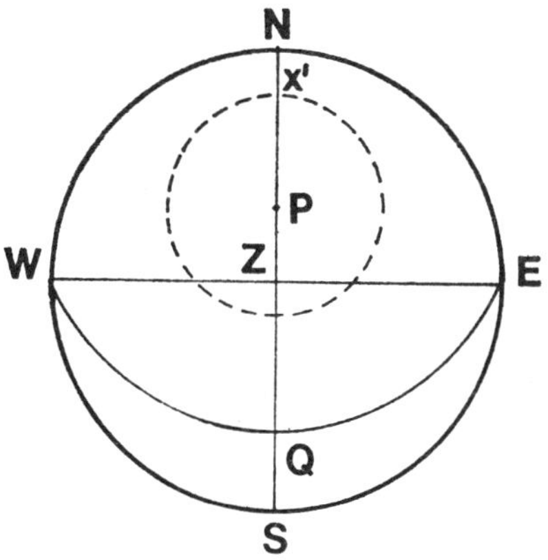

Fig.8

Dec Capella		45°	57·4′ N	(QX)
		90		
Polar Dist		44°	02·6′ N	(PX^1)
Sext Alt		12°	30′ N	
I.E	(+)		02	
Obs Alt		12	32	
Dip	(−)		05·3′	
App Alt		12	26·7	
Main Corrn	(−)		4·3	
True Alt		12	22·4 N	(X^1N)
Polar Dist		44	02·6 N	(PX^1)
Lat at Lower Mer Passage		56°	25·0′ N	(QZ)

To Find Latitude Using an Unknown Circumpolar Star.

The latitude of a place on land can be found by taking the mean of the true meridian altitudes above and below the pole of any unknown circumpolar star.

Figures 9 and 10 show the upper and lower transits of a star at X and X^1. Figure 9 is drawn on the plane of the observer's rational horizon and Figure 10 on the plane of the observer's meridian.

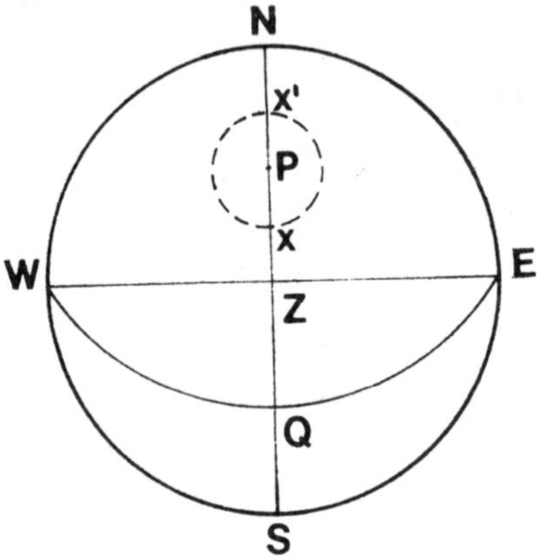

Fig.9

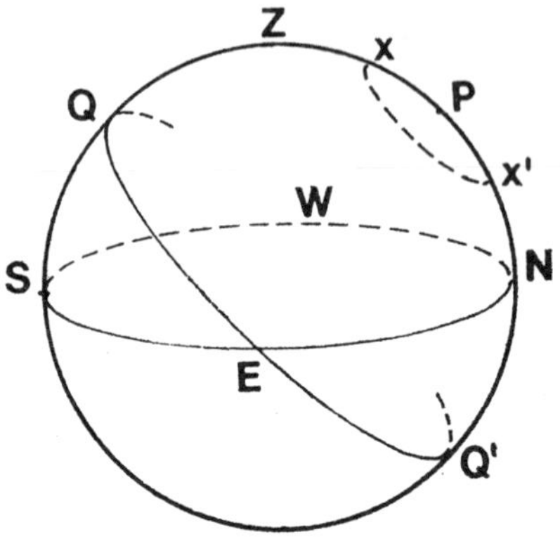

Fig.10

X	Star at Upper Meridian Passage
X^1	Star at Lower Meridian Passage
NX	Meridian altitude Above the Pole
NX^1	Meridian altitude Below the Pole
NP	Altitude of the elevated pole
QZ	Latitude of the observer
PZ	Co. Lat. of the observer
WQE	Equator
SZN	Observer's meridian
NESW	Observer's rational horizon

Refer to Figure 9 and the Legend above:

Now
$$NP + PZ = 90^\circ$$
$$QZ + PZ = 90^\circ$$
$$NP = QZ = \text{Lat}$$
$$NX + NX^1 = (NP + PX) + (NP - PX^1)$$
$$= 2\ NP$$
$$= 2 \times \text{Lat}$$

So
$$\frac{NX + NX^1}{2} = \text{Latitude}$$

or
$$\text{Latitude of Observer} = \frac{\text{Meridian Alt above Pole} + \text{Meridian Alt Below Pole}}{2}$$

Examples and Questions for Exercise.

In examples 1–6 calculate the LMT of upper and lower meridian passage of the star in question. State if the star is visible at lower transit and if so calculate the approximate lower meridian altitude for the given latitude.

1. Acrux, January 8th at ship in D.R. 52° 00′ S 163° 40′ E
2. Regulus, December 22nd at ship in D.R. 24° 30′ S 138° 46′ W
3. Vega, September 20th at ship in D.R. 60° 00′ N 176° 49′ E
4. Atria, June 22nd at ship in D.R. 38° 40′ S 17° 45′ W
5. Antares, June 17th at ship in D.R. 38° 00′ N 177° 40′ E
6. Aldebaran, September 19th at ship in D.R. 56° 30′ N 154° 20′ W
7. Re-calculate the time of upper meridian passage for questions 1–6 above using the Time of Meridian Passage Aries.
8. Explain the conditions necessary for circumpolarity and use diagrams to illustrate your answer.
9. Comment on the conditions necessary in order to obtain an acceptable meridian altitude of a star.
10. Prove that, $(180° - \text{SHA*}) \pm \text{Long} = \text{GHA} ♈$ at lower meridian passage.

TO DETERMINE WHAT STARS ARE NEAR THE MERIDIAN.

It may sometimes be required to find what stars are near the meridian so that those suitable for meridian altitude or ex-meridian sights may be selected. The Planet Diagram given in the Almanac can be used to find the Mean time of transit of any star for any date during the year, by simply reading the graph of Sidereal Hour Angle. Another method is to find the S.H.A. of the observer's meridian (360°—L.H.A. Aries) and compare this with the star list.

Star are tabulated in the main star list in the *Nautical Almanac* in the order of their sidereal hour angle which increases to the westward from zero at the First Point of Aries, so that stars having a greater S.H.A. than the S.H.A. of the observer are to the westward of the observer's upper meridian but to the eastward of his lower, or ante-meridian.

It will be remembered that stars are below the horizon when their meridian zenith distance exceeds 90°, that is, Latitude +Declination; also, that the brightness of a star is indicated by its magnitude. The brighter the star the smaller its magnitude number.

Example I.

Supposing you are not familiar with the stars, and wish to take an observation, find the names of the stars of magnitude 2·0 or brighter that are within one hour East and West of your meridian, above the pole and above the horizon, at 06h 30m Mean time at ship on January 9, the Latitude being 36° 30′ N., and the Longitude 48° 18′ W. Find also the approximate local hour angle of each of these stars and also whether to the North or South of your zenith when passing the meridian.

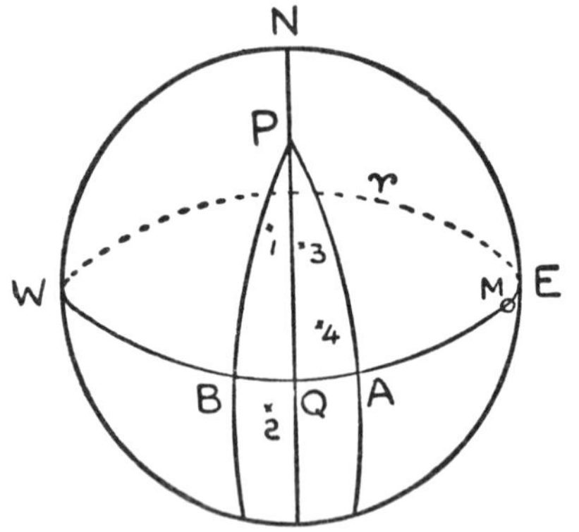

In figure:
N E S W the rational horizon
P the pole, Z the zenith
M the Mean sun
♈ the First Point of Aries
♈*EQ* the S.H.A. Observer's meridian.

The meridians *PA* and *PB* are 15° East and 15° West, and indicate respectively the limits of hour angle as prescribed in the question. The stars are represented within those meridians and are plotted approximately as numbered, according to their declinations.

S.H.A. Observer's meridian=360°—L.H.A. Aries.

	d	h	m
L.M.T.	9	06	30
Long. W. +		3	13
G.M.T.	9	09	43

G.H.A. Aries	243°	23·7′
Incre.	10	46·8
G.H.A. Aries	254	10·5
Long. W.	− 48	18
L.H.A. Aries	205	52·5
	360	00
S.H.A. Merid.	154	07·5

The required stars must have S.H.A's within 15° greater or less than the S.H.A. of observer. From the star list pick out the stars of the required magnitude whose S.H.A's come between 139° 07′ and 169° 07′ and write them down in the order in which they cross the meridian, as shown below.

Name	1. *Alioth*	2. *Spica*	3. *Alkaid*	4. *Arcturus.*
Magnitude	1·7	1·2	1·9	0·2
Declination	56° N.	11° S.	50° N.	20° N.
S.H.A. Merid.	154° 07·5′	154° 07·5′	154° 07·5′	154° 07·5′
S.H.A. Star	166 57·1	159 15·1	153 31·7	146 33·7
L.H.A. Star	12 49·6 W.	5 07·6 W.	0 35·8 E.	7 33·8 E.
	Has crossed N.	Has crossed S.	Will cross N.	Will cross S.

The stars *Hadar* and *Rigil Kent* are omitted because their declination being of a different name to the latitude, and the sum in each case exceeding 90°, we know that they are below the horizon.

Example II.

On December 27, in Lat. 38° 42′ S., Long. 138° 15′ W., find the names of the stars not less bright than the first magnitude that are within 1½ hours East of the meridian above the pole and above the horizon. Time by chronometer 28d 09h 04m 00s which was correct G.M.T. Find the approximate local hour angle of these stars, and state if they cross the meridian to the North or South of the zenith.

S.H.A. Obs. Meridian=360°−L.H.A. Aries.

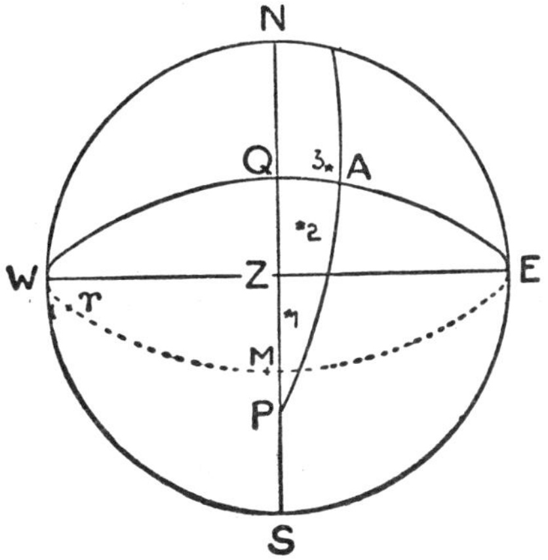

December			
G.H.A. 28d 09h	231°	19·6′	
Incre. 04d 00h	1	00·2	
G.H.A. Aries	232	19·8	
Long. W.	−138	15	
L.H.A. Aries	94	04·8	
	360	00	
S.H.A. Obs. Merid.	265	55·2	Limits of S.H.A.
1½ hours East	− 22	30	
East limit S.H.A.	243	25·2	

	Canopus	*Sirius*	*Procyon*
Magnitude	—0·7	—1·6	0·5
Declination	53° S.	17° S.	5° N.
S.H.A. Obs.	265 55·2	265 55·2′	265 55·2
S.H.A. Star	264 13·8	259 09·5	245 42·3
L.H.A. Star	1 41·4 E.	6 45·7 E.	20 12·9 E.
	Will cross S.	**Will cross N.**	**Will cross N.**

Notice the magnitudes of Canopus and Sirius. The minus sign prefixed to them indicates that they are brighter than magnitude 0, and, therefore, both are brighter than magnitude 1·0. These are the only two stars with a minus magnitude.

When the magnitude of a star is written 1·0—1·4 it means that its magnitude varies between 1·0 and 1·4. If the magnitude required in any example comes in between the two given magnitudes write the star down and mark it "magnitude variable 1·0—1·4," or whatever it may be.

Example III.

On June 24, at 05h 35m local Mean time, in Lat. 54° 30′ S., Long. 32° 10′ E., find the names of the stars up to magnitude of 2·5 that are within one hour East and West of the meridian above and below the pole and above the horizon. Find also the approximate local hour angle of each star, and state if to the North or South of the zenith at Transit.

S.H.A. Observer's meridian=360°—L.H.A. Aries.

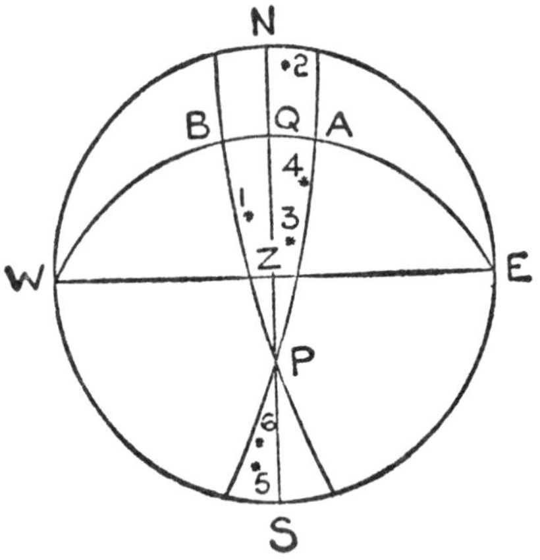

L.M.T.		24d	05h	35m
Long E.	—		2	09
G.M.T.		24	03	26
G.H.A. ♈		316°	45·9	
Incre		6	31·1	
G.H.A. ♈		323	17·0	
Long. E.	+	32	10	
L.H.A. ♈		355	27	
		360	00	
S.H.A. Obs.		4	33	(upper)
		180	00	
S.H.A. Obs.		184	33	(lower)

"Upper Meridian" limits of S.H.A. 19° 33′ to 349° 33′.

Maximum declination to be visible 35° 30′ N. This is equal to the co-latitude but of opposite name to the latitude.

	1. *Fomalhaut*	2. *Alpheratz*	3. *Ankaa*	4. *Diphda*
	Mag. 1·3	Mag. 2·2	Mag. 2·4	Mag. 2·2
S.H.A. Obs.	4° 33′	364° 33′	364° 33′	364° 33′
S.H.A. Star	16 09·5	358 26·3	353 56·6	349 37
L.H.A. Star	11 36·5 W.	6 06·7 E.	10 36·4 E.	14 56 E.

All crossing North of Observer.

"Lower Meridian" limits of S.H.A. 199° 33′ to 169° 33′.

Minimum declination 35° 30′ S.=maximum Polar distance (*PS*) to be visible. The minimum declination is also equal to the co-latitude but of the same name as the latitude.

	5. *Acrux* (Mag. 1·1)	6. *Gacrux* (Mag. 1·6)
S.H.A. (lower)	184° 33′	184° 33′
S.H.A. Star	173 55·3	172 46·8
L.H.A. Star	10 37·7 W.	11 46·2 W.

Both crossing South of Observer.

Note.—Although stars (5) and (6) have sidereal hour angles less than the S.H.A. of the observer's lower meridian they bear South and West from the observer and have yet to come to the ante-meridian as their apparent daily motion, due to the earth's rotation, is counter clockwise round the South pole of the figure and their lower passage is from West to East.

Examples for Exercise.

Supposing you are not familiar with the stars and wish to take an observation, find, the names from the *Nautical Almanac* of the stars above, and below the pole if any, but above the horizon; give also the approximate local hour angle of each of the stars in the following examples, and state also whether they will be to the North or South of your zenith when passing the meridian.

	Date	Time	Lat.	Long.	Mag. up to	Limits of L.H.A.
1.	January 10	L.M.T. 03h 50m	40° 00′ S.	43° 20′ W.	2·5	1½hrs. E. and W.
2.	Sept. 19	G.M.T. 16h 24m	19 40 S.	0 30 E.	2·0	1h 45m E. and W.
3.	June 15	L.M.T. 02h 00m	54 00 S.	108 00 W.	2·1	1½ h. East
4.	June 18	G.M.T. 02h 35m	50 00 S.	38 05 E.	2·4	3h. East
5.	Sept. 23	L.M.T. 0h 30m	35 20 S.	155 00 W.	2·2	1h. E. and W.
6.	Sept. 20	G.M.T. 14h 40m	32 00 S.	135 00 W.	1·0	1½h East.

STAR IDENTIFICATION.

The easiest way for the star-gazer to begin is to ask someone who knows them to point out a few of the principal stars suitable for navigation, and, afterwards, to keep up a nodding acquaintanceship with them each night. Star identification at sea is then a simple matter.

The chief officer on the evening and morning watches night after night cannot fail to become familiar with the groups of suitable stars above the horizon and their position in the sky at dusk and dawn. The third officer will soon learn to know when and where to look for particular stars during the first watch, and the second officer will no doubt have a few in sight which he favours during the middle watch for position finding and azimuths. Allowance can readily be made each night for the slow automatic changes in the positions of the stars, relative to the horizon, and to the meridian, caused by the ship gradually changing her Latitude and to the fact that stars rise, culminate and set four minutes earlier each day.

The four diagrams given here are from *Brown's Nautical Almanac* and *Star Atlas*, to facilitate the recognition of stars. The maps are drawn on the gnomonic projection in which the meridians are straight lines radiating from the pole, spaced one hour or 15° apart to represent one hour of right ascension, and the parallels of declination are circles.

To read off the position of α Dubhe in the Great Bear, for example, we note that its right ascension is 11 hours and declination 62° N.

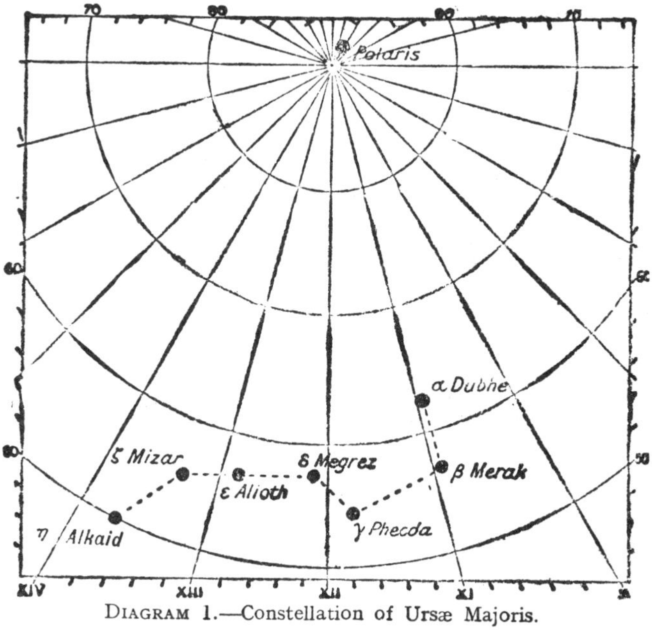

DIAGRAM 1.—Constellation of Ursæ Majoris.
Relative Positions of the Seven Principal Stars in the Constellation Ursæ Majoris, and of the Pole Star.

Amongst the most remarkable constellations which are the glory of our northern heavens, Ursæ Majoris, the Great Bear, is conspicuous (Diagram 1). The Great Bear consists of seven principal stars not differing greatly in magnitude. These stars are of sufficient brightness to be of some use in navigation, but they have an even more important mission to fulfil in enabling us to identify α Ursæ Minoris, otherwise known as Polaris, or the Pole Star. This star is so near the North Pole that we may deduce the Latitude from its altitude at any time when the star is visible, without waiting for it to approach the meridian.

To Find Polaris.

Draw a line from β to α of the Great Bear (Diagram 1) and produce it. This line will lead to the Pole Star.

To Find the Constellation Cassiopeiæ.

Cassiopeiæ is on the opposite side of the pole to the Great Bear, so that when one bears East the other bears West. It is situated at about the same distance from the pole as the Great Bear, and is easily recognized from its resemblance to a chair, of which α, β, the two brightest stars, form the base of the seat, β being the westernmost star when the constellation is above pole, and therefore the first to pass the meridian.

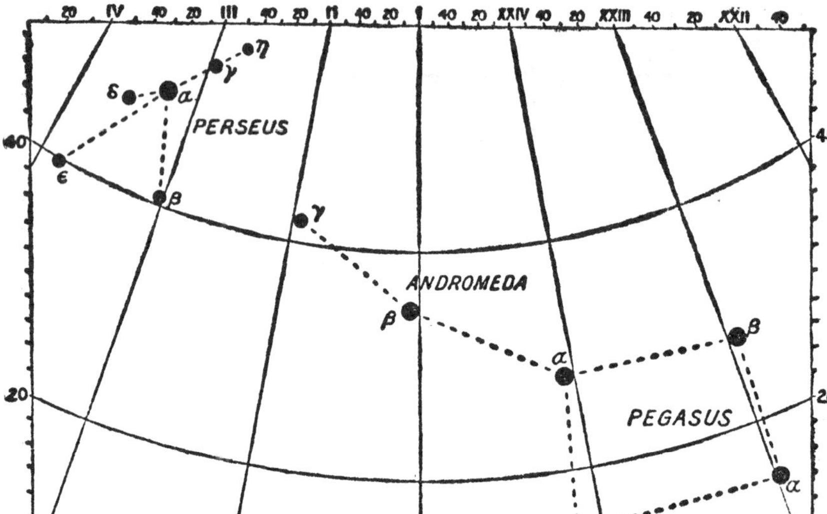

DIAGRAM 2.—The Constellations Perseus, Andromedæ and Pegasus.

To Find the Constellation Pegasus.

This constellation, with Perseus and Andromedæ, forms a configuration easy to remember (Diagram 2). A line from Polaris through β Cassiopeiæ, and produced about the same distance, terminates near the four stars which form "the great square of Pegasus." The stars α, β are those which form the outer side of the square, α, or Markab, being that which is nearer to the equator.

To Find Andromedae.

The star at the angle of the square of Pegasus opposite to α Pegasi is sometimes called δ Pegasi, but it is known also as α Andromedæ, and if this diagonal is continued further it indicates β, γ, two of the principal stars in Andromedæ.

To Find Perseus.

The above line produced still further passes through the chief star in Perseus, situated in the middle of an arc formed by the stars β, δ, α, γ and η Persei.

To Find Capella (α Aurigæ).

From the bow of Perseus proceed two rows of small stars, one of which, curving to the East, extends to Aurigæ, which takes the shape of a large irregular pentagon. Capella, the brightest star (magnitude 0·2), is well marked by a small isosceles triangle, formed by three small stars in close proximity.

To Find the Pleiades.

The second row of stars proceeding from Perseus is directed towards the well-known group of small stars called the Pleiades.

To Find Arcturus (α Bootis).

Continue the curved tail of the Great Bear about twice the length of the constellation, and it will terminate near the bright star Arcturus (magnitude 0·2).

To Find Vega (α Lyræ).

A line from Capella, midway between Polaris and Cassiopeiæ, and continued something more than an equal distance beyond the Pole, leads to Vega (magnitude 0·1).

To Find Altair (α Aquilæ).

The Eagle lies to the southward of Lyra. It contains three bright stars in the same line, the centre one of which is Altair (magnitude 0·9).

To Find α Coronæ Borealis.

The Northern Crown lies between Vega and Arcturus. It is composed of several stars in the shape of a crown, the principal of which is known as α Coronæ or Alphacca. The brightness of this star has suffered some diminution in recent years and its magnitude is now 2·3. It lies upon a line drawn from Vega to Arcturus, about one-third of the length of the line from Arcturus.

To Find Hamal (α Arietis).

A line drawn from the neighbourhood of α Persei (Diagram 2) through β Persei, and produced twice their distance apart, points out α Arietis (magnitude 2·2).

To Find Aldebaran (α Tauri).

A line drawn from Polaris between Capella and α Persei passes no conspicuous star until it comes to Aldebaran (magnitude 1·1). This is a particularly easy star to determine since it is situated near the Pleiades, in a companion cluster called the Hyades, which have the shape of a well-marked V. Aldebaran is situated on the left, at the bottom of the cluster.

To Find Gemini (Castor and Pollux).

To the eastward of Aldebaran, about the distance of that star from Capella, are four stars in a line. These stars form the feet of the Twins. In the head are two bright stars, Castor (magnitude 1·6), and Pollux (magnitude 1·2).

Castor and Aldebaran form with Capella an isosceles triangle, which has Capella at the vertex. Pollux is a little to the South of Castor.

To Find Regulus (α Leonis).

A line drawn from α to β Ursæ Majoris, the Pointers, in the opposite direction to Polaris, and produced about twice the length of Ursæ Majoris will pass between Regulus and β Leonis. The magnitude of Regulus (1·2) will serve to distinguish it from β Leonis (magnitude 2·2).

To Find Cor Caroli (α Canum Venaticorum) and thence Spica, or α Virginis.

A line drawn from Dubhe (Diagram 1) through the opposite angle of the square, and produced nearly twice their distance apart, leads to Cor Caroli. The magnitude of this star (2·8) is not such as to give it any great importance in itself, but it is useful in helping to identify the bright star Spica (magnitude 1·2).

A line drawn from ε Ursæ Majoris, the star in the tail which is nearest to the square, through Cor Caroli, and produced about same distance, will terminate near Spica.

To Find α and β Libræ.

To the east of Spica are two stars, α Libræ (magnitude 2·9) and β Libræ (magnitude 2·8), β, the most distant from Spica, being about as far from Spica as is the latter from Arcturus.

To Find Antares (α Scorpii).

The Scorpion is to the South-East of Libra. In the head are five small stars, the centre one of which is β^1 Scorpii (magnitude 2·7); Antares (magnitude 1·3) forms, with Vega and Arcturus, a right-angled triangle, of which Vega is at the right angle.

To Find Fomalhaut (α Piscis Australis).

A line drawn from β to α Pegasi (Diagram 2), produced three times the distance, will terminate near Fomalhaut (magnitude 1·8).

To Find α Orionis (Betelguese) and β Orionis (Rigel).

The constellation Orion is one of great beauty, and its appearance should be familiar to every seaman (Diagram 3). The stars, α β, γ, κ, have the form of a great quadrilateral of which Betelguese (magnitude 1·0 to 1·4) is in the North-East angle. The star Rigel (magnitude 0·8) is in the opposite foot of the constellation. In the middle of the quadrilateral are three stars of the second magnitude, disposed in an oblique line. These are said to form the " belt of Orion."

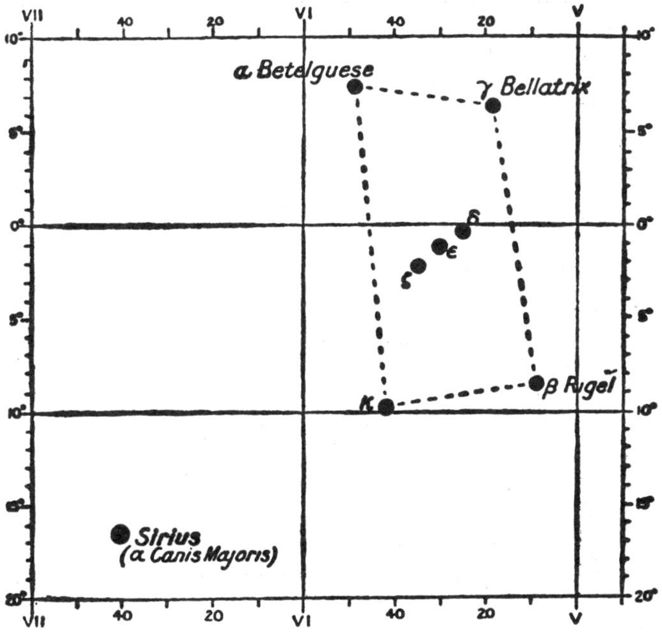

DIAGRAM 3.—Constellation of Orion.
Relative Position of the Seven Principal Stars in the Constellation Orion, showing also the Position of Sirius (α Canis Majoris).

To Find α Canis Majoris (Sirius).

A line drawn through the belt of Orion (Diagram 3) to the left, continued for about the distance which separates Rigel from Betelguese, will terminate near Sirius (magnitude −1·4).

To Find α Canis Minoris (Procyon).

Procyon (magnitude 0·5) lies midway between Gemini and Sirius. Or it may be regarded as forming the eastern angle of a triangle which has Sirius and Betelguese at the other angles.

To Find α Columbæ and thence Canopus (α Argus).

A line drawn from Procyon to Sirius and produced nearly the same distance passes near the first star (magnitude 2·7). Again a line drawn from Rigel through α Columbæ, and produced half the same distance, will terminate near Canopus (magnitude −1·0).

To Find the Centaur and the Cross.

These constellations lie to the South of Virgo and Libra. The various stars which compose them may easily be made out from Diagram 4 without any formal rule. The most southern star of the Cross is α (magnitude 1·0); the most northern is γ (magnitude 1·6). β Crucis (magnitude 1·5) and β Centauri (magnitude 0·8) are adjacent to each other.

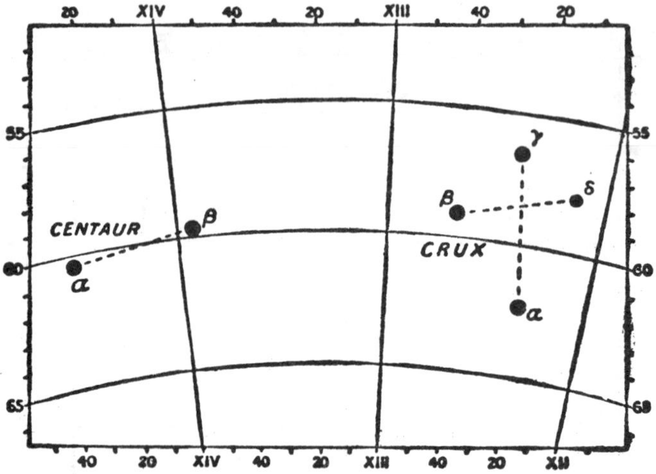

DIAGRAM 4.—The Constellations Centaur and Crux.

To Find Achernar (α Eridani).

A line drawn from Fomalhaut to Canopus is bisected by Achrenar (magnitude 0·5.)

Identification of Stars by Means of Azimuth Tables.

The foregoing directions will, it is hoped, enable the observer to locate the principal stars that are of service in navigation. Their use, however, is obviously limited to clear nights, when the outline of the various constellations can be made out, and are of little service in the case of an altitude "snapped during the momentary appearance of a star through a rift in the clouds, although it is in gloomy and overcast weather that star observations are often of paramount importance." A star may occasionally be identified by means of its altitude or azimuth, computed roughly.

For stars near the meridian the altitude, computed approximately in a few seconds from known values of Latitude and declination, is all that is required. A bright star bearing nearly due North or South, as the case may be, and having the altitude so obtained, may be accepted without much risk of error as the star required.

For stars observed on larger bearings it is necessary to know the altitude and azimuth of the star approximately before identification can be established.

The position of the observer's zenith is plotted on a star chart using declination for the latitude and right ascension equal to the local hour angle of Aries. From this position the star may be identified by laying off its zenith distance in the direction of its azimuth.

The following method of identifying a star by calculation and reference to the star list in the *Nautical Almanac* may not be quite obsolete yet. In the *PZX* triangle of nautical astronomy, *PZ* the co-latitude is known, *ZX* is the complement of the altitude observed by sextant and angle *Z* is found from the compass bearing of the star. Using these three factors calculate *PX* the star's polar distance, and angle *P* the L.H.A. star as shown in the following example.

Example.—On January 10, D.R. Latitude 51° N., Longitude 30° 30′ W., the altitude of an "unknown" bright star was observed to be 42° 15′, bearing 239° (T) G.M.T. by chronometer 02h 02m 00s. What star was it?

Given: $PZ = 39° 00', ZX = 47° 45', Z = 121°$.

Find: *PX* thence the declination.

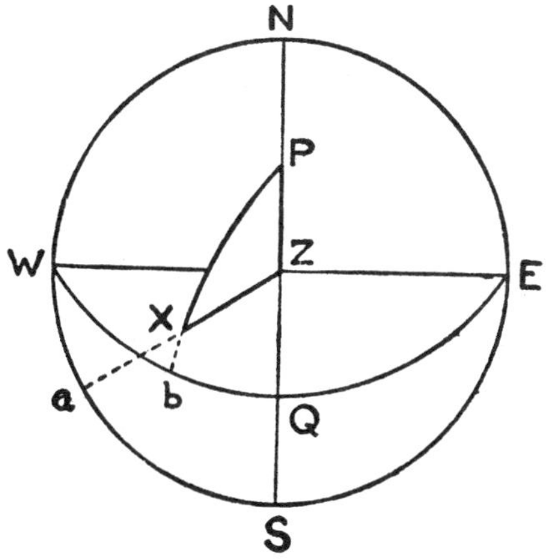

hav PX = hav Z sin PZ sin ZX + hav ($PZ \sim ZX$)

Z	121° 00′	log hav	9·87939
PZ	39 00	log sin	9·79887
ZX	47 45	log sin	9·86936
		log hav	9·54762
		nat hav	·35288
(*PZ*∼*ZX*)	8 45	nat hav	·00582
PX	73 35	nat hav	·35870
	90 00		
Decl.	16 25 N.		

(2) Find *P* and with the G.M.T. get G.H.A. Aries.

S.H.A. Star = G.H.A. Star − G.H.A. Aries.

$$\text{hav } P = \frac{\text{hav } ZX - \text{hav } (PZ \sim PX)}{\sin PZ \sin Px}$$

ZX	47° 45′	nat hav	·16382	G.M.T. 10d 02h 02m 00s
(*PZ*∼*PX*)	34 45	nat hav	·08918	G.H.A. Aries 139° 33·4′
		nat hav	·07464	
		log hav	8·87298	
PZ	39 00	cosec	0·20113	
PX	73 45	cosec	0·01771	
L.H.A.	41 09·5	log hav	9·09182	
Long. W.	+30 30			
G.H.A. Star	71 39·5			
G.H.A. Aries	139 33·4			
S.H.A. Star	292 06·1	(Approx.)		The star whose S.H.A. and Decl. agree most closely with this will be the star required.
Decl.	16 25 N.	(Approx.)		*Ans.—Aldebaran*

LATITUDE BY MERIDIAN ALTITUDE (MOON).

The G.M.T. of moon's transit across the observer's meridian is required in order to find the moon's declination and horizontal parallax at the time of observation. The L.M.T. of transit of the moon over the meridian of Greenwich is given in the *Nautical Almanac* at the foot of the daily pages in the column headed "Moon, Mer. Pass. Upper". If the observer is not on the meridian of Greenwich, a correction for Longitude must be applied to the L.M.T. of transit at Green wich to findthe local time of the moon's meridian passage at ship.

The moon passes the meridian of Greenwich later each day by the number of minutes indicated by taking the difference of the times tabulated from one day to the next day. In the interval between two such transits the moon passes over 360° of Longitude, therefore, the Mean time of transit over any meridian of *West* Longitude is later than that over the meridian of Greenwich by $\frac{\text{Long. in time}}{24} \times \text{diff.}$ The Mean time of transit over any meridian of *East* Longitude is *earlier* than that over the meridian of Greenwich by $\frac{\text{Long. in time}}{24} \times \text{diff.}$ This is the correction to be applied to the tabulated time of meridian passage at Greenwich for the given date. Use the difference between the day in question and the following date when in West Longitude, but use the difference between the preceeding date when in East Longitude.

When the moon does not pass the upper meridian of Greenwich on the given date at ship, take the meridian passage for the preceeding day when in West Longitude, and for the following day when in East Longitude.

Example I.

On June 22, in D.R. Latitude 39° 40′ S., Longitude 60° 18′ W., the sextant meridian altitude of the moon's lower limb was 45° 22′. Index error 1′ off the arc, height of eye 34 feet. Required the Latitude.

		d	h	m
Mer. Pass. at Greenwich		22	16	32
Corr. for W. Long.	+			8·5
L.M.T. Mer. Pass. ship		22	16	40·5
Long. in time			4	01·2
G.M.T. Mer. Pass. ship		22	20	41·7

diff. $51 \times \frac{4}{24} = 8{\cdot}5$ m.

Declination 22d 20h		4°	09·8′ N.
d correction (11·2)	−		7·9
Corr. Decl.		4	01·9 N.
Hor. Parlx. 58·9			

Sext. alt.		45°	22′
Index error	+		1
		45	23
Dip.	−		5·7
		45	17·3
Corr.	+		50·2
H.P. (58·9)	+		6·1
		46	13·6
True alt.		90	00
Zen. dist.		43	46·4 S.
Declination		4	01·9 N.
Latitude		39	44·5 S.

The altitude of the moon's lower limb has been corrected above using the correction tables given in the *Nautical Almanac*.

Example II.

On September 23, in D.R. Latitude 30° 12′ N., Longitude 100° 49·5′ E., the sextant meridian altitude of the moon's Upper Limb was 47° 00′. Index error 2′ off the arc; height of eye 25 feet. Required the Latitude.

	d	h	m
Mer. Pass. at Greenwich	23	20	56
Corr. for E. Long.	—		13·4
L.M.T. Mer. Pass. ship	23	20	42·6
Long. in time	—	6	·43·3
G.M.T. Mer. Pass. ship	23	13	59·3

Declination	12° 32·1′ S.
Hor. Parlx.	55·7′

Diff. $48 \times \frac{6 \cdot 7}{24} = 13 \cdot 4$ mins.

Sext. alt.		47°	00′
Index error	+		2
		47	02
Dip	—		4·9
		46	57·1
Main corr.	+		49·1
H.P. (U) corr.	+		2·8
		47	49
Upper limb cor.	—		30
True alt.		47	19
		90	00
Zen. dist.		42	41 N.
Declination		12	32·1 S.
Latitude		30	08·9 N.

It should be noticed that the constant, 30′, is subtracted *after* the other corrections have been applied to the altitude of the moon's upper limb.

Examples for Exercise.

(1) On January 6, in D.R. Latitude 30° 20′ S., Longitude 47° 30′ E., the observed meridian altitude of the moon's upper limb was 41° 27′. Height of eye 35 feet. Required the Latitude.

(2) On December 21, in D.R. position 48° 00′ N., 43° 15′ W., the sextant meridian altitude of the moon's lower limb was 55° 10′. Index error the 1′ off arc; height of eye 30 feet. Find the Latitude.

(3) On June 23, in D.R. Lat. 28° 47′ S., Long. 84° 36′ E., the sextant meridian altitude of the moon's upper limb was 59° 50′. Index error 1′ on the arc; height of eye 35 feet. Required the Latitude.

(4) On September 19, in D.R. Lat. 41° 55′ N., Long. 52° 06′ W., the sextant meridian altitude of the moon's lower limb was 28° 24′. Index error 2′ off the arc; height of eye 28 feet. Required the Latitude.

(5) On June 24, in D.R. Lat. 50° 15′ N., Long. 45° 00′ W., the observed altitude of the moon's lower limb was 33° 35′. Height of eye 40 feet. Required the Latitude.

LATITUDE BY MERIDIAN ALTITUDE (PLANET).

The Mean time of transit of a planet over the meridian of Greenwich is given in the daily pages of the *Nautical Almanac* in the column headed Mer. Pass., opposite the name of the planet. The times are given for the middle day of the three days on each page, so if the date required does not happen to be the one tabulated, find the difference in time of meridian passage in one day and apply this to the time nearest the required date. The L.M.T. of transit of a planet over any other meridian is found by applying a correction for Longitude to the time of transit at Greenwich. If the Longitude is East, find the difference in times of transit between the date of the question and the preceding date, but the following date if the Longitude is West, and multiply this by the Longitude in time divided by 24. The correction is added if the time of transit increases between the date of the question and the second date, but subtracted if it decreases. To the L.M.T. of transit over the observer's meridian, apply the Longitude in time to get the G.M.T. at the time of observation.

Example I.

On June 16, in D.R. Lat. 27° 37′ S., Long. 167° 21′ W., the sextant meridian altitude of Venus was 47° 12′, index error 1′ on the arc, height of eye 35 feet. Required the Latitude.

June		d	h	m
Mer. Pass.		16	09	26
Corr. for W. Long.	+			0·3
L.M.T. Mer. Pass.		16	09	26·3
Long. in time	+		11	09·4
G.M.T. Mer. Pass.		16	20	35·7

Declination 15° 16·5′ N.

	h	m
Mer. Pass. 16th	9	26
„ „ 19th	9	28
3)		2
Diff. in 1 day		0·6

$$\text{Corr.} = 0{\cdot}6 \times \frac{11}{24}$$
$= 0{\cdot}3$mins.

Sext. Alt.	47°	12
In ex error	—	1
Dip	—	5·7
Correction	—	0·9
True alt.	47	04·4
	90	00
Zen. dist.	42	55·6 S.
Declination	15	16·5 N.
Latitude	27	39·1 S.

Example II.

On January 10, in D.R. Lat. 44° 05′ N., Long. 130° 47′ E., compute the altitude to be set on the sextant for the meridian altitude observation of Venus, index error 1′ on the arc, height of eye 34 feet.

		d	h	m
Mer. Pass.		10	13	52·3
Corr. E. Long.	+			1·5
L.M.T. Mer. Pass.		10	13	53·8
Long. in time	—		8	43·1
G.M.T. Mer. Pass.		10	05	10·7

Declination 13° 11·3′ S.

	h	m
Mer. Pass. 11th	13	48
„ „ 8th	14	01
3)		13
Diff. in 1 day		4·3
Diff. in 8·7h.		1·5

D.R. Lat.	44°	05′ N.
Decl.	13	11·3 S.
Zen. Dist.	57	16·3
	90	00
True alt.	32	43·7
Index error	+	1
Dip	+	5·7
Corr.	+	1·5
Sext. alt.	32	51·9

Examples for Exercise.

(1) On September 20, in D.R. Lat. 31° 25′ N., Long. 178° 37′ W., the observed meridian altitude of Saturn was 36° 51′, height of eye 36 feet. Required the Latitude.

(2) On January 10, in D.R. Lat. 39° 20′ N., Long. 157° 20′ W., the observed meridian altitude of Jupiter was 40° 37′, height of eye 30 feet. Required the Latitude.

(3) On June 20, in D.R. Lat. 26° 12′ S., Long. 71° 30′ E., index error 1′ on the arc, height of eye 35 feet, compute the altitude to be set on the sextant for the meridian altitude observation of Mars.

(4) On December 22, in D.R. Lat. 46° 25′ N., Long. 18° 12′ W., compute the sextant meridian altitude of Mars, index error 1′ on the arc, height of eye 36 feet.

(5) On January 8, in Longitude 132° 15′ E., the observed meridian altitude of Venus South of observer, was 58° 15′; height of eye 40 feet. Required the Latitude.

CHAPTER VII.

EX-MERIDIANS AND POSITION LINE METHODS.

LATITUDE BY ALTITUDE OF THE POLE STAR.

The altitude of the celestial pole is equal to the latitude of the observer, and if "Polaris" were exactly at the pole of the heavens its true altitude would be the latitude. It is situated, however, about 56′ away from the pole and slowly approaching it, but the star describes such a small daily circle in the sky that the latitude can be computed from an altitude of the Pole Star at any time. Polaris changes its azimuth from about N.2° E. to N2° W. during its daily circuit round the pole of the heavens and bears true North on the two occasions, each day, when it crosses the meridian above and below the pole. See Pole Star Tables in Appendix.

$$\text{Latitude} = \text{True altitude} - 1^\circ + a_0 + a_1 + a_2.$$

The table is entered with the local hour angle Aries and the a° correction is interpolated mentally. The a_1 and a_2 corrections are taken from the tables nearest the L.H.A. Aries, abreast of the approximate Latitude and of the Month, respectively.

Example.

On September 20, at 22h 00m 00s local Mean time in Longitude 152° 30′ E., the sextant altitude of Polaris was 36° 44′, index error 1·5′ on the arc, height of eye 39 feet. Required the Latitude.

	d	h	m	s
L.M.T.	20	22	00	00
Long. E.	—	10	10	00
G.M.T.	20	11	50	00

L.H.A.♈ = G.H.A.♈ + E. Long.

G.H.A.♈ 20d 11h	163°	49·8′
Incre. 50m 00s	12	32·1
G.H.A. Aries	176	21·9
Long. E.	+152	30
L.H.A.	328	51·9

True Az. 001°
P.L. 091°–271° T.

Sext. alt.	36°	44′
Index error	—	1·5
Obs. alt.	36	42·5
Dip	—	6·1
	36	36·4
Corr.	—	1·3
True alt.	36	35·1
	−1	
	35	35·1
a	+	31·4
a	+	0·5
a	+	0·7
Latitude	36	07·7 N.

The Pole Star Tables are based on the formula:—

$$\text{Lat.} = \text{Alt.} - \overbrace{\text{P. dist} \cos \text{L.H.A. star}}^{\text{2nd term}} + \overbrace{\tfrac{1}{2} P\ \text{dist}^2 \sin^2 \text{L.H.A. star} \tan (\text{lat.})}^{\text{3rd term}}$$

The correction a_0 gives the value of the 2nd and 3rd terms, using a mean latitude of 50° and a mean S.H.A. and Declination of Polaris. A constant, 58·8′, is always added.

The corrections a_1 and a^2 are adjustments to the 3rd and 2nd terms, respectively when the ship's Latitude and the position of Polaris differ from the mean values, used. A constant (0·6′) is added to each to make them always positive, and to make the sum of the contants equal to one degree.

Examples for Exercise.

(1) On January 12, local Mean time 03h 16m at ship in Longitude 32° 00′ W., the sextant altitude of the Pole star (α *Ursae Minoris*) out of the meridian was 45° 19′, height of eye 35 feet, index error 1·5′ off the arc. Required the Latitude.

(2) On June 18, at 23h 39m Mean time at ship in Longitude 147° 20′ W., the sextant altitude of Polaris off the meridian was 34° 58′, index error 1·0′ on the arc, height of eye 27 feet. Required the Latitude.

(3) On June 23, in Longitude 165° 18′ E., the observed altitude of Polaris out of the meridian was 51° 15′, G.M.T. 22d 17h 02m 24s., height of eye 20 feet. Required the Latitude.

(4) On September 19, at G.M.T. 20d 00h 36m 31s in Longitude 30° 40′ W., the observed altitude of the Pole star out of the meridian was 61° 09′, height of eye 30 feet. Required the Latitude.

(5) On December 24 in Longitude 149° 30′ W., when a chronometer (corrected) indicated Mean time at Greenwich 15h 17m 40s, the sextant altitude of Polaris 42° 20′, index error 1·4′ off the arc, height of eye 26 feet. Required the Latitude.

(6) On January 10, in Longitude 20° 10′ E., when a chronometer (corrected) indicated Mean time at Greenwich January 9d 22h 53m 40s, the sextant altitude of the Pole star was 36° 02′, index error 1′ on the arc, height of eye 24 feet. Required the Latitude.

EX-MERIDIAN PROBLEMS.

The "Ex-Meridian" problem is given in Volume I., mainly in its application to the sun. We shall now proceed to give examples of all bodies worked by the "Ex-Meridian" and the "Intercept" methods, and show how to fix the ship's position by combining ex-meridian sights with position lines obtained from the observations of other bodies by the several methods described in Volume I.

EX-MERIDIAN (STAR).

Example I.

On June 15, DR position 33° 07′ N., 138° 30′ W., the observed altitude of the star *Fomalhaut* near the meridian was 26° 48·2′ South of the observer, height of eye 36 feet, time by chronometer 15h 13m 12s, which was 10m 08s fast of Mean time at Greenwich. Required the position line and the latitude where the position line crosses the DR longitude.

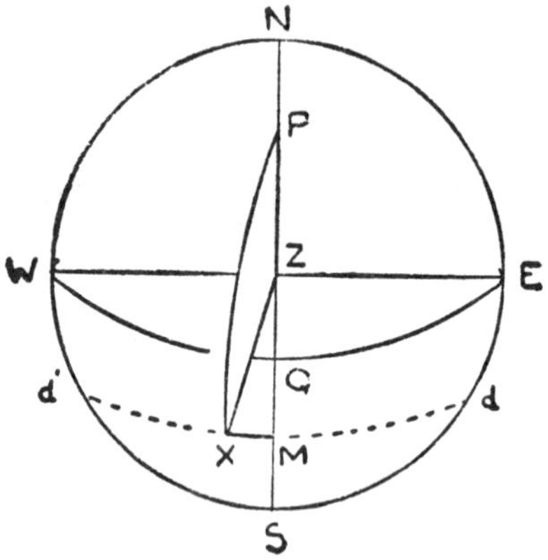

W Q E the equinoctial
W Z E the prime vertical
N Z S the observer's meridian
d M X d′ the parallel of declination
P the pole, *Z* the zenith, *X* the star at the time of observation, position exaggeration to show the *P Z X* triangle
M the position of the star when it was on the meridian
∠*Z P X* is the local hour angle of *X*
P Z the co-latitude
Z Q the latitude, which is the quantity required
P X the polar distance, which in this illustration is (90° +decl.)
Z X the zenith distance of the body at the time of observation
Z M the zenith distance of the body when it was on the meridian, that is its MZD.

Z Q the latitude is equal to (*Z M*−*Q M*), that is (MZD−decl.) and the purpose of the calculation is to find *Z M* the meridian zenith distance.

Now, the least zenith distance of a body for the day occurs when it is on the observer's meridian, so that the zenith distance at any other time must be greater than its MZD and therefore *ZX* is slightly greater than *ZM*, and the operation of finding the smaller arc *ZM* is called "The Reduction to the Meridian".

The "Reduction" is a small quantity which is the difference between the MZD and the observed zenith distance, or, arc *ZX*−arc *ZM*.

The body is so close to the meridian that *PX* is practically equal to *PM*, and if (*P M*−*P Z*) is equal to *Z M* then (*P X*−*P Z*) is assumed to be also equal to *Z M*, the meridian zenith distance, which we are out to find in triangle *P Z X*.

A D.R. latitude has to be introduced into the formula so that the latitude found by calculation must consequently be approximate, but when the body is suitably placed the result obtained is a very close approximation to the actual latitude.

If the calculated latitude should differ much from the D.R. latitude the triangle should be reworked, using the new latitude to obtain a nearer approximation, but a second working, however, should not be necessary.

Order of Work.

(1) Find the local hour angle of the body
(2) Correct the observed altitude.
(3) Apply the formula to find $(PZ \sim PX)$ the meridian zenith distance, using PX, PZ, $\angle P$ and a D.R. co-latitude PZ.
(4) Apply the declination to the M.Z.D. and get the latitude at sights.
(5) Apply the sine formula in the PZX triangle and get the azimuth of the body for a position line.

		d	h	m	s
Chron.		15	15	13	12
Fast	−			10	08
G.M.T.		15	15	03	04

G.H.A. ♈ 15d 15h	128°	23·2′
Incre. 03m 04s	0	46·1
G.H.A. ♈	129	09·3
S.H.A.*	16	09·5
G.H.A.*	145	18·8
Long. W.	−138	30·0
L.H.A.*	6	48·8 (P)

Obs. Alt.		26°	48·2′
Dip	−		5·8
Corr.	−		1·9
True Alt.		26	40·5
		90	00
ZX		63	19·5 N.
Star's Decl.		29°	50·3′ S.
		90	00
Polar Dist.		119	50·3 (PX)
D.R. Lat.		33°	07′ N.
		90	00
Approx. co-lat.		56	53 (PZ)

Hav $(PZ \sim PX) = $ Hav $ZX -$ Hav P Sin PZ Sin PX, for the M.Z.D.
Sin $Z =$ Sin P Sin PX cosec ZX, for the azimuth.

P	6°	48·8′	log hav	7·54790	sin	9·07421
PZ	56	53	log sin	9·92302		
PX	119	50·3	log sin	9·93824	sin	9·93824
			log hav	7·40916		
			nat hav	0·00257		
ZX	63	19·5	nat hav	0·27554	cosec	0·04887
Mer. zen. dist.	62	59·8	nat hav	0·27297		
Declination	29	50·3				
					sin	9·06132
Latitude by obs.	33	09·5 N.				
Latitude by D.R.	33	07 N.			Azimuth S. 6° 37′ W.	
D. Lat.		2·5 N.			Position Line 096·6°—276·3° T.	

The Position Line.

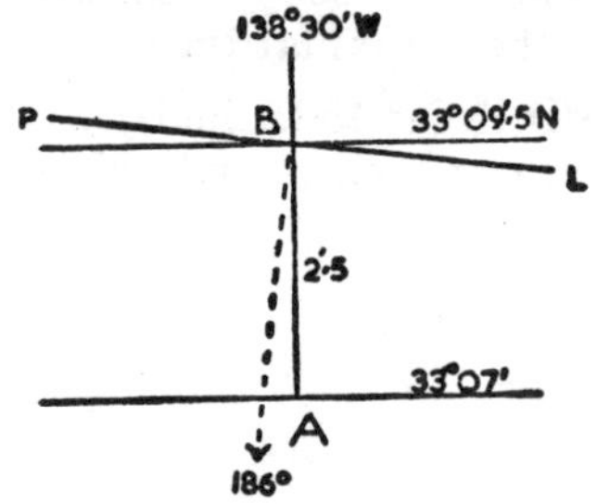

In the figure:

A is the D.R. position, Lat. 33° 07′ N., Long. 138° 30′ W.

AB is the D. Lat. 2·5′ N.

B is the amended D.R. position, 33° 09·5′ N. by observation, 138° 30′ W. by D.R., through which to draw the position line *PL*.

Another method of calculating the latitude by ex-meridian is by means of right angled spherical triangles. It was a popular method before the haversine formula was evolved and it possesses the merit of being independent of an estimated latitude which is an uncertain factor in the reduction formula.

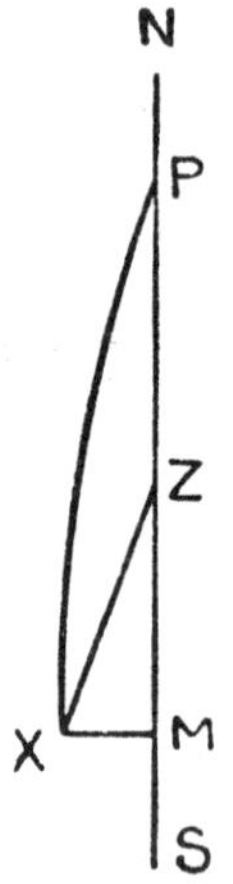

PZMX shows the essential parts from the figure illustrating the foregoing example.

$\angle M=90°$, we know the hour angle, the polar distance and the zenith distance of Fomalhaut from which to find *PZ* the co-latitude as follows.

In triangle PMX

(i) Sin *comp* $P=$tan *comp* PX tan PM
cos P $=$cot PX tan PM
tan $PM=$cos P tan PX—
tan $PM=$cos hour angle cot decl.

(ii) sin $XM=$cos *comp* P cos *comp* PX
sin $XM=$sin hour angle cos decl.

In triangle ZMX

(iii) sin *comp* $ZX=$cos XM cos ZM
sin alt sec $XM=$cos ZM
then arc $PM-$arc $ZM=$arc PZ, and $90°-PZ=$latitude

Using the values given in the example we have

L.H.A.	6° 48·8′	cos	9·99693	sin	9·07421
Decl.	29 50·3	cot	10·24140	cos	9·93824
	59 59·2	Tan	10·23833	sin	9·01245
	180				
				XM	5° 54·5′
**PM*	120 00·8				

Alt·	26° 40·5	sin	9·65218
XM	5 54·5	sec	0.00231
ZM	63 10·3	cos	9·65449
PM	120 00·8		
Co-lat	56 50·5	Lat.	33° 09·5′ N.

*When the polar distance exceeds 90°, arc *PM* is also greater than 90° and, by the rule of signs, the arc taken from the Tables is then subtracted from 180° to yield arc *PM*.

EX-MERIDIAN (PLANET).

Example II.

On January 9, in DR 42° 20′ N., Long. 179° 30′ E., the sextant altitude of Jupiter near the meridian was 37° 22′, index error 1·2′ on the arc, height of eye 26 feet, time by chronometer which was correct Mean time at Greenwich 9d 18h 58m 57s. Required the latitude where the position line crosses the DR longitude. The vessel then steamed 075° T., for four hours at 10 knots, a current setting 130° T., at 2·5 knots. Required a position through which to draw the transferred position line.

LHA = GHA + E. Long.

GMT Jan. 9d 18h 58m 57s			Sext. alt.		37° 22·0′
GHA Jupiter 9d 18h	170°	46·2′	Index error	—	1·2
Incre. 58m 57s	14	44·3	Dip.	—	4·9
v corr. (2·2)		2·1	Corr.	—	1·3
GHA	185	32·6	True alt.		37 14·6
Long. E.	+ 179	30			90 00
LHA Jupiter	5	02·6 (*P*)	Zen. dist.		52 45·4

Decl. 10° 09·1′ S.

The formula

$$\text{hav}\,(PZ \sim PX) = \text{hav}\,ZX - \text{hav}\,P \sin PZ \sin PX$$

may be modified to

$$\text{hav}\,(PZ \sim PX) = \text{hav}\,ZX - \text{hav}\,P \cos \text{lat.} \cos \text{decl.}$$

thus saving two subtractions from 90°. When cos decl. is substituted the azimuth formula becomes sin *Az*=sin LHA cos decl cosec zen dist.

P	5° 02·6′	hav	7·28680	sin	8·94403
Lat.	42 20 N.	cos	9·86879		
Decl.	10 09·1 S.	cos	9·99315	cos	9·99315
		hav	7·14874		
		nat hav	0·00141		
ZX	52 45·4	nat hav	0·19739	cosec	10·09905
(*PZ*∼*PX*)	52 33·1	nat hav	0·19598	sin	9·03623
M.Z.D.	52 33·1 N.			Azimuth S. 6·1° W. or 186° T.	
Decl.	10 09·1 S.			P.L. 096°—276° T.	
Obs. Lat.	42 24·0 N.				
D.R. Lat.	42 20·0 N.				
D. Lat.	4·0 N.				

Run *B* to *C*			D. Lat.	Dep.
Course	075° T.	40	10·4′ N.	38·6′ E.
Current	130° T.	10	6·4 S.	7·7 E.
			4·0 N.	46·3 E.
			D. Long.	60·5 E.

	Lat.	Long.
A	42° 20′ N.	179° 30′ E.
A to *B*	4 N.	0 00
B	42 24	179 30 E.
B to *C*	4 N.	1 00·5 E.
C	42 28 N.	180 30·5 E.
		360 00
		179 29·5 W.

Transfer the position line to Lat. 42° 28′ N., Long. 179° 29·5′ W.

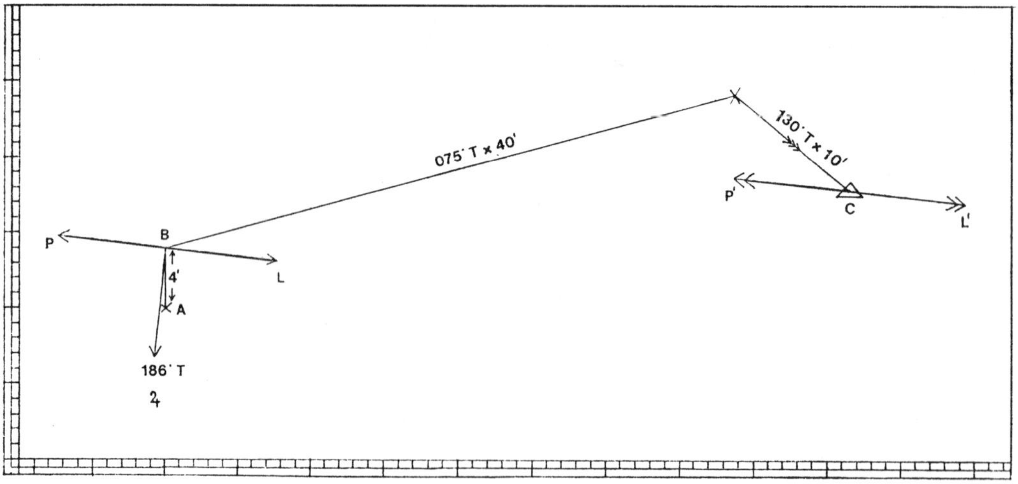

In the figure:

A is the D.R. position.

A B the D. Lat. 4′ N. got from the ex-meridian of Jupiter.

B is a Latitude by observation, but a D.R. Longitude position through which to draw the position line *PL*.

C is the estimated point through which to draw the transferred position line *P′L′*, its accuracy being dependent on the correctness of the stated course and distance made good between *B* and *C*.

EX-MERIDIAN (MOON).

Example III.

(i) On December 21d 22h 01m 51s GMT, position by DR 39° 10′ N., 22° 30′ W., the observed altitude of the moon's upper limb was 64° 31′, height of eye 13·5 m. Required the Latitude and the position line.

(ii) At GMT 21d 22h 03m 02s DR Lat. 39° 10′ N. Long. 22° 30′ W., the observed altitude of Betelguese was 35° 28′ East of the meridian. Height of eye 13·5 m. Required the position line through the DR Long. and the ship's position as determined by combining both sights.

LHA = GHA − W. Long.

GMT Dec. 21d 22h 01m 51s

GHA Moon 21d 22h	19° 48·2′	Obs alt Moon UL	64° 31·0′
Incre 01m 51s	0 26·5	Dip	− 6·5
v corr (13·6)	0·3		64 24·5
GHA Moon	20 15·0	1st Corr.	+ 35·1
Long W	−22 30	HP (U)	+ 3·0
LHA Moon	357 45		65 02·6
			− 30
Decl 13° 37·0′ N.			
d (6·7) ·2		True Alt	64 32·6
Dec 13 37·2 N.		True Zen Dist	25 27·4

Hor Parlx 54·2

We might apply the "Ex-Meridian" formula in this example by way of a change.

Sin ½ red=hav LHA cos. Lat.∼cos Decl. cosec *MZD.*

NOTE.—MZD=lat+decl when they are of opposite name, and lat∼decl when of the same name.

EHA	2° 15′	hav	6·58600
Lat.	39 10 N.	cos	9·88948
Decl.	13 37·2 N.	cos	9·98761
Approx. MZD	25 32·8	cosec	0·36527
½ red	2·3	sin	6·82836
Reduction	00° 04·6′		
Obs. Zen. Dist.	25 27·4		
MZD	25 22·8 N.		
Dec.	13 37·2 N.		
Lat.	39 00·0 N. by obs.		
Long.	22 30 W. by D.R.		

ABC Tables

A	20·6 S.
B	6·16 N.
C	14·44 S.

Azimuth 175° T.
P.L. 085°—265° T.

Example III.

(ii) Longitude by Betelguese.

G.M.T. Dec. 21d 22h 03m 02s.

G.H.A. Star=G.H.A. Aries+S.H.A. Star.

G.H.A. Aries	59°	57·7′	Obs. alt.	35°	28′
Incre.		45·6	Dip	—	6·5
S.H.A. Betelguese	271	45·3		35	21·5
G.H.A. ,,	332	28·6	Corr.	—	1·4
			True alt.	35	20·1
				90	
			True Zen. dist.	54	39·9

$$\text{hav } P = \frac{\text{hav } ZX - \text{hav } (L \sim D)}{\cos \text{Lat.} \cos \text{Decl.}}$$

ZX	54°	39·9′	nat hav	·21082		
(L—D)	31	46·1	nat hav	·07491		
			nat hav	·13591		
			log hav	9·13326	ABC Tables	
Lat.	39	10 N.	log sec	10·11052	*A*	·68 S.
Decl.	7	23·9 N.	log sec	10·00363	*B*	·16 N.
P	49	43·5	log hav	9·24741	*C*	·52 S.
L.H.A.	310°	16·5′			Azimuth=S. 68° E.=112° T.	
G.H.A.	332	28·6			P.L. 022°—202° T.	
Obs. Long.	22	12·1 W.				
D.R. Long.	22	30 W.				
D. Long.		17·9 E.				
Dep.		13·9 E.				

Both Sights Combined.

In the squared paper diagram:

A is an assumed position.

AB is D. Lat. 10′ S. from the moon's ex-meridian.

B is a Latitude by observation but a D.R. Longitude through which to draw position line *PL*.

AC is Dep. 13·9 miles East found by observation of Betelguese.

C is a Longitude by observation but a D.R. Latitude through which to draw position line *P′L′*.

F is the ship's position by observation.

To Transfer *A* to *F*.

	Lat.		Long.
A	39° 10′ N.		22° 30′ W.
A to *F*, D. Lat.	9·2 S.	(Dep. 10 miles)=D. Long.	12·8 E.
F	39 00·8 N.		22 17·2 W.

Ship's position Lat. 39° 00·8′ N., Long. 22° 17·2′ W.

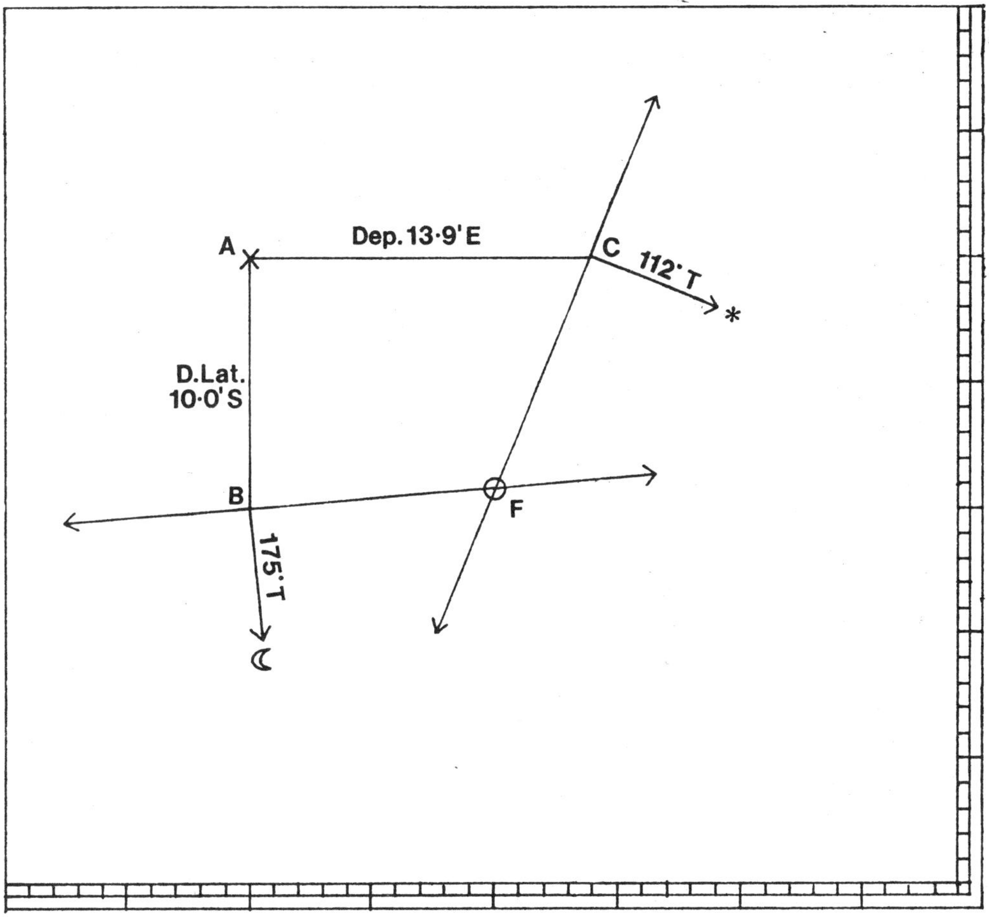
A
Dep. 13·9' E
C
112° T
D.Lat.
10·0' S
B
F
175° T

EX-MERIDIAN POSITION LINE. INTERCEPT METHOD.

Example IV.

(i) December 28, about 2320 ship time, D.R. position Lat. 13° 50′ N., Long. 86° 36′ E., the sextant altitude of Canopus near the meridian was 23° 06′ South of the observer, eye 21 feet, index error 1·4 on arc. G.M.T. 28d 17h 34m 07s. Find the intercept and the position line.

(ii) At G.M.T. 28d 17h 34m 42s the same observer found the sextant altitude of Regulus to be 27° 29·5′. Find the intercept and the direction of the position line, also fix the ship's position by combining both sights.

(i) Canopus.

L.H.A.=G.H.A. ♈ +S.H.A. +E. Long.

G.M.T. Dec. 28d 17h 34m 07s.

G.H.A. ♈ 28d 17h	351°	39·3′	Sext. Alt.	23°	06′
Incre. 34m 07s.	8	33·2	Index Error	—	1·4
G.H.A. ♈	000	12·5		23	04·6
S.H.A.*	264	13·8	Dip	—	4·4
G.H.A.*	264	26·3		23	00·2
Long. E.	+ 86	36	Corr.	—	2·3
L.H.A.*	351	02·3	True Alt.	22	57·9
				90	
Angle P	8	57·7			
			Zen. Dist. (Obs.)	67	02·1

Hav ZX=hav P sin PZ sin PZ +hav ($PZ \sim PX$)

or Hav ZX=hav P cos lat cos decl +hav ($L+D$)

and Sin az=sin P cos decl cosec ZX

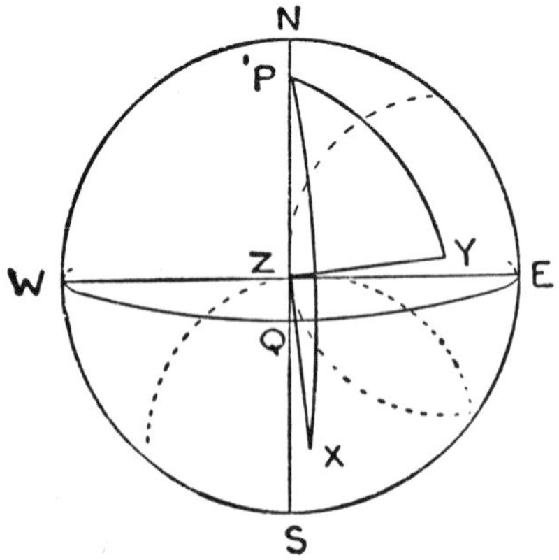

P	8°	57·7′	hav	7·78558	sin	9·19250
Lat	13	50 N.	cos	9·98722		
Decl.	52	40·5 S.	cos	9·78271	cos	9·78271
			hav	7·55551		
			nat hav	0·00360		
(L+ D)	66	30·5	nat hav	0·30069		
ZX	66	57·4	nat hav	0·30429	cosec	0·03612
Calc. zen. dist.	66°	57·4′			sin	9·01133
Obs. zen. dist.	67	02·1				
Intercept, away		4·7				Az. S. 5° 53·5′ E. P.L. 084°–264° T.

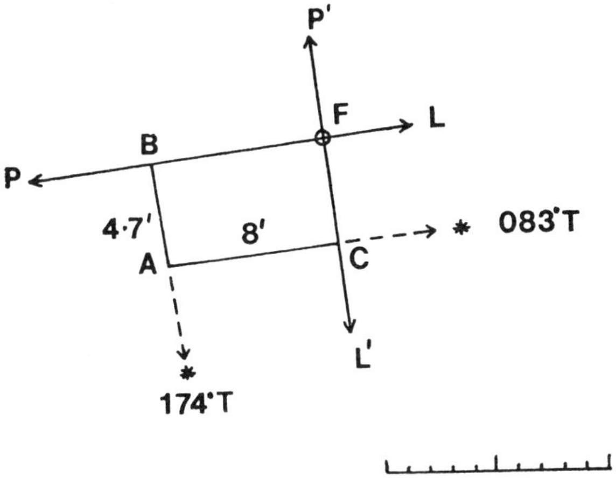
P'
F
L
B
P
4·7'
8'
083°T
A
C
L'
174°T

(ii) The observation of Regulus when worked up gives calculated zenith distance 62° 46·2′, observed zenith distance 62° 38·2′, Intercept 8′ towards azimuth 083° T., P.L. 173°—353° T., which the student may verify for himself. The combined position lines are then plotted as shown on the squared paper diagram.

A is the D.R. position

AB the intercept 4·7′ away from Canopus and *PL* the position line drawn through *B* the intercept terminal point.

AC the intercept 8′ towards Regulus and *P′L′* the position line drawn through *C*, the intercept terminal point.

F is the ship's position—

A	13° 50′ N.		86 36′ E.
A to *F*, D. Lat.	5·5 N.	Dep. 7 miles=D. Long.	7·3 E.
F	13 55·5 N.		86 43·3 E.

Ship's position Lat. 13° 55·5′ N., Long. 86° 43·3′ E.

A Forenoon Position Line combined with an Ex-Meridian Position Line. (Sun).

Example V.

(i) On December 24, at 0838 at ship in D.R. position, Lat. 24° 43′ N., Long. 72° 57′ W., sights were taken for a position line. The sextant altitude of the sun's lower limb was 20° 46′, index error 1·5′ off the arc, eye 48 feet, G.M.T. 24d 13h 27m 55s. Required a position through which to draw the position line.

(ii) At 1138 hours on the same day an ex-meridian sight was taken, the ship's course being 330° T., speed 11 knots since 0838 when the position was assumed to be Lat. 24° 51′ N., Long. 73° 07′ W. Find the new position for working up the ex-meridian sight and also transfer the position line.

(iii) On December 24, at 1138 hours, position by D.R. Lat. 25° 19·5′ N., Long. 73° 25′ W., the observed altitude of the sun's lower limb was 40° 11·5′ South of observer, no index error, eye 48 feet, G.M.T. 24d 16h 15m 14s. Required the sun's true azimuth and the intercept, also the ship's position as determined by the position lines.

First work up the morning sights for a position line as follows:—

(i) L.H.A. Sun=G.H.A. Sun—W. Long.

G.M.T. Dec. 24d 13h 27m 55s.

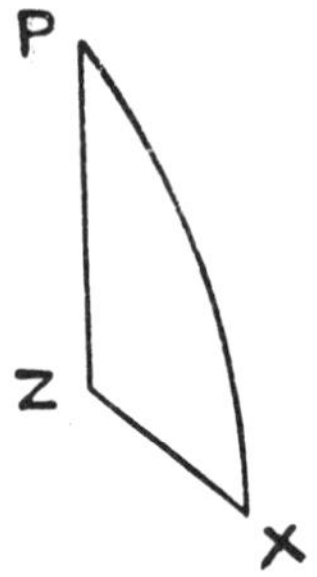

G.H.A. Sun 24d 13h	15° 07·6′	Sext. alt.	20° 46′
Incre. 27m 55s	6 58·8	Index error	+ 1·5
G.H.A.	22 06·4		20 47·5
Long. W.	− 72 57	Dip	− 6·7
L.H.A. Sun	309 09·4		20 40·8
		Corr.	+ 13·7
Angle *P*	50 50·6 E.	True Alt.	20 54·5
			90 00
		True zen. dist.	69 05·5

hav ZX = hav P cos lat cos decl + hav $(L \sim D)$

P	50° 50·6′	hav	9·26548	Azimuth from ABC Tables		
Lat	24 43 N.	cos	9·95827			
Decl	23 25·4 S.	cos	9·96265			
					A	·37 S.
		hav	9·18640		*B*	·56 S.
		nat hav	0·15360		*C*	·93 S.
$(L+D)$	48 08·4	nat hav	0·16634			
				Azimuth S. 49° E.		
Calc ZX	68 53·5	nat hav	0·31994	Position Line 041°—121° T.		
True ZX	69 05·5					
Intercept	12·0 away					

Plotting the Position Line and Run.

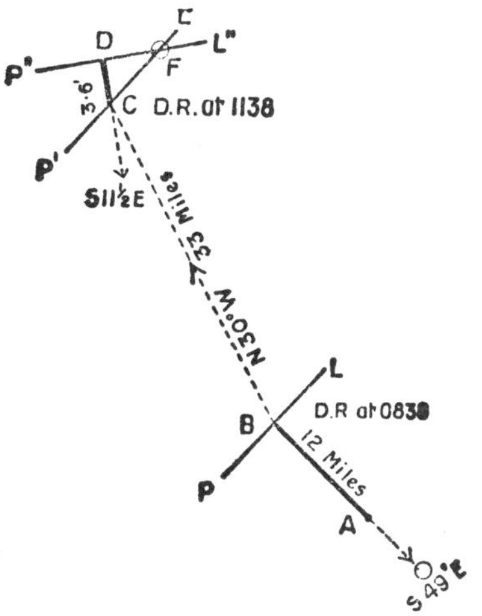

(1) Plot the D.R. position at 0838 and name it *A*. Lay off the sun's true azimuth S. 49° E. from *A* and measure the intercept *AB*, 12 miles away. Through *B* draw the position line *PL* at right angles to the sun's bearing. The ship is somewhere on *PL* and *B* is the new D.R. position of which we have to find the latitude and longitude.

(2) Lay off the run *BC* N. 30° W., 33 miles, and transfer *PL* parallel to itself through *C*. Find the latitude and longitude of *C*. The ship is somewhere on the transferred position line *P′L′* at the end of the run, not necessarily at *C* but somewhere near it.

(3) The ex-meridian sight is then worked up, using position *C* as D.R., and the intercept and sun's true azimuth found for a second position line.

(i) and (ii)

Traverse Table.

INTERCEPT *A B* AWAY
N. 49° W., 12 miles.
D. Lat. 7·9 N.
Dep. 9·1 W., D. Long. 10′ W.

RUN *B C*
N. 30 W., 33 miles
D. Lat. 28·6 N.
Dep. 16·5 W., D. Long. 18·2′ W.

	Lat.		Long.
D.R. Posn. *A*	24° 43′ N.		72° 57′ W.
D. Lat.	7·9 N.	D. Long.	10·0 W.
Posn. *B*	24 50·9 N.		73 07·0 W.
D. Lat.	28·6 N.	D. Long.	18·2 W.
Posn. *C*	25 19·5 N.		73 25·2 W.

(iii)

L.H.A. Sun = G.H.A. Sun − W. Long.

G.M.T. Dec. 24d 16h 15m 14s.

G.H.A. Sun 24d 16h	60° 06·6′	Obs. alt.	40° 11·5′
Incre. 15m 14s	3 48·5	Dip	− 6·7
G.H.A.	63 55·1		40 04·8
Long. W.	− 73 25·2	Corr.	+ 15·1
L.H.A. Sun	350 29·9	True alt.	40 19·9
			90 00
Angle *P*	9 30·1 (E)	True zen. dist.	49 40·1

hav ZX = hav P cos lat cos decl + hav ($L \sim D$)

P	9° 30·1′	hav 7·83630		
Lat.	25 19·5 N.	cos 9·95612		
Decl.	23 25·3 S.	cos 9·96265		
		hav 7·75507		
		nat hav 0·00569	Azimuth from ABC Tables	
($L+D$)	48 44·8	nat hav 0·17031	*A*	2·83 S.
Calc. zen. dist.	49° 36·5		*B*	2·62 S.
True zen. dist.	49 40·1	0·17600		
Intercept	3·6 away		*C*	5·45 S.

Position Line 078·5—258·5° T. True Azimuth S. 11·5° E.

The above figure is on a reduced scale to illustrate the charting up to position *C*, but the triangle *CDF* is too small to enable us to measure the **D. Lat** and **Dep.** between *C* and *F* from the figure. Indeed, in practice, it is better to work up position *C* by the Traverse Table and leave only the transferred position line and the position line at second observation to be laid off on a scale large enough to transfer *C* to *F* by measurement. We shall now do so on squared paper.

(1) Mark a spot *C* and through it draw *P′L′* the transferred position line N. 41° E., which is right angles to the azimuth S. 49° E. given in part (i).

(2) From *C* lay off the azimuth at ex-meridian observation, S. 11½° E., measure the intercept *CD*, 3·6 miles away, and draw the position line *P″L″* which cuts *P′L′* at *F* and gives us the position required.

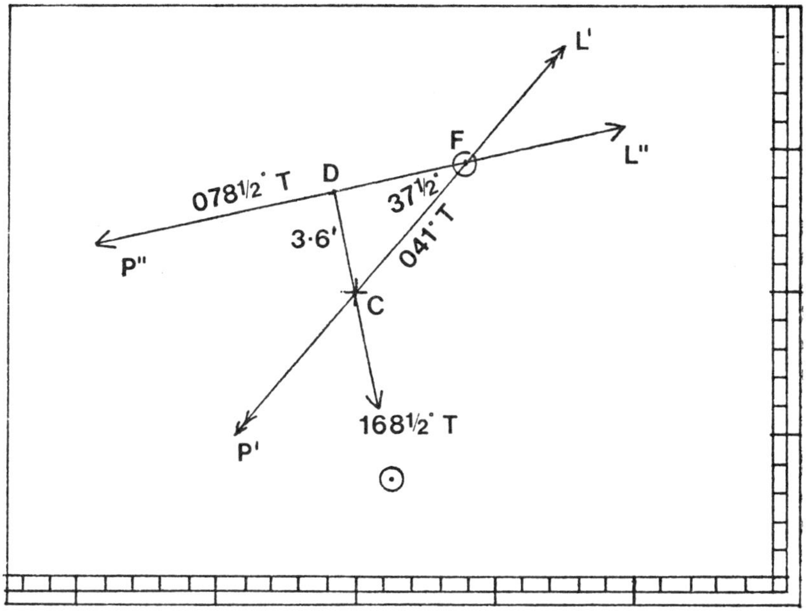

(3) Count the number of squares (two to the mile) vertically between *C* and *F*. These are 4·5 ∴ D. Lat. is 4·5′ N. Count the number of squares horizontally between *C* and *F*. There are 4, which is 4 miles of departure to be turned into **D. Long.**

Lat. 25°, Dep 4 miles gives D. Long. 4·4′ E.

C Lat.	25° 19·5′ N.	Long.	73° 25·2′ W.
D. Lat.	4·5 N.	D. Long.	4·4 E.
F Lat.	25 24·0 N.	Long.	73 20·8 W.

	Lat.	Long.
Answers.—At 1138, position by D.R.	25° 19·5′ N.	73° 25·2′ W.
,, ,, obs.	25 24·0 N.	73 20·8 W.

To Shift Position C to F by Traverse Table.

In triangle *CDF* ∠*D*=90°, *CD*=3·6 miles. ∠*F* is the angle between the two position lines, which is always equal to the difference between the two azimuths: in this case the azimuths are S. 49° E. and S. 11½° E. so ∠*F*=37½°.

The length of *CF* is found by inspection from the Traverse Tables; angle 37¼°, *CD* 3·6 miles as **Dep.** gives *CF* as 5·9 miles in the **Dist.** column.

The direction from *C* to *F* is along the position line *P′L′* at right angles to the azimuth at 1st observation; the azimuth was S. 49° E., therefore *CF* is N. 41° E., distance 5·9 miles to be treated as a course and distance and applied to *C*.

Traverse Table.
C TO *F*
∠*F* 37½°, *CD* 3·6 (Dep.)
Gives *CF* 5·9 (dist.)
CF is N. 41° E., 5·9 miles
D. Lat. 4·5 N.
Dep. 3·9 E., D. Long. 4·3 E.

	Lat.		Long.
C	25° 19·5′ N.		73° 25·2′ W.
CF, D. Lat.	4·5 N.	D. Long.	4·3 E.
F	25 24·0 N.		73 20·9 W.

At 1138, position by obs., Lat. 25° 24′ N., Long. 73° 21′ W.

DEGREE OF DEPENDENCE.

The position line at 1138 in the example we have worked is right. The trans, ferred position line (*P'L'*) is at least right in direction but it is drawn N. 30° W., 33- miles from *B* and the ship may not have made that course and distance exactly so that *P'L'* may be just a little ahead or behind where it should be. This however, is just one of the differences between theory and practice. The practical navigator learns to know from experience how much reliance to place upon the D.R., and during a three or four hour's straight run his reckoning is not likely to be more than a mile in error. The accuracy of the position by observation depends upon the ship making good the course and distance allowed between successive sights, and any error in this factor will probably be accentuated when the position lines cut each other at a small angle. The best "fix" is obtained when their angle of intersection is 90°, just as in a cross bearing "fix" of two shore objects.

Example VI.

(i) On June 15, at 1220 ship time, D.R. position Lat. 41° 21′ S., Long. 154° 16′E., the sextant altitude of sun's lower limb was 24° 49′ north of observer, index error 1′ off the arc, eye 32 feet, time by chronometer 2h 08m 16s which was 1m 10s fast on G.M.T. Required the position line and the Latitude and Longitude of the intercept terminal point.

(ii) On the same day at 1550 hours sights were taken for a position line, the ship having steamed 212° T., 10 knots during the $3\frac{1}{2}$ hours since 1220 when the ship's position was assumed to be Lat. 41° 30·2′ S., Long. 154° 17·2′ E. Required the D.R. position for working the sights.

(iii) At 1550 hours, given D.R. position Lat. 41° 59·9′ S., Long. 153° 52·7′ E., the sun's azimuth 304° T., intercept 8′ towards. Required the ship's position, assuming the sun's azimuth for the transferred position line to be 354° T.

The ex-meridian sight in question (i) gives:—

June	d	h	m	s
Chron.	15	02	08	16
Fast —			1	10
G.M.T.	15	02	07	06

G.H.A. sun 15d 02h	209°	57·4′
Incre. 07m 06s	1	46·5
G.H.A.	211	43·9
Long. E.	+154	16
L.H.A. Sun	5	59·9

Sext. Alt.	24°	49′
Index Error+		1
	24	50
Dip —		5·5
	24	44·5
Corr. +		13·9
True alt.	24	58·4
	90	00
True zen dist	65	01·6

P	5° 59·9′		hav	7·43736
Lat.	41 21 S.		cos	9·87546
Decl.	23 16·9 N.		cos	9·96312
			hav	7·27594
			nat hav	0·00188
(*L*+*D*)	64 37·9		nat hav	0·28579
Calc. zen. dist.	64° 52·3′		nat hav	0·28767
True zen. dist.	65 01·6			
Intercept	9·3′ away			

Azimuth Tables	
A	8·37 N.
B	4·12 N.
C	12·49 N.
Azimuth N. 6·1° W.=354° T.	
P.L. 084°—264° T.	

(i) In the figure, position *A* is the assumed position at the time of taking the ex-meridian sight at 1220; *AB* is the intercept 9·3 miles away, azimuth 354° T., and *B* is the D.R. position through which to draw the position line *PL.*

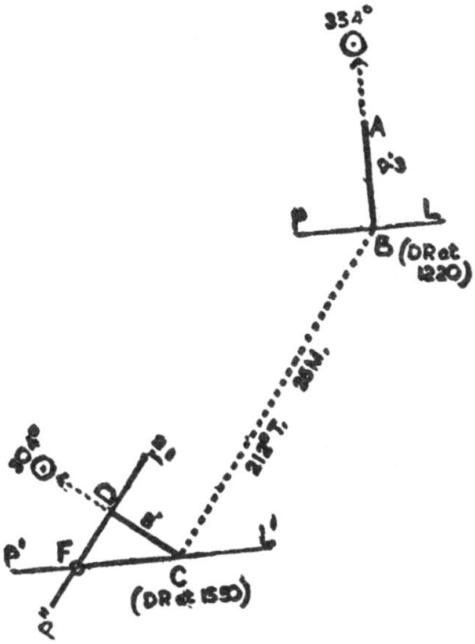

(ii) The run *BC*, 212° T., 35 miles, is laid off from *B*, and *C* now becomes the position at afternoon sights through which to draw the transferred position line *P′L′*. The ship is somewhere on this line at the end of the run.

Traverse Table.

		Lat.	Long.
(i) INTERCEPT *A B*			
174°, 9·3 miles	*A*	41° 21′ S.	154° 16′ E.
D. Lat. 9·2 S.	*A B*	9·2 S.	1·2E.
Dep. 1·0, D. Long. 1·3 E.			
	B	41 30·2 S.	154 17·2 E.
(ii) RUN *B C*	*B C*	29·7 S.	24·5 W.
212° W., 35 miles			
D. Lat. 29·7 S.	*C*	41 59·9 S.	153 52·7 E.
Dep. 18·5, D. Long. 24·5 W.			

(iii) The plotting of the position lines at *C* is best done on squared paper.

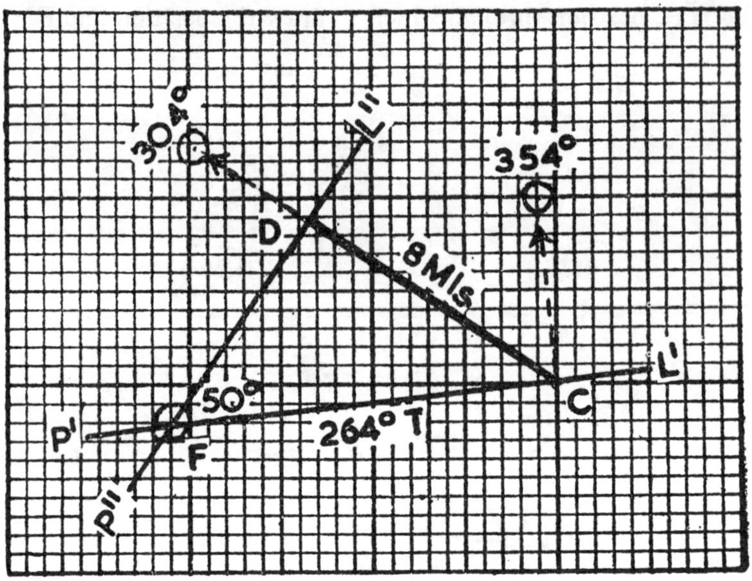

	Lat.		Long.
C	41° 59·9′ S.		153° 52·7′ E.
D. Lat.	1·1 S.	Dep. 10·5=D. Long.	14·0 W.
F	42 01 S.		153 38·7 E.

Answers—

(i) Azimuth 354° T., intercept 9·3 away, Lat. 41° 30·2′ S., Long. 154° 17·2′ E.
(ii) Lat. 41° 59·9′ S., Long. 153° 52·7′ E.
(iii) Lat. 42° 01′ S., Long. 153° 38·7′ E.

SUN
Examples for Exercise.

(1) A Forenoon Sight combined with an Ex-Meridian of the Sun.

(i) June 16, at 0935 ship's time, D.R. position 34° 48′ S., 163° 44′ E., sun's azimuth 036° T., intercept 10 miles towards. Required the position of the intercept terminal point for drawing in the position line.

(ii) On the same day at 1135 L.M.T. an ex-meridian sight was taken, ship's run 245° T., speed 12 knots since 0935. Required the latitude and longitude through which to draw the transferred position line.

(iii) June 16, about 1135 at ship, D.R. position 34° 50′ S., 163° 24·6′ E., sextant altitude of the sun's lower limb 31° 20′ North of observer, index error 1·4′ on the arc, eye 22 feet, time by chronometer 00h 42m 34s, which was fast 5m 21s on G.M.T. Required the sun's true azimuth and intercept, also the ship's position by observation as determined by position lines, assuming that the sun's true azimuth for the transferred position line to be 036° T.

(2) A Forenoon Position Line combined with an Ex-Meridian Position line.

(i) June 17, at 1000 L.M.T., D.R. position 48° 54′ N., 127° 52′ W., the azimuth of the sun was 127° T., intercept 10′ away. Required the latitude and longitude of the intercept terminal point through which to draw the position line.

(ii) On the same day about 1220 L.M.T., ex-meridian sights were taken, the ship having sailed 213° T., 40 miles since the morning sights were taken. Required the latitude and longitude through which to draw the transferred position line.

(iii) June 17, at 1220 ship's time, D.R. position 48° 26·5′ N., 128° 37·4′ W., observed altitude sun's lower limb 64° 23′, eye 24 feet, time by chronometer 20h 59m 44s which was 5m 16s fast on G.M.T. Required the intercept and the sun's true azimuth, also the position by observation, assuming the sun's azimuth for the transferred position line to be 127° T.

(3) An Ex-Meridian Position Line combined with an Afternoon Sight.

(i) June 22, about 1220 L.M.T., D.R. position 46° 01′ N., 35° 12′ E., sextant altitude sun's lower limb 66° 47′, index error 1·5′ off the arc, eye 30 feet, time by chronometer 22d 09h 50m 47s which was 12m 59s slow of G.M.T. Required the sun's true azimuth and intercept, also the latitude and longitude of the point through which to draw the position line.

(ii) On the same day at 1550 L.M.T., the ship having steamed 056° T., speed 12 knots since 1220, find the D.R. position for the transferred position line.

(iii) In the afternoon at 1550 the sun's azimuth was 265° T., intercept 5′ away, and D.R. position 46° 23·3′ N., 36° 01·7′ E. Required the ship's position by observation, assuming the sun's azimuth for the transferred position line to be 194° T.

(4) An Ex-Meridian Position Line combined with an Afternoon Position Line.

(i) January 7, about 10 minutes before the sun came to the meridian, D.R. position 56° 18′ S., 67° 22′ W., sextant altitude sun's lower limb 56° 01′ index error 1′ off the arc, eye 25 feet, G.M.T. 16h 25m 49s. Required the sun's true azimuth, intercept, and the latitude and longitude of the point through which to draw the position line.

(ii) From D.R. position 56° 07′ S., 67° 20·5′ W., ship steamed 145° T., 46 miles. Required the latitude and longitude for transferring the position line.

(iii) An afternoon sight worked for D.R. 56° 44·7′ S., 66° 32·4′ W., gave an intercept 8 miles away from sun's azimuth 296° T. Required the latitude and longitude of the observation position, assuming the azimuth of the sun for the transferred position line to be 004° T.

STARS, PLANETS, MOON.

Examples for Exercise

1. On January 10, D.R. position 14° 20′ S., 31° 22′ W., the observed altitude of the star Acrux near the meridian was 41° 34·5′, height of eye 26 feet, time by chronometer 07h 20m 10s which was 8m 06s slow of Mean time at Greenwich. Required the position line and the Latitude in which it crosses the D.R. Longitude.

2. (i) On June 17, D.R. position 48° 34′ N., 155° 30′ E., the observed altitude of the star Altair near the meridian was 50° 09′, height of eye 41 feet, G.M.T. 16d 16h 01m 26s. Required the position line and a position through which to draw it.

   (ii) If the ship steamed 035° T., 35 miles, find the Latitude and Longitude for the transferred position line.

3. (i) On October 12, D.R. position 36° 28′ N., 127° 41′ W., the sextant altitude of Sirius near the meridian was 36° 45′, index error 2′ on the arc, height of eye 45 feet, chronometer time 14h 06m 10s which was 11m 14s slow of G.M.T. Required the position line and the intercept terminal point.

   (ii) If the ship steamed 220° T. at 12 knots and current set 280° T., at 2 knots, find the position through which to draw the position line at the end of a 4 hours run.

4. June 19, D.R. position 38° 25′ S., 82° 16′ W., at G.M.T. 19d 11h 54m 09s the sextant altitude of Mars near the meridian was 50° 04′, index error 1′ off the arc, height of eye 40 feet. Find the latitude and the position line.

5. September 23, in D.R. Lat. 30° 20′ N., Long. 100° 50′ E., the sextant altitude of the moon's upper limb near the meridian was 46° 45′, index error 2′ off the arc, height of eye 30 feet, G.M.T. 23d 14h 14m 18s. Find the latitude in which the position line crosses the D.R. Longitude, also the position line.

6. January 8, in D.R. Lat. 36° 28′ S., Long. 127° 41′ W., the sextant altitude of the moon's lower limb was 42° 42′, index error 0·4′ on the arc, height of eye 52 feet, G.M.T. 8d 10h 25m 02s. Find the latitude in which the position line crosses the D.R. Longitude.

7. (i) June 16 at ship about 0650 hours L.M.T., D.R. Lat. 40° 06′ S., Long. 158° 11′ E., the observed altitude of Mars near the meridian was 49° 28′, height of eye 52 feet, when the chronometer which was 3m 12s fast of mean time at Greenwich showed 8h 10m 00s. Find the position line and a position through which it passes.

   (ii) The ship then steamed 140° C. (variation 14° E., deviation 6° E.) to counteract a current setting 090° M., drift 4 miles, the distance by log being 30 miles.

   (iii) At the end of the above run it was about 0840 hours L.M.T. on the 16th and the observed altitude of the sun's upper limb was then 11° 44′; and the same chronometer showed 10h 12m 12s. Find the ship's position at the time of the second observation.

8. (i) September 21, in D.R. Lat. 12° 18′ N., Long. 62° 12′ E., the sextant altitude of the moon's lower limb was 59° 41′, index error 1′ off the arc, height of eye 45 feet. G.M.T. 21d 15h 10m 10s. Find the position line and the latitude in which it crosses the D.R. Longitude.

(ii) If the ship then steamed 275° T., 45 miles, current setting 225° T., 6 miles, find a position through which to draw the transferred position line.

9. (i) September 23, D.R. position 30° 20′ N., 100° 50′ E., G.M.T. 23d 14h 14m 18s. the sextant altitude of the moon's upper limb near the meridian was 46° 45′, index error 2′ off the arc, height of eye 30 feet. Find the position line and the intercept terminal point.

(ii) Ship then ran 130° T., 48 miles, when the same sextant gave the altitude of the moon's upper limb 15° 13′, eye 30 feet, G.M.T. 23d 18h 20m 06s. Find the position at the time of taking the second observation.

10. (i) On September 23, G.M.T. 23d 03h 31m 50s, D.R. position 0° 24′ S., 26° 42′ W., the sextant altitude of star Achernar, near the meridian was 33° 05′, index error 1′ on the arc, height of eye 31 feet. Find the position line and a position through which to draw it.

(ii) At the same instant the observed altitude of Rigel was 39° 26′ (no index error), height of eye 31 feet. Find the intercept terminal point and plot the position lines to fix the ship's position.

11. (i) September 20, morning twilight at ship in D.R. position 56° 15′ N., 143° 25′ E., the observed altitude of Mars near the meridian was 51° 43′, height of eye 32 feet, a chronometer which had no error on mean time at Greenwich showed 6h 46m 40s. Find the position line and a position through which to draw it.

(ii) At the same instant the observed altitude of the star Pollux East of the meridian, was 43° 54′. Find the ship's position.

12. September 21, D.R. Lat. 36° 20′ N., Long. 18° 45′ W., G.M.T. 21d 19h 15m 09s, the observed altitude of Jupiter was 16° 47′, height of eye 60 feet.

At the same instant the observed altitude of the moon's upper limb was 33° 20′. Find the ship's position by combining both sights.

SIMULTANEOUS AND DOUBLE ALTITUDES

Examples for Exercise

1. January 9, in D.R. Lat. 32° 12′ S., Long. 80° 38′ E., G.M.T. 9d 14h 32m 12s, eye 35 feet, the observed altitude of Betelguese was 34° 34·5′ and Canopus was 50° 00′, both stars East of the meridian. Find the ship's position.

2. December 28, G.M.T. 20h 20m 06s, D.R. Lat. 11° 15′ N., Long. 32° 18′ W., height of eye 36 feet, the observed altitude of Fomalhaut was 42° 11·5′ and observed altitude of Deneb was 29° 26′, both stars West of the meridian. Find the ship's position.

3. June 15, about 0930 hrs. in D.R. Lat. 32° 04′ N., Long. 57° 18′ W., the observed altitude of Venus was 72° 45′ and the observed altitude of the Sun's lower limb was 55° 52′ when a chronometer which was slow 2m 06s on G.M.T. showed 1h 18m 01s. Height of eye 40 feet. Find the ship's position.

4. September 22, at ship, morning twilight, D.R. Lat. 42° 40′ N., Long. 171° 12′ E., chronometer showed 5h 53m 12s, no error, the observed altitude of Mars was 59° 48′, and when chronometer showed 5h 54m 20s the observed altitude of Procyon was 42° 01′, height of eye 55 feet. Find the ship's position.

5. September 26, about 0530 hrs. at ship in D.R. Lat. 13° 58′ N., Long. 78° 12′ W., the observed altitude of Sirius was 56° 33′ at G.M.T. 10h 42m 03s and the observed altitude of Mars was 62° 48′ at G.M.T. 10h 44m 00s, height of eye 30 feet. Find the ship's position.

6. December 21, at ship in D.R. Lat. 14° 03′ S., Long. 88° 05′ W., the observed altitude of the moon's L.L. was 61° 21′ and observed altitude of Adhara was 31° 21′, height of eye 45 feet G.M.T. 22d 02h 41m 43s. Find the ship's position.

7. September 23, G.M.T. 20h 07m 48s in D.R. Lat. 0° 24′ S., Long. 26° 42′ W., height of eye 39 feet, the observed altitude of Fomalhaut was 20° 34′ and the observed altitude of Saturn was 62° 05′. Find the ship's position.

8. December 28, at morning twilight, D.R. Lat. 35° 30′ S., Long. 80° 06′ W., a chronometer which was 2m 12s. slow on G.M.T. showed 9h 35m 50s., when the sextant altitude of Canopus was 42° 15′ and the sextant altitude of the moon's L.L. was 27° 50·5′. Index error 1′ off the arc, eye 40 feet. Find the ship's position.

9. January 8, in D.R. Lat. 25° 20′ N., Long. 155° 40′ E., the observed altitude of Capella was 47° 01′ at G.M.T. 08h 37m 04s, height of eye 28 feet. The ship then ran 100° T., 34 miles until G.M.T. 11h 37m 28s when the observed altitude of Capella was 69° 26′. Find the ship's position at second observation.

10. January 7, about 1400 hours at ship in D.R. Lat. 12° 04′ N., Long. 57° 18′ W., G.M.T. 7d 17h 54m 11s, the observed altitude of Venus was 64° 14′ and at the same time the observed altitude of the sun's L.L. was 44° 47′, eye 40 feet. Find the ship's position.

11. December 22, about 2015 hours at ship, D.R. Lat. 43° 08′ S., Long. 83° 25′ E., a chronometer showed 2h 49m 34s and was 8m 14s fast of G.M.T. when the observed altitude of Mars was 27° 53′ and at the same time Fomalhaut was observed to be 48° 05′, height of eye 30 feet. Find the ship's position.

12. September 20, in D.R. Lat. 24° 28′ S., Long. 156° 10′ E., about 1420 hours at ship, the observed altitude of Jupiter was 76° 27′, height of eye 45 feet, G.M.T. 20d 03h 58m 20.s Ship ran 060° T., 44 miles until G.M.T. 20d 07h 38m 06s when the observed altitude of Jupiter was 32° 33′. Find the ship's position at second observation.

13. June 24, in D.R. Lat. 46° 00′ N., Long. 22° 18′ W., the observed altitude of the moon's upper limb was 22° 02′, height of eye 60 feet, G.M.T. 24d 16h 15m 02s; the ship then ran 120° T., 48 miles until G.M.T. 24d 20h 49m 07s when the observed altitude of the moon's lower limb was 36° 28′. Find the ship's position at second observation.

14. January 10, in D.R. Lat. 35° 30′ N., Long. 151° 05′ W., the observed altitude of Spica was 41° 43′, G.M.T. 10d 15h 11m 27s, eye 65 feet; the ship then ran 230° T., 34 miles and the observed altitude of Regulus was then 28° 33′, G.M.T. 10d 17h 08m 56s. Find the ship's position at the second observation.

15. September 23, G.M.T. 23d 14h 10m 26s in D.R. Lat. 33° 40′ S., Long. 104° 12′ E., the observed altitude of the moon's L.L. was 67° 38′, near the meridian, height of eye 52 feet. At G.M.T. 23d 14h 10m 42s the observed altitude of Shaula was 45° 00′. Find the ship's position by combining the position lines.

CHAPTER VIII.

MISCELLANEOUS PROBLEMS ON PASSAGE FROM GLASGOW TO THE WEST INDIES

The following series of examples for exercise have been made up for a vessel on passage from the Clyde to Mona Passage. The day to day navigation has been arranged progressively to include virtually all the questions of practical application that have been referred to in the text, the consecutive problems being set so as to be as independent of each other as possible, although, in some cases where one question runs into the next, it may be necessary to read the preceding one in order to get the sense of a question.

1. SWINGING SHIP FOR COMPASS DEVIATION.

September 15.—The ship was swung for deviation at the Tail of the Bank, Firth of Clyde, Lat. 55° 58′ N., Long. 4° 50′ W., and the following bearings of the sun were observed. GHA 30° 57·5′; Declination 3° 00′ N.; Variation 16° W.

Ship's Head Compass	Compass Bearing Sun	G M T
000° C	236° C	1436
045°	247°	1440
090°	251°	1444
135°	249·5	1448
180°	243°	1452
225°	234°	1456
270°	233·5	1500
315°	238·5	1504

Find the true azimuth of the sun from ABC Tables, thence the deviation for each direction on which the ship's head was steadied, then draw a curve of deviations on squared paper.

Ans.—Dev. 1·5° E ; 8·3° W ; 11·5° W ; 1·3° W ; 8·7° E ; 10·3° E ; 6·3° E.

2. A TRAVERSE.

September 15.

At 1230 hrs. GMT, dropped pilot and took departure from off Cloch Point (Lat. 55° 56½′ N., Long. 4° 52·6′ W.) bearing 098° T., distant 5 cables, set course 210° C. (variation 16° W., deviation 6° E.), log at zero.

At 1300 hrs., altered course 201° C. (Var. 16° W., Dev. 4° E.), log 4½.

At 1359 hrs., Cumbrae Lighthouse (Lat. 55° 43·2′ N., Long. 4° 58′ W.) bore four points on the port bow, log 13¼, and at 1402 hrs. it was abeam, log 13¾.

Find the position on the beam bearing, the course and distance made good from the point of departure, and the set and drift, if any.

Ans.—Beam position Lat. 55° 43·3′ N., Long. 4° 58·9′ W.; course 193° T., distance 13·8 miles; set and drift nil.

3. BOW AND BEAM BEARING.

From the abeam position (Lat. 55° 43·3′ N., Long. 4° 58·9′ W.) off Cumbrae Lighthouse altered course 198° C. (error 14° W.) log 13¾.

At 1650 hrs. Ailsa Craig Lighthouse (Lat. 55° 15¼′ N., Long. 5° 06¼′ W.) was 45° on the starboard bow, log 39, and at 1710 hrs. it was abeam, log 42. Allow leeway 4° for a NW gale and a current setting 223° T. at 3 knots.

Find the beam position and the bearing and distance from Ailsa Craig at 1710 hrs.

Ans.—Beam position Lat. 55° 15′ N., Long. 5° 00·2′ W., bearing 094° T., distant 3·4 miles.

4. SET AND DRIFT OF CURRENT.

At 1710 hrs. in Lat. 55° 15′ N., Long. 5° 00¼′ W., with Ailsa Craig abeam, altered course 213° T., log 42, leeway 5°, wind N.W.

At 1857 hrs., altered course 185° T., log 58, ship making 5° leeway with wind on starboard beam.

At 2400 hrs., log 94.

Find the DR position and if the position by observation was Lat. 54° 19′ N., Long. 5° 06′ W.

Find the set and drift experienced since 1710 hrs.

Ans.—Set 144° T., drift 7·3 miles in 6·8 hrs. Rate 1·07 knots.

September 16.

5. D.R. POSITION AT NOON.

At 0000 hrs. (Lat. 54° 19′ N., Long. 5° 06′ W.) altered course 203° C. (Var. 18° W., Dev. 4° E.), log 94.

At noon, log 202.

Find the DR position at noon allowing for a current setting 144° T., 1·07 knots.

Ans.—Position at noon, Lat. 52° 21·7′ N., Long. 5° 21·6′ W.

6. POSITION BY BEARING.

At noon, log set at zero.

At 1600 hrs., log 40, Smalls Lighthouse (Lat. 51° 43′ N., Long. 5° 40′ W.) bore 110° T., distant 3 miles. Find the ship's position and the course and distance made good, and the average speed since noon.

Ans.—Position Lat. 51° 44′ N., 5° 44½′ W., course 201° T.; distance 40½ miles; speed 10·2 knots.

7. GREAT CIRCLE AND MERCATOR SAILINGS.

Find the initial course and distance on the great circle track from the position off the Smalls (Lat. 51° 44′ N., Long. 5° 44½′ W.) to Lat. 36° 00′ N., Long. 34° 33′ W. Find also the latitudes of points *C*, *D* and *E* every 10° along the track from Long. 10° W.

Find also the mercatorial courses and distances between these points along the track.

Ans.—Course 243° 44½′ T., distance 1546 miles; vertex Lat. 56° 15½′ N., Long. 26° 24′ E., Point *C*, Lat. 50° 18½′ N., Long. 10° W.; Point *D*, Lat. 45° 55′ N., Long. 20° W.; Point *E*, Lat. 39° 38½ N., Long. 30° W.

Mercator courses and distances 242° T., 182 miles; 237° T., 480 miles; 229·5° T., 580 miles; 224·5° T., 307 miles.

8. SETTING COURSE.

At 1600 hrs., in Lat. 51° 34′ N., Long. 5° 44½′ W., being of the Smalls, log. 40, course was set from chart 242° (Var. 18° W., Dev. 10° E.), allow 3° leeway for moderate N.W. gale. Find the compass course to steer.

Ans.—Course 253° C.

September 19.

9. FIX BY POSITION LINE AND RADIO BEARING.

At about 0745 hrs., ship time, log 198, having steered by compass 253° (error 8° W., allow leeway 3° for moderate N.W. gale) from Lat. 51° 44′ N., Long. 5° 44½′ W. when the log read 40, at 1600 hrs.

The observed altitude of the sun's lower limb was then 17° 42′, eye 50 feet. Chronometer showed 8h 19m 45s being fast 2m 30s on G.M.T. Required the position line.

At the same time Mizzen Head Ro D/F station (Lat. 51° 27′ N., Long. 9° 49·5′ W.) bore 330° (corrected). Find by plotting the ship's position by observation at 0745 hrs., and the set and drift of the current experienced since 1600 hrs. on the 18th, question 8.

(Chart S.W. Coast of Ireland No. 1123 used for plotting.)

Ans.—Fix at 0745 hrs., Lat. 50° 44′ N., Long. 9° 11′ W.; set 034° T., drift 17 miles. Intercept 4·2′ towards Az. 109·6° T.

10. STEAMING TIME AND SPEED.

Continued on the same course as above until noon the log then read **243.** Find the D.R. position at noon allowing for a similar set and drift of current; find also the course and distance made good, the steaming time and the average speed since 1600 hrs.

Ans.—September 19 noon, D.R. Lat. 50° 26·8′ N., Long. 10° 09′ W.; Course 245° T.; distance 183 miles; steaming time 20h 17m; average speed 8·9 knots.

11. MERCATOR SAILING.

Find the mercatorial course and distance from the above noon position (Lat. 50° 27′ N., Long. 10° 09′ W.) to the next point *D* on the great circle track (Lat. 45° 55′ N., Long. 20° 00′ W.). Find also the compass course (Var. 19° W., Dev. 12° E.); log reset at noon to zero.

Ans.—Course 236° T., 243° C.; distance 480 miles.

12. A RUN AND FIX BY A.M. SIGHTS AND LATITUDE AT NOON.

About 0810 hrs. ship time, steered 245° C. (error 7° W., leeway 2° for N.W. wind), log 190 miles from noon position of the 19th (Lat. 50° 27′ N., Long. 10° 09′ W.).

The observed altitude of the sun's L.L. was then 21° 32′, eye 50 feet. Chronometer time 9h 01m 11s which was 2m 32s fast on G.M.T. Required the D.R. position, also the position by sights.

At noon, log 226. Observed meridian altitude of the sun's L.L. 42° 38′, eye 50 feet. Find the noon position combining both sights; the set and drift if any; the course and distance made good during the day, the steaming time and the average speed.

Ans.—D.R. at 0810, Lat. 48° 41′ N., Long. 14° 11·5′ W.; intercept terminal point Lat. 48° 37·6′ N., Long. 14° 00′ W., intercept 8·3′ towards. Position line 024°—204° T.

September 20, noon Lat. 48° 24·3′ N., Long. 14° 40·5′ W.

Course made good 235° T., distance 214 miles. Set 068° T., drift 10 miles= 0·42 knots; steaming time 24h 18m, average speed 8·8 knots. Log reset to zero.

13. RUN AND FIX BY STAR POSITION LINE.

Steering 244° C. (error 8° W.) from noon position Lat. 48° 24′ N., Long. 14° 40·5′ W.

At 1530 hrs. log 34, altered course 241° C. to allow 3° leeway for S.E. wind and sea.

About 1835 hrs. apparent time at ship, log 68, D.R. Lat. 47° 46′ N., Long. 16° 05′ W., the observed altitude of Altair was 47° 53·5′ East of meridian, eye 50 feet, G.M.T. 19h 38m 05s.

Almost simultaneously the observed altitude of Alphecca was 49° 47·5′ West of meridian, G.M.T. 19h 38m 12s. Find the ship's position and the set and drift experienced since noon.

Ans.—Lat. 47° 39′ N., Long. 15° 57·6′ W.; set 145° T., drift 8·6 miles=1·3 knots.

September 21.

14. FIX BY STAR POSITION LINES.

From Lat. 47° 39′ N., Long. 15° 57·6′ W., on September 20, at 1835 hrs., when log showed 68, steered 241° C. (error 10° W., allow 3° for S.E. wind and sea.)

About 0715 hrs. apparent time at ship, log 200, the observed altitude of Betelguese was 47° 30·5′ West of the meridian, G.M.T. 21d 08h 26m 24s, and the observed altitude of Mirfak was 52° 22′ West of meridian, G.M.T. 21d 08h 27m 08s, eye 50 feet.

Find the ship's position, the set and drift of current, the course and distance made good, the steaming time and the average speed between evening and morning star sights.

Ans.—D.R. Lat. 46° 21·4′ N., Long. 18° 34′ W.; Obs. Lat. 46° 30·7′ N., Long. 18° 21·5′ W.; course 235° T.; dist. 120 miles; set 044° T., drift 13 miles; steaming time 12h 49m; average speed 9·3 knots.

September 20.

15. POSITION LINE BY PLANET.

About 0715 hrs. altered course 256° C. (error 9° W.), log 200.

About 1000 hrs. log 224, D.R. Lat. 46° 21′ N., Long. 18° 53·5′ W., the observed altitude of Jupiter was 11° 00′ East of the meridian, G.M.T. 21d 11h 15m 28s, eye 50 feet. Find the position line and the intercept terminal point.

Ans.—Lat. 46° 15·4′ N., Long. 18° 38·8′ W.; P.L. 029°—209° T.

16. SUN MERIDIAN ALTITUDE′ SET AND DRIFT.

At 1000 hrs. altered course 254° C. (error 7° W.) log 224; at noon log 240·5. The observed meridian altitude of the sun's lower limb was 44° 32′, eye 50 feet. Required the latitude by observation at noon and the longitude by combining the 1000 hrs. sight of Jupiter.

Required also, the course and distance made good, the set and drift of the current, and the steaming time and the average speed since noon on the 20th.

Ans.—Noon D.R. Lat. 46° 09′ N., Long. 19° 00·5′ W.; Obs. Lat. 46° 06·5′ N. Long. 19° 02·5′ W.; course 233° T.; 224 miles; set 105° T.; drift 23 miles steaming time 24h 17m; average speed 9·2 knots.

17. MERCATOR SAILING.

Find the course and distance by Mercator principle from the noon position on 21st, Lat. 46° 06·5′ N., Long. 19° 02·5′ W., to Point '*E*', Lat. 39° 39′ N., Long. 30° W., on the original great circle, as in Question 6.

Ans.—Course 231° 15′ T., distance 619 miles.

18. DAY'S WORK.

At noon (21st) course set 244° C. (error 12° W.), log reset at zero, noon position as above, Lat. 46° 06·5′ N., Long. 19° 02·5′ W.

September 22.

At noon (22nd), log 252. Find the estimated position allowing for a current estimated to have set East (mag.), Var. 20° W., drift 0·8 knots. Find also the estimated course and distance made good, the steaming time and the average speed.

Ans.—E.P. Lat. 43° 38′ N., Long. 23° 17·5′ W.; course 230·5° T.; distance 233·5 miles; steaming time 24h 17m; average speed 9·6 knots

September 23.

19. FIX BY STAR AND MARS.

At noon (22nd) the log was reset to zero. Continued on same course 244° C. (Error 12° W.) until 0400 hrs. on the 23rd, log 164. The observed altitude of Mars was 66° 19′ near the meridian, G.M.T. 23d 05h 45m 20s, eye 50 feet. At the same instant the observed altitude of Pollux was 44° 28′ East of meridian.

Find the ship's position and the set and drift experienced since the last observation at noon on 21st, Question 16, viz., Lat. 46° 06·5′ N., Long. 19° 02·5′ W.

Ans.—D.R. Lat. 41° 57′ N., Long. 26° 14′ W.; Obs. Lat. 42° 07′ N., Long. 26° 24′ W.; D.R. from 21st noon, Lat. 41° 50′ N., Long. 26° 38′ W.; set 031° T.; drift 19 miles in 40·5 hours.

20. FIX BY POSITION LINES OF SUN AND SATURN.

September 23, at 1600 hrs., having steamed 244° C. (Error 17° W.) for 130 miles from 0400 position (Lat. 42° 07′ N., Long. 26° 24′ W.) the observed altitude of the sun's L.L. was 22° 00′, G.M.T. 23d 17h 46m 56s, eye 50 feet.

About 1800 hours having steamed a further 22 miles on the same course, the observed altitude of Saturn was 26° 55′, G.M.T. 23d 19h 50m 42s. Find the ship's position at second observation.

Ans.—1st D.R. Lat. 40° 38·3′ N., Long. 28° 30′ W.; 2nd D.R. Lat. 40° 24′ N., Long. 28° 48′ W.; Sun azimuth 250° T., Intercept 2·7′ away. Saturn Azimuth 193° T., intercept 4·2′ towards Fix. Lat. 40° 19·5′ N., Long. 28° 46′ W.

September 24.

21. RIGEL AND SUN.

September 24, about 0612 hrs. in D.R. Lat. 38° 47′ N., Long. 30° 22′ W., the observed altitude of Rigel was 39° 42′, G.M.T. 24d 08h 22m 14s, height of eye 50 feet. The ship then steamed 232° T., for 31 miles through a current estimated to have set 130° T., at ¾ of a knot, until 0900 hrs. when the observed altitude of the sun's lower limb was 32° 51′, G.M.T. 24d 10h 53m 47s.

Find the ship's position at second observation.

Ans.—Obs. Lat. 38° 27·3′ N., Long. 30° 58·5′ W.

22. COURSE AND DISTANCE.

Find the true course and distance from Lat. 38° 27′ N., Long. 30° 58′ W., to Lat. 36° 00′ N., Long. 34° 33′ W.

Ans.—Course 229·5° T., distance 226 miles.

23. A GREAT CIRCLE.

Find the initial course and distance on the great circle track from Lat. 36° 00′ N., Long. 34° 33′ W., to a position in the Mona Passage, Lat. 18° 30′ N., Long. 67° 45′ W. Find also the position of the vertex and the latitudes of points every 10° from longitude 50° W.

Ans.—Course 248° T., distance 2043 miles; Vertex Lat. 41° 24′ N., Long. 00° 03′ W.; points 29° 34′ N., in 50° W.; 23° 49′ N. in 60° W.

September 25.

24. MARS AND SIRIUS.

September 25, about 0530 hrs., in D.R. Lat. 36° 00′ N., Long. 34° 33′ W., at G.M.T. 25d 07h 23m 48s the observed altitude of Mars West of the meridian was 61° 36′ and of Sirius 35° 24′ East of the meridian, eye 50 feet.

Required the position.

Ans.—Lat. 35° 56·5′ N., Long. 34° 39′ W.

25. VENUS—RUN—VENUS.

At 0530 hrs. on 25th in Lat. 35° 57′ N., Long. 34° 39′ W., altered course 281° C. (Var. 22° W., Dev. 11° W.) log 85.

At about 1110 hrs. on the same day log 147 G.M.T. 25d 13h 32m 55s the observed altitude of Venus was 59° 54′, eye 50 feet, and again about 1500 hours when log read 184 the observed altitude of Venus was 28° 09′, G.M.T. 25d 17h 23m 00s.

Required the ship's position at second observation.

Ans.—Mer. Alt. Lat. 35° 32·6′ N.; 2nd D.R. Lat. 35° 18·7′ N., Long. 36° 32′ W. Intercept 3·8′ away. Azimuth 254° T. Fix Lat. 35° 18·7′ N., Long. 36° 26′ W.

September 26.

26. ALTAIR AND MOON.

At about 1500 hrs. on 25th in Lat. 35° 19′ N., Long. 36° 26′ W., altered course 280° C. (Error 32° W.).

At about 1900 hrs. on 26th G.M.T. 26d 21h 36m 20s the observed altitude of Altair was 63° 47·5′ and at the same instant the observed altitude of the moon's lower limb was 19° 32′, eye 50 feet.

The distance by log from 1500 hrs. was 310 miles and the current was estimated to have set 150° T., 11 miles.

Find the ship's position.

Ans.—D.R. Lat. 33° 13·4′ N., Long. 42° 07′ W.; Obs. Lat. 33° 23·5′ N., Long. 42° 11′ W.

27. AN S.O.S. CALL.

At about 1900 hrs. L.M.T., September 26, in Lat. 33° 24′ N., Long. 42° 11′ W., altered course to 270° C. (Error 30° W.) log 90.

At G.M.T. 27d 03h 56m 38s log 185, an S.O.S. was picked up from a ship in Lat. 30° N., Long. 60° W., hove to in a hurricane moving north-easterly. Find the compass course to reach her, and if your maximum speed is 12·5 knots, find the L.A.T. of arrival (Var. 20° W., Dev. 8° E.).

Ans.—Course 271° C., distance 845 miles L.A.T. 29d 19h 40m.

September 27

28. POLE STAR.

At about 0105 hours L.M.T. In D.R. Lat. 32° 14′ N., Long. 43° 49′ W., G.M.T. 04h 00m 20s, the observed altitude of the Pole Star was 33° 16′, eye 50 feet. Find the latitude.

Ans.—Lat. 32° 12·2′ N., Long. 43° 49′ W., bearing North.

September 28.

29 MARS AND SIRIUS.

At about 0530 hours on September 28, having steamed 373 miles from last position (32° 12′ N., 43° 49′ W.) course 273° C. (Error 11° W.) G.M.T. 28d 08h 55m 56s. observed the altitude of Mars 68° 22′ West of meridian and of Sirius 40° 59′ East of meridian. Required the ship's position by D.R. and by observation.

Ans.—D.R. Lat. 31° 20′ N., Long. 51° 04′ W. Obs. Lat. 31° 16′ N., Long. 51° 09′ W.

30. D.R. POSITION AND MERCATOR SAILING.

Find the D.R, position at Noon on the 28th having steered 262° T., at speed 13 knots from 0530 position (31° 16′ N., 51° 09′ W.). Find also the course and distance from your noon position to the latest position of the vessel in distress (Lat. 30° 25′ N., Long. 59° 40′ W.) also the L.A.T. of arrival.

Ans.—Noon D.R. Lat. 31° 04·4′ N., Long. 52° 45′ W., 264° T., 360 miles; L.A.T. 29d 15h 13m.

31. RESETTING COURSE.

At G.M.T. 28d 22h 40m 00s, log 91, received radio message from disabled ship that no assistance would be required. Find the D.R. position and thence the course and distance to a position 310° T., 15 miles from Mona Island light (Lat. 18° 04′ N. Long. 67° 56′ W.), also the L.AT. of arrival assuming an average speed of 10·5 knots.

Ans.—D.R. Lat. 30° 55′ N., Long. 54° 31′ W.; course 225° T., distance 1081 miles, L.A.T. Oct. 3d 01h 13m.

September 29.

32. D.R. POSITIONS

At G.M.T. 28d 22h 50m, altered course 224° T., speed 10 knots, D.R. position Lat. 30° 55′ N., Long. 54° 33′ W.

About 0300 hours Sept. 29 at ship when G.M.T. was 29d 06h 55m, engines were stopped to effect repairs, wind East force 5, estimated leeway 1 knot, estimated current setting 300° T., 2·5 knots, for the period under repair.

About 1340 hours, at G.M.T. 29d 17h 25m, repairs completed, find the estimated position. Resumed full speed 10 knots, course 224° T. (Variation 17° W., Deviation 10° E.) give the compass course.

Ans.—Lat. 30° 07′ N., Long. 56° 07′ W.; course 231° C.

September 29.

33. STAR POSITION LINE.

At about 1750 hours, G.M.T. 29d 21h 40m 10s, having steamed 224° T., 45 miles since getting under way at 1340 hours when the D.R. position was Lat. 30° 07′ N., Long. 56° 07′ W., the observed altitude of Antares was 27° 19′, height of eye 50 feet, and bearing 216° by compass. Find the Longitude in which the position line crosses the D.R. Latitude, and also the error of the compass.

Ans.—D.R. Lat. 29° 34·6′ N., Long. 56° 33·4′ W.; P/L 120-300° T., error 6° W.

34. STARS MERIDIAN PASSAGE.

Find the Mean time at ship to the nearest minute when Vega will be on the meridian, September 29 in Lat. 29° 35′ N., Long. 56° 45′ W., and find the altitude to set on the sextant, no index error, eye 50 feet.

Ans.—18h 03m, 80° 57′ north of observer.

EXAMINATION PAPERS

PAST EXAMINATION PAPERS SUITABLE FOR CLASS 4—CLASS 1 MASTER MARINER EXAMINATION CANDIDATES.

PRACTICAL NAVIGATION.

Paper 1. (*3 hours.*)

1. From the following data find the position of the ship:—Date at ship September 26, about 23h in DR Lat. 38° 10′ N, Long. 32° 20′ W.

   1st Obs.—Observed altitude Moon's lower limb 51° 04′, near the meridian, chronometer time 1h 35m 14s.

   2nd Obs.—Observed altitude star Aldebaran 24° 58′ when the same chronometer showed 1h 36m 08s. HE 12 m. Chronometer slow on GMT 1m 06s.

2. Find the initial course and the distance on the great circle track from Cape Clear (Lat. 51° 25′ N, Long. 9° 29′ W) to New York (Lat. 40° 43′ N, Long. 74° 07′ W) and the latitude and longitude of the vertex.

3. On Sept. 19, about 2010h at ship in DR Lat. 47° 00′ N, Long. 37° 00′ W, the sextant altitude of Polaris was 47° 04′ when a chronometer which was 2m 40s slow of GMT showed 10h 40m 00s. HE 9 m. Index error 1′ on the arc. Required the Latitude and position line.

4. On September 26, in Lat. 49° 50′ N, Long. 22° 00′ W, the star Bellatrix bore 091° C. at 23h 46m LMT. Required the true azimuth and error of the compass. If the variation is 20° W, what will be the deviation

5. Find the LMT on January 11 when the star Menkar will be on the meridian in Lat. 19° 15′ S, Long. 47° 15′ W. State whether the star will be N or S of the observer.

Paper 2. (*3 hours.*)

1. On December 25, about 1400h at ship in DR Lat. 46° 28′ N, Long. 179° 58′ W, the observed altitude of the sun's lower limb was 14° 32′, height of eye 14 m, GMT 2h 05m 54s.

   Ship then ran 292° T. at 10 knots through a current estimated to set 180° T at 2 knots until next day at GMT 22h 06m 00s when a point of land in Lat. 47° 03′ N, Long. 176° 13′ E bore 050° T. Find the position of the ship at the time of 2nd observation.

2. From the following data find the Latitude and position line:—GMT September 19d 00h 55m 00s, observed altitude of Polaris 51° 22′, height of eye 7·7 m. DR Lat. 50° 40′ N, Long. 28° 30′ W.

3. On September 26 in DR Lat. 45° 20′ S, Long. 132° 15′ W, required the LMT of meridian passage of the Moon.

The sextant meridian altitude of the moon's upper limb was 42° 48′. index error 2′ on the arc, height of eye 36 feet. Find the latitude and the position line.

4. The position of the vertex of the great circle from Port Jackson to Valparaiso is Lat. 60° 54′ S., Long. 140° 40′ W. Calculate the latitudes at which the great circle track crosses the meridian of 159° 20′ E., and thence every 10th degree to destination, and plot the points on a mercator's chart.

Port Jackson—Lat. 33° 52′ S., Long. 151° 16′ E.

Valparaiso—Lat. 33° 01′ S., Long. 71° 52′ W.

Paper 3. (3 *hours.*)

1. On September 27, about 1600h L.M.T. in D.R. position Lat. 52° 20′ N., Long. 5° 40′ W., the observed altitude of the sun's lower limb was 15° 09′, height of eye 50 feet, G.M.T. 16h 21m 50s. The ship then ran 359° T., for 16 miles until G.M.T. 27d 18h 41m 47s when the observed altitude Altair was 44° 22′. Find the ship's position at the time of the 2nd observation.

2. On November 27, at 0830h A.T.S., it was ascertained that the ship was on the position line 050°-230° T., passing through D.R. Lat. 50° 10′ N., Long. 27° 30′ W., the log registering 97.

At 1140h A.T.S. an ex-meridian observation of the sun gave Latitude 50° 35′ N., log 25.

In the interval the ship steered 322° T., to allow for a current estimated to set 090° T., at 1 knot.

Find by calculation, or by plotting, the position of the ship at 1140 hours.

3. The ship's head by compass being in turn on the eight principal points of the compass, and the compass bearing of a distant object on each course being as stated below, find the magnetic bearing of the object and construct, on squared paper, a deviation curve.

Compass Course	N.	N.E.	E.	S.E.	S.	S.W.	W.	N.W.
Compass bearing	255°C.	244°C.	236°C.	230°C.	214°C.	211°C.	229°C.	251°C.

From the curve ascertain the deviation for the following compass courses. N.N.E., E.N.E., N.N.W., and W.N.W.

4. On September 20, the sextant meridian altitude of the star Rigel, south of the observer, was 44° 53′. Index error 2′ on the arc, height of eye 37 feet. Required the latitude.

Paper 4. (3 *hours.*)

1. On September 26, about 1815h at ship in D.R. Lat. 35° 20′ N., Long. 36° 22′ W., the observed altitude of the moon's lower limb, east of the meridian, was 12° 48′, G.M.T. 26d 20h 40m 28s. Height of eye 35 feet. Find the position line and a position through which to draw it.

(*b*) At G.M.T. 26d 20h 41m 04s, the observed altitude of Nunki near the meridian was 28° 22′, height of eye 35 feet. Find the Latitude and position line and also the position by observation.

2. On September 16, at 0100h a point of land in Lat. 44° 20′ S., Long. 100° 08′ E., bore 4 points on the port bow and at 0200 it was abeam; variation 6° E., deviation 0°. Ship steamed the following courses during the day:—

Noon to 2200: Course 160° C., Dev. 2° E., Speed 12·4 knots.
2200 to 0600: Course 152° C., Dev. 0°, Speed 12·25 knots.
0600 to Noon: Course 146° C., Dev. 2° E., Speed 13·5 knots.

Find the course and distance made good during the 24 hours and the position of the ship at noon next day.

3. On January 7, at 05h 04m 15s G.M.T. in D.R. Lat. 34° 50′ N., Long. 24° 16′ E., the sextant altitude of Polaris was 34° 00′, index error 2′ on the arc, height of eye 28 feet. Required the latitude and position line.

4. Required the difference between the great circle distance and the Mercator distance from Lat. 48° 25′ N., Long. 123° 15′ W., to Lat. 21° 20′ N., Long. 157° 48′ W.

Paper 5. (3 *hours.*)

1. On January 25 at L.M.T. 0920, log 17, the ship was found to be on a position line 160-340° T., D.R. position Lat. 49° 25′ N., Long. 179° 50′ E. Ship steamed on course 080° T., till noon, during which time a current was estimated to set 125° T., at 1·5 knots. At noon, log 45, the latitude by meridian altitude was found to be 49° 34′ N. Find by calculation, or by plotting, the ship's position at noon.

2. The position of the vertex of the great circle from *A* Lat. 48° 26′ N., Long. 161° 24′ E. to *B*, Lat. 37° 54′ N., Long. 142° 46′ W., being Lat. 48° 46·4′ N., Long. 170° 14·7′ E. Calculate the latitudes at which the great circle track crosses the meridian of 170° E. and each 10th degree of longitude from it towards the destination and plot the track on the Mercator's chart provided.

3. On September 28 G.M.T. 27d 23h 17m 42s in D.R. Lat. 40° 56′ S., the observed altitude of Procyon, east of the meridian was 23° 09′, height of eye 42 feet. Find the longitude in which the position line crosses the D.R. Latitude.

4. On September 20 L.M.T. 05h 04m 00s at ship in Longitude 152° 30′ E., the sextant altitude of Polaris was 36° 46′, index error 2′ on the arc, height of eye 39 feet. Required the latitude and position line.

Paper 6. (3 *hours.*)

1. Find the initial and final courses and the distance on the great circle track from Sydney (Lat. 33° 52′ S., Long. 151° 18′ E.) to San Francisco (Lat. 37° 42′ N., Long. 123° W.) and the position of the vertex of the great circle.

2. Find the true course and distance by use of meridional parts between (*A*) Lat. 43° 08′ N., Long. 5° 56′ E., and (*B*) Lat. 39° 29′ N., Long. 0° 24′ W.

3. At midnight the ship bore 300° T., distant 20 miles from Lundy Island Light in Lat. 51° 10′ N., Long. 4° 40′ W. She then sailed 168° T., 35 miles, 230° T., 68 miles, and a current set 045° T., 15 miles. Find the position arrived at.

4. Cape Amber (Lat. 11° 57′ S., Long. 49° 18′ E.) bore 248° C., at 1800 hours, error 8° W., distant 20 miles. Stellar obs. at midnight placed the ship in Lat. 12° 34′ S., Long. 50° 12′ E.

Noon to 2000: Course 101° C., Dev. 3° W., Var. 5° W., Speed 10 knots.
2000 to noon: Course 156° C., Dev. 5° W., Var. 5° W., Speed 9·5 knots.

Find the set and drift experienced between 1800 and midnight, and allowing for a similar set and drift throughout the day, find the course and distance made good from noon to noon, also the position at noon.

Paper 7. (3 *hours.*)

1. On January 8, D.R. Lat. 46° 20′ N., Long. 19° 15′ W., find the L.A.T. of meridian passage of Venus, and compute the altitude to set on the sextant, index error 1′ on the arc, height of eye 36 feet. If the sextant meridian altitude was observed to be 30° 18′, find the latitude.

2. On January 5, in D.R. Lat. 31° 35′ S., Long. 110° 08′ E., the obs. mer. alt, of the moon's upper limb was 39° 43′, height of eye 35 feet. Find the latitude and position line.

3. On September 19, at L.M.T. 05h 22m in D.R. Lat. 22° 00′ N., Long. 65° 48′ W., the sextant altitude of the Pole star was 22° 41′, index error 1′ off the arc, height of eye 36 feet. Required the latitude and position line.

4. On October 30, in D.R. Lat. 51° 10′ S., Long. 00° 20′ W., at 1510 A.T.S., log 83, the ship was found by observation to be on a position line 063°-243° T. At 21h 55m A.T.S. an ex-meridian observation of the moon gave Lat. 51° 50′ S., log 37. Ship steamed 130° T., between observations. Find by calculation or by plotting, the position of the ship at 21h 55m.

5. On December 24, in D.R. Lat. 12° 20′ S., Long. 45° 00′ W., the following observations were taken:

(*a*) Observed altitude Jupiter 34° 08′ east of meridian, G.M.T. 24d 08h 20m 10s.

(*b*) Observed altitude Denebola near the meridian 62° 38′, G.M.T. 24d 08h 21m 20s, height of eye 40 feet.

Find the ship's position by means of position lines.

Paper 8. (3 *hours.*)

1. On September 25, in D.R. Lat. 16° 50′ N., Long. 113° 48′ E., the sextant meridian altitude of the moon's upper limb was 67° 31′, index error 2′ off the arc, height of eye 25 feet. Required the latitude.

2. On January 12, in D.R. Lat. 59° 18′ N., Long. 85° 30′ W., find the G.M.T. of transit of Betelguese, and compute the altitude to set on a sextant having an index error of 2′ off the arc; eye 40 feet.

3. On September 27, in D.R. Lat. 48° 52′ S., Long. 45° 10′ E., the observed altitude of the sun's upper limb was 28° 38′ when the chronometer, which was 5m 18s slow on G.M.T. showed 5h 39m 32s. The ship then ran 320° T., 46 miles until noon, when the observed altitude of the sun's lower limb was 43° 07′, height of eye 35 feet. Find the ship's position at noon.

4. Steering 107° C. (error 17° W.), Portland Bill (Lat. 50° 31′ N., Long. 2° 27′ W.) bore four points on the port bow and after steaming 9 miles the Bill was abeam. Find the position of the ship on the beam bearing making allowance for a current setting 225° T., 2 miles, and leeway for a N.W. gale 10°.

5. If the position of the vertex of the great circle from Tristan da Cunna to Java Head is Lat. 42° 37·3′ S., Long. 22° 38·6′ E., calculate the latitudes of the points where the great circle crosses the meridians of 10° W. and each 10th degree from it to the destination, and plot the points on a Mercator's chart.
Tristan da Cunha. Lat. 37° 02′ S., Long. 12° 17′ W.
Java Head. Lat. 6° 47′ S., Long. 105° 13′ E.

Paper 9.—Chartwork.

3 hours.

English Channel Practice Chart.

1. From a position with Bishop Rock bearing 300° T., and Lizard Light bearing 045° M., find the courses and distances to a position 180° T., 6 miles, from Helwick Shoal Lightvessel. Alter course when midway between Longships and Sevenstones Lightvessel, and pass 12 miles off Lundy Island North Light, the deviations being as follows:—Deviation on 1st course 3° E. On 2nd course Longships Light changed from white to red bearing 188° Compass. On 3rd course the Pole Star bore 353° C.

2. After steaming for 3 hours at 10 knots on the 2nd course the Longships Light changed from white to red and Trevose Head Light bore 103° C. Find the set and drift experienced and the Latitude and Longitude of both positions.

3. Lundy Island North Light bore 055° C., and after steaming $2\frac{1}{2}$ hours at 10 knots it bore 121° C. Find the ship's position on the second bearing making allowance for the set and drift found in Question 2.

4. Find the course by compass to counteract the tide setting W.S.W. Magnetic, 3 knots, from a position 290° T., from Bishop Rock to a position 225° T. from the Smalls so as just to skirt the 50 fathom line. If the ship is logging 12 knots how long will she take to reach the final position? (Smalls Lt. 51° 43′ N., 5° 40′ W.)

Paper 10.—Practical Navigation.

3 hours.

1. On June 24, in D.R. Lat. 46° 22′ N., Long. 24° 20′ W., G.M.T. 24d 16h 16m 20s, the sextant altitude of the Moon's upper limb was 20° 48′ east of the meridian; index error 1′ on the arc; height of eye 55 feet. Following this observation

the ship ran 309° T., for 42 miles when a second altitude of the moon's lower limb was 36° 05′ west of the meridian at G.M.T. 24d 20h 50m 15s. Find the ship's position at the second observation.

2. Find the true course and distance by Mercator's Sailing from Cape Palmas, Lat. 4° 24′ N., Long. 7° 46′ W., to St. Paul de Loando, Lat. 8° 48′ S., Long. 13° 08′ E.

3. On January 11, in D.R. Lat. 42° 10′ N., Long. 157° 20′ W., find the local mean time of meridian passage of Jupiter and compute the sextant meridian altitude; index error 1·4′ on the arc; height of eye 30 feet. If the sextant meridian altitude was found to be 37° 47′ what was the latitude of the observer?

4. A lighthouse in Lat. 47° 19′ N., Long. 3° 14′ W., bore 45° on the port bow at 1735, log 24. At 1935 it was abeam, log 42. The course was 120° C., error 10° E., wind South force 7, leeway 5°, tide set 250° T., at 3 knots.
Find the ship's position on the beam bearings.

5. On December 27, in D.R. lat. 25° 20′ N., Long. 73° 26′ W., the sextant altitude of the sun's lower limb, near the meridian, was 40° 24′; index error 3′ off the arc; height of eye 38 feet; G.M.T. 27d 16h 21m 15s. Required the latitude and position line.

Paper 11.—Practical Navigation.

3 hours.

1. On September 19, in D.R. Lat. 10° 00′ S., Long. 60° 00′ E., required the time of transit of Capella to the nearest minute of G.M.T., also the sextant meridian altitude if index error is 1′ off the arc, and height of eye is 40 feet.

2. On September 29, Chronometer B was 2m 11s slow of A, and C was 4m 11s fast of A. Again on October 11, B was 2m 14s slow of A, and C was 4m 30s fast of A. If the latter's daily rate was 1·6s gaining, find the daily rates of B and C.

3. The ship took her departure from Lat. 47° 30′ N., Long. 179° 25′ W., at noon and steered the following courses: Noon to midnight—Course 176° C., 16·5 knots, Dev. 3° E., Var. 15° E. Midnight to noon—Course 202° C., 15·3 knots, Dev. 5° E., Var. 15° E.

At 2200 hours her position by observation was Lat. 44° 55′ N., Long. 180° E. Find the set and drift of the current. Find also the ship's position at noon the next day and the course and distance made good during the twenty-four hours, making due allowance for the current just found.

4. On January 10, in D.R. Lat. 11° 25′ N., Long. 37° 42′ W., the observed altitude of Fomalhaut, west of the meridian was 33° 46′, G.M.T. 20h 45m 48s; and the observed altitude of Deneb, west of the meridian was 19° 41′, G.M.T. 20h 46m 36s. Height of eye 40 feet. Required the position lines and the ship's position.

5. June 22, in Lat. 50° N., Long. 10° W., the following bearings were taken of the sun.

Ship's head Compass	Compass Bearing	A.T.S.	Ship's head Compass	Compass Bearing	A.T.S.
000° C	109°	0820	180° C	089°	0836
045°	104°	0824	225°	089°	0840
090°	102°	0828	270°	099°	0844
135°	096°	0832	315°	101°	0848

Variation 6° E. Draw a curve on squared paper and from it convert 040° T. into the equivalent compass course.

Paper 12.—Practical Navigation.

3 *hours.*

1. At 2000 hours, on February 17, Finisterre Light, in Lat. 42° 53′ N., Long. 9° 15′ W., was on the starboard beam distant 10 miles. Ship steamed 030° C. (Var. 20° W., Dev. 10° E.) until noon next day; speed 9 knots. Find the position of the ship at noon on February 18.

2. On September 23, in D.R. Lat. 42° 00′ N., Long. 52° 00′ W., the sextant meridian altitude of the moon's lower limb was 35° 53′; index error 2′ off the arc, height of eye 28 feet. Required the latitude.

3. Find the initial and final courses, also the distance from A, Lat. 36° 35′ N., Long. 160° W., to B, in Lat. 26° 25′ S., Long. 80° W. Give also the position of the vertex and of successive points every 10° on longitude apart from 150° W.

4. On January 7, in D.R. Lat. 45° 10′ N., Long. 20° 44′ W., the star Capella bore 100° C., G.M.T. 7d 21h 20m 00s. Required the true azimuth and the deviation for the direction of the ship's head if the variation was 18° W.

5. On September 24, in D.R. Lat. 42° 17′ S., Long. 76° 43′ E., the following observations were taken:

    (*a*) Sextant altitude of moon's lower limb, east of the meridian, 38° 30′; G.M.T. 24d 13h 23m 22s.

    (*b*) Sextant altitude of star Rasalhague, west of the meridian, 32° 55·5′; G.M.T. 24d 13h 24m 48s; index error 1·8′ on the arc; height of eye 37 feet.

    Find the ship's position by means of position lines.

Paper 13.—Practical Navigation.

3 *hours.*

1. On January 12, at evening twilight, D.R. Lat. 46° 00′ N., Long. 37° 26′ W., the sextant altitude of Polaris was 46° 57′; index error 1′ on the arc; height of eye 30 feet; the chronometer which was 1m 42s fast on G.M.T. showed 7h 43m 15s. Required the latitude and position line.

2. Thursday at noon, a point of land in Lat. 47° 30′ N., Long. 179° 25′ W., bore 322° T., distant 12 miles. During the next twenty-four hours ship steered 210° C. (Dev. nil): at 11 knots. Variation during the first 10 hours 12° E., and for the remainder of the day 10° E.

Friday at 0800 position by observation Lat. 44° 40′ N., Long. 177° 00′ E. Find current for the 24 hours, the course and distance made good and the position at noon on Friday.

3. Find the G.M.T., to the nearest minute, when Dubhe will transit the observer's meridian on June 15, in Lat. 40° N., Long. 40° 10′ W.

4. On January 6, in D.R. Lat. 35° 20′ S., Long. 47° 30′ E., the observed meridian altitude of the moon's upper limb was 36° 24′, height of eye 50 feet. Required the latitude.

5. On January 10, in D.R. Lat. 26° 15′ N., Long. 37° 42′ W., the following observations were taken:

(*a*) G.M.T. 10d 20h 43m 56s; observed altitude Capella 42° 29′.

(*b*) G.M.T. 10d 20h 45m 48s; observed altitude Deneb 29° 40′.

Height of eye 40 feet.

Find the ship's position by means of position lines.

Paper 14.—Practical Navigation.

1. Having taken the following compass bearings of a distant object, find its magnetic bearing and thence the deviations on those points on which the bearings were observed.

Ship's head Compass	Compass Bearing	Ship's head Compass	Compass Bearing
000° C	270° C	180° C	248° C
045°	256°	225°	256°
090°	244°	270°	266°
135°	241°	315°	270°

Find the magnetic bearing of the object and construct a curve of deviations on squared paper.

Ascertain from the curve the deviations on the following courses:

Compass Courses 045° C, 098° C, 200° C, 312° C.

Magnetic Courses 011° M, 087° M, 217° M, 337° M.

2. On September 21, in D.R. Lat. 33° 00′ N., Long. 24° 16′ E., the sextant altitude of Polaris was 33° 48′; index error 1·4′ on the arc; height of eye 8·3 m.; G.M.T. 21d 03h 43m 24s. Required the latitude.

3. On December 24, in D.R. Lat. 23° 50′ N., Long. 32° 20′ W., G.M.T. 24d 22h 26m 00s, the moon bore 106° C., variation 20° W. Required the true azimuth and the deviation for the direction of the ship's head.

4. Find by calculation on Mercator's principle the true course and distance from *A* 20° 00′ S., 178° 00′ W., to *B* 39° 52′ S., 147° 45′ E.

5. On January 12, in D.R. Lat. 30° 10′ S., Long. 17° 40′ W., height of eye 50 feet, the following observations were taken:—

(*a*) G.M.T. 12d 05h 54m 12s. observed altitude Acrux near the meridian 57° 18′.

(*b*) G.M.T. 12d 05h 55m 40s, observed altitude Antares east of the meridian 34° 14′.

Find the ship's position.

Paper 15.—Practical Navigation.

3 hours.

1. May 1, at 0130 hours in Lat. 49° 30′ N., Long. 4° 50′ W., course set 220° C. (Dev. 10° E., Var. 15° W.) log reading 142. 0400 hrs. altered course to North (Dev. 7° W.) log 162. 0730 hrs. altered course to 220° C., log 180. Course maintained until 1200 hrs. log 216.

   Leeway 5° throughout for a westerly wind. Current from 0130 to 0600 hrs. 080° T., 2 knots, and thereafter till noon, 260° T., 2·5 knots. Find D.R. position at noon.

2. January 7, morning twilight at ship in D.R. Lat. 32° 46′ N., Long. 151° 12′ W., the observed altitude of Jupiter near the meridian was 46° 42′ when a chronometer which was correct on G.M.T. showed 5h 17m 10s; height of eye 45 feet. Required the position line and the latitude in which it crosses the D.R. longitude.

3. On October 20, Chronometer *A* was 32m 19s fast of G.M.T. and losing 2·5s daily. On the same date *B* was 1m 03s slow of *A*. If on November 30, *B* was 17s fast of *A*, find *B*'s daily rate.

4. June 15, at 0530 L.A.T., in D.R. Lat. 17° 25′ N., Long. 62° 42′ E., the sun rose bearing 070° C. Find the true amplitude and the deviation for the direction of the ship's head if the variation was 3° E.

5. January 7, D.R. Lat. 39° 03′ S., Long. 41° 45′ E., cloudy weather, at G.M.T. 7d 17h 39m 15s the obs, alt, of a bright star whose identity was unknown was 42° 52′, bearing 075·5° T. Height of eye 45 feet. Establish the identity of the star.

CHAPTER IX.

MARINE SURVEYING.

Marine surveying is the art of portraying on paper in the form of a chart or plan the physical characteristics of any bay, harbour, or other locality of which the survey is to be made.

While the position and details of permanent land marks near the coastline form an important part of the work of marine surveying, it more particularly deals with the bed of the sea, shoal patches, islands, rocks, and any other features there may be, together with the set of the tides, depth of water, and nature of the bottom.

The first part of the work is the selection of the site and the laying off of a suitable base line on the land, by means of which the positions of other objects may be ascertained. The base line should be situated on a level surface if possible, and its two ends marked by definite fixed objects. If no desirable objects are already in position—that is, if the existing objects are not suitable—the purpose can be served by erecting two flag poles in convenient places, the distance between them being very accurately measured. In modern surveys this is usually carried out by Electronic Distance Measure (EDM). What that distance should be would depend on the area of the survey, but a length should be chosen with a view to saving labour in the future calculations which will have to be made. The true direction of this base line will also have to be found with great accuracy. The exact latitude and longitude of the marks at each end of it will have to be carefully calculated.

All the permanent landmarks which do not naturally offer a conspicuous point for observation should be marked, possibly with whitewash. The same process should also be adopted in the case of rocks and islands, a flag stuck firmly in the ground being useful on a level surface.

The high-water mark should then be carefully denoted by means of a succession of flags placed along it, or by otherwise marking it, the flags being most closely spaced where the greatest curvature or irregularity occurs in the beach or shore, this being done to facilitate the correct drawing in of the coastline.

Having marked (where necessary) all the objects which are to be shown on the chart or plan, it is advisable to give each one an appropriate name, as this will greatly assist the memory during the progress of the work.

The next thing to be done is to carefully ascertain the position of all these objects so that they may be denoted on the chart or plan in their correct places. This is done by a process called "triangulation". Each end of the base line is made a station for observations, and from these stations angles are measured between the base line and the objects to be fixed, the angles being very carefully recorded. These stations are called "triangulation stations". By means of the angles which have been measured triangles are formed, the base line forming one side of the triangle. As the position and true direction of the base line are known, also its exact length, the positions of the other objects can be easily deduced from it. Triangulation stations are also chosen in other known positions from which angles are measured and bearings taken to find the exact position of other objects and to verify the position of objects which have been previously fixed. A new base line may be found when a new triangulation station is established.

Angles are also measured and bearings taken from the object whose position is required to be fixed.

Having obtained sets of triangles, the next performance is to calculate the length of the sides of the triangles where necessary, and lay off the exact position of the base line (or lines) on cartridge paper.

In modern surveys the lengths of the sides of the triangles are often measured direct by EDM and in these cases the process is known as, "trilateration".

The first point chosen is then plotted off on its true bearings from both ends of the base line, its position being checked by means of the angles previously measured and recorded, also by the calculated length of the side of the triangle thus formed. The positions of other objects are then fixed by means of angles which have been measured between them and other known marks, the sides of the triangles which have been formed again providing a useful check. When all the triangles are laid off they should agree in fixing the position of each object as being in one definite place. If there are any discrepancies, an error has been made. It should be investigated and put right so that there shall be no doubt that the exact positions have been determined.

The coastline must be put in by means of plotting off the positions of the flags which were stuck in the ground along the high-water mark. A line drawn through them will be a good representation of it. The nature of the shore should be carefully noted, and sketches should be made of any important or unusual features. An aerial photographic survey would be helpful when drawing in the contour of the coast.

The set and drift of the tides, the depth of water, and the nature of the bottom form one of the most important parts of the survey, the value of the chart or plan depending greatly on their accuracy. Where echo sounders are not used the leadline must be carefully marked and, for the older non-metric charts, feet are the best unit of measurement up to the first five fathoms. The whole of the soundings being taken in feet if the water is not very deep. In the case of surveys for metric charts the metric leadline is used and soundings are taken in metres and decimetres. The soundings should be marked, as nearly as possible, in parallel straight lines, this being done by choosing or placing two fixed objects on shore and keeping the boat in line with them as the casts are taken. The exact time of each cast must also be taken so that the soundings may be reduced to chart datum when its level is decided upon. The distance between the position of one sounding and the next one to it will depend upon the locality of the place, the depth of water, the nature of the bottom, and the scale to which the chart or plan is to be constructed. Should a shoal patch be discovered, care must be taken to fix its position very accurately. This is best done by means of horizontal angles between objects ashore, which angles must be carefully observed from the boat.

Today most surveys are obtained by the use of echo sounders which, for accurate results, should first be calibrated by means of a "bar check" before survey commences. The other precautions and procedures as related to leadline surveys still apply.

A "tide pole" should be erected in a suitable and sheltered position which is always accessible to the sea, so that it may show the correct level of the water at all times. It should be graduated to spaces of, say, 1 decimetre, and very clearly marked

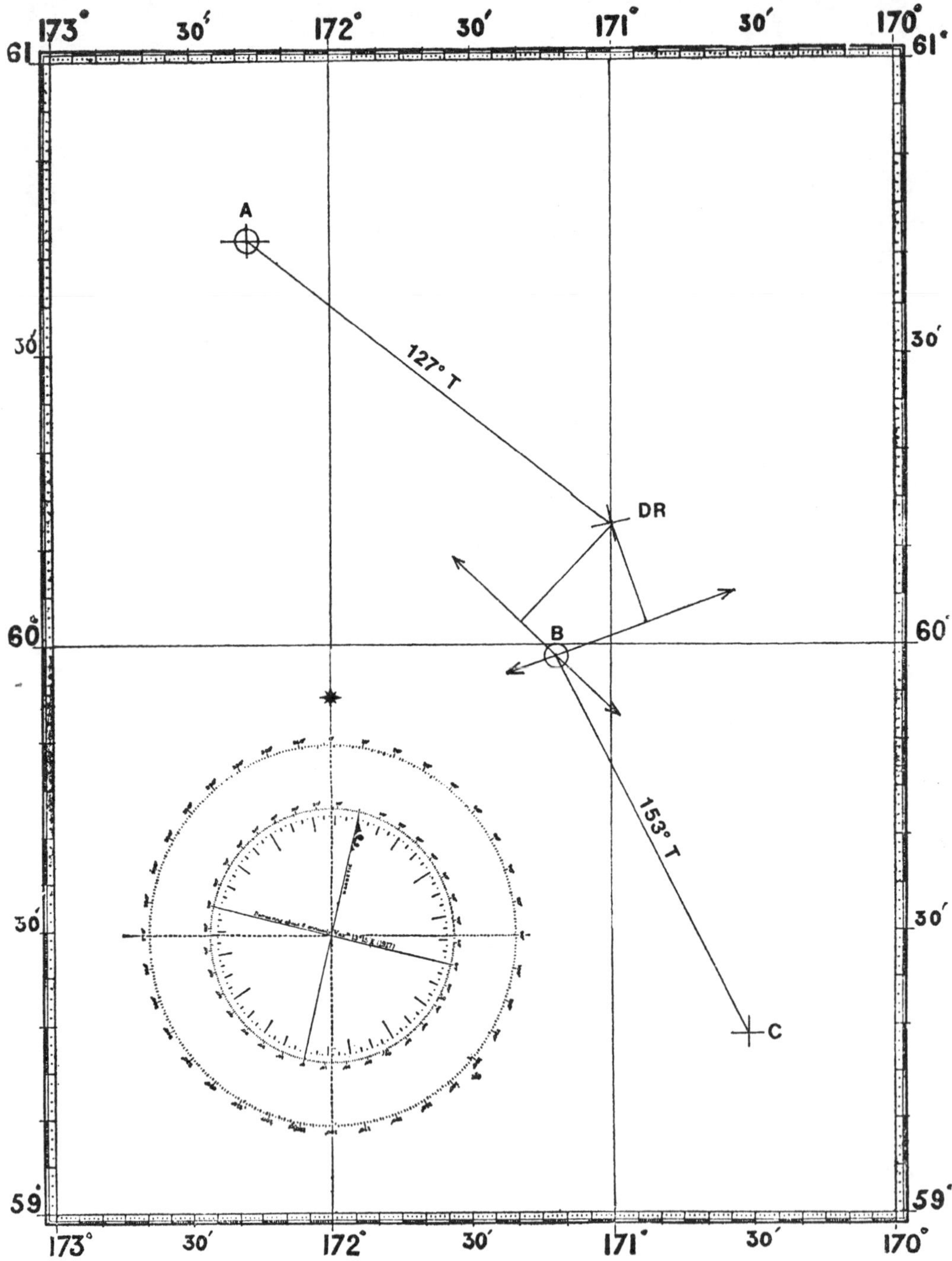
173°
30′
172°
30′
171°
30′
170°
61°
30′
60°
30′
59°
A
127° T
DR
B
153° T
C

so that the height of the water-level may be read without having to closely approach the pole. This height should be recorded at regular intervals of time, and special attention paid to the times of high and low water, most particularly on the dates of the new and full moon, so that the datum of the port" may be decided upon.

Allowance should be made for wind, atmospheric pressure, and other influences likely to affect the height of the tide. As the height of the water-level can be recorded at any time of the day by the above arrangements, it becomes a simple matter to reduce all the soundings to chart-datum level. On new editions of British Charts this is the level of the Lowest Astronomical Tide (LAT) which was chosen so as to ensure that the tide will rarely fall below this level and that the charted depth of water is the least the mariner can expect under normal conditions.

The height of all objects on the land should be marked on the chart or plan in feet above the level of mean high-water spring tides. These can generally be ascertained by means of the vertical angles which they subtend, allowance being made for the height of the tide. A compass should be engraved showing the variation.

CONSTRUCTION OF A MERCATOR CHART.

Example—Construct a chart on Mercator projection extending from Lat. 59° N. to 61° N. and from Long. 170° W. to 173° W. Scale 4 inches to 1 degree of longitude. Divide the scales of latitude and longitude into minutes. Insert a compass, Admiralty pattern, divided into degrees and showing variation 13° 15′ E.

Refer to the figure when reading the following explanation. The original drawing was drawn to a 4-inch scale, but has been reduced in size to fit the page of this book. Plot the following work on the chart:—

A ship at position *A*, Lat. 60° 42′ N., Long. 172° 17′ W., sailed 127°T, 48 miles. From this DR position, the azimuth of a star was 160°T, intercept 10·5 miles towards. Lay off the position line.

At the same time the azimuth of another star was 224°T, intercept 14 miles towards. Lay off the position line and find ship's position.

(*Ans.* Lat. 59° 59′ N., Long. 171° 11′ W.)

Find the course and distance from this position *B* to *C* in Lat. 59° 19′ N., Long. 170° 30′ W.

(*Ans.* Course 153°T, Dist. 45 miles).

Construction.—A horizontal line is drawn to represent the parallel of Latitude 59° N. and a perpendicular drawn to represent the meridian of 170° West Longitude. This meridian must be drawn exactly perpendicular by geometrical methods and then checked with a protractor.

Longitude Scale.—The parallel should then be divided into spaces for each degree of longitude; in this example 1 degree of longitude is 4 inches. Each degree should then be divided into 10′ spaces and subdivided into 5′ and finally into 1′ spaces. This must be done as accurately as possible because the accuracy of the whole chart depends upon the scale of longitude. The left hand marginal meridian may now be drawn exactly parallel to the right hand meridian.

Latitude Scale.—The spacing of the parallels of latitude depends on the meridional difference of latitude between each one, and increases with the latitude: the higher the latitude the greater the increase.

Write down each half degree of latitude and against each write its meridional parts taken from *Norie's Tables*. The D.M.P. will be the number of minutes of longitude separating each successive parellel as follows:—

Lat.	Mer. Parts	D.M.P.	Sphere
59°	4409·19		
59½	4467·82	58·57	between 59° and 59½°
60	4527·37	59·45	,, 59½ ,, 60
60½	4587·83	60·36	,, 60 ,, 60½
61	4649·23	61·29	,, 60½ ,, 61

Take 58·57 parts from the longitude scale and lay it off upwards on the meridian. This fixes the spot for 59½°. Then 59·45 parts from scale is laid off from 59½° to fix the spot for 60°, and 60·36 parts for 60° to 60½° and 61·29 parts for 60½° to 61°. The parallels of latitude may then be drawn. It is desirable, however, to lay off the D.M.P. on the marginal meridians on each side of the chart with the same opening of the compasses and to draw in the parallels with a straight edge, as parallel rulers are not likely to be accurate enough for the purpose. It will be obvious that great care should be taken when lining in the parallels and meridians (the graticule) so that they may be accurately parallel and perpendicular, otherwise, the chart will be twisted.

Another Method of determining by calculation the spacing between the parallels is—

D. Lat. scale = D. Long. scale × nat secant mid lat.

59° and 59½°	= 2 × nat sec	59° 15′	= 2 × 1·956	= 3·912	inches
59½ ,, 60	= 2 ,,	59 45	= 2 × 1·975	= 3·970	,,
60 ,, 60½	= 2 ,,	60 15	= 2 × 2·015	= 4·030	,,
60½ ,, 61	= 2 ,,	60 45	= 2 × 2·047	= 4·094	,,

The half degree spacing may then be divided off into minutes of latitude.

The Compass.—This may be drawn in with a protractor. The Admiralty pattern compass has the outer circle graduated true from 0° to 359° clockwise, and the inner circle graduated magnetic from 0° to 359°, with due allowance made for the angle of variation for the year, or, as given in the example.

1. Construct a chart on Mercator projection extending from Lat. 49° 55′ N. to Lat. 50° 45′ N., and from Long. 0° 00′ to Long. 2° 30′ W. Scale 6 ins. to 1 degree of longitude. Construct a compass, Admiralty pattern showing 12° W. variation. Divide margin scales to minutes.

	Lat.	Long.
St. Catherines	50° 34′ 30″ N.	1° 17′ 47″ W.
Portland Bill	50 31 18 N.	2 27 18 W.
Anvil Point	50 35 30 N.	1 57 30 W.

Portland Bill bore 291° T., and Anvil Point bore 047° T. Ship then steered 104° M. Find ship's position when St. Catherines is abeam and the distance off.

2. Construct a Mercator chart extending from Lat. 54° 40′ N., to Lat. 55° 40′ N. and from Long. 7° 40′ W. to Long. 9° 20′ W. Scale 7·5 ins. to 1 degree of longitude. Construct a compass showing 20° 40′ variation. Plot the following positions:—

A	Lat.	55° 17′ N.	Long.	8° 15′ W.
B	,,	55 13¾ N.	,,	7 59 W.
C	,,	55 08 N.	,,	8 15½ W.
D	,,	55 01 N.	,,	8 33 W.
E	,,	54 59 N.	,,	8 07 W.
F	,,	54 51 N.	,,	8 30 W.
G	,,	54 46 N.	,,	8 34 W.

When *A* and *B* were in transit the distance from *C* by vertical angle was 22 miles. Ship then sailed 230° M., 18 miles, when the distance from *C* was found to be 30 miles with *D* and *E* in transit. Find the set and drift of the current, also the course and distance made good by the ship.

Vessel stood to the south-east. The horizontal angle between *D* and *F* was 79° and at the same time the vertical angle of Hill *G* (1502 ft.) was 1° 18′ 40″ off the arc and 1° 23′ 00″ on the arc. Find the ship's position.

3. Construct a Mercator chart, scale 3½ ins. to 1 degree of longitude, extending from Lat. 53° 00′ N. to Lat. 54° 30′ N. and from Long. 3° W. to Long. 6° W. No variation. Plot the following positions:—

A	Lat.	54° 02′ N.	Long.	4° 50′ W.
B	,,	54 03 N.	,,	4 37 W.
C	,,	55 31 N.	,,	3 31 W.

At 02 p.m. *A* bore 311° M. and *B* bore 022° M. Set course 104° T., speed 15 knots. At 04·40 p.m., altered course to 172° T. At 05·50 p.m., *C* bore 126° M. and at 06·10 p.m. it bore 047° M. Find ship's position and the set and drift of current.

4. Construct a chart on Mercator's projection extending from Lat. 48° 20′ N. to Lat. 49° 10′ N and from Long 164° 20′ E. to Long. 165° 50′ E. Scale 8 ins. to 1 degree of longitude. Insert a true compass.

Project the following position lines and find the ship's position at the time of taking the 2nd observation:—

1st obs.—Position line 007°-187° T. drawn through Lat. 48° 55′ N., Long. 165° 35′ E.

2nd obs.—Position line 052°-232° T. drawn through Lat. 48° 25′ N., Long. 164° 38′ E.

Between 1st and 2nd observations the ship sailed 240° T., for 15 miles, 260° T., 13 miles, 310° T., 7 miles and 250° T., for 10 miles.

5. Construct a plan chart 4·8 equal to 1 mile of latitude. Divide latitude and Longitude scales to seconds.

An observation spot (Lat. 58° 11′ 00″ N., Long. 6° 19′ 22″ W.) bore 029° T., and 7 minutes later it bore 347° T., and 10 minutes later it bore 303° T. Required the course made good, and if the gyro course steered was 096°, speed 6 knots, find the set and drift of the tide experienced.

If a point in Lat. 58° 10′ 47″, Long. 6° 18′ 00″ N. then bore 081° T., find the course to bring the point 5 cables off when it comes abeam, and the distance to steam.

6. Construct a plan chart, scale of latitude 2·8 ins. to the mile, also show the scale of longitude, and a true compass divided into degrees.

Plot position *A* about 3 ins. from the bottom left hand corner of the paper.

Lighthouse	*A*	Lat.	50° 39′ 42″ N.,	Long.	1° 35′ 25″ W.
,,	*B*	,,	50 42 27 N.,	,,	1 32 56 W.

Find the course to steer and the distance from a position with *A*, 4 cables on starboard beam to a position with *B*, 3 cables on the port beam.

From the beam position of *B*, alter course to fetch an anchorage with *B* bearing 253½° and *C* (Lat. 50° 42′ 30″ N., Long. 1° 29′ 54″ W.) bearing 200° T. Required the distance of the anchorage from *B* and *C*.

7. Construct a chart extending from Lat. 60° 22′ N. to Lat. 60° 33′ N., and from Long. 27° 55′ E. to Long. 28° 25′ E., scale 1 mile of middle latitude equal to 4 ins. Divide margins to minutes, compass to degrees, variation 0° 40′ W.

Plot	*A*	Lat.	60°	24′	30″	N.,	Long.	27°	58′	40″	E.
,,	*B*	,,	60	28	40	N.,	,,	28	06	00	E.
,,	*C*	,,	60	26	50	N.,	,,	28	22	30	E.

Fix the centre of a shoal by sextant angles as follows:—*A* 54° *B* 96° *C*. Find this position, also its magnetic bearing from each of the points, *A*, *B* and *C*

8. Construct a chart, scale 7 ins. to 1 minute of longitude, variation 14° 15′ W., extending from Lat. 45° 30′ N., to Lat. 45° 44′ N., and from Long. 1° 00′ W. to Long. 1° 25′ W. A lighthouse in Lat. 45° 41′ 45″ N., Long. 1° 14′ 3″ W.

Showing a	white sector	between	309° M.	and	297° M.
,,	red	,,	297° M.	,,	245° M.
,,	white	,,	245° M.	,,	236° M.

Another light in Lat. 45° 35′ 12″ N., Long. 1° 10′ 30″ W.

Showing	white light	between	338° M.	and	332° M.
,,	green	,,	332° M.	,,	312° M.
,,	white	,,	312° M.	,,	208° M.
,,	red	,,	338° M.	,,	025° M.

All bearings are from the light.

A buoy is moored at the intersection of the northern limits of the white lights. Find its latitude and longitude and its magnetic bearing and distance from the second light.

Answers.

1. Lat. 50° 26′ N., Long. 1° 18′ W. Beam distance 8½ miles.

2. Set 262° M., drift 5 miles. Course 236° M., distance 22 miles. Position Lat. 54° 55′ N., Long. 8° 44′ W.

3. Lat. 53° 28½′ N., Long. 3° 35½′ W. Set 306° T., drift 4½ miles.

4. Lat. 48° 21′ N., Long. 164° 29′ E.

5. Course 085° T. Set 187° T. Drift 3·5 cables in 17 minutes. Course 111° T. Distance 8·3 cables.

6. Course 042° T. Distance 3 miles 1 cable. Distance 2 miles 1·5 cables from *B* and 6·6 cables from *C*.

7. Lat. 60° 24′ 20″ N., Long. 28° 11′ 45″ E. 094° M., 6·4 miles from *A*. 148° M. 5·3 miles from *B*. 244° M. 5·8 miles from *C*.

8. Lat. 45° 43′ 15″ N., Long. 1° 18′ 40″ W. Bearing 338° M. Distance 10 miles.

GNOMONIC CHART.

The eye, in the gnomonic projection, is conceived to be at the centre of the sphere and the area to be mapped out is projected on a tangent plane in contact with the sphere. Lines are drawn from the centre of the sphere through various points on its surface until they meet the tangent plane in spots which are projections of these points.

A simple case may illustrate the principle. Let *O*, in Fig. 1, represent the centre

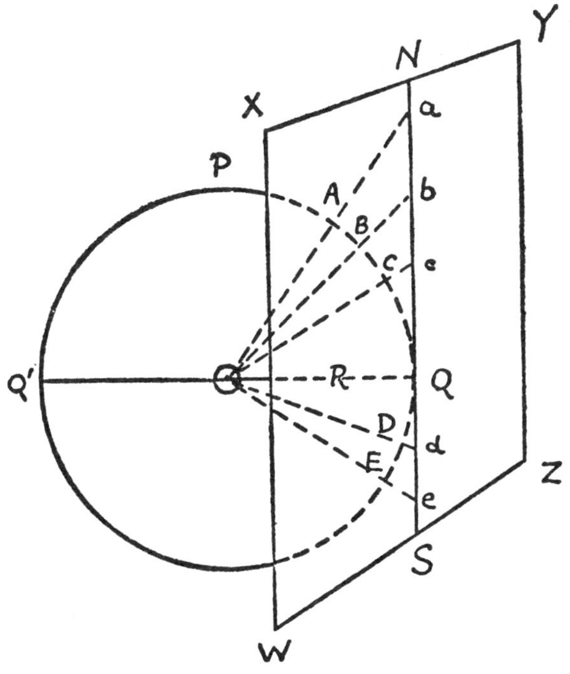

FIG. 1.

of the Earth and *QPQ'* be any meridian. *WXYZ* is a plane sheet of paper touching the sphere at *Q*. *Q* is called the tangential point. Let *A*, *B*, *C*, *D* and *E* be points on the meridian. Produce their respective radii *OA*, *OB*, *OC*, *OD* and *OE* until they meet the tangent plane at *a*, *b*, *c*, *d* and *e* respectively, then these spots will be projected along the line *NQS* which now represents the gnomonic projection of the meridian. But *QPQ'* is any meridian and all meridians, therefore, appear as straight lines on the chart.

The angles *QOA*, *QOB*, *QOC*., etc., represent the latitude of *A*, *B* and *C* respectively. *OQ* is the radius of the sphere = *R*, $\angle OQN = 90°$, therefore $Qa = R$ tan Lat. *A*, $Qb = R$ tan Lat. *B*, etc.

All Great Circles project as straight lines on the Gnomonic Chart.—A great circle is one whose plane passes through the centre of the sphere. Imagine, as in Fig. 2, a sphere resting on a plane surface at *A*. Let *ABC* and *ADE* be arcs of two great circles. The tangential point is projected on the plane at *A*, *OB* produced projects *B* at *b*, and, similarly, *C* is projected at *c*. Thus the projection of the great circle arc *ABC* is the straight line *Abc*.

Again, the points *A*, *D* and *E* on the great circle *ADA'* are similarly projected on

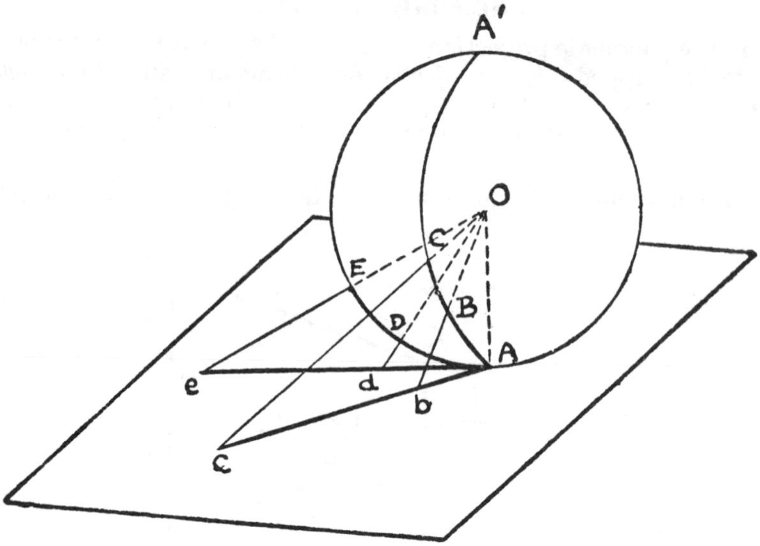

FIG. 2.

the tangent plane by producing their respective radii *OD* and *OE*. Now all the radii *OA*, *OD* and *OE* lie in the plane of the great circle and the points *A*, *d* and *e* lie on this plane and also on the tangent plane and they must, therefore, be on the line of intersection of the two planes.

Thus the projection of this great circle is a straight line, as indeed are all great circles on the gnomonic projection.

It may be noted that the spherical angle *CAE* on the sphere is equal to the plane angle *cAe* on the tangent plane.

Parallels of Latitude Project as Conic Sections.—Let *ABCDEFG*, in Fig. 3 be a parallel of latitude and *O* the centre of the sphere, then the projections of points on

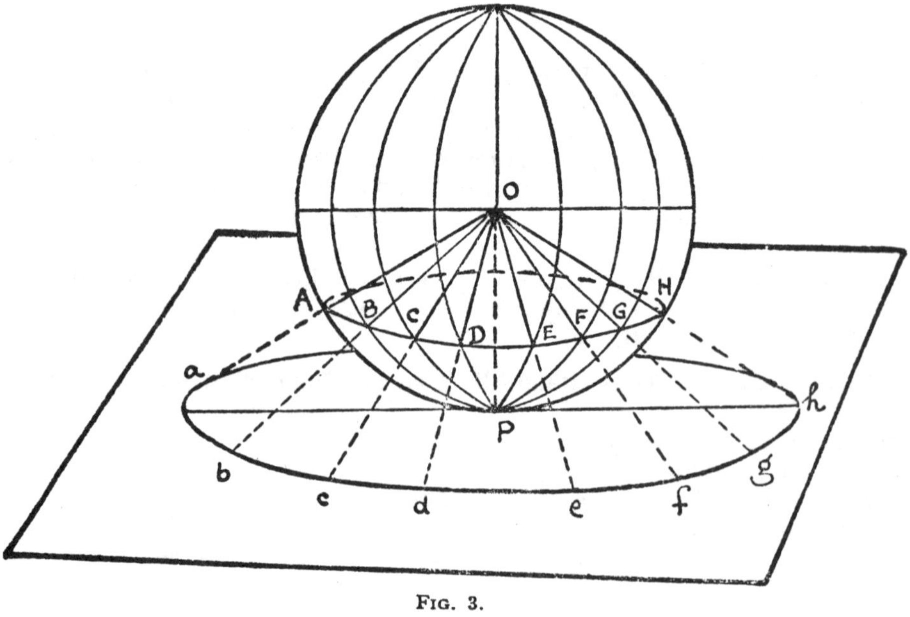

FIG. 3.

the parallel, such as *A, B, C, D, E, F, G,* and *H,* will trace out the curve *a, b, c, d, e, f, g* and *h* on the tangent plane.

But the radii *OA, OB, OC,* etc., lie on a right circular cone whose vertex is *O* and its base the parallel of latitude, hence this curve is a section of a cone projected on the tangent plane. This curve is usually a conic section and may be an ellipse, a parabola or hyperbola (see Fig. 4).

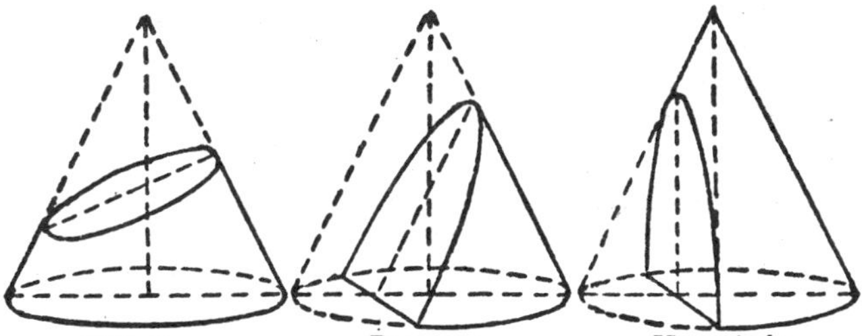

FIG. 4.

An ellipse is a conic section or curve made by a plane passing obliquely through the opposite sides of a cone.

A parabola is the curve formed by cutting the cone with the plane of the section parallel to the side of the cone.

A hyperbola is the curve formed by the intersecting plane making a greater angle with the base than the angle made by the side of the cone.

Should, however, the tangent plane be parallel to the base of the cone the section becomes a circle. This happens only when the plane touches the Earth sphere at the

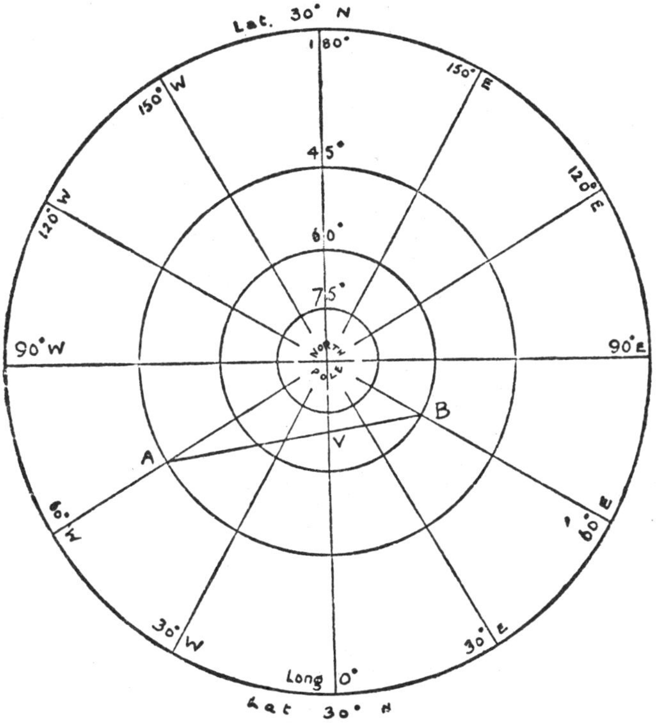

FIG. 5.

pole and, in this particular case called the "polar projection", the parallels of latitude are circles and the meridians are diameters, as in Fig. 5.

In triangle OPh we have P the pole, $\angle OPh = 90°$, OH the radius of earth $= R$, $\angle POH$ is the co-latitude of H, therefore, the radius of the parallel of latitude $= R$ tan $POH = R$ cot Lat. Thus the radii of the circles representing parallels of latitude on a polar gnomonic chart are proportional to the cotangent of their latitudes. But the cotangent increases quickly with decrease of the angle and finally becomes infinite, so that cot lat. 0° = infinity, and thus the polar chart fails to represent the tropical regions. Gnomonic charts cannot represent satisfactorily areas which are more than 75° from the tangential point, and consequently considerably less than a hemisphere can be shown on this projection. Refer back so Fig. 3.

The simplest gnomonic chart to construct is the polar projection (Fig. 5). The tangential point is the pole. The radius of latitude 45° is the unit of measure so that cot 45° $= R$ units and the radii of the other parallels is R cot latitude respectively. The parallels of latitude are circles and the meridians are diameters.

In drawing this figure we made the radius of 45° = 2 inches. Radius 75° = 2 tan 75° = 2 × ·267 = ·543, similarly radius 60° = 1·154, radius 30° = 3·46. The meridians were laid off with a protractor.

The great circle from A Lat. 45° N., Long. 60° W. to B Lat. 60° N., Long. 60° E., is represented by the straight line AB. A perpendicular from the pole to AB gives the position of vertex. The positions of successive points can be read from the chart,

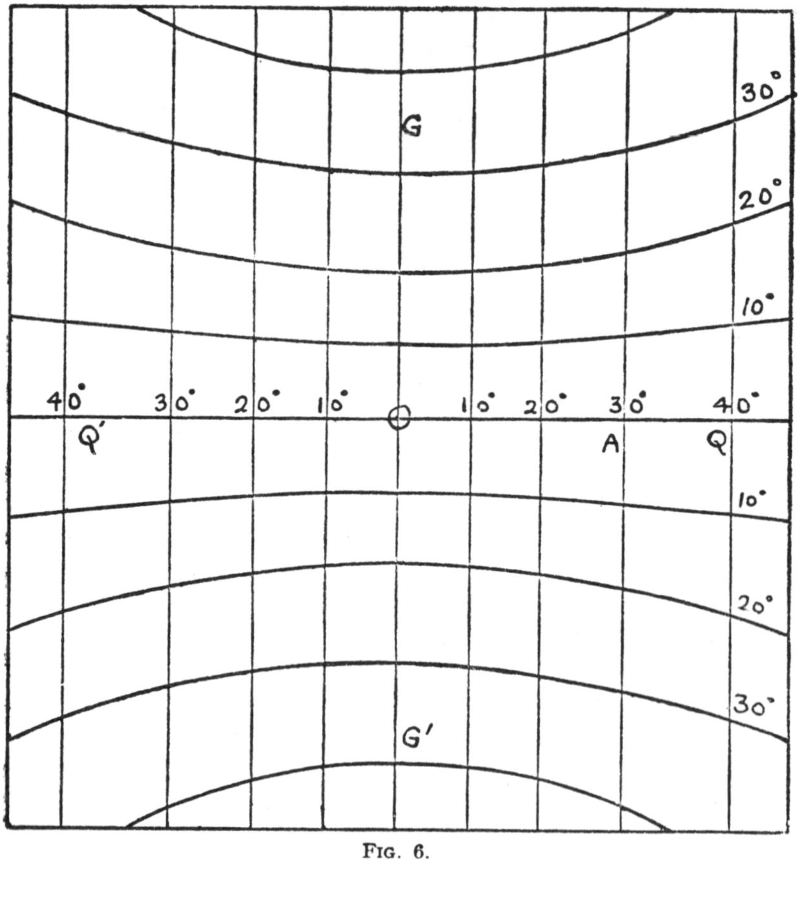

FIG. 6.

and courses measured with a protractor, the degree of accuracy being limited only by the scale of the chart.

Figure 6 represents an equatorial gnomonic projection the tangential point being at *O*. This network map is called a graticule. The parallels of latitude are conic sections, and the meridians are parallel straight lines but not equidistant from each other. If GOG^1 be the principal meridian and QOQ^1 the equator, the distance of each meridian from the principal meridian is given by, *R* tan difference of longitude, where R=the scale selected for a D. Long. of 45°. Thus, if D. Long. 45° were made equal to 3 ins., then *OA* (the D. Long. corresponding to 30°) equals 3 tan 30° $= 3 \times \cdot 58 = 1\cdot 74$ ins., and similarly for the spacing of the respective meridians having D. Long. 10°, 20°, etc. The parallels are not so easily drawn. Spots on each parallel corresponding to their latitudes, and their differences of longitude from the principal meridian, are calculated, the spots plotted and curves drawn through them. The equatorial projection is, however, of little use in navigation as the advantages of great circle sailing are confined to high latitudes.

When the tangential point is between the pole and the equator the projection of the meridians and parallels is somewhat complicated, and requires the calculation of a series of points through which to draw the respective meridians and the parallels of latitude. See gnomonic chart facing page 69.

Although hydrographic surveying is no longer included as a separate subject in the new Extra Master Examination, certain aspects of the subject are included in the Marine Environmental Studies Paper and also in maritime studies degree courses. The following questions, which include examples from past Extra Master examination papers, are here included for students wishing to acquire a deeper understanding of the subject either for examination or other purposes. The student is referred to the comprehensive chapter on "Surveying" contained in the *Admiralty Manual of Navigation* and the *Admiralty Manual of Hydrographic Surveying* for more detailed information.

Questions for Exercise

1. Regarding the use of echo sounding devices, what precautions would you take before making a survey?
2. Describe the buoyage system and lights around the British Isles giving distinctive features.
3. Give a brief description of the method adopted in the triangulation for a survey and explain with figure how you would extend the base line.
4. Describe the method of finding the true bearing of a triangulation station.
5. How would you make tidal observations when surveying a small bay?
6. What is meant by "running a line of soundings" and what are the precautions to be taken before plotting it?
7. Describe how soundings are plotted on a "field plan" and also the method adopted to indicate soundings in shallow water.
8. Describe a survey of a bay, or harbour, with special reference to hidden dangers and the system of observations.
9. What is put on a fair chart before sending it to the Hydrographer?
10. On which side of a river is the deepest water found when it curves? Why are shoals formed at river entrances?

11. Describe the tests you would make before using a station pointer for surveying.
12. Describe the methods of coast lining. Enumerate the instruments used and describe those which are not in general use in navigation.
13. Describe the theodolite and its adjustments.
14. Show that in surveying when obtaining distance by an angle of depression the correction for refraction is 100D/4 in seconds of arc, where D is the distance in miles.
15. The position of two stations and the distance between them has been established in a triangulation survey. Explain how the reliability of the position of the third station can be built up by: (*a*) intersection, (*b*) resection, and (*c*) trilateration.

CHAPTER X.

MATHEMATICAL FORMULAE AND THEIR APPLICATION TO SOME PROBLEMS. IN NAVIGATION, NAUTICAL AND SPHERICAL ASTRONOMY.

THIS chapter is devoted to recording very briefly, but in a more or less progressive order, proofs of the fundamental trigonometrical formulae usually employed in solving navigational problems, practical as well as academic. It is really a continuation of Chapter VI, Volume I, the contents of which the reader will, no doubt, be familiar with. But it is expected that students seeking after more comprehensive and elaborate investigations will also refer to standard works on the subject such as *Goodwin's Plane and Spherical Trigonometry, The Admiralty Manuals on Navigation,* and school textbooks on algebra and geometry, our purpose here being merely to lead up to the solution of some types of theoretical questions as indicated by the syllabus and the experience of extra masters' examinations.

PROOFS OF FORMULAE.

Prove $\frac{a}{b} = \frac{\sin A}{\sin B}$ . . . . (1)

In figure CD is perpendicular to AB

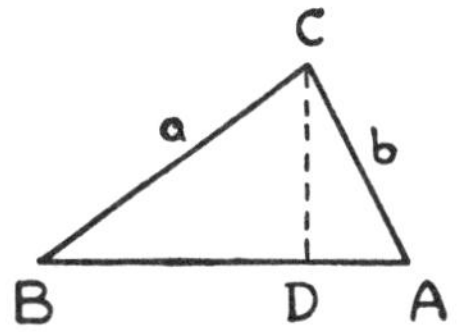

$$CD = a \sin B = b \sin A$$

$$\therefore \frac{a}{b} = \frac{\sin A}{\sin B}$$

Similarly, by moving the lettering A, B, C one place, we get

$$\frac{c}{b} = \frac{\sin C}{\sin B} \text{ and } \frac{a}{c} = \frac{\sin A}{\sin C} \quad \ldots \quad (1)$$

Prove $c = a \cos B + b \cos A$ . . . . (2)

From above figure $c = BD + DA = a \cos B + b \cos A$

Similarly, it may be shown that

$$a = b \cos C + c \cos B \quad \ldots \quad (2)$$
$$b = a \cos C + c \cos A \quad \ldots \quad (2)$$

Prove $a^2 = b^2 + c^2 - 2bc \cos A$ . . . . (8)

In $\Delta\ ABC$, CD is perpendicular to AB.

Let $AD = x$, then $BD = (c - x)$ in fig. (i) and $(c + x)$ in fig. (ii).

$$CD^2 = a^2 - (c - x)^2 = b^2 - x^2$$
$$a^2 - c^2 + 2cx - x^2 = b^2 - x^2$$
$$a^2 = b^2 + c^2 - 2cx$$
$$a^2 = b^2 + c^2 - 2cb \cos A \quad . . . . (3)$$

or, writing it in another form

$$\text{Cos } A = \frac{b^2 + c^2 - a^2}{2bc} \quad . . . . (4)$$

(i)

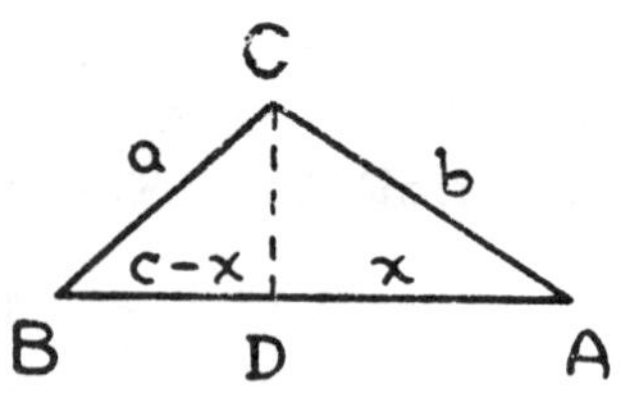

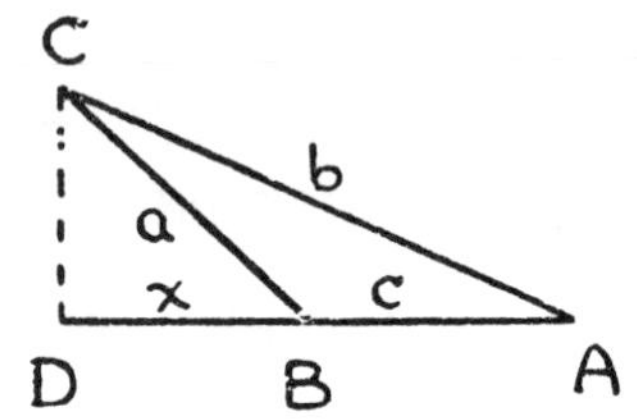

(ii)

Similarly, it may be shown that

$$\text{Cos } B = \frac{a^2 + c^2 - b^2}{2ac} \quad . . . . (4)$$

$$\text{Cos } C = \frac{a^2 + b^2 - c^2}{2ab} \quad . . . . (4)$$

MULTIPLE ANGLES.

Prove $\sin (A + B) = \sin A, \cos B + \cos A, \sin B$ . . . . (5)

,, $\cos (A + B) = \cos A \cos B - \sin A \sin B$ . . . . (6)

,, $\tan (A + B) = \dfrac{\tan A + \tan B}{1 - \tan A \tan B}$ . . . . (7)

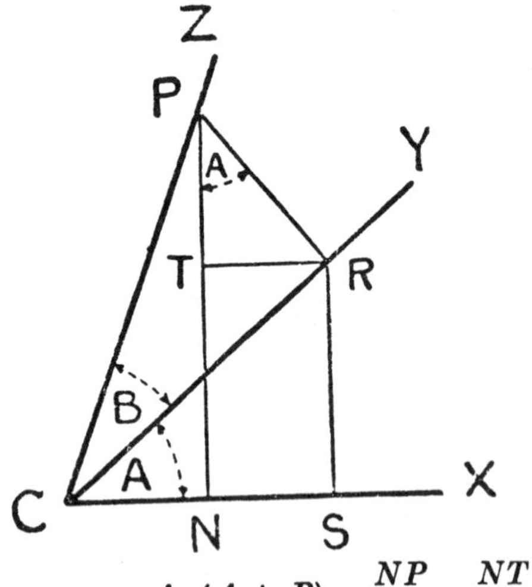

Let $\angle YCX = A$ and $\angle ZCY = B$, then $\angle ZCX = (A + B)$.

On CZ (the line that includes the angle $(A + B)$), take any point P, and from this point drop a perpendicular on CY (PR), make PN and RS perpendicular to CX, and RT perpendicular to PN.

Because TR is parallel to CX, and CY cuts both at the same angle, therefore angle $TRC = RCX = A$.

As $\angle TPR$ and $\angle TRC$ are both complements of $\angle TRP$, we have

$$\angle TPR = \angle TRC = \angle A.$$

$$\sin (A + B) = \frac{NP}{CP} = \frac{NT + TP}{CP} = \frac{SR + TP}{CP} = \frac{SR}{CP} + \frac{TP}{CP}$$

By multiplying and dividing $\frac{SR}{CP}$ by CR, and $\frac{TP}{CP}$ by PR, we obtain

$$\sin (A + B) = \frac{SR \times CR}{CR \times CP} + \frac{TP \times PR}{PR \times CP}$$

$$\therefore \sin (A + B) = \sin A, \cos B + \cos A, \sin B \quad \ldots \ldots \quad (5)$$

$$\cos (A + B) = \frac{CN}{CP} = \frac{CS - NS}{CP} = \frac{CS - TR}{CP} = \frac{CS}{CP} - \frac{TR}{CP}$$

Multiplying and dividing by CR and PR as before, we obtain:—

$$\cos (A + B) = \frac{CS \times CR}{CR \times CP} - \frac{TR \times PR}{PR \times CP}$$

i.e., $$\cos (A + B) = \cos A, \cos B - \sin A, \sin B \quad \ldots \ldots \quad (6)$$

Triangles CRS and PTR are equiangular because

$$\angle RCS = \angle RPT = \angle A$$
$$\angle RSC = \angle PTR = 90°$$
$$\angle CRS = \angle PRT$$

and the corresponding sides are proportional ;

$\therefore$ $PT : PR = CS : CR$, which may be written as

$$\frac{PT}{CS} = \frac{PR}{CR} = \tan B, \text{ and from fig. } \frac{TR}{PT} = \tan A$$

$$\tan (A + B) = \frac{PN}{CN} = \frac{RS + PT}{CS - TR}. \text{ Dividing by } CS \text{ we get}$$

$$\tan (A + B) = \frac{\frac{RS}{CS} + \frac{PT}{CS}}{\frac{CS}{CS} - \frac{TR}{CS}} = \frac{\frac{RS}{CS} + \frac{PT}{CS}}{1 - \frac{TR}{PT} \times \frac{PT}{CS}} \quad \text{using } PT \text{ as a link line}$$

$$\tan (A + B) = \frac{\tan A + \tan B}{1 - \tan A \tan B} \quad \ldots \ldots \quad (7)$$

Some may prefer to combine (5) and (6) to get (7), as follows :—

$$\tan (A + B) = \frac{\sin (A + B)}{\cos (A + B)} = \frac{\sin A \cos B + \cos A \sin B}{\cos A \cos B - \sin A \sin B}$$

Divide throughout by $\cos A \cos B$ to get tangents

$$\tan (A + B) = \frac{\frac{\sin A \cos B}{\cos A \cos B} + \frac{\cos A \sin B}{\cos A \cos B}}{\frac{\cos A \cos B}{\cos A \cos B} - \frac{\sin A \sin B}{\cos A \cos B}}$$

$$\tan (A + B) = \frac{\tan A + \tan B}{1 - \tan A \tan B} \quad \ldots \ldots \quad (7)$$

If $\tan B = A$, we may write

$$\tan (A + A) = \frac{\tan A + \tan A}{1 - \tan A \tan A}$$

$$\therefore \tan 2A = \frac{2 \tan A}{1 - \tan^2 A}$$

Prove $\sin (A - B) = \sin A \cos B - \cos A \sin B$ . . . . (8)

„ $\cos (A - B) = \cos A \cos B + \sin A \sin B$ . . . . (9)

„ $\tan (A - B) = \dfrac{\tan A - \tan B}{1 + \tan A \tan B}$ . . . . (10)

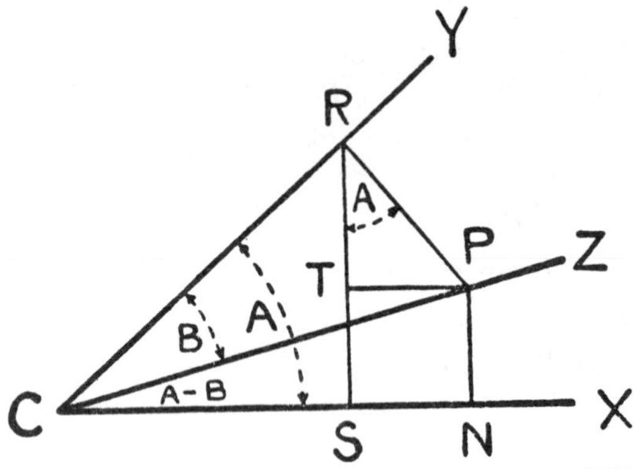

Let $\angle XCY = A$ and $\angle ZCY = B$, then $\angle XCZ = (A - B)$.

From any point P on CZ the line that includes the angle $(A - B)$ make PR perpendicular to CY. Make RS and PN perpendicular to CX, and PT perpendicular to RS.

$\angle TRP$ and $\angle XCR$ are both complements of $\angle CRS$

$\therefore \quad \angle TRP = \angle XCR = \angle A.$

$$\text{Now, } \sin (A - B) = \frac{NP}{CP} = \frac{SR - RT}{CP} = \frac{SR}{CP} - \frac{RT}{CP}$$

By multiplying and dividing each fraction by CR and RP respectively we obtain :—

$$\sin (A - B) = \frac{SR \times CR}{CR \times CP} - \frac{RT \times RP}{RP \times CP}$$

$\therefore \sin (A - B) = \sin A \ \cos B - \cos A \ \sin B$ . . . . (8)

Also, $$\cos (A - B) = \frac{CN}{CP} = \frac{CS + SN}{CP} = \frac{CS + TP}{CP} = \frac{CS}{CP} + \frac{TP}{CP}$$

by introducing CR and PR as link lines

$$\cos (A - B) = \frac{CS \times CR}{CR \times CP} + \frac{TP \times RP}{RP \times CP}$$

$\therefore \cos (A - B) = \cos A \ \cos B + \sin A \ \sin B$ . . . . (9)

$$\tan (A - B) = \frac{PN}{CN} = \frac{RS - RT}{CS + PT}$$

„ $$= \frac{\dfrac{RS}{CS} - \dfrac{RT}{CS}}{\dfrac{CS}{CS} + \dfrac{PT}{RT} \times \dfrac{RT}{CS}}$$ dividing by CS and using RT as a link line

Remembering that in the similar triangles RPT and CRS

$\dfrac{RT}{RP} = \dfrac{CS}{RC}$ and by transposing we get

$\dfrac{RT}{CS} = \dfrac{RP}{RC} = \tan B$, so that the above becomes

$$\tan (A - B) = \frac{\tan A - \tan B}{1 + \tan A \tan B}$$ . . . . (10)

or it may be deduced from (8) and (9)

$$\tan(A - B) = \frac{\sin(A - B)}{\cos(A - B)}$$

$$\text{,,} \qquad = \frac{\sin A \cos B - \cos A \sin B}{\cos A \cos B + \sin A \sin B}$$

Dividing by $\cos A \cos B$

$$\tan(A - B) = \frac{\dfrac{\sin A \cos B}{\cos A \cos B} - \dfrac{\cos A \sin B}{\cos A \cos B}}{\dfrac{\cos A \cos B}{\cos A \cos B} + \dfrac{\sin A \sin B}{\cos A \cos B}}$$

$$\tan(A - B) = \frac{\tan A - \tan B}{1 + \tan A \tan B} \quad \ldots \ldots \quad (10)$$

SUM AND DIFFERENCE OF TWO ANGLES.

From (5) and (8) the ratios of the sum and difference of angles may be established.

Prove $\sin A + \sin B = 2 \sin \tfrac{1}{2}(A + B) \cos \tfrac{1}{2}(A - B) \quad \ldots \ldots \quad (11)$

,, $\sin A - \sin B = 2 \cos \tfrac{1}{2}(A + B) \sin \tfrac{1}{2}(A - B) \quad \ldots \ldots \quad (12)$

By (5) $\sin(A + B) = \sin A \cos B + \cos A \sin B$.
By (8) $\sin(A - B) = \sin A \cos B - \cos A \sin B$.

By adding and subtracting we get

$$\sin(A + B) + \sin(A - B) = 2 \sin A \cos B$$
$$\sin(A + B) - \sin(A - B) = 2 \cos A \sin B.$$

Let $(A + B) = S$, and $(A - B) = D$

then $S + D = 2A$, and $A = \tfrac{1}{2}(S + D)$

$S - D = 2B$, and $B = \tfrac{1}{2}(S - D)$

and by substituting S and D in above we get

$$\sin S + \sin D = 2 \sin \tfrac{1}{2}(S + D) \cos \tfrac{1}{2}(S - D)$$
$$\sin A - \sin D = 2 \cos \tfrac{1}{2}(S + D) \sin \tfrac{1}{2}(S - D)$$

but S and D represent the sum and difference of any two angles so that the general expression may be written

$$\sin A + \sin B = 2 \sin \tfrac{1}{2}(A + B) \cos \tfrac{1}{2}(A - B) \quad \ldots \ldots \quad (11)$$
$$\sin A - \sin B = 2 \cos \tfrac{1}{2}(A + B) \sin \tfrac{1}{2}(A - B) \quad \ldots \ldots \quad (12)$$

Prove $\cos A + \cos B = 2 \cos \tfrac{1}{2}(A + B) \cos \tfrac{1}{2}(A - B) \quad \ldots \ldots \quad (13)$

,, $\cos B - \cos A = 2 \sin \tfrac{1}{2}(A + B) \sin \tfrac{1}{2}(A - B) \quad \ldots \ldots \quad (14)$

by (6) $\cos(A + B) = \cos A \cos B - \sin A \sin B$
by (9) $\cos(A - B) = \cos A \cos B + \sin A \sin B$
add $\cos(A + B) + \cos(A - B) = 2 \cos A \cos B$
subtract $\cos(A - B) - \cos(A + B) = 2 \sin A \sin B$

If $(A + B) = S$ and $(A - B) = D$

then $\tfrac{1}{2}(S + D) = A$ and $\tfrac{1}{2}(S - D) = B$

and by substituting in above we get

$$\cos S + \cos D = 2 \cos \tfrac{1}{2}(S + D) \cos \tfrac{1}{2}(S - D)$$
$$\cos D - \cos S = 2 \sin \tfrac{1}{2}(S + D) \sin \tfrac{1}{2}(S - D)$$

or, by substituting A and B we get the general expression

$$\cos A + \cos B = 2 \cos \tfrac{1}{2}(A + B) \cos \tfrac{1}{2}(A - B) \quad . \quad . \quad . \quad . \quad . \quad (13)$$

$$\cos B - \cos A = 2 \sin \tfrac{1}{2}(A + B) \sin \tfrac{1}{2}(A - B) \quad . \quad . \quad . \quad . \quad . \quad (14)$$

Prove $\sin A = 2 \sin \frac{A}{2} \cos \frac{A}{2}$ . . . . (15)

Substituting $\frac{A}{2}$ for A in (5) we may write

$$\sin\left(\frac{A}{2} + \frac{A}{2}\right) = \sin \frac{A}{2} \cos \frac{A}{2} + \cos \frac{A}{2} \sin \frac{A}{2}$$

$$\therefore \sin A = 2 \sin \frac{A}{2} \cos \frac{A}{2} \quad . \quad . \quad . \quad . \quad . \quad . \quad . \quad . \quad . \quad . \quad . \quad (15)$$

similarly $\quad \sin B = 2 \sin \frac{B}{2} \cos \frac{B}{2}$ . . . . (15)

$$\sin C = 2 \sin \frac{C}{2} \cos \frac{C}{2} \quad . \quad . \quad . \quad . \quad . \quad . \quad . \quad . \quad . \quad . \quad . \quad . \quad (15)$$

or $\quad \sin 2A = 2 \sin A \cos A$

When dealing with spherical triangles we may substitute $(A + B)$ for A, or $(a + b)$ for A and write for angles and sides respectively in (15) :—

for angles $\sin (A + B) = 2 \sin \frac{1}{2}(A + B) \cos \frac{1}{2}(A + B)$ . . . . . . (16)

for sides $\sin (a + b) = 2 \sin \frac{1}{2}(a + b) \cos \frac{1}{2}(a + b)$ . . . . (17)

Prove $\cos A = 1 - 2 \sin^2 \frac{A}{2}$ . . . . (18)

by (6) $\quad \cos\left(\frac{A}{2} + \frac{A}{2}\right) = \cos \frac{A}{2} \cos \frac{A}{2} - \sin \frac{A}{2} \sin \frac{A}{2}$

$$\text{,,} \quad = \cos^2 \frac{A}{2} - \sin^2 \frac{A}{2}$$

$$\text{,,} \quad = 1 - \sin^2 \frac{A}{2} - \sin^2 \frac{A}{2}$$

$$\therefore \cos A = 1 - 2 \sin^2 \frac{A}{2}$$

$$\text{or } 2 \sin^2 \frac{A}{2} = 1 - \cos A = \text{versine } A \quad . \quad . \quad . \quad . \quad . \quad . \quad . \quad . \quad . \quad (18)$$

Prove $\cos A = 2 \cos^2 \frac{A}{2} - 1$ . . . . (19)

by (6) $\quad \cos\left(\frac{A}{2} + \frac{A}{2}\right) = \cos \frac{A}{2} \cos \frac{A}{2} - \sin \frac{A}{2} \sin \frac{A}{2}$

$$= \cos^2 \frac{A}{2} - \sin^2 \frac{A}{2}$$

$$= \cos^2 A - \left(1 - \cos^2 \frac{A}{2}\right)$$

$$\therefore \cos A = 2 \cos^2 \frac{A}{2} - 1$$

$$\text{or } 1 + \cos A = 2 \cos^2 \frac{A}{2} \quad . \quad . \quad . \quad . \quad . \quad . \quad . \quad . \quad . \quad . \quad . \quad . \quad . \quad (19)$$

Similarly, $\quad \cos 2A = 2 \cos^2 A - 1$

Prove $\cos\dfrac{A}{2} = \sqrt{\dfrac{s(s-a)}{bc}}$ where $s = \dfrac{a+b+c}{2}$ . . . . (20)

by (19) $\qquad 2\cos^2\dfrac{A}{2} = \cos A + 1$

by (4) $\qquad$,, $\quad = \dfrac{b^2+c^2-a^2}{2bc} + 1$

,, $\quad = \dfrac{b^2+2bc+c^2-a^2}{2bc}$

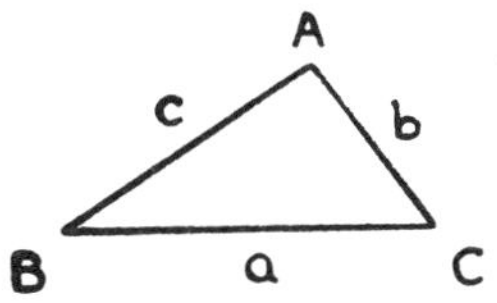

,, $\quad = \dfrac{(b+c)^2-a^2}{2bc}$

,, $\quad = \dfrac{(b+c+a)(b+c-a)}{2bc}$

But if $2s = a+b+c$
then $2(s-a) = b+c-a$ because $2s - 2a = a+b+c-2a$
and $2(s-b) = a+c-b$,, $2s-2b = a+b+c-2b$
and $2(s-c) = a+b-c$,, $2s-2c = a+b+c-2c$

$\therefore\ 2\cos^2\dfrac{A}{2} = \dfrac{2s\,.\,2(s-a)}{2bc}$ the 2's cancel,

and $\qquad \cos\dfrac{A}{2} = \sqrt{\dfrac{s\,.\,(s-a)}{bc}}$ . . . . (20)

Prove $\sin\dfrac{A}{2} = \sqrt{\dfrac{(s-b)(s-c)}{bc}}$ where $s = \dfrac{a+b+c}{2}$ . . . . (21)

by (18) $\qquad 2\sin^2\dfrac{A}{2} = 1 - \cos A$

by (4) $\qquad$,, $\quad = 1 - \dfrac{b^2+c^2-a^2}{2bc}$

,, $\quad = \dfrac{2bc-b^2-c^2+a^2}{2bc}$

,, $\quad = \dfrac{a^2-(b^2-2bc+c^2)}{2bc}$

,, $\quad = \dfrac{a^2-(b-c)^2}{2bc}$

,, $\quad = \dfrac{(a+b-c)(a-b+c)}{2bc}$

$\therefore\ 2\sin^2\dfrac{A}{2} = \dfrac{2(s-c)\,2(s-b)}{2bc}$

$\therefore\ \sin\dfrac{A}{2} = \sqrt{\dfrac{(s-b)(s-c)}{bc}}$ . . . . (21)

But $2\sin^2\dfrac{A}{2} = 1 - \cos A = \text{vers } A$

$\therefore\ \sin^2\dfrac{A}{2} = \dfrac{\text{vers } A}{2}$ = haversine A, so the above expression may be written.

$$\text{Havs. } A = \frac{(s-b)(s-c)}{bc} \quad . . . . (22)$$

Prove $\dfrac{a+b}{a-b} = \dfrac{\tan \frac{1}{2}(A+B)}{\tan \frac{1}{2}(A-B)}$ . . . . (23)

by (1) $\dfrac{a}{b} = \dfrac{\sin A}{\sin B}$ and by $\dfrac{\text{Componendo}}{\text{Dividendo}}$

$\therefore \dfrac{a+b}{a-b} = \dfrac{\sin A + \sin B}{\sin A - \sin B}$

by (11)
by (12) $\quad$,, $= \dfrac{2 \sin \frac{1}{2}(A+B) \cos \frac{1}{2}(A-B)}{2 \cos \frac{1}{2}(A+B) \sin \frac{1}{2}(A-B)}$

,, $= \tan \frac{1}{2}(A+B) \cot \frac{1}{2}(A-B)$

or $\dfrac{a+b}{a-b} = \dfrac{\tan \frac{1}{2}(A+B)}{\tan \frac{1}{2}(A-B)}$ . . . . (23)

Prove $\sin^2 A - \sin^2 B = \sin(A+B) \sin(A-B)$ . . . . (24)

$\sin^2 A - \sin^2 B = (\sin A + \sin B)(\sin A - \sin B)$ . . . . . difference of squares.
,, $= 2 \sin \frac{1}{2}(A+B) \cos \frac{1}{2}(A-B) \cdot 2 \cos \frac{1}{2}(A+B) \sin \frac{1}{2}(A-B)$ by (11)(12)
,, $= 2 \sin \frac{1}{2}(A+B) \cos \frac{1}{2}(A+B) . 2 \sin \frac{1}{2}(A-B) \cos \frac{1}{2}(A-B)$ rearrange
,, $= \sin(A+B) \sin(A-B)$ . . . . by (16) . . . . . (24)

Various other expressions may be deduced from the foregoing, for example—

$\cos 3A = \cos(2A + A)$

by (6) ,, $= \cos 2A \cos A - \sin 2A \sin A$
by (19) and (15) ,, $= (2\cos^2 A - 1)\cos A - (2 \sin A \cos A) \sin A$
,, $= 2\cos^3 A - \cos A - 2 \sin^2 A \cos A$
,, $= 2\cos^3 A - \cos A - 2(1 - \cos^2 A)\cos A$
,, $= 2\cos^3 A - \cos A - 2\cos A + 2\cos^3 A$

$\cos 3A = 4\cos^3 A - 3\cos A$ . . . . (25)

In a similar way it can be proved that $\sin 3A = 3 \sin A - 4 \sin^3 A$ . . (26)

A THREE BEARING FORMULA.

Given three bearings from a fixed point to find the angle at which a given ratio line will cross them.

In figure, let A be a fixed point, AB, AC and AD three lines of bearings drawn from it, α and β the angles between the bearings, BCD a line drawn across the bearings the ratio of BC to CD being as p is to q, expressed either in units of distance or of time.

It is required to find $\angle C$, which will give the angle between the ratio line and the middle bearing.

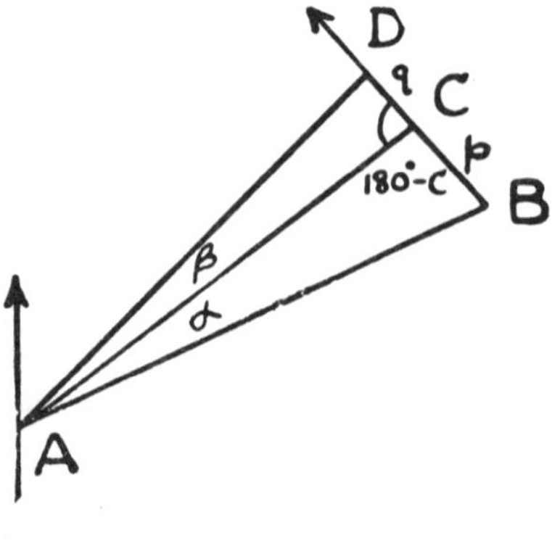

In $\triangle$BAC $\dfrac{AC}{\sin B} = \dfrac{p}{\sin \alpha}$ $\quad \therefore AC = p \dfrac{\sin B}{\sin \alpha}$

In $\triangle$DAC $\dfrac{AC}{\sin D} = \dfrac{q}{\sin \beta}$ $\quad \therefore AC = q \dfrac{\sin D}{\sin \beta}.$

$\therefore p \dfrac{\sin B}{\sin \alpha} = q \dfrac{\sin D}{\sin \beta}$

From figure, $\angle B = C - a \therefore \sin B = \sin (C-a)$

" $\angle D = 180° - C - \beta = 180° - (C + \beta) \therefore \sin D = \sin (C + \beta)$

$$A\,C = \frac{p \sin B}{\sin a} = \frac{q \sin D}{\sin \beta}$$

$$\therefore \frac{p \sin (C - a)}{\sin a} = \frac{q \sin (C + \beta)}{\sin \beta}$$

$$p \frac{\sin C \cos a - \cos C \sin a}{\sin a} = q \frac{\sin C \cos \beta + \cos C \sin \beta}{\sin \beta}$$

$$p (\sin C \cot a - \cos C) = q (\sin C \cot \beta + \cos C)$$

divide both sides by $\sin C$.

$$\therefore p (\cot a - \cot C) = q (\cot \beta + \cot C)$$

$$p \cot a - p \cot C = q \cot \beta + q \cot C$$

$$p \cot a - q \cot \beta = (p + q) \cot C$$

$$\text{or } \cot C = \frac{p \cot a - q \cot \beta}{p + q} \quad . \quad . \quad . \quad . \quad . \quad . \quad . \qquad (27)$$

NOTE.—$\angle C$ is on the opposite side to p and a, but should $\cot C$ come out negative take $\angle A\,C\,B$ instead of $\angle A\,C\,D$.

NOTES

THE COSINE FORMULA.

Prove $\cos A = \dfrac{\cos a - \cos b \cos c}{\sin b \sin c}$ . . . . (28)

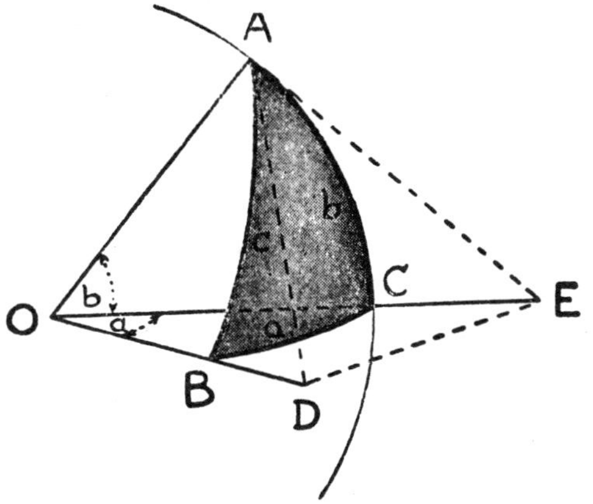

In figure, let O be the centre of a sphere, and ABC a spherical triangle on its surface, so that the radii OB, OC and OA are equal.

AE is a tangent to the great circle AC at A.
AD ,, ,, ,, AB ,,

and the plane $\angle DAE$ = spherical $\angle BAC$.

Produce OB and OC to meet the tangents at D and E. Join DE.

$\angle AOD$ measures the arc AB, side c.
$\angle DOE$,, ,, BC, side a.
$\angle AOE$,, ,, AC, side b.

by (3) In $\triangle ODE$ $DE^2 = OD^2 + OE^2 - 2\,OD\,.\,OE \cos a$.
In $\triangle ADE$ $DE^2 = AD^2 + AE^2 - 2\,AD\,.\,AE \cos A$.

Subtract, $0 = \underbrace{OD^2 - AD^2}_{OA^2} + \underbrace{OE^2 - AE^2}_{OA^2} + 2\,AD\,.\,AE \cos A - 2\,OD.OE \cos a$

$0 = OA^2 + OA^2 +$ do. $-$ do.

AD being tangent to circle AB, of which OA is a radius
$\therefore \angle OAD = 90°$ and $AO^2 = OD^2 - AD^2$.
AE being tangent to circle AC of which OA is a radius
$\therefore \angle OAE = 90°$ and $AO^2 = OE^2 - AE^2$.

$$\therefore 2\,OD\,.\,OE \cos a = 2\,OA^2 + 2\,AD\,.\,AE \cos A$$

$$\cos a = \frac{OA}{OE}\cdot\frac{OA}{OD} + \frac{AD}{OD}\cdot\frac{AE}{OE}\cos A$$

$$\cos a = \cos b \cos c + \sin c \sin b \cos A$$

or, $\cos A = \dfrac{\cos a - \cos b \cos c}{\sin b \sin c}$ . . . . (28)

similarly, $\cos B = \dfrac{\cos b - \cos c \cos a}{\sin c \sin a}$ . . . . (28)

and $\cos C = \dfrac{\cos c - \cos a \cos b}{\sin a \sin b}$ . . . . (28)

THE SINE FORMULA.

Prove $\frac{\sin a}{\sin A} = \frac{\sin b}{\sin B}$ . . . . (29)

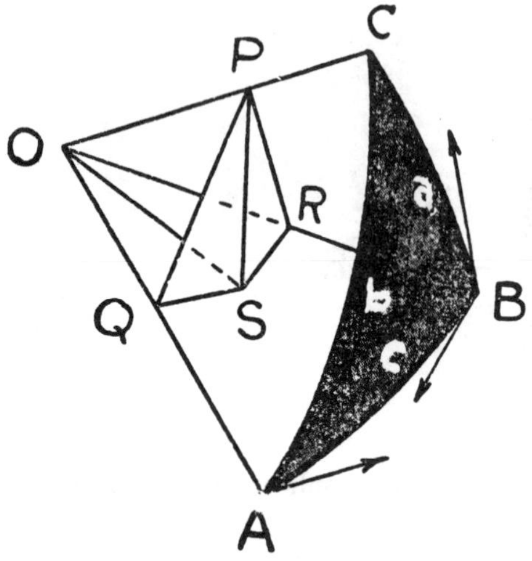

ABC is a spherical triangle, *O* the centre of the sphere, *OA*, *OB* and *OC* are radii forming the three inclined planes *OCB*, *OCA* and *OBA*.

$\angle COB$ measures the side *a*, $\angle COA$ the side *b* and $\angle BOA$ the side *c*.

Construction.—Take any point *P* in *OC*, draw *PS* perpendicular to the plane *OAB* and from *S* draw *SR* and *SQ* perpendicular to *OB* and *OA* respectively; join *PR*, *PQ*, *OS*.

Since *PS* is perpendicular to the plane *OAB*, it makes right angles with every straight line meeting it in that plane, hence angles *PSR*, *PSO* and *PSQ* are each 90°.

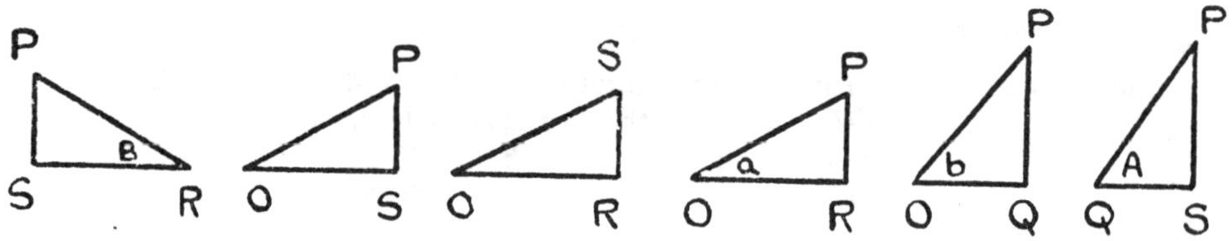

In triangles *PSR*, *PSO*, *RSO* and *PRO* in succession we have the following connections:—

$$PR^2 = PS^2 + SR^2 = PO^2 - OS^2 + SR^2 = PO^2 - OR^2 = PR^2 \therefore \angle PRO = 90°$$

In $\triangle$ *PRO*, $R=90°$, $PR = OP \sin POR = OP \sin a$

In $\triangle$ *PSR*, $S=90°$, $PS = PR \sin PRS = PR \sin B$

$= OP \sin a \sin B$

In $\triangle$ *PQO*, $Q=90°$, $PQ = OP \sin POQ = OP \sin b$ I

In $\triangle$ *PSQ*, $S=90°$, $PS = PQ \sin PQS = PQ \sin A$

$= OP \sin b \sin A$ II

$\therefore OP \sin a \sin B = OP \sin b \sin A$

and $\frac{\sin a}{\sin A} = \frac{\sin b}{\sin B}$ . . . . (29)

THE FOUR ADJACENT PARTS FORMULA.

Prove $\sin A \cot B = \sin c \cot b - \cos c \cos A$

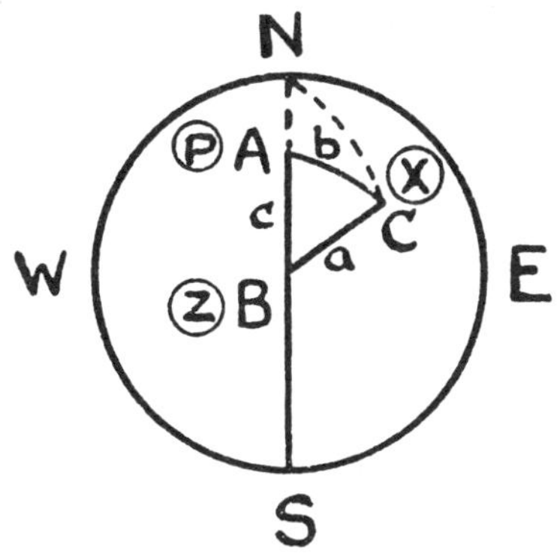

In figure, given $\Delta\ ABC$. Draw in great circle NC

If $\angle BAC = A$, then $\angle NAC = 180° - A$ so that $\cos NAC = \cos(180° - A) = -\cos A$, side $NB = 90°$ so NA and AB are complementary
and $\cos NA = \sin AB = \sin c$
and $\sin NA = \cos AB = \cos c$
In $\Delta\ NBC$, $NB = 90°$,
$\therefore \sin$ "comp" $NC = \cos B \cos$ "comp" a

$$\cos NC = \cos B \sin a \qquad \text{I}$$

In $\Delta\ ABC$, $\dfrac{\sin a}{\sin A} = \dfrac{\sin b}{\sin B}$ $\therefore \sin a = \dfrac{\sin A \sin b}{\sin B}$ which

substitute in I so that

$$\cos NC = \cos B \times \frac{\sin A \sin b}{\sin B} = \cot B \sin A \sin b \quad . \quad . \quad . \quad . \quad \text{II}$$

In $\Delta\ NAC$, $\cos NC = \cos NA \cos b + \sin NA \sin b \cos(180° - A)$. III

Eq. II. = Eq. III.

$$\cot B \sin A \sin b = \cos NA \cos b + \sin NA \sin b(-\cos A)$$

dividing by $\sin b$ we get

$$\cot B \sin A = \sin c \cot b - \cos c \cos A \quad . \quad . \quad . \quad . \quad . \quad . \quad (80)$$

It may be noticed that the $\Delta\ ABC$ is just the PZX triangle of nautical astronomy and that the four part formula is the basis of the ABC Azimuth Tables, so by substitution the formula

$$\cot B \sin A = \sin c \cot b - \cos c \cos A$$

becomes

$$\cot Z \sin P = \sin PZ \cot PX - \cos PZ \cos P$$

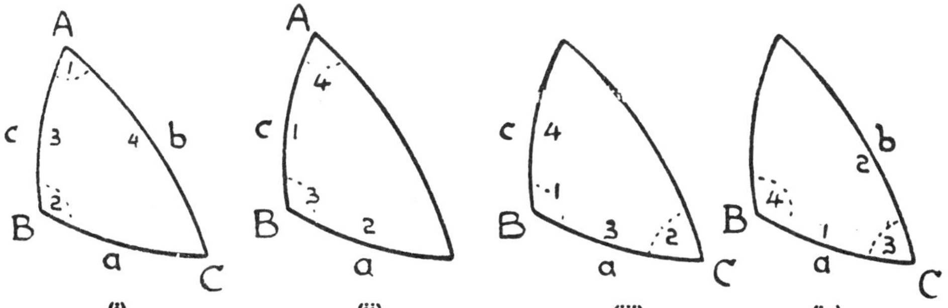

(i) (ii) (iii) (iv)

This formula is rather difficult to remember, but it will be noted that the four parts occur in their order round the triangle and that there is a symmetry and sequence of the ratios and sides that help one to visualise the connection between them.

We have drawn four triangles leaving out the letters of the two parts not required, and have moved the given parts round one place to the left, as marked in the successive triangles, and have numbered the parts in the order in which they appear in the formula. If we can visualise and apply the formula in (i) we can write it down from symmetry in the other cases by simply moving our numbers round as shown in (ii), (iii) and (iv.)

$$\text{(i)}\quad \overset{1}{\sin A}\ \overset{2}{\cot B} = \overset{3}{\sin c}\ \overset{4}{\cot b} - \overset{3}{\cos c}\ \overset{1}{\cos A}$$

$$\text{(ii)}\quad \sin c \cot a = \sin B \cot A + \cos B \cos c$$

Note that the sign is + when the left side of the formula contains two sides.

$$\text{(iii)}\quad \overset{1}{\sin B}\ \overset{2}{\cot C} = \overset{3}{\sin a}\ \overset{4}{\cot c} - \overset{3}{\cos a}\ \overset{1}{\cos B}$$

$$\text{(iv)}\quad \sin a \cot b = \sin C \cot B + \cos C \cos a$$

Reference is made on page 119 to Norie's ABC Azimuth Table formula, but no special advantage seems to be gained in the modification of the direct formula by dividing each term by sin *PZ* sin *P* as previously shown.

The following example will show the application of the formula without the camouflage adopted in Norie's Tables.

Given L.H.A. 45°, Decl. 15° N., Lat. 40° N., required the Azimuth.

$$\cot Z \sin P = \underset{A}{\sin PZ \cot PX} - \underset{B}{\cos PZ \cos P}$$

PZ 50°	sin	9·88425			*P* 3h	cos	9·80807
PX 75°	cot	9·42804			- - -	cos	9·84949
	log	9·31230				log	9·65756
Nat No. A +		·2053					
Nat No. B −		·4545					
+		·2492	log	9·39653			
P		3h	cosec	0·15052			
Z		70 35½	cot	9·54705			

Azimuth S. 70° 35½′ W.

The quantity *A* has a + sign when *PX* is less than 90°

THE HAVERSINE FORMULA.

Prove $\text{hav}\,A = \dfrac{\text{hav}\,a - \text{hav}\,(b-c)}{\sin b \sin c}$ . . . . (31)

by (28) $$\cos A = \frac{\cos a - \cos b \cos c}{\sin b \sin c}$$

$$1 - \cos A = 1 - \frac{\cos a - \cos b \cos c}{\sin b \sin c}$$

by (18) $$\text{vers}\,A = \frac{(\sin b \sin c + \cos b \cos c) - \cos a}{\sin b \sin c}$$

$$\sin b \sin c \,\text{vers}\,A = \cos(b-c) - \cos a$$

add and subtract unity and rearrange.

" $= 1 - \cos a - [1 - \cos(b-c)]$

" $= \text{vers}\,a - \text{vers}\,(b-c)$

Divide throughout by 2.

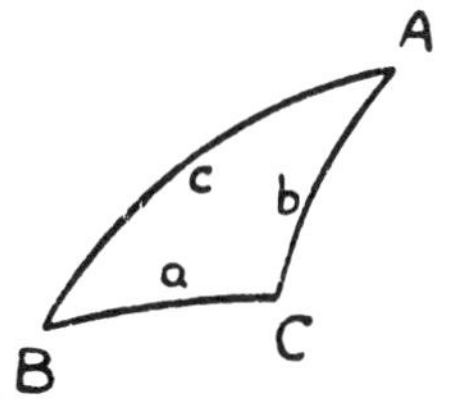

$$\sin b \sin c \frac{\text{vers}\,A}{2} = \frac{\text{vers}\,a}{2} - \frac{\text{vers}\,(b-c)}{2}$$

$$\therefore \text{hav}\,A = \frac{\text{hav}\,a - \text{hav}\,(b-c)}{\sin b \sin c} \quad . \quad . \quad . \quad (31)$$

This form is used when three sides are given to find an angle but it may be rewritten in the form

$$\text{hav}\,a = \sin b \sin c \,\text{hav}\,A + \text{hav}\,(b-c) \quad . \quad . \quad . \quad (31)$$

when given two sides and included angle, to find the third side, and proved as follows:

In (18) it was shown that $\cos A = 1 - 2\,\text{hav}\,A$ so by symmetry the angle $\cos(b-c) = 1 - 2\,\text{hav}(b-c)$

by (28) $\cos a = \cos b \cos c + \sin b \sin c \cos A$

" $= \cos b \cos c + \sin b \sin c\,(1 - 2\,\text{hav}\,A)$

" $= (\cos b \cos c + \sin b \sin c) - 2 \sin b \sin c \,\text{hav}\,A$

" $= \cos(b-c) - 2 \sin b \sin c \,\text{hav}\,A$

$$\cancel{1} - \cancel{2}\,\text{hav}\,a = \cancel{1} - \cancel{2}\,\text{hav}\,(b-c) - \cancel{2} \sin b \sin c \,\text{hav}\,A$$

$$\text{hav}\,a = \text{hav}\,(b-c) + \sin b \sin c \,\text{hav}\,A$$

$$\text{or hav}\,A = \frac{\text{hav}\,a - \text{hav}\,(b-c)}{\sin b \sin c} \quad . \quad . \quad . \quad . \quad . \quad . \quad . \quad (31)$$

In the *PZX* triangle of nautical astronomy,

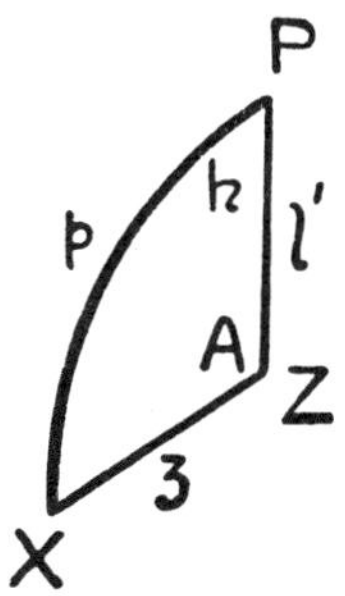

where $\angle P$ = hour angle
$\angle A$ = azimuth
z = zenith distance
p = polar distance
l' = co-latitude

the formula would read

$$\text{hav}\,P = \frac{\text{hav}\,z - \text{hav}\,(p - l')}{\sin p \sin l'} \quad . \quad . \quad . \quad (31)$$

the steps of the proof being exactly as above.

ALL HAVERSINE FORMULA.

$$\text{Prove hav } A = \frac{\text{hav } a - \text{hav }(b-c)}{\text{hav }(b+c) - \text{hav }(b-c)} \quad \ldots\ldots \quad (32)$$

by (28)

$$\cos a = \cos b \cos c + \sin b \sin c \cos A$$
$$\text{"} = \cos b \cos c + \sin b \sin c\,(1 - 2 \text{ hav } A)$$
$$\text{"} = (\cos b \cos c + \sin b \sin c) - 2 \sin b \sin c \text{ hav } A$$
$$\text{"} = \cos (b-c) - 2 \sin b \sin c \text{ hav } A$$
$$\not{1} - \not{2} \text{ hav } a = \not{1} - \not{2} \text{ hav }(b-c) - \not{2} \sin b \sin c \text{ hav } A$$
$$\text{hav } a = \text{hav }(b-c) + \sin b \sin c \text{ hav } A$$
$$\text{hav } a = \text{hav }(b-c) + [\text{hav }(b+c) - \text{hav }(b-c)] \text{ hav } A$$
$$\therefore \text{hav } A = \frac{\text{hav } a - \text{hav }(b-c)}{\text{hav }(b+c) - \text{hav }(b-c)} \quad \ldots\ldots \quad (32)$$

It may be noted that the first five lines in the transformation are the same steps as in proof (31) and that it is left to convert $\sin b \sin c$ into haversines as follows:—

By proofs (13) and (14) 4th line we get

$$2 \sin A \sin B = \cos (A - B) - \cos (A + B)$$
$$\text{so that } \sin b \sin c = \tfrac{1}{2}\{\cos (b-c) - \cos (b+c)\}$$
$$\text{"} = \tfrac{1}{2}\{1 - 2 \text{ hav }(b-c) - [1 - 2 \text{ hav }(b+c)]\}$$

$\therefore \sin b \sin c = \text{hav }(b+c) - \text{hav }(b-c)$ as substituted above.

In the *PZX* triangle the formula would be written

$$\text{hav } P = \frac{\text{hav } z - \text{hav }(p-l)}{\text{hav }(p+l) - \text{hav }(p-l)}$$

but if the complementary parts, latitude (L) and declination (D) be substituted for l and p respectively, due attention must be paid to the angle ($L \pm D$), viz., one named north and the other south use ($L + D$), but when they have the same names use ($L - D$). The formula is then modified to meet the condition $1 - \text{hav } A = \text{hav }(180° - A)$ into

$$\text{hav } P = \frac{\text{hav } z - \text{hav }(L \pm D)}{1 - \{\text{hav}(L + D) + \text{hav }(L - D)\}} \quad \ldots\ldots \quad (32)$$

The transformation is as follows:—

$$1 - \text{hav } A = 1 - \frac{1 - \cos A}{2} = \frac{2 - (1 - \cos A)}{2} = \frac{1 + \cos A}{2}$$
$$= \frac{1 - \cos (180° - A)}{2} = \text{hav }(180° - A)$$

again

$$\text{hav }(p + l) = \text{hav }(90° - D + 90° - L)$$
$$\text{"} = \text{hav }[180° - (L + D)] = 1 - \text{hav }(L + D)$$
$$\text{hav }(p - l) = \text{hav }(90° - D - 90° + L) = \text{hav }(L - D)$$
$$\therefore \text{hav}(p + l) - \text{hav}(p - l) = 1 - \text{hav }(L + D) - \text{hav }(L - D)$$
$$\text{"} = 1 - [\text{hav }(L + D) + \text{hav }(L - D)] \text{ as substituted.}$$

A THREE SIDES FORMULA.

Prove $\text{hav } A = \dfrac{\sin (s - b) \sin (s - c)}{\sin b \sin c}$ where $s = \dfrac{a + b + c}{2}$ . . . . . (33)

by (28) $$1 - \cos A = 1 - \frac{\cos a - \cos b \cos c}{\sin b \sin c}$$

$$\text{,,} \quad = \frac{\cos b \cos c + \sin b \sin c - \cos a}{\sin b \sin c}$$

by (9) $$\text{,,} \quad = \frac{\cos (b - c) - \cos a}{\sin b \sin c}$$

by (14) $$\text{vers } A = \frac{2 \sin \frac{1}{2} (a + b - c) \sin \frac{1}{2} (a + c - b)}{\sin b \sin c}$$

If $2s = a + b + c$, then $2s - 2c = a + b + c - 2c$

$$2 (s - c) = a + b - c$$

$$\therefore (s - c) = \tfrac{1}{2} (a + b - c)$$

Similarly, $$(s - b) = \tfrac{1}{2} (a + c - b)$$

Substituting in above we get

$$\frac{\text{vers } A}{2} = \frac{\sin (s - c) \sin (s - b)}{\sin b \sin c}$$

$$\therefore \text{hav } A = \frac{\sin (s - b) \sin (s - c)}{\sin b \sin c} \quad . \; . \; . \; . \; . \; . \; . \quad (33)$$

This may also be written as

$$\sin \frac{A}{2} = \sqrt{\frac{\sin (s - b) \sin (s - c)}{\sin b \sin c}} \quad . \; . \; . \; . \; . \; . \quad (33)$$

because $$\frac{\text{vers } A}{2} = \frac{1 - \cos A}{2} = \frac{2 \sin^2 A/2}{2} = \sin^2 \frac{A}{2}$$

In the *PZX* triangle this would read

$$\text{hav } P = \frac{\sin (s - p) \sin (s - l')}{\sin p \sin l'} \quad . \; . \; . \; . \; . \; . \; . \quad (33)$$

where $$s = \frac{z + p + l'}{2}$$

A THREE SIDES FORMULA.

Prove $\cos\frac{A}{2} = \sqrt{\frac{\sin s \sin (s - a)}{\sin b \sin c}}$ where $s = \frac{a + b + c}{2}$ . . . . (34)

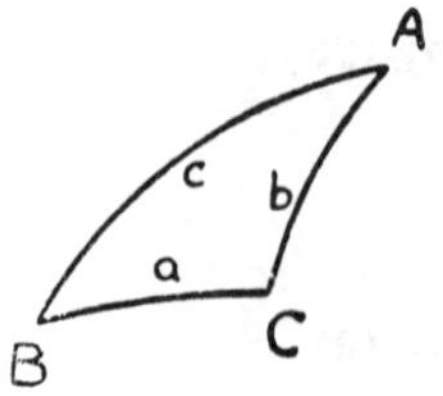

by (28) $1 + \cos A = 1 + \frac{\cos a - \cos b \cos c}{\sin b \sin c}$

" $= \frac{\cos a - (\cos b \cos c - \sin b \sin c)}{\sin b \sin c}$

by (6) " $= \frac{\cos a - \cos (b + c)}{\sin b \sin c}$

by (14) $2\cos^2\frac{A}{2} = \frac{2 \sin \frac{1}{2}(a + b + c) \sin \frac{1}{2}(b + c - a)}{\sin b \sin c}$

$\cos^2\frac{A}{2} = \frac{\sin s \sin (s - a)}{\sin b \sin c}$

$\therefore \cos\frac{A}{2} = \sqrt{\frac{\sin s \sin (s - a)}{\sin b \sin c}}$ . . . . (34)

This would read in the PZX triangle

$$\cos\frac{P}{2} = \sqrt{\frac{\sin s \sin (s - z)}{\sin p \sin l'}} \text{ where } s = \frac{p + l' + z}{2}$$

A THREE SIDES FORMULA.

Prove $\tan\frac{A}{2} = \sqrt{\frac{\sin (s - b) \sin (s - c)}{\sin s \sin (s - a)}}$ where $s = \frac{a + b + c}{2}$ . . . . . (35)

by (33) and (34) $\tan\frac{A}{2} = \frac{\sin\frac{A}{2}}{\cos\frac{A}{2}} = \frac{\sqrt{\frac{\sin (s - b) \sin (s - c)}{\sin b \sin c}}}{\sqrt{\frac{\sin s \sin (s - a)}{\sin b \sin c}}}$

$\therefore \tan\frac{A}{2} = \sqrt{\frac{\sin (s - b) \sin (s - c)}{\sin s \sin (s - a)}}$ . . . . . . (35)

Similarly, $\cot\frac{A}{2} = \sqrt{\frac{\sin s \sin (s - a)}{\sin (s - b) \sin (s - c)}}$ . . . . . . (36)

THE SINE FORMULA.

Prove $\frac{\sin a}{\sin b} = \frac{\sin A}{\sin B}$ . . . . (37)

by (15) $\sin A = 2 \sin\frac{A}{2} \cos\frac{A}{2}$

by (33) and (34) $\sin A = 2 \sqrt{\frac{\sin (s - b) \sin (s - c)}{\sin b \sin c}} \sqrt{\frac{\sin s \sin (s - a)}{\sin b \sin c}}$

(i.) $\sin A = 2 \frac{\sqrt{\sin s \sin (s - a) \sin (s - b) \sin (s - c)}}{\sin b \sin c}$

Similarly, (ii.) $\sin B = 2\frac{\sqrt{\sin s \sin (s-a) \sin (s-b) \sin (s-c)}}{\sin a \sin c}$

Dividing (i.) by (ii.) we get

$$\frac{\sin A}{\sin B} = \frac{\sin a}{\sin b} \quad \dots\dots\dots\dots \quad (37)$$

See also geometrical proof (29).

NAPIER'S ANALOGIES.

Prove $\tan \frac{1}{2}(A+B) = \frac{\cos \frac{1}{2}(a-b)}{\cos \frac{1}{2}(a+b)} \cot \frac{C}{2}$ . . . . (38)

by (28) (i.) $\cos A = \frac{\cos a - \cos b \cos c}{\sin b \sin c}$

(ii.) $\cos c = \cos a \cos b + \sin a \sin b \cos C$

Substitute (ii.) for $\cos c$ in (i.)

$$\cos A = \frac{\cos a - \cos b (\cos a \cos b + \sin a \sin b \cos C)}{\sin b \sin c}$$

$$,, \quad = \frac{\cos a - \cos a \cos^2 b + \sin a \sin b \cos b \cos C}{\sin b \sin c}$$

$$,, \quad = \frac{\cos a (1 - \cos^2 b) - \sin a \sin b \cos b \cos C}{\sin b \sin c}$$

$$,, \quad = \frac{\cos a \sin^2 b - \sin a \sin b \cos b \cos C}{\sin b \sin c}$$

$$\cos A = \frac{\cos a \sin b - \sin a \cos b \cos C}{\sin c}$$

By substituting B for A and b for a, we may write down from symmetry

$$\cos B = \frac{\cos b \sin a - \sin b \cos a \cos C}{\sin c}$$

then adding and grouping the quantities we get

$\cos A + \cos B = (\cos a \sin b, + \sin a \cos b) - \cos C (\sin a \cos b + \cos a \sin b)$

$$\cos A + \cos B = \frac{\sin (a+b) - \sin (a+b) \cos C}{\sin c}$$

$$,, \quad = \frac{\sin (a+b) \; [1 - \cos C]}{\sin c}$$

by (18) $\cos A + \cos B = \frac{\sin (a+b) \, 2 \sin^2 \frac{C}{2}}{\sin c}$ . . . . (iii.)

But by (29) $\frac{\sin A}{\sin C} = \frac{\sin a}{\sin c}$ $\therefore \sin A = \frac{\sin a \sin C}{\sin c}$

and $\sin B = \frac{\sin b \sin C}{\sin c}$

Add, $\sin A + \sin B = \frac{\sin C (\sin a + \sin b)}{\sin c}$ . . . . . . (iv.)

Divide (iv.) by (iii.).

$$\frac{\sin A + \sin B}{\cos A + \cos B} = \frac{\sin C\,(\sin a + \sin b)}{2\sin^2\frac{C}{2}\sin(a+b)}$$

by (11) ,, $$= \frac{\sin C}{2\sin^2\frac{C}{2}} \times \frac{2\sin\frac{1}{2}(a+b)\cos\frac{1}{2}(a-b)}{\sin(a+b)}$$

by (15) $\sin C = 2\sin\frac{C}{2}\cos\frac{C}{2}$, and expanding $\sin(a+b)$ by (17) we get

$$\frac{\sin A + \sin B}{\cos A + \cos B} = \frac{2\sin\frac{C}{2}\cos\frac{C}{2}}{2\sin^2\frac{C}{2}} \times \frac{2\sin\frac{1}{2}(a+b)\cos\frac{1}{2}(a-b)}{2\sin\frac{1}{2}(a+b)\cos\frac{1}{2}(a+b)}$$

the right hand side cancels to

by (11)
by (13) $$\frac{2\sin\frac{1}{2}(A+B)\cos\frac{1}{2}(A-B)}{2\cos\frac{1}{2}(A+B)\cos\frac{1}{2}(A-B)} = \frac{\cos\frac{C}{2}}{\sin\frac{C}{2}} \times \frac{\cos\frac{1}{2}(a-b)}{\cos\frac{1}{2}(a+b)}$$

$$\tan\tfrac{1}{2}(A+B) = \frac{\cos\frac{1}{2}(a-b)}{\cos\frac{1}{2}(a+b)}\cot\frac{C}{2} \quad \ldots\ldots \quad (38)$$

Prove $\tan\frac{1}{2}(A-B) = \dfrac{\sin\frac{1}{2}(a-b)}{\sin\frac{1}{2}(a+b)}\cot\dfrac{C}{2}$ . . . . (39)

by (12)
by (13) $$\frac{\sin A - \sin B}{\cos A + \cos B} = \frac{2\cos\frac{1}{2}(A+B)\sin\frac{1}{2}(A-B)}{2\cos\frac{1}{2}(A+B)\cos\frac{1}{2}(A-B)} = \tan\tfrac{1}{2}(A-B)$$

$$\therefore \tan\tfrac{1}{2}(A-B) = \frac{\sin A - \sin B}{\cos A + \cos B}$$

by (29)
by (iii.) in (38)

$$\therefore \tan\tfrac{1}{2}(A-B) = \frac{\dfrac{\sin a\sin C}{\sin c} - \dfrac{\sin b\sin C}{\sin c}}{\dfrac{\sin(a+b)\,2\sin^2\frac{C}{2}}{\sin c}}$$

sin c cancels and numerator becomes

$$\therefore \tan\tfrac{1}{2}(A-B) = \frac{\sin C\,(\sin a - \sin b)}{2\sin^2\frac{C}{2}\sin(a+b)}$$

by (15), (12) and (17)

$$\therefore \tan\tfrac{1}{2}(A-B) = \frac{2\sin\frac{C}{2}\cos\frac{C}{2}}{2\sin^2\frac{C}{2}} \times \frac{2\cos\frac{1}{2}(a+b)\sin\frac{1}{2}(a-b)}{2\cos\frac{1}{2}(a+b)\sin\frac{1}{2}(a+b)}$$

which cancels to

$$\tan \tfrac{1}{2}(A - B) = \frac{\sin \tfrac{1}{2}(a - b)}{\sin \tfrac{1}{2}(a + b)} \cot \frac{C}{2} \quad \ldots \ldots \quad (39)$$

$$\tan \tfrac{1}{2}(A + B) = \frac{\cos \tfrac{1}{2}(a - b)}{\cos \tfrac{1}{2}(a + b)} \cot \frac{C}{2} \quad \ldots \ldots \quad (38)$$

Napier's Analogies (38) and (39) when combined are applied to finding the two angles when given two sides with their included angle, as in great circle sailing to find the courses, and in time azimuth examples.

One of the following two analogies may be applied to find a side when the other two sides with their opposite angles are given.

$$\frac{\tan \tfrac{1}{2} c}{\tan \tfrac{1}{2}(a + b)} = \frac{\cos \tfrac{1}{2}(A + B)}{\cos \tfrac{1}{2}(A - B)} \quad \ldots \ldots \quad (40)$$

$$\frac{\tan \tfrac{1}{2} c}{\tan \tfrac{1}{2}(a - b)} = \frac{\sin \tfrac{1}{2}(A + B)}{\sin \tfrac{1}{2}(A - B)} \quad \ldots \ldots \quad (41)$$

Given three angles to find a side

$$\text{hav } a = \frac{\cos S \cos (S - A)}{\sin B \sin C} \quad \ldots \ldots \quad (42)$$

where S=half sum of the three angles.

RIGHT ANGLED SPHERICAL TRIANGLES.

ABC is a spherical triangle, *O* the centre of the sphere, $\angle ACB = 90°$ so the plane *AOC* is perpendicular to the plane *BOC*. $OA = OC = OB$ radii of the sphere.

Construction.—Make *DE* perpendicular to *OC* and *EF* perpendicular to *OA* $\angle DEF = 90°$ being at the intersection of two perpendicular planes, $\angle DEF = \angle C$.

Join *DF* and prove $\angle OFD = 90°$.

$\angle EFO = 90° \quad \therefore \; OF^2 + FE^2 = OE^2$. Add DE^2

$OF^2 + (FE^2 + DE^2) = OE^2 + DE^2$

$OF^2 + DF^2 = OD^2$

$\therefore \; \angle DFO = 90°$

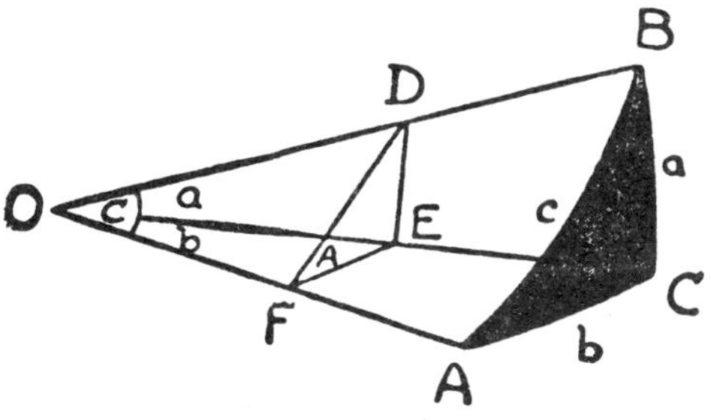

$\angle BOC$	measures	side a
$\angle COA$	,,	,, b
$\angle AOB$	,,	,, c
$\angle EFD$	,,	$\angle A$
$\angle DEF$	,,	$\angle C$

The triangles referred to are as named in the following proofs, the middle letters being at the right angled corner, there being so many of them that one is easily confused.

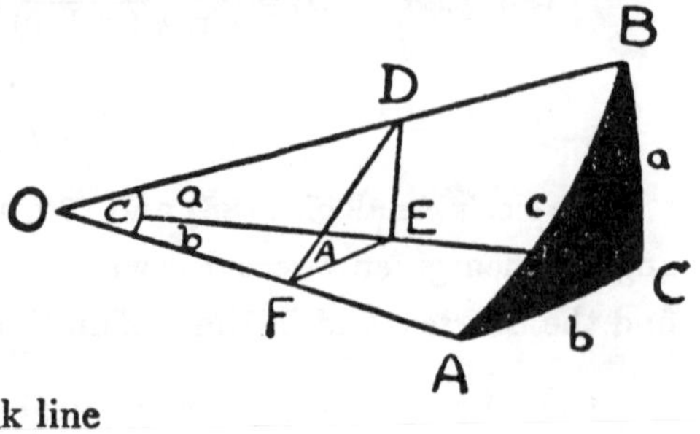

(1) $\angle DFO$, $\angle EFO$, $\angle DEF = 90°$

$\dfrac{OF}{OD} = \dfrac{OF}{OE} \cdot \dfrac{OE}{OD}$, using OE as a link line

$\therefore$ $\cos FOD = \cos EOF \cos DOE$

or $\cos c = \cos b \cos a$.

(2) $\angle DEO$, $\angle DEF$, $\angle DFO = 90°$

$\dfrac{DE}{OD} = \dfrac{DE}{DF} \cdot \dfrac{DF}{OD}$, using DF as a link line

$\sin a = \sin A \sin c$ . . . . or $\sin A = \dfrac{\sin a}{\sin c}$

(3) Similarly it can be shown, by making FE and FD perpendicular to OC and OB respectively,

$$\angle FDE = B$$

$\sin b = \sin B \sin c$ . . . . or $\sin B = \dfrac{\sin b}{\sin c}$

(4) $\angle EFO$, $\angle DEF$, $\angle DFO = 90°$

$\dfrac{EF}{OF} = \dfrac{EF}{DF} \cdot \dfrac{DF}{OF}$

$\therefore$ $\tan b = \cos A \tan c$ . . . . or $\cos A = \dfrac{\tan b}{\tan c}$

(5) Similarly $\tan a = \cos B \tan c$ . . . . or $\cos B = \dfrac{\tan a}{\tan c}$

(6) $\angle DEF$, $\angle OFE$, $\angle DEO = 90°$

$\dfrac{ED}{EF} = \dfrac{OE}{EF} \cdot \dfrac{ED}{OE}$

$\therefore$ $\tan A = \operatorname{cosec} b \tan a$ . . . . or $\tan A = \dfrac{\tan a}{\sin b}$

(7) Similarly $\tan B = \operatorname{cosec} a \tan b$ . . . . or $\tan B = \dfrac{\tan b}{\sin a}$

(8) By (2) $\sin a = \sin A \sin c$

By (5) $\tan a = \cos B \tan c$

Divide (2) by (5) $\cos a = \dfrac{\sin A \cos c}{\cos B}$

or $\cos B = \dfrac{\sin A \cos c}{\cos a}$ but by (1) $\cos b = \dfrac{\cos c}{\cos a}$

$\therefore$ $\cos B = \sin A \cos b$ . . . . or $\sin A = \dfrac{\cos B}{\cos b}$

(9) By (3) $\sin b = \sin B \sin c$

By (4) $\tan b = \cos A \tan c$

Divide (8) by (4) $\cos b = \dfrac{\sin B \cos c}{\cos A}$

or $\cos A = \dfrac{\sin B \cos c}{\cos b}$ but by (1) $\cos a = \dfrac{\cos c}{\cos b}$

$\therefore \cos A = \sin B \cos a$ . . . . or $\sin B = \dfrac{\cos A}{\cos a}$

(10) Multiply (6) by (7).

$$\tan A \tan B = \frac{\tan a}{\sin b} \cdot \frac{\tan b}{\sin a}$$

$$" \qquad = \frac{1}{\cos a \cos b} = \frac{1}{\cos c,} \quad \ldots\ldots\ldots \text{ by (1)}$$

or $\cot A \cot B = \cos c$ by using reciprocals.

It is interesting to compare these formulæ with the corresponding ones for plane triangles and to note the similarity.

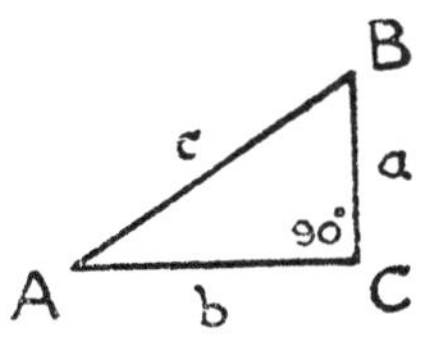

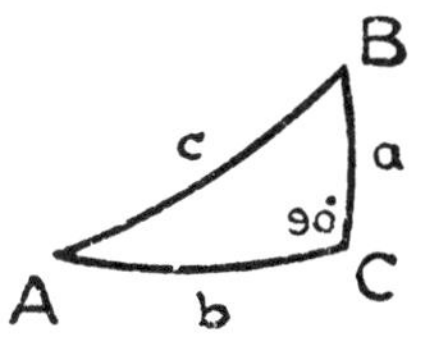

In plane right triangles.

$\sin A = \dfrac{a}{c}$ $\qquad \sin B = \dfrac{b}{c}$

$\cos A = \dfrac{b}{c}$ $\qquad \cos B = \dfrac{a}{c}$

$\tan A = \dfrac{a}{b}$ $\qquad \tan B = \dfrac{b}{a}$

$\sin A = \cos B$ $\qquad \sin B = \cos A$

$c^2 = a^2 + b^2$

$1 = \cot A \cot B$

In spherical right triangles.

(2) $\sin A = \dfrac{\sin a}{\sin c}$ $\qquad$ (3) $\sin B = \dfrac{\sin b}{\sin c}$

(4) $\cos A = \dfrac{\tan b}{\tan c}$ $\qquad$ (5) $\cos B = \dfrac{\tan a}{\tan c}$

(6) $\tan A = \dfrac{\tan a}{\sin b}$ $\qquad$ (7) $\tan B = \dfrac{\tan b}{\sin a}$

(8) $\sin A = \dfrac{\cos B}{\cos b}$ $\qquad$ (9) $\sin B = \dfrac{\cos A}{\cos a}$

(1) $\cos c = \cos a \cos b$

(10) $\cos c = \cot A \cot B$

These ten formulae for right angled triangles are grouped under the two rules referred to on page 112, as Napier's Rules for Circular Parts.

THE TRIGONOMETRICAL RATIOS FOR SMALL ANGLES.

If a small angle θ is expressed in circular measure, we have

$$\sin\theta = \theta \text{ radians}; \quad \cos\theta = 1; \quad \tan\theta = \theta \text{ radians}$$

$$1 \text{ radian} = 57^\circ\ 17'\ 45'' = 3437{\cdot}75' = 206265''$$

so that $\sin 1' = \dfrac{1}{3438}$ radian, $\sin 1'' = \dfrac{1}{206265}$ radian.

When angle θ is very small the chord and the subtended arc of the circle are very nearly equal, and if θ be indefinitely diminished we can make $\dfrac{\theta''}{\sin\theta''}$ differ from unity by less than any assignable quantity, however small, so that we may write

$$\sin\theta'' = \theta''\sin 1''$$
$$\text{or } \sin\theta' = \theta'\sin 1' \quad \ldots\ldots \quad (48)$$

This approximation is frequently made use of in spherical astronomy, remembering that angles are frequently expressed in terms of time,

$$24 \text{ hours} = 360^\circ;\ 1\text{h} = 15^\circ;\ 1\text{m} = 15';\ 1\text{s} = 15''$$

and approximately,

$$\sin 1\text{m} = \sin 15' = 15\sin 1'$$
$$\text{or } \sin 1\text{s} = \sin 15'' = 15\sin 1'' \quad \ldots\ldots \quad (43)$$

THE POLAR TRIANGLE.

Every spherical triangle has its polar triangle formed by the great circles on the sphere joining the poles of its respective three sides. The pole of a great circle is that point on the sphere 90° removed from every part of the great circle.

In figure, given triangle ABC with A^1, B^1, C^1, the respective poles of the arcs CB AC, and BA. Then $A^1B^1C^1$ is the polar triangle of ABC, and A^1 is 90° from every spot on CB. Produce AC and AB to meet the polar arc C^1B^1 at x and y, then A is the pole of arc C^1B^1 and arc xy=angle A, because Ax and Ay=90°

B^1 is the pole of arc AC and is 90° removed from it.

C^1 " AB " " "

so that $B^1x = 90^\circ$ and $C^1y = 90^\circ \therefore B^1x + C^1y = 180^\circ$

But $B^1C^1 = C^1y + yB^1$

and $xy = B^1x - yB^1$

add, $B^1C^1 + xy = B^1x + C^1y = 180^\circ$ as above, but B^1C^1 is a^1, the side opposite angle A' of the polar triangle $A^1B^1C^1$, and xy = angle A of the ABC triangle

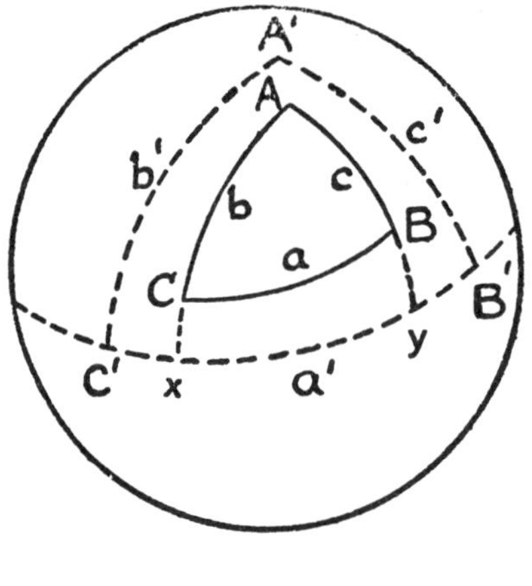

$\therefore B^1C^1 + xy = a^1 + A = 180°$. Similarly, if $A^1B^1C^1$ be the polar triangle of any spherical triangle ABC, with sides lettered $a^1b^1c^1$ correspondingly, we deduce that—

$180° - A$	of any triangle	=	side	a^1	of its polar triangle
$180° - B$	„	=	„	b^1	„
$180° - C$	„	=	„	c^1	„

And conversely—

$180° - A^1$	of polar triangle	=	side	a	of the ABC triangle.
$180° - B^1$	„	=	„	b	„
$180° - C^1$	„	=	„	c	„

The formulae of a spherical triangle apply also to its polar triangle, for example, in the fundamental formula $\cos a = \cos b \cos c + \sin b \sin c \cos A$.

If $\pi = 180°$, then the polar equivalent is
$\cos(\pi - A^1) = \cos(\pi - B^1)\cos(\pi - C^1) + \sin(\pi - B^1)\sin(\pi - C^1)\cos(\pi - a^1)$
or, simply, $-\cos A^1 = \cos B^1 \cos C^1 - \sin B^1 \sin C^1 \cos a^1$.

Thus the angles and sides of the triangle may be converted into the corresponding sides and angles of its polar triangle, and the usual formulae applied to the solution of the triangle.

NOTE.—Cos of an angle = − cos of its supplement
Sin „ = + sin „ „

POLAR TRIANGLE.

Example.—In a spherical triangle ABC, $\angle A = 57° 26' 45''$, $\angle B = 94° 0' 45''$, $\angle C = 71° 51' 30''$. Find sides a and b.

Subtract each angle from 180° and name them a^1, b^1, c^1 respectively, then apply the appropriate spherical formula. The result will be the supplement of the side required. Suppose we apply the following established formula—

$$\text{hav } A^1 = \frac{\sin(s - b^1)\sin(s - c^1)}{\sin b^1 \sin c^1}$$

$$\text{hav } B^1 = \frac{\sin(s - a^1)\sin(s - c^1)}{\sin a^1 \sin c^1}$$

where $s = \frac{1}{2}(a^1 + b^1 + c^1)$

						(A^1)	(B^1)
$(180° - A)$	a^1	122°	33′	15″	cosec	. . . .	. 0·07423
$(180° - B)$	b^1	85	59	15	cosec	0·00107	
$(180° - C)$	c^1	108	08	30	cosec	0·02214	0·02214
		2)316	41	00			
s		158	20	30			
$(s - a^1)$		35	47	15	sin	. . . .	. 9·76700
$(s - b^1)$		72	21	15	sin	9·97907	
$(s - c^1)$		50	12	00	sin	9·88552	9·88552
A^1		123	00		hav	9·88780	9·74889
		180					
Side a		57	00				

B^1	96°	59′	45″
	180		
side b	83	00	15

The triangle may be solved direct from the three given angles, and thus avoid supplementary parts, by the formula—

$$\text{hav } a = \frac{\cos S \times \cos (S - A)}{\sin B \sin C}$$

$$\text{hav } b = \frac{\cos S \times \cos (S - B^1)}{\sin A \sin C}$$

$$\text{where } S = \tfrac{1}{2}(A + B + C)$$

				(a)	(b)
A	57° 26′ 45″	cosec	. .	0·07428
B	94 00 45	cosec	0·00112	. . .
C	71 51 30	cosec	0·02214	0·02214
	2)223 19 00			
S	111 39 30	cos	9·56711	9·56711
$(S - A)$	54° 12′ 45″	cos	9·76698	. . .
$(S - B)$	17 38 45	cos	. . .	9·97907
Side a	57 00 00	hav	9·35735	9·64255
			side b	83° 00½′

1. Initial course of a great circle track in South hemisphere is N. 66° 13′ 45″ E., the final course is N. 43° 20′ 40″ E., the difference of longitude is 112°. Required the great circle distance between the two positions and their latitudes.

Ans. Lat. 42° 46¾′ S., Long. 11° 50¼′ S., Dist. 97° 27′ = 5847 miles.

2. In spherical triangle ABC, find the value of the sides, if $\angle A = 57° \, 26' \, 45''$, $\angle B = 94° \, 00' \, 45''$, $\angle C = 71° \, 51' \, 30''$.

Ans. a 57° 00′, b 83° 00½′, c 71° 01′.

SPHERICAL EXCESS.

A geodetic survey is one which covers a considerable area of the Earth's surface and takes into account the curvature of the sphere.

The distances between observation stations in a geodetic survey may exceed 40 or 50 miles and the triangles are therefore spherical. The angles measured by theodolite at the triangulation stations are spherical angles and the sum of the three angles of the triangle will exceed 180 degrees. The excess of angles $A + B + C$ over 180° is called the *spherical excess* and is usually denoted by E, so that $E = A + B + C - 180°$.

The amount of spherical excess depends upon the area of the triangle and also upon the latitudes of the observation stations; it is approximately equal to 1 second of arc for each 75·5 square miles area of the triangle.

The quantity E may be computed more accurately from the formula—

$$E = \frac{\text{area of triangle}}{R^2 \sin 1''} = \frac{\frac{1}{2} b c \sin A}{R^2 \sin 1''}$$

where b and c are the sides of the plane triangle and A the included angle, R the mean radius of the Earth. The mean radius is accepted as being 20,889,000 feet and $\log R = 7{\cdot}319917$.

Example.—If side $b = 20$ miles (105,600 ft.), side $c = 40$ miles (211,208 ft.). $\angle A = 60°$, the approximate method gives $E = 4{\cdot}5''$, and, by the accurate formula, $E = 4{\cdot}56''$.

The spherical excess varies as the area of the triangle, the area $= E\,r^2$ where r is the radius of the sphere.

The area of a spherical triangle varies as its spherical excess, so that for any two triangles $A\ B\ C$ and $A^1B^1C^1$ on the same sphere.

$$\frac{\text{area of } A\ B\ C}{\text{area of } A^1B^1C^1} = \frac{E}{E^1}$$

E is usually expressed in circular measure.

A radian $= 57°\ 17'\ 45'' = 3438' = 206265''$ so that, in circular measure, $1' = \frac{1}{3438}$ and $1'' = \frac{1}{206265}$

Example.—If the spherical excess be $1'$ and the radius of the sphere 9 inches, find the area of the triangle, and also of the sphere.

$$\text{Area of triangle} = E\,r^2 = \frac{1}{3438} \times 9^2 = {\cdot}0236 \text{ sq. ins.}$$

$$\text{Area of sphere} = 4\,\pi\,r^2 = \frac{4 \times 22 \times 9^2}{7} = 1018{\cdot}3 \text{ sq. inches.}$$

Example.—The surface area of a sphere is 4096 square feet. An equilateral spherical triangle whose angles are each $64°\ 51'\ 40''$ is described on the sphere. Find the area of the triangle.

$$A + B + C = 3 \times 64°\ 51'\ 40'' = 194°\ 35'$$
$$E = 194°\ 35' - 180° = 14°\ 35'$$
$$\text{Area of sphere} = 4\pi\,r^2 = 4096 \text{ sq. ft.}$$
$$\text{from which } r^2 = 325{\cdot}8 \text{ ft.}$$
$$E \text{ in circular measure} = \frac{14°\ 35'}{57°\ 18'} = \frac{875}{3438} \text{ in minutes of arc.}$$
$$\text{Area of triangle} = E\,r^2 = \frac{875 \times 325{\cdot}8}{3438} = 82{\cdot}96 \text{ sq. ft.}$$

A right-angled isosceles spherical triangle has an area equal to one-ninth the whole area of the sphere. Find the hypotenuse of the triangle.

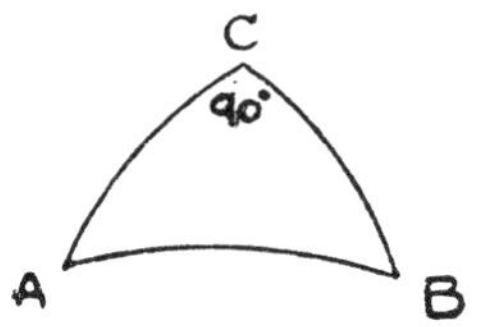

Given $\angle A = \angle B$ and $\angle C = 90°$

$E = A + B + C - 180° = 2A + 90° - 180° = 2A - 90°$

Area of sphere $= 4\,\pi\,r^2$

Area of triangle $= E\,r^2 = \frac{4}{9}\,\pi\,r^2$

$\therefore\ E = \frac{4\pi}{9} = \frac{4 \times 3{\cdot}1416}{9} = 1{\cdot}4$ radians

or $E = 1{\cdot}4 \times 57{\cdot}3° = 80{\cdot}22$

$E = 2A - 90\ \therefore\ 2A = 90 + 80°\ 22'\ \therefore\ A = 85°\ 11'$

In triangle ABC—

$$\cos AB = \cot A \cot B = \cot^2 85° \ 11'$$
from which hypotenuse $AB = 89° \ 34\frac{1}{2}'$

Example.—In a spherical triangle the arc of spherical excess is 1″. What proportion does the area of the triangle bear to that of the sphere? If the radius of the sphere is 3960 miles, give also the area of the triangle.

$$\text{Here } E = 1'' = \frac{1}{206265}$$

$$\text{Ratio required} = \frac{\text{Area of triangle}}{\text{area of sphere}} = \frac{E\,r^2}{4\,\pi\,r^2} = \frac{E}{4\pi}$$

$$\text{,,} \quad = \frac{1}{206265 \times 4 \times 3{\cdot}1416} = \frac{1}{2592008}$$

Ratio is as 1 to 2592008

$$\text{Area of triangle} = E\,r^2 = \frac{3960^2}{206265} = 76{\cdot}01 \text{ sq. miles.}$$

SUMMARY OF FORMULAE FOR REFERENCE.

			page
1.	Proofs of Formulae	$\frac{a}{b} = \frac{\sin A}{\sin B}$	225
2.		$c = a \cos B + b \cos A$	
3.		$a^2 = b^2 + c^2 - 2\,bc \cos A$	
4.		$\cos A = \frac{b^2 + c^2 - a^2}{2\,bc}$	
5.	Multiple Angles	$\sin (A + B) = \sin A \cos B + \cos A \sin B$	226
6.		$\cos (A + B) = \cos A \cos B - \sin A \sin B$	
7.		$\tan (A + B) = \frac{\tan A + \tan B}{1 - \tan A \tan B}$	
8.		$\sin (A - B) = \sin A \cos B - \cos A \sin B$	
9.		$\cos (A - B) = \cos A \cos B + \sin A \sin B$	
10.		$\tan (A - B) = \frac{\tan A - \tan B}{1 + \tan A \tan B}$	
11.	Sun and Difference of 2 Angles	$\sin A + \sin B = 2 \sin \frac{1}{2} (A + B) \cos \frac{1}{2} (A - B)$	229
12.		$\sin A - \sin B = 2 \cos \frac{1}{2} (A + B) \sin \frac{1}{2} (A - B)$	
13.		$\cos A + \cos B = 2 \cos \frac{1}{2} (A + B) \cos \frac{1}{2} (A - B)$	
14.		$\cos B - \cos A = 2 \sin \frac{1}{2} (A + B) \sin \frac{1}{2} (A - B)$	
15.		$\sin A = 2 \sin \frac{A}{2} \cos \frac{A}{2}$	
16.		$\sin (A + B) = 2 \sin \frac{1}{2} (A + B) \cos \frac{1}{2} (A + B)$	
17.		$\sin (a + b) = 2 \sin \frac{1}{2} (a + b) \cos \frac{1}{2} (a + b)$ (16) and (17) apply to spherical triangles.	
18.		$\cos A = 1 - 2 \sin^2 \frac{A}{2} = 1 - 2 \text{ hav } A$	
19.		$\cos A = 2 \cos^2 \frac{A}{2} - 1$, and $\cos 2\,A = 2 \cos^2 A - 1$	
20.		$\cos \frac{A}{2} = \sqrt{\frac{s\,(s - a)}{bc}}$, where $s = \frac{1}{2} (a + b + c)$	
21.		$\sin \frac{A}{2} = \sqrt{\frac{(s - b)\,(s - c)}{bc}}$	
22.		$\text{hav } A = \frac{(s - b)\,(s - c)}{bc}$	
23.		$\frac{a + b}{a - b} = \frac{\tan \frac{1}{2} (A + B)}{\tan \frac{1}{2} (A - B)}$	
24.		$\sin^2 A - \sin^2 B = \sin (A + B) \sin (A - B)$	
25.		$\cos 3\,A = 4 \cos^3 A - 3 \cos A$	
26.		$\sin 3\,A = 3 \sin A - 4 \sin^3 A$	232

		page
27. Three Bearing Formula	$\cot C = \dfrac{p \cot a - q \cot \beta}{p + q}$	232
28. Cosine formula	$\cos A = \dfrac{\cos a - \cos b \cos c}{\sin b \sin c}$	235
29. Sine formula	$\dfrac{\sin a}{\sin A} = \dfrac{\sin b}{\sin B}$	236
30. Four Adealout Parts formula	$\cot B \sin A = \cot b \sin c - \cos c \cos A$ $\sin a \cot b = \sin C \cot B + \cos C \cos a$	237 238
31. Haversine Formula	$\text{hav } A = \dfrac{\text{hav } a - \text{hav } (b - c)}{\sin b \sin c}$	239
32. All Haversine Formula	$\text{hav } A = \dfrac{\text{hav } a - \text{hav } (b - c)}{\text{hav } (b + c) - \text{hav } (b - c)}$	240
33. Three Sides Formula	$\text{hav } A = \dfrac{\sin (s - b) \sin (s - c)}{\sin b \sin c}$	241
34.	$\cos \dfrac{A}{2} = \sqrt{\dfrac{\sin s \sin (s - a)}{\sin b \sin c}}$	242
35.	$\tan \dfrac{A}{2} = \sqrt{\dfrac{\sin (s - b) \sin (s - c)}{\sin s \sin (s - a)}}$	
36.	$\cot \dfrac{A}{2} = \sqrt{\dfrac{\sin s \sin (s - a)}{\sin (s - b) \sin (s - c)}}$	
37. Sine Formula	$\dfrac{\sin a}{\sin b} = \dfrac{\sin A}{\sin B}$	242
38. Napier's Analogies	$\tan \frac{1}{2} (A + B) = \dfrac{\cos \frac{1}{2} (a - b)}{\cos \frac{1}{2} (a + b)} \cot \dfrac{C}{2}$	243
39.	$\tan \frac{1}{2} (A - B) = \dfrac{\sin \frac{1}{2} (a - b)}{\sin \frac{1}{2} (a + b)} \cot \dfrac{C}{2}$	
40.	$\dfrac{\tan \frac{1}{2} c}{\tan \frac{1}{2} (a + b)} = \dfrac{\cos \frac{1}{2} (A + B)}{\cos \frac{1}{2} (A - B)}$	
41.	$\dfrac{\tan \frac{1}{2} c}{\tan \frac{1}{2} (a - b)} = \dfrac{\sin \frac{1}{2} (A + B)}{\sin \frac{1}{2} (A - B)}$	
	$\text{hav } a = \dfrac{\cos S \cos (S - A)}{\sin B \sin C}$. $S = \frac{1}{2}(A + B + C)$	
43. Small Angle Ratios	$\sin \theta' = \theta' \sin'$ $\sin 1^s = \sin 15'' = 15 \sin 1''$	248
44. Spherical Excess	Spherical excess $= A + B + C - 180 = E$	250
45. Area of Spherical Triangle	area of spherical triangle $= Er^2$	251
46. Polar Triangle	In Polar triangle $\cos a = \dfrac{\cos B^1 \cos C^1 + \cos A^1}{\sin B^1 \sin C^1}$	249

CHAPTER XI.

NUMEROUS examples, also exercises with hints as to their solution, are given in this chapter, many of which are similar in type to the questions set in the extra master's papers on mathematics and navigation.

There are five sections: I—Problems introducing algebraic equations. II—Plane triangles dealing with heights, distances and relative bearings. III—Speed and relative velocities. IV—Various examples. V—Theoretical problems in nautical astronomy.

It is suggested that the student should go through the worked examples, then re-work them without this assistance. When working the exercises he should read the question, draw an appropriate figure and attempt a solution before referring to the hint. It is to be understood that the methods adopted, or suggested, may not be the only way of solving particular problems and the student may probably be able to evolve more elegant solutions in some cases.

I.—ALGEBRA.

Quadratic Equations.

A quadratic equation is one which contains the square of the unknown quantity, and when it contains the first power as well as the second power of the unknown it is called an "adfected" quadratic.

A complete quadratic equation is solved in the following steps:—

(1) Arrange the quantities so that the terms involving the unknown quantity are on the left side and the known quantities to the right.

(2) Make the coefficient of x^2 unity and positive by dividing throughout by the coefficient of x^2.

(3) Add to each side of the equation the square of half the coefficient of x.

(4) Extract the square root of each side and solve the resulting simple equation.

Example (i.)

Solve the equation $\frac{5}{6}x^2 - \frac{1}{2}x + \frac{3}{4} = 8 - \frac{2}{3}x - x^2 + \frac{278}{12}$

clearing of fractions, common denominator 12.

$$10\,x^2 - 6\,x + 9 = 96 - 8\,x - 12x^2 + 278$$

Transposing and reducing we get $22\,x^2 + 2\,x = 360$

Dividing by 22.

$$x^2 + \frac{2}{22}\,x = \frac{360}{22}$$

adding $\left(\frac{1}{22}\right)^2$ to both sides $x^2 + \frac{2}{22}x + \left(\frac{1}{22}\right)^2 = \frac{360}{22} + \left(\frac{1}{22}\right)^2$

Extracting the square root

$$x + \frac{1}{22} = \pm\sqrt{\frac{360}{22} + \left(\frac{1}{22}\right)^2}$$

$$,, \quad = \pm\sqrt{\frac{7921}{(22)^2}} = \pm\,\frac{89}{22}$$

$$\therefore \quad x = -\frac{1}{22} + \frac{89}{22} \quad \text{or} \; -\frac{1}{22} - \frac{89}{22}$$

$$x = \quad 4 \quad \text{or} \; -\frac{45}{11}$$

Instead of going through the process of completing the square we may solve many quadratics by the general formula $x = \dfrac{-\,b \pm \sqrt{(b^2 - 4ac)}}{2a}$

where a is the coefficient of x^2, the first term,

,, b is the coefficient of the middle term,

,, c is the third term.

In the foregoing example we have

$$22x^2 + 2x = 360$$

$$\text{or } 22x^2 + 2x - 360 = 0.$$

$a = 22$, $b = 2$, $c = -360$, and introducing these values in the general formula we have

$$x = \frac{-\,2 \pm \sqrt{(4 - 4 \times 22\,(-\,360)}}{2 \times 22}$$

$$x = \frac{-\,2 \pm \sqrt{31684}}{44}$$

$$x = \frac{-\,2 \pm 178}{44} = \frac{176}{44} \quad \text{or} \; -\,\frac{180}{44}$$

$$\therefore \quad x = 4 \text{ or } -\,\frac{45}{11}$$

Example (ii.)

A journey of 209 miles would be made by a train in 16 minutes less than the time actually taken, if the speed were increased by one mile per hour; find the speed in miles per hour.

Let x = original speed in m.p.h.

then $x + 1$ = increased speed in m.p.h.

$\therefore \dfrac{209}{x}$ = original time to perform journey

and $\dfrac{209}{x+1}$ = reduced time in hours to perform journey,

hence $\dfrac{209}{x} = \dfrac{209}{x+1} + \dfrac{16}{60}$

when reduced this gives the quadratic $4x^2 + 4x - 3135 = 0$

introducing the general formula $x = \dfrac{-b \pm \sqrt{(b^2 - 4ac)}}{2a}$,

we have $a = 4,\ b = 4,\ c = -3135$

$$\therefore x = \frac{-4 \pm \sqrt{(16 - 4 \times 4(-3135)}}{2 \times 4}$$

$$x = \frac{-4 \pm \sqrt{50176}}{8} = \frac{-4 \pm 224}{8}$$

$$\therefore x = \frac{220}{8} = 27\tfrac{1}{2} \text{ m.p.h.} \quad \text{Answer.}$$

It may be noticed that $4x^2 + 4x - 3135 = 0$

factorises into $(2x - 55)(2x + 57) = 0$

$\therefore 2x = 55$ or -57

$x = 27\frac{1}{2}$ or $-28\frac{1}{2}$, which does not apply

$\therefore$ speed $x = 27\frac{1}{2}$ m.p.h.

Example (iii.)

A and *B* are two ports 2,640 miles apart. Two vessels start simultaneously from *A* and *B* and one reaches port 4 days 4 hours and the other 6 days after they meet each other en route. Find their respective speeds.

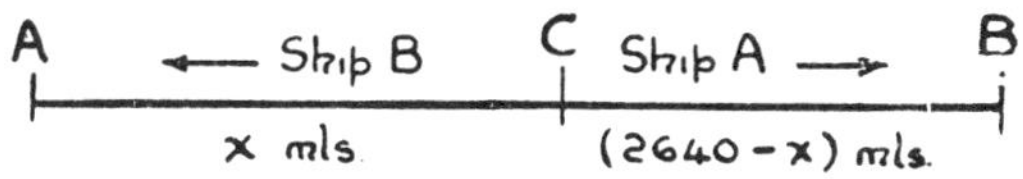

Let *A* and *B* be the ports and *C* the passing point of the ships x miles from *A*, then ship *B* goes x miles in 144 hours, and ship *A* goes $(2{,}640 - x)$ miles in 100 hours.

A's speed $= \dfrac{2640 - x}{100}$, *B*'s speed $= \dfrac{x}{144}$.

But the time taken by ship *A* in going from *A* to *C* is equal to the time taken by ship *B* in going from *B* to *C*, and time = distance ÷ speed.

$$\frac{\text{Distance}}{\text{Speed}} = \frac{x}{\dfrac{2640 - x}{100}} = \frac{2640 - x}{\dfrac{x}{144}}$$

$$i.e., \quad \frac{100x}{2640 - x} = \frac{144(2640 - x)}{x}$$

$$\therefore 25x^2 = 36(2640 - x)^2$$

Extract square root

$$5\,x = 6\,(2640 - x)$$
$$\text{and } x = 1440 \text{ miles.}$$
$$A\text{'s speed} = \frac{2640 - 1440}{100} = 12 \text{ knots}$$
$$B\text{'s speed} = \frac{1440}{144} = 10 \text{ knots}$$

Example (iv.)

Two ports *A* and *B* are 1,560 miles apart, steamer *Y* leaves *A* two days after steamer *X* and after 2 days passes a position that *X* passed 24 hours previously.

Y then reduced speed by 1 knot and arrived at *B* at the same time as *X*. Find the speeds of *X* and *Y*.

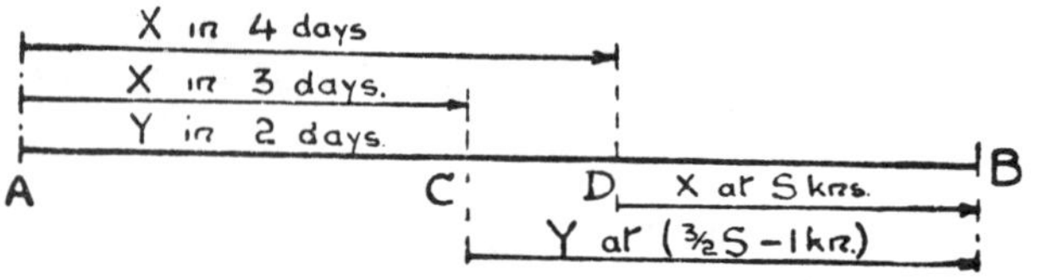

Let *S* knots be the speed of *X*.
X takes 3 days to go from *A* to *C*.
Y „ 2 „ „ „

$\therefore$ Y's speed $= \frac{3}{2}S$ knots

and Y's reduced speed $= \left(\frac{3}{2}S - 1\right)$ knots

X does $24\,S$ miles per day.
X does $72\,S$ miles in 3 days from *A* to *C*
and *X* does $96\,S$ miles in 4 days from *A* to *D*

$\therefore$ $AC = 72\,S$ miles and $CB = (1560 - 72\,S)$ miles
$AD = 96\,S$ „ „ $DB = (1560 - 96\,S)$ miles

when ship *X* is at *D* ship *Y* is at C.

Time taken by *X* to go *DB* miles is equal to the time taken by *Y* to go *CB* miles at her reduced speed, and the time is distance ÷ speed.

$$\therefore \frac{(1560 - 96\,S)}{S} = \frac{(1560 - 72\,S)}{\frac{3}{2}S - 1}$$

$$1560\,S - 72\,S^2 = 2340\,S - 144\,S^2 - 1560 + 96\,S$$
$$72\,S^2 - 876\,S + 1560 = 0$$
$$6\,S^2 - 73\,S + 130 = 0$$
$$(6\,S - 13)(S - 10) = 0$$
$$S = \frac{13}{6} \text{ or } 10$$

$\therefore$ X's speed $= 2\frac{1}{6}$ knots or 10 knots
Y's speed $= 3\frac{1}{4}$ „ or 15 „

EXAMPLES FOR EXERCISE.

I. ALGEBRA.

Solve for x—

1. $12x - 210 = 205 - 3x^2 + 5$.
2. $2x^2 + 34 = 20x + 2$.

Solve for x and y—

3. $5x + y = 3, 2x^2 - 3xy - y^2 = 1$.
4. $x^2 + xy = 15, y^2 + xy = 10$.

5. Two vessels leave ports A and B at the same time. One reaches B in 9 days and the other reaches A in 4 days after they meet. If the places A and B are 3000 miles apart, find the speed of each vessel.

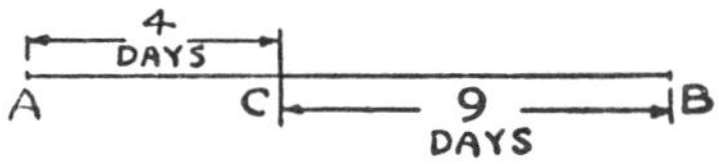

If $x = A$'s speed then

distance $CB = (9 \times 24\ x)$ miles

„ $AC = AB - CB = 3000 - (9 \times 24\ x)$ miles

B's speed $= \dfrac{AC}{4 \times 24}$ knots.

The time taken to meet each other being the same

we get $\dfrac{CB}{B\text{'s speed}} = \dfrac{AC}{A\text{'s speed}}$

so put in the above values and solve for x.

6. The distance between two ports is 800 miles. A steamer outward bound goes half-way at full speed with a current. She then goes at reduced speed without a current. On the return journey the steamer does the first half at reduced speed and then, the current having turned against her, she completes the journey at full speed.

Outward she takes 3 days 11 hours, and homeward she takes 4 days 04 hours. If the vessel can steam the total distance in 3 days 18 hours without a current at full speed, find (*a*) the vessel's reduced speed, (*b*) speed of current, (*c*) her full speed.

1600 miles in 90 hours at full speed = 17·77 knots.

Let x = current, y = reduced speed.

Outward—400 miles at $(17{\cdot}77 + x)$ knots and 400 miles at y knots.

Time taken $\dfrac{400}{17{\cdot}77 + x} + \dfrac{400}{y} = 83$ hours . . . . (1)

Homeward—400 miles at y knots and 400 miles at $(17{\cdot}77 - x)$ knots.

Time taken $\dfrac{400}{17{\cdot}77 - x} + \dfrac{400}{y} = 100$ hours . . . . (2)

Subtract equation (1) from (2) and solve for x and y.

7. A convoy 1 mile in length is steaming at a uniform rate of $7\frac{1}{2}$ knots. A destroyer proceeds along the length of the convoy from rear to front and notes the time taken to do so. She then steams from front to rear again noting the time to do so. If the sum of the two times was 10 mins. 40 secs., find her speed.

Let S knots be the required speed of destroyer.

Distance = 1 mile.

Speed with convoy $= S - 7{\cdot}5$ knots.

„ against „ $= S + 7{\cdot}5$ „

$\therefore \dfrac{1}{S - 7{\cdot}5} + \dfrac{1}{S + 7{\cdot}5} = 10\text{m } 40\text{s}$

Solve for S.

8. A steamer making a passage of 2000 miles, does half the distance at full speed, then owing to bad weather reduces to half speed. After covering quarter of the remaining distance she is able to increase speed by 3 knots. The passage takes 1 day 19 hrs. 45 mins. longer than at full speed. Find her full speed.

Let $2x$ knots = full speed, then x knots = half speed and $(x + 3)$ knots = final speed, then—

$$\frac{1000}{2x} + \frac{250}{x} + \frac{750}{x+3} = \frac{2000}{2x} + 43\tfrac{3}{4}$$

Solve for $2x$.

9. Two ships A and B leave ports 4320 miles apart at the same time and steam towards each other. A steams 60 miles more per day than B. They meet in a number of days equal to one-thirtieth of B's miles per day. Find the number of days.

10. A ship leaves port A at 0800 and is due at port B at 1800, distance between the ports being 120 miles. At 0900 another ship leaves B and is due at A at 1700. Show graphically, (*a*) when and where the two vessels will meet, (*b*) when they will be 10 miles apart.

(*a*) Draw a horizontal line AB = 120 miles to any convenient scale. Erect perpendiculars AC and BD, and divide them AC and BD into a scale of time from 0800 to 1800 hours.

Let A, at the bottom left-hand corner, be the point ot origin, viz. 0 miles and 08 hours. Draw a line from A to 18 hours on BD.

Mark 09 hours on BD and 17 hours on AC, join the two points by a straight line. The graphs intersect at X, which indicates where and when they meet.

(*b*) With 10 miles from the scale of distance, fit dividers horizontal between the two graphs and read off the corresponding times.

11. Ship A leaves a port steaming at 10 knots. Three hours later B leaves at 15 knots on a course inclined to A's course at an angle $X°$ such that $\cos X = \frac{4}{5}$. If they are y miles apart after z hours from B's time of leaving, show that $y^2 = 900 + 85z^2 - 120z$; also find distance apart when they are closest to each other.

12. A ship going South at 12·75 knots has a current which adds 2·1 knots to her speed. If the current were reversed her speed would be reduced by 1·875 knots. Find rate and direction of current.

13. A and B are two stations 300 miles apart. Two trains start simultaneously from A and B, each to the opposite station. The train from A reaches B in 9 hours, and the train from B reaches A 4 hours after they pass. Find the rate at which each train travels.

14. A tank can be filled by two pipes running together in $22\frac{1}{2}$ minutes: the larger pipe would fill the tank in 24 minutes less than the smaller one. Find the time taken by each.

15. A tank is holed and will fill in 2 hours. One pump cannot keep the water under, so a second pump is started when the tank is three-quarters full. In an hour the tank is half full. If the second pump takes 70 minutes longer than the first pump to empty the tank under normal conditions, find time taken by each.

In 1 hour, 2 pumps empty $\frac{1}{4}$ tank + leak, but the leak fills $\frac{1}{2}$ tank in 1 hour, therefore the two pumps are capable of emptying $\frac{3}{4}$ tank in 1 hour with no leak.

Let 1st pump empty tank in x hours, *i.e.* $\frac{1}{x}$ tank per hour

„ 2nd „ „ $(x + \frac{7}{6})$ hours *i.e.* $\frac{1}{x + \frac{7}{6}}$ tank per hour

$$\therefore \frac{3}{4} = \frac{1}{x} + \frac{1}{x + \frac{7}{6}}$$

Solve this equation for x.

16. Two pumps A and B are put on a ballast tank. A can empty the tank in 30 minutes less than B. The tank springs a leak which would fill it in 40 minutes. Pumps A and B were tried separately but could not reduce the water. At noon the tank was half full and both pumps were started; at 1 p.m. pump A broke down, but resumed work at 1h 30m p.m. and at 1h 40m p.m. the tank was a quarter full. Find how long each pump would take to empty the tank if there was no leak.

Into tank by leak	
at 12·00 - - -	$\frac{1}{2}$ tank
at 12·40 - - -	1 ,,
at 1·20 - - -	1 ,,
at 1·40 - - -	$\frac{1}{2}$,,
	3 tanks
Left in at 1·40 - -	$\frac{1}{4}$,,
Pumped out - - -	$2\frac{3}{4}$ tanks.

A pumps for 70 mins.
B ,, ,, 100 ,,

If x mins. = time for A to empty the tank and $x + 30$ mins. the time B takes, then together they would take

$$\frac{70}{x} + \frac{100}{x+30} = 2\tfrac{3}{4} \text{ (tanks)}$$

17. Two pumps working together empty a ballast tank in 3 hours, pump A working at half speed during 30 mins. of the time. If B takes 30 mins. longer than A normally to empty the tank, find the time taken by each separately.

If A takes x hours alone at normal capacity
then B takes $(x + \frac{1}{2})$ hours at normal capacity

A empties $\frac{1}{x}$ tank per hour

B ,, $\frac{1}{x+\frac{1}{2}}$ tank per hour

A ,, $\frac{5}{2x}$ tank in $2\frac{1}{2}$ hours

and $\frac{1}{4x}$ tank in $\frac{1}{2}$ hour at $\frac{1}{2}$ speed

B empties $\frac{3}{x+\frac{1}{2}} = \frac{6}{2x+1}$ tank in 3 hours

$$\therefore \quad \frac{5}{2x} + \frac{1}{4x} + \frac{6}{2x+1} = 1$$

Solve for x

18. When a flywheel of circumference $14\frac{2}{3}$ ft. takes 1 sec. more per revolution, a spot on the rim travels $2\frac{3}{4}$ miles less per hour. Find the circumferential speed of the flywheel.

Circumference $= \frac{44}{3}$ ft.

If t (secs.) = normal time per revolution
$t + 1$ (secs.) = new time per revolution

Circumferential speed $= \frac{44}{3}\,t$ (feet per sec.)

Revs. per hour $\frac{3600}{t}$ and $\frac{3600}{t+1}$

I $\quad \frac{44}{3} \times \frac{3600}{t}$ = speed in feet per hour

II $\quad \frac{44}{3} \times \frac{3600}{t+1}$ = new speed per hour

Eq. 1 — Eq. 2 = 14,520 ft. ($2\frac{3}{4}$ miles)
Solve for t.

19. A ship steaming at 10 knots burns 30 tons of coal per day. In addition to her fuel bill her expenses are £25 per day, she has to make a voyage of 4800 miles. Find by any method her most economical speed for the voyage if the cost of coal is 10s. per ton.

Let x = the economical speed.

Then $\frac{4800 \text{ miles}}{24x} = \frac{200}{x}$ = number of days.

Consumption varies as the cube of the speed, therefore new consumption at x knots is $\frac{30x^3}{10^3}$

Cost of coal for the voyage at x knots

$= £\left(\frac{30x^3}{10^3} \times \frac{1}{2}\right) \times \frac{200}{x} = £3x^2$

Other expenses at £25 per day $= 25 \times \frac{200}{x} = \frac{£5000}{x}$

Total cost $= 3x^2 + \frac{5000}{x} = y$, it is required to find the minimum value of y which we can do by means of a graph.

If x = knots	8	$8\frac{1}{2}$	9	$9\frac{1}{2}$	10	$10\frac{1}{2}$
then $3x^2$ =	192	216·7	243	270·7	300	330·7
and $\frac{5000}{x}$ =	625	588·2	555·5	527·3	500	476·5
$\therefore 3x^2 + \frac{5000}{x}$ =	£817	804·9	798·5	798	800	807·2

From graph, x = 9·4 knots, y = £797·5

How to draw the graph.—Write down some values of x on each side of the 10 knots and under each write the corresponding values of $3x^2$, $\frac{5000}{x}$ and $3x^2 + \frac{5000}{x}$.

On graph paper (use page 153) mark off the base line from 8 to $10\frac{1}{2}$ knots (abscissa) say 10 spaces per knot, and on the vertical line (ordinate) the values of y on each side of the approximate minimum value, say, from £790 to £820, one space per pound.

Mark the spots corresponding to the x's and y's and draw a smooth curve through them. The turn of the curve gives the minimum value, £797·5, and the corresponding speed, 9·4 knots.

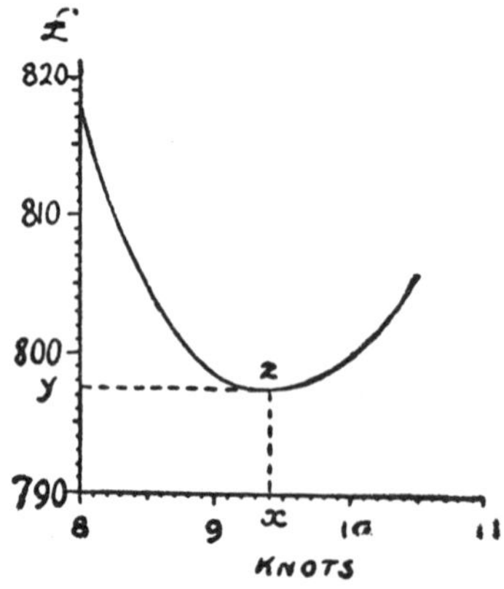

The speed may be found by means of the calculus. In this case the consumption x, and the cost y, vary together, but the cost is to be a minimum and equates to zero.

$$\text{thus } y = 3x^2 + \frac{5000}{x} = 0$$
$$\text{or } 3x^2 + 5000x^{-1} = 0$$
$$\text{differentiate } x,\ 3 \times 2x + 5000(-1x^{-2}) = 0$$
$$6x - 5000x^{-2} = 0$$
$$\text{or } 6x - \frac{5000}{x^2} = 0$$
$$\therefore 6x^3 = 5000$$
$$x = \sqrt[3]{\frac{5000}{6}} = 9{\cdot}4 \text{ knots.}$$

The process of differentiating is to diminish the power of x by 1 and at the same time multiplying x by its power. Thus if y and x vary together and $y = x^2$, then $\frac{dy}{dx} = 2x$. We might tabulate some examples to see how they appear. Put them in two columns, the values y in one and the corresponding values for $\frac{dy}{dx}$ in the other.

y	$\frac{dy}{dx}$
x^3	$3x^2$
$5x^4$	$5 \times 4x^3 = 20x^3$
x^{-1}	$-1 \times x^{-2} = \frac{-1}{x^2}$
$3x^{-2}$	$3 \times -2x^{-3} = \frac{-6}{x^3}$

Here are two similar examples for practice in the calculus as in previous question.

1. Consumption 30 tons per day at 10 knots, coal at £1 per ton; other expenses £36 per day, distance to go 5000 miles, find the most economical speed. *Ans.*—8·44 kns.
2. Consumption 40 tons per day at 12 knots, coal at 15s. per ton; other expenses £50 per day, distance to go 3600 miles, required the most economical speed. *Ans.*—11·29 kns.

TRIGONOMETRICAL EQUATIONS.

Find the value of angle A when

$$\frac{1}{1 - \sin^2 A} - \frac{1}{2}\sqrt{1 + \frac{\sin^2 A}{\cos^2 A}} = 8{\cdot}06$$

The first step is to express the equation in terms of one function.

$$\frac{1}{\cos^2 A} - \frac{1}{2}\sqrt{1 + \tan^2 A} = 8{\cdot}06$$
$$\sec^2 A - \tfrac{1}{2}\sec A = 8{\cdot}06$$

Complete the square by adding to each side of the equation the square of half the coefficient of sec. A.

$$\sec^2 A - \tfrac{1}{2}\sec A + (\tfrac{1}{4})^2 = 8{\cdot}06 + \tfrac{1}{4}^2$$
$$(\sec A - {\cdot}25)^2 = 8{\cdot}1225$$

Extract the square root of both sides

$$\sec A - {\cdot}25 = \pm 2{\cdot}85$$
$$\sec A = 3{\cdot}10 \text{ or } - 2{\cdot}6$$

Referring to a table of natural secants, we find

$$\text{when } \sec A = 3{\cdot}10,\ \angle A = 71^\circ\ 11' \text{ or } 288^\circ\ 49'$$
$$\text{,, } \sec A = -2{\cdot}6,\ \angle A = 67^\circ\ 23'$$

but having a negative sign the angle must be in the second or third quadrants, that is

$$(180^\circ - 67^\circ\ 23') = 112^\circ\ 37'$$
$$(180^\circ + 67^\circ\ 23') = 247^\circ\ 23'.$$

Find the values of the angles in the following equations:—

1. $\sin^2 A + \dfrac{1}{\sqrt{1 + \cot^2 A}} + \dfrac{1}{1 + \cot^2 A} = {\cdot}837$

2. $\cot^2 A - \dfrac{\cos A}{\sin A} = 17$

3. $\dfrac{1}{\operatorname{cosec}^2 A} - \sqrt{1 - \dfrac{1}{1 + \tan^2 A}} = {\cdot}8525$

   NOTE.—In this example sin A has two values, 1·55 and $-$ ·55, but sin A cannot exceed unity as the value 1·55 does not apply trigonometrically and $-$ ·55 indicates that the angle is in the 3rd or 4th quadrants.

4. $\tan^2 A + \dfrac{4 \sin A}{\sqrt{1 - \sin^2 A}} - \dfrac{2}{\cot A} = 179$

5. $2 \cos \theta = 3 \sec \theta - 2$
6. $4\sqrt{3} \cot \theta = 7 \operatorname{cosec} \theta - 4 \sin \theta$
7. $\tan^2 \theta - \sin^2 \theta = 2{\cdot}25$
8. Find value of cos 15° from the expression $\cos 4A = \tfrac{1}{2}$
9. Find the greatest angle in a plane triangle whose sides are $x^2 + x + 1,\ 2x + 1, \text{ and } x^2 - 1.$

IDENTITIES.

Identities differ from equations in that they hold good for any angle whereas an equation can only be satisfied by introducing certain values for the angle.

Prove the following identities:—

1. $\operatorname{Sin}^4 A + \cos^4 A = 1 - \tfrac{1}{2}\sin^2 2A.$
2. $\sin 3A = 4 \sin A \sin (60^\circ + A) \sin (60^\circ - A)$
3. $\tan 2A - \sin A \sec A = \tan A \sec 2A$
4. $\tan A \tan 2A = \sec 2A - 1$
5. $\tan^2 A + \cot^2 A = 2 + 4 \cot^2 2A$
6. $(\sin A - \sin B)^2 + (\cos A - \cos B)^2 = 2 \operatorname{vers} (A - B)$

AN ARTIFICE FOR SOLVING AN AWKWARD EQUATION.

Given an equation in the form of $3 \cos B + 4 \sin B = 4$, to find $\angle B$.

i. Let $R \sin (A + B) = 4$, then $\sin (A + B) = \frac{4}{R}$ Expanding we get $R \sin A \cos B + R \cos A \sin B = 4$.

ii. Note that $R \sin A =$ coeff. 3 $\therefore \sin A = \frac{3}{R}$

iii. Note that $R \cos A =$ coeff. 4 $\therefore \cos A = \frac{4}{R}$

Square and add ii and iii and we get $R^2 (\sin^2 A + \cos^2 A) = 9 + 16$.

$$R^2 \times 1 = 25 \quad \therefore R = 5$$

From i. $\sin (A + B) = \frac{4}{R} = \frac{4}{5} = \cdot 8. \quad \therefore A + B = 53^\circ \; 8'$

From ii. $\sin A = \frac{3}{R} = \frac{3}{5} = \cdot 6. \quad \therefore A = 36 \; 52$

$\angle B \quad 16 \; 16$

Verification from Tables

3 nat cos $16^\circ\ 16' = \cdot 96 \times 3 = 2\cdot 88$
4 nat sin $16 \quad 16 = \cdot 28 \times 4 = 1\cdot 12$

$\therefore 3 \cos 16^\circ\ 16' + 4 \sin 16^\circ\ 16' = 4$

$\therefore B = 16^\circ\ 16'$ satisfies the equation.

Examples for Exercise—

10. $5 \cos x + 2 \sin x = 2$, find $\angle x$
11. $2 \cos B + 3 \sin B = 3$, find $\angle B$
12. $\sqrt{3} \sin \theta - \cos \theta = \sqrt{2}$, finds $\angle \theta$

NOTES

II.—TRIANGLES.

DEALING WITH HEIGHTS, DISTANCES AND BEARINGS.

Example 1.

A cliff 250 feet high has a flagstaff 50 feet high on top of it. To an observer on the same plane as the base of the cliff the flagstaff subtends an angle the same as at a point 10 feet up from the base of the cliff. Find the distance of the observer from the base of the cliff.

In figure, $AB = 10$ feet, $AC = 250$ feet, $AD = 300$ feet, $\angle AOB = DOC = x$. Let $AO = d$ the distance. Let $\angle AOD = a$, and $\angle AOC = b$, then $\angle DOC = (a - b) = x$.

$$\text{By (10) } \tan x = \tan(a-b) = \frac{\tan a - \tan b}{1 + \tan a \tan b}$$

$$\text{,, ,, } = \frac{\frac{300}{d} - \frac{250}{d}}{1 + \frac{300 \times 250}{d^2}}$$

$$\text{but } \tan x = \frac{10}{d} = \frac{50\,d}{d^2 + 75{,}000}$$

$$\text{from which } 5d^2 = d^2 + 75{,}000$$

$$\therefore \quad d = \sqrt{\frac{75{,}000}{4}} = 136{\cdot}93$$

Answer. Distance 136·93 feet.

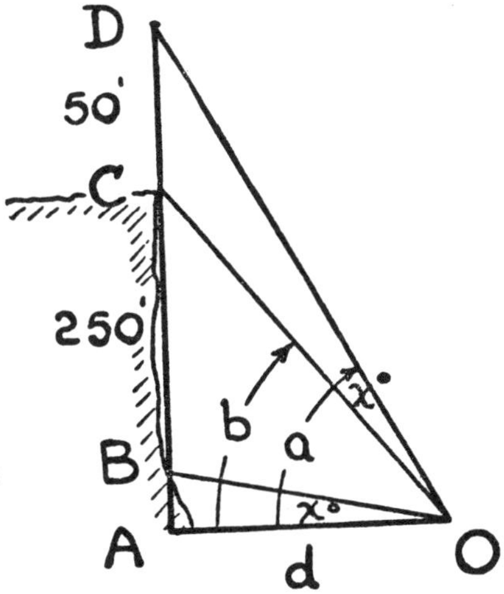

Example 2.

A cliff is 420 feet high. From a boat a conspicuous object 48 feet high on top of the cliff subtends an angle of $x°$. After going 180 feet towards the cliff the object subtends the same angle x. Find distance of observer from the base of the cliff at each observation.

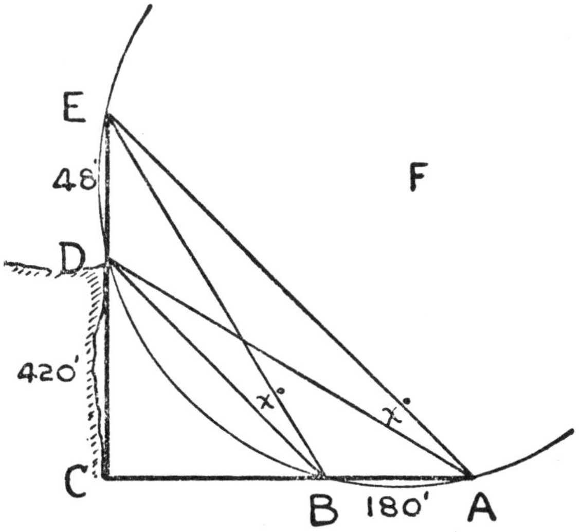

In figure, let A and B be the two positions of the boat. $AB = 180$ feet, CD the cliff 420 feet, $D\,E$ the object 48 feet, F is the centre of the circle passing through A, B, D, and E.

$\angle EAD = \angle EBD = x°$.

It is required to find CB and CA.

$$CD \times CE = CB \times CA$$

$$468 \times 420 = CB\,(CB + 180\,\text{ft.})$$

$$196560 = CB^2 + 180\,CB.$$

Complete the square and solve for CB

$$CB^2 + 180\ CB + (90)^2 = 196560 + 90^2$$

$$CB + 90 = \pm\sqrt{204660} = \pm 452{\cdot}4$$

$$CB = \pm 452{\cdot}4 - 90 = 362{\cdot}4 \text{ ft.}$$

$$AC = (CB + 180) = (362{\cdot}4 + 180) = 542{\cdot}4 \text{ ft.}$$

Answer. 362·4 ft. and 542·4 ft.

Example 3.

The angle of elevation of a balloon from a station due South of it is 45° 35′ and from another station 725 feet due West of the former, the elevation is 40° 22′. Find the height of the balloon.

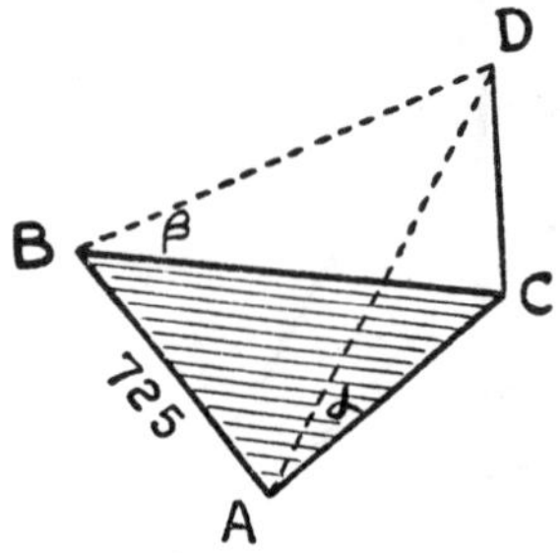

In figure, D is the balloon, CD its height above the ground at C. Station A is due south from C, $\angle DAC$ is the vertical angle at $A = \alpha = 45°\ 35'$. Station B is due west of A, so that $\angle CAB$ is 90° and $AB = 725$ feet. The vertical angle at B is $\angle DBC = \beta = 40°\ 22'$. BCD is a vertical triangle, right angled at C. BCA is a horizontal, or ground triangle, right angled at A.

From figure:—

$$CD = CA \tan\alpha = CB \tan\beta$$

$$\therefore \frac{\tan\alpha}{\tan\beta} = \frac{CB}{CA} = \operatorname{cosec} CBA = \frac{\tan 45°\ 35'}{\tan 40°\ 22'} = \frac{\log 0{\cdot}008844}{\log 9{\cdot}929452}$$

$$\angle CBA = \quad 56°\ 24' \operatorname{cosec} 10{\cdot}079392$$

$$DC = BC \tan\beta$$

$$\text{,,} \quad = AB \sec CBA \tan\beta$$

$$\text{,,} \quad = 725 \sec.\ 56°\ 24' \tan 40°\ 22'$$

,, $= 1118$ feet, the height of the balloon. *Ans.*

Example 4.

Three ships start from the same point steering in a north-easterly direction. A steams 8 knots, B 10 knots, C 12 knots. After 2 hours steaming they arrive on the same meridian, B to the North of A and to the South of C; B's distance from A and C is in the ratio of 4 to 3. Find the courses steered.

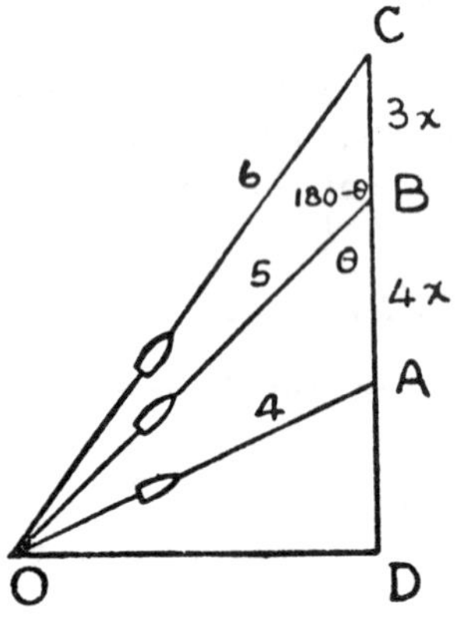

The distances steamed by ships A, B and C are, respectively 16, 20 and 24 miles, or, in the ratio of 4, 5 and 6. If $AB = 4x$ then, from question, $BC = 3x$. It is required to find the value of x and then solve the triangles to find angles A, B and C, the respective courses of the ships.

Call $\angle OBA$, θ, then $\angle OBC = (180° - \theta)$

Apply the general formula (3) for two sides and included angle in triangles OBA and OBC.

In $\triangle OBA$ by (3) $OA^2 = BO^2 + BA^2 - 2BO \cdot BA \cos\theta$

$$4^2 = 5^2 + (4x)^2 - 2 \times 5 \times 4x \times \cos\theta$$
$$16 = 25 + 16x^2 - 40x\cos\theta$$
$$\cos\theta = \frac{16x^2 + 9}{40x} \quad . \quad . \quad . \quad . \quad . \quad . \quad . \quad . \quad . \quad . \quad (i.)$$

In $\triangle OBC$ by (3) $OC^2 = BO^2 + BC^2 - 2BO.BC \cos(180° - \theta)$

$$6^2 = 5^2 + (3x)^2 - 2 \times 5 \times 3x \cos(180° - \theta)$$
$$36 = 25 + 9x^2 - 30x \times -\cos\theta$$
$$36 = 25 + 9x^2 + 30x\cos\theta$$
$$\cos\theta = \frac{-9x^2 + 11}{30x} \quad . \quad . \quad . \quad . \quad . \quad . \quad . \quad . \quad . \quad . \quad (ii.)$$

Eq. i = Eq. ii, and multiplying each side by $10x$ we get

$$\frac{16x^2 + 9}{4} = \frac{-9x^2 + 11}{3} \text{ from which}$$

$$84x^2 = 17 \text{ and } x = \sqrt{\frac{17}{84}} = \cdot 4498 \quad \therefore 3x = 1\cdot 3494 \text{ and } 4x = 1\cdot 7992.$$

In $\triangle COB$ the sides being in the ratio of 6, 5 and 1·3494 to find $\angle C$

by (22) $\text{hav } C = \dfrac{(s - 6)(s - 1\cdot 349)}{6 \times 1\cdot 349} \quad . \quad . \quad . \quad . \quad . \quad . \quad . \quad . \quad s = 6\cdot 174$

$$\text{,,} \quad = \frac{\cdot 174 \times 4\cdot 825}{8\cdot 094}$$

From which $\angle C$ is 37° 34′ 30″.

$\triangle OCD$, $OD = 6$ nat sin $37° 34\tfrac{1}{2}' = 6 \times \cdot 6098 = 3\cdot 6588$

$\triangle OBD$, $\sin\theta = \dfrac{3\cdot 6588}{5} = \cdot 73174 \therefore \theta = 47° 2'$

$\triangle OAD$, $\sin A = \dfrac{3\cdot 6588}{4} = \cdot 9147 \therefore A = 66° 10'$

Answer. Courses, A N 66° 10′ E.
B N 47° 02′ E.
C N 37° 34½′ E.

Example 5.

A ship steering 270°, speed 24 knots, sights a light bearing 295°. After an interval of 40 minutes it bears 328° and after another interval of 20 minutes it bears 360°, a current of unknown strength was setting 137°. Find the course made good and the speed of the current.

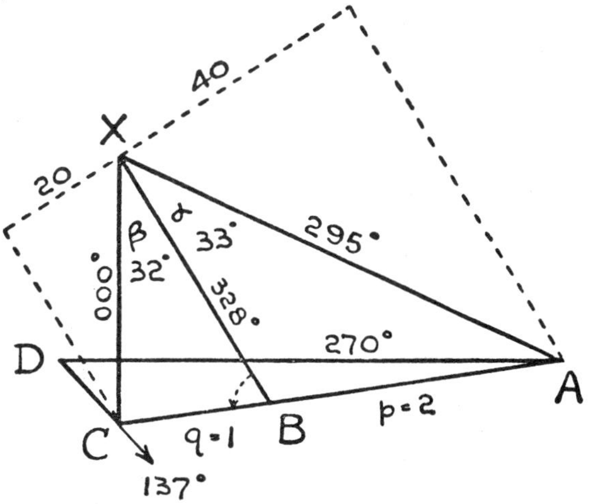

XA, XB and XC are the three bearings. The dotted lines are constructional as described on page 21, chap. II, p = 40 minutes, q = 20 minutes or in the ratio of 2 to 1, $\alpha = 33°$, $\beta = 32°$. *ABC* is the course made good. Find $\angle B$.

by (27) $$\text{nat} \cot B = \frac{p \cot \alpha - q \cot \beta}{p + q}$$

$$\text{,,} \quad = \frac{2 \cot 33° - 1 \cot 32°}{3}$$

$$\text{,,} \quad = \cdot 4932$$

$$\therefore \angle B = 63° \ 45'$$

Direction of XB = 328° 00′
$\angle B$ = 63 45 to the left

ABC the course = 264 15

or S. 84° 15′ W.

AD is the course and distance steered, 270°, 24 miles, *DC* the direction of current 137°, $\angle D = 47°$ $\angle A = 5° \ 45'$, $\angle C = 127° \ 15'$. In triangle *ACD* find *DC* the drift of the current in one hour.

$$\frac{DC}{\sin A} = \frac{DA}{\sin C} \quad \therefore \ DC = 24 \sin 5° \ 45' \operatorname{cosec} 127° \ 15'$$

$$DC = 3 \cdot 02 \text{ knots.}$$

Answer. Course S. 84° 15′ W., current 3·02 knots.

II. Plane Triangles.

Examples 1 to 11 are exercises in vertical angles somewhat similar to Worked Examples 1 and 2.

1. From a boat on the same horizontal plane as the base of a cliff 360 ft. high, the angle subtended by a tower 50 feet high on the top of the cliff is 2° 54′ 27″. Find the distance of the boat from the base of the cliff.

2. A flagstaff 60 ft. high on top of a cliff subtends the same angle at an observer 200 ft. from the base of the cliff as a tree 12 ft. high at the base of the cliff. Find height of the cliff.

3. An observer on a vessel's bridge, height of eye 40 ft., observes the vertical angle of *B*'s mast, 120 ft. from truck to water-line, to be 11° 18′ 36″. Find the distance between the two vessels.

4. A tower 80 ft. high stands on top of a cliff 600 ft. high. From a boat the tower subtends the same angle as a house 20 ft. high at the foot of the cliff. Find distance of the boat from the foot of the cliff.

See Example 1.

5. From the bridge of a vessel the vertical angle subtended by a cliff 100 ft. high is 9° 30′. If the angle subtended between the cliff top and its reflection is 18° 54′, how far off is the cliff?

Draw a figure such that $BC\ (q) = CD\ (p) = 100$ ft.
$\angle DAC = 9° \ 30' \ (\alpha)$, $\angle CAB = 9° \ 24' \ (\beta)$

Apply (27) $\cot C = \dfrac{p \cot \alpha - q \cot \beta}{p + q}$, $C = 88° \ 8\frac{1}{2}'$

In triangle *B A C*, having found *C* as above, find *A C* by formula (1), then make *A E* perpendicular to *B C D* and find *A E* the horizontal distance of the bridge from the cliff.

6. An observer, whose height of eye was 42 ft., found the vertical angle of the upper half of a flagpole to be 30° while the lower half subtended 33°. Find the distance of the observer from the pole.

NOTE.—Construct a figure so that $A B = 42$ ft., the height of eye, and $C E$ the flagpole, then $\angle D B C = 33°$ (β), $D B E = 30°$ (α), D being halfway up the flagstaff, $C D = D E = x$, $\angle C D B = \theta$.

$$\cot \theta = \frac{x \cot \alpha - x \cot \beta}{x + x} = \frac{\cot 30° - \cot 33°}{2}$$

and $A C$ the distance $= 42 \tan B C A$.

7. A ship's deck cargo is 9 ft. high. A 6 ft. man standing on the main deck finds the angle of elevation to be $2d$ when the top of the cargo is in line with the masthead. He moves back until the angle is d and he then sees the upper one-third of the mast. Bottom of deck cargo and heel of mast are in the same horizontal plane. If $\cos 2d = \frac{3}{5}$ find the height of the mast.

NOTE.—Construct a figure so that $A B = 6$ ft. $B C = 3$ ft. and D the masthead the height of which is h feet. $C D = (h - 9)$.

If $\angle E = 2d$ and $\angle F = d$, draw $C G$ parallel to the deck line $B E F$, to represent the top of the deck cargo. $\angle D G C = 2d$, $\angle H G C = d$ and as $H C$ is unknown call it a ft.

The mast $h = \frac{1}{3} h + a + 9$ ft. $\therefore a = (\frac{2}{3} h - 9)$, $H D = \frac{1}{3} h$.

Given $\cos 2d = \frac{3}{5} = \frac{\text{adj}}{\text{hyp}}$ therefore the opposite side is 4, from which $\cot 2d = \frac{3}{4} = 53° 8'$

$$d = 26° 34', \text{ nat } \cot d = 2$$

$\triangle G D C$, $C G = (h - 9) \cot 2d$

$\triangle H G C$, $C G = (\frac{2}{3} h - 9) \cot d$

from which $(h - 9)\frac{3}{4} = (\frac{2}{3} h - 9)\, 2$

Solve for h.

8. An observer 20 ft. above the level of a lake observes the angle of elevation of a tower to be 45° and the angle of depression of its exact reflection to be 60°. Find the height of the tower.

NOTE.—Construct a figure so that A is the observer, $A B$ the level of eye, $C D$ the level of lake.

$C A = D B = 20$ ft. $\angle B A E = 45°$, $\angle B A F = 60°$

If h be the height of tower and $A B = C D = d$, then $d = (h - 20) \cot 45° = (h + 20) \cot 60°$.

Solve for h.

9. An observer 7 ft. above the deck observes the angles subtended between the deck and the top of a conning tower 28 ft. and between the conning tower and the masthead 30 ft. to be equal, tower and mast in same plane perpendicular to the deck. Find distance of observer from the tower.

NOTE.—Make a rough figure such that A is the observer, $D B = 28$ ft. (q), $B C = 30$ ft. (p), the vertical angles $D A B$ and $B A C$ are equal (θ).

Apply (27) $\cot A B D = \frac{p \cot \theta - q \cot \theta}{p + q}$

from which $\tan B = 29 \tan \theta$.

If x be the length of base line on deck and also drawn as $A E$ 7 ft. higher, then $\tan B = \frac{x}{21}$ and $\tan d = \frac{21}{x}$ where d is $\angle B A E$ and $(\theta - d)$ is $\angle E A D$, $\tan(\theta - d) = 7/x$.

Apply (10) $\tan(\theta - d) = \frac{\tan \theta - \tan d}{1 + \tan \theta \tan d}$

And solve the resulting equation.

10. An observer on deck observes the vertical sextant angle of masthead to derrick table to be 26° 33′ 26″. Another observer 36 ft. above the first finds the angle to be 28° 00′ 35″. The eye of first observer is on a level with the derrick table and the mast is perpendicular to the deck. Required distance of observer from the mast.

NOTE.—Draw a rough figure. Draw a horizontal line through the upper observer so that his angle is $(\alpha + \beta)$. Call the lower angle θ and total height of mast H, then the bit above the table is $(H - 36)$.

Let d be the length of base line so that $\tan \alpha = \frac{36}{d}$ and $\tan \beta = \frac{H - 36}{d}$

Apply (7) $\tan (\alpha + \beta) = \frac{\tan \alpha + \tan \beta}{1 - \tan \alpha \tan \beta} = \frac{H d}{d^2 - 36 H + 36^2}$

when the right hand side is cleared up.

Now substitute $d \tan \theta$ for H and solve the resulting quadratic.

11. When steering East two buoys were observed in line bearing North. After going 1 cable the horizontal angle subtended by the two buoys is θ°, and after going a further 3 cables the horizontal angle is again θ°. Find the course from a point 3 cables beyond where the second angle was observed to pass midway between the two buoys. The distance between the buoys is twice the distance between the yacht and the nearest buoy when they were in transit.

This problem may be treated as a vertical angle problem similar to Example 2, page 247, by assuming C to be 1st position of yacht, D and E the two buoys, so that $C D = d$, the unknown distance and $D E = 2 d$.

$\angle D B E = \angle D A E = \theta$, $C B = 1$ cable, $B A = A F = 3$ cables.

Join F to G midway between D and E then $C D \times C E = C B \times C A$ or $d \times 3 d = 1 \times 4$

$\therefore d = 1{\cdot}1547 \tan F = \frac{2 d}{7}$

12. The angles of elevation of the top and bottom of a flagstaff on the top of a tower are A and B respectively. Prove that the height of the tower is equal to $\frac{x \cos A \sin B}{\sin (A - B)}$ where x is the height of the flagstaff.

13. Three stations P, M and O in same horizontal plane. P and O subtended an angle of 90° at M. R is a point midway between P and O and on a line joining them. There is a flagstaff at M. If x, y and z be the vertical angles subtended by the flagstaff at P, R and O respectively, show that $\tan^2 y = 2 \tan x \tan z \sin 2 P$, where P is the angle $M P O$ in the $M P O$ triangle.

Exercises 14 to 17 are in two planes similar in type to Worked Example 3.

14. A balloon is observed from two stations A and B 7560 ft. apart, B being due North of A. The angle of elevation at B was 27° 13′ and at A 46° 48′, the balloon bearing due East from A, find its distance from B.

15. At two stations A and B the following simultaneous observations were got. At A a balloon bore N. 3° E., alt. 47° 39′. At B it bore N. 76° E. Suppose B to be N. 40° W., 9120 ft. from A, find the height of the balloon and its distance in feet from A.

16. Two observers A and B are 1200 ft. apart. A observes the vertical sextant angle of the top of a tower to be 8° 17′, and the horizontal angle between B and the tower to be 67° 30′. B observes the horizontal angle between A and the tower to be 49° 15′. Find the height of the tower.

17. From a ship a mountain bore 000° and its vertical angle was 15° 20′. The ship then steamed due East for 5 miles and its vertical angle was now 11° 25′. Find the height of the mountain.

18. An observer finds that the angle between the mastheads of two masts of equal height is 120° from a point midway between the masts and on a line joining them. Walking 60 ft. backwards and perpendicular to the line joining the two masts the angle is now 81° 35′ 16″. Find the distance between the masts.

Construct a figure such that $A\ D$ and $A^1\ D^1$ represent the masts and B is midway between A and A^1.

$\angle D\ B\ D^1 = 120°$. Join $D\ D^1$ with E midway between. Let $D\ E = d$ the half distance required, and h the height of the masts. $\angle D\ B\ E = 60°$, $\angle D\ E\ B = 90°$.

Make $B\ C = 60$ ft., so that $\angle A\ B\ C = 90°$, then $A\ B\ A^1\ C$ is a ground triangle. Join $D\ C$ and D^1C. $\angle D\ C\ D^1 = 81° 35' 16''$.

$B\ E$ is a vertical line and $B\ E\ D$ is a vertical triangle.

$B\ C$ is a horizontal line, and $C\ E\ D$ is a slant triangle in which $\angle E\ C\ D = 40° 47' 38''$, $\angle D\ E\ C = 90°$.

Let $\angle E\ C\ D = \theta$, side $A\ C = a$, and $C\ D = b$.

1. $\triangle\ B\ A\ D$. $h = d \tan 30°$. $h^2 = d^2 \tan^2 30°$.
2. $\triangle\ C\ E\ D$. $b = d \operatorname{cosec} \theta$. $b^2 = d^2 \operatorname{cosec}^2 \theta$.
3. $\triangle\ D\ C\ A$. $\angle D\ A\ C = 90°$. $a^2 = b^2 - h^2$.
4. $\triangle\ A\ B\ C$. $d^2 = a^2 - 60^2 = b^2 - h^2 - 3600$ and by submitting 1 and 2 for

b^2 and h^2 we get $d^2 = d^2 \operatorname{cosec}^2 \theta - d^2 \tan 30° - 3600$ which simplifies to
$d^2(\operatorname{cosec}^2\theta - 1 - \tan^2 30°) = 3600$, but $\operatorname{cosec}^2\theta - 1 = \cot^2\theta$ $\therefore$ $d^2(\cot^2\theta - \tan^2 30) = 3600$
Solve for d.

Exercises 19 to 33 are examples of running fixes involved with an unknown quantity which has either to be found or eliminated as in Worked Example 4.

19. Three ships A, B, and C leave the same point steering to the eastwards, A and C at equal speeds and B at 12 knots. Find the courses they must steer so that after one hour A is 3 miles due North of B, and B is 5 miles due North of C.

20. Three ships start simultaneously from the same position steering in a north-westerly direction, A steams 8 knots, B at 10 knots, C at 14 knots. After one hour steaming the ships are on the same meridian, B being midway between A and C. Find the courses steered.

21. A vessel observes a point of land abeam and at the same time the horizontal angle between it and a lighthouse is 42°. After steaming 3 miles the horizontal angle is still 42°, and after steaming another 4 miles the lighthouse is abeam. Find the distance between the point and the lighthouse.

22. A ship steering 000° sees a lighthouse ahead and a beacon on the starboard bow, the horizontal angle between them being θ°. After 54 minutes the angle is 2θ and 34 minutes later it is $(90° - \theta)$, speed 10 knots. Find the distances off the beacon when it is abeam.

If A, C, D and E be the successive positions of ship, B the beacon and the angles named in terms of θ, then $A = \theta°$, $C = 2\ \theta°$, $D = 90° - 3$, $E = 90°$, $\angle E\ B\ D = \theta$, $\angle D\ B\ A = (90° - 2\ \theta)$, $\angle E\ B\ A = (90° - \theta)$, $\angle C\ B\ A, = \theta$ $\angle D\ B\ C = (90° - 3\ \theta)$

Given $D\ C = 5{\cdot}67$ miles, $A\ C = C\ B = 9$ miles.

$B\ E = 9 \sin 2\ \theta = 9\ (2 \sin \theta \cos \theta) = 18 \sin \theta \cos \theta$

$$\triangle\ D\ B\ C\ \frac{5{\cdot}67}{9} = \frac{\sin(90 - 3\ \theta)}{\sin(90 + \theta)} = \frac{\cos 3\ \theta}{\cos \theta} = \frac{4 \cos^3 \theta - 3 \cos \theta}{\cos \theta} = 4 \cos^2 \theta - 3^*$$

from which $\theta = 17° 42\frac{1}{2}'$ and $B\ E = 9 \sin 2\ \theta$.

* (See Eq. (25) for this transformation.)

23. A ship observes a beacon on the starboard bow and after making good 6 miles observes the bearing to have trebled.

After making good another 2 miles the beacon was abeam. Find distance off when abeam.

If A, B and C be the three successive positions of the ship and D the beacon, the $\angle A = \theta$, $\angle B = 3\,\theta$, $\angle C = 90°$, $B\,C = 2$ miles.
$A\,B = 6$ miles, $C\,D = 2 \tan 3\theta = 8 \tan\,\theta$ $\therefore \tan 3\,\theta = 4 \tan \theta$.

Apply Eq. (7) $\tan (2\,\theta + \theta) = \dfrac{\tan 2\,\theta + \tan\,\theta}{1 - \tan 2\,\theta \tan\,\theta}$ which resolves itself into $\tan\,\theta = \sqrt{\dfrac{1}{11}}$

24. A vessel steering 270° sights a point of land A bearing 000°. After two hours on the same course a point P (29·985 miles due West of A) bore 311°. After another 2 hours on the same course P bore 030°. Find ship's distance from P at nearest approach.

Construct a figure so that B, C and D are the three successive positions of the ship from point of land P, then $A\,P = 29{\cdot}985$ miles.

Let $\angle B\,P\,C = \alpha$, $\angle C\,P\,D = \beta = 79°$, $\angle C = 41°$, $B\,C = p$, $C\,D = q$, $p = q$

Apply (27) $\cot C = \dfrac{p \cot \alpha - q \cot \beta}{p + q}$

$$\cot 41° = \frac{\cot \alpha - \cot 79°}{2}$$

from which $\alpha = 21°\ 50'$

Make $P\,E$ perpendicular to $D\,C$. E is the point of nearest approach.

25. Two beacons are on same meridian. A ship steaming 290° sights the North beacon right ahead distant 12 miles and the horizontal angle between the beacons to be $A°$. After proceeding 6 miles the angle is $d°$ and after another 3 miles it is $2d°$. Find angle A.

If X and Y be the beacons, X to North of Y, A the ship's 1st position where the horizontal angle is $A°$, B the 2nd position where the angle is d and C the 3rd position where the angle is $2\,d$. C is the centre of a circle passing through X, Y and B.

$C\,Y = C\,B = 3$ miles. $\angle X = \angle Y = 70°$.

$\triangle\ C\,Y\,A$, given $C\,Y$, $C\,A$, $\angle C$ Find $\angle A$.

26. Ship steering North, a light bearing 100° T. After sailing 2 miles the light bore 120° T. Ship then increased speed and after 20 minutes the light bore 140° T. Find vessel's speed between the second and third bearings.

NOTE.—Draw figure so that A, B, and C are the three successive positions of the ship and D the position of the light.

In $\triangle\ A\,B\,D$ all the angles are known and side $A\,B$. Find side $B\,D$.

In $\triangle\ B\,D\,C$ all the angles are known and now side $B\,D$. Find $B\,C$ the distance ship has sailed in 20 minutes.

27. When steering North the maximum horizontal angle between two buoys was 44°, and after steaming 2 miles the buoys are in line bearing 158°. Find distance between the buoys.

Construct a figure such that A is the position when the hor. angle $X\,A\,Y = 44°$, X and Y being the buoys. Make $A\,B$ due North, 2 miles and $\angle A\,B\,Y$ 22°.

The hor. angle being maximum a circle passes through X, Y, and A of which $A\,B$ is a tangent and $X\,Y$ a chord.

The angle between the tangent and chord is equal to the angle in the opposite segment.

i.e. $\angle B\,A\,X = \angle X\,Y\,A = \theta$.

In $\triangle\ A\,B\,X$, $\angle\ \theta + (\theta + 44) + 22{\cdot} = 180$. $\therefore\ \theta = 57°$.

The solution is now obvious.

28. Two points of land in transit bearing 320° T. from a ship. After making 9·5 miles on a westerly course the horizontal angle becomes a maximum at 22°. If the two points are 4 miles apart, find the course made good by the ship.

Construct a figure. Make angles at X and Y equal to the complement of 22° the horizontal angle, the lines intersect at C which is the centre of the circle of position passing through X, Y and the ship. Join CY and CX and make CD perpendicular to XY, $DX = D'Y = 2$ miles. Draw tangent from A meeting circle at E. Join EY, EX and EC $\angle XEY = 22°$, $\angle XCY = 44°$, $\angle DCX = 22°$.

Calculate $CX = r$ the radius of the circle.

$\tan CAE = r/AE$, $CA = AE$ Sec. CAE.

Because AE is a tangent and AXY a secant to the circle, $AE^2 = AX(AX + XY)$ by Geometry, from which $AX = 7{\cdot}71$.

In triangle ACD, given three sides, find $\angle CAD$. The course required may now be found.

29. A vessel steering to the westward sights two beacons 3 miles apart and in transit bearing 008° T. She steams 6 miles and the maximum horizontal angle subtended by the two beacons is then 29° 18′. Find the course steered.

30. From a vessel steering north-easterly a beacon bears 000° and after 40 minutes it bears 315°; shortly afterwards it bears 270° and again in another 20 minutes it bears 225° T. Find the ship's course.

Draw figure. Let A, B, C and D be the respective positions of the ship when the bearings are 000°, 315°, 270° and 225°, the difference between each bearing being 45°. If P be the point then $\angle$s $APB = BPC = CPD = NPD = 45°$. $\angle PBD = (45° + A)$, $AB = 2$ units, $BC = x$ units, $CD = 1$ unit of time intervals.

$$\text{By Geometry—}\quad \frac{CP}{PA} = \frac{CB}{BA} = \frac{x}{2} = \tan A \text{ in } \triangle APC$$

$$\text{and}\quad \frac{PD}{PB} = \frac{DC}{CB} = \frac{1}{x} = \tan(45° + A) \text{ in } \triangle PBD$$

$$\text{By (7)}\quad \tan(45 + A) = \frac{\tan 45 + \tan A}{1 - \tan 45 \tan A} = \frac{1 + \frac{x}{2}}{1 - \frac{x}{2}} = \frac{2 + x}{2 - x} = \frac{1}{x}$$

on solving the quadratic equation, $x = {\cdot}5615$, from which $\angle A$ is found.

31. Three buoys, A, B, C are in line, B being midway between A and C. A ship sails 6 miles from B and the angle between A and B is then 20°, and between B and C 30°. Find distance from A to C.

Draw a figure. If D be the ship, then $\angle ADB = 20°$, $\angle BDC = 30°$, $AB = BC = x$.

$x = c \sin 20 \operatorname{cosec} ABD = a \sin 30 \operatorname{cosec} CBD$

$$\therefore\ \frac{c}{a} = \frac{\sin 30}{\sin 20} = \frac{5}{3{\cdot}42} \text{ call side } c \text{ 5 and side } a \text{ 3·42}$$

Substitute this ratio for sides, use tan formula to find $\angle$s A and C, thence find X, the half distance.

32. A, B and C are three buoys in line on a bearing 087° T. AB is 4 miles BC is 2 miles. A ship steams from A northwards 8 miles so as to subtend equal angles between AB and BC. Find the ship's distance from each buoy.

If S be the ship's position, then SB bisects $\angle ASC$ and $\frac{AB}{BC} = \frac{AS}{SC} = \frac{4}{2} = \frac{2}{1}$ $\therefore 2\,SC = AS$, $SC = 4$

$\triangle ASC$ three sides given. Find $\angle A$ and $\angle S$.

In $\triangle ABS$, with $\angle$s A and ASB and side AB. Find SB.

33. Three buoys A, B and C are in line bearing 080° T. A ship steams 5 miles to the northward from B and A, and C are then distant 8 and 6 miles respectively, and the sextant angles between A and B and B and C are equal. Find the distance between the buoys.

If S be the ship distant from A, B and C, 8, 5 and 6 miles respectively, and $AB = x$ miles, $BC = y$ miles then $\frac{x}{y} = \frac{8}{6}$ and $x = \frac{4y}{3}$. $\angle ASC$ is bisected $= 2\theta$,

$$\cos\theta = \frac{AS^2 + BS^2 - x^2}{2\,AS\,.\,BS} = \frac{CS^2 + BS^2 - y^2}{2\,CS\,.\,BS}$$

Substitute values and get $\frac{89 - x^2}{80} = \frac{61 - y^2}{60}$, solve this quadratic for x and y.

Exercises 34 to 37 are of similar type to Worked Example 4 which is sometimes referred to in chartwork as the three bearing problem.

34. A ship steering East through a current setting 204°, of unknown strength, sights a lighthouse bearing 030°; after sailing 4 miles it bore 000°; and after sailing another 4 miles it bore 340°. Find the course made good, also the drift of the current.

35. A ship steers north-easterly from a position A, then alters course 120° to port, finishing 3 miles, 270° from A; she steams altogether $4\frac{1}{2}$ miles. Find the distance on each course.

36. A vessel steaming at a uniform speed is observed from the shore to bear 025° T.; after an interval of 15 minutes she bears 350° T.; and after a further interval of 20 minutes she bears 280° T. Find her course.

37. Steering 100° T. through a current of unknown strength, setting 315° T., a lighthouse is sighted bearing 036° T., and after sailing 30 minutes it bore 010° T, and after sailing another 20 minutes it bore 348° T. Ship's speed 10 knots. Find course and distance made good, also drift of current.

38. A ship heading East observes a buoy B right ahead and another buoy C on her port bow subtending an angle θ° with B. Ship proceeds x miles due East and is then 3 miles due South of C and x miles from B. At the same time from a vessel at anchor due South of C the buoys B and C subtend an angle 2θ°. If the distance AC be 5 miles, find the distance between C and the anchored ship.

III.—RELATIVE VELOCITIES.

Example 6.

A cruiser receives information that a vessel is bearing N.W. from her, distant 36 miles and steering 240° T., speed 18 knots. How must the cruiser steer to intercept the vessel in shortest possible time at 24 knots, and how long will she take to do so?

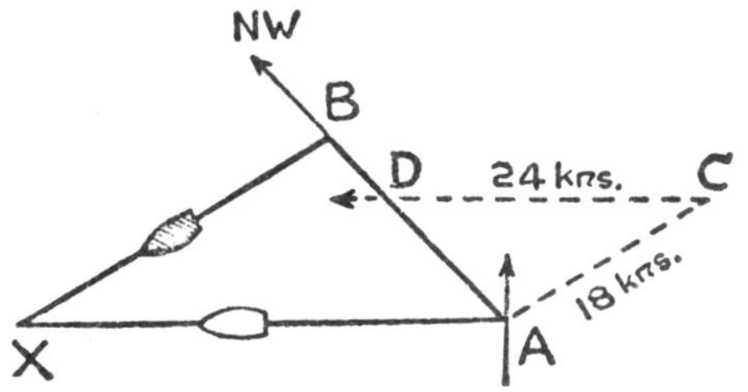

In relative velocity problems it usually simplifies the work if we assume the other ship to be stopped and treat her speed and course (reversed) as if it were the set and drift of a current.

In figure, let A be the cruiser and B the vessel, N.W. 36 miles from A. From A, lay off AC 060° T., 18 knots. With centre C and radius 24 knots describe an arc cutting AB at D. Then CD is the direction A must steer to reach B, and AD is the closing speed. Actually, ship A steams along AX which is parallel to CD, and B along BX, 240° T., the ships meeting at X. ACD is a triangle of velocity—given AC, CD and $\angle A = 105°$. Find $\angle D$, also AD.

by (1) $$\frac{\sin D}{18 \text{ kns.}} = \frac{\sin 105°}{24 \text{ kns.}} \text{ from which } D = 46° \ 26' \ 16''$$

A to B is N.	45°	00′	00″ W.
or	315	00	00
$\angle D = \angle A$	46	26	16
A to X	268	33	44
$\angle C$ =	28°	33′	44″

$$\frac{AD}{\sin C} = \frac{24}{\sin 105°} \text{ from which } AD = 11{\cdot}88 \text{ kns.}$$

But the distance from A to B is 36 miles.

$$\text{Time} = \frac{\text{Distance}}{\text{Speed}} = \frac{36 \text{ mls.}}{11{\cdot}88 \text{ kns.}} = 3\text{h } 1{\cdot}8\text{m.}$$

Answer.—Course 268° 33′ 44″, time 3h 1·8m.

Example 7.

Two ships *A* and *B* on patrol steering North, speed 10 knots. Ship *B* one mile on *A*'s port beam is ordered to increase speed to 12 knots and take station 2 cables astern of *A*. Find *B*'s course and time to get into position.

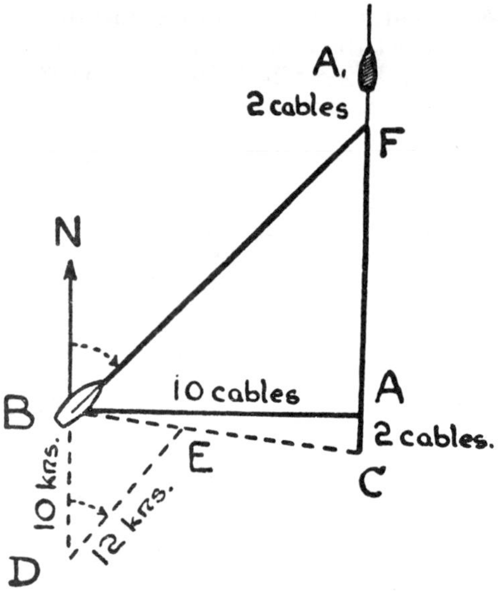

In figure, *A* and *B* are the two ships, *B* has to cross over to *C*. $AB = 10$ cables, $AC = 2$ cables.

Assume ship *A* to be stopped and apply his course (reversed), and speed, to ship *B* and read the question as follows:—Find the course *B* must steer to reach *C in order to counteract a current setting South, 10 knots*, *B*'s speed being 12 knots.

Construction:—Join *BC*. Lay off *BD* from *B*, South 10 knots, and with centre *D* and radius 12 knots describe an arc cutting *BC* at *E*. Join *DE*.

Then *BDE* is a triangle of velocity in which *DE* is the direction *B* must steer to get into station and *BE* is the effective or closing speed.

First find the distance from *B* to *C*.

$\cot B = \frac{10}{2} = 5$

$\therefore B = 11° \ 19'$

$\angle DBE = 78° \ 41'$

$BC = AB \sec B$

$\quad" = 10 \sec 11° \ 19'$

$\quad" = 10{\cdot}2$ cables.

In $\triangle BDE$.

By (1) $\frac{\sin E}{10} = \frac{\sin 78° \ 41'}{12}$ from which $E = 54° \ 48\tfrac{1}{2}'$, $\angle D = 46° \ 30\tfrac{1}{2}'$,

but BF is parallel to DE $\therefore$ $\angle NBF = \angle D = 046° \ 30\tfrac{1}{2}'$ course.

Again, $\frac{BE}{\sin D} = \frac{12}{\sin 78° \ 41'}$, from which $BE = 8{\cdot}88$ knots.

$$\text{Time} = \frac{\text{Distance}}{\text{Speed}} = \frac{BC}{BE} = \frac{1{\cdot}02 \text{ mls.}}{8{\cdot}88 \text{ kns.}} = 7 \text{ minutes.}$$

Answer. *B*'s course is 046° $30\tfrac{1}{2}'$, time to take station, 7 minutes. When *B* reaches *F*, ship *A* is at A_1.

Example 8.

Ship A steering North at 12 knots sees ship B 4 points on her starboard bow, 10 minutes later B bears 3 points on the same bow and 5 minutes later she is $2\frac{1}{2}$ points on the bow, if B's speed is 15 knots find her course.

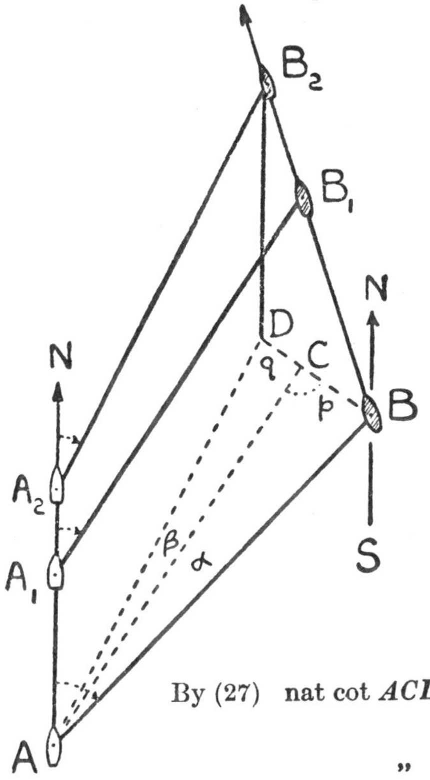

It is obvious from the question that B is passing ahead of A. Assume ship A to be stopped at position A in figure and imagine her speed and course (reversed) to be transferred to ship B so that the relative bearings of the two ships will be maintained as indicated by the dotted constructional lines on the figure where AB, AC and AD would be the 4, 3, and $2\frac{1}{2}$ point bearings respectively, $a = 11° \ 15'$, $\beta = 5° \ 37{\cdot}5'$, $p = 10$ minutes, $q =$ 5 minutes, a ratio of 2 to 1. It is required to find $\angle C$, which is the angle between the middle bearing and the ratio line BCD. BCD would be the course steered by B if A were actually stopped, but at the moment it only serves as a line of direction as a step towards finding B's course.

By (27) $\text{nat cot } ACD = \dfrac{p \cot a - q \cot \beta}{p+q}$

" $= \dfrac{2 \cot 11° \, 15' - 1 \cot 5° \, 37{\cdot}5'}{3} = -{\cdot}0329$

$\angle ACD = 88° \, 7'$

but this is really $\angle ACB$ the supplement of $\angle ACD$ as the sign is negative.

$\angle ABC = 180° - (88° \, 7' + 11° \, 15') = 80° \, 38'$

To find B's course.

From A lay off A's course North so that $AA_1 = 2$ miles, $A_1A_2 = 1$ mile. From A_1 lay off the 3 point bearing and from A_2 the $2\frac{1}{2}$ point bearing. From D lay off A's course North cutting the $2\frac{1}{2}$ bearing at B_2. Join B to B_2 and this will give the direction of B's track, the successive relative positions of the ships being A and B, A_1 and B_1, A_2 and B_2.

In triangle BDB_2 it is required to find $\angle DBB_2$.

We know $DB_2 = 15$ minutes at 12 kns. $= 3$ miles $= AA_2$

$BB_2 = 15$ minutes at 15 kns $= 3\frac{3}{4}$,,

and we can get $\angle B_2DB$ which is equal to $DBS = 80° \, 38' + 45° = 125° \, 38'$.

by (1) $\dfrac{\sin B}{3} = \dfrac{\sin 125° \, 38'}{3{\cdot}75}$ from which $\angle DBB_2 = 40° \, 33' \, 36''$.

$\angle B_2 = 180° - (40° \, 33' \, 36'' + 125° \, 38') = 13° \, 48' \, 24''$.

Answer. B's course N. 13° 48′ 24″ W. or 346° 11′ 36″ T.

Example 9.

Ship *B* is 18 miles North of *A* and steering E.S.E.; *A* is steering East and both are steaming at 15 knots. Find (1) speed of approach; (2) distance of nearest approach and time to reach it; (3) the interval of time they will be within 10 miles of each other.

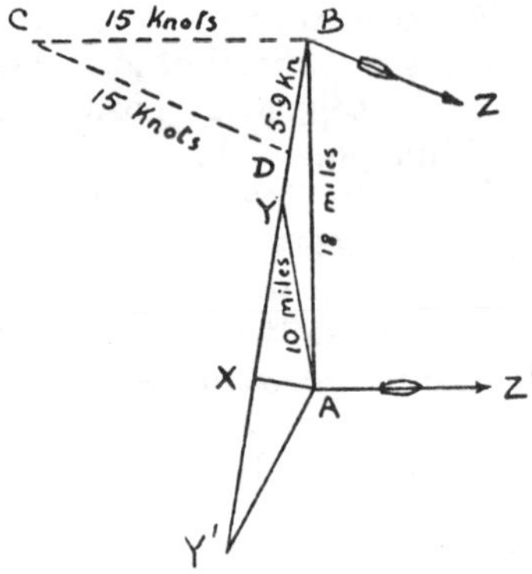

Construction.—1. Let *A* and *B* be the relative positions of the two ships. *AB*= 18 miles from any convenient scale.

Lay off *BC*, West, 15 knots = *A*'s course (reversed) and speed.

Lay off *CD*, E.S.E., 15 knots = *B*'s course and speed.

Join *BD*, which indicates the relative bearing line, and the length of *BD* (5·9 knots) their speed of approach.

2. Drop *AX* perpendicular to *BD* produced, then *X* indicates the relative position of nearest approach but not the actual position. The length of *AX* gives the distance on the same scale as *AB* (3·5 miles) and *BX* gives the distance to go (17·5 miles) at approaching speed (5·9 knots).

$$\text{Time to reach } X = \frac{\text{Dist.}}{\text{Speed}} = \frac{17{\cdot}5}{5{\cdot}9} = 2{\cdot}97 \text{ hours.}$$

The distance each ship would have to go on their respective courses to reach *X* = 2·97 hrs. at 15 knots = 44·55 miles.

The actual position of *X* relatively to *A* and *B* may be found by laying off from *B*, *BZ*, E.S.E. 44·55 miles and from *A*, AZ^1 East 44·55 miles The distance ZZ^1 will measure 3·5 miles the distance of nearest approach.

3. Centre *A*, radius 10 miles, describe an arc cutting *BX* at *Y* and Y^1. Then YY^1 is the distance they cover (18·7 miles) when within 10 miles range of each other.

$$\text{Time} = \frac{\text{Dist.}}{\text{Speed}} = \frac{18{\cdot}7}{5{\cdot}9} = 3\text{h } 10\text{m.}$$

Relative velocity problems can be solved by constructing a figure to scale with compasses and protractor, the answers so found being near enough in practice as the data are only approximate, no allowance being made for the turning circle, delay in gaining speed, etc. The larger the scale of distances the more accurate will be the results. The facts can be laid off quickly on a prepared diagram, a manoeuvring board, which is just a compass rose, or protractor showing concentric circles spaced equally apart, the spaces being marked off in tenths for convenience of measurement. These spaces may be called miles, or knots, depending on whether distance, or speed, are the factors under consideration.

III. Relative Velocities.

1. A warship receives information that an enemy ship is bearing N.N.W. from him and distant 30 miles, steering W.S.W. at 20 knots. If the warship's maximum speed is 35 knots, how long will it take him to intercept the enemy, and what course must he steer to do so?

1. Lay off *AB*, N.N.W., the direction of enemy.
2. Lay off *AC*, E.N.E. 20 knots.
3. Centre *C*, radius 35 knots, describe an arc cutting *AB* at *D*. *CD* gives the direction warship must steer (302°) and *AD* (28·2 knots) his closing speed

2. *A* and *B* are two destroyers steaming North at 15 knots; *B* being on *A*'s port beam distant 5 miles has orders to take up a position 2 miles astern of *A*, and to be in this position in 15 minutes. Find the course that *B* must steer and her speed to carry out the order.

1. Mark spot *A*. Lay off *AB*, 5 miles on *A*'s port beam and *AC* 2 miles astern of *A*; *BC* measures 5·4 miles and this is the distance *B* has to go in 15 minutes. *B*'s speed is therefore 21·6 knots.

2. Lay off *BD*, South 15 knots. Centre *D* and radius 21·6 knots cut *BC*, produced if need be, at *E*. *DE* is the course *B* must steer.

3. Two ships steering 338° T., at 12 knots, ship *B* is one mile on *A*'s port quarter and is ordered to increase speed to 16 knots and take station one mile on *A*'s starboard beam. Find *B*'s course and closing speed.

4. Two ships *A* and *B* steering the same course; *B* 2 cables astern of *A* is ordered to increase speed to 12 knots and take station one mile on *A*'s starboard beam, *A*'s speed is 10 knots. Find *B*'s alteration of course and the time she would take to get into position.

5. Two vessels patrolling steering 300° T., at 10 knots. Ship *B* is four points on *A*'s starboard quarter distant one mile, and is ordered to increase to 15 knots and take station one mile on *A*'s port beam. Required the course to steer and the time to get into position.

6. *A* and *B* are two steamers in a convoy steering N.N.W. at 10 knots. *B* being two miles on A's starboard beam is ordered to take a position 1 mile ahead of *A* and to increase speed to 12 knots. Find the course that *B* must steer and the time it will take to reach her new position ahead of *A*.

7. Two vessels *A* and *B* on the same meridian; *B* is 10 miles North of *A*. *B* steams S.E. at 10 knots and *A* steams East at 20 knots. Find how long it will be until they are within nearest approach of each other.
(See Worked Example 9).

8. A liner steering 180° at 20 knots to pass 2 miles due East of a lighthouse at 1600. At 1530 a 12-knot launch starts out from the lighthouse to intercept the liner in the shortest possible time. Find the course the launch must steer and give the time of intercepting.

1. Mark a spot *B*. Lay off *BA* due North, 10 miles. *A* gives the liner's position at 1530 and *B* is to be her position at 1600.

2. Lay off *BC*, due West, 2 miles. *C* is the lighthouse. Join *CA*; this gives the course and distance launch would have to go if liner were stopped at *A*.

3. Lay off *CD*, South, 20 knots. With centre *C* and radius 12 knots, describe the arc of a circle towards *A*. Through *D* draw a line parallel to *CA* cutting the 12 knot circle at *E*. Join *CE*; then *CE* is the direction launch must steer, viz. (031°) and *DE* is the closing speed (30·8 knots).

4. Where *CE* cuts *BA* is the meeting point. Call this spot *X*. By measurement *AX* is 6·6 miles and at 20 knots the liner will be there in 19·8 minutes, viz. at 15h 49·8m. *CX* is 4 miles by measurement and is the distance the launch has to go, viz. 4 miles at 12 knots = 20 minutes.

9. *A* and *B* are steering the same course 220° at same speed, 12 knots, *B* is one mile on *A*'s starboard beam. Ship *A* sights a destroyer bearing 000°, distant 6 miles, and *B* alters course 22° to starboard. Find the course of the destroyer at 25 knots to pass midway between *A* and *B*.

1. Mark spot *A*, and spot *B* 1 mile on *A*'s starboard beam, *A* heading 220°; mark also spot *X* the position of the destroyer bearing 000°, 6 miles from *A*.

2. Bisect *AB* at *Y*, being midway between *A* and *B*, the destroyer's objective if *A* and *B* were stopped.

3 Lay off *YC* = 231°, viz. 220° + half the alteration of *B*'s course.

4. From X lay off XD, the mean course YC (reversed) and speed viz. 051°, 11·8 knots. With centre D and radius destroyer's speed, 25 knots describe an arc to cut XY at E. Join DE. DE gives the destroyer's course (205°) and XE her closing speed (14·8 knots) and XY (5·6 miles) the distance she has to go.

10. A vessel steering West, 10 knots, sights a cruiser bearing 020° T., distant 5·5 miles as measured by the flash and report of a gun fired from the cruiser. Half an hour later the cruiser bears 350° T., distant 2·5 miles as found by the same method. Find the course and speed of the cruiser.

1. Mark a spot A to represent the vessel and C the cruiser 020°, 5·5 miles from A.

2. Lay off AA^1, West 5 miles, and A^1 C^1 350°, 2·5 miles. Join CC^1. CC^1 is the direction of the course (250°) and measures 7·85 miles in 30 minutes.

11. A's course 000°, speed 10 knots; B's speed 12 knots. At 1 p.m., B bore from A 066°; at 2 p.m.—046°, at 2.15 p.m.—025°. Find B's course.

Similar to Worked Example 8.

12. A's course 180°, speed 12 knots; B's speed 16 knots. B bears from A 140°, and 30 minutes later B bears 160° and 15 minutes later 170°. Find B's course.

IV.—VARIOUS EXAMPLES.

Example 10.

A vessel sights a beacon 48 feet high bearing 295° and subtending a vertical angle of 0° 6′ 20″. At the same instant another beacon of the same height subtends the same vertical angle and bears 320. If the distance between the beam bearings to port is one mile, find the vessel's course.

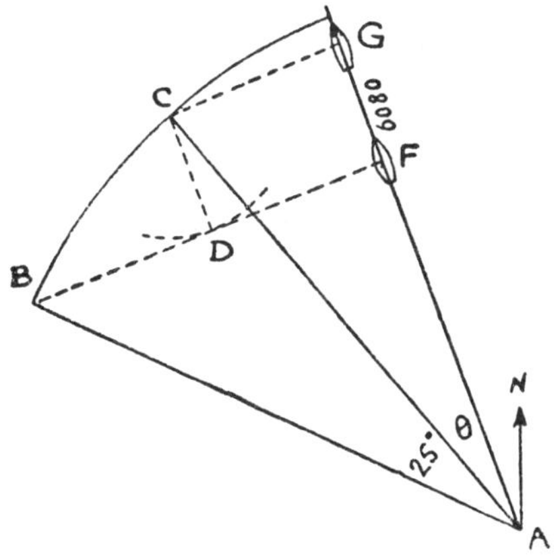

Construction:—Let A be the position of the ship. $AB = AC$ the distance of the beacons B and $C = 48 \cot 0° \, 6' \, 20'' =$ 26046 feet.

With centre A and radius this distance from any convenient scale, describe an arc passing through the beacons B and C, lay off the bearing of B (295°) and of C (320°), $\angle BAC = 25°$. Join BC.

With centre C and radius 1 mile describe an arc. Draw BD a tangent to this arc and produce it.

Draw CD and AF perpendicular to the tangent BD, and make CG parallel to BD. $FG = 6080$ feet given, AF is the vessel's track.

In $\triangle\ AGC \quad \angle G = 90°. \quad AG = AC \cos \theta.$

In $\triangle\ AFB \quad \angle F\ 90°. \quad AF = AB \cos (25° + \theta)$

Subtracting we get $GF = 26046 \{ \cos \theta - \cos (25° + \theta) \}$

By (14) $\cos \theta - \cos (25° + \theta) = 2 \sin \frac{1}{2} (25 + \theta + \theta) \sin \frac{1}{2} (25° + \theta - \theta)$

$\therefore\ 6080 = 26046 \times 2 \sin 12° \, 30' \times \sin (12° \, 30' + \theta).$

$$\sin (12° \, 30' + \theta) = \frac{6080 \times \operatorname{cosec} 12° \, 30'}{2 \times 26046}$$

„	=	32°	38′
		12	30
$\angle\ \theta$	=	20	08
Bearing of C is		320	00
A to F the course		340	08 *Answer*

Example 11.

A and B are two points 5 miles apart, B being 090° from A. A vessel on an easterly course observes A bearing 000° and, after sailing 3 miles, observes B bearing 030° and after sailing a further 8 miles B was observed bearing 330°. Find the ship's course.

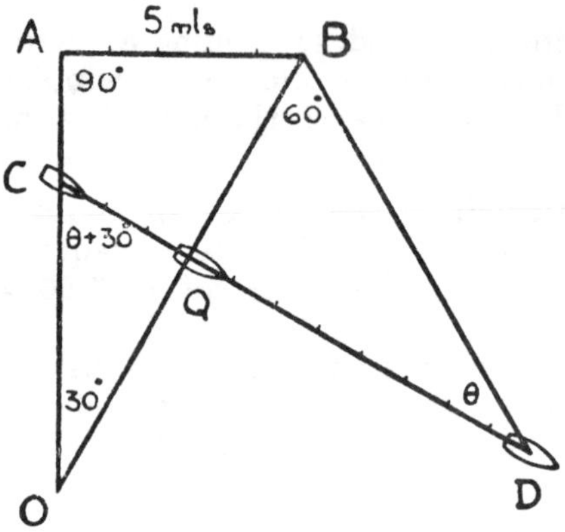

In figure, let A and B be the two points, $AB = 5$ miles, $CQ = 3$ miles, $QD = 8$ miles, $\angle O = 30°$, $\angle OBD = 60°$, $\angle A = 90°$, $\angle OCQ = (\theta + 30°)$, $\angle D = \theta$.

In $\triangle OAB$.

$OB = 5 \text{ cosec } 30° = 5 \times 2 = 10$ miles.

$\therefore OQ + QB = OB = 10$ miles. . . (i.)

In $\triangle OCQ$ by (1) $\dfrac{OQ}{\sin(\theta + 30)} = \dfrac{3}{\sin 30}$

$\therefore OQ = \dfrac{3 \sin(\theta + 30)}{\sin 30} = 6 \sin(\theta + 30)$ (ii).

In $\triangle BQD$. $\dfrac{QB}{\sin \theta} = \dfrac{8}{\sin 60}$

$\therefore QB = \dfrac{8 \sin \theta}{\sin 60} = \dfrac{16}{\sqrt{3}} \sin \theta$ . . (iii).

(i), (ii), (iii) $OQ + QB = 10 = 6 \sin(\theta + 30) + \dfrac{16}{\sqrt{3}} \sin \theta$

by (5) $\therefore 10 = 6(\sin \theta \cos 30 + \cos \theta \sin 30) + \dfrac{16}{\sqrt{3}} \sin \theta$

$$10 = 6\left(\frac{\sqrt{3}}{2} \sin \theta + \tfrac{1}{2} \cos \theta\right) + \frac{16}{\sqrt{3}} \sin \theta$$

$$10 = \frac{25}{\sqrt{3}} \sin \theta + 3 \cos \theta$$

We might adopt the artifice explained on page 245 to solve this equation for θ. Re-arranging the equation.

$$3 \cos \theta + \frac{25}{\sqrt{3}} \sin \theta = 10$$

Let $R \sin(A + \theta) = 10$, then $R \sin A \cos \theta + \cos A \sin \theta = 10$.

From which $R \sin A = 3$ and $R \cos A = \dfrac{25}{\sqrt{3}}$

Square and add, $R^2 = \dfrac{625}{3} + 9 = \dfrac{652}{3} = 217{\cdot}3$, $\therefore R = 14{\cdot}74$

$\text{Sin}(A + \theta) = \dfrac{10}{R} = \dfrac{10}{14{\cdot}74}$ or $\text{cosec}(A + \theta) = \dfrac{14{\cdot}74}{10} = 1{\cdot}474$, $A + \theta = 42°\ 43'$

$\text{Sin } A = \dfrac{3}{R} = \dfrac{3}{14{\cdot}74}$ or cosec $A = 4{\cdot}913$ . . $A = 11\ \ 44$

$\theta = 30\ \ 59$

The direction B to D is		150° 00′
	$\angle \theta$	30 59
The course C to D is		119 01

Example **12.**

The apparent wind is 260° to a vessel steering 220° at 10 knots. When she altered course to 260° the apparent wind was 15° on the starboard bow. Find the direction and velocity of the true wind.

In figure, AC is 260°, 10 miles and CD is the apparent wind 15° on the starboard bow. AB is 220°, 10 miles, and BD is the apparent wind from 260°, or 40° on the starboard bow. The apparent winds meet at D. Join AD which gives the direction of the true wind and its velocity per hour.

Join CB, then $\triangle\ ABC$ is isosceles and $\angle ACB = \angle ABC = 70°$
But $\angle ACD = 15°$, $\therefore\ \angle BCD = 55°$ and $\angle BDC = 15°$.

Make AE perpendicular to CB, then $CB = 2CE = 2\,(AC \sin 20°) = 20 \sin 20°$.

In $\triangle$BCD by (1)

$$BD = CB \sin BCD \operatorname{cosec} BDC$$
$$\quad = 20 \sin 20 \sin 55° \operatorname{cosec} 15°$$
$$\quad = 21{\cdot}65$$

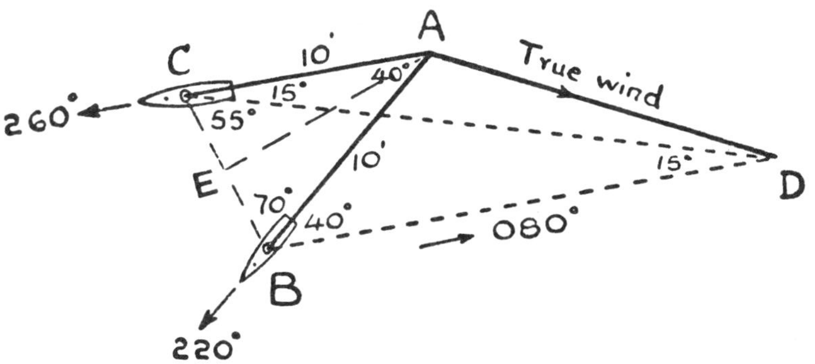

In $\triangle$ABD $\qquad \angle B = 40°,\ (A + D) = 140°$

by (23)
$$\frac{\tan \frac{1}{2}(A - D)}{\tan \frac{1}{2}(A + D)} = \frac{BD - BA}{DB + BA} = \frac{11{\cdot}65}{31{\cdot}65}$$

$$\therefore\ \tan \tfrac{1}{2}(A - D) = \frac{\tan 70° \times 11{\cdot}65}{31{\cdot}65}$$

from which $\frac{1}{2}(A - D) = 45°\ 19\frac{1}{2}'$
$\qquad\qquad\quad \frac{1}{2}(A + D) = 70\ \ 00$

$\angle BAD$	$=115$	$19\frac{1}{2}$
A to B is	220	00
A to D is	104	$40\frac{1}{2}$
+	180	00

$\therefore$ the direction of the true wind is from 284 $40\frac{1}{2}$

In $\triangle$DAB
$$\frac{AD}{AB} = \frac{\sin ABD}{\sin ADB}$$
$$\frac{AD}{10} = \frac{\sin 40°}{\sin 24°\ 40\frac{1}{2}'}$$
$$AD = 15{\cdot}4 \text{ miles}$$

Answer. True wind 284° 40′
Velocity 15·4 m.p.h.

Example **18.**

At noon a vessel steaming south-easterly at 10 knots sights a lighthouse in transi with a peak bearing South true and, when the lighthouse bears 270°, sights a beaco: bearing 180°. A cross bearing shows each to be 4 miles distant. At 2 p.m. afte continuing on the same course, the vessel anchors with the beacon and peak in transi bearing 270°. Find the course made good during the two hours.

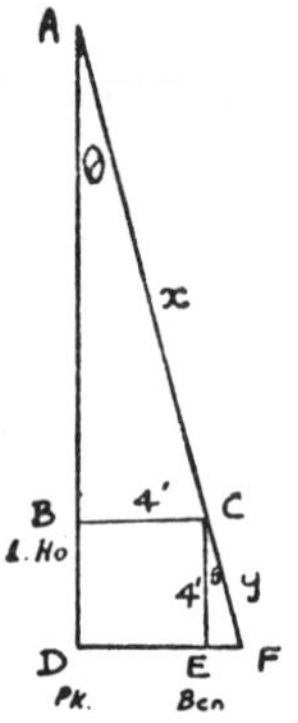

Given ABD a meridian, B the lighthouse, D th peak, $\angle A = \theta$ to be found. ACF the ship's track such that $AC = x$, $CF = y$ and $x + y = 20$ miles When ship is at C, B bears 270°, 4 miles, and beacon E bears 180°, 4 miles. When she reaches F, E and D ar in transit. $\angle ECF = \theta$.

$$\triangle ABC \qquad \frac{x}{4} = \text{cosec } \theta = \frac{1}{\sin \theta} \quad . \quad . \quad . \quad . \quad (1)$$

$$\triangle CEF \qquad \frac{y}{4} = \sec \theta = \frac{1}{\cos \theta} \quad . \quad . \quad . \quad . \quad (2)$$

(1) and (2) add $\frac{x+y}{4} = \frac{20}{4} = 5 = \frac{1}{\sin \theta} + \frac{1}{\cos \theta} = \frac{\cos \theta + \sin \theta}{\sin \theta \cos \theta}$

$$5 \sin \theta \cos \theta = \sin \theta + \cos \theta$$

Square each side. $25 \sin^2 \theta \cos^2 \theta = \sin^2 \theta + \cos^2 \theta + 2 \sin \theta \cos \theta$

NOTE. * $\quad 25 \frac{\sin^2 2\theta}{4} = 1 + \sin 2\theta$

$$\frac{25}{4} \sin^2 2\theta - \sin 2\theta = 1$$

Divide throughout by $\frac{25}{4}$, complete the square and solve for 2θ.

$$\sin^2 2\theta - \frac{4}{25} \sin 2\theta + \left(\frac{2}{25}\right)^2 = \frac{4}{25} + \frac{4}{625}$$

$$\therefore \sin 2\theta - \frac{2}{25} = \pm \sqrt{\frac{104}{625}}$$

$$\sin 2\theta = \pm \frac{10{\cdot}2}{25} + \frac{2}{25} = \frac{12{\cdot}2}{25} = + {\cdot}488$$

$2\theta = 29° \ 12'$, $\theta = 14° \ 36'$ Course S. 14° 36′ E. = 165° 24′.

NOTE.*—This transformation is derived from (15) $\sin 2\theta = 2 \sin \theta \cos \theta$. Square both sides, $\sin^2 2\theta = 4 \sin^2 \theta \cos^2 \theta$ $\therefore \sin^2 \theta \cos^2 \theta = \sin^2 2\theta/4$.

IV.—Various Examples.

1. From a vessel two towers, A and B, subtend a horizontal angle of 24° and A is distant 6 miles. After steaming a short distance the hor. angle was 36° and A is distant 4 miles. The distances from B to the points of observation are as 5 is to 6. Find the distance of B at the 1st and 2nd observations.

Construct a figure such that D is the 1st position of the vessel and $\angle ADB = 24°$ and $DA = 6$ miles. Make C the 2nd position and $\angle ACB = 36°$, $CA = 4$ miles. $DB = 5X$, $CB = 6X$. Join AB.

$BA^2 = AC^2 + BC^2 - 2\ AC \cdot CB \cos 36° = BD^2 + DA^2 - 2\ BD \cdot DA \cos 24°$

$\therefore\ 16 + 36X^2 - 48\ X \times \cdot 809 = 25\ X^2 + 36 - 60X \times \cdot 913$ from which $X = \cdot 8055$ thence find distances BC and BD.

2. A ship observes the horizontal angle between a conspicuous object and a lighthouse bearing East to be 24°. She sails 132° T., 3 miles, and finds the lighthouse bearing North and the hor. angle the same as before. Find the distance between the object and the lighthouse.

Draw a figure so that A is the 1st position of ship, B the 2nd position and C the lighthouse, then $\triangle ACB$ is right angled at C. Bisect AB at D, then D is the centre of a circle passing through A, C and B, $DA = 1\frac{1}{2}$ miles.

If E be the object, then it is also on the circumference of the circle because $\angle CAE = \angle CBE = 24°$ Join DC and DE, radii of circle each being $1\frac{1}{2}$ miles. $\angle CDE = 48°$, double $\angle CAE$ which is 24° and stands on the same chord CE. Drop DF perpendicular to CE, it bisects $\angle CDE$ and also the chord CE, so that $CE = 2\ CF$ the distance required. $CF = CD \sin CDF$.

3. A survey vessel observes two points A and B the same distance off, subtending an angle of 86°, A bearing 310° T. After steaming due South for 7 cables the angle now subtended by A and B is 43°. Find bearing and distance from A to B.

4. A vessel steering East, 15 knots, has the apparent wind S.E., and when she alters course two points to starboard the apparent wind is S.E. by S. Find the direction and velocity of the true wind.

5. A ship steering 000° has the apparent wind 4 points on her starboard bow. She alters course to 045° and the apparent wind is now 20° on her starboard bow. Find the true direction of the wind and its velocity if the ship's speed were 12 knots.

See Worked Example 12.

6. A cubical box of 4 feet sides is laid against a wall. A ladder is also laid against the wall just touching the edge of the box. The ladder is 16 feet long. Find the height of the wall.

Draw figure and note that the example is similar to Worked Example 13.

V. SPHERICAL TRIANGLES.

Example 14.

A and *B* are two observers 2640 miles apart in latitude 24° N., *B* being to the westward of *A*. The sun's zenith distance on *A's* meridian was 25° 16′. Find the bearing of the sun from *B*.

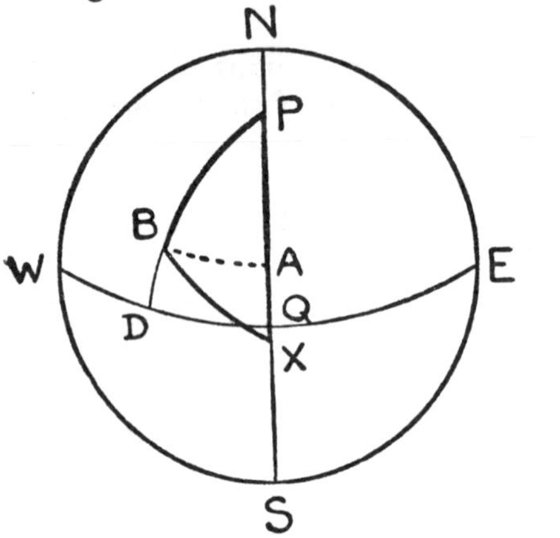

Let *A* be the zenith of observer *A*, and *B* that of observer *B*.

X the sun on the meridian of *A* so that *AX* is 25° 16′ and *PX* = 91° 16′. *AB* is the arc of the parallel of latitude, joining the two observers, 2640 miles; it is the departure between them and has to be converted into *D* Long. (*QD*) and transferred to ∠*P*. *D* Long. = 2640 sec 24° = 2890′ = 48° 10′ = ∠*P*. *BX* is a great circle between *B* and *X*.

In △PBX.
Given $PB = 66°$, $PX = 91°\ 16'$, $\angle P = 48°\ 10'$. Find ∠*X*.

by (38) $$\tan \tfrac{1}{2}(B + X) = \frac{\cos \frac{1}{2}(PB - PX)}{\cos \frac{1}{2}(PB + PX)} \cot \frac{P}{2}$$

by (39) $$\tan \tfrac{1}{2}(B - X) = \frac{\sin \frac{1}{2}(PB - PX)}{\sin \frac{1}{2}(PB + PX)} \cot \frac{P}{2}$$

from which $\frac{1}{2}(B + X) = 84°\ 50\frac{1}{2}'$
$\frac{1}{2}(B - X) = 26\ \ 31\frac{1}{2}$
∠*PBX*. 111 22
∴ Sun bears from *B* 111° 22′ T. *Answer.*

Example 15.

In north latitude a star is on the prime vertical $4^h\ 10^m$ after passing the meridian, and $2^h\ 32^m$ later it sets. Find the latitude.

In figure, $\angle ZPX = 4^h\ 10^m = P$
$\angle ZPY = 6\ \ 42 = P_1$
$\angle PZX = 90°$, side $ZY = 90°$
Find *PZ*.

In △PZX. $\cos P = \cot PX \tan PZ$
In △PZY. $-\cos P_1 = \cot PY \cot PZ$
but $PX = PY$ and dividing we get

$$\frac{\cos P}{-\cos P_1} = \tan^2 PZ$$

$$\therefore \cot \text{lat.} = \sqrt{\cos P \sec P_1}$$

P	$4^h\ 10^m$	cos	9·664406
P_1	6 42	− sec	10·739367
			2)20·403773
Answer. Lat.	32° $8\frac{1}{2}'$ N.	cot	10·201886

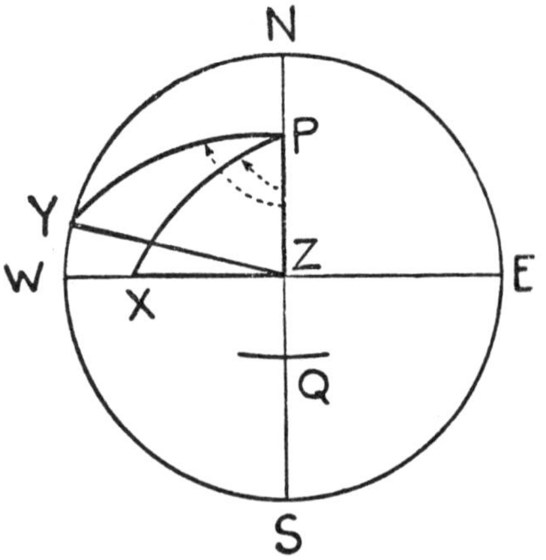

Example **16.**

Two stars have the same altitude at the same instant, A bears 060°, Decl 25° N., and B bears 090°, Decl 15° N. Find the latitude.

In figure. A and B are the two stars.

$$PA = 65°, \quad PB = 75°, \quad \angle PZA = 60°$$
$$\angle PZB = 90°, \quad ZA = ZB$$

In $\triangle$PZB. $\cos PB = \cos PZ \cos ZB = \cos 75°$ . . . . (i.)

In $\triangle$PZA. by (28) $\cos 60° = \dfrac{\cos 65° - \cos PZ \cos ZA}{\sin PZ \sin ZA}$

$\frac{1}{2} = \dfrac{\cos 65° - \cos 75°}{\sin PZ \sin ZA}$ . . . . (ii.)

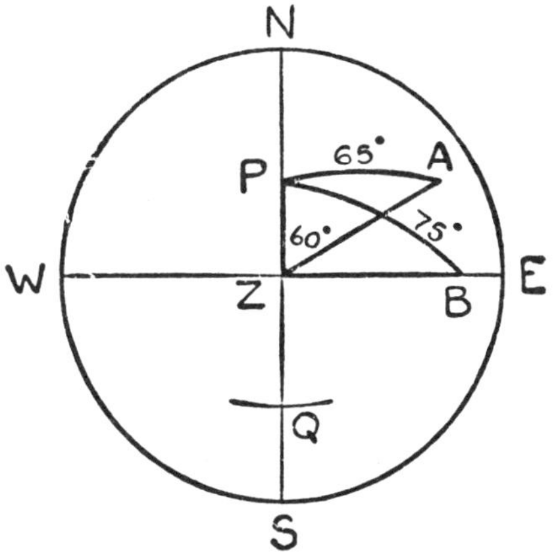

From (ii.) $\sin PZ \sin ZA = 2(\cos 65° - \cos 75°) =$ ·327598

„ (i.) $\cos PZ \cos ZA = \cos 75° =$ ·258819

add $\cos PZ \cos ZA + \sin PZ \sin ZA = \cos(PZ - ZA) =$ ·586417 (54° 06′)

subt. $\cos PZ \cos ZA - \sin PZ \sin ZA = \cos(PZ + ZA) = -$ ·068779 (93 57)

Nat cos − ·068779 = 86° 3′, but having a negative sign $(PZ + ZA) = 180° - 86° 3' = 93° 57'$

From which	$(PZ - ZA)$	54°	06′	54°	06′
„	$(PZ + ZA)$	93	57	93	57
	2)	148	03	2) 39	51
	PZ	74	1½	or PZ 19	55½

Answer. Lat. = 15° 58½′ N. or 70° 4½′ N.

Example **17.**

In what latitude will a star of Decl 45° N. be on the prime vertical while having an hour angle of half that of another star on the prime vertical whose Decl is 20°N.?

In figure, let x and y be the two stars bearing West, such that $\angle ZPY = \theta$ and $\angle ZPX = 2\theta$, $PX = 70°$, $PY = 45°$.

In $\triangle$PZY. $\cos\theta = \tan PZ \cot PY$ . . . . (i.)

$\therefore$ $\tan PZ = \cos\theta \tan PY$

$\triangle$PZX. $\cos 2\theta = \tan PZ \cot PX$ . . . . (ii.)

Dividing (ii.) by (i.) $\dfrac{\cos 2\theta}{\cos\theta} = \dfrac{\cot PX}{\cot PY} = \dfrac{\cot 70°}{\cot 45°} = ·364$

But $$\frac{\cos 2\theta}{\cos\theta} = \frac{2\cos^2\theta - 1}{\cos\theta} = \cdot 364$$

$$2\cos^2\theta - \cdot 364\cos\theta = 1$$

$$\cos^2\theta - \cdot 182\cos\theta = \cdot 5$$

Complete the square by adding half the coefficient of cos θ to each side of the equation and we get

$$\cos^2\theta - \cdot 182\cos\theta + \cdot 091^2 = \cdot 5 + \cdot 091^2$$

$$(\cos\theta - \cdot 091)^2 = \cdot 5 + \cdot 091^2 = \cdot 5083$$

$$\cos\theta - \cdot 091 = \pm\sqrt{\cdot 5083} = \pm\cdot 7130$$

$$\cos\theta = +\cdot 7130 + \cdot 091 = +\cdot 804$$

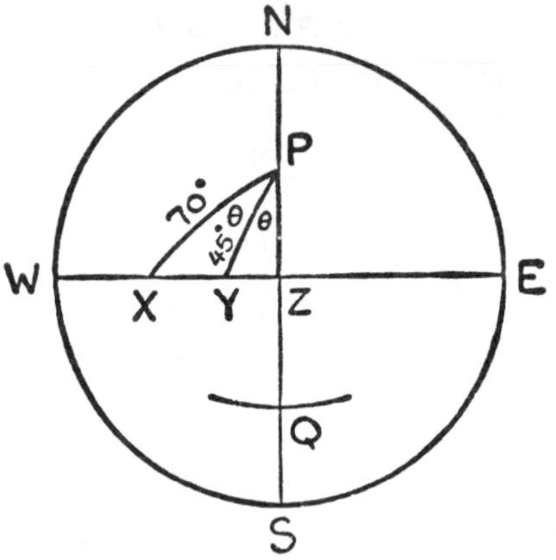

From Eq. (i.)

$$\tan PZ = \cos\theta \tan PY$$

$$\text{nat cot Lat} = \cos\theta \tan 45$$

$$,, \quad = \cdot 804 \times 1 = \cdot 804$$

$$\text{Lat} = 51° \ \ 12' \text{ N.}$$ *Answer.*

Example 18.

A star is above the horizon to an observer in 60° N. for twice the period it is to an observer in 30° S. Find the star's declination.

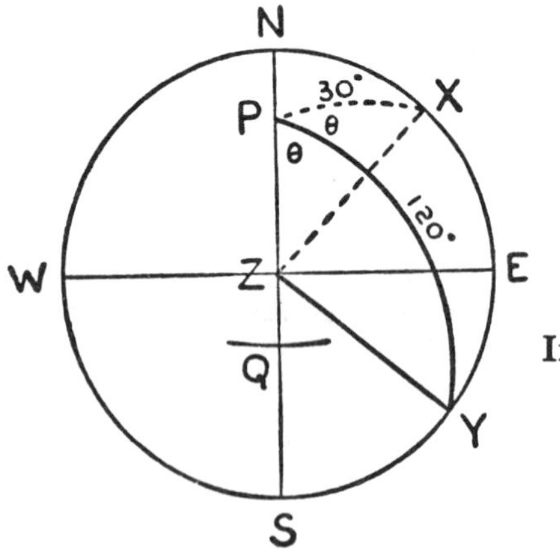

This problem is somewhat similar to Example 15 if we substitute declination for latitude in the question so that Lat 60° N. = Decl 60° N., making $PX = 30°$. Lat 30° S. = Decl 30° S., making $PY = 120°$ and hour angles θ and 2θ. Find PZ the colatitude which will be the polar distance of the star. It is merely a question of thinking of the observers as two stars X and Y, and the star as an observer Z.

In quadrantal $\triangle$PZX—$\cos 2\theta = \cot PX \cot PZ$

,, $\triangle$PZY—$\cos\theta = \cot PY \cot PZ$

$$\therefore \frac{\cos 2\theta}{\cos\theta} = \frac{\cot 30°}{\cot 120°}$$

$$= \cot 30° \tan 60° = \sqrt{3} \times -\sqrt{3} = -3$$

(Note that cot 120° is negative.)

By (19) we may write . . . $\cos 2\theta = 2\cos^2\theta - 1$

so that $\dfrac{\cos 2\theta}{\cos\theta} = \dfrac{2\cos^2\theta - 1}{\cos\theta} = -3$

$\therefore \quad 2\cos^2\theta + 3\cos\theta - 1 = 0$

Applying the quadratic formula mentioned on page 236, viz.:—

$$x = \frac{-b \pm \sqrt{(b^2 - 4ac)}}{2a}$$

$$\cos\theta = \frac{-3 \pm \sqrt{(3^2 - 4 \times 2 \times (-1))}}{2 \times 2} = \frac{-3 \pm \sqrt{17}}{4} = \frac{-3 \pm 4{\cdot}12}{4}$$

$$\cos\theta = \frac{1{\cdot}12}{4} = {\cdot}28 \quad \therefore \theta = 73° \; 41\tfrac{3}{4}'.$$

The negative value of $\cos\theta$ does not apply.

We already noted, $\cot PZ = \cos\theta \tan PY$

,, $= \cos 73° \; 41\tfrac{3}{4}' \tan 120°$

from which $PZ = 64° \; 04' =$ P. dist of star

Answer. Decl of star $= 25° \; 56'$N.

Example 19.

A circumpolar star has maximum azimuth 42° from an observer A in North latitude. From another observer B, 10° South from A, the maximum azimuth is 32°. Find the latitude.

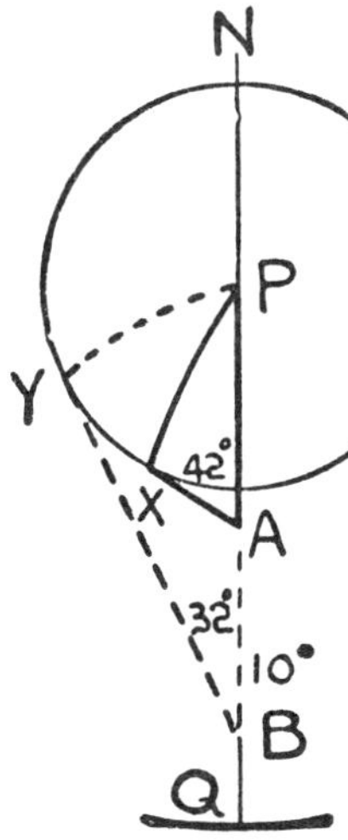

In figure, let A be the position of one observer and B that of the other, so that $AB = 10°$.

X is the position of the star at maximum azimuth 42°.
Y is the position of the star at maximum azimuth 32°.
Angles X and $Y = 90°$.

Required to find PA the co-latitude.

In $\triangle$PXA. $\sin PX = \sin PA \sin A$.
In $\triangle$PYB. $\sin PY = \sin PB \sin B$.
But $PX = PY$ and $PB = (PA + 10°)$

$\therefore \sin PA \sin 42 = \sin 32 \sin (PA + 10°)$

by (5)

$\sin PA \sin 42 = \sin 32 [\sin PA \cos 10 + \cos PA \sin 10]$

$$\frac{\sin 42}{\sin 32} = \frac{\sin PA \cos 10 + \cos PA \sin 10°}{\sin PA}$$

,, $= \cos 10 + \cot PA \sin 10$

$$\therefore \cot PA = \frac{\sin 42}{\sin 32 \sin 10} - \frac{\cos 10}{\sin 10}$$

Nat $\cot PA = 7{\cdot}271617 - 5{\cdot}671280 = 1{\cdot}600337$

$\therefore PA = 32°$.

Answer. Lat of $A = 58°$N.
Lat of $B = 48°$N.

Example 20.

A star, declination 43° N., which crossed the meridian North of the observer and reached its maximum azimuth at the same time as another star, declination 57° N., has an hour angle of 6 hours West. The azimuth of the first star is twice that of the second. Find the observer's latitude.

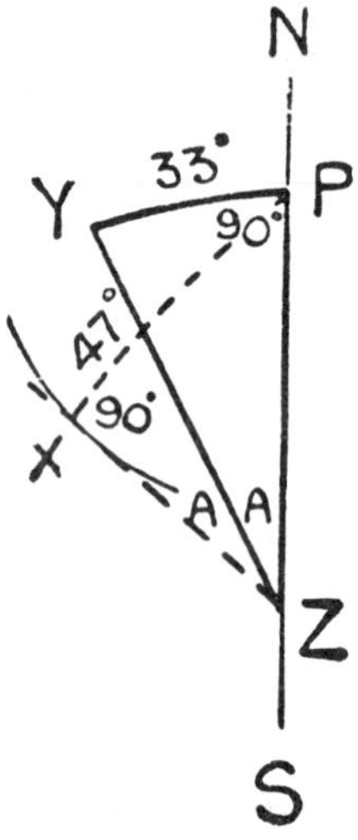

In figure, Z is the zenith, P the pole, X the star having maximum azimuth, $\angle X = 90°$, $PX = 47°$, Y the star whose hour angle is 6^h, $\angle P = 90°$, and PY 33°.

Let $\angle PZY = A$, then $\angle PZX = 2A$.

Find angle A, thence the latitude.

In △PZX. $\sin PX = \sin PZ \sin 2A$

$\therefore \sin PZ = \sin PX \operatorname{cosec} 2A$

In △PZY. $\sin PZ = \tan PY \cot A$

$\therefore \sin PX \operatorname{cosec} 2A = \tan PY \cot A$

or, $\dfrac{\sin PX}{\tan PY} = \cot A \sin 2A$

$$= \frac{\cos A \; 2 \sin A \cos A}{\sin A} \quad . \quad . \quad \text{by (15)}$$

$$\frac{\sin PX}{\tan PY} = 2\cos^2 A = \frac{\sin 47°}{\tan 33°}$$

$\cos^2 A = \frac{1}{2}(\sin 47° \cot 33°)$ from which

$A = 41°\ 22\frac{1}{2}'$.

In △PZY.

$\sin PZ = \tan PY \cot A$

$\cos \text{lat} = \tan 33 \cot 41°\ 22\frac{1}{2}'$

Answer. Lat $= 42°\ 30'$ N.

Example 21.

Two stars x and y are observed on the prime vertical at the same instant, x Decl 40°, y Decl 32° 27′, the hour angle of y is three times that of x. Find the latitude.

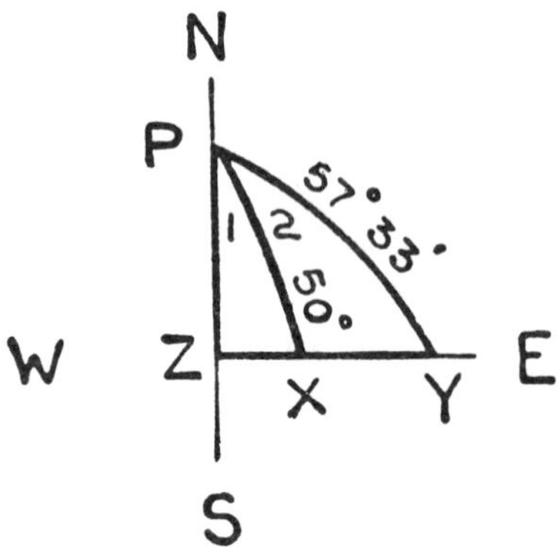

In figure, X and Y are the two stars. If $\angle ZPX = P$, then $\angle ZPY = 3P$, $PX = 50°$, $PY = 57°\ 33'$. Find angle P, thence PZ the co-latitude.

In △ZPY. $\cos 3P = \cot PY \tan PZ$

In △ZPX. $\cos P = \cot PX \tan PZ$

Divide $\dfrac{\cos 3P}{\cos P} = \dfrac{\cot PY}{\cot PX} = \dfrac{\cot 57°\ 33'}{\cot 50°} = \cdot 7578$

by (25) $\dfrac{\cos 3P}{\cos P} = \dfrac{4\cos^3 P - 3\cos P}{\cos P} = \cdot 7578$

$\cos P$ cancels, $4\cos^2 P - 3 = \cdot 7578$

$\cos P = \sqrt{\cdot 93945}$

and $P = 56^m\ 59^s$

In △PZX. $\cos P = \cot PX \tan PZ$

$\tan PZ = \cos 56^m\ 59^s \tan 50°$

Answer Lat 40° 53′.

Example 22.

A circumpolar star with hour angle 6 hours East has an altitude of 50° 9′, and later on reaching its maximum azimuth for the same position of the observer it had an altitude of 60° 5′. Find the mean solar interval.

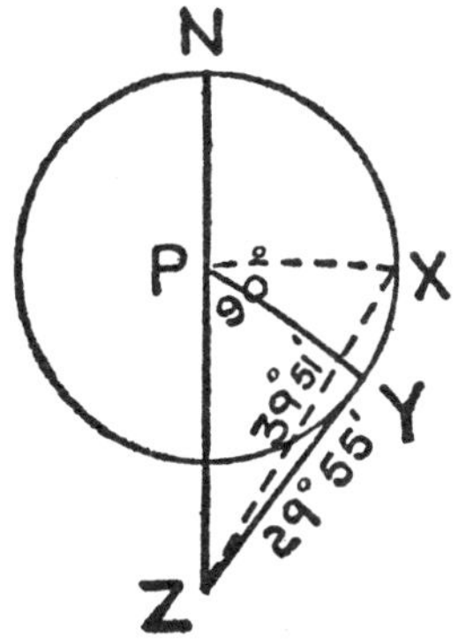

In figure, let X and Y be the two positions of the star, P the pole, Z the zenith.

$\triangle PZX$. Given $\angle ZPX = 90°$, $ZX = 39° 51'$

$\triangle PZY$. ,, $\angle ZYP = 90°$, $ZY = 29° 55'$

$PX = PY$, PZ common.

In $\triangle$PZX. $\cos ZX = \cos PZ \cos PX$

$\therefore \cos PZ = \cos ZX \sec PX$

In $\triangle$PZY. $\cos PZ = \cos ZY \cos PY$

equating, $\cos ZX \sec PX = \cos ZY \cos PY$

but $PX = PY$, $\sec^2 PX = \cos ZY \sec ZX$

$\sec PX = \sqrt{\cos 29° 55' \sec 39° 51'}$

from which $PX = 19° 45' 21'' = PY$

In $\triangle$PZY. $\sin PY = \cot P \tan ZY$

$\cot P = \sin PY \cot ZY$

,, $= \sin 19° 45' 21'' \cot 29° 55'$

		h.	m	s.
from which	$\angle ZPY =$	3	58	17
	$\angle ZPX =$	6	00	00
	$\angle ZPY =$	3	58	17
Sidereal interval,	$\angle XPY =$	2	01	43
Retardation from Tables			—	19·8
Answer. Mean solar interval		2	01	23·2

Example 23.

What is the latitude of an observer who observes that stars Alioth and Deneb are at their maximum azimuth on opposite sides of the meridian at the same instant?

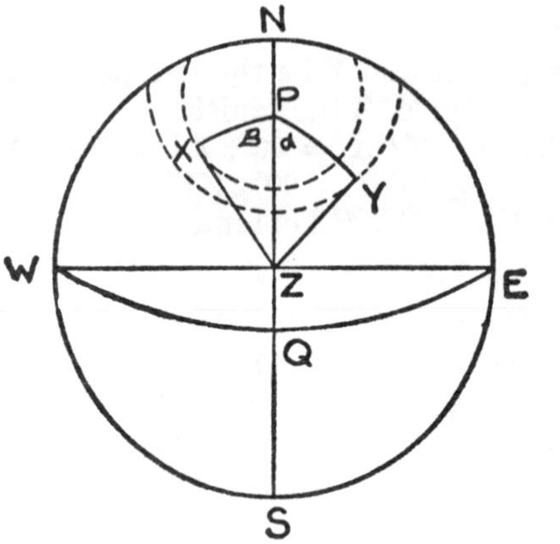

In figure, Z is the observer.

PY = P. Dist. of Deneb Decl. 45° 1½′ N.

PX = P. Dist. of Alioth Decl. 56° 21′ N.

$\angle XPY = \theta =$ difference of star's right ascension, viz., Y 20h 39m 00s, X 12h 50m 55s ∴ θ = 7h 48m 5s.

If α be the H. ang. of Y and β the H. Ang. of X. then $\alpha + \beta = \theta$ and $\beta = \theta - \alpha$ Angles Y and $X = 90°$. Find value of α.

I. $\cos \alpha = \tan PY \cot PZ$

$\cos \beta = \cos(\theta - \alpha) = \tan PX \cot PZ.$

II. But $\cos(\theta - \alpha) = \cos\theta \cos\alpha + \sin\theta \sin\alpha = \tan P \times \cot PZ.$

Divide Eq. II. by Eq. I. and cancel and we get—

$$\cos\theta + \sin\theta \tan a = \tan PX \cot PY$$

$$\therefore \tan\alpha = \frac{\tan PX \cot PY - \cos\theta}{\sin\theta} \text{ from which}$$

$$\alpha = 51° \ 31\tfrac{1}{2}', \quad \beta = 65° \ 29\tfrac{3}{14}'.$$

NOTE.—θ is over 90° so that $-\cos\theta$ becomes $+\cos\theta$ in the equation. Solve one of the triangles for PZ.

Answer. Lat of observer 81° 48½′ N.

Example 24.

Two observers on the same meridian observed altitudes of the same star at the same instant. A's altitude was 41° 09′ and B's 30° 59′, and the azimuth of the star from the same pole was $2\theta°$ at A and $\theta°$ at B. If B's latitude was 20° S., find A's latitude.

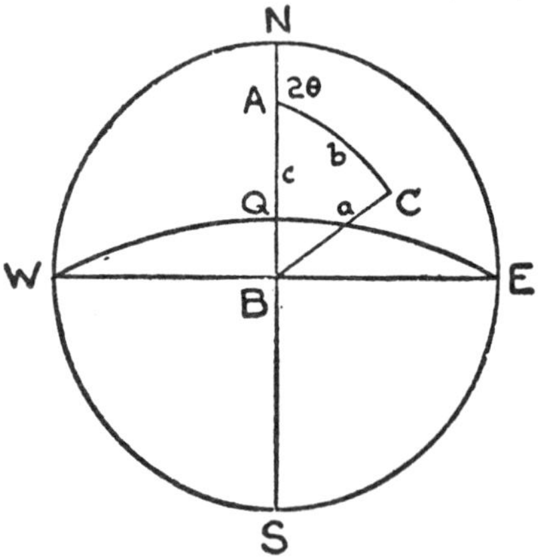

In figure, A and B are the two positions, and C the star, then side b is the zenith distance of star from A and side a its zenith distance from B, $b = 48° \ 51'$, $a = 59° \ 01'$.

$\angle NAC$ = azimuth of star from $A = 2\theta$

$\angle NBC$ = azimuth of star from $B = \theta$

Find value of θ.

$$\frac{\sin a}{\sin b} = \frac{\sin A}{\sin B} = \frac{\sin(180° - 2\theta)}{\sin\theta} = \frac{\sin 2\theta}{\sin\theta} = \frac{2\sin\theta\cos\theta}{\sin\theta} = 2\cos\theta \quad . \quad (15)$$

$$\therefore 2\cos\theta = \frac{\sin a}{\sin b} = \frac{\sin 59° \ 01'}{\sin 48° \ 51'} \quad \therefore \underline{\underline{\theta = 55° \ 18'}}$$

By Napier's Analogy (40).

$\tan\frac{c}{2} = \tan\frac{1}{2}(a + b)\cos\frac{1}{2}(A + B)\sec\frac{1}{2}(A - B)$ from which $c = 65° 24'$ and is the D. Lat. of A north of B.

Answers. B's Lat. 20° S.
A's Lat. 45° 24′ N.

Example 25.

If the meridian zenith distance of a body were three times the latitude, and the hour angle when rising were 5h 46m 16s, find the latitude and indicate the names of the latitude and declination.

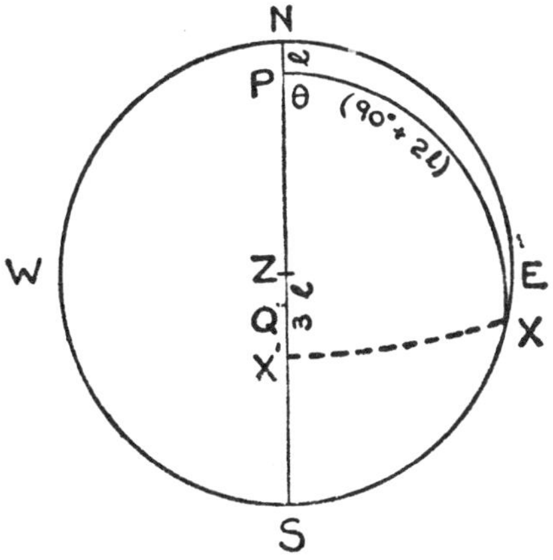

Given NP the Lat. $= l$, and $ZX^1 = 3l$, $\theta = $ 5h 46m 16s.

X the star at rising, X^1 when on meridian, $\angle N = 90°$ then $PX = PX^1 = (90° + 2l)$

$$\begin{aligned}\text{Cos}\,(180° - \theta) &= -\cos\theta \\ \text{,,} &= \tan l \cot(90° + 2l) \\ \text{,,} &= -\tan l \tan 2l \\ \text{but } \tan 2l &= \tan(l + l) \\ \text{,,} &= \frac{2\tan l}{1 - \tan^2 l} \quad (7)\end{aligned}$$

$\therefore \cos\theta = \frac{2\tan^2 l}{1 - \tan^2 l}$ or $\sec\theta = \frac{1 - \tan^2 l}{2\tan^2 l}$ using reciprocal for convenience so that

$$2\sec\theta = \frac{1}{\tan^2 l} - \frac{\tan^2 l}{\tan^2 l} = \cot^2 l - 1$$

and $\cot l = \sqrt{2\sec\theta + 1}$ from which $l = 9° 40\frac{1}{2}'$

Answer. Lat. 9° 40½′ N. when Decl. is S. but Lat. S. when Decl. is N.

Example 26.

Find the time of true sunset in Lat. 30° 40′ N., Long. 3° 30′ W., and from thence deduce the duration of twilight on the longest day.

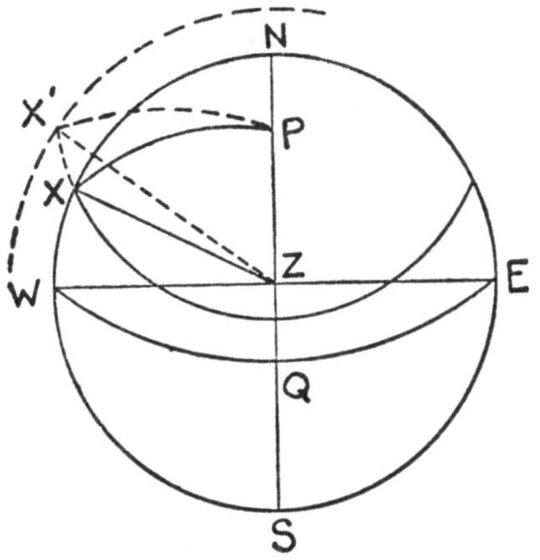

In figure—

Given $PX = PX^1 =$ P. Dist. of Sun on longest day 66° 33′

$PZ =$ co-latitude

$ZX = 90°$. $ZX^1 = 108°$

Find $\angle ZPX$ the H. Ang.

The dotted circle represents the 18° arc of twilight. Twilight lasts during the period the sun takes to travel from X to X^1 and $\angle ZPX^1$ is the hour angle at end of twilight, and $\angle ZPX$ is the hour angle at sunset.

In $\angle PZX^1$. Given PZ, ZX^1, PX^1. Find $\angle ZPX^1$ by haversine formula. In $\triangle PZX$ find $\angle ZPX$. Difference of the two hour angles is the duration of the setting twilight.

Hour angle of sunset	= 6h 59 38s
Hour angle at end of twilight	= 8 41 50
Duration of twilight	1 42 12

Example 27.

If the celestial position of a star is Lat. 4° 52′ 41″ S., Long. 246° 42′, and assuming the obliquity of ecliptic to be 23° 28′, find the star's right ascension and declination.

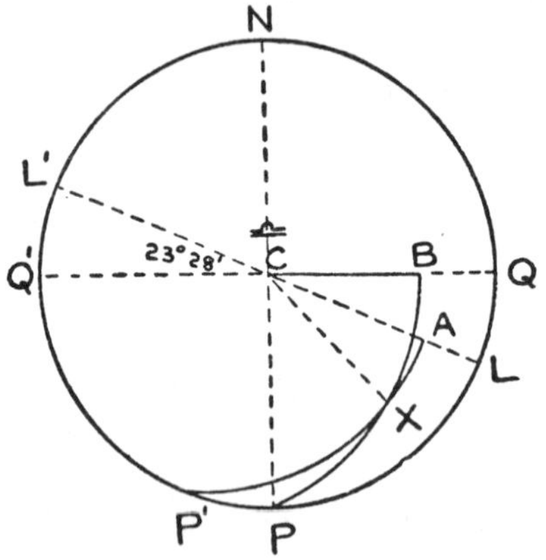

In figure—Q^1Q the equinoctial.
L^1L the ecliptic.
$\angle QCL$ the obliquity 23° 28′.
P the pole of equinoctial.
P^1 the pole of ecliptic.

If C be the point of Libra, the longitude of which is 180° from Aries, then the longitude of star X 246° 42′ is 180° + CA and CA = 66° 42′.

AX is the Lat. of X = 4° 52′ 41″ S. (position exaggerated to show up in figure).

BX the Decl. of X.

12h + CB is the R.A.'s of X. Join CX by a great circle.

In $\triangle CAX$ Given CA, AX and $\angle A = 90°$. Find CX and $\angle XCA$.
In $\triangle CBX$. Given CX, $\angle XCB$ and $\angle B = 90°$. Find CB and BX.
(12h + CB) = R.A. 16h 15m 44s. (BX) Decl. 26° 15.6′ S.

V.—Spherical Triangles.

1. In North latitude a star, Decl. 8° N., bears South, and another star, Decl. 19° N., bears West, the arc of the celestial concave between them is 47°. Find the latitude.

2. An observer in Lat. 23° 10′ N., Long. 20° W., observes the Sun on his meridian, Decl. 23 °10′ N. At the same time another observer, whose great circle distance is 1760 miles from the first observer, notes the Sun's bearing to be 045°. Find the latitude and longitude of the second observer.

Draw figure. Let A be observer whose latitude is known and equal to declination of Sun. B the other observer so that $\angle B = 45°$, AB his great circle distance from A, $\angle APB$ his D. Long. from A and PB his co-latitude. Find $\angle P$ by sine formula and thence PB by Napier's Analogy.

3. A and B are in the same North latitude. When the Sun is on A's meridian its bearing from B is 119° 30′, declination 4° S. The difference of longitude between A and B is equal to the true altitude of the Sun at B. Find the latitude.

This question is the same type as Example 14, the figures being somewhat similar. Refer to the figure and note that the question gives
$\angle B = 119° 30'$, $PX = 94°$, $\angle P = \theta°$, side $BX = (90° - \theta)$ from which find value of θ.

In triangle PBX.

$$\frac{\sin\theta}{\sin(90^\circ-\theta)} = \frac{\sin\theta}{\cos\theta} = \tan\theta = \frac{\sin B}{\sin PX} = \frac{\sin 119^\circ 30'}{\sin 94^\circ} \text{ whence } \theta = 41^\circ 06'$$

Again by Napier's Analogy . . . . . (40)

$\tan\frac{PB}{2} = \tan\frac{1}{2}(PX + BX)\cos\frac{1}{2}(\theta + B)\sec\frac{1}{2}(\theta - B)$ which gives PB the co-latitude of A and B.

4. If the hour angle of a body is 6h 30m at rising and 5h 16m when it reaches the prime vertical, find the latitude of observer and declination of the body.

See Example 15.

5. In North latitude a star was on the horizon at 22h 00m S.M.T., on the prime vertical at 0h 10m S.M.T., and on the meridian at 4h 0m S.M.T. Find the star's declination.

Convert the mean solar intervals into sidereal intervals. Construct figure. Let A be the star on the horizon, B on the prime vertical and C on the meridian.

In $\triangle ZPA$, $ZA = 90^\circ$, use $\angle ZPA$ and PZ and write the equation for PA.

In $\triangle ZPB$, $\angle Z = 90^\circ$, use $\angle ZPB$, and PZ and equate for PB. Divide one equation by the other, PZ cancels and $\frac{\sin ZPB}{\cos APZ} = \tan^2 PB$ the polar distance.

6. The difference between the longest and shortest days is 5h 5m 24s., the greatest declination of the Sun being 23° 27′; find the latitude.

See example on page 376 Vol. I.

Figure on plane of horizon. Let Z and Y be the Sun at setting on shortest and longest days respectively. Then $\angle XPY$ is half the difference of time = 2h 32m 42s. But $\angle ZPX + \angle XPY + \angle YPN = 12\text{h}$ and $\angle ZPX = \angle YPN$ since X and Y are the positions of the Sun at extreme North and South declination. $\angle NPY$ is therefore 4h 43m 29s. Combine this angle with P. Dist. and find NP the latitude.

7. The rising amplitude of a star was E. 16° 53′ N. (true) and later its true altitude was 20° 25′ 35″ when bearing East (true). Find the latitude, North hemisphere.

Draw figure. Let X be the star on the horizon and Y when it is on the prime vertical.

In $\triangle ZXY$. Given ZY, $\angle Z$, side $ZX = 90^\circ$. Find XY.

In $\triangle PZX$. $\cos PX = \sin PZ \cos PZX$ I

In $\triangle PZY$. $\cos PY = \cos PZ \cos ZY$ II

$PX = PY$. Divide I by II $\tan PZ = \cos ZY \sec PZX$.

8. A star has maximum azimuth to an observer in Lat. 21° N., Long. 33° W. At the same time he finds that this star and another 15° W. of him and on the equinoctial differ in azimuth by 67° 6′. Required the declination of the first star and its altitude to an observer on the equator in Long. 48° W.

Construct figure. If B is the star on equator, and A the other star, then $\triangle PZB$ is quadrantal, $PB = 90^\circ$. Given PZ and $\angle P$ find $\angle Z$ and ZB.

In $\triangle PZA$ with $\angle A = 90^\circ$, PZ and $\angle PZA$ find PA and $\angle ZPA$.

In $\triangle PAB$ with PA, $\angle APB$, find BA the zenith distance of star A from observer B.

9. The altitude of a star when due South was 47° and when due West 18°, required the latitude.

Construct figure. If X be the star when on the meridian and Y when on the prime vertical, solve triangle ZXY for side XY. $\triangle PXY$ is isosceles. Draw PA perpendicular to XY. $\angle A = 90^\circ$

In $\triangle PAX$, given $AX = \frac{1}{2}XY$ also ZX, find PX, then $PX - 43^\circ =$ co-lat.

10. Betelgeuse R.A. 5h 51m 10s., Decl. 7° 23′ 30″ N.
Bellatrix R.A. 5 21 10, Decl. 6 16 54 N.

are both on the prime vertical to the westward at the same time, required the latituae.

If X and Y be the positions of the stars when on the prime vertical, then in $\triangle PXY$, given PX and PY their polar distances and $\angle XPY$ the difference of their right ascensions, calculate angles X and Y by tan formula.

In $\triangle PZY$, $Z = 90°$, find PZ the co-lat.

11. Given apparent time at ship 8h 52m a.m., Sun's Decl. 3° 10′ S., Alt. 29° 39′. Find the latitude.

Figure on plane of horizon.
Draw in XY perpendicular to meridian.
Solve $\triangle PXY$, for PY and XY.
Solve $\triangle ZXY$, for ZY.
$PY - ZY$ = co-lat.

12. Given Lat. 30° N., Alt. 22° 8·2′, Az. S. 76° 25′ E. Find the hour angle and declination of the body.

13. Given *R.A. 4h 31m 40s, *Decl. 16° 21·6′ N
*Az. 252° T., G.H.A. Aries 124° 26·5′.

If the star's meridian zenith distance when bearing South was 23° 13′ 28″, find the longitude.

In $\triangle PZX$, given PZ, PX, $\angle Z$, find X by the sine formula and ZX by Napier's Analogy (40), then $\angle P$ by sine formula, thence the longitude.

14. Given Vega R.A. 18h 34m 28s, Decl. 38° 42′ 09″ N. Az. N. 50° E.
Spica R.A. 13 21 18, Decl. 10 46 06 S. Az. S. 50° W.

Find the latitude of observer.

If X and Y be the respective stars, then in $\triangle PXY$, given PX, PY, $\angle XPY$ find $\angle$s X and Y.
In $\triangle PZX$, find PZ by the sine formula.

15. In North latitude, the difference of azimuth between Elnath and Betelguese was 20°, both stars to the eastward, the true altitude of Betelguese was 24°, find the altitude of Elnath.

Betelguese	R.A.	5h 51m 28s,	Decl.	7° 23′ 45″	N.
Elnath	R.A.	5 21 56,	Decl.	28 32 48	N.

Construct figure. If X be Betelguese and Y be Elnath, join XY by a great circle.
In $\triangle XPY$, given PX, PY, $\angle XPY$, find XY.
In $\triangle XZY$, given ZX, XY and $\angle Z$, find $\angle ZYX$ then find ZY by Napier's Analogy.

16. Find the celestial latitude and longitude of a star whose R.A. is 16h 25m 5s, and Decl. 26° 16·7′ S., obliquity 23° 27′.
See Example 25.

17. A star R.A. 2h 59m 36s., Decl. 21° 27′ 45″ N., find its celestial latitude and longitude.

CHAPTER XII.

NAUTICAL AND SPHERICAL ASTRONOMY.

(I.) The arc of a small circle subtending a given angle at the pole is proportional to the cosine of the angular distance of the small circle from the parallel great circle.

This property will be recognised on referring to the figure, in which P is the pole of the great circle WQE, and XY the arc of a small circle KL parallel to it, then PA and PB are secondaries of the circle WQE and $KXYL$.

$$\therefore \text{ arc } XY = \text{arc } AB \cos AX$$
$$\text{or dep.} = d \text{ long} \cos \text{Lat.}$$

if WQE be the earth's equator. The student will recognise this as the formula for parallel sailing proved in Volume I., page 153.

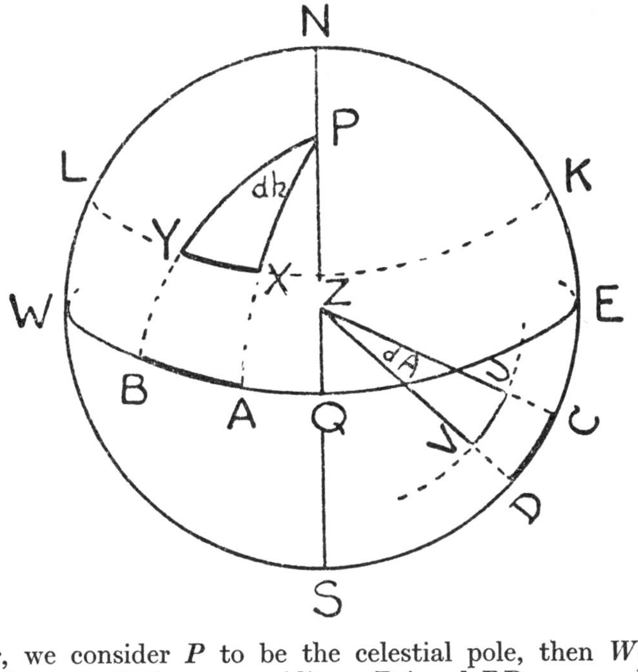

If, however, we consider P to be the celestial pole, then WQE will be the equinoctial and the secondaries, or meridians, PA and PB are now hour circles and XY is an arc of a parallel of declination.

The time taken by a body to travel from X to Y is measured by the angle APB or arc AB, this interval of time, or difference in hour angle, is written as dh, and as before

$$\text{arc } XY = \text{arc } AB \cos AX$$
$$\text{or } XY = dh \cos \text{Decl.}$$

Again, Z the zenith is the pole of the horizon $NESW$. With centre Z and radius ZU, describe a small circle of equal altitude of which UV is an arc, then $\angle UZV$ or arc CD measures the difference of azimuth, between U and V, written dA, and, as before, we may write

$$\text{arc } UV = \text{arc } CD \cos CU$$
$$UV = dA \cos \text{alt}$$

DIP.

(II.) To Find the Distance of the Sea Horizon.—

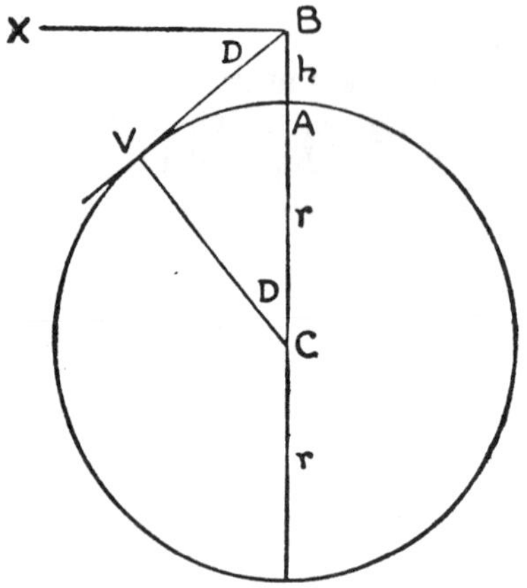

In figure:

BX is the plane of the sensible horizon
BV the tangent to the visible horizon
$\angle XBV$ the dip $= D$
AB the height of eye $= h$
AC the radius of the earth $= r = 3960$ miles.
h is small compared with r the earth's radius and therefore the arc AV is small and equals BV nearly.

$$BV^2 = BC^2 - CV^2$$
$$= (r+h)^2 - r^2$$
$$= r^2 + 2rh + h^2 - r^2$$
$$= 2rh + h^2, \text{ but } h \text{ is small so that } h^2 \text{ may be neglected}$$
$$= 2rh \text{ nearly. } h \text{ in mls.} = h/5280 \text{ ft.}$$

$$BV = \sqrt{2rh} = \sqrt{\frac{2 \times 3960 \times h}{5280}} = \sqrt{1{\cdot}5h}$$

Example.—Calculate the distance of the sea horizon for height of eye 100 feet, and radius of the earth 3960 statute miles, or 3444 nautical miles.

Dist. $= \sqrt{2rh} = \sqrt{1{\cdot}5 \times 100} = 12{\cdot}2$ statute miles
$12{\cdot}2$ (statute) $\div 1{\cdot}15 = 10{\cdot}6$ nautical miles $= 1{\cdot}06\sqrt{h}$.

This calculated distance does not take into account the uncertain effect of refraction which increases the distance by an amount variously estimated at from $\frac{1}{10}$ to $\frac{1}{15}$ of itself, so if we add $\frac{1}{12}$ of $10{\cdot}6$ we get $\cdot 88 + 10{\cdot}6 = 11{\cdot}48$ miles as the distance of the sea horizon. This agrees closely with the approximate formula result given in Volume I, page 245, viz.

$$\text{Distance} = 1{\cdot}15\sqrt{\text{height of eye}} = 1{\cdot}15\sqrt{100} = 11{\cdot}5 \text{ miles.}$$

(III.) To Find the Angle of Dip, and Area of Visible Horizon.

In the figure, angle XBV the dip is equal to the angle BCV at the centre of the earth, because both angles are complements of $\angle VBC$. Find dip for 100 feet.

$$\tan D = \frac{BV}{CV} = \frac{\sqrt{2rh}}{r} = \sqrt{\frac{1{\cdot}5 \times 100}{3960}} = 10'\ 38''$$

About one-twelfth of the calculated amount should be deducted for the uncertain effect of refraction.

	D =	10′ 38″
Deduct for refraction		53
Dip for 100 feet	=	9 45

Nautical Tables give 9·8′ dip for 100 feet.

The angle of dip may be got approximately from the formula $\cdot 98\sqrt{\text{height of eye}}$.
In the above example the dip $= \cdot 98\sqrt{100} = 9{\cdot}8'$

Area within the Visible Horizon $= \frac{3\pi h}{2}$ approximately.

Example.—Find the area visible from a height of 100 feet.

Area $= \frac{3\pi h}{2} = \frac{3 \times 22 \times 100}{2 \times 7} = 471{\cdot}4$ sq. miles (statute) and $471{\cdot}4$ (statute) $\div\ 1{\cdot}15 = 410$ sq. miles (nautical).

Another approximate method is to assume the area within the visible horizon to be that of a plane circle, the radius of the circle being the distance of the sea horizon, thus:—

Distance $= 1{\cdot}15\ \sqrt{100}$ ft. $= 11{\cdot}5$ miles (nautical)

Area of circle $= \pi r^2 = \frac{22 \times 11{\cdot}5^2}{7} = 415$ sq. miles (nautical)

Example.—Find the area contained by the visible horizon when the eye is 64 feet above sea level.

Answer. 304 sq. miles (statute)
263 sq. miles (nautical)

(IV.) **The Dip of a Shore Horizon.**

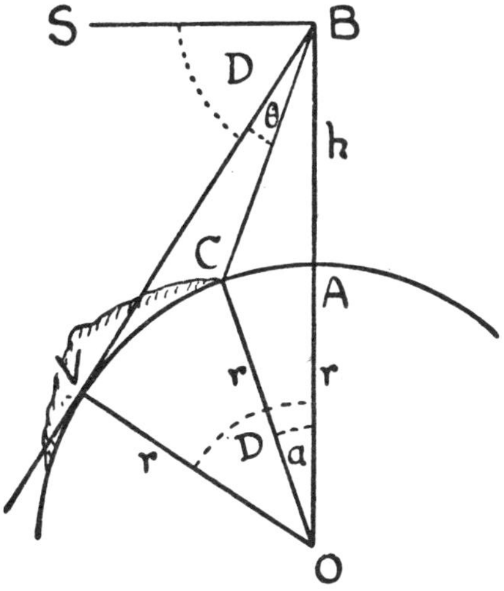

In the figure BV represents the direction of the sea horizon for height of eye $AB = h$, and $r =$ radius of the earth. Let the true dip $\angle SBV = D$, then, because $\angle SBV = \angle BOV$ both being complements of $\angle VBO$, $\angle BOV = D$. the apparent dip.

When land intervenes between the sea horizon and the observer the altitude of a body is measured from the waterline of the shore at C, and the dip of the shore horizon is $\angle SBC = (D + \theta)$, $\angle\theta$ being the correction to be added to the true dip as taken from the dip table.

The arc AC is the distance of the shore and, being very small compared with the radius of the earth, the arc $AC = \angle AOC = a$.

$$\angle BCO = 180° - (CBO + a) = 180° - CBO - a$$
$$\text{,,} \quad = 180° - (90° - D - \theta) - a$$
$$\text{,,} \quad = 180° - 90° + D + \theta - a$$
$$\angle BCO = 90° + (D + \theta - a) \text{ and } \angle CBO = 90° - (D + \theta)$$

In $\triangle BCO$.
$$\frac{OB}{OC} = \frac{r+h}{r} = \frac{\sin BCO}{\sin CBO} = \frac{\sin\{90° + (D + \theta - a)\}}{\sin\{90° - (D + \theta)\}}$$
$$= \frac{\cos(D + \theta - a)}{\cos(D + \theta)} \quad \ldots \quad 1$$

From $\triangle BVO$.
$$\frac{OB}{OV} = \frac{r+h}{r} = \sec D = \frac{1}{\cos D} \quad \ldots \quad 2$$

From 1 and 2. $\dfrac{\cos(D + \theta - a)}{\cos(D + \theta)} = \dfrac{1}{\cos D}$

$$\therefore \cos(D + \theta - a) = \frac{\cos(D + \theta)}{\cos D} = \frac{\cos D \cos\theta - \sin D \sin\theta}{\cos D}$$

or $\cos\{(D - a) + \theta\} = \cos\theta - \tan D \sin\theta$

$\cos(D - a)\cos\theta - \sin(D - a)\sin\theta = \cos\theta - \tan D \sin\theta$

See footnote for this transformation.

Dividing both sides by $\cos\theta$ we get

$\cos(D - a) - \sin(D - a)\tan\theta = 1 - \tan D \tan\theta.$

By re-arranging and multiplying both sides by -1 we get

$1 - \cos(D - a) = \tan D \tan\theta - \sin(D - a)\tan\theta = \tan\theta\{\tan D - \sin(D - a)\}$

$$\therefore \tan\theta = \frac{1 - \cos(D - a)}{\tan D - \sin(D - a)} = \frac{2\sin^2\left(\frac{D - a}{2}\right)}{\tan D - \sin(D - a)}$$

But all these angles are small, therefore the equation may be reduced to

$$\theta = \frac{2\left(\frac{D - a}{2}\right)^2}{D-(D-a)} = \frac{(D - a)^2}{2a} = \frac{(\text{dip} - \text{dist.})^2}{2\text{ dist.}}$$

NOTE.—Let the compound angle $(D + \theta - a)$ be written $(D - a) + \theta$, then, if $(D - a) = A$ and $\theta = B$, it takes the form of the equation $\cos(A + B) = \cos A \cos B - \sin A \sin B$ from whence by substitution $\cos\{(D - a) + \theta\} = \cos(D - a)\cos\theta - \sin(D - a)\sin\theta$.

The *Admiralty Manual* formula for the dip of a shore horizon is $\cdot565\,\frac{h}{a} + \cdot423a$, where h is the height of eye in feet, and a is the offshore distance in miles.

Example.—Distance of shore horizon 3 miles, eye 49 feet, calculate the dip.

The distance of the sea horizon for 49 feet is $1\cdot15\sqrt{49} = 8$ miles (see *Norie's Table*). The offshore distance is only 2 miles so that the shore horizon is well within the range of the sea horizon.

Dip of sea horizon $= \cdot98\sqrt{49} = 6\cdot9$ (see *Norie's Table*; Refr. allowed for).

The apparent dip is calculated from Eq. III, page 280.

$$\text{Tan } D^1 = \sqrt{\frac{1\cdot5 \times 49}{3960}} = \sqrt{\frac{49}{2640}} = \begin{array}{r} \log\ 1\cdot690196 \\ \log\ 3\cdot421604 \\ 2)\overline{18\cdot268592} \end{array}$$

apparent dip $D = 7'\,45''$ or $7\cdot7'$ $\quad \tan\ \underline{9\cdot134296}$

$$\theta^1 = \frac{(D - a)^2}{2a} = \frac{(7\cdot7 - 3)^2}{2 \times 3} = \frac{4\cdot7 \times 4\cdot7}{6} = \frac{22}{6} = 3\cdot7'$$

True dip of sea horizon	6·9′	(from Dip Table)
Add θ as above	3·7	(∠*VBC* in fig.)
Dip of shore horizon	10·6	(∠*SBC* in fig.)

The above example by *Admiralty Manual* method.

$$\text{Dip of shore horizon} = \cdot565 \times \frac{49\text{ ft.}}{3\text{ miles}} + \cdot423 \times 3\text{ miles}$$

$$= 9\cdot276 + 1\cdot269$$

$$= 10\cdot545'$$

Example.—

Distance of shore horizon 3 miles, eye 36 feet, calculate the dip.

Answers.	*Guide*	formula	8·1′
	Admiralty	„	08·0′
	Norie's Table	„	08·2′

Example.—

Offshore distance 1·5 miles, eye 25 feet, calculate the dip.

Answer. As per *Guide* 10·1′; *Admiralty* 10′; *Norie* 10′.

(V) Effect of Dip and Refraction on Times of Rising and Setting.

The effect of the height of eye is to depress the visible horizon and to cause celestial bodies to rise sooner, and set later, than they would if there were no dip.

Refraction raises the position of celestial bodies and, like dip, accelerates their times of rising and retards their times of setting.

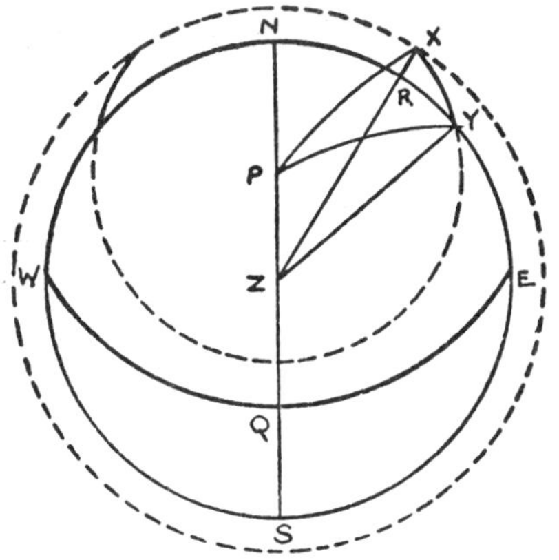

In figure, *NESW* represents the rational horizon and the dotted circle the sea horizon.

If *X* be the position of a body at visible rising and *Y* its position at theoretical rising, then $\angle XPY$ is the interval of time taken by the body to travel along its diurnal path from *X* to *Y*, and $\angle XPY$ is the difference between the hour angles *ZPY* and *ZPX*.

In quadrantal triangle *ZPY*. $PZ =$ co-lat. $PY =$ P. dist. and *ZX* the zenith distance $= 90°$, so by Napier's Rules cos H. ang. $=$ tan lat. tan decl., which gives the hour angle of theoretical rising without taking into account the effect of dip and refraction. These are the times given in *Norie's Tables.*

In triangle *PZX*, we have the same co-latitude and polar distance, *PZ* and *PX*, but *ZX*, the zenith distance, is equal to the arc *ZR* + arc *RX*, which is 90° plus the dip and refraction combined, or to each one separately if the effect of only one of the factors is being considered. The three sides of the triangle *ZPX* being known, the hour angle of the body when on the sea horizon is calculated from the formula

$$\text{hav } ZPX = \frac{\text{hav } ZX - \text{hav } (L - D)}{\cos L \cos D}$$

Example.—Find the acceleration of the rising of a celestial body, due to the effect of dip and refraction, in Lat. 50° N., from the top of a hill 900 feet above sea level, declination of body 20° N.

Dip $= \cdot 98\sqrt{900} = \cdot 98 \times 30 = 29\cdot 4'$

Refr. $= 33'$ $\therefore$ $RX = 29\cdot 4 + 33 = 62\cdot 4'$, $ZX = 91°\ 02\cdot 4'$

ZX	91° 2·4'		nat hav	·50905
$L-D$	30 00		„	·06699
			„	·44206
			hav	9·64548
Lat.	50 00		sec	0·19193
Decl.	20 00		sec	0·02701
$\angle ZPX$	7h 50·5m		hav	9·86442
$\angle ZPY$	7 43·0 (taken from Table)			
Interval	07·5			

Thus visible time of rising occurs 7·5 minutes earlier than the theoretical time.

This interval may be calculated by means of the "rate of change of altitude" formula (XX).

$dh = dz$ cosec azimuth sec. lat. where dh is the interval of time in minutes corresponding to a change of altitude dz in minutes of arc which, in this case, is 62·4'.

The azimuth at rising may be got from the Amplitude Table in *Norie*.

$$dh^{m} = \frac{62\cdot 4'}{15} \text{ cosec } 57\cdot 8° \text{ sec } 50°$$

dz	62·4'	log	1·79518
amp.	57·8°	cosec	0·07253
lat.	50°	sec	0·19193
			2·05964
	15'	log	1·17609—
dh	7·6	log	0·88355

H. ang. from Table	7h 43m
Interval	+ 07·6
H. ang. of visible rising	7 50·6

Had dip (29·4') and refraction (33') been worked out separately, the interval for the height of eye (900 feet) would have been 3·6 minutes, and for refraction 4·0 minutes, the combined effect being 7·6m as above.

Exercises.

1. Given height of eye 400 feet, Lat. 56° N., Decl. 22° N., find acceleration of the rising of the body due to refraction and dip. (8·4 minutes)

2. In Lat. 56° N., eye 20 feet, Decl. of Sun 10° N., required the interval between visible and theoretical sunset. (6·7 minutes)

3. Lat. 30° N., Decl. 40°, eye 100 feet, required the interval between visible and theoretical rising of the body due to dip and refraction. (4·9 minutes)

(VI.) Visible and Theoretical Sunrise.

Visible sunrise and sunset is taken to be the moment when the Sun's upper limb is in contact with the sea horizon.

The time of theoretical sunrise and sunset is the instant when the Sun's centre is on the rational horizon. These are the times given in *Norie's Tables*, no allowance being made for dip, refraction and semi-diameter.

Example.—Find the time interval between visible and theoretical sunrise on the longest day in Lat. 45° N., eye 20 feet.

Lat. 45° N., Decl. 23° 27′ N. gives Amp. E. 34·3° N. = Az. N. 55·7° E.

Dip	− 4·4′	Obs. alt.	0° 00′	
Ref.	− 33·0	Corr.	− 53·4	
S. diam.	− 16·0	True alt.	− 0 53·4	(*RX* in fig.)
			90 00	
Total corr.	− 53·4			
		Zen. dist.	90 53·4	(*ZX* in fig.)

Refer to the figure in previous section.

$$\text{hav}\, ZPX = \frac{\text{hav}\, ZX - \text{hav}\, (L - D)}{\cos L \cos D}$$

ZX	90° 53·4′	nat hav	·50778
L−D	21 33·0	,,	·03495
		,,	·47283
		hav	9·67469
Lat.	45 00 N.	sec	0·15051
Decl.	23 27 N.	sec	0·03744
∠*ZPX*	7h 49m	hav	9·86264
∠*ZPY*	7 43 (from Table)		
Interval	06		

Thus the Sun rises 6 minutes before, and sets 6 minutes after, the times of theoretical rising and setting.

The same result is got by using the formula, $dh = dz$ cosec az. sec. lat.

$$dh^{m} = \frac{53{\cdot}4'}{15} \text{ cosec } 55{\cdot}7° \text{ sec. } 45°$$

RX	53·4′	log	1·72754
Az.	55·7°	cosec	0·08297
Lat.	45·0°	sec	0·15051
			1·96102
	15	log	1·17609
Interval	6·1 min.	log	0·78493

Example.—Find the time taken at sunrise between contact of U.L. and L.L. with the sea horizon in Lat. 60° on March 21st.

$$dh = dz \text{ cosec az. sec lat}$$

$$,, = \frac{32'}{15} \text{ cosec } 90° \text{ sec } 60°$$

$$,, = \frac{32}{15} \times 1 \times 2$$

$$,, = \frac{64}{15} = 4{\cdot}26 \text{ minutes}$$

NOTE.—The average diameter of the Sun is about 32′. The declination is 0° on March 21 and the Sun rises bearing due East or azimuth N. 90° E., cosec 90° = 1, sec. 60° = 2.

A question, of the following type, associated with the rising and setting of the Sun, is sometimes given in examination papers.

In South latitude, S.M.T. of sunrise 6h 56m bearing by compass S. 78·5° E., S.M.T. of sunset 16h 34m., bearing by compass S.48·5 W., variation 10° E. Find the deviation of the compass, the Equation of Time, the declination of the Sun and the latitude of the observer.

1. The error of the compass is half the difference of the compass bearings at rising and setting.

Error in this case = 30°/2 = 15° E., named East because it has to be applied to the right of S. 78·5° E. to get the mean compass bearing.

Compass bearing	S. 78·5° E.
Error	15·0 E.
True bearing	S. 63·5 E.

Dev. = Error 15° E − Var. 10° E. = 5° E.

2. The equation of time is half the difference between the hour angles of the mean Sun at rising and setting.

H.A.M.S. rising	05h 04m
,, setting	04 34
Difference	30

Eq. of time is 15m 00s to be subtracted from apparent time.

Eq. of time is − from apparent time when it has to be subtracted from the H.A.M.S. at rising to get the mean of the hour angles.

L.M.T. sunset	16h	34m	00s
Eq. of Time +		15	00
L.A.T.	16	49	00

3. We now have azimuth 63·5° and hour angle 4h 49m 00s at rising or setting, and with these values in the quadrantal triangle *PZX* we have, by Napier's Rules—

Sin. lat. = cot H.A. tan az. and cos dec = cosec H.A. sin az. from which

Lat. = 39° 56½′ N. Decl. = 20° 00′ S.

Exercise.—In N. Latitude the Sun rose at 4h 37·6m S.M.T. bearing by compass N. 60° E., and set at 18h 51·6m S.M.T. bearing N. 80° W. by compass. If the variation is 20° W., find the deviation of the compass, the equation of time, the declination of the Sun and the latitude of the observer.

Answer. Dev. 30° E., Eq. time 15·4m—from App. time. Decl. 11° 5′ N., Lat. 56° 47′ N.

(VII.) Time of Moonrise and Moonset.

The remarks regarding the rising and setting of the Sun do not apply equally to the Moon owing to the larger parllax of the latter body. The effect of dip and refraction is to accelerate the time of visible sunrise or moonrise, whilst parallax retards the appearance of the bodies which in the case of the Moon averages about 57′ as against 1·5′ for the sun.

The effect of refraction minus parallax on the Sun is +33′−1·5′=+31·5′; so that the Sun's centre is a full diameter above the visible horizon when his centre is on the rational horizon. With the Moon, however, the combined effect is Refr. +33′ Parlx. −57′=−24′, so that the Moon's centre is about 24′ below the visible horizon at the time of theoretical moonrise, her upper limb would then be almost appearing to an observer whose eye is 40 or 50 feet above sea level.

Calculate the time of moonrise on January 4, in Lat. 40° N., Long. 45° W.

Mer. pass.	−	4d	23h	16m	diff. 57	Enter *Nautical Tables* (Times of Rising and Setting) with Lat. 40° N., Decl. 19° N. (approx.) and get H. ang. 7h 07m.
Corr. for 3h Long.	+			7		
L.M.T. mer. pass		4	23	23		
Long. W.	−		03	00		
G.M.T. mer. pass.		5	02	23		
Approx. H.A.	−		7	07		
Approx. G.M.T. of moonrise	−	4	19	16		Decl. 19° 27′ N. (correct)

Again enter *Nautical Tables* with Lat. 40° N., and Decl. 19° 27′ N., and get

Hour angle	−	07h	09m
Diff. 57m × 7·1h ÷ 24h = correction	− +		17
Intervals between moonrise and mer. pass.	−	07	26
G.M.T. of mer. pass. as above	−	26	23
G.M.T. of moonrise	−	18	57

The first part of the work is the same as the moon's meridian passage. The first entry in Nautical Tables to get the hour angle of moonrise is an approximation because the declination at the time of transit is used. The second entry with the declination at moonrise is more correct. The acceleration of the solar interval due to the moon's proper motion is evident. The same operation is performed to get the time of moonset by adding the hour angle to the time of transit.

The above information can be obtained from the moonrise column in the daily pages of the *Nautical Almanac* and the Lat. and Long. correction tables for sunrise, moonrise, etc., as follows:—

Jan. 4—Lat. 40° N., L.M.T. of moonrise at Gr.		15h	50m
Correction for Long. (Diff. 1h 02m)	− +		7
L.M.T. moonrise at ship	−	15	57
Long. in time	− +	3	00
G.M.T. moonrise at ship	−	18	57

(VIII) THEORY OF REFRACTION.

Refraction depends upon two theorems in optics. (I) The angles of refraction and incidence are in the same plane, which plane passes through the normal to the refracting surfaces. (II.) The sine of the angle of incidence has to the sine of the angle of refraction a constant ratio for the same medium.

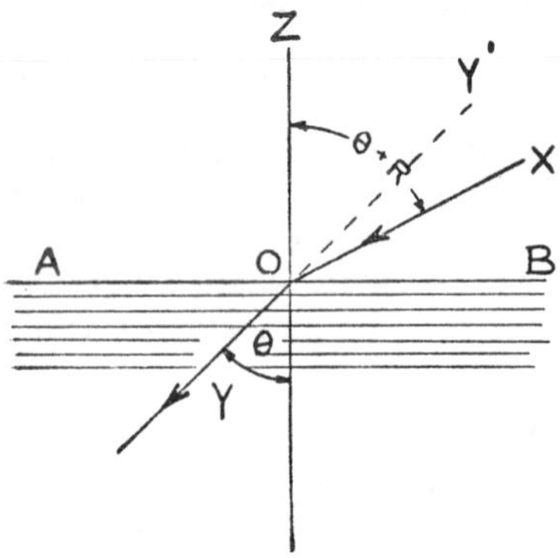

In figure: *AB* represents a surface separating two media of different density, *OZ* being the normal to that surface, *XO* a ray of light refracted in the direction of *OY* on entering the denser medium.

If R = mean refraction in seconds

θ = angle of refraction

then $(\theta + R) = \angle ZOX$ the angle of incidence; it is also the true zenith distance of the body *X*, and $\angle ZOY$, or θ, is its apparent zenith distance.

If μ be the relative index of refraction which is a constant ratio, then by theorem **II.**

$$\mu \sin \theta = \sin (\theta + R)$$

$$,, \quad = \sin \theta \cos R + \cos \theta \sin R,$$ but R is small so that $\cos R = 1$, and $\sin R$ may be written $R'' \sin 1''$ . . . . by (43)

$\therefore \mu \sin \theta = \sin \theta + \cos \theta \, R'' \sin 1''$, or $R'' \cos \theta \sin 1'' = \mu \sin \theta - \sin \theta$

$$= \sin \theta (\mu - 1)$$

$$R'' = \frac{\sin \theta (\mu - 1)}{\cos \theta \sin 1''} = \tan \theta \frac{(\mu - 1)}{\sin 1''}$$

but $\dfrac{\mu - 1''}{\sin 1''}$ is a constant and for atmospheric refraction it is **57·538″.**

∴ Refraction in seconds = tan zen. dist. × 57·538″

Example.—Required the refraction for apparent altitude **37° 30′.**

R'' = 57·538 cot alt = 57·538 cot 37° 30′

	log	1·759954
	cot	0·115020
R'' =	log	1·874974 = 74·985″

Refraction = 1′ 14·98″ for alt 37° 30′. *Norie's Table* gives 1′ 14″

(IX.) DIURNAL PARALLAX OF THE MOON.

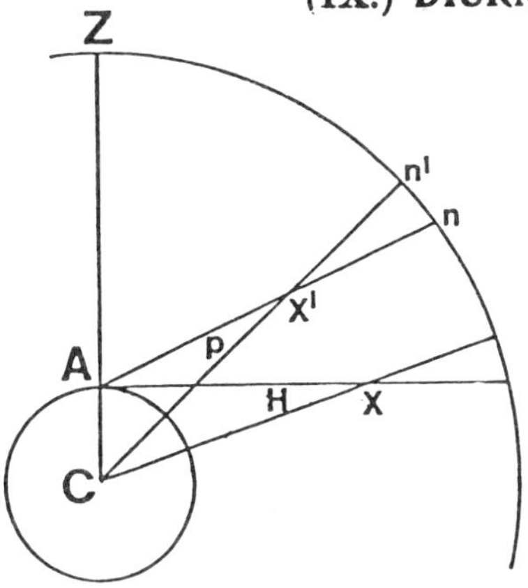

In figure:

C the centre of the Earth

A the observer, **Z** his Zenith

X **a** body in the sensible horizon CX

X the body in altitude

From the observer at A the body X' would be projected on the celestial sphere at n, and from the centre of the Earth it would be projected at n'. The arc nn' is the parallax in altitude $= \angle nX'n' = \angle AX'C$ or simply p, which we want to find.

Let $\angle AXC = H$, the horizontal **parallax** as given in the *Nautical Almanac*.

App. Alt. = Sext. Alt. corrected for Index Error and Dip so:

In triangle ACX'

by (1) $$\frac{AC}{CX'} = \frac{\sin p}{\sin CAX'} = \frac{\sin p}{\sin ZAX'} = \frac{\sin p}{\cos \text{App. Alt.}}$$

In triangle ACX $$\frac{AC}{CX} = \sin H$$

But $CX' =$ **CX** nearly, so that $$\frac{AC}{CX'} = \frac{AC}{CX} = \frac{\sin p}{\cos \text{App. Alt.}} = \sin H$$

$$\therefore \sin p = \sin H \cos \text{alt}$$

The angles p and H being small we may write

by $$\sin p' = p' \sin 1' \text{ and } \sin h' = H' \sin 1'$$
$$\therefore p' \sin 1' = H' \sin 1' \cos \text{App. Alt.}$$
$$\text{or } p = H \cos \text{App. Alt.}$$

i.e., parallax in alt = hor parallax cos App. Alt. as in the case of correcting altitudes of the Moon.

Example.—Calculate the parallax in altitude having given Earth's radius 3960 miles, Moon's distance 238,840 miles, Moon's altitude 34° 48'.

From figure: $$\frac{\sin p}{AC} = \frac{\sin CAX'}{CX'}$$

$$\therefore \sin p = \frac{AC \cos \text{alt}}{CX'}$$

AC	3960	log	3·59770
alt	34° 48'	cos	9·91442
			3·51212
CX'	238840	log	5·37810 −
p =	46' 48"	sin	8·13402

(X.) AUGMENTATION OF THE MOON'S SEMI-DIAMETER.

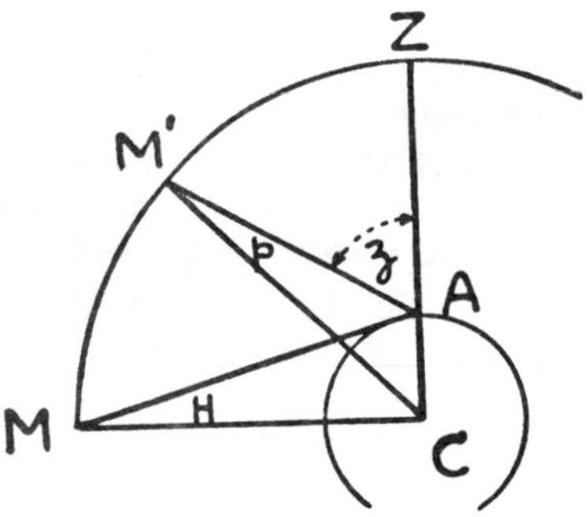

In figure:

A the observer, *Z* the zenith

M the Moon in horizon for which the semi-diameter is tabulated in the *Nautical Almanac.*

M' the Moon in altitude

If *S* represents the semi-diameter for *M*, then $(S + a)$, will represent the augmented semi-diameter for *M'*

The semi-diameters are inversely proportional to the distances *CM* and *AM'*. and as *AM'* is always less than *CM* when the Moon is above the horizon the semi-diameter for *M'* is greater than for *M*.

p = parallax in alt. $= H \sin z$, as proved in (IX.).

$$\frac{S+a}{S} = \frac{M'C}{M'A} = \frac{\sin M'AC}{\sin M'CA} = \frac{\sin ZAM'}{\sin (ZAM' - p)} = \frac{\sin z}{\sin (z - p)}$$

$$\frac{S+a}{S} = 1 + \frac{a}{S} \quad \therefore \quad \frac{a}{S} = \frac{\sin z}{\sin (z-p)} - 1$$

$$,, = \frac{\sin z - \sin (z-p)}{\sin (z-p)}$$

$$,, = \frac{2 \cos \frac{1}{2} (z + z - p) \sin \frac{1}{2} (z - z + p)}{\sin (z-p)}$$

$$a = 2\, S \cos (z - \tfrac{1}{2} p) \sin \tfrac{1}{2} p \operatorname{cosec} (z - p).$$

Example.—Given Moon's Hor. Par. 58′ 57″ (H), App. Z. Dist. 48° (z), Semi-diam. 16′ 4″ (S), to find the augmentation.

by (IX.) $p = H \sin z$

$,, = 3537'' \times$ Nat. sin 48°

$,, = 3537 \times \cdot 7431 = 2629'' = 43'\ 49''$

$\therefore \quad \frac{1}{2}p = 21'\ 54''$, $(z - \frac{1}{2}p) = 47°\ 38'\ 06''$, $2S = 1928''$

$2S$			1928″	log	3·285107
$(z - \frac{1}{2}p)$	47°	38′	06″	cos	9·828564
$\frac{1}{2}p$	0	21	54	sin	7·804173
$(z - p)$	47	16	11	cosec	0·133975
Augmentation, a =			11·27″	log	1·051819

(XI.) SEXTANT.

An error of collimation is caused when the axis of the telescope is not parallel to the plane of the sextant.

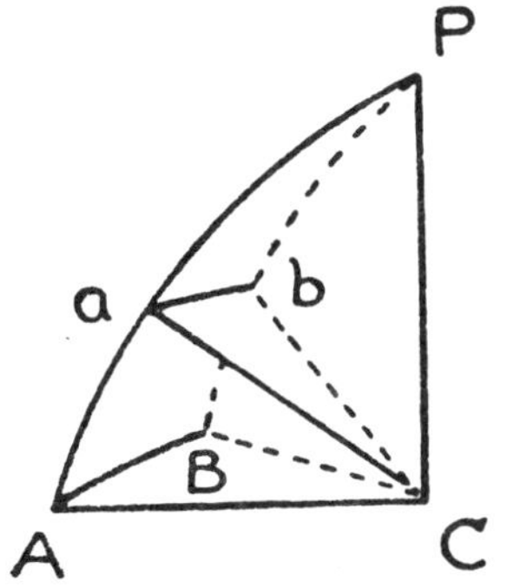

In figure, let C be the centre of a sphere, Aa the inclination of the telescope, greatly exaggerated to show up the figure, the inclination is always small.

$\angle aCb$ the true angular distance between two objects and $\angle ACB$ the measured distance between them.

$$\cos Aa = \frac{\text{arc } ab}{\text{arc } AB} = \frac{\text{chord } ab}{\text{chord } AB}$$

$$\cos Aa = \frac{2 \sin \frac{1}{2} aCb}{2 \sin \frac{1}{2} ACB}$$

$$\cos Aa = \frac{\sin \frac{1}{2} \text{ true dist.}}{\sin \frac{1}{2} \text{ measured dist.}}$$

The student may notice that $\cos Aa = \frac{ab}{AB}$ is equivalent to the parallel sailing formula $\cos \text{lat} = \frac{\text{dep.}}{\text{D. Long.}}$

Example.—If the line of collimation be inclined 42′ to the plane of the sextant, what is the error in an observed angle of 117° 22′ 50″?

$$\begin{aligned} \sin \tfrac{1}{2} \text{ true dist.} &= \cos Aa \sin \tfrac{1}{2} \text{ measured dist.} \\ \text{,,} \quad &= \cos 42' \sin 58° \ 41' \ 25'' \\ \text{,,} \quad &= 9{\cdot}999968 + 9{\cdot}931646 = 9{\cdot}931614 \\ \tfrac{1}{2} \text{ true dist.} &= 58° \ 41' \ 00'' \end{aligned}$$

True dist.	117	22	00
Obs. dist.	117	22	50
Error in sextant angle			50″

Example.—Calculate the error in an observed sextant angle of 90° if the line of collimation be inclined 1°. *Answer.* 1′ 4″.

(XII.) TO FIND THE ERROR OF LATITUDE DUE TO A SMALL ERROR IN ALTITUDE.

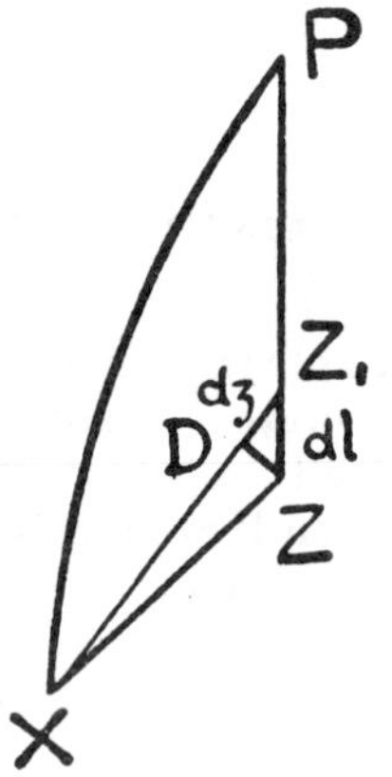

In figure, let Z be the correct zenith, Z_1 the faulty one, then ZZ_1 will be the error in latitude *dl.*

Make ZD perpendicular to Z_1X, then $DZ_1 = dz$ the error in zenith distance arising from *dl.*

Triangle DZZ_1 being very small, is practically a right angled plane triangle.

$$\therefore\ Z_1Z = DZ_1 \sec ZZ'D$$
$$dl = dz \sec \text{Az}$$

or, error in lat. = error in alt. × sec Az.

This indicates that observations for latitude should be taken near the meridian because the value of the secant increases with the azimuth from sec 0° = 1 to infinity.

(XIII.) TO FIND THE ERROR IN HOUR ANGLE DUE TO A SMALL ERROR IN LATITUDE.

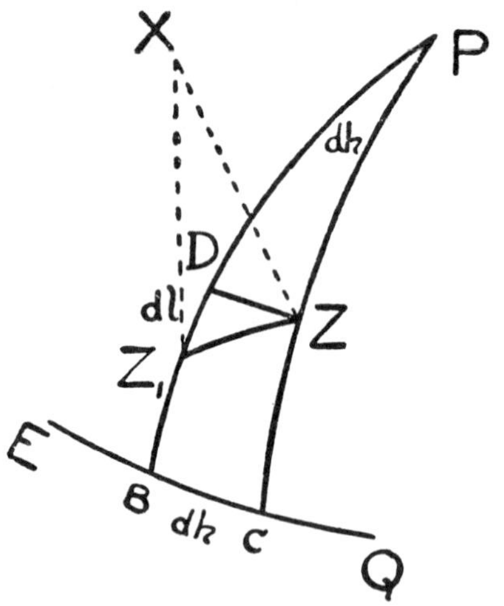

In figure, let Z be the correct zenith, Z_1 the erroneous zenith. With centre P and radius PZ describe an arc ZD, then Z_1D = error in latitude (dl) and ZPZ_1 = error in hour angle (dh), DZ being an arc of a parallel, and BC the corresponding arc of equator, $BC = \angle ZPZ_1$ (dh), and $DZ = BC \cos CZ = dh \cos$ lat, by (I.) $\angle PZD = 90°$ $\therefore$ $\angle PZX$ is the complement of $\angle DZX$

$\angle XZZ_1 = 90°$ $\therefore$ DZZ_1 is the complement of DZX

$\therefore$ $DZZ_1 = PZX$ the azimuth of X.

In $\triangle DZZ_1$ $DZ = DZ_1 \cot DZZ_1$

$dh \cos \text{lat.} = dl \cot Az$

$\therefore$ $dh = dl \sec \text{lat.} \cot Az$ (in arc).

Error in H. Ang. = $\frac{1}{15}$ Error in lat. sec lat. cot Az (in time).

Example.—Sights for hour angle were worked with Lat. 40° instead of 40° 10′, the azimuth of the body being 160°. Required the resulting error in the calculated hour angle.

dl	10′	log	1·000000
Lat.	40°	sec	0·115746
Az	20°	cot	0·438934
15)	35·8′	log	1·554680
Error in H. Ang.	$2^m 22{\cdot}8^s$		

(XIV.) TO FIND THE ERROR IN HOUR ANGLE DUE TO A SMALL ERROR IN ALTITUDE.

The figure is drawn on the plane of the meridian, Z is the zenith, NWS the horizon, ZC and ZD are two vertical circles so that $\angle CZD = \text{arc } CD = dA$, the change of azimuth between X and Y.

XV is the arc of a small circle of equal altitude
by (I.) $\therefore\ VX = DC \cos CX = dA \sin ZX$ . . . . (a)

P is the pole, QWQ the equinoctial, X and Y are two successive positions of a body travelling along its diurnal arc KL. The secondaries PA and PB are meridians through X and Y, so that

arc $AB = \angle APB = dh$ the change of hour angle between X and Y.
by (I.) $\therefore\ XY = AB \cos AX = dh \sin PX$ . . . . (b)

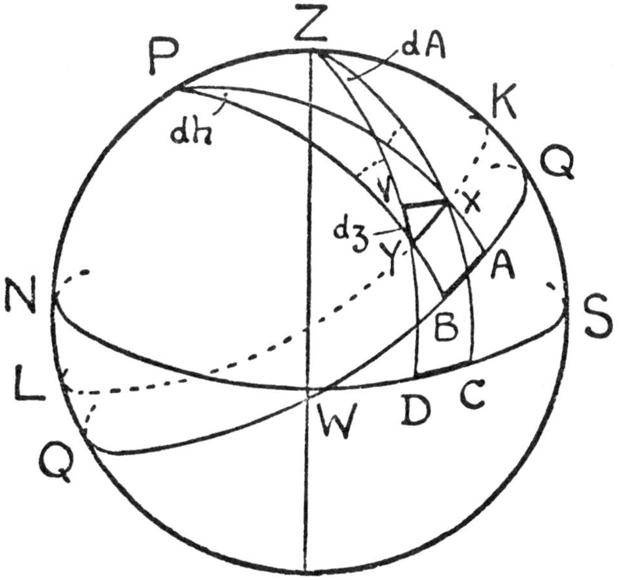

VY is the change of zenith distance, or of altitude, written dz, corresponding to the interval dh while the body moved from X to Y.

Triangle XVY being small may be considered to be plane, right angled at V, and X is very close to Y so that $\angle PXZ = \angle PYZ =$ complement of $\angle VYX$ as $\angle PYX = 90°$.

In △XVY $\qquad VY = XY \cos VYX$

substitute (b) for XY, $\qquad dz = dh \sin PX \sin PXZ$.

In △PZX
by (37) $\qquad \dfrac{\sin PXZ}{\sin PZ} = \dfrac{\sin PZX}{\sin PX}$

$$\therefore\ \sin PX \sin PXZ = \sin PZ \sin PZX.$$

Substituting this in the above we get

$$dz = dh \sin PZ \sin PZX$$

expressed in arc $\qquad dz' = dh' \cos \text{lat.} \sin \text{azimuth}$

or $dh' = dz' \sec \text{lat.} \operatorname{cosec} \text{azimuth}$.

Divide the result by 15 to convert minutes of arc into minutes of time.

(XV.) MAXIMUM RATE OF CHANGE OF AZIMUTH.

Equation (*a*) and (*b*) in (XIV.) when combined give the maximum rate of change of azimuth of a body. This occurs when it is crossing the meridian, the zenith distance of the body being then its meridian zenith distance, that is (lat. ± decl.) by (*a*) and (*b*) where *dA* and *dh* vary together

$$dA \sin ZX = dh \sin PX$$

$$dA = \frac{dh^{m} \sin PX}{\sin ZX} = \frac{15' \cos \text{decl.}}{\sin (\text{Lat.} \pm \text{decl.})} \text{ per minute.}$$

Example.—Calculate the maximum rate of change in azimuth of Vega in Lat. 50° N.

	15′	log	1·176091
decl.	38° 43′ N.	cos	9·892233
(L. – D.)	11 17	cosec	0·708496
dA	59·8′	log	1·776820

Rate of change 59·8′ per minute.

(XVI.) DISPLACEMENT IN LONGITUDE OF A POSITION LINE DUE TO AN ERROR IN HOUR ANGLE.

The equation *dz* = *dh* cos lat. sin Az. enables us to show that an error in the G.M.T. used in computing the hour angle of a body when calculating the intercept produces an equal error in longitude when expressed in minutes of arc. In the figure (XIV.) *ZX* is the shorter zenith distance and *ZY* the longer one, *VY* being the *dz*, or intercept. If we were to project the information of the figure on to a chart the projection would appear as follows, the bearing of the body *X* being south-eastward, *Z* the correct spot through which to draw the position line *PL* corresponding to the shorter zenith distance *ZX*, and *Z′* will be the wrong spot corresponding to the longer zenith distance, then the intercept *ZZ′* or *VY* in figure under (XIV.) = *dz*. In the projection it is the zenith we move towards or away from the body whereas, in the figure, it is the body that has moved away from the zenith, the difference, however, is merely a relative one.

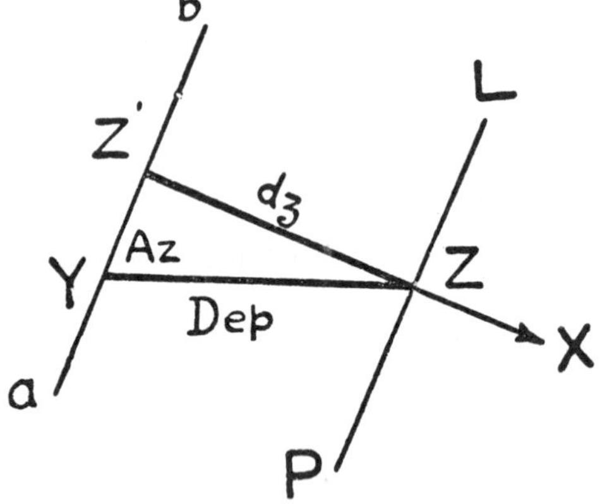

ZY is the east-west displacement of the position line = departure.

ZY = *dz* cosec *Y* = *dz* cosec Az, because *Z′ZY* is the complement of ∠*Y* and also of the bearing of the body *X*. ∴ ∠*Y* = azimuth. But between **Z** and *Y* the

D. Long. = Dep. sec lat.
= *ZY* sec lat.
= *dz* cosec Az. sec lat.

$$= dh \cos \text{lat.} \sin \text{Az.} \operatorname{cosec} \text{Az.} \sec \text{lat.} = dh \frac{\cos \text{lat.} \sin \text{Az.}}{\cos \text{lat.} \sin \text{Az.}}$$

∴ D. Long.′ = *dh*′.

The reciprocals of the latitude and azimuth cancel out.

The displacement of the position line in longitude is, therefore, equal to the errors in hour angle in minutes of arc.

It is usually easier to rework, or replot, position lines, which have been laid off wrongly owing to an error in the course steered or in the distance run between sights, than to investigate a formula which will yield the resulting error. The student should refer to the *Admiralty Manuals of Navigation* for information regarding errors in position lines and fixes.

(XVII.) ERROR IN AZIMUTH OR HOUR ANGLE.

We may deduce from the figure in (XIV.) the error in azimuth resulting from a small error in hour angle.

Since $\angle PXY$ and $\angle ZXV$ are right angles and $\angle PXV$ is common to both, the angle $VXY = \angle PXZ$, which is known as the angle of position.

In $\triangle XVY$ $\quad VX = \cos VXY$

by (*a*) and (*b*) $\quad dA \sin ZX = dh \sin PX \cos PXZ$

$$dA = dh \sin PX \operatorname{cosec} ZX \cos PXZ$$

$$dA = dh \cos \text{decl.} \sec \text{alt.} \cos PXZ$$

(XVIII.) SMALL ERRORS IN LATITUDE, HOUR ANGLE AND ALTITUDE COMBINED.

Admiral Sir H. E. Purey-Cust, formerly Hydrographer to the Admiralty, has mentioned in the *Nautical Magazine* of August, 1933, that the reactions on each other of small errors of observation in altitude, hour angle and latitude can be very simply and clearly demonstrated by reference to the position line, as follows :—

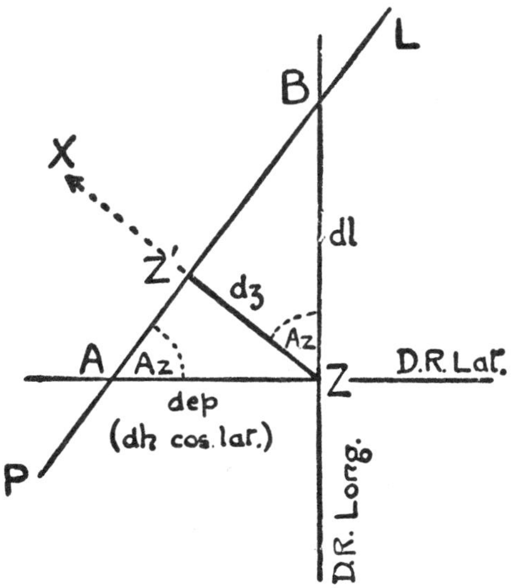

In figure:

If Z be the D.R. position,

ZZ^1 an intercept or small error in altitude (dz),

PL a position line cutting the D.R. parallel of latitude at A, and the D.R. meridian of longitude at B, then

ZB is the error in latitude (dl) due to dz,

and ZA is the error of departure due to dz;

but dep. = D. long. cos lat.

or ZA = error in H. ang. cos lat.

but by (XVI.) D. Long. = dh

$ZA = dh$ cos lat.

$\angle BZZ'$ is the azimuth of body X, and $\angle BZZ' = \angle BAZ$, both being complements of $\angle B$.

In $\triangle$AZZ′.

$$ZZ' = ZA \sin A$$
$$\text{or } dz = dh \cos \text{lat} \sin Az \quad \ldots\ldots\ldots\ldots \quad (1)$$
$$\therefore \text{ error in alt.} = \text{error in H.A. cos lat sin Az.}$$

By transposing we have

$$\text{error in H.A.} = \text{error in alt. sec lat. cosec Az}$$
$$\text{or } dh = dz \sec \text{lat} \operatorname{cosec} Az \quad \ldots\ldots\ldots\ldots \quad (2)$$

In $\triangle$ BZZ′.

$$ZZ' = ZB \cos BZZ'$$
$$\text{or } dz = dl \cos Az \quad \ldots\ldots\ldots\ldots \quad (3)$$
$$\therefore \text{ error in alt.} = \text{error in lat. cos Az}$$

by transposing we have

$$\text{error in lat.} = \text{error in alt sec Az}$$
$$\text{or } dl = dz \sec Az \quad \ldots\ldots\ldots\ldots \quad (4)$$

In $\triangle$BZA.

$$ZB = ZA \tan A$$
$$\text{or } dl = dh \cos \text{lat} \tan Az \quad \ldots\ldots\ldots\ldots \quad (5)$$
$$\therefore \text{ error in lat.} = \text{error in H.A. cos lat. tan Az.}$$

By transposing we have

$$\text{error in H.A.} = \text{error in lat. sec lat. cot Az}$$
$$\text{or } dh = dl \sec \text{lat.} \cot. Az \quad \ldots\ldots\ldots\ldots \quad (6)$$

The Admiral points out that everything depends mainly on the size of the azimuth and has little to do with the size of the hour angle.

When calculating latitude from altitude (4), or from hour angle (5), the smaller the azimuth the better as in case (4), sec Az will be unity, and in case (5), tan Az will be zero when the body is on the meridan, that is azimuth 0° or 180°.

When calculating hour angle from altitude (2), and from latitude (6), the larger the azimuth the better, as, in case (2) cosec Az will be unity, and in case (6) cot Az will be zero when the body is on the prime vertical, that is azimuth 90° or 270°.

When calculating altitude from hour angle (1), and from latitude (3), it does not matter much whether the azimuth is large or small as in both cases (1) and (3) the sine and cosine of the azimuth are involved and the value of those functions is always less than unity.

While not laying down any strict rule as to what are small and what are large azimuths, it is convenient to consider azimuths under 45° as small and calculate latitude; azimuths over 45° as large and calculate hour angle.

(XIX.) TRIGONOMETRICAL DIFFERENTIAL COEFFICIENTS.

These coefficients are very convenient in determining the mathematical error in a quantity arising from a small error of observation in one of the factors used in a computation; thus, for example, an error in the altitude when working up longitude sights will produce a corresponding error in the resulting hour angle and it may save a long recalculation if we could determine readily the amount of this error.

We give here the standard forms of the six trigonometrical differential coefficients although those of the sine and cosine are the most commonly used in navigational problems.

Trigonometrical function.	Differential coefficient.	Differential
1. $\sin x$	$\cos x$	$\cos x \, . \, dx$
2. $\cos x$	$-\sin x$	$-\sin x \, . \, dx$
3. $\tan x$	$\sec^2 x$	$\sec^2 x \, . \, dx$
4. $\cot x$	$-\text{cosec}^2 x$	$-\text{cosec}^2 x \, . \, dx$
5. $\sec x$	$\sec x \tan x$	$(\sec x \tan x) \, dx$
6. $\text{cosec}\, x$	$-\text{cosec}\, x \cot x$	$-(\text{cosec}\, x \cot x) \, dx$

It will be noted that the differentials of cos x, cot x, cosec x are negative because these functions decrease as x increases.

The symbol d does not represent a quantity but simply indicates that dx is a small " increment " or " increase " in the value of x. Note that the rate of change of sin x with x is cos x, etc.

When quantities attain maximum or minimum values their differential coefficient is zero, the values of x, etc., which give these maximum and minimum values are therefore obtained by equating the differential coefficient to zero.

We are really concerned only with the application of the differentials to navigation, but perhaps, we should give here the proofs of these rules for the benefit of those who may wish to know their origin.

(1) sin x.

If $y = \sin x$ then $y + \delta y = \sin (x + \delta x)$ and by subtraction

$$\delta y = \sin (x + \delta x) - \sin x$$

by (12) page 233

$$\delta y = 2 \cos \left(\frac{x + \delta x + x}{2}\right) \sin \left(\frac{x + \delta x - x}{2}\right)$$

$$\delta y = 2 \cos \left(x + \frac{\delta x}{2}\right) \sin \frac{\delta x}{2}$$

divide by δx so that

$$\frac{\delta y}{\delta x} = \frac{2 \cos \left(x + \frac{\delta x}{2}\right) \sin \frac{\delta x}{2}}{\delta x}$$

$$\frac{\delta y}{\delta x} = \cos \left(x + \frac{\delta x}{2}\right) \left(\sin \frac{\delta x}{2} \Big/ \frac{\delta x}{2}\right)$$

now let δx tend to zero, then $\frac{\delta y}{\delta x}$ tends to $\frac{dy}{dx}$, and $\sin \frac{\delta x}{2} \Big/ \frac{\delta x}{2}$ tends to 1

and we have $\frac{dy}{dx} = \cos x$ but $y = \sin x$

$$\therefore d \sin x = \cos x \, . \, d\, x$$

(2) cos x.

If $y = \cos x$ then $y + \delta y = \cos (x + \delta x)$ and by subtraction

$$\delta y = \cos (x + \delta x) - \cos x$$

by (14) page 233

$$\delta y = 2 \sin \left(\frac{x + \delta x + x}{2}\right) \sin \left(\frac{x - \delta x - x}{2}\right)$$

$$\delta y = 2 \sin \left(x + \frac{\delta x}{2}\right) \sin \left(- \frac{\delta x}{2}\right)$$

then $$\frac{\delta y}{\delta x} = -\frac{2\sin\left(x+\frac{\delta x}{2}\right)\sin\frac{\delta x}{2}}{\delta x}$$ since $\text{Sin}(-A) = -\sin A$

$$\frac{\delta y}{\delta x} = -\sin\left(x+\frac{\delta x}{2}\right)\left(\sin\frac{\delta x}{2}\Big/\frac{\delta x}{2}\right)$$

now let δx tend to zero, etc., as in (1) and we have

$$\frac{dy}{dx} = -\sin x \qquad \text{but } y = \cos x$$

$$\therefore d\cos x = -\sin x \,.\, dx$$

(3) tan x.

If $y = \tan x$ then $y + \delta y = \tan(x + \delta x)$ and by subtraction

$$\delta y = \tan(x+\delta x) - \tan x$$

$$\delta y = \frac{\sin(x+\delta x)}{\cos(x+\delta x)} - \frac{\sin x}{\cos x}$$

$$\delta y = \frac{\sin(x+\delta x)\cos x - \cos(x+\delta x)\sin x}{\cos(x+\delta x)\cos x}$$

by (8) where $x + \delta x = A$, $x = B$

$$\delta y = \frac{\sin(x+\delta x - x)}{\cos(x+\delta x)\cos x}$$

$$\delta y = \frac{\sin\delta x}{\cos(x+\delta x)\cos x}$$

then $$\frac{\delta y}{\delta x} = \frac{\sin\delta x}{\delta x}\times\frac{1}{\cos(x+\delta x)}\times\frac{1}{\cos x}$$

now let δx tend to zero, etc., as in (1) then we have

$$\frac{dy}{dx} = 1\,.\,\frac{1}{\cos x}\,.\,\frac{1}{\cos x} = \frac{1}{\cos^2 x} = \sec^2 x, \text{ but } y = \tan x$$

$$\therefore d\tan x = \sec^2 x \,.\, dx$$

(6) cosec x.

If $y = \operatorname{cosec} x$ then, $y + \delta y = \operatorname{cosec}(x + \delta x)$ and by subtraction

$$\delta y = \operatorname{cosec}(x+\delta x) - \operatorname{cosec} x$$

$$\delta y = \frac{1}{\sin(x+\delta x)} - \frac{1}{\sin x}$$

$$\delta y = \frac{\sin x - \sin(x+\delta x)}{\sin(x+\delta x)\sin x}$$

by (12) where $x = A$, $(x+\delta x) = B$

$$\delta y = \frac{2\cos\left(\frac{x+x+\delta x}{2}\right)\sin\left(\frac{x-x-\delta x}{2}\right)}{\sin(x+\delta x)\sin x}$$

$$\delta y = \frac{2\cos\left(x+\frac{\delta x}{2}\right)\sin\left(-\frac{\delta x}{2}\right)}{\sin(x+\delta x)\sin x}$$

then $$\frac{\delta y}{\delta x} = -\frac{2\cos\left(x+\frac{\delta x}{2}\right)}{\sin(x+\delta x)\sin x}\,.\,\frac{\sin\frac{\delta x}{2}}{\delta x}$$

$$\frac{\delta y}{\delta x} = - \frac{\cos\left(x + \frac{\delta x}{2}\right)}{\sin(x + \delta x)} \times \frac{1}{\sin x} \times \left(\sin \frac{\delta x}{2} \Big/ \frac{\delta x}{2}\right)$$

Again let δx tend to zero, etc., as in (1),

then $\frac{dy}{dx} = - \frac{\cos x}{\sin x} \cdot \frac{1}{\sin x} \cdot 1 = - \cot x \operatorname{cosec} x$ but $y = \operatorname{cosec} x$

$\therefore d \operatorname{cosec} x = - \cot x \operatorname{cosec} x \cdot dx$

The differential coefficients for cot x and sec x are obtained in the same way as those for tan x and cosec x respectively.

(XX.) RATE OF CHANGE OF ALTITUDE.

It is required to find the error in hour angle arising from a small error in altitude.

The hour angle and zenith distance, that is h and z, are the two variable quantities, so we write down the fundamental formula and then re-write it, introducing the appropriate differential coefficient, remembering to write zero against any factor that does not contain a variable.

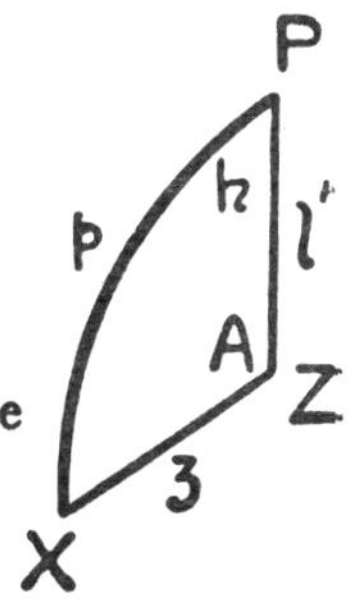

$$\cos z = \cos p \cos l' + \sin p \sin l' \cos h$$

differentials, $- \sin z \,.\, dz = o + \sin p \sin l' (- \sin h \,.\, dh)$

$$- \sin z \,.\, dz = - \sin p \sin l' \sin h \,.\, dh$$

$$dh = \frac{\sin z \,.\, dz}{\sin p \sin l' \sin h}$$

by (29) from figure, $\frac{\sin z}{\sin p} = \frac{\sin h}{\sin A}$ and by substitution in above we have

$$dh = \frac{\sin h \,.\, dz}{\sin A \sin l' \sin h} \text{ but } \sin h \text{ cancels}$$

$$\therefore \; dh = \frac{dz}{\sin A \sin l'} = \frac{dz}{\sin \mathrm{Az} \cos \mathrm{lat.}}$$

Thus if $dz = 1'$ the resulting

$$dh = \frac{1'}{\sin \mathrm{Az} \cos \mathrm{lat.}} = \operatorname{cosec} \mathrm{Az} \sec \mathrm{lat.},$$

and if z be expressed in minutes of arc, h will also be in minutes of arc to be divided by 15′ to convert it into time, as otherwise proved in (XIV.).

Incidentally, the formula $dh = \frac{dz}{\cos \mathrm{lat.} \sin \mathrm{Az}}$ indicates that when the azimuth is greatest, that is $\sin 90° = 1$, its maximum value, the error in hour angle is least, so that sights for longitude should be taken when the body is near the prime vertical as a small error in altitude will then produce a comparatively small error in time.

By transposing the equation we get $dz = dh \sin \mathrm{Az} \cos \mathrm{lat.}$, and if $dh = 1^{m} = 15'$ we may write $dz = 15 \sin \mathrm{Az} \cos \mathrm{lat.}$, which expresses the rate of change of altitude of a body in 1 minute of time. Thus the rate of change of altitude depends on the azimuth for when the azimuth $= 0°$, $\sin 0° = 0$ and the rate of change of altitude disappears for the moment because the body is just crossing the meridian, but when the azimuth is 90°, $\sin 90° = 1$, its maximum value and so the equation tells us that a body changes in altitude most rapidly when on the prime vertical, whatever its declination may be.

(XXI.) MAXIMUM ALTITUDE.

The change of altitude of a star is due to the Earth's rotation on its axis. If the Earth did not rotate, and an observer remained in the same latitude, the fixed stars would retain their same altitudes but the Sun, Moon and planets would have a slow change of altitude depending on their rate of change of declination, an increasing altitude when they travel towards the observer and, obviously, a decreasing altitude when they move away from him.

Again, assuming the Earth and celestial bodies to be fixed in space but the observer moving along a meridian, then, as he travelled towards a body, his horizon would be falling from under it and its altitude would increase; and conversely, when travelling away from the body its altitude would decrease, or, one may think of it in terms of zenith distance, the geographical position of the body being fixed and the observer's zenith moving with him, either towards or away from the body.

When a body is moving rapidly in declination, as the Moon for instance, the maximum altitude may occur a little before or after the time the body is on the meridian, before when the body is moving away from the observer, that is polar distance increasing, and after its meridian passage when the polar distance is decreasing, that is when the body is travelling towards the observer.

This change in altitude continues until the rate of change of altitude due to the proper motion of the body is equal to the rate of change of altitude due to the Earth's rotation on her axis.

(XXII.) THE TIME INTERVAL IN MAXIMUM ALTITUDES.

$$\cos z = \cos p \cos l' + \sin p \sin l' \cos h$$

l' is the only constant, h is required and differentiating with respect to h we have

$$-\sin z \frac{dz}{dh} = -\sin p \frac{dp}{dh} \cos l' + \cos p \frac{dp}{dh} \sin l' \cos h - \sin p \sin l \sin h$$

but $dz/dh = 0$ since the zenith distance is to be a minimum, and $\cos h = \cos 0° = 1$; by substituting we get

$$0 = -\frac{dp}{dh} \sin p \cos l' + \frac{dp}{dh} \cos p \sin l' \times 1 - \sin p \sin l' \sin h$$

$$\sin p \sin l' \sin h = -(\sin p \cos l' - \cos p \sin l') \frac{dp}{dh}$$

$$\sin h = h'' \sin 1'' = -\frac{\sin (p - l')}{\sin p \sin l'} \frac{dp}{dh}$$

$$\therefore h \sec = -\frac{1}{15 \sin 1''} \times \frac{\sin (p - l')}{\sin p \sin l'} \times \frac{dp}{dh}$$

By substituting Lat. and Decl. we get

$$h = -\frac{1}{15 \sin 1''} \sin (L \pm D) \sec L \sec D \frac{dp}{dh}$$

$(L + D)$ when Lat and Decl. are of opposite names

$(L - D)$ when of same name.

$$15 \sin 1'' = \log 5{\cdot}861666, \text{ and } \frac{dp}{dh} = \frac{\text{change of decl. per hour}}{\text{rate of Earth's rotation per hour } (15°)}$$

The maximum altitude may occur before, or after, the meridian passage of the body so that h seconds may have to be subtracted from, or added to, the time of meridian passage, but this can be decided algebraically by placing a plus, or minus, sign after dp/dh.

$\frac{dp}{dh}$ — when the body is approaching the observer

$\frac{dp}{dh}$ + when the body is receding from the observer

in other words, — when "closing" and + when "opening".

This sign when combined with the — sign which precedes the formula indicates whether h is to be added to, or subtracted from, the time of passage.

Example.—Dec. 22, Lat. 50° 22′ N., Long. 30° W. Required the interval before or after meridian passage when the moon will attain her maximum altitude.

Mer. Pass. Dec.	22d	21h	24m	Dec. –	16°	05·2′ N.	diff. 5·1′ N. in 1h.
Corr. for Long. 30°	+		04	Cor. –		2·4 N.	
L.M.T. – –	22	21	28	D. –	16	07·6 N.	
Long. W. –		2	00	L. –	50	22 N.	
G.M.T. – –	22	23	28	(L.—D.)	34	14·4	

$$\frac{dp}{dh} = \frac{\text{change of decl. per hour}}{\text{minutes of arc per hour}} = \frac{5\cdot1}{900} = \frac{1}{176\cdot5}$$

and the foregoing formula may be written as

$$h \sec = \frac{-1}{15 \sin 1''} \sin (\text{L.} \sim \text{D.}) \sec \text{D.} \sec \text{L.} \frac{dp}{dh} -$$

where h=Hour angle, L. the Latitude, D. the Declination.

L.	50° 22′	N.	sec	0·19527
D.	16 07·6	N.	sec	0·01743
(L.∼D.)	34 14·4		sin	9·75025
				9·96295
$\frac{dp}{dh}$	$\frac{1}{176\cdot5}$		log	2·24675
				7·71620
	15 sin 1″		log	5·86167
h	71·5s.		log	1·85453

The time interval is + 1m 11·1s.

The body is "closing", therefore $\frac{dp}{dh}$ is minus and this minus, combined with the minus sign which precedes the formula, makes the quantity plus to the time of meridian passage.

The maximum altitude occurs 1m 11·1s after the Moon's transit.

Example—Find the interval between the meridian passage and the maximum altitude of the Sun in Lat. 55° N. on March 23rd. Given Decl. 0° 56′ N., Sun going North at the rate of 1′ per hour; course 230°, speed 20 knots.

$$\textit{Answer}: \frac{dp}{dh} = \frac{13\cdot9}{873} = \frac{1}{62\cdot8} h = +3\text{m } 39\text{s}$$

The time of maximum altitude, combined with equal altitudes of a body taken when the hour angles *E* and *W* are small (not more than about 40 minutes), provides a method of finding longitude.

Example.—Jan. 12, D.R. Lat. 10° 30′ N., Long. 32° 30′ E., the Sun had equal altitudes when the G.M. times by chronometer were 09h 27m 30s and 10h 26m 50s. The ship's course was 060°, and speed 15 knots during the interval between sights. Find the longitude.

1st G.M.T. – – –	09h	27m	30s
2nd ,, – – –	10	26	50
Interval – – –	00	59	20
$\frac{1}{2}$ interval – – –	00	29	40
1st G.M.T. – – –	09	27	30
Middle G.M.T. of max. alt.	09	57	10

The quantity *dp* has to be adjusted for the course and speed of the ship, when the course is towards North or South the rate of change of altitude is due to the change in declination per hour plus or minus the ship's D. Lat. per hour.

Add them when ship and body are approaching each other, or "closing".
Subtract them when ship and body are receding from each other, or "opening".
In this example:—

Course 060°, 15 knots, D. Lat. 7·5′ N., Dep. 13·0 E., D. Long. 13·2′ E.
Decl. 21° 41·6′ S.—Sun going N. 0·4′ per hour
Ship going N. 7·5 ,,
Ship is leaving Sun at rate of 7·1 per hour = *dp*

To make sure whether the body and ship are opening or closing, it is desirable to make a rough sketch showing their relative positions.

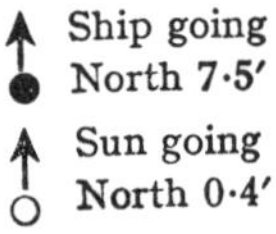

In this case they are "opening", the ship is running away from the Sun, so that dp=(change of Decl.—D. Lat.) per hour.

The rate of change of altitude is also due to the Earth's rotation per hour (900′), plus or minus the ship's D. Long. per hour.

Add D. Long. when ship is making easting.
Subtract D. Long, when ship is making westing.

In this example the ship is making easting so that dh=(900′ +D. Long. 13·2′ E.) = 913·2′. These adjustments give $\frac{dp}{dh} = \frac{7\cdot1}{913\cdot2}$

$$h=-\frac{1}{15 \sin 1''} \sin (L+D) \sec L \sec D \frac{dp}{dh}+$$

A +sign should be placed after $\frac{dp}{dh}$ because the ship and body are "opening" out, or receding from each other.

L	10° 30′	N.	sec	10·00733
D	21 41·6	S.	sec	10·03191
$L+D$	32 11·6		sin	9·72655
dp	7·1		log	0·85126
				0·61705
dh	913·2		log	2·96057—
				7·65648
15 sin	1″		log	5·86167—
h=—	62·35s.		log	1·79481

L.A.T. at noon		12h	00m	00
h	—		1	02
L.A.T. at max. alt.		11	58	58
Eq. of time	+		8	14
L.M.T. at max. alt.		12	07	12
G.M.T. ,,		9	57	10
Longitude		2	10	02
		32°	30·5′	E.

The *Admiralty Manual* formula for time of maximum altitude is

$$h \text{ sec} = 15{\cdot}28 \ (\tan L \pm \tan D) \times \left(\frac{1 \pm 2X}{900}\right) Y$$

$(\tan L \pm \tan D)$ + for contrary names, — for like names

$\left(1 \pm \frac{2X}{900}\right)$, + for ship going West, — for ship going East

X = D. Long. of ship per hour.

$Y = dp$ = decl. per hour combined with ship's D. Lat. per hour.

In the foregoing example, course 060°, speed 15 knots, D. Lat. 7·5′ N., Dep. 13·0 E., D. Long. 13·2′ E., change of Decl. 0·4′ per hour, Decl. 21° 41·6′ S., Lat. 10° 30′ N., X=13·2′ and Y=7·1′.

$$\left(1 - \frac{2X}{900}\right) Y = \left(\frac{900 - 26{\cdot}4}{900}\right) 6{\cdot}5 = \frac{874 \times 7{\cdot}1}{900} = 6{\cdot}3$$

15·28 tan L + 15·28 tan D = 8·911 as follows:—

Constant	15·28	log	1·18412		log	1·18412
L	10° 30′	tan	9·26797	D 21° 41·6′	tan	9·59968
	2·832	log	0·45209	6·079	log	0·78380
	6·079					
	8·911	then h=8·91×6·9=61·5 sec.				

The maximum altitude occurs 61·5s before the time of meridian passage because the ship and body are receding. The other formula gave 62·3 seconds.

Exercises

1. Sept. 20, D.R. Lat. 20° N., Long. 152° 30′ E. the Sun had equal altitudes when the G.M. times by chronometer were 01h 05m 02s and 02h 26m 00s, ship's course during the interval 190°, speed 20 knots, required the longitude.

Answer. $\frac{dp}{dh} = \frac{18{\cdot}8}{896{\cdot}3}$, h = +98s Long. 152° 27·5′ E.

2. Oct. 11, D.R. Lat. 2° 10′ N., Long. 66° 05′ E., Sun had equal altitudes when chronometer times were 06h 58m 53s and 07h 54m 57s, error on G.M.T. 5m 09s fast. Course during interval between sights, 325°, speed 18 knots, find the longitude.

Answer. $\frac{dp}{dh} = \frac{15{\cdot}6}{889{\cdot}7}$, h = − 3·8s, Long. 66° 08′ E.

3. Sept. 19, D.R. Lat. 30° 30′ N., Long. 31° 30′ W., G.M. times of Sun's equal altitudes, 13h 32m 30s and 14h 31m 50s, course during interval 200°, speed 20 knots, find the longitude.

Answer. $\frac{dp}{dh} = \frac{17{\cdot}8}{892}$, h = +154s, Long. 31° 25·5′ W.

4. A ship is steaming 180°, at 20 knots, in Lat. 50° N., Long. 0°. Find when the Sun will dip if the declination is 16° 4′ 30″ S. and travelling South at the rate of 45″ per hour.

Answer. 07m 15s past noon.

5. A ship steaming North at 16 knots in Lat. 37° N. If the Sun's declination is 4° 33·4′ N., and travelling North at the rate of 1′ per hour, at what time will the Sun dip?

Answer. 02m 34s before noon.

6. In Lat. 42° N., Decl. 10° 11′ N., the Sun to an observer on board a ship steaming on the meridian dipped 03m 21s before noon. Was the ship steaming North or South and what was her speed?

NOTE. $dp = \dfrac{dh\ 15 \sin 1'' \times h}{\sin (\text{L.} - \text{D.}) \sec \text{L} \sec \text{D}}$

Answer. Steaming North, 17·3 knots.

(XXIII.) EQUATION OF EQUAL ALTITUDES.

The frequent daily transmission by W.T. of G.M.T. ensures a greater degree of accuracy in determining the error of a ship's chronometer when at sea than was possible by means of time balls, or by delicate astronomical observations made either when in port or at sea, before the development of wireless telegraphy. The method of determining the error of chronometer by equal altitudes of the same body observed before and after it has crossed the meridian offers an example of the application of a fundamental principle in nautical astronomy and, as such, it may still be given as a test of theoretical knowledge.

If the Sun's declination did not alter, the mean of the chronometer times at which the Sun has the same altitude on opposite sides of the meridian would be the chronometer time of apparent noon. But the Sun's declination does alter, and, for the same altitude, the E.H.A. is not quite the same as the W.H.A., so a correction, called the "Equation of Equal Altitudes," must be applied to the mean of the chronometer times to get the correct middle time. The difference between the chronometer time of the apparent noon thus found, and the Greenwich time of apparent noon, is the error of chronometer on G.M.T.

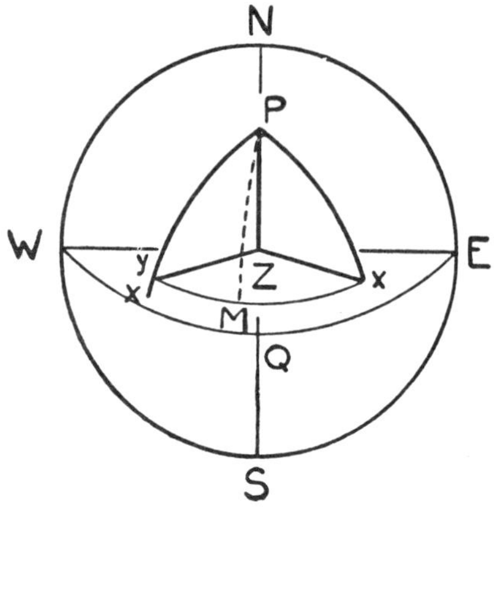

In figure, X and Y are two positions of the Sun having equal altitudes east and west of the meridian so that $ZX = ZY$, but if the polar distance PY be shorter than PX, giving an hour angle west greater than the hour angle east for the same altitude due to the change of declination, then the middle meridian PM will be west of the observer's meridian, but if the polar distance were increasing PM would be to the eastward of the observer's meridian.

We can apply the differential calculus to determine an expression for the equation of equal altitudes.

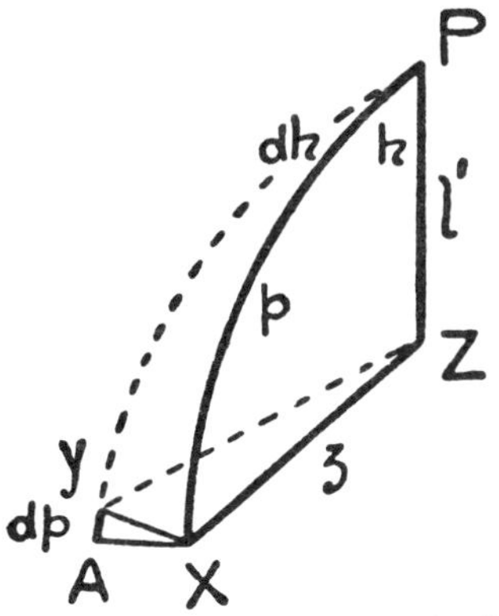

In figure, let $ZX=ZY$

$AY=dp$ the change of declination during the time elapsed between observations.

$\angle XPA=dh$ the corresponding increment in hour angle.

Then in triangle PZX

$$\cos z=\cos p\cos l'+\sin p\sin l'\cos h$$

p and h are the variables.

$$0=-\sin p\,.\,dp\cos l'+\cos p\,.\,dp\sin l'\cos h$$
$$+\sin p\sin l'(-\sin h\,.\,dh)$$
$$\sin p\sin l'\sin h\,.\,dh=dp(\cos p\sin l'\cos h-\sin p\cos l')$$
$$dh=\frac{dp(\cos p\sin l'\cos h-\sin p\cos l')}{\sin p\sin l'\sin h}$$
$$dh=-dp(\cot p\cot h\quad-\cot l'\operatorname{cosec} h)$$
$$dh=dp(\tan \text{lat.}\operatorname{cosec} h\quad-\tan \text{decl.}\cot h)$$
$$dh=dp(\quad A\quad-\quad B\quad)$$

The sign $-dp$ is introduced because the hour angle (h) and polar distance (p) vary inversely. Half the time elapsed between the observations gives h, and dp is the change of declination in the half elapsed time.

The parts A and B will be $-$ or $+$, depending on whether $\cot p$ is positive (less than 90°) or negative (greater than 90°). dh is the correction to apply to the mean chronometer time and is $+$ when polar distance is increasing and $-$ when it is decreasing.

Example.—September 22, Lat. 30° 30′ N., Long. 75° W., the sun had equal altitudes when the chronometer showed 12h 46m 57s and 20h 22m 45s, find the chronometer error on G.M.T.

	d	h	m	s
L.A.T. -	22	12	00	00
Long. -		5	00	00
G.A.T. -	22	17	00	00

Eq. of time +7m 12s
Dec. 0° 19·6′ N. Hrly diff.=·9′
=54″

(change of decl. per hour × ½ interval)=dp=54″ × 3·8h=205″

	h	m	s
1st chron. time	12	46	57
2nd „ „ -	20	22	45
Interval - -	07	35	48
½ interval - -	03	47	54
1st chron. time	12	46	57
Mean chron. time	16	34	51

	h	m	s
L.A.T. (Noon) -	12	00	00
Eq. of time - +		07	12
L.M.T. (Noon) -	12	07	12
Long. W. -	5	00	00
G.M.T. (Noon) -	17	07	12

$dh=dp$ tan lat. cosec $h-dp$ tan decl. cot h

dp	205″	log	2·31218		log	2·31218
Lat.	30° 30′	tan	9·77015	Dec. 0° 19·6′	tan	7·75488
h	3h 47m 54s	cosec	0·07653		cot	9·81293
A	144·2″	log	2·15886	0·8″	log	9·87991

B −0·8 (−*B* because P. dist. is less than 90°)

15)143·4

dh 9·6s.

	h	m	s	
Mean chron. time - - -	16	34	51	
Eq. of eq. alts. - - -	+		10	(plus because P. dist. is increasing).
Chron. time (noon) - -	16	35	01	
G.M.T. (noon) - - -	17	7	12	
Chron. slow of G.M.T. - -		32	11	

Star.

Stars do not change their declination so that the mean of the chronometer times when they have equal altitudes E. and W. of the meridian will be the chronometer time of the star's meridian passage. The problem resolves itself into finding G.M.T. of the star's upper transit as in Chapter VI.

Example III, page 125, may be written as follows:—December 22, Lat. 43° 20′ S., Long. 175° 30′ E., Sirius had equal altitudes E. and W. of the meridian when the chronometer times were 10h 35m 12s and 15h 13m 20s, required the error of the chronometer.

The working up to G.M.T. of upper passage is exactly the same as in the worked examples, the additional work being:—

	h	m	s
1st chron. time -	10	35	12
2nd „ - -	15	13	20
Interval - -	4	38	08
½ elapsed time -	2	19	04
1st chron. time -	10	35	12
Middle chron. time of star's passage	12	54	16

	h	m	s
G.M.T. of star pass.	13	03	00
Chron. time of pass.	12	54	16
Chron. slow of G.M.T.		08	44

Given Lat. 27° 10′ N., Long. 56° 17′ E., chron. times when Sun had equal altitudes 04h 22m 17s and 12h 09m 52s, Sun's decl. 16° 48′ N., decreasing ·65′ per hour. eq. of time +5m 49s. Required the equation of equal altitude.

Answer. +04·2s.

Given Lat. 33° 02′ S., Long. 71° 42′ W., chron. times when Sun had equal altitudes 02h 16m 12s and 07h 20m 42s, Decl. 18° 02·7′ S. decreasing 39″ per hour, eq. of time +6m 11s, find the equation of equal altitudes.

Answer. +4·22s.

(XXIV.) EX-MERIDIAN—REDUCTION FORMULA.

The meridian zenith distance of a body is the least zenith distance possible and is smaller than the zenith distance obtained from observations taken a few minutes before, or after, the body has crossed the meridian by a small quantity called the reduction.

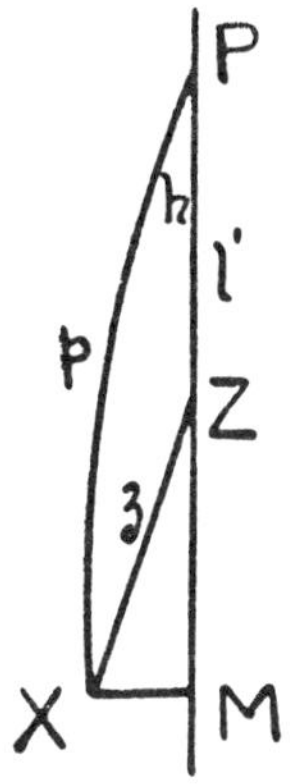

In figure: If z be the observed zenith distance (ZX) R the reduction, then ZM the M.Z.D. $= (z - R)$

by (28) $$\cos h = \frac{\cos z - \cos p \cos l'}{\sin p \sin l'} = \frac{\cos z - \cos p \sin l}{\sin p \cos l} \quad . \quad . \quad \text{I.}$$

by substituting lat. for co-lat.

Now, when the body is on the meridian the hour angle $= 0$ and $\cos h = 1$.

$$1 = \frac{\cos (z - R) - \cos p \sin l}{\sin p \cos l} \quad . \quad . \quad . \quad . \quad . \quad . \quad . \quad . \quad . \quad . \quad . \quad \text{II.}$$

Subtracting equation I. from II.

$$1 - \cos h = \frac{\cos (z - R) - \cos p \sin l - \cos z + \cos p \sin l}{\sin p \cos l}$$

$$1 - \cos h = \frac{\cos (z - R) - \cos z}{\sin p \cos l}$$

by (18) and (14)

$$2 \sin^2 \frac{h}{2} = \frac{2 \sin \frac{1}{2} (z + z - R) \sin \frac{1}{2} (z - z + R)}{\sin p \cos l}$$

$$\sin^2 \frac{h}{2} = \frac{\sin \left(z - \frac{R}{2}\right) \sin \frac{R}{2}}{\sin p \cos l}$$

but R is small so that $\left(z - \frac{R}{2}\right) = (z - R)$ nearly and by transposing we get

$$\sin \frac{R}{2} = \frac{\sin p \cos l \sin^2 \frac{h}{2}}{\sin (z - R)}$$

i.e., sin ½ Reduction = cos decl. cos lat. havs H. ang. cosec (M.Z.D.)

$(z - R) =$ M.Z.D. $=$ (lat. $\pm$ decl.) nearly.

This expression involves the latitude, the quantity we wish to find, and as it is only known approximately, it may be necessary to repeat the calculation using the latitude last found.

Example, given D.R. Lat. 13° 50′ N., Decl. 52° 39·2′ S., H. ang. $37^m\ 4^s$, zen. dist. 67° 2·7′ N., calculate the reduction to the meridian.

Lat.	13°	50′	N.	cos	9·987217
Decl.	52	39·2	S.	cos	9·782930
M.Z.D.	66	29·2		cosec	0·037646
H. Ang.	37^m	4^s		hav	7·814590
½ Red.		14·4′		sin	7·622383
Reduction		28·8			
Z. Dist.	67	02·7			
M.Z.D.	66	33·9	N.		
Decl.	52	39·2	S.		
Lat.	13	54·7	N.		

(XXV.) NORIE'S EX-MERIDIAN TABLES.

The reduction formula

$$\sin\frac{R}{2} = \frac{\sin p \cos l \sin^2\frac{h}{2}}{\sin(z - R)}$$

may be written as follows, and h being small, $\sin^2\frac{h}{2} = \frac{\sin^2 h}{4}$ nearly.

$$\frac{R''}{2}\sin 1'' = \frac{\cos d \cos l}{\sin(d \pm l)} \times \frac{\sin^2 h,}{4}$$

$$R = \frac{\cos d \cos l}{\sin(d \pm l)} \times \frac{2\sin^2 h}{4\sin 1''}$$

$$R = \frac{\cos d \cos l}{\sin(d \pm l)} \times \frac{\sin^2 15'}{2\sin 1''}, h^2$$

the quantity $\frac{\sin^2 15'}{2\sin 1''}$ is constant = 1·9635

15′ sin	7·639816		1″	sin	4·685575
sin² 15′	15·279632		2	log	0·301030
2 sin 1″	4·986605	. . . .			4·986605
1·9635 log	0·293027				

The expression as given in Norie's Ex-Meridian Tables is

$$R = 1{\cdot}9635 \cos d \cos l \operatorname{cosec}(d \pm l) \times h^2$$
$$R = A \times h^2$$
$$R = \text{Table I.} \times \text{Table II.}$$

A is the first part of the above expression worked up for each degree of latitude and declination up to 60° and 63° respectively, it is also the change of altitude per minute when the body is near the meridian.

$R = A \times h^2 = A$ multiplied by the minutes in hour angle squared.

(XXVI.) POLE STAR.

In figure, if HO be the horizon, Z the zenith and X the pole star on its diurnal path round the pole, then PX = P. dist. (p), $\angle ZPX$ = H. ang. (h), ZX = zenith distance (z), ZP = co-latitude (l').

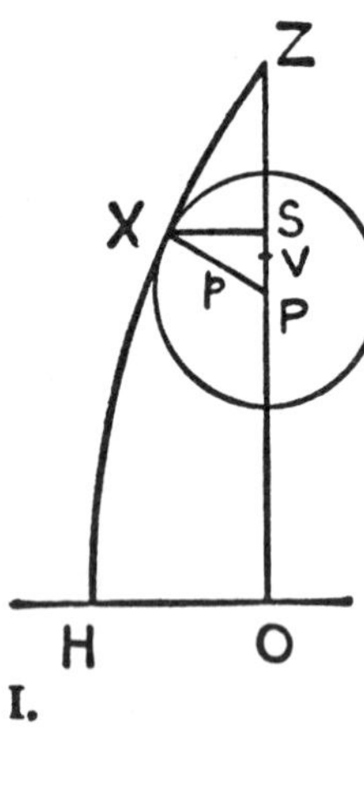

I.

Construction.—Make XS perpendicular to PZ. Centre Z and radius ZX, describe an arc cutting meridian at V, then $PZ = ZV - SV + PS$ and PS and SV are parts of corrections given in the Pole Star Tables.

In $\angle PXS$

$$PS = p\cos h \text{ (1st part of correction)}$$
$$ZX = (ZS + SV)$$
$$\therefore \cos ZX = \cos(ZS + SV)$$
$$= \cos ZS \cos SV - \sin ZS \sin SV \text{ by (6)}$$

but SV is small, $\therefore \cos SV = 1$ nearly

and $\sin SV = SV'' \sin 1''$ by (43)

so that $\cos ZX = \cos ZS - \sin ZS\,(SV'' \sin 1'')$.

II. In $\triangle ZXS$ $$\cos ZX = \cos ZS \cos SX$$

I. = II. $$\cos ZS - \sin ZS\,(SV'' \sin 1'') = \cos ZS \cos SX$$

$$\sin ZS\,(SV'' \sin 1'') = \cos ZS - \cos ZS \cos SX$$

$$\text{,,} \qquad = \cos ZS\,(1 - \cos SX)$$

$$SV'' \sin 1'' = \cot ZS\ 2\sin^2 \frac{SX}{2} \text{ by (18)}$$

But
$$\sin \frac{SX}{2} = \frac{SX''}{2} \sin 1''$$

$$\sin^2 \frac{SX}{2} = \frac{SX^2}{4} \sin^2 1''$$

$$2\sin^2 \frac{SX}{2} = \frac{SX^2}{2} \sin^2 1''$$

$$\text{but } SX = p \sin h$$

$$\therefore\ 2\sin^2 \frac{SX}{2} = \frac{p^2 \sin^2 h}{2} \sin^2 1'' \quad \ldots\ldots\ldots \quad (a)$$

$$\cot ZS = \tan SO = \tan HX = \tan \text{alt.} \quad \ldots\ldots\ldots \quad (b$$

Substituting (a) and (b) in above, viz.

$$SV'' \sin 1'' = \cot ZS \,.\, 2\sin^2 \frac{SX}{2}$$

$$\text{becomes } SV'' = \tan \text{alt.} \frac{p^2 \sin^2 h}{2} \sin^2 1''$$

The Pole Star Tables were based on the formula:—

$$\text{Lat.} = \text{Alt.} - p \cos h + \tfrac{1}{2} p^2 \sin^2 h \tan \text{alt.} \sin^2 1''$$

but since 1952 the tables have been compiled using a mean Latitude of 50° in the first instance, and the formula for the correction is:—

$$\text{Lat.} = \text{True Alt.} - p \cos h + \tfrac{1}{2} p^2 \sin^2 h \tan \text{Lat.}$$

The value of correction a_0 is calculated from the formula using the assumed Latitude and mean position of the star, and a_1 and a_2 are the adjustments necessary when the observer and the star positions differ from the mean positions. The corrections are further adjusted to avoid negative values by adding constants 58·8′, 0·6′ and 0·6′ to a_0, a_1 and a_2 respectively.

SUMMARY OF FORMULAE, I TO XXVI.

I. **On any Sphere and between the same Secondaries.**

$$\frac{\text{Arc of a small circle}}{\text{Arc of parallel great circle}} = \cos\theta.$$

Where θ is the angular distance of the small circle from the great circle.

II. **Distance of sea horizon** $= \sqrt{2rh} = \sqrt{1{\cdot}5h}$ statute miles (no refraction):
" $= 1{\cdot}15\sqrt{h}$ in nautical miles including allowance for effect of refraction. h = height of eye, r = radius of earth.

III. **Angle of dip** $= \frac{\sqrt{2rh}}{r}$ without refraction.
" $= {\cdot}98\sqrt{h}$ allowing for refraction.

IV. **Dip of shore horizon** $= \frac{(D-a)^2}{2a}$

Where D = dip for height of eye.
a = distance of observer from the shore.

Admiralty Manual formula. ${\cdot}565\frac{h}{a} + {\cdot}423a.$

Where h = height of eye in feet and a = offshore distance in miles.

V. **Effect of Dip and Refraction on times of rising and setting.**

dh = Time in minutes $= \frac{RX}{15}$ cosec Az sec lat.

Where RX = dip and refraction in minutes of arc and dh is the interval between visible and theoretical times of rising.

VI. and VII. Sunrise and Moonrise.

VIII. **Refraction** $(R)'' = \tan\theta\,\frac{(\mu - 1)}{\sin 1''}$

Where θ = zenith distance, R = refraction.

$\frac{\mu - 1}{\sin 1''} = 57{\cdot}538'$ the constant of atmospheric refraction.

IX. **Diurnal parallax** = hor. parallax cos alt.

X. **Augmentation of Moon's Semi-diameter.**

Aug. $= 2s\,(\cos z - \tfrac{1}{2}p)\sin\tfrac{1}{2}p\,\text{cosec}\,(z - p).$
Where s = semi-diameter, z = zenith distance, p = parallax in altitude.

XI. **Sextant. Error of collimation** (Aa)

$$\cos Aa = \frac{\sin\frac{1}{2}\text{ true distance}}{\sin\frac{1}{2}\text{ apparent distance}}$$

XII., XIII., XIV., also XVI., XVII., XVIII.

Errors due to Small Errors in Azimuth, Latitude and Hour Angle.

1. $dz = dh$ cos lat sin Az.
2. $dl = dh$ cos lat tan Az.
3. $dz = dl$ cos Az.

Where dz = error in azimuth, dh = error in hour angle.
dl = error in latitude.

XV. Maximum Rate of Change in Azimuth (dA)

$$dA \text{ per minute} = \frac{15' \cos \text{decl.}}{\sin (\text{lat} \pm \text{decl.})}$$

XIX. Trigonometrical Coefficients.

XX. Rate of change of Altitude (dh)

$$dh = dz \operatorname{cosec} Az \sec \text{lat.}$$

XXI. Maximum Altitude.

XXII. Time Interval in Maximum Meridian Altitude.

$$h \sec = \frac{1}{15 \sin 1''} \sin (L \pm D) \sec L \sec D \frac{dp}{dh}$$

Where h = hour angle, L = Lat., D = Decl.

$$\frac{dp}{dh} = \frac{\text{change of decl. per hour}}{\text{minutes of arc per hour}}$$

$\frac{dp}{dh}$ has to be adjusted for course and speed of ship.

Admiralty Manual Formula.

$$h \sec = 15{\cdot}28 (\tan L \pm \tan D) \left(1 \pm \frac{2X}{900}\right) Y.$$

($\tan L \pm \tan D$), + for contrary names, − for like names.

$\left(1 \pm \frac{2X}{900}\right)$, + for ship going west, − for ship going east.

X = D. Long. of ship per hour.

Y = Decl. per hour combined with ship's D. Lat. per hour.

XXIII. Equation of Equal Altitudes.

$$dh = dp (\cot L \operatorname{cosec} h \pm h \tan D \cot h).$$

+ when polar distance greater than 90°

− „ „ less than 90°

$+ dh$ „ „ is increasing

$- dh$ „ „ is decreasing.

L = Lat., D = decl., h = ½ interval between times of observation.

dp = change of decl. during the ½ interval.

XXIV. Reduction to Meridian (R).

$$\sin \tfrac{1}{2} R = \cos L \cos D \operatorname{hav} h \operatorname{cosec} (\text{M.Z.D.})$$

M.Z.D. = ($L \pm D$) nearly. L = Lat., D = decl., h = H. ang.

XXV. Reduction to Meridian ("Norie's Table")

$$R = 1{\cdot}9635 \cos L \cos D \operatorname{cosec} (L \pm D) \times h^2.$$

XXVI. Pole Star.

Lat. $= p \cos h + \frac{1}{2} \tan \text{alt} (p^2 \sin^2 h) \sin^2 1''$, where p = P. Dist., h = H. ang.

Lat. = True Alt. $- p \cos h + \frac{1}{2} p^2 \sin^2 h \tan$ (latitude).

CHAPTER XIII.

MENSURATION.

MENSURATION assumes a knowledge of algebra, geometry and a little trigonometry. We have assumed that knowledge. A brief reference is made in Volume I to the mensuration of a few simple figures and we now wish to extend the subject to include types of questions set in the extra master's examinations, such examples where given in this chapter being marked with an asterisk.

THE TRIANGLE.

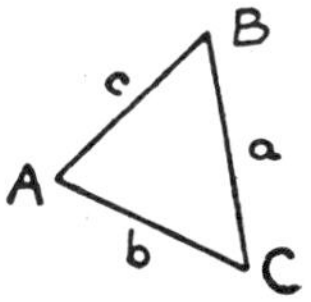

Three sides given.

$$Area = \sqrt{(s - a)(s - b)(s - c)} \text{ where } s = \tfrac{1}{2}(a + b + c)$$

Two sides and included angle given.

$$Area = \tfrac{1}{2}bc \sin A$$

Three angles and a side given.

$$Area = \frac{a^2 \sin B \sin C}{2 \sin A}$$

I.—QUADRILATERALS.

A parallelogram is a quadrilateral whose pairs of opposite sides are parallel. It is particularised as a square, a rectangle, rhombus, rhomboid and trapezium according to the shape of the figure.

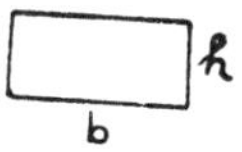

A rectangle has all its angles right angles, but its sides are not all equal.

$$Area = \text{length} \times \text{breadth} = bh$$

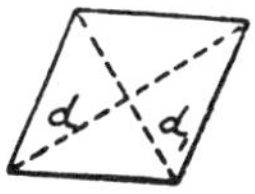

A rhombus has all its sides equal but angles oblique. Its diagonals bisect each other at right angles.

$$Area = \tfrac{1}{2}d \times d_1 \text{ where } d \text{ and } d_1 \text{ are the diagonals}$$

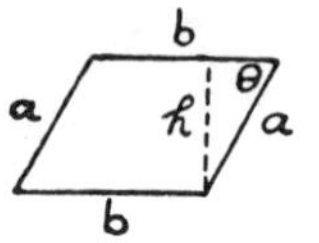

A rhomboid is an oblique parallelogram.

(i) *Area* = base × height = bh.

(ii) *Area* = $ab \sin \theta$, when two sides are given with the angle θ included between them. This will be obvious from the figure where $h = a \sin \theta$, and, substituting this in (i), we get area = $ab \sin \theta$.

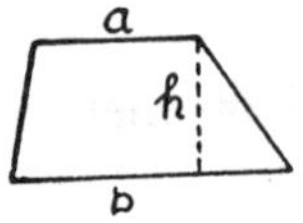

A trapezium has two sides parallel.

$$Area = \tfrac{1}{2}(a + b)h$$

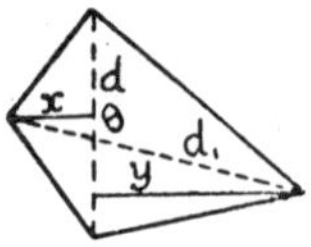

Any quadrilateral.—The offsets are the perpendiculars from two corners on to a diagonal, as x and y in figure.

(i) *Area* = the areas of the two triangles formed by the diagonal.

$$Area = \tfrac{1}{2}dx + \tfrac{1}{2}dy = \tfrac{1}{2}d(x + y).$$

(ii) *Area* = $\tfrac{1}{2}dd_1 \sin \theta$, where θ is the angle at the intersection of the two diagonals d and d_1.

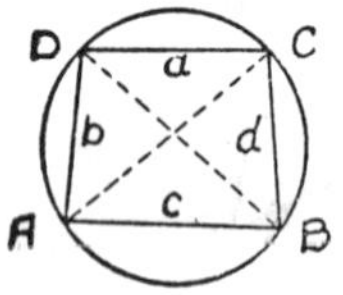

A quadrilateral having its opposite angles supplementary can be inscribed in a circle in which case its

(i) *Area*$^2 = (s - a)(s - b)(s - c)(s - d)$, where s is its semiperimeter and a, b, c, d its sides.

(ii) $Radius^2 = \dfrac{(ab + cd)(ad + bc)(ac + bd)}{16\ area^2}$

(iii) $Diagonal\ BD^2 = \dfrac{(ab + cd)(ac + bd)}{(ad + bc)}$

(iv) $Diagonal\ AC^2 = \dfrac{(ad + bc)(ac + bd)}{(ab + cd)}$

EXERCISES—QUADRILATERALS.

1. Find the number of square feet in a carpet 14 ft. 6 ins. long, 4 ft. 9 ins. broad.

2. Required the number of square yards in a floor $16\frac{1}{2}$ yds. long, $10\frac{1}{5}$ yds. broad.

3 Find the area of a rectangle base 21 ft., diagonal 35 ft.

4. Find the area of a rectangle of base 36 ft., and angle between the base and diagonal 32° 25′.

5. A rectangle, base 14·4 yds., angle between base and diagonal 35° 40′; required its area.

6. Given a parallelogram, length 45 ft., breadth 24 ft.; find its area.

7. Find the area of a rhombus whose sides are 42 ft. 6 ins. and acute angles 53° 20′.

8. In a rhomboid its adjacent sides are 18 ft. and 25 ft. 6 ins., and the angle contained by them is 58°; find its area.

9. The diagonals of a square courtyard measure 36 ft. 8 ins.; find its area.

10. How many square yards are contained in a rectangular field whose diagonals are each 96 ft. and contain an angle of 30°.

11. A rhombus, diagonals 75 ft. and 60 ft.; find its area.

12. Find the area of a trapezoid whose parallel sides are 41 ft. and 24·5 ft. and perpendicular height 43 ft.

13. A trapezoid whose parallel sides are 30 ft. and 40 ft. and area 525 sq. ft.; find its perpendicular breadth.

14. How many square yards are contained in a court, the diagonals of which are 180 ft. and 210 ft. and the angle contained between them is 30°?

15. The four sides of a trapezium inscribed in a circle are 40, 75, 55 and 60 ft.; find its area.

16. The diagonal of a quadrilateral is 60 yds. and the offsets 12·6 yds. and 11·4 yds.; find its area.

17. The diagonal of a trapezium is 40 units, and the offsets 21·6 units and 13 units; find its area.

* 18. *ABCD* are four buoys, 6, 8, 7 and 9 miles apart respectively. A ship is anchored in a position such that she is equidistant from the buoys; find the distance.

(See note under quadrilateral.)

* 19. A circular area is marked by four buoys, *A*, *B*, *C* and *D*. *A* and *B* are the northern buoys two miles apart, *C* and *D* the southern, *A* and *C* are to the westward and one mile apart. At *A*, *C* and *D* subtend an angle of 37°. At *B*, *A* and *D* subtend an angle of 65°. If *B* bears 072° T. from *A*, find the bearing of *C* and *D* from *A*.

NOTE.—Draw a figure with the buoys in their relative positions as given in the question. The quadrilateral is cyclic, draw the diagonals and mark the angles that are equal to each other. In $\triangle ABC$, Given $\angle ABC = 28°$, $AB = 2$, $AC = 1$, find $\angle ACB$. $\angle ADB = \angle ACB$. All the angles are now known. Note that the opposite angles of a cyclic quadrilateral are supplementary so that $\angle C = (180° - \angle B)$.

* 20. Four beacons *A*, *B*, *C* and *D* mark a circular area, *C* and *D* being north of *A* and *B*. At *A*, *C* and *D* subtend a horizontal angle of 32° and the vertical angle of *D* (82 ft. high) is 0° 19′ 59″. The ship steams eastward to *B*; *A* and *D* then subtend an angle of 80° and the vertical angle of *D* is 0° 27′ 27″. Find distance between *C* and *D*.

NOTE.—Calculate the distances by vertical angle. $AD = 14{,}106$ ft., $DB = 10{,}269$ ft. The quadrilateral is cyclic. Draw in the diagonals. In $\triangle ADC$ given AD, $\angle DAC$, $\angle DCA$. Solve for DC.

21. Four buoys moored round the circumference of a circle; *B* is 12 cables to northward of *A* and 17 cables to north-westward of *C*; *D* is 10 cables northward of *C* and 16 cables north-eastward of *A*. Find distance from *A* to *C*.

NOTE.—Construct rough figure. $\angle ABC = \angle ADC = \theta$ on same arc AC. $AC^2 = AB^2 + BC^2 - 2\,AB\,.\,BC \cos\theta = AD^2 + DC^2 - 2\,AD\,.\,DC \cos.\ \theta$. Substitute the values of the sides and find $\theta = 28° 58'$. Then find AC.

II.—THE CIRCLE.

$$\textit{Circumference} = 2\pi r$$
$$\textit{Area} = \pi r^2$$
$$\textit{Area} = \frac{\text{circumference}^2}{4\pi}$$

ANNULUS.

To find the area of a circular annulus or ring.

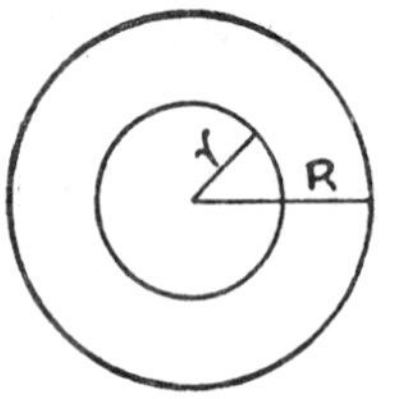

Area of annulus = area of the outer circle of radius R, less the area of the inner ring of radius r.

Area of annulus $= \pi R^2 - \pi r^2 = \pi(R^2 - r^2) = \pi(R + r)(R - r)$.

If C be the circumference of the outer circle and c the circumference of the inner circle, then

(ii) *Area of annulus* $= \dfrac{(C + c)(C - c)}{4\pi}$

Exercises.

1. Find area of a circular annulus, the diameter of the containing circles being 30 and 40 ft.

2. The diameters of a ring are 12 and 10; find the area of the annulus.

3. The circumferences of the containing circles of an annulus are 30 and 40 units; find its area.

4. Find the area of a circular annulus the circumferences of which are 62·8 and 37·7.

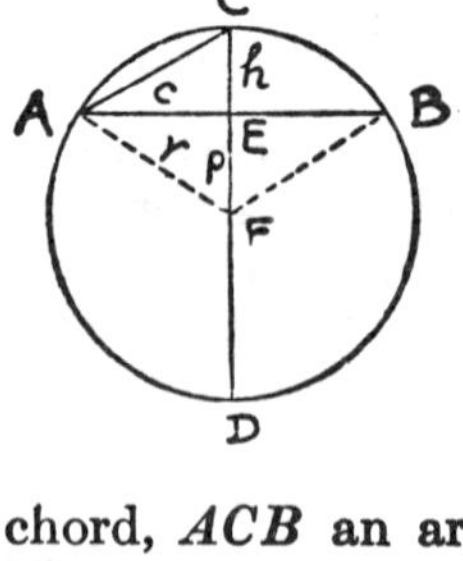

In above figure, AEB is a chord, ACB an arc of the circle, both chord and arc subtend $\angle AFB$.

AC is the half chord of $\angle AFB$.

The area $ACBE$ enclosed by the arc and the chord is a segment of the circle and the area $AFBC$ is a sector.

Area of a segment = area of sector − area of triangle thus area $AEBC$ = area $AFBC$ − area $AFBE$.

CE is the height (h) of the segment.

EF, or p, is the perpendicular, and r is the radius of the circle.

$AE^2 = \frac{1}{4}AB^2 = CE.ED = h(r+p)$

thus if radius = 9, $h = 2$, p = 7, then

$AE^2 = h(r+p) = 2(9+7) = 32$ and
$AC^2 = AE^2 + EC^2 = 32 + 4 = 36$.

If chord $AB = 2c$, then $AE = c$ and $p = r - h$.

From figure, $r^2 = c^2 + p^2 = c^2 + (r-h)^2$

$$r^2 = c^2 + r^2 - 2rh + h^2$$

$\therefore\ 2rh = c^2 + h^2$ and $r = \dfrac{c^2+h^2}{2h}$ = radius . . . . . . (i)

Half chord $AC^2 = c^2 + h^2 = 2rh$

$\therefore$ Half chord of angle $= \sqrt{2rh}$ . . . . (ii)

Example.—Given chord = 20, perpendicular 12; find radius of circle.

$$r^2 = (c^2 + p^2) = (100 + 144) \quad \therefore\ r = \sqrt{244} = 15{\cdot}62.$$

Example.—Given height of segment 2, chord 15; find the diameter of the circle.

$$r = \frac{c^2+h^2}{2h} = \frac{7{\cdot}5^2 + 2^2}{2 \times 2} = \frac{60{\cdot}25}{4} = 15{\cdot}06.$$

Diameter = 30·12.

Example.—Given height 6, perpendicular 24; find the number of degrees in the arc of the segment.

In figure, $r = p + h = 30$.

$$\text{Cos } AFE = \frac{p}{r} = \frac{24}{30} = {\cdot}8 \quad \therefore\ AFE = 36^\circ\ 52'.$$

Angle $AFB = 73^\circ\ 44'$.

Exercises.

5. Given chord 12, perpendicular 10; find radius of circle.
6. Chord 36, perpendicular 25; find radius.
7. Chord 24, height 10; find radius.
8. Chord 24, height 4; find radius.
9. Given chord 40, radius 60; required the number of degrees in the arc.
10. Chord 36, radius 54; find the arc of the segment.
11. Height 12, radius 56; find the angle subtended by the arc.
12. Chord 40, height 5; find the angle subtended by the arc.
13. Given the chord of half the angle of a segment to be 20, the height of which is 2; find the radius and thence the arc.

SECTORS.

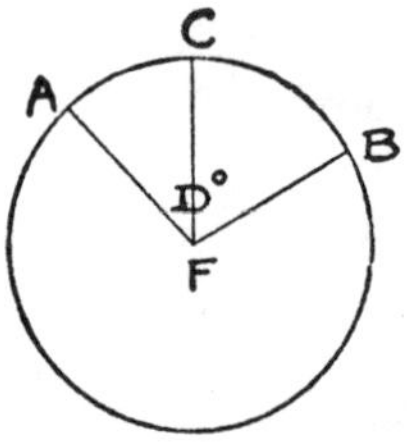

In any circle

$$\frac{\text{length of arc}}{\text{length of circumference}} = \frac{D^\circ}{360^\circ}$$

where D° is the angle AFB at the centre of the circle subtended by the arc AB in figure; thus

$$\frac{\text{arc } AB}{2\pi r} = \frac{D^\circ}{360^\circ} \quad \therefore \text{ arc} = \frac{2\pi r D^\circ}{360^\circ}$$

AFC and CFB are sectors. The areas of sectors in the same circle are proportional to their angles, thus the

$$\frac{\text{area } AFC}{\text{area } CFB} = \frac{\text{angle } AFC}{\text{angle } CFB}$$

from which it follows

$$\frac{\text{area of sector}}{\text{area of circle}} = \frac{D^\circ}{360^\circ} = \frac{\text{arc}}{\text{circumference}}$$

as above, where D° is the angle of the sector.

(i) *Area of sector* $= \dfrac{\text{area of circle} \times D^\circ}{360^\circ} = \dfrac{\pi r^2 D^\circ}{360^\circ}$

again $\dfrac{\text{sector}}{\text{circle}} = \dfrac{\text{arc}}{\text{circumference}}$, or, $\dfrac{\text{sector}}{\pi r^2} = \dfrac{\text{arc}}{2\pi r}$

(ii) *Area of sector* $= \dfrac{\text{arc} \times r}{2}$

$\dfrac{\text{Area of sector}}{\text{Area of circle}} = \dfrac{\text{angle of sector}}{360^\circ}$, $\quad \dfrac{\text{sector}}{\pi r^2} = \dfrac{\theta}{2\pi}$

where 2π is the radian measure of the circumference of the circle ($2\pi = 360^\circ$) and θ is the angle of the sector expressed in radian measure. ($R = 57{\cdot}3^\circ$) approximately, thus $\theta = \dfrac{D^\circ}{57{\cdot}3^\circ}$.

(iii) *Area of sector* $= \dfrac{r^2\theta}{2}$

SEGMENTS.

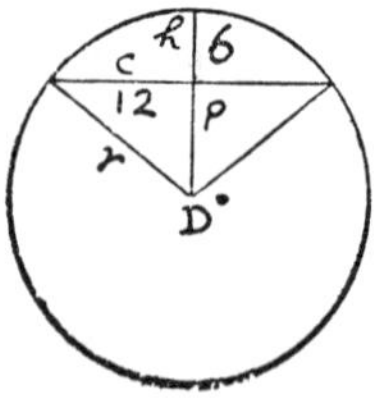

In figure,

Area of a segment = area of sector − area of triangle but area of sector $= \frac{1}{2}r^2\theta$, and area of triangle $= \frac{1}{2}r^2 \sin D^\circ$

Area of segment = sector − triangle

" $= \frac{1}{2}r^2\theta - \frac{1}{2}r^2 \sin D^\circ$

" $= \frac{1}{2}r^2(\theta - \text{nat} \sin D^\circ)$

where $\theta = \dfrac{D^\circ}{57{\cdot}3^\circ}$

Example.—Given the chord of a segment = 24 and its height 6; find its area.

1st.—Find radius. $r = \dfrac{c^2 + h^2}{2h} = \dfrac{144 + 36}{2 \times 6} = \dfrac{180}{12} = 15$

$\therefore p = 15 - 6 = 9.$

2nd.—Find $D°$. $\text{Sin } \dfrac{D}{2} = \dfrac{c}{r} = \dfrac{12}{15} = \cdot 8.$

$\dfrac{D}{2} = 53° \; 8' \; \therefore \; D = 106° \; 16' = 106{\cdot}3°$ nearly.

3rd.—Area segment = area sector — area triangle; find area of sector.

By (i) sector $= \dfrac{\pi r^2 D°}{360°} = \dfrac{15^2 \times 22 \times 106{\cdot}3°}{7 \times 360°} = \underline{\underline{208{\cdot}7}}$ when worked out by logs.

or by (ii) sector $= \dfrac{r^2 D°}{2} = \dfrac{15^2}{2} \times \dfrac{106{\cdot}3°}{57{\cdot}3°} = \underline{\underline{208{\cdot}8}}$ when worked out by logs.

or by (iii) sector $= \dfrac{\text{length of arc} \times r}{2}$

To find the length of the arc

$$\frac{\text{arc}}{r} = \frac{\text{angle}}{57{\cdot}3°}, \text{ arc} = \frac{15 \times 106{\cdot}3}{57{\cdot}3} = 27{\cdot}82$$

$$\text{sector} = \frac{27{\cdot}82 \times 15}{2} = \underline{\underline{208{\cdot}8.}}$$

The area of triangle $= \frac{1}{2}$ base $\times \perp^r$ height
$= \frac{1}{2}\, cp = 6 \times 9 = 54.$

Area of whole triangle $= 2 \times 54 = 108.$

Segment = sector — triangle = $208{\cdot}8 - 108 = 100{\cdot}8$ sq. units.

The area of the sector has been worked out in three different ways to show the application of the formulæ. The area of the segment may be found direct, viz., area $= \frac{1}{2}r^2(\theta - \text{nat sin } D°)$; having found $r = 15$ and $D° = 106{\cdot}3°$ as above, then

$$\text{area} = \frac{15 \times 15}{2}\left(\frac{106{\cdot}3°}{57{\cdot}3} - \cdot 96\right) = 100{\cdot}7 \text{ when worked out by logs.}$$

Nat sine 106° 16′ = ·96 from tables.

Exercises.

14. The chord of circle = 40, and height of the segment = 4; find area of segment.

15. Given chord 24, height of segment 9; find its area.

16. The diameter of a circle = 50, and a chord = 30; find the area of the segment.

17. Diameter 20, chord 16; find the area of this segment of the circle.

18. If the arc of a segment contains 90°, and the diameter of the circle is 12 ft., find area of the segment.

19. Diameter of circle = 50, height of the segment 18; find area of segment.

20. Diameter of circle 10 ft., arc of the segment 280°; find its area.

III.—SPECIFIC GRAVITY.

Frequently examination questions in mensuration involve the volumes and the weights of bodies, introducing their specific gravity.

The specific gravity of a body is the ratio of the weight of a given volume of it, to the weight of an equal volume of fresh water at 4° Centigrade. Fresh water is the standard, or unit, of specific gravity to which other bodies are referred.

One cubic foot of fresh water weighs 1,000 ounces ($62\frac{1}{2}$ lbs.).

If a cubic foot of a substance weighs 1,000 ounces, or $62\frac{1}{2}$ lbs., then its specific gravity = 1.

If the weight of any given volume of a substance weighs the same as an equal volume of water, its specific gravity is 1, and its density, that is, its weight per cubic foot, is 1,000 ounces.

But should a cubic inch of the substance weigh five times more than a cubic inch of water, its specific gravity is 5. If a log of wood floats with half its volume immersed in fresh water, its specific gravity is half that of the water, viz., ·5.

If the average weight of a ship per cubic foot weighs less than a cubic foot of the water which supports her, she will float because her specific gravity is less than that of the water, the water, of course, may be sea water, having a specific gravity of 1·025. If weights be placed on board the ship her specific gravity will be increased until such time as her average weight per cubic foot will be equal to the weight of a cubic foot of the water, her specific gravity will then be the same as the water, she will have reached the limits of her buoyancy and, thereafter, if more weights be added, her specific gravity will be greater than that of the water and she will sink. Fresh water will support a weight of $62\frac{1}{2}$ lbs. per cubic foot; average sea water supports 64 lbs. per cubic foot. When the density of the ship, that is her weight per cubic foot, exceeds these weights she will not float.

To find the Specific Gravity of a Substance.

The substance is weighed in air, and again in water, and the apparent loss of weight noted.

If S = the specific gravity of the substance,
s = ,, ,, ,, ,, ,, water,
W = weight of substance in air,
w = weight of an equal volume of fresh water,

then $W - w$ = the apparent loss of weight when the substance is weighed in water, and we get the equation.

$$\frac{S}{s} = \frac{W}{W - w}.$$

Example.—Find the specific gravity of a piece of steel which weighed 78·5 lbs in air and 68·5 lbs. in fresh water by means of a hydrostatic balance.

$$\frac{S}{s} = \frac{W}{W - w}. \quad \text{Specific gravity of F.W.} = 1.$$

$$S = \frac{sW}{W - w} = \frac{1 \times 78{\cdot}5}{10} = \underline{7{\cdot}85} \text{ specific gravity of steel.}$$

Example.—Find specific gravity of a body which weighed 10 lbs. in air and 6 lbs. when suspended in sea water of specific gravity 1·025.

$$S = \frac{sW}{W - w} = \frac{1 \cdot 025 \times 10}{4} = \underline{\underline{2 \cdot 562}} \text{ specific gravity.}$$

1. Find specific gravity of a substance which weighed 20 lbs. in air and $13\frac{1}{2}$ lbs. in fresh water.

2. A substance weighed 33 ounces in air and 29 ounces in water; find its specific gravity.

To find the Specific Gravity of a Body lighter than Water.

When a body is lighter than water it floats; it does not displace its own volume of water but only its own weight of water, *i.e.* its displacement.

To determine its specific gravity it has to be wholly submerged by attaching a weight to it, and then to eliminate the effect of the weight on the compound mass, that is, the substance and weight combined.

Let S = the specific gravity of the substance,
s = „ „ „ „ „ water,
W = weight of the substance in air,
w_1 = weight of sinker in water,
W_1 = weight of sinker and substance together in water,

then $$\frac{S}{s} = \frac{W}{(W + w_1) - W_1}.$$

Example.—A piece of cork weighs 2·5 lbs. in air, a sinker weighs 9·1 lbs. in fresh water. The cork and sinker together weigh 1·2 lbs. in water. Find specific gravity of cork.

$$\frac{S}{s} = \frac{W}{(W + w_1) - W_1}. \qquad s = \text{specific gravity } 1.$$

$$S = \frac{25}{25 + 91 - 12} = \frac{25}{104} = \underline{\underline{\cdot 243}} \text{ specific gravity.}$$

3. A substance in air weighs 10 gms.; a sinker in water weighs 91 gms. Substance and sinker together in water weigh 61 gms. Find specific gravity of the substance.

4. A piece of wood weighs 42·6 lbs. in air, a piece of iron weighs 40·7 lbs. in water; when tied together they weigh 33·3 lbs. in water. Find specific gravity of the wood.

Given the Volume and Specific Gravity of a Body to find its Weight.

V = volume of the body in cubic feet,
W = weight of the body,
S = specific gravity of the body,

then $W = VS \times 62\frac{1}{2}$ lbs. (one cubic foot of F.W. weighs $62\frac{1}{2}$ lbs.).

Example.—Find the weight of a block of marble if specific gravity 2·7, length 6 ft., breadth 5 ft., thickness 18 inches.

$$W = VS \times 62\tfrac{1}{2}$$
$$= \frac{6 \times 5 \times 3 \times 27 \times 125}{2 \times 10 \times 2} = \frac{30375}{4} = \underline{\underline{7593{\cdot}75}} \text{ lbs.}$$

NOTE.—$2{\cdot}7 = \frac{27}{10}$, 18 ins. $= \frac{3}{2}$ ft.

5. Find weight of a log 24 ft. × 3 ft. × 1 ft., specific gravity ·925.

6. A square pillar of specific gravity 2·7 measures 63 ft. × 12 ft. × 12 ft.; find its weight.

7. The area of the end of a 10 ft. log is 93·582 sq. ins. and its specific gravity is ·8; find its weight.

To find the Volume of a Body of known Specific Gravity.

A re-arrangement of the preceding formula will suffice, viz.,

$$W = VS \times 62\tfrac{1}{2} \text{ lbs.} \quad \therefore \; V = \frac{W}{62\frac{1}{2}S}.$$

Example.—Find the volume of a stone weighing 8 tons, specific gravity 2·52.

$$V = \frac{W}{62\frac{1}{2}S}$$
$$V = 8 \text{ tons} \times 2240 \text{ lbs.} \div (62\tfrac{1}{2} \times 2{\cdot}52) = 113{\cdot}7 \text{ cub. ft.}$$

8. How many cubic feet are in a ton of timber of specific gravity ·550?

9. A block of stone weighs 112 lbs. and its specific gravity is 2·52; find its volume.

* 10. A ring is 3 ins. in diameter, the internal radius is 10 ins., specific gravity of the iron is 7·5; find weight of the ring.

* 11. Find the weight of the greatest circular band 10 ins. wide that can be cut from a square iron sheet of 3 ft. 6 ins. side and 1 in. thick; specific gravity of iron is 7·75.

* 12. A hollow metal tube 6 ft. long, open at both ends, external diameter 2 ft., displaces 4 cubic ft. of fresh water when completely submerged; find thickness of the metal.

IV.—POLYGONS.

Rectilineal figures having several sides are named polygons and are distinguished by the number of their sides. They are called regular polygons when their sides and their angles are equal.

A triangle, for example, has 3 sides, a quadrilateral 4, a pentagon 5, a hexagon 6, heptagon 7, octagon 8, nonagon 9, decagon 10, undecagon 11, and a dodecagon 12.

The central angle of the figure is 360° divided by the number of sides.

Example.—Given a regular pentagon of side 6 ins. to find its area.

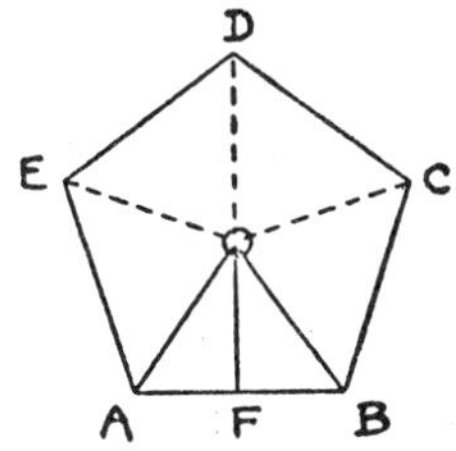

The pentagon is made up of five equal triangles, AOB, BOC, etc.

$$\angle AOB = \frac{360°}{5} = 72°.$$

Make OF perpendicular to AB, it bisects $\angle AOB$, also side AB. Find OF.

$$OF = AF \cot AOF = 3 \cot 36° = 3 \times 1{\cdot}376 = 4{\cdot}128 \text{ ins.}$$

$$\text{Area } AOB = \tfrac{1}{2}AB \, . \, OF = 3 \times 4{\cdot}128 = 12{\cdot}384 \text{ sq. ins.}$$

$$\text{Area of pentagon} = 5 \times 12{\cdot}384 = \underline{\underline{61{\cdot}92}} \text{ sq. ins.}$$

Similarly with a hexagon or other regular polygonic figures.

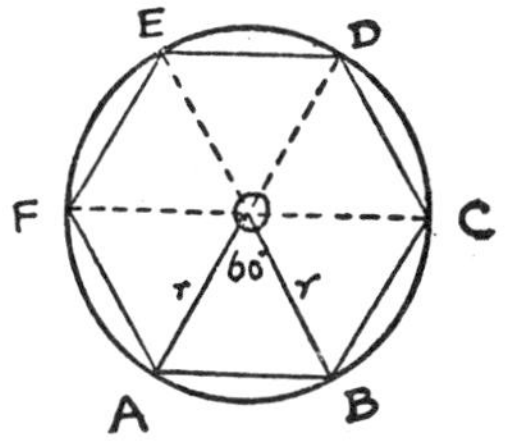

If the radius of a circle be given as 6 inches and it is required to find the area of the inscribed hexagon, then the central angle $AOB = \frac{360°}{6} = 60°$, and $AO = OB = r$ the radius of the circle.

$$\text{Area of } AOB = \tfrac{1}{2}AO \, . \, OB \sin AOB$$

$$\text{,, \quad ,,} \quad = \tfrac{1}{2}r^2 \sin 60°$$

$$\text{,, \quad ,,} \quad = \tfrac{1}{2}\, 36 \times {\cdot}866 = 15{\cdot}588 \text{ sq. ins.}$$

$$\text{Area of hexagon} = 6 \times 15{\cdot}588 = \underline{\underline{93{\cdot}528}} \text{ sq. ins.}$$

*Example.**—An octagonal log of wood 16 ft. long, diagonal 18 ins., floats in fresh water with three sides immersed; find (*a*) depth of log in the water; (*b*) the volume of water displaced if floating in salt water of density 1020.

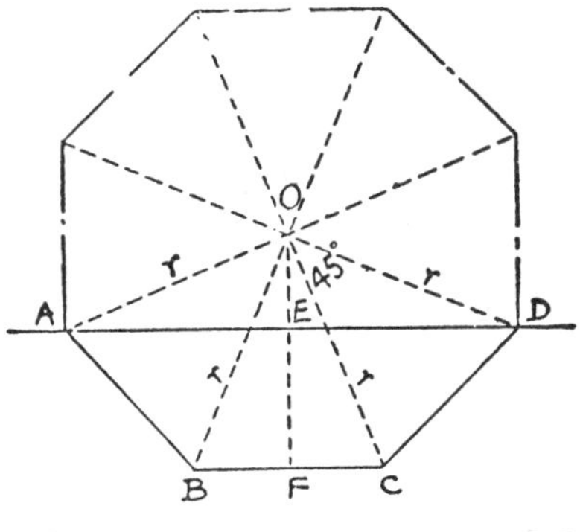

In figure the eight triangles are equal, $r = 9$ ins., central angle $= 45°$, $\angle AOD = 135°$.

$$\text{Area of one triangle} \quad = \tfrac{1}{2}\, r \, . \, r \sin 45° \quad = \tfrac{1}{2}\, 81 \times \frac{1}{\sqrt{2}} = \frac{81}{2\sqrt{2}}$$

$$\text{Area of triangle } AOD = \tfrac{1}{2}\, r \, . \, r \sin 135° = \tfrac{1}{2}\, 81 \times \frac{1}{\sqrt{2}} = \frac{81}{2\sqrt{2}}$$

The triangles are all of equal area and the figure $AODCB = 8$ triangles.

Immersed area $ABCD =$ fig. $AODCB -$ triangle $AODE$

" " $= 8$ triangles $- 1$ triangle

" " $= 2$ triangles $= \dfrac{2 \times 81}{2\sqrt{2}} = \dfrac{81\sqrt{2}}{2}$ sq. ins.

$= \dfrac{81\sqrt{2}}{2 \times 144}$ sq. ft.

Immersed volume = length of log $\times$ area of immersed end.

" " $= 16 \times \dfrac{81\sqrt{2}}{2 \times 144} = 6{\cdot}354$ cub. ft. in F.W.

The volume varies inversely as the density of the liquid, so that the immersed volume $= \dfrac{6{\cdot}354}{1{\cdot}020} = 6{\cdot}23$ cub. ft. in salt water.

The draught of water $= EF = OF - OE$

" $= 9 \cos 22\frac{1}{2}° - 9 \cos 67\frac{1}{2}°$

" $= 9 (\cos 22\frac{1}{2} - \cos 67\frac{1}{2})$*

" $= 18 \sin 45 \sin 22\frac{1}{2}$*

" $= 4{\cdot}86$ ins.

Answer (*a*) 4·86 ins.; (*b*) 6·23 cub. ft.

* See $\cos A - \cos B$, formula (14), page 233 for this transformation.

Exercises.

1. Find the area of a regular pentagon whose side is 30 ft.

2. Find the area of a heptagon each side of which is 15 ft.

* 3. Find the area of a circle inscribed in a rhombus of 80·902 sq. ins. area, the angle at the base being 54°.

* 4. Find the perimeter of a hexagon circumscribing a circle of 2-inch radius.

* 5. A pentagonal log, each side 18 ins., is cut at an angle of 45°; find the area of the section.

(NOTE.—Area of section = area of end $\times$ secant angle of cut.)

* 6. Find the cross sectional area of a hexagonal log 20 ins. each side and cut at an angle of 50°.

* Find the weight of a coil of wire one-sixteenth inch in diameter, wound in the form of a screw having 12 turns, the pitch being one-fifth of an inch. The diameter of the coil is 2 ins. and the specific gravity of the metal is 7·6.

NOTE.—Diam. of helix $= 2'' + \frac{1}{16} = 2\frac{1}{16}''$
Pitch $\frac{1}{5}''$.

The circumference, pitch and length of one turn of wire may be illustrated by a right-angled triangle from which—

$$\begin{aligned} \text{Length}^2 &= \text{pitch}^2 + \text{circumference}^2 \\ \text{,,} \quad &= (\tfrac{1}{5})^2 + (\pi\tfrac{33}{16})^2 \\ \text{Length of one turn} &= 6{\cdot}48 \\ \text{Length of 12 turns} &= 77{\cdot}81 \\ \text{Weight of metal} &= \text{vol. of wire} \times \text{specific gravity} \\ &= \pi r^2 l \times 7{\cdot}6 = 1{\cdot}05 \text{ ozs.} \end{aligned}$$

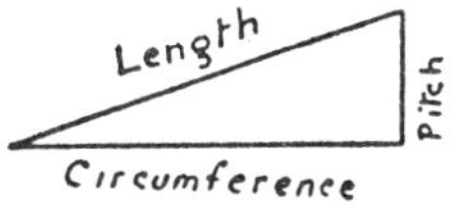

* 7. Nine buoys are moored equidistant from each other round the circumference of a circle the area of which is 32 square miles; find the distance apart of the buoys.

* 8. Seven buoys equidistant from each other are moored round the circumference of a circle, the area of which is 23 square miles; find the distance between the buoys.

* 9. An octagonal log 12 ft. long, specific gravity ·87, greatest diagonal 16 ins. Find the volume of the immersed portion of the log when floating in sea water of density 1023.

* 10. Find the weight of the largest hexagonal log that can be cut from a cylindrical log 3 ft. in diameter, volume 85 cub. ft., specific gravity of the timber ·82.

* 11. Find the length of a helix having 64 turns, pitch $\frac{1}{8}$ in., diameter of helix 1 in.

* 12. A metal plate 3 ins. × 8 ins. × 1 in. is made into a wire 200 ft. long; find diameter of the wire.

* 13. A light shown on a disc 10 ins. away, diameter of disc 24 ins., light in line with centre of the disc. Four inches further away from the disc is a second disc 48 ins. in diameter, the centres of the two discs being in line. Find the area of the annular of the second disc.

V.—THE SPHERE.

In figure, r is the radius of the sphere, $ABCD$ is a zone, r_1 and r_2 the radii of its two ends, GF its thickness (k in formula).

BEC is a segment or spherical cap, r_2 the radius of its base, FE its height (h). AHD is also a segment greater than a hemisphere, its height being HG, and r_1 the radius of its base.

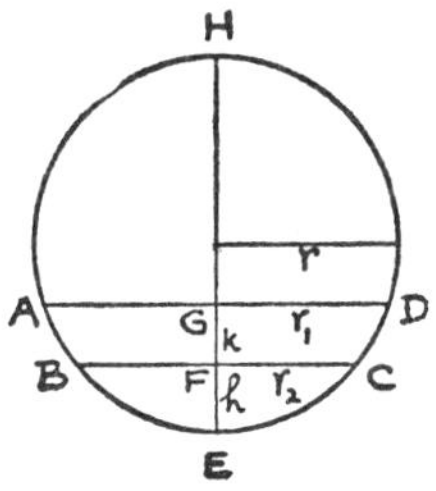

Sphere.

Surface area $= 4\pi r^2$ sq. units.

$$\textit{Volume} = \frac{4\pi r^3}{3} \text{ cub. units.}$$

Segment.

Curved surface $= 2\pi rh$ sq. units, where r is the radius of the sphere.

Volume $= \pi h^2(r - \frac{1}{3}h)$ cub. units, where r is the radius of the sphere.

Volume $= \frac{\pi h}{6}(3r_1^2 + h^2)$, where r_1 is the radius of base of segment.

Zone.

Curved surface $= 2\pi rk$ sq. units, where $r =$ radius of sphere, and k the thickness of the zone.

Volume $= \frac{\pi k}{6}\left\{3(r_1^2 + r_2^2) + k^2\right\}$ cub. units, where r_1 and r_2 are the radii of ends of zone.

Volume of a sphere having a hole bored through its centre.

$$\text{Volume} = \frac{4}{3}\pi(R^2 - r^2)^{\frac{3}{2}}$$

where R is the radius of the sphere and r the radius of the hole.

Example.—The surface area of a sphere is 616 sq. ft.; find its radius.

$$\text{Area} = 4\pi r^2 = 616 \therefore r^2 = \frac{616}{4\pi}, \quad r = \underline{\underline{7 \text{ ins.}}}$$

Example.—Find volume of a sphere of diameter 2 ft. 8 ins.

$$\text{Vol.} = \frac{4}{3}\pi r^3 = \frac{4}{3} \times \frac{22}{7} \times \frac{16^3}{1728} = \underline{\underline{9{\cdot}93}} \text{ cu. ft.}$$

Exercises.

1. Find the area and the volume of a sphere of 5 ins. diameter.

2. Find area and volume of a globe 12 ins. diameter, also area of its circumscribing square.

Example.—Find the diameter of the greatest possible sphere that can be turned out of a square block of wood, if the weight of wood removed after turning weighs 217·24 lbs. and 1 cub. ft. of the wood weighs 57 lbs.

Vol. of block $= L \times B \times D = (2r)^3 = 8r^3$.

Vol. of sphere $= \frac{4\pi r^3}{3}$.

Wt. of turnings = Block − Sphere $= 8r^3 - \frac{4\pi r^3}{3} = \frac{217{\cdot}24}{57}$.

Solve for r and get 1 ft.

Diameter of sphere = 2 ft.

Segment of Sphere.

Example.—Find (*a*) the surface area, (*b*) the volume of a segment of a sphere of diameter 6 ins., height of segment 2 ins.

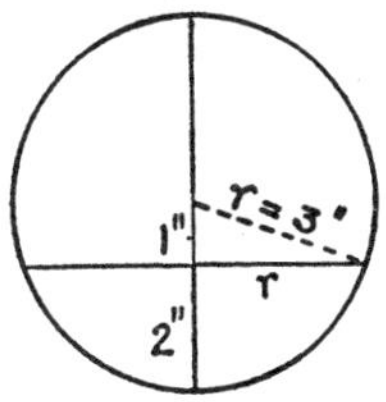

Area of segment $= 2\pi rh = \dfrac{2 \times 22 \times 3 \times 2}{7} = 37{\cdot}7$ sq. ins.

Volume of segment $= \dfrac{\pi h}{6}(3r_1^2 + h^2) = \dfrac{22 \times 2}{7 \times 6}(3 \times 8 + 4)$

$= \underline{\underline{29{\cdot}33}}$ cub. ins.

Note.—$r_1^2 = 3^2 - 1^2 = 8$.

The other formula might be used, viz. :—

$$\text{Vol. of segment} = \pi h^2(r - \tfrac{1}{3}h) = \frac{22}{7} \times \frac{4}{1}\left(\frac{9-2}{3}\right) = 29{\cdot}33 \text{ cub. ins.}$$

The volume of the segment greater than a hemisphere when calculated is 83·31 cub. ins., its height being 4 ins. The volumes of the two segments together make up the volume of the whole sphere, viz. (29·33 + 83·81 = 113·14 cub. ins.) which is the same as found from $\dfrac{4\pi r^3}{3}$.

Exercises.

3. Find surface area of a segment, height 5 ins., diameter of sphere 25 ins.
4. Height of segment 3 ft. 6 ins., diameter of sphere 10 ft.; find area of segment.
5. Height of segment 8 ins., radius of base of segment 14 ins.; find its volume.
6. Diameter of sphere 5 ft., height of segment 2 ft.; find volume of segment.
7. Diameter of sphere 10 ft., height of segment 2 ft.; find volume of segment.

Zone.

Example.—The radii of the circular ends of a zone on a sphere are 12 ins. and 5 ins.; find the volume of the zone if its thickness is 3 ins.

Vol. of zone $= \dfrac{\pi k}{6}\left\{3(r_1^2 + r_2^2) + k^2\right\}$ where k is its thickness and r_1 r_2 the radii of its ends.

,, ,, $= \dfrac{22}{7} \times \dfrac{3}{6}\left\{3(144 + 25) + 9\right\}$

,, ,, $= \underline{\underline{132}}$ cub. ins.

Example.—From a sphere of 10-in. diameter a parallel zone is cut, the ends of which are 4 ins. and 3 ins. on each side of the centre. Find the volume of the zone.

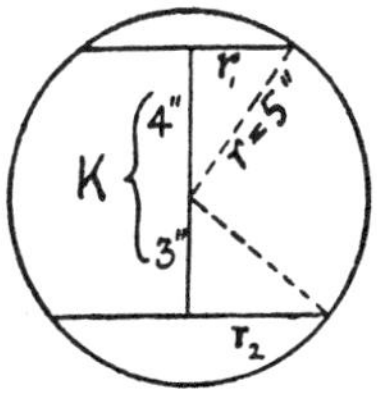

First find the radii of the ends of the zone.

$r_1^2 = 5^2 - 4^2 = 9$, $r_2^2 = 5^2 - 3^2 = 16$, $k = 7$.

$$\text{Vol.} = \frac{22}{7} \times \frac{7}{6}\left\{3(9 + 16) + 49\right\} = \underline{\underline{454\tfrac{2}{3}}} \text{ cub. ins.}$$

Exercises.

8. Find volume of a spherical zone, the diameters of its ends being **10** and **12** ins., and its height **2** ins.

9. Spherical zone, diameters **8** and **12** ins., thickness **10** ins.; find its volume.

10. Find volume of a middle zone of a sphere, its thickness being **8** ft. and the diameters of its ends **6** ft.

Volume and Density.

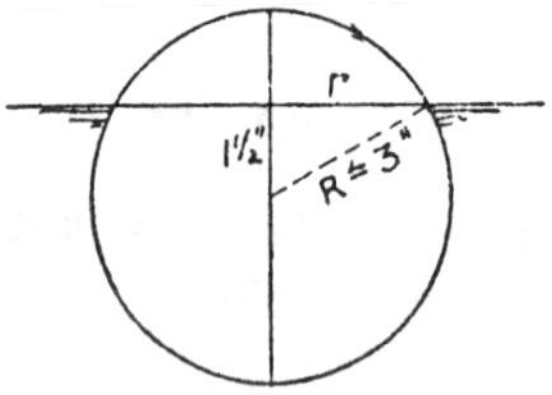

*Example.**—A sphere **6** ins. diameter floats in fresh water with three-quarters of its diameter immersed; find the density of this sphere.

1st. Find volume of segment.

2nd. ,, ,, ,, sphere.

Then vol. of segment ÷ vol. of sphere = specific gravity of sphere.

Given height of segment $h = 4\frac{1}{2}$ ins., radius of sphere $R = 3$ ins., radius of segment r_1.

So that $r_1^2 = 3^2 - 1\frac{1}{2}^2 = 6\frac{3}{4}$.

$$\text{Vol. of segment} = \frac{\pi h}{6}(3r_1^2 + h^2) = \pi\frac{4\frac{1}{2}}{6}(3 \times 6\tfrac{3}{4} + 20\tfrac{1}{4}) = \frac{243\pi}{8}$$

$$\text{Vol. of sphere} = \frac{4\pi R^3}{3} = \pi \times \frac{4 \times 3 \times 3 \times 3}{3} = 36\pi$$

$$\frac{\text{Vol. of segment}}{\text{Vol. of sphere}} = \frac{243\pi}{8} \times \frac{1}{36\pi} = \frac{27}{32} = \cdot 8437$$

Sp. gr. of sphere = ·8437, or density 843·7 oz. per cub. ft.

Example.—*A hollow iron sphere of internal diameter **4** ins. weighs **760** lbs., specific gravity of iron **7·6**; find thickness of the iron.

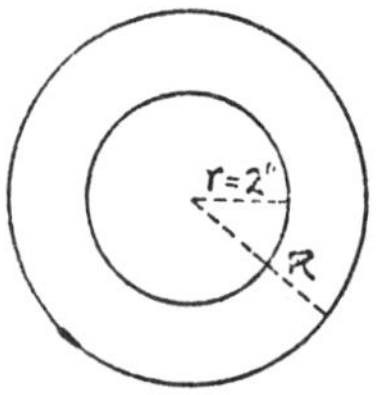

NOTE.—A cub. ft. F.W. weighs 62·5 lbs. = $\frac{1000}{16}$ ozs.

1 cub. ft. = 1,728 cu. ins.

∴ A cub. inch of F.W. = $\frac{1000}{16 \times 1728}$ lbs. = w.

Let W represent the weight of substance (760 lbs.) and w the weight of unit volume of water $\left(\frac{1000}{16 \times 1728}\right)$ lbs., then as

Weight = vol. × sp. gr. × wt. of unit vol. of water

we have

$$\text{Vol.} = \frac{W}{\text{sp. gr.} \times w} = \frac{760 \times 1728 \times 16}{7 \cdot 6 \times 1000} = \frac{13824}{5} \text{ cub. ins.}$$

$$\text{Vol. of whole sphere} = \frac{4\pi R^3}{3}$$

$$\text{Vol. of hollow sphere} = \frac{4\pi r^3}{3}$$

$$\text{Vol. of shell} = \frac{4\pi R^3}{3} - \frac{4\pi r^3}{3} = \frac{4\pi}{3}(R^3 - r^3) = \frac{W}{\text{sp. gr.} \times w} = \frac{13824}{5}$$

from which $$R^3 - 8 = \frac{13824 \times 3 \times 7}{5 \times 4 \times 22} = 659 \cdot 7$$

$$\therefore R = \sqrt[3]{667 \cdot 2} = 8 \cdot 73 \text{ ins.}$$

Thickness of metal $= R - r = 6 \cdot 73$ ins.

*Example.**—A hollow aluminium sphere 15-in. diameter floats with one-third of its diameter immersed in liquid of density 932·5; specific gravity of metal 2·5. Find the thickness of the metal.

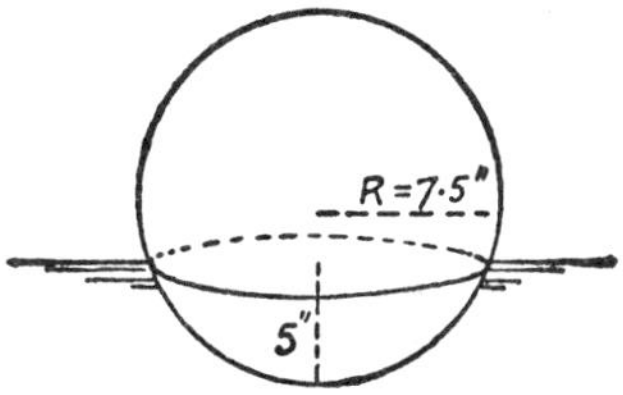

In figure $R = 7 \cdot 5$ ins., $h = 5$ ins.

(1) Vol. of sphere $= \dfrac{4\pi R^3}{3}$.

(2) Weight of liquid = volume × density $= \pi h^2(R - \frac{1}{3}h) \times 932 \cdot 5$.

(3) Vol. of shell $= \dfrac{4\pi}{3}(R^3 - r^3) = \dfrac{\text{Weight}}{\text{density}} = \dfrac{\pi h^2(R - \frac{1}{3}h)932 \cdot 5}{2500}$

Vol. of hollow part = vol. of sphere — vol. of shell

$$\frac{4\pi r^3}{3} = \frac{4\pi R^3}{3} - \frac{\pi h^2(R - \frac{1}{3}h)932 \cdot 5}{2500}$$

π cancels, divide throughout by $\frac{4}{3}$, rearrange and we get

$$R^3 - r^3 = \frac{25}{1} \times \frac{17 \cdot 5}{3} \times \frac{932 \cdot 5}{2500}$$

$$= 40 \cdot 8 \text{ by logs.}$$

$$r^3 = R^3 - 40 \cdot 8 = 421 \cdot 9 - 40 \cdot 8 = 381 \cdot 1$$

$$r = 7 \cdot 25$$

$$R = 7 \cdot 50$$

$R - r = \cdot 25$ in. the thickness of metal.

Exercises.

* 11. A hollow aluminium sphere has an external diameter of 16 ins. and floats at half its diameter in water weighing 1015 ounces per cub. ft. If specific gravity of the metal is 2·6, find the thickness of the shell.

* 12. A solid iron sphere of 6-in. diameter has a hole 1-in. diameter bored through its centre. Find the weight of the sphere after the hole is bored, specific gravity of the iron 7·8.

* 13. A hollow metal sphere, external diameter 14 ins., floats with upper edge just flush with the surface of the water of density 1023. If the specific gravity of the metal is 8·6, find its thickness.

* 14. A hollow metal sphere, external diameter 12 ins., internal diameter 10 ins., when full of water weighs 136 lbs. and when full of spirit it weighs 124 lbs. Find specific gravity of the sphere and of the spirit.

VI.—THE CONE AND PRISM.

THEOREMS.

1. The solid content, or volume, of prisms, and cylinders, of the same base **area** and height, are equal to each other.

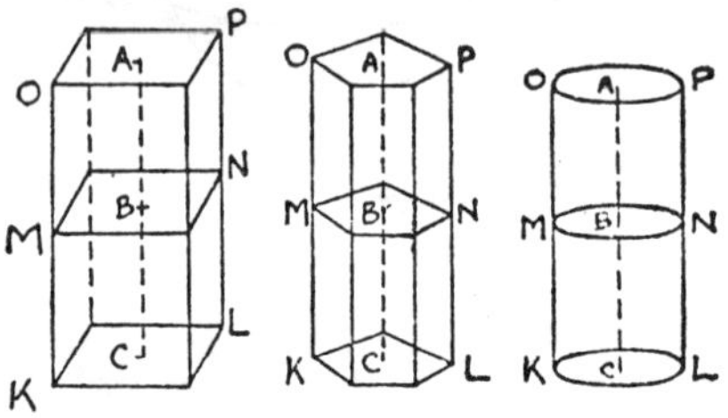

If ABC represent the height or length of the figures represented by l units in each case, and the base area KL are also equal, then the volume of each is equal to base times l.

2. Similar prisms and cylinders are to each other as the cubes of their heights, or of any other linear dimensions, like to like.

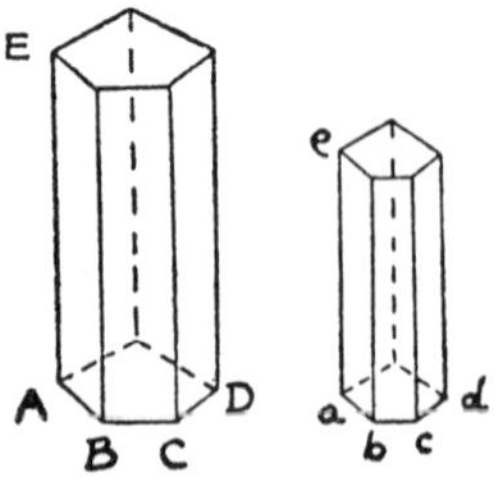

If the two pentagonal prisms in figure are similar, then

$$\frac{\text{Vol. } DE}{\text{Vol. } de} = \frac{AB^3}{ab^3} = \frac{AE^3}{ae^3}$$

3. The ends of the frustum of a right pyramid, or of a right cone, are to each other as the squares of their distances from the vertex.

Thus, $$\frac{\text{area } ab}{\text{area } AB} = \frac{Vp^2}{VP^2}$$

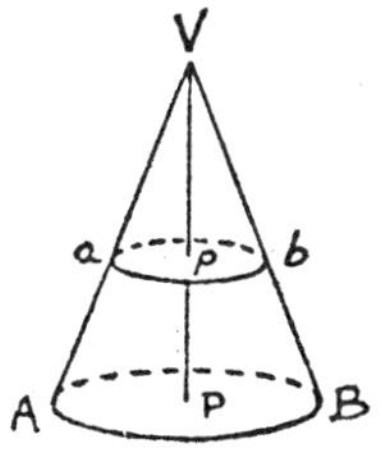

4. Every pyramid, whatever figure it may be, is the third part of a prism having the same base and height.

5. A right cone is the third part of a cylinder having the same base and height; it is also two-thirds of a hemisphere having the same radius.

6. All spheres are to each other as the cubes of their diameters.

7. The volume of a sphere is two-thirds of its circumscribing cylinder. If the volume of such a sphere be 320 cub. units, and x the volume of the cylinder, then

$$\frac{2x}{3} = 320. \quad x = 480 \text{ cub. units.}$$

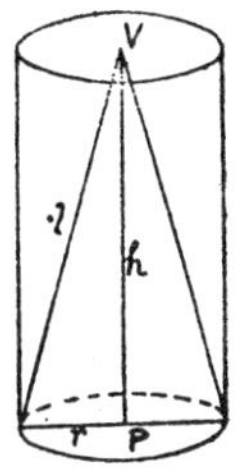

The cylinder and cone are somewhat similar in character.

The figure represents a cone inside a cylinder, the height and base of each being the same.

VP or h = perpendicular height of cone and cylinder, l the slant height of cone and r the radius of the base.

Cylinder.

Curved area of cylinder $= 2\pi r \times h$.

Area of its ends $= 2\pi r^2$.

Area of whole surface $= 2\pi r(h + r)$ sq. units.

Cone.

Curved area of cone $= \pi rl$.

Area of its base $= \pi r^2$.

Area of whole surface $= \pi r(l + r)$ sq. units.

Vol. of cylinder $= \pi r^2 h$ cu. units.

The volume of a cone is one-third that of a cylinder on the same base and of the same height.

Vol. of cone $= \dfrac{\pi r^2 h}{3}$.

Example.—Required the surface area of a cylinder and also of a cone of the same height 20 ft., and same base of 5 ft. radius.

Cylinder.

$$\text{Vol.} = \pi r^2 h = \frac{22}{7} \times 5^2 \times 20 = 1571{\cdot}43 \text{ cub. ft.}$$

Cone.

$$\text{Vol.} = \tfrac{1}{3} \text{ of cylinder} = \frac{1571{\cdot}43}{3} = 523{\cdot}8 \text{ cub. ft.}$$

Cylinder.

$$\text{Area} = 2\pi r(h + r) = \frac{2 \times 22 \times 5 \times 25}{7} = 785{\cdot}77 \text{ sq. ft.}$$

Cone.

Slant height of cone . $l = \sqrt{h^2 + r^2} = 20{\cdot}62.$

$$\text{Area} = \pi r(l + r) = \frac{22 \times 5 \times 25{\cdot}62}{7} = 402{\cdot}6 \text{ sq. ft.}$$

Show that the volume of a cone, a hemisphere and a cylinder of the same base and height are in the ratio of 1, 2 and 3.

In this case $r = l$ and πr^3 cancels and as it is common to each fraction it may be regarded as unity.

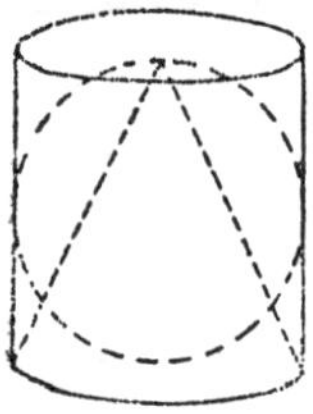

Vol. of cone: $\frac{\pi r^2 l}{3} = \frac{\pi r^3}{3} = \frac{1}{3} = 1$

Vol. of hemisphere: $\frac{2\pi r^3}{3} = \frac{2\pi r^3}{3} = \frac{2}{3} = 2$

Vol. of cylinder: $\pi r^2 l = \pi r^3 = \frac{1}{1} = 3$

Multiply each fraction by 3 and we have the ratio of 1, 2 and 3.

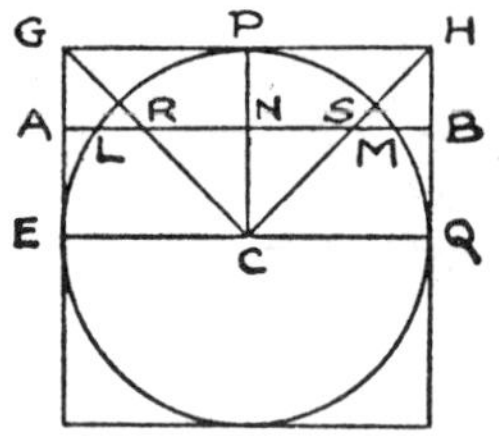

The figure represents a sphere within a cylinder, or cube, the diameter of the sphere being equal to the height of the cylinder.

Frustum $EQML$ = cylinder $EQBA$ — cone CRS, all having the same height NC.

Segment LPM = cylinder $AGHB$ — conic frustum $GRSH$, all having the same height PN.

Exercises.

1. Find the number of gallons of fresh water a cylindrical tank, height 10 ft., diameter of base 6 ft., would hold.

2. A conical vessel apex down measures 28 ins. across the top and is $4\frac{2}{3}$ ft. deep; find the weight of fresh water it would hold.

3. Find the volume of a cylinder, height 25 ins., diameter of base 15 ins.

4. Find volume of a cone, perpendicular height 30 ft., diameter of base 3 ft.

5. Cone of base diameter 3 ft. 4 ins. and slant side 16 ft.; find its volume.

Example.—Two similar cones, one 24 ins. in height, and volume 2772 cub. ins., the other 15 ins. high; find the diameter of the latter.

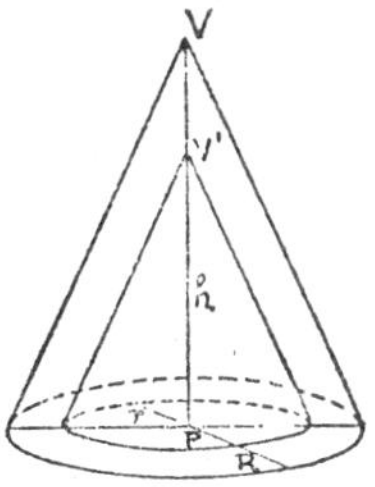

Similar cones are proportional to their heights and radii, thus, in figure, let h and H be the height of the two cones and r and R their radii, then $\frac{r}{R} = \frac{h}{H}$ $\therefore$ $r = \frac{hR}{H}$, but h and H are known, find R.

Vol. $= \frac{\pi R^2 H}{3} = 2772$ cub. ins., from which $R = \frac{\sqrt{441}}{2} = \frac{21}{2}$. Substitute this value for R, and $r = 6{\cdot}5625$. Diameter of smaller cone 13·125 ins.

Example.—Find the volume and surface area of a rod having hemispherical ends. The cylindrical part of the rod is 11 ins. long, and 2 ins. diameter.

Total vol. = vol. of cylinder + vol. of two hemispheres.

,, ,, $= \pi r^2 h + \frac{4\pi r^3}{3}$.

,, ,, = 38·76 cub. ins. when worked out.

Total area = area of cylinder + area of sphere

,, ,, $= 2\pi rh + 4\pi r^2$.

,, ,, = 81·71 sq. ins. when worked out.

*Example.**—A right circular cone has the angle at the apex equal to half the angle at the base. A sphere is placed inside so that when it touches the sides of the cone it is resting on the base. The surface area of the sphere is 59·72 sq. ins.; find the volume of the space left vacant in the cone.

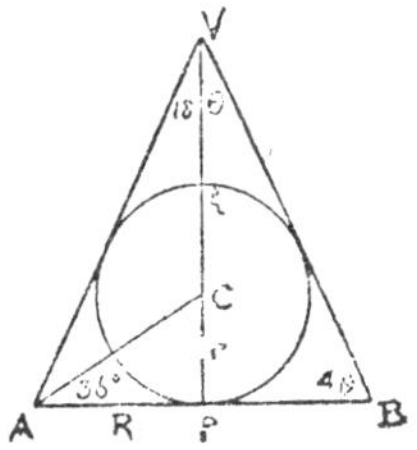

In figure, $h = VP$ the height of cone, R = radius of base of cone, r = radius of sphere.

Let $\angle A$ and $\angle B$ each = say 4θ, then $\angle V = 2\theta$.

but $V + A + B = 180° = 10\theta$, $\therefore$ $\theta = 18°$.

$\angle PVA = \theta = 18°$, $\angle VAP = 4\theta = 72°$,

$\angle CAP = 2\theta = 36°$.

Vol. of space = vol. of cone — vol. of sphere

,, ,, $= \frac{\pi R^2 h}{3} - \frac{4\pi r^3}{3}$

,, ,, $= \frac{1}{3}\pi(R^2 h - 4r^3)$.

We require to find a value for R^2h.

$R = r \cot 36°$, $\therefore R^2 = r^2 \cot^2 36°$.

$VP = h = R \cot 18° = r \cot 36° \cot 18°$.

$R^2h = (r^2 \cot^2 36°)(r \cot 36° \cot 18°) = r^3 \cot^3 36° \cot 18°$.

Substitute this value for R^2h in the above.

Vol. of space $= \frac{1}{3}\pi(r^3 \cot^3 36° \cot 18° - 4r^3)$

" " $= \frac{1}{3}\pi r^3(\cot^3 36 \cot 18 - 4)$

" " $= 43{\cdot}66$ cub. ins. when worked out by logs.

NOTE.—The vol. of sphere works out at 43·4 cub. ins. and vol. of cone works out at 87·06 cub. ins. when the two volumes are worked out separately.

Exercises.

*** 6.** A sphere whose radius is $3\frac{1}{2}$ ins. is placed inside a cone whose base radius is $4\frac{1}{2}$ ins. and slant height 12 ins.; find the volume of that part of the sphere projecting beyond the cone.

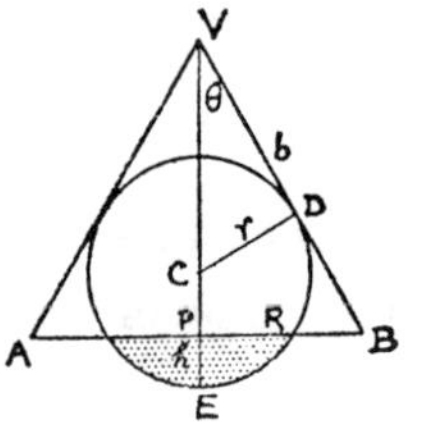

From figure. $CD = r = 3\frac{1}{2}$, $PB = R = 4\frac{1}{2}$, $VB = l = 12$, $VP =$ height of cone, $PE = h =$ height of spherical cap.

In $\triangle VPB$, find $\angle\theta$ and VP.

In $\triangle VCD$, find VC and thence h.

Vol. of cap $= \pi h^2(r - \frac{1}{3}h)$.

Exercises.

*** 7.** A right cone whose base and height are equal to that of a hemisphere, has a volume of 28 cub. ins.; find the area of the hemisphere.

Draw a figure. Write down the formula for volume of cone and area of hemisphere. Note that radius of hemisphere = height of cone, that is $r = h$, and by substitution get area $= 2\sqrt[3]{9\pi V^2}$, where $V =$ volume of cone.

*** 8.** Find the volume of the largest sphere which can be cut from a cone of volume 80 cub. ins., angle at apex 40°.

NOTE.—Refer to figure. $\angle AVP = 20°$, $\angle A = 70°$, $\angle CAP = 35°$. Note that $h = R \cot 20°$. Vol. of cone $= \frac{1}{3}\pi R^2h$. Find R, then r in terms of R. Vol. of sphere $= \frac{4}{3}\pi r^3$.

*** 9.** A right cone, height 12 ins., base diameter 10 ins., find (*a*) diameter of sphere having the same volume, (*b*) and the same surface area as the cone.

PYRAMIDS.

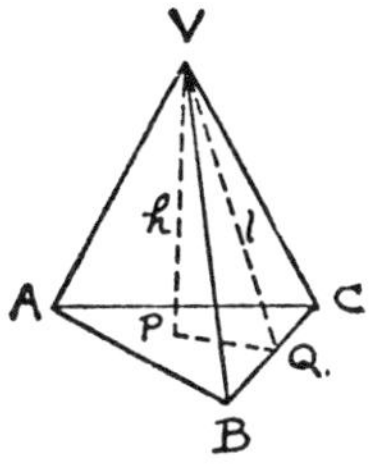

(i)

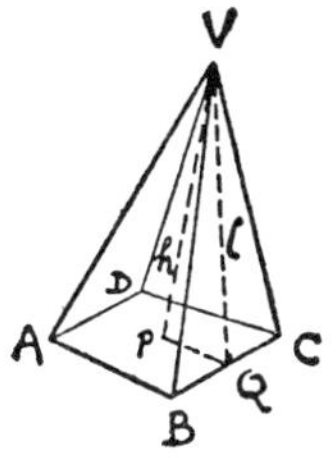

(ii)

The base of a pyramid may be any rectilineal figure. The sides, or faces, are triangles, the number of triangles depending upon the number of edges contained by the base.

Figure (i) represents a pyramid having a triangular base, and (ii) one having a rectilineal base of four sides. V is the vertex, VP is perpendicular to the base and is the height (h) of the pyramid. Bisect the edge BC at Q. Join VQ, then VQ is the slant height (l) of the face VBC and it bisects BC perpendicularly. $\angle VPQ = 90°$. Any line on the base drawn from P is at right angles to VP.

It will be obvious from the figure that the whole surface area of a pyramid is equal to the area of all the triangles plus area of the base.

A right pyramid is one standing on a regular base having any number of equal sides.

The slant surface area $= \Delta$s $BVC + BVA + AVC$ (fig. i).
" " " $= (\frac{1}{2}BC + \frac{1}{2}AC + \frac{1}{2}AB) \times$ slant height.
" " " $= \frac{1}{2}$ perimeter of base $\times$ slant height.

This applies to a right pyramid standing on a regular base, but in the case of an oblique pyramid the area of each triangle would need to be found separately.

The area of the base will depend upon its shape.

The volume $= \frac{1}{3}$(area of base) $\times$ (perp. height) for all pyramids.

Example.—Find the slant surface area of a pyramid the slant height of which is 10 ft., and each side of its triangular base is 18 ins.

If s be the semi-perimeter of base and l the slant height, then the

$$\text{slant area} = s \times l = \frac{3 \times 1{\cdot}5 \text{ ft.}}{2} \times 10 \text{ ft.} = 22{\cdot}5 \text{ sq. ft.}$$

Example.—What is the surface of a square pyramid, a side of its base being 5 ft. and slant height 12 ft.?

Area of base $= 5 \times 5$	. . . .	$=$ 25	sq. ft.
$l = 12$, $s = 10$, . .	slant area $s.l$	$=$ 120	"
Whole area	. . . .	145	"

Example.—A regular pentagonal pyramid, whose slant height is 10 ft., each side of base 1 ft. 8 ins.; find its surface area.

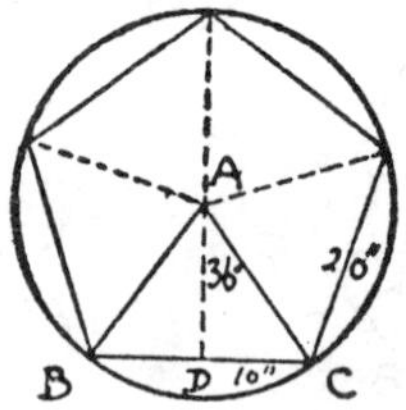

To find area of sides. Each side of the pentagon is 1 ft. 8 ins.

$$s = \tfrac{1}{2}(5 \times 1 \text{ ft. } 8 \text{ ins.}) = 4 \text{ ft. } 2 \text{ ins.}$$

Slant area $= s \times l = 4$ ft. 2 ins. $\times$ 10 ft. $= 41{\cdot}735$ sq. ft.

To find area of base. The figure shows the pentagonal base on which the pyramid stands. The area of the base will be equal to the sum of five triangles equal to ABC.

$\angle A = 72°$, AD bisects $\angle A$ and also BC perpendicularly, $DC = 10$ ins., $\angle DAC = 36°$.

$AD = 10 \cot 36° = 13{\cdot}7$ ins.

Area of $ABC = \frac{1}{2}$ base $\times$ perp. height $= 10 \times 13{\cdot}7 = 137$ sq. ins.

Area of base $= 5 \times 137$ . . .	$=$	4·76 sq. ft.
Area of sides as above	$=$	41·73 ,,
Area of whole pyramid . . .	$=$	46·49 ,,

Example.—Find volume of a regular triangular pyramid, sides of base 6 ft., height 60 ft.

$$\text{Area of base} = \sqrt{s(s-a)(s-b)(s-c)}$$
$$= \sqrt{9 \times 3 \times 3 \times 3} = 9\sqrt{3} = 15{\cdot}588 \text{ sq. ft.}$$

Vol. of pyramid $= \frac{1}{3}$ area of base $\times$ height

,, $= \frac{1}{3} \times 15{\cdot}588 \times 60 = 311{\cdot}76$ cub. ft.

Example.—A square pyramid, slant height 25 ft., side of base 30 ft.; find volume.

Refer to figure under pyramids. Note that in this question $BQ = QP = 15$ ft.; $l = 25$ ft.; find h.

$$h = \sqrt{l^2 - PQ^2} = \sqrt{25^2 - 15^2} = \sqrt{400} = 20 \text{ ft.}$$

Vol. of pyramid $= \frac{1}{3}$ area of base $\times$ height

,, $= \frac{1}{3} \times 30 \times 30 \times 20 = 6000$ cub. ft.

FRUSTUMS.

A frustum is the lower part of a pyramid, or cone, sliced off parallel to the base.

The solid lined parts in figures (i) and (ii) are the frustums of the pyramid and cone respectively.

The ends of a right pyramid frustum are similar figures and its sides, or faces, are trapeziums, and if the base be a regular polygon the trapeziums are all equal.

The ends of a cone frustum are circles of radius r_1 and r_2. In both cases pP represents the perpendicular distance between the base end and top end of the frustum, and is called its thickness, usually indicated in equations by letter k.

The areas of the ends of a frustum (pyramid or cone) are proportional to the squares of their heights, thus in the pyramid

$$\frac{\text{Area } abcd}{\text{Area } ABCD} = \frac{Vp}{VP}$$

and in the cone

$$\frac{\text{Area } ab}{\text{Area } AB} = \frac{Vp}{VP}$$

Pyramid.

The area of the slant surface of the frustum of a right pyramid in square units is n times the area of the figure $ABba$, where n is the number of equal sides, that is,

$$area = \frac{l}{2}(AB + ab)n,$$

or, in words,

$$area = \tfrac{1}{2}(\text{sum of perimeter of ends}) \times (\text{slant height}).$$

The volume of the frustum $= \frac{1}{3}k(E_1 + \sqrt{E_1E_2} + E_2)$ cub. units,

where k is the thickness of the frustum,

E_1 the area of one end,

E_2 the area of the other end.

CONE.

Area of frustum.

$$\textit{Curved area} = \pi(r_1 + r_2)l \text{ sq. units.}$$

where r_1 and r_2 are the radii of the ends and l the slant height.

Volume of frustum.

$$\textit{Vol.} = \tfrac{1}{3}k(E_1 + \sqrt{E_1E_2} + E_2) \text{ cub. units,}$$

where E_1 and E_2 are the area of its ends and k the thickness of the frustum.

This is the same as for a pyramid, but the ends of a cone frustum are circles, the areas of which are given by πr_1^2 and πr_2^2, so by substituting these symbols for E_1 and E_2 in the above equation we get

$$\textit{Vol.} = \frac{\pi}{3}k(r_1^2 + r_1r_2 + r_2^2) \text{ cub. units.}$$

Example.—A square pyramid, the sides of a frustum are 20 ins. and 4 ins., thickness 15 ins.; find the area of its slant surface and its volume.

To find the volume of the frustum.

$$\text{Vol.} = \tfrac{1}{3}k(E_1 + \sqrt{E_1E_2} + E_2)$$

$$\text{Area of } E_1 = 20 \times 20 = 400 \text{ sq. ins.}$$

$$\text{Area of } E_2 = 4 \times 4 = 16 \text{ sq. ins.}$$

$$k = 15 \text{ ins.}$$

$$\text{Vol.} = \frac{15}{3}(400 + 16 + \sqrt{400 \times 16})$$

$$\text{,,} \quad = 5(416 + 80)$$

$$\text{,,} \quad = 2480 \text{ cu. ins. or } \underline{\underline{1{\cdot}43}} \text{ cub. ft.}$$

To find the area of the frustum.

The thickness (k) is given so we have first to find the slant height (l) or qQ.

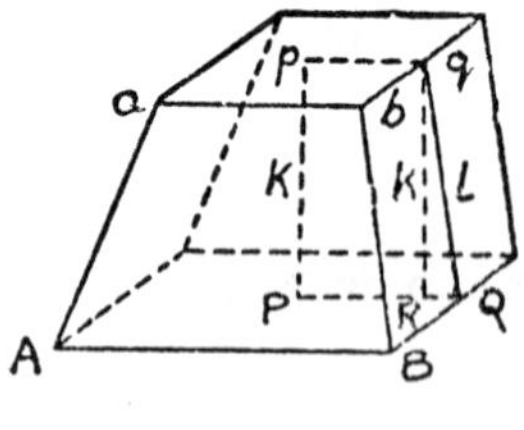

In figure

$$PQ = 10,$$

$$pq = 2 = PR,$$

$$qR = pP = 15 \text{ ins.} = k.$$

$$RQ = PQ - PR = 10 - 2 = 8 \text{ ins.}$$

$$qQ^2 = qR^2 + RQ^2 = 15^2 + 8^2 = 289. \quad \therefore\ qQ = 17.$$

$$\text{Area} = \frac{l}{2}(AB + ab)n = \frac{17}{2}(20 + 4)4 = 816 \text{ sq. ins.}$$

Area of slant surface $=$ $\underline{\underline{\text{5 sq. ft. 96 sq. ins.}}}$

*Example.**—A cone of height 26 ins. is made up of three materials of specific gravity 2·5, 8·0 and 8·5 respectively. The weight of each material is the same, the lightest being on top and the heaviest below; find the thickness of each.

We have here a cone made up of two frustums surmounted by a cone, each of different density but of equal weight. The volume of the whole cone is made up of the volume of the top cone and the volumes of the two frustums, and their relative volumes must vary inversely as their densities because the weight of material is the same.

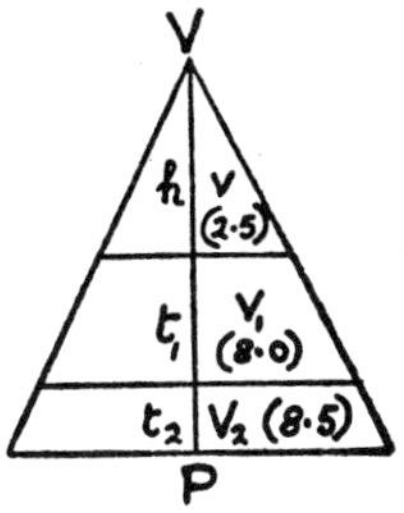

Let $VP = H$ (26 ins.), the height of the whole cone and V its volume.

h = height of the top cone and v (2·5) its volume.

t_1 and t_2 the thickness of the frustums and v_1 and v_2 their volumes, viz. (8·0) and (8·5).

There are really three similar cones of heights h, $(h + t_1)$ and H, to find h, t_1 and t_2.

$$\frac{v_1}{v} = \frac{25}{80} \quad \therefore \ v_1 = \frac{25v}{80} \qquad \frac{v_2}{v} = \frac{25}{85} \quad \therefore \ v_2 = \frac{25v}{85}$$

$$V = v + v_1 + v_2 = \frac{v}{1} + \frac{25v}{80} + \frac{25v}{85} = \frac{10925v}{6800} = 1{\cdot}607v.$$

The volumes of similar cones vary as the cube of their heights, so for the top cone

$$\frac{h^3}{H^3} = \frac{v}{V} \quad \therefore \ h^3 = \frac{H^3v}{V} = \frac{26^3v}{1{\cdot}607v} \quad \therefore \ h = \underline{22{\cdot}2} \text{ ins.}$$

For the top cone plus frustum t_1

$$\frac{(h + t_1)^3}{H^3} = \frac{v + v_1}{V} = \frac{v + \dfrac{25v}{80}}{1{\cdot}607v} = \frac{105}{80} \times \frac{1}{1{\cdot}607} = \frac{10500}{12856}$$

$$\therefore \ h + t_1 = 26\sqrt[3]{\frac{10500}{12856}} = 24{\cdot}3$$

$$\therefore \quad t_1 = 24{\cdot}3 - 22{\cdot}2 = \underline{2{\cdot}1} \text{ ins.}$$

$$t_2 = VP - (h + t_1) = 26 - 24{\cdot}3 = \underline{1{\cdot}7} \text{ ins.}$$

Answer :—

Height of top cone	22·2 ins.
Thickness of middle frustum . .	2·1 „
Thickness of bottom frustum . .	1·7 „

Exercises.

10. Find the slant surface area of an upright triangular pyramid, the slant height being 20 ft., and each side of the base 3 ft.

11. Find the area of the frustum of a square pyramid whose slant height is 10 ft., and each side of the base end 8 ft. 4 ins., and the top end 2 ft. 2 ins.

12. Required the number of cubic feet in a tapered log of timber the ends of which are square, one end being 15 ins. square, the other end 6 ins. square, and length of log 24 ft.

13. Find the volume of a pentagonal frustum, thickness 5 ft., sides of base 18 ins., sides of top 6 ins.

14. Given a square pyramid, slant height 2 ft. 3 ins., the sides of its ends being 24 and 14 ins.; find its surface area and its volume.

(NOTE.—The slant height has to be turned into the thickness k in order to get the volume.)

15. Required the surface area of a cone, slant height 50 ft., base diameter $8\frac{1}{2}$ ft.

16. The frustum of a cone, slant height $12\frac{1}{2}$ ft., circumferences of the two ends 6 ft. and 8·4 ft.; find its surface area.

17. A cone, of thickness $10\frac{1}{2}$ ft., circumference of base 9 ft.; find its volume.

18. Find the content of a conic frustum, thickness 18 units, greatest diameter 8 units, and least diameter 4 units.

19. Required the solid content of the frustum of a cone, thickness 25 ins., circumference of base 20 ins., and of top end 10 ins.

20. A cask which measured 40 ins. in length was cut across at the middle of its length, bung diameter 28 ins., head diameter 20 ins.; find how many gallons of fresh water the tub would hold. (NOTE.—Its shape is the frustum of a cone.)

*** 21.** A pyramid of square base having equilateral triangular sides of 12 ins. is placed in a lathe and the largest possible right cone made from it; find the weight of the turnings, the specific gravity of the wood being ·5.

*** 22.** A regular pentagonal prism of slant height 5 ft., bottom sides 20 ins., top sides 14 ins.; find its slant area.

*** 23.** The areas of the two ends of a frustum are 154 and 616 sq. ins. The area of the curved surface is equal to the surface area of a sphere of radius $10\frac{1}{2}$ ins.; find the slant height of the frustum.

*** 24.** A right cone is cut parallel to the base in such a way that the volume of the small cone so formed is half the volume of the remaining frustum; find the proportion of their heights.

*** 25.** A right cone is cut by a plane parallel to the base so that the volume of the cone cut off is half as large again as the remaining frustum; find the proportion in which the cutting plane divides the altitude of the original cone.

*** 26.** The frustum of a cone and a similar right cone each displace the same amount of water. Each is 30 ins. high and the cone has a base diameter of 12 ins., while the diameter of the frustum is 14 ins. If the specific gravity of the cone is 0·9, find the specific gravity of the frustum.

NOTE.—Find volume of cone and frustum.
Vol. of frustum × specific gravity = vol. of cone × specific gravity.
r_1 of frustum = 7″, and r_2 = 1″.

*** 27.** A conical buoy is floating apex up and weighs 3 tons, drawing 2 ft. of water. (35 cub. ft. water to 1 ton.) If the base radius of the buoy is 5 ft., find the radius of the waterplane and the height of the buoy above water.

VII.—THE ELLIPSE.

Definitions.

1. An ellipse is a closed curve such that, if lines be drawn from any point on the curve to two given points on the major axis, the sum of the two straight lines will always be the same.

2. **Foci.**—In figure, $SP + PH = SP^1 + P^1H = AA^1$ and this condition is constant for all positions of P on the curve. The fixed points S and H are called the foci.

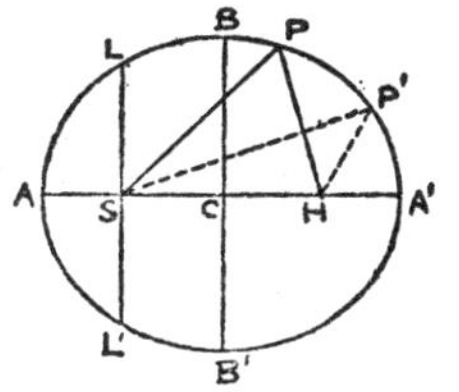

3. **Eccentricity.**—The centre of the ellipse is at **C** midway between the foci so that $CS = CH$. The eccentricity of the ellipse, usually represented by $e, = \frac{CS}{CA}$,

4. **Axes.**—AA^1 is the major axis. BB^1 is the minor axis drawn through C at right angles to the axis major. $CA = CA^1$. The major axis is usually referred to as $2a$ and the minor axis as $2b$, so that $CA = a$ and $CB = b$.

The simplest way to draw an ellipse is to make a loop of thread equal to the major axis desired. Slip the loop over two pins pushed into the paper. Hold the point of a pencil in the bight of the thread as at P and, keeping the thread tight, sweep in the circumference P, A^1, B^1, A, B. The pins mark the foci S and H.

5. **Ordinates.**—Any straight line drawn perpendicular to the major axis and terminated both ways by the curve is called an ordinate.

6. **Latus Rectum.**—The ordinate which passes through either foci, such as LSL^1 in figure is called the latus rectum, usually denoted by $2l$.

7. **Vertices.**—The points A and A^1 where the major axis meets the curve are named the principal vertices or apsides.

In figure, if PC continued be any diameter then P is its vertex, DCD^1 its conjugate diameter and FHF^1 its parameter.

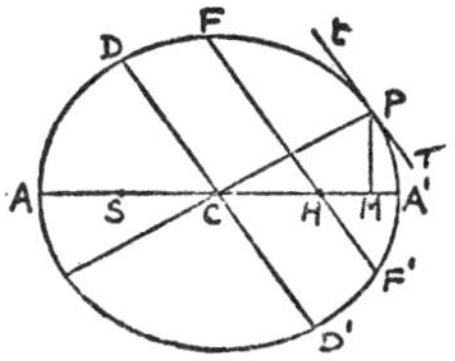

8. **Conjugate Diameter.**—tT is a tangent at P. The diameter DCD^1 drawn through the centre of the ellipse and parallel to the tangent is known as the conjugate diameter.

9. **Parameter.**—The line drawn parallel to the tangent and through one of the foci, such as FHF^1, is called the parameter.

10. **Ordinates.**—A line drawn from a point on the curve and perpendicular to either axis is called the ordinate of that axis. Thus PM is the ordinate of P to the major axis, and the segments AM and A^1M are its abscissae.

11. **The Directrix** is a line outside the ellipse drawn perpendicular to the axis major. In figure the tangent LX meets the major axis produced at X and KXK^1 is the direction drawn through X perpendicular to CA.

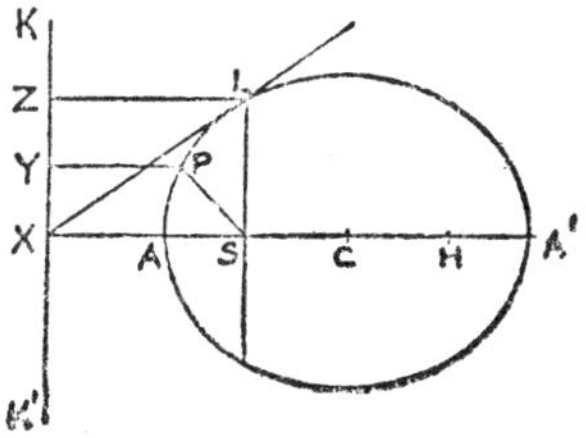

THEOREMS.

Theorem 1.—The fixed line KK^1 is the directrix, and, in the foregoing figure, the property of the ellipse is such that if A and P and L be any points on the curve and AX, PY and LZ are drawn perpendicular to the directrix then the eccentricity $SC = \frac{SA}{AX} = \frac{SP}{PY} = \frac{SL}{LZ}$, the ratio of the distances being constant.

From the above we have $SA = e.AX$, $SP = e.PY$ and $SL = e.LZ$, but from figure, $SA = CA - CS$ and $AX = CX - CA$, from which $SA = e.AX$ becomes $CA - CS = e(CX - CA)$ I

Again $SA^1 = CA^1 + CS$ and $A^1X = CX + CA^1$ and so $SA^1 = e.A^1X$ becomes
$CA^1 + CS = e(CX + CA^1)$ . . . II

bring down Eq. I.................. $CA - CS = e(CX - CA)$ . . . I

and by subtraction...................... $CS = e.CA$ . . . . (1)

$$\therefore e = \frac{CS}{CA} \text{ and if } CA = 1, \text{ then } e = CS \quad . \quad . \quad . \quad (2)$$

By adding Eq. I and II we get $CA = e.CX$.

$$\therefore e = \frac{CA}{CX} = \frac{1}{CX} \text{ if } CA = 1.$$

From above—(1) $CS = e.CA$.
(2) $CA = e.CX$.

Combining (1) and (2) we get $CS.CX = CA^2$ . . . . . . (3)

Theorem 2.—To prove $SP + PH = AA^1$ or $2a$

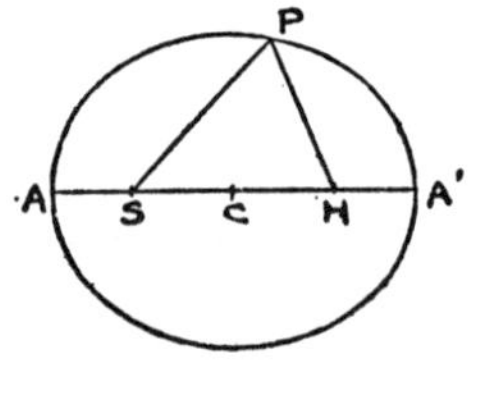

By Def. $SP + PH = AS + AH = AS + AS + SH = 2AS + SH$

$\underline{SP + PH} = A^1S + A^1H = A^1H + HS + A^1H = \underline{2A^1H + SH}$

Add $2(SP + PH) = 2(AS + SH + A^1H) = 2AA^1$.

$\therefore SP + PH = AA^1$ or $2a$ the major axis.

Theorem 3.—To prove $CB^2 = AS.AH = b^2$.

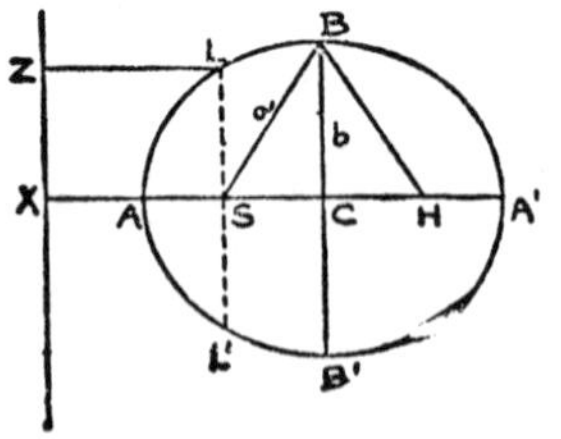

If the point P coincides with B, which is the vertex of the minor axis, as in figure, then $SB + BH = AA^1$, but $SB = BH = CA = a$.

In triangle SCB,

$CB^2 = SB^2 - SC^2 = CA^2 - SC^2 =$
$(CA - SC)(CA + SC) = AS.AH$.

Theorem **4.**—The latus rectum is a third proportional to the major and minor axes.

By Theorem 1, $$e = \frac{SL}{LZ} = \frac{SL}{SX} = \frac{CS}{CA}$$

$$\therefore SL \,.\, CA = CS \,.\, SX = CS(CX - CS) = CS \,.\, CX - CS^2$$
$$= CA^2 - CS^2 = CB^2. \quad \text{(See Theorem I.)}$$
$$SL = \frac{CB^2}{CA} = \frac{b^2}{a} = l \text{ the semi latus rectum.}$$
$$\therefore \quad LSL^1 = 2l = 2\left(\frac{b^2}{a}\right).$$

Theorem 5.—The eccentricity of an ellipse is always less than unity. A circle is a particular form of ellipse, its axes being equal, so that $a = b$ and its eccentricity $e = o$.

By Def. 3, eccentricity $e = \dfrac{CS}{CA} \therefore CS^2 = e^2 \,.\, CA^2 = e^2 \,.\, a^2$.

In figure, $SB = AC = a$ . . . Theorem 3.

$$b^2 = BC^2 = SB^2 - SC^2 = a^2 - e^2a^2 = a^2(1 - e^2) = b^2$$
$$1 - e^2 = \frac{b^2}{a^2} \therefore \; e^2 = 1 - \frac{b^2}{a^2} = \frac{a^2 - b^2}{a^2}.$$

again $\quad SA = CA - CS = a - ea = a(1 - e)$

and $\quad SA^1 = CA^1 + CS = a + ea = a(1 + e)$

The semi latus rectum $SL = l = \dfrac{b^2}{a}$ as in Theorem 4.

" $\quad = \dfrac{a^2(1 - e^2)}{a}$ as above

" $\quad = a(1 - e^2)$.

Theorem 6.—$SP(1 + e \cos ASP) = l$ and is constant for all positions of P on the curve

By Theorem 5 we have $1 - e^2 = \dfrac{b^2}{a}$.

From figure, if P were at B then $SB = SP = a$, $SB = BH$

$\therefore SB + BH = SP + PH = 2SP = 2a$, which is constant for all positions of P.

$$\text{Cos } ASP = \cos BSA = \cos BSC = \frac{SC}{SB} = \frac{SC}{a} = e$$
$$\therefore (1 + e \cos ASP) = 1 + e^2 = \frac{b^2}{a}$$
$$\therefore SP(1 + e \cos ASP) = a \times \frac{b^2}{a^2} = \frac{b^2}{a} = l = \text{constant as in Theorem 4.}$$

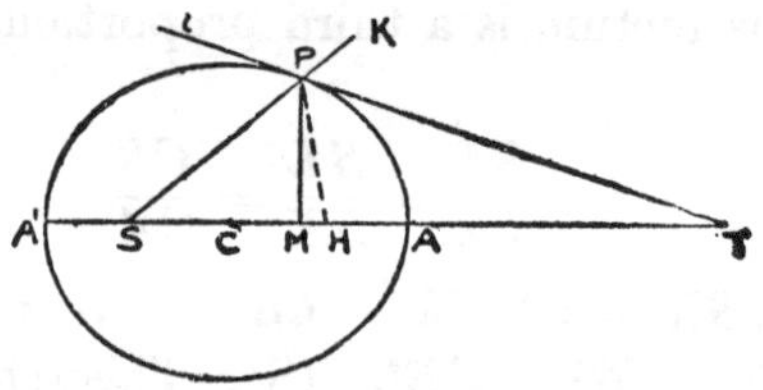

Theorem 7.—In figure, if Tt be the tangent at P meeting the major axis produced at T, and PM a semi-ordinate, then $\frac{CT}{CA} = \frac{CA}{CM}$, thus CA is a mean proportional of CT and CM, and Tt bisects $\angle HPK$.

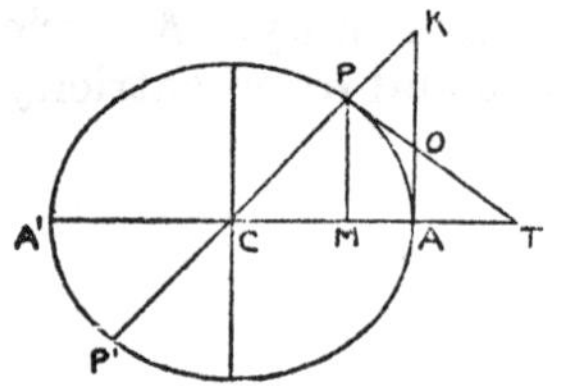

Theorem 8.—In figure, if P be any point, PT a tangent meeting the major axis produced at T, and AK be a tangent at the principal vertex meeting diameter P^1P produced at K, then $\Delta CAK = \Delta CPT$.

Take away the common part $CPOA$ and we have $\Delta PKO = \Delta OAT$; or, take away the common triangle CMP and the trapezium $PKAM = \Delta PMT$.

Mensuration of the Ellipse.

$$\textit{Surface area} = \pi a \,.\, b$$

$$\textit{Volume} = \pi ab \times \text{thickness}$$

$$\textit{Circumference} = 2\pi\sqrt{\frac{a^2 + b^2}{2}}$$

where a is the semi major axis
and b the semi minor axis.

Exercises.

Find the area of the following ellipses :—

1. Axes 5 and 10.
2. Axes 12 and 16.
3. Axes 6 and 7.

Example.—Required the length of the circumference of an ellipse whose axes are 10 and 12.

$$\text{Circumference} = \pi\sqrt{\frac{a^2 + b^2}{2}} = \frac{22}{7}\sqrt{\frac{244}{2}} = 34{\cdot}7 \text{ units.}$$

Find the circumference of the following ellipses :—

4. Axes 6 and 8.
5. Axes 4 and 6.
6. Find the weight of an elliptical steel plate 6 ft. by 4 ft. and 1 in. thick, specific gravity 7·8.
7. Find the weight of an elliptical steel plate $\frac{1}{2}$ in. thick, major axis 12 ft., minor axis 8 ft., specific gravity 7·6.

*Example.**—Given an ellipse, major axis 15 ft., minor axis 9 ft. The focal points S and H are joined by straight lines to P on the curve and PH is perpendicular to the major axis; find the area of triangle SPH.

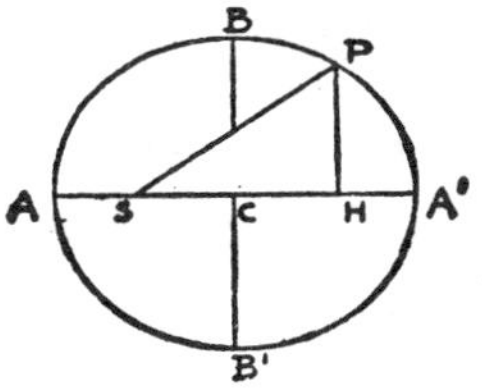

In figure, find the sides of triangle SPH, thence its area.
Given AA^1 the major axis $= 2a = 15$ ft. $\therefore a = 7{\cdot}5$ ft.
Given BB^1 the minor axis $= 2b = 9$ ft. $\therefore b = 4{\cdot}5$ ft.

PH is half the latus rectum.

By Theorem 4. $PH = \dfrac{BC^2}{CA} = \dfrac{b^2}{a} \quad \dfrac{(4{\cdot}5)^2}{7{\cdot}5} = \dfrac{81}{30} = 2{\cdot}7$ ft.

By Theorem 2. $SP + PH = 2a = 15 \therefore SP = 15 - 2{\cdot}7 = 12{\cdot}8$ ft.
$\angle H = 90°$. $SH^2 = SP^2 - PH^2 = (12{\cdot}3)^2 - (2{\cdot}7)^2 = 144 \therefore SH = 12$ ft.
Area $\triangle SPH = \frac{1}{2}$ base $\times$ perp. ht. $= 6 \times 2{\cdot}7 = \underline{16{\cdot}2}$ sq. ft.

---

*Example.**—A square of 3 ins. has circumscribed about it an ellipse whose foci are in the sides of the square; find area of the ellipse.

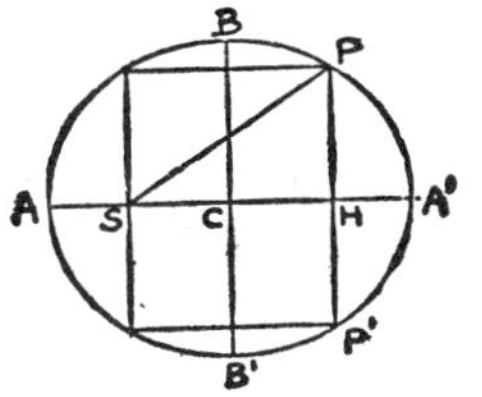

The area of an ellipse $= \pi ab$; find a and b, given the sides of the square 3 ins.

$$PH = 1\tfrac{1}{2},\ PP^1 = 3,\ SH = 3.$$

$$SP^2 = PH^2 + SH^2 = \frac{9}{4} + 9 = \frac{45}{4} \therefore SP = 3{\cdot}354.$$

By Theorem 2. $AA^1 = 2a = SP + PH = 4{\cdot}854 \therefore a = 2{\cdot}427.$

PH is the semi latus rectum $= \dfrac{b^2}{a}$ by Theorem 4

$\therefore b^2 = a \times PH = 2{\cdot}427 \times 1{\cdot}5 = 3{\cdot}6405 \therefore b = 1{\cdot}908.$

Area $= \pi ab = \dfrac{22}{7} \times 2{\cdot}427 \times 1{\cdot}908 = 14{\cdot}55$ sq. ins.

*Example.**—The total surface area of a solid cylindrical frustum is 94·875 sq. ft. The axis is 6 ft., and the frustum is cut at an angle of 60° to the horizontal; find the diameter of the cylinder.

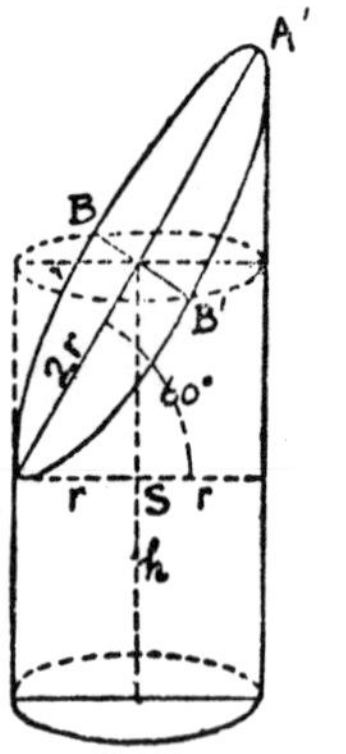

In figure, the height of cylinder = 6 ft. (h). ABA^1B^1 represents the cylindrical cut, $\angle A^1AS = 60°$. If r = radius of the cylinder, then the semi minor axis $\frac{1}{2}(BB^1) = b = r$ the radius of cylinder, and the semi major axis $\frac{1}{2}(A^1A) = a = 2r$.

$$AA^1 = 2a = 2r \sec 60° = 2r \times 2 \quad \therefore a = 2r.$$

The total area of the frustum, 94·875 sq. ft., is made up of

area of bottom		πr^2
area of ellipse	$\pi ab =$	$2\pi r^2$
area of curved surface . .	$2\pi rh =$	$12\pi r$

that is,

$$\pi r^2 + 2\pi r^2 + 12\pi r = 94{\cdot}875$$
$$\pi(3r^2 + 12r) = \text{,,}$$
$$3\pi(r^2 + 4r) = \text{,,}$$

$$\therefore r^2 + 4r = \frac{94{\cdot}875 \times 7}{3 \times 22} = 10{\cdot}0625$$

$$r^2 + 4r - 10{\cdot}0625 = 0$$

On solving this quadratic equation we get

$$r = \frac{3{\cdot}5}{2} \quad \therefore \text{ diameter of cylinder is } \underline{\underline{3{\cdot}5}} \text{ ft.}$$

*Example.**—A ship is holed 9 ft. below the waterline; the hole is elliptical, its axes being 12 ins. and 8 ins. A pump takes the water away at the rate of 7 tons per minute. The ship's hold measures 30 ft. by 20 ft. Find how long it will take the water to reach a height of 6 ft. in the hold, the density of the water being 1024.

The volume of water entering through a hole at any depth below the waterline is given by the formula $V = A\sqrt{2gh}$ cub. ft. per sec., where V = volume, A = area of hole in sq. ft., $g = 32$, h = depth below waterline.

$$\text{Inflow of water in feet per min.} = \frac{\pi ab\sqrt{2gh} \times 60}{144}$$

$$\text{,, ,, ,, ,,} = \frac{22}{7} \times 6 \times 4 \times \sqrt{2 \times 32 \times 9} \times \frac{60}{144}$$

$$\text{,, ,, ,, ,,} = \frac{22 \times 6 \times 4 \times 3 \times 8 \times 60}{7 \times 144}$$

$$\text{,, ,, ,, ,,} = 754{\cdot}3 \text{ cub. ft. per minute.}$$

Pump discharges 7 tons per min. = 7 × 35 = 245 cub. ft. per minute.

Water gains on pump 754 − 245 = 509 cub. ft. per min.

$$\text{Time} = \frac{\text{Vol. of hold}}{\text{Inflow of water}} = \frac{30 \times 20 \times 6}{509} = \underline{\underline{7{\cdot}1}} \text{ minutes.}$$

NOTE.—Sea water of density 1024 = 35 cub. ft. per ton.

Exercises.

*8. A tank measuring 28 ft. × 16 ft. × 12 ft. is filled through a pipe 10 ins. in diameter in 20 minutes; find the rate of flow in feet per second.

$$\frac{\text{Capacity of tank}}{\text{Time to fill}} = \text{cub. ft. per sec.} \quad \text{Rate of flow} = \frac{\text{cub. ft. per sec.}}{\text{area of pipe}}$$

*9. A vessel is holed 16 ft. below the waterline; the hole is a rhombus in shape, the diagonals being 12 and 18 ins. The hold measures 60 ft. × 48 ft. and the centre of buoyancy is ·75. The pumps are capable of discharging $20\frac{4}{7}$ tons per minute. How long will it take the water to rise 8 ft. in the hold? Density of water 1024.

*10. An elliptical tank is 9 ft. high, the axes of the base are 12 ft. and 8 ft., the axes of the top are 9 ft. and 6 ft.; find how many gallons of fresh water it can hold.

KEPLER'S LAWS.

The only important application of the ellipse in nautical astronomy is the motion of the Sun in the ecliptic as enunciated in Kepler's Laws.

Assuming the Earth to travel in its orbit at a uniform speed, then it would be apparent to an observer on the Earth that he is passing the Sun faster in December when he is nearest to the Sun, than in June when he is furthest from it; that is, at an apparently decreasing speed during the first half of the year and at an increasing speed during the other half. Conversely, on reversing the relative conditions and assuming the theory that the Sun travels on the ecliptic with the Earth remaining fixed, then the Sun's motion in longitude appears to be faster in December than in June.

Kepler's First Law.—The Earth's orbit is an ellipse having the Sun in one focus.

Kepler's Second Law.—The radius vector joining the Earth and Sun traces out equal areas in equal times about the Sun.

Kepler's Third Law.—If T is the sidereal period of a planet and d its distance from the Sun, then $T^2 \div d^3$ is the same for all planets.

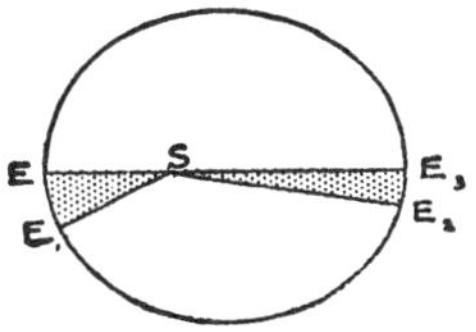

If S be the Sun and E, E_1, E_2, E_3 successive positions of the Earth in its orbit, then area ESE_1 = area E_2SE_3 when the time taken by the Earth to travel from E to E_1 is equal to the time taken to travel from E_2 to E_3. But in nautical astronomy we assume the Sun and Earth to change places so that S becomes the Earth and E, E_1, E_2, E_3 successive positions of the Sun, in above figure.

Assume the ellipse in the following figure to represent the ecliptic, S the Earth, then CS is the eccentricity $= e = \frac{CS}{CA}$, using the same lettering as adopted throughout this chapter.

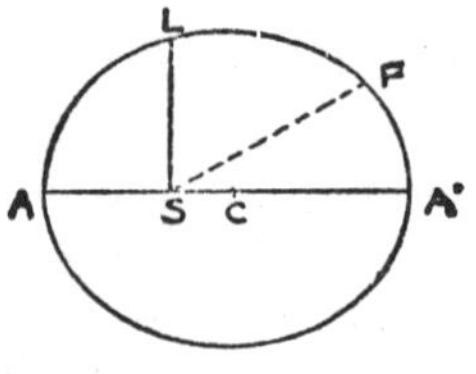

The eccentricity of the Earth's orbit may be verified approximately by a comparison of the measurable changes in the Sun's diameter. When the Sun is at A on 31st December, the angular diameter is greatest (32′ 36″ = 1956″) because the Sun is then at its nearest distance (AS) to the Earth. The Sun is then in perigee and the Earth in perihelion. Again, when the Sun is at A^1 on 30th June, the angular diameter is least (31′ 32″ = 1892″) for then the distance (A^1S) is greatest. The Sun is then in apogee and the Earth in aphelion.

By Def. $e = \frac{CS}{CA} = \frac{CS}{a}$ $\therefore$ $CS = a.e.$

$$\text{Dist. } SA = CA - CS = a - a.e = a(1 - e)$$

$$\text{Dist. } SA^1 = CA^1 + CS = a + a.e = a(1 + e)$$

See also Theorem 5.

The diameter of the sun varies inversely as the distance, thus

$$\frac{\text{diam. at } A}{\text{diam. at } A_1} = \frac{SA^1}{SA} = \frac{a(1+e)}{a(1-e)} = \frac{1+e}{1-e} = \frac{1956}{1892} = \frac{489}{473}$$

$$\therefore e = \frac{489 - 473}{489 + 473} = \frac{16}{962} = \frac{8}{481} = \frac{1}{60}$$

which is approximately the eccentricity of the Earth and, e being very small, its orbit is almost circular.

Verification of Kepler's First Law.

In any ellipse it was shown by Theorem 6 that $SP(1 + e \cos ASP) = l$ = constant for all positions of P on the curve.

Suppose P to be at L, then $\angle ASP = 90°$, $\cos 90° = O$, and SP then becomes SL because $e \cos 90° = O$.

$\therefore SP(1 + e \cos ASP) = SP \times 1 = SL = l.$ Theorem 6.

Suppose P to be at A, then $\angle ASP = 0°$, $\cos 0° = 1$, so that $e \cos ASP = e$ and $SP = SA = a(1 - e)$ as given previously.

$\therefore SP(1 + e \cos ASP) = a(1 - e)(1 - e) = a(1 - e^2) = l.$ Theorem 5.

The truth of the equation being constant for all positions of P on the curve when $\angle ASP$ is known, it follows from the results of numerous observations, that if P represents the Sun travelling on the ecliptic and S the Earth at a focus, that the ecliptic is an ellipse and Kepler's First Law is confirmed.

Verification of Kepler's Second Law.

The result of observation indicates that the increase in the Sun's longitude in a day at different times of the year is inversely proportional to the square of its distance. This fact offers a method of verifying Kepler's Second Law.

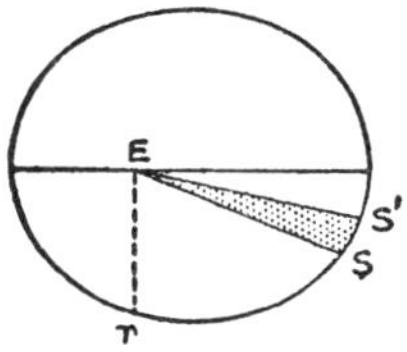

If S and S^1 be positions of the Sun S on two successive days, then the arc SS^1 is the change of longitude in a day and the sector SES^1 is the area swept out in one day by the radius vector.

$$\text{Area } SES^1 = \tfrac{1}{2}\, ES \,.\, ES^1 \sin SES.$$

The angle SES^1 being small, about 1° per day, the sector SES^1 is practically equal to triangle SES^1, so we may assume $ES = ES^1$ and $\sin SES^1 =$ the angle SES^1 in circular measure without appreciably affecting the result, so the equation becomes

$$\text{Area } SES^1 = \tfrac{1}{2}\, ES^2 \times SES^1.$$

But, as stated, the change of longitude $\angle SES^1$ varies inversely as ES^2, thus $ES^2 \times SES^1$ is constant. It follows, therefore, that area SES^1 is also constant, which confirms Kepler's Second Law, viz., the area described by the radius vector in a day is constant. Also that the area swept out in any number of days is proportional to the number of days and, the general statement, that equal areas are traced out in equal times.

EXAMINATION PAPERS

PAST EXAMINATION PAPERS FOR THE EXTRA MASTER EXAMINATION.

Paper 1.—Mathematics.

4 hours.

1. Find the centre of gravity of a triangle and its perpendicular distance from the longest side, the length of the sides being 8, 7 and 5 yards.
2. A hexagonal log, sides 12 inches, is cut through at an angle of 45°. Find the area of the section.
3. Find angle A having given
$$\cot^2 A + \operatorname{cosec} A = 3.$$
4. An iron sphere 12″ in diameter and 1″ thick, find its weight in lbs., the specific gravity being 7·3.
5. The Sun rose at 7h a.m. apparent time, and its meridian zenith distance was twice the latitude; required the latitude.
6. A tank can be filled by two pipes running together in $22\frac{1}{2}$ minutes; but when running separately the larger pipe would fill the tank in 24 minutes less than the smaller one; find the time taken by each.

Paper 2.—Mathematics.

1. Two ships on the same parallel of latitude are 20 miles apart. They are in touch by W/T and know they are on converging courses. A is steering S. 29° E. at 10 knots and B S. 40° W., at 9 knots. If the visibility is 5 miles, for how long will they be able to signal each other visually?
2. An elliptical tank 9 feet high and base 12 feet by 8 feet and top 9 feet by 6 feet; find the number of gallons of fresh water it can hold.
3. Solve the equation $3x + 7y = 27$, $5x + 2y = 16$.
4. Prove that if two circles intersect each other the angle subtended by any line which cuts the point of intersection and is terminated by the circumference is constant at the other point of intersection.
5. Prove that $\cos A = \dfrac{\cos a - \cos b \cos c}{\sin b \sin c}$
6. Find the length of a helix round a cylinder 12 inches in diameter and 10 feet long, there being 10 complete turns.

Paper 3.—Mathematics.

1. Find the value of ·378 to an infinite geometric progression.
2. Find the value of A which satisfies the equation
$$\operatorname{cosec}^2 A + \frac{1}{\sin A} + \frac{\sqrt{1 + \tan^2 A}}{\tan A} = 19$$

3. A ballast tank which is holed takes 2 hours to fill. A pump is put to work but fails to take effect. When the tank is three-quarters full a second pump is started, and the combined efforts of the pumps reduce the water to a half of the tank in one hour. If the second pump normally takes 70 minutes longer than the first to empty the tank, calculate how long each pump will take.

4. Find the weight of $13\frac{1}{2}$ fathoms of chain having a circular link. The outside diameter of the link is 4·5″, diameter of its cross section 1″, specific gravity of iron 7·8.

5. At a certain place, when the Sun's declination was 10° N., it rose an hour later than when it was 20° N., required the latitude.

6. Solve the equation

$$3x - 4y = 5$$
$$3x^2 - xy + 3y^2 = 27$$

Paper 4.—Mathematics.

1. (i) The sum of two numbers is 216. Find these numbers if $\frac{1}{4}$ of one number exceeds $\frac{1}{3}$ of the other by 5.

   (ii) Two ships A and B leave ports 4320 miles apart at the same instant. A does a certain number of miles daily and B this number + 60 miles daily. If the number of days before the ships meet is $\frac{1}{30}$ of the speed of B in miles per day, find the number of days until they meet.

2. What is an arithmetical mean? If the angles of a quadrilateral are in arithmetical progression and the larger and smaller angles are in the radio of 17 to 3, find the value of the angles.

3. What is the latitude of an observer who observes that the stars Alioth and Deneb are at their maximum azimuth, one east and one west of his meridian at the same instant?

4. In triangle ABC, if $C = 90°$ prove

   (i) $$\frac{\cos 2B - \cos 2A}{\sin 2A} = \tan A - \tan B$$

   (ii) Find value of θ if $4\sqrt{3} \cot \theta = 7 \operatorname{cosec} \theta - 4 \sin \theta$.

5. The surface area of a solid circular cylinder frustum is 100 sq. ft., its axis is 4 ft. and the angle of the cut section is 60° with the horizontal. Find diameter of the frustum.

Paper 5.—Mathematics.

1. A sphere 14″ in diameter is partially submerged in F.W. and displaces 1078 cubic inches. When a spherical sector (solid angle 39°) is removed and replaced by some other material the sphere just submerges. Find the specific gravity of the material.

2. If a fly-wheel, circumference $14\frac{2}{3}$ feet, takes one second more per revolution, a spot on the rim will do $2\frac{2}{3}$ miles less per hour. Find circumferential speed of the fly-wheel.

3. A vessel steers westwards on a G.C. from 25° N., 120° W., to 58° N. If the D. Long. = G.C. Dist., find Long. of arrival and the final course.

4. (i) Prove formula for arithmetical progression to the nth term.
   (ii) Two masts 120 ft. apart have a wire stay between them, secured at an equal height below each truck. From the mid-point the wire rises 1 inch in the 1st fathom, 2 more inches in the 2nd fathom, 3 more in the 3rd fathom, and so on. Find the height of wire on mast above the mid-point.

5. (i) Prove $\tan A \tan 2A = \sec 2A - 1$.
   (ii) $2 \cos \theta = 3 \sec \theta + 2$.

6. Celestial Lat. 4° 52′ 41·5″ S., celestial Long. 246° 42′, is the position of a star obliquity of ecliptic 23° 28′. Find star's R.A. and Decl.

Paper 6.—Mathematics.

1. In a right angled triangle ABC the hypotenuse is trisected at D and E. Prove that the sum of the squares on the sides of the triangle CDE is equal to two-thirds AB^2.

2. Draw figure showing a spherical triangle as it would appear on a sphere and also the polar triangle. Deduce their relationship and give an example of where used in spherical trigonometry.

3. Find the greatest angle in a plane triangle whose sides are $x^2 + x + 1$, $2x + 1$ and $x^2 - 1$.

4. (i) Prove $(\sin A - \sin B)^2 + (\cos A - \cos B)^2 = 2 \text{ vers } (A - B)$.
   (ii) Solve $\tan^2 A - \sin^2 A = 2{\cdot}25$.

5. Find the celestial Lat. and Long. of a star Decl. 26° 16·7′ S., R.A. 16h 25m 5s.

6. A ship sailing 270° sights a point bearing 000°. After steaming 2 hours a point B due west of A and distant 29·985 miles bore 311°, and after steaming another 2 hours B bore 030°. Find ship's speed.

Paper 7.—Navigation (Extra Master).

3 *hours.*

1. Calculate the maximum rate of change of azimuth of the star α Capella on June 21st in Lat. 55° 30′ N.

2. Explain the principle of the gyro compass and describe the errors.

3. What is meant by the equation of equal altitudes? Deduce the equation.

4. What are tides? Draw a figure showing the more important factors, neglecting the diurnal rotation of the earth.

5. In a certain place where the declination is twice the latitude, the Sun rose at 6h 15m a.m. What is the latitude?

6. Calculate the true angular distance between two bodies if the line of collimation be inclined 1° to the plane of the sextant and the observed angle were 150°.

Paper 8.—Navigation.

1. Prove the formula,
   Dip of sea horizon $= \cdot 98\sqrt{\text{height of eye in feet above sea level}}$.
2. Prove the conclusion that an error in hour angle due to an error in altitude is least when the body is on the prime vertical.
3. Explain the principle of the sounding tube used in the Kelvin sounding machine. What precautions must be taken with the boxwood scale?
4. Find the time of visible moonrise on Jan. 9, in Lat. 34° N., Long. 144° W.
5. Write briefly what the Harmonic Tidal Constants are and what the letters and suffixes denote.
6. A point in Lat. 50° 44′ N., Long. 0° 15′ E., bore 45° on port bow at 22h 20m and at 23h 27½m it was abeam. Course 195°, speed 8 knots, wind north, leeway 4°, tidal stream west (magnetic) 2½ knots, var. 14° W. Find the distance off on beam bearing and the ship's position.

Paper 9.—Navigation.

1. Investigate the formula governing the rate of change of altitude of a heavenly body and show that such change is greatest when the body is on the prime vertical.
2. Explain the principle of the sextant vernier, and show how an arc cut to 10′ can be read to 10″.
3. Describe in detail the 1st correction in Norie's Ex-meridian Table.
4. Investigate the conclusion that the error in longitude of a position line is equal to the error in the intercept caused by an incorrect G.M.T.
5. What is "precession" in a gyro compass? Give the rules governing precession of a freely mounted gyro.
6. A ship steered 050° for 1000 miles from Lat. 50° S., Long. 70° E.; required the position reached.

Paper 10.—Navigation.

1. Sept. 24th, D.R. 10° 35′ S., 67° 15′ E., course 325°, speed 12 knots. The Sun had equal altitudes when the chronometer times were 06h 56m 20s, and 07h 46m 24s, but slow 02m 20s on G.M.T. Find the longitude.
2. At noon when the Sun's declination was 15° N. a post cast a shadow x feet long, and when the Sun was 15° S. the shadow was $4x$ feet in length. Find the latitude.
3. When observing simultaneous altitudes for position lines in Lat. 35° S. (approx.) the index error + 3′ was applied the wrong way to the first altitude. If the Sun's true bearing at 1st obs. was 160° and at 2nd obs. 220°, required the errors of latitude and longitude.

4. At what height above the earth's surface will the time of sunrise be accelerated by 4 minutes on March 21st in Lat. 64° 40′ N.

5. Give a brief description of wireless navigational instruments and methods.

6. Prove that the distance of the sea horizon is $1{\cdot}06\ \sqrt{h}$. Under what conditions will it be $1{\cdot}15\ \sqrt{h}$?

Paper 11.—Navigation.

1. Dec. 23rd, in Lat. 25° 40′ S., Long. 179° 10′ W., sext. alt. Vega 18° 14′, G.M.T. 24d 02h 41m 47s. Ship steamed 245° for 80 miles. Sext. alt. Aldebaran 20° 50′, G.M.T. 24d 06h 41m 41s. Index error 2′ off arc, eye 37 feet. Find ship's position.
2. Dec. 24th, at 03h 30m G.M.T. in Lat. 22° 40′ S., Long. 45° 10′ W. Eye 33 feet, index error 1′ on arc, variation 18° W., deviation 17° E. Calculate the sextant altitude and bearing by compass of Rigel.
3. June 16th, D.R. position Lat. 40° 10′ N., Long. 30° 40′ W., Altair had equal altitudes when chronometer showed 01h 50m 41s star East, and 07h 09m 49s star West. Find the error of chronometer.
4. Two stars have equal declinations but opposite names and differ by 12 hours right ascension. Show that at any instant at a given place the rates of change of altitude of the two stars is numerically equal.
5. Calculate the Moon's augmentation having given the apparent altitude of the Moon's centre 51°, horizontal parallax 54′ 36″ and semi-diameter 14′ 54″.

Paper 12.—Navigation.

1. Find the G.M.T. of astronomical sunset and duration of twilight in Lat. 30° 40′ N., Long. 3° 30′ W., on the longest day in the year.
2. A man cast away in the North Atlantic has no *Nautical Almanac*. His height above sea level is 6 feet and he observes a star rising. Later the star has an altitude of 20° 30′ 30″ and bears E.S.E. magnetic. If the star's amplitude was E. ½ S. magnetic and the variation was 22½ W., find the latitude of the observer.
3. South latitude, cloudy weather in a partial clearance two stars were seen on the same bearing west of the meridian. Star *X*, R.A. 20h 00m 00s, Decl. 24° 45′ 30″, true alt. 13° 45′. Star *Y*, R.A. 20h 45m 00s, Decl. 36° 53′ 00″, hour angle 6h 00m west of the meridian. Required the latitude.
4. Find the terrestrial positions of Jupiter (J.) and Saturn (S.) on Jan. 10th, 00h 00m G.M.T. Then find the great circle distance from J. to S., also the initial course. Find also the Mercator course from J. to S. Explain why these two courses differ.
5. The Sun rose at 04h 06m and set at 20h 04m ship mean time. Find the equation of time.
6. In North latitude, the semi-diameter 16′ was not applied to the Sun's lower limb at 1st observation, the azimuth being 065°. It was applied however, to the altitude at the 2nd observation when the azimuth was 140°. Find the resulting error in latitude.

ANSWERS

ANSWERS

Plane; Parallel; Mercator and Traverse Sailings. Page 8.

1. 3725 miles to Lat. 18° 05′ N. or 1554 to Lat. 18° 05′ S.
2. Latitudes 34° 41·9′ and 34° 07·1 N. or S., distance 201 miles.
3. Latitude 61° 23′ N.
4. Latitude 51° 22′ S., distance 983·8 miles.
5. Latitude 20° 36·5′ S., distance 1427 miles.
6. Distance 421·7 miles.
7. Longitude 62° 30·3′ W.
8. Course 209° 45·4′ T., distance 400·8 miles.
9. Speed 10·29 knots.
10. Course 350° T., speed 15·25 knots.
11. Lat. 39° 46·8′ N., Long. 167° 37·2′ W.
12. Lat. 39° 17·2′ S., Long. 87° 57·5′ E.
13. Course 074° 56·3′ T., distance 2761 miles, time of arrival May 20d 04h 19m.
14. Course 161° T., distance 186 miles. Time of arrival G.M.T. 09h 30m on the 12th.

Answers to Four Point Bearings with Leeway and Current.

Bow and Beam Bearings with Current. Page 10

	Course and distance made good		Beam distance	Position	
1.	220° T.	10 miles	10 miles	50° 48·5′ N.	1° 10·5′ E.
2.	091·5	7·9 miles	10·2 miles	50 45·3 N.	1 04 E.
3.	104	6·5 miles	7·7 miles	50 36·3 N.	0 14·5 E.
4.	210	7·2 miles	8·8 miles	50 54·5 N.	1 21·5 E.
5.	169·5	6·7 miles	9·2 miles	51 05·3 N.	1 36·3 E.
6.	338	6·7 miles	8·5 miles	51 28 N.	1 37·2 E.
7.	Brg. 060	5 miles from Flamborough Head		54 09·5 N.	0 02·7 E.
8.	,, 218	9·9 miles from Needles Light		50 31·9 N.	1 44·9 W.
9.	46° 53·5′ N.	175° 56·6′ E.			
10.	349° 50′ T.	15·25 knots.			
11.	284° 48′ T.	24h 30m.	15·98 knots.		
12.	298 miles	12·67 knots.			
13.	51° 19′ N.	9° 53′ W.			

CHAPTER II.

Deviation Curves. Page 15.

1. Bearing 236° Magnetic.
   Deviations—21° W. 10° W. 2° W. 4° E. 20° E. 23° E. 5° E. 17° W.
   Courses—315° C. 149° C. 307° C. 135° C.
   Courses—007° M. 062° M. 225° M. 267° M.
   Bearings—137° M. 329° M.
2. Bearing 172° Magnetic.
   Deviations—12° E. 18° E. 10° E. 6° W. 10° W. 14° W. 9° W. 3° W.
   Courses—009½° C. 294½° C.
   Courses—072° M. 224° M. 326° M. 009° M.
   Bearings—036° M. 306° M.
3. Bearing—North Magnetic.
   Deviations—3° W. 17° E. 20° E. 15° E. 3° E. 9° W. 21° W. 22° W.
   Courses—097° M. 254° M.
4. Deviations—3·1° W. 9·6° W. 10·0° W. 5·5° W. 3·1° E. 9·6° E. 10·2° E. 2·9° E.

Day's Work. Page 37.

1. Lat. 48° 40½′ N., Long. 177° 07′ E. 302° T. 190 miles.
2. Current 320° T. 2·5 knots.
   Steer 243° T., 258° C.
   Speed 12·3 knots.
3. Lat. 48° 01·2′ N., Long. 5° 18′ W.
4. Lat. 48° 58′ N., Long. 5° 50′ W.
5. Lat. 44° 07′ N., Long. 176° 18′ E. D.R. position at noon. Course 224° T. 268·5 miles.
   Set and drift 292° T., 14 miles in 20 hrs. = 16·8 miles in 24 hrs.
6. Lat. 55° 15′ N., Long. 5° 01′ W. 094° 3 miles.
7. Set and drift 065° T., 30 miles Course 100° T. 95 miles.
8. Lat. 19° 42′ S., Long. 51° 08·3′ W.
9. Lat. 29° 53′ S., Long. 23° 21′ W. 2nd obs.
   Lat. 29° 53′ S., Long. 23° 17·6′ W., at noon.
10. D.R. at 8 p.m. Lat. 24° 15·3′ S., Long. 4° 23·3′ E.
    At noon. Lat. 22° 33′ S., Long. 4° 27·5′ E.
    Set and drift in 8 hrs. 198° T., 15·5 miles.
    Course 028·5° T. 167·5 miles. Speed 7·0 knots.

Radar Plotting. Page 77

1. Course 122° T., Speed 11 knots. CPA 1·7 miles. 025° Relative.
2. Course 080° T., Speed 6·4 knots. A/C 62° to Port.
3. Course 222° T., Speed 4·2 knots.
4. Set 252° T., Rate 3·8 knots.
5. Reduce speed to 5·5 knots. Co. 347° T., spd. 17·5 knots. Aspect Red 46° or 314° Relative.
6. "X" has stopped, CPA 1·2 miles. "Y" has altered course 5 points to Port, CPA 1·2 miles.
7. 1200, Target bears 030° Relative, Range 7 miles.
   1209, Target bears 030° Relative, Range 5·5 miles. Collision in 33 minutes.
   Co. 345° T., speed 6 knots.
   1220 Target bears 287° Relative, Range 3·3 miles. CPA 105° × 3·3 miles.
8. 1800, "X" bears 340° Relative at 8 miles;
   1800, "Y" bears 050° Relative at 7 miles.
   1806, "X" bears 317° Relative at 4 miles (decreasing). CPA 2·7 miles, 270° Relative in 4 mins.
   Co. 270° T., speed 26·5 knots.
   1806, "Y" bears 049° Relative (drawing forward), Range 5 miles decreasing. CPA 0·3 miles, 322° Relative in 15 mins. Target altered course 46° to starboard.
   Expected bearing at 1812, 345° Relative at 2·7 miles.
9. No action required; CPA 3 miles, 250° Relative. Target has altered course 61° to starboard.
10. No, Target has reduced speed from 19 knots to 8·5 knots. CPA 2 miles, 230° Relative. Co. 335° T.
11. "X" Course 018° T., speed 12 knots, crosses 2·9 miles ahead.
    "Y" Course 268° T., speed 10 knots, crosses 3·5 miles ahead.
12. CPA 153° T., 2·8 miles, about 3 mins. after last plot. Target has altered course 37° to starboard. Original course and speed 045° T., 10 knots. Final course 072° T., speed 10 knots.
13. "X" Course 172° T., speed 10 knots, has stopped in the interval between 3rd and 4th observations·
    "Y" Course 250° T., speed 20 knots, has made no alteration.
    "Z" Course 352° T., speed 15 knots, has altered course approx. 80° to starboard presumably to avoid collision with "Y".
    No action necessary by own ship.
14. Set 314° T., drift 1 mile. Alter course 16° to starboard.
15. (*a*) Alter course 34° to starboard.
    (*b*) Set 150° T., 4 knots.
    (*c*) Course 146° T., speed 10 knots, she will cross 1·6 miles ahead before I pass the Lt. Vessel.

CHAPTER III.

Great Circle Sailing. Page 87.

1. Initial course 249° 50′ T., final course 218° 26′ T., distance 3433·7 miles, latitude of vertex 52° 44′ N., longitude of vertex 20° 28′ E.
   Position of points on Great Circle:—Long. 15° 12′ W., Lat. 46° 53′ N. Long. 25° 12′ W., Lat. 42° 34′ N. Long. 35° 12′ W., Lat. 36° 33′ N. Long. 45° 12′ W., Lat. 28° 26′ N. Long. 55° 12′ W., Lat. 18° 1′ N.
2. Initial course 309° 12·7′ T., final course 272° 51·5′ T., distance 2269·4 miles, latitude of vertex 51° 43·5′ N., longitude of vertex 166° 21·5′ E.
   Position of points:—Long. 150° W., Lat. 42° 31·4′ N. Long. 160° W., Lat. 46° 32′ N. Long. 170° W., Lat. 49° 15·5′ N. Long. 180°, Lat. 50° 55·4′ N.
3. Initial course 060° 00′ T., final course 046° 28′ T., distance 3913·9 miles, latitude of vertex 44° 8′ S., longitude of vertex 20° 23′ W.
   Position of points:—Long. 35° 30′ E., Lat. 28° 33′ S. Long. 45° 30′ E., Lat. 21° 37′ S. Long. 55° 30′ E., Lat. 13° 19′ S. Long. 65° 30′ E., Lat. 3° 59′ S. Long. 75° 30′ E., Lat. 5° 41′ N.
4. Initial course 093° 16·5′ T., final course 056° 04·5′ T., distance 4975 miles, latitude of vertex 34° 30′ S., longitude of vertex 24° 16½′ E.
   Position of points:—Long. 34° 16½′ E., Lat. 34° 5½′ S. Long. 44° 16½′ E., Lat. 32° 51½′ S. Long. 54° 16½′ E., Lat. 30° 45½′ S. Long. 64° 16½′ E., Lat. 27° 46′ S. Long. 74° 16½′ E., Lat. 23° 50′ S. Long. 84° 16½′ E., Lat. 18° 58′ S. Long. 94° 16½′ E., Lat. 13° 13½′ S. Long. 104° 16½′ E., Lat. 6° 48½′ S.
5. 049° 48′ T., 091° 23′ T., 2809¾ miles. Vertex 49° 46′ N., 7° 00½′ W. Points 55° W., 38° 20½′ N., 45° W., 42° 58′ N.; 35° W., 46° 13½′ N.; 25° W., 48° 21′ N.; 15° W., 49° 29½′ N.
6. 237° 49′ T., 217° 05′ T., 2317 miles. Vertex 55° 49½′ N., 83° 11′ W. Points 135° W., 42° 19′ N.; 145° W., 34° 49½′ N.; 155° W., 24° 41′ N.
7. 121° 25·3′ T., 108° 03·3′ T., 5939 miles. Vertex 36° 41½′ S., 40° 45½′ W. Points 140° W., 6° 49½′ N.; 120° W., 7° 55½′ S.; 100° W., 20° 51½′ S.; 80° W., 29° 59½′ S.
8. 071° 19·5′ T., 116° 58′ T., 3154·5 miles. 43° 51½′ N., 46° 26′ W. Points 60° W., 43° 3′ N.; 50° W., 43° 48′ N.; 40° W., 43° 40½′ N.; 30° W., 42° 40′ N.; 20° W., 40° 42½′ N.; 10° W., 37° 42½′ N.
9. 068° 18·3′ T., 111° 41·7′ T., 3141½ miles. Vertex 42° 02′ N., 42° 30′ W. Points 67° 30′ W., 39° 15′ N.; 55° 00′ W., 41° 21′ N.; 30° 00′ W., 41° 21′ N.; 17° 30′ W., 39° 15′ N.
10. 209° 45′ T., 219° 56′ T., 3619 miles. Vertex 60° 48½′ S., 6° 05′ W. Points 85° 00′ E., 1° 56′ N.; 80° E., 6° 59′ S.; 75° E., 15° 31′ S.; 70° E., 23° 18′ S.; 65° E., 30° 08′ S.; 60° E., 35° 58′ S.

Composite Sailing. Page 92.

1. Initial course 241° 12·2′ T., final course 304° 29·5′ T., distance on first great circle 1358·25 miles' distance on second great circle 1725·42 miles, Longitude where track meets the maximum Latitude 111° 51′ 10″ W., Longitude where track leaves the maximum Latitude 145° 29′ 58″ W., distance along parallel of maximum Latitude 1157·94 miles, total distance 4241·61 miles.

First point in Long.	80°	W. is 31°	51′	10″	from vertex, and is in Lat.	50°	29′	58″	S.
Second point in Long.	90	W. is 21	51	10	from vertex, and is in Lat.	52	58	7	S.
Third point in Long.	100	W. is 11	51	10	from vertex, and is in Lat.	54	25	4	S.
Fourth point in Long.	110	W. is 1	51	10	from vertex, and is in Lat.	54	59	9	S.

The points in Longitudes 120°, 130° and 140° W. are on the parallel of maximum latitude.

Eighth point in Long.	150°	W. is 4°	30′	2″	from vertex, and is in Lat.	54°	55′	0″	S.
Ninth point in Long.	160	W. is 14	30	2	from vertex, and is in Lat.	54	7	26	S.
Tenth point in Long.	170	W. is 24	30	2	from vertex, and is in Lat.	52	25	19	S.
Eleventh point in Long.	180	W. is 34	30	2	from vertex, and is in Lat.	49	38	51	S.

2. Initial course 060° 30·1′ T., final course 116° 30·3′ T., distance on first great circle 2067·22 miles, distance on second great circle 1794·8 miles, longitude in which the track meets maximum Latitude 174° 51′ 51″ W., longitude in which the track leaves maximum Lat. 161° 48′ 00″ W., distance along parallel of maximum Latitude 554·27 miles, total distance 4416·29 miles.

First point in Long.	150°	E. is 35°	8′	9″	from vertex, and is in Lat.	39°	16′	34″	N.
Second point in Long.	160	E. is 25	8	9	from vertex, and is in Lat.	42	9	17	N.
Third point in Long.	170	E. is 15	8	8	from vertex, and is in Lat.	43	59	20	N.
Fourth point in Long.	180	E. is 5	8	9	from vertex, and is in Lat.	44	53	5	N.

The fifth point in Longitude 170° W. is on the parallel of maximum latitude.

Sixth point in Long.	160°	W. is 1°	48′	0″	from vertex, and is in Lat.	44°	59′	9″	N.
Seventh point in Long.	150	W. is 11	48	0	from vertex, and is in Lat.	44	23	17	N.
Eighth point in Long.	140	W. is 21	48	0	from vertex, and is in Lat.	42	52	4	N.
Ninth point in Long.	130	W. is 31	48	0	from vertex, and is in Lat.	40	21	39	N.

3. 133° 04·5′ T. East, 2329 mls.+1921 mls.=4250 mls. Vertex 56° 10′ S., 124° 41′ W. Pts. 170° W., 46° 22′ S.; 160° W., 50° 36′ S.; 150° W., 53° 26½′ S.; 140° W., 55° 12′ S.; 130° W., 56° 3′ S.; 120° W., 56° 10′ S.

4. 113° 29·5′ T., 050° 03′ T., 1284 mls.+2679½ mls.+1149 mls.=5112½ mls. Vertex 158° 25′ W. Vertex 128° 37′ W. Pts. 180°, 47° 56½′ S.; 170° W., 49° 25′ S.; 160° W., 49° 59½′ S.; 4th, 5th, 6th pts. in 50° S.; 120° W.; 49° 41′ S.; 110° W., 48° 28½′ S.; 100° W., 46° 17½′ S.; 90° W., 42° 57½′ S.; 80° W., 38° 14′ S.

Gnomonic Chart. Page 92.

1. Belle Isle to The Naze (Composite Track).

Long.	Lat.
Long. 50° W.	Lat. 54° 20′ N.
,, 45 W.	,, 56 00 N.
,, 40 W.	,, 57 25 N.
,, 35 W.	,, 58 30 N.
,, 30 W.	,, 59 20 N.
,, 5 W.	,, 59 42 N.
,, 0 W.	,, 59 10 N.
,, 5 E.	,, 58 20 N.

2. Sable Island to Lisbon (Great Circle).

Long. 50° W.	Course 087° T.
,, 40 W.	,, 093 T.
,, 30 W.	,, 100 T.
,, 20 W.	,, 107 T.
,, 10 W.	,, 115 T.

CHAPTER III.

Great Circle Sailing. Page 97.

No.	Initial Course	Final Course	Dist. miles	Vertex		Points
	° ′	° ′		° ′	° ′	° ′ ° ′ ° ′
1	288 45·3 T.	307 50 T.	3359	38 08 S.	49 33 E.	31 11½ S. 27 00 S. 21 42 S. 15 20½ S. 8 06½ S.
2	064 58	092 02·7	1411	51 25 N.	12 15 W.	47 57½ N. 49 08 N. 50 2½ N. 50 43 N. 51 09½ N. 51 23 N. 51 23¾ N.
3	293 36	247 52	3323	42 14 N.	41 54½ W.	
4	079 38	055 35	2962	37 30½ S.	73 47 W.	
5	083 21·7	123 22	2466	48 46 N.	170 16 E.	48 46 N. 48 21½ N. 47 03 N. 44 45 N. 41 17½ N.
6	101 12	059 41	4224	35 47 S.	45 29 E.	
7	114 15·5	040 33·2	5764	50 31 S.	157 5½ W.	48 11½ S. 50 29 S. 49 14½ S. 44 4½ S. 33 24 S. 15 10 S.
8	110 48·5	040 14·5	3516	58 56½ S.	43 00 W.	
9	113 52	076 08	3654	34 35 S.	90 33 E.	29 17 S. 33 35 S. 34 30 S. 32 6½ S.
10	292 56	254 00	1825	53 33½ N.	35 29 W.	51 10 N. 53 00 N. 53 33½ N. 52 54 N.
11	235 34	272 50	4930	35 1 S.	24 56 E.	21 52 S.
12	295 07·5	253 23·5	2948	42 50½ N.	45 8¾ W.	37 10 N. 40 01 N. 41 50 N. 42 43½ N. 42 44½ N. 41 52 N.
13	113 30	050 03	1284 2679½ 1149	158 25 W.	128 37 W.	47 56½ S. 49 25 S. 49 59½ S. 49 41 S. 48 28½ S. 46 17 S. 42 57½ S. 38 14 S.
14	133 04·5	090 00	2329 1921	56 10 S.	124 41 W.	46 22 S. 50 36 S. 53 26 S. 55 12 S. 56 03 S. 56 10 S.
15	139 50·5	066 38·5	2590½ 866½ 624	118 03 W.	97 15 W.	
16	142 23	067 14·5	2911½ 840½ 519½	113 50½ W. 96 32 W.		

CHAPTER VI.

Stars on the Meridian. Page 147.

		Upper				Lower				Altitude
		d	h	m	s	d	h	m	s	
1.	L.M.T.	8	05	17	03	8	17	15	05	24° 52′
2.	,,	22	04	03	51	22	16	01	53	Not visible
3.	,,	20	18	40	54	20	06	42	52	8 45
4.	,,	22	22	41	44	22	10	43	42	17 37
5.	,,	17	22	50	17	17	10	52	15	Not visible
6.	,,	19	04	41	32	19	16	39	34	Not visible

Stars Near the Meridian. Page 151.

1.	S.H.A. Observer's Meridian 193° 13′				
	Regulus	has crossed meridian	North	L.H.A.	15° 15′
	Denebola	will cross meridian	North	E'ly H.A.	9° 57′
	Acrux	,, ,, ,,	South	,,	19° 18′
	Gacrux	,, ,, ,,	South	,,	20° 26′
2.	S.H.A. Meridian 115° 26′				
	Rigil Kent	has crossed	South	L.H.A.	25° 22·4′
	Antares	will cross meridian	South	E'ly H.A.	2° 09′
	Atria	,, ,, ,,	South	,,	6° 30′
	Shaula	,, ,, ,,	South	,,	18° 08′
3.	S.H.A. Meridian 67° 51′				
	Altair	will cross meridian	North	E'ly H.A.	5° 03′
	Peacock	,, ,, ,,	South	,,	13° 28′
	Canopus	crossed lower meridian	South	,,	16° 57′
4.	S.H.A. Meridian 17° 20′				
	Fomalhaut	will cross meridian	North	E'ly H.A.	1° 10′
	Ankaa	,, ,, ,,	North	,,	23° 23′
	Achernar	,, ,, ,,	South	,,	41° 22′
	Avior	crossed lower meridian	South	E'ly H.A.	37° 15′
	Suhail	,, ,, ,,	South	,,	26° 03′
	Miaplacidus	,, ,, ,,	South	,,	24° 29′
5.	S.H.A. Meridian 350° 43′				
	Alpheratz	has crossed meridian	North	L.H.A.	7° 43′
	Diphda	will cross meridian	North	E'ly H.A.	1° 06′
	Achernar	,, ,, ,,	South	,,	14° 46′
	Acrux	crossed lower meridian	South	,,	3° 13′
	Gacrux	,, ,, ,,	South	,,	2° 04′
6.	S.H.A. Meridian 276° 01′				
	Betelguese	will cross meridian	North	E'ly H.A.	4° 15′
	Canopus	,, ,, ,,	South	,,	11° 46′
	Sirius	,, ,, ,,	North	,,	16° 5′

Meridian Altitudes. Page 161.

Moon.

	G.M.T.	Decl.	Altitude	Latitude
1.	5d 20h 56m	17° 55·4′ N.	41° 48·3′	30° 16·3′ S.
2.	21 23 36·5	13 47·8 N.	55 50·9	47 56·9 N.
3.	23 11 32·6	1 12·9 N.	59 56·5	28 50·6 S.
4.	19 21 06	18 41·8 S.	29 26·1	41 52·1 N.
5.	24 21 22	5 15·4 S.	34 33	50 11·6 N.

PLANETS. Page 163.

	G.M.T.	Decl.	Altitude	Latitude
1.	21d 05h 11m	21° 53·7′ S.	36° 43·9′	31° 22·4′ S.
2.	10 17 03·3	10° 11·0′ S.	40° 30·6′	39° 18·4′ N.
3.	20 01 59·7	1° 33·2′ N.	62· 22′	—
4.	22 22 03·8	18° 28·1′ N.	62° 10·4′	—
5.	8 05 12	13° 37·6′ S.	58° 08·3′	18° 14·1′ N.

CHAPTER VII

Latitude by Pole Star. Page 165.

1.	Latitude 45° 50·9′ N.	Azimuth 359·1° T.	4.	Latitude60° 29·6′ N.	Azimuth 001·5° T.
2.	,, 35° 24·3′ N.	,, 001° T.	5.	,, 43° 00·0′ N.	,, 359·3° T.
3.	,, 50° 40·0′ N.	,, 001·2° T.	6.	,, 35° 49·2′ N.	,, 358·8° T.

SUN. Position Lines and Ex-Meridian Sights. Page 182.

1. (i) At 0935 D.R. Lat. 34° 48′ S., Long. 163° 44′ E.
(ii) At 1135 D.R. Lat. 34° 50′ S., Long. 163° 51·2′ E.
(iii) L.H.A. 352° 37·4′. Decl. 23° 19·4′ N. Calc. *Z*. Dist. 58° 34·6′. True *Z*. Dist. 58° 31·5′. Intercept 3·1′ towards True Az. 008° T. *C* to *F* 306° T., 6·6 miles, Obs. Posn. 34° 46·1′ S., 163° 18·1′ E.
2. (i) At 1000, D.R. Lat. 49° 00′ N., Long. 128° 04·2′ W.
(ii) At 1220, D.R. Lat. 48° 26·5′ N., Long. 128° 37·4′ W.
(iii) L.H.A. 4° 50·6′. Decl. 23° 23·2′ N. Calc. *Z*. Dist. 25° 20·9′. True *Z*. Dist. 25° 26·3′, Intercept 5·4′ away. Az. 190·4° T., Obs. Posn. 48° 31·3′ N., 128° 32·0′ W.
3. (i) At 1220 L.H.A. 5° 42′. Decl. 23° 26·5′ N. Calc. *Z*. Dist. 23° 02·5′. True *Z*. Dist. 23° 01·3′. Intercept 1·2′ towards. Az. 193·5° T. Intercept Terminal Point 45° 54·8′ N., 35° 11·6′ E.
(ii) At 1550 D.R. Lat. 46° 23·3′ N., Long. 36° 01·7′ E.
(iii) *C* to *F* 104° T., 5·3 miles. Obs. Posn. 46° 22′ N., 39° 09·2′ E.
4. (i) At 1150 L.H.A. 357° 30·8′, Decl. 22° 22·6′ S., Calc. *Z*. Dist. 33° 58·3′. True *Z*. Dist. 33° 47·3′. Intercept 11′ towards. Azimuth 004° T. Int. Terminal Pt. Lat. 56° 07′ S., Long. 67° 20·5′ W.
(ii) D.R. Position Lat. 56° 44·7′ S., Long. 66° 32·4′ W.
(iii) *C* to *F* 096° T., 8·6 miles. Obs. Posn. Lat. 56° 45·3′ S., Long. 66° 16·6′ W.

Stars' Planets and Moon. Page 184.

1. Ex-meridian L.H.A. 3° 56·7′, M.Z.D. 48° 26·7′, Lat. 14° 25·1′ S., Azimuth 182·3° T., Posn. Line 092°-272° T.
2. Ex-meridian L.H.A. 003° 04·7′, M.Z.D. 39° 53′, Lat. 48° 36·6′ N., Azimuth 184·6° T., Posn. Line 094·6°-274·6° T. Intercept method 4·5′ away. (i) 48° 38·5′ N., 155° 30·6′ E. (ii) 49° 06·4′ N., 156° 00·6′ E.
3. Intercept 3·7 towards Azimuth 188° T. L.H.A. 6° 28·8′. Posn. Line 098°-278° T. (i) 36° 24·4′ N., 127° 41·5′ W. (ii) 35° 49′ N., 128° 29·5′ W.
4. Ex-meridian L.H.A. 354° 38·4′; M.Z.D. 39° 43·5′; Lat. 38° 19·7′ S., Azimuth 008° T., P.L. 098°-278° T.
5. Ex-meridian L.H.A. 3° 32·9′; M.Z.D. 42° 48′; Lat. 30° 17·7′ N. Azimuth 185° T. P.L. 095°-275° T.
6. L.H.A. 352° 54·2′; M.Z.D. 46° 26·2′; Lat. 36° 31·1′ S. Azimuth 010° T. P.L. 100°-280° T. or Intercept method 3·2′ away.
7. Intercept method L.H.A. Mars 357° 01·4′; Obs. Z.D. 40° 39·7′, Intercept 3·2′ away, Azimuth 004·5° T.; P.L. 094·5-274·5° T. L.H.A. Sun 310° 38·9′; Obs. Z.D. 78° 44·6′; Intercept 3·2′ away, Azimuth 04·5° T., P.L. 135°-315° T. 2nd D.R. position 40° 38·3′ S., 158° 29·2′ E. Fix 40° 38′ S., 158° 22·7′ E.
8. Intercept method; L.H.A. 2° 25·1′, Intercept 3·1′ away, Azimuth 185° T., P.L. 095°-275° T. (i) Lat. 12° 21′ N., Long. 62° 12·3′ E. (ii) Lat. 12° 20·7′ N., Long. 61° 22·3′ E.
9. Intercept method L.H.A. 3° 32·9′, T.Z.D. 42° 56·2′, Intercept 2·4′ towards, Az. 185° T. P.L. 095°-275° T. Int. Pt. 30° 17·7′ N., 100° 49·7′ E. 2nd D.R. Lat. 29° 47′ N., Long. 101° 32·2′ E. L.H.A. 63° 45·1′. T.Z.D. 74° 15·4′. Int. 3·5′ away. Azimuth 246° T. P.L. 156°-336° T. Fix 29° 46·5′ N., 101° 36·8′ E.
10. Intercept method L.H.A. 3° 41·2′. Int. 4·3′ towards. Az. 182° T. P.L. 092°-272° T. Int. Pt. 0° 28·2′ S., 26° 42·2′ W. (i) L.H.A. 309° 35·9′. Int. 8·3′ towards. Az. 100° T. P.L. 010°-190° T. Int. Pt. 0° 25·4′ S., 26° 34′ W. Fix 0° 28·6′ S., 26° 34·5′ W.
11. Intercept method Mars L.H.A. 5° 22·6′, T.Z.D. 38° 33·1′, Intercept 4·5′ away. Azimuth 188° T. P.L. 098°-278° T. Pollux L.H.A. 307° 33′, T.Z.D. 46° 12·5′, Intercept 7·0′ towards. Azimuth 105° T. P.L. 015°-195° T. Fix 56° 18·3′ N., 143° 40′ E.
12. Intercept method. Jupiter L.H.A. 59° 18·6′, T.Z.D. 73° 23·7′, Intercept 2·9′ towards, Azimuth 241° T. P.L. 151°-331° T., Moon L.H.A. 340° 35·2′, T.Z.D. 56° 17′. Intercept 3·3′ towards, Azimuth 157° T., P.L. 067°-247° T. Fix Lat. 36° 16′ N., Long. 18° 45·7′ W.

Simultaneous and Double Altitudes. Page 186.

		LHA	Intercept	Bearing	2nd DR and Fix
1.	(a)	319° 04·3′	1·6 Towards	052° T	32° 14·4′ S 80° 42·6′ E
	(b)	311° 32·4′	4·5 ,,	135°	
2.	(a)	25° 40·3′	2·0 ,,	210·5°	11° 14·1′ N 32° 20·4′ W
	(b)	59° 30·8′	1·0 ,,	315·5°	
3.	(a)	001° 16′	4·9 Away	184°	38° 08·7′ N 57° 14·2′ W
	(b)	322° 39·7′	2·7 Towards	095°	
4.	(a)	21° 13·1′	3·0 Away	222°	42° 40·5′ N 171° 17′ E
	(b)	325° 35·8′	2·4 Towards	134°	
5.	(a)	346° 12·7′	2·0 Away	156°	13° 58·6′ N 78° 15·7′ W
	(b)	28° 07·1′	3·7 Towards	283°	
6.	(a)	0° 00·1′	3·5 ,,	000°	13° 59·5′ S 88° 07·3′ W
	(b)	298° 14·3′	3·5 Away	116°	
7.	(a)	293° 33·7′	2·9 Towards	122°	0° 23·2′ S 26° 38·2′ W
	(b)	18° 21·3′	3·0 Away	218°	
8.	(a)	64° 59·5′	2·6 Towards	228°	35° 29·5′ S 80° 11′ W
	(b)	36° 50·9′	3·0 ,,	319°	
9.	(a)	313° 55·1′	4·7 ,,	048°	2nd DR 25° 17·2′ N 156° 21′ E
	(b)	0° 51·5′	1·4 ,,	000°	Fix 25° 18·6′ N 156° 19·6′ E
10.	(a)	0° 00′	4·5 Away	180°	12° 08·5′ N 57° 24′ W
	(b)	29° 40′	0·4 Towards	220°	
11.	(a)	350° 46·3′	3·2 Away	010°	43° 10·8′ S 83° 21·7′ E
	(b)	50° 33·1′	2·4 Towards	271°	
12.	(a)	3° 43′	6·0 Away	344°	2nd DR 24° 11·6′ S 156° 53·5′ E
	(b)	59° 30·3′	5·6 Towards	272°	Fix 24° 13·4′ S 156° 43·3′ E
13.	(a)	308° 53·7′	6·8 ,,	123°	2nd DR 45° 32·3′ N 21° 10′ W
	(b)	16° 07·8′	4·7 ,,	200°	Fix 45° 28·3′ N 21° 13·7′ W
14.	(a)	345° 39·9′	3·1 Away	161°	2nd DR 35° 11′ N 151° 38·2′ W
	(b)	63° 46·2′	3·5 Towards	265°	Fix 35° 09·8′ N 151° 42·3′ W
15.	(a)	5° 58·8′	2·2 Away	344°	33° 43′ S 104° 08·6′ E
	(b)	56° 05·4′	3·7 Towards	250°	

EXAMINATION PAPERS CLASS 4—CLASS 1. Page 201.

Paper 1.

1. (a) Intercept 3·9′ towards. Azimuth 185° T. Lat. 38° 06′ N.
   (b) 2·3′ towards. 088° T. Long. 32° 17′ W.
2. Initial Co. 282° 35½′ T., Final Co. 233° 25¼′ T. Distance 2674 miles. Vertex Lat. 52° 30½′ N., Long. 25° 26′ W.
3. Lat. 46° 54·4′ N. P.L. 091·4°-271·4° T.
4. Azimuth 094·5° T. Error 3·5° E. Dev. 23·5° E.
5. L.M.T. 19h 37m. North of observer.

Paper 2.

1. L.H.A. 31° 26·6′; T.Z.D. 75° 22′; Intercept 3·8′ towards; Azimuth 210° T. 2nd D.R. Lat. 47° 00′ N. Long. 175° 29′ E. Fix Lat. 46° 50·5′ N., Long. 175° 51′ E.
2. Lat. 50° 38′ N. P.L. 091°-271° T.
3. L.M.T. 26d 23h 21m; Lat. 45° 23·3′ S. P.L. 090-270° T.
4. Points C Long. 159° 20′ E. Lat. 41° 56′ S.
   ,, D ,, 169 20 E. ,, 49 07 S.
   ,, E ,, 179 20 E. ,, 54 00 S.
   ,, F ,, 170 40 W. ,, 57 16 S.
   ,, G ,, 160 40 W. ,, 59 22 S.
   ,, H ,, 150 40 W. ,, 60 32 S.

There are other six points on other side of vertex, their polar angles are the same as the above, they must also be in, respectively, the same latitudes.

Paper 3.

1. Intercept 4·8′ away; Sun Az. 246° T. Lat. 52° 35′ N.
   ,, 3·4′ towards; Star Az. 157° T. Long. 5° 31′ W.
2. Lat. 50° 35·3′ N., Long. 27° 43·3′ W.
3. Bearing 233·7° M. Dev. 21·3° W., 10·3° W., 2·3° W., 3·7° E.
   ,, 19·7° E., 22·7° E., 4·7° E., 17·3° W.
   Compass courses deviations (i) 15·5° W. (ii) 6·5° W. (iii) 19° W. (iv) 6° W.
4. Lat. 37° 01′ N.

Paper 4.

1. (*a*) L.H.A. 287° 28·6′; Intercept 3·5′ towards; Az. 101° T.; Int. Terminal Pt. 35° 19·3′ N., 36° 18′ W.; P.L. 011°-191° T.
   (*b*) L.H.A. 355° 51·7′; Int. 2·0′ towards; Az. 176° T., P.L. 086°-266° T.
   Fix. Lat. 35° 18·5′ N., Long. 36° 18′ W.
2. Course 161° T., Dist. 301 miles. Lat. 46° 23′ S., Long. 101° 09·3′ E.
3. Lat. 34° 47·1′ N., P.L. 090°-270° T.
4. G.C. Dist. 2317 miles. Mercator (spheroid) 2335 miles, difference 18; Mercator (sphere) 2330 miles, difference 13 miles.

Paper 5.

1. Lat. 49° 34′ N., Long. 179° 26·5′ W.
2. Points A 170° E. Lat. 48° 46·5′ N.
   ,, B 180 E. ,, 48 21·5 N.
   ,, C 170 W. ,, 47 02·8 N.
   ,, D 160 W. ,, 44 44 N.
   ,, E 150 W. ,, 41 15·8 N.
3. Long. 65° 32·2′ E. P.L. 150°-330° T.
4. Lat. 35° 47·3′ N. P.L. 089°-269° T.

Paper 6.

1. Initial course 055° 34′ T. Final course 059° 57′ T. Dist. 6417 miles. Vertex Lat. 46° 46·5′ N., Long. 79° 35·5′ W.
2. Course 232° 38·8′ T. Distance 361·1 miles (spheroid).
   ,, 232° 27·5′ T. ,, 359·4 ,, (sphere).
3. Lat. 50° 12·7′ N., Long. 6° 01′ W.
4. Course 143° T. Dist. 236 miles; Set. 201° T., Drift 15·6 miles; Noon position Lat. 14° 38′ S., Long. 51° 06′ E.

Paper 7.

1. Mer. Pass. 8d 13h 54m L.A.T., Sextant Alt. 30° 16′; Lat. 46° 18′ N.
2. Lat. 31° 38·2′ S., P.L. 090°-270° T.
3. Lat. 21° 57·5′ N., P.L. 089°-269° T.
4. Lat. 51° 50′ S., Long. 00° 30′ E.
5. Jupiter L.H.A. 302° 01·6′; Int. 4·1′ towards; Az. 103° T. Denebola L.H.A. 355° 56·9′; Int. 2·8′ away; Az. 009° T. Lat. 12° 23·3′ S., Long. 44° 56·6′ W.

Paper 8.

1. 25d 14h 37m 12s Lat. 16° 52·4′ N.
2. G.M.T. 13d 04h 06·3m. Altitude 38° 11·2′.
3. Lat. 48° 13′ S., Long. 44° 29′ E.
4. Lat. 50° 20·5′ N., Long. 2° 27′ W.
5. Points C Long. 10° W. Lat. 37° 43·6′ S.
   ,, D ,, 0 E. ,, 40 20·5 S.
   ,, E ,, 10 E. ,, 41 55·3 S.
   ,, F ,, 20 E. ,, 42 35·5 S.
   ,, G ,, 30 E. ,, 42 23 S.
   ,, H ,, 40 E. ,, 41 17·5 S.
   ,, J ,, 50 E. ,, 39 15·5 S.
   ,, K ,, 60 E. ,, 36 11 S.
   ,, L ,, 70 E. ,, 31 56·3 S.
   ,, M ,, 80 E. ,, 26 24 S.
   ,, N ,, 90 E. ,, 19 30·5 S.
   ,, O ,, 100 E. ,, 11 23·3 S.

Paper 9. Chartwork.

1. Courses—350° C., 34 miles; 026° C., 90 miles; 080° C., 17 miles.
2. Lat. 50° 31′ N., Long. 5° 34′ W. Set 005° T., Drift 7·5 miles. Lat. 50° 38′ N., Long. 5° 37′ W.
3. Lat. 51° 23′ N., Long. 5° 06′ W.
4. Course 025° C. Effective speed 9·4 knots. Time 11½ hours.

Paper 10.

1. (*a*) Intercept 6·1′ towards; Azimuth 121° T.; 2nd D.R. position 46° 45·3′ N., 25° 00′ W.
   (*b*) Intercept 5·0′ towards; Azimuth 195° T.; Fix Lat. 46° 39′ N., Long. 24° 55′ W.
2. Course 122° 10·6′ T. Distance 1487·2 miles (spheroid). or Course 122° 21′ T. Distance 1480 miles (sphere).
3. Mer. Pass. 11d 06h 30m L.M.T.; Sext. alt. 37° 45′; Lat. 42° 08′ N.
4. Co. made good 144° T., dist. 15·4 miles. Beam dist. 18·6 miles. Lat. 47° 05′ N., Long. 3° 31′ W.
5. Lat. 25° 23·5′ N. P/L 080-260° T.

Paper 11.

1. G.M.T. 19d 01h 24m. Altitude 34° 09·3′.
2. B 1·35s gaining. C 3·18s gaining.
3. Current to 2200 hours 073° T., 17 miles; Course 203·5° T., distance 342 miles. Lat. 42° 13·6′ N., Long. 177° 29′ E.
4. (*a*) Intercept 3·6′ away; Azimuth 222° T.
   (*b*) Intercept 5·0′ towards; Azimuth 314° T.
   Fix: Lat. 11° 31′ N., Long. 27° 43·2′ W.
   Longitude method: Lat. 11° 30·8′ N. Long. 37° 43·3′ W.
5. Course 043° C.

Paper 12.

1. Lat. 45° 11·7′ N., Long. 8° 19·5′ W.
2. Lat. 42° 01·8′ N.
3. G.C. Initial Co. 117° 01¾′ T., Final Co. 126° 59¾′ T., Dist. 5883·8 miles. Vertex 44° 20′ N., 159° 26¼′ E. Lats. of points 31° 49½′ N., 25° 39′ N., 18° 00½′ N., 0° 33′ S., 10° 09¼′ S., 18° 56¼′ S.
4. Az. 076·5° T., Dev. 5·5′ W.
5. (*a*) Intercept 1·9′ towards Az. 063° T. } Lat. 42° 18·2′ S.
   (*b*) ,, 2·0′ away Az. 343° T. } Long. 76° 46·8′ E.
   or
   (*a*) Long. 76° 46′ E. P/L 153° 333° T. } Lat. 42° 18′ S.
   (*b*) Long. 76° 50·8′ E. P/L 075° 253° T. } Long. 76° 46·6′ E.

Paper 13.

1. Lat. 45° 57·3′ N. P/L 090·5°-270·5° T.
2. 0800 D.R. Lat. 44° 35·5′ N., Long. 177° 17·8′ E. Current 293° T., 12 miles in 20 hours. Noon Lat. 44° 07·4′ N., Long. 176° 17′ E. Course 224·5° T., distance 268·5 miles.
3. G.M.T. 15d 20h 07m 28s.
4. Lat. 35° 17·3′ S.
5. (*a*) Int. 2·9′ towards Az. 050° T. }
   (*b*) Int. 4·1′ towards Az. 309° T. } Lat. 26° 20·5′ N., Long. 37° 43′ W.
   or
   (*a*) Long. 37° 37·6′ W. P/L 140-320° T. }
   (*b*) Long. 37° 47·8′ W. P/L 039-219° T. } Lat. 26° 20·6′ N., Long. 37° 42·8′ W.

Paper 14.

1. Brg. 256° M., Dev. 14° W., 0°, 12° E., 15° E., 11° E., 0°, 10° W., 14° W. Compass courses dev. 0°, 14° E., 6° E., 14° W. Mag. courses dev. 8° W., 9° E., 3° E., 14·5° W.
2. Lat. 33° 04·7′ N.
3. Az. 090° T., Dev. 4° E.
4. Course 236° 07′ T., Distance 2138 miles (spheroid) or
   ,, 255° 59′ T., ,, 2131 ,, (sphere).
5. (*a*) Intercept 1·8′ away Az. 177° T. }
   (*b*) ,, 3·5′ towards Az. 103° T. } Lat. 30° 08′ S., 17° 35′ W.
   or
   (*a*) Ex.-Merid. Lat. 30° 08·3′ S., P.L. 087°-267° T. } Lat. 30° 08′ S.
   (*b*) Long. 17° 35·1′ W. P.L. 013°-193° T. } Long. 17° 34·7′ W.

Paper 15

1. Lat. 48° 58′ N., Long. 5° 50′ W.
2. Lat. 32° 49′ N., P.L. 100°-280° T.
3. B 55 secs. losing.
4. Amp. E. 4·5° N., Dea. 6·5° W.
5. Sirius.

CHAPTER XI.

I.—Algebra. Page 259.

1. $x=10$ or -14.
2. $x=8$ or 2.
3. $x=2$ or $\frac{5}{8}$, $y=7$ or $-\frac{1}{8}$.
4. $x=\pm 3$, $y=\pm 2$.
5. A's speed $8\frac{1}{3}$ knots, B's speed 12·5 knots.
6. Current 5·96 knots, reduced speed 6·05 knots, full speed 17·77 knots.
7. Destroyer's speed 15 knots.
8. Full speed 10 knots.
9. 8 days.
10. Meeting place halfway. Time 13·00. They are 10 miles apart at 12·38 and 13·22.
11. Closest distance 29·29 miles.
12. Current sets S. 39° 56′ E. or W., rate 2·615 knots.
13. 20 and 30 m.p.h.
14. 36 and 60 minutes.
15. First pump takes 2h 12·3m. Second pump takes 3h 22·3m.
16. Time 47·8 and 77·8 minutes.
17. Time $5\frac{1}{2}$ hours and 6 hours.
18. Speed 21·57 f.p.s.

Trigonometrical Equations. Page 264 and 265.

1. A = 26° 18′ or 153° 42′.
2. A = 12° 8′ or 192° 8′.
3. A = 213° 23′ or 326° 37′.
4. A = 85° 24′ or 265° 24′. 93° 58′ or 273° 58′.
5. θ = 34° 37′ or 325° 23′.
6. θ = 30° or 330°.
7. θ = 60° or 120°.
8. Cos 15° = ·966.
9. Angle 120°.
10. x = —46° 24′ or 313° 30′.
11. B = 22° $37\frac{1}{4}$′.
12. θ = 75° or 165°.

II.—Plane Triangles. Page 270.

1. Distance 820 feet or 180 feet.
2. Cliff 371 feet high.
3. Distance 605·3 feet.
4. Distance 369 feet.
5. Distance 600·2 feet.
6. Distance 80·62 feet.
7. Height of mast 19·3 feet.
8. Height of tower 74·62 feet.
9. Distance 131·15 feet.
10. Distance 176·4 or 121·8 feet.
11. Course N. 71° $44\frac{1}{2}$′ W.
12. and 13. Theoretical proofs.
14. Distance 9705 feet, bearing from B, S. 28° $52\frac{1}{2}$′ E.
15. Distance 12724 feet, height 9403 feet.
16. Height of tower 148·1 feet.
17. Height of mountain 1·493 miles.
18. Distance between masts 119·44 feet.

19. Courses A, N. 71° 31′ E.; B, N. 85° 13′ E.; C, S. 71° 31′ E.
20. ,, A, N. 86° 07′ W.; B, N. 52° 27′ W.; C, N. 34° 45′ W.
21. Distance 7·87 miles.
22. Distance 5·216 miles.
23. Distance 2·412 miles.
24. Distance 10·42 miles.
25. Angle A 9° 41′.
26. Speed 9·192 knots.
27. Distance ·632 miles.
28. Course S. 83° 33′ W., Angle 56° 27′.
29. Course N. 42° 26′ W.
30. Course N. 15° 41′ E.
31. Distance 6 miles.
32. Distance from B 4·9 miles; from C 4 miles.
33. Distance A B 5·537 miles; B C 4·153 miles.
34. Course 116·9°, drift 3·62 miles.
35. Distances 1·11 miles, 3·39 miles.
36. Course S. 58° W.
37. Course N. 86° 26½′ W., distance 6·376 miles, drift 2·61 miles in 50 minutes.
38. Distance 4·167 miles.

III.—Relative Velocities. Page 280.

1. Course N. 57° 21′ W., time 62·7 minutes.
2. Course N. 70° 21′ E., speed 21·2 knots.
3. Course N. 1° 38′ E., speed 6·95 knots.
4. Alter 23° 53′ to starboard, time 12·4 minutes.
5. Course N. 89½° W., time 13·9 minutes.
6. Course N. 37° 41′ W., time 38 minutes.
7. Speed of approach 14·8 knots. Distance to go 5 miles. Time to reach position 19·8 minutes.
8. Course N. 31° E., time 15h 49·8m.
9. Course 205°, time 22·8 minutes.
10. Course S. 70° W., speed 15·7 knots.
11. Ratio line ∠C 30° 49′. Direction of ratio line 257°. Course N. 49° W.
12. Ratio line 246½°. ∠C 93° 23′. Course S. 22° W.

IV.—Various Examples. Page 287.

1. Distances 4·02 and 4·82 miles.
2. Distance 1·22 miles.
3. Brg. N. 3° W., distance 9·548 cables.
4. True wind South, 15·27 m.p.h.
5. True wind 081·9°, 14·14 m.p.h.
6. Height 15·08 feet.

V.—Spherical Triangles. Page 296.

1. Lat. 26° 51′ 20″ N.
2. Lat. 3° 08′ N., Long. 42° 08′ W.
3. Lat. 24° 8½′ N.
4. Lat. 39° 35½′, Decl. 8° 58′.
5. Decl. 2° 44′ 27″.
6. Lat. 37° 01′ N.
7. Lat. 39° 46′ N.
8. Decl. 25° N., Alt. 58° 38′.
9. Lat. 58° 14′ N.
10. Lat. 10° 53½′ N.
11. Lat. 38° 57½′ N. or 48° 14½′ S.
12. Hour angle 4h 16m 50s, Decl. 0° 0′.
13. Long. 11° 58′ W.
14. Lat. 11° 20′ N.
15. Alt. 38° 23½′.
16. Lat. 4° 33′ 42″ S., Long. 248° 46′ 46″.
17. Lat. 4° 15′ 30″ N., Long. 48° 37′ 06″.

CHAPTER XII.

I.—Quadrilaterals. Page 334.

1. 68 feet 126 sq. inches.
2. 168·3 sq. yards.
3. 588 sq. feet.
4. 823 sq. feet.
5. 148·8 sq. yards.
6. 120 sq. yards.
7. 1448·8 sq. feet.
8. 389·3 sq. feet.
9. 74·69 sq. yards.
10. 256 sq. yards.
11. 2250 sq. feet.
12. 1408·2 sq. feet.
13. 15 feet.
14. 1050 sq. yards.
15. 3146·4 sq. feet.
16. 720 sq. yards.
17. 692 units.
18. 5·363 miles.
19. C bears S. 25° 52½′ E. from A.
    D bears S. 62° 52½′ E. from A.
20. 7417 feet.
21. 8·718 cables.

II.—Circle. Page 336.

1. 550 sq. feet.
2. 34·5 units.
3. 55·7 units.
4. 200·6 units.
5. 11·66 units.
6. 30·8 units.
7. 12·2 units.
8. 20 units.
9. 38° 56′.
10. 38° 56′.
11. 76° 25′.
12. 56° 08′.
13. 22° 57′.
14. 107·5 sq. units.
15. 159·1 sq. units.
16. 102·2 sq. units.
17. 44·7 sq. units.
18. 10·27 sq. units.
19. 636·4 units.
20. 73·4 sq. feet.

III.—Specific Gravity. Page 341.

1. Specific gravity 3·077.
2. Specific gravity 8·25.
3. Specific gravity ·25.
4. Specific gravity ·852.
5. 4162·5 lbs.
6. 683·2 tons.
7. 27·06 lbs.
8. 65·16 cub. feet.
9. ·71 cub. feet.
10. 138·55 lbs.
11. 281·8 lbs.
12. 1·356 inches.

IV.—Polygons. Page 344.

1. 1548·5 sq. feet.
2. 817·6 sq. feet.
3. 51·41 sq. inches.
4. 13·85 inches.
5. 788·3 sq. inches.
6. 1357 sq. inches.
7. 2·188 miles.
8. 2·348 miles.
9. 12·83 cub. feet.
10. Weight 1 ton, 12 cwt., 0 oz., 13 lbs.
11. 201·2 inches.
12. ·1128 inches.
13. Area 922·87 sq. inches.

V.—The Sphere. Page 346.

1. Area 78·54 sq. inches, vol. 65·45 cub. inches.
2. Area 452·4 sq. inches, vol. 905·1 cub. feet.
3. 392·7 sq. inches.
4. 110 sq. feet.
5. 2732 cub. feet.
6. 23·0384 cub. feet.
7. 54·45 cub. feet.
8. 195·8 cub. inches.
9. 1340·4 cub. inches.
10. 494·3 cub. inches.
11. ·56 inches.
12. 30·6 lbs.
13. ·29 inches.
14. Metal specific gravity 8·40, spirit ·366.

VI.—Cones. Page 352.

1. 1847·87 gallons.
2. 428 lbs.
3. 2·55 cub. feet.
4. 70·68 cub. feet.
5. 42·28 cub. feet.
6. 26·9 cub. inches.
7. 56·2 sq. inches.
8. 39·98 cub. inches.
9. Diameter (*a*) 8·434; (*b*) 9·486 inches.
10. 90 feet.
11. 110 feet.
12. 19·5 cub. feet.
13. 9·32 cub. feet.
14. 19·61 sq. feet, 6882 cub. inches.
15. 667·6 sq. feet.
16. 90 sq. feet.
17. 22·56 cub. feet.
18. 527·8 cub. units.
19. 464·2 cub. inches.
20. 5·31 cub. feet, 33·18 gallons.
21. 31·07 ounces.
22. 2700 sq. inches.
23. 21 inches.
24. $h = {}^{\cdot}6933\ H$.
25. As 721 is to 134.
26. Specific gravity ·57.
27. Radius 3·1 feet, height 3·26 feet above water.

VII.—Ellipse. Page 364.

1. 39·27.
2. 150·8.
3. 33·0.
4. 22·2.
5. 160.
6. 766·1 lbs.
7. 1492¼ lbs.
8. 8·21 f.p.s.
9. 24 minutes.
10. Vol. 530·36 cub. ft., 3314·7 galls. F.W.

EXAMINATION PAPERS. EXTRA MASTER. Page 371.

Paper 1.

1. 1·44 yds. from longest side.
2. Area of end section 374·1 sq. ins.
   Area of slant section 529·1 sq. ins.
3. $A = 39° 52'$.
4. Weight 100·7 lbs.
5. Lat. 26° 58'.
6. 36m and 60m.

Paper 2.

1. 40·4 minutes.
2. 3270·5 gallons.
3. $x = 2,\ y = 3$.
6. 32·97 feet.

Paper 3.

1. $\frac{375}{990}$

   NOTE.—Sum to infinity $= \frac{a}{1 - r}$. a = first term, rz = common ratio. In this question ·3 does not recur and the progression is $\frac{78}{1000}$ as the 1st term, and $\frac{1}{100} = r$ ... the sum $S - \cdot 3 = \frac{a}{1 - r}$.
2. $A = 16° 44'$ or $163° 16'$, $190° 32'$ or $349° 28'$.
3. 2h 12·3m and 3h 22·3m.
4. Weight of one link 2·438 lbs.
   Weight of chain 946 lbs.
5. Lat. 52° 27' N.
6. $x = 3$, or $-\frac{137}{9}$; $y = 1$, or $-\frac{38}{3}$.

Paper 4.

1. (i) 132 and 84. (ii) 8 days.
2. 27°, 69°, 111°, 153°.
3. Lat. 31° 48½' N.
4. $\theta = 30°$ or $330°$.
5. Diameter 8·926 feet.

Paper 5.

1. Specific gravity 9·45.
2. One rev. 1·5 secs., speed 9·8 f.p.s.
3. Long. 166° 24' W., Dist. 2784 mls.
4. 100 inches.
5. (ii) $\theta = 145° 30'$ or $325° 30'$.
6. R.A. 16h 22m 50s, Decl. 16° 38' 36" N.

Paper 6.

3. 120°—apply formula $\cos A = \frac{b^2 + c^2 - a^2}{2bc}$
4. (ii) $A = 60°, 120°, 240°$ or $300°$.
5. Lat. 47° 8' 22" S., Long. 237° 57' 50".
6. Speed 9 knots. Distance 10·43 miles.

Paper 7.

1. Rate of change 62·7' per minute.
5. Lat. 10° 5½' N., Decl. 20° 11' S.
6. True dist. 149° 56' 8".

Paper 8.

4. Moonrise 14h 59m.
6. Lat. 50° 47·3' N., Long. 00° 3·2' W.,
   Dist. 12·16 miles.

Paper 9.

6. Lat. 39° 17·2' S., Long. 78° 01' E.

Paper 10.

1. dp/dh 10·8/893. Long. 67° 12¾' E.
2. Lats. 61° 47' N., or 28° 13' N.
3. D. Lat. 4·44' apply to North, D. Long. 6·45' apply to the West.
4. 686 feet.

Paper 11.

1. Lat. 26° 11½' S., Long. 179° 26' E.
2. Alt. 71° 28½'. Az. N. 74° 27' E.
3. 4m 13·5s fast.
5. Augmentation 11·22".

Paper 12.

1. G.M.T. 21d 19h 15m 14s. Duration 1h 42m 12s.
2. Lat. 39° 45½' N.
3. Lat. 54° 26' S.
4. G.C.Co. S. 71° 51' E. Distance 3442·5 miles. Rhumb Co. S. 67° 49½' E. Distance 3443·0 miles.
5. Eq. of time + 5m.
6. Error 10·67' to add.

EXTRACTS FROM THE ADMIRALTY TIDE TABLES

INDEX

VOLUME 1 EUROPEAN WATERS

Index to Standard Ports

(Page numbers refer to the Full Edition of ATT Vol. 1, not this book.)

TABLE V

(ATT Vol. 1)

TIDAL LEVELS—DEFINITIONS AND NOTES

(*a*) The levels are referred to *CHART DATUM*, which is the same as the zero of the tidal predictions in all cases. By international agreement, Chart Datum is defined as a level so low that the tide will not frequently fall below it. In the United Kingdom, this level is normally approximately the level of Lowest Astronomical Tide.

(*b*) *H.A.T.* (*Highest Astronomical Tide*). *L.A.T.* (*Lowest Astronomical Tide*). The highest and lowest levels respectively which can be predicted to occur under average meteorological conditions and under any combination of astronomical conditions; these levels will not be reached every year. H.A.T. and L.A.T. are not the extreme levels which can be reached, as storm surges (see page vii) may cause considerably higher and lower levels to occur.

(*c*) *M.H.W.S.* (*Mean High Water Springs*). *M.L.W.S.* (*Mean Low Water Springs*). The height of mean high water springs is the average, throughout a year when the average maximum declination of the moon is $23\frac{1}{2}°$, of the heights of *two* successive high waters during those periods of 24 hrs (approximately once a fortnight) when the range of the tide is greatest. The height of mean low water springs is the average height obtained by the two successive low waters during the same periods.

(*d*) *M.H.W.N.* (*Mean High Water Neaps*). *M.L.W.N.* (*Mean Low Water Neaps*). The height of mean high water neaps is the average, throughout a year as defined in (*b*) above, of the heights of two successive high waters during those periods (approximately once a fortnight) when the range of the tide is least. The height of mean low water neaps is the average height obtained from the two successive low waters during the same periods.

(*e*) The values of M.H.W.S., M.H.W.N., M.L.W.N. and M.L.W.S. vary from year to year in a cycle of approximately 18·6 years. In general the levels are computed from at least a year's predictions and are adjusted for the long period variations to give values which are the average over the whole cycle. The values of Lowest Astronomical Tide (L.A.T.) and Highest Astronomical Tide (H.A.T.) are determined by inspection over a span of years.

(*f*) *M.T.L.* (*Mean Tide Level*). Mean Tide Level, as given above, is the mean of the heights of M.H.W.S., M.H.W.N., M.L.W.S. and M.L.W.N.

(*g*) *Abbreviations for authorities:*

B.	L'Administration des Ponts et Chaussées, Belgium.
D.	Danish Meteorological Institute.
F.	Service Hydrographique et Océanographique de la Marine, France.
G.	German Hydrographic Institute.
H.A.	Local Harbour Authority.
Hyd.	Hydrographer of the Navy.
I.	Lighthouse and Harbour Administration, Iceland.
It.	Instituto Idografico, Genova, Italy.
N.	Department van Waterstaat, Netherlands.
Nor.	Norwegian Hydrographic Service.
P.	Ministerio de Marinha, Lisboa, Portugal.
R.	Russian sources.
I.O.S.	Institute of Oceangraphic Sciences.

(*h*) *Abbreviations for methods of predicting:*

H. By tide-predicting machine or electronic computer using harmonic constants.

H.C. As above but with additional shallow water corrections.

Diff. By variable differences on another port.

(*j*) *Years of observations.* Wherever possible tidal predictions for Standard Ports are based on the analysis of at least one complete year's observations. Where a period of less than one year has been used, the dates are given in italics.

TABLE V (*cont.*)

TIDAL LEVELS IN METRES AT STANDARD PORTS

(with data concerning predictions, etc.)

Standard Port	L.A.T. (*b*)	M.L.W.S. (*c*)	M.L.W.N. (*d*)	M.T.L. (*f*)	M.H.W.N. (*d*)	M.H.W.S. (*c*)	H.A.T. (*b*)	Authority for (*g*) Observations	Constants	Predictions	Method of Predicting (*h*)	Years of Observations (*j*)
Plymouth (Devonport)	0·0	+0·8	+2·2	+3·2	+4·4	+5·5	+6·1	Hyd.	I.O.S.	Hyd.	H.	1961–62
Portland	−0·1	+0·2	+0·7	+1·1	+1·4	+2·1	+2·7	Hyd.	I.O.S.	Hyd.	H.C.	1961–62
Southampton	0·0	+0·5	+1·8	+2·6	+3·7	+4·5	+4·9	H.A.	I.O.S.	I.O.S.	H.C.	1957
Portsmouth	0·0	+0·6	+1·8	+2·7	+3·8	+4·7	+5·2	Hyd.	I.O.S.	Hyd.	H.C.	1961–62
Shoreham	0·0	+0·7	+1·9	+3·5	+5·0	+6·2	+6·7	H.A.	I.O.S.	I.O.S.	H.	1959
Dover	0·0	+0·8	+2·0	+3·7	+5·3	+6·7	+7·3	H.A	I.O.S.	I.O.S.	H.C.	1975
Margate	−0·1	+0·5	+1·4	+2·6	+3·9	+4·8	+5·2	H.A.	I.O.S.	I.O.S.	H.	1969–70
Sheerness	−0·1	+0·6	+1·5	+3·1	+4·8	+5·7	+6·4	H.A.	I.O.S.	I.O.S.	H.	1969–70
Chatham	−0·4	+0·4	+1·4	+3·2	+4·9	+6·0	+6·7	H.A.	I.O.S.	Hyd.	H.	1971
London Bridge	−0·2	+0·5	+1·6	+3·7	+5·8	+7·1	+7·9	H.A.	I.O.S.	I.O.S.	H.C.	1969–70
Walton-on-the-Naze	−0·1	+0·4	+1·1	+2·3	+3·4	+4·2	+4·7	H.A.	I.O.S.	I.O.S.	H.	1969–70
Harwich	−0·1	+0·4	+1·1	+2·2	+3·4	+4·0	+4·4	H.A.	I.O.S.	I.O.S.	H.C.	1961
Lowestoft	0·0	+0·5	+1·0	+1·5	+2·1	+2·4	+2·8	H.A.	I.O.S.	I.O.S.	H.	1959
Immingham	−0·1	+0·9	+2·6	+4·1	+5·8	+7·3	+8·0	H.A.	I.O.S.	I.O.S.	H.C.	1977
Hull	−0·1	+0·8	+2·4	+4·1	+5·8	+7·3	+8·3	H.A.	I.O.S.	I.O.S.	H.C.	1977
River Tees	0·0	+0·9	+2·0	+3·2	+4·3	+5·5	+6·1	H.A.	I.O.S.	I.O.S.	H.	1948
River Tyne	0·0	+0·7	+1·8	+2·9	+3·9	+5·0	+5·7	H.A.	I.O.S.	I.O.S.	H.	1962
Leith	0·0	+0·8	+2·1	+3·3	+4·5	+5·6	+6·1	H.A.	I.O.S.	I.O.S.	H.C.	1956
Rosyth	0·0	+0·8	+2·2	+3·4	+4·7	+5·8	+6·4	Hyd.	I.O.S.	Hyd.	H.C.	1966
Aberdeen	0·0	+0·6	+1·6	+2·5	+3·4	+4·3	+4·8	H.A.	I.O.S.	I.O.S.	H.	1964
Lerwick	0·0	+0·5	+0·9	+1·3	+1·6	+2·2	+2·5	H.A.	I.O.S.	Hyd.	H.	1967–71
Ullapool	0·0	+0·7	+2·1	+3·0	+3·9	+5·2	+5·8	H.A.	I.O.S.	Hyd.	H.	1966
Oban	0·0	+0·7	+1·8	+2·3	+2·9	+4·0	+4·5	Hyd.	I.O.S.	Hyd.	H.	1971–72
Greenock	−0·1	+0·4	+1·0	+1·9	+2·9	+3·4	+4·1	H.A.	I.O.S.	Hyd.	H.C.	1948
Glasgow	−0·1	+0·7	+1·6	+2·8	+4·0	+4·8	+5·4	H.A.	I.O.S.	I.O.S.	H.C.	1948
Liverpool	−0·2	+0·9	+2·9	+5·2	+7·4	+9·3	+10·3	H.A.	I.O.S.	I.O.S.	H.C.	1970
Holyhead	0·0	+0·7	+2·0	+3·2	+4·5	+5·7	+6·5	H.A.	I.O.S.	I.O.S.	H.	1959–60
Milford Haven	−0·1	+0·7	+2·5	+3·8	+5·2	+7·0	+8·0	Hyd.	I.O.S.	I.O.S.	H.	1961
Swansea	0·0	+1·0	+3·2	+5·3	+7·3	+9·6	+10·5	H.A.	I.O.S.	I.O.S.	H.C.	1961
Cardiff	0·0	+1·2	+3·8	+6·7	+9·4	+12·3	+13·4	H.A.	—	I.O.S.	Diff.	1956
Port of Bristol (Avonmouth)	0·0	+0·9	+3·5	+6·9	+10·0	+13·2	+14·6	H.A.	I.O.S.	I.O.S.	H.C.	1970

TABLE V (*cont.*)
TIDAL LEVELS IN METRES AT STANDARD PORTS
(with data concerning predictions, etc.)

Standard Port	L.A.T. (*b*)	M.L.W.S. (*c*)	M.L.W.N. (*d*)	M.T.L. (*f*)	M.H.W.N. (*d*)	M.H.W.S. (*c*)	H.A.T. (*b*)	Authority for (*g*) Observations	Constants	Predictions	Method of Predicting (*h*)	Years of Observations (*j*)
Dublin	0·0	+0·5	+1·5	+2·4	+3·4	+4·1	+4·7	H.A.	I.O.S.	I.O.S.	H.	1948–49
Belfast	0·0	+0·4	+1·1	+2·0	+3·0	+3·5	+3·9	H.A.	I.O.S.	I.O.S.	H.	1947–48
Londonderry	0·0	+0·4	+1·0	+1·5	+2·0	+2·7	+3·2	H.A.	I.O.S.	I.O.S.	H.	1935–36
Galway	−0·1	+0·6	+2·0	+2·9	+3·9	+5·1	+5·7	H.A.	I.O.S.	I.O.S.	H.	1965
Tarbert Island	0·0	+0·5	+1·7	+2·8	+3·8	+5·0	+5·5	H.A.	I.O.S.	I.O.S.	H.	1966
Cobh	−0·1	+0·5	+1·3	+2·3	+3·3	+4·1	+4·6	Hyd.	I.O.S.	I.O.S.	H.	1906
Reykjavik	−0·4	+0·2	+1·3	+2·1	+2·9	+4·0	+4·6	I.	I.	Hyd.	H.	1963
Port of Kem'	+0·1	+0·4	+0·6	+1·1	+1·6	+1·9	+2·0	R.	R.	Hyd.	H.	1910
Ostrov Yekaterininskaya	−0·1	+0·5	+1·3	+2·1	+3·0	+3·7	+4·2	R.	G.	G.	H.	1906–07
Narvik	+0·1	+0·5	+1·2	+1·8	+2·4	+3·2	+3·7	—	—	Nor.	—	—
Bergen	−0·1	+0·2	+0·5	+0·8	+1·1	+1·4	+1·6	—	—	Nor.	—	—
Esbjerg	−0·4	−0·1	+0·2	+0·8	+1·4	+1·6	+1·9	D.	D.	D.	H.	—
Helgoland	−0·3	0·0	+0·4	+1·3	+2·3	+2·6	+3·0	—	—	G.	—	—
Cuxhaven	−0·4	−0·1	+0·3	+1·5	+2·8	+3·2	+3·5	—	—	G.	—	—
Wilhelmshaven	−0·4	0·0	+0·6	+2·0	+3·7	+4·2	+4·5	—	—	G.	—	—
Hoek van Holland	−0·2	+0·4	+0·3	+1·2	+1·8	+2·3	+2·6	—	—	N.	—	—
Vlissingen (Flushing)	−0·4	+0·5	+1·0	+2·6	+4·0	+4·9	+5·3	—	—	N.	—	—
Antwerp (Prosperpolder)	−0·6	0·0	+0·9	+2·6	+4·3	+5·6	+5·9	B.	I.O.S.	I.O.S.	H.C.	1972
Dunkerque	+0·2	+0·6	+1·5	+3·1	+4·8	+5·8	+6·2	—	—	F.	—	—
Calais	+0·4	+1·0	+2·2	+4·1	+6·0	+7·2	+7·6	—	—	F.	—	—
Boulogne	+0·2	+1·1	+2·8	+5·0	+7·2	+8·9	+9·5	—	—	F.	—	—
Dieppe	+0·2	+0·7	+2·5	+4·9	+7·2	+9·3	+10·0	—	—	F.	—	—
Le Havre	+0·3	+1·1	+2·8	+4·5	+6·5	+7·8	+8·3	—	—	F.	—	—
Cherbourg	+0·5	+1·1	+2·5	+3·5	+5·0	+6·3	+7·0	—	—	F.	—	—
St. Helier	0·0	+1·3	+4·1	+6·2	+8·1	+11·1	+12·3	H.A.	I.O.S.	I.O.S.	H.	1952
Brest	+0·5	+1·4	+3·0	+4·4	+5·9	+7·5	+8·3	—	—	F.	—	—
Pointe de Grave	+0·5	+1·0	+2·1	+3·1	+4·3	+5·3	+6·0	—	—	F.	—	—
Lisbon	+0·0	+0·5	+1·4	+2·2	+3·0	+3·8	+4·2	—	—	P.	—	—
Gibraltar	−0·1	+0·1	+0·3	+0·5	+0·7	+1·0	+1·1	Hyd.	I.O.S.	Hyd.	H.	1961–62
Venezia (Venice)	−0·1	+0·1	+0·4	+0·5	+0·6	+0·8	+1·0	—	—	It.	—	—

Volume 1 EUROPEAN WATERS—Part I (Standard Ports)

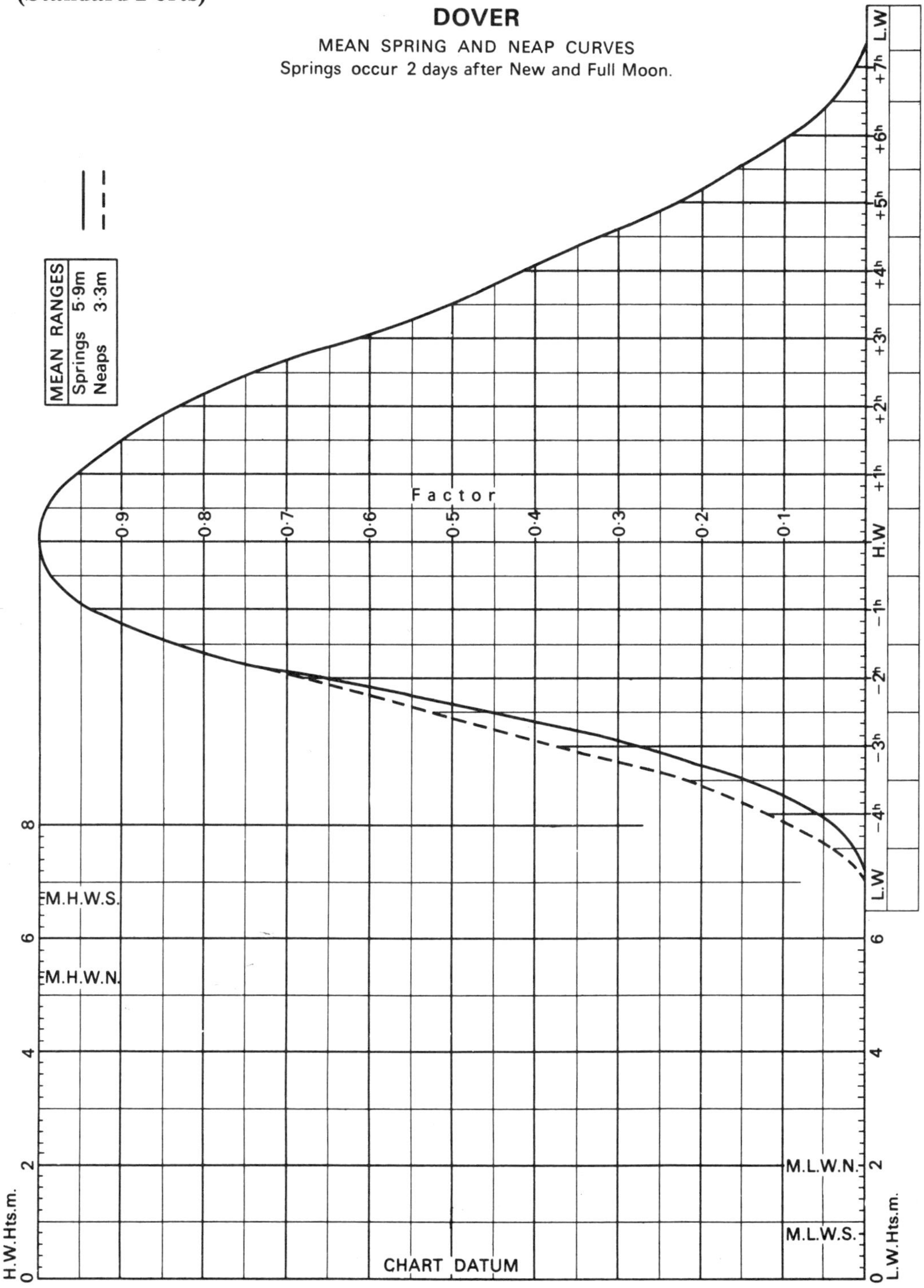

ENGLAND, SOUTH COAST - DOVER

LAT 51°07'N LONG 1°19'E

TIME ZONE GMT — TIMES AND HEIGHTS OF HIGH AND LOW WATERS — YEAR 1982

JANUARY

Day	TIME	M	TIME	M	TIME	M	TIME	M
1 F	0209	6.2	0934	1.5	1430	5.9	2146	1.6
2 SA	0256	6.1	1016	1.6	1527	5.7	2231	1.8
3 SU	0352	6.0	1106	1.7	1633	5.6	2327	1.9
4 M	0458	5.9	1210	1.8	1743	5.6		
5 TU	0038	2.0	0607	5.9	1326	1.7	1850	5.7
6 W	0157	1.9	0713	6.0	1439	1.5	1954	5.9
7 TH	0308	1.7	0815	6.2	1548	1.3	2056	6.2
8 F	0414	1.4	0914	6.4	1651	1.1	2153	6.4
9 SA	0516	1.2	1010	6.6	1751	0.9	2247	6.6
10 SU	0614	1.0	1102	6.7	1849	0.8	2336	6.8
11 M	0709	0.8	1153	6.7	1942	0.8		
12 TU	0021	6.8	0801	0.7	1239	6.7	2032	0.8
13 W	0103	6.8	0849	0.7	1323	6.5	2115	1.0
14 TH	0144	6.7	0934	0.9	1406	6.3	2155	1.3
15 F	0226	6.5	1014	1.2	1453	6.0	2231	1.6
16 SA	0312	6.3	1057	1.5	1543	5.7	2309	1.9
17 SU	0406	6.0	1140	1.8	1642	5.5	2351	2.2
18 M	0508	5.7	1232	2.0	1756	5.3		
19 TU	0046	2.4	0621	5.5	1335	2.2	1914	5.3
20 W	0201	2.4	0735	5.5	1451	2.1	2019	5.4
21 TH	0318	2.2	0836	5.6	1549	1.9	2111	5.7
22 F	0412	1.9	0927	5.8	1634	1.6	2155	5.9
23 SA	0455	1.6	1007	6.0	1715	1.4	2231	6.2
24 SU	0539	1.4	1041	6.1	1757	1.2	2304	6.3
25 M	0619	1.2	1112	6.2	1838	1.1	2333	6.4
26 TU	0700	1.1	1142	6.3	1916	1.0		
27 W	0005	6.5	0737	1.0	1215	6.4	1948	1.0
28 TH	0039	6.6	0809	1.0	1252	6.4	2018	1.1
29 F	0116	6.5	0840	1.0	1331	6.3	2049	1.1
30 SA	0152	6.4	0912	1.1	1412	6.2	2124	1.2
31 SU	0233	6.3	0950	1.2	1500	6.0	2204	1.4

FEBRUARY

Day	TIME	M	TIME	M	TIME	M	TIME	M
1 M	0322	6.1	1035	1.4	1557	5.8	2254	1.7
2 TU	0421	5.9	1130	1.6	1702	5.6	2357	1.9
3 W	0529	5.8	1245	1.8	1815	5.5		
4 TH	0119	2.0	0643	5.7	1409	1.8	1933	5.6
5 F	0244	1.8	0759	5.8	1532	1.5	2049	5.8
6 SA	0403	1.5	0910	6.0	1645	1.2	2152	6.2
7 SU	0509	1.2	1009	6.3	1751	1.0	2241	6.4
8 M	0611	0.9	1058	6.5	1850	0.8	2325	6.7
9 TU	0706	0.7	1143	6.6	1940	0.7		
10 W	0004	6.8	0754	0.6	1224	6.6	2020	0.7
11 TH	0043	6.8	0834	0.6	1302	6.5	2056	0.8
12 F	0121	6.7	0910	0.8	1341	6.4	2124	1.1
13 SA	0201	6.6	0941	1.0	1420	6.1	2146	1.4
14 SU	0242	6.4	1010	1.3	1504	5.9	2213	1.7
15 M	0325	6.0	1044	1.7	1553	5.5	2248	2.0
16 TU	0417	5.7	1125	2.0	1654	5.2	2337	2.3
17 W	0525	5.3	1219	2.3	1819	5.0		
18 TH	0041	2.5	0653	5.1	1333	2.4	1942	5.1
19 F	0211	2.4	0806	5.2	1500	2.2	2042	5.3
20 SA	0334	2.1	0901	5.4	1602	1.8	2127	5.7
21 SU	0428	1.7	0943	5.7	1651	1.5	2203	5.9
22 M	0516	1.3	1014	5.9	1737	1.2	2234	6.2
23 TU	0601	1.1	1044	6.2	1821	1.0	2305	6.4
24 W	0643	0.9	1116	6.4	1900	0.9	2340	6.6
25 TH	0721	0.8	1153	6.5	1933	0.8		
26 F	0017	6.7	0752	0.8	1232	6.6	1958	0.8
27 SA	0055	6.7	0819	0.8	1312	6.5	2027	0.8
28 SU	0131	6.6	0851	0.8	1352	6.4	2104	0.9

MARCH

Day	TIME	M	TIME	M	TIME	M	TIME	M
1 M	0212	6.4	0929	1.0	1437	6.2	2145	1.2
2 TU	0258	6.2	1012	1.3	1531	5.9	2233	1.5
3 W	0355	5.9	1106	1.6	1634	5.6	2334	1.8
4 TH	0505	5.6	1222	1.9	1751	5.3		
5 F	0100	1.9	0628	5.4	1357	1.9	1923	5.4
6 SA	0234	1.8	0801	5.5	1528	1.6	2047	5.7
7 SU	0357	1.4	0918	5.9	1644	1.2	2145	6.0
8 M	0506	1.0	1009	6.2	1747	0.9	2228	6.3
9 TU	0605	0.8	1051	6.4	1842	0.7	2306	6.6
10 W	0656	0.6	1127	6.5	1924	0.6	2344	6.7
11 TH	0737	0.5	1203	6.5	1959	0.7		
12 F	0021	6.7	0811	0.6	1239	6.5	2025	0.8
13 SA	0057	6.7	0837	0.7	1314	6.4	2044	1.0
14 SU	0133	6.5	0903	0.9	1349	6.2	2107	1.2
15 M	0208	6.3	0929	1.2	1425	5.9	2135	1.4
16 TU	0243	6.0	1002	1.5	1504	5.6	2210	1.8
17 W	0325	5.6	1041	1.9	1553	5.2	2255	2.1
18 TH	0420	5.1	1130	2.2	1706	4.9	2354	2.4
19 F	0556	4.8	1235	2.4	1849	4.8		
20 SA	0113	2.4	0727	4.9	1402	2.3	1958	5.1
21 SU	0250	2.2	0825	5.2	1522	1.9	2047	5.5
22 M	0356	1.7	0907	5.5	1620	1.5	2125	5.8
23 TU	0448	1.3	0939	5.9	1711	1.2	2159	6.2
24 W	0536	1.0	1013	6.2	1757	0.9	2234	6.5
25 TH	0621	0.8	1049	6.5	1838	0.8	2312	6.7
26 F	0659	0.6	1129	6.6	1909	0.7	2350	6.8
27 SA	0728	0.6	1210	6.7	1937	0.6		
28 SU	0031	6.8	0757	0.5	1252	6.6	2009	0.6
29 M	0112	6.6	0832	0.6	1335	6.5	2049	0.7
30 TU	0154	6.4	0912	0.8	1422	6.2	2131	1.0
31 W	0242	6.1	0959	1.2	1515	5.9	2221	1.3

APRIL

Day	TIME	M	TIME	M	TIME	M	TIME	M
1 TH	0342	5.8	1058	1.6	1619	5.6	2329	1.7
2 F	0454	5.4	1219	1.9	1739	5.3		
3 SA	0057	1.8	0628	5.3	1354	1.8	1919	5.3
4 SU	0230	1.6	0809	5.5	1521	1.5	2036	5.7
5 M	0349	1.2	0912	5.9	1630	1.1	2127	6.0
6 TU	0454	0.9	0957	6.1	1730	0.9	2207	6.3
7 W	0549	0.7	1033	6.3	1818	0.7	2244	6.5
8 TH	0634	0.6	1105	6.4	1857	0.7	2319	6.6
9 F	0710	0.6	1140	6.5	1926	0.8	2356	6.7
10 SA	0737	0.7	1215	6.4	1947	0.9		
11 SU	0032	6.6	0801	0.8	1249	6.3	2008	0.9
12 M	0104	6.4	0827	0.9	1320	6.2	2036	1.1
13 TU	0134	6.2	0857	1.1	1351	6.0	2108	1.3
14 W	0202	5.9	0931	1.4	1423	5.7	2143	1.6
15 TH	0236	5.6	1009	1.7	1505	5.4	2226	1.9
16 F	0325	5.2	1054	2.0	1607	5.1	2319	2.2
17 SA	0449	4.8	1153	2.3	1740	4.9		
18 SU	0028	2.3	0631	4.8	1310	2.3	1902	5.1
19 M	0159	2.1	0735	5.1	1433	2.0	1958	5.5
20 TU	0314	1.7	0823	5.5	1539	1.6	2043	5.9
21 W	0412	1.2	0904	5.9	1633	1.2	2124	6.3
22 TH	0502	0.9	0942	6.3	1722	0.9	2203	6.5
23 F	0549	0.7	1021	6.5	1805	0.8	2242	6.7
24 SA	0629	0.6	1104	6.7	1841	0.6	2325	6.8
25 SU	0703	0.5	1149	6.8	1916	0.6		
26 M	0008	6.8	0738	0.5	1235	6.7	1955	0.6
27 TU	0053	6.6	0819	0.6	1324	6.5	2039	0.7
28 W	0142	6.4	0904	0.8	1415	6.3	2127	0.9
29 TH	0236	6.1	0956	1.2	1508	6.0	2223	1.3
30 F	0336	5.8	1101	1.5	1609	5.7	2333	1.5

HEIGHTS IN METRES

ENGLAND, SOUTH COAST - DOVER

LAT 51°07'N LONG 1°19'E

TIME ZONE GMT — TIMES AND HEIGHTS OF HIGH AND LOW WATERS — YEAR 1982

MAY

Day	TIME	M	Day	TIME	M
1 SA	0447	5.5	16 SU	0407	5.1
	1219	1.7		1118	2.0
	1723	5.4		1648	5.3
				2351	2.0
2 SU	0055	1.6	17 M	0536	5.0
	0625	5.4		1224	2.1
	1342	1.7		1805	5.3
	1857	5.5			
3 M	0218	1.4	18 TU	0109	1.9
	0754	5.6		0646	5.3
	1500	1.4		1341	1.9
	2009	5.7		1909	5.6
4 TU	0329	1.1	19 W	0225	1.6
	0850	5.8		0740	5.6
	1604	1.2		1450	1.6
	2058	6.0		2001	6.0
5 W	0430	0.9	20 TH	0328	1.3
	0932	6.0		0827	6.0
	1701	1.0		1549	1.3
	2141	6.2		2047	6.3
6 TH	0522	0.8	21 F	0423	1.0
	1007	6.2		0912	6.3
	1746	1.0		1642	1.1
	2217	6.4		2132	6.6
7 F	0604	0.8	22 SA	0513	0.8
	1041	6.3		0956	6.5
	1821	1.0		1730	0.9
	2255	6.5		2216	6.7
8 SA	0635	0.9	23 SU	0558	0.7
	1118	6.4		1042	6.7
	1846	1.0		1815	0.7
	2332	6.5		2302	6.8
9 SU	0700	0.9	24 M	0642	0.6
	1153	6.4		1132	6.7
	1910	1.0		1900	0.6
				2351	6.8
10 M	0008	6.5	25 TU	0727	0.6
	0727	0.9		1224	6.7
	1228	6.3		1947	0.6
	1940	1.0			
11 TU	0039	6.3	26 W	0043	6.6
	0758	0.9		0815	0.7
	1257	6.2		1316	6.6
	2012	1.1		2036	0.7
12 W	0106	6.1	27 TH	0137	6.4
	0832	1.1		0905	0.9
	1326	6.0		1406	6.4
	2047	1.3		2129	0.9
13 TH	0133	5.9	28 F	0230	6.2
	0905	1.3		1002	1.1
	1357	5.8		1457	6.1
	2124	1.5		2227	1.1
14 F	0206	5.7	29 SA	0327	5.9
	0943	1.6		1102	1.4
	1437	5.6		1552	5.9
	2203	1.8		2330	1.3
15 SA	0254	5.4	30 SU	0431	5.6
	1024	1.8		1207	1.6
	1534	5.4		1658	5.7
	2251	1.9			
			31 M	0038	1.4
				0553	5.5
				1317	1.6
				1817	5.6

JUNE

Day	TIME	M	Day	TIME	M
1 TU	0151	1.4	16 W	0021	1.8
	0714	5.6		0557	5.4
	1427	1.6		1248	1.9
	1928	5.8		1819	5.7
2 W	0300	1.3	17 TH	0133	1.6
	0812	5.7		0657	5.6
	1531	1.4		1359	1.8
	2023	6.0		1919	6.0
3 TH	0359	1.2	18 F	0240	1.4
	0858	5.9		0752	5.9
	1624	1.3		1505	1.5
	2110	6.1		2012	6.2
4 F	0448	1.1	19 SA	0342	1.2
	0939	6.0		0844	6.2
	1708	1.3		1606	1.3
	2152	6.3		2104	6.4
5 SA	0527	1.1	20 SU	0440	1.0
	1019	6.2		0936	6.4
	1742	1.3		1704	1.1
	2233	6.4		2155	6.6
6 SU	0600	1.1	21 M	0536	0.8
	1057	6.3		1028	6.6
	1811	1.2		1758	0.9
	2311	6.4		2247	6.7
7 M	0628	1.1	22 TU	0631	0.8
	1134	6.3		1122	6.7
	1842	1.1		1852	0.7
	2347	6.3		2340	6.7
8 TU	0700	1.0	23 W	0724	0.7
	1208	6.3		1215	6.7
	1917	1.1		1945	0.6
9 W	0019	6.2	24 TH	0035	6.6
	0735	1.0		0818	0.7
	1239	6.2		1304	6.7
	1954	1.1		2037	0.6
10 TH	0046	6.1	25 F	0127	6.5
	0811	1.1		0910	0.8
	1309	6.2		1351	6.5
	2030	1.2		2129	0.7
11 F	0116	6.0	26 SA	0216	6.3
	0846	1.3		0959	1.0
	1341	6.0		1437	6.4
	2105	1.4		2220	0.9
12 SA	0152	5.8	27 SU	0307	6.1
	0921	1.4		1049	1.3
	1420	5.9		1527	6.1
	2143	1.5		2312	1.2
13 SU	0237	5.6	28 M	0402	5.8
	0959	1.6		1142	1.5
	1507	5.8		1624	5.9
	2226	1.7			
14 M	0338	5.4	29 TU	0008	1.4
	1044	1.8		0506	5.6
	1607	5.6		1236	1.7
	2316	1.8		1729	5.8
15 TU	0449	5.4	30 W	0110	1.6
	1140	1.9		0621	5.5
	1715	5.6		1340	1.9
				1842	5.7

JULY

Day	TIME	M	Day	TIME	M
1 TH	0216	1.6	16 F	0042	1.7
	0728	5.6		0614	5.6
	1446	1.9		1312	1.9
	1947	5.8		1838	5.8
2 F	0319	1.6	17 SA	0155	1.6
	0825	5.7		0719	5.7
	1543	1.8		1426	1.7
	2042	5.9		1942	6.0
3 SA	0412	1.6	18 SU	0308	1.5
	0914	5.9		0823	6.0
	1630	1.6		1539	1.5
	2129	6.0		2044	6.2
4 SU	0452	1.5	19 M	0417	1.2
	0957	6.0		0925	6.2
	1708	1.5		1647	1.2
	2213	6.1		2143	6.4
5 M	0527	1.4	20 TU	0523	1.1
	1038	6.2		1021	6.5
	1744	1.3		1749	1.0
	2252	6.2		2240	6.5
6 TU	0603	1.2	21 W	0627	0.9
	1115	6.3		1113	6.6
	1822	1.2		1848	0.8
	2327	6.2		2333	6.6
7 W	0641	1.1	22 TH	0726	0.8
	1149	6.3		1201	6.8
	1900	1.1		1942	0.6
	2358	6.2			
8 TH	0717	1.1	23 F	0024	6.7
	1219	6.3		0818	0.7
	1938	1.1		1246	6.8
				2033	0.5
9 F	0028	6.2	24 SA	0110	6.6
	0754	1.1		0903	0.8
	1250	6.3		1330	6.7
	2015	1.2		2119	0.6
10 SA	0100	6.1	25 SU	0154	6.4
	0827	1.2		0945	0.9
	1323	6.3		1412	6.6
	2049	1.2		2202	0.8
11 SU	0137	6.0	26 M	0239	6.2
	0900	1.3		1023	1.2
	1359	6.2		1457	6.4
	2122	1.3		2242	1.1
12 M	0218	5.9	27 TU	0327	6.0
	0935	1.5		1059	1.6
	1440	6.1		1546	6.1
	2159	1.4		2325	1.5
13 TU	0307	5.8	28 W	0421	5.7
	1014	1.6		1140	1.9
	1529	5.9		1644	5.8
	2242	1.5			
14 W	0406	5.7	29 TH	0012	1.8
	1102	1.8		0527	5.4
	1628	5.8		1229	2.1
	2336	1.6		1754	5.6
15 TH	0509	5.6	30 F	0112	2.0
	1201	1.9		0646	5.3
	1732	5.8		1337	2.3
				1912	5.5
			31 SA	0226	2.1
				0755	5.4
				1457	2.2
				2018	5.6

AUGUST

Day	TIME	M	Day	TIME	M
1 SU	0331	2.0	16 M	0249	1.7
	0851	5.7		0816	5.7
	1555	2.0		1527	1.6
	2111	5.7		2040	5.9
2 M	0420	1.8	17 TU	0410	1.4
	0938	5.9		0924	6.1
	1641	1.7		1640	1.3
	2156	5.9		2145	6.2
3 TU	0502	1.5	18 W	0520	1.1
	1019	6.1		1017	6.4
	1723	1.4		1743	0.9
	2234	6.0		2237	6.5
4 W	0543	1.3	19 TH	0624	0.9
	1052	6.3		1102	6.7
	1804	1.2		1843	0.7
	2305	6.1		2323	6.7
5 TH	0622	1.2	20 F	0720	0.7
	1123	6.4		1144	6.8
	1845	1.1		1935	0.5
	2333	6.2			
6 F	0702	1.1	21 SA	0005	6.7
	1154	6.5		0805	0.7
	1924	1.0		1225	6.9
				2019	0.5
7 SA	0004	6.3	22 SU	0046	6.7
	0737	1.1		0843	0.7
	1227	6.5		1304	6.8
	1958	1.0		2057	0.6
8 SU	0039	6.3	23 M	0126	6.5
	0808	1.1		0915	1.0
	1300	6.5		1344	6.7
	2029	1.0		2132	0.8
9 M	0116	6.3	24 TU	0206	6.3
	0837	1.2		0942	1.3
	1334	6.4		1425	6.5
	2058	1.1		2202	1.2
10 TU	0152	6.2	25 W	0249	6.0
	0908	1.3		1007	1.6
	1411	6.3		1508	6.2
	2132	1.2		2234	1.6
11 W	0234	6.0	26 TH	0336	5.7
	0946	1.4		1040	1.9
	1453	6.1		1559	5.8
	2213	1.4		2313	1.9
12 TH	0325	5.9	27 F	0435	5.4
	1030	1.6		1125	2.2
	1546	5.9		1705	5.4
	2301	1.6			
13 F	0427	5.7	28 SA	0004	2.2
	1125	1.9		0558	5.1
	1651	5.8		1225	2.5
				1835	5.2
14 SA	0005	1.8	29 SU	0114	2.4
	0536	5.5		0723	5.2
	1238	2.0		1357	2.5
	1804	5.6		1952	5.2
15 SU	0124	1.8	30 M	0243	2.3
	0655	5.5		0826	5.4
	1402	1.9		1521	2.2
	1924	5.7		2051	5.5
			31 TU	0346	2.0
				0915	5.7
				1614	1.8
				2136	5.7

HEIGHTS IN METRES

ENGLAND, SOUTH COAST - DOVER

LAT 51°07'N LONG 1°19'E

TIME ZONE GMT — TIMES AND HEIGHTS OF HIGH AND LOW WATERS — YEAR 1982

SEPTEMBER

Day	TIME	M	TIME	M	TIME	M	TIME	M
1 W	0435	1.7	0952	6.0	1701	1.4	2209	5.9
2 TH	0519	1.4	1023	6.3	1744	1.2	2235	6.2
3 F	0603	1.2	1052	6.5	1827	1.0	2305	6.4
4 SA	0643	1.1	1125	6.6	1904	0.9	2337	6.5
5 SU	0717	1.0	1158	6.7	1937	0.9		
6 M	0012	6.6	0744	1.0	1234	6.7	2004	0.9
7 TU	0049	6.5	0811	1.0	1307	6.6	2033	0.9
8 W	0127	6.4	0844	1.1	1342	6.5	2107	1.1
9 TH	0206	6.2	0922	1.3	1423	6.2	2148	1.3
10 F	0254	6.0	1007	1.6	1515	6.0	2237	1.6
11 SA	0356	5.7	1104	1.9	1623	5.7	2344	1.9
12 SU	0511	5.4	1221	2.1	1746	5.4		
13 M	0113	2.0	0645	5.4	1354	2.0	1927	5.5
14 TU	0247	1.8	0819	5.7	1522	1.6	2051	5.9
15 W	0407	1.4	0919	6.1	1634	1.2	2146	6.2
16 TH	0513	1.1	1006	6.5	1736	0.8	2228	6.5
17 F	0612	0.8	1045	6.7	1831	0.6	2306	6.7
18 SA	0702	0.7	1122	6.9	1917	0.5	2343	6.7
19 SU	0741	0.7	1200	6.9	1954	0.6		
20 M	0019	6.7	0812	0.8	1238	6.9	2026	0.7
21 TU	0056	6.5	0834	1.0	1314	6.8	2051	0.9
22 W	0133	6.3	0857	1.3	1349	6.5	2119	1.2
23 TH	0209	6.1	0924	1.6	1426	6.2	2149	1.6
24 F	0249	5.8	0957	1.9	1508	5.7	2227	2.0
25 SA	0338	5.4	1041	2.2	1604	5.3	2315	2.3
26 SU	0452	5.0	1139	2.5	1743	4.9		
27 M	0019	2.5	0635	5.0	1259	2.6	1919	5.0
28 TU	0147	2.5	0748	5.2	1436	2.4	2020	5.2
29 W	0305	2.2	0839	5.6	1541	1.9	2103	5.6
30 TH	0402	1.7	0917	6.0	1631	1.5	2134	5.9

OCTOBER

Day	TIME	M	TIME	M	TIME	M	TIME	M
1 F	0449	1.4	0949	6.3	1716	1.1	2203	6.2
2 SA	0534	1.1	1020	6.6	1800	1.0	2234	6.5
3 SU	0615	1.0	1054	6.8	1838	0.9	2309	6.7
4 M	0649	0.9	1129	6.9	1910	0.8	2347	6.8
5 TU	0716	0.9	1205	6.9	1937	0.8		
6 W	0025	6.7	0747	0.9	1242	6.8	2009	0.9
7 TH	0104	6.6	0823	1.0	1320	6.6	2047	1.0
8 F	0147	6.3	0905	1.2	1404	6.3	2131	1.3
9 SA	0237	6.0	0953	1.5	1500	5.9	2226	1.7
10 SU	0341	5.7	1055	1.9	1613	5.6	2340	2.0
11 M	0458	5.4	1218	2.0	1744	5.4		
12 TU	0113	2.1	0641	5.4	1354	1.9	1940	5.5
13 W	0244	1.7	0809	5.7	1517	1.4	2049	5.9
14 TH	0357	1.3	0904	6.1	1623	1.0	2136	6.3
15 F	0458	1.0	0946	6.5	1719	0.8	2213	6.5
16 SA	0551	0.9	1023	6.7	1810	0.7	2245	6.6
17 SU	0634	0.8	1059	6.9	1849	0.7	2319	6.7
18 M	0707	0.9	1136	6.9	1920	0.8	2356	6.7
19 TU	0730	1.0	1211	6.9	1945	0.9		
20 W	0031	6.6	0752	1.1	1246	6.7	2012	1.0
21 TH	0103	6.4	0820	1.3	1317	6.4	2042	1.3
22 F	0135	6.2	0853	1.5	1348	6.1	2115	1.6
23 SA	0209	5.9	0928	1.8	1422	5.8	2152	1.9
24 SU	0250	5.6	1010	2.1	1510	5.3	2237	2.3
25 M	0350	5.2	1102	2.4	1633	4.9	2336	2.5
26 TU	0522	5.0	1212	2.5	1819	4.9		
27 W	0052	2.6	0648	5.2	1341	2.4	1928	5.1
28 TH	0213	2.3	0747	5.5	1456	1.9	2016	5.5
29 F	0318	1.9	0832	5.9	1552	1.5	2053	5.9
30 SA	0412	1.5	0910	6.3	1641	1.2	2129	6.3
31 SU	0458	1.2	0946	6.6	1726	0.9	2204	6.6

NOVEMBER

Day	TIME	M	TIME	M	TIME	M	TIME	M
1 M	0542	1.0	1023	6.8	1807	0.8	2244	6.8
2 TU	0618	0.9	1102	6.9	1841	0.8	2323	6.9
3 W	0652	0.9	1142	6.9	1913	0.8		
4 TH	0005	6.8	0728	0.9	1222	6.8	1951	0.8
5 F	0050	6.7	0811	0.9	1307	6.6	2034	1.0
6 SA	0140	6.5	0857	1.2	1358	6.3	2124	1.4
7 SU	0233	6.2	0952	1.4	1458	5.9	2224	1.7
8 M	0334	5.8	1058	1.7	1609	5.6	2342	2.0
9 TU	0444	5.6	1218	1.8	1739	5.5		
10 W	0107	2.0	0617	5.5	1342	1.7	1924	5.6
11 TH	0227	1.7	0741	5.8	1458	1.3	2026	5.9
12 F	0335	1.4	0836	6.1	1600	1.1	2112	6.2
13 SA	0431	1.2	0919	6.4	1655	0.9	2149	6.4
14 SU	0520	1.1	0957	6.6	1740	0.9	2223	6.5
15 M	0600	1.1	1035	6.7	1815	1.0	2258	6.6
16 TU	0628	1.2	1112	6.8	1843	1.0	2334	6.6
17 W	0652	1.2	1149	6.7	1910	1.0		
18 TH	0010	6.6	0721	1.2	1222	6.6	1940	1.1
19 F	0042	6.5	0755	1.3	1252	6.4	2013	1.3
20 SA	0112	6.3	0830	1.5	1321	6.1	2047	1.5
21 SU	0144	6.1	0907	1.7	1354	5.8	2124	1.8
22 M	0222	5.9	0946	1.9	1439	5.5	2204	2.1
23 TU	0312	5.6	1033	2.2	1543	5.2	2255	2.3
24 W	0421	5.4	1132	2.3	1712	5.1		
25 TH	0000	2.4	0540	5.4	1246	2.2	1827	5.2
26 F	0116	2.3	0648	5.6	1401	2.0	1924	5.5
27 SA	0226	2.0	0741	5.9	1504	1.6	2012	5.9
28 SU	0327	1.7	0829	6.2	1600	1.3	2056	6.2
29 M	0420	1.4	0911	6.5	1649	1.1	2138	6.5
30 TU	0509	1.2	0955	6.7	1736	0.9	2221	6.7

DECEMBER

Day	TIME	M	TIME	M	TIME	M	TIME	M
1 W	0554	1.0	1038	6.8	1817	0.8	2306	6.8
2 TH	0636	0.9	1123	6.9	1859	0.8	2356	6.8
3 F	0720	0.9	1211	6.8	1942	0.9		
4 SA	0045	6.8	0808	0.9	1303	6.6	2033	1.0
5 SU	0135	6.6	0858	1.0	1357	6.4	2127	1.3
6 M	0225	6.4	0955	1.2	1453	6.1	2227	1.6
7 TU	0319	6.1	1058	1.4	1553	5.8	2334	1.8
8 W	0420	5.9	1205	1.6	1708	5.6		
9 TH	0045	1.9	0533	5.7	1317	1.6	1838	5.6
10 F	0155	1.8	0652	5.8	1429	1.5	1947	5.7
11 SA	0301	1.7	0757	6.0	1531	1.3	2039	5.9
12 SU	0359	1.5	0847	6.2	1624	1.3	2122	6.1
13 M	0447	1.5	0932	6.3	1708	1.2	2202	6.3
14 TU	0525	1.4	1014	6.4	1743	1.2	2241	6.4
15 W	0556	1.4	1054	6.5	1812	1.2	2319	6.5
16 TH	0627	1.3	1132	6.5	1843	1.2	2354	6.5
17 F	0700	1.2	1205	6.4	1917	1.2		
18 SA	0027	6.5	0737	1.3	1235	6.3	1952	1.2
19 SU	0056	6.4	0813	1.4	1304	6.1	2026	1.4
20 M	0127	6.3	0850	1.5	1337	6.0	2101	1.6
21 TU	0202	6.1	0925	1.7	1418	5.8	2136	1.8
22 W	0244	6.0	1004	1.8	1508	5.5	2217	2.0
23 TH	0336	5.8	1052	2.0	1616	5.3	2309	2.2
24 F	0440	5.6	1151	2.0	1726	5.3		
25 SA	0014	2.2	0546	5.6	1300	2.0	1829	5.4
26 SU	0126	2.2	0648	5.8	1409	1.8	1927	5.7
27 M	0236	2.0	0745	6.0	1515	1.5	2022	6.0
28 TU	0341	1.7	0839	6.2	1616	1.3	2114	6.2
29 W	0441	1.4	0929	6.4	1711	1.1	2206	6.5
30 TH	0536	1.1	1021	6.6	1804	0.9	2258	6.7
31 F	0628	1.0	1113	6.7	1855	0.9	2349	6.8

HEIGHTS IN METRES

ENGLAND, SOUTH COAST - PORTSMOUTH

LAT 50°48′N LONG 1°07′W

ME ZONE GMT TIMES AND HEIGHTS OF HIGH AND LOW WATERS YEAR 1982

MAY

	TIME	M		TIME	M
13	0205	4.2	28	0247	4.4
	0718	1.2		0810	1.0
TH	1440	4.1	F	1534	4.4
	1937	1.5		2038	1.4
14	0241	4.0	29	0345	4.2
	0757	1.4		0910	1.2
F	1521	3.9	SA	1640	4.3
	2023	1.7		2148	1.6
15	0325	3.8	30	0453	4.0
	0847	1.6		1021	1.4
SA	1612	3.8	SU	1753	4.1
	2124	1.9		2305	1.7

JUNE

	TIME	M		TIME	M
13	0303	3.9	28	0423	4.1
	0820	1.3		0951	1.3
SU	1546	4.0	M	1715	4.3
	2050	1.6		2228	1.6
14	0351	3.8	29	0530	4.0
	0912	1.4		1058	1.5
M	1638	4.0	TU	1820	4.2
	2149	1.7		2337	1.7
15	0449	3.8	30	0642	3.9
	1014	1.4		1207	1.5
TU	1739	4.0	W	1923	4.1
	2255	1.7			

JULY

	TIME	M		TIME	M
13	0326	4.1	28	0445	4.1
	0844	1.1		1011	1.5
TU	1608	4.2	W	1727	4.2
	2112	1.4		2246	1.7
14	0415	4.0	29	0551	3.8
	0937	1.2		1118	1.7
W	1659	4.2	TH	1829	4.0
	2210	1.5		2356	1.8
15	0515	3.9	30	0706	3.7
	1040	1.4		1230	1.8
TH	1800	4.1	F	1934	3.9
	2318	1.5			

AUGUST

	TIME	M		TIME	M
13	0447	4.0	28	0615	3.6
	1009	1.4		1143	2.1
F	1727	4.1	SA	1845	3.7
	2246	1.6			
14	0559	3.9	29	0026	2.0
	1127	1.6		0740	3.6
SA	1840	4.0	SU	1305	2.1
				2001	3.8
15	0008	1.6	30	0140	1.9
	0723	3.9		0852	3.7
SU	1249	1.6	M	1411	1.9
	1956	4.1		2105	3.9

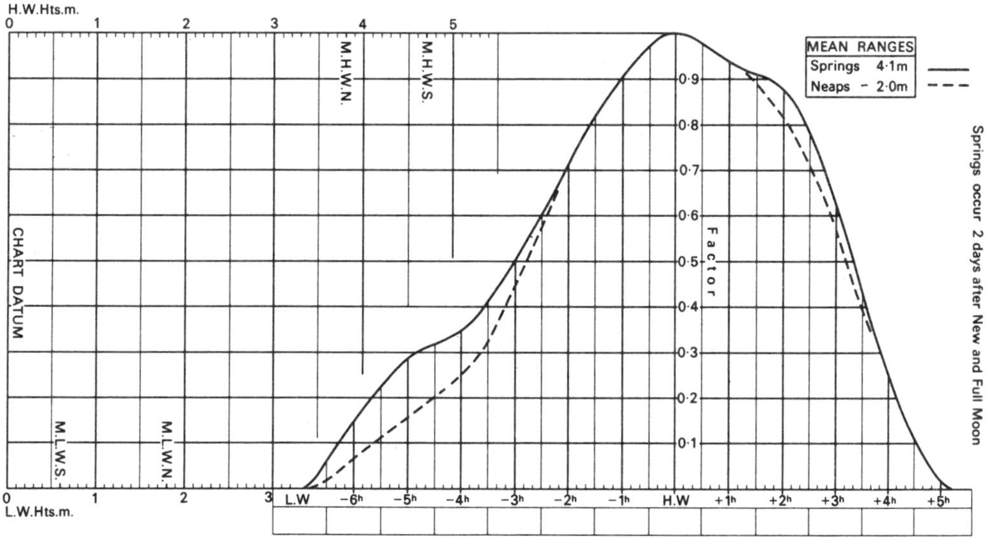

PORTSMOUTH
MEAN SPRING AND NEAP CURVES
Springs occur 2 days after New and Full Moon

SHEERNESS

MEAN SPRING AND NEAP CURVES

Springs occur 2 days after New and Full Moon.

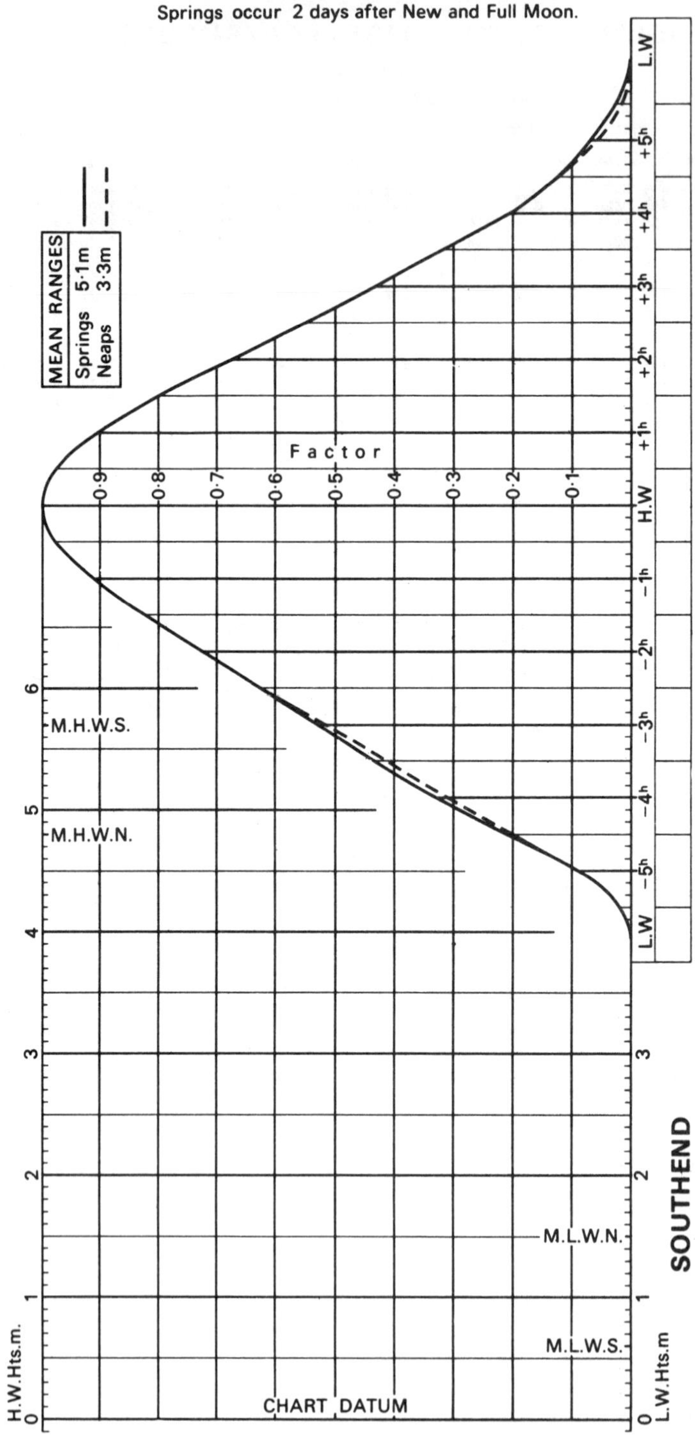

SOUTHEND

To obtain predictions for SOUTHEND the following corrections should be applied to the predictions for SHEERNESS:—

All Times	−5 mins.
H.W. Heights	Nil.
L.W. Heights	−0·1m

ENGLAND, EAST COAST - SHEERNESS

LAT 51°27'N LONG 0°45'E

TIME ZONE GMT — TIMES AND HEIGHTS OF HIGH AND LOW WATERS — YEAR 1982

JANUARY

Day	TIME	M	Day	TIME	M
1	0349	5.4	16	0452	5.4
	1002	1.0		1111	1.0
F	1623	5.3	SA	1726	5.3
	2212	1.3		2304	1.5
2	0430	5.3	17	0539	5.2
	1042	1.0		1151	1.1
SA	1709	5.1	SU	1817	5.0
	2255	1.4		2351	1.6
3	0516	5.1	18	0632	5.0
	1130	1.1		1246	1.3
SU	1804	5.0	M	1916	4.8
	2350	1.5			
4	0614	5.0	19	0057	1.7
	1232	1.1		0738	4.7
M	1913	4.9	TU	1358	1.3
				2023	4.7
5	0059	1.5	20	0223	1.7
	0724	5.0		0853	4.7
TU	1342	1.1	W	1505	1.3
	2026	5.0		2132	4.8
6	0212	1.4	21	0336	1.6
	0842	5.1		1004	4.7
W	1454	1.0	TH	1602	1.2
	2135	5.2		2231	4.9
7	0325	1.2	22	0433	1.3
	0952	5.3		1059	4.9
TH	1609	0.9	F	1648	1.1
	2237	5.5		2316	5.1
8	0440	1.0	23	0518	1.1
	1055	5.5		1142	5.1
F	1720	0.8	SA	1729	1.0
	2333	5.7		2356	5.3
9	0550	0.8	24	0558	0.9
	1153	5.7		1219	5.3
SA	1819	0.7	SU	1808	0.9
10	0024	5.8	25	0031	5.5
	0649	0.6		0638	0.8
SU	1245	5.9	M	1255	5.5
	1912	0.7		1849	0.8
11	0112	5.9	26	0107	5.6
	0741	0.4		0719	0.7
M	1335	6.0	TU	1331	5.6
	1957	0.7		1928	0.7
12	0158	5.9	27	0145	5.7
	0829	0.3		0801	0.6
TU	1423	6.1	W	1409	5.7
	2039	0.8		2009	0.7
13	0243	5.9	28	0222	5.7
	0914	0.4		0842	0.6
W	1510	6.0	TH	1447	5.7
	2118	0.9		2047	0.8
14	0325	5.8	29	0258	5.7
	0956	0.5		0921	0.6
TH	1555	5.8	F	1525	5.7
	2153	1.1		2122	0.9
15	0409	5.6	30	0334	5.6
	1034	0.7		0955	0.7
F	1640	5.5	SA	1604	5.5
	2227	1.3		2155	1.1
			31	0412	5.5
				1026	0.8
			SU	1647	5.3
				2230	1.2

FEBRUARY

Day	TIME	M	Day	TIME	M
1	0452	5.3	16	0542	5.0
	1101	0.9		1142	1.2
M	1736	5.1	TU	1814	4.8
	2315	1.3		2353	1.6
2	0543	5.2	17	0638	4.7
	1149	1.1		1243	1.5
TU	1836	4.9	W	1913	4.6
3	0014	1.4	18	0107	1.7
	0648	5.0		0751	4.4
W	1256	1.2	TH	1404	1.6
	1951	4.9		2027	4.5
4	0133	1.5	19	0244	1.7
	0811	4.9		0919	4.4
TH	1420	1.2	F	1518	1.5
	2107	5.0		2146	4.6
5	0301	1.3	20	0357	1.4
	0932	5.0		1031	4.7
F	1549	1.2	SA	1617	1.3
	2217	5.2		2247	4.9
6	0430	1.1	21	0452	1.2
	1045	5.3		1119	5.0
SA	1709	1.0	SU	1705	1.1
	2319	5.5		2332	5.2
7	0546	0.8	22	0539	0.9
	1146	5.6		1200	5.3
SU	1810	0.8	M	1750	0.9
8	0012	5.7	23	0012	5.4
	0643	0.5		0622	0.8
M	1239	5.8	TU	1236	5.5
	1900	0.7		1832	0.8
9	0100	5.8	24	0050	5.6
	0733	0.3		0706	0.6
TU	1326	6.0	W	1314	5.7
	1942	0.7		1914	0.7
10	0145	5.9	25	0127	5.8
	0818	0.2		0748	0.5
W	1411	6.0	TH	1351	5.8
	2023	0.7		1955	0.6
11	0226	5.9	26	0204	5.9
	0858	0.3		0839	0.4
TH	1451	6.0	F	1429	5.9
	2100	0.8		2034	0.7
12	0305	5.9	27	0239	5.9
	0935	0.4		0907	0.4
F	1532	5.8	SA	1507	5.8
	2132	0.9		2108	0.8
13	0343	5.7	28	0314	5.8
	1006	0.7		0939	0.6
SA	1610	5.6	SU	1545	5.7
	2159	1.1		2138	0.9
14	0420	5.5			
	1031	0.9			
SU	1647	5.3			
	2226	1.2			
15	0458	5.3			
	1059	1.0			
M	1726	5.1			
	2301	1.4			

MARCH

Day	TIME	M	Day	TIME	M
1	0352	5.7	16	0421	5.3
	1006	0.7		1020	1.0
M	1624	5.4	TU	1641	5.2
	2209	1.0		2224	1.2
2	0433	5.5	17	0501	5.0
	1035	0.9		1057	1.2
TU	1711	5.2	W	1722	4.9
	2251	1.2		2308	1.4
3	0522	5.3	18	0550	4.7
	1120	1.1		1150	1.5
W	1807	4.9	TH	1812	4.6
	2349	1.3			
4	0628	5.0	19	0012	1.6
	1228	1.4		0656	4.4
TH	1921	4.8	F	1306	1.7
				1923	4.4
5	0112	1.4	20	0144	1.7
	0755	4.8		0827	4.3
F	1404	1.5	SA	1434	1.7
	2046	4.8		2054	4.4
6	0254	1.3	21	0314	1.5
	0924	5.0		0953	4.6
SA	1542	1.4	SU	1542	1.4
	2203	5.1		2210	4.7
7	0430	1.0	22	0417	1.2
	1040	5.3		1048	5.0
SU	1701	1.1	M	1637	1.2
	2308	5.4		2302	5.1
8	0540	0.6	23	0509	0.9
	1139	5.6		1132	5.3
M	1758	0.9	TU	1725	0.9
				2344	5.4
9	0000	5.6	24	0557	0.7
	0634	0.4		1211	5.6
TU	1228	5.9	W	1811	0.8
	1845	0.7			
10	0046	5.8	25	0024	5.6
	0719	0.2		0643	0.5
W	1312	6.0	TH	1250	5.8
	1926	0.6		1855	0.7
11	0127	5.9	26	0102	5.8
	0758	0.2		0727	0.4
TH	1352	6.0	F	1328	6.0
	2002	0.6		1937	0.6
12	0205	5.9	27	0140	6.0
	0834	0.3		0809	0.4
F	1429	6.0	SA	1406	6.0
	2036	0.6		2016	0.6
13	0242	5.9	28	0216	6.0
	0905	0.4		0847	0.4
SA	1504	5.8	SU	1444	6.0
	2105	0.8		2051	0.6
14	0315	5.7	29	0253	6.0
	0932	0.7		0919	0.5
SU	1536	5.6	M	1522	5.8
	2131	0.9		2124	0.7
15	0349	5.5	30	0334	5.9
	0955	0.9		0949	0.7
M	1607	5.4	TU	1603	5.5
	2153	1.1		2159	0.9
			31	0417	5.6
				1021	1.0
			W	1649	5.2
				2242	

APRIL

Day	TIME	M	Day	TIME	M
1	0511	5.3	16	0513	4.8
	1109	1.3		1113	1.5
TH	1749	5.0	F	1727	4.7
	2343	1.2		2333	1.5
2	0621	5.0	17	0612	4.5
	1219	1.5		1219	1.7
F	1903	4.8	SA	1831	4.5
3	0112	1.3	18	0055	1.6
	0748	4.9		0734	4.4
SA	1358	1.6	SU	1345	1.7
	2027	4.8		1957	4.4
4	0257	1.1	19	0225	1.4
	0915	5.1		0904	4.6
SU	1534	1.4	M	1500	1.5
	2146	5.1		2122	4.7
5	0421	0.8	20	0332	1.1
	1027	5.5		1007	5.0
M	1647	1.1	TU	1559	1.2
	2251	5.4		2221	5.0
6	0525	0.5	21	0430	0.9
	1123	5.8		1057	5.4
TU	1740	0.9	W	1651	1.0
	2343	5.6		2309	5.4
7	0614	0.3	22	0523	0.6
	1211	5.9		1140	5.7
W	1824	0.7	TH	1740	0.8
				2353	5.6
8	0027	5.7	23	0614	0.5
	0655	0.3		1221	5.9
TH	1252	5.9	F	1828	0.7
	1902	0.6			
9	0106	5.8	24	0034	5.9
	0731	0.3		0700	0.4
F	1328	5.9	SA	1302	6.0
	1935	0.6		1913	0.6
10	0141	5.8	25	0113	6.0
	0802	0.4		0744	0.4
SA	1402	5.9	SU	1342	6.1
	2009	0.6		1957	0.5
11	0215	5.8	26	0154	6.1
	0832	0.5		0825	0.4
SU	1434	5.8	M	1422	6.0
	2039	0.7		2036	0.5
12	0247	5.7	27	0234	6.1
	0858	0.7		0903	0.6
M	1503	5.6	TU	1504	5.8
	2104	0.8		2115	0.6
13	0318	5.5	28	0319	6.0
	0922	0.9		0938	0.8
TU	1532	5.4	W	1548	5.6
	2128	1.0		2157	0.8
14	0352	5.3	29	0410	5.7
	0949	1.1		1017	1.1
W	1603	5.2	TH	1637	5.3
	2157	1.1		2245	0.9
15	0428	5.1	30	0508	5.4
	1026	1.3		1108	1.4
TH	1641	5.0	F	1736	5.1
	2237	1.3		2349	1.1

HEIGHTS IN METRES

ENGLAND, EAST COAST - SHEERNESS

LAT 51°27'N LONG 0°45'E

TIME ZONE GMT

TIMES AND HEIGHTS OF HIGH AND LOW WATERS

YEAR 1982

MAY

	TIME	M		TIME	M
1	0617	5.2	16	0540	4.7
	1217	1.6		1143	1.6
SA	1846	4.9	SU	1753	4.7
2	0116	1.1	17	0017	1.4
	0737	5.1		0648	4.6
SU	1347	1.6	M	1256	1.7
	2006	5.0		1904	4.6
3	0244	0.9	18	0135	1.3
	0857	5.3		0809	4.8
M	1511	1.4	TU	1411	1.5
	2122	5.2		2026	4.7
4	0400	0.6	19	0244	1.1
	1006	5.6		0919	5.1
TU	1621	1.1	W	1512	1.3
	2227	5.4		2135	5.0
5	0501	0.5	20	0343	0.8
	1101	5.8		1016	5.4
W	1715	0.9	TH	1610	1.0
	2319	5.6		2230	5.3
6	0547	0.4	21	0442	0.6
	1149	5.8		1106	5.7
TH	1757	0.8	F	1705	0.8
				2319	5.6
7	0003	5.6	22	0540	0.5
	0625	0.5		1153	5.9
F	1228	5.8	SA	1800	0.7
	1834	0.7			
8	0041	5.7	23	0005	5.9
	0657	0.5		0634	0.5
SA	1303	5.8	SU	1236	6.0
	1907	0.6		1852	0.6
9	0116	5.7	24	0050	6.0
	0728	0.6		0721	0.4
SU	1334	5.7	M	1321	6.0
	1941	0.6		1940	0.5
10	0148	5.7	25	0135	6.1
	0758	0.6		0806	0.5
M	1404	5.7	TU	1405	6.0
	2012	0.7		2026	0.5
11	0220	5.6	26	0222	6.1
	0827	0.8		0849	0.6
TU	1433	5.6	W	1450	5.8
	2042	0.8		2112	0.5
12	0254	5.5	27	0312	6.0
	0856	0.9		0931	0.8
W	1503	5.5	TH	1536	5.7
	2108	0.9		2200	0.6
13	0328	5.3	28	0404	5.9
	0927	1.1		1014	1.1
TH	1535	5.3	F	1628	5.5
	2139	1.0		2251	0.7
14	0406	5.1	29	0501	5.6
	1002	1.3		1104	1.3
F	1613	5.1	SA	1723	5.3
	2217	1.2		2351	0.8
15	0448	4.9	30	0604	5.4
	1045	1.5		1204	1.5
SA	1657	4.9	SU	1827	5.2
	2308	1.3			
			31	0103	0.9
				0714	5.3
			M	1319	1.5
				1937	5.2

JUNE

	TIME	M		TIME	M
1	0218	0.8	16	0050	1.1
	0826	5.4		0720	4.9
TU	1436	1.4	W	1319	1.5
	2050	5.2		1934	4.9
2	0327	0.7	17	0157	1.0
	0935	5.5		0832	5.1
W	1545	1.2	TH	1425	1.3
	2156	5.4		2047	5.0
3	0426	0.6	18	0301	0.9
	1033	5.6		0936	5.3
TH	1641	1.1	F	1528	1.1
	2251	5.4		2150	5.3
4	0512	0.6	19	0403	0.7
	1120	5.6		1034	5.6
F	1726	1.0	SA	1631	1.0
	2337	5.5		2248	5.5
5	0550	0.7	20	0509	0.6
	1201	5.6		1126	5.8
SA	1805	0.9	SU	1736	0.8
				2342	5.8
6	0017	5.5	21	0611	0.6
	0622	0.7		1217	5.9
SU	1236	5.6	M	1835	0.6
	1841	0.8			
7	0053	5.5	22	0034	6.0
	0655	0.7		0704	0.6
M	1307	5.6	TU	1304	5.9
	1914	0.7		1930	0.5
8	0126	5.5	23	0124	6.1
	0727	0.7		0754	0.6
TU	1338	5.6	W	1352	5.9
	1948	0.7		2020	0.4
9	0159	5.5	24	0215	6.1
	0801	0.8		0839	0.6
W	1409	5.6	TH	1439	5.9
	2022	0.8		2110	0.3
10	0234	5.5	25	0304	6.1
	0834	0.9		0924	0.8
TH	1443	5.5	F	1527	5.8
	2056	0.9		2157	0.4
11	0310	5.4	26	0356	6.0
	0908	1.0		1006	1.0
F	1518	5.4	SA	1614	5.7
	2129	1.0		2247	0.5
12	0348	5.3	27	0448	5.8
	0945	1.2		1049	1.2
SA	1555	5.2	SU	1705	5.5
	2207	1.0		2337	0.7
13	0430	5.1	28	0542	5.6
	1026	1.3		1137	1.4
SU	1637	5.1	M	1800	5.4
	2252	1.1			
14	0516	5.0	29	0032	0.8
	1113	1.5		0642	5.4
M	1725	4.9	TU	1236	1.5
	2347	1.2		1900	5.3
15	0611	4.9	30	0135	0.9
	1212	1.5		0747	5.3
TU	1822	4.8	W	1347	1.5
				2008	5.2

JULY

	TIME	M		TIME	M
1	0240	0.9	16	0109	1.1
	0854	5.2		0748	5.0
TH	1500	1.4	F	1338	1.4
	2118	5.1		2004	5.0
2	0341	0.9	17	0219	1.0
	0956	5.3		0900	5.1
F	1603	1.3	SA	1451	1.3
	2220	5.2		2118	5.2
3	0431	0.9	18	0334	1.0
	1051	5.3		1006	5.3
SA	1654	1.1	SU	1606	1.1
	2312	5.2		2226	5.4
4	0513	0.9	19	0448	0.9
	1134	5.3		1106	5.5
SU	1737	1.0	M	1719	0.9
	2356	5.3		2327	5.7
5	0550	0.9	20	0556	0.8
	1212	5.4		1201	5.7
M	1817	0.9	TU	1825	0.6
6	0034	5.3	21	0024	5.9
	0625	0.9		0652	0.7
TU	1245	5.4	W	1253	5.8
	1853	0.8		1923	0.4
7	0107	5.4	22	0116	6.1
	0702	0.8		0741	0.6
W	1319	5.5	TH	1341	5.9
	1930	0.8		2013	0.3
8	0142	5.5	23	0205	6.2
	0738	0.8		0827	0.6
TH	1352	5.6	F	1427	5.9
	2006	0.8		2101	0.2
9	0218	5.5	24	0253	6.2
	0816	0.9		0910	0.7
F	1429	5.5	SA	1512	5.9
	2044	0.8		2146	0.3
10	0254	5.5	25	0339	6.0
	0853	0.9		0949	0.9
SA	1504	5.5	SU	1556	5.8
	2122	0.8		2228	0.5
11	0332	5.5	26	0426	5.8
	0931	1.0		1027	1.1
SU	1541	5.4	M	1640	5.7
	2159	0.9		2309	0.7
12	0410	5.4	27	0512	5.6
	1006	1.2		1102	1.3
M	1617	5.3	TU	1726	5.5
	2237	1.0		2349	0.9
13	0452	5.2	28	0601	5.3
	1045	1.3		1144	1.5
TU	1659	5.2	W	1817	5.2
	2318	1.0			
14	0540	5.1	29	0036	1.1
	1132	1.4		0656	5.1
W	1747	5.1	TH	1242	1.6
				1919	5.0
15	0008	1.1	30	0140	1.3
	0638	5.0		0801	4.9
TH	1229	1.4	F	1402	1.6
	1849	5.0		2032	4.9
			31	0249	1.3
				0911	4.9
			SA	1519	1.5
				2148	4.9

AUGUST

	TIME	M		TIME	M
1	0348	1.3	16	0314	1.3
	1016	4.9		0943	5.1
SU	1623	1.3	M	1552	1.2
	2248	5.0		2214	5.3
2	0438	1.2	17	0435	1.1
	1108	5.1		1051	5.4
M	1713	1.2	TU	1713	0.9
	2336	5.1		2320	5.6
3	0522	1.1	18	0544	0.9
	1149	5.2		1150	5.6
TU	1756	1.0	W	1819	0.6
4	0014	5.3	19	0017	5.9
	0603	1.0		0639	0.8
W	1225	5.3	TH	1241	5.8
	1834	0.9		1913	0.3
5	0049	5.4	20	0106	6.1
	0641	0.9		0727	0.7
TH	1300	5.5	F	1326	5.9
	1912	0.8		2001	0.2
6	0124	5.5	21	0152	6.2
	0719	0.8		0811	0.6
F	1335	5.6	SA	1409	6.0
	1951	0.7		2044	0.2
7	0159	5.6	22	0236	6.1
	0758	0.8		0850	0.7
SA	1411	5.7	SU	1451	6.0
	2030	0.7		2125	0.3
8	0234	5.7	23	0317	6.0
	0836	0.8		0925	0.9
SU	1446	5.7	M	1531	5.9
	2108	0.7		2200	0.6
9	0311	5.7	24	0356	5.8
	0912	0.9		0956	1.1
M	1521	5.6	TU	1609	5.7
	2143	0.8		2230	0.9
10	0348	5.6	25	0435	5.5
	0945	[illegible]		1023	1.3
TU	1555	5.5	W	1648	5.4
	2216	0.9		2258	1.1
11	0427	5.4	26	0515	5.2
	1017	1.2		1055	1.4
W	1633	5.4	TH	1732	5.2
	2247	1.0		2333	1.3
12	0509	5.2	27	0600	4.9
	1057	1.3		1140	1.6
TH	1718	5.2	F	1825	4.8
	2327	1.1			
13	0603	5.0	28	0029	1.6
	1147	1.4		0656	4.7
F	1815	5.1	SA	1250	1.8
				1935	4.6
14	0025	1.2	29	0148	1.7
	0710	4.9		0809	4.5
SA	1257	1.5	SU	1430	1.7
	1933	4.9		2107	4.6
15	0145	1.3	30	0305	1.6
	0829	4.9		0934	4.6
SU	1425	1.4	M	1549	1.5
	2057	5.0		2221	4.8
			31	0406	1.4
				1037	4.9
			TU	1645	1.3
				2312	5.1

HEIGHTS IN METRES

ENGLAND, EAST COAST - SHEERNESS

LAT 51°27′N LONG 0°45′E

TIME ZONE **GMT** TIMES AND HEIGHTS OF HIGH AND LOW WATERS YEAR **1982**

SEPTEMBER

Day	TIME	M	Day	TIME	M
1 W	0455	1.2	16 TH	0532	1.0
	1123	5.1		1136	5.6
	1732	1.1		1808	0.5
	2350	5.3			
2 TH	0539	1.1	17 F	0004	5.9
	1201	5.3		0624	0.8
	1811	0.9		1224	5.8
				1857	0.3
3 F	0025	5.5	18 SA	0050	6.1
	0618	0.9		0707	0.7
	1236	5.5		1306	5.9
	1850	0.8		1940	0.3
4 SA	0100	5.6	19 SU	0133	6.1
	0657	0.8		0745	0.7
	1312	5.7		1347	6.0
	1930	0.7		2019	0.3
5 SU	0135	5.8	20 M	0212	6.0
	0737	0.8		0822	0.7
	1347	5.8		1425	6.0
	2009	0.6		2053	0.5
6 M	0211	5.8	21 TU	0249	5.9
	0815	0.8		0854	0.8
	1420	5.8		1501	5.9
	2047	0.6		2122	0.8
7 TU	0246	5.8	22 W	0322	5.7
	0850	0.8		0922	1.0
	1456	5.8		1535	5.6
	2121	0.7		2146	1.0
8 W	0322	5.7	23 TH	0355	5.4
	0921	1.0		0945	1.2
	1531	5.7		1610	5.4
	2149	0.9		2210	1.2
9 TH	0400	5.5	24 F	0428	5.1
	0950	1.1		1010	1.4
	1609	5.6		1648	5.1
	2216	1.1		2242	1.5
10 F	0442	5.2	25 SA	0506	4.9
	1028	1.3		1051	1.6
	1654	5.3		1736	4.8
	2255	1.3		2332	1.7
11 SA	0533	5.0	26 SU	0556	4.6
	1120	1.4		1151	1.8
	1753	5.0		1841	4.5
	2354	1.5			
12 SU	0642	4.8	27 M	0045	1.9
	1234	1.5		0704	4.4
	1914	4.9		1326	1.9
				2013	4.4
13 M	0124	1.6	28 TU	0218	1.9
	0806	4.8		0837	4.4
	1416	1.5		1505	1.7
	2047	5.0		2143	4.6
14 TU	0307	1.5	29 W	0329	1.6
	0929	5.0		0957	4.7
	1552	1.1		1609	1.3
	2207	5.3		2237	5.0
15 W	0428	1.2	30 TH	0423	1.3
	1038	5.3		1048	5.1
	1709	0.8		1657	1.1
	2312	5.7		2319	5.3

OCTOBER

Day	TIME	M	Day	TIME	M
1 F	0508	1.1	16 SA	0601	0.9
	1129	5.4		1203	5.8
	1739	0.9		1835	0.4
	2356	5.6			
2 SA	0549	0.9	17 SU	0029	6.0
	1207	5.6		0641	0.8
	1819	0.7		1243	5.9
				1912	0.5
3 SU	0031	5.8	18 M	0109	6.0
	0629	0.8		0717	0.7
	1242	5.8		1320	5.9
	1900	0.6		1945	0.5
4 M	0107	5.9	19 TU	0144	5.9
	0710	0.7		0751	0.7
	1317	5.9		1355	5.9
	1941	0.6		2015	0.7
5 TU	0144	6.0	20 W	0216	5.8
	0748	0.7		0822	0.8
	1352	6.0		1430	5.8
	2019	0.6		2043	0.9
6 W	0220	5.9	21 TH	0247	5.6
	0825	0.8		0849	1.0
	1429	6.0		1503	5.6
	2053	0.7		2107	1.1
7 TH	0257	5.8	22 F	0317	5.4
	0858	0.9		0912	1.2
	1507	5.9		1536	5.4
	2122	0.9		2132	1.3
8 F	0336	5.5	23 SA	0348	5.2
	0932	1.1		0938	1.3
	1550	5.7		1613	5.1
	2155	1.2		2203	1.5
9 SA	0420	5.2	24 SU	0421	5.0
	1013	1.2		1013	1.5
	1640	5.4		1657	4.8
	2237	1.4		2248	1.7
10 SU	0513	4.9	25 M	0506	4.7
	1109	1.4		1106	1.7
	1744	5.0		1754	4.5
	2342	1.7		2353	1.9
11 M	0624	4.7	26 TU	0607	4.5
	1232	1.5		1227	1.8
	1909	4.9		1912	4.4
12 TU	0117	1.8	27 W	0119	2.0
	0749	4.8		0731	4.4
	1420	1.4		1405	1.7
	2039	5.0		2044	4.5
13 W	0257	1.6	28 TH	0239	1.7
	0912	5.0		0901	4.6
	1548	1.0		1517	1.4
	2156	5.4		2150	4.9
14 TH	0414	1.3	29 F	0339	1.4
	1021	5.4		1003	5.0
	1657	0.6		1610	1.1
	2257	5.8		2238	5.3
15 F	0513	1.0	30 SA	0428	1.2
	1116	5.7		1049	5.3
	1750	0.4		1658	0.8
	2347	5.9		2319	5.6
			31 SU	0513	0.9
				1130	5.6
				1744	0.7
				2358	5.8

NOVEMBER

Day	TIME	M	Day	TIME	M
1 M	0558	0.8	16 TU	0042	5.8
	1210	5.8		0646	0.8
	1829	0.6		1255	5.7
				1909	0.7
2 TU	0038	6.0	17 W	0116	5.7
	0642	0.7		0720	0.8
	1248	6.0		1330	5.7
	1912	0.5		1937	0.8
3 W	0117	6.0	18 TH	0147	5.7
	0724	0.7		0752	0.8
	1327	6.1		1402	5.7
	1952	0.6		2006	0.9
4 TH	0155	5.9	19 F	0216	5.6
	0805	0.7		0822	0.9
	1408	6.1		1436	5.6
	2030	0.7		2034	1.0
5 F	0236	5.8	20 SA	0246	5.5
	0846	0.8		0849	1.0
	1451	6.0		1511	5.4
	2105	0.9		2104	1.2
6 SA	0318	5.6	21 SU	0317	5.3
	0927	1.0		0915	1.2
	1539	5.7		1548	5.2
	2145	1.2		2136	1.4
7 SU	0406	5.3	22 M	0352	6.1
	1013	1.1		0950	1.3
	1634	5.4		1628	5.0
	2231	1.5		2217	1.6
8 M	0502	5.0	23 TU	0434	4.9
	1113	1.3		1038	1.4
	1740	5.2		1718	4.7
	2336	1.8		2312	1.7
9 TU	0610	4.9	24 W	0526	4.7
	1238	1.3		1143	1.5
	1859	5.0		1819	4.6
10 W	0103	1.8	25 TH	0021	1.8
	0728	4.9		0632	4.6
	1411	1.2		1303	1.5
	2020	5.2		1938	4.6
11 TH	0233	1.7	26 F	0138	1.8
	0847	5.1		0755	4.6
	1528	0.9		1418	1.3
	2134	5.4		2053	4.9
12 F	0348	1.4	27 SA	0244	1.5
	0955	5.4		0907	4.9
	1633	0.6		1518	1.1
	2233	5.7		2152	5.2
13 SA	0447	1.1	28 SU	0342	1.3
	1051	5.6		1003	5.2
	1725	0.6		1613	0.8
	2323	5.8		2241	5.5
14 SU	0533	1.0	29 M	0435	1.0
	1137	5.7		1052	5.5
	1805	0.6		1706	0.7
				2327	5.8
15 M	0005	5.8	30 TU	0527	0.9
	0612	0.9		1139	5.8
	1218	5.7		1758	0.6
	1838	0.7			

DECEMBER

Day	TIME	M	Day	TIME	M
1 W	0011	5.9	16 TH	0050	5.6
	0618	0.8		0656	0.8
	1224	5.9		1309	5.5
	1848	0.6		1906	0.8
2 TH	0055	5.9	17 F	0121	5.6
	0707	0.7		0730	0.8
	1309	6.1		1344	5.5
	1933	0.6		1940	0.9
3 F	0138	5.9	18 SA	0152	5.6
	0755	0.7		0802	0.8
	1355	6.1		1418	5.5
	2016	0.7		2012	0.9
4 SA	0222	5.8	19 SU	0225	5.5
	0842	0.7		0834	0.9
	1443	6.0		1453	5.4
	2058	0.9		2044	1.1
5 SU	0308	5.6	20 M	0258	5.4
	0931	0.8		0905	1.0
	1535	5.8		1529	5.3
	2142	1.2		2118	1.2
6 M	0357	5.5	21 TU	0334	5.3
	1021	0.9		0939	1.1
	1630	5.6		1609	5.2
	2228	1.4		2156	1.3
7 TU	0451	5.3	22 W	0412	5.1
	1118	1.0		1020	1.2
	1730	5.4		1651	5.0
	2323	1.6		2240	1.5
8 W	0551	5.1	23 TH	0455	5.0
	1227	1.1		1111	1.2
	1838	5.2		1740	4.9
				2333	1.6
9 TH	0032	1.7	24 F	0547	4.8
	0659	5.1		1211	1.3
	1341	1.0		1842	4.8
	1949	5.2			
10 F	0152	1.7	25 SA	0038	1.6
	0811	5.1		0652	4.8
	1453	0.9		1319	1.2
	2100	5.3		1954	4.8
11 SA	0307	1.5	26 SU	0147	1.6
	0921	5.3		0808	4.8
	1556	0.8		1426	1.1
	2203	5.4		2103	5.1
12 SU	0412	1.3	27 M	0254	1.4
	1021	5.4		0918	5.0
	1648	0.8		1529	0.9
	2255	5.5		2204	5.3
13 M	0502	1.1	28 TU	0359	1.2
	1113	5.4		1019	5.3
	1730	0.8		1633	0.8
	2340	5.5		2259	5.6
14 TU	0544	1.0	29 W	0501	1.0
	1157	5.5		1115	5.6
	1803	0.8		1734	0.7
				2350	5.7
15 W	0018	5.5	30 TH	0601	0.8
	0621	0.9		1207	5.8
	1234	5.5		1831	0.7
	1834	0.8			
			31 F	0039	5.8
				0659	0.6
				1257	6.0
				1923	0.7

HEIGHTS IN METRES

SCOTLAND, WEST COAST - GREENOCK

LAT 55°57'N LONG 4°46'W

TIME ZONE **GMT** TIMES AND HEIGHTS OF HIGH AND LOW WATERS YEAR **1982**

MAY

	TIME	M		TIME	M
1	0525	3.2	**16**	0510	3.0
	1215	0.7		1141	0.9
SA	1854	2.8	SU	1754	2.8
2	0109	0.9	**17**	0001	1.2
	0634	3.0		0607	2.9
SU	1340	0.6	M	1252	0.9
	2058	2.7		1857	2.8
3	0222	0.8	**18**	0118	1.2
	0824	2.9		0726	2.9
M	1446	0.5	TU	1356	0.7
	2203	2.9		2021	2.8
4	0320	0.6	**19**	0225	1.1
	0940	3.0		0853	3.0
TU	1536	0.3	W	1449	0.6
	2246	3.0		2128	3.0
5	0405	0.5	**20**	0317	0.9
	1030	3.1		0952	3.1
W	1618	0.3	TH	1535	0.4
	2323	3.1		2217	3.1

JUNE

	TIME	M		TIME	M
1	0149	0.8	**16**	0020	1.1
	0738	2.9		0635	3.0
TU	1409	0.5	W	1300	0.7
	2124	2.8		1922	2.9
2	0247	0.7	**17**	0128	1.1
	0858	2.9		0756	3.0
W	1503	0.5	TH	1401	0.6
	2211	2.9		2036	3.0
3	0336	0.6	**18**	0233	0.9
	0954	3.0		0909	3.1
TH	1548	0.5	F	1457	0.5
	2251	3.0		2138	3.1
4	0418	0.6	**19**	0328	0.7
	1041	3.1		1008	3.3
F	1629	0.5	SA	1548	0.4
	2329	3.1		2229	3.3
5	0456	0.5	**20**	0416	0.5
	1125	3.2		1103	3.3
SA	1708	0.5	SU	1637	0.3
				2320	3.4

JULY

	TIME	M		TIME	M
1	0208	0.8	**16**	0039	1.0
	0807	2.9		0711	3.1
TH	1424	0.7	F	1320	0.7
	2130	2.8		1948	3.0
2	0304	0.8	**17**	0156	1.0
	0914	2.9		0833	3.1
F	1518	0.7	SA	1428	0.6
	2219	3.0		2102	3.1
3	0352	0.7	**18**	0304	0.8
	1009	3.0		0945	3.1
SA	1605	0.7	SU	1529	0.6
	2301	3.1		2205	3.2
4	0436	0.6	**19**	0401	0.6
	1057	3.0		1047	3.2
SU	1648	0.7	M	1623	0.5
	2340	3.2		2301	3.4
5	0516	0.5	**20**	0451	0.4
	1142	3.0		1147	3.2
M	1728	0.6	TU	1714	0.4
				2354	3.5

AUGUST

	TIME	M		TIME	M
1	0329	0.8	**16**	0253	0.8
	0939	2.8		0938	3.0
SU	1545	0.9	M	1521	0.7
	2236	3.0		2152	3.1
2	0417	0.7	**17**	0352	0.5
	1035	2.9		1046	3.0
M	1631	0.8	TU	1615	0.5
	2319	3.1		2252	3.3
3	0458	0.6	**18**	0442	0.3
	1123	2.9		1145	3.1
TU	1711	0.7	W	1704	0.4
	2357	3.2		2345	3.4
4	0535	0.4	**19**	0528	0.2
	1207	2.9		1239	3.1
W	1745	0.6	TH	1750	0.4
5	0031	3.2	**20**	0037	3.6
	0608	0.3		0612	0.1
TH	1247	2.9	F	1330	3.1
	1816	0.6		1835	0.4

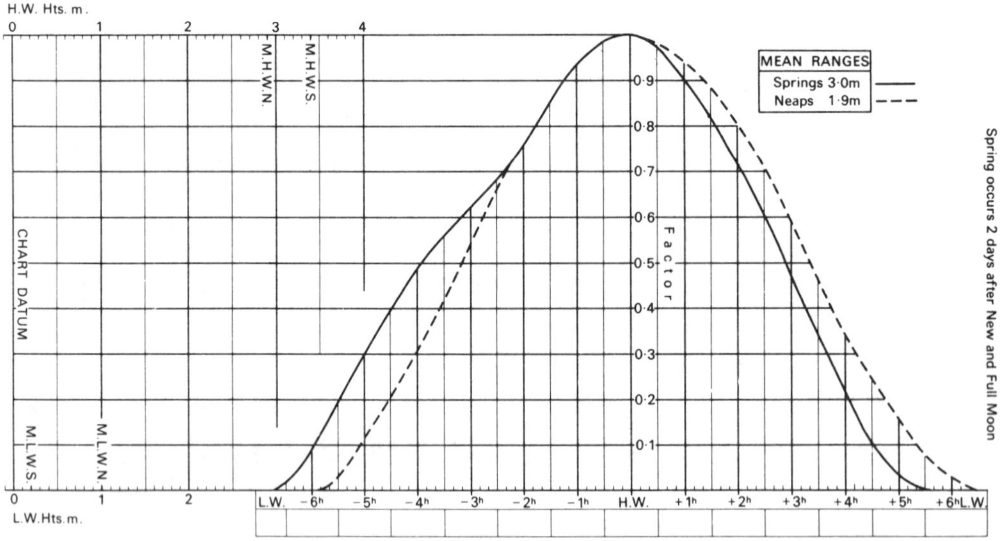

GREENOCK
MEAN SPRING AND NEAP CURVES
Spring occurs 2 days after New and Full Moon

ENGLAND, WEST COAST - PORT OF BRISTOL (AVONMOUTH)

LAT 51°30'N LONG 2°43'W

TIME ZONE GMT TIMES AND HEIGHTS OF HIGH AND LOW WATERS YEAR 1982

SEPTEMBER

		TIME	M			TIME	M
11	SA	0608	3.1	26	SU	0003	9.5
		1210	10.4			0615	3.8
		1834	3.6			1235	9.1
						1848	4.3
12	SU	0100	10.1	27	M	0114	8.8
		0709	3.7			0716	4.6
		1340	10.0			1426	8.8
		2004	4.0			2040	4.7
13	M	0233	10.0	28	TU	0310	9.0
		0905	3.8			0938	4.5
		1511	10.3			1600	9.6
		2200	3.4			2226	3.8
14	TU	0400	10.7	29	W	0427	9.8
		1038	3.0			1051	3.5
		1634	11.2			1655	10.6
		2327	2.4			2323	2.8

OCTOBER

		TIME	M			TIME	M
11	M	0055	9.9	26	TU	0015	8.8
		0700	3.9			0628	4.4
		1328	9.9			1300	8.7
		1959	3.9			1919	4.6
12	TU	0220	9.9	27	W	0154	8.6
		0853	3.8			0751	4.7
		1457	10.3			1500	9.1
		2145	3.2			2122	4.3
13	W	0348	10.6	28	TH	0331	9.4
		1021	2.9			0953	4.1
		1620	11.3			1603	10.2
		2306	2.2			2233	3.3
14	TH	0459	11.7	29	F	0428	10.5
		1139	1.8			1058	3.1
		1723	12.3			1652	11.3
						2329	2.5

NOVEMBER

		TIME	M			TIME	M
11	TH	0321	10.8	26	F	0218	9.6
		0952	2.8			0826	4.1
		1553	11.3			1503	10.2
		2230	2.2			2121	3.5
12	F	0433	11.7	27	SA	0334	10.4
		1104	2.0			0953	3.4
		1657	12.2			1604	11.2
		2340	1.6			2231	2.7
13	SA	0526	12.5	28	SU	0431	11.5
		1207	1.4			1104	2.7
		1747	12.8			1658	12.2
						2339	2.1
14	SU	0035	1.2	29	M	0525	12.4
		0612	13.0			1212	2.0
		1257	1.2			1750	13.0
		1829	13.1				

DECEMBER

		TIME	M			TIME	M
11	SA	0353	11.3	26	SU	0237	10.5
		1016	2.5			0853	3.6
		1621	11.7			1518	11.1
		2249	2.2			2142	3.0
12	SU	0452	11.9	27	M	0349	11.3
		1120	2.1			1020	3.1
		1716	12.2			1623	11.9
		2351	2.0			2258	2.4
13	M	0543	12.4	28	TU	0451	12.1
		1218	1.8			1136	2.4
		1803	12.5			1722	12.6
14	TU	0042	1.7	29	W	0011	1.9
		0625	12.7			0550	12.8
		1304	1.6			1248	1.8
		1845	12.7			1819	13.1

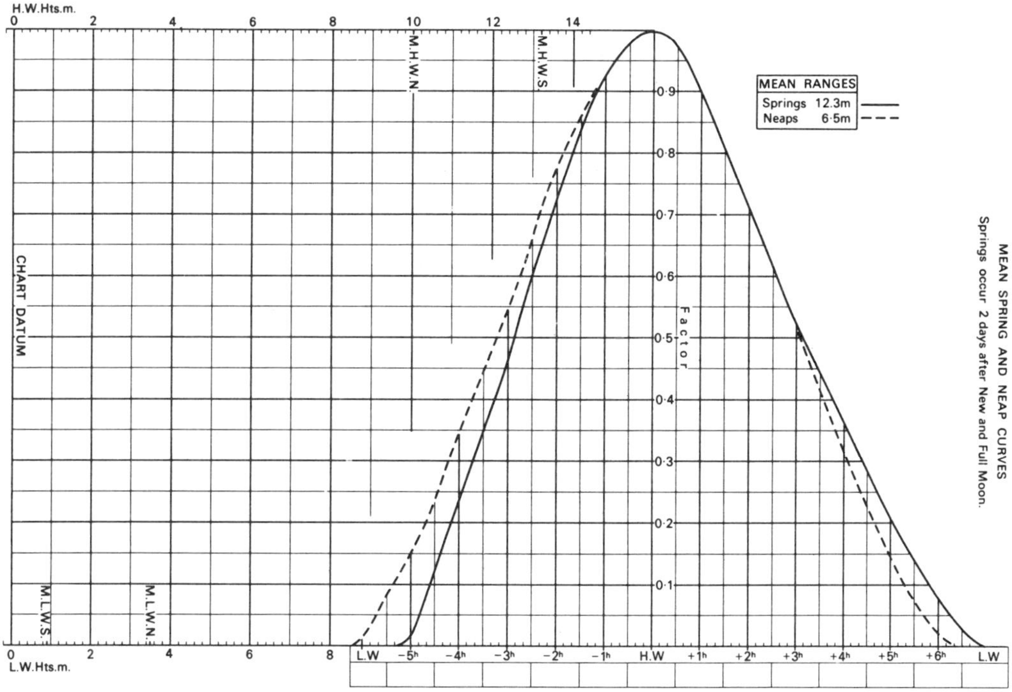

AVONMOUTH

MEAN SPRING AND NEAP CURVES

Springs occur 2 days after New and Full Moon.

IRELAND, NORTH COAST - LONDONDERRY

LAT 55°00'N LONG 7°19'W

TIME ZONE GMT — TIMES AND HEIGHTS OF HIGH AND LOW WATERS — YEAR **1982**

MAY

Day	TIME	M	Day	TIME	M
7	0131	0.4	22	0113	0.4
	0748	2.5		0720	2.8
F	1409	0.4	SA	1335	0.3
	2001	2.5		1942	2.6
8	0209	0.4	23	0201	0.3
	0820	2.5		0805	2.8
SA	1443	0.4	SU	1419	0.2
	2034	2.6		2023	2.7
9	0246	0.4	24	0247	0.3
	0851	2.4		0850	2.8
SU	1515	0.4	M	1503	0.2
	2105	2.6		2105	2.7
10	0321	0.5	25	0335	0.3
	0921	2.4		0935	2.7
M	1546	0.5	TU	1545	0.3
	2136	2.5		2148	2.7
11	0353	0.6	26	0423	0.4
	0949	2.3		1021	2.5
TU	1617	0.5	W	1628	0.4
	2209	2.5		2233	2.6

JUNE

Day	TIME	M	Day	TIME	M
7	0227	0.6	22	0237	0.4
	0830	2.3		0837	2.7
M	1450	0.5	TU	1446	0.3
	2044	2.5		2054	2.8
8	0304	0.6	23	0328	0.3
	0901	2.3		0924	2.7
TU	1524	0.5	W	1531	0.3
	2118	2.6		2139	2.8
9	0339	0.6	24	0416	0.4
	0932	2.2		1012	2.5
W	1556	0.5	TH	1614	0.4
	2152	2.5		2224	2.7
10	0414	0.7	25	0506	0.5
	1006	2.2		1059	2.4
TH	1631	0.6	F	1659	0.5
	2227	2.5		2313	2.6
11	0452	0.7	26	0600	0.6
	1044	2.1		1153	2.2
F	1709	0.7	SA	1746	0.6
	2308	2.4			

JULY

Day	TIME	M	Day	TIME	M
7	0250	0.7	22	0318	0.3
	0844	2.3		0910	2.7
W	1503	0.5	TH	1515	0.3
	2100	2.6		2127	2.9
8	0325	0.7	23	0404	0.4
	0918	2.3		0955	2.6
TH	1536	0.5	F	1557	0.3
	2134	2.6		2210	2.9
9	0400	0.7	24	0449	0.4
	0950	2.3		1038	2.5
F	1612	0.5	SA	1640	0.4
	2209	2.6		2255	2.7
10	0435	0.7	25	0536	0.5
	1026	2.2		1125	2.3
SA	1647	0.6	SU	1722	0.5
	2247	2.6		2342	2.5
11	0513	0.7	26	0625	0.7
	1105	2.1		1214	2.1
SU	1725	0.7	M	1805	0.7
	2329	2.5			

AUGUST

Day	TIME	M	Day	TIME	M
7	0338	0.6	22	0424	0.4
	0931	2.4		1012	2.6
SA	1548	0.5	SU	1614	0.3
	2148	2.8		2228	2.8
8	0412	0.6	23	0504	0.5
	1003	2.4		1051	2.4
SU	1621	0.5	M	1654	0.5
	2223	2.8		2306	2.5
9	0448	0.6	24	0544	0.7
	1040	2.3		1130	2.3
M	1658	0.6	TU	1732	0.6
	2304	2.7		2347	2.3
10	0527	0.7	25	0628	0.9
	1118	2.2		1215	2.1
TU	1737	0.7	W	1815	0.8
	2349	2.5			
11	0612	0.7	26	0035	2.1
	1204	2.1		0723	1.0
W	1825	0.8	TH	1313	2.0
				1909	1.0

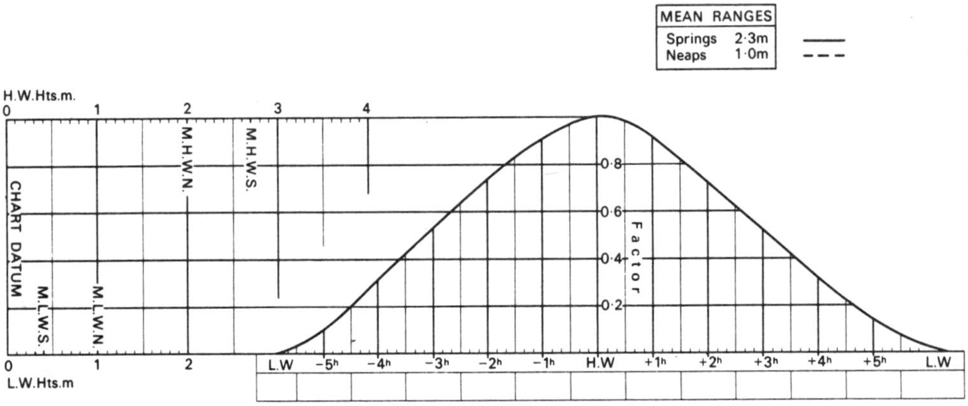

LONDONDERRY
MEAN SPRING AND NEAP CURVES
Springs occur 1 day after New and Full Moon.

ST HELIER

MEAN SPRING AND NEAP CURVES

Springs occur 2 days after New and Full Moon.

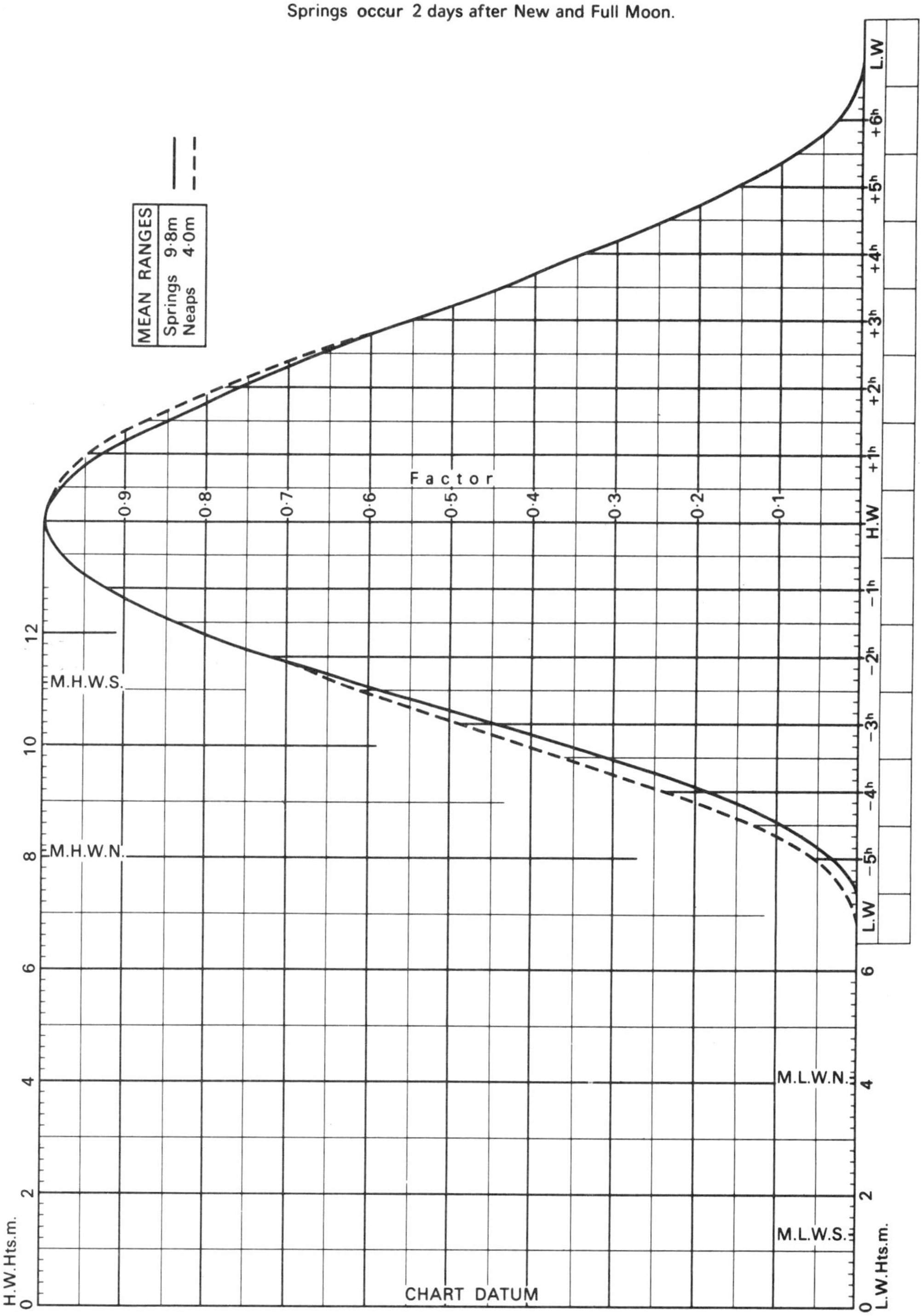

CHANNEL ISLANDS - ST. HELIER

LAT 49°11'N LONG 2°07'W

TIME ZONE GMT — TIMES AND HEIGHTS OF HIGH AND LOW WATERS — YEAR 1982

JANUARY

Day	TIME	M	Day	TIME	M
1 F	0403 0948 1627 2210	3.0 9.6 2.9 9.3	16 SA	0502 1045 1727 2316	3.2 9.2 3.3 8.7
2 SA	0440 1033 1709 2301	3.4 9.3 3.3 9.0	17 SU	0543 1134 1812	3.7 8.7 3.8
3 SU	0530 1129 1807	3.7 9.0 3.5	18 M	0011 0638 1236 1917	8.2 4.1 8.2 4.1
4 M	0005 0639 1242 1923	8.8 3.8 8.8 3.5	19 TU	0121 0749 1351 2030	8.0 4.1 8.1 4.1
5 TU	0123 0804 1401 2042	8.8 3.6 9.0 3.2	20 W	0236 0901 1504 2136	8.2 3.9 8.2 3.8
6 W	0240 0919 1514 2153	9.2 3.1 9.5 2.7	21 TH	0339 1004 1606 2234	8.5 3.5 8.6 3.4
7 TH	0346 1027 1619 2257	9.8 2.5 10.1 2.1	22 F	0434 1059 1658 2325	9.0 3.1 9.0 3.0
8 F	0447 1129 1718 2357	10.5 1.8 10.6 1.6	23 SA	0520 1150 1744	9.5 2.6 9.4
9 SA	0543 1228 1812	11.0 1.3 11.0	24 SU	0011 0603 1236 1824	2.6 9.9 2.2 9.8
10 SU	0053 0634 1324 1903	1.2 11.3 0.9 11.2	25 M	0056 0641 1319 1900	2.3 10.2 1.9 10.1
11 M	0145 0721 1413 1951	1.1 11.4 0.8 11.2	26 TU	0135 0716 1358 1935	2.0 10.5 1.7 10.3
12 TU	0233 0806 1458 2033	1.1 11.3 1.0 10.9	27 W	0212 0749 1433 2008	1.9 10.7 1.6 10.4
13 W	0315 0847 1539 2114	1.5 10.9 1.4 10.4	28 TH	0247 0822 1507 2040	1.9 10.7 1.7 10.5
14 TH	0353 0927 1616 2152	2.0 10.4 2.0 9.8	29 F	0318 0854 1538 2112	2.0 10.6 1.9 10.3
15 F	0427 1004 1651 2231	2.6 9.8 2.7 9.2	30 SA	0349 0929 1610 2150	2.3 10.4 2.2 10.0
			31 SU	0423 1009 1647 2233	2.7 10.0 2.6 9.6

FEBRUARY

Day	TIME	M	Day	TIME	M
1 M	0504 1058 1734 2327	3.1 9.5 3.1 9.1	16 TU	0534 1130 1800	3.7 8.4 4.0
2 TU	0603 1200 1843	3.5 9.0 3.5	17 W	0005 0636 1239 1913	8.0 4.1 7.8 4.4
3 W	0041 0726 1324 2011	8.8 3.7 8.7 3.5	18 TH	0133 0759 1415 2042	7.8 4.3 7.7 4.4
4 TH	0209 0856 1454 2135	8.8 3.5 8.9 3.2	19 F	0301 0924 1538 2159	8.0 4.1 8.0 4.0
5 F	0332 1014 1610 2247	9.2 2.9 9.5 2.6	20 SA	0407 1031 1637 2259	8.5 3.5 8.6 3.4
6 SA	0438 1123 1713 2351	9.9 2.2 10.1 2.0	21 SU	0458 1127 1723 2350	9.1 2.9 9.2 2.8
7 SU	0536 1224 1807	10.6 1.5 10.7	22 M	0542 1215 1803	9.7 2.3 9.8
8 M	0046 0627 1316 1855	1.4 11.1 1.0 11.1	23 TU	0035 0619 1257 1839	2.2 10.3 1.8 10.3
9 TU	0135 0712 1402 1937	1.1 11.3 0.8 11.2	24 W	0114 0655 1337 1913	1.8 10.8 1.4 10.8
10 W	0219 0752 1443 2016	1.0 11.3 0.8 11.0	25 TH	0152 0728 1413 1945	1.4 11.1 1.1 11.0
11 TH	0257 0829 1518 2050	1.2 11.1 1.2 10.7	26 F	0227 0802 1447 2019	1.3 11.3 1.1 11.1
12 F	0328 0901 1549 2122	1.6 10.7 1.7 10.2	27 SA	0301 0834 1519 2053	1.4 11.2 1.3 11.0
13 SA	0356 0934 1614 2153	2.1 10.2 2.3 9.7	28 SU	0334 0910 1553 2128	1.7 10.9 1.7 10.6
14 SU	0423 1006 1640 2227	2.7 9.6 2.9 9.1			
15 M	0452 1042 1712 2308	3.2 9.0 3.5 8.6			

MARCH

Day	TIME	M	Day	TIME	M
1 M	0407 0948 1628 2209	2.1 10.4 2.3 10.0	16 TU	0416 1003 1630 2221	2.8 9.3 3.2 8.9
2 TU	0447 1033 1713 2259	2.7 9.7 2.9 9.3	17 W	0452 1041 1709 2305	3.4 8.5 3.9 8.2
3 W	0540 1132 1818	3.4 8.9 3.6	18 TH	0543 1136 1810	4.0 7.8 4.4
4 TH	0008 0702 1300 1951	8.6 3.8 8.3 3.9	19 F	0024 0702 1320 1945	7.7 4.5 7.4 4.7
5 F	0151 0843 1449 2127	8.3 3.8 8.4 3.6	20 SA	0220 0839 1504 2122	7.7 4.4 7.7 4.4
6 SA	0325 1010 1607 2242	8.8 3.2 9.1 2.9	21 SU	0336 1000 1607 2230	8.2 3.9 8.3 3.7
7 SU	0433 1119 1708 2343	9.6 2.3 9.9 2.2	22 M	0428 1058 1652 2322	8.9 3.1 9.1 2.9
8 M	0526 1214 1756	10.3 1.6 10.5	23 TU	0511 1146 1732	9.7 2.4 9.9
9 TU	0035 0612 1302 1839	1.6 10.9 1.1 11.0	24 W	0005 0549 1228 1808	2.2 10.4 1.7 10.6
10 W	0119 0653 1344 1917	1.2 11.2 0.8 11.1	25 TH	0046 0625 1309 1845	1.6 11.0 1.2 11.2
11 TH	0158 0730 1420 1951	1.1 11.3 0.9 11.1	26 F	0127 0702 1347 1920	1.2 11.5 0.8 11.5
12 F	0232 0804 1451 2022	1.2 11.1 1.1 10.8	27 SA	0205 0738 1425 1957	0.9 11.7 0.7 11.6
13 SA	0300 0833 1517 2050	1.4 10.8 1.6 10.5	28 SU	0242 0815 1501 2032	1.0 11.6 0.9 11.4
14 SU	0325 0903 1539 2118	1.8 10.4 2.1 10.1	29 M	0318 0851 1538 2110	1.3 11.2 1.4 10.9
15 M	0349 0931 1603 2148	2.3 9.9 2.6 9.5	30 TU	0355 0931 1616 2150	1.8 10.6 2.2 10.2
			31 W	0437 1017 1702 2241	2.6 9.7 3.0 9.3

APRIL

Day	TIME	M	Day	TIME	M
1 TH	0530 1116 1805 2353	3.4 8.8 3.8 8.4	16 F	0511 1057 1730 2333	3.9 7.9 4.4 7.8
2 F	0652 1255 1942	3.9 8.1 4.2	17 SA	0617 1225 1853	4.4 7.4 4.7
3 SA	0144 0837 1446 2121	8.2 3.9 8.3 3.8	18 SU	0127 0751 1418 2036	7.6 4.5 7.6 4.5
4 SU	0315 1004 1559 2233	8.7 3.2 9.0 3.1	19 M	0251 0917 1525 2150	8.1 4.0 8.3 3.9
5 M	0419 1106 1651 2327	9.4 2.4 9.8 2.4	20 TU	0346 1017 1613 2242	8.8 3.3 9.1 3.1
6 TU	0508 1156 1736	10.1 1.8 10.4	21 W	0431 1106 1654 2329	9.7 2.5 10.0 2.3
7 W	0012 0550 1238 1814	1.8 10.7 1.3 10.7	22 TH	0513 1151 1734	10.4 1.7 10.8
8 TH	0053 0628 1317 1850	1.4 11.0 1.1 10.9	23 F	0014 0554 1236 1815	1.6 11.1 1.1 11.4
9 F	0130 0703 1351 1923	1.3 11.0 1.2 10.9	24 SA	0057 0635 1320 1855	1.1 11.6 0.8 11.7
10 SA	0202 0735 1420 1952	1.3 10.9 1.4 10.8	25 SU	0141 0716 1402 1934	0.8 11.8 0.7 11.8
11 SU	0230 0805 1446 2020	1.5 10.7 1.7 10.5	26 M	0223 0757 1444 2015	0.8 11.6 0.9 11.5
12 M	0256 0834 1510 2049	1.8 10.4 2.1 10.2	27 TU	0305 0837 1525 2056	1.1 11.2 1.4 11.0
13 TU	0322 0903 1534 2117	2.2 9.9 2.6 9.7	28 W	0348 0921 1609 2141	1.7 10.5 2.2 10.2
14 W	0350 0932 1602 2149	2.7 9.3 3.2 9.1	29 TH	0433 1010 1658 2234	2.5 9.6 3.0 9.3
15 TH	0426 1007 1638 2228	3.3 8.6 3.8 8.5	30 F	0529 1115 1801 2347	3.2 8.7 3.8 8.5

HEIGHTS IN METRES

CHANNEL ISLANDS - ST. HELIER

LAT 49°11'N LONG 2°07'W

TIME ZONE GMT — TIMES AND HEIGHTS OF HIGH AND LOW WATERS — YEAR 1982

MAY

Day	TIME	M	Day	TIME	M
1 SA	0646 1250 1931	3.8 8.2 4.2	16 SU	0546 1144 1812	4.1 7.8 4.5
2 SU	0128 0823 1426 2101	8.3 3.8 8.4 3.9	17 M	0029 0702 1317 1941	7.9 4.3 7.8 4.4
3 M	0251 0942 1534 2209	8.6 3.3 8.9 3.3	18 TU	0154 0823 1430 2058	8.2 3.9 8.4 3.9
4 TU	0350 1040 1624 2259	9.2 2.7 9.5 2.6	19 W	0257 0929 1527 2157	8.8 3.3 9.1 3.1
5 W	0438 1126 1706 2343	9.8 2.1 10.0 2.1	20 TH	0349 1023 1614 2249	9.6 2.6 10.0 2.4
6 TH	0520 1207 1744	10.2 1.8 10.4	21 F	0437 1113 1701 2339	10.3 1.9 10.7 1.7
7 F	0021 0558 1243 1819	1.8 10.5 1.6 10.6	22 SA	0525 1204 1747	11.0 1.3 11.3
8 SA	0057 0635 1319 1853	1.6 10.6 1.6 10.6	23 SU	0029 0611 1255 1832	1.2 11.4 0.9 11.6
9 SU	0131 0709 1349 1926	1.6 10.5 1.8 10.6	24 M	0120 0657 1344 1919	0.9 11.6 0.8 11.7
10 M	0202 0741 1419 1955	1.7 10.4 2.0 10.4	25 TU	0209 0744 1430 2004	0.8 11.5 1.0 11.4
11 TU	0233 0812 1447 2026	1.9 10.1 2.3 10.1	26 W	0256 0830 1517 2049	1.1 11.1 1.4 10.9
12 W	0304 0842 1515 2056	2.2 9.7 2.7 9.7	27 TH	0343 0918 1603 2136	1.6 10.4 2.1 10.2
13 TH	0335 0914 1545 2129	2.7 9.3 3.1 9.2	28 F	0431 1009 1652 2230	2.2 9.6 2.9 9.5
14 F	0410 0949 1620 2209	3.2 8.7 3.7 8.7	29 SA	0525 1111 1751 2334	2.9 8.9 3.6 8.8
15 SA	0451 1035 1706 2305	3.7 8.2 4.2 8.2	30 SU	0631 1227 1904	3.5 8.4 3.9
			31 M	0055 0749 1347 2023	8.5 3.6 8.4 3.9

JUNE

Day	TIME	M	Day	TIME	M
1 TU	0211 0903 1453 2129	8.6 3.4 8.7 3.5	16 W	0053 0727 1331 2004	8.5 3.7 8.6 3.8
2 W	0312 1000 1546 2220	8.9 3.1 9.1 3.1	17 TH	0204 0839 1439 2112	8.8 3.3 9.1 3.2
3 TH	0403 1048 1631 2305	9.3 2.7 9.5 2.6	18 F	0307 0942 1538 2212	9.4 2.7 9.8 2.6
4 F	0448 1129 1712 2346	9.7 2.4 9.8 2.3	19 SA	0404 1041 1633 2311	10.0 2.1 10.4 1.9
5 SA	0529 1208 1750	9.9 2.2 10.1	20 SU	0459 1139 1725	10.6 1.6 11.0
6 SU	0025 0608 1246 1827	2.1 10.0 2.1 10.2	21 M	0008 0554 1235 1817	1.4 11.0 1.2 11.3
7 M	0103 0646 1321 1903	2.0 10.0 2.1 10.3	22 TU	0104 0646 1330 1907	1.1 11.3 1.0 11.5
8 TU	0140 0721 1357 1937	2.0 10.0 2.2 10.2	23 W	0159 0735 1422 1955	0.9 11.3 1.1 11.3
9 W	0216 0755 1430 2011	2.0 9.9 2.4 10.0	24 TH	0250 0825 1510 2043	1.0 11.0 1.4 11.0
10 TH	0251 0829 1503 2043	2.2 9.6 2.6 9.8	25 F	0336 0912 1556 2128	1.3 10.5 1.9 10.5
11 F	0325 0901 1535 2117	2.5 9.3 3.0 9.5	26 SA	0423 0959 1641 2216	1.8 9.9 2.5 9.8
12 SA	0359 0936 1609 2155	2.9 9.0 3.4 9.1	27 SU	0509 1049 1727 2308	2.5 9.3 3.1 9.2
13 SU	0435 1019 1647 2241	3.3 8.7 3.7 8.7	28 M	0600 1146 1822	3.1 8.7 3.7
14 M	0519 1111 1737 2340	3.6 8.4 4.0 8.5	29 TU	0008 0700 1252 1927	8.7 3.5 8.4 3.9
15 TU	0615 1218 1846	3.8 8.3 4.1	30 W	0114 0806 1359 2033	8.5 3.7 8.4 3.8

JULY

Day	TIME	M	Day	TIME	M
1 TH	0222 0908 1500 2131	8.5 3.6 8.6 3.5	16 F	0114 0754 1355 2032	8.8 3.4 8.9 3.4
2 F	0321 1002 1552 2223	8.7 3.3 8.9 3.2	17 SA	0230 0908 1507 2145	9.1 3.0 9.4 2.9
3 SA	0414 1049 1640 2309	9.0 3.0 9.3 2.8	18 SU	0342 1017 1613 2251	9.6 2.5 10.0 2.3
4 SU	0502 1134 1725 2356	9.3 2.8 9.6 2.5	19 M	0445 1122 1712 2356	10.1 2.0 10.6 1.7
5 M	0546 1217 1805	9.5 2.6 9.9	20 TU	0544 1224 1807	10.7 1.5 11.1
6 TU	0039 0628 1259 1845	2.3 9.7 2.4 10.1	21 W	0056 0638 1320 1859	1.2 11.1 1.2 11.4
7 W	0123 0706 1340 1923	2.1 9.8 2.3 10.2	22 TH	0151 0728 1412 1947	0.8 11.2 1.0 11.4
8 TH	0202 0742 1416 1957	2.0 9.9 2.3 10.2	23 F	0240 0815 1458 2030	0.8 11.1 1.2 11.2
9 F	0239 0816 1451 2030	2.1 9.8 2.4 10.1	24 SA	0324 0857 1539 2111	1.0 10.8 1.5 10.8
10 SA	0312 0847 1524 2103	2.2 9.7 2.6 9.9	25 SU	0404 0938 1619 2152	1.5 10.3 2.1 10.2
11 SU	0345 0919 1555 2136	2.5 9.6 2.9 9.7	26 M	0442 1017 1655 2231	2.1 9.7 2.7 9.6
12 M	0416 0956 1627 2214	2.8 9.3 3.2 9.4	27 TU	0519 1059 1733 2316	2.8 9.0 3.3 9.0
13 TU	0452 1038 1706 2301	3.1 9.1 3.5 9.1	28 W	0600 1149 1821	3.5 8.5 3.8
14 W	0537 1132 1801	3.3 8.9 3.7	29 TH	0011 0655 1252 1924	8.4 3.9 8.1 4.0
15 TH	0001 0638 1239 1914	8.8 3.5 8.8 3.7	30 F	0121 0802 1406 2034	8.1 4.1 8.1 4.0
			31 SA	0237 0911 1515 2139	8.1 4.0 8.3 3.7

AUGUST

Day	TIME	M	Day	TIME	M
1 SU	0343 1012 1613 2238	8.4 3.7 8.8 3.3	16 M	0332 1007 1603 2245	9.1 3.0 9.6 2.6
2 M	0441 1105 1704 2332	8.8 3.3 9.2 2.9	17 TU	0441 1115 1705 2350	9.8 2.3 10.3 1.8
3 TU	0529 1154 1747	9.2 2.9 9.7	18 W	0539 1215 1758	10.5 1.6 11.0
4 W	0019 0611 1239 1828	2.5 9.6 2.5 10.0	19 TH	0048 0629 1310 1848	1.2 11.1 1.2 11.4
5 TH	0104 0649 1321 1904	2.1 9.9 2.2 10.3	20 F	0138 0714 1357 1931	0.7 11.3 0.9 11.5
6 F	0144 0724 1359 1938	1.8 10.1 2.0 10.5	21 SA	0223 0757 1439 2011	0.6 11.3 1.0 11.4
7 SA	0220 0755 1434 2011	1.7 10.3 2.0 10.5	22 SU	0303 0833 1515 2047	0.8 11.0 1.3 11.1
8 SU	0254 0826 1505 2040	1.7 10.3 2.1 10.5	23 M	0338 0908 1548 2121	1.3 10.6 1.8 10.5
9 M	0325 0857 1535 2112	1.9 10.3 2.3 10.3	24 TU	0407 0941 1616 2153	2.0 10.0 2.4 9.9
10 TU	0355 0931 1604 2148	2.2 10.1 2.6 10.0	25 W	0434 1013 1645 2230	2.7 9.4 3.0 9.2
11 W	0426 1007 1640 2228	2.6 9.7 3.0 9.6	26 TH	0504 1052 1722 2312	3.4 8.7 3.6 8.5
12 TH	0505 1054 1726 2322	3.0 9.3 3.4 9.1	27 F	0544 1143 1815	4.0 8.2 4.1
13 F	0601 1156 1836	3.4 8.9 3.7	28 SA	0015 0649 1304 1934	7.9 4.4 7.8 4.3
14 SA	0035 0720 1319 2005	8.7 3.7 8.6 3.7	29 SU	0151 0816 1439 2100	7.6 4.5 7.9 4.2
15 SU	0208 0849 1450 2131	8.6 3.5 8.9 3.3	30 M	0319 0938 1550 2212	7.9 4.2 8.4 3.7
			31 TU	0421 1041 1642 2309	8.5 3.6 9.0 3.1

HEIGHTS IN METRES

CHANNEL ISLANDS - ST. HELIER

LAT 49°11'N LONG 2°07'W

TIME ZONE GMT — TIMES AND HEIGHTS OF HIGH AND LOW WATERS — YEAR 1982

SEPTEMBER

Day	TIME	M	Day	TIME	M
1 W	0508	9.1	16 TH	0527	10.6
	1132	3.0		1205	1.7
	1726	9.6		1746	11.0
	2357	2.5			
2 TH	0549	9.7	17 F	0034	1.1
	1217	2.5		0614	11.1
	1804	10.2		1252	1.2
				1829	11.4
3 F	0041	2.0	18 SA	0119	0.8
	0624	10.2		0655	11.4
	1257	2.0		1335	0.9
	1839	10.6		1909	11.5
4 SA	0119	1.6	19 SU	0159	0.7
	0657	10.6		0731	11.4
	1335	1.7		1413	1.0
	1913	10.9		1945	11.4
5 SU	0155	1.4	20 M	0234	1.0
	0728	10.9		0805	11.1
	1409	1.5		1446	1.3
	1944	11.1		2018	11.1
6 M	0229	1.3	21 TU	0305	1.4
	0801	11.0		0836	10.7
	1443	1.6		1514	1.7
	2016	11.1		2049	10.6
7 TU	0301	1.5	22 W	0331	2.0
	0832	10.9		0904	10.2
	1514	1.8		1539	2.3
	2049	10.9		2118	10.0
8 W	0332	1.8	23 TH	0353	2.7
	0905	10.7		0934	9.6
	1545	2.2		1606	2.8
	2124	10.5		2149	9.3
9 TH	0403	2.3	24 F	0419	3.3
	0942	10.2		1006	9.0
	1619	2.7		1638	3.4
	2203	9.9		2226	8.6
10 F	0442	2.9	25 SA	0454	3.9
	1026	9.5		1048	8.3
	1705	3.3		1726	4.1
	2254	9.1		2318	7.9
11 SA	0537	3.6	26 SU	0550	4.5
	1126	8.8		1201	7.7
	1814	3.8		1839	4.5
12 SU	0010	8.4	27 M	0100	7.4
	0700	4.0		0723	4.8
	1259	8.3		1402	7.7
	1954	4.0		2019	4.5
13 M	0201	8.3	28 TU	0250	7.7
	0842	3.9		0904	4.5
	1444	8.6		1521	8.2
	2129	3.5		2143	4.0
14 TU	0332	8.9	29 W	0353	8.3
	1006	3.2		1014	3.9
	1559	9.4		1614	8.9
	2244	2.6		2241	3.3
15 W	0437	9.8	30 TH	0438	9.1
	1111	2.4		1104	3.1
	1657	10.3		1655	9.6
	2343	1.8		2327	2.6

OCTOBER

Day	TIME	M	Day	TIME	M
1 F	0516	9.8	16 SA	0012	1.3
	1147	2.5		0551	11.0
	1733	10.3		1229	1.4
				1805	11.2
2 SA	0008	2.0	17 SU	0053	1.1
	0551	10.4		0628	11.2
	1227	1.9		1309	1.2
	1808	10.8		1843	11.3
3 SU	0048	1.5	18 M	0131	1.1
	0627	11.0		0703	11.2
	1304	1.5		1344	1.2
	1842	11.2		1919	11.2
4 M	0126	1.2	19 TU	0204	1.3
	0700	11.3		0735	11.0
	1342	1.2		1415	1.5
	1917	11.5		1949	10.9
5 TU	0202	1.1	20 W	0233	1.7
	0734	11.5		0805	10.7
	1418	1.2		1443	1.8
	1951	11.5		2020	10.5
6 W	0237	1.2	21 TH	0257	2.2
	0808	11.4		0834	10.3
	1453	1.5		1510	2.2
	2027	11.2		2049	10.0
7 TH	0312	1.6	22 F	0321	2.7
	0844	11.0		0903	9.8
	1529	1.9		1538	2.8
	2104	10.7		2119	9.4
8 F	0349	2.2	23 SA	0348	3.2
	0922	10.4		0935	9.2
	1607	2.6		1610	3.3
	2146	9.9		2155	8.7
9 SA	0431	3.0	24 SU	0421	3.9
	1009	9.6		1013	8.6
	1657	3.3		1654	3.9
	2240	9.0		2241	8.0
10 SU	0527	3.7	25 M	0512	4.5
	1111	8.7		1115	7.9
	1810	3.9		1758	4.5
11 M	0004	8.3	26 TU	0008	7.5
	0655	4.2		0632	4.9
	1253	8.2		1309	7.7
	1952	4.0		1931	4.6
12 TU	0204	8.3	27 W	0204	7.6
	0840	4.1		0818	4.7
	1439	8.6		1437	8.1
	2127	3.5		2101	4.2
13 W	0327	8.9	28 TH	0312	8.2
	0959	3.3		0935	4.1
	1548	9.4		1534	8.7
	2234	2.6		2202	3.5
14 TH	0423	9.8	29 F	0359	9.0
	1058	2.5		1026	3.3
	1640	10.2		1617	9.5
	2326	1.9		2248	2.8
15 F	0509	10.5	30 SA	0438	9.8
	1146	1.8		1109	2.6
	1725	10.8		1655	10.2
				2330	2.1
			31 SU	0516	10.5
				1151	1.9
				1734	10.8

NOVEMBER

Day	TIME	M	Day	TIME	M
1 M	0012	1.5	16 TU	0100	1.6
	0554	11.1		0636	10.8
	1232	1.4		1314	1.6
	1812	11.3		1853	10.7
2 TU	0055	1.2	17 W	0134	1.8
	0632	11.5		0710	10.7
	1314	1.1		1348	1.7
	1852	11.6		1927	10.5
3 W	0135	1.0	18 TH	0204	2.0
	0710	11.7		0741	10.5
	1357	1.1		1419	1.9
	1931	11.6		1958	10.2
4 TH	0218	1.1	19 F	0233	2.3
	0749	11.6		0812	10.2
	1439	1.3		1449	2.3
	2012	11.3		2030	9.8
5 F	0258	1.5	20 SA	0300	2.7
	0830	11.1		0843	9.9
	1521	1.7		1519	2.7
	2054	10.7		2101	9.4
6 SA	0341	2.1	21 SU	0331	3.2
	0912	10.5		0917	9.4
	1604	2.4		1555	3.1
	2142	9.9		2136	8.9
7 SU	0427	2.9	22 M	0404	3.6
	1002	9.6		0955	8.9
	1658	3.1		1634	3.7
	2240	9.0		2220	8.3
8 M	0526	3.7	23 TU	0448	4.2
	1108	8.8		1047	8.3
	1808	3.7		1726	4.1
				2323	7.9
9 TU	0004	8.4	24 W	0549	4.6
	0649	4.2		1204	8.0
	1242	8.4		1838	4.4
	1941	3.9			
10 W	0147	8.4	25 TH	0053	7.8
	0825	4.0		0716	4.6
	1416	8.7		1331	8.1
	2108	3.4		2001	4.2
11 TH	0304	8.9	26 F	0211	8.2
	0938	3.4		0839	4.2
	1524	9.3		1437	8.5
	2212	2.8		2110	3.7
12 F	0359	9.6	27 SA	0308	8.8
	1033	2.7		0938	3.5
	1614	9.9		1529	9.2
	2301	2.2		2203	3.0
13 SA	0444	10.1	28 SU	0355	9.6
	1119	2.2		1028	2.8
	1659	10.4		1617	9.9
	2344	1.8		2251	2.3
14 SU	0525	10.5	29 M	0440	10.3
	1200	1.8		1115	2.1
	1740	10.7		1702	10.6
				2339	1.7
15 M	0024	1.6	30 TU	0525	11.0
	0601	10.7		1204	1.6
	1239	1.6		1747	11.1
	1818	10.7			

DECEMBER

Day	TIME	M	Day	TIME	M
1 W	0027	1.3	16 TH	0109	2.2
	0608	11.4		0650	10.3
	1252	1.2		1327	2.0
	1834	11.4		1910	10.1
2 TH	0116	1.1	17 F	0144	2.2
	0653	11.6		0726	10.3
	1341	1.0		1402	2.0
	1919	11.4		1945	10.0
3 F	0204	1.1	18 SA	0216	2.4
	0738	11.5		0759	10.2
	1429	1.1		1437	2.2
	2005	11.2		2018	9.8
4 SA	0250	1.4	19 SU	0249	2.6
	0823	11.2		0832	10.0
	1517	1.5		1510	2.4
	2051	10.7		2051	9.5
5 SU	0338	1.9	20 M	0319	2.9
	0910	10.6		0905	9.7
	1604	2.0		1543	2.8
	2141	10.1		2124	9.2
6 M	0426	2.6	21 TU	0352	3.2
	1000	9.9		0941	9.3
	1655	2.6		1617	3.2
	2237	9.3		2202	8.9
7 TU	0520	3.3	22 W	0428	3.6
	1059	9.2		1021	8.9
	1757	3.2		1657	3.6
	2346	8.7		2247	8.5
8 W	0628	3.8	23 TH	0512	4.0
	1214	8.7		1112	8.5
	1913	3.6		1746	3.9
				2344	8.3
9 TH	0107	8.5	24 F	0611	4.2
	0748	3.9		1217	8.3
	1334	8.6		1852	4.0
	2030	3.5			
10 F	0223	8.7	25 SA	0056	8.3
	0900	3.6		0728	4.1
	1444	8.9		1330	8.5
	2135	3.2		2006	3.8
11 SA	0322	9.0	26 SU	0208	8.6
	0957	3.2		0843	3.7
	1542	9.3		1439	8.9
	2227	2.8		2114	3.3
12 SU	0412	9.5	27 M	0311	9.2
	1047	2.7		0946	3.1
	1630	9.6		1539	9.5
	2312	2.5		2214	2.7
13 M	0455	9.8	28 TU	0407	9.9
	1129	2.4		1044	2.5
	1713	9.9		1635	10.1
	2353	2.3		2312	2.1
14 TU	0536	10.1	29 W	0501	10.6
	1210	2.1		1142	1.9
	1754	10.0		1730	10.7
15 W	0032	2.2	30 TH	0008	1.6
	0614	10.3		0553	11.1
	1249	2.0		1238	1.3
	1834	10.1		1822	11.1
			31 F	0103	1.2
				0643	11.4
				1333	1.0
				1913	11.3

HEIGHTS IN METRES

WILHELMSHAVEN

MEAN SPRING AND NEAP CURVES

Spring occurs 3 days after New and Full Moon.

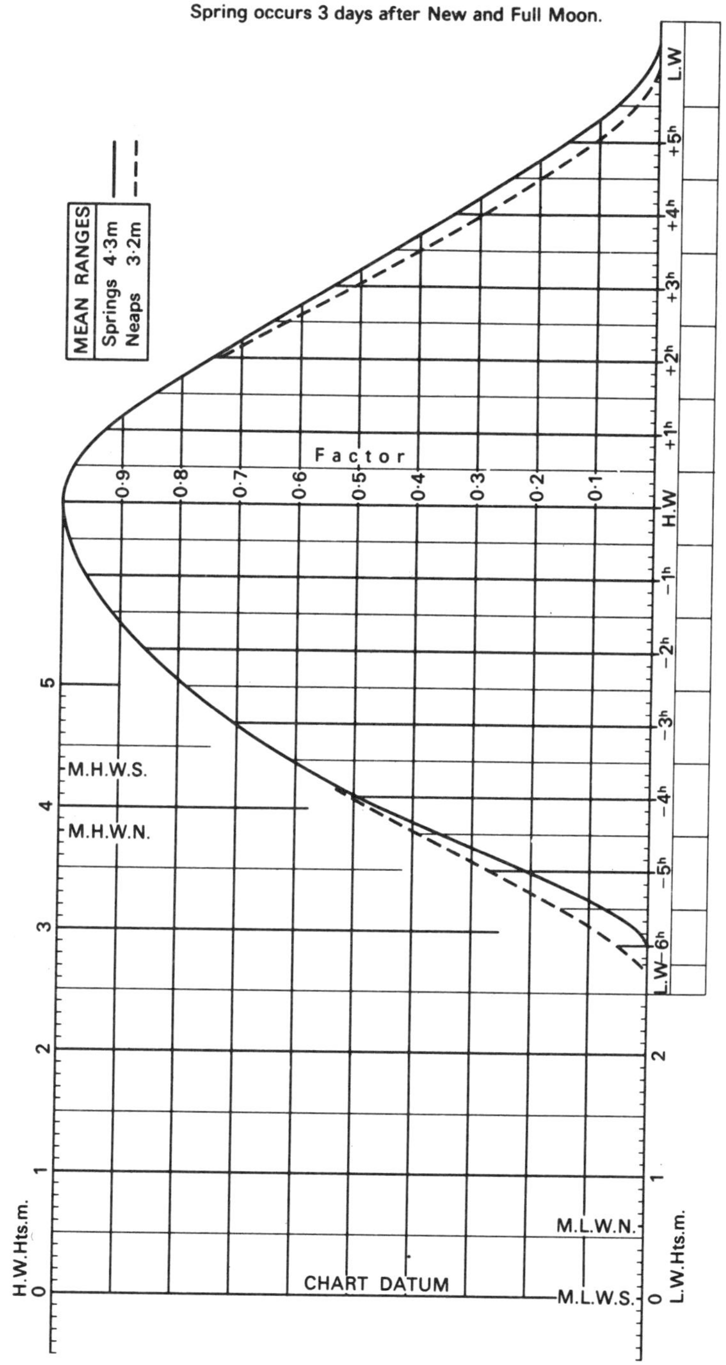

GERMANY - WILHELMSHAVEN

LAT 53°31'N LONG 8°09'E

TIME ZONE -0100 TIMES AND HEIGHTS OF HIGH AND LOW WATERS YEAR 1982

JANUARY						FEBRUARY					
	TIME	M		TIME	M		TIME	M		TIME	M
1	0408	4.1	16	0517	4.2	1	0523	4.1	16	0558	3.8
	1025	0.2		1132	0.1		1136	0.1		1156	0.2
F	1640	3.8	SA	1749	3.7	M	1752	3.8	TU	1818	3.5
	2234	0.3		2338	0.3		2347	0.3			
2	0452	4.1	17	0559	4.0	2	0605	4.1	17	0009	0.3
	1108	0.2		1209	0.2		1215	0.3		0644	3.6
SA	1726	3.7	SU	1830	3.6	TU	1837	3.7	W	1240	0.5
	2319	0.4								1912	3.4
3	0538	4.1	18	0017	0.3	3	0035	0.4	18	0110	0.5
	1154	0.3		0645	3.9		0700	4.0		0751	3.5
SU	1815	3.7	M	1252	0.3	W	1311	0.4	TH	1348	0.6
				1917	3.5		1939	3.7		2029	3.5
11	0133	4.3	26	0138	4.1	11	0256	4.3	26	0233	4.1
	0753	0.0		0751	0.0		0915	-0.2		0848	-0.2
M	1406	4.2	TU	1403	3.9	TH	1529	3.9	F	1501	3.9
	2015	0.1		2006	0.1		2127	-0.1		2101	-0.2

MARCH						APRIL					
	TIME	M		TIME	M		TIME	M		TIME	M
1	0428	4.2	16	0449	3.9	1	0541	3.9	16	0523	3.6
	1042	-0.1		1049	0.1		1137	0.2		1108	0.5
M	1653	3.9	TU	1701	3.7	TH	1759	3.8	F	1739	3.7
	2252	0.0		2257	0.1					2343	0.4
2	0508	4.1	17	0518	3.8	2	0002	0.1	17	0615	3.5
	1117	0.0		1111	0.3		0640	3.7		1202	0.7
TU	1730	3.9	W	1730	3.7	F	1233	0.4	SA	1844	3.6
	2326	0.1		2327	0.3		1907	3.8			
3	0548	4.0	18	0556	3.6	3	0116	0.3	18	0052	0.5
	1152	0.2		1144	0.5		0801	3.6		0731	3.4
W	1812	3.8	TH	1817	3.5	SA	1357	0.5	SU	1323	0.7
							2035	3.8		2008	3.7
11	0157	4.2	26	0134	4.1	11	0250	4.1	26	0230	4.3
	0813	-0.3		0747	-0.3		0855	-0.2		0842	-0.2
TH	1429	3.9	F	1404	4.0	SU	1508	4.0	M	1455	4.1
	2027	-0.2		2006	-0.2		2108	-0.2		2101	-0.2

VOL. 1 EUROPEAN WATERS—PART II (Secondary Ports)

Seasonal Changes in Mean Level and Harmonic Constants (For worked examples and exercises.)

No.		Jan. 1	Feb. 1	Mar. 1	Apr. 1	May 1	June 1	July 1	Aug. 1	Sep. 1	Oct. 1	Nov. 1	Dec. 1	Jan. 1
1–60								Negligible						
61–63a		+0·1	0·0	−0·1	−0·1	−0·1	0·0	0·0	0·0	0·0	0·0	+0·1	+0·1	+0·1
64–130								Negligible						
308–390		+0·1	0·0	0·0	0·0	−0·1	−0·1	−0·1	0·0	0·0	0·0	0·0	+0·1	+0·1
391–398		+0·1	0·0	−0·1	−0·1	−0·1	−0·1	0·0	0·0	0·0	0·0	+0·1	+0·1	+0·1
399–407		+0·2	+0·1	0·0	−0·1	−0·1	−0·1	−0·1	−0·1	0·0	0·0	+0·1	+0·2	+0·2
408–414		+0·1	0·0	−0·1	−0·1	−0·1	−0·1	0·0	0·0	0·0	0·0	+0·1	+0·1	+0·1
415–444		0·0	0·0	0·0	−0·1	−0·1	0·0	0·0	0·0	0·0	0·0	+0·1	+0·1	0·0
445–464		0·0	0·0	−0·1	−0·1	0·0	0·0	0·0	0·0	0·0	0·0	+0·1	+0·1	0·0
466–478		+0·1	0·0	0·0	−0·1	−0·1	−0·1	0·0	0·0	0·0	0·0	+0·1	+0·1	+0·1
479–512		0·0	0·0	0·0	0·0	0·0	0·0	−0·1	0·0	0·0	0·0	+0·1	+0·1	0·0
513–534		0·0	0·0	0·0	0·0	−0·1	−0·1	0·0	0·0	0·0	0·0	0·0	0·0	0·0
535–548		+0·1	0·0	0·0	−0·1	−0·1	−0·1	0·0	0·0	0·0	0·0	+0·1	+0·1	+0·1
601–644		0·0	0·0	−0·1	−0·1	−0·1	0·0	0·0	0·0	0·0	0·0	+0·1	+0·1	0·0
651–672		+0·1	0·0	0·0	−0·1	−0·1	−0·1	0·0	0·0	0·0	0·0	0·0	+0·1	+0·1
681–712		+0·1	0·0	−0·1	−0·1	0·0	0·0	0·0	−0·1	−0·1	0·0	+0·1	+0·1	+0·1
713–719								Negligible						
721–737		+0·1	0·0	0·0	−0·1	−0·1	0·0	0·0	0·0	0·0	0·0	0·0	+0·1	+0·1
741–766		0·0	0·0	0·0	−0·1	−0·1	−0·1	−0·1	0·0	0·0	+0·1	+0·1	+0·1	0·0
65	M_2	1·40	1·40	1·42	1·43	1·42	1·39	1·39	1·41	1·45	1·47	1·45	1·42	1·40
89	M_2	2·28	2·24	2·22	2·22	2·25	2·25	2·23	2·19	2·18	2·21	2·26	2·29	2·28
103	M_2	1·70	1·65	1·60	1·59	1·61	1·63	1·64	1·62	1·61	1·63	1·68	1·71	1·70
108	M_2	2·00	2·00	2·02	2·03	2·01	1·97	1·97	2·01	2·06	2·08	2·07	2·03	2·00
109	M_2	2·18	2·17	2·14	2·12	2·13	2·16	2·18	2·17	2·14	2·12	2·13	2·16	2·18
110	M_2	2·01	2·02	2·05	2·08	2·09	2·07	2·05	2·03	2·03	2·04	2·03	2·02	2·01
404	M_2	1·19	1·18	1·17	1·18	1·18	1·18	1·17	1·15	1·14	1·15	1·17	1·18	1·19
440	M_2	3·08	3·07	3·05	3·04	3·04	3·04	3·05	3·07	3·08	3·09	3·09	3·09	3·08
	S_2	0·99	1·09	1·13	1·08	0·98	0·92	0·94	0·98	0·98	0·93	0·87	0·89	0·99
441	M_2	3·19	3·17	3·17	3·16	3·15	3·14	3·15	3·18	3·22	3·25	3·24	3·22	3·19
	S_2	1·05	1·04	1·02	1·02	1·04	1·06	1·07	1·05	1·01	1·00	1·01	1·04	1·05
452	M_2	3·13	3·13	3·12	3·12	3·15	3·18	3·18	3·15	3·10	3·07	3·08	3·12	3·13
496	M_2	2·21	2·20	2·20	2·21	2·23	2·25	2·27	2·28	2·28	2·27	2·25	2·23	2·21
513a	M_2	3·77	3·82	3·89	3·94	3·97	4·00	3·97	3·91	3·84	3·78	3·75	3·75	3·77
523	M_2	4·30	4·26	4·23	4·25	4·31	4·35	4·31	4·21	4·13	4·13	4·21	4·28	4·30
751	M_2	1·42	1·42	1·43	1·43	1·41	1·39	1·39	1·42	1·45	1·47	1·46	1·43	1·42

Volume 1 EUROPEAN WATERS—Part II (Secondary Ports)

ENGLAND, SCOTLAND AND IRELAND

No.	PLACE	Lat. N.	Long. W.	TIME DIFFERENCES High Water (Zone G.M.T.)		Low Water		HEIGHT DIFFERENCES (IN METRES) MHWS	MHWN	MLWN	MLWS	M.L. Z_0 m.
65	**PORTSMOUTH**	(see page 14)		**0000** and **1200**	**0600** and **1800**	**0500** and **1700**	**1100** and **2300**	**4·7**	**3·8**	**1·8**	**0·6**	
51	Ventnor	50 36	1 12	−0025	−0030	−0025	−0030	−0·8	−0·6	−0·2	+0·2	2·33
53	Sandown	50 39	1 09	0000	+0005	+0010	+0025	−0·6	−0·5	−0·2	0·0	2·41
54	Brading	50 42	1 06	−0010	+0005	+0020	0000	−1·6	−1·5	−1·4	−0·6	⊙
58	Ryde	50 44	1 07	−0010	+0010	−0005	−0010	−0·2	−0·1	0·0	+0·1	2·76
65	**PORTSMOUTH**	(see page 14)		**1000** and **2200**	**0500** and **1700**	**0000** and **1200**	**0600** and **1800**	**4·7**	**3·8**	**1·8**	**0·6**	**2·84**
	Chichester Harbour											
68	Entrance	50 47	0 56	+0005	−0010	+0015	+0020	+0·2	+0·2	0·0	+0·1	2·83
68a	Bosham	50 50	0 52	+0020	+0005	⊙	⊙	+0·4	+0·2	⊙	⊙	⊙
68b	Itchenor	50 48	0 52	+0015	0000	⊙	⊙	+0·4	+0·2	0·0	0·0	⊙
68c	Dell Quay	50 49	0 49	+0025	+0010	⊙	⊙	+0·4	+0·2	⊙	⊙	⊙
69	Selsey Bill	50 43	0 47	−0005	−0005	+0035	+0035	+0·6	+0·6	0·0	0·0	2·94
70	Nab Tower	50 40	0 57	0000	+0015	+0015	+0015	−0·2	0·0	+0·2	0·0	2·58
108	**SHEERNESS**	(see page 30)		**0600** and **1800**	**1100** and **2300**	**0100** and **1300**	**0800** and **2000**	**5·7**	**4·8**	**1·5**	**0·6**	
	River Thames											
110	Southend	51 31	0 43	−0005	−0005	−0005	−0005	0·0	0·0	−0·1	−0·1	3·02
110a	Thames Haven	51 30	0 31	+0010	+0010	0000	+0010	+0·5	+0·4	−0·1	−0·1	3·15
111	Tilbury	51 28	0 22	+0010	+0030	+0005	+0035	+0·8	+0·6	0·0	−0·1	3·33
404	**GREENOCK**	(see page 94)		**0000** and **1200**	**0600** and **1800**	**0000** and **1200**	**0600** and **1800**	**3·4**	**2·9**	**1·0**	**0·4**	**2·00**
	River Clyde											
405	Port Glasgow	55 56	4 41	+0011	+0005	+0010	+0018	+0·2	+0·1	0·0	0·0	⊙
406	Bowling	55 56	4 29	+0032	+0016	+0043	+0107	+0·6	+0·5	+0·3	+0·1	⊙
408	Brodick Bay	55 35	5 08	0000	0000	+0005	+0005	−0·2	−0·2	0·0	0·0	1·86
409	Lamlash	55 32	5 07	−0016	−0036	−0024	−0004	−0·2	−0·2	⊙	⊙	⊙
410	Ardrossan	55 38	4 49	−0020	−0020	−0030	−0010	−0·2	−0·2	−0·1	−0·1	⊙
411	Irvine	55 36	4 41	−0020	−0020	−0030	−0010	−0·3	−0·3	−0·1	0·0	⊙
523	**PORT OF BRISTOL (AVONMOUTH)**	(see page 122)		**0600** and **1800**	**1100** and **2300**	**0300** and **1500**	**0800** and **2000**	**13·2**	**10·0**	**3·5**	**0·9**	
513	Barry	51 23	3 16	−0030	−0015	−0125	−0030	−1·8	−1·3	+0·2	0·0	6·11
514	**CARDIFF**	51 27	3 09	STANDARD PORT					See Table V			6·45
515	Newport	51 33	2 59	−0020	−0010	0000	−0020	−1·1	−1·0	−0·6	−0·7	*w*
	River Wye											
516	Chepstow	51 39	2 40	+0020	+0020	⊙	⊙	⊙	⊙	⊙	⊙	⊙
658	**LONDONDERRY**	(see page 134)		**0200** and **0400**	**0800** and **2000**	**1100** and **2300**	**0500** and **1700**	**2·7**	**2·0**	**1·0**	**0·4**	
652	Portrush	55 12	6 40	−0140	−0140	−0120	−0200	−0·6	−0·6	+0·1	0·0	1·37
653	Coleraine	55 08	6 40	−0108	−0126	−0037	−0101	−0·6	−0·4	−0·3	−0·1	⊙
	Lough Foyle											
654	Warren Point	55 12	6 57	−0121	−0139	−0132	−0156	−0·4	−0·2	⊙	⊙	⊙
655	Moville	55 11	7 03	−0046	−0058	−0052	−0108	−0·4	−0·2	−0·2	−0·1	⊙
656	Quigley's Point	55 07	7 11	−0025	−0040	−0040	−0025	−0·4	−0·3	−0·3	−0·2	⊙
657	Calmore Point	55 03	7 15	−0010	−0030	−0040	−0020	−0·3	−0·3	−0·2	−0·1	⊙

		N.	E.																						
83	Newhaven	50 47	0 04	−0010	−0015	0000	0000	+0·4	+0·2	0·0	−0·2	3·61	323	2·26	010	0·73	107	0·07	331	0·01	322	0·020	⊙	⊙	
84	Eastbourne	50 46	0 17	−0005	−0010	+0015	+0020	+1·1	+0·6	+0·2	+0·1	3·77	325	2·48	006	0·80	091	0·04	345	0·03	307	0·022	⊙	⊙	
89	**DOVER**	(see page 22)		**0000 and 1200**	**0600 and 1800**	**0100 and 1300**	**0700 and 1900**	**6·7**	**5·3**	**2·0**	**0·8**														
85	Hastings	50 51	0 35	0000	−0010	−0030	−0030	+0·8	+0·5	+0·1	−0·1	3·85	323	2·47	017	0·89	095	0·08	223	0·01	302	0·035	272	0·001	
86	Rye (Approaches)	50 55	0 47	+0005	−0010	⊙	⊙	+1·0	+0·7	⊙	⊙	⊙	⊙	⊙	⊙	⊙	⊙	⊙	⊙	⊙	⊙	⊙	⊙	⊙	
86a	Rye (Harbour)	50 56	0 46	+0005	−0010	⊙	⊙	−1·4	−1·7	⊙	⊙	1·97	328	2·19	015	0·60	149	0·08	031	0·03	327	0·076	⊙	⊙	*l*
87	Dungeness	50 54	0 58	−0010	−0015	−0020	−0010	+1·0	+0·5	+0·2	−0·1	4·03	326	2·43	023	0·94	023	0·10	182	0·09	279	0·041	⊙	⊙	*l*
88	Folkestone	51 05	1 12	−0020	−0005	−0010	−0010	+0·4	+0·4	0·0	−0·1	3·74	332	2·45	032	0·79	055	0·05	241	0·02	282	0·057	⊙	⊙	
89	**DOVER**	51 07	1 19	STANDARD PORT				See Table V				3·70	332	*v*	023	0·71	046	0·06	172	0·06	278	0·054	188	0·006	*v*
98	Deal	51 13	1 25	+0010	+0020	+0010	+0005	−0·6	−0·3	0·0	0·0	3·13	336	2·07	027	0·63	020	0·06	175	0·07	274	0·044	187	0·009	*i x*
102	Ramsgate	51 20	1 25	+0020	+0020	−0007	−0007	−1·8	−1·5	−0·8	−0·4	2·56	339	1·86	030	0·56	012	0·08	182	0·09	281	0·038	191	0·007	
103	**MARGATE**	(see page 26)		**0100 and 1300**	**0700 and 1900**	**0100 and 1300**	**0700 and 1900**	**4·8**	**3·9**	**1·4**	**0·5**														
102a	Broadstairs	51 21	1 27	−0020	−0008	+0007	+0010	−0·2	−0·2	−0·1	−0·1	⊙	⊙	⊙	⊙	⊙	⊙	⊙	⊙	⊙	⊙	⊙	⊙	⊙	
103	**MARGATE**	51 24	1 23	STANDARD PORT				See Table V				2·67	342	*v*	034	0·47	008	0·10	184	0·12	339	0·017	⊙	⊙	*v*
104	Herne Bay	51 23	1 07	+0034	+0022	+0015	+0032	+0·4	+0·4	0·0	0·0	2·73	349	1·80	048	0·54	346	0·09	171	0·12	004	0·013	040	0·007	
105	Whitstable Approaches	51 22	1 02	+0042	+0029	+0025	+0050	+0·6	+0·6	+0·1	0·0	⊙	⊙	⊙	⊙	⊙	⊙	⊙	⊙	⊙	⊙	⊙	⊙	⊙	
108	**SHEERNESS**	(see page 30)		**0200 and 1400**	**0800 and 2000**	**0200 and 1400**	**0700 and 1900**	**5·7**	**4·8**	**1·5**	**0·6**														
	River Swale																								
106	Grovehurst Jetty	51 22	0 46	−0007	0000	0000	+0016	0·0	0·0	0·0	−0·1	⊙	⊙	⊙	⊙	⊙	⊙	⊙	⊙	⊙	⊙	⊙	⊙	⊙	
	River Medway																								
108	**SHEERNESS**	51 27	0 45	STANDARD PORT				See Table V				3·07	355	*v*	052	0·58	013	0·12	190	0·13	027	0·023	049	0·006	★
108a	Bee Ness	51 25	0 39	+0002	+0002	0000	+0005	+0·2	+0·1	0·0	0·0	⊙	⊙	⊙	⊙	⊙	⊙	⊙	⊙	⊙	⊙	⊙	⊙	⊙	
108b	Bartlett Creek	51 23	0 38	+0016	+0008	⊙	⊙	+0·1	0·0	⊙	⊙	⊙	⊙	⊙	⊙	⊙	⊙	⊙	⊙	⊙	⊙	⊙	⊙	⊙	
108c	Darnett Ness	51 24	0 36	+0004	+0004	0000	+0010	+0·2	+0·1	0·0	−0·1	⊙	⊙	⊙	⊙	⊙	⊙	⊙	⊙	⊙	⊙	⊙	⊙	⊙	

⊙ No data.
★ See notes on page 362.
c For intermediate heights, use harmonic constants and N.P.159.
i Constants inferred.
j For intermediate heights, see page xxiii.
k For intermediate heights, use Standard Curve for Shoreham on page 22.
l Constants from 15 days' observations.
v Owing to large seasonal variations, see Table VI.
x M.L. inferred.

ENGLAND, EAST COAST

No.	PLACE	Lat. N.	Long. E.	Time Differences (Zone G.M.T.) High Water	High Water	Low Water	Low Water	Height Differences (in metres) MHWS	MHWN	MLWN	MLWS	M.L. Z_0 m.	Harmonic Constants (Zone G.M.T.) M_2 g°	M_2 H.m.	S_2 g°	S_2 H.m.	K_1 g°	K_1 H.m.	O_1 g°	O_1 H.m.	S.W. Corrections ¼ diurnal f_4	¼ diurnal F_4	⅙ diurnal f_6	⅙ diurnal F_6	
108	**SHEERNESS**	(see page 30)		0200 and 1400	0800 and 2000	0200 and 1400	0700 and 1900	5·7	4·8	1·5	0·6														
109	**CHATHAM LOCK APPROACHES**			STANDARD PORT				See Table V				2·95	357	2·15	054	0·63	015	0·11	188	0·14	030	0·027	060	0·008	
109a	Upnor	51 25	0 32	+0015	+0015	+0015	+0025	+0·2	+0·2	−0·1	−0·1	⊙	⊙	⊙	⊙	⊙	⊙	⊙	⊙	⊙	⊙	⊙	⊙	⊙	
109b	Rochester (Strood Pier)	51 24	0 30	+0018	+0018	+0018	+0028	+0·2	+0·2	−0·2	−0·3	2·91	000	2·19	056	0·59	011	0·10	200	0·16	030	0·033	⊙	⊙	
109c	Wouldham	51 22	0 27	+0030	+0025	+0035	+0120	−0·2	−0·3	−1·0	−0·3	⊙	⊙	⊙	⊙	⊙	⊙	⊙	⊙	⊙	⊙	⊙	⊙	⊙	
109d	New Hythe	51 19	0 28	+0035	+0035	+0220	+0240	−1·6	−1·7	−1·2	−0·3	⊙	⊙	⊙	⊙	⊙	⊙	⊙	⊙	⊙	⊙	⊙	⊙	⊙	
109e	Allington Lock	51 17	0 30	+0050	+0035	⊙	⊙	−2·1	−2·2	−1·3	−0·4	⊙	⊙	⊙	⊙	⊙	⊙	⊙	⊙	⊙	⊙	⊙	⊙	⊙	
108	**SHEERNESS**	(see page 30)		0600 and 1800	1100 and 2300	0100 and 1300	0800 and 2000	5·7	4·8	1·5	0·6														
	River Thames																								
110	Southend	51 31	0 43	−0005	−0005	−0005	−0005	0·0	0·0	−0·1	−0·1	3·02	354	*v*	049	0·59	011	0·11	188	0·13	021	0·022	050	0·006	* *v*
110a	Thames Haven	51 30	0 31	+0010	+0010	0000	+0010	+0·5	+0·4	−0·1	−0·1	3·15	000	2·17	056	0·63	014	0·12	192	0·13	012	0·021	057	0·005	*
111	Tilbury	51 28	0 22	+0010	+0030	+0005	+0035	+0·8	+0·6	0·0	−0·1	3·33	009	2·32	069	0·66	021	0·13	197	0·14	007	0·016	053	0·005	*
113	**LONDON BRIDGE**	(see page 38)		0300 and 1500	0900 and 2100	0400 and 1600	1100 and 2300	7·1	5·8	1·6	0·5														
112	Woolwich (Gallion's Point)	51 30	0 05	−0020	−0025	−0035	−0045	−0·1	+0·1	−0·1	0·0	*w*	021	2·53	084	0·74	027	0·13	202	0·14	358	0·010	053	0·007	**wx*
		N.	W.																						
113	**LONDON BRIDGE**	51 30	0 05	STANDARD PORT				See Table V				*w*	034	2·50	100	0·71	035	0·13	208	0·14	311	0·016	048	0·008	**w*
114	Chelsea Bridge	51 29	0 09	+0020	+0015	+0055	+0100	−0·8	−0·7	−0·4	−0·1	⊙	⊙	⊙	⊙	⊙	⊙	⊙	⊙	⊙	⊙	⊙	⊙	⊙	*
115	Barnes Bridge	51 28	0 15	+0045	+0040	+0220	+0210	−1·6	−1·7	−0·5	0·0	⊙	⊙	⊙	⊙	⊙	⊙	⊙	⊙	⊙	⊙	⊙	⊙	⊙	*
116	Richmond Lock	51 28	0 19	+0100	+0055	+0325	+0305	−2·1	−2·2	−0·3	+0·2	⊙	⊙	⊙	⊙	⊙	⊙	⊙	⊙	⊙	⊙	⊙	⊙	⊙	*
108	**SHEERNESS**	(see page 30)		0200 and 1400	0700 and 1900	0100 and 1300	0700 and 1900	5·7	4·8	1·5	0·6														
		N.	E.																						
116a	*Thames estuary* Shivering Sand Tower	51 30	1 05	−0025	−0019	−0008	−0026	−0·6	−0·6	−0·1	−0·1	2·75	347	1·80	040	0·52	010	0·10	185	0·13	010	0·024	032	0·007	*
103	**MARGATE**	(see page 26)		0100 and 1300	0700 and 1900	0100 and 1300	0700 and 1900	4·8	3·9	1·4	0·5														
117	S.E. Longsand	51 32	1 21	−0006	−0003	−0004	−0004	0·0	+0·1	0·0	−0·1	2·5	338	1·69	032	0·49	351	0·08	167	0·12	000	0·032	043	0·004	* *x*

FRANCE, NORTH COAST

350

No.	PLACE	Lat. N.	Long. W.	TIME DIFFERENCES (Zone −0100) High Water		Low Water		HEIGHT DIFFERENCES (IN METRES) MHWS	MHWN	MLWN	MLWS	M.L. Z_0 m.	HARMONIC CONSTANTS (Zone −0100) M_2 g°	M_2 H.m.	S_2 g°	S_2 H.m.	K_1 g°	K_1 H.m.	O_1 g°	O_1 H.m.	S.W. CORRECTIONS ¼ diurnal f_4	F_4	⅙ diurnal f_6	F_6	
1605	**ST. HELIER**	(see page 222)		**0800** and **2000**	**0100** and **1300**	**0200** and **1400**	**0700** and **1900**	11·1	8·1	4·1	1·3														
1609	Iles Chausey	48 52	1 49	+0048	+0044	+0104	+0058	+1·8	+1·7	+0·8	+0·6	7·50	218	3·94	258	1·46	111	0·09	355	0·09	266	0·035	177	0·002	i
1610	Dielette	49 33	1 52	+0119	+0116	+0115	+0120	−1·6	−0·9	−0·5	−0·1	5·51	223	3·07	263	1·13	111	0·09	355	0·09	300	0·026	034	0·002	i
1611	Carteret	49 22	1 47	+0110	+0100	+0120	+0115	−0·1	−0·3	0·0	+0·1	6·30	223	3·50	263	1·30	111	0·09	355	0·09	293	0·020	073	0·001	i
1612	Granville	48 50	1 36	+0049	+0040	+0115	+0053	+1·7	+1·5	+0·5	+0·1	7·21	218	4·16	258	1·54	111	0·09	355	0·09	323	0·025	347	0·003	i
1613	Cancale	48 40	1 51	+0050	+0035	+0115	+0100	+2·2	+2·0	+1·0	+0·7	7·76	216	4·12	256	1·53	111	0·09	355	0·09	326	0·024	345	0·003	i
1614	St. Malo	F 48 38	2 02	+0044	+0034	+0105	+0050	+1·0	+1·0	+0·3	+0·1	6·85	207	3·69	257	1·44	111	0·09	355	0·09	286	0·020	⊙	⊙	
1615	Erquy	48 38	2 28	+0040	+0030	+0035	+0032	+0·1	+0·4	0·0	0·0	6·40	204	3·58	244	1·32	111	0·09	355	0·09	302	0·016	041	0·001	i
1616	Dahouet	48 35	2 34	+0038	+0031	+0027	+0036	+0·1	+0·4	−0·2	−0·1	⊙	⊙	⊙	⊙	⊙	⊙	⊙	⊙	⊙	⊙	⊙	⊙	⊙	
1617	Le Légué	48 32	2 44	+0045	+0030	+0035	+0031	+0·1	+0·4	0·0	0·0	5·6	205	3·61	245	1·34	111	0·09	355	0·09	345	0·027	004	0·004	ix
1618	Binic	48 36	2 49	+0045	+0030	+0035	+0031	+0·1	+0·4	0·0	0·0	5·6	205	3·61	245	1·34	111	0·09	355	0·09	345	0·027	004	0·004	ix
1619	Portrieux	48 38	2 49	+0045	+0030	+0030	+0030	+0·1	+0·4	0·0	0·0	6·38	203	3·61	243	1·34	111	0·09	355	0·09	298	0·017	043	0·001	i
1620	Paimpol	48 47	3 02	+0038	+0025	+0025	+0021	−0·8	−0·3	−0·9	−0·8	5·52	202	3·58	242	1·32	102	0·10	009	0·08	295	0·013	180	0·001	i
1621	Ile de Bréhat	48 51	3 00	+0040	+0020	+0010	+0015	−0·7	−0·1	−0·5	−0·2	5·85	198	3·36	238	1·24	102	0·10	009	0·08	321	0·008	072	0·001	i
1622	Les Heaux de Brehat	F 48 55	3 05	+0030	+0031	−0011	+0042	−1·3	−0·6	−0·7	−0·3	5·51	195	3·25	244	1·15	097	0·09	346	0·08	060	0·003	⊙	⊙	i
1623	Lezardrieux	48 47	3 06	+0038	+0026	+0015	+0020	−1·1	−0·6	−0·7	−0·4	5·57	193	3·15	241	1·22	102	0·10	009	0·08	290	0·006	⊙	⊙	
1624	Plougrescant	48 51	3 13	+0004	−0005	−0017	−0013	−1·5	−0·7	−0·6	0·0	5·55	⊙	⊙	⊙	⊙	⊙	⊙	⊙	⊙	⊙	⊙	⊙	⊙	
1625	Tréguier	48 47	3 13	+0012	−0004	−0018	−0007	−1·4	−0·7	−0·8	−0·4	5·46	185	3·14	225	1·16	090	0·06	346	0·06	318	0·018	166	0·001	i
1626	Ploumanac'h	48 50	3 29	+0005	0000	−0025	−0015	−2·2	−1·1	−0·7	−0·4	5·15	181	2·85	221	1·05	090	0·06	346	0·06	194	0·017	140	0·003	i
1638	**BREST**	(see page 226)		**0500** and **1700**	**1000** and **2200**	**1100** and **2300**	**0400** and **1600**	7·5	5·9	3·0	1·4														
1628	Trebeurden	48 46	3 35	+0111	+0107	+0120	+0100	+1·6	+1·3	+0·5	−0·1	5·28	175	2·81	215	1·04	090	0·06	346	0·06	259	0·025	158	0·005	i
1629	*Rade de Morlaix* Morlaix (Chateau du Taureau)	F 48 41	3 53	+0114	+0056	+0114	+0046	+1·5	+1·1	+0·5	−0·1	5·20	173	2·77	213	1·03	090	0·06	346	0·06	⊙	⊙	⊙	⊙	i
1630	Roscoff	48 43	3 58	+0105	+0055	+0110	+0045	+1·4	+1·1	+0·5	−0·1	5·17	172	2·77	212	1·03	090	0·06	346	0·06	264	0·017	173	0·002	i
1631	Ile de Batz	48 44	4 00	+0105	+0045	+0105	+0045	+1·4	+1·1	+0·5	0·0	⊙	⊙	⊙	⊙	⊙	⊙	⊙	⊙	⊙	⊙	⊙	⊙	⊙	
1632	*L'Aberwrac'h* Fort Cézon	48 37	4 36	+0035	+0020	+0035	+0015	+0·5	+0·2	−0·1	−0·3	4·52	154	2·45	194	0·90	090	0·06	346	0·06	254	0·016	199	0·003	i
1633	L'Aberbenoit	48 35	4 38	+0020	+0025	+0035	+0035	+0·6	+0·5	+0·1	−0·2	4·70	⊙	⊙	⊙	⊙	⊙	⊙	⊙	⊙	⊙	⊙	⊙	⊙	
1638	**BREST**	(see page 226)		**0600** and **1800**	**0000** and **1200**	**0000** and **1200**	**0600** and **1800**	7·5	5·9	3·0	1·4														
1634	*Ushant (Ouessant)* Baie de Lampaul	48 27	5 06	+0005	0000	−0005	−0005	0·0	−0·1	0·0	+0·1	4·45	139	2·06	180	0·83	093	0·07	338	0·06	⊙	⊙	⊙	⊙	

1635	Molene		48 24	4 58	+0010	+0010	+0015	+0015	−0·1	−0·1	−0·2	−0·2	4·30	⊙	⊙	⊙	⊙	⊙	⊙	⊙	⊙	⊙	⊙	⊙	⊙	
1636	Le Conquet		48 22	4 47	0000	0000	+0010	0000	−0·3	−0·3	−0·1	0·0	4·26	141	2·02	181	0·73	090	0·07	342	0·07	281	0·017	⊙	⊙	
1637	La Penfeld		48 23	4 30	−0005	−0005	0000	−0005	+0·1	0·0	+0·2	+0·2	⊙	⊙	⊙	⊙	⊙	⊙	⊙	⊙	⊙	⊙	⊙	⊙	⊙	
1638	**BREST**		48 23	4 29	STANDARD PORT					See Table V			*w*	138	2·02	178	0·74	090	0·06	346	0·06	258	0·011	⊙	⊙	*w*
1639	Camaret		48 16	4 36	−0015	−0015	−0015	−0020	−0·5	−0·5	−0·2	−0·1	4·14	132	1·96	170	0·72	089	0·07	335	0·06	270	0·007	⊙	⊙	
1640	Douarnenez		48 06	4 19	−0010	−0015	−0015	−0020	−0·5	−0·4	−0·2	0·0	4·17	133	1·91	170	0·64	085	0·07	339	0·07	266	0·012	⊙	⊙	
1641	Ile de Sein		48 02	4 51	−0010	−0010	−0010	−0015	−1·1	−0·9	−0·5	−0·3	3·78	133	1·90	173	0·70	090	0·06	346	0·06	341	0·020	351	0·006	*i*
1643	Audierne		48 01	4 33	−0040	−0020	−0040	−0020	−2·3	−1·8	−0·9	−0·5	3·08	122	1·57	162	0·58	090	0·06	346	0·06	330	0·013	135	0·007	*i*
1644	Guilvinec		47 48	4 17	−0035	−0010	−0035	−0020	−2·4	−1·9	−1·0	−0·5	3·03	⊙	⊙	⊙	⊙	⊙	⊙	⊙	⊙	⊙	⊙	⊙	⊙	
	Pont l'Abbe River																									
1645	Loctudy		47 50	4 10	−0030	−0010	−0030	−0025	−2·6	−2·1	−1·2	−0·9	2·76	123	1·53	163	0·57	090	0·06	346	0·06	151	0·006	180	0·007	*i*
	Odet River																									
1646	Benodet		47 53	4 07	−0025	−0010	−0040	−0015	−2·6	−2·2	−1·2	−0·8	2·75	125	1·53	165	0·57	090	0·06	346	0·06	180	0·023	171	0·012	*i*
1647	Corniguel		47 58	4 06	+0010	+0015	−0015	−0010	−2·6	−2·1	−1·4	−1·1	2·66	⊙	⊙	⊙	⊙	⊙	⊙	⊙	⊙	⊙	⊙	⊙	⊙	
1648	Concarneau		47 52	3 54	−0035	−0005	−0030	−0020	−2·5	−2·0	−1·0	−0·6	2·93	122	1·50	162	0·55	090	0·06	346	0·06	339	0·038	174	0·020	*i*
	Iles de Glenan																									
1649	Penfret		47 44	3 57	−0025	−0010	−0030	−0020	−2·6	−2·0	−1·2	−0·6	2·87	125	1·53	165	0·57	090	0·06	346	0·06	331	0·006	134	0·008	*i*
1650	Port Louis		47 42	3 21	−0025	−0010	−0020	−0020	−2·4	−1·9	−0·9	−0·5	3·04	127	1·62	167	0·57	090	0·06	346	0·06	191	0·032	192	0·008	*i*
1651	Lorient		47 45	3 21	−0025	+0005	−0025	−0015	−2·4	−1·9	−1·0	−0·5	3·04	130	1·50	163	0·55	092	0·06	345	0·06	189	0·026	⊙	⊙	
	Ile de Groix																									
1652	Port Tudy	F	47 39	3 27	−0035	−0005	−0030	−0025	−2·4	−1·9	−0·9	−0·5	3·06	129	1·49	163	0·55	093	0·06	346	0·06	196	0·028	⊙	⊙	
1653	Port-Haliguen		47 29	3 06	−0020	0000	−0025	−0015	−2·3	−1·9	−1·0	−0·8	2·98	126	1·68	166	0·62	090	0·06	346	0·06	191	0·043	180	0·012	*i*
	Belle-Ile																									
1654	Le Palais		47 21	3 09	−0030	−0015	−0030	−0025	−2·3	−1·9	−1·0	−0·6	3·01	⊙	⊙	⊙	⊙	⊙	⊙	⊙	⊙	⊙	⊙	⊙	⊙	
	Crac'h River																									
1655	La Trinite		47 35	3 01	−0020	+0010	−0025	−0015	−2·1	−1·7	−0·9	−0·7	3·09	127	1·68	167	0·62	090	0·06	346	0·06	219	0·030	135	0·008	*i*
	Morbihan																									
1656	Port-Navalo		47 33	2 55	−0005	+0025	−0010	−0005	−2·5	−2·0	−1·1	−0·7	2·90	135	1·53	175	0·57	090	0·06	346	0·06	202	0·043	⊙	⊙	*i*
1657	Auray		47 40	2 59	+0005	+0040	⊙	⊙	−2·2	−1·8	−1·1	−0·6	3·1	147	1·64	187	0·61	090	0·06	346	0·06	237	0·064	109	0·009	*ix*
1658	Vannes		47 40	2 46	+0200	+0200	⊙	⊙	−2·8	−1·9	−0·9	−0·1	2·7	207	1·39	247	0·51	090	0·06	346	0·06	307	0·070	000	0·003	*ix*
1659	Ile de Hoedic		47 20	2 52	−0025	−0005	−0030	−0020	−2·3	−1·9	−1·1	−0·8	2·95	125	1·64	165	0·61	093	0·06	347	0·07	166	0·036	169	0·008	*i*
1660	Pénerf		47 31	2 37	−0020	−0010	−0025	−0020	−2·0	−1·7	−1·0	−0·7	3·09	128	1·75	168	0·65	093	0·06	347	0·07	180	0·017	112	0·001	*i*
1661	Le Croisic		47 18	2 31	−0030	+0030	−0015	−0010	−2·3	−1·8	−1·1	−0·8	2·97	134	1·59	170	0·57	109	0·08	340	0·07	199	0·055	359	0·007	
	La Loire																									
1662	Le Pouliguen		47 17	2 25	−0040	+0015	0000	−0020	−2·1	−1·8	−1·2	−0·8	2·95	⊙	⊙	⊙	⊙	⊙	⊙	⊙	⊙	⊙	⊙	⊙	⊙	
1663	Le Grand-Charpentier		47 13	2 19	−0030	+0010	−0020	−0020	−2·2	−2·0	−1·3	−0·9	3·0	126	1·72	166	0·63	093	0·06	347	0·07	204	0·045	052	0·007	*ix*
1664	St. Nazaire	F	47 16	2 12	−0025	+0030	−0005	−0010	−2·0	−1·7	−1·1	−0·8	3·06	136	1·71	170	0·60	093	0·06	347	0·07	205	0·045	⊙	⊙	
1665	Paimboeuf		47 17	2 02	−0005	+0015	+0120	+0030	−1·9	−1·6	−1·3	−0·5	3·2	153	1·72	193	0·63	093	0·06	347	0·07	270	0·052	131	0·010	*ix*
1666	Le Pellerin		47 12	1 45	+0100	+0140	+0300	+0210	−1·7	−1·4	−1·2	+0·1	3·32	172	1·93	206	0·68	107	0·04	001	0·05	252	0·079	⊙	⊙	
1667	Nantes (Chantenay)		47 12	1 35	+0140	+0155	+0330	+0245	−1·7	−1·3	−1·1	+0·2	3·74	195	1·59	229	0·56	217	0·04	111	0·05	243	0·116	⊙	⊙	

⊙ No data.
F Tides predicted in French Tide Tables.
i Constants inferred.
w Owing to large fortnightly variations, see Table VIa.
x M.L. inferred.

SEASONAL CHANGES IN MEAN LEVEL

No.	Jan. 1	Feb. 1	Mar. 1	Apr. 1	May 1	June 1	July 1	Aug. 1	Sep. 1	Oct. 1	Nov. 1	Dec. 1	Jan. 1
1605–1626	0·0	−0·1	−0·1	0·0	0·0	0·0	0·0	0·0	0·0	+0·1	+0·1	+0·1	0·0
1628–1643							Negligible						
1645–1667	0·0	0·0	0·0	0·0	0·0	0·0	−0·1	−0·1	0·0	0·0	0·0	+0·1	0·0

GEOGRAPHICAL INDEX

N.B. Specimen Page only—page numbers refer to ATT Volume 1.

Volume 2 ATLANTIC AND INDIAN OCEANS

TABLE V Tidal Levels at Standard Ports

(with data concerning predictions, etc.)

Standard Port	L.A.T.	M.L.W.S.	M.L.W.N.	M.S.L.	M.H.W.N.	M.H.W.S.	H.A.T.	Authority for (*b*) Observations	Authority for (*b*) Constants	Authority for (*b*) Predictions	Years of tidal observations (*c*)
	m.	m.	m.	m.	m.	m.	m.				
Puerto Gallegos	−0·4	+1·6	+3·8	+6·7	+9·5	+12·0	+13·2	Ar.	Ar.	Ar.	—
Puerto Madryn	−0·3	+0·3	+1·5	+3·0	+4·5	+5·3	+6·1	Ar.	Ar.	Ar.	1969 (1)
Puerto Belgrano (Bahia Blanca)	−0·1	+0·5	+1·0	+2·4	+3·7	+4·1	+4·3	Ar.	Ar.	Ar.	1905–37 (5)
Rio De Janeiro	−0·1	+0·1	+0·4	+0·7	+0·9	+1·2	+1·4	B.	B.	B.	1953 (1)
Georgetown (Guyana)	−0·1	+0·3	+1·0	+1·6	+2·2	+2·9	+3·4	H.A.	I.O.S.	I.O.S.	1974 (1)
Bermuda	−0·1	+0·2	+0·4	+0·7	+0·9	+1·2	+1·5	Hyd.	I.O.S.	I.O.S.	1853, 1920 (1)
Charleston	−0·5	−0·2	+0·1	+0·8	+1·4	+1·7	+2·1	U.S.	U.S.	U.S.	1948 (1)
Sandy Hook (New York Bay)	−0·4	−0·2	+0·2	+0·7	+1·2	+1·6	+2·0	U.S.	U.S.	U.S.	1940 (3)
Boston	−0·7	−0·2	+0·2	+1·5	+2·7	+3·2	+3·7	U.S.	U.S.	U.S.	1940 (5)
Saint John N.B.	−0·3	+0·6	+1·7	+4·3	+6·9	+8·0	+8·9	C.	C.	C.	(*d*)
Halifax	+0·1	+0·4	+0·7	+1·2	+1·7	+2·1	+2·4	C.	C.	C.	(*d*)
Pointe-au-Père	−0·1	+0·5	+1·4	+2·3	+3·1	+4·2	+4·9	C.	C.	C.	(*d*)
Quebec	−0·1	+0·3	+0·6	+2·6	+4·0	+5·2	+6·1	C.	C.	C.	(*d*)
Harrington	+0·1	+0·3	+0·7	+1·1	+1·4	+1·8	+2·3	C.	C.	C.	(*d*)
Argentia	0·0	+0·3	+0·7	+1·2	+1·7	+2·2	+2·7	C.	C.	C.	(*d*)
St. John's Harbour (Newfoundland)	0·0	+0·2	+0·5	+0·8	+0·9	+1·3	+1·6	C.	C.	C.	(*d*)
Ponta Delgada	0·0	+0·3	+0·7	+1·0	+1·3	+1·7	+2·0	P.	P.	P.	1939 (1)
Casablanca	+0·4	+0·7	+1·5	+2·1	+2·8	+3·6	+4·0	F.	F.	F.	1958 (1)
Dakar	0·0	+0·3	+0·7	+1·0	+1·3	+1·6	+1·9	F.	F.	F.	1965 (7)
Freetown (Sierra Leone)	0·0	+0·4	+1·0	+1·7	+2·3	+3·0	+3·4	H.A.	I.O.S.	I.O.S.	1926–27 (1)
Takoradi	−0·1	+0·2	+0·6	+1·0	+1·2	+1·5	+1·9	Gh.	I.O.S.	I.O.S.	1927–30 (2)
Bonny Town	0·0	+0·4	+0·9	+1·5	+1·9	+2·3	+2·7	H.A.	I.O.S.	I.O.S.	1958 (1)
Cape Town	+0·1	+0·3	+0·8	+1·0	+1·3	+1·8	+2·1	H.A.	S.A	S.A.	(*d*)
Durban	0·0	+0·2	+0·8	+1·1	+1·3	+2·0	+2·3	H.A.	S.A.	S.A.	(*d*)
Beira	+0·2	+0·9	+2·7	+3·5	+4·2	+6·5	+7·1	P.	P.	P.	1952, 3 (1)
Dar Es Salaam	−0·5	−0·1	+1·0	+1·5	+2·1	+3·2	+3·7	H.A.	I.O.S.	I.O.S.	1929–30 (1)
Zanzibar	−0·2	+0·2	+1·5	+2·1	+2·7	+3·9	+4·5	H.A.	I.O.S.	I.O.S.	1925–27 (1)
Kilindini	−0·1	+0·3	+1·3	+1·9	+2·4	+3·5	+4·0	H.A.	I.O.S.	I.O.S.	1975, 76 (1)
Suez	0·0	+0·4	+0·7	+1·1	+1·6	+1·9	+2·3	I.	I.	I.	1897–1904 (7)
Mina Salman	0·0	+0·6	+1·0	+1·5	+1·9	+2·4	+2·7	Hyd.	I.O.S.	Hyd.	1976 (1)
Bhavnagar	+0·3	+1·4	+3·5	+6·1	+8·3	+10·2	+12·2	I.	I.	I.	—
Bombay	−0·2	+0·8	+1·9	+2·5	+3·3	+4·4	+5·1	I.	I.	I.	1951–54 (4)
Colombo	−0·1	+0·1	+0·3	+0·4	+0·5	+0·7	+0·9	I.	I.	I.	1884–90 (6)
Trincomalee	0·0	+0·1	+0·3	+0·4	+0·5	+0·7	+0·9	I.	I.	I.	1890–95 (6)
Madras	−0·1	+0·1	+0·4	+0·6	+0·8	+1·1	+1·5	I.	I.	I.	1952–65 (6)
Sagar Roads (Hooghly river)	+0·2	+0·9	+2·2	+3·0	+3·8	+5·2	+5·9	H.A.	I.	I.	1948–64 (9)
Pussur River Ent.	−0·3	+0·6	+1·3	+1·8	+2·2	+3·0	+3·8	Pa.	I.O.S.	I.O.S.	1976 (1)
Chittagong	−0·1	+0·5	+1·0	+2·1	+3·0	+4·0	+5·0	Pa.	I.O.S.	I.O.S.	1978 (1)
Bassein River Ent. (Thamihla Kyun)	−0·2	+0·4	+1·1	+1·4	+1·8	+2·4	+2·8	I.	I.	I.	1895–99 (5)
Elephant Point (Rangoon river)	+0·1	+0·8	+2·5	+3·7	+4·9	+6·6	+7·3	I.	I.	I.	1884–88, 1930 (6)
Mergui	−0·8	+0·1	+2·0	+2·8	+3·6	+5·4	+6·2	I.	I.	I.	1889–94 (5)
Penang	+0·1	+0·6	+1·3	+1·6	+1·8	+2·5	+2·9	H.A.	I.O.S.	I.O.S.	1953 (1)
Pelabuhan Kelang	−0·1	+0·7	+2·3	+2·9	+3·6	+4·8	+5·4	H.A.	I.O.S.	I.O.S.	1911–12 (1)
Port Dickson	−0·1	+0·3	+1·1	+1·5	+1·9	+2·8	+3·4	I.T.S.	I.T.S.	I.O.S.	1979 (1)
Kuala Batu Pahat	−0·2	+0·3	+1·1	+1·5	+1·9	+2·7	+3·3	I.T.S.	I.T.S.	I.O.S.	1979 (1)
Singapore (Victoria Dock)	−0·3	+0·6	+1·1	+1·6	+2·1	+2·8	+3·3	H.A.	I.O.S.	H.A.	1954 (3)

The above levels, in metres, are referred to *CHART DATUM*, which is the same as the zero of the tidal predictions in all cases.

For Notes (*b*), (*c*) and (*d*) see ATT.

TABLE V—*continued*

Standard Port	L.A.T.	M.L.W.S.	M.L.W.N.	M.S.L.	M.H.W.N.	M.H.W.S.	H.A.T.	Authority for (*b*)			
								Observations	Constants	Predictions	Years of tidal observations (*c*)
	m.	m.	m.	m.	m.	m.	m.				
Cape Horn (Orange Bay)	+0·1	+0·5	+0·9	+1·3	+1·7	+2·2	+2·7	U.S.	U.S.	U.S.	1882–83 (1)
Stanley	0·0	+0·4	+0·7	+1·0	+1·4	+1·6	+2·1	H.A.	Hyd.	Hyd.	1964 (3 mths.)
Buenos Aires (Rio de la Plata)	−0·1	+0·3	+0·5	+0·8	+1·0	+1·3	+1·7	Ar.	Ar.	Ar.	1970 (1)
Port of Spain	0·0	+0·4	+0·5	+0·7	+0·9	+1·1	+1·3	H.A.	I.O.S.	I.O.S.	1938 (1)
Maracaibo Appr. Malecon	0·0	+0·2	+0·5	+0·8	+1·2	+1·3	+1·7	H.A.	U.S.	U.S.	—
Colon	−0·2	0·0	0·0	+0·1	+0·1	+0·3	+0·5	U.S.	U.S.	U.S.	1950 (1)
Galveston	−0·4	0·0	+0·3	+0·2	+0·4	+0·4	+0·5	U.S.	U.S.	U.S.	1908, 23, 32 (3)
Pictou	−0·1	+0·5	+0·9	+1·1	+1·5	+1·7	+2·0	C.	C.	C.	(*d*)
Aden	−0·2	+0·5	+1·2	+1·3	+1·8	+2·0	+2·6	I.	I.	I.O.S.	1951–55 (4)
Muscat	−0·1	+0·8	+1·6	+1·9	+2·4	+2·7	+3·2	I.	I.	I.	—
Musay'id outer channel entr.	+0·3	+0·7	+1·0	+1·1	+1·2	+1·9	+2·5	Q.	I.O.S.	I.O.S.	1959 (1)
Mina Saud	−0·2	+0·4	+1·2	+1·2	+1·6	+2·0	+2·4	H.A.	I.O.S.	I.O.S.	1966–67 (1)
Mina Al Ahmadi	−0·2	+0·4	+1·2	+1·5	+2·0	+2·4	+2·9	A.I.	Hyd.	I.O.S.	1952 (2 mths.)
Shatt Al Arab Bar	−0·5	+0·4	+1·3	+1·7	+2·4	+3·0	+3·5	H.A.	Rob.	I.O.S.	1918–19 (1)
Khawr-e Musa Bar	0·0	+0·9	+1·9	+2·3	+2·9	+3·5	+4·3	Ir.	I.O.S.	I.O.S.	1960, 1963 (2)
Jazireh Ye Khark	+0·1	+0·5	+1·1	+1·2	+1·4	+2·0	+2·6	Ir.	I.O.S.	I.O.S.	1958, 1962 (2)
Karachi	−0·4	+0·4	+1·1	+1·6	+2·4	+2·7	+3·2	Pa.	I.O.S.	I.O.S.	1950, 1953 (2)
Cochin	0·0	+0·3	+0·6	+0·6	+0·8	+0·9	+1·2	I.	I.	I.	1886–92 (6)
Air Musi (Outer Bar)	−0·3	+0·9	△	+1·9	△	+3·1	+4·4	N.	N.	Hyd.	1954 (2)
Horsburgh Lt. Ho.	−0·3	+0·6	+1·3	+1·5	+2·1	+2·2	+2·8	I.T.S.	I.T.S.	I.O.S.	1979 (1)
Karang Djamuang	−0·2	+0·4	△	+1·1	△	+1·6	+2·2	N.	N.	Hyd.	—
Mui Vung Tau	−0·4	+0·9	+2·2	+2·4	+3·3	+3·5	+4·1	F.	F.	F.	—

Notes:

(*a*) The above levels, in metres, are referred to *CHART DATUM*, which is the same as the zero of the tidal predictions in all cases.

(*b*) *Abbreviations:*

- A.I. Anglo-Iranian Oil Company.
- Ar. Ministerio de Marina, Buenos Aires.
- B. Brazilian Government.
- C. Canadian Hydrographic Service, Ottawa.
- F. Service Hydrographique et Océanographique, Brest.
- Gh. Survey of Ghana.
- H.A. Local Harbour Authority.
- Hyd. Hydrographer of the Navy.
- I. Geodetic Branch, Survey of India.
- Ir. Iranian Oil Exploration and Producing Company.
- I.O.S. Institute of Oceanographic Sciences.
- I.T.S. International Tidal Survey.
- N. Netherlands Government.
- P. Ministerio de Marinha, Lisbon.
- Pa. Pakistan Navy.
- Q. Qatar Petroleum Company Limited.
- Rob. Messrs. Edward Roberts & Son.
- S.A. Hydrographer, South African Navy.
- U.S. United States National Ocean Survey.

(*c*) The years between which the observations were obtained are given, with the number of complete year's observations in brackets.

(*d*) Constants from the latest available year.

(*e*) All predictions in this volume are calculated by the harmonic method, except Sagar Roads, Elephant Point and Chittagong where the harmonic method with shallow water corrections is used. Predictions for Quebec have been prepared using constants obtained from the analysis of High and Low waters over a 19-year period.

(*f*) △ Tide is usually diurnal.

(*g*) For definitions of tidal levels see page 104.

Volume 2 ATLANTIC AND INDIAN OCEANS—Part I (Standard Ports)

CANADA, BAY OF FUNDY - SAINT JOHN, N.B.

LAT 45°16'N LONG 66°04'W

TIME ZONE +0400 TIMES AND HEIGHTS OF HIGH AND LOW WATERS YEAR 1982

JANUARY

Day	TIME	M	Day	TIME	M
1	0320	7.3	16	0450	7.4
	0930	1.6		1100	1.5
F	1540	7.4	SA	1710	7.3
	2155	1.2		2320	1.4
2	0410	7.3	17	0540	7.2
	1025	1.6		1155	1.6
SA	1620	7.3	SU	1805	7.0
	2255	1.3			
3	0500	7.3	18	0025	1.6
	1115	1.6		0630	7.2
SU	1715	7.2	M	1250	1.6
	2345	1.4		1915	6.9
4	0555	7.3	19	0115	1.7
	1220	1.6		0735	7.2
M	1820	7.1	TU	1355	1.6
				2005	6.9
5	0050	1.4	20	0200	1.7
	0700	7.4		0815	7.3
TU	1315	1.4	W	1440	1.6
	1930	7.1		2050	6.9
6	0155	1.3	21	0300	1.7
	0805	7.7		0905	7.3
W	1425	1.2	TH	1525	1.4
	2040	7.3		2140	7.0
7	0245	1.2	22	0345	1.7
	0905	7.9		0955	7.5
TH	1525	0.9	F	1610	1.3
	2130	7.5		2220	7.1
8	0400	0.9	23	0420	1.6
	1005	8.1		1030	7.6
F	1620	0.5	SA	1645	1.1
	2240	7.7		2305	7.3

FEBRUARY

Day	TIME	M	Day	TIME	M
1	0435	7.4	16	0545	7.1
	1050	1.4		1205	1.7
M	1655	7.1	TU	1815	6.8
	2320	1.4			
2	0530	7.3	17	0020	2.0
	1155	1.5		0635	7.0
TU	1805	7.0	W	1300	1.9
				1910	6.7
3	0015	1.5	18	0115	2.0
	0640	7.4		0740	7.0
W	1305	1.4	TH	1350	1.8
	1905	6.9		2000	6.7
4	0130	1.6	19	0215	2.0
	0745	7.4		0830	7.1
TH	1405	1.3	F	1455	1.7
	2025	7.1		2100	6.8
5	0230	1.4	20	0315	1.9
	0855	7.7		0920	7.3
F	1515	1.0	SA	1530	1.5
	2120	7.3		2150	7.0
6	0335	1.1	21	0350	1.6
	0945	8.0		1010	7.5
SA	1615	0.7	SU	1625	1.2
	2220	7.6		2230	7.2
7	0445	0.9	22	0445	1.4
	1050	8.2		1050	7.7
SU	1700	0.4	M	1705	1.0
	2320	7.8		2305	7.4
8	0530	0.7	23	0515	1.2
	1135	8.3		1120	7.8
M	1800	0.3	TU	1740	0.9
				2350	7.6

MARCH

Day	TIME	M	Day	TIME	M
1	0315	7.7	16	0410	7.3
	0940	1.0		1025	1.5
M	1540	7.4	TU	1635	6.9
	2200	1.2		2240	1.9
2	0400	7.6	17	0450	7.0
	1035	1.2		1105	1.8
TU	1635	7.1	W	1720	6.7
	2255	1.5		2335	2.1
3	0500	7.4	18	0550	6.9
	1130	1.4		1210	2.0
W	1750	6.9	TH	1815	6.6
4	0010	1.7	19	0030	2.3
	0620	7.3		[illegible]	6.8
TH	1250	1.5	F	[illegible]	2.0
	1900	6.9		1925	6.6
5	0115	1.7	20	0130	2.2
	0735	7.3		0745	6.9
F	1355	1.4	SA	1405	1.9
	2010	7.0		2020	6.7
6	0220	1.6	21	0220	2.0
	0840	7.5		0835	7.1
SA	1505	1.1	SU	1450	1.7
	2120	7.3		2100	7.0
7	0330	1.2	22	0315	1.8
	0945	7.8		0925	7.3
SU	1605	0.8	M	1540	1.4
	2215	7.6		2150	7.3
8	0420	0.9	23	0405	1.5
	1035	8.0		1010	7.6
M	1650	0.5	TU	1625	1.1
	2300	7.8		2240	7.6

APRIL

Day	TIME	M	Day	TIME	M
1	0455	7.4	16	0500	7.0
	1120	1.4		1125	1.9
TH	1735	6.9	F	1735	6.7
	2355	1.8		2345	2.3
2	0600	7.3	17	0550	6.9
	1230	1.5		1215	2.0
F	1845	6.9	SA	1830	6.7
3	0100	1.8	18	0040	2.3
	0715	7.3		0650	6.9
SA	1350	1.4	SU	1315	2.0
	2005	7.1		1925	6.8
4	0215	1.6	19	0140	2.1
	0830	7.5		0745	7.1
SU	1445	1.2	M	1405	1.7
	2115	7.4		2015	7.1
5	0320	1.3	20	0240	1.8
	0935	7.7		0850	7.3
M	1555	0.9	TU	1505	1.5
	2200	7.7		2110	7.4
6	0410	1.0	21	0320	1.5
	1020	8.0		0940	7.6
TU	1630	0.7	W	1555	1.2
	2255	7.9		2155	7.7
7	0500	0.7	22	0410	1.1
	1105	8.0		1025	7.8
W	1715	0.6	TH	1640	0.9
	2335	8.0		2245	8.0
8	0545	0.6	23	0500	0.8
	1150	8.1		1115	7.9
TH	1810	0.6	F	1725	0.8
				2325	8.2

BURMA - BASSEIN RIVER ENTRANCE (THAMIHLA KYUN)

LAT 15°52'N LONG 94°16'E

TIME ZONE -0630 TIMES AND HEIGHTS OF HIGH AND LOW WATERS YEAR 1982

SEPTEMBER

Day	TIME	M	Day	TIME	M
9	0053	2.5	24	0113	2.2
	0704	0.8		0726	1.1
TH	1300	2.3	F	1319	2.0
	1920	0.7		1933	1.0
10	0138	2.4	25	0158	2.1
	0757	1.0		0819	1.3
F	1348	2.2	SA	1405	1.8
	2013	0.9		2030	1.2
11	0237	2.2	26	0308	1.9
	0912	1.1		0956	1.4
SA	1456	2.0	SU	1548	1.7
	2129	1.0		2210	1.3
12	0406	2.1	27	0513	1.9
	1054	1.2		1222	1.3
SU	1644	1.8	M	1822	1.7
	2311	1.1			
13	0603	2.1	28	0008	1.3
	1242	1.1		0655	2.0
M	1843	1.9	TU	1337	1.2
				1928	1.8
14	0049	1.0	29	0121	1.1
	0727	2.3		0744	2.2
TU	1355	0.9	W	1413	1.0
	1955	2.1		2006	2.0
15	0159	0.8	30	0206	1.0
	0822	2.5		0818	2.3
W	1447	0.7	TH	1443	0.8
	2046	2.2		2039	2.2

OCTOBER

Day	TIME	M	Day	TIME	M
9	0120	2.4	24	0123	2.2
	0749	1.0		0749	1.2
SA	1337	2.1	SU	1335	1.8
	1957	1.0		1947	1.2
10	0222	2.2	25	0220	2.0
	0910	1.1		0907	1.3
SU	1457	1.9	M	1503	1.7
	2121	1.2		2111	1.3
11	0357	2.1	26	0400	1.9
	1054	1.2		1104	1.3
M	1704	1.9	TU	1732	1.7
	2311	1.2		2309	1.3
12	0556	2.1	27	0551	2.0
	1235	1.1		1238	1.2
TU	1849	2.0	W	1852	1.9
13	0049	1.1	28	0041	1.2
	0714	2.3		0655	2.1
W	1340	0.9	TH	1326	1.0
	1949	2.2		1934	2.1
14	0152	0.9	29	0134	1.0
	0805	2.4		0737	2.2
TH	1426	0.7	F	1401	0.8
	2033	2.4		2009	2.3
15	0240	0.8	30	0216	0.9
	0846	2.5		0812	2.4
F	1504	0.6	SA	1433	0.7
	2110	2.5		2042	2.4
			31	0253	0.7
				0847	2.4
			SU	1505	0.5
				2115	2.6

NOVEMBER

Day	TIME	M	Day	TIME	M
9	0348	2.1	24	0250	2.0
	1040	1.1		0945	1.2
TU	1702	1.9	W	1610	1.7
	2301	1.2		2156	1.3
10	0529	2.1	25	0420	1.9
	1207	1.0		1111	1.1
W	1832	2.0	TH	1746	1.8
				2336	1.2
11	0034	1.1	26	0543	2.0
	0645	2.2		1218	1.0
TH	1310	0.9	F	1846	2.0
	1928	2.2			
12	0135	1.0	27	0049	1.1
	0738	2.3		0643	2.1
F	1357	0.7	SA	1307	0.8
	2011	2.4		1930	2.2
13	0223	0.8	28	0141	0.9
	0819	2.3		0731	2.2
SA	1434	0.6	SU	1351	0.6
	2046	2.5		2009	2.4
14	0301	0.7	29	0226	0.7
	0854	2.3		0813	2.3
SU	1507	0.6	M	1430	0.5
	2117	2.5		2049	2.6
15	0336	0.6	30	0310	0.6
	0927	2.3		0856	2.3
M	1538	0.6	TU	1511	0.4
	2146	2.6		2128	2.7

DECEMBER

Day	TIME	M	Day	TIME	M
9	0438	2.0	24	0301	1.9
	1120	0.9		0949	0.9
TH	1750	2.0	F	1621	1.8
				2226	1.1
10	0000	1.1	25	0421	1.8
	0558	2.0		1102	0.8
F	1227	0.8	SA	1742	1.9
	1855	2.1		2354	1.0
11	0110	1.0	26	0540	1.8
	0702	2.0		1210	0.7
SA	1319	0.8	SU	1845	2.1
	1941	2.2			
12	0202	0.8	27	0103	0.8
	0749	2.0		0648	1.9
SU	1401	0.7	M	1309	0.5
	2019	2.3		1937	2.2
13	0243	0.7	28	0201	0.6
	0829	2.0		0744	2.0
M	1437	0.6	TU	1401	0.4
	2053	2.4		2025	2.4
14	0319	0.6	29	0251	0.5
	0903	2.0		0834	2.1
TU	1511	0.5	W	1450	0.3
	2125	2.4		2111	2.5
15	0352	0.6	30	0339	0.4
	0935	2.0		0924	2.1
W	1543	0.5	TH	1536	0.2
	2156	2.4		2157	2.6
			31	0426	0.3
				1013	2.1
			F	1623	0.2
				2242	2.6

HEIGHTS IN METRES

Volume 2 ATLANTIC AND INDIAN OCEANS—Part II (Secondary Ports)

UNITED STATES: CANADA, BAY OF FUNDY

No.	Place	Lat. N.	Long. W.	Time Differences MHW	Time Differences MLW	Height Differences (in metres) MHWS	MHWN	MLWN	MLWS	M.L. Z_0 m.
				(Zone +0500)						
2864	SAINT JOHN, N.B.	see page 51				**8·0**	**6·9**	**1·7**	**0·6**	
2837	Blue Hill Harbour	44 24	68 34	−0055	−0045	−4·6	−4·0	−1·5	−0·8	1·52
2838	Bar Harbour	44 23	68 12	−0111	−0111	−4·6	−4·0	−1·4	−0·8	1·58
2840	Prospect Harbour	44 24	68 01	−0105	−0055	−4·5	−3·9	−1·4	−0·8	1·58
2842	Jonesport	44 32	67 36	−0105	−0055	−4·3	−3·6	−1·4	−0·8	1·77
2843	Cutler, Little River	44 39	67 13	−0110	−0100	−3·5	−3·0	−1·4	−0·9	2·07
2845	West Quoddy Head	44 49	66 59	−0110	−0100	−2·0	−1·6	−0·4	0·0	3·29
	Canada			**(Zone +0400)**						
	BAY OF FUNDY									
	Grand Manan Island									
2847	Outer Wood Island	44 36	66 48	−0029	−0028	−2·5	−2·2	−0·2	+0·1	3·09
2849	North Head	44 46	66 45	−0006	−0004	−1·5	−1·2	−0·3	+0·1	3·56
2850	Lubec (U.S.)	44 52	66 59	+0005	+0015	−1·4	−1·1	−0·3	0·0	3·57
2853	Welshpool	44 53	66 57	+0005	+0010	−1·3	−1·0	−0·2	+0·1	3·72

Seasonal Changes in Mean Level

No.	Jan. 1	Feb. 1	Mar. 1	Apr. 1	May 1	June 1	July 1	Aug. 1	Sep. 1	Oct. 1	Nov. 1	Dec.1	Jan. 1
2847–2902	0·0	−0·1	−0·1	0·0	0·0	+0·1	0·0	0·0	0·0	0·0	0·0	0·0	0·0

INDIA, EAST COAST: BANGLADESH

No.	Place		Lat. N.	Long. E.	Time Differences MHW	Time Differences MLW	Height Differences (in metres) MHWS	MHWN	MLWN	MLWS	M.L. Z_0 m.
					(Zone −0530)						
4512	**CHITTAGONG**		see page 168				**4·0**	**3·0**	**1·0**	**0·5**	
4507	Sandwip Island	B	22 30	91 25	+0320	+0320	+1·4	+1·1	+0·8	0·0	*w*
4507a	Ilsa Ghat	B	22 48	90 39	+0630	+0630	−1·2	−0·7	+0·4	+0·3	*w*
4508	Chandpur		23 15	90 35	+5000	⊙	−2·3	⊙	⊙	⊙	⊙
4509	Narayanganj		23 27	90 32	+0700	⊙	−3·1	⊙	⊙	⊙	⊙
	Kharnaphuli River										
4510	Norman Point		22 11	91 49	−0119	−0135	+0·6	+0·5	+0·9	+0·3	2·32
4512	CHITTAGONG		22 20	91 50	Standard Port			See Table V			*w*
4514	Kutubdia Island		21 52	91 50	−0152	−0222	−0·3	−0·3	+0·4	−0·1	2·04
4515	Cox's Bazaar		21 26	91 59	−0250	−0340	−0·6	−0·5	+0·3	−0·1	1·9
4539	**BASSEIN RIVER ENTRANCE**		see page 171				**2·4**	**1·8**	**1·1**	**0·4**	
4517	St. Martin's Island		20 37	92 19	−0056	−0101	+0·8	+0·5	+0·2	+0·1	1·83
	Burma				**(Zone −0630)**						
4520	Akyab	I	20 08	92 54	+0004	+0001	0·0	−0·1	−0·3	−0·3	1·27
4525	Kyauk Pyu		19 26	93 33	−0020	−0033	+0·6	+0·3	0·0	−0·2	1·62
4526	Searle Point		18 54	93 37	−0017	−0015	+0·4	+0·3	0·0	0·0	1·62
4529	Sagu Island		18 48	93 59	−0013	−0016	+0·4	+0·3	−0·3	−0·3	1·46

Seasonal Changes in Mean Level

No.	Jan. 1	Feb. 1	Mar. 1	Apr. 1	May 1	June 1	July 1	Aug. 1	Sep. 1	Oct. 1	Nov. 1	Dec. 1	Jan. 1
4510–4515	−0·3	−0·4	−0·4	−0·3	−0·1	+0·2	+0·4	+0·5	+0·4	+0·2	0·0	−0·2	−0·3
4517–4532	−0·2	−0·3	−0·3	−0·2	−0·1	+0·1	+0·2	+0·3	+0·2	+0·1	+0·1	0·0	−0·2
4534–4539	−0·1	−0·2	−0·3	−0·2	0·0	+0·1	+0·1	+0·1	+0·1	+0·1	+0·1	+0·1	−0·1
4541	−0·4	−0·5	−0·5	−0·5	−0·4	−0·1	+0·3	+0·7	+0·8	+0·6	+0·2	−0·2	−0·4

			Harmonic Constants								S.W. Corrections				
		M.L. Z_0	M_2		S_2		K_1		O_1		$\frac{1}{4}$-diurnal		$\frac{1}{6}$-diurnal		
No.	Place	m.	g°	H.m.	g°	H.m.	g°	H.m.	g°	H.m.	f_4	F_1	f_6	F_6	
2853	Welshpool	3·72	344	2·60	018	0·37	136	0·15	120	0·11	237	0·009	⊙	⊙	
3092*	Quebec	*w*	186	1·73	226	0·42	270	0·23	247	0·22	272	0·095	056	0·013	*w*
4574*	Elephant Point	*w*	099	1·80	142	0·73	020	0·23	000	0·10	247	0·032	129	0·012	*w*
6259	Port Hedland	4·12	305	1·68	014	1·02	292	0·24	273	0·15	157	0·014	330	0·005	
6319	Daly River	3·7	139	1·9	194	0·9	336	0·6	310	0·3	⊙	⊙	⊙	⊙	

⊙ No data.
? Dries out except for river water.
n See notes on page 362.
c For intermediate heights, use harmonic constants and N.P. 159.
l Constants from 15 days' observations.
v Owing to large seasonal variations, see Table VI.
w Owing to large fortnightly variations, see Table VIa.
x M.L. inferred.

Table VIa FORTNIGHTLY SHALLOW WATER CORRECTIONS

Investigations have shown that at certain ports with large Shallow Water Effects the tidal constituent with a period of one cycle per fortnight (MSf) can be significant. The normal time and height difference allow for this effect but at certain ports predictions obtained by the use of the Admiralty Method of Tidal Prediction N.P. 159 can be improved by a simple additional adjustment. Corrected values of M.L. are given below for the ports concerned, for the day of Springs and for each day before or after springs. The day of springs should be found by inspection of the daily predictions for the Standard Port of the area concerned. The corrected value should be entered as M.L. on N.P. 159. The seasonal correction is then applied to this value as usual.

Corrected M.L.									
			Days before or after springs						
Port No.	Average M.L.	At Springs	1	2	3	4	5	6	7
3092*	2·59	2·74	2·73	2·68	2·62	2·56	2·50	2·45	2·44
4547*	3·66	3·72	3·71	3·70	3·67	3·65	3·62	3·61	3·60

* *Important Note:* Port No. 3092 Assume time in question 7 days after Springs.
Port No. 4547 Assume time in question at Springs.

Table VII TIDAL ANGLES AND FACTORS YEAR 1982

JANUARY

	M2		S2		K1		O1	
DAY	A	F	A	F	A	F	A	F
1	128	0·99	001	0·80	001	1·33	126	0·92
2	152	1·03	001	0·81	001	1·35	150	0·96
3	176	1·07	002	0·81	001	1·37	176	1·00
4	202	1·11	003	0·81	001	1·39	202	1·04
5	228	1·15	003	0·81	001	1·41	229	1·08
6	255	1·19	004	0·81	000	1·42	257	1·11
7	282	1·22	005	0·81	000	1·42	285	1·13
8	310	1·23	005	0·82	000	1·42	314	1·15
9	338	1·23	006	0·82	359	1·41	343	1·15
10	005	1·22	007	0·82	358	1·40	012	1·14
11	033	1·20	007	0·83	358	1·38	040	1·12
12	060	1·16	008	0·83	357	1·35	069	1·09
13	087	1·12	008	0·84	356	1·31	096	1·05
14	113	1·07	009	0·84	356	1·28	123	1·00
15	138	1·02	009	0·85	355	1·24	149	0·96
16	162	0·97	010	0·85	354	1·20	174	0·91
17	185	0·93	010	0·86	354	1·17	198	0·87
18	207	0·90	011	0·87	353	1·14	221	0·84
19	229	0·88	011	0·87	353	1·12	244	0·83
20	250	0·87	012	0·88	352	1·11	266	0·82
21	272	0·87	012	0·88	352	1·10	289	0·82
22	293	0·88	012	0·89	351	1·10	312	0·83
23	315	0·89	013	0·90	350	1·10	335	0·84
24	338	0·91	013	0·91	350	1·11	358	0·85
25	000	0·92	014	0·91	349	1·12	021	0·87
26	023	0·94	014	0·92	348	1·14	045	0·88
27	046	0·96	014	0·93	347	1·15	069	0·90
28	069	0·98	014	0·94	347	1·16	094	0·92
29	093	1·01	015	0·94	346	1·18	118	0·95
30	117	1·04	015	0·95	346	1·19	143	0·97
31	142	1·07	015	0·96	345	1·20	169	1·00

FEBRUARY

	M2		S2		K1		O1	
DAY	A	F	A	F	A	F	A	F
1	167	1·10	015	0·97	345	1·21	195	1·03
2	193	1·13	015	0·98	344	1·22	222	1·06
3	220	1·15	015	0·99	344	1·22	250	1·08
4	246	1·17	016	0·99	344	1·21	277	1·10
5	273	1·18	016	1·00	343	1·20	305	1·11
6	301	1·18	016	1·01	342	1·19	334	1·11
7	328	1·17	016	1·02	342	1·16	002	1·10
8	355	1·16	016	1·03	341	1·13	030	1·09
9	022	1·13	016	1·04	340	1·10	057	1·06
10	048	1·09	016	1·04	339	1·06	085	1·03
11	073	1·05	016	1·05	339	1·02	111	0·99
12	098	1·01	016	1·06	338	0·98	137	0·95
13	122	0·97	016	1·07	337	0·94	162	0·92
14	145	0·94	016	1·08	336	0·91	186	0·88
15	168	0·91	016	1·08	336	0·87	210	0·85
16	190	0·89	016	1·09	335	0·85	233	0·83
17	211	0·88	015	1·10	334	0·83	255	0·83
18	233	0·88	015	1·11	333	0·82	278	0·83
19	254	0·89	015	1·11	332	0·82	300	0·83
20	276	0·90	015	1·12	330	0·82	323	0·85
21	299	0·92	015	1·13	329	0·82	346	0·87
22	322	0·95	015	1·14	327	0·83	010	0·89
23	345	0·97	015	1·14	325	0·85	035	0·92
24	009	1·00	014	1·15	324	0·86	059	0·94
25	033	1·02	014	1·16	322	0·87	084	0·97
26	057	1·05	014	1·16	321	0·88	110	0·99
27	082	1·07	014	1·17	320	0·89	136	1·01
28	108	1·10	014	1·18	319	0·89	162	1·03

MARCH

	M2		S2		K1		O1	
DAY	A	F	A	F	A	F	A	F
1	134	1·11	013	1·18	318	0·90	189	1·05
2	160	1·13	013	1·19	317	0·89	216	1·07
3	186	1·14	013	1·19	316	0·89	244	1·08
4	213	1·14	013	1·20	315	0·88	271	1·08
5	239	1·14	012	1·20	314	0·86	299	1·08
6	266	1·13	012	1·21	313	0·85	326	1·07
7	292	1·12	012	1·21	312	0·82	353	1·06
8	318	1·10	012	1·22	310	0·80	021	1·04
9	344	1·08	011	1·22	309	0·77	047	1·02
10	009	1·05	011	1·23	307	0·74	074	1·00
11	034	1·02	011	1·23	305	0·70	100	0·97
12	059	0·99	010	1·23	303	0·67	125	0·94
13	082	0·96	010	1·24	301	0·64	150	0·91
14	106	0·93	010	1·24	299	0·61	174	0·88
15	128	0·91	009	1·24	297	0·58	198	0·86
16	150	0·89	009	1·25	294	0·57	221	0·84
17	172	0·88	009	1·25	292	0·56	243	0·83
18	193	0·88	008	1·25	290	0·55	266	0·83
19	215	0·88	008	1·25	288	0·56	288	0·84
20	237	0·90	008	1·25	286	0·57	311	0·86
21	259	0·93	007	1·26	284	0·59	334	0·88
22	282	0·96	007	1·26	282	0·62	358	0·91
23	306	1·00	006	1·26	280	0·64	023	0·95
24	330	1·03	006	1·26	278	0·67	048	0·98
25	355	1·07	006	1·26	276	0·69	074	1·01
26	020	1·10	005	1·26	274	0·72	100	1·04
27	046	1·12	005	1·26	272	0·73	127	1·07
28	073	1·14	005	1·26	271	0·75	155	1·09
29	099	1·15	004	1·26	269	0·76	182	1·10
30	126	1·16	004	1·26	268	0·76	210	1·10
31	153	1·15	003	1·26	266	0·76	238	1·10

APRIL

	M2		S2		K1		O1	
DAY	A	F	A	F	A	F	A	F
1	179	1·14	003	1·26	265	0·76	265	1·09
2	206	1·12	003	1·26	263	0·75	293	1·07
3	232	1·10	002	1·25	262	0·74	320	1·05
4	258	1·08	002	1·25	260	0·72	347	1·03
5	283	1·05	001	1·25	258	0·71	013	1·00
6	308	1·03	001	1·25	256	0·70	039	0·98
7	332	1·00	001	1·25	253	0·68	064	0·96
8	356	0·98	000	1·24	250	0·67	089	0·93
9	020	0·96	000	1·24	248	0·66	114	0·91
10	043	0·94	000	1·24	245	0·65	138	0·89
11	066	0·92	359	1·23	242	0·64	162	0·87
12	089	0·90	359	1·23	240	0·64	186	0·86
13	111	0·89	358	1·23	237	0·64	209	0·84
14	133	0·88	358	1·22	236	0·65	232	0·84
15	154	0·87	358	1·22	234	0·66	254	0·83
16	176	0·88	357	1·21	234	0·68	277	0·84
17	197	0·90	357	1·21	233	0·71	299	0·86
18	219	0·93	356	1·20	233	0·74	322	0·89
19	242	0·97	356	1·20	232	0·77	346	0·92
20	266	1·01	356	1·19	232	0·81	011	0·96
21	290	1·05	355	1·19	231	0·84	036	1·01
22	316	1·10	355	1·18	231	0·88	062	1·05
23	342	1·14	355	1·17	230	0·91	090	1·09
24	009	1·17	354	1·17	229	0·94	117	1·12
25	036	1·19	354	1·16	228	0·97	145	1·14
26	063	1·20	354	1·16	227	0·98	174	1·15
27	091	1·20	353	1·15	226	1·00	202	1·15
28	118	1·19	353	1·14	225	1·00	231	1·14
29	145	1·17	353	1·13	224	1·00	259	1·12
30	172	1·14	352	1·13	223	0·99	287	1·09

Table VII—*continued*

MAY

	M2		S2		K1		O1	
DAY	A	F	A	F	A	F	A	F
1	198	1·10	352	1·12	222	0·98	314	1·06
2	224	1·07	352	1·11	220	0·97	340	1·02
3	249	1·03	351	1·11	219	0·96	006	0·99
4	273	0·99	351	1·10	218	0·95	032	0·95
5	297	0·97	351	1·09	216	0·94	056	0·93
6	320	0·94	350	1·08	215	0·93	080	0·90
7	343	0·92	350	1·07	213	0·92	104	0·89
8	005	0·91	350	1·07	212	0·92	128	0·87
9	028	0·90	350	1·06	210	0·92	151	0·86
10	050	0·89	349	1·05	209	0·92	175	0·85
11	072	0·88	349	1·04	208	0·93	198	0·85
12	094	0·88	349	1·03	207	0·94	221	0·84
13	116	0·87	349	1·02	207	0·95	243	0·84
14	137	0·88	349	1·01	206	0·97	266	0·84
15	159	0·89	348	1·00	206	1·00	288	0·86
16	181	0·92	348	1·00	206	1·02	311	0·89
17	203	0·96	348	0·99	206	1·06	334	0·92
18	227	1·01	348	0·98	207	1·09	359	0·97
19	251	1·06	348	0·97	207	1·13	024	1·02
20	276	1·11	348	0·96	206	1·16	051	1·07
21	303	1.16	348	0·95	206	1·20	078	1·11
22	330	1·20	347	0·94	206	1·23	106	1·15
23	357	1·22	347	0·93	205	1·25	135	1·18
24	025	1·24	347	0·92	204	1·27	164	1·19
25	053	1·24	347	0·91	203	1·27	193	1·19
26	081	1·23	347	0·90	203	1·28	222	1·18
27	109	1·20	347	0·90	202	1·27	250	1·16
28	136	1·17	347	0·89	201	1·26	279	1·12
29	163	1·12	347	0·88	200	1·24	306	1·08
30	189	1·08	347	0·87	199	1·22	333	1·04
31	214	1·03	347	0·86	198	1·20	359	0·99

JUNE

	M2		S2		K1		O1	
DAY	A	F	A	F	A	F	A	F
1	238	0·98	347	0·85	197	1·19	024	0·95
2	261	0·95	348	0·84	197	1·17	048	0·91
3	284	0·92	348	0·83	196	1·15	072	0·89
4	306	0·90	348	0·83	195	1·14	095	0·87
5	328	0·89	348	0·82	194	1·13	118	0·86
6	350	0·88	348	0·81	193	1·12	141	0·85
7	012	0·88	349	0·80	193	1·12	164	0·85
8	034	0·88	349	0·79	192	1·13	187	0·85
9	056	0·88	349	0·79	192	1·13	210	0·85
10	078	0·88	349	0·78	191	1·15	233	0·85
11	100	0·88	350	0·77	191	1·16	256	0·85
12	121	0·89	350	0·76	191	1·18	278	0·86
13	143	0·92	350	0·76	191	1·20	301	0·89
14	165	0·95	351	0·75	191	1·23	324	0·92
15	189	1·00	351	0·75	191	1·26	349	0·96
16	213	1·05	352	0·74	191	1·29	014	1·01
17	238	1·10	352	0·73	190	1·32	040	1·06
18	264	1·15	352	0·73	190	1·35	067	1·11
19	291	1·19	353	0·72	190	1·37	095	1·16
20	319	1·23	353	0·72	189	1·39	124	1·19
21	347	1·25	354	0·71	188	1·40	153	1·21
22	015	1·25	355	0·71	188	1·41	182	1·21
23	043	1·25	355	0·71	187	1·40	211	1·21
24	071	1·22	356	0·70	186	1·39	240	1·19
25	099	1·19	356	0·70	185	1·37	269	1·15
26	126	1·15	357	0·70	185	1·35	297	1·11
27	152	1·09	358	0·70	184	1·32	324	1·06
28	178	1·04	358	0·70	183	1·29	350	1·01
29	202	0·99	359	0·69	183	1·26	016	0·96
30	225	0·95	359	0·69	182	1·24	040	0·92

JULY

	M2		S2		K1		O1	
DAY	A	F	A	F	A	F	A	F
1	248	0·91	000	0·69	182	1·21	063	0·89
2	270	0·89	001	0·69	181	1·19	086	0·86
3	291	0·88	001	0·69	181	1·18	109	0·85
4	313	0·87	002	0·69	180	1·17	132	0·85
5	335	0·87	003	0·69	180	1·16	154	0·85
6	357	0·88	003	0·70	179	1·16	178	0·85
7	019	0·88	004	0·70	179	1·16	201	0·86
8	041	0·89	004	0·70	178	1·17	224	0·86
9	063	0·89	005	0·70	178	1·18	247	0·87
10	085	0·91	006	0·71	177	1·20	270	0·88
11	107	0·93	006	0·71	177	1·21	293	0·90
12	130	0·96	007	0·71	177	1·23	316	0·93
13	153	1·00	007	0·72	176	1·26	340	0·97
14	177	1·05	008	0·72	176	1·28	005	1·02
15	201	1·10	008	0.73	176	1·30	031	1·07
16	227	1·14	009	0·73	175	1·32	057	1·11
17	253	1·19	009	0·74	175	1·34	085	1·15
18	281	1·22	010	0·74	174	1·35	113	1·19
19	308	1·24	010	0·75	173	1·35	142	1·21
20	336	1·24	010	0·76	172	1·35	170	1·21
21	004	1·23	011	0·76	172	1·34	199	1·20
22	032	1·21	011	0·77	171	1·32	228	1·18
23	060	1·18	011	0·78	170	1·30	257	1·15
24	087	1·13	012	0·78	169	1·27	285	1·10
25	113	1·08	012	0·79	169	1·24	313	1·05
26	139	1·02	012	0·80	168	1·20	340	1·00
27	164	0·97	013	0·81	168	1·16	005	0·95
28	188	0·92	013	0·81	167	1·13	030	0·90
29	211	0·89	013	0·82	167	1·10	054	0·87
30	233	0·86	013	0·83	167	1·07	077	0·84
31	255	0·85	013	0·84	166	1·05	100	0·83

AUGUST

	M2		S2		K1		O1	
DAY	A	F	A	F	A	F	A	F
1	277	0·85	013	0·85	166	1·04	123	0·83
2	299	0·85	013	0·86	165	1·03	146	0·83
3	321	0·87	013	0·87	165	1·02	169	0·85
4	343	0·88	014	0·87	164	1·02	192	0·86
5	005	0·90	014	0·88	163	1·03	216	0·88
6	028	0·91	014	0·89	162	1·03	239	0·89
7	050	0·94	014	0·90	161	1·04	263	0·91
8	073	0·96	014	0·91	160	1·05	286	0·94
9	096	0·99	014	0·92	159	1·07	310	0·97
10	119	1·03	013	0·93	158	1·08	334	1·00
11	143	1·07	013	0·94	157	1·09	359	1·04
12	167	1·11	013	0·95	157	1·10	024	1·09
13	192	1·15	013	0·96	156	1·12	050	1·13
14	218	1·18	013	0·97	155	1·12	077	1·16
15	244	1·21	013	0·98	154	1·13	104	1·19
16	271	1·23	013	0·98	154	1·13	132	1·20
17	298	1·23	013	0·99	153	1·12	160	1·20
18	325	1·22	012	1·00	152	1·11	188	1·19
19	352	1;19	012	1·01	151	1·09	216	1·17
20	019	1·16	012	1·02	150	1·06	244	1·13
21	046	1·11	012	1·03	149	1·03	272	1·09
22	073	1·06	012	1·04	148	0·99	300	1·04
23	098	1·01	011	1·05	147	0·95	326	0·99
24	123	0·95	011	1·06	146	0·91	352	0·94
25	148	0·91	011	1·06	146	0·88	018	0·89
26	171	0·87	010	1·07	145	0·84	042	0·85
27	194	0·84	010	1·08	144	0·81	066	0·82
28	216	0·82	010	1·09	144	0·79	089	0·81
29	238	0·82	010	1·10	143	0·77	112	0·80
30	260	0·82	009	1·10	142	0·76	135	0·81
31	283	0·84	009	1·11	141	0·76	158	0·83

Table VII—*continued*

SEPTEMBER

DAY	M2 A	M2 F	S2 A	S2 F	K1 A	K1 F	O1 A	O1 F
1	305	0·86	009	1·12	139	0·76	182	0·85
2	328	0·89	008	1·13	137	0·76	206	0·87
3	351	0·92	008	1·14	135	0·77	230	0·90
4	014	0·95	008	1·14	133	0·78	254	0·93
5	038	0·98	007	1·15	132	0·79	279	0·97
6	061	1·02	007	1·16	130	0·80	303	1·00
7	085	1·06	007	1·16	128	0·81	328	1·04
8	110	1·09	006	1·17	127	0·82	354	1·08
9	134	1·13	006	1·18	125	0·83	019	1·11
10	159	1·16	005	1·18	124	0·84	045	1·14
11	185	1·18	005	1·19	123	0·84	072	1·17
12	210	1·20	005	1·19	122	0·84	098	1·18
13	236	1·21	004	1·20	120	0·84	125	1·19
14	262	1·20	004	1·20	119	0·83	152	1·19
15	289	1·19	003	1·21	118	0·82	179	1·17
16	315	1·17	003	1·22	116	0·81	206	1·15
17	341	1·13	003	1·22	114	0·79	233	1·12
18	007	1·09	002	1·22	113	0·76	260	1·08
19	032	1·04	002	1·23	111	0·73	287	1·03
20	057	0·99	001	1·23	109	0·70	313	0·98
21	082	0·95	001	1·24	107	0·67	339	0·93
22	106	0·90	001	1·24	105	0·64	004	0·89
23	130	0·86	000	1·24	103	0·61	028	0·85
24	153	0·83	000	1·25	101	0·59	052	0·82
25	175	0·81	359	1·25	099	0·58	076	0·80
26	197	0·80	359	1·25	097	0·57	099	0·79
27	220	0·81	359	1·26	095	0·57	122	0·80
28	242	0·82	358	1·26	093	0·57	146	0·81
29	265	0·84	358	1·26	090	0·59	169	0·83
30	288	0·88	357	1·26	088	0·61	193	0·87

OCTOBER

DAY	M2 A	M2 F	S2 A	S2 F	K1 A	K1 F	O1 A	O1 F
1	312	0·91	357	1·26	086	0·63	218	0·90
2	336	0·96	357	1·26	084	0·66	243	0·95
3	000	1·00	356	1·26	082	0·68	268	0·99
4	025	1·04	356	1·27	080	0·71	294	1·03
5	050	1·08	355	1·27	078	0·73	320	1·07
6	075	1·12	355	1·27	077	0·76	347	1·11
7	101	1·15	354	1·27	075	0·78	013	1·14
8	127	1·18	354	1·27	074	0·79	040	1·17
9	152	1·19	354	1·27	072	0·80	067	1·18
10	178	1·20	353	1·27	071	0·81	093	1·19
11	204	1·19	353	1·26	070	0·81	120	1·18
12	229	1·18	352	1·26	068	0·81	147	1·17
13	255	1·16	352	1·26	067	0·81	173	1·15
14	280	1·13	352	1·26	065	0·80	199	1·12
15	305	1·10	351	1·26	064	0·79	225	1·09
16	330	1·06	351	1·26	062	0·78	251	1·06
17	354	1·03	351	1·25	060	0·77	276	1·02
18	018	0·99	350	1·25	057	0·76	301	0·98
19	042	0·95	350	1·25	055	0·75	326	0·94
20	066	0·91	349	1·24	053	0·74	351	0·90
21	089	0·88	349	1·24	050	0·73	015	0·87
22	112	0·85	349	1·24	048	0·73	039	0·84
23	134	0·83	348	1·23	047	0·73	062	0·82
24	156	0·81	348	1·23	045	0·73	085	0·81
25	178	0·81	348	1·22	044	0·74	109	0·81
26	201	0·82	347	1·22	043	0·76	132	0·81
27	223	0·84	347	1·21	043	0·79	155	0·84
28	246	0·87	347	1·21	042	0·82	179	0·87
29	270	0·91	347	1·20	041	0·85	204	0·91
30	295	0·96	346	1·20	041	0·89	230	0·96
31	320	1·01	346	1·19	040	0·93	256	1·01

NOVEMBER

DAY	M2 A	M2 F	S2 A	S2 F	K1 A	K1 F	O1 A	O1 F
1	346	1·06	346	1·19	039	0·97	283	1·06
2	012	1·11	345	1·18	039	1·00	310	1·10
3	038	1·15	345	1·18	038	1·03	337	1·14
4	065	1·18	345	1·17	037	1·06	005	1·18
5	092	1·20	345	1·16	036	1·08	033	1·20
6	119	1·21	344	1·15	035	1·10	060	1·20
7	145	1·21	344	1·15	034	1·10	088	1·20
8	171	1·19	344	1·14	033	1·11	115	1·19
9	197	1·17	344	1·13	033	1·11	142	1·17
10	222	1·14	344	1·13	032	1·10	168	1·13
11	247	1·10	344	1·12	031	1·09	194	1·10
12	271	1·07	343	1·11	030	1·09	219	1·06
13	295	1·03	343	1·10	029	1·08	244	1·03
14	319	1·00	343	1·09	028	1·07	268	0·99
15	342	0·97	343	1·09	027	1·06	293	0·96
16	005	0·94	343	1·08	025	1·05	316	0·94
17	027	0·91	343	1·07	024	1·04	340	0·91
18	050	0·89	343	1·06	023	1·04	004	0·89
19	073	0·87	343	1·05	022	1·04	027	0·87
20	095	0·85	343	1·04	021	1·04	051	0·85
21	117	0·84	343	1·03	021	1·04	074	0·84
22	139	0·83	343	1·03	020	1·06	097	0·83
23	161	0·84	343	1·02	020	1·07	119	0·84
24	183	0·86	343	1·01	020	1·10	143	0·86
25	206	0·89	343	1·00	019	1·13	166	0·89
26	229	0·93	343	0·99	019	1·16	191	0·93
27	253	0·98	343	0·98	019	1·20	216	0·98
28	279	1·03	343	0·97	019	1·24	242	1·03
29	305	1·08	343	0·96	018	1·27	269	1·08
30	332	1·13	343	0·95	018	1·31	297	1·13

DECEMBER

DAY	M2 A	M2 F	S2 A	S2 F	K1 A	K1 F	O1 A	O1 F
1	359	1·17	343	0·95	017	1·34	325	1·17
2	026	1·20	344	0·94	016	1·36	354	1·21
3	054	1·22	344	0·93	016	1·38	023	1·22
4	082	1·23	344	0·92	015	1·39	051	1·23
5	109	1·22	344	0·91	014	1·39	080	1·22
6	136	1·19	345	0·90	014	1·39	108	1·20
7	163	1·16	345	0·89	013	1·38	135	1·17
8	189	1·12	345	0·89	013	1·37	162	1·13
9	214	1·08	346	0·88	012	1·35	188	1·08
10	238	1·03	346	0·87	012	1·33	214	1·04
11	262	0·99	347	0·86	011	1·31	238	1·00
12	285	0·96	347	0·86	011	1·29	262	0·96
13	307	0·93	348	0·85	010	1·27	286	0·94
14	329	0·91	348	0·84	010	1·26	309	0·92
15	352	0·90	349	0·84	009	1·25	332	0·90
16	014	0·89	349	0·83	009	1·23	355	0·89
17	036	0·88	350	0·82	008	1·23	018	0·88
18	058	0·87	350	0·82	008	1·22	041	0·87
19	080	0·86	351	0·81	007	1·23	064	0·87
20	102	0·86	352	0·81	007	1·23	087	0·87
21	123	0·87	352	0·80	006	1·24	110	0·87
22	145	0·88	353	0·80	006	1·26	133	0·89
23	167	0·91	354	0·79	006	1·28	156	0·92
24	190	0·95	354	0·79	006	1·31	180	0·96
25	214	1·00	355	0·79	005	1·34	204	1·01
26	239	1·05	356	0·79	005	1·37	230	1·06
27	265	1·11	356	0·78	004	1·40	257	1·11
28	291	1·16	357	0·78	004	1·43	285	1·16
29	319	1·20	358	0·78	003	1·46	313	1·20
30	347	1·22	359	0·78	002	1·47	342	1·23
31	015	1·24	359	0·78	002	1·48	011	1·25

Volume 3 PACIFIC OCEAN AND ADJACENT SEAS—Part I (Standard Ports)

Canada, British Columbia—Prince Rupert

LAT 54°19′N LONG 130°20′W

TIME ZONE +0800 TIMES AND HEIGHTS OF HIGH AND LOW WATERS YEAR 1982

MAY

	TIME	M		TIME	M
1	0150	2.8	16	0110	3.1
	0755	5.5		0655	4.8
SA	1430	1.7	SU	1335	2.0
	2110	5.4		2015	5.0
2	0310	2.6	17	0225	2.9
	0900	5.5		0820	4.9
SU	1535	1.7	M	1450	2.0
	2210	5.7		2115	5.3
3	0415	2.2	18	0335	2.5
	1015	5.6		0930	5.1
M	1630	1.6	TU	1550	1.9
	2300	6.0		2205	5.6

JUNE

	TIME	M		TIME	M
1	0355	2.0	16	0250	2.3
	0955	5.3		0850	4.9
TU	1555	2.0	W	1445	2.1
	2215	6.0		2125	5.8
2	0440	1.7	17	0355	1.8
	1055	5.4		0955	5.1
W	1655	2.0	TH	1545	2.1
	2310	6.1		2200	6.2
3	0530	1.4	18	0445	1.3
	1140	5.5		1100	5.4
TH	1740	2.0	F	1650	2.1
	2345	6.2		2255	6.6

JULY

	TIME	M		TIME	M
1	0405	1.8	16	0310	1.8
	1015	5.0		0930	4.9
TH	1605	2.5	F	1510	2.5
	2225	5.9		2135	6.1
2	0505	1.6	17	0420	1.3
	1125	5.1		1040	5.2
F	1705	2.5	SA	1625	2.4
	2305	6.0		2230	6.4
3	0550	1.3	18	0520	0.9
	1200	5.3		1145	5.5
SA	1750	2.5	SU	1725	2.3
	2345	6.1		2320	6.8

AUGUST

	TIME	M		TIME	M
1	0515	1.5	16	0510	1.0
	1155	5.1		1140	5.5
SU	1720	2.7	M	1705	2.3
	2325	5.9		2315	6.6
2	0615	1.3	17	0600	0.7
	1230	5.3		1220	5.9
M	1810	2.5	TU	1810	2.0
3	0015	6.0	18	0005	6.9
	0650	1.1		0655	0.3
TU	1310	5.5	W	1305	6.2
	1845	2.3		1850	1.6

Volume 3 PACIFIC OCEAN AND ADJACENT SEAS—Part II (Secondary Ports)

No.	Place			MHW	MLW	MHWS	MHWN	MLWN	MLWS	M.L.
8850	**PRINCE RUPERT**	see page 207		MHW	MLW.	**6·5**	**5·2**	**2·5**	**1·1**	Z_0 m.
9020	Port Hardy	50 43	127 29	−0025	−0030	−1·9	−1·4	−0·4	+0·1	2·94
9022	Bay	50 51	127 51	−0030	−0035	−2·1	−1·6	−0·6	−0·1	2·75
	QUEEN CHARLOTTE SOUND									
8936	Port Blackney	52 19	128 21	−0030	−0035	−2·0	−1·5	−0·6	−0·2	2·77
8937	Bellabella C	52 10	128 08	−0030	−0030	−1·9	−1·4	−0·6	−0·1	2·85
	Cousins Inlet									
8939	Ocean Falls	52 21	127 41	−0041	−0035	−1·8	−1·3	−0·5	−0·1	2·93
8942	Bella Coola	52 23	126 48	−0025	−0030	−1·7	−1·3	−0·6	−0·2	2·90
8943	Namu Harbour	51 52	127 52	−0036	−0036	−2·2	−1·7	−0·8	−0·3	2·59
8944	Addenbroke Island	51 36	127 49	−0038	−0036	−2·2	−1·7	−0·6	−0·1	2·70
	Rivers Inlet									
8946	Wadhams	51 31	127 31	−0035	−0040	−2·2	−1·7	−0·7	−0·2	2·65
8947	Kildala	51 42	127 21	−0047	−0041	−2·2	−1·7	−0·7	−0·2	2·7

Seasonal Changes in Mean Level

No.	Jan. 1	Feb. 1	Mar. 1	Apr. 1	May 1	June 1	July 1	Aug. 1	Sep. 1	Oct. 1	Nov. 1	Dec. 1	Jan. 1
8850	+0·1	+0·1	0·0	0·0	0·0	0·0	−0·1	−0·1	−0·1	0·0	0·0	+0·1	+0·1
8871–8881	+0·1	+0·1	+0·1	0·0	−0·1	−0·1	−0·1	−0·1	−0·1	0·0	0·0	+0·1	+0·1
8884–8927	+0·1	0·0	0·0	−0·1	−0·1	0·0	0·0	−0·1	0·0	0·0	+0·1	+0·1	+0·1
8928–8952	+0·1	+0·1	0·0	0·0	−0·1	−0·1	−0·1	−0·1	−0·1	0·0	0·0	+0·1	+0·1
9050	0·0	+0·1	+0·1	0·0	0·0	0·0	0·0	0·0	−0·1	−0·1	−0·1	0·0	0·0
8850	+0·1	+0·1	0·0	0·0	0·0	0·0	−0·1	−0·1	−0·1	0·0	0·0	+0·1	+0·1
8960–9022	+0·1	+0·1	0·0	0·0	−0·1	−0·1	−0·1	−0·1	−0·1	0·0	0·0	+0·1	+0·1
9023–9057	0·0	+0·1	+0·1	0·0	0·0	0·0	0·0	0·0	−0·1	−0·1	−0·1	0·0	0·0
9133	+0·1	+0·1	0·0	−0·1	−0·1	0·0	0·0	0·0	0·0	0·0	0·0	0·0	+0·1

ADMIRALTY TIDAL PREDICTION FORM—NP 204

INSTRUCTIONS FOR SECONDARY PORTS

Box No.

Complete heading (A.T.T. Part II).

1- 4. Enter HW/LW times and heights at Standard Port (A.T.T. Part I).

5. Calculate range at Standard Port (Difference between heights in boxes 3 and 4).

6. Enter Seasonal Change at Standard Port with sign changed (A.T.T. Part II).

7-10.* Calculate Differences (A.T.T. Part II). Interpolate/extrapolate.

11. Enter Seasonal Change at Secondary Port (A.T.T. Part II).

12-15.† Calculate Secondary HW/LW (i.e. box 1 adjusted by box 7; box 2 by box 8; box 3 by boxes 6, 9 and 11; box 4 by boxes 6, 10 and 11).

16.‡ Calculate Duration at Secondary Port (Difference between boxes 12 and 13) (Not required for A.T.T. Vol. 1).

TO FIND HEIGHT AT GIVEN TIME

First enter boxes 1 to 16.

‡ On appropriate Standard Port Curve diagram, plot heights of H.W. and L.W. occurring either side of required time and join by sloping line.

Enter H.W. time and sufficient others to embrace required time.

From required time proceed vertically to curves using Range at Standard Port (box 5) to interpolate between Springs and Neaps. Do NOT extrapolate.

Proceed horizontally to sloping line and thence vertically to Height scale.

Read off height above Chart Datum.

TO FIND TIME FOR A GIVEN HEIGHT ABOVE CHART DATUM

First enter boxes 1 to 16.

‡ On appropriate Standard Port Curve diagram, plot heights of H.W. and L.W. occuring either side of required event and join by sloping line.

Enter H.W. time and those for half-tidal cycle covering required event.

From required height, proceed vertically to sloping line, thence horizontally to curves, using Range at Standard Port (box 5) to interpolate between Springs and Neaps. Do NOT extrapolate.

Proceed vertically to Time scale.

Read off time.

INSTRUCTIONS FOR STANDARD PORTS

As above but boxes 6 to 15 are not required.

* A diagram to assist the interpolation of differences is on page 5.

† When calculating boxes 12 to 15 great care should be taken with the SIGNS of the contributions.

‡ In A.T.T. Vols. 2 and 3, use curves on pages 46 and 47. If the Duration is outside 5-7 hours, use Simplified Harmonic Method of Tidal Prediction (N.P. 159) or calculator equivalent).

NOTE For the area Swanage to Selsey see instructions in A.T.T. Vol. 1 and pages 8 to 11.

---

* *Note:* Do not attempt to extrapolate additional curves. In ATT Vol. 1 use the Spring curve for Spring Ranges and greater, and the Neap curve for Neap Ranges and less. For intermediate ranges interpolate between curves using the Range at the Standard Port (Box 5) as argument. In ATT Vols. 2 and 3 if duration is outside those shown in Table II use NP 159.

ADMIRALTY TIDAL PREDICTION FORM—NP 204

STANDARD PORT..............................TIME/HEIGHT REQUIRED...................

SECONDARY PORT...........................DATE..............TIME ZONE..............

	TIME		HEIGHT		
STANDARD PORT	HW	LW	HW	LW	RANGE
	1	2	3	4	5
Seasonal change	Standard Port		6	6	
DIFFERENCES	7*	8*	9*	10*	
Seasonal change	Secondary port		11	11	
SECONDARY PORT	12	13	14	15	
Duration	16				

STANDARD PORT..............................TIME/HEIGHT REQUIRED...................

SECONDARY PORT...........................DATE..............TIME ZONE..............

	TIME		HEIGHT		
STANDARD PORT	HW	LW	HW	LW	RANGE
	1	2	3	4	5
Seasonal change	Standard Port		6	6	
DIFFERENCES	7*	8*	9*	10*	
Seasonal change	Secondary port		11	11	
SECONDARY PORT	12	13	14	15	
Duration	16				

SECONDARY PORT HEIGHT DIFFERENCE INTERPOLATION

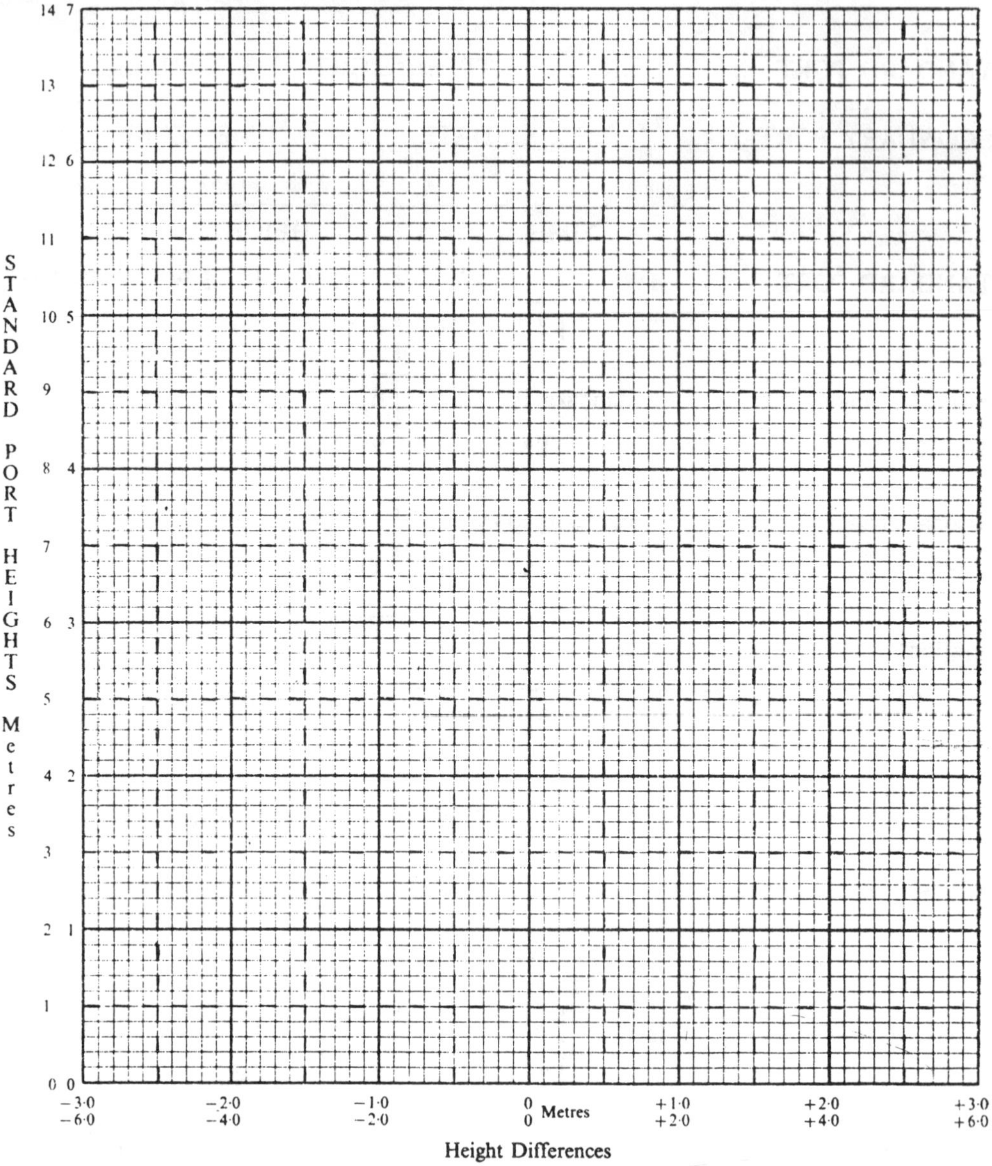

Height Difference Interpolation for High or Low Water

Plot two points using the left-hand and bottom scales for the two high or low water heights and their associated differences. Join these two points by a straight line and extend it to the edge of the diagram. From the intersection of this line with the horizontal line representing the predicted Standard Port height follow the vertical line to the bottom scale and read off the required height difference.

Notes: 1. Various scales are given to cover the ranges likely to be encountered. Further scales can be constructed if required. Care must be taken to ensure that the same scales are used for both plotting and reading off the results.

2. Where only one HW and one LW height value is tabulated in Part II, interpolation for both HW and LW must be made between these two values and extrapolating if necessary.

Pages 816–856 Reproduced by kind permission of the Hydrographer of the Navy.

Admiralty Method of Tidal Prediction — NP 159

Forms A, B & C for Predictions using Harmonic Constants and the Table of Tidal Angles and Factors in ATT.

FORM A

Place .... Date ....

Lat .... Long .... Zone Time ....

TABLE

			M_2		S_2		K_1		O_1		M.L.	
	α°		012		000		180		192			
from A.T.T., part II	g°	H										Mean level (Zo)
from A.T.T., table VII	A°	F										Seasonal correction
	α+g+A	H×F	=m°	=M	=s°	=S	=k°	=K	=o°	=O	$=A_0$	ADD.

logs:

h_2=
H_2=

PLOTTING CIRCLE

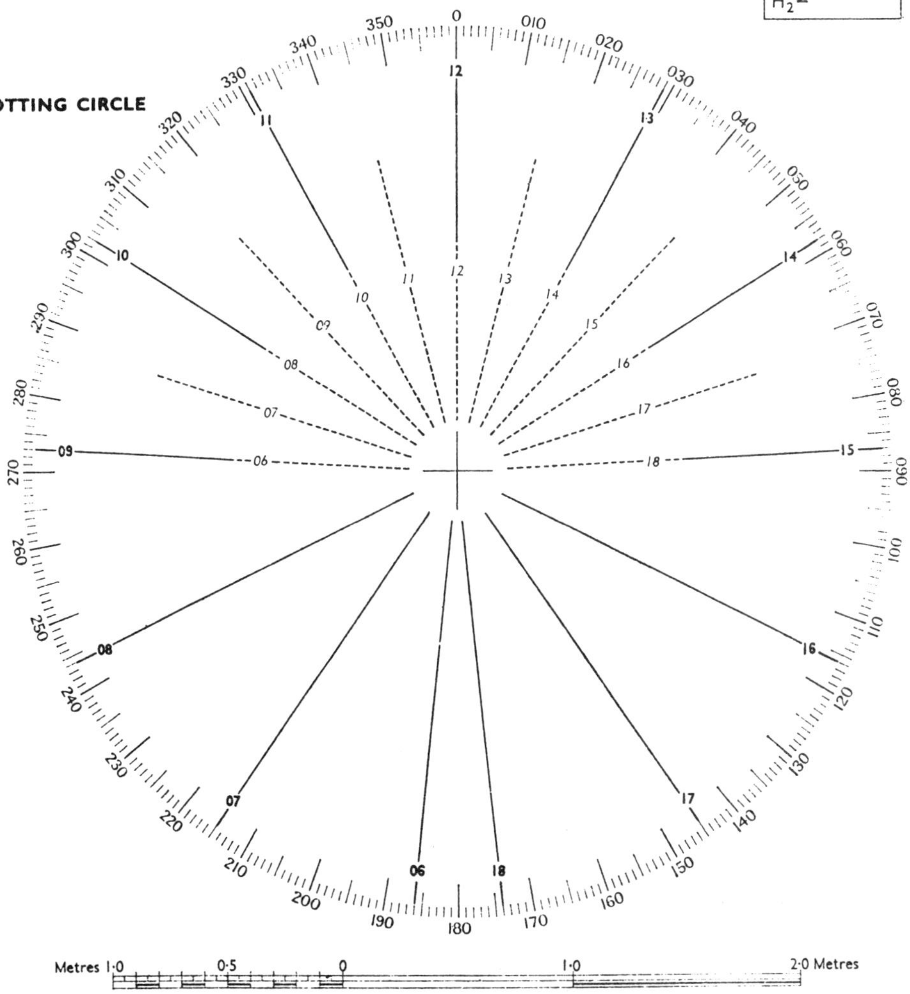

FORM B

Place .... Date ....

Lat .... Long .... Zone Time ....

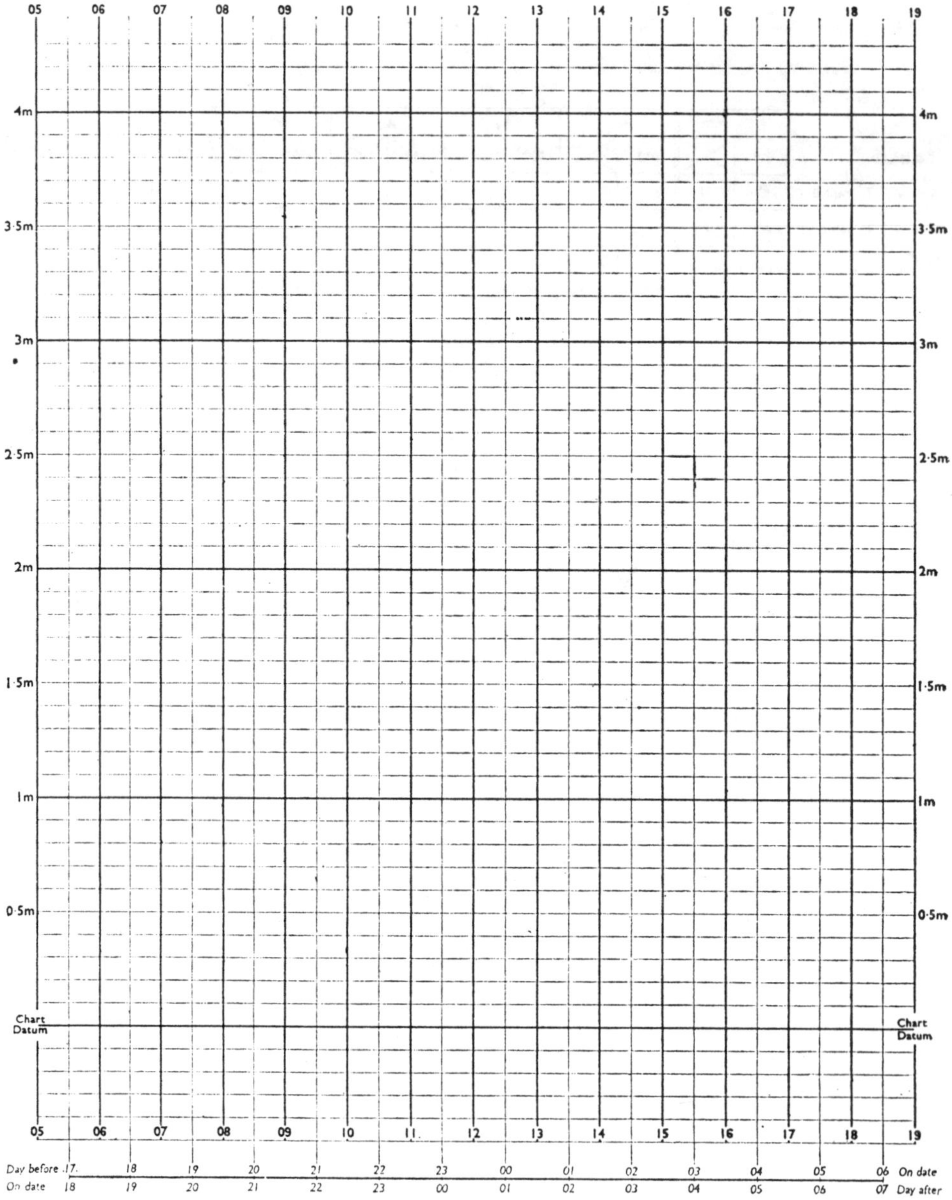

Times at top and bottom of main framework relate to continuous curve (drawn first)
Lower set of times (in italics) relate to pecked curve (drawn second).

FORM C

Place .... Date ....

Lat .... Long .... Zone Time ....

TABLE

$h_2 =$	$2h_2 =$	$f_4 =$	$h_4 = f_4 + 2h_2 =$	$3h_2 =$	$f_6 =$	$h_6 = f_6 + 3h_2 =$
$H_2 =$	$H_2^2 =$	$F_4 =$	$H_4 = F_4 \times H_2^2 =$	$H_2^3 =$	$F_6 =$	$H_6 = F_6 \times H_2^3 =$

logs: ....

PLOTTING CIRCLE

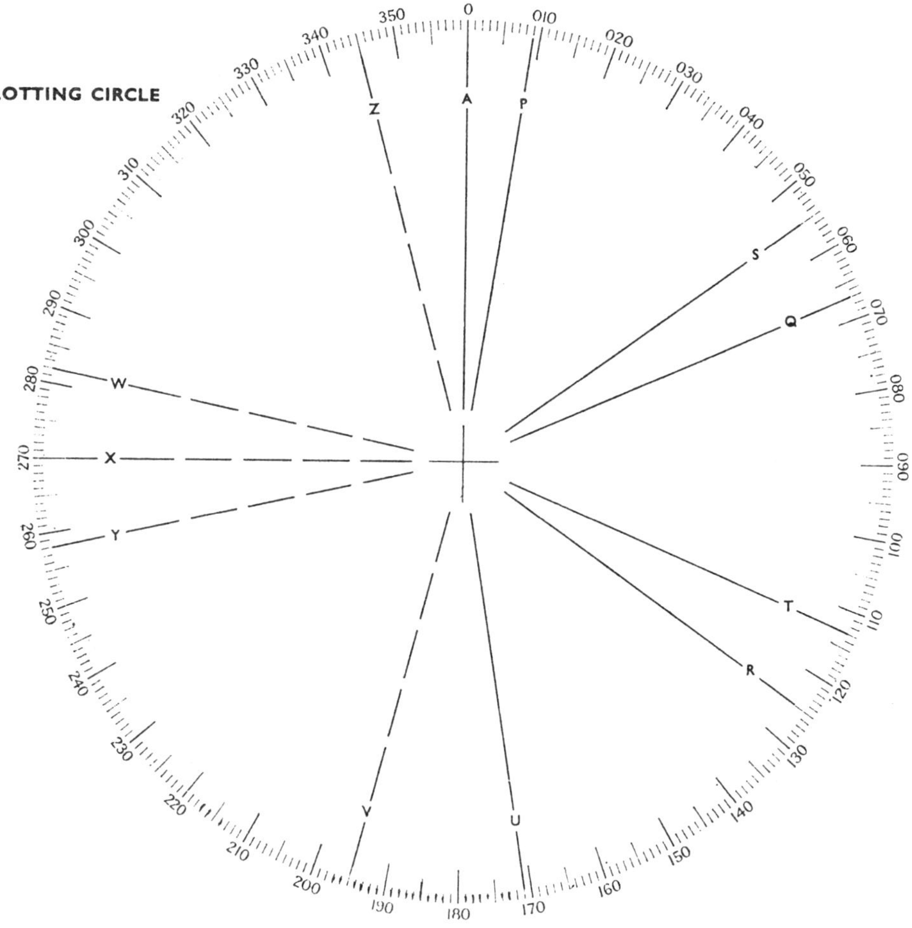

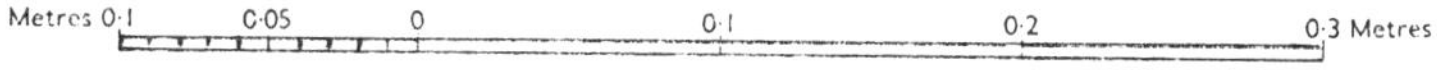

TABLE

Time	0600	0700	0800	0900	1000	1100	1200	1300	1400	1500	1600	1700	1800
QD	P	Q	R	–P	–Q	–R	A	S	T	U	–S	–T	–U
6-D	V	W	–V	–W	–A	X	A	–X	–A	Y	Z	–Y	–Z
Sum													

TABLE I

Conversion Table: Metres to Feet

Conversion Factor 1 Foot = 0·3048 m

Metres	Feet	Metres	Feet	Metres	Feet	Metres	Feet	Metres	Feet	Metres	Feet	Metres	Feet	Metres	Feet
0·05	0·2	4·05	13·3	8·05	26·4	12·05	39·5	16·05	52·7	20·05	65·8	24·05	78·9	28·05	92·0
0·10	0·3	4·10	13·5	8·10	26·6	12·10	39·7	16·10	52·8	20·10	65·9	24·10	79·1	28·10	92·2
0·15	0·5	4·15	13·6	8·15	26·7	12·15	39·9	16·15	53·0	20·15	66·1	24·15	79·2	28·15	92·4
0·20	0·7	4·20	13·8	8·20	26·9	12·20	40·0	16·20	53·1	20·20	66·3	24·20	79·4	28·20	92·5
0·25	0·8	4·25	13·9	8·25	27·1	12·25	40·2	16·25	53·3	20·25	66·4	24·25	79·6	28·25	92·7
0·30	1·0	4·30	14·1	8·30	27·2	12·30	40·4	16·30	53·5	20·30	66·6	24·30	79·7	28·30	92·8
0·35	1·1	4·35	14·3	8·35	27·4	12·35	40·5	16·35	53·6	20·35	66·8	24·35	79·9	28·35	93·0
0·40	1·3	4·40	14·4	8·40	27·6	12·40	40·7	16·40	53·8	20·40	66·9	24·40	80·1	28·40	93·2
0·45	1·5	4·45	14·6	8·45	27·7	12·45	40·8	16·45	54·0	20·45	67·1	24·45	80·2	28·45	93·3
0·50	1·6	4·50	14·8	8·50	27·9	12·50	41·0	16·50	54·1	20·50	67·3	24·50	80·4	28·50	93·5
0·55	1·8	4·55	14·9	8·55	28·1	12·55	41·2	16·55	54·3	20·55	67·4	24·55	80·5	28·55	93·7
0·60	2·0	4·60	15·1	8·60	28·2	12·60	41·3	16·60	54·5	20·60	67·6	24·60	80·7	28·60	93·8
0·65	2·1	4·65	15·3	8·65	28·4	12·65	41·5	16·65	54·6	20·65	67·7	24·65	80·9	28·65	94·0
0·70	2·3	4·70	15·4	8·70	28·5	12·70	41·7	16·70	54·8	20·70	67·9	24·70	81·0	28·70	94·2
0·75	2·5	4·75	15·6	8·75	28·7	12·75	41·8	16·75	55·0	20·75	68·1	24·75	81·2	28·75	94·3
0·80	2·6	4·80	15·7	8·80	28·9	12·80	42·0	16·80	55·1	20·80	68·2	24·80	81·4	28·80	94·5
0·85	2·8	4·85	15·9	8·85	29·0	12·85	42·2	16·85	55·3	20·85	68·4	24·85	81·5	28·85	94·7
0·90	3·0	4·90	16·1	8·90	29·2	12·90	42·3	16·90	55·4	20·90	68·6	24·90	81·7	28·90	94·8
0·95	3·1	4·95	16·2	8·95	29·4	12·95	42·5	16·95	55·6	20·95	68·7	24·95	81·9	28·95	95·0
1·00	3·3	5·00	16·4	9·00	29·5	13·00	42·7	17·00	55·8	21·00	68·9	25·00	82·0	29·00	95·1
1·05	3·4	5·05	16·6	9·05	29·7	13·05	42·8	17·05	55·9	21·05	69·1	25·05	82·2	29·05	95·3
1·10	3·6	5·10	16·7	9·10	29·9	13·10	43·0	17·10	56·1	21·10	69·2	25·10	82·3	29·10	95·5
1·15	3·8	5·15	16·9	9·15	30·0	13·15	43·1	17·15	56·3	21·15	69·4	25·15	82·5	29·15	95·6
1·20	3·9	5·20	17·1	9·20	30·2	13·20	43·3	17·20	56·4	21·20	69·6	25·20	82·7	29·20	95·8
1·25	4·1	5·25	17·2	9·25	30·3	13·25	43·5	17·25	56·6	21·25	69·7	25·25	82·8	29·25	96·0
1·30	4·3	5·30	17·4	9·30	30·5	13·30	43·6	17·30	56·8	21·30	69·9	25·30	83·0	29·30	96·1
1·35	4·4	5·35	17·6	9·35	30·7	13·35	43·8	17·35	56·9	21·35	70·0	25·35	83·2	29·35	96·3
1·40	4·6	5·40	17·7	9·40	30·8	13·40	44·0	17·40	57·1	21·40	70·2	25·40	83·3	29·40	96·5
1·45	4·8	5·45	17·9	9·45	31·0	13·45	44·1	17·45	57·3	21·45	70·4	25·45	83·5	29·45	96·6
1·50	4·9	5·50	18·0	9·50	31·2	13·50	44·3	17·50	57·4	21·50	70·5	25·50	83·7	29·50	96·8
1·55	5·1	5·55	18·2	9·55	31·3	13·55	44·5	17·55	57·6	21·55	70·7	25·55	83·8	29·55	96·9
1·60	5·2	5·60	18·4	9·60	31·5	13·60	44·6	17·60	57·7	21·60	70·9	25·60	84·0	29·60	97·1
1·65	5·4	5·65	18·5	9·65	31·7	13·65	44·8	17·65	57·9	21·65	71·0	25·65	84·2	29·65	97·3
1·70	5·6	5·70	18·7	9·70	31·8	13·70	44·9	17·70	58·1	21·70	71·2	25·70	84·3	29·70	97·4
1·75	5·7	5·75	18·9	9·75	32·0	13·75	45·1	17·75	58·2	21·75	71·4	25·75	84·5	29·75	97·6
1·80	5·9	5·80	19·0	9·80	32·2	13·80	45·3	17·80	58·4	21·80	71·5	25·80	84·6	29·80	97·8
1·85	6·1	5·85	19·2	9·85	32·3	13·85	45·4	17·85	58·6	21·85	71·7	25·85	84·8	29·85	97·9
1·90	6·2	5·90	19·4	9·90	32·5	13·90	45·6	17·90	58·7	21·90	71·9	25·90	85·0	29·90	98·1
1·95	6·4	5·95	19·5	9·95	32·6	13·95	45·8	17·95	58·9	21·95	72·0	25·95	85·1	29·95	98·3
2·00	6·6	6·00	19·7	10·00	32·8	14·00	45·9	18·00	59·1	22·00	72·2	26·00	85·3	30·00	98·4
2·05	6·7	6·05	19·8	10·05	33·0	14·05	46·1	18·05	59·2	22·05	72·3	26·05	85·5	30·05	98·6
2·10	6·9	6·10	20·0	10·10	33·1	14·10	46·3	18·10	59·4	22·10	72·5	26·10	85·6	30·10	98·8
2·15	7·1	6·15	20·2	10·15	33·3	14·15	46·4	18·15	59·5	22·15	72·7	26·15	85·8	30·15	98·9
2·20	7·2	6·20	20·3	10·20	33·5	14·20	46·6	18·20	59·7	22·20	72·8	26·20	86·0	30·20	99·1
2·25	7·4	6·25	20·5	10·25	33·6	14·25	46·8	18·25	59·9	22·25	73·0	26·25	86·1	30·25	99·2
2·30	7·5	6·30	20·7	10·30	33·8	14·30	46·9	18·30	60·0	22·30	73·2	26·30	86·3	30·30	99·4
2·35	7·7	6·35	20·8	10·35	34·0	14·35	47·1	18·35	60·2	22·35	73·3	26·35	86·5	30·35	99·6
2·40	7·9	6·40	21·0	10·40	34·1	14·40	47·2	18·40	60·4	22·40	73·5	26·40	86·6	30·40	99·7
2·45	8·0	6·45	21·2	10·45	34·3	14·45	47·4	18·45	60·5	22·45	73·7	26·45	86·8	30·45	99·9
2·50	8·2	6·50	21·3	10·50	34·4	14·50	47·6	18·50	60·7	22·50	73·8	26·50	86·9	30·50	100·1
2·55	8·4	6·55	21·5	10·55	34·6	14·55	47·7	18·55	60·9	22·55	74·0	26·55	87·1	30·55	100·2
2·60	8·5	6·60	21·7	10·60	34·8	14·60	47·9	18·60	61·0	22·60	74·1	26·60	87·3	30·60	100·4
2·65	8·7	6·65	21·8	10·65	34·9	14·65	48·1	18·65	61·2	22·65	74·3	26·65	87·4	30·65	100·6
2·70	8·9	6·70	22·0	10·70	35·1	14·70	48·2	18·70	61·4	22·70	74·5	26·70	87·6	30·70	100·7
2·75	9·0	6·75	22·1	10·75	35·3	14·75	48·4	18·75	61·5	22·75	74·6	26·75	87·8	30·75	100·9
2·80	9·2	6·80	22·3	10·80	35·4	14·80	48·6	18·80	61·7	22·80	74·8	26·80	87·9	30·80	101·0
2·85	9·4	6·85	22·5	10·85	35·6	14·85	48·7	18·85	61·8	22·85	75·0	26·85	88·1	30·85	101·2
2·90	9·5	6·90	22·6	10·90	35·8	14·90	48·9	18·90	62·0	22·90	75·1	26·90	88·3	30·90	101·4
2·95	9·7	6·95	22·8	10·95	35·9	14·95	49·0	18·95	62·2	22·95	75·3	26·95	88·4	30·95	101·5
3·00	9·8	7·00	23·0	11·00	36·1	15·00	49·2	19·00	62·3	23·00	75·5	27·00	88·6	31·00	101·7

Metres	Feet	Metres	Feet	Metres	Feet	Metres	Feet	Metres	Feet	Metres	Feet	Metres	Feet	Metres	Feet
3·05	10·0	7·05	23·1	11·05	36·3	15·05	49·4	19·05	62·5	23·05	75·6	27·05	88·7	31·05	101·9
3·10	10·2	7·10	23·3	11·10	36·4	15·10	49·5	19·10	62·7	23·10	75·8	27·10	88·9	31·10	102·0
3·15	10·3	7·15	23·5	11·15	36·6	15·15	49·7	19·15	62·8	23·15	76·0	27·15	89·1	31·15	102·2
3·20	10·5	7·20	23·6	11·20	36·7	15·20	49·9	19·20	63·0	23·20	76·1	27·20	89·2	31·20	102·4
3·25	10·7	7·25	23·8	11·25	36·9	15·25	50·0	19·25	63·2	23·25	76·3	27·25	89·4	31·25	102·5
3·30	10·8	7·30	24·0	11·30	37·1	15·30	50·2	19·30	63·3	23·30	76·4	27·30	89·6	31·30	102·7
3·35	11·0	7·35	24·1	11·35	37·2	15·35	50·4	19·35	63·5	23·35	76·6	27·35	89·7	31·35	102·9
3·40	11·2	7·40	24·3	11·40	37·4	15·40	50·5	19·40	63·6	23·40	76·8	27·40	89·9	31·40	103·0
3·45	11·3	7·45	24·4	11·45	37·6	15·45	50·7	19·45	63·8	23·45	76·9	27·45	90·1	31·45	103·2
3·50	11·5	7·50	24·6	11·50	37·7	15·50	50·9	19·50	64·0	23·50	77·1	27·50	90·2	31·50	103·3
3·55	11·6	7·55	24·8	11·55	37·9	13·55	51·0	19·55	64·1	23·55	77·3	27·55	90·4	31·55	103·5
3·60	11·8	7·60	24·9	11·60	38·1	15·60	51·2	19·60	64·3	23·60	77·4	27·60	90·6	31·60	103·7
3·65	12·0	7·65	25·1	11·65	38·2	15·65	51·3	19·65	64·5	23·65	77·6	27·65	90·7	31·65	103·8
3·70	12·1	7·70	25·3	11·70	38·4	15·70	51·5	19·70	64·6	23·70	77·8	27·70	90·9	31·70	104·0
3·75	12·3	7·75	25·4	11·75	38·5	15·75	51·7	19·75	64·8	23·75	77·9	27·75	91·0	31·75	104·2
3·80	12·5	7·80	25·6	11·80	38·7	15·80	51·8	19·80	65·0	23·80	78·1	27·80	91·2	31·80	104·3
3·85	12·6	7·85	25·8	11·85	38·9	15·85	52·0	19·85	65·1	23·85	78·2	27·85	91·4	31·85	104·5
3·90	12·8	7·90	25·9	11·90	39·0	15·90	52·2	19·90	65·3	23·90	78·4	27·90	91·5	31·90	104·7
3·95	13·0	7·95	26·1	11·95	39·2	15·95	52·3	19·95	65·5	23·95	78·6	27·95	91·7	31·95	104·8
4·00	13·1	8·00	26·2	12·00	39·4	16·00	52·5	20·00	65·6	24·00	78·7	28·00	91·9	32·00	105·0

ADMIRALTY METHOD OF TIDAL PREDICTION—NP 159

Calculator Box Diagram

A

Port	
A.T.T. No.	
Date	
Time Zone	

	Mean Level	
1	Zo (Part II) or (Tab VIa)	
2	Seasonal Corr. (Part II)	
3	Sum = M.L.	

		M_2	S_2	K_1	O_1
4	A1 (Tab VII)				
5	A2 (Tab VII)				
6	A1 − A2				
7	360. n				
8	(A1 − A2) + 360.n = p				
9	p/24				
0	A1 (I.4.)				
1	g. (Part II)				
2	A1 + g				
3	F2 (Tab VII)				
4	F1 (Tab VII)				
5	F2 − F1 = P				
6	P/24				

* n = 0 or smallest integer necessary to make 1.8 > 600° in M_2 and S_2 columns and > 300° in K_1 and O_1 columns.

† R sin r = sum of H.Ft sin Θ for M_2 and S_2
R cos r = sum of H.Ft cos Θ for M_2 and S_2

‡ H.Ft cos Θ(K_1)

§ H.Ft cos Θ(O_1)

B

17	Time = T				
	H.C.	M_2	S_2	K_1	O_1
18	p/24 (I.9)				
19	p/24 × T				
20	(A1 + g) (I.12)				
21	(A1 + g) − p.T/24 = Θ				
22	Sin Θ			✕	✕
23	Cos Θ				
24	P/24 (I.16)				
25	P/24 × T				
26	F1(I.14)				
27	F1 + P.T/24 = Ft				
28	H. (Part II)				
29	H × Ft.				
30	(H × Ft) Sin Θ				
31	(H × Ft) Cos Θ			‡	
				§	
†32	R sin r : R cos r			⟶	
33	r : R			M.L. (I.3)	
34	2r : R^2			✕	
35	f_4 : F_4 (Part II)			✕	
36	2r + f_4 = d_4 : R^2. F_4 = D_4			D_4 cos d_4	
37	3r : R^3			✕	
38	f_6 : F_6 (Part II)			✕	
39	3r + f_6 = d_6 : R^3. F_6 = D_6			D_6 cos d_6	
40		Sum lines 30-39 = Height			

VOL. 1 EUROPEAN WATERS

Tidal Curves—Swanage to Christchurch—Ryde to Selsey—Lymington to Cowes

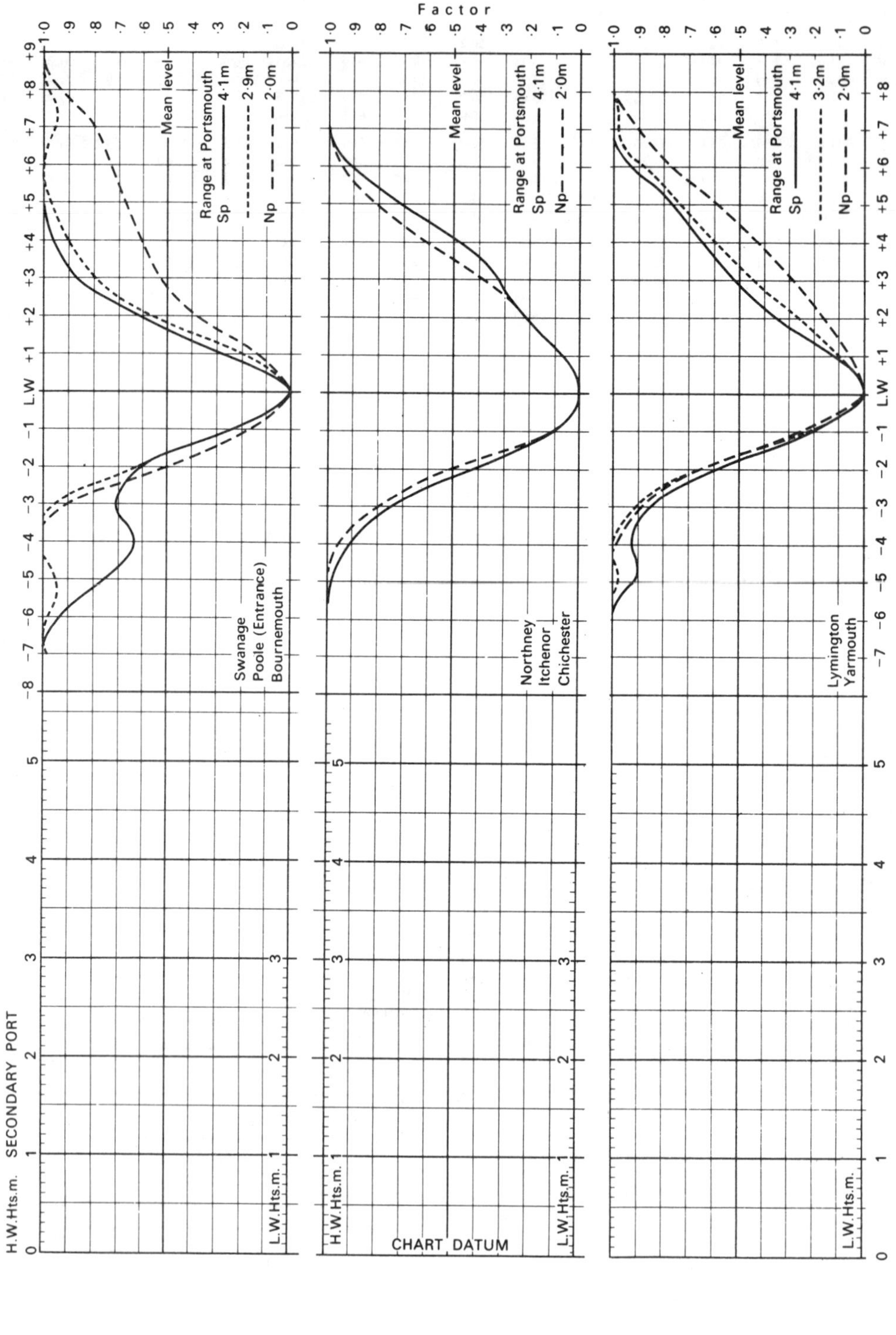

VOLS. 2 AND 3 ATLANTIC, INDIAN AND PACIFIC OCEANS

FOR FINDING THE HEIGHT OF THE TIDE AT TIMES BETWEEN HIGH AND LOW WATER

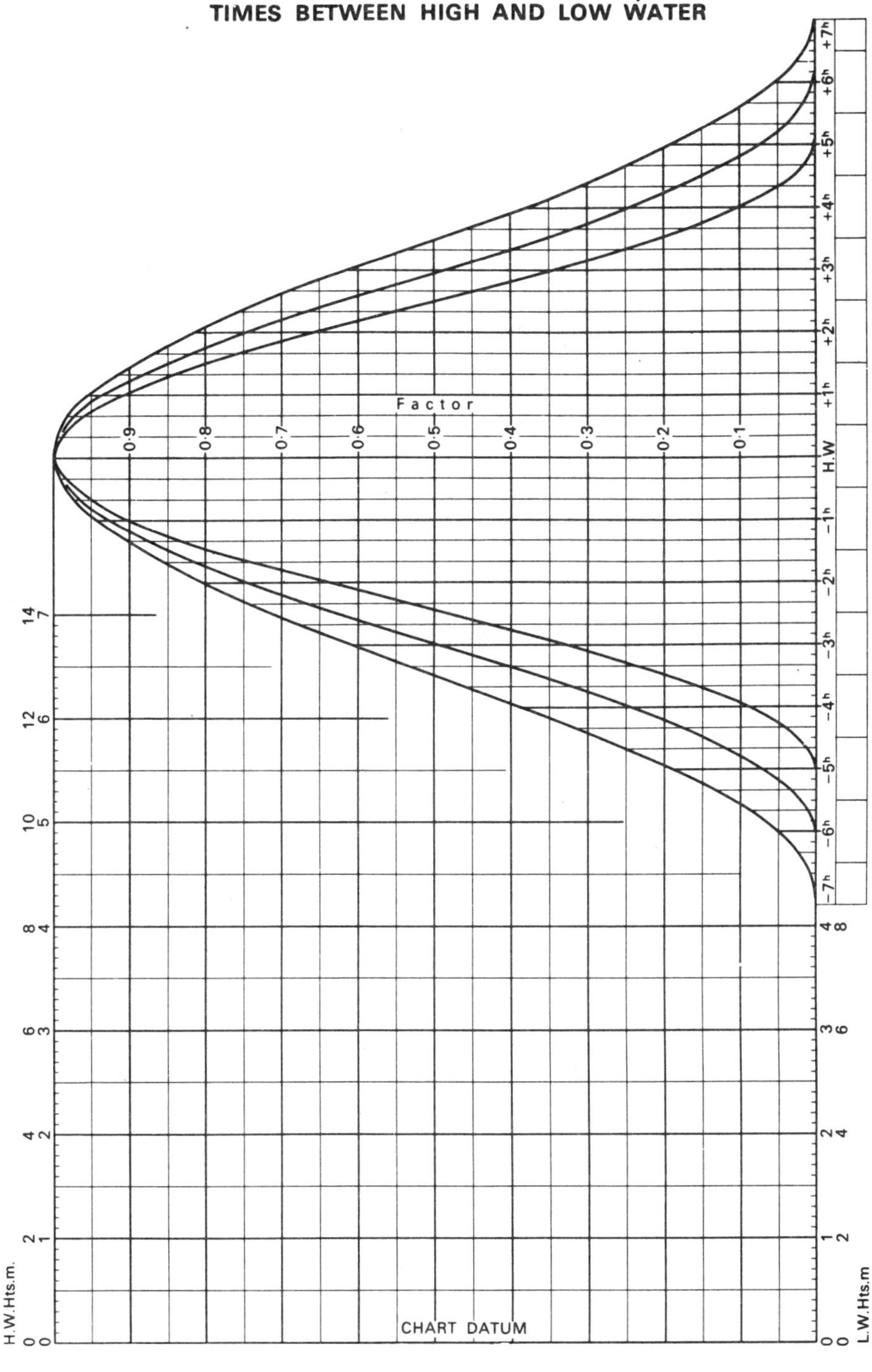

TABLE II

Multiplication Table

FACTOR

	RANGE AT STANDARD PORT																														
1·00	1·2	1·4	1·6	1·8	2·0	2·2	2·4	2·6	2·8	3·0	3·2	3·4	3·6	3·8	4·0	4·2	4·4	4·6	4·8	5·0	5·2	5·4	5·6	5·8	6·0	6·2	6·4	6·6	6·8	7·0	1·00
0·98	1·2	1·4	1·6	1·8	2·0	2·2	2·4	2·5	2·7	2·9	3·1	3·3	3·5	3·7	3·9	4·1	4·3	4·5	4·7	4·9	5·1	5·3	5·5	5·7	5·9	6·1	6·3	6·5	6·7	6·9	0·98
0·96	1·2	1·3	1·5	1·7	1·9	2·1	2·3	2·5	2·7	2·9	3·1	3·3	3·5	3·6	3·8	4·0	4·2	4·4	4·6	4·8	5·0	5·2	5·4	5·6	5·8	6·0	6·1	6·3	6·5	6·7	0·96
0·94	1·1	1·3	1·5	1·7	1·9	2·1	2·3	2·4	2·6	2·8	3·0	3·2	3·4	3·6	3·8	3·9	4·1	4·3	4·5	4·7	4·9	5·1	5·3	5·5	5·6	5·8	6·0	6·2	6·4	6·6	0·94
0·92	1·1	1·3	1·5	1·7	1·8	2·0	2·2	2·4	2·6	2·8	2·9	3·1	3·3	3·5	3·7	3·9	4·0	4·2	4·4	4·6	4·8	5·0	5·2	5·3	5·5	5·7	5·9	6·1	6·3	6·4	0·92
0·90	1·1	1·3	1·4	1·6	1·8	2·0	2·2	2·3	2·5	2·7	2·9	3·1	3·2	3·4	3·6	3·8	4·0	4·1	4·3	4·5	4·7	4·9	5·0	5·2	5·4	5·6	5·8	5·9	6·1	6·3	0·90
0·88	1·1	1·2	1·4	1·6	1·8	1·9	2·1	2·3	2·5	2·6	2·8	3·0	3·2	3·3	3·5	3·7	3·9	4·0	4·2	4·4	4·6	4·8	4·9	5·1	5·3	5·5	5·6	5·8	6·0	6·2	0·88
0·86	1·0	1·2	1·4	1·5	1·7	1·9	2·1	2·2	2·4	2·6	2·8	2·9	3·1	3·3	3·4	3·6	3·8	4·0	4·1	4·3	4·5	4·6	4·8	5·0	5·2	5·3	5·5	5·7	5·8	6·0	0·86
0·84	1·0	1·2	1·3	1·5	1·7	1·8	2·0	2·2	2·4	2·5	2·7	2·9	3·0	3·2	3·4	3·5	3·7	3·9	4·0	4·2	4·4	4·5	4·7	4·9	5·0	5·2	5·4	5·5	5·7	5·9	0·84
0·82	1·0	1·1	1·3	1·5	1·6	1·8	2·0	2·1	2·3	2·5	2·6	2·8	3·0	3·1	3·3	3·4	3·6	3·8	3·9	4·1	4·3	4·4	4·6	4·8	4·9	5·1	5·2	5·4	5·6	5·7	0·82
0·80	1·0	1·1	1·3	1·4	1·6	1·8	1·9	2·1	2·2	2·4	2·6	2·7	2·9	3·0	3·2	3·4	3·5	3·7	3·8	4·0	4·2	4·3	4·5	4·6	4·8	5·0	5·1	5·3	5·4	5·6	0·80
0·78	0·9	1·1	1·2	1·4	1·6	1·7	1·9	2·0	2·2	2·3	2·5	2·7	2·8	3·0	3·1	3·3	3·4	3·6	3·7	3·9	4·1	4·2	4·4	4·5	4·7	4·8	5·0	5·1	5·3	5·5	0·78
0·76	0·9	1·1	1·2	1·4	1·5	1·7	1·8	2·0	2·1	2·3	2·4	2·6	2·7	2·9	3·0	3·2	3·3	3·5	3·6	3·8	4·0	4·1	4·3	4·4	4·6	4·7	4·9	5·0	5·2	5·3	0·76
0·74	0·9	1·0	1·2	1·3	1·5	1·6	1·8	1·9	2·1	2·2	2·4	2·5	2·7	2·8	3·0	3·1	3·3	3·4	3·6	3·7	3·8	4·0	4·1	4·3	4·4	4·6	4·7	4·9	5·0	5·2	0·74
0·72	0·9	1·0	1·2	1·3	1·4	1·6	1·7	1·9	2·0	2·2	2·3	2·4	2·6	2·7	2·9	3·0	3·2	3·3	3·5	3·6	3·7	3·9	4·0	4·2	4·3	4·5	4·6	4·8	4·9	5·0	0·72
0·70	0·8	1·0	1·1	1·3	1·4	1·5	1·7	1·8	2·0	2·1	2·2	2·4	2·5	2·7	2·8	2·9	3·1	3·2	3·4	3·5	3·6	3·8	3·9	4·1	4·2	4·3	4·5	4·6	4·8	4·9	0·70
0·68	0·8	1·0	1·1	1·2	1·4	1·5	1·6	1·8	1·9	2·0	2·2	2·3	2·4	2·6	2·7	2·9	3·0	3·1	3·3	3·4	3·5	3·7	3·8	3·9	4·1	4·2	4·4	4·5	4·6	4·8	0·68
0·66	0·8	0·9	1·1	1·2	1·3	1·5	1·6	1·7	1·8	2·0	2·1	2·2	2·4	2·5	2·6	2·8	2·9	3·0	3·2	3·3	3·4	3·6	3·7	3·8	4·0	4·1	4·2	4·4	4·5	4·6	0·66
0·64	0·8	0·9	1·0	1·1	1·3	1·4	1·5	1·7	1·8	1·9	2·0	2·2	2·3	2·4	2·6	2·7	2·8	2·9	3·1	3·2	3·3	3·5	3·6	3·7	3·8	4·0	4·1	4·2	4·4	4·5	0·64
0·62	0·7	0·9	1·0	1·1	1·2	1·4	1·5	1·6	1·7	1·9	2·0	2·1	2·2	2·4	2·5	2·6	2·7	2·9	3·0	3·1	3·2	3·3	3·5	3·6	3·7	3·8	4·0	4·1	4·2	4·3	0·62
0·60	0·7	0·8	1·0	1·1	1·2	1·3	1·4	1·6	1·7	1·8	1·9	2·0	2·2	2·3	2·4	2·5	2·6	2·8	2·9	3·0	3·1	3·2	3·4	3·5	3·6	3·7	3·8	4·0	4·1	4·2	0·60
0·58	0·7	0·8	0·9	1·0	1·2	1·3	1·4	1·5	1·6	1·7	1·9	2·0	2·1	2·2	2·3	2·4	2·6	2·7	2·8	2·9	3·0	3·1	3·2	3·4	3·5	3·6	3·7	3·8	3·9	4·1	0·58
0·56	0·7	0·8	0·9	1·0	1·1	1·2	1·3	1·5	1·6	1·7	1·8	1·9	2·0	2·1	2·2	2·4	2·5	2·6	2·7	2·8	2·9	3·0	3·1	3·2	3·4	3·5	3·6	3·7	3·8	3·9	0·56
0·54	0·6	0·8	0·9	1·0	1·1	1·2	1·3	1·4	1·5	1·6	1·7	1·8	1·9	2·1	2·2	2·3	2·4	2·5	2·6	2·7	2·8	2·9	3·0	3·1	3·2	3·3	3·5	3·6	3·7	3·8	0·54
0·52	0·6	0·7	0·8	0·9	1·0	1·1	1·2	1·4	1·5	1·6	1·7	1·8	1·9	2·0	2·1	2·2	2·3	2·4	2·5	2·6	2·7	2·8	2·9	3·0	3·1	3·2	3·3	3·4	3·5	3·6	0·52
0·50	0·6	0·7	0·8	0·9	1·0	1·1	1·2	1·3	1·4	1·5	1·6	1·7	1·8	1·9	2·0	2·1	2·2	2·3	2·4	2·5	2·6	2·7	2·8	2·9	3·0	3·1	3·2	3·3	3·4	3·5	0·50
0·48	0·6	0·7	0·8	0·9	1·0	1·1	1·2	1·2	1·3	1·4	1·5	1·6	1·7	1·8	1·9	2·0	2·1	2·2	2·3	2·4	2·5	2·6	2·7	2·8	2·9	3·0	3·1	3·2	3·3	3·4	0·48
0·46	0·6	0·6	0·7	0·8	0·9	1·0	1·1	1·2	1·3	1·4	1·5	1·6	1·7	1·7	1·8	1·9	2·0	2·1	2·2	2·3	2·4	2·5	2·6	2·7	2·8	2·9	2·9	3·0	3·1	3·2	0·46
0·44	0·5	0·6	0·7	0·8	0·9	1·0	1·1	1·1	1·2	1·3	1·4	1·5	1·6	1·7	1·8	1·8	1·9	2·0	2·1	2·2	2·3	2·4	2·5	2·6	2·6	2·7	2·8	2·9	3·0	3·1	0·44
0·42	0·5	0·6	0·7	0·8	0·8	0·9	1·0	1·1	1·2	1·3	1·3	1·4	1·5	1·6	1·7	1·8	1·8	1·9	2·0	2·1	2·2	2·3	2·4	2·4	2·5	2·6	2·7	2·8	2·9	2·9	0·42
0·40	0·5	0·6	0·6	0·7	0·8	0·9	1·0	1·0	1·1	1·2	1·3	1·4	1·4	1·5	1·6	1·7	1·8	1·8	1·9	2·0	2·1	2·2	2·2	2·3	2·4	2·5	2·6	2·6	2·7	2·8	0·40
0·38	0·5	0·5	0·6	0·7	0·8	0·8	0·9	1·0	1·1	1·1	1·2	1·3	1·4	1·4	1·5	1·6	1·7	1·7	1·8	1·9	2·0	2·1	2·1	2·2	2·3	2·4	2·4	2·5	2·6	2·7	0·38
0·36	0·4	0·5	0·6	0·6	0·7	0·8	0·9	0·9	1·0	1·1	1·2	1·2	1·3	1·4	1·4	1·5	1·6	1·7	1·7	1·8	1·9	1·9	2·0	2·1	2·2	2·2	2·3	2·4	2·4	2·5	0·36
0·34	0·4	0·5	0·5	0·6	0·7	0·7	0·8	0·9	1·0	1·0	1·1	1·2	1·2	1·3	1·4	1·4	1·5	1·6	1·6	1·7	1·8	1·8	1·9	2·0	2·0	2·1	2·2	2·2	2·3	2·4	0·34
0·32	0·4	0·4	0·5	0·6	0·6	0·7	0·8	0·8	0·9	1·0	1·0	1·1	1·2	1·2	1·3	1·3	1·4	1·5	1·5	1·6	1·7	1·7	1·8	1·9	1·9	2·0	2·0	2·1	2·2	2·2	0·32
0·30	0·4	0·4	0·5	0·5	0·6	0·7	0·7	0·8	0·8	0·9	1·0	1·0	1·1	1·1	1·2	1·3	1·3	1·4	1·4	1·5	1·6	1·6	1·7	1·7	1·8	1·9	1·9	2·0	2·0	2·1	0·30
0·28	0·3	0·4	0·4	0·5	0·6	0·6	0·7	0·7	0·8	0·8	0·9	1·0	1·0	1·1	1·1	1·2	1·2	1·3	1·3	1·4	1·5	1·5	1·6	1·6	1·7	1·7	1·8	1·8	1·9	2·0	0·28
0·26	0·3	0·4	0·4	0·5	0·5	0·6	0·6	0·7	0·7	0·8	0·8	0·9	0·9	1·0	1·0	1·1	1·1	1·2	1·2	1·3	1·4	1·4	1·5	1·5	1·6	1·6	1·7	1·7	1·8	1·8	0·26
0·24	0·3	0·3	0·4	0·4	0·5	0·5	0·6	0·6	0·7	0·7	0·8	0·8	0·9	0·9	1·0	1·0	1·1	1·1	1·2	1·2	1·2	1·3	1·3	1·4	1·4	1·5	1·5	1·6	1·6	1·7	0·24
0·22	0·3	0·3	0·4	0·4	0·4	0·5	0·5	0·6	0·6	0·7	0·7	0·7	0·8	0·8	0·9	0·9	1·0	1·0	1·1	1·1	1·1	1·2	1·2	1·3	1·3	1·4	1·4	1·5	1·5	1·5	0·22
0·20	0·2	0·3	0·3	0·4	0·4	0·4	0·5	0·5	0·6	0·6	0·6	0·7	0·7	0·8	0·8	0·8	0·9	0·9	1·0	1·0	1·0	1·1	1·1	1·2	1·2	1·2	1·3	1·3	1·4	1·4	0·20
0·18	0·2	0·3	0·3	0·3	0·4	0·4	0·4	0·5	0·5	0·5	0·6	0·6	0·6	0·7	0·7	0·8	0·8	0·8	0·9	0·9	0·9	1·0	1·0	1·0	1·1	1·1	1·2	1·2	1·2	1·3	0·18
0·16	0·2	0·2	0·3	0·3	0·3	0·4	0·4	0·4	0·4	0·5	0·5	0·5	0·6	0·6	0·6	0·7	0·7	0·7	0·8	0·8	0·8	0·9	0·9	0·9	0·9	1·0	1·0	1·1	1·1	1·1	0·16
0·14	0·2	0·2	0·2	0·3	0·3	0·3	0·3	0·4	0·4	0·4	0·4	0·5	0·5	0·5	0·6	0·6	0·6	0·6	0·7	0·7	0·7	0·8	0·8	0·8	0·8	0·9	0·9	0·9	1·0	1·0	0·14
0·12	0·1	0·2	0·2	0·2	0·2	0·3	0·3	0·3	0·3	0·4	0·4	0·4	0·4	0·5	0·5	0·5	0·5	0·6	0·6	0·6	0·6	0·6	0·7	0·7	0·7	0·7	0·8	0·8	0·8	0·8	0·12
0·10	0·1	0·1	0·2	0·2	0·2	0·2	0·2	0·3	0·3	0·3	0·3	0·3	0·4	0·4	0·4	0·4	0·4	0·5	0·5	0·5	0·5	0·5	0·6	0·6	0·6	0·6	0·6	0·7	0·7	0·7	0·10
0·08	0·1	0·1	0·1	0·1	0·2	0·2	0·2	0·2	0·2	0·2	0·3	0·3	0·3	0·3	0·3	0·3	0·4	0·4	0·4	0·4	0·4	0·4	0·4	0·5	0·5	0·5	0·5	0·5	0·5	0·6	0·08
0·06	0·1	0·1	0·1	0·1	0·1	0·1	0·1	0·2	0·2	0·2	0·2	0·2	0·2	0·2	0·2	0·3	0·3	0·3	0·3	0·3	0·3	0·3	0·3	0·3	0·4	0·4	0·4	0·4	0·4	0·4	0·06
0·04	0·0	0·1	0·1	0·1	0·1	0·1	0·1	0·1	0·1	0·1	0·1	0·1	0·1	0·2	0·2	0·2	0·2	0·2	0·2	0·2	0·2	0·2	0·2	0·2	0·2	0·2	0·3	0·3	0·3	0·3	0·04
0·02	0·0	0·0	0·0	0·0	0·0	0·0	0·0	0·1	0·1	0·1	0·1	0·1	0·1	0·1	0·1	0·1	0·1	0·1	0·1	0·1	0·1	0·1	0·1	0·1	0·1	0·1	0·1	0·1	0·1	0·1	0·02

FACTOR

TABLE II—*continued*

FACTOR

	RANGE AT STANDARD PORT																														
1·00	7·2	7·4	7·6	7·8	8·0	8·2	8·4	8·6	8·8	9·0	9·2	9·4	9·6	9·8	10·0	10·2	10·4	10·6	10·8	11·0	11·2	11·4	11·6	11·8	12·0	13·0	14·0	15·0	16·0	17·0	1·00
0·98	7·1	7·3	7·4	7·6	7·8	8·0	8·2	8·4	8·6	8·8	9·0	9·2	9·4	9·6	9·8	10·0	10·2	10·4	10·6	10·8	11·0	11·2	11·4	11·6	11·8	12·7	13·7	14·7	15·7	16·7	0·98
0·96	6·9	7·1	7·3	7·5	7·7	7·9	8·1	8·3	8·4	8·6	8·8	9·0	9·2	9·4	9·6	9·8	10·0	10·2	10·4	10·6	10·8	10·9	11·1	11·3	11·5	12·5	13·4	14·4	15·4	16·3	0·96
0·94	6·8	7·0	7·1	7·3	7·5	7·7	7·9	8·1	8·3	8·5	8·6	8·8	9·0	9·2	9·4	9·6	9·8	10·0	10·2	10·3	10·5	10·7	10·9	11·1	11·3	12·2	13·2	14·1	15·0	16·0	0·94
0·92	6·6	6·8	7·0	7·2	7·4	7·5	7·7	7·9	8·1	8·3	8·5	8·6	8·8	9·0	9·2	9·4	9·6	9·8	9·9	10·1	10·3	10·5	10·7	10·9	11·0	12·0	12·9	13·8	14·7	15·6	0·92
0·90	6·5	6·7	6·8	7·0	7·2	7·4	7·6	7·7	7·9	8·1	8·3	8·5	8·6	8·8	9·0	9·2	9·4	9·5	9·7	9·9	10·1	10·3	10·4	10·6	10·8	11·7	12·6	13·5	14·4	15·3	0·90
0·88	6·3	6·5	6·7	6·9	7·0	7·2	7·4	7·6	7·7	7·9	8·1	8·3	8·4	8·6	8·8	9·0	9·2	9·3	9·5	9·7	9·9	10·0	10·2	10·4	10·6	11·4	12·3	13·2	14·1	15·0	0·88
0·86	6·2	6·4	6·5	6·7	6·9	7·1	7·2	7·4	7·6	7·7	7·9	8·1	8·3	8·4	8·6	8·8	8·9	9·1	9·3	9·5	9·6	9·8	10·0	10·1	10·3	11·2	12·0	12·9	13·8	14·6	0·86
0·84	6·0	6·2	6·4	6·6	6·7	6·9	7·1	7·2	7·4	7·6	7·7	7·9	8·1	8·2	8·4	8·6	8·7	8·9	9·1	9·2	9·4	9·6	9·7	9·9	10·1	10·9	11·8	12·6	13·4	14·3	0·84
0·82	5·9	6·1	6·2	6·4	6·6	6·7	6·9	7·1	7·2	7·4	7·5	7·7	7·9	8·0	8·2	8·4	8·5	8·7	8·9	9·0	9·2	9·3	9·5	9·7	9·8	10·7	11·5	12·3	13·1	13·9	0·82
0·80	5·8	5·9	6·1	6·2	6·4	6·6	6·7	6·9	7·0	7·2	7·4	7·5	7·7	7·8	8·0	8·2	8·3	8·5	8·6	8·8	9·0	9·1	9·3	9·4	9·6	10·4	11·2	12·0	12·8	13·6	0·80
0·78	5·6	5·8	5·9	6·1	6·2	6·4	6·6	6·7	6·9	7·0	7·2	7·3	7·5	7·6	7·8	8·0	8·1	8·3	8·4	8·6	8·7	8·9	9·0	9·2	9·4	10·1	10·9	11·7	12·5	13·3	0·78
0·76	5·5	5·6	5·8	5·9	6·1	6·2	6·4	6·5	6·7	6·8	7·0	7·1	7·3	7·4	7·6	7·8	7·9	8·1	8·2	8·4	8·5	8·7	8·8	9·0	9·1	9·9	10·6	11·4	12·2	12·9	0·76
0·74	5·3	5·5	5·6	5·8	5·9	6·1	6·2	6·4	6·5	6·7	6·8	7·0	7·1	7·3	7·4	7·5	7·7	7·8	8·0	8·1	8·3	8·4	8·6	8·7	8·9	9·6	10·4	11·1	11·8	12·6	0·74
0·72	5·2	5·3	5·5	5·6	5·8	5·9	6·0	6·2	6·3	6·5	6·6	6·8	6·9	7·1	7·2	7·3	7·5	7·6	7·8	7·9	8·1	8·2	8·4	8·5	8·6	9·4	10·1	10·8	11·5	12·2	0·72
0·70	5·0	5·2	5·3	5·5	5·6	5·7	5·9	6·0	6·2	6·3	6·4	6·6	6·7	6·9	7·0	7·1	7·3	7·4	7·6	7·7	7·8	8·0	8·1	8·3	8·4	9·1	9·8	10·5	11·2	11·9	0·70
0·68	4·9	5·0	5·2	5·3	5·4	5·6	5·7	5·8	6·0	6·1	6·3	6·4	6·5	6·7	6·8	6·9	7·1	7·2	7·3	7·5	7·6	7·8	7·9	8·0	8·2	8·8	9·5	10·2	10·9	11·6	0·68
0·66	4·8	4·9	5·0	5·1	5·3	5·4	5·5	5·7	5·8	5·9	6·1	6·2	6·3	6·5	6·6	6·7	6·9	7·0	7·1	7·3	7·4	7·5	7·7	7·8	7·9	8·6	9·2	9·9	10·6	11·2	0·66
0·64	4·6	4·7	4·9	5·0	5·1	5·2	5·4	5·5	5·6	5·8	5·9	6·0	6·1	6·3	6·4	6·5	6·7	6·8	6·9	7·0	7·2	7·3	7·4	7·6	7·7	8·3	9·0	9·6	10·2	10·9	0·64
0·62	4·5	4·6	4·7	4·8	5·0	5·1	5·2	5·3	5·5	5·6	5·7	5·8	6·0	6·1	6·2	6·3	6·4	6·6	6·7	6·8	6·9	7·1	7·2	7·3	7·4	8·1	8·7	9·3	9·9	10·5	0·62
0·60	4·3	4·4	4·6	4·7	4·8	4·9	5·0	5·2	5·3	5·4	5·5	5·6	5·8	5·9	6·0	6·1	6·2	6·4	6·5	6·6	6·7	6·8	7·0	7·1	7·2	7·8	8·4	9·0	9·6	10·2	0·60
0·58	4·2	4·3	4·4	4·5	4·6	4·8	4·9	5·0	5·1	5·2	5·3	5·5	5·6	5·7	5·8	5·9	6·0	6·1	6·3	6·4	6·5	6·6	6·7	6·8	7·0	7·5	8·1	8·7	9·3	9·9	0·58
0·56	4·0	4·1	4·3	4·4	4·5	4·6	4·7	4·8	4·9	5·0	5·2	5·3	5·4	5·5	5·6	5·7	5·8	5·9	6·0	6·2	6·3	6·4	6·5	6·6	6·7	7·3	7·8	8·4	9·0	9·5	0·56
0·54	3·9	4·0	4·1	4·2	4·3	4·4	4·5	4·6	4·8	4·9	5·0	5·1	5·2	5·3	5·4	5·5	5·6	5·7	5·8	5·9	6·0	6·2	6·3	6·4	6·5	7·0	7·6	8·1	8·6	9·2	0·54
0·52	3·7	3·8	4·0	4·1	4·2	4·3	4·4	4·5	4·6	4·7	4·8	4·9	5·0	5·1	5·2	5·3	5·4	5·5	5·6	5·7	5·8	5·9	6·0	6·1	6·2	6·8	7·3	7·8	8·3	8·8	0·52
0·50	3·6	3·7	3·8	3·9	4·0	4·1	4·2	4·3	4·4	4·5	4·6	4·7	4·8	4·9	5·0	5·1	5·2	5·3	5·4	5·5	5·6	5·7	5·8	5·9	6·0	6·5	7·0	7·5	8·0	8·5	0·50
0·48	3·5	3·6	3·6	3·7	3·8	3·9	4·0	4·1	4·2	4·3	4·4	4·5	4·6	4·7	4·8	4·9	5·0	5·1	5·2	5·3	5·4	5·5	5·6	5·7	5·8	6·2	6·7	7·2	7·7	8·2	0·48
0·46	3·3	3·4	3·5	3·6	3·7	3·8	3·9	4·0	4·0	4·1	4·2	4·3	4·4	4·5	4·6	4·7	4·8	4·9	5·0	5·1	5·2	5·2	5·3	5·4	5·5	6·0	6·4	6·9	7·4	7·8	0·46
0·44	3·2	3·3	3·3	3·4	3·5	3·6	3·7	3·8	3·9	4·0	4·0	4·1	4·2	4·3	4·4	4·5	4·6	4·7	4·8	4·8	4·9	5·0	5·1	5·2	5·3	5·7	6·2	6·6	7·0	7·5	0·44
0·42	3·0	3·1	3·2	3·3	3·4	3·4	3·5	3·6	3·7	3·8	3·9	3·9	4·0	4·1	4·2	4·3	4·4	4·5	4·5	4·6	4·7	4·8	4·9	5·0	5·0	5·5	5·9	6·3	6·7	7·1	0·42
0·40	2·9	3·0	3·0	3·1	3·2	3·3	3·4	3·4	3·5	3·6	3·7	3·8	3·8	3·9	4·0	4·1	4·2	4·2	4·3	4·4	4·5	4·6	4·6	4·7	4·8	5·2	5·6	6·0	6·4	6·8	0·40
0·38	2·7	2·8	2·9	3·0	3·0	3·1	3·2	3·3	3·3	3·4	3·5	3·6	3·6	3·7	3·8	3·9	4·0	4·0	4·1	4·2	4·3	4·3	4·4	4·5	4·6	4·9	5·3	5·7	6·1	6·5	0·38
0·36	2·6	2·7	2·7	2·8	2·9	3·0	3·0	3·1	3·2	3·2	3·3	3·4	3·5	3·5	3·6	3·7	3·7	3·8	3·9	4·0	4·0	4·1	4·2	4·2	4·3	4·7	5·0	5·4	5·8	6·1	0·36
0·34	2·4	2·5	2·6	2·7	2·7	2·8	2·9	2·9	3·0	3·1	3·1	3·2	3·3	3·3	3·4	3·5	3·5	3·6	3·7	3·7	3·8	3·9	3·9	4·0	4·1	4·4	4·8	5·1	5·4	5·8	0·34
0·32	2·3	2·4	2·4	2·5	2·6	2·6	2·7	2·8	2·8	2·9	2·9	3·0	3·1	3·1	3·2	3·3	3·3	3·4	3·5	3·5	3·6	3·6	3·7	3·8	3·8	4·2	4·5	4·8	5·1	5·4	0·32
0·30	2·2	2·2	2·3	2·3	2·4	2·5	2·5	2·6	2·6	2·7	2·8	2·8	2·9	2·9	3·0	3·1	3·1	3·2	3·2	3·3	3·4	3·4	3·5	3·5	3·6	3·9	4·2	4·5	4·8	5·1	0·30
0·28	2·0	2·1	2·1	2·2	2·2	2·3	2·4	2·4	2·5	2·5	2·6	2·6	2·7	2·7	2·8	2·9	2·9	3·0	3·0	3·1	3·1	3·2	3·2	3·3	3·4	3·6	3·9	4·2	4·5	4·8	0·28
0·26	1·9	1·9	2·0	2·0	2·1	2·1	2·2	2·2	2·3	2·3	2·4	2·4	2·5	2·5	2·6	2·7	2·7	2·8	2·8	2·9	2·9	3·0	3·0	3·1	3·1	3·4	3·6	3·9	4·2	4·4	0·26
0·24	1·7	1·8	1·8	1·9	1·9	2·0	2·0	2·1	2·1	2·2	2·2	2·3	2·3	2·4	2·4	2·4	2·5	2·5	2·6	2·6	2·7	2·7	2·8	2·8	2·9	3·1	3·4	3·6	3·8	4·1	0·24
0·22	1·6	1·6	1·7	1·7	1·8	1·8	1·8	1·9	1·9	2·0	2·0	2·1	2·1	2·2	2·2	2·2	2·3	2·3	2·4	2·4	2·5	2·5	2·6	2·6	2·6	2·9	3·1	3·3	3·5	3·7	0·22
0·20	1·4	1·5	1·5	1·6	1·6	1·6	1·7	1·7	1·8	1·8	1·8	1·9	1·9	2·0	2·0	2·0	2·1	2·1	2·2	2·2	2·2	2·3	2·3	2·4	2·4	2·6	2·8	3·0	3·2	3·4	0·20
0·18	1·3	1·3	1·4	1·4	1·4	1·5	1·5	1·5	1·6	1·6	1·7	1·7	1·7	1·8	1·8	1·8	1·9	1·9	1·9	2·0	2·0	2·1	2·1	2·1	2·2	2·3	2·5	2·7	2·9	3·1	0·18
0·16	1·2	1·2	1·2	1·2	1·3	1·3	1·3	1·4	1·4	1·4	1·5	1·5	1·5	1·6	1·6	1·6	1·7	1·7	1·7	1·8	1·8	1·8	1·9	1·9	1·9	2·1	2·2	2·4	2·6	2·7	0·16
0·14	1·0	1·0	1·1	1·1	1·1	1·1	1·2	1·2	1·2	1·3	1·3	1·3	1·3	1·4	1·4	1·4	1·5	1·5	1·5	1·5	1·6	1·6	1·6	1·7	1·7	1·8	2·0	2·1	2·2	2·4	0·14
0·12	0·9	0·9	0·9	0·9	1·0	1·0	1·0	1·0	1·1	1·1	1·1	1·1	1·2	1·2	1·2	1·2	1·2	1·3	1·3	1·3	1·3	1·4	1·4	1·4	1·4	1·6	1·7	1·8	1·9	2·0	0·12
0·10	0·7	0·7	0·8	0·8	0·8	0·8	0·8	0·9	0·9	0·9	0·9	0·9	1·0	1·0	1·0	1·0	1·0	1·1	1·1	1·1	1·1	1·1	1·2	1·2	1·2	1·3	1·4	1·5	1·6	1·7	0·10
0·08	0·6	0·6	0·6	0·6	0·6	0·7	0·7	0·7	0·7	0·7	0·7	0·8	0·8	0·8	0·8	0·8	0·8	0·8	0·9	0·9	0·9	0·9	0·9	0·9	1·0	1·0	1·1	1·2	1·3	1·4	0·08
0·06	0·4	0·4	0·5	0·5	0·5	0·5	0·5	0·5	0·5	0·5	0·6	0·6	0·6	0·6	0·6	0·6	0·6	0·6	0·6	0·7	0·7	0·7	0·7	0·7	0·7	0·8	0·8	0·9	1·0	1·0	0·06
0·04	0·3	0·3	0·3	0·3	0·3	0·3	0·3	0·3	0·4	0·4	0·4	0·4	0·4	0·4	0·4	0·4	0·4	0·4	0·4	0·4	0·4	0·5	0·5	0·5	0·5	0·5	0·6	0·6	0·6	0·7	0·04
0·02	0·1	0·1	0·2	0·2	0·2	0·2	0·2	0·2	0·2	0·2	0·2	0·2	0·2	0·2	0·2	0·2	0·2	0·2	0·2	0·2	0·2	0·2	0·2	0·2	0·2	0·3	0·3	0·3	0·3	0·3	0·02

FACTOR

ADMIRALTY TIDAL STREAM ATLAS (EXTRACTS)

INDEX

Page No.

4	BEFORE HW SHEERNESS
	2h 35m before HW Dover

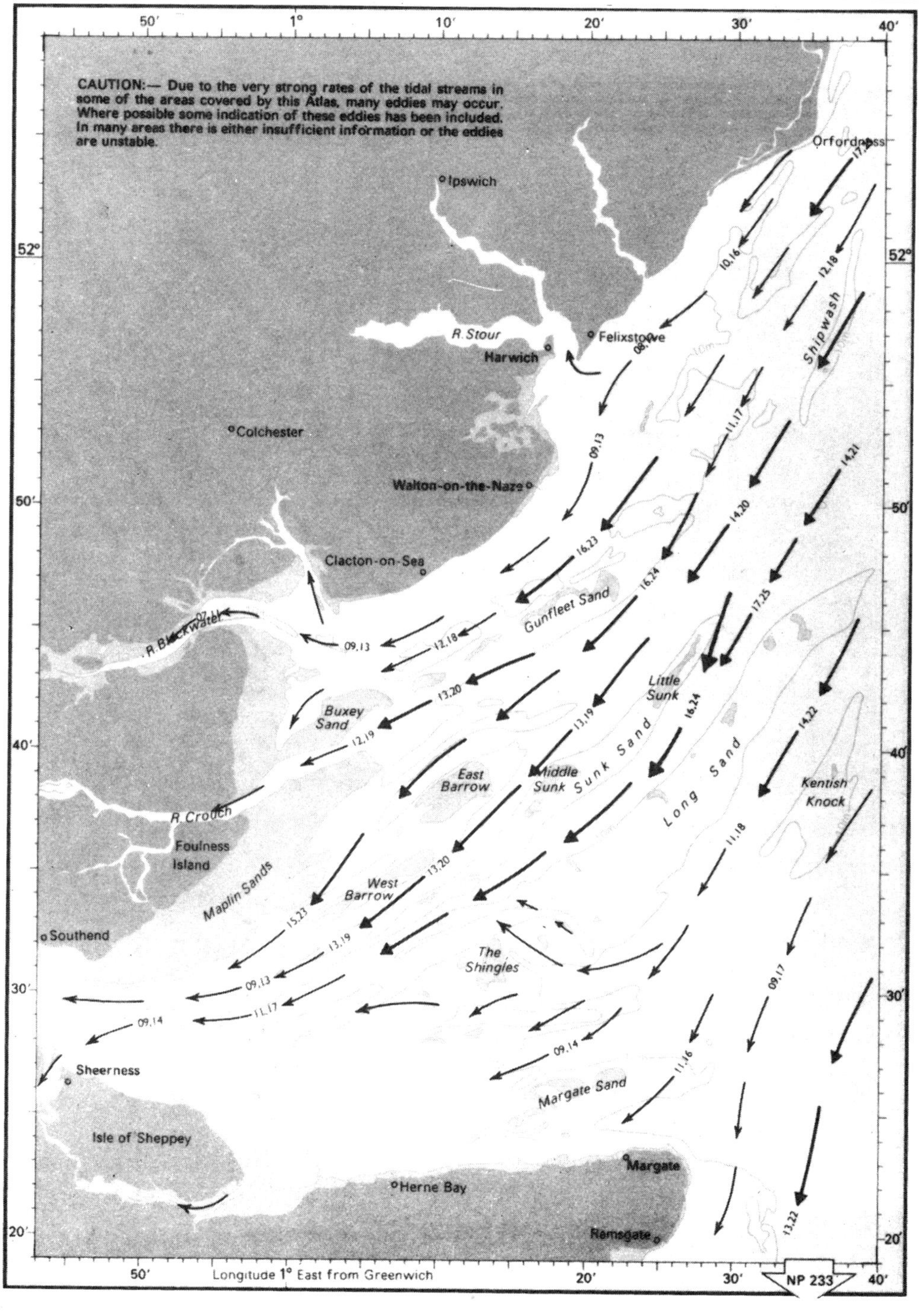

3 **BEFORE HW SHEERNESS**

1h 35m before HW Dover

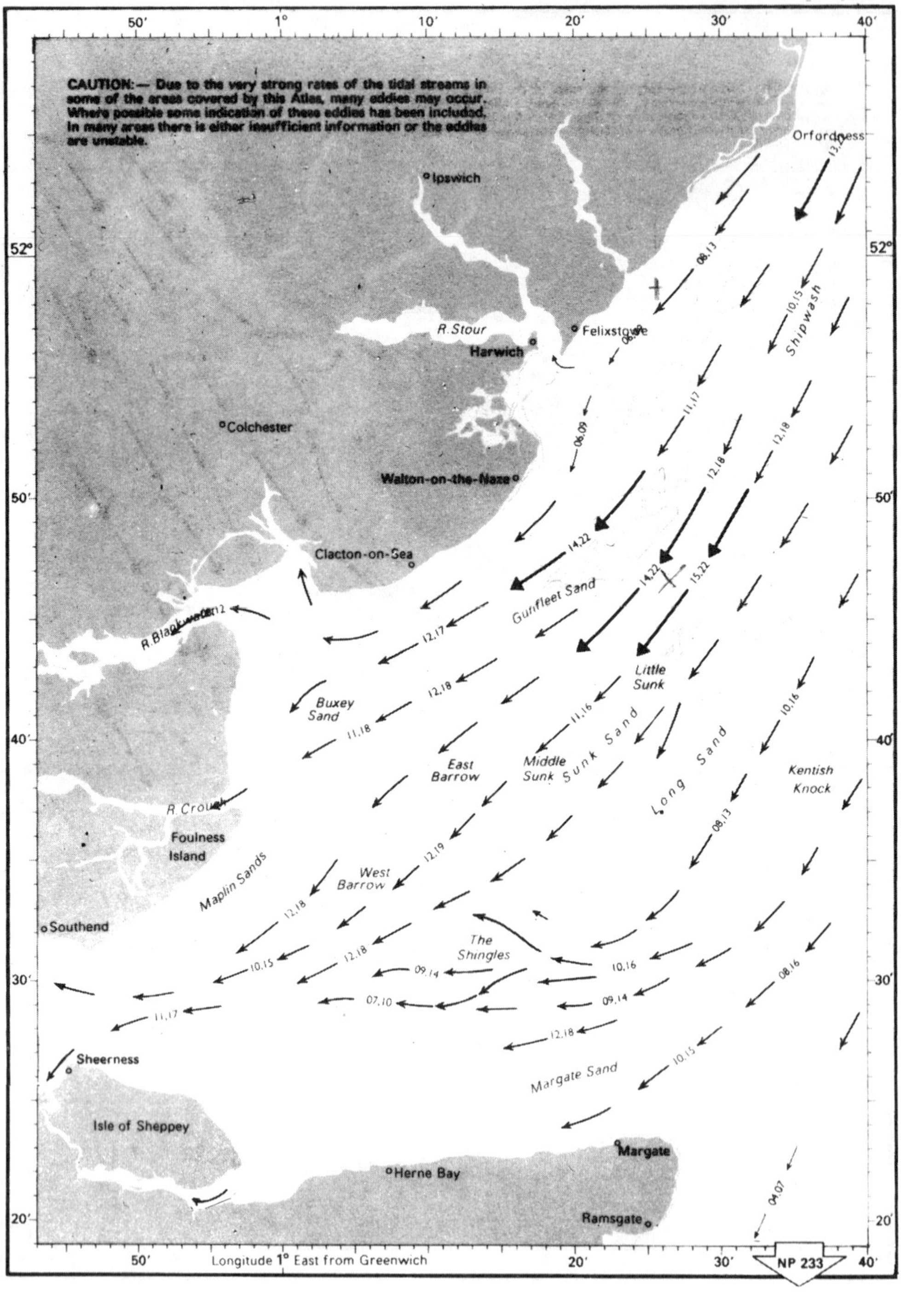

2 BEFORE HW SHEERNESS

0h 35m before HW Dover

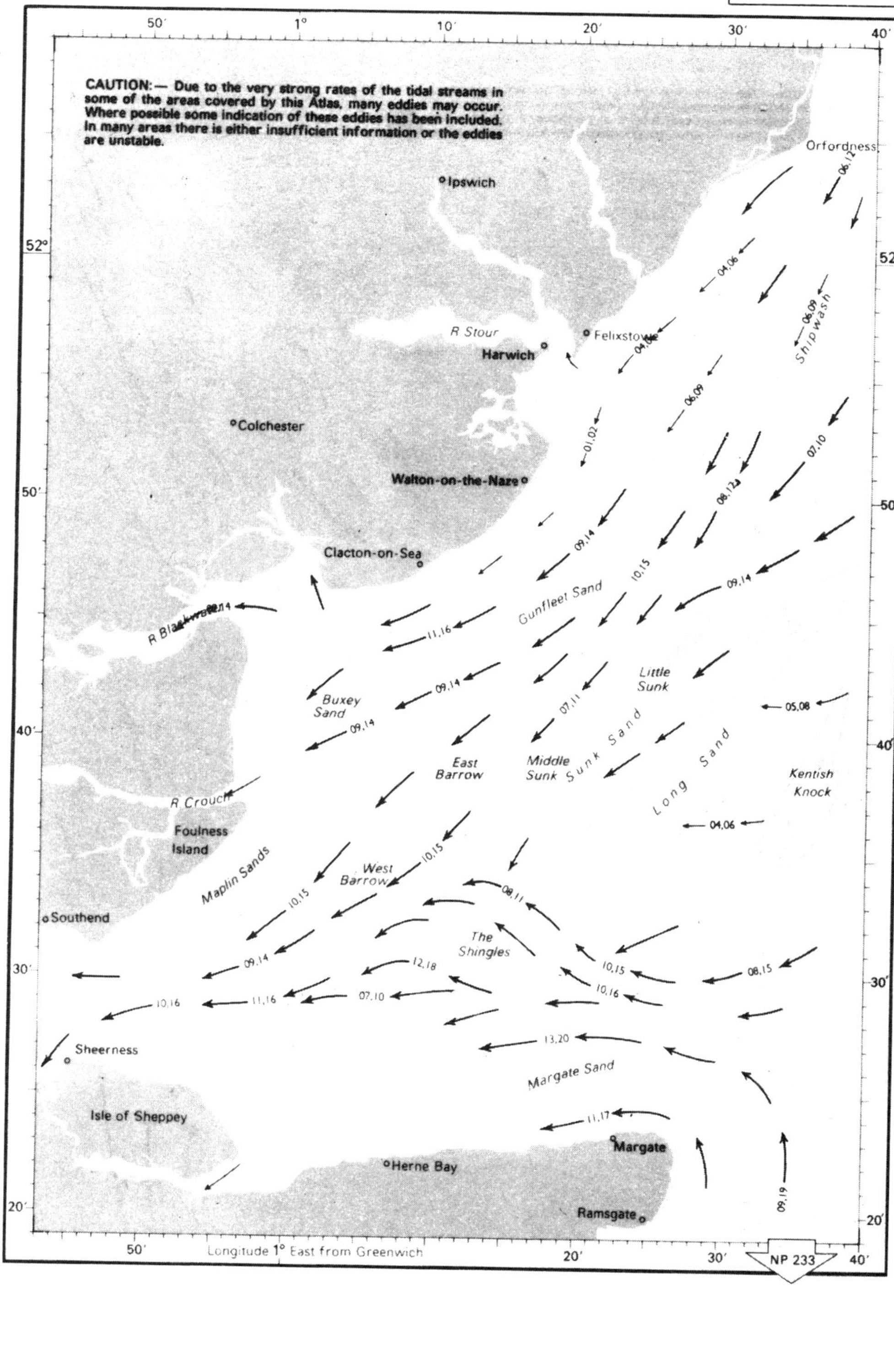

HW SHEERNESS

1h 25m after HW Dover

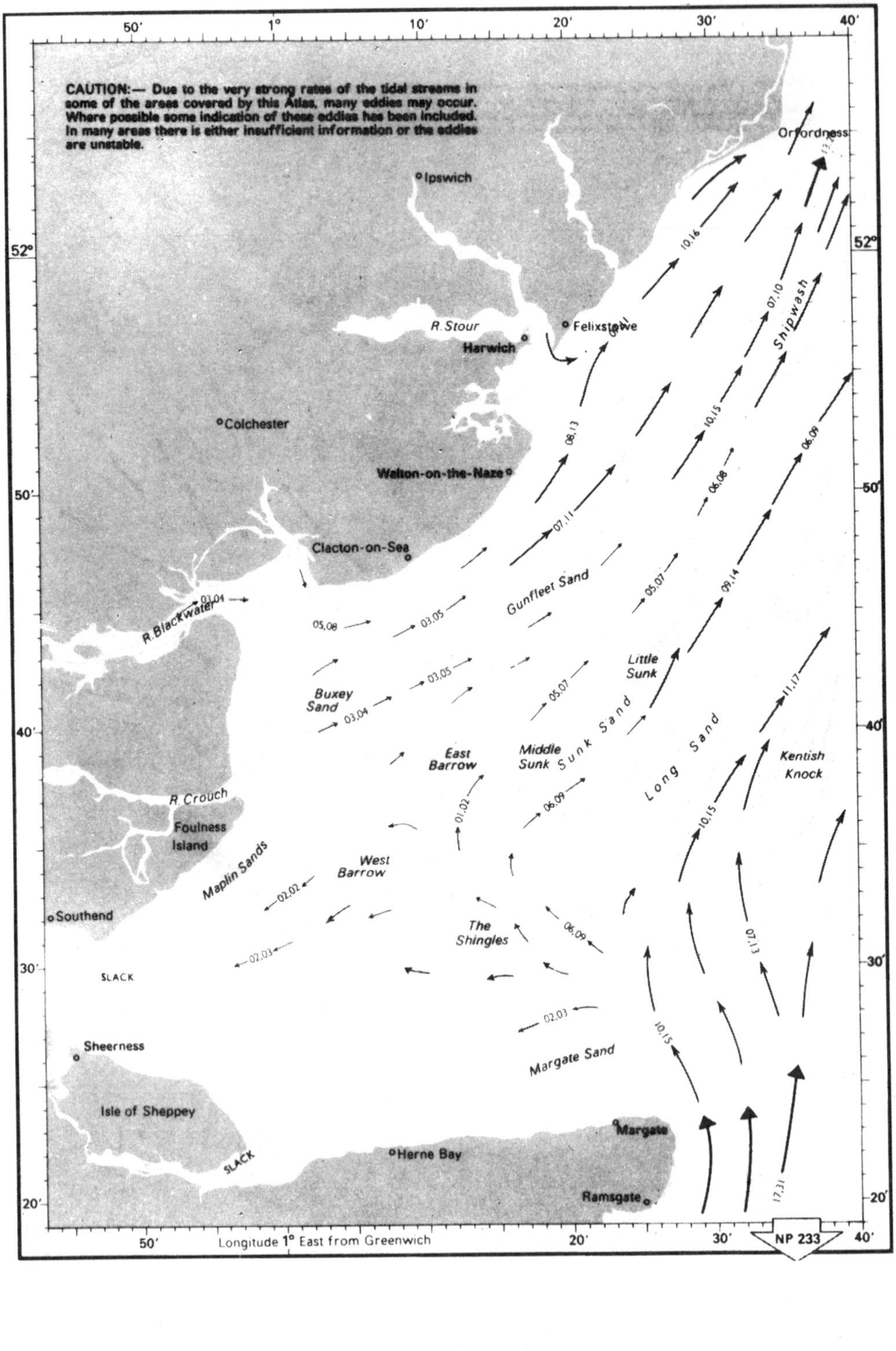

2	AFTER HW SHEERNESS
3h 25m after HW Dover	

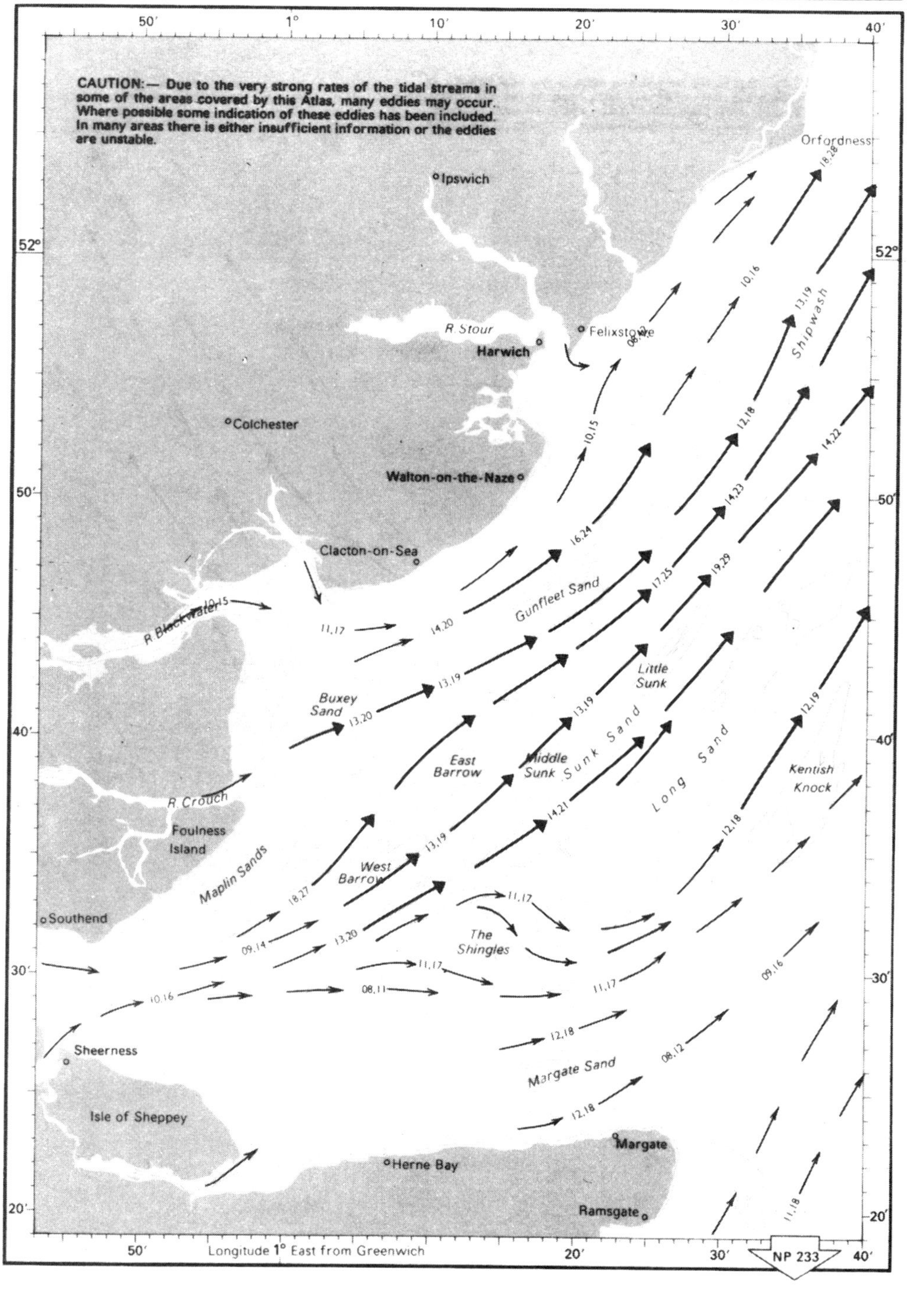

3	AFTER HW SHEERNESS
4h 25m after HW Dover	

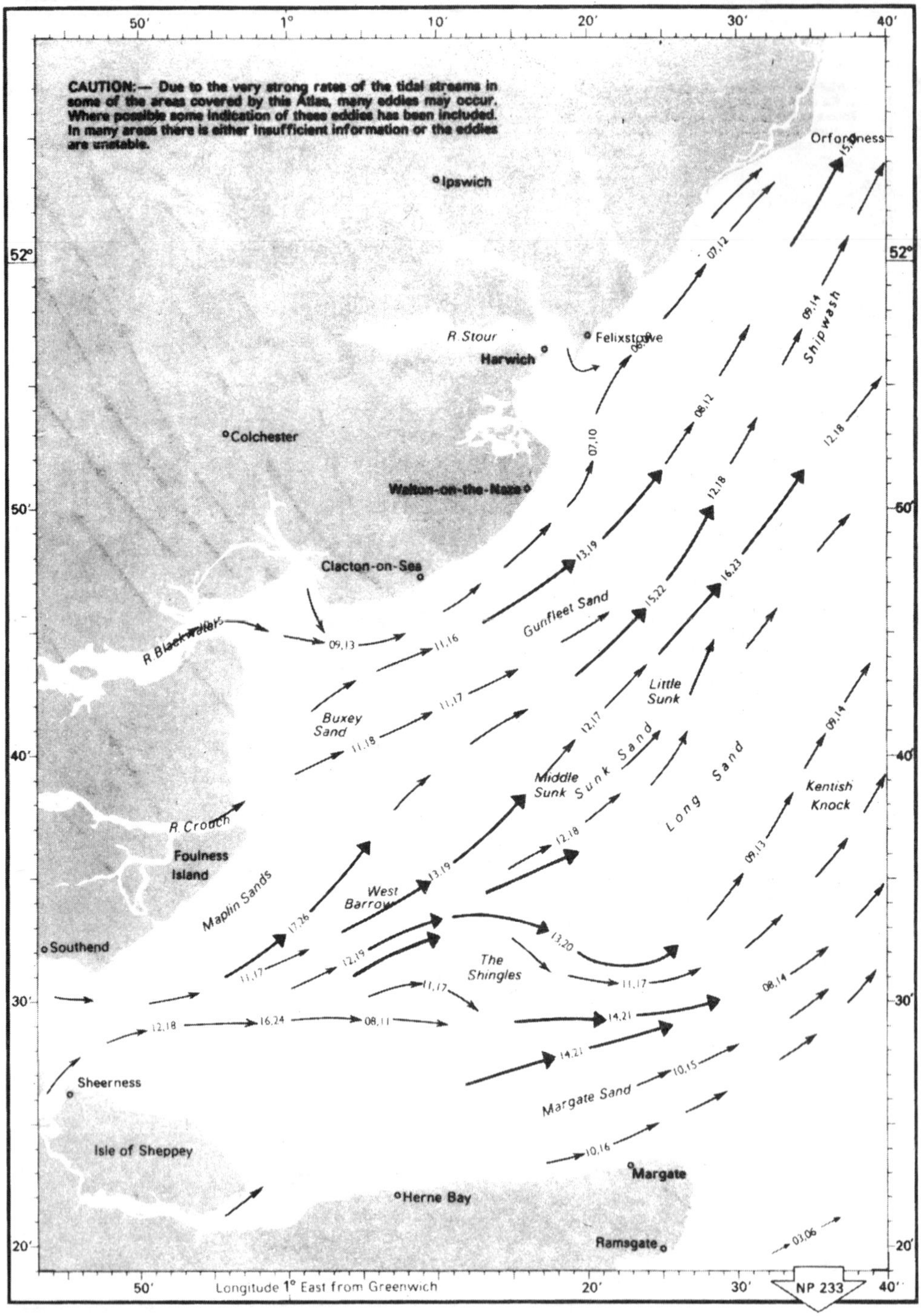

6	AFTER HW SHEERNESS
5h before HW Dover	

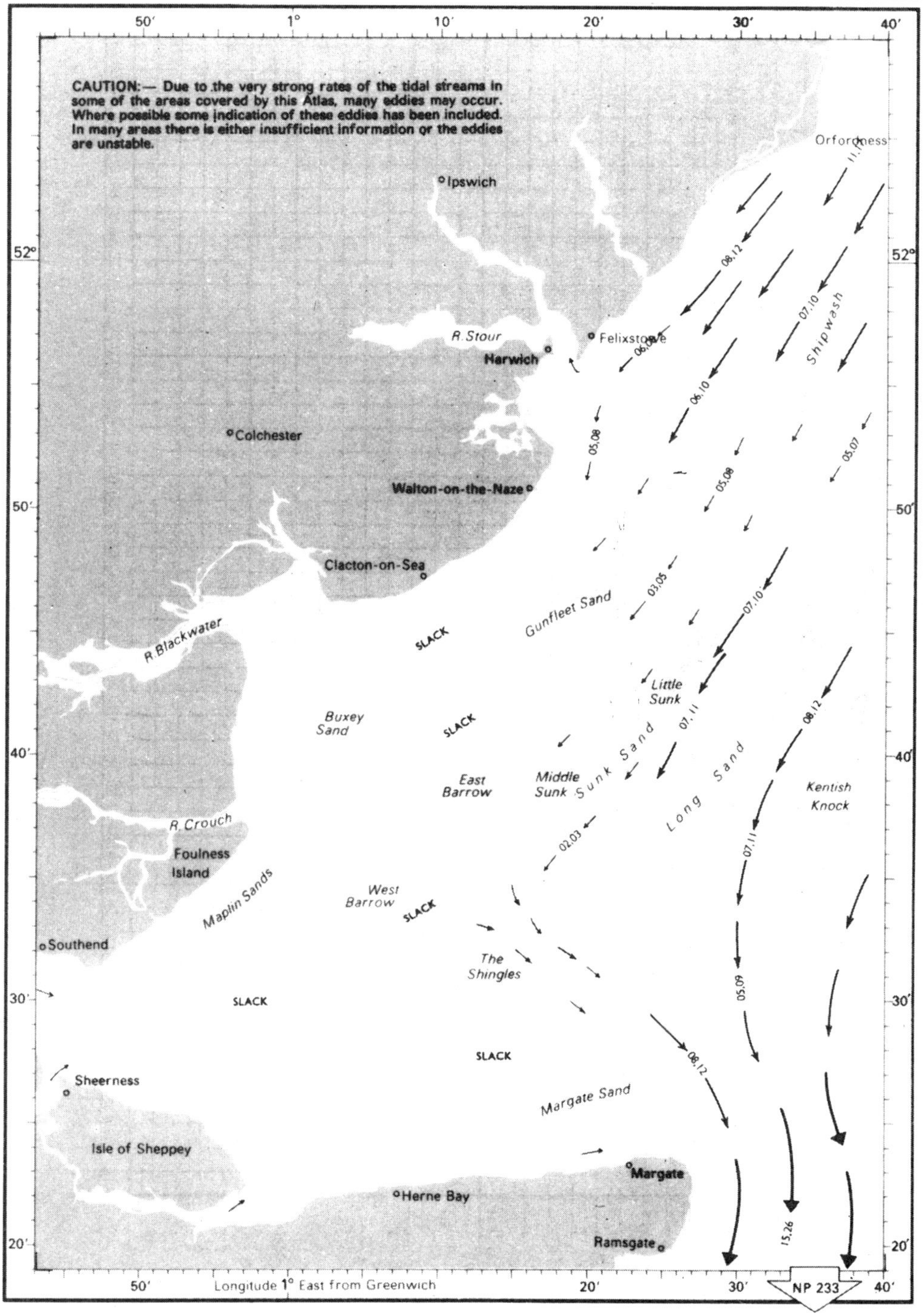

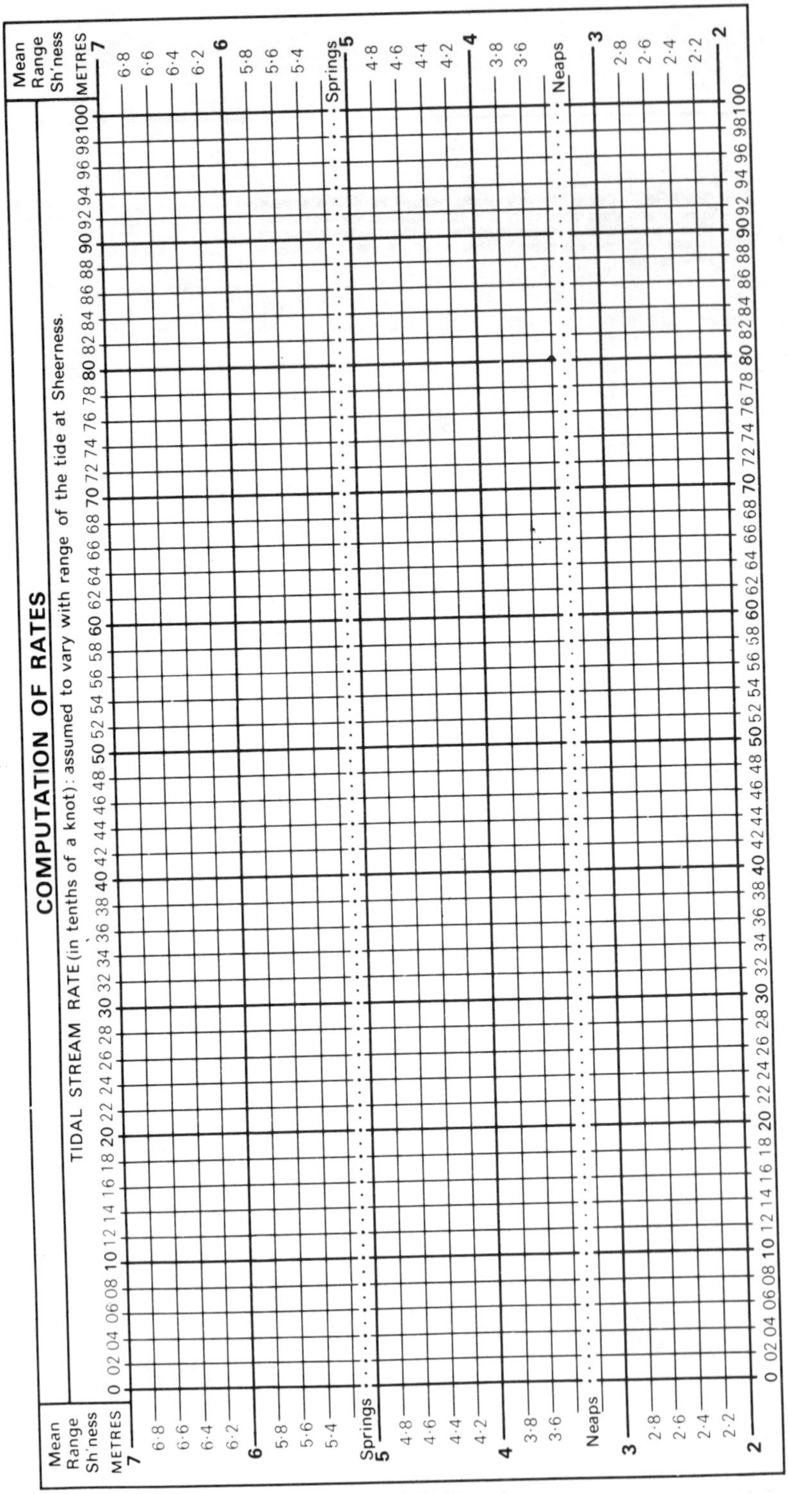
COMPUTATION OF RATES
TIDAL STREAM RATE (in tenths of a knot): assumed to vary with range of the tide at Sheerness.
Mean Range Sh'ness METRES
Springs
Neaps

ELEMENTS FROM THE

ADMIRALTY

NAUTICAL ALMANAC

JANUARY 4, 5, 6 (SAT., SUN., MON.)

G.M.T. d	h	ARIES G.H.A. ° ′	VENUS −4·3 G.H.A. ° ′	VENUS Dec. ° ′	MARS +1·7 G.H.A. ° ′	MARS Dec. ° ′	JUPITER −1·5 G.H.A. ° ′	JUPITER Dec. ° ′	SATURN +0·7 G.H.A. ° ′	SATURN Dec. ° ′
4	00	103 05·8	145 07·7	S14 39·3	216 33·3	S21 38·6	255 43·9	S 9 56·3	204 06·4	S21 43·1
	01	118 08·3	160 10·0	38·6	231 33·9	39·0	270 46·0	56·4	219 08·6	43·1
	02	133 10·7	175 12·2	38·0	246 34·5	39·3	285 48·2	56·5	234 10·7	43·1
	03	148 13·2	190 14·4	37·3	261 35·2	39·6	300 50·4	56·6	249 12·9	43·1
	04	163 15·7	205 16·7	36·7	276 35·8	39·9	315 52·6	56·7	264 15·1	43·2
	05	178 18·1	220 18·9	36·1	291 36·4	40·2	330 54·7	56·8	279 17·2	43·2
	06	193 20·6	235 21·2	S14 35·4	306 37·0	S21 40·5	345 56·9	S 9 56·9	294 19·4	S21 43·2
	07	208 23·1	250 23·4	34·8	321 37·6	40·8	0 59·1	57·0	309 21·5	43·2
S	08	223 25·5	265 25·7	34·1	336 38·2	41·1	16 01·3	57·1	324 23·7	43·3
A	09	238 28·0	280 28·0	33·5	351 38·9	41·4	31 03·4	57·2	339 25·9	43·3
T	10	253 30·5	295 30·2	32·9	6 39·5	41·7	46 05·6	57·3	354 28·0	43·3
U	11	268 32·9	310 32·5	32·2	21 40·1	42·0	61 07·8	57·4	9 30·2	43·3
R	12	283 35·4	325 34·8	S14 31·6	36 40·7	S21 42·3	76 10·0	S 9 57·5	24 32·4	S21 43·3
D	13	298 37·8	340 37·1	30·9	51 41·3	42·6	91 12·2	57·6	39 34·5	43·4
A	14	313 40·3	355 39·3	30·3	66 41·9	42·9	106 14·3	57·7	54 36·7	43·4
Y	15	328 42·8	10 41·6	29·7	81 42·6	43·2	121 16·5	57·8	69 38·9	43·4
	16	343 45·2	25 43·9	29·0	96 43·2	43·5	136 13·7	57·9	84 41·0	43·4
	17	358 47·7	40 46·2	28·4	111 43·8	43·8	151 20·9	58·0	99 43·2	43·5
	18	13 50·2	55 48·5	S14 27·8	126 44·4	S21 44·1	166 23·0	S 9 58·1	114 45·4	S21 43·5
	19	28 52·6	70 50·8	27·1	141 45·0	44·4	181 25·2	58·2	129 47·5	43·5
	20	43 55·1	85 53·1	26·5	156 45·6	44·7	196 27·4	58·3	144 49·7	43·5
	21	58 57·6	100 55·4	25·9	171 46·2	44·9	211 29·6	58·4	159 51·9	43·5
	22	74 00·0	115 57·8	25·2	186 46·9	45·2	226 31·8	58·5	174 54·0	43·6
	23	89 02·5	131 00·1	24·6	201 47·5	45·5	241 33·9	58·6	189 56·2	43·6
5	00	104 05·0	146 02·4	S14 24·0	216 48·1	S21 45·8	256 36·1	S 9 58·7	204 58·3	S21 43·6
	01	119 07·4	161 04·7	23·3	231 48·7	46·1	271 38·3	58·8	220 00·5	43·6
	02	134 09·9	176 07·1	22·7	246 49·3	46·4	286 40·5	58·9	235 02·7	43·6
	03	149 12·3	191 09·4	22·1	261 49·9	46·7	301 42·7	58·9	250 04·8	43·7
	04	164 14·8	206 11·8	21·5	276 50·5	47·0	316 44·8	59·0	265 07·0	43·7
	05	179 17·3	221 14·1	20·8	291 51·1	47·3	331 47·0	59·1	280 09·2	43·7
	06	194 19·7	236 16·5	S14 20·2	306 51·8	S21 47·6	346 49·2	S 9 59·2	295 11·3	S21 43·7
	07	209 22·2	251 18·8	19·6	321 52·4	47·9	1 51·4	59·3	310 13·5	43·8
	08	224 24·7	266 21·2	19·0	336 53·0	48·2	16 53·6	59·4	325 15·7	43·8
S	09	239 27·1	281 23·5	18·3	351 53·6	48·5	31 55·8	59·5	340 17·8	43·8
U	10	254 29·6	296 25·9	17·7	6 54·2	48·8	46 57·9	59·6	355 20·0	43·8
N	11	269 32·1	311 28·3	17·1	21 54·8	49·1	62 00·1	59·7	10 22·2	43·8
D	12	284 34·5	326 30·7	S14 16·5	36 55·4	S21 49·4	77 02·3	S 9 59·8	25 24·3	S21 43·9
A	13	299 37·0	341 33·0	15·8	51 56·0	49·7	92 04·5	9 59·9	40 26·5	43·9
Y	14	314 39·4	356 35·4	15·2	66 56·7	49·9	107 06·7	10 00·0	55 28·7	43·9
	15	329 41·9	11 37·8	14·6	81 57·3	50·2	122 08·9	00·1	70 30·8	43·9
	16	344 44·4	26 40·2	14·0	96 57·9	50·5	137 11·0	00·2	85 33·0	43·9
	17	359 46·8	41 42·6	13·4	111 58·5	50·8	152 13·2	00·3	100 35·2	44·0
	18	14 49·3	56 45·0	S14 12·7	126 59·1	S21 51·1	167 15·4	S10 00·4	115 37·3	S21 44·0
	19	29 51·8	71 47·4	12·1	141 59·7	51·4	182 17·6	00·5	130 39·5	44·0
	20	44 54·2	86 49·8	11·5	157 00·3	51·7	197 19·8	00·6	145 41·7	44·0
	21	59 56·7	101 52·2	10·9	172 00·9	52·0	212 22·0	00·7	160 43·8	44·0
	22	74 59·2	116 54·6	10·3	187 01·5	52·3	227 24·2	00·8	175 46·0	44·1
	23	90 01·6	131 57·1	09·7	202 02·1	52·5	242 26·3	00·9	190 48·2	44·1
6	00	105 04·1	146 59·5	S14 09·1	217 02·7	S21 52·8	257 28·5	S10 01·0	205 50·3	S21 44·1
	01	120 06·6	162 01·9	08·4	232 03·4	53·1	272 30·7	01·0	220 52·5	44·1
	02	135 09·0	177 04·4	07·8	247 04·0	53·4	287 32·9	01·1	235 54·7	44·2
	03	150 11·5	192 06·8	07·2	262 04·6	53·7	302 35·1	01·2	250 56·8	44·2
	04	165 13·9	207 09·2	06·6	277 05·2	54·0	317 37·3	01·3	265 59·0	44·2
	05	180 16·4	222 11·7	06·0	292 05·8	54·3	332 39·5	01·4	281 01·2	44·2
	06	195 18·9	237 14·1	S14 05·4	307 06·4	S21 54·6	347 41·6	S10 01·5	296 03·3	S21 44·2
	07	210 21·3	252 16·6	04·8	322 07·0	54·8	2 43·8	01·6	311 05·5	44·3
	08	225 23·8	267 19·1	04·2	337 07·6	55·1	17 46·0	01·7	326 07·7	44·3
M	09	240 26·3	282 21·5	03·6	352 08·2	55·4	32 48·2	01·8	341 09·8	44·3
O	10	255 28·7	297 24·0	03·0	7 08·8	55·7	47 50·4	01·9	356 12·0	44·3
N	11	270 31·2	312 26·5	02·4	22 09·4	56·0	62 52·6	02·0	11 14·2	44·3
D	12	285 33·7	327 28·9	S14 01·8	37 10·0	S21 56·3	77 54·8	S10 02·1	26 16·3	S21 44·4
A	13	300 36·1	342 31·4	01·2	52 10·6	56·5	92 57·0	02·2	41 18·5	44·4
Y	14	315 38·6	357 33·9	00·6	67 11·3	56·8	107 59·1	02·3	56 20·7	44·4
	15	330 41·1	12 36·4	14 00·0	82 11·9	57·1	123 01·3	02·4	71 22·8	44·4
	16	345 43·5	27 38·9	13 59·4	97 12·5	57·4	138 03·5	02·5	86 25·0	44·4
	17	0 46·0	42 41·4	58·8	112 13·1	57·7	153 05·7	02·5	101 27·2	44·5
	18	15 48·4	57 43·9	S13 58·2	127 13·7	S21 58·0	168 07·9	S10 02·6	116 29·4	S21 44·5
	19	30 50·9	72 46·4	57·6	142 14·3	58·2	183 10·1	02·7	131 31·5	44·5
	20	45 53·4	87 48·9	57·0	157 14·9	58·5	198 12·3	02·8	146 33·7	44·5
	21	60 55·8	102 51·4	56·4	172 15·5	58·8	213 14·5	02·9	161 35·9	44·5
	22	75 58·3	117 54·0	55·8	187 16·1	59·1	228 16·7	03·0	176 38·0	44·6
	23	91 00·8	132 56·5	55·2	202 16·7	59·4	243 18·9	03·1	191 40·2	44·6
Mer. Pass.		17 00·9 (h m)	v 2·4	d 0·6	v 0·6	d 0·3	v 2·2	d 0·1	v 2·2	d 0·0

STARS

Name	S.H.A. ° ′	Dec. ° ′
Acamar	315 49·7	S 40 28·6
Achernar	335 57·7	S 57 27·3
Acrux	173 55·7	S 62 51·8
Adhara	255 44·8	S 28 55·0
Aldebaran	291 36·9	N 16 25·5
Alioth	166 57·2	N 56 10·9
Alkaid	153 31·8	N 49 31·1
Al Na'ir	28 36·3	S 47 10·0
Alnilam	276 28·3	S 1 13·8
Alphard	218 36·7	S 8 28·7
Alphecca	126 46·5	N 26 51·2
Alpheratz	358 26·7	N 28 51·7
Altair	62 49·2	N 8 45·5
Ankaa	353 56·8	S 42 32·2
Antares	113 17·6	S 26 20·3
Arcturus	146 33·8	N 19 23·9
Atria	108 57·3	S 68 56·9
Avior	234 34·4	S 59 22·5
Bellatrix	279 16·3	N 6 18·7
Betelgeuse	271 46·0	N 7 23·9
Canopus	264 14·1	S 52 40·5
Capella	281 35·5	N 45 57·4
Deneb	50 00·2	N 45 08·0
Denebola	183 16·0	N 14 48·2
Diphda	349 37·7	S 18 13·1
Dubhe	194 42·4	N 61 58·3
Elnath	279 04·9	N 28 34·3
Eltanin	91 06·0	N 51 29·6
Enif	34 28·2	N 9 41·0
Fomalhaut	16 10·1	S 29 50·3
Gacrux	172 47·2	S 56 52·5
Gienah	176 35·1	S 17 18·5
Hadar	149 47·1	S 60 10·0
Hamal	328 47·6	N 23 16·0
Kaus Aust.	84 39·4	S 34 24·2
Kochab	137 18·9	N 74 19·4
Markab	14 20·0	N 14 58·9
Menkar	314 58·4	N 3 55·5
Menkent	148 56·8	S 36 09·7
Miaplacidus	221 47·7	S 69 32·6
Mirfak	309 39·7	N 49 42·9
Nunki	76 50·3	S 26 20·9
Peacock	54 25·4	S 56 52·3
Pollux	244 18·2	N 28 07·5
Procyon	245 42·9	N 5 19·8
Rasalhague	96 45·4	N 12 35·4
Regulus	208 27·6	N 12 10·2
Rigel	281 51·8	S 8 15·1
Rigil Kent.	140 48·9	S 60 39·5
Sabik	103 00·6	S 15 40·4
Schedar	350 28·0	N 56 18·8
Shaula	97 18·8	S 37 04·3
Sirius	259 10·1	S 16 39·6
Spica	159 15·2	S 10 56·6
Suhail	223 22·7	S 43 15·8
Vega	81 07·6	N 38 44·7
Zuben'ubi	137 51·7	S 15 52·0

	S.H.A. ° ′	Mer. Pass. h m
Venus	41 57·4	14 14
Mars	112 43·1	9 32
Jupiter	152 31·1	6 53
Saturn	100 53·3	10 19

JANUARY 4, 5, 6 (SAT., SUN., MON.)

G.M.T. d h	SUN G.H.A. ° ′	SUN Dec. ° ′	MOON G.H.A. ° ′	v ′	MOON Dec. ° ′	d ′	H.P. ′
4 00	178 49·9	S 22 47·7	24 09·6	7·6	N19 42·8	0·3	57·7
01	193 49·6	47·5	38 36·2	7·6	19 43·1	0·1	57·7
02	208 49·3	47·2	53 02·8	7·5	19 43·2	0·1	57·7
03	223 49·0	46·9	67 29·3	7·5	19 43·3	0·1	57·8
04	238 48·7	46·7	81 55·8	7·5	19 43·2	0·2	57·8
05	253 48·5	46·4	96 22·3	7·4	19 43·0	0·3	57·8
06	268 48·2	S 22 46·2	110 48·7	7·4	N19 42·7	0·5	57·9
07	283 47·9	45·9	125 15·1	7·4	19 42·2	0·5	57·9
S 08	298 47·6	45·7	139 41·5	7·3	19 41·7	0·7	57·9
A 09	313 47·3	45·4	154 07·8	7·3	19 41·0	0·8	57·9
T 10	328 47·0	45·2	168 34·1	7·2	19 40·2	0·9	58·0
U 11	343 46·7	44·9	183 00·3	7·2	19 39·3	1·1	58·0
R 12	358 46·5	S 22 44·6	197 26·5	7·2	N19 38·2	1·1	58·0
D 13	13 46·2	44·4	211 52·7	7·2	19 37·1	1·3	58·1
A 14	28 45·9	44·1	226 18·9	7·1	19 35·8	1·4	58·1
Y 15	43 45·6	43·9	240 45·0	7·1	19 34·4	1·6	58·1
16	58 45·3	43·6	255 11·1	7·1	19 32·8	1·6	58·2
17	73 45·0	43·3	269 37·2	7·0	19 31·2	1·8	58·2
18	88 44·8	S 22 43·1	284 03·2	7·0	N19 29·4	1·9	58·2
19	103 44·5	42·8	298 29·2	7·0	19 27·5	2·0	58·2
20	118 44·2	42·5	312 55·2	7·0	19 25·5	2·1	58·3
21	133 43·9	42·3	327 21·2	7·0	19 23·4	2·3	58·3
22	148 43·6	42·0	341 47·2	6·9	19 21·1	2·4	58·3
23	163 43·4	41·7	356 13·1	6·9	19 18·7	2·5	58·4
5 00	178 43·1	S 22 41·5	10 39·0	6·9	N19 16·2	2·7	58·4
01	193 42·8	41·2	25 04·9	6·9	19 13·5	2·7	58·4
02	208 42·5	40·9	39 30·8	6·9	19 10·8	2·9	58·4
03	223 42·2	40·7	53 56·7	6·8	19 07·9	3·0	58·5
04	238 41·9	40·4	68 22·5	6·8	19 04·9	3·1	58·5
05	253 41·7	40·1	82 48·3	6·9	19 01·8	3·3	58·5
06	268 41·4	S 22 39·9	97 14·2	6·8	N18 58·5	3·4	58·6
07	283 41·1	39·6	111 40·0	6·8	18 55·1	3·5	58·6
08	298 40·8	39·3	126 05·8	6·8	18 51·6	3·6	58·6
S 09	313 40·5	39·0	140 31·6	6·8	18 48·0	3·7	58·6
U 10	328 40·3	38·8	154 57·4	6·8	18 44·3	3·9	58·7
N 11	343 40·0	38·5	169 23·2	6·7	18 40·4	4·0	58·7
D 12	358 39·7	S 22 38·2	183 48·9	6·8	N18 36·4	4·1	58·7
A 13	13 39·4	37·9	198 14·7	6·8	18 32·3	4·2	58·7
Y 14	28 39·2	37·6	212 40·5	6·7	18 28·1	4·4	58·8
15	43 38·9	37·4	227 06·2	6·8	18 23·7	4·5	58·8
16	58 38·6	37·1	241 32·0	5·8	18 19·2	4·6	58·8
17	73 38·3	36·8	255 57·8	6·7	18 14·6	4·7	58·8
18	88 38·0	S 22 36·5	270 23·5	6·8	N18 09·9	4·8	58·9
19	103 37·8	36·2	284 49·3	6·8	18 05·1	5·0	58·9
20	118 37·5	36·0	299 15·1	6·8	18 00·1	5·0	58·9
21	133 37·2	35·7	313 40·9	6·8	17 55·1	5·2	58·9
22	148 36·9	35·4	328 06·7	6·7	17 49·9	5·3	59·0
23	163 36·7	35·1	342 32·4	6·8	17 44·6	5·4	59·0
6 00	178 36·4	S 22 34·8	356 58·2	6·9	N17 39·2	5·6	59·0
01	193 36·1	34·5	11 24·1	6·8	17 33·6	5·6	59·0
02	208 35·8	34·2	25 49·9	6·8	17 28·0	5·8	59·1
03	223 35·5	33·9	40 15·7	6·8	17 22·2	5·9	59·1
04	238 35·3	33·7	54 41·5	6·9	17 16·3	6·0	59·1
05	253 35·0	33·4	69 07·4	6·9	17 10·3	6·1	59·1
06	268 34·7	S 22 33·1	83 33·3	6·9	N17 04·2	6·2	59·2
07	283 34·4	32·8	97 59·2	6·8	16 58·0	6·4	59·2
08	298 34·2	32·5	112 25·0	7·0	16 51·6	6·4	59·2
M 09	313 33·9	32·2	126 51·0	6·9	16 45·2	6·6	59·2
O 10	328 33·6	31·9	141 16·9	6·9	16 38·6	6·7	59·2
N 11	343 33·3	31·6	155 42·8	7·0	16 31·9	6·8	59·3
D 12	358 33·1	S 22 31·3	170 08·8	7·0	N16 25·1	6·9	59·3
A 13	13 32·8	31·0	184 34·8	7·0	16 18·2	7·0	59·3
Y 14	28 32·5	30·7	199 00·8	7·0	16 11·2	7·1	59·3
15	43 32·2	30·4	213 26·8	7·1	16 04·1	7·2	59·3
16	58 32·0	30·1	227 52·9	7·0	15 56·9	7·3	59·4
17	73 31·7	29·8	242 18·9	7·1	15 49·6	7·4	59·4
18	88 31·4	S 22 29·5	256 45·0	7·1	N15 42·2	7·6	59·4
19	103 31·2	29·2	271 11·1	7·2	15 34·6	7·6	59·4
20	118 30·9	28·9	285 37·3	7·2	15 27·0	7·7	59·4
21	133 30·6	28·6	300 03·5	7·2	15 19·3	7·9	59·4
22	148 30·3	28·3	314 29·7	7·2	15 11·4	7·9	59·5
23	163 30·1	28·0	328 55·9	7·2	15 03·5	8·0	59·5
	S.D. 16·3	d 0·3	S.D. 15·8		16·0		16·2

Lat. °	Twilight Naut. h m	Twilight Civil h m	Sunrise h m	Moonrise 4 h m	Moonrise 5 h m	Moonrise 6 h m	Moonrise 7 h m
N 72	08 20	10 31	■	□	□	14 39	16 57
N 70	08 02	09 43	■	11 43	13 25	15 22	17 19
68	07 47	09 12	11 32	12 48	14 10	15 50	17 37
66	07 35	08 50	10 20	13 23	14 39	16 12	17 51
64	07 25	08 31	09 45	13 49	15 02	16 29	18 03
62	07 16	08 16	09 19	14 09	15 19	16 43	18 13
60	07 08	08 04	09 00	14 25	15 34	16 54	18 21
N 58	07 01	07 53	08 44	14 39	15 47	17 05	18 28
56	06 55	07 43	08 30	14 51	15 57	17 13	18 35
54	06 49	07 35	08 18	15 01	16 07	17 21	18 40
52	06 44	07 27	08 07	15 10	16 15	17 28	18 46
50	06 39	07 20	07 58	15 18	16 23	17 35	18 50
45	06 28	07 05	07 38	15 36	16 39	17 48	19 00
N 40	06 18	06 52	07 22	15 50	16 52	17 59	19 09
35	06 09	06 41	07 08	16 03	17 04	18 09	19 16
30	06 01	06 30	06 57	16 13	17 14	18 17	19 22
20	05 45	06 12	06 36	16 31	17 31	18 32	19 35
N 10	05 29	05 55	06 18	16 47	17 45	18 44	19 43
0	05 13	05 39	06 02	17 02	17 59	18 56	19 51
S 10	04 55	05 22	05 45	17 17	18 13	19 07	20 00
20	04 33	05 02	05 26	17 33	18 28	19 20	20 10
30	04 05	04 38	05 05	17 51	18 44	19 34	20 21
35	03 47	04 23	04 52	18 02	18 54	19 42	20 27
40	03 25	04 06	04 38	18 14	19 05	19 52	20 34
45	02 56	03 44	04 21	18 29	19 19	20 03	20 42
S 50	02 13	03 16	03 59	18 46	19 34	20 16	20 52
52	01 48	03 01	03 49	18 55	19 42	20 22	20 56
54	01 11	02 45	03 37	19 04	19 50	20 29	21 01
56	////	02 24	03 24	19 14	19 59	20 36	21 07
58	////	01 57	03 08	19 26	20 10	20 45	21 13
S 60	////	01 17	02 49	19 40	20 22	20 55	21 20

Lat. °	Sunset h m	Twilight Civil h m	Twilight Naut. h m	Moonset 4 h m	Moonset 5 h m	Moonset 6 h m	Moonset 7 h m
N 72	■	13 40	15 51	□	□	10 56	10 26
N 70	■	14 27	16 09	09 54	10 11	10 12	10 11
68	12 38	14 58	16 23	08 49	09 25	09 43	09 53
66	13 50	15 21	16 35	08 13	08 55	09 21	09 37
64	14 25	15 39	16 46	07 48	08 33	09 03	09 25
62	14 51	15 54	16 54	07 28	08 14	08 49	09 14
60	15 11	16 06	17 02	07 11	07 59	08 36	09 05
N 58	15 27	16 17	17 09	06 58	07 46	08 26	08 57
56	15 41	16 27	17 15	06 46	07 35	08 16	08 50
54	15 53	16 36	17 21	06 35	07 26	08 08	08 43
52	16 03	16 44	17 27	06 26	07 17	08 00	08 37
50	16 13	16 51	17 32	06 18	07 09	07 54	08 32
45	16 33	17 06	17 43	06 00	06 52	07 39	08 21
N 40	16 49	17 19	17 53	05 46	06 39	07 27	08 11
35	17 02	17 30	18 02	05 33	06 27	07 17	08 03
30	17 14	17 40	18 10	05 22	06 17	07 08	07 56
20	17 34	17 58	18 26	05 04	05 59	06 53	07 44
N 10	17 52	18 15	18 41	04 48	05 44	06 39	07 33
0	18 09	18 32	18 57	04 33	05 29	06 26	07 22
S 10	18 26	18 49	19 16	04 18	05 15	06 13	07 12
20	18 44	19 09	19 38	04 02	04 59	05 59	07 01
30	19 05	19 33	20 05	03 43	04 41	05 43	06 48
35	19 18	19 47	20 23	03 32	04 31	05 34	06 41
40	19 32	20 05	20 46	03 20	04 19	05 24	06 32
45	19 49	20 26	21 14	03 05	04 05	05 11	06 22
S 50	20 11	20 54	21 56	02 47	03 48	04 56	06 10
52	20 21	21 08	22 21	02 39	03 39	04 49	06 04
54	20 33	21 25	22 56	02 30	03 30	04 41	05 58
56	20 46	21 45	////	02 19	03 20	04 32	05 51
58	21 02	22 12	////	02 07	03 09	04 21	05 43
S 60	21 20	22 51	////	01 53	02 55	04 10	05 34

Day	SUN Eqn. of Time 00h m s	SUN Eqn. of Time 12h m s	SUN Mer. Pass. h m	MOON Mer. Pass. Upper h m	MOON Mer. Pass. Lower h m	MOON Age d	MOON Phase
4	04 40	04 54	12 05	23 16	10 47	14	
5	05 07	05 21	12 05	24 13	11 44	15	○
6	05 34	05 47	12 06	00 13	12 41	16	

JANUARY 7, 8, 9 (TUES., WED., THURS.)

G.M.T. d	h	ARIES G.H.A. ° ′	VENUS 4·2 G.H.A. ° ′	VENUS Dec. ° ′	MARS +1·7 G.H.A. ° ′	MARS Dec. ° ′	JUPITER −1·5 G.H.A. ° ′	JUPITER Dec. ° ′	SATURN +0·7 G.H.A. ° ′	SATURN Dec. ° ′
7	00	106 03·2	147 59·0	S13 54·6	217 17·3	S21 59·6	258 21·0	S10 03·2	206 42·4	S21 44·6
	01	121 05·7	163 01·6	54·0	232 17·9	21 59·9	273 23·2	03·3	221 44·5	44·6
	02	136 08·2	178 04·1	53·4	247 18·5	22 00·2	288 25·4	03·4	236 46·7	44·6
	03	151 10·6	193 06·6	· · 52·8	262 19·1	· · 00·5	303 27·6	· · 03·5	251 48·9	· · 44·7
	04	166 13·1	208 09·2	52·2	277 19·7	00·8	318 29·8	03·6	266 51·0	44·7
	05	181 15·6	223 11·7	51·6	292 20·3	01·0	333 32·0	03·7	281 53·2	44·7
	06	196 18·0	238 14·3	S13 51·0	307 20·9	S22 01·3	348 34·2	S10 03·8	296 55·4	S21 44·7
	07	211 20·5	253 16·8	50·4	322 21·5	01·6	3 36·4	03·8	311 57·5	44·7
TUESDAY	08	226 22·9	268 19·4	49·8	337 22·1	01·9	18 38·6	03·9	326 59·7	44·8
	09	241 25·4	283 22·0	· · 49·3	352 22·7	· · 02·1	33 40·8	· · 04·0	342 01·9	· · 44·8
	10	256 27·9	298 24·5	48·7	7 23·3	02·4	48 43·0	04·1	357 04·0	44·8
	11	271 30·3	313 27·1	48·1	22 23·9	02·7	63 45·2	04·2	12 06·2	44·8
	12	286 32·8	328 29·7	S13 47·5	37 24·5	S22 03·0	78 47·4	S10 04·3	27 08·4	S21 44·8
	13	301 35·3	343 32·3	46·9	52 25·1	03·2	93 49·6	04·4	42 10·6	44·9
	14	316 37·7	358 34·9	46·3	67 25·7	03·5	108 51·7	04·5	57 12·7	44·9
	15	331 40·2	13 37·5	· · 45·7	82 26·3	· · 03·8	123 53·9	· · 04·6	72 14·9	· · 44·9
	16	346 42·7	28 40·1	45·2	97 26·9	04·1	138 56·1	04·7	87 17·1	44·9
	17	1 45·1	43 42·7	44·6	112 27·5	04·3	153 58·3	04·8	102 19·2	44·9
	18	16 47·6	58 45·3	S13 44·0	127 28·2	S22 04·6	169 00·5	S10 04·8	117 21·4	S21 45·0
	19	31 50·1	73 47·9	43·4	142 28·8	04·9	184 02·7	04·9	132 23·6	45·0
	20	46 52·5	88 50·5	42·9	157 29·4	05·2	199 04·9	05·0	147 25·7	45·0
	21	61 55·0	103 53·1	· · 42·3	172 30·0	· · 05·4	214 07·1	· · 05·1	162 27·9	· · 45·0
	22	76 57·4	118 55·7	41·7	187 30·6	05·7	229 09·3	05·2	177 30·1	45·0
	23	91 59·9	133 58·4	41·1	202 31·2	06·0	244 11·5	05·3	192 32·3	45·1
8	00	107 02·4	149 01·0	S13 40·5	217 31·8	S22 06·2	259 13·7	S10 05·4	207 34·4	S21 45·1
	01	122 04·8	164 03·6	40·0	232 32·4	06·5	274 15·9	05·5	222 36·6	45·1
	02	137 07·3	179 06·3	39·4	247 33·0	06·8	289 18·1	05·6	237 38·8	45·1
	03	152 09·8	194 08·9	· · 38·8	262 33·6	· · 07·1	304 20·3	· · 05·7	252 40·9	· · 45·1
	04	167 12·2	209 11·6	38·3	277 34·2	07·3	319 22·5	05·7	267 43·1	45·2
	05	182 14·7	224 14·2	37·7	292 34·8	07·6	334 24·7	05·8	282 45·3	45·2
	06	197 17·2	239 16·9	S13 37·1	307 35·4	S22 07·9	349 26·9	S10 05·9	297 47·4	S21 45·2
WEDNESDAY	07	212 19·6	254 19·6	36·5	322 36·0	08·1	4 29·1	06·0	312 49·6	45·2
	08	227 22·1	269 22·2	36·0	337 36·6	08·4	19 31·3	06·1	327 51·8	45·2
	09	242 24·5	284 24·9	· · 35·4	352 37·1	· · 08·7	34 33·5	· · 06·2	342 53·9	· · 45·3
	10	257 27·0	299 27·6	34·8	7 37·7	08·9	49 35·7	06·3	357 56·1	45·3
	11	272 29·5	314 30·2	34·3	22 38·3	09·2	64 37·9	06·4	12 58·3	45·3
	12	287 31·9	329 32·9	S13 33·7	37 38·9	S22 09·5	79 40·1	S10 06·5	28 00·5	S21 45·3
	13	302 34·4	344 35·6	33·2	52 39·5	09·7	94 42·3	06·5	43 02·6	45·3
	14	317 36·9	359 38·3	32·6	67 40·1	10·0	109 44·5	06·6	58 04·8	45·4
	15	332 39·3	14 41·0	· · 32·0	82 40·7	· 10·3	124 46·7	· 06·7	73 07·0	· 45·4
	16	347 41·8	29 43·7	31·5	97 41·3	10·5	139 48·9	06·8	88 09·1	45·4
	17	2 44·3	44 46·4	30·9	112 41·9	10·8	154 51·1	06·9	103 11·3	45·4
	18	17 46·7	59 49·1	S13 30·3	127 42·5	S22 11·1	169 53·3	S10 07·0	118 13·5	S21 45·4
	19	32 49·2	74 51·8	29·8	142 43·1	11·3	184 55·5	07·1	133 15·7	45·5
	20	47 51·7	89 54·6	29·2	157 43·7	11·6	199 57·7	07·2	148 17·8	45·5
	21	62 54·1	104 57·3	· · 28·7	172 44·3	· 11·9	214 59·9	07·3	163 20·0	· · 45·5
	22	77 56·6	120 00·0	28·1	187 44·9	12·1	230 02·1	07·3	178 22·2	45·5
	23	92 59·0	135 02·7	27·6	202 45·5	12·4	245 04·3	07·4	193 24·3	45·5
9	00	108 01·5	150 05·5	S13 27·0	217 46·1	S22 12·7	260 06·5	S10 07·5	208 26·5	S21 45·6
	01	123 04·0	165 08·2	26·5	232 46·7	12·9	275 08·7	07·6	223 28·7	45·6
	02	138 06·4	180 11·0	25·9	247 47·3	13·2	290 10·9	07·7	238 30·8	45·6
	03	153 08·9	195 13·7	· · 25·4	262 47·9	· · 13·4	305 13·1	· 07·8	253 33·0	· · 45·6
	04	168 11·4	210 16·5	24·8	277 48·5	13·7	320 15·3	07·9	268 35·2	45·6
	05	183 13·8	225 19·2	24·3	292 49·1	14·0	335 17·5	08·0	283 37·4	45·7
	06	198 16·3	240 22·0	S13 23·7	307 49·7	S22 14·2	350 19·7	S10 08·0	298 39·5	S21 45·7
	07	213 18·8	255 24·7	23·2	322 50·3	14·5	5 21·9	08·1	313 41·7	45·7
THURSDAY	08	228 21·2	270 27·5	22·6	337 50·9	14·8	20 24·1	08·2	328 43·9	45·7
	09	243 23·7	285 30·3	· · 22·1	352 51·5	· · 15·0	35 26·3	· · 08·3	343 46·0	· · 45·7
	10	258 26·2	300 33·1	21·5	7 52·0	15·3	50 28·5	08·4	358 48·2	45·8
	11	273 28·6	315 35·9	21·0	22 52·6	15·5	65 30·8	08·5	13 50·4	45·8
	12	288 31·1	330 38·6	S13 20·4	37 53·2	S22 15·8	80 33·0	S10 08·6	28 52·6	S21 45·8
	13	303 33·5	345 41·4	19·9	52 53·8	16·1	95 35·2	08·7	43 54·7	45·8
	14	318 36·0	0 44·2	19·4	67 54·4	16·3	110 37·4	08·7	58 56·9	45·8
	15	333 38·5	15 47·0	· · 18·8	82 55·0	· · 16·6	125 39·6	· · 08·8	73 59·1	· · 45·9
	16	348 40·9	30 49·8	18·3	97 55·6	16·8	140 41·8	08·9	89 01·3	45·9
	17	3 43·4	45 52·6	17·7	112 56·2	17·1	155 44·0	09·0	104 03·4	45·9
	18	18 45·9	60 55·5	S13 17·2	127 56·8	S22 17·3	170 46·2	S10 09·1	119 05·6	S21 45·9
	19	33 48·3	75 58·3	16·7	142 57·4	17·6	185 48·4	09·2	134 07·8	45·9
	20	48 50·8	91 01·1	16·1	157 58·0	17·9	200 50·6	09·3	149 09·9	46·0
	21	63 53·3	106 03·9	· · 15·6	172 58·6	· · 18·1	215 52·8	· · 09·3	164 12·1	· · 46·0
	22	78 55·7	121 06·8	15·1	187 59·2	18·4	230 55·0	09·4	179 14·3	46·0
	23	93 58·2	136 09·6	14·5	202 59·7	18·6	245 57·2	09·5	194 16·5	46·0
Mer. Pass.		h m 16 49·1	*v* 2·7	*d* 0·6	*v* 0·6	*d* 0·3	*v* 2·2	*d* 0·1	*v* 2·2	*d* 0·0

STARS Name	S.H.A. ° ′	Dec. ° ′
Acamar	315 49·7	S 40 28·6
Achernar	335 57·7	S 57 27·3
Acrux	173 55·6	S 62 51·8
Adhara	255 44·8	S 28 55·0
Aldebaran	291 36·9	N 16 25·5
Alioth	166 57·2	N 56 10·9
Alkaid	153 31·8	N 49 31·0
Al Na'ir	28 36·3	S 47 10·0
Alnilam	276 28·3	S 1 13·8
Alphard	218 36·7	S 8 28·7
Alphecca	126 46·5	N 26 51·2
Alpheratz	358 26·7	N 28 51·7
Altair	62 49·2	N 8 45·5
Ankaa	353 56·8	S 42 32·2
Antares	113 17·6	S 26 20·3
Arcturus	146 33·8	N 19 23·9
Atria	108 57·3	S 68 56·9
Avior	234 34·4	S 59 22·5
Bellatrix	279 16·3	N 6 18·7
Betelgeuse	271 46·0	N 7 23·9
Canopus	264 14·1	S 52 40·5
Capella	281 35·5	N 45 57·4
Deneb	50 00·3	N 45 08·0
Denebola	183 16·0	N 14 48·2
Diphda	349 37·7	S 18 13·1
Dubhe	194 42·4	N 61 58·3
Elnath	279 04·9	N 28 34·3
Eltanin	91 06·0	N 51 29·6
Enif	34 28·2	N 9 41·0
Fomalhaut	16 10·1	S 29 50·8
Gacrux	172 47·2	S 56 52·5
Gienah	176 35·1	S 17 18·5
Hadar	149 47·1	S 60 10·0
Hamal	328 47·7	N 23 16·0
Kaus Aust.	84 39·4	S 34 24·2
Kochab	137 18·8	N 74 19·4
Markab	14 20·0	N 14 58·9
Menkar	314 58·4	N 3 55·5
Menkent	148 56·8	S 36 09·7
Miaplacidus	221 47·6	S 69 32·6
Mirfak	309 39·7	N 49 42·9
Nunki	76 50·2	S 26 20·9
Peacock	54 25·4	S 56 52·2
Pollux	244 18·2	N 28 07·5
Procyon	245 42·9	N 5 19·8
Rasalhague	96 45·4	N 12 35·4
Regulus	208 27·5	N 12 10·2
Rigel	281 51·8	S 8 15·1
Rigil Kent.	140 48·8	S 60 39·5
Sabik	103 00·6	S 15 40·4
Schedar	350 28·0	N 56 18·8
Shaula	97 18·8	S 37 04·3
Sirius	259 10·1	S 16 39·6
Spica	159 15·2	S 10 56·6
Suhail	223 22·6	S 43 15·8
Vega	81 07·6	N 38 44·7
Zuben'ubi	137 51·7	S 15 52·0

	S.H.A. ° ′	Mer. Pass. h m
Venus	41 58·6	14 01
Mars	110 29·4	9 29
Jupiter	152 11·3	6 42
Saturn	100 32·0	10 08

JANUARY 7, 8, 9 (TUES., WED., THURS.)

G.M.T. d h	SUN G.H.A. ° ′	SUN Dec. ° ′	MOON G.H.A. ° ′	v ′	MOON Dec. ° ′	d ′	H.P. ′
7 00	178 29·8	S22 27·7	343 22·1	7·3	N14 55·5	8·2	59·5
01	193 29·5	27·4	357 48·4	7·3	14 47·3	8·2	59·5
02	208 29·2	27·1	12 14·7	7·3	14 39·1	8·3	59·5
03	223 29·0	· · 26·8	26 41·0	7·4	14 30·8	8·5	59·5
04	238 28·7	26·5	41 07·4	7·4	14 22·3	8·5	59·5
05	253 28·4	26·2	55 33·8	7·4	14 13·8	8·6	59·6
06	268 28·2	S22 25·9	70 00·2	7·4	N14 05·2	8·7	59·6
07	283 27·9	25·5	84 26·6	7·5	13 56·5	8·8	59·6
T 08	298 27·6	25·2	98 53·1	7·5	13 47·7	8·9	59·6
U 09	313 27·3	· · 24·9	113 19·6	7·6	13 38·8	9·0	59·6
E 10	328 27·1	24·6	127 46·2	7·5	13 29·8	9·0	59·6
S 11	343 26·8	24·3	142 12·7	7·6	13 20·8	9·2	59·6
D 12	358 26·5	S22 24·0	156 39·3	7·7	N13 11·6	9·2	59·7
A 13	13 26·3	23·7	171 06·0	7·7	13 02·4	9·3	59·7
Y 14	28 26·0	23·3	185 32·7	7·7	12 53·1	9·4	59·7
15	43 25·7	· · 23·0	199 59·4	7·7	12 43·7	9·5	59·7
16	58 25·5	22·7	214 26·1	7·8	12 34·2	9·6	59·7
17	73 25·2	22·4	228 52·9	7·8	12 24·6	9·6	59·7
18	88 24·9	S22 22·1	243 19·7	7·8	N12 15·0	9·8	59·7
19	103 24·7	21·8	257 46·5	7·9	12 05·2	9·8	59·7
20	118 24·4	21·4	272 13·4	7·9	11 55·4	9·8	59·7
21	133 24·1	· · 21·1	286 40·3	7·9	11 45·6	10·0	59·7
22	148 23·9	20·8	301 07·2	8·0	11 35·6	10·0	59·8
23	163 23·6	20·5	315 34·2	8·0	11 25·6	10·1	59·8
8 00	178 23·3	S22 20·1	330 01·2	8·1	N11 15·5	10·2	59·8
01	193 23·1	19·8	344 28·3	8·1	11 05·3	10·3	59·8
02	208 22·8	19·5	358 55·4	8·1	10 55·0	10·3	59·8
03	223 22·5	· · 19·2	13 22·5	8·1	10 44·7	10·4	59·8
04	238 22·3	18·8	27 49·6	8·2	10 34·3	10·4	59·8
05	253 22·0	18·5	42 16·8	8·2	10 23·9	10·5	59·8
06	268 21·7	S22 18·2	56 44·0	8·3	N10 13·4	10·6	59·8
W 07	283 21·5	17·9	71 11·3	8·3	10 02·8	10·6	59·8
E 08	298 21·2	17·5	85 38·6	8·3	9 52·2	10·7	59·8
D 09	313 20·9	· · 17·2	100 05·9	8·4	9 41·5	10·8	59·8
N 10	328 20·7	16·9	114 33·3	8·3	9 30·7	10·8	59·8
E 11	343 20·4	16·5	129 00·6	8·5	9 19·9	10·9	59·8
S 12	358 20·1	S22 16·2	143 28·1	8·4	N 9 09·0	10·9	59·8
D 13	13 19·9	15·9	157 55·5	8·5	8 58·1	11·0	59·8
A 14	28 19·6	15·5	172 23·0	8·5	8 47·1	11·1	59·8
Y 15	43 19·3	· · 15·2	186 50·5	8·6	8 36·0	11·1	59·8
16	58 19·1	14·9	201 18·1	8·6	8 24·9	11·1	59·9
17	73 18·8	14·5	215 45·7	8·6	8 13·8	11·2	59·9
18	88 18·6	S22 14·2	230 13·3	8·7	N 8 02·6	11·2	59·9
19	103 18·3	13·9	244 41·0	8·7	7 51·4	11·3	59·9
20	118 18·0	13·5	259 08·7	8·7	7 40·1	11·4	59·9
21	133 17·8	· · 13·2	273 36·4	8·7	7 28·7	11·3	59·9
22	148 17·5	12·8	288 04·1	8·8	7 17·4	11·5	59·9
23	163 17·2	12·5	302 31·9	8·9	7 05·9	11·4	59·9
9 00	178 17·0	S22 12·1	316 59·8	8·8	N 6 54·5	11·5	59·9
01	193 16·7	11·8	331 27·6	8·9	6 43·0	11·6	59·9
02	208 16·5	11·5	345 55·5	8·9	6 31·4	11·6	59·9
03	223 16·2	· · 11·1	0 23·4	8·9	6 19·8	11·6	59·9
04	238 15·9	10·8	14 51·3	9·0	6 08·2	11·6	59·9
05	253 15·7	10·4	29 19·3	9·0	5 56·6	11·7	59·9
06	268 15·4	S22 10·1	43 47·3	9·0	N 5 44·9	11·8	59·9
07	283 15·1	09·7	58 15·3	9·0	5 33·1	11·7	59·9
T 08	298 14·9	09·4	72 43·3	9·1	5 21·4	11·8	59·9
H 09	313 14·6	· · 09·0	87 11·4	9·1	5 09·6	11·8	59·8
U 10	328 14·4	08·7	101 39·5	9·2	4 57·8	11·8	59·8
R 11	343 14·1	08·3	116 07·7	9·1	4 46·0	11·9	59·8
S 12	358 13·9	S22 08·0	130 35·8	9·2	N 4 34·1	11·9	59·8
D 13	13 13·6	07·6	145 04·0	9·2	4 22·2	11·9	59·8
A 14	28 13·3	07·3	159 32·2	9·2	4 10·3	11·9	59·8
Y 15	43 13·1	· · 06·9	174 00·4	9·3	3 58·4	12·0	59·8
16	58 12·8	06·6	188 28·7	9·3	3 46·4	12·0	59·8
17	73 12·6	06·2	202 57·0	9·3	3 34·4	12·0	59·8
18	88 12·3	S22 05·9	217 25·3	9·3	N 3 22·4	12·0	59·8
19	103 12·0	05·5	231 53·6	9·3	3 10·4	12·0	59·8
20	118 11·8	05·2	246 21·9	9·4	2 58·4	12·0	59·8
21	133 11·5	· · 04·8	260 50·3	9·4	2 46·4	12·1	59·8
22	148 11·3	04·4	275 18·7	9·4	2 34·3	12·0	59·8
23	163 11·0	04·1	289 47·1	9·5	2 22·3	12·1	59·8
	S.D. 16·3	d 0·3	S.D. 16·3		16·3		16·3

Lat. °	Twilight Naut. h m	Twilight Civil h m	Sunrise h m	Moonrise 7 h m	Moonrise 8 h m	Moonrise 9 h m	Moonrise 10 h m
N 72	08 15	10 20	■	16 57	19 03	21 05	23 04
N 70	07 58	09 37	■	17 19	19 15	21 09	23 00
68	07 44	09 08	11 11	17 37	19 25	21 12	22 57
66	07 33	08 46	10 13	17 51	19 33	21 14	22 54
64	07 23	08 29	09 40	18 03	19 39	21 16	22 52
62	07 14	08 14	09 16	18 13	19 45	21 18	22 50
60	07 07	08 02	08 57	18 21	19 50	21 20	22 49
N 58	07 00	07 51	08 42	18 28	19 55	21 21	22 47
56	06 54	07 42	08 28	18 35	19 58	21 23	22 46
54	06 48	07 34	08 16	18 40	20 02	21 24	22 45
52	06 43	07 26	08 06	18 46	20 05	21 25	22 44
50	06 38	07 19	07 57	18 50	20 08	21 26	22 43
45	06 28	07 04	07 38	19 00	20 14	21 28	22 41
N 40	06 18	06 52	07 22	19 09	20 19	21 30	22 39
35	06 09	06 41	07 08	19 16	20 24	21 31	22 38
30	06 01	06 31	06 57	19 22	20 28	21 32	22 37
20	05 46	06 13	06 37	19 33	20 34	21 35	22 34
N 10	05 31	05 57	06 19	19 43	20 40	21 37	22 33
0	05 15	05 40	06 03	19 51	20 46	21 39	22 31
S 10	04 56	05 23	05 46	20 00	20 51	21 41	22 29
20	04 35	05 04	05 28	20 10	20 57	21 43	22 27
30	04 07	04 40	05 07	20 21	21 04	21 45	22 25
35	03 50	04 26	04 55	20 27	21 08	21 46	22 24
40	03 28	04 09	04 41	20 34	21 12	21 48	22 23
45	03 00	03 47	04 24	20 42	21 17	21 50	22 21
S 50	02 18	03 20	04 03	20 52	21 23	21 52	22 20
52	01 55	03 06	03 53	20 56	21 26	21 53	22 19
54	01 21	02 50	03 41	21 01	21 29	21 54	22 18
56	////	02 30	03 28	21 07	21 32	21 55	22 17
58	////	02 04	03 13	21 13	21 36	21 57	22 16
S 60	////	01 27	02 55	21 20	21 40	21 58	22 15

Lat. °	Sunset h m	Twilight Civil h m	Twilight Naut. h m	Moonset 7 h m	Moonset 8 h m	Moonset 9 h m	Moonset 10 h m
N 72	■	13 53	15 58	10 36	10 23	10 13	10 04
N 70	■	14 36	16 15	10 11	10 09	10 07	10 04
68	13 02	15 05	16 29	09 53	09 58	10 01	10 04
66	14 00	15 27	16 41	09 37	09 49	09 57	10 04
64	14 33	15 44	16 50	09 25	09 41	09 53	10 04
62	14 57	15 59	16 59	09 14	09 34	09 49	10 04
60	15 16	16 11	17 06	09 05	09 28	09 47	10 04
N 58	15 32	16 22	17 13	08 57	09 22	09 44	10 04
56	15 45	16 31	17 19	08 50	09 17	09 41	10 04
54	15 57	16 40	17 25	08 43	09 13	09 39	10 04
52	16 07	16 47	17 30	08 37	09 09	09 37	10 04
50	16 16	16 54	17 35	08 32	09 06	09 36	10 04
45	16 36	17 09	17 46	08 21	08 58	09 32	10 04
N 40	16 51	17 22	17 55	08 11	08 52	09 29	10 04
35	17 05	17 32	18 04	08 03	08 46	09 26	10 04
30	17 16	17 42	18 12	07 56	08 41	09 23	10 04
20	17 36	18 00	18 27	07 44	08 32	09 19	10 04
N 10	17 54	18 16	18 43	07 33	08 25	09 15	10 04
0	18 10	18 33	18 59	07 22	08 17	09 11	10 03
S 10	18 27	18 50	19 17	07 12	08 10	09 07	10 03
20	18 45	19 09	19 38	07 01	08 02	09 03	10 03
30	19 06	19 33	20 05	06 48	07 53	08 59	10 03
35	19 18	19 47	20 23	06 41	07 48	08 56	10 03
40	19 32	20 04	20 45	06 32	07 42	08 53	10 03
45	19 49	20 25	21 13	06 22	07 35	08 49	10 03
S 50	20 10	20 53	21 54	06 10	07 27	08 45	10 02
52	20 20	21 06	22 17	06 04	07 23	08 43	10 02
54	20 31	21 22	22 49	05 58	07 19	08 41	10 02
56	20 44	21 42	////	05 51	07 14	08 38	10 02
58	20 59	22 07	////	05 43	07 09	08 35	10 02
S 60	21 17	22 43	////	05 34	07 03	08 32	10 02

Day	SUN Eqn. of Time 00h m s	SUN Eqn. of Time 12h m s	SUN Mer. Pass. h m	MOON Mer. Pass. Upper h m	MOON Mer. Pass. Lower h m	MOON Age d	MOON Phase
7	06 00	06 13	12 06	01 09	13 37	17	
8	06 26	06 39	12 07	02 04	14 32	18	○
9	06 52	07 04	12 07	02 58	15 25	19	

JANUARY 10, 11, 12 (FRI., SAT., SUN.)

G.M.T.		ARIES	VENUS −4·1		MARS +1·7		JUPITER −1·5		SATURN +0·7	
d	h	G.H.A. ° ′	G.H.A. ° ′	Dec. ° ′	G.H.A. ° ′	Dec. ° ′	G.H.A. ° ′	Dec. ° ′	G.H.A. ° ′	Dec. ° ′
10	00	109 00·6	151 12·4	S13 14·0	218 00·3	S22 18·9	260 59·4	S10 09·6	209 18·6	S21 46·0
	01	124 03·1	166 15·3	13·5	233 00·9	19·1	276 01·6	09·7	224 20·8	46·1
	02	139 05·6	181 18·1	12·9	248 01·5	19·4	291 03·9	09·8	239 23·0	46·1
	03	154 08·0	196 21·0	·· 12·4	263 02·1	·· 19·6	306 06·1	· 09·9	254 25·2	·· 46·1
	04	169 10·5	211 23·8	11·9	278 02·7	19·9	321 08·3	09·9	269 27·3	46·1
	05	184 13·0	226 26·7	11·4	293 03·3	20·1	336 10·5	10·0	284 29·5	46·1
	06	199 15·4	241 29·6	S13 10·8	308 03·9	S22 20·4	351 12·7	S10 10·1	299 31·7	S21 46·2
	07	214 17·9	256 32·4	10·3	323 04·5	20·7	6 14·9	10·2	314 33·8	46·2
	08	229 20·4	271 35·3	09·8	338 05·1	20·9	21 17·1	10·3	329 36·0	46·2
F	09	244 22·8	286 38·2	·· 09·3	353 05·7	· 21·2	36 19·3	·· 10·4	344 38·2	· 46·2
R	10	259 25·3	301 41·1	08·7	8 06·2	21·4	51 21·5	10·5	359 40·4	46·2
I	11	274 27·8	316 44·0	08·2	23 06·8	21·7	66 23·7	10·5	14 42·5	46·2
D	12	289 30·2	331 46·9	S13 07·7	38 07·4	S22 21·9	81 26·0	S10 10·6	29 44·7	S21 46·3
A	13	304 32·7	346 49·7	07·2	53 08·0	22·2	96 28·2	10·7	44 46·9	46·3
Y	14	319 35·1	1 52·7	06·7	68 08·6	22·4	111 30·4	10·8	59 49·1	46·3
	15	334 37·6	16 55·6	·· 06·2	83 09·2	·· 22·7	126 32·6	·· 10·9	74 51·2	· 46·3
	16	349 40·1	31 58·5	05·6	98 09·8	22·9	141 34·8	11·0	89 53·4	46·3
	17	4 42·5	47 01·4	05·1	113 10·4	23·2	156 37·0	11·0	104 55·6	46·4
	18	19 45·0	62 04·3	S13 04·6	128 11·0	S22 23·4	171 39·2	S10 11·1	119 57·8	S21 46·4
	19	34 47·5	77 07·2	04·1	143 11·5	23·7	186 41·4	11·2	134 59·9	46·4
	20	49 49·9	92 10·1	03·6	158 12·1	23·9	201 43·7	11·3	150 02·1	46·4
	21	64 52·4	107 13·1	·· 03·1	173 12·7	· 24·2	216 45·9	·· 11·4	165 04·3	·· 46·4
	22	79 54·9	122 16·0	02·6	188 13·3	24·4	231 48·1	11·5	180 06·4	46·5
	23	94 57·3	137 18·9	02·0	203 13·9	24·7	246 50·3	11·5	195 08·6	46·5
11	00	109 59·8	152 21·9	S13 01·5	218 14·5	S22 24·9	261 52·5	S10 11·6	210 10·8	S21 46·5
	01	125 02·3	167 24·8	01·0	233 15·1	25·1	276 54·7	11·7	225 13·0	46·5
	02	140 04·7	182 27·8	00·5	248 15·7	25·4	291 56·9	11·8	240 15·1	46·5
	03	155 07·2	197 30·7	13 00·0	263 16·2	·· 25·6	306 59·2	· 11·9	255 17·3	46·6
	04	170 09·6	212 33·7	12 59·5	278 16·8	25·9	322 01·4	12·0	270 19·5	46·6
	05	185 12·1	227 36·7	59·0	293 17·4	26·1	337 03·6	12·0	285 21·7	46·6
	06	200 14·6	242 39·6	S12 58·5	308 18·0	S22 26·4	352 05·8	S10 12·1	300 23·8	S21 46·6
	07	215 17·0	257 42·6	58·0	323 18·6	26·6	7 08·0	12·2	315 26·0	46·6
S	08	230 19·5	272 45·6	57·5	338 19·2	26·9	22 10·2	12·3	330 28·2	46·6
A	09	245 22·0	287 48·6	· 57·0	353 19·8	·· 27·1	37 12·4	· 12·4	345 30·4	· 46·7
T	10	260 24·4	302 51·6	56·5	8 20·3	27·4	52 14·7	12·4	0 32·5	46·7
U	11	275 26·9	317 54·6	56·0	23 20·9	27·6	67 16·9	12·5	15 34·7	46·7
R	12	290 29·4	332 57·6	S12 55·5	38 21·5	S22 27·8	82 19·1	S10 12·6	30 36·9	S21 46·7
D	13	305 31·8	348 00·6	55·0	53 22·1	28·1	97 21·3	12·7	45 39·1	46·7
A	14	320 34·3	3 03·6	54·5	68 22·7	28·3	112 23·5	12·8	60 41·2	46·8
Y	15	335 36·8	18 06·6	·· 54·0	83 23·3	· 28·6	127 25·7	· 12·9	75 43·4	· 46·8
	16	350 39·2	33 09·6	53·6	98 23·9	28·8	142 28·0	12·9	90 45·6	46·8
	17	5 41·7	48 12·6	53·1	113 24·4	29·0	157 30·2	13·0	105 47·8	46·8
	18	20 44·1	63 15·6	S12 52·6	128 25·0	S22 29·3	172 32·4	S10 13·1	120 49·9	S21 46·8
	19	35 46·6	78 18·6	52·1	143 25·6	29·5	187 34·6	13·2	135 52·1	46·9
	20	50 49·1	93 21·7	51·6	158 26·2	29·8	202 36·8	13·3	150 54·3	46·9
	21	65 51·5	108 24·7	·· 51·1	173 26·8	· 30·0	217 39·1	·· 13·3	165 56·5	· 46·9
	22	80 54·0	123 27·7	50·6	188 27·4	30·2	232 41·3	13·4	180 58·7	46·9
	23	95 56·5	138 30·8	50·1	203 27·9	30·5	247 43·5	13·5	196 00·8	46·9
12	00	110 58·9	153 33·8	S12 49·7	218 28·5	S22 30·7	262 45·7	S10 13·6	211 03·0	S21 46·9
	01	126 01·4	168 36·9	49·2	233 29·1	31·0	277 47·9	13·7	226 05·2	47·0
	02	141 03·9	183 39·9	48·7	248 29·7	31·2	292 50·2	13·8	241 07·4	47·0
	03	156 06·3	198 43·0	·· 48·2	263 30·3	·· 31·4	307 52·4	·· 13·8	256 09·5	·· 47·0
	04	171 08·8	213 46·1	47·7	278 30·8	31·7	322 54·6	13·9	271 11·7	47·0
	05	186 11·2	228 49·1	47·3	293 31·4	31·9	337 56·8	14·0	286 13·9	47·0
	06	201 13·7	243 52·2	S12 46·8	308 32·0	S22 32·1	352 59·0	S10 14·1	301 16·1	S21 47·1
	07	216 16·2	258 55·3	46·3	323 32·6	32·4	8 01·3	14·2	316 18·2	47·1
	08	231 18·6	273 58·4	45·8	338 33·2	32·6	23 03·5	14·2	331 20·4	47·1
S	09	246 21·1	289 01·4	· 45·4	353 33·8	·· 32·9	38 05·7	·· 14·3	346 22·6	·· 47·1
U	10	261 23·6	304 04·5	44·9	8 34·3	33·1	53 07·9	14·4	1 24·8	47·1
N	11	276 26·0	319 07·6	44·4	23 34·9	33·3	68 10·2	14·5	16 26·9	47·1
D	12	291 28·5	334 10·7	S12 43·9	38 35·5	S22 33·6	83 12·4	S10 14·6	31 29·1	S21 47·2
A	13	306 31·0	349 13·8	43·5	53 36·1	33·8	98 14·6	14·6	46 31·3	47·2
Y	14	321 33·4	4 16·9	43·0	68 36·7	34·0	113 16·8	14·7	61 33·5	47·2
	15	336 35·9	19 20·1	·· 42·5	83 37·2	·· 34·3	128 19·0	· 14·8	76 35·7	·· 47·2
	16	351 38·4	34 23·2	42·1	98 37·8	34·5	143 21·3	14·9	91 37·8	47·2
	17	6 40·8	49 26·3	41·6	113 38·4	34·7	158 23·5	15·0	106 40·0	47·3
	18	21 43·3	64 29·4	S12 41·1	128 39·0	S22 35·0	173 25·7	S10 15·0	121 42·2	S21 47·3
	19	36 45·7	79 32·5	40·7	143 39·6	35·2	188 27·9	15·1	136 44·4	47·3
	20	51 48·2	94 35·7	40·2	158 40·1	35·4	203 30·2	15·2	151 46·5	47·3
	21	66 50·7	109 38·8	·· 39·7	173 40·7	·· 35·6	218 32·4	·· 15·3	166 48·7	· 47·3
	22	81 53·1	124 41·9	39·3	188 41·3	35·9	233 34·6	15·3	181 50·9	47·3
	23	96 55·6	139 45·1	38·8	203 41·9	36·1	248 36·8	15·4	196 53·1	47·4
Mer. Pass.		16 37·3 (h m)	v 3·0	d 0·5	v 0·6	d 0·2	v 2·2	d 0·1	v 2·2	d 0·0

STARS

Name	S.H.A. ° ′	Dec. ° ′
Acamar	315 49·7	S 40 28·6
Achernar	335 57·7	S 57 27·3
Acrux	173 55·6	S 62 51·8
Adhara	255 44·8	S 28 55·0
Aldebaran	291 36·9	N 16 25·5
Alioth	166 57·2	N 56 10·9
Alkaid	153 31·7	N 49 31·0
Al Na'ir	28 36·3	S 47 10·0
Alnilam	276 28·3	S 1 13·8
Alphard	218 36·7	S 8 28·7
Alphecca	126 46·5	N 26 51·2
Alpheratz	358 26·7	N 28 51·7
Altair	62 49·1	N 8 45·5
Ankaa	353 56·9	S 42 32·2
Antares	113 17·6	S 26 20·3
Arcturus	146 33·8	N 19 23·9
Atria	108 57·2	S 68 56·9
Avior	234 34·4	S 59 22·5
Bellatrix	279 16·4	N 6 18·6
Betelgeuse	271 46·0	N 7 23·9
Canopus	264 14·1	S 52 40·6
Capella	281 35·5	N 45 57·4
Deneb	50 00·3	N 45 08·0
Denebola	183 15·9	N 14 48·2
Diphda	349 37·7	S 18 13·1
Dubhe	194 42·3	N 61 58·3
Elnath	279 04·9	N 28 34·3
Eltanin	91 06·0	N 51 29·6
Enif	34 28·2	N 9 41·0
Fomalhaut	16 10·1	S 29 50·8
Gacrux	172 47·1	S 56 52·5
Gienah	176 35·0	S 17 18·6
Hadar	149 47·0	S 60 10·0
Hamal	328 47·7	N 23 16·0
Kaus Aust.	84 39·4	S 34 24·2
Kochab	137 18·8	N 74 19·3
Markab	14 20·0	N 14 58·9
Menkar	314 58·4	N 3 55·5
Menkent	148 56·7	S 36 09·7
Miaplacidus	221 47·6	S 69 32·7
Mirfak	309 39·7	N 49 43·0
Nunki	76 50·2	S 26 20·9
Peacock	54 25·4	S 56 52·2
Pollux	244 18·2	N 28 07·5
Procyon	245 42·9	N 5 19·8
Rasalhague	96 45·4	N 12 35·4
Regulus	208 27·5	N 12 10·2
Rigel	281 51·8	S 8 15·1
Rigil Kent.	140 48·8	S 60 39·5
Sabik	103 00·6	S 15 40·4
Schedar	350 28·1	N 56 18·8
Shaula	97 18·8	S 37 04·3
Sirius	259 10·0	S 16 39·6
Spica	159 15·1	S 10 56·6
Suhail	223 22·6	S 43 15·8
Vega	81 07·6	N 38 44·7
Zuben'ubi	137 51·6	S 15 52·0

	S.H.A. ° ′	Mer. Pass. h m
Venus	42 22·1	13 48
Mars	108 14·7	9 27
Jupiter	151 52·7	6 31
Saturn	100 11·0	9 58

JANUARY 10, 11, 12 (FRI., SAT., SUN.)

G.M.T. d h	SUN G.H.A. ° ′	SUN Dec. ° ′	MOON G.H.A. ° ′	v ′	MOON Dec. ° ′	d ′	H.P. ′
10 00	178 10·8	S22 03·7	304 15·6	9·4	N 2 10·2	12·1	59·8
01	193 10·5	03·4	318 44·0	9·5	1 58·1	12·1	59·8
02	208 10·2	03·0	333 12·5	9·5	1 46·0	12·1	59·8
03	223 10·0	·· 02·6	347 41·0	9·5	1 33·9	12·1	59·8
04	238 09·7	02·3	2 09·5	9·5	1 21·8	12·1	59·7
05	253 09·5	01·9	16 38·0	9·5	1 09·7	12·1	59·7
06	268 09·2	S22 01·5	31 06·5	9·6	N 0 57·6	12·1	59·7
07	283 09·0	01·2	45 35·1	9·5	0 45·5	12·1	59·7
08	298 08·7	00·8	60 03·6	9·6	0 33·4	12·1	59·7
F 09	313 08·5	·· 00·4	74 32·2	9·6	0 21·3	12·1	59·7
R 10	328 08·2	22 00·1	89 00·8	9·6	N 0 09·2	12·1	59·7
I 11	343 08·0	21 59·7	103 29·4	9·6	S 0 02·9	12·1	59·7
D 12	358 07·7	S21 59·3	117 58·0	9·7	S 0 15·0	12·1	59·7
A 13	13 07·4	59·0	132 26·7	9·6	0 27·1	12·1	59·7
Y 14	28 07·2	58·6	146 55·3	9·7	0 39·2	12·0	59·7
15	43 06·9	·· 58·2	161 24·0	9·6	0 51·2	12·1	59·6
16	58 06·7	57·9	175 52·6	9·7	1 03·3	12·0	59·6
17	73 06·4	57·5	190 21·3	9·7	1 15·3	12·1	59·6
18	88 06·2	S21 57·1	204 50·0	9·7	S 1 27·4	12·0	59·6
19	103 05·9	56·7	219 18·7	9·7	1 39·4	12·0	59·6
20	118 05·7	56·4	233 47·4	9·7	1 51·4	12·0	59·6
21	133 05·4	·· 56·0	248 16·1	9·7	2 03·4	12·0	59·6
22	148 05·2	55·6	262 44·8	9·7	2 15·4	11·9	59·6
23	163 04·9	55·2	277 13·5	9·7	2 27·3	12·0	59·6
11 00	178 04·7	S21 54·9	291 42·2	9·8	S 2 39·3	11·9	59·5
01	193 04·4	54·5	306 11·0	9·7	2 51·2	11·9	59·5
02	208 04·2	54·1	320 39·7	9·7	3 03·1	11·9	59·5
03	223 03·9	·· 53·7	335 08·4	9·8	3 15·0	11·8	59·5
04	238 03·7	53·3	349 37·2	9·7	3 26·8	11·8	59·5
05	253 03·4	53·0	4 05·9	9·7	3 38·6	11·8	59·5
06	268 03·2	S21 52·6	18 34·6	9·8	S 3 50·4	11·8	59·5
S 07	283 02·9	52·2	33 03·4	9·7	4 02·2	11·7	59·5
08	298 02·7	51·8	47 32·1	9·8	4 13·9	11·7	59·4
A 09	313 02·4	·· 51·4	62 00·9	9·7	4 25·6	11·7	59·4
T 10	328 02·2	51·0	76 29·6	9·7	4 37·3	11·6	59·4
U 11	343 01·9	50·7	90 58·3	9·8	4 48·9	11·7	59·4
R 12	358 01·7	S21 50·3	105 27·1	9·7	S 5 00·6	11·5	59·4
D 13	13 01·4	49·9	119 55·8	9·7	5 12·1	11·6	59·4
A 14	28 01·2	49·5	134 24·5	9·8	5 23·7	11·5	59·4
Y 15	43 00·9	·· 49·1	148 53·3	9·7	5 35·2	11·4	59·3
16	58 00·7	48·7	163 22·0	9·7	5 46·6	11·5	59·3
17	73 00·4	48·3	177 50·7	9·7	5 58·1	11·4	59·3
18	88 00·2	S21 47·9	192 19·4	9·7	S 6 09·5	11·3	59·3
19	103 00·0	47·5	206 48·1	9·7	6 20·8	11·3	59·3
20	117 59·7	47·1	221 16·8	9·7	6 32·1	11·2	59·3
21	132 59·5	·· 46·7	235 45·5	9·7	6 43·3	11·3	59·3
22	147 59·2	46·4	250 14·2	9·6	6 54·6	11·1	59·2
23	162 59·0	46·0	264 42·8	9·7	7 05·7	11·1	59·2
12 00	177 58·7	S21 45·6	279 11·5	9·6	S 7 16·8	11·1	59·2
01	192 58·5	45·2	293 40·1	9·7	7 27·9	11·0	59·2
02	207 58·2	44·8	308 08·8	9·6	7 38·9	11·0	59·2
03	222 58·0	·· 44·4	322 37·4	9·6	7 49·9	10·9	59·2
04	237 57·8	44·0	337 06·0	9·7	8 00·8	10·9	59·1
05	252 57·5	43·6	351 34·7	9·5	8 11·7	10·8	59·1
06	267 57·3	S21 43·2	6 03·2	9·6	S 8 22·5	10·8	59·1
07	282 57·0	42·8	20 31·8	9·6	8 33·3	10·7	59·1
08	297 56·8	42·4	35 00·4	9·6	8 44·0	10·6	59·1
S 09	312 56·5	· 42·0	49 29·0	9·5	8 54·6	10·6	59·1
U 10	327 56·3	41·6	63 57·5	9·5	9 05·2	10·6	59·1
N 11	342 56·1	41·2	78 26·0	9·6	9 15·8	10·4	59·0
D 12	357 55·8	S21 40·8	92 54·6	9·5	S 9 26·2	10·5	59·0
A 13	12 55·6	40·4	107 23·1	9·5	9 36·7	10·3	59·0
Y 14	27 55·3	39·9	121 51·6	9·4	9 47·0	10·3	59·0
15	42 55·1	·· 39·5	136 20·0	9·5	9 57·3	10·2	59·0
16	57 54·8	39·1	150 48·5	9·4	10 07·5	10·2	59·0
17	72 54·6	38·7	165 16·9	9·4	10 17·7	10·1	58·9
18	87 54·4	S21 38·3	179 45·3	9·5	S10 27·8	10·0	58·9
19	102 54·1	37·9	194 13·8	9·3	10 37·8	10·0	58·9
20	117 53·9	37·5	208 42·1	9·4	10 47·8	9·9	58·9
21	132 53·6	·· 37·1	223 10·5	9·4	10 57·7	9·8	58·9
22	147 53·4	36·7	237 38·9	9·3	11 07·5	9·8	58·8
23	162 53·2	36·3	252 07·2	9·3	11 17·3	9·7	58·8
	S.D. 16·3	d 0·4	S.D. 16·3		16·2		16·1

Lat. °	Twilight Naut. h m	Twilight Civil h m	Sunrise h m	Moonrise 10 h m	Moonrise 11 h m	Moonrise 12 h m	Moonrise 13 h m
N 72	08 10	10 09	▬	23 04	25 03	01 03	03 05
N 70	07 54	09 30	▬	23 00	24 51	00 51	02 43
68	07 40	09 02	10 55	22 57	24 41	00 41	02 26
66	07 29	08 42	10 05	22 54	24 34	00 34	02 12
64	07 20	08 25	09 35	22 52	24 27	00 27	02 01
62	07 12	08 11	09 12	22 50	24 22	00 22	01 51
60	07 05	07 59	08 54	22 49	24 17	00 17	01 43
N 58	06 58	07 49	08 39	22 47	24 12	00 12	01 36
56	06 52	07 40	08 26	22 46	24 09	00 09	01 29
54	06 47	07 32	08 14	22 45	24 05	00 05	01 24
52	06 42	07 25	08 04	22 44	24 02	00 02	01 19
50	06 38	07 18	07 56	22 43	23 59	25 14	01 14
45	06 27	07 04	07 37	22 41	23 53	25 04	01 04
N 40	06 18	06 51	07 21	22 39	23 48	24 56	00 56
35	06 09	06 41	07 08	22 38	23 44	24 49	00 49
30	06 02	06 31	06 57	22 37	23 40	24 43	00 43
20	05 46	06 14	06 37	22 34	23 34	24 32	00 32
N 10	05 32	05 57	06 20	22 33	23 28	24 23	00 23
0	05 16	05 42	06 04	22 31	23 23	24 15	00 15
S 10	04 58	05 25	05 48	22 29	23 17	24 06	00 06
20	04 37	05 06	05 30	22 27	23 12	23 57	24 45
30	04 10	04 43	05 10	22 25	23 06	23 47	24 31
35	03 53	04 29	04 58	22 24	23 02	23 41	24 23
40	03 32	04 12	04 44	22 23	22 58	23 35	24 14
45	03 04	03 51	04 27	22 21	22 53	23 27	24 04
S 50	02 24	03 24	04 07	22 20	22 48	23 18	23 51
52	02 02	03 11	03 57	22 19	22 45	23 14	23 46
54	01 31	02 55	03 46	22 18	22 43	23 09	23 39
56	00 26	02 36	03 33	22 17	22 40	23 04	23 32
58	////	02 12	03 19	22 16	22 36	22 58	23 24
S 60	////	01 39	03 01	22 15	22 32	22 52	23 15

Lat. °	Sunset h m	Twilight Civil h m	Twilight Naut. h m	Moonset 10 h m	Moonset 11 h m	Moonset 12 h m	Moonset 13 h m
N 72	▬	14 07	16 06	10 04	09 54	09 44	09 32
N 70	▬	14 46	16 22	10 04	10 01	09 58	09 55
68	13 21	15 13	16 35	10 04	10 06	10 09	10 14
66	14 10	15 34	16 46	10 04	10 11	10 18	10 29
64	14 41	15 50	16 56	10 04	10 15	10 26	10 41
62	15 04	16 04	17 04	10 04	10 18	10 33	10 51
60	15 22	16 16	17 11	10 04	10 21	10 39	11 00
N 58	15 37	16 26	17 17	10 04	10 23	10 44	11 08
56	15 50	16 36	17 23	10 04	10 26	10 49	11 15
54	16 02	16 44	17 29	10 04	10 28	10 53	11 21
52	16 11	16 51	17 34	10 04	10 30	10 57	11 27
50	16 20	16 58	17 38	10 04	10 32	11 00	11 32
45	16 39	17 12	17 49	10 04	10 35	11 08	11 43
N 40	16 54	17 24	17 58	10 04	10 39	11 14	11 52
35	17 07	17 35	18 06	10 04	10 41	11 20	12 00
30	17 19	17 45	18 14	10 04	10 44	11 25	12 07
20	17 38	18 02	18 29	10 04	10 48	11 33	12 19
N 10	17 55	18 18	18 44	10 04	10 52	11 40	12 30
0	18 11	18 34	19 00	10 03	10 55	11 47	12 40
S 10	18 28	18 51	19 17	10 03	10 59	11 54	12 50
20	18 45	19 09	19 38	10 03	11 03	12 02	13 01
30	19 06	19 33	20 05	10 03	11 07	12 10	13 13
35	19 17	19 47	20 22	10 03	11 09	12 15	13 20
40	19 31	20 03	20 43	10 03	11 12	12 21	13 28
45	19 48	20 24	21 11	10 03	11 15	12 27	13 37
S 50	20 08	20 51	21 50	10 02	11 19	12 35	13 49
52	20 18	21 04	22 12	10 02	11 21	12 38	13 54
54	20 29	21 19	22 41	10 02	11 23	12 42	14 00
56	20 41	21 38	23 38	10 02	11 25	12 47	14 07
58	20 56	22 01	////	10 02	11 27	12 52	14 14
S 60	21 13	22 34	////	10 02	11 30	12 57	14 22

Day	SUN Eqn. of Time 00h m s	SUN Eqn. of Time 12h m s	SUN Mer. Pass. h m	MOON Mer. Pass. Upper h m	MOON Mer. Pass. Lower h m	Age d	Phase
10	07 16	07 29	12 07	03 51	16 17	20	◑
11	07 41	07 53	12 08	04 43	17 09	21	
12	08 05	08 16	12 08	05 35	18 01	22	

JUNE 15, 16, 17 (SUN., MON., TUES.)

G.M.T. d h	ARIES G.H.A. ° ′	VENUS −3·5 G.H.A. ° ′	VENUS Dec. ° ′	MARS +0·4 G.H.A. ° ′	MARS Dec. ° ′	JUPITER −1·8 G.H.A. ° ′	JUPITER Dec. ° ′	SATURN +0·2 G.H.A. ° ′	SATURN Dec. ° ′
15 00	262 46·3	218 37·6	N14 38·5	256 51·6	N 0 11·4	62 07·6	S 7 13·8	1 01·0	S21 48·2
01	277 48·7	233 37·2	39·4	271 52·4	12·0	77 10·1	13·8	16 03·7	48·2
02	292 51·2	248 36·8	40·2	286 53·2	12·7	92 12·6	13·8	31 06·4	48·2
03	307 53·7	263 36·4	·· 41·1	301 54·1	·· 13·4	107 15·1	·· 13·8	46 09·0	·· 48·2
04	322 56·1	278 36·0	42·0	316 54·9	14·1	122 17·6	13·8	61 11·7	48·2
05	337 58·6	293 35·6	42·8	331 55·8	14·7	137 20·1	13·8	76 14·3	48·2
06	353 01·0	308 35·1	N14 43·7	346 56·6	N 0 15·4	152 22·6	S 7 13·8	91 17·0	S21 48·2
07	8 03·5	323 34·7	44·5	1 57·4	16·1	167 25·1	13·8	106 19·7	48·2
08	23 06·0	338 34·3	45·4	16 58·3	16·8	182 27·6	13·8	121 22·3	48·2
09	38 08·4	353 33·9	·· 46·3	31 59·1	·· 17·4	197 30·1	·· 13·8	136 25·0	·· 48·2
10	53 10·9	8 33·5	47·1	46 59·9	18·1	212 32·6	13·8	151 27·7	48·1
11	68 13·4	23 33·1	48·0	62 00·8	18·8	227 35·1	13·8	166 30·3	48·1
SUNDAY 12	83 15·8	38 32·7	N14 48·8	77 01·6	N 0 19·5	242 37·6	S 7 13·8	181 33·0	S21 48·1
13	98 18·3	53 32·3	49·7	92 02·4	20·1	257 40·1	13·8	196 35·6	48·1
14	113 20·8	68 31·8	50·5	107 03·3	20·8	272 42·6	13·8	211 38·3	48·1
15	128 23·2	83 31·4	·· 51·4	122 04·1	·· 21·5	287 45·1	·· 13·8	226 41·0	·· 48·1
16	143 25·7	98 31·0	52·3	137 04·9	22·2	302 47·5	13·8	241 43·6	48·1
17	158 28·1	113 30·6	53·1	152 05·8	22·8	317 50·0	13·8	256 46·3	48·1
18	173 30·6	128 30·2	N14 54·0	167 06·6	N 0 23·5	332 52·5	S 7 13·8	271 49·0	S21 48·1
19	188 33·1	143 29·8	54·8	182 07·4	24·2	347 55·0	13·8	286 51·6	48·1
20	203 35·5	158 29·3	55·7	197 08·3	24·9	2 57·5	13·8	301 54·3	48·1
21	218 38·0	173 28·9	·· 56·5	212 09·1	·· 25·5	18 00·0	·· 13·8	316 56·9	·· 48·1
22	233 40·5	188 28·5	57·4	227 09·9	26·2	33 02·5	13·8	331 59·6	48·1
23	248 42·9	203 28·1	58·2	242 10·8	26·9	48 05·0	13·8	347 02·3	48·1
16 00	263 45·4	218 27·7	N14 59·1	257 11·6	N 0 27·6	63 07·5	S 7 13·8	2 04·9	S21 48·0
01	278 47·9	233 27·2	14 59·9	272 12·5	28·2	78 10·0	13·8	17 07·6	48·0
02	293 50·3	248 26·8	15 00·8	287 13·3	28·9	93 12·5	13·8	32 10·3	48·0
03	308 52·8	263 26·4	·· 01·6	302 14·1	·· 29·6	108 14·9	·· 13·8	47 12·9	·· 48·0
04	323 55·3	278 26·0	02·5	317 15·0	30·3	123 17·4	13·8	62 15·6	48·0
05	338 57·7	293 25·6	03·3	332 15·8	30·9	138 19·9	13·8	77 18·2	48·0
06	354 00·2	308 25·1	N15 04·2	347 16·6	N 0 31·6	153 22·4	S 7 13·8	92 20·9	S21 48·0
07	9 02·6	323 24·7	05·0	2 17·5	32·3	168 24·9	13·8	107 23·6	48·0
08	24 05·1	338 24·3	05·9	17 18·3	33·0	183 27·4	13·8	122 26·2	48·0
09	39 07·6	353 23·9	·· 06·7	32 19·1	·· 33·6	198 29·9	·· 13·8	137 28·9	·· 48·0
10	54 10·0	8 23·4	07·6	47 20·0	34·3	213 32·4	13·8	152 31·6	48·0
11	69 12·5	23 23·0	08·4	62 20·8	35·0	228 34·8	13·8	167 34·2	48·0
MONDAY 12	84 15·0	38 22·6	N15 09·3	77 21·7	N 0 35·7	243 37·3	S 7 13·8	182 36·9	S21 48·0
13	99 17·4	53 22·2	10·1	92 22·5	36·3	258 39·8	13·8	197 39·5	48·0
14	114 19·9	68 21·7	11·0	107 23·3	37·0	273 42·3	13·8	212 42·2	47·9
15	129 22·4	83 21·3	·· 11·8	122 24·2	·· 37·7	288 44·8	·· 13·8	227 44·9	·· 47·9
16	144 24·8	98 20·9	12·6	137 25·0	38·3	303 47·3	13·8	242 47·5	47·9
17	159 27·3	113 20·5	13·5	152 25·8	39·0	318 49·8	13·8	257 50·2	47·9
18	174 29·8	128 20·0	N15 14·3	167 26·7	N 0 39·7	333 52·2	S 7 13·8	272 52·9	S21 47·9
19	189 32·2	143 19·6	15·2	182 27·5	40·4	348 54·7	13·8	287 55·5	47·9
20	204 34·7	158 19·2	16·0	197 28·4	41·0	3 57·2	13·8	302 58·2	47·9
21	219 37·1	173 18·7	·· 16·9	212 29·2	·· 41·7	18 59·7	·· 13·8	318 00·8	·· 47·9
22	234 39·6	188 18·3	17·7	227 30·0	42·4	34 02·2	13·8	333 03·5	47·9
23	249 42·1	203 17·9	18·5	242 30·9	43·1	49 04·7	13·9	348 06·2	47·9
17 00	264 44·5	218 17·5	N15 19·4	257 31·7	N 0 43·7	64 07·2	S 7 13·9	3 08·8	S21 47·9
01	279 47·0	233 17·0	20·2	272 32·6	44·4	79 09·6	13·9	18 11·5	47·9
02	294 49·5	248 16·6	21·1	287 33·4	45·1	94 12·1	13·9	33 14·2	47·9
03	309 51·9	263 16·2	·· 21·9	302 34·2	·· 45·7	109 14·6	·· 13·9	48 16·8	·· 47·9
04	324 54·4	278 15·7	22·7	317 35·1	46·4	124 17·1	13·9	63 19·5	47·8
05	339 56·9	293 15·3	23·6	332 35·9	47·1	139 19·6	13·9	78 22·1	47·8
06	354 59·3	308 14·9	N15 24·4	347 36·8	N 0 47·8	154 22·0	S 7 13·9	93 24·8	S21 47·8
07	10 01·8	323 14·4	25·2	2 37·6	48·4	169 24·5	13·9	108 27·5	47·8
08	25 04·3	338 14·0	26·1	17 38·4	49·1	184 27·0	13·9	123 30·1	47·8
09	40 06·7	353 13·6	·· 26·9	32 39·3	·· 49·8	199 29·5	·· 13·9	138 32·8	·· 47·8
10	55 09·2	8 13·1	27·7	47 40·1	50·4	214 32·0	13·9	153 35·5	47·8
11	70 11·6	23 12·7	28·6	62 41·0	51·1	229 34·4	13·9	168 38·1	47·8
TUESDAY 12	85 14·1	38 12·2	N15 29·4	77 41·8	N 0 51·8	244 36·9	S 7 13·9	183 40·8	S21 47·8
13	100 16·6	53 11·8	30·2	92 42·6	52·5	259 39·4	13·9	198 43·4	47·8
14	115 19·0	68 11·4	31·1	107 43·5	53·1	274 41·9	13·9	213 46·1	47·8
15	130 21·5	83 10·9	·· 31·9	122 44·3	·· 53·8	289 44·4	·· 13·9	228 48·8	·· 47·8
16	145 24·0	98 10·5	32·7	137 45·2	54·5	304 46·8	13·9	243 51·4	47·8
17	160 26·4	113 10·1	33·6	152 46·0	55·1	319 49·3	14·0	258 54·1	47·8
18	175 28·9	128 09·6	N15 34·4	167 46·8	N 0 55·8	334 51·8	S 7 14·0	273 56·8	S21 47·7
19	190 31·4	143 09·2	35·2	182 47·7	56·5	349 54·3	14·0	288 59·4	47·7
20	205 33·8	158 08·7	36·0	197 48·5	57·2	4 56·8	14·0	304 02·1	47·7
21	220 36·3	173 08·3	·· 36·9	212 49·4	·· 57·8	19 59·2	·· 14·0	319 04·7	·· 47·7
22	235 38·8	188 07·8	37·7	227 50·2	58·5	35 01·7	14·0	334 07·4	47·7
23	250 41·2	203 07·4	38·5	242 51·0	59·2	50 04·2	14·0	349 10·1	47·7
Mer. Pass.	6h 23m·9	v −0·4	d 0·8	v 0·8	d 0·7	v 2·5	d 0·0	v 2·7	d 0·0

STARS

Name	S.H.A. ° ′	Dec. ° ′
Acamar	315 50·1	S 40 28·1
Achernar	335 57·9	S 57 26·6
Acrux	173 55·3	S 62 52·5
Adhara	255 45·4	S 28 55·1
Aldebaran	291 37·2	N 16 25·5
Alioth	166 56·8	N 56 11·3
Alkaid	153 31·2	N 49 31·4
Al Na'ir	28 35·4	S 47 09·4
Alnilam	276 28·7	S 1 13·8
Alphard	218 36·9	S 8 28·8
Alphecca	126 45·7	N 26 51·4
Alpheratz	358 26·3	N 28 51·5
Altair	62 48·3	N 8 45·6
Ankaa	353 56·6	S 42 31·6
Antares	113 16·6	S 26 20·4
Arcturus	146 33·2	N 19 24·0
Atria	108 55·0	S 68 57·2
Avior	234 35·4	S 59 22·9
Bellatrix	279 16·7	N 6 18·7
Betelgeuse	271 46·4	N 7 23·9
Canopus	264 15·0	S 52 40·5
Capella	281 36·0	N 45 57·3
Deneb	49 59·3	N 45 07·8
Denebola	183 15·8	N 14 48·3
Diphda	349 37·5	S 18 12·7
Dubhe	194 42·5	N 61 58·7
Elnath	279 05·3	N 28 34·3
Eltanin	91 04·8	N 51 29·7
Enif	34 27·5	N 9 41·1
Fomalhaut	16 09·5	S 29 50·3
Gacrux	172 46·8	S 56 53·2
Gienah	176 34·8	S 17 18·8
Hadar	149 46·1	S 60 10·6
Hamal	328 47·7	N 23 15·9
Kaus Aust.	84 38·2	S 34 24·2
Kochab	137 17·4	N 74 19·8
Markab	14 19·5	N 14 58·9
Menkar	314 58·6	N 3 55·6
Menkent	148 56·1	S 36 10·1
Miaplacidus	221 49·0	S 69 33·1
Mirfak	309 40·0	N 49 42·7
Nunki	76 49·2	S 26 20·8
Peacock	54 23·9	S 56 51·9
Pollux	244 18·6	N 28 07·6
Procyon	245 43·3	N 5 19·8
Rasalhague	96 44·5	N 12 35·5
Regulus	208 27·6	N 12 10·2
Rigel	281 52·1	S 8 15·0
Rigil Kent.	140 47·7	S 60 40·1
Sabik	102 59·6	S 15 40·4
Schedar	350 27·8	N 56 18·3
Shaula	97 17·6	S 37 04·4
Sirius	259 10·5	S 16 39·7
Spica	159 14·7	S 10 56·8
Suhail	223 23·1	S 43 16·2
Vega	81 06·5	N 38 44·7
Zuben'ubi	137 50·9	S 15 52·2

	S.H.A. ° ′	Mer. Pass. h m
Venus	314 42·3	9 26
Mars	353 26·2	6 51
Jupiter	159 22·1	19 44
Saturn	98 19·5	23 47

JUNE 15, 16, 17 (SUN., MON., TUES.)

G.M.T.	SUN G.H.A.	SUN Dec.	MOON G.H.A.	v	MOON Dec.	d	H.P.
d h	° ′	° ′	° ′	′	° ′	′	′
15 00	179 57·7	N23 16·7	208 24·7	11·5	N17 06·1	4·7	55·2
01	194 57·5	16·8	222 55·2	11·4	17 10·8	4·6	55·2
02	209 57·4	16·9	237 25·6	11·4	17 15·4	4·5	55·3
03	224 57·3	·· 17·0	251 56·0	11·3	17 19·9	4·5	55·3
04	239 57·1	17·1	266 26·3	11·2	17 24·4	4·4	55·3
05	254 57·0	17·2	280 56·5	11·3	17 28·8	4·3	55·3
06	269 56·9	N23 17·4	295 26·8	11·1	N17 33·1	4·2	55·3
07	284 56·7	17·5	309 56·9	11·1	17 37·3	4·1	55·4
08	299 56·6	17·6	324 27·0	11·1	17 41·4	4·0	55·4
S 09	314 56·5	·· 17·7	338 57·1	11·0	17 45·4	4·0	55·4
U 10	329 56·3	17·8	353 27·1	11·0	17 49·4	3·8	55·4
N 11	344 56·2	17·9	7 57·1	10·9	17 53·2	3·8	55·4
D 12	359 56·1	N23 18·0	22 27·0	10·9	N17 57·0	3·7	55·5
A 13	14 55·9	18·1	36 56·9	10·8	18 00·7	3·6	55·5
Y 14	29 55·8	18·2	51 26·7	10·7	18 04·3	3·5	55·5
15	44 55·7	·· 18·4	65 56·4	10·8	18 07·8	3·5	55·5
16	59 55·5	18·5	80 26·2	10·6	18 11·3	3·3	55·5
17	74 55·4	18·6	94 55·8	10·6	18 14·6	3·2	55·6
18	89 55·3	N23 18·7	109 25·4	10·6	N18 17·8	3·2	55·6
19	104 55·1	18·8	123 55·0	10·6	18 21·0	3·1	55·6
20	119 55·0	18·9	138 24·6	10·4	18 24·1	2·9	55·6
21	134 54·9	·· 19·0	152 54·0	10·5	18 27·0	2·9	55·7
22	149 54·7	19·1	167 23·5	10·4	18 29·9	2·8	55·7
23	164 54·6	19·2	181 52·9	10·3	18 32·7	2·7	55·7
16 00	179 54·5	N23 19·3	196 22·2	10·3	N18 35·4	2·6	55·7
01	194 54·3	19·4	210 51·5	10·3	18 38·0	2·5	55·7
02	209 54·2	19·5	225 20·8	10·2	18 40·5	2·4	55·8
03	224 54·1	·· 19·6	239 50·0	10·2	18 42·9	2·3	55·8
04	239 53·9	19·7	254 19·2	10·1	18 45·2	2·3	55·8
05	254 53·8	19·8	268 48·3	10·1	18 47·5	2·1	55·8
06	269 53·7	N23 19·9	283 17·4	10·0	N18 49·6	2·0	55·9
07	284 53·5	20·0	297 46·4	10·0	18 51·6	2·0	55·9
08	299 53·4	20·1	312 15·4	10·0	18 53·6	1·8	55·9
M 09	314 53·2	·· 20·2	326 44·4	9·9	18 55·4	1·7	55·9
O 10	329 53·1	20·3	341 13·3	9·9	18 57·1	1·7	55·9
N 11	344 53·0	20·4	355 42·2	9·9	18 58·8	1·5	56·0
D 12	359 52·8	N23 20·5	10 11·1	9·8	N19 00·3	1·4	56·0
A 13	14 52·7	20·6	24 39·9	9·7	19 01·7	1·4	56·0
Y 14	29 52·6	20·7	39 08·6	9·8	19 03·1	1·2	56·0
15	44 52·4	·· 20·7	53 37·4	9·7	19 04·3	1·1	56·1
16	59 52·3	20·8	68 06·1	9·7	19 05·4	1·0	56·1
17	74 52·2	20·9	82 34·8	9·6	19 06·4	1·0	56·1
18	89 52·0	N23 21·0	97 03·4	9·6	N19 07·4	0·8	56·1
19	104 51·9	21·1	111 32·0	9·5	19 08·2	0·7	56·1
20	119 51·8	21·2	126 00·5	9·6	19 08·9	0·7	56·2
21	134 51·6	·· 21·3	140 29·1	9·5	19 09·6	0·5	56·2
22	149 51·5	21·4	154 57·6	9·4	19 10·1	0·4	56·2
23	164 51·4	21·4	169 26·0	9·5	19 10·5	0·3	56·2
17 00	179 51·2	N23 21·5	183 54·5	9·4	N19 10·8	0·2	56·3
01	194 51·1	21·6	198 22·9	9·4	19 11·0	0·1	56·3
02	209 51·0	21·7	212 51·3	9·3	19 11·1	0·0	56·3
03	224 50·8	·· 21·8	227 19·6	9·3	19 11·1	0·1	56·3
04	239 50·7	21·9	241 47·9	9·3	19 11·0	0·2	56·3
05	254 50·5	21·9	256 16·2	9·3	19 10·8	0·3	56·4
06	269 50·4	N23 22·0	270 44·5	9·3	N19 10·5	0·4	56·4
07	284 50·3	22·1	285 12·8	9·2	19 10·1	0·6	56·4
T 08	299 50·1	22·2	299 41·0	9·2	19 09·5	0·6	56·4
U 09	314 50·0	·· 22·3	314 09·2	9·2	19 08·9	0·7	56·5
E 10	329 49·9	22·3	328 37·4	9·1	19 08·2	0·9	56·5
S 11	344 49·7	22·4	343 05·5	9·2	19 07·3	0·9	56·5
D 12	359 49·6	N23 22·5	357 33·7	9·1	N19 06·4	1·1	56·5
A 13	14 49·5	22·6	12 01·8	9·1	19 05·3	1·1	56·6
Y 14	29 49·3	22·6	26 29·9	9·0	19 04·2	1·3	56·6
15	44 49·2	·· 22·7	40 57·9	9·1	19 02·9	1·4	56·6
16	59 49·1	22·8	55 26·0	9·0	19 01·5	1·5	56·6
17	74 48·9	22·9	69 54·0	9·1	19 00·0	1·6	56·6
18	89 48·8	N23 22·9	84 22·1	9·0	N18 58·4	1·7	56·7
19	104 48·6	23·0	98 50·1	9·0	18 56·7	1·8	56·7
20	119 48·5	23·1	113 18·1	8·9	18 54·9	1·9	56·7
21	134 48·4	·· 23·2	127 46·0	9·0	18 53·0	2·0	56·7
22	149 48·2	23·2	142 14·0	9·0	18 51·0	2·1	56·8
23	164 48·1	23·3	156 42·0	8·9	18 48·9	2·3	56·8
	S.D. 15·8	*d* 0·1	S.D.		15·1	15·3	15·4

Lat.	Twilight Naut.	Twilight Civil	Sun-rise	Moonrise 15	Moonrise 16	Moonrise 17	Moonrise 18
°	h m	h m	h m	h m	h m	h m	h m
N 72	□	□	□	□	□	□	□
N 70	□	□	□	00 18	00 28	00 59	02 11
68	□	□	□	00 50	01 13	01 53	02 57
66	□	□	□	01 14	01 42	02 26	03 28
64	////	////	01 33	01 33	02 05	02 50	03 50
62	////	////	02 10	01 48	02 23	03 09	04 09
60	////	00 52	02 36	02 01	02 37	03 25	04 23
N 58	////	01 41	02 56	02 12	02 50	03 38	04 36
56	////	02 11	03 13	02 22	03 01	03 49	04 47
54	00 48	02 33	03 27	02 31	03 11	03 59	04 57
52	01 33	02 51	03 39	02 38	03 19	04 08	05 05
50	02 00	03 06	03 50	02 45	03 27	04 16	05 13
45	02 46	03 35	04 13	03 00	03 43	04 33	05 29
N 40	03 16	03 58	04 30	03 13	03 57	04 47	05 43
35	03 39	04 16	04 45	03 23	04 09	04 59	05 54
30	03 58	04 31	04 58	03 32	04 19	05 10	06 04
20	04 27	04 56	05 20	03 48	04 36	05 27	06 22
N 10	04 49	05 16	05 39	04 02	04 51	05 43	06 36
0	05 08	05 34	05 57	04 15	05 05	05 58	06 51
S 10	05 25	05 51	06 14	04 28	05 20	06 12	07 05
20	05 41	06 09	06 33	04 43	05 35	06 28	07 19
30	05 58	06 27	06 54	04 59	05 53	06 46	07 37
35	06 06	06 38	07 06	05 08	06 03	06 56	07 47
40	06 16	06 50	07 20	05 19	06 14	07 08	07 58
45	06 26	07 03	07 37	05 32	06 28	07 22	08 11
S 50	06 38	07 19	07 58	05 47	06 45	07 39	08 28
52	06 43	07 27	08 08	05 55	06 53	07 47	08 35
54	06 48	07 35	08 19	06 03	07 02	07 56	08 44
56	06 55	07 44	08 31	06 12	07 12	08 07	08 53
58	07 01	07 54	08 46	06 22	07 23	08 18	09 04
S 60	07 09	08 06	09 03	06 34	07 37	08 32	09 17

Lat.	Sun-set	Twilight Civil	Twilight Naut.	Moonset 15	Moonset 16	Moonset 17	Moonset 18
°	h m	h m	h m	h m	h m	h m	h m
N 72	□	□	□	□	□	□	□
N 70	□	□	□	21 12	22 28	23 07	23 20
68	□	□	□	20 28	21 34	22 20	22 46
66	□	□	□	19 58	21 01	21 49	22 21
64	22 29	////	////	19 36	20 37	21 26	22 02
62	21 51	////	////	19 18	20 18	21 08	21 46
60	21 25	23 10	////	19 04	20 03	20 53	21 33
N 58	21 05	22 20	////	18 51	19 49	20 40	21 22
56	20 48	21 51	////	18 40	19 38	20 29	21 12
54	20 34	21 29	23 14	18 31	19 28	20 19	21 03
52	20 22	21 11	22 29	18 22	19 19	20 10	20 55
50	20 11	20 56	22 01	18 15	19 11	20 02	20 48
45	19 49	20 26	21 16	17 59	18 54	19 46	20 33
N 40	19 31	20 03	20 45	17 46	18 40	19 32	20 20
35	19 16	19 45	20 22	17 34	18 28	19 20	20 09
30	19 03	19 30	20 03	17 25	18 18	19 10	20 00
20	18 41	19 05	19 34	17 08	18 00	18 53	19 44
N 10	18 22	18 45	19 12	16 53	17 45	18 37	19 29
0	18 04	18 27	18 53	16 39	17 30	18 23	19 16
S 10	17 47	18 10	18 36	16 25	17 16	18 08	19 03
20	17 28	17 52	18 20	16 11	17 00	17 53	18 48
30	17 07	17 34	18 03	15 54	16 42	17 35	18 32
35	16 55	17 23	17 55	15 44	16 32	17 25	18 22
40	16 41	17 11	17 45	15 33	16 20	17 13	18 11
45	16 24	16 58	17 35	15 20	16 06	16 59	17 58
S 50	16 03	16 42	17 23	15 04	15 49	16 42	17 42
52	15 53	16 34	17 18	14 56	15 41	16 34	17 35
54	15 42	16 26	17 12	14 48	15 32	16 25	17 26
56	15 30	16 17	17 06	14 39	15 22	16 15	17 17
58	15 15	16 07	16 59	14 28	15 10	16 03	17 07
S 60	14 58	15 55	16 52	14 16	14 57	15 50	16 54

Day	SUN Eqn. of Time 00h	SUN Eqn. of Time 12h	SUN Mer. Pass.	MOON Mer. Pass. Upper	MOON Mer. Pass. Lower	Age	Phase
	m s	m s	h m	h m	h m	d	
15	00 09	00 15	12 00	10 27	22 52	28	
16	00 22	00 28	12 00	11 18	23 44	29	●
17	00 35	00 41	12 01	12 10	24 37	00	

JUNE 18, 19, 20 (WED., THURS., FRI.)

G.M.T.		ARIES	VENUS −3·4		MARS +0·4		JUPITER −1·8		SATURN +0·2	
d	h	G.H.A. ° ′	G.H.A. ° ′	Dec. ° ′	G.H.A. ° ′	Dec. ° ′	G.H.A. ° ′	Dec. ° ′	G.H.A. ° ′	Dec. °
18	00	265 43·7	218 07·0	N15 39·4	257 51·9	N 0 59·8	65 06·7	S 7 14·0	4 12·7	S21 47·7
	01	280 46·1	233 06·5	40·2	272 52·7	1 00·5	80 09·1	14·0	19 15·4	47·7
	02	295 48·6	248 06·1	41·0	287 53·6	01·2	95 11·6	14·0	34 18·1	47·7
	03	310 51·1	263 05·6	· · 41·8	302 54·4	· · 01·8	110 14·1	· · 14·0	49 20·7	· · 47·7
	04	325 53·5	278 05·2	42·7	317 55·3	02·5	125 16·6	14·0	64 23·4	47·7
	05	340 56·0	293 04·7	43·5	332 56·1	03·2	140 19·0	14·0	79 26·0	47·7
	06	355 58·5	308 04·3	N15 44·3	347 56·9	N 1 03·9	155 21·5	S 7 14·1	94 28·7	S21 47·7
W	07	11 00·9	323 03·9	45·1	2 57·8	04·5	170 24·0	14·1	109 31·4	47·7
E	08	26 03·4	338 03·4	45·9	17 58·6	05·2	185 26·5	14·1	124 34·0	47·6
D	09	41 05·9	353 03·0	· · 46·8	32 59·5	· · 05·9	200 28·9	· · 14·1	139 36·7	· · 47·6
N	10	56 08·3	8 02·5	47·6	48 00·3	06·5	215 31·4	14·1	154 39·4	47·6
E	11	71 10·8	23 02·1	48·4	63 01·2	07·2	230 33·9	14·1	169 42·0	47·6
S	12	86 13·2	38 01·6	N15 49·2	78 02·0	N 1 07·9	245 36·4	S 7 14·1	184 44·7	S21 47·6
D	13	101 15·7	53 01·2	50·0	93 02·8	08·5	260 38·8	14·1	199 47·3	47·6
A	14	116 18·2	68 00·7	50·9	108 03·7	09·2	275 41·3	14·1	214 50·0	47·6
Y	15	131 20·6	83 00·3	· · 51·7	123 04·5	· · 09·9	290 43·8	· · 14·1	229 52·7	· · 47·6
	16	146 23·1	97 59·8	52·5	138 05·4	10·5	305 46·2	14·1	244 55·3	47·6
	17	161 25·6	112 59·4	53·3	153 06·2	11·2	320 48·7	14·1	259 58·0	47·6
	18	176 28·0	127 58·9	N15 54·1	168 07·1	N 1 11·9	335 51·2	S 7 14·2	275 00·6	S21 47·6
	19	191 30·5	142 58·5	54·9	183 07·9	12·6	350 53·7	14·2	290 03·3	47·6
	20	206 33·0	157 58·0	55·8	198 08·8	13·2	5 56·1	14·2	305 06·0	47·6
	21	221 35·4	172 57·6	· · 56·6	213 09·6	· · 13·9	20 58·6	· · 14·2	320 08·6	· · 47·6
	22	236 37·9	187 57·1	57·4	228 10·4	14·6	36 01·1	14·2	335 11·3	47·5
	23	251 40·4	202 56·7	58·2	243 11·3	15·2	51 03·5	14·2	350 14·0	47·5
19	00	266 42·8	217 56·2	N15 59·0	258 12·1	N 1 15·9	66 06·0	S 7 14·2	5 16·6	S21 47·5
	01	281 45·3	232 55·7	15 59·8	273 13·0	16·6	81 08·5	14·2	20 19·3	47·5
	02	296 47·7	247 55·3	16 00·6	288 13·8	17·2	96 10·9	14·2	35 21·9	47·5
	03	311 50·2	262 54·8	· · 01·4	303 14·7	· · 17·9	111 13·4	· · 14·2	50 24·6	· · 47·5
	04	326 52·7	277 54·4	02·2	318 15·5	18·6	126 15·9	14·3	65 27·3	47·5
	05	341 55·1	292 53·9	03·1	333 16·4	19·2	141 18·3	14·3	80 29·9	47·5
	06	356 57·6	307 53·5	N16 03·9	348 17·2	N 1 19·9	156 20·8	S 7 14·3	95 32·6	S21 47·5
	07	12 00·1	322 53·0	04·7	3 18·0	20·6	171 23·3	14·3	110 35·2	47·5
T	08	27 02·5	337 52·6	05·5	18 18·9	21·2	186 25·7	14·3	125 37·9	47·5
H	09	42 05·0	352 52·1	· · 06·3	33 19·7	· · 21·9	201 28·2	· · 14·3	140 40·6	· · 47·5
U	10	57 07·5	7 51·6	07·1	48 20·6	22·6	216 30·7	14·3	155 43·2	47·5
R	11	72 09·9	22 51·2	07·9	63 21·4	23·2	231 33·1	14·3	170 45·9	47·5
S	12	87 12·4	37 50·7	N16 08·7	78 22·3	N 1 23·9	246 35·6	S 7 14·3	185 48·6	S21 47·4
D	13	102 14·9	52 50·3	09·5	93 23·1	24·6	261 38·1	14·4	200 51·2	47·4
A	14	117 17·3	67 49·8	10·3	108 24·0	25·2	276 40·5	14·4	215 53·9	47·4
Y	15	132 19·8	82 49·3	· · 11·1	123 24·8	· · 25·9	291 43·0	· · 14·4	230 56·5	· · 47·4
	16	147 22·2	97 48·9	11·9	138 25·7	26·6	306 45·5	14·4	245 59·2	47·4
	17	162 24·7	112 48·4	12·7	153 26·5	27·2	321 47·9	14·4	261 01·9	47·4
	18	177 27·2	127 47·9	N16 13·5	168 27·3	N 1 27·9	336 50·4	S 7 14·4	276 04·5	S21 47·4
	19	192 29·6	142 47·5	14·3	183 28·2	28·6	351 52·9	14·4	291 07·2	47·4
	20	207 32·1	157 47·0	15·1	198 29·0	29·2	6 55·3	14·4	306 09·8	47·4
	21	222 34·6	172 46·6	· · 15·9	213 29·9	· · 29·9	21 57·8	· · 14·5	321 12·5	· · 47·4
	22	237 37·0	187 46·1	16·7	228 30·7	30·6	37 00·3	14·5	336 15·2	47·4
	23	252 39·5	202 45·6	17·5	243 31·6	31·2	52 02·7	14·5	351 17·8	47·4
20	00	267 42·0	217 45·2	N16 18·3	258 32·4	N 1 31·9	67 05·2	S 7 14·5	6 20·5	S21 47·4
	01	282 44·4	232 44·7	19·1	273 33·3	32·6	82 07·6	14·5	21 23·1	47·4
	02	297 46·9	247 44·2	19·9	288 34·1	33·2	97 10·1	14·5	36 25·8	47·3
	03	312 49·4	262 43·8	· · 20·7	303 35·0	· · 33·9	112 12·6	· · 14·5	51 28·5	· · 47·3
	04	327 51·8	277 43·3	21·5	318 35·8	34·6	127 15·0	14·6	66 31·1	47·3
	05	342 54·3	292 42·8	22·3	333 36·7	35·2	142 17·5	14·6	81 33·8	47·3
	06	357 56·7	307 42·3	N16 23·1	348 37·5	N 1 35·9	157 19·9	S 7 14·6	96 36·4	S21 47·3
	07	12 59·2	322 41·9	23·9	3 38·4	36·6	172 22·4	14·6	111 39·1	47·3
	08	28 01·7	337 41·4	24·7	18 39·2	37·2	187 24·9	14·6	126 41·8	47·3
F	09	43 04·1	352 40·9	· · 25·5	33 40·1	· · 37·9	202 27·3	· · 14·6	141 44·4	· · 47·3
R	10	58 06·6	7 40·5	26·3	48 40·9	38·5	217 29·8	14·6	156 47·1	47·3
I	11	73 09·1	22 40·0	27·0	63 41·8	39·2	232 32·2	14·7	171 49·8	47·3
D	12	88 11·5	37 39·5	N16 27·8	78 42·6	N 1 39·9	247 34·7	S 7 14·7	186 52·4	S21 47·3
A	13	103 14·0	52 39·1	28·6	93 43·5	40·5	262 37·2	14·7	201 55·1	47·3
Y	14	118 16·5	67 38·6	29·4	108 44·3	41·2	277 39·6	14·7	216 57·7	47·3
	15	133 18·9	82 38·1	· · 30·2	123 45·2	· · 41·9	292 42·1	· · 14·7	232 00·4	· · 47·3
	16	148 21·4	97 37·6	31·0	138 46·0	42·5	307 44·5	14·7	247 03·1	47·2
	17	163 23·8	112 37·2	31·8	153 46·9	43·2	322 47·0	14·7	262 05·7	47·2
	18	178 26·3	127 36·7	N16 32·6	168 47·7	N 1 43·9	337 49·4	S 7 14·8	277 08·4	S21 47·2
	19	193 28·8	142 36·2	33·4	183 48·6	44·5	352 51·9	14·8	292 11·0	47·2
	20	208 31·2	157 35·7	34·1	198 49·4	45·2	7 54·4	14·8	307 13·7	47·2
	21	223 33·7	172 35·3	· · 34·9	213 50·3	· · 45·9	22 56·8	· · 14·8	322 16·4	· · 47·2
	22	238 36·2	187 34·8	35·7	228 51·1	46·5	37 59·3	14·8	337 19·0	47·2
	23	253 38·6	202 34·3	36·5	243 52·0	47·2	53 01·7	14·8	352 21·7	47·2
Mer. Pass.		h m 6 12·1	v −0·5	d 0·8	v 0·8	d 0·7	v 2·5	d 0·0	v 2·7	d 0·0

STARS		
Name	S.H.A. ° ′	Dec. ° ′
Acamar	315 50·0	S 40 28·1
Achernar	335 57·9	S 57 26·6
Acrux	173 55·3	S 62 52·5
Adhara	255 45·4	S 28 55·1
Aldebaran	291 37·2	N 16 25·5
Alioth	166 56·8	N 56 11·3
Alkaid	153 31·2	N 49 31·4
Al Na'ir	28 35·4	S 47 09·4
Alnilam	276 28·7	S 1 13·8
Alphard	218 36·9	S 8 28·8
Alphecca	126 45·7	N 26 51·4
Alpheratz	358 26·3	N 28 51·6
Altair	62 48·2	N 8 45·6
Ankaa	353 56·6	S 42 31·6
Antares	113 16·6	S 26 20·4
Arcturus	146 33·2	N 19 24·0
Atria	108 55·0	S 68 57·2
Avior	234 35·4	S 59 22·8
Bellatrix	279 16·7	N 6 18·7
Betelgeuse	271 46·4	N 7 23·9
Canopus	264 15·0	S 52 40·5
Capella	281 36·0	N 45 57·3
Deneb	49 59·3	N 45 07·8
Denebola	183 15·8	N 14 48·3
Diphda	349 37·4	S 18 12·7
Dubhe	194 42·6	N 61 58·7
Elnath	279 05·3	N 28 34·3
Eltanin	91 04·8	N 51 29·7
Enif	34 27·5	N 9 41·1
Fomalhaut	16 09·5	S 29 50·3
Gacrux	172 46·8	S 56 53·2
Gienah	176 34·8	S 17 18·8
Hadar	149 46·1	S 60 10·6
Hamal	328 47·6	N 23 15·9
Kaus Aust.	84 38·2	S 34 24·2
Kochab	137 17·4	N 74 19·8
Markab	14 19·4	N 14 58·9
Menkar	314 58·5	N 3 55·6
Menkent	148 56·1	S 36 10·1
Miaplacidus	221 49·0	S 69 33·1
Mirfak	309 40·0	N 49 42·7
Nunki	76 49·2	S 26 20·8
Peacock	54 23·9	S 56 51·9
Pollux	244 18·6	N 28 07·6
Procyon	245 43·3	N 5 19·8
Rasalhague	96 44·5	N 12 35·5
Regulus	208 27·6	N 12 10·2
Rigel	281 52·1	S 8 15·0
Rigil Kent.	140 47·7	S 60 40·1
Sabik	102 59·6	S 15 40·4
Schedar	350 27·7	N 56 18·3
Shaula	97 17·6	S 37 04·4
Sirius	259 10·5	S 16 39·6
Spica	159 14·7	S 10 56·8
Suhail	223 23·1	S 43 16·2
Vega	81 06·5	N 38 44·7
Zuben'ubi	137 50·9	S 15 52·2

	S.H.A. ° ′	Mer. Pass. h m
Venus	311 13·4	9 28
Mars	351 29·3	6 47
Jupiter	159 23·2	19 32
Saturn	98 33·8	23 35

JUNE 18, 19, 20 (WED., THURS., FRI.)

G.M.T.		SUN G.H.A.	SUN Dec.	MOON G.H.A.	v	MOON Dec.	d	H.P.
d	h	° ′	° ′	° ′	′	° ′	′	
18	00	179 48·0	N23 23·4	171 09·9	8·9	N18 46·6	2·3	56·8
	01	194 47·8	23·4	185 37·8	8·9	18 44·3	2·5	56·8
	02	209 47·7	23·5	200 05·7	9·0	18 41·8	2·5	56·8
	03	224 47·6	· · 23·6	214 33·7	8·9	18 39·3	2·7	56·9
	04	239 47·4	23·6	229 01·6	8·8	18 36·6	2·7	56·9
	05	254 47·3	23·7	243 29·4	8·9	18 33·9	2·9	56·9
	06	269 47·1	N23 23·7	257 57·3	8·9	N18 31·0	3·0	56·9
W	07	284 47·0	23·8	272 25·2	8·9	18 28·0	3·1	57·0
E	08	299 46·9	23·9	286 53·1	8·9	18 24·9	3·2	57·0
D	09	314 46·7	· · 23·9	301 21·0	8·8	18 21·7	3·3	57·0
N	10	329 46·6	24·0	315 48·8	8·9	18 18·4	3·4	57·0
E	11	344 46·5	24·1	330 16·7	8·9	18 15·0	3·5	57·0
S	12	359 46·3	N23 24·1	344 44·6	8·8	N18 11·5	3·6	57·1
D	13	14 46·2	24·2	359 12·4	8·9	18 07·9	3·8	57·1
A	14	29 46·1	24·2	13 40·3	8·8	18 04·1	3·8	57·1
Y	15	44 45·9	· · 24·3	28 08·1	8·9	18 00·3	3·9	57·1
	16	59 45·8	24·3	42 36·0	8·9	17 56·4	4·1	57·2
	17	74 45·6	24·4	57 03·9	8·8	17 52·3	4·1	57·2
	18	89 45·5	N23 24·5	71 31·7	8·9	N17 48·2	4·2	57·2
	19	104 45·4	24·5	85 59·6	8·8	17 44·0	4·4	57·2
	20	119 45·2	24·6	100 27·4	8·9	17 39·6	4·4	57·2
	21	134 45·1	· · 24·6	114 55·3	8·9	17 35·2	4·6	57·3
	22	149 45·0	24·7	129 23·2	8·9	17 30·6	4·7	57·3
	23	164 44·8	24·7	143 51·1	8·9	17 25·9	4·7	57·3
19	00	179 44·7	N23 24·8	158 19·0	8·9	N17 21·2	4·9	57·3
	01	194 44·6	24·8	172 46·9	8·9	17 16·3	5·0	57·3
	02	209 44·4	24·9	187 14·8	8·9	17 11·3	5·0	57·4
	03	224 44·3	· · 24·9	201 42·7	8·9	17 06·3	5·2	57·4
	04	239 44·1	25·0	216 10·6	8·9	17 01·1	5·3	57·4
	05	254 44·0	25·0	230 38·5	9·0	16 55·8	5·3	57·4
	06	269 43·9	N23 25·1	245 06·5	8·9	N16 50·5	5·5	57·4
	07	284 43·7	25·1	259 34·4	9·0	16 45·0	5·6	57·5
T	08	299 43·6	25·1	274 02·4	8·9	16 39·4	5·6	57·5
H	09	314 43·5	· · 25·2	288 30·3	9·0	16 33·8	5·8	57·5
U	10	329 43·3	25·2	302 58·3	9·0	16 28·0	5·8	57·5
R	11	344 43·2	25·3	317 26·3	9·0	16 22·2	6·0	57·5
S	12	359 43·0	N23 25·3	331 54·3	9·0	N16 16·2	6·0	57·6
D	13	14 42·9	25·4	346 22·3	9·0	16 10·2	6·2	57·6
A	14	29 42·8	25·4	0 50·3	9·1	16 04·0	6·2	57·6
Y	15	44 42·6	· · 25·4	15 18·4	9·0	15 57·8	6·4	57·6
	16	59 42·5	25·5	29 46·4	9·1	15 51·4	6·4	57·7
	17	74 42·4	25·5	44 14·5	9·1	15 45·0	6·5	57·7
	18	89 42·2	N23 25·6	58 42·6	9·1	N15 38·5	6·6	57·7
	19	104 42·1	25·6	73 10·7	9·1	15 31·9	6·8	57·7
	20	119 42·0	25·6	87 38·8	9·2	15 25·1	6·8	57·7
	21	134 41·8	· · 25·7	102 07·0	9·1	15 18·3	6·9	57·7
	22	149 41·7	25·7	116 35·1	9·2	15 11·4	6·9	57·8
	23	164 41·5	25·7	131 03·3	9·2	15 04·5	7·1	57·8
20	00	179 41·4	N23 25·8	145 31·5	9·2	N14 57·4	7·2	57·8
	01	194 41·3	25·8	159 59·7	9·2	14 50·2	7·2	57·8
	02	209 41·1	25·8	174 27·9	9·2	14 43·0	7·4	57·8
	03	224 41·0	· · 25·9	188 56·1	9·3	14 35·6	7·4	57·9
	04	239 40·9	25·9	203 24·4	9·3	14 28·2	7·5	57·9
	05	254 40·7	25·9	217 52·7	9·3	14 20·7	7·6	57·9
	06	269 40·6	N23 25·9	232 21·0	9·3	N14 13·1	7·7	57·9
	07	284 40·5	26·0	246 49·3	9·3	14 05·4	7·8	57·9
	08	299 40·3	26·0	261 17·6	9·4	13 57·6	7·8	58·0
F	09	314 40·2	· · 26·0	275 46·0	9·3	13 49·8	7·9	58·0
R	10	329 40·0	26·1	290 14·3	9·4	13 41·9	8·1	58·0
I	11	344 39·9	26·1	304 42·7	9·4	13 33·8	8·1	58·0
D	12	359 39·8	N23 26·1	319 11·1	9·5	N13 25·7	8·1	58·0
A	13	14 39·6	26·1	333 39·6	9·4	13 17·6	8·3	58·0
Y	14	29 39·5	26·2	348 08·0	9·5	13 09·3	8·3	58·1
	15	44 39·4	· · 26·2	2 36·5	9·5	13 01·0	8·4	58·1
	16	59 39·2	26·2	17 05·0	9·5	12 52·6	8·5	58·1
	17	74 39·1	26·2	31 33·5	9·5	12 44·1	8·6	58·1
	18	89 39·0	N23 26·2	46 02·0	9·6	N12 35·5	8·6	58·1
	19	104 38·8	26·3	60 30·6	9·5	12 26·9	8·8	58·2
	20	119 38·7	26·3	74 59·1	9·6	12 18·1	8·7	58·2
	21	134 38·5	· · 26·3	89 27·7	9·6	12 09·4	8·9	58·2
	22	149 38·4	26·3	103 56·3	9·7	12 00·5	8·9	58·2
	23	164 38·3	26·3	118 25·0	9·6	11 51·6	9·1	58·2
		S.D. 15·8	d 0·0	S.D.		15·5	15·7	15·8

Lat.	Twilight Naut.	Twilight Civil	Sunrise	Moonrise 18	Moonrise 19	Moonrise 20	Moonrise 21
°	h m	h m	h m	h m	h m	h m	h m
N 72	□	□	□	□	02 52	05 04	07 05
N 70	□	□	□	02 11	03 48	05 35	07 22
68	□	□	□	02 57	04 22	05 57	07 36
66	□	□	□	03 28	04 46	06 14	07 48
64	////	////	01 31	03 50	05 05	06 28	07 57
62	////	////	02 09	04 09	05 20	06 40	08 05
60	////	00 49	02 36	04 23	05 33	06 50	08 12
N 58	////	01 40	02 56	04 36	05 44	06 59	08 13
56	////	02 10	03 13	04 47	05 54	07 07	08 24
54	00 45	02 32	03 27	04 57	06 02	07 14	08 29
52	01 32	02 50	03 39	05 05	06 10	07 20	08 33
50	02 00	03 06	03 50	05 13	06 17	07 25	08 37
45	02 46	03 35	04 13	05 29	06 31	07 37	08 45
N 40	03 16	03 58	04 31	05 43	06 43	07 47	08 53
35	03 39	04 16	04 46	05 54	06 53	07 55	08 59
30	03 58	04 31	04 59	06 04	07 02	08 03	09 04
20	04 27	04 56	05 21	06 22	07 18	08 15	09 13
N 10	04 50	05 17	05 40	06 36	07 31	08 26	09 21
0	05 09	05 35	05 57	06 51	07 44	08 37	09 29
S 10	05 26	05 52	06 15	07 05	07 56	08 47	09 36
20	05 42	06 09	06 33	07 19	08 10	08 58	09 44
30	05 58	06 28	06 55	07 37	08 25	09 11	09 54
35	06 07	06 39	07 07	07 47	08 34	09 18	09 59
40	06 17	06 51	07 21	07 58	08 44	09 26	10 05
45	06 27	07 04	07 38	08 11	08 56	09 36	10 12
S 50	06 39	07 20	07 59	08 28	09 10	09 47	10 20
52	06 44	07 28	08 09	08 35	09 17	09 52	10 24
54	06 50	07 36	08 20	08 44	09 24	09 58	10 28
56	06 56	07 45	08 33	08 53	09 33	10 05	10 32
58	07 03	07 55	08 47	09 04	09 42	10 12	10 38
S 60	07 10	08 07	09 05	09 17	09 53	10 20	10 43

Lat.	Sunset	Twilight Civil	Twilight Naut.	Moonset 18	Moonset 19	Moonset 20	Moonset 21
°	h m	h m	h m	h m	h m	h m	h m
N 72	□	□	□	□	{00 17} {23 56}	23 45	23 36
N 70	□	□	□	23 20	23 24	23 25	23 25
68	□	□	□	22 46	23 01	23 10	23 16
66	□	□	□	22 21	22 43	22 57	23 08
64	22 32	////	////	22 02	22 28	22 47	23 01
62	21 53	////	////	21 46	22 15	22 38	22 56
60	21 27	23 13	////	21 33	22 05	22 30	22 51
N 58	21 06	22 22	////	21 22	21 55	22 23	22 47
56	20 50	21 52	////	21 12	21 47	22 17	22 43
54	20 35	21 30	23 17	21 03	21 40	22 11	22 39
52	20 23	21 12	22 30	20 55	21 33	22 07	22 36
50	20 12	20 57	22 02	20 48	21 27	22 02	22 33
45	19 50	20 27	21 17	20 33	21 15	21 52	22 27
N 40	19 32	20 04	20 46	20 20	21 04	21 44	22 22
35	19 17	19 46	20 23	20 09	20 55	21 37	22 17
30	19 03	19 31	20 04	20 00	20 47	21 31	22 13
20	18 41	19 06	19 35	19 44	20 33	21 20	22 06
N 10	18 22	18 45	19 13	19 29	20 21	21 11	21 59
0	18 05	18 27	18 54	19 16	20 09	21 02	21 53
S 10	17 47	18 10	18 37	19 03	19 58	20 53	21 47
20	17 29	17 53	18 20	18 48	19 45	20 43	21 41
30	17 08	17 34	18 04	18 32	19 31	20 32	21 33
35	16 55	17 23	17 55	18 22	19 23	20 25	21 29
40	16 41	17 12	17 46	18 11	19 13	20 18	21 24
45	16 24	16 58	17 35	17 58	19 02	20 09	21 19
S 50	16 03	16 42	17 23	17 42	18 49	19 59	21 12
52	15 53	16 34	17 18	17 35	18 42	19 54	21 08
54	15 42	16 26	17 12	17 26	18 35	19 49	21 05
56	15 30	16 17	17 06	17 17	18 27	19 43	21 01
58	15 15	16 07	16 59	17 07	18 18	19 36	20 57
S 60	14 57	15 55	16 52	16 54	18 08	19 28	20 52

Day	SUN Eqn. of Time 00h	SUN Eqn. of Time 12h	SUN Mer. Pass.	MOON Mer. Pass. Upper	MOON Mer. Pass. Lower	Age	Phase
	m s	m s	h m	h m	h m	d	
18	00 48	00 54	12 01	13 03	00 37	01	●
19	01 01	01 08	12 01	13 56	01 30	02	
20	01 14	01 21	12 01	14 49	02 23	03	

JUNE 21, 22, 23 (SAT., SUN., MON.)

G.M.T. d h	ARIES G.H.A. ° ′	VENUS −3·4 G.H.A. ° ′	VENUS Dec. ° ′	MARS +0·4 G.H.A. ° ′	MARS Dec. ° ′	JUPITER −1·8 G.H.A. ° ′	JUPITER Dec. ° ′	SATURN +0·2 G.H.A. ° ′	SATURN Dec. ° ′
21 00	268 41·1	217 33·8	N16 37·3	258 52·8	N 1 47·9	68 04·2	S 7 14·8	7 24·3	S21 47·2
01	283 43·6	232 33·4	38·1	273 53·7	48·5	83 06·6	14·9	22 27·0	47·2
02	298 46·0	247 32·9	38·8	288 54·5	49·2	98 09·1	14·9	37 29·7	47·2
03	313 48·5	262 32·4	· · 39·6	303 55·4	· · 49·8	113 11·5	· · 14·9	52 32·3	· · 47·2
04	328 51·0	277 31·9	40·4	318 56·2	50·5	128 14·0	14·9	67 35·0	47·2
05	343 53·4	292 31·4	41·2	333 57·1	51·2	143 16·4	14·9	82 37·7	47·1
06	358 55·9	307 31·0	N16 42·0	348 57·9	N 1 51·8	158 18·9	S 7 14·9	97 40·3	S21 47·1
07	13 58·3	322 30·5	42·7	3 58·8	52·5	173 21·4	15·0	112 43·0	47·1
S 08	29 00·8	337 30·0	43·5	18 59·6	53·2	188 23·8	15·0	127 45·6	47·1
A 09	44 03·3	352 29·5	· · 44·3	34 00·5	· · 53·8	203 26·3	· · 15·0	142 48·3	· · 47·1
T 10	59 05·7	7 29·0	45·1	49 01·3	54·5	218 28·7	15·0	157 51·0	47·1
U 11	74 08·2	22 28·5	45·9	64 02·2	55·1	233 31·2	15·0	172 53·6	47·1
R 12	89 10·7	37 28·1	N16 46·6	79 03·0	N 1 55·8	248 33·6	S 7 15·0	187 56·3	S21 47·1
D 13	104 13·1	52 27·6	47·4	94 03·9	56·5	263 36·1	15·1	202 58·9	47·1
A 14	119 15·6	67 27·1	48·2	109 04·7	57·1	278 38·5	15·1	218 01·6	47·1
Y 15	134 18·1	82 26·6	· · 49·0	124 05·6	· · 57·8	293 41·0	· · 15·1	233 04·2	· · 47·1
16	149 20·5	97 26·1	49·7	139 06·4	58·4	308 43·4	15·1	248 06·9	47·1
17	164 23·0	112 25·6	50·5	154 07·3	59·1	323 45·9	15·1	263 09·6	47·1
18	179 25·5	127 25·1	N16 51·3	169 08·1	N 1 59·8	338 48·3	S 7 15·1	278 12·2	S21 47·1
19	194 27·9	142 24·7	52·0	184 09·0	2 00·4	353 50·8	15·2	293 14·9	47·0
20	209 30·4	157 24·2	52·8	199 09·8	01·1	8 53·2	15·2	308 17·5	47·0
21	224 32·8	172 23·7	· · 53·6	214 10·7	· · 01·8	23 55·7	· · 15·2	323 20·2	· · 47·0
22	239 35·3	187 23·2	54·4	229 11·5	02·4	38 58·1	15·2	338 22·9	47·0
23	254 37·8	202 22·7	55·1	244 12·4	03·1	54 00·6	15·2	353 25·5	47·0
22 00	269 40·2	217 22·2	N16 55·9	259 13·3	N 2 03·7	69 03·0	S 7 15·3	8 28·2	S21 47·0
01	284 42·7	232 21·7	56·7	274 14·1	04·4	84 05·5	15·3	23 30·8	47·0
02	299 45·2	247 21·2	57·4	289 15·0	05·1	99 07·9	15·3	38 33·5	47·0
03	314 47·6	262 20·7	· · 58·2	304 15·8	· · 05·7	114 10·3	· · 15·3	53 36·2	· · 47·0
04	329 50·1	277 20·3	59·0	319 16·7	06·4	129 12·8	15·3	68 38·8	47·0
05	344 52·6	292 19·8	16 59·7	334 17·5	07·0	144 15·2	15·4	83 41·5	47·0
06	359 55·0	307 19·3	N17 00·5	349 18·4	N 2 07·7	159 17·7	S 7 15·4	98 44·1	S21 47·0
07	14 57·5	322 18·8	01·2	4 19·2	08·4	174 20·1	15·4	113 46·8	47·0
08	29 59·9	337 18·3	02·0	19 20·1	09·0	189 22·6	15·4	128 49·5	47·0
S 09	45 02·4	352 17·8	· · 02·8	34 20·9	· · 09·7	204 25·0	· · 15·4	143 52·1	· · 47·0
U 10	60 04·9	7 17·3	03·5	49 21·8	10·3	219 27·5	15·5	158 54·8	46·9
N 11	75 07·3	22 16·8	04·3	64 22·7	11·0	234 29·9	15·5	173 57·4	46·9
D 12	90 09·8	37 16·3	N17 05·1	79 23·5	N 2 11·7	249 32·4	S 7 15·5	189 00·1	S21 46·9
A 13	105 12·3	52 15·8	05·8	94 24·4	12·3	264 34·8	15·5	204 02·8	46·9
Y 14	120 14·7	67 15·3	06·6	109 25·2	13·0	279 37·2	15·5	219 05·4	46·9
15	135 17·2	82 14·8	· · 07·3	124 26·1	· · 13·6	294 39·7	· · 15·6	234 08·1	· · 46·9
16	150 19·7	97 14·3	08·1	139 26·9	14·3	309 42·1	15·6	249 10·7	46·9
17	165 22·1	112 13·8	08·8	154 27·8	15·0	324 44·6	15·6	264 13·4	46·9
18	180 24·6	127 13·3	N17 09·6	169 28·6	N 2 15·6	339 47·0	S 7 15·6	279 16·0	S21 46·9
19	195 27·1	142 12·8	10·4	184 29·5	16·3	354 49·5	15·6	294 18·7	46·9
20	210 29·5	157 12·3	11·1	199 30·4	16·9	9 51·9	15·7	309 21·4	46·9
21	225 32·0	172 11·8	· · 11·9	214 31·2	· · 17·6	24 54·3	· · 15·7	324 24·0	· · 46·9
22	240 34·4	187 11·3	12·6	229 32·1	18·2	39 56·8	15·7	339 26·7	46·9
23	255 36·9	202 10·8	13·4	244 32·9	18·9	54 59·2	15·7	354 29·3	46·9
23 00	270 39·4	217 10·3	N17 14·1	259 33·8	N 2 19·6	70 01·7	S 7 15·7	9 32·0	S21 46·9
01	285 41·8	232 09·8	14·9	274 34·6	20·2	85 04·1	15·8	24 34·7	46·8
02	300 44·3	247 09·3	15·6	289 35·5	20·9	100 06·5	15·8	39 37·3	46·8
03	315 46·8	262 08·8	· · 16·4	304 36·4	· · 21·5	115 09·0	· · 15·8	54 40·0	· · 46·8
04	330 49·2	277 08·3	17·1	319 37·2	22·2	130 11·4	15·8	69 42·6	46·8
05	345 51·7	292 07·8	17·9	334 38·1	22·9	145 13·9	15·9	84 45·3	46·8
06	0 54·2	307 07·3	N17 18·6	349 38·9	N 2 23·5	160 16·3	S 7 15·9	99 48·0	S21 46·8
07	15 56·6	322 06·8	19·4	4 39·8	24·2	175 18·7	15·9	114 50·6	46·8
08	30 59·1	337 06·3	20·1	19 40·6	24·8	190 21·2	15·9	129 53·3	46·8
M 09	46 01·6	352 05·8	· · 20·9	34 41·5	· · 25·5	205 23·6	· · 15·9	144 55·9	· · 46·8
O 10	61 04·0	7 05·3	21·6	49 42·4	26·1	220 26·1	16·0	159 58·6	46·8
N 11	76 06·5	22 04·8	22·4	64 43·2	26·8	235 28·5	16·0	175 01·2	46·8
D 12	91 08·9	37 04·3	N17 23·1	79 44·1	N 2 27·5	250 30·9	S 7 16·0	190 03·9	S21 46·8
A 13	106 11·4	52 03·8	23·8	94 44·9	28·1	265 33·4	16·0	205 06·6	46·8
Y 14	121 13·9	67 03·3	24·6	109 45·8	28·8	280 35·8	16·1	220 09·2	46·8
15	136 16·3	82 02·8	· · 25·3	124 46·6	· · 29·4	295 38·2	· · 16·1	235 11·9	· · 46·7
16	151 18·8	97 02·2	26·1	139 47·5	30·1	310 40·7	16·1	250 14·5	46·7
17	166 21·3	112 01·7	26·8	154 48·4	30·7	325 43·1	16·1	265 17·2	46·7
18	181 23·7	127 01·2	N17 27·6	169 49·2	N 2 31·4	340 45·6	S 7 16·2	280 19·9	S21 46·7
19	196 26·2	142 00·7	28·3	184 50·1	32·1	355 48·0	16·2	295 22·5	46·7
20	211 28·7	157 00·2	29·0	199 50·9	32·7	10 50·4	16·2	310 25·2	46·7
21	226 31·1	171 59·7	· · 29·8	214 51·8	· · 33·4	25 52·9	· · 16·2	325 27·8	· · 46·7
22	241 33·6	186 59·2	30·5	229 52·7	34·0	40 55·3	16·2	340 30·5	46·7
23	256 36·0	201 58·7	31·2	244 53·5	34·7	55 57·7	16·3	355 33·1	46·7
Mer. Pass.	6 00·3 (h m)	v −0·5	d 0·8	v 0·9	d 0·7	v 2·4	d 0·0	v 2·7	d 0·0

STARS Name	S.H.A. ° ′	Dec. ° ′
Acamar	315 50·0	S 40 28·1
Achernar	335 57·8	S 57 26·6
Acrux	173 55·3	S 62 52·5
Adhara	255 45·4	S 28 55·1
Aldebaran	291 37·1	N 16 25·5
Alioth	166 56·9	N 56 11·4
Alkaid	153 31·2	N 49 31·5
Al Na'ir	28 35·4	S 47 09·4
Alnilam	276 28·6	S 1 13·8
Alphard	218 36·9	S 8 28·8
Alphecca	126 45·7	N 26 51·4
Alpheratz	358 26·3	N 28 51·6
Altair	62 48·2	N 8 45·6
Ankaa	353 56·5	S 42 31·6
Antares	113 16·6	S 26 20·4
Arcturus	146 33·2	N 19 24·0
Atria	108 55·0	S 68 57·2
Avior	234 35·4	S 59 22·8
Bellatrix	279 16·7	N 6 18·7
Betelgeuse	271 46·4	N 7 23·9
Canopus	264 15·0	S 52 40·5
Capella	281 36·0	N 45 57·3
Deneb	49 59·2	N 45 07·9
Denebola	183 15·8	N 14 48·3
Diphda	349 37·4	S 18 12·7
Dubhe	194 42·6	N 61 58·7
Elnath	279 05·3	N 28 34·3
Eltanin	91 04·8	N 51 29·7
Enif	34 27·5	N 9 41·1
Fomalhaut	16 09·4	S 29 50·3
Gacrux	172 46·8	S 56 53·2
Gienah	176 34·8	S 17 18·8
Hadar	149 46·2	S 60 10·6
Hamal	328 47·6	N 23 15·9
Kaus Aust.	84 38·2	S 34 24·2
Kochab	137 17·5	N 74 19·8
Markab	14 19·4	N 14 58·9
Menkar	314 58·5	N 3 55·6
Menkent	148 56·1	S 36 10·1
Miaplacidus	221 49·0	S 69 33·1
Mirfak	309 39·9	N 49 42·7
Nunki	76 49·1	S 26 20·8
Peacock	54 23·8	S 56 51·9
Pollux	244 18·6	N 28 07·6
Procyon	245 43·3	N 5 19·8
Rasalhague	96 44·5	N 12 35·5
Regulus	208 27·6	N 12 10·2
Rigel	281 52·1	S 8 15·0
Rigil Kent.	140 47·7	S 60 40·1
Sabik	102 59·6	S 15 40·4
Schedar	350 27·7	N 56 18·3
Shaula	97 17·6	S 37 04·4
Sirius	259 10·5	S 16 39·6
Spica	159 14·7	S 10 56·8
Suhail	223 23·1	S 43 16·1
Vega	81 06·5	N 38 44·8
Zuben'ubi	137 51·0	S 15 52·2

	S.H.A. ° ′	Mer. Pass. h m
Venus	307 42·0	9 31
Mars	349 33·1	6 43
Jupiter	159 22·8	19 21
Saturn	98 48·0	23 22

JUNE 21, 22, 23 (SAT., SUN., MON.)

	G.M.T. d h	SUN G.H.A. ° ′	SUN Dec. ° ′	MOON G.H.A. ° ′	v ′	MOON Dec. ° ′	d ′	H.P. ′
	21 00	179 38·1	N23 26·4	132 53·6	9·7	N11 42·5	9·0	58·2
	01	194 38·0	26·4	147 22·3	9·7	11 33·5	9·2	58·3
	02	209 37·9	26·4	161 51·0	9·7	11 24·3	9·2	58·3
	03	224 37·7	·· 26·4	176 19·7	9·7	11 15·1	9·3	58·3
	04	239 37·6	26·4	190 48·4	9·7	11 05·8	9·3	58·3
	05	254 37·5	26·4	205 17·1	9·8	10 56·5	9·4	58·3
	06	269 37·3	N23 26·4	219 45·9	9·8	N10 47·1	9·5	58·3
S	07	284 37·2	26·4	234 14·7	9·8	10 37·6	9·6	58·4
A	08	299 37·0	26·5	248 43·5	9·8	10 28·0	9·6	58·4
T	09	314 36·9	·· 26·5	263 12·3	9·9	10 18·4	9·6	58·4
U	10	329 36·8	26·5	277 41·2	9·8	10 08·8	9·8	58·4
R	11	344 36·6	26·5	292 10·0	9·9	9 59·0	9·8	58·4
D	12	359 36·5	N23 26·5	306 38·9	9·9	N 9 49·2	9·8	58·4
A	13	14 36·4	26·5	321 07·8	9·9	9 39·4	9·9	58·4
Y	14	29 36·2	26·5	335 36·7	9·9	9 29·5	10·0	58·5
	15	44 36·1	·· 26·5	350 05·6	10·0	9 19·5	10·0	58·5
	16	59 35·9	26·5	4 34·6	9·9	9 09·5	10·1	58·5
	17	74 35·8	26·5	19 03·5	10·0	8 59·4	10·1	58·5
	18	89 35·7	N23 26·5	33 32·5	10·0	N 8 49·3	10·2	58·5
	19	104 35·5	26·5	48 01·5	10·0	8 39·1	10·3	58·5
	20	119 35·4	26·5	62 30·5	10·0	8 28·8	10·2	58·6
	21	134 35·3	·· 26·5	76 59·5	10·1	8 18·6	10·4	58·6
	22	149 35·1	26·5	91 28·6	10·0	8 08·2	10·4	58·6
	23	164 35·0	26·5	105 57·6	10·1	7 57·8	10·4	58·6
	22 00	179 34·9	N23 26·5	120 26·7	10·1	N 7 47·4	10·5	58·6
	01	194 34·7	26·5	134 55·8	10·1	7 36·9	10·6	58·6
	02	209 34·6	26·5	149 24·9	10·1	7 26·3	10·5	58·6
	03	224 34·4	·· 26·5	163 54·0	10·1	7 15·8	10·7	58·7
	04	239 34·3	26·5	178 23·1	10·1	7 05·1	10·6	58·7
	05	254 34·2	26·5	192 52·2	10·1	6 54·5	10·8	58·7
	06	269 34·0	N23 26·5	207 21·3	10·2	N 6 43·7	10·7	58·7
	07	284 33·9	26·5	221 50·5	10·2	6 33·0	10·8	58·7
	08	299 33·8	26·5	236 19·7	10·1	6 22·2	10·8	58·7
S	09	314 33·6	·· 26·5	250 48·8	10·2	6 11·4	10·9	58·7
U	10	329 33·5	26·5	265 18·0	10·2	6 00·5	10·9	58·8
N	11	344 33·4	26·5	279 47·2	10·2	5 49·6	11·0	58·8
D	12	359 33·2	N23 26·5	294 16·4	10·2	N 5 38·6	11·0	58·8
A	13	14 33·1	26·4	308 45·6	10·2	5 27·6	11·0	58·8
Y	14	29 32·9	26·4	323 14·8	10·2	5 16·6	11·0	58·8
	15	44 32·8	·· 26·4	337 44·0	10·3	5 05·6	11·1	58·8
	16	59 32·7	26·4	352 13·3	10·2	4 54·5	11·1	58·8
	17	74 32·5	26·4	6 42·5	10·2	4 43·4	11·2	58·8
	18	89 32·4	N23 26·4	21 11·7	10·3	N 4 32·2	11·2	58·9
	19	104 32·3	26·4	35 41·0	10·2	4 21·0	11·2	58·9
	20	119 32·1	26·4	50 10·2	10·3	4 09·8	11·2	58·9
	21	134 32·0	·· 26·3	64 39·5	10·2	3 58·6	11·3	58·9
	22	149 31·9	26·3	79 08·7	10·3	3 47·3	11·2	58·9
	23	164 31·7	26·3	93 38·0	10·2	3 36·1	11·4	58·9
	23 00	179 31·6	N23 26·3	108 07·2	10·3	N 3 24·7	11·3	58·9
	01	194 31·5	26·3	122 36·5	10·3	3 13·4	11·3	58·9
	02	209 31·3	26·2	137 05·8	10·2	3 02·1	11·4	59·0
	03	224 31·2	·· 26·2	151 35·0	10·3	2 50·7	11·4	59·0
	04	239 31·1	26·2	166 04·3	10·3	2 39·3	11·4	59·0
	05	254 30·9	26·2	180 33·5	10·3	2 27·9	11·4	59·0
	06	269 30·8	N23 26·2	195 02·8	10·2	N 2 16·5	11·5	59·0
	07	284 30·6	26·1	209 32·0	10·3	2 05·0	11·5	59·0
M	08	299 30·5	26·1	224 01·3	10·2	1 53·5	11·4	59·0
O	09	314 30·4	·· 26·1	238 30·5	10·3	1 42·1	11·5	59·0
N	10	329 30·2	26·1	252 59·8	10·2	1 30·6	11·5	59·0
D	11	344 30·1	26·0	267 29·0	10·2	1 19·1	11·5	59·1
A	12	359 30·0	N23 26·0	281 58·2	10·3	N 1 07·6	11·5	59·1
Y	13	14 29·8	26·0	296 27·5	10·2	0 56·1	11·6	59·1
	14	29 29·7	26·0	310 56·7	10·2	0 44·5	11·5	59·1
	15	44 29·6	·· 25·9	325 25·9	10·2	0 33·0	11·6	59·1
	16	59 29·4	25·9	339 55·1	10·2	0 21·4	11·5	59·1
	17	74 29·3	25·9	354 24·3	10·1	N 0 09·9	11·6	59·1
	18	89 29·2	N23 25·8	8 53·4	10·2	S 0 01·7	11·5	59·1
	19	104 29·0	25·8	23 22·6	10·2	0 13·2	11·6	59·1
	20	119 28·9	25·8	37 51·8	10·1	0 24·8	11·6	59·1
	21	134 28·8	·· 25·7	52 20·9	10·2	0 36·4	11·5	59·2
	22	149 28·6	25·7	66 50·1	10·1	0 47·9	11·6	59·2
	23	164 28·5	25·7	81 19·2	10·1	0 59·5	11·5	59·2
		S.D. 15·8	d 0·0	S.D.		15·9	16·0	16·1

Lat. °	Twilight Naut. h m	Twilight Civil h m	Sunrise h m	Moonrise 21 h m	Moonrise 22 h m	Moonrise 23 h m	Moonrise 24 h m
N 72	□	□	□	07 05	09 01	10 55	12 51
N 70	□	□	□	07 22	09 10	10 57	12 45
68	□	□	□	07 36	09 17	10 59	12 41
66	□	□	□	07 48	09 23	11 00	12 37
64	////	////	01 31	07 57	09 28	11 01	12 34
62	////	////	02 09	08 05	09 33	11 01	12 31
60	////	00 49	02 36	08 12	09 37	11 02	12 29
N 58	////	01 40	02 56	08 18	09 40	11 03	12 27
56	////	02 10	03 13	08 24	09 43	11 04	12 25
54	00 45	02 33	03 27	08 29	09 46	11 04	12 23
52	01 32	02 51	03 40	08 33	09 48	11 05	12 22
50	02 00	03 06	03 51	08 37	09 50	11 05	12 20
45	02 46	03 36	04 13	08 45	09 55	11 06	12 17
N 40	03 17	03 58	04 31	08 53	09 59	11 07	12 15
35	03 40	04 17	04 46	08 59	10 03	11 07	12 13
30	03 59	04 32	04 59	09 04	10 06	11 08	12 11
20	04 28	04 57	05 22	09 13	10 11	11 09	12 08
N 10	04 50	05 17	05 40	09 21	10 16	11 10	12 05
0	05 09	05 36	05 58	09 29	10 20	11 11	12 02
S 10	05 26	05 53	06 16	09 36	10 24	11 12	12 00
20	05 43	06 10	06 34	09 44	10 29	11 13	11 57
30	05 59	06 29	06 55	09 54	10 34	11 14	11 54
35	06 08	06 40	07 08	09 59	10 37	11 15	11 52
40	06 17	06 51	07 22	10 05	10 41	11 16	11 50
45	06 28	07 05	07 39	10 12	10 45	11 17	11 48
S 50	06 40	07 21	08 00	10 20	10 50	11 18	11 46
52	06 45	07 29	08 10	10 24	10 52	11 18	11 44
54	06 50	07 37	08 21	10 28	10 54	11 19	11 43
56	06 57	07 46	08 33	10 32	10 57	11 19	11 42
58	07 03	07 56	08 48	10 38	11 00	11 20	11 40
S 60	07 11	08 08	09 06	10 43	11 03	11 21	11 38

Lat. °	Sunset h m	Twilight Civil h m	Twilight Naut. h m	Moonset 21 h m	Moonset 22 h m	Moonset 23 h m	Moonset 24 h m
N 72	□	□	□	23 36	23 28	23 20	23 13
N 70	□	□	□	23 25	23 24	23 22	23 21
68	□	□	□	23 16	23 20	23 23	23 27
66	□	□	□	23 08	23 17	23 24	23 33
64	22 32	////	////	23 01	23 14	23 25	23 37
62	21 54	////	////	22 56	23 12	23 26	23 41
60	21 28	23 14	////	22 51	23 10	23 27	23 45
N 58	21 07	22 23	////	22 47	23 08	23 27	23 48
56	20 50	21 53	////	22 43	23 06	23 28	23 51
54	20 36	21 31	23 18	22 39	23 04	23 29	23 54
52	20 24	21 13	22 31	22 36	23 03	23 29	23 56
50	20 13	20 58	22 03	22 33	23 02	23 30	23 58
45	19 50	20 28	21 18	22 27	22 59	23 30	24 03
N 40	19 32	20 05	20 47	22 22	22 57	23 31	24 07
35	19 17	19 47	20 24	22 17	22 55	23 32	24 10
30	19 04	19 32	20 05	22 13	22 53	23 33	24 13
20	18 42	19 07	19 36	22 06	22 50	23 34	24 18
N 10	18 23	18 46	19 13	21 59	22 47	23 34	24 23
0	18 05	18 28	18 54	21 53	22 44	23 35	24 27
S 10	17 48	18 11	18 37	21 47	22 42	23 36	24 31
20	17 30	17 53	18 21	21 41	22 39	23 37	24 36
30	17 08	17 35	18 04	21 33	22 35	23 38	24 41
35	16 56	17 24	17 56	21 29	22 33	23 38	24 44
40	16 42	17 12	17 46	21 24	22 31	23 39	24 47
45	16 25	16 59	17 36	21 19	22 29	23 40	24 51
S 50	16 04	16 43	17 24	21 12	22 26	23 40	24 56
52	15 54	16 35	17 19	21 08	22 24	23 41	24 58
54	15 43	16 27	17 13	21 05	22 23	23 41	25 01
56	15 30	16 18	17 07	21 01	22 21	23 42	25 03
58	15 16	16 07	17 00	20 57	22 19	23 42	25 06
S 60	14 58	15 56	16 52	20 52	22 17	23 43	25 09

Day	SUN Eqn. of Time 00^h m s	SUN Eqn. of Time 12^h m s	SUN Mer. Pass. h m	MOON Mer. Pass. Upper h m	MOON Mer. Pass. Lower h m	MOON Age d	Phase
21	01 27	01 34	12 02	15 41	03 15	04	◑
22	01 40	01 47	12 02	16 32	04 07	05	
23	01 53	02 00	12 02	17 23	04 58	06	

JUNE 24, 25, 26 (TUES., WED., THURS.)

G.M.T. d h	ARIES G.H.A.	VENUS −3·4 G.H.A.	VENUS Dec.	MARS +0·3 G.H.A.	MARS Dec.	JUPITER −1·7 G.H.A.	JUPITER Dec.	SATURN +0·3 G.H.A.	SATURN Dec.	STARS Name	S.H.A.	Dec.
24 00	271 38·5	216 58·2	N17 32·0	259 54·4	N 2 35·3	71 00·2	S 7 16·3	10 35·8	S21 46·7	Acamar	315 50·0	S 40 28·1
01	286 41·0	231 57·6	32·7	274 55·2	36·0	86 02·6	16·3	25 38·5	46·7	Achernar	335 57·8	S 57 26·6
02	301 43·4	246 57·1	33·5	289 56·1	36·6	101 05·0	16·3	40 41·1	46·7	Acrux	173 55·4	S 62 52·5
03	316 45·9	261 56·6	· · 34·2	304 56·9	· · 37·3	116 07·5	· · 16·4	55 43·8	· · 46·7	Adhara	255 45·4	S 28 55·0
04	331 48·4	276 56·1	34·9	319 57·8	38·0	131 09·9	16·4	70 46·4	46·7	Aldebaran	291 37·1	N 16 25·5
05	346 50·8	291 55·6	35·7	334 58·7	38·6	146 12·3	16·4	85 49·1	46·6			
06	1 53·3	306 55·1	N17 36·4	349 59·5	N 2 39·3	161 14·8	S 7 16·4	100 51·7	S21 46·6	Alioth	166 56·9	N 56 11·4
07	16 55·8	321 54·5	37·1	5 00·4	39·9	176 17·2	16·5	115 54·4	46·6	Alkaid	153 31·3	N 49 31·5
T 08	31 58·2	336 54·0	37·8	20 01·2	40·6	191 19·6	16·5	130 57·1	46·6	Al Na'ir	28 35·4	S 47 09·4
U 09	47 00·7	351 53·5	· · 38·6	35 02·1	· · 41·2	206 22·1	· · 16·5	145 59·7	· · 46·6	Alnilam	276 28·6	S 1 13·7
E 10	62 03·2	6 53·0	39·3	50 03·0	41·9	221 24·5	16·5	161 02·4	46·6	Alphard	218 36·9	S 8 28·8
S 11	77 05·6	21 52·5	40·0	65 03·8	42·5	236 26·9	16·6	176 05·0	46·6			
D 12	92 08·1	36 52·0	N17 40·8	80 04·7	N 2 43·2	251 29·3	S 7 16·6	191 07·7	S21 46·6	Alphecca	126 45·7	N 26 51·4
A 13	107 10·5	51 51·4	41·5	95 05·6	43·8	266 31·8	16·6	206 10·3	46·6	Alpheratz	358 26·2	N 28 51·6
Y 14	122 13·0	66 50·9	42·2	110 06·4	44·5	281 34·2	16·6	221 13·0	46·6	Altair	62 48·2	N 8 45·6
15	137 15·5	81 50·4	· · 42·9	125 07·3	· · 45·2	296 36·6	· · 16·7	236 15·7	· · 46·6	Ankaa	353 56·5	S 42 31·6
16	152 17·9	96 49·9	43·7	140 08·1	45·8	311 39·1	16·7	251 18·3	46·6	Antares	113 16·6	S 26 20·4
17	167 20·4	111 49·4	44·4	155 09·0	46·5	326 41·5	16·7	266 21·0	46·6			
18	182 22·9	126 48·8	N17 45·1	170 09·9	N 2 47·1	341 43·9	S 7 16·7	281 23·6	S21 46·6	Arcturus	146 33·3	N 19 24·0
19	197 25·3	141 48·3	45·8	185 10·7	47·8	356 46·4	16·8	296 26·3	46·6	Atria	108 55·0	S 68 57·3
20	212 27·8	156 47·8	46·6	200 11·6	48·4	11 48·8	16·8	311 28·9	46·5	Avior	234 35·4	S 59 22·8
21	227 30·3	171 47·3	· · 47·3	215 12·4	· · 49·1	26 51·2	· · 16·8	326 31·6	· · 46·5	Bellatrix	279 16·7	N 6 18·7
22	242 32·7	186 46·7	48·0	230 13·3	49·7	41 53·6	16·9	341 34·3	46·5	Betelgeuse	271 46·4	N 7 23·9
23	257 35·2	201 46·2	48·7	245 14·2	50·4	56 56·1	16·9	356 36·9	46·5			
25 00	272 37·7	216 45·7	N17 49·4	260 15·0	N 2 51·0	71 58·5	S 7 16·9	11 39·6	S21 46·5	Canopus	264 15·0	S 52 40·5
01	287 40·1	231 45·2	50·2	275 15·9	51·7	87 00·9	16·9	26 42·2	46·5	Capella	281 36·0	N 45 57·3
02	302 42·6	246 44·6	50·9	290 16·8	52·3	102 03·3	17·0	41 44·9	46·5	Deneb	49 59·2	N 45 07·9
03	317 45·0	261 44·1	· · 51·6	305 17·6	· · 53·0	117 05·8	· · 17·0	56 47·5	· · 46·5	Denebola	183 15·8	N 14 48·3
04	332 47·5	276 43·6	52·3	320 18·5	53·6	132 08·2	17·0	71 50·2	46·5	Diphda	349 37·4	S 18 12·7
05	347 50·0	291 43·1	53·0	335 19·3	54·3	147 10·6	17·0	86 52·8	46·5			
06	2 52·4	306 42·5	N17 53·7	350 20·2	N 2 55·0	162 13·0	S 7 17·1	101 55·5	S21 46·5	Dubhe	194 42·6	N 61 58·7
W 07	17 54·9	321 42·0	54·5	5 21·1	55·6	177 15·5	17·1	116 58·2	46·5	Elnath	279 05·2	N 28 34·3
E 08	32 57·4	336 41·5	55·2	20 21·9	56·3	192 17·9	17·1	132 00·8	46·5	Eltanin	91 04·8	N 51 29·8
D 09	47 59·8	351 41·0	· · 55·9	35 22·8	· · 56·9	207 20·3	· · 17·2	147 03·5	· · 46·5	Enif	34 27·5	N 9 41·1
N 10	63 02·3	6 40·4	56·6	50 23·7	57·6	222 22·7	17·2	162 06·1	46·5	Fomalhaut	16 09·4	S 29 50·3
E 11	78 04·8	21 39·9	57·3	65 24·5	58·2	237 25·2	17·2	177 08·8	46·4			
S 12	93 07·2	36 39·4	N17 58·0	80 25·4	N 2 58·9	252 27·6	S 7 17·2	192 11·4	S21 46·4	Gacrux	172 46·8	S 56 53·2
D 13	108 09·7	51 38·8	58·7	95 26·2	2 59·5	267 30·0	17·3	207 14·1	46·4	Gienah	176 34·8	S 17 18·8
A 14	123 12·1	66 38·3	17 59·4	110 27·1	3 00·2	282 32·4	17·3	222 16·8	46·4	Hadar	149 46·2	S 60 10·7
Y 15	138 14·6	81 37·8	18 00·1	125 28·0	· · 00·8	297 34·9	· · 17·3	237 19·4	· · 46·4	Hamal	328 47·6	N 23 15·9
16	153 17·1	96 37·2	00·9	140 28·8	01·5	312 37·3	17·4	252 22·1	46·4	Kaus Aust.	84 38·2	S 34 24·2
17	168 19·5	111 36·7	01·6	155 29·7	02·1	327 39·7	17·4	267 24·7	46·4			
18	183 22·0	126 36·2	N18 02·3	170 30·6	N 3 02·8	342 42·1	S 7 17·4	282 27·4	S21 46·4	Kochab	137 17·5	N 74 19·8
19	198 24·5	141 35·6	03·0	185 31·4	03·4	357 44·5	17·4	297 30·0	46·4	Markab	14 19·4	N 14 58·9
20	213 26·9	156 35·1	03·7	200 32·3	04·1	12 47·0	17·5	312 32·7	46·4	Menkar	314 58·5	N 3 55·6
21	228 29·4	171 34·6	· · 04·4	215 33·2	· · 04·7	27 49·4	· · 17·5	327 35·3	· · 46·4	Menkent	148 56·1	S 36 10·1
22	243 31·9	186 34·0	05·1	230 34·0	05·4	42 51·8	17·5	342 38·0	46·4	Miaplacidus	221 49·0	S 69 33·1
23	258 34·3	201 33·5	05·8	245 34·9	06·0	57 54·2	17·6	357 40·7	46·4			
26 00	273 36·8	216 32·9	N18 06·5	260 35·8	N 3 06·7	72 56·6	S 7 17·6	12 43·3	S21 46·4	Mirfak	309 39·9	N 49 42·7
01	288 39·3	231 32·4	07·2	275 36·6	07·3	87 59·1	17·6	27 46·0	46·3	Nunki	76 49·1	S 26 20·8
02	303 41·7	246 31·9	07·9	290 37·5	08·0	103 01·5	17·6	42 48·6	46·3	Peacock	54 23·8	S 56 51·9
03	318 44·2	261 31·3	· · 08·6	305 38·4	· · 08·6	118 03·9	· · 17·7	57 51·3	· · 46·3	Pollux	244 18·6	N 28 07·6
04	333 46·6	276 30·8	09·3	320 39·2	09·3	133 06·3	17·7	72 53·9	46·3	Procyon	245 43·5	N 5 19·8
05	348 49·1	291 30·3	10·0	335 40·1	09·9	148 08·7	17·7	87 56·6	46·3			
06	3 51·6	306 29·7	N18 10·7	350 41·0	N 3 10·6	163 11·2	S 7 17·8	102 59·2	S21 46·3	Rasalhague	96 44·5	N 12 35·5
T 07	18 54·0	321 29·2	11·4	5 41·8	11·2	178 13·6	17·8	118 01·9	46·3	Regulus	208 27·6	N 12 10·2
H 08	33 56·5	336 28·6	12·1	20 42·7	11·9	193 16·0	17·8	133 04·6	46·3	Rigel	281 52·1	S 8 15·0
U 09	48 59·0	351 28·1	· · 12·8	35 43·6	· · 12·5	208 18·4	· · 17·9	148 07·2	· · 46·3	Rigil Kent.	140 47·8	S 60 40·1
R 10	64 01·4	6 27·6	13·5	50 44·4	13·2	223 20·8	17·9	163 09·9	46·3	Sabik	102 59·6	S 15 40·4
S 11	79 03·9	21 27·0	14·2	65 45·3	13·8	238 23·2	17·9	178 12·5	46·3			
D 12	94 06·4	36 26·5	N18 14·9	80 46·2	N 3 14·5	253 25·7	S 7 18·0	193 15·2	S21 46·3	Schedar	350 27·7	N 56 18·4
A 13	109 08·8	51 25·9	15·6	95 47·0	15·1	268 28·1	18·0	208 17·8	46·3	Shaula	97 17·6	S 37 04·4
Y 14	124 11·3	66 25·4	16·3	110 47·9	15·8	283 30·5	18·0	223 20·5	46·3	Sirius	259 10·5	S 16 39·6
15	139 13·8	81 24·8	· · 16·9	125 48·8	· · 16·4	298 32·9	· · 18·0	238 23·1	· · 46·3	Spica	159 14·7	S 10 56·8
16	154 16·2	96 24·3	17·6	140 49·6	17·1	313 35·3	18·1	253 25·8	46·2	Suhail	223 23·1	S 43 16·1
17	169 18·7	111 23·8	18·3	155 50·5	17·7	328 37·7	18·1	268 28·5	46·2			
18	184 21·1	126 23·2	N18 19·0	170 51·4	N 3 18·4	343 40·2	S 7 18·1	283 31·1	S21 46·2	Vega	81 06·5	N 38 44·8
19	199 23·6	141 22·7	19·7	185 52·2	19·0	358 42·6	18·2	298 33·8	46·2	Zuben'ubi	137 51·0	S 15 52·2
20	214 26·1	156 22·1	20·4	200 53·1	19·7	13 45·0	18·2	313 36·4	46·2			
21	229 28·5	171 21·6	· · 21·1	215 54·0	· · 20·3	28 47·4	· · 18·2	328 39·1	· · 46·2			
22	244 31·0	186 21·0	21·8	230 54·8	21·0	43 49·8	18·3	343 41·7	46·2			
23	259 33·5	201 20·5	22·5	245 55·7	21·6	58 52·2	18·3	358 44·4	46·2			
Mer. Pass.	5 48·5 (h m)	v −0·5	d 0·7	v 0·9	d 0·7	v 2·4	d 0·0	v 2·7	d 0·0			

	S.H.A. ° ′	Mer. Pass. h m
Venus	304 08·0	9 33
Mars	347 37·3	6 39
Jupiter	159 20·8	19 09
Saturn	99 01·9	23 09

JUNE 24, 25, 26 (TUES., WED., THURS.)

G.M.T. d h	SUN G.H.A. ° ′	SUN Dec. ° ′	MOON G.H.A. ° ′	v ′	MOON Dec. ° ′	d ′	H.P. ′
24 00	179 28·4	N23 25·6	95 48·3	10·1	S 1 11·0	11·6	59·2
01	194 28·2	25·6	110 17·4	10·1	1 22·6	11·5	59·2
02	209 28·1	25·6	124 46·5	10·0	1 34·1	11·6	59·2
03	224 28·0	·· 25·5	139 15·5	10·1	1 45·7	11·5	59·2
04	239 27·8	25·5	153 44·6	10·0	1 57·2	11·6	59·2
05	254 27·7	25·4	168 13·6	10·0	2 08·8	11·5	59·2
06	269 27·5	N23 25·4	182 42·6	10·0	S 2 20·3	11·5	59·2
07	284 27·4	25·4	197 11·6	10·0	2 31·8	11·5	59·2
T 08	299 27·3	25·3	211 40·6	10·0	2 43·3	11·5	59·2
U 09	314 27·1	· 25·3	226 09·6	9·9	2 54·8	11·5	59·3
E 10	329 27·0	25·2	240 38·5	9·9	3 06·3	11·4	59·3
S 11	344 26·9	25·2	255 07·4	9·9	3 17·7	11·5	59·3
D 12	359 26·7	N23 25·2	269 36·3	9·9	S 3 29·2	11·4	59·3
A 13	14 26·6	25·1	284 05·2	9·8	3 40·6	11·4	59·3
Y 14	29 26·5	25·1	298 34·0	9·8	3 52·0	11·4	59·3
15	44 26·3	·· 25·0	313 02·8	9·8	4 03·4	11·4	59·3
16	59 26·2	25·0	327 31·6	9·8	4 14·8	11·3	59·3
17	74 26·1	24·9	342 00·4	9·8	4 26·1	11·3	59·3
18	89 25·9	N23 24·9	356 29·2	9·7	S 4 37·4	11·3	59·3
19	104 25·8	24·8	10 57·9	9·7	4 48·7	11·3	59·3
20	119 25·7	24·8	25 26·6	9·7	5 00·0	11·2	59·3
21	134 25·5	·· 24·7	39 55·3	9·6	5 11·2	11·3	59·3
22	149 25·4	24·7	54 23·9	9·6	5 22·5	11·2	59·3
23	164 25·3	24·6	68 52·5	9·6	5 33·7	11·1	59·3
25 00	179 25·1	N23 24·6	83 21·1	9·6	S 5 44·8	11·2	59·4
01	194 25·0	24·5	97 49·7	9·5	5 56·0	11·1	59·4
02	209 24·9	24·5	112 18·2	9·5	6 07·1	11·0	59·4
03	224 24·7	·· 24·4	126 46·7	9·5	6 18·1	11·1	59·4
04	239 24·6	24·3	141 15·2	9·4	6 29·2	11·0	59·4
05	254 24·5	24·3	155 43·6	9·4	6 40·2	10·9	59·4
06	269 24·3	N23 24·2	170 12·0	9·4	S 6 51·1	10·9	59·4
W 07	284 24·2	24·2	184 40·4	9·4	7 02·0	10·9	59·4
E 08	299 24·1	24·1	199 08·8	9·3	7 12·9	10·9	59·4
D 09	314 23·9	·· 24·1	213 37·1	9·2	7 23·8	10·8	59·4
N 10	329 23·8	24·0	228 05·3	9·3	7 34·6	10·7	59·4
E 11	344 23·7	23·9	242 33·6	9·2	7 45·3	10·7	59·4
S 12	359 23·5	N23 23·9	257 01·8	9·2	S 7 56·0	10·7	59·4
D 13	14 23·4	23·8	271 30·0	9·1	8 06·7	10·6	59·4
A 14	29 23·3	23·8	285 58·1	9·1	8 17·3	10·6	59·4
Y 15	44 23·1	· 23·7	300 26·2	9·0	8 27·9	10·5	59·4
16	59 23·0	23·6	314 54·2	9·1	8 38·4	10·5	59·4
17	74 22·9	23·6	329 22·3	9·0	8 48·9	10·5	59·4
18	89 22·7	N23 23·5	343 50·3	8·9	S 8 59·4	10·3	59·4
19	104 22·6	23·4	358 18·2	8·9	9 09·7	10·4	59·4
20	119 22·5	23·4	12 46·1	8·9	9 20·1	10·2	59·4
21	134 22·3	·· 23·3	27 14·0	8·8	9 30·3	10·2	59·4
22	149 22·2	23·2	41 41·8	8·8	9 40·5	10·2	59·4
23	164 22·1	23·2	56 09·6	8·8	9 50·7	10·1	59·4
26 00	179 21·9	N23 23·1	70 37·4	8·7	S10 00·8	10·0	59·4
01	194 21·8	23·0	85 05·1	8·7	10 10·8	10·0	59·4
02	209 21·7	22·9	99 32·8	8·6	10 20·8	9·9	59·4
03	224 21·6	· 22·9	114 00·4	8·6	10 30·7	9·9	59·4
04	239 21·4	22·8	128 28·0	8·6	10 40·6	9·8	59·4
05	254 21·3	22·7	142 55·6	8·5	10 50·4	9·7	59·5
06	269 21·2	N23 22·7	157 23·1	8·5	S11 00·1	9·7	59·5
T 07	284 21·0	22·6	171 50·6	8·4	11 09·8	9·6	59·5
H 08	299 20·9	22·5	186 18·0	8·4	11 19·4	9·5	59·5
U 09	314 20·8	· 22·4	200 45·4	8·3	11 28·9	9·5	59·5
R 10	329 20·6	22·4	215 12·7	8·4	11 38·4	9·3	59·5
S 11	344 20·5	22·3	229 40·1	8·2	11 47·7	9·4	59·5
D 12	359 20·4	N23 22·2	244 07·3	8·2	S11 57·1	9·2	59·5
A 13	14 20·2	22·1	258 34·5	8·2	12 06·3	9·2	59·4
Y 14	29 20·1	22·0	273 01·7	8·2	12 15·5	9·1	59·4
15	44 20·0	· 22·0	287 28·9	8·1	12 24·6	9·0	59·4
16	59 19·8	21·9	301 56·0	8·0	12 33·6	9·0	59·4
17	74 19·7	21·8	316 23·0	8·1	12 42·6	8·8	59·4
18	89 19·6	N23 21·7	330 50·1	7·9	S12 51·4	8·8	59·4
19	104 19·5	21·6	345 17·0	8·0	13 00·2	8·7	59·4
20	119 19·3	21·5	359 44·0	7·9	13 08·9	8·6	59·4
21	134 19·2	· 21·5	14 10·9	7·8	13 17·5	8·6	59·4
22	149 19·1	21·4	28 37·7	7·8	13 26·1	8·4	59·4
23	164 18·9	21·3	43 04·5	7·8	13 34·5	8·4	59·4
	S.D. 15·8	d 0·1	S.D. 16·2		16·2		16·2

Lat. °	Twilight Naut. h m	Twilight Civil h m	Sunrise h m	Moonrise 24 h m	Moonrise 25 h m	Moonrise 26 h m	Moonrise 27 h m
N 72	□	□	□	12 51	14 49	16 52	19 08
N 70	□	□	□	12 45	14 35	16 28	18 22
68	□	□	□	12 41	14 24	16 09	17 53
66	□	□	□	12 37	14 15	15 54	17 30
64	////	////	01 32	12 34	14 08	15 42	17 13
62	////	////	02 10	12 31	14 01	15 31	16 59
60	////	00 51	02 37	12 29	13 56	15 22	16 47
N 58	////	01 42	02 57	12 27	13 51	15 15	16 36
56	////	02 12	03 14	12 25	13 47	15 08	16 27
54	00 47	02 34	03 28	12 23	13 43	15 02	16 19
52	01 34	02 52	03 41	12 22	13 39	14 56	16 12
50	02 01	03 07	03 51	12 20	13 36	14 52	16 06
45	02 47	03 37	04 14	12 17	13 29	14 41	15 52
N 40	03 18	03 59	04 32	12 15	13 23	14 32	15 41
35	03 41	04 17	04 47	12 13	13 19	14 25	15 31
30	03 59	04 33	05 00	12 11	13 14	14 18	15 22
20	04 28	04 58	05 22	12 08	13 07	14 07	15 08
N 10	04 51	05 18	05 41	12 05	13 00	13 57	14 55
0	05 10	05 36	05 59	12 02	12 54	13 48	14 44
S 10	05 27	05 53	06 16	12 00	12 49	13 39	14 32
20	05 43	06 11	06 35	11 57	12 42	13 29	14 19
30	06 00	06 30	06 56	11 54	12 35	13 19	14 05
35	06 08	06 40	07 08	11 52	12 31	13 12	13 57
40	06 18	06 52	07 22	11 50	12 27	13 05	13 48
45	06 28	07 05	07 39	11 48	12 21	12 57	13 37
S 50	06 40	07 21	08 00	11 46	12 15	12 47	13 24
52	06 45	07 29	08 10	11 44	12 12	12 43	13 18
54	06 51	07 37	08 21	11 43	12 09	12 38	13 11
56	06 57	07 46	08 34	11 42	12 05	12 32	13 04
58	07 04	07 57	08 48	11 40	12 02	12 26	12 56
S 60	07 11	08 08	09 06	11 38	11 57	12 19	12 46

Lat. °	Sunset h m	Twilight Civil h m	Twilight Naut. h m	Moonset 24 h m	Moonset 25 h m	Moonset 26 h m	Moonset 27 h m
N 72	□	□	□	23 13	23 04	22 53	22 35
N 70	□	□	□	23 21	23 19	23 19	23 21
68	□	□	□	23 27	23 32	23 39	23 51
66	□	□	□	23 33	23 42	23 55	24 14
64	22 32	////	////	23 37	23 51	24 08	00 08
62	21 54	////	////	23 41	23 59	24 19	00 19
60	21 28	23 13	////	23 45	24 05	00 05	00 29
N 58	21 07	22 23	////	23 48	24 11	00 11	00 37
56	20 51	21 53	////	23 51	24 16	00 16	00 45
54	20 36	21 31	23 17	23 54	24 20	00 20	00 51
52	20 24	21 13	22 31	23 56	24 25	00 25	00 57
50	20 13	20 58	22 03	23 58	24 28	00 28	01 03
45	19 51	20 28	21 18	24 03	00 03	00 37	01 14
N 40	19 33	20 06	20 47	24 07	00 07	00 44	01 24
35	19 18	19 47	20 24	24 10	00 10	00 50	01 33
30	19 05	19 32	20 05	24 13	00 13	00 55	01 40
20	18 43	19 07	19 37	24 18	00 18	01 04	01 53
N 10	18 24	18 47	19 14	24 23	00 23	01 12	02 04
0	18 06	18 29	18 55	24 27	00 27	01 20	02 14
S 10	17 49	18 11	18 38	24 31	00 31	01 27	02 25
20	17 30	17 54	18 22	24 36	00 36	01 35	02 36
30	17 09	17 35	18 05	24 41	00 41	01 45	02 49
35	16 57	17 25	17 57	24 44	00 44	01 50	02 56
40	16 42	17 13	17 47	24 47	00 47	01 56	03 05
45	16 26	16 59	17 37	24 51	00 51	02 03	03 15
S 50	16 05	16 44	17 25	24 56	00 56	02 12	03 27
52	15 55	16 36	17 20	24 58	00 58	02 16	03 32
54	15 44	16 28	17 14	25 01	01 01	02 20	03 38
56	15 31	16 19	17 08	25 03	01 03	02 25	03 45
58	15 17	16 09	17 01	25 06	01 06	02 30	03 53
S 60	14 59	15 57	16 53	25 09	01 09	02 36	04 02

Day	SUN Eqn. of Time 00h m s	SUN Eqn. of Time 12h m s	SUN Mer. Pass. h m	MOON Mer. Pass. Upper h m	MOON Mer. Pass. Lower h m	Age d	Phase
24	02 06	02 13	12 02	18 15	05 49	07	
25	02 19	02 26	12 02	19 07	06 41	08	
26	02 32	02 38	12 03	20 01	07 34	09	

SEPTEMBER 19, 20, 21 (FRI., SAT., SUN.)

G.M.T.	ARIES	VENUS −3·4		MARS −0·9		JUPITER −1·3		SATURN +0·7	
	G.H.A.	G.H.A.	Dec.	G.H.A.	Dec.	G.H.A.	Dec.	G.H.A.	Dec.
d h	° ′	° ′	° ′	° ′	° ′	° ′	° ′	° ′	° ′
19 00	357 23·6	193 43·8	N 8 24·7	299 44·2	N18 05·6	147 02·1	S11 17·4	98 34·0	S21 52·8
01	12 26·0	208 43·3	23·5	314 45·9	05·8	162 04·1	17·6	113 36·4	52·8
02	27 28·5	223 42·9	22·4	329 47·7	06·0	177 06·1	17·8	128 38·8	52·9
03	42 31·0	238 42·5	.. 21·2	344 49·5	.. 06·2	192 08·1	.. 18·0	143 41·1	.. 52·9
04	57 33·4	253 42·0	20·1	359 51·2	06·4	207 10·1	18·1	158 43·5	52·9
05	72 35·9	268 41·6	18·9	14 53·0	06·6	222 12·1	18·3	173 45·8	52·9
06	87 38·4	283 41·1	N 8 17·8	29 54·8	N18 06·8	237 14·1	S11 18·5	188 48·2	S21 52·9
07	102 40·8	298 40·7	16·6	44 56·6	07·0	252 16·1	18·6	203 50·5	52·9
08	117 43·3	313 40·2	15·5	59 58·3	07·2	267 18·1	18·8	218 52·9	53·0
F 09	132 45·8	328 39·8	.. 14·3	75 00·1	.. 07·3	282 20·1	.. 19·0	233 55·2	.. 53·0
R 10	147 48·2	343 39·3	13·2	90 01·9	07·5	297 22·1	19·2	248 57·6	53·0
I 11	162 50·7	358 38·9	12·0	105 03·7	07·7	312 24·1	19·3	263 59·9	53·0
D 12	177 53·1	13 38·5	N 8 10·9	120 05·5	N18 07·9	327 26·1	S11 19·5	279 02·3	S21 53·0
A 13	192 55·6	28 38·0	09·7	135 07·2	08·1	342 28·1	19·7	294 04·6	53·0
Y 14	207 58·1	43 37·6	08·6	150 09·0	08·3	357 30·1	19·8	309 07·0	53·0
15	223 00·5	58 37·1	.. 07·4	165 10·8	.. 08·5	12 32·1	.. 20·0	324 09·3	.. 53·1
16	238 03·0	73 36·7	06·3	180 12·6	08·7	27 34·1	20·2	339 11·7	53·1
17	253 05·5	88 36·2	05·1	195 14·4	08·9	42 36·1	20·4	354 14·0	53·1
18	268 07·9	103 35·8	N 8 04·0	210 16·2	N18 09·1	57 38·1	S11 20·5	9 16·4	S21 53·1
19	283 10·4	118 35·4	02·8	225 18·0	09·3	72 40·1	20·7	24 18·7	53·1
20	298 12·9	133 34·9	01·6	240 19·7	09·5	87 42·1	20·9	39 21·1	53·1
21	313 15·3	148 34·5	8 00·5	255 21·5	.. 09·7	102 44·1	.. 21·1	54 23·4	.. 53·2
22	328 17·8	163 34·0	7 59·3	270 23·3	09·9	117 46·1	21·2	69 25·8	53·2
23	343 20·2	178 33·6	58·2	285 25·1	10·0	132 48·1	21·4	84 28·1	53·2
20 00	358 22·7	193 33·2	N 7 57·0	300 26·9	N18 10·2	147 50·1	S11 21·6	99 30·5	S21 53·2
01	13 25·2	203 32·7	55·9	315 28·7	10·4	162 52·1	21·8	114 32·8	53·2
02	28 27·6	223 32·3	54·7	330 30·5	10·6	177 54·0	21·9	129 35·2	53·2
03	43 30·1	238 31·8	.. 53·5	345 32·3	.. 10·8	192 56·0	22·1	144 37·5	.. 53·2
04	58 32·6	253 31·4	52·4	0 34·1	11·0	207 58·0	22·3	159 39·9	53·3
05	73 35·0	268 31·0	51·2	15 35·9	11·2	223 00·0	22·4	174 42·2	53·3
06	88 37·5	283 30·5	N 7 50·1	30 37·7	N18 11·4	238 02·0	S11 22·6	189 44·6	S21 53·3
07	103 40·0	298 30·1	48·9	45 39·5	11·6	253 04·0	22·8	204 46·9	53·3
S 08	118 42·4	313 29·7	47·7	60 41·3	11·8	268 06·0	23·0	219 49·3	53·3
A 09	133 44·9	328 29·2	.. 46·6	75 43·1	.. 12·0	283 08·0	.. 23·1	234 51·6	.. 53·3
T 10	148 47·4	343 28·8	45·4	90 44·9	12·1	298 10·0	23·3	249 54·0	53·4
U 11	163 49·8	358 28·3	44·2	105 46·7	12·3	313 12·0	23·5	264 56·3	53·4
R 12	178 52·3	13 27·9	N 7 43·1	120 48·5	N18 12·5	328 14·0	S11 23·7	279 58·7	S21 53·4
D 13	193 54·7	28 27·5	41·9	135 50·3	12·7	343 16·0	23·8	295 01·0	53·4
A 14	208 57·2	43 27·0	40·8	150 52·2	12·9	358 18·0	24·0	310 03·4	53·4
Y 15	223 59·7	58 26·6	.. 39·6	165 54·0	.. 13·1	13 20·0	.. 24·2	325 05·7	.. 53·4
16	239 02·1	73 26·2	38·4	180 55·8	13·3	28 22·0	24·3	340 08·1	53·5
17	254 04·6	88 25·7	37·3	195 57·6	13·5	43 24·0	24·5	355 10·4	53·5
18	269 07·1	103 25·3	N 7 36·1	210 59·4	N18 13·6	58 26·0	S11 24·7	10 12·8	S21 53·5
19	284 09·5	118 24·9	34·9	226 01·2	13·8	73 28·0	24·9	25 15·1	53·5
20	299 12·0	133 24·4	33·8	241 03·0	14·0	88 30·0	25·0	40 17·5	53·5
21	314 14·5	148 24·0	. 32·6	256 04·9	.. 14·2	103 32·0	.. 25·2	55 19·8	.. 53·5
22	329 16·9	163 23·5	31·4	271 06·7	14·4	118 34·0	25·4	70 22·1	53·5
23	344 19·4	178 23·1	30·3	286 08·5	14·6	133 36·0	25·6	85 24·5	53·6
21 00	359 21·8	193 22·7	N 7 29·1	301 10·3	N18 14·8	148 38·0	S11 25·7	100 26·8	S21 53·6
01	14 24·3	208 22·2	27·9	316 12·2	15·0	163 40·0	25·9	115 29·2	53·6
02	29 26·8	223 21·8	26·8	331 14·0	15·1	178 42·0	26·1	130 31·5	53·6
03	44 29·2	238 21·4	.. 25·6	346 15·8	.. 15·3	193 44·0	.. 26·2	145 33·9	.. 53·6
04	59 31·7	253 20·9	24·4	1 17·6	15·5	208 45·9	26·4	160 36·2	53·6
05	74 34·2	268 20·5	23·3	16 19·5	15·7	223 47·9	26·6	175 38·6	53·7
06	89 36·6	283 20·1	N 7 22·1	31 21·3	N18 15·9	238 49·9	S11 26·8	190 40·9	S21 53·7
07	104 39·1	298 19·6	20·9	46 23·1	16·1	253 51·9	26·9	205 43·3	53·7
08	119 41·6	313 19·2	19·7	61 25·0	16·3	268 53·9	27·1	220 45·6	53·7
S 09	134 44·0	328 18·8	. 18·6	76 26·8	.. 16·4	283 55·9	.. 27·3	235 47·9	.. 53·7
U 10	149 46·5	343 18·4	17·4	91 28·6	16·6	298 57·9	27·5	250 50·3	53·7
N 11	164 49·0	358 17·9	16·2	106 30·5	16·8	313 59·9	27·6	265 52·6	53·8
D 12	179 51·4	13 17·5	N 7 15·1	121 32·3	N18 17·0	329 01·9	S11 27·8	280 55·0	S21 53·8
A 13	194 53·9	28 17·1	13·9	136 34·1	17·2	344 03·9	28·0	295 57·3	53·8
Y 14	209 56·3	43 16·6	12·7	151 36·0	17·4	359 05·9	28·2	310 59·7	53·8
15	224 58·8	58 16·2	.. 11·5	166 37·8	.. 17·5	14 07·9	.. 28·3	326 02·0	.. 53·8
16	240 01·3	73 15·8	10·4	181 39·7	17·7	29 09·9	28·5	341 04·3	53·8
17	255 03·7	88 15·3	09·2	196 41·5	17·9	44 11·9	28·7	356 06·7	53·9
18	270 06·2	103 14·9	N 7 08·0	211 43·3	N18 18·1	59 13·9	S11 28·8	11 09·0	S21 53·9
19	285 08·7	118 14·5	06·9	226 45·2	18·3	74 15·8	29·0	26 11·4	53·9
20	300 11·1	133 14·0	05·7	241 47·0	18·5	89 17·8	29·2	41 13·7	53·9
21	315 13·6	148 13·6	.. 04·5	256 48·9	.. 18·6	104 19·8	.. 29·4	56 16·1	. 53·9
22	330 16·1	163 13·2	03·3	271 50·7	18·8	119 21·8	29·5	71 18·4	53·9
23	345 18·5	178 12·8	02·2	286 52·6	19·0	134 23·8	29·7	86 20·7	54·0
Mer. Pass.	h m 0 06·5	v −0·4	d 1·2	v 1·8	d 0·2	v 2·0	d 0·2	v 2·3	d 0·0

STARS

Name	S.H.A. ° ′	Dec. ° ′
Acamar	315 49·3	S 40 28·0
Achernar	335 56·9	S 57 26·6
Acrux	173 55·9	S 62 52·3
Adhara	255 45·0	S 28 54·8
Aldebaran	291 36·5	N 16 25·6
Alioth	166 57·4	N 56 11·1
Alkaid	153 31·7	N 49 31·3
Al Na'ir	28 35·0	S 47 09·6
Alnilam	276 28·1	S 1 13·6
Alphard	218 36·8	S 8 28·7
Alphecca	126 46·1	N 26 51·5
Alpheratz	358 25·7	N 28 51·9
Altair	62 48·2	N 8 45·8
Ankaa	353 55·9	S 42 31·6
Antares	113 16·9	S 26 20·4
Arcturus	146 33·6	N 19 24·0
Atria	108 55·8	S 68 57·4
Avior	234 35·2	S 59 22·4
Bellatrix	279 16·1	N 6 18·8
Betelgeuse	271 45·9	N 7 24·0
Canopus	264 14·5	S 52 40·2
Capella	281 35·2	N 45 57·2
Deneb	49 59·2	N 45 08·3
Denebola	183 15·9	N 14 48·2
Diphda	349 36·9	S 18 12·6
Dubhe	194 42·8	N 61 58·4
Elnath	279 04·6	N 28 34·3
Eltanin	91 05·3	N 51 30·0
Enif	34 27·2	N 9 41·4
Fomalhaut	16 09·0	S 29 50·3
Gacrux	172 47·3	S 56 52·9
Gienah	176 35·0	S 17 18·7
Hadar	149 46·8	S 60 10·5
Hamal	328 47·0	N 23 16·1
Kaus Aust.	84 38·3	S 34 24·3
Kochab	137 19·1	N 74 19·8
Markab	14 19·0	N 14 59·2
Menkar	314 57·9	N 3 55·8
Menkent	148 56·4	S 36 10·0
Miaplacidus	221 49·1	S 69 32·7
Mirfak	309 39·0	N 49 42·8
Nunki	76 49·2	S 26 20·9
Peacock	54 23·7	S 56 52·1
Pollux	244 18·2	N 28 07·5
Procyon	245 42·9	N 5 19·9
Rasalhague	96 44·7	N 12 35·6
Regulus	208 27·6	N 12 10·2
Rigel	281 51·6	S 8 14·8
Rigil Kent.	140 48·4	S 60 40·0
Sabik	102 59·8	S 15 40·4
Schedar	350 26·9	N 56 18·7
Shaula	97 17·8	S 37 04·5
Sirius	259 10·1	S 16 39·4
Spica	159 14·9	S 10 56·7
Suhail	223 23·0	S 43 15·8
Vega	81 06·7	N 38 45·1
Zuben'ubi	137 51·2	S 15 52·2

	S.H.A. ° ′	Mer. Pass. h m
Venus	195 10·5	11 06
Mars	302 04·2	3 58
Jupiter	149 27·4	14 07
Saturn	101 07·8	17 19

SEPTEMBER 19, 20, 21 (FRI., SAT., SUN.)

G.M.T.	SUN G.H.A.	SUN Dec.	MOON G.H.A.	*v*	MOON Dec.	*d*	H.P.
d h	° ′	° ′	° ′	′	° ′	′	′
19 00	181 28·6	N 1 46·1	107 22·5	6·9	S18 12·2	2·5	58·9
01	196 28·8	45·1	121 48·4	6·8	18 14·7	2·5	58·8
02	211 29·0	44·1	136 14·2	6·9	18 17·2	2·4	58·8
03	226 29·2	· 43·2	150 40·1	6·9	18 19·6	2·2	58·8
04	241 29·5	42·2	165 06·0	7·0	18 21·8	2·1	58·7
05	256 29·7	41·2	179 32·0	7·0	18 23·9	2·0	58·7
06	271 29·9	N 1 40·3	193 58·0	7·0	S18 25·9	1·9	58·7
07	286 30·1	39·3	208 24·0	7·0	18 27·8	1·7	58·6
08	301 30·3	38·3	222 50·0	7·0	18 29·5	1·7	58·6
F 09	316 30·6	· · 37·4	237 16·0	7·1	18 31·2	1·5	58·6
R 10	331 30·8	36·4	251 42·1	7·2	18 32·7	1·4	58·5
I 11	346 31·0	35·4	266 08·3	7·1	18 34·1	1·3	58·5
D 12	1 31·2	N 1 34·4	280 34·4	7·2	S18 35·4	1·2	58·5
A 13	16 31·4	33·5	295 00·6	7·2	18 36·6	1·0	58·4
Y 14	31 31·7	32·5	309 26·8	7·3	18 37·6	1·0	58·4
15	46 31·9	· 31·5	323 53·1	7·2	18 38·6	0·8	58·4
16	61 32·1	30·6	338 19·3	7·4	18 39·4	0·7	58·4
17	76 32·3	29·6	352 45·7	7·3	18 40·1	0·6	58·3
18	91 32·6	N 1 28·6	7 12·0	7·4	S18 40·7	0·5	58·3
19	106 32·8	27·7	21 38·4	7·5	18 41·2	0·3	58·3
20	121 33·0	26·7	36 04·9	7·4	18 41·5	0·3	58·2
21	136 33·2	· · 25·7	50 31·3	7·5	18 41·8	0·1	58·2
22	151 33·4	24·7	64 57·8	7·6	18 41·9	0·0	58·2
23	166 33·7	23·8	79 24·4	7·6	18 41·9	0·1	58·1
20 00	181 33·9	N 1 22·8	93 51·0	7·6	S18 41·8	0·2	58·1
01	196 34·1	21·8	108 17·6	7·7	18 41·6	0·3	58·1
02	211 34·3	20·9	122 44·3	7·7	18 41·3	0·5	58·0
03	226 34·6	· · 19·9	137 11·0	7·8	18 40·8	0·5	58·0
04	241 34·8	18·9	151 37·8	7·8	18 40·3	0·7	58·0
05	256 35·0	18·0	166 04·6	7·8	18 39·6	0·8	57·9
06	271 35·2	N 1 17·0	180 31·4	7·9	S18 38·8	0·8	57·9
07	286 35·4	16·0	194 58·3	8·0	18 38·0	1·0	57·9
S 08	301 35·7	15·0	209 25·3	8·0	18 37·0	1·1	57·8
A 09	316 35·9	· · 14·1	223 52·3	8·0	18 35·9	1·2	57·8
T 10	331 36·1	13·1	238 19·3	8·1	18 34·7	1·4	57·8
U 11	346 36·3	12·1	252 46·4	8·1	18 33·3	1·4	57·7
R 12	1 36·5	N 1 11·2	267 13·5	8·2	S18 31·9	1·5	57·7
D 13	16 36·8	10·2	281 40·7	8·2	18 30·4	1·7	57·7
A 14	31 37·0	09·2	296 07·9	8·3	18 28·7	1·7	57·6
Y 15	46 37·2	· · 08·2	310 35·2	8·3	18 27·0	1·9	57·6
16	61 37·4	07·3	325 02·5	8·4	18 25·1	1·9	57·6
17	76 37·7	06·3	339 29·9	8·4	18 23·2	2·1	57·5
18	91 37·9	N 1 05·3	353 57·3	8·5	S18 21·1	2·2	57·5
19	106 38·1	04·4	8 24·8	8·5	18 18·9	2·3	57·5
20	121 38·3	03·4	22 52·3	8·6	18 16·6	2·3	57·5
21	136 38·5	· · 02·4	37 19·9	8·6	18 14·3	2·5	57·4
22	151 38·8	01·4	51 47·5	8·7	18 11·8	2·6	57·4
23	166 39·0	1 00·5	66 15·2	8·8	18 09·2	2·7	57·4
21 00	181 39·2	N 0 59·5	80 43·0	8·8	S18 06·5	2·8	57·3
01	196 39·4	58·5	95 10·8	8·8	18 03·7	2·9	57·3
02	211 39·6	57·6	109 38·6	8·9	18 00·8	2·9	57·3
03	226 39·9	· · 56·6	124 06·5	9·0	17 57·9	3·1	57·2
04	241 40·1	55·6	138 34·5	9·0	17 54·8	3·2	57·2
05	256 40·3	54·6	153 02·5	9·1	17 51·6	3·3	57·2
06	271 40·5	N 0 53·7	167 30·6	9·1	S17 48·3	3·3	57·1
07	286 40·7	52·7	181 58·7	9·2	17 45·0	3·5	57·1
08	301 41·0	51·7	196 26·9	9·2	17 41·5	3·6	57·1
S 09	316 41·2	· · 50·8	210 55·1	9·3	17 37·9	3·6	57·1
U 10	331 41·4	49·8	225 23·4	9·4	17 34·3	3·8	57·0
N 11	346 41·6	48·8	239 51·8	9·4	17 30·5	3·8	57·0
D 12	1 41·8	N 0 47·8	254 20·2	9·5	S17 26·7	4·0	57·0
A 13	16 42·1	46·9	268 48·7	9·5	17 22·7	4·0	56·9
Y 14	31 42·3	45·9	283 17·2	9·6	17 18·7	4·1	56·9
15	46 42·5	· · 44·9	297 45·8	9·6	17 14·6	4·2	56·9
16	61 42·7	44·0	312 14·4	9·7	17 10·4	4·3	56·9
17	76 43·0	43·0	326 43·1	9·8	17 06·1	4·4	56·8
18	91 43·2	N 0 42·0	341 11·9	9·8	S17 01·7	4·5	56·8
19	106 43·4	41·0	355 40·7	9·9	16 57·2	4·6	56·8
20	121 43·6	40·1	10 09·6	9·9	16 52·6	4·6	56·7
21	136 43·8	· · 39·1	24 38·5	10·0	16 48·0	4·8	56·7
22	151 44·1	38·1	39 07·5	10·1	16 43·2	4·8	56·7
23	166 44·3	37·1	53 36·6	10·1	16 38·4	4·9	56·7
	S.D. 16·0	*d* 1·0	S.D. 15·9		15·7		15·5

Lat.	Twilight Naut.	Twilight Civil	Sunrise	Moonrise 19	Moonrise 20	Moonrise 21	Moonrise 22
°	h m	h m	h m	h m	h m	h m	h m
N 72	02 42	04 18	05 27	■	■	18 02	17 46
N 70	03 05	04 28	05 30	16 04	16 50	17 07	17 13
68	03 23	04 36	05 32	15 14	16 05	16 33	16 49
66	03 37	04 42	05 34	14 43	15 35	16 09	16 31
64	03 48	04 48	05 36	14 20	15 13	15 50	16 16
62	03 58	04 52	05 37	14 01	14 55	15 34	16 03
60	04 06	04 57	05 38	13 46	14 40	15 21	15 52
N 58	04 12	05 00	05 39	13 33	14 27	15 10	15 43
56	04 18	05 03	05 40	13 22	14 16	15 00	15 35
54	04 23	05 06	05 41	13 12	14 06	14 51	15 27
52	04 28	05 08	05 42	13 03	13 58	14 43	15 21
50	04 32	05 10	05 42	12 55	13 50	14 36	15 15
45	04 40	05 15	05 44	12 39	13 34	14 21	15 02
N 40	04 46	05 18	05 45	12 25	13 20	14 08	14 51
35	04 51	05 21	05 46	12 13	13 09	13 58	14 42
30	04 55	05 23	05 47	12 03	12 59	13 49	14 34
20	05 01	05 26	05 48	11 46	12 41	13 33	14 20
N 10	05 04	05 28	05 49	11 31	12 26	13 19	14 08
0	05 06	05 30	05 50	11 17	12 12	13 06	13 56
S 10	05 06	05 30	05 51	11 03	11 58	12 52	13 45
20	05 04	05 30	05 52	10 48	11 43	12 38	13 33
30	05 01	05 29	05 53	10 30	11 26	12 22	13 19
35	04 59	05 28	05 53	10 20	11 16	12 13	13 11
40	04 55	05 27	05 54	10 09	11 04	12 02	13 01
45	04 51	05 25	05 54	09 56	10 51	11 50	12 51
S 50	04 45	05 22	05 54	09 39	10 35	11 35	12 37
52	04 42	05 21	05 55	09 32	10 27	11 27	12 31
54	04 39	05 20	05 55	09 23	10 18	11 20	12 25
56	04 35	05 18	05 55	09 13	10 09	11 11	12 17
58	04 30	05 16	05 55	09 03	09 58	11 01	12 08
S 60	04 25	05 14	05 56	08 50	09 45	10 49	11 59

Lat.	Sunset	Twilight Civil	Twilight Naut.	Moonset 19	Moonset 20	Moonset 21	Moonset 22
°	h m	h m	h m	h m	h m	h m	h m
N 72	18 18	19 27	21 00	■	■	20 42	22 43
N 70	18 15	19 17	20 38	18 55	20 04	21 37	23 15
68	18 13	19 09	20 21	19 45	20 49	22 10	23 38
66	18 11	19 03	20 08	20 16	21 19	22 34	23 55
64	18 10	18 58	19 57	20 40	21 41	22 53	24 10
62	18 09	18 53	19 48	20 58	21 59	23 08	24 22
60	18 08	18 50	19 40	21 14	22 13	23 21	24 32
N 58	18 07	18 46	19 33	21 26	22 26	23 32	24 41
56	18 06	18 43	19 28	21 38	22 37	23 41	24 49
54	18 05	18 40	19 23	21 48	22 46	23 50	24 56
52	18 04	18 38	19 18	21 56	22 55	23 57	25 02
50	18 04	18 36	19 14	22 04	23 02	24 04	00 04
45	18 03	18 32	19 06	22 21	23 18	24 18	00 18
N 40	18 01	18 28	19 00	22 35	23 32	24 30	00 30
35	18 01	18 26	18 55	22 46	23 43	24 41	00 41
30	18 00	18 24	18 52	22 56	23 53	24 49	00 49
20	17 59	18 21	18 46	23 14	24 10	00 10	01 05
N 10	17 58	18 19	18 43	23 29	24 24	00 24	01 18
0	17 57	18 18	18 42	23 43	24 38	00 38	01 30
S 10	17 56	18 17	18 41	23 58	24 52	00 52	01 43
20	17 55	18 18	18 43	24 13	00 13	01 06	01 56
30	17 55	18 19	18 47	24 30	00 30	01 23	02 11
35	17 55	18 20	18 49	24 40	00 40	01 33	02 19
40	17 54	18 21	18 53	24 52	00 52	01 44	02 29
45	17 54	18 23	18 58	00 07	01 05	01 56	02 41
S 50	17 54	18 26	19 04	00 23	01 22	02 12	02 54
52	17 53	18 27	19 07	00 30	01 29	02 19	03 01
54	17 53	18 29	19 11	00 39	01 38	02 28	03 08
56	17 53	18 30	19 14	00 48	01 48	02 37	03 16
58	17 53	18 32	19 19	00 59	01 59	02 47	03 25
S 60	17 53	18 34	19 24	01 11	02 11	02 59	03 35

Day	SUN Eqn. of Time 00h	SUN Eqn. of Time 12h	SUN Mer. Pass.	MOON Mer. Pass. Upper	MOON Mer. Pass. Lower	Age	Phase
	m s	m s	h m	h m	h m	d	
19	05 54	06 04	11 54	17 30	05 02	06	◐
20	06 15	06 26	11 54	18 25	05 58	07	
21	06 36	06 47	11 53	19 18	06 52	08	

SEPTEMBER 22, 23, 24 (MON., TUES., WED.)

G.M.T.	ARIES	VENUS −3·4		MARS −1·0		JUPITER −1·3		SATURN +0·7		STARS		
d h	G.H.A. ° ′	G.H.A. ° ′	Dec. ° ′	G.H.A. ° ′	Dec. ° ′	G.H.A. ° ′	Dec. ° ′	G.H.A. ° ′	Dec. ° ′	Name	S.H.A. ° ′	Dec. ° ′
22 00	0 21·0	193 12·3	N 7 01·0	301 54·4	N18 19·2	149 25·8	S11 29·9	101 23·1	S21 54·0	Acamar	315 49·3	S 40 28·0
01	15 23·5	208 11·9	6 59·8	316 56·3	19·4	164 27·8	30·1	116 25·4	54·0	Achernar	335 56·9	S 57 26·6
02	30 25·9	223 11·5	58·6	331 58·1	19·5	179 29·8	30·2	131 27·8	54·0	Acrux	173 55·9	S 62 52·2
03	45 28·4	238 11·0	·· 57·4	347 00·0	·· 19·7	194 31·8	·· 30·4	146 30·1	·· 54·0	Adhara	255 44·9	S 28 54·8
04	60 30·8	253 10·6	56·3	2 01·8	19·9	209 33·8	30·6	161 32·5	54·0	Aldebaran	291 36·5	N 16 25·6
05	75 33·3	268 10·2	55·1	17 03·7	20·1	224 35·8	30·8	176 34·8	54·1			
06	90 35·8	283 09·8	N 6 53·9	32 05·6	N18 20·3	239 37·8	S11 30·9	191 37·1	S21 54·1	Alioth	166 57·4	N 56 11·1
07	105 38·2	298 09·3	52·7	47 07·4	20·4	254 39·8	31·1	206 39·5	54·1	Alkaid	153 31·8	N 49 31·3
08	120 40·7	313 08·9	51·6	62 09·3	20·6	269 41·7	31·3	221 41·8	54·1	Al Na'ir	28 35·0	S 47 09·6
M 09	135 43·2	328 08·5	·· 50·4	77 11·1	·· 20·8	284 43·7	·· 31·4	236 44·2	·· 54·1	Alnilam	276 28·1	S 1 13·6
O 10	150 45·6	343 08·1	49·2	92 13·0	21·0	299 45·7	31·6	251 46·5	54·1	Alphard	218 36·7	S 8 28·7
N 11	165 48·1	358 07·6	48·0	107 14·9	21·2	314 47·7	31·8	266 48·8	54·2			
D 12	180 50·6	13 07·2	N 6 46·8	122 16·7	N18 21·3	329 49·7	S11 32·0	281 51·2	S21 54·2	Alphecca	126 46·1	N 26 51·5
A 13	195 53·0	28 06·8	45·7	137 18·6	21·5	344 51·7	32·1	296 53·5	54·2	Alpheratz	358 25·7	N 28 51·9
Y 14	210 55·5	43 06·4	44·5	152 20·5	21·7	359 53·7	32·3	311 55·9	54·2	Altair	62 48·2	N 8 45·8
15	225 57·9	58 05·9	·· 43·3	167 22·3	·· 21·9	14 55·7	·· 32·5	326 58·2	·· 54·2	Ankaa	353 55·9	S 42 31·6
16	241 00·4	73 05·5	42·1	182 24·2	22·0	29 57·7	32·7	342 00·5	54·2	Antares	113 16·9	S 26 20·4
17	256 02·9	88 05·1	40·9	197 26·1	22·2	44 59·7	32·8	357 02·9	54·2			
18	271 05·3	103 04·7	N 6 39·7	212 27·9	N18 22·4	60 01·7	S11 33·0	12 05·2	S21 54·3	Arcturus	146 33·6	N 19 24·0
19	286 07·8	118 04·2	38·6	227 29·8	22·6	75 03·6	33·2	27 07·5	54·3	Atria	108 55·8	S 68 57·4
20	301 10·3	133 03·8	37·4	242 31·7	22·8	90 05·6	33·4	42 09·9	54·3	Avior	234 35·2	S 59 22·4
21	316 12·7	148 03·4	·· 36·2	257 33·6	·· 22·9	105 07·6	·· 33·5	57 12·2	·· 54·3	Bellatrix	279 16·1	N 6 18·8
22	331 15·2	163 03·0	35·0	272 35·4	23·1	120 09·6	33·7	72 14·6	54·3	Betelgeuse	271 45·8	N 7 24·0
23	346 17·7	178 02·5	33·8	287 37·3	23·3	135 11·6	33·9	87 16·9	54·3			
23 00	1 20·1	193 02·1	N 6 32·6	302 39·2	N18 23·5	150 13·6	S11 34·1	102 19·2	S21 54·4	Canopus	264 14·5	S 52 40·2
01	16 22·6	208 01·7	31·5	317 41·1	23·6	165 15·6	34·2	117 21·6	54·4	Capella	281 35·2	N 45 57·2
02	31 25·1	223 01·3	30·3	332 43·0	23·8	180 17·6	34·4	132 23·9	54·4	Deneb	49 59·2	N 45 08·3
03	46 27·5	238 00·8	·· 29·1	347 44·9	·· 24·0	195 19·6	·· 34·6	147 26·3	·· 54·4	Denebola	183 15·9	N 14 48·2
04	61 30·0	253 00·4	27·9	2 46·7	24·2	210 21·6	34·8	162 28·6	54·4	Diphda	349 36·8	S 18 12·6
05	76 32·4	268 00·0	26·7	17 48·6	24·3	225 23·5	34·9	177 30·9	54·4			
06	91 34·9	282 59·6	N 6 25·5	32 50·5	N18 24·5	240 25·5	S11 35·1	192 33·3	S21 54·5	Dubhe	194 42·8	N 61 58·4
07	106 37·4	297 59·1	24·3	47 52·4	24·7	255 27·5	35·3	207 35·6	54·5	Elnath	279 04·6	N 28 34·3
T 08	121 39·8	312 58·7	23·1	62 54·3	24·9	270 29·5	35·4	222 37·9	54·5	Eltanin	91 05·3	N 51 30·0
U 09	136 42·3	327 58·3	·· 22·0	77 56·2	·· 25·0	285 31·5	·· 35·6	237 40·3	·· 54·5	Enif	34 27·3	N 9 41·4
E 10	151 44·8	342 57·9	20·8	92 58·1	25·2	300 33·5	35·8	252 42·6	54·5	Fomalhaut	16 09·0	S 29 50·3
S 11	166 47·2	357 57·5	19·6	108 00·0	25·4	315 35·5	36·0	267 44·9	54·5			
D 12	181 49·7	12 57·0	N 6 18·4	123 01·9	N18 25·5	330 37·5	S11 36·1	282 47·3	S21 54·6	Gacrux	172 47·3	S 56 52·9
A 13	196 52·2	27 56·6	17·2	138 03·8	25·7	345 39·4	36·3	297 49·6	54·6	Gienah	176 35·0	S 17 18·7
Y 14	211 54·6	42 56·2	16·0	153 05·7	25·9	0 41·4	36·5	312 51·9	54·6	Hadar	149 46·8	S 60 10·5
15	226 57·1	57 55·8	·· 14·8	168 07·5	·· 26·1	15 43·4	·· 36·7	327 54·3	·· 54·6	Hamal	328 46·9	N 23 16·1
16	241 59·5	72 55·4	13·6	183 09·4	26·2	30 45·4	36·8	342 56·6	54·6	Kaus Aust.	84 38·3	S 34.24·3
17	257 02·0	87 54·9	12·5	198 11·3	26·4	45 47·4	37·0	357 59·0	54·6			
18	272 04·5	102 54·5	N 6 11·3	213 13·3	N18 26·6	60 49·4	S11 37·2	13 01·3	S21 54·7	Kochab	137 19·2	N 74 19·7
19	287 06·9	117 54·1	10·1	228 15·2	26·8	75 51·4	37·4	28 03·6	54·7	Markab	14 19·0	N 14 59·2
20	302 09·4	132 53·7	08·9	243 17·1	26·9	90 53·4	37·5	43 06·0	54·7	Menkar	314 57·9	N 3 55·8
21	317 11·9	147 53·3	·· 07·7	258 19·0	·· 27·1	105 55·3	·· 37·7	58 08·3	·· 54·7	Menkent	148 56·4	S 36 10·0
22	332 14·3	162 52·8	06·5	273 20·9	27·3	120 57·3	37·9	73 10·6	54·7	Miapiacidus	221 49·0	S 69 32·7
23	347 16·8	177 52·4	05·3	288 22·8	27·4	135 59·3	38·1	88 13·0	54·7			
24 00	2 19·3	192 52·0	N 6 04·1	303 24·7	N18 27·6	151 01·3	S11 38·2	103 15·3	S21 54·8	Mirfak	309 39·0	N 49 42·8
01	17 21·7	207 51·6	02·9	318 26·6	27·8	166 03·3	38·4	118 17·6	54·8	Nunki	76 49·2	S 26 20·9
02	32 24·2	222 51·2	01·7	333 28·5	28·0	181 05·3	38·6	133 20·0	54·8	Peacock	54 23·7	S 56 52·2
03	47 26·7	237 50·7	6 00·5	348 30·4	·· 28·1	196 07·3	·· 38·8	148 22·3	·· 54·8	Pollux	244 18·2	N 28 07·5
04	62 29·1	252 50·3	5 59·3	3 32·3	28·3	211 09·3	38·9	163 24·6	54·8	Procyon	245 42·9	N 5 19·9
05	77 31·6	267 49·9	58·1	18 34·3	28·5	226 11·2	39·1	178 27·0	54·8			
06	92 34·0	282 49·5	N 5 57·0	33 36·2	N18 28·6	241 13·2	S11 39·3	193 29·3	S21 54·9	Rasalhague	96 44·7	N 12 35·6
W 07	107 36·5	297 49·1	55·8	48 38·1	28·8	256 15·2	39·5	208 31·6	54·9	Regulus	208 27·6	N 12 10·2
E 08	122 39·0	312 48·7	54·6	63 40·0	29·0	271 17·2	39·6	223 34·0	54·9	Rigel	281 51·6	S 8 14·8
D 09	137 41·4	327 48·2	·· 53·4	78 41·9	·· 29·1	286 19·2	·· 39·8	238 36·3	·· 54·9	Rigil Kent.	140 48·4	S 60 40·0
N 10	152 43·9	342 47·8	52·2	93 43·9	29·3	301 21·2	40·0	253 38·6	54·9	Sabik	102 59·8	S 15 40·4
E 11	167 46·4	357 47·4	51·0	108 45·8	29·5	316 23·2	40·2	268 41·0	54·9			
S 12	182 48·8	12 47·0	N 5 49·8	123 47·7	N18 29·6	331 25·1	S11 40·3	283 43·3	S21 55·0	Schedar	350 26·9	N 56 18·7
D 13	197 51·3	27 46·6	48·6	138 49·6	29·8	346 27·1	40·5	298 45·6	55·0	Shaula	97 17·9	S 37 04·5
A 14	212 53·8	42 46·2	47·4	153 51·6	30·0	1 29·1	40·7	313 48·0	55·0	Sirius	259 10·1	S 16 39·4
Y 15	227 56·2	57 45·7	·· 46·2	168 53·5	·· 30·2	16 31·1	·· 40·9	328 50·3	·· 55·0	Spica	159 14·9	S 10 56·7
16	242 58·7	72 45·3	45·0	183 55·4	30·3	31 33·1	41·0	343 52·6	55·0	Suhail	223 23·0	S 43 15·8
17	258 01·2	87 44·9	43·8	198 57·3	30·5	46 35·1	41·2	358 54·9	55·0			
18	273 03·6	102 44·5	N 5 42·6	213 59·3	N18 30·7	61 37·1	S11 41·4	13 57·3	S21 55·1	Vega	81 06·8	N 38 45·1
19	288 06·1	117 44·1	41·4	229 01·2	30·8	76 39·0	41·6	28 59·6	55·1	Zuben'ubi	137 51·2	S 15 52·2
20	303 08·5	132 43·7	40·2	244 03·1	31·0	91 41·0	41·7	44 01·9	55·1			
21	318 11·0	147 43·2	·· 39·0	259 05·1	·· 31·2	106 43·0	·· 41·9	59 04·3	·· 55·1			
22	333 13·5	162 42·8	37·8	274 07·0	31·3	121 45·0	42·1	74 06·6	55·1			
23	348 15·9	177 42·4	36·6	289 09·0	31·5	136 47·0	42·3	89 08·9	55·1			
Mer. Pass.	23 50·7 (h m)	v −0·4	d 1·2	v 1·9	d 0·2	v 2·0	d 0·2	v 2·3	d 0·0			

	S.H.A. ° ′	Mer. Pass. h m
Venus	191 42·0	11 08
Mars	301 19·1	3 49
Jupiter	148 53·5	13 57
Saturn	100 59·1	17 08

SEPTEMBER 22, 23, 24 (MON., TUES., WED.)

G.M.T. d h	SUN G.H.A. ° ′	SUN Dec. ° ′	MOON G.H.A. ° ′	v ′	MOON Dec. ° ′	d ′	H.P. ′
22 00	181 44·5	N 0 36·2	68 05·7	10·2	S16 33·5	5·0	56·6
01	196 44·7	35·2	82 34·9	10·2	16 28·5	5·1	56·6
02	211 44·9	34·2	97 04·1	10·3	16 23·4	5·1	56·6
03	226 45·2	· · 33·3	111 33·4	10·4	16 18·3	5·3	56·5
04	241 45·4	32·3	126 02·8	10·4	16 13·0	5·3	56·5
05	256 45·6	31·3	140 32·2	10·4	16 07·7	5·4	56·5
06	271 45·8	N 0 30·3	155 01·6	10·6	S16 02·3	5·4	56·5
07	286 46·0	29·4	169 31·2	10·5	15 56·9	5·6	56·4
08	301 46·3	28·4	184 00·7	10·7	15 51·3	5·6	56·4
M 09	316 46·5	· · 27·4	198 30·4	10·7	15 45·7	5·7	56·4
O 10	331 46·7	26·4	213 00·1	10·8	15 40·0	5·8	56·4
N 11	346 46·9	25·5	227 29·9	10·8	15 34·2	5·9	56·3
D 12	1 47·1	N 0 24·5	241 59·7	10·9	S15 28·3	5·9	56·3
A 13	16 47·4	23·5	256 29·6	10·9	15 22·4	6·0	56·3
Y 14	31 47·6	22·6	270 59·5	11·0	15 16·4	6·1	56·3
15	46 47·8	· · 21·6	285 29·5	11·1	15 10·3	6·2	56·2
16	61 48·0	20·6	299 59·6	11·1	15 04·1	6·2	56·2
17	76 48·2	19·6	314 29·7	11·1	14 57·9	6·3	56·2
18	91 48·5	N 0 18·7	328 59·8	11·3	S14 51·6	6·4	56·2
19	106 48·7	17·7	343 30·1	11·3	14 45·2	6·4	56·1
20	121 48·9	16·7	358 00·4	11·3	14 38·8	6·5	56·1
21	136 49·1	· · 15·7	12 30·7	11·4	14 32·3	6·6	56·1
22	151 49·3	14·8	27 01·1	11·5	14 25·7	6·6	56·1
23	166 49·6	13·8	41 31·6	11·5	14 19·1	6·7	56·0
23 00	181 49·8	N 0 12·8	56 02·1	11·6	S14 12·4	6·8	56·0
01	196 50·0	11·8	70 32·7	11·6	14 05·6	6·8	56·0
02	211 50·2	10·9	85 03·3	11·7	13 58·8	6·9	56·0
03	226 50·4	· · 09·9	99 34·0	11·7	13 51·9	7·0	55·9
04	241 50·7	08·9	114 04·7	11·8	13 44·9	7·0	55·9
05	256 50·9	08·0	128 35·5	11·9	13 37·9	7·1	55·9
06	271 51·1	N 0 07·0	143 06·4	11·9	S13 30·8	7·1	55·9
07	286 51·3	06·0	157 37·3	11·9	13 23·7	7·2	55·8
T 08	301 51·5	05·0	172 08·2	12·1	13 16·5	7·3	55·8
U 09	316 51·7	· · 04·1	186 39·3	12·0	13 09·2	7·3	55·8
E 10	331 52·0	03·1	201 10·3	12·2	13 01·9	7·4	55·8
S 11	346 52·2	02·1	215 41·5	12·1	12 54·5	7·4	55·8
D 12	1 52·4	N 0 01·1	230 12·6	12·3	S12 47·1	7·5	55·7
A 13	16 52·6	N 0 00·2	244 43·9	12·3	12 39·6	7·5	55·7
Y 14	31 52·8	S 0 00·8	259 15·2	12·3	12 32·1	7·6	55·7
15	46 53·1	· · 01·8	273 46·5	12·4	12 24·5	7·7	55·7
16	61 53·3	02·8	288 17·9	12·4	12 16·8	7·7	55·6
17	76 53·5	03·7	302 49·3	12·5	12 09·1	7·7	55·6
18	91 53·7	S 0 04·7	317 20·8	12·6	S12 01·4	7·8	55·6
19	106 53·9	05·7	331 52·4	12·5	11 53·6	7·9	55·6
20	121 54·2	06·6	346 23·9	12·7	11 45·7	7·9	55·6
21	136 54·4	· · 07·6	0 55·6	12·7	11 37·8	8·0	55·5
22	151 54·6	08·6	15 27·3	12·7	11 29·8	8·0	55·5
23	166 54·8	09·6	29 59·0	12·8	11 21·8	8·0	55·5
24 00	181 55·0	S 0 10·5	44 30·8	12·9	S11 13·8	8·1	55·5
01	196 55·2	11·5	59 02·7	12·9	11 05·7	8·2	55·5
02	211 55·5	12·5	73 34·6	12·9	10 57·5	8·2	55·4
03	226 55·7	· · 13·5	88 06·5	13·0	10 49·3	8·2	55·4
04	241 55·9	14·4	102 38·5	13·0	10 41·1	8·3	55·4
05	256 56·1	15·4	117 10·5	13·1	10 32·8	8·3	55·4
06	271 56·3	S 0 16·4	131 42·6	13·1	S10 24·5	8·4	55·4
W 07	286 56·6	17·4	146 14·7	13·2	10 16·1	8·4	55·3
E 08	301 56·8	18·3	160 46·9	13·2	10 07·7	8·4	55·3
D 09	316 57·0	· · 19·3	175 19·1	13·3	9 59·3	8·5	55·3
N 10	331 57·2	20·3	189 51·4	13·3	9 50·8	8·5	55·3
E 11	346 57·4	21·3	204 23·7	13·3	9 42·3	8·6	55·3
S 12	1 57·6	S 0 22·2	218 56·0	13·4	S 9 33·7	8·6	55·2
D 13	16 57·9	23·2	233 28·4	13·4	9 25·1	8·6	55·2
A 14	31 58·1	24·2	248 00·8	13·5	9 16·5	8·7	55·2
Y 15	46 58·3	· · 25·1	262 33·3	13·5	9 07·8	8·7	55·2
16	61 58·5	26·1	277 05·8	13·6	3 59·1	8·8	55·2
17	76 58·7	27·1	291 38·4	13·6	8 50·3	8·8	55·1
18	91 59·0	S 0 28·1	306 11·0	13·6	S 8 41·5	8·8	55·1
19	106 59·2	29·0	320 43·6	13·7	8 32·7	8·8	55·1
20	121 59·4	30·0	335 16·3	13·7	8 23·9	8·9	55·1
21	136 59·6	· · 31·0	349 49·0	13·8	8 15·0	8·9	55·1
22	151 59·8	32·0	4 21·8	13·8	8 06·1	9·0	55·1
23	167 00·0	32·9	18 54·6	13·8	7 57·1	8·9	55·0
	S.D. 16·0	d 1·0	S.D. 15·3		15·2		15·1

Lat. °	Twilight Naut. h m	Twilight Civil h m	Sunrise h m	Moonrise 22 h m	23 h m	24 h m	25 h m
N 72	03 01	04 32	05 41	17 46	17 37	17 29	17 23
N 70	03 21	04 40	05 42	17 13	17 16	17 16	17 16
68	03 36	04 47	05 43	16 49	16 59	17 05	17 10
66	03 48	04 52	05 44	16 31	16 46	16 56	17 05
64	03 58	04 57	05 44	16 16	16 34	16 49	17 01
62	04 07	05 01	05 45	16 03	16 25	16 42	16 57
60	04 14	05 04	05 45	15 52	16 17	16 37	16 54
N 58	04 20	05 07	05 46	15 43	16 09	16 32	16 51
56	04 25	05 09	05 46	15 35	16 03	16 27	16 48
54	04 29	05 11	05 46	15 27	15 57	16 23	16 46
52	04 33	05 13	05 47	15 21	15 52	16 19	16 44
50	04 37	05 15	05 47	15 15	15 47	16 16	16 42
45	04 44	05 18	05 48	15 02	15 37	16 09	16 38
N 40	04 49	05 21	05 48	14 51	15 29	16 03	16 34
35	04 53	05 23	05 48	14 42	15 21	15 58	16 31
30	04 57	05 25	05 48	14 34	15 15	15 53	16 28
20	05 01	05 27	05 49	14 20	15 04	15 45	16 24
N 10	05 04	05 28	05 49	14 08	14 54	15 38	16 20
0	05 05	05 29	05 49	13 56	14 45	15 31	16 16
S 10	05 04	05 28	05 49	13 45	14 36	15 25	16 12
20	05 02	05 27	05 49	13 33	14 26	15 17	16 08
30	04 57	05 25	05 49	13 19	14 15	15 09	16 03
35	04 54	05 24	05 49	13 11	14 08	15 05	16 00
40	04 50	05 22	05 49	13 01	14 01	14 59	15 57
45	04 45	05 19	05 48	12 51	13 52	14 53	15 54
S 50	04 38	05 16	05 48	12 37	13 41	14 46	15 49
52	04 34	05 14	05 48	12 31	13 37	14 42	15 47
54	04 31	05 12	05 47	12 25	13 31	14 39	15 45
56	04 26	05 10	05 47	12 17	13 25	14 34	15 43
58	04 21	05 08	05 47	12 08	13 19	14 30	15 40
S 60	04 15	05 05	05 47	11 59	13 11	14 24	15 37

Lat. °	Sunset h m	Twilight Civil h m	Twilight Naut. h m	Moonset 22 h m	23 h m	24 h m	25 h m
N 72	18 02	19 10	20 39	22 43	24 32	00 32	02 14
N 70	18 01	19 03	20 21	23 15	24 51	00 51	02 25
68	18 01	18 56	20 06	23 38	25 07	01 07	02 35
66	18 00	18 51	19 54	23 55	25 19	01 19	02 42
64	17 59	18 47	19 45	24 10	00 10	01 30	02 49
62	17 59	18 43	19 37	24 22	00 22	01 38	02 54
60	17 58	18 40	19 30	24 32	00 32	01 46	02 59
N 58	17 58	18 38	19 24	24 41	00 41	01 52	03 03
56	17 58	18 35	19 19	24 49	00 49	01 58	03 07
54	17 58	18 33	19 15	24 56	00 56	02 03	03 10
52	17 57	18 31	19 11	25 02	01 02	02 08	03 14
50	17 57	18 29	19 07	00 04	01 08	02 12	03 16
45	17 57	18 26	19 00	00 18	01 20	02 21	03 22
N 40	17 56	18 23	18 55	00 30	01 30	02 29	03 27
35	17 56	18 21	18 51	00 41	01 38	02 35	03 32
30	17 56	18 20	18 48	00 49	01 46	02 41	03 35
20	17 56	18 18	18 43	01 05	01 58	02 51	03 42
N 10	17 56	18 17	18 41	01 18	02 10	02 59	03 47
0	17 56	18 16	18 40	01 30	02 20	03 07	03 53
S 10	17 56	18 17	18 41	01 43	02 30	03 15	03 58
20	17 56	18 18	18 44	01 56	02 41	03 24	04 04
30	17 56	18 20	18 48	02 11	02 54	03 33	04 10
35	17 57	18 22	18 52	02 19	03 01	03 39	04 14
40	17 57	18 24	18 56	02 29	03 09	03 45	04 18
45	17 58	18 27	19 01	02 41	03 19	03 52	04 22
S 50	17 58	18 31	19 08	02 54	03 30	04 01	04 28
52	17 58	18 32	19 12	03 01	03 35	04 05	04 31
54	17 59	18 34	19 16	03 08	03 41	04 09	04 33
56	17 59	18 36	19 21	03 16	03 48	04 14	04 36
58	17 59	18 39	19 26	03 25	03 55	04 19	04 40
S 60	18 00	18 41	19 32	03 35	04 03	04 25	04 44

Day	SUN Eqn. of Time 00h m s	SUN Eqn. of Time 12h m s	SUN Mer. Pass. h m	MOON Mer. Pass. Upper h m	MOON Mer. Pass. Lower h m	Age d	Phase
22	06 58	07 08	11 53	20 08	07 43	09	○
23	07 19	07 29	11 53	20 56	08 32	10	
24	07 40	07 50	11 52	21 42	09 19	11	

SEPTEMBER 25, 26, 27 (THURS., FRI., SAT.)

G.M.T.		ARIES	VENUS −3·4		MARS −1·0		JUPITER −1·3		SATURN +0·7	
d	h	G.H.A. ° ′	G.H.A. ° ′	Dec. ° ′	G.H.A. ° ′	Dec. ° ′	G.H.A. ° ′	Dec. ° ′	G.H.A. ° ′	Dec. ° ′
25	00	3 18·4	192 42·0	N 5 35·4	304 10·9	N18 31·7	151 49·0	S11 42·4	104 11·3	S21 55·2
	01	18 20·9	207 41·6	34·2	319 12·8	31·8	166 51·0	42·6	119 13·6	55·2
	02	33 23·3	222 41·2	33·0	334 14·8	32·0	181 52·9	42·8	134 15·9	55·2
	03	48 25·8	237 40·7	·· 31·8	349 16·7	·· 32·2	196 54·9	·· 43·0	149 18·2	·· 55·2
	04	63 28·3	252 40·3	30·6	4 18·7	32·3	211 56·9	43·1	164 20·6	55·2
	05	78 30·7	267 39·9	29·4	19 20·6	32·5	226 58·9	43·3	179 22·9	55·2
	06	93 33·2	282 39·5	N 5 28·2	34 22·6	N18 32·6	242 00·9	S11 43·5	194 25·2	S21 55·3
	07	108 35·7	297 39·1	27·0	49 24·5	32·8	257 02·9	43·7	209 27·6	55·3
T	08	123 38·1	312 38·7	25·8	64 26·5	33·0	272 04·8	43·8	224 29·9	55·3
H	09	138 40·6	327 38·3	·· 24·6	79 28·4	·· 33·1	287 06·8	·· 44·0	239 32·2	·· 55·3
U	10	153 43·0	342 37·9	23·4	94 30·4	33·3	302 08·8	44·2	254 34·5	55·3
R	11	168 45·5	357 37·4	22·2	109 32·3	33·5	317 10·8	44·4	269 36·9	55·3
S	12	183 48·0	12 37·0	N 5 21·0	124 34·3	N18 33·6	332 12·8	S11 44·5	284 39·2	S21 55·4
D	13	198 50·4	27 36·6	19·8	139 36·2	33·8	347 14·8	44·7	299 41·5	55·4
A	14	213 52·9	42 36·2	18·6	154 38·2	34·0	2 16·7	44·9	314 43·9	55·4
Y	15	228 55·4	57 35·8	·· 17·4	169 40·1	·· 34·1	17 18·7	·· 45·1	329 46·2	·· 55·4
	16	243 57·8	72 35·4	16·2	184 42·1	34·3	32 20·7	45·2	344 48·5	55·4
	17	259 00·3	87 35·0	15·0	199 44·1	34·4	47 22·7	45·4	359 50·8	55·4
	18	274 02·8	102 34·6	N 5 13·8	214 46·0	N18 34·6	62 24·7	S11 45·6	14 53·2	S21 55·5
	19	289 05·2	117 34·1	12·6	229 48·0	34·8	77 26·7	45·8	29 55·5	55·5
	20	304 07·7	132 33·7	11·3	244 49·9	34·9	92 28·6	45·9	44 57·8	55·5
	21	319 10·1	147 33·3	·· 10·1	259 51·9	·· 35·1	107 30·6	·· 46·1	60 00·1	·· 55·5
	22	334 12·6	162 32·9	08·9	274 53·9	35·3	122 32·6	46·3	75 02·5	55·5
	23	349 15·1	177 32·5	07·7	289 55·8	35·4	137 34·6	46·5	90 04·8	55·5
26	00	4 17·5	192 32·1	N 5 06·5	304 57·8	N18 35·6	152 36·6	S11 46·6	105 07·1	S21 55·6
	01	19 20·0	207 31·7	05·3	319 59·8	35·7	167 38·6	46·8	120 09·4	55·6
	02	34 22·5	222 31·3	04·1	335 01·8	35·9	182 40·5	47·0	135 11·8	55·6
	03	49 24·9	237 30·9	·· 02·9	350 03·7	·· 36·1	197 42·5	·· 47·2	150 14·1	·· 55·6
	04	64 27·4	252 30·5	01·7	5 05·7	36·2	212 44·5	47·3	165 16·4	55·6
	05	79 29·9	267 30·0	5 00·5	20 07·7	36·4	227 46·5	47·5	180 18·7	55·6
	06	94 32·3	282 29·6	N 4 59·3	35 09·7	N18 36·5	242 48·5	S11 47·7	195 21·1	S21 55·7
	07	109 34·8	297 29·2	58·1	50 11·6	36·7	257 50·4	47·9	210 23·4	55·7
	08	124 37·3	312 28·8	56·9	65 13·6	36·9	272 52·4	48·0	225 25·7	55·7
F	09	139 39·7	327 28·4	·· 55·7	80 15·6	·· 37·0	287 54·4	·· 48·2	240 28·0	·· 55·7
R	10	154 42·2	342 28·0	54·4	95 17·6	37·2	302 56·4	48·4	255 30·4	55·7
I	11	169 44·6	357 27·6	53·2	110 19·6	37·3	317 58·4	48·6	270 32·7	55·7
D	12	184 47·1	12 27·2	N 4 52·0	125 21·6	N18 37·5	333 00·4	S11 48·7	285 35·0	S21 55·8
A	13	199 49·6	27 26·8	50·8	140 23·5	37·6	348 02·3	48·9	300 37·3	55·8
Y	14	214 52·0	42 26·4	49·6	155 25·5	37·8	3 04·3	49·1	315 39·7	55·8
	15	229 54·5	57 26·0	·· 48·4	170 27·5	·· 38·0	18 06·3	·· 49·3	330 42·0	·· 55·8
	16	244 57·0	72 25·5	47·2	185 29·5	38·1	33 08·3	49·4	345 44·3	55·8
	17	259 59·4	87 25·1	46·0	200 31·5	38·3	48 10·3	49·6	0 46·6	55·9
	18	275 01·9	102 24·7	N 4 44·8	215 33·5	N18 38·4	63 12·2	S11 49·8	15 49·0	S21 55·9
	19	290 04·4	117 24·3	43·5	230 35·5	38·6	78 14·2	50·0	30 51·3	55·9
	20	305 06·8	132 23·9	42·3	245 37·5	38·7	93 16·2	50·1	45 53·6	55·9
	21	320 09·3	147 23·5	·· 41·1	260 39·5	·· 38·9	108 18·2	·· 50·3	60 55·9	·· 55·9
	22	335 11·7	162 23·1	39·9	275 41·5	39·1	123 20·2	50·5	75 58·2	55·9
	23	350 14·2	177 22·7	38·7	290 43·5	39·2	138 22·1	50·7	91 00·6	56·0
27	00	5 16·7	192 22·3	N 4 37·5	305 45·5	N18 39·4	153 24·1	S11 50·8	106 02·9	S21 56·0
	01	20 19·1	207 21·9	36·3	320 47·5	39·5	168 26·1	51·0	121 05·2	56·0
	02	35 21·6	222 21·5	35·1	335 49·5	39·7	183 28·1	51·2	136 07·5	56·0
	03	50 24·1	237 21·1	·· 33·8	350 51·5	·· 39·8	198 30·1	·· 51·4	151 09·9	·· 56·0
	04	65 26·5	252 20·7	32·6	5 53·5	40·0	213 32·0	51·5	166 12·2	56·0
	05	80 29·0	267 20·2	31·4	20 55·5	40·2	228 34·0	51·7	181 14·5	56·1
	06	95 31·5	282 19·8	N 4 30·2	35 57·5	N18 40·3	243 36·0	S11 51·9	196 16·8	S21 56·1
	07	110 33·9	297 19·4	29·0	50 59·5	40·5	258 38·0	52·1	211 19·1	56·1
S	08	125 36·4	312 19·0	27·8	66 01·6	40·6	273 40·0	52·2	226 21·5	56·1
A	09	140 38·9	327 18·6	·· 26·6	81 03·6	·· 40·8	288 41·9	·· 52·4	241 23·8	·· 56·1
T	10	155 41·3	342 18·2	25·3	96 05·6	40·9	303 43·9	52·6	256 26·1	56·1
U	11	170 43·8	357 17·8	24·1	111 07·6	41·1	318 45·9	52·8	271 28·4	56·2
R	12	185 46·2	12 17·4	N 4 22·9	126 09·6	N18 41·2	333 47·9	S11 52·9	286 30·7	S21 56·2
D	13	200 48·7	27 17·0	21·7	141 11·6	41·4	348 49·9	53·1	301 33·1	56·2
A	14	215 51·2	42 16·6	20·5	156 13·7	41·5	3 51·8	53·3	316 35·4	56·2
Y	15	230 53·6	57 16·2	·· 19·3	171 15·7	·· 41·7	18 53·8	·· 53·5	331 37·7	·· 56·2
	16	245 56·1	72 15·8	18·0	186 17·7	41·8	33 55·8	53·6	346 40·0	56·3
	17	260 58·6	87 15·4	16·8	201 19·7	42·0	48 57·8	53·8	1 42·3	56·3
	18	276 01·0	102 15·0	N 4 15·6	216 21·8	N18 42·1	63 59·7	S11 54·0	16 44·7	S21 56·3
	19	291 03·5	117 14·6	14·4	231 23·8	42·3	79 01·7	54·2	31 47·0	56·3
	20	306 06·0	132 14·2	13·2	246 25·8	42·4	94 03·7	54·3	46 49·3	56·3
	21	321 08·4	147 13·8	·· 12·0	261 27·8	·· 42·6	109 05·7	· 54·5	61 51·6	·· 56·3
	22	336 10·9	162 13·4	10·7	276 29·9	42·7	124 07·7	54·7	76 53·9	56·4
	23	351 13·4	177 12·9	09·5	291 31·9	42·9	139 09·6	54·9	91 56·3	56·4
Mer. Pass.		23 38·9 (h m)	v −0·4	d 1·2	v 2·0	d 0·2	v 2·0	d 0·2	v 2·3	d 0·0

STARS

Name	S.H.A. ° ′	Dec. ° ′
Acamar	315 49·3	S 40 28·0
Achernar	335 56·9	S 57 26·6
Acrux	173 55·9	S 62 52·2
Adhara	255 44·9	S 28 54·8
Aldebaran	291 36·5	N 16 25·6
Alioth	166 57·4	N 56 11·1
Alkaid	153 31·8	N 49 31·3
Al Na'ir	28 35·0	S 47 09·6
Alnilam	276 28·1	S 1 13·6
Alphard	218 36·7	S 8 28·7
Alphecca	126 46·1	N 26 51·5
Alpheratz	358 25·7	N 28 51·9
Altair	62 48·2	N 8 45·8
Ankaa	353 55·9	S 42 31·6
Antares	113 16·9	S 26 20·4
Arcturus	146 33·6	N 19 24·0
Atria	108 55·8	S 68 57·4
Avior	234 35·1	S 59 22·4
Bellatrix	279 16·1	N 6 18·8
Betelgeuse	271 45·8	N 7 24·0
Canopus	264 14·4	S 52 40·2
Capella	281 35·2	N 45 57·2
Deneb	49 59·2	N 45 08·3
Denebola	183 15·9	N 14 48·2
Diphda	349 36·8	S 18 12·6
Dubhe	194 42·8	N 61 58·3
Elnath	279 04·6	N 28 34·3
Eltanin	91 05·3	N 51 30·0
Enif	34 27·3	N 9 41·4
Fomalhaut	16 09·0	S 29 50·3
Gacrux	172 47·3	S 56 52·9
Gienah	176 35·0	S 17 18·7
Hadar	149 46·8	S 60 10·5
Hamal	328 46·9	N 23 16·1
Kaus Aust.	84 38·4	S 34 24·3
Kochab	137 19·2	N 74 19·7
Markab	14 19·0	N 14 59·2
Menkar	314 57·9	N 3 55·8
Menkent	148 56·5	S 36 10·0
Miaplacidus	221 49·0	S 69 32·7
Mirfak	309 39·0	N 49 42·8
Nunki	76 49·2	S 26 20·9
Peacock	54 23·7	S 56 52·2
Pollux	244 18·2	N 28 07·5
Procyon	245 42·9	N 5 19·9
Rasalhague	96 44·7	N 12 35·6
Regulus	208 27·6	N 12 10·2
Rigel	281 51·5	S 8 14·8
Rigil Kent.	140 48·5	S 60 40·0
Sabik	102 59·8	S 15 40·4
Schedar	350 26·9	N 56 18·7
Shaula	97 17·9	S 37 04·5
Sirius	259 10·0	S 16 39·4
Spica	159 14·9	S 10 56·7
Suhail	223 23·0	S 43 15·8
Vega	81 06·8	N 38 45·1
Zuben'ubi	137 51·3	S 15 52·2

	S.H.A. ° ′	Mer. Pass. h m
Venus	188 14·6	11 10
Mars	300 40·3	3 40
Jupiter	148 19·1	13 48
Saturn	100 49·6	16 57

SEPTEMBER 25, 26, 27 (THURS., FRI., SAT.)

	G.M.T.	SUN G.H.A.	SUN Dec.	MOON G.H.A.	*v*	MOON Dec.	*d*	H.P.
	d h	° ′	° ′	° ′		° ′	′	′
	25 00	182 00·3	S 0 33·9	33 27·4	13·9	S 7 48·2	9·0	55·0
	01	197 00·5	34·9	48 00·3	13·9	7 39·2	9·1	55·0
	02	212 00·7	35·9	62 33·2	13·9	7 30·1	9·0	55·0
	03	227 00·9	· · 36·8	77 06·1	14·0	7 21·1	9·1	55·0
	04	242 01·1	37·8	91 39·1	14·0	7 12·0	9·1	55·0
	05	257 01·3	38·8	106 12·1	14·1	7 02·9	9·2	54·9
	06	272 01·6	S 0 39·8	120 45·2	14·1	S 6 53·7	9·1	54·9
	07	287 01·8	40·7	135 18·3	14·1	6 44·6	9·2	54·9
T	08	302 02·0	41·7	149 51·4	14·1	6 35·4	9·2	54·9
H	09	317 02·2	· · 42·7	164 24·5	14·2	6 26·2	9·3	54·9
U	10	332 02·4	43·7	178 57·7	14·2	6 16·9	9·2	54·9
R	11	347 02·6	44·6	193 30·9	14·3	6 07·7	9·3	54·8
S	12	2 02·9	S 0 45·6	208 04·2	14·2	S 5 58·4	9·3	54·8
D	13	17 03·1	46·6	222 37·4	14·3	5 49·1	9·3	54·8
A	14	32 03·3	47·6	237 10·7	14·4	5 39·8	9·4	54·8
Y	15	47 03·5	· · 48·5	251 44·1	14·3	5 30·4	9·3	54·8
	16	62 03·7	49·5	266 17·4	14·4	5 21·1	9·4	54·8
	17	77 03·9	50·5	280 50·8	14·4	5 11·7	9·4	54·8
	18	92 04·2	S 0 51·5	295 24·2	14·5	S 5 02·3	9·4	54·7
	19	107 04·4	52·4	309 57·7	14·5	4 52·9	9·4	54·7
	20	122 04·6	53·4	324 31·2	14·5	4 43·5	9·5	54·7
	21	137 04·8	· · 54·4	339 04·7	14·5	4 34·0	9·4	54·7
	22	152 05·0	55·4	353 38·2	14·6	4 24·6	9·5	54·7
	23	167 05·2	56·3	8 11·8	14·5	4 15·1	9·5	54·7
	26 00	182 05·5	S 0 57·3	22 45·3	14·6	S 4 05·6	9·5	54·7
	01	197 05·7	58·3	37 18·9	14·7	3 56·1	9·5	54·6
	02	212 05·9	0 59·2	51 52·6	14·6	3 46·6	9·5	54·6
	03	227 06·1	1 00·2	66 26·2	14·7	3 37·1	9·6	54·6
	04	242 06·3	01·2	80 59·9	14·7	3 27·5	9·5	54·6
	05	257 06·5	02·2	95 33·6	14·7	3 18·0	9·5	54·6
	06	272 06·7	S 1 03·1	110 07·3	14·7	S 3 08·5	9·6	54·6
	07	287 07·0	04·1	124 41·0	14·8	2 58·9	9·6	54·6
	08	302 07·2	05·1	139 14·8	14·7	2 49·3	9·6	54·6
F	09	317 07·4	· · 06·1	153 48·5	14·8	2 39·7	9·5	54·5
R	10	332 07·6	07·0	168 22·3	14·8	2 30·2	9·6	54·5
I	11	347 07·8	08·0	182 56·1	14·9	2 20·6	9·6	54·5
D	12	2 08·0	S 1 09·0	197 30·0	14·8	S 2 11·0	9·6	54·5
A	13	17 08·2	10·0	212 03·8	14·9	2 01·4	9·7	54·5
Y	14	32 08·5	10·9	226 37·7	14·9	1 51·7	9·6	54·5
	15	47 08·7	· · 11·9	241 11·6	14·9	1 42·1	9·6	54·5
	16	62 08·9	12·9	255 45·5	14·9	1 32·5	9·6	54·5
	17	77 09·1	13·9	270 19·4	14·9	1 22·9	9·6	54·4
	18	92 09·3	S 1 14·8	284 53·3	14·9	S 1 13·3	9·6	54·4
	19	107 09·5	15·8	299 27·2	15·0	1 03·7	9·6	54·4
	20	122 09·7	16·8	314 01·2	14·9	0 54·1	9·7	54·4
	21	137 10·0	· · 17·8	328 35·1	15·0	0 44·4	9·6	54·4
	22	152 10·2	18·7	343 09·1	15·0	0 34·8	9·6	54·4
	23	167 10·4	19·7	357 43·1	15·0	0 25·2	9·6	54·4
	27 00	182 10·6	S 1 20·7	12 17·1	15·0	S 0 15·6	9·6	54·4
	01	197 10·8	21·6	26 51·1	15·1	S 0 06·0	9·6	54·4
	02	212 11·0	22·6	41 25·2	15·0	N 0 03·6	9·7	54·3
	03	227 11·2	· · 23·6	55 59·2	15·0	0 13·3	9·6	54·3
	04	242 11·5	24·6	70 33·2	15·1	0 22·9	9·6	54·3
	05	257 11·7	25·5	85 07·3	15·0	0 32·5	9·6	54·3
	06	272 11·9	S 1 26·5	99 41·3	15·1	N 0 42·1	9·5	54·3
	07	287 12·1	27·5	114 15·4	15·1	0 51·6	9·6	54·3
S	08	302 12·3	28·5	128 49·5	15·1	1 01·2	9·6	54·3
A	09	317 12·5	· · 29·4	143 23·6	15·0	1 10·8	9·6	54·3
T	10	332 12·7	30·4	157 57·6	15·1	1 20·4	9·5	54·3
U	11	347 12·9	31·4	172 31·7	15·1	1 29·9	9·6	54·3
R	12	2 13·2	S 1 32·4	187 05·8	15·1	N 1 39·5	9·5	54·3
D	13	17 13·4	33·3	201 39·9	15·1	1 49·0	9·5	54·2
A	14	32 13·6	34·3	216 14·0	15·1	1 58·5	9·5	54·2
Y	15	47 13·8	· · 35·3	230 48·1	15·1	2 08·0	9·5	54·2
	16	62 14·0	36·3	245 22·2	15·2	2 17·5	9·5	54·2
	17	77 14·2	37·2	259 56·4	15·1	2 27·0	9·5	54·2
	18	92 14·4	S 1 38·2	274 30·5	15·1	N 2 36·5	9·5	54·2
	19	107 14·6	39·2	289 04·6	15·1	2 46·0	9·4	54·2
	20	122 14·8	40·1	303 38·7	15·1	2 55·4	9·5	54·2
	21	137 15·1	· · 41·1	318 12·8	15·1	3 04·9	9·4	54·2
	22	152 15·3	42·1	332 46·9	15·1	3 14·3	9·4	54·2
	23	167 15·5	43·1	347 21·0	15·1	3 23·7	9·4	54·2
		S.D. 16·0	*d* 1·0	S.D. 14·9		14·9		14·8

Lat.	Twilight Naut.	Twilight Civil	Sun-rise	Moonrise 25	26	27	28
°	h m	h m	h m	h m	h m	h m	h m
N 72	03 19	04 46	05 54	17 23	17 17	17 11	17 05
N 70	03 36	04 53	05 54	17 16	17 15	17 14	17 14
68	03 49	04 58	05 54	17 10	17 13	17 17	17 21
66	04 00	05 02	05 53	17 05	17 12	17 19	17 27
64	04 08	05 06	05 53	17 01	17 11	17 21	17 32
62	04 15	05 09	05 53	16 57	17 10	17 23	17 36
60	04 21	05 11	05 52	16 54	17 09	17 24	17 40
N 58	04 27	05 13	05 52	16 51	17 08	17 26	17 43
56	04 31	05 15	05 52	16 48	17 08	17 27	17 46
54	04 35	05 17	05 52	16 46	17 07	17 28	17 49
52	04 38	05 18	05 52	16 44	17 07	17 29	17 51
50	04 41	05 19	05 51	16 42	17 06	17 30	17 54
45	04 48	05 22	05 51	16 38	17 05	17 32	17 59
N 40	04 52	05 24	05 51	16 34	17 04	17 33	18 03
35	04 56	05 25	05 50	16 31	17 03	17 35	18 06
30	04 59	05 26	05 50	16 28	17 02	17 36	18 09
20	05 02	05 28	05 50	16 24	17 01	17 38	18 15
N 10	05 04	05 28	05 49	16 20	17 00	17 40	18 20
0	05 04	05 28	05 48	16 16	16 59	17 42	18 25
S 10	05 02	05 26	05 47	16 12	16 58	17 44	18 29
20	04 59	05 24	05 47	16 08	16 57	17 46	18 34
30	04 53	05 21	05 45	16 03	16 56	17 48	18 40
35	04 50	05 19	05 44	16 00	16 55	17 49	18 43
40	04 45	05 17	05 44	15 57	16 54	17 51	18 47
45	04 39	05 13	05 43	15 54	16 53	17 53	18 51
S 50	04 31	05 09	05 41	15 49	16 52	17 55	18 57
52	04 27	05 07	05 41	15 47	16 52	17 56	18 59
54	04 23	05 05	05 40	15 45	16 51	17 57	19 02
56	04 18	05 02	05 39	15 43	16 51	17 58	19 05
58	04 12	04 59	05 38	15 40	16 50	17 59	19 08
S 60	04 06	04 56	05 37	15 37	16 49	18 01	19 12

Lat.	Sun-set	Twilight Civil	Twilight Naut.	Moonset 25	26	27	28
°	h m	h m	h m	h m	h m	h m	h m
N 72	17 47	18 54	20 20	02 14	03 51	05 27	07 01
N 70	17 47	18 48	20 04	02 25	03 57	05 26	06 55
68	17 48	18 43	19 51	02 35	04 01	05 25	06 49
66	17 48	18 39	19 41	02 42	04 04	05 25	06 45
64	17 49	18 36	19 33	02 49	04 07	05 24	06 41
62	17 49	18 33	19 26	02 54	04 10	05 24	06 38
60	17 49	18 31	19 20	02 59	04 12	05 24	06 35
N 58	17 50	18 29	19 15	03 03	04 14	05 23	06 32
56	17 50	18 27	19 11	03 07	04 15	05 23	06 30
54	17 50	18 25	19 07	03 10	04 17	05 23	06 28
52	17 50	18 24	19 03	03 14	04 18	05 23	06 26
50	17 51	18 23	19 00	03 16	04 20	05 22	06 25
45	17 51	18 20	18 54	03 22	04 23	05 22	06 21
N 40	17 51	18 18	18 50	03 27	04 25	05 22	06 18
35	17 52	18 17	18 47	03 32	04 27	05 21	06 15
30	17 52	18 16	18 44	03 35	04 29	05 21	06 13
20	17 53	18 15	18 41	03 42	04 32	05 21	06 09
N 10	17 54	18 15	18 39	03 47	04 34	05 20	06 06
0	17 55	18 15	18 39	03 53	04 37	05 20	06 02
S 10	17 56	18 17	18 41	03 58	04 39	05 19	05 59
20	17 57	18 19	18 44	04 04	04 42	05 19	05 56
30	17 58	18 22	18 50	04 10	04 45	05 18	05 52
35	17 59	18 24	18 54	04 14	04 46	05 18	05 49
40	18 00	18 27	18 59	04 18	04 48	05 18	05 47
45	18 01	18 31	19 05	04 22	04 50	05 17	05 44
S 50	18 03	18 35	19 13	04 28	04 53	05 17	05 40
52	18 03	18 37	19 17	04 31	04 54	05 16	05 39
54	18 04	18 40	19 22	04 33	04 55	05 16	05 37
56	18 05	18 42	19 27	04 36	04 57	05 16	05 35
58	18 06	18 45	19 33	04 40	04 58	05 16	05 33
S 60	18 07	18 49	19 40	04 44	05 00	05 15	05 30

Day	SUN Eqn. of Time 00ʰ	SUN Eqn. of Time 12ʰ	SUN Mer. Pass.	MOON Mer. Pass. Upper	MOON Mer. Pass. Lower	Age	Phase
	m s	m s	h m	h m	h m	d	
25	08 01	08 11	11 52	22 26	10 04	12	
26	08 21	08 32	11 51	23 09	10 48	13	○
27	08 42	08 52	11 51	23 52	11 31	14	

SEPTEMBER 28, 29, 30 (SUN., MON., TUES.)

	G.M.T.	ARIES	VENUS −3·4		MARS −1·1		JUPITER −1·3		SATURN +0·7		STARS		
	d h	G.H.A. ° ′	G.H.A. ° ′	Dec. ° ′	G.H.A. ° ′	Dec. ° ′	G.H.A. ° ′	Dec. ° ′	G.H.A. ° ′	Dec. ° ′	Name	S.H.A. ° ′	Dec. ° ′
	28 00	6 15·8	192 12·5	N 4 08·3	306 33·9	N18 43·1	154 11·6	S11 55·0	106 58·6	S21 56·4	Acamar	315 49·3	S 40 28·0
	01	21 18·3	207 12·1	07·1	321 36·0	43·2	169 13·6	55·2	122 00·9	56·4	Achernar	335 56·9	S 57 26·6
	02	36 20·7	222 11·7	05·9	336 38·0	43·4	184 15·6	55·4	137 03·2	56·4	Acrux	173 55·9	S 62 52·2
	03	51 23·2	237 11·3	· · 04·6	351 40·0	· · 43·5	199 17·5	· · 55·6	152 05·5	· · 56·4	Adhara	255 44·9	S 28 54·8
	04	66 25·7	252 10·9	03·4	6 42·1	43·7	214 19·5	55·7	167 07·8	56·5	Aldebaran	291 36·5	N 16 25·6
	05	81 28·1	267 10·5	02·2	21 44·1	43·8	229 21·5	55·9	182 10·2	56·5			
	06	96 30·6	282 10·1	N 4 01·0	36 46·2	N18 44·0	244 23·5	S11 56·1	197 12·5	S21 56·5	Alioth	166 57·4	N 56 11·1
	07	111 33·1	297 09·7	3 59·8	51 48·2	44·1	259 25·4	56·3	212 14·8	56·5	Alkaid	153 31·8	N 49 31·3
	08	126 35·5	312 09·3	58·5	66 50·3	44·2	274 27·4	56·4	227 17·1	56·5	Al Na'ir	28 35·0	S 47 09·6
SUNDAY	09	141 38·0	327 08·9	· · 57·[illegible]	81 52·3	· · 44·4	289 29·4	· · 56·6	242 19·4	· · 56·5	Alnilam	276 28·1	S 1 13·6
	10	156 40·5	342 08·5	56·1	96 54·3	44·5	304 31·4	56·8	257 21·7	56·6	Alphard	218 36·7	S 8 28·7
	11	171 42·9	357 08·1	54·9	111 56·4	44·7	319 33·4	57·0	272 24·1	56·6			
	12	186 45·4	12 07·7	N 3 53·7	126 58·4	N18 44·8	334 35·3	S11 57·1	287 26·4	S21 56·6	Alphecca	126 46·1	N 26 51·5
	13	201 47·8	27 07·3	52·4	142 00·5	45·0	349 37·3	57·3	302 28·7	56·6	Alpheratz	358 25·7	N 28 51·9
	14	216 50·3	42 06·9	51·2	157 02·6	45·1	4 39·3	57·5	317 31·0	56·6	Altair	62 48·3	N 8 45·8
	15	231 52·8	57 06·5	· · 50·0	172 04·6	· · 45·3	19 41·3	· · 57·7	332 33·3	· · 56·6	Ankaa	353 55·9	S 42 31·7
	16	246 55·2	72 06·1	48·8	187 06·7	45·4	34 43·2	57·9	347 35·6	56·7	Antares	113 16·9	S 26 20·4
	17	261 57·7	87 05·7	47·5	202 08·7	45·6	49 45·2	58·0	2 38·0	56·7			
	18	277 00·2	102 05·3	N 3 46·3	217 10·8	N18 45·7	64 47·2	S11 58·2	17 40·3	S21 56·7	Arcturus	146 33·6	N 19 24·0
	19	292 02·6	117 04·9	45·1	232 12·8	45·9	79 49·2	58·4	32 42·6	56·7	Atria	108 55·9	S 68 57·4
	20	307 05·1	132 04·5	43·9	247 14·9	46·0	94 51·1	58·6	47 44·9	56·7	Avior	234 35·1	S 59 22·4
	21	322 07·6	147 04·1	· · 42·7	262 17·0	· · 46·2	109 53·1	· · 58·7	62 47·2	· · 56·8	Bellatrix	279 16·1	N 6 18·8
	22	337 10·0	162 03·7	41·4	277 19·0	46·3	124 55·1	58·9	77 49·5	56·8	Betelgeuse	271 45·8	N 7 24·0
	23	352 12·5	177 03·3	40·2	292 21·1	46·5	139 57·1	59·1	92 51·8	56·8			
	29 00	7 15·0	192 02·9	N 3 39·0	307 23·2	N18 46·6	154 59·0	S11 59·3	107 54·2	S21 56·8	Canopus	264 14·4	S 52 40·2
	01	22 17·4	207 02·5	37·8	322 25·2	46·8	170 01·0	59·4	122 56·5	56·8	Capella	281 35·1	N 45 57·3
	02	37 19·9	222 02·1	36·5	337 27·3	46·9	185 03·0	59·6	137 58·8	56·8	Deneb	49 59·3	N 45 08·3
	03	52 22·3	237 01·7	· · 35·3	352 29·4	· · 47·0	200 05·0	11 59·8	153 01·1	· · 56·9	Denebola	183 15·9	N 14 48·2
	04	67 24·8	252 01·3	34·1	7 31·4	47·2	215 06·9	12 00·0	168 03·4	56·9	Diphda	349 36·8	S 18 12·6
	05	82 27·3	267 00·9	32·9	22 33·5	47·3	230 08·9	00·1	183 05·7	56·9			
	06	97 29·7	282 00·5	N 3 31·6	37 35·6	N18 47·5	245 10·9	S12 00·3	198 08·0	S21 56·9	Dubhe	194 42·8	N 61 58·3
	07	112 32·2	297 00·1	30·4	52 37·7	47·6	260 12·9	00·5	213 10·4	56·9	Elnath	279 04·6	N 28 34·3
	08	127 34·7	311 59·7	29·2	67 39·7	47·8	275 14·8	00·7	228 12·7	56·9	Eltanin	91 05·3	N 51 30·0
MONDAY	09	142 37·1	326 59·3	· · 28·0	82 41·8	· · 47·9	290 16·8	· · 00·8	243 15·0	· · 57·0	Enif	34 27·3	N 9 41·4
	10	157 39·6	341 58·9	26·7	97 43·9	48·1	305 18·8	01·0	258 17·3	57·0	Fomalhaut	16 09·0	S 29 50·4
	11	172 42·1	356 58·5	25·5	112 46·0	48·2	320 20·8	01·2	273 19·6	57·0			
	12	187 44·5	11 58·1	N 3 24·3	127 48·1	N18 48·3	335 22·7	S12 01·4	288 21·9	S21 57·0	Gacrux	172 47·3	S 56 52·9
	13	202 47·0	26 57·7	23·0	142 50·1	48·5	350 24·7	01·5	303 24·2	57·0	Gienah	176 35·0	S 17 18·7
	14	217 49·4	41 57·3	21·8	157 52·2	48·6	5 26·7	01·7	318 26·5	57·0	Hadar	149 46·8	S 60 10·5
	15	232 51·9	56 56·9	· · 20·6	172 54·3	· · 48·8	20 28·7	· · 01·9	333 28·9	· · 57·1	Hamal	328 46·9	N 23 16·1
	16	247 54·4	71 56·5	19·4	187 56·4	48·9	35 30·6	02·1	348 31·2	57·1	Kaus Aust.	84 38·4	S 34 24·3
	17	262 56·8	86 56·1	18·1	202 58·5	49·1	50 32·6	02·3	3 33·5	57·1			
	18	277 59·3	101 55·7	N 3 16·9	218 00·6	N18 49·2	65 34·6	S12 02·4	18 35·3	S21 57·1	Kochab	137 19·2	N 74 19·7
	19	293 01·8	116 55·3	15·7	233 02·7	49·3	80 36·6	02·6	33 38·1	57·1	Markab	14 19·0	N 14 59·2
	20	308 04·2	131 54·9	14·5	248 04·8	49·5	95 38·5	02·8	48 40·4	57·2	Menkar	314 57·9	N 3 55·8
	21	323 06·7	146 54·5	· · 13·2	263 06·9	· · 49·6	110 40·5	· · 03·0	63 42·7	· · 57·2	Menkent	148 56·5	S 36 10·0
	22	338 09·2	161 54·1	12·0	278 09·0	49·8	125 42·5	03·1	78 45·0	57·2	Miaplacidus	221 49·0	S 69 32·7
	23	353 11·6	176 53·7	10·8	293 11·1	49·9	140 44·4	03·3	93 47·4	57·2			
	30 00	8 14·1	191 53·3	N 3 09·5	308 13·2	N18 50·1	155 46·4	S12 03·5	108 49·7	S21 57·2	Mirfak	309 38·9	N 49 42·8
	01	23 16·6	206 52·9	08·3	323 15·3	50·2	170 48·4	03·7	123 52·0	57·2	Nunki	76 49·3	S 26 20·9
	02	38 19·0	221 52·5	07·1	338 17·4	50·3	185 50·4	03·8	138 54·3	57·3	Peacock	54 23·8	S 56 52·2
	03	53 21·5	236 52·1	· · 05·9	353 19·5	· · 50·5	200 52·3	· · 04·0	153 56·6	· · 57·3	Pollux	244 18·1	N 28 07·5
	04	68 23·9	251 51·7	04·6	8 21·6	50·6	215 54·3	04·2	168 58·9	57·3	Procyon	245 42·9	N 5 19·9
	05	83 26·4	266 51·3	03·4	23 23·7	50·8	230 56·3	04·4	184 01·2	57·3			
	06	98 28·9	281 50·9	N 3 02·2	38 25·8	N18 50·9	245 58·3	S12 04·5	199 03·5	S21 57·3	Rasalhague	96 44·7	N 12 35·6
	07	113 31·3	296 50·5	3 00·9	53 27·9	51·0	261 00·2	04·7	214 05·8	57·3	Regulus	208 27·5	N 12 10·2
TUESDAY	08	128 33·8	311 50·1	2 59·7	68 30·0	51·2	276 02·2	04·9	229 08·1	57·4	Rigel	281 51·5	S 8 14·8
	09	143 36·3	326 49·7	· · 58·5	83 32·1	· · 51·3	291 04·2	· · 05·1	244 10·5	· · 57·4	Rigil Kent.	140 48·5	S 60 40·0
	10	158 38·7	341 49·3	57·2	98 34·2	51·5	306 06·1	05·2	259 12·8	57·4	Sabik	102 59·9	S 15 40·4
	11	173 41·2	356 48·9	56·0	113 36·3	51·6	321 08·1	05·4	274 15·1	57·4			
	12	188 43·7	11 48·5	N 2 54·8	128 38·5	N18 51·7	336 10·1	S12 05·6	289 17·4	S21 57·4	Schedar	350 26·9	N 56 18·8
	13	203 46·1	26 48·1	53·5	143 40·6	51·9	351 12·1	05·8	304 19·7	57·4	Shaula	97 17·9	S 37 04·5
	14	218 48·6	41 47·7	52·3	158 42·7	52·0	6 14·0	05·9	319 22·0	57·5	Sirius	259 10·0	S 16 39·4
	15	233 51·0	56 47·3	· · 51·1	173 44·8	· · 52·1	21 16·0	· · 06·1	334 24·3	· · 57·5	Spica	159 14·9	S 10 56·7
	16	248 53·5	71 46·9	49·9	188 46·9	52·3	36 18·0	06·3	349 26·6	57·5	Suhail	223 23·0	S 43 15·8
	17	263 56·0	86 46·5	48·6	203 49·1	52·4	51 19·9	06·5	4 28·9	57·5			
	18	278 58·4	101 46·1	N 2 47·4	218 51·2	N18 52·6	66 21·9	S12 06·6	19 31·2	S21 57·5	Vega	81 06·8	N 38 45·1
	19	294 00·9	116 45·7	46·2	233 53·3	52·7	81 23·9	06·8	34 33·5	57·6	Zuben'ubi	137 51·3	S 15 52·2
	20	309 03·4	131 45·3	44·9	248 55·4	52·8	96 25·9	07·0	49 35·8	57·6		S.H.A. ° ′	Mer. Pass. h m
	21	324 05·8	146 44·9	· · 43·7	263 57·6	· · 53·0	111 27·8	· · 07·2	64 38·2	· · 57·6			
	22	339 08·3	161 44·5	42·5	278 59·7	53·1	126 29·8	07·4	79 40·5	57·6	Venus	184 47·9	11 12
	23	354 10·8	176 44·1	41·2	294 01·8	53·2	141 31·8	07·5	94 42·8	57·6	Mars	300 08·2	3 30
	Mer. Pass.	h m 23 27·2	v −0·4	d 1·2	v 2·1	d 0·1	v 2·0	d 0·2	v 2·3	d 0·0	Jupiter	147 44·0	13 38
											Saturn	100 39·2	16 46

SEPTEMBER 28, 29, 30 (SUN., MON., TUES.)

G.M.T. d		h	SUN G.H.A. ° ′	SUN Dec. ° ′	MOON G.H.A. ° ′	v ′	MOON Dec. ° ′	d ′	H.P.
28		00	182 15·7	S 1 44·0	1 55·1	15·2	N 3 33·1	9·4	54·2
		01	197 15·9	45·0	16 29·3	15·1	3 42·5	9·3	54·2
		02	212 16·1	46·0	31 03·4	15·1	3 51·8	9·3	54·1
		03	227 16·3	· · 47·0	45 37·5	15·1	4 01·1	9·4	54·1
		04	242 16·5	47·9	60 11·6	15·1	4 10·5	9·3	54·1
		05	257 16·8	48·9	74 45·7	15·0	4 19·8	9·2	54·1
		06	272 17·0	S 1 49·9	89 19·7	15·1	N 4 29·0	9·3	54·1
		07	287 17·2	50·9	103 53·8	15·1	4 38·3	9·2	54·1
	S	08	302 17·4	51·8	118 27·9	15·1	4 47·5	9·2	54·1
	U	09	317 17·6	· · 52·8	133 02·0	15·0	4 56·7	9·2	54·1
	N	10	332 17·8	53·8	147 36·0	15·1	5 05·9	9·2	54·1
	D	11	347 18·0	54·7	162 10·1	15·1	5 15·1	9·1	54·1
	A	12	2 18·2	S 1 55·7	176 44·2	15·0	N 5 24·2	9·2	54·1
	Y	13	17 18·4	56·7	191 18·2	15·0	5 33·4	9·0	54·1
		14	32 18·6	57·7	205 52·2	15·1	5 42·4	9·1	54·1
		15	47 18·9	· · 58·6	220 26·3	15·0	5 51·5	9·1	54·1
		16	62 19·1	1 59·6	235 00·3	15·0	6 00·6	9·0	54·1
		17	77 19·3	2 00·6	249 34·3	15·0	6 09·6	9·0	54·1
		18	92 19·5	S 2 01·6	264 08·3	15·0	N 6 18·6	8·9	54·1
		19	107 19·7	02·5	278 42·3	14·9	6 27·5	9·0	54·0
		20	122 19·9	03·5	293 16·2	15·0	6 36·5	8·9	54·0
		21	137 20·1	· · 04·5	307 50·2	14·9	6 45·4	8·9	54·0
		22	152 20·3	05·5	322 24·1	15·0	6 54·3	8·8	54·0
		23	167 20·5	06·4	336 58·1	14·9	7 03·1	8·8	54·0
29		00	182 20·8	S 2 07·4	351 32·0	14·9	N 7 11·9	8·8	54·0
		01	197 21·0	08·4	6 05·9	14·9	7 20·7	8·8	54·0
		02	212 21·2	09·3	20 39·8	14·9	7 29·5	8·7	54·0
		03	227 21·4	· · 10·3	35 13·7	14·9	7 38·2	8·7	54·0
		04	242 21·6	11·3	49 47·6	14·8	7 46·9	8·6	54·0
		05	257 21·8	12·3	64 21·4	14·8	7 55·5	8·6	54·0
		06	272 22·0	S 2 13·2	78 55·2	14·9	N 8 04·1	8·6	54·0
		07	287 22·2	14·2	93 29·1	14·8	8 12·7	8·6	54·0
		08	302 22·4	15·2	108 02·9	14·8	8 21·3	8·5	54·0
	M	09	317 22·6	· · 16·2	122 36·7	14·7	8 29·8	8·5	54·0
	O	10	332 22·8	17·1	137 10·4	14·8	8 38·3	8·4	54·0
	N	11	347 23·0	18·1	151 44·2	14·7	8 46·7	8·4	54·0
	D	12	2 23·3	S 2 19·1	166 17·9	14·7	N 8 55·1	8·4	54·0
	A	13	17 23·5	20·0	180 51·6	14·7	9 03·5	8·3	54·0
	Y	14	32 23·7	21·0	195 25·3	14·7	9 11·8	8·3	54·0
		15	47 23·9	· · 22·0	209 59·0	14·6	9 20·1	8·3	54·0
		16	62 24·1	23·0	224 32·6	14·7	9 28·4	8·2	54·0
		17	77 24·3	23·9	239 06·3	14·6	9 36·6	8·1	54·0
		18	92 24·5	S 2 24·9	253 39·9	14·6	N 9 44·7	8·2	54·0
		19	107 24·7	25·9	268 13·5	14·5	9 52·9	6·0	54·0
		20	122 24·9	26·9	282 47·0	14·6	10 00·9	8·1	54·0
		21	137 25·1	· · 27·8	297 20·6	14·5	10 09·0	8·0	54·0
		22	152 25·3	28·8	311 54·1	14·5	10 17·0	7·9	54·0
		23	167 25·5	29·8	326 27·6	14·5	10 24·9	8·0	54·0
30		00	182 25·7	S 2 30·7	341 01·1	14·5	N10 32·9	7·8	54·0
		01	197 25·9	31·7	355 34·6	14·4	10 40·7	7·9	54·0
		02	212 26·2	32·7	10 08·0	14·4	10 48·6	7·7	54·0
		03	227 26·4	· 33·7	24 41·4	14·4	10 56·3	7·8	54·0
		04	242 26·6	34·6	39 14·8	14·4	11 04·1	7·7	54·0
		05	257 26·8	35·6	53 48·2	14·3	11 11·8	7·6	54·0
		06	272 27·0	S 2 36·6	68 21·5	14·3	N11 19·4	7·6	54·0
		07	287 27·2	37·5	82 54·8	14·3	11 27·0	7·5	54·0
	T	08	302 27·4	38·5	97 28·1	14·2	11 34·5	7·5	54·0
	U	09	317 27·6	· · 39·5	112 01·3	14·3	11 42·0	7·5	54·0
	E	10	332 27·8	40·5	126 34·6	14·2	11 49·5	7·3	54·0
	S	11	347 28·0	41·4	141 07·8	14·2	11 56·8	7·4	54·0
	D	12	2 28·2	S 2 42·4	155 41·0	14·1	N12 04·2	7·3	54·0
	A	13	17 28·4	43·4	170 14·1	14·1	12 11·5	7·2	54·0
	Y	14	32 28·6	44·3	184 47·2	14·1	12 18·7	7·2	54·0
		15	47 28·8	· · 45·3	199 20·3	14·1	12 25·9	7·2	54·0
		16	62 29·0	46·3	213 53·4	14·0	12 33·1	7·0	54·0
		17	77 29·2	47·3	228 26·4	14·1	12 40·1	7·1	54·0
		18	92 29·4	S 2 48·2	242 59·5	13·9	N12 47·2	6·9	54·0
		19	107 29·6	49·2	257 32·4	14·0	12 54·1	7·0	54·0
		20	122 29·8	50·2	272 05·4	13·9	13 01·1	6·8	54·0
		21	137 30·0	· · 51·1	286 38·3	13·9	13 07·9	6·8	54·0
		22	152 30·3	52·1	301 11·2	13·9	13 14·7	6·8	54·0
		23	167 30·5	53·1	315 44·1	13·8	13 21·5	6·7	54·0
			S.D. 16·0	d 1·0	S.D. 14·7		14·7		14·7

Lat. °	Twilight Naut. h m	Twilight Civil h m	Sunrise h m	Moonrise 28 h m	Moonrise 29 h m	Moonrise 30 h m	Moonrise 1 h m
N 72	03 36	05 00	06 08	17 05	16 59	16 53	16 45
N 70	03 50	05 05	06 06	17 14	17 14	17 15	17 18
68	04 01	05 09	06 04	17 21	17 26	17 32	17 43
66	04 10	05 12	06 03	17 27	17 35	17 46	18 01
64	04 18	05 14	06 02	17 32	17 44	17 58	18 17
62	04 24	05 16	06 01	17 36	17 51	18 08	18 29
60	04 29	05 18	05 59	17 40	17 57	18 17	18 40
N 58	04 34	05 20	05 59	17 43	18 02	18 24	18 50
56	04 37	05 21	05 58	17 46	18 07	18 31	18 58
54	04 41	05 22	05 57	17 49	18 12	18 37	19 06
52	04 44	05 23	05 57	17 51	18 15	18 42	19 12
50	04 46	05 24	05 56	17 54	18 19	18 47	19 18
45	04 51	05 26	05 55	17 59	18 27	18 57	19 31
N 40	04 55	05 27	05 54	18 03	18 33	19 06	19 42
35	04 58	05 27	05 53	18 06	18 39	19 14	19 51
30	05 00	05 28	05 52	18 09	18 44	19 20	19 59
20	05 03	05 28	05 50	18 15	18 53	19 32	20 13
N 10	05 03	05 28	05 49	18 20	19 00	19 42	20 25
0	05 03	05 27	05 47	18 25	19 08	19 52	20 37
S 10	05 00	05 25	05 46	18 29	19 15	20 01	20 49
20	04 56	05 22	05 44	18 34	19 23	20 12	21 01
30	04 50	05 18	05 42	18 40	19 32	20 24	21 15
35	04 45	05 15	05 40	18 43	19 37	20 30	21 24
40	04 40	05 12	05 39	18 47	19 43	20 38	21 33
45	04 33	05 07	05 37	18 51	19 50	20 48	21 44
S 50	04 24	05 02	05 35	18 57	19 58	20 59	21 58
52	04 19	05 00	05 34	18 59	20 02	21 04	22 04
54	04 15	04 57	05 33	19 02	20 06	21 09	22 11
56	04 09	04 54	05 31	19 05	20 11	21 16	22 19
58	04 03	04 50	05 30	19 08	20 16	21 23	22 28
S 60	03 55	04 46	05 28	19 12	20 22	21 31	22 38

Lat. °	Sunset h m	Twilight Civil h m	Twilight Naut. h m	Moonset 28 h m	Moonset 29 h m	Moonset 30 h m	Moonset 1 h m
N 72	17 31	18 38	20 02	07 01	08 36	10 14	11 54
N 70	17 34	18 34	19 48	06 55	08 23	09 53	11 22
68	17 35	18 31	19 37	06 49	08 13	09 36	10 58
66	17 37	18 28	19 29	06 45	08 04	09 23	10 40
64	17 38	18 25	19 22	06 41	07 57	09 12	10 26
62	17 39	18 23	19 16	06 38	07 51	09 03	10 13
60	17 40	18 22	19 11	06 35	07 45	08 55	10 03
N 58	17 41	18 20	19 06	06 32	07 41	08 48	09 54
56	17 42	18 19	19 03	06 30	07 36	08 42	09 46
54	17 43	18 18	18 59	06 28	07 33	08 37	09 39
52	17 43	18 17	18 56	06 26	07 29	08 32	09 33
50	17 44	18 16	18 54	06 25	07 26	08 27	09 27
45	17 46	18 15	18 49	06 21	07 20	08 18	09 15
N 40	17 47	18 13	18 45	06 18	07 14	08 10	09 05
35	17 48	18 13	18 42	06 15	07 09	08 03	08 56
30	17 49	18 12	18 40	06 13	07 05	07 57	08 49
20	17 50	18 12	18 38	06 09	06 58	07 46	08 36
N 10	17 52	18 13	18 37	06 06	06 51	07 37	08 24
0	17 54	18 14	18 38	06 02	06 45	07 29	08 13
S 10	17 55	18 17	18 41	05 59	06 39	07 20	08 03
20	17 57	18 20	18 45	05 56	06 33	07 11	07 51
30	18 00	18 24	18 52	05 52	06 26	07 01	07 39
35	18 01	18 27	18 56	05 49	06 22	06 55	07 31
40	18 03	18 30	19 02	05 47	06 17	06 48	07 23
45	18 05	18 34	19 09	05 44	06 11	06 41	07 13
S 50	18 07	18 40	19 18	05 40	06 05	06 31	07 01
52	18 08	18 42	19 23	05 39	06 02	06 27	06 55
54	18 10	18 45	19 28	05 37	05 59	06 22	06 49
56	18 11	18 48	19 34	05 35	05 55	06 17	06 42
58	18 12	18 52	19 41	05 33	05 51	06 11	06 35
S 60	18 14	18 56	19 48	05 30	05 46	06 05	06 26

Day	SUN Eqn. of Time 00h m s	SUN Eqn. of Time 12h m s	SUN Mer. Pass. h m	MOON Mer. Pass. Upper h m	MOON Mer. Pass. Lower h m	MOON Age d	MOON Phase
28	09 02	09 13	11 51	24 35	12 13	15	○
29	09 23	09 33	11 50	00 35	12 56	16	
30	09 43	09 52	11 50	01 18	13 40	17	

OCTOBER 10, 11, 12 (FRI., SAT., SUN.)

G.M.T.	ARIES	VENUS −3·4		MARS −1·4		JUPITER −1·2		SATURN +0·8		STARS		
d h	G.H.A. ° ′	G.H.A. ° ′	Dec. ° ′	G.H.A. ° ′	Dec. ° ′	G.H.A. ° ′	Dec. ° ′	G.H.A. ° ′	Dec. ° ′	Name	S.H.A. ° ′	Dec. ° ′
10 00	18 05·5	190 18·4	S 1 48·9	317 20·1	N19 18·0	163 37·5	S12 45·9	117 59·8	S22 01·6	Acamar	315 49·2	S 40 28·0
01	33 07·9	205 18·0	50·1	332 22·6	18·1	178 39·4	46·1	133 02·1	01·6	Achernar	335 56·8	S 57 26·7
02	48 10·4	220 17·6	51·4	347 25·0	18·2	193 41·4	46·3	148 04·4	01·7	Acrux	173 55·9	S 62 52·2
03	63 12·9	235 17·2	·· 52·6	2 27·5	·· 18·3	208 43·3	·· 46·5	163 06·6	·· 01·7	Adhara	255 44·8	S 28 54·8
04	78 15·3	250 16·8	53·9	17 30·0	18·4	223 45·3	46·6	178 08·9	01·7	Aldebaran	291 36·4	N 16 25·6
05	93 17·8	265 16·4	55·1	32 32·5	18·5	238 47·2	46·8	193 11·2	01·7			
06	108 20·3	280 16·0	S 1 56·4	47 34·9	N19 18·5	253 49·2	S12 47·0	208 13·5	S22 01·7	Alioth	166 57·4	N 56 11·0
07	123 22·7	295 15·6	57·6	62 37·4	18·6	268 51·1	47·2	223 15·7	01·8	Alkaid	153 31·8	N 49 31·3
08	138 25·2	310 15·2	1 58·9	77 39·9	18·7	283 53·1	47·3	238 18·0	01·8	Al Na'ir	28 35·1	S 47 09·7
F 09	153 27·6	325 14·8	2 00·1	92 42·4	·· 18·8	298 55·0	·· 47·5	253 20·3	·· 01·8	Alnilam	276 28·0	S 1 13·6
R 10	168 30·1	340 14·4	01·4	107 44·9	18·9	313 57·0	47·7	268 22·5	01·8	Alphard	218 36·7	S 8 28·7
I 11	183 32·6	355 14·0	02·6	122 47·3	19·0	328 58·9	47·9	283 24·8	01·8			
D 12	198 35·0	10 13·6	S 2 03·9	137 49·8	N19 19·1	344 00·9	S12 48·0	298 27·1	S22 01·8	Alphecca	126 46·1	N 26 51·4
A 13	213 37·5	25 13·2	05·1	152 52·3	19·2	359 02·8	48·2	313 29·4	01·9	Alpheratz	358 25·7	N 28 52·0
Y 14	228 40·0	40 12·8	06·4	167 54·8	19·3	14 04·8	48·4	328 31·6	01·9	Altair	62 48·3	N 8 45·8
15	243 42·4	55 12·4	·· 07·6	182 57·3	·· 19·3	29 06·8	·· 48·6	343 33·9	·· 01·9	Ankaa	353 55·9	S 42 31·7
16	258 44·9	70 12·0	08·9	197 59·8	19·4	44 08·7	48·8	358 36·2	01·9	Antares	113 17·0	S 26 20·4
17	273 47·4	85 11·6	10·1	213 02·3	19·5	59 10·7	48·9	13 38·5	01·9			
18	288 49·8	100 11·2	S 2 11·4	228 04·8	N19 19·6	74 12·6	S12 49·1	28 40·7	S22 02·0	Arcturus	146 33·6	N 19 24·0
19	303 52·3	115 10·8	12·6	243 07·3	19·7	89 14·6	49·3	43 43·0	02·0	Atria	108 56·0	S 68 57·4
20	318 54·8	130 10·4	13·9	258 09·8	19·8	104 16·5	49·5	58 45·3	02·0	Avior	234 35·0	S 59 22·4
21	333 57·2	145 10·0	·· 15·1	273 12·3	·· 19·9	119 18·5	·· 49·6	73 47·5	·· 02·0	Bellatrix	279 16·0	N 6 18·8
22	348 59·7	160 09·6	16·4	288 14·8	20·0	134 20·4	49·8	88 49·8	02·0	Betelgeuse	271 45·7	N 7 24·0
23	4 02·1	175 09·2	17·6	303 17·3	20·0	149 22·4	50·0	103 52·1	02·1			
11 00	19 04·6	190 08·8	S 2 18·9	318 19·8	N19 20·1	164 24·3	S12 50·2	118 54·4	S22 02·1	Canopus	264 14·3	S 52 40·2
01	34 07·1	205 08·4	20·1	333 22·3	20·2	179 26·3	50·3	133 56·6	02·1	Capella	281 35·0	N 45 57·3
02	49 09·5	220 08·0	21·4	348 24·8	20·3	194 28·2	50·5	148 58·9	02·1	Deneb	49 59·3	N 45 08·4
03	64 12·0	235 07·6	·· 22·6	3 27·3	·· 20·4	209 30·2	·· 50·7	164 01·2	·· 02·1	Denebola	183 15·9	N 14 48·2
04	79 14·5	250 07·2	23·9	18 29·8	20·5	224 32·1	50·9	179 03·4	02·1	Diphda	349 36·8	S 18 12·6
05	94 16·9	265 06·8	25·1	33 32·3	20·6	239 34·1	51·1	194 05·7	02·2			
06	109 19·4	280 06·4	S 2 26·4	48 34·8	N19 20·6	254 36·0	S12 51·2	209 08·0	S22 02·2	Dubhe	194 42·7	N 61 58·3
07	124 21·9	295 06·0	27·6	63 37·4	20·7	269 38·0	51·4	224 10·3	02·2	Elnath	279 04·5	N 28 34·3
S 08	139 24·3	310 05·6	28·9	78 39·9	20·8	284 39·9	51·6	239 12·5	02·2	Eltanin	91 05·4	N 51 30·0
A 09	154 26·8	325 05·2	·· 30·1	93 42·4	·· 20·9	299 41·9	·· 51·8	254 14·8	·· 02·2	Enif	34 27·3	N 9 41·4
T 10	169 29·2	340 04·8	31·4	108 44·9	21·0	314 43·8	51·9	269 17·1	02·3	Fomalhaut	16 09·0	S 29 50·4
U 11	184 31·7	355 04·4	32·6	123 47·4	21·1	329 45·8	52·1	284 19·3	02·3			
R 12	199 34·2	10 04·0	S 2 33·9	138 50·0	N19 21·2	344 47·7	S12 52·3	299 21·6	S22 02·3	Gacrux	172 47·3	S 56 52·8
D 13	214 36·6	25 03·6	35·1	153 52·5	21·2	359 49·7	52·5	314 23·9	02·3	Gienah	176 35·0	S 17 18·7
A 14	229 39·1	40 03·2	36·3	168 55·0	21·3	14 51·6	52·6	329 26·1	02·3	Hadar	149 46·9	S 60 10·5
Y 15	244 41·6	55 02·8	·· 37·6	183 57·6	·· 21·4	29 53·6	·· 52·8	344 28·4	·· 02·4	Hamal	328 46·9	N 23 16·2
16	259 44·0	70 02·4	38·8	199 00·1	21·5	44 55·5	53·0	359 30·7	02·4	Kaus Aust.	84 38·4	S 34 24·3
17	274 46·5	85 02·0	40·1	214 02·6	21·6	59 57·5	53·2	14 32·9	02·4			
18	289 49·0	100 01·6	S 2 41·3	229 05·2	N19 21·7	74 59·4	S12 53·4	29 35·2	S22 02·4	Kochab	137 19·4	N 74 19·7
19	304 51·4	115 01·2	42·6	244 07·7	21·7	90 01·4	53·5	44 37·5	02·4	Markab	14 19·0	N 14 59·2
20	319 53·9	130 00·8	43·8	259 10·2	21·8	105 03·3	53·7	59 39·7	02·4	Menkar	314 57·8	N 3 55·8
21	334 56·4	145 00·4	·· 45·1	274 12·8	·· 21·9	120 05·3	·· 53·9	74 42·0	·· 02·5	Menkent	148 56·5	S 36 10·0
22	349 58·8	160 00·0	46·3	289 15·3	22·0	135 07·2	54·1	89 44·3	02·5	Miaplacidus	221 48·8	S 69 32·6
23	5 01·3	174 59·5	47·6	304 17·9	22·1	150 09·2	54·2	104 46·5	02·5			
12 00	20 03·7	189 59·1	S 2 48·8	319 20·4	N19 22·1	165 11·1	S12 54·4	119 48·8	S22 02·5	Mirfak	309 38·8	N 49 42·8
01	35 06·2	204 58·7	50·1	334 22·9	22·2	180 13·1	54·6	134 51·1	02·5	Nunki	76 49·3	S 26 20·9
02	50 08·7	219 58·3	51·3	349 25·5	22·3	195 15·0	54·8	149 53·3	02·6	Peacock	54 23·9	S 56 52·2
03	65 11·1	234 57·9	·· 52·6	4 28·0	·· 22·4	210 17·0	·· 55·0	164 55·6	·· 02·6	Pollux	244 18·0	N 28 07·5
04	80 13·6	249 57·5	53·8	19 30·6	22·5	225 18·9	55·1	179 57·9	02·6	Procyon	245 42·8	N 5 19·9
05	95 16·1	264 57·1	55·1	34 33·1	22·6	240 20·9	55·3	195 00·2	02·6			
06	110 18·5	279 56·7	S 2 56·3	49 35·7	N19 22·6	255 22·8	S12 55·5	210 02·4	S22 02·6	Rasalhague	96 44·8	N 12 35·6
07	125 21·0	294 56·3	57·6	64 38·3	22·7	270 24·8	55·7	225 04·7	02·7	Regulus	208 27·5	N 12 10·1
08	140 23·5	309 55·9	2 58·8	79 40·8	22·8	285 26·7	55·8	240 07·0	02·7	Rigel	281 51·4	S 8 14·8
09	155 25·9	324 55·5	3 00·0	94 43·4	·· 22·9	300 28·7	·· 56·0	255 09·2	·· 02·7	Rigil Kent.	140 48·5	S 60 39·9
10	170 28·4	339 55·1	01·3	109 45·9	23·0	315 30·6	56·2	270 11·5	02·7	Sabik	102 59·9	S 15 40·4
11	185 30·9	354 54·7	02·5	124 48·5	23·0	330 32·6	56·4	285 13·7	02·7			
12	200 33·3	9 54·3	S 3 03·8	139 51·1	N19 23·1	345 34·5	S12 56·5	300 16·0	S22 02·7	Schedar	350 26·8	N 56 18·8
13	215 35·8	24 53·9	05·0	154 53·6	23·2	0 36·5	56·7	315 18·3	02·8	Shaula	97 17·9	S 37 04·5
14	230 38·2	39 53·5	06·3	169 56·2	23·3	15 38·4	56·9	330 20·5	02·8	Sirius	259 09·9	S 16 39·4
15	245 40·7	54 53·1	·· 07·5	184 58·8	·· 23·3	30 40·4	·· 57·1	345 22·6	·· 02·8	Spica	159 14·9	S 10 56·7
16	260 43·2	69 52·7	08·8	200 01·3	23·4	45 42·3	57·3	0 25·1	02·8	Suhail	223 22·9	S 43 15·8
17	275 45·6	84 52·3	10·0	215 03·9	23·5	60 44·3	57·4	15 27·3	02·8			
18	290 48·1	99 51·9	S 3 11·3	230 06·5	N19 23·6	75 46·2	S12 57·6	30 29·6	S22 02·9	Vega	81 06·9	N 38 45·1
19	305 50·6	114 51·5	12·5	245 09·1	23·7	90 48·2	57·8	45 31·9	02·9	Zuben'ubi	137 51·3	S 15 52·1
20	320 53·0	129 51·0	13·8	260 11·6	23·7	105 50·1	58·0	60 34·1	02·9		S.H.A. ° ′	Mer. Pass. h m
21	335 55·5	144 50·6	·· 15·0	275 14·2	·· 23·8	120 52·1	·· 58·1	75 36·4	·· 02·9			
22	350 58·0	159 50·2	16·3	290 16·8	23·9	135 54·0	58·3	90 38·7	02·9	Venus	171 04·2	11 20
23	6 00·4	174 49·8	17·5	305 19·4	24·0	150 56·0	58·5	105 40·9	03·0	Mars	299 15·2	2 46
r. Pass.	h m 22 40·0	v −0·4	d 1·2	v 2·5	d 0·1	v 2·0	d 0·2	v 2·3	d 0·0	Jupiter	145 19·7	13 01
										Saturn	99 49·8	16 02

OCTOBER 10, 11, 12 (FRI., SAT., SUN.)

G.M.T. d h		SUN G.H.A.	SUN Dec.	MOON G.H.A.	v	MOON Dec.	d	H.P.
		° ′	° ′	° ′	′	° ′	′	′
10 00		183 11·1	S 6 21·9	221 37·0	8·7	N 5 42·4	11·1	59·9
01		198 11·3	22·8	236 04·7	8·7	5 31·3	11·1	60·0
02		213 11·5	23·8	250 32·4	8·7	5 20·2	11·2	60·0
03		228 11·6	· · 24·7	265 00·1	8·7	5 09·0	11·2	60·0
04		243 11·8	25·7	279 27·8	8·6	4 57·8	11·3	60·1
05		258 12·0	26·6	293 55·4	8·7	4 46·5	11·3	60·1
06		273 12·1	S 6 27·6	308 23·1	8·6	N 4 35·2	11·4	60·1
07		288 12·3	28·5	322 50·7	8·5	4 23·8	11·4	60·2
08		303 12·5	29·5	337 18·2	8·6	4 12·4	11·5	60·2
09	F	318 12·6	· · 30·4	351 45·8	8·5	4 00·9	11·5	60·2
10	R	333 12·8	31·4	6 13·3	8·5	3 49·4	11·5	60·3
11	I	348 13·0	32·3	20 40·8	8·5	3 37·9	11·5	60·3
12	D	3 13·1	S 6 33·3	35 08·3	8·5	N 3 26·4	11·6	60·3
13	A	18 13·3	34·2	49 35·8	8·4	3 14·8	11·7	60·4
14	Y	33 13·5	35·2	64 03·2	8·4	3 03·1	11·6	60·4
15		48 13·6	36·1	78 30·6	8·4	2 51·5	11·7	60·4
16		63 13·8	37·1	92 58·0	8·3	2 39·8	11·7	60·5
17		78 14·0	38·0	107 25·3	8·4	2 28·1	11·8	60·5
18		93 14·1	S 6 39·0	121 52·7	8·3	N 2 16·3	11·7	60·5
19		108 14·3	39·9	136 20·0	8·2	2 04·6	11·8	60·5
20		123 14·5	40·9	150 47·2	8·3	1 52·8	11·8	60·6
21		138 14·6	· · 41·8	165 14·5	8·2	1 41·0	11·9	60·6
22		153 14·8	42·8	179 41·7	8·2	1 29·1	11·8	60·6
23		168 15·0	43·7	194 08·9	8·2	1 17·3	11·9	60·7
11 00		183 15·1	S 6 44·7	208 36·1	8·1	N 1 05·4	11·9	60·7
01		198 15·3	45·6	223 03·2	8·1	0 53·5	11·9	60·7
02		213 15·4	46·5	237 30·3	8·1	0 41·6	12·0	60·7
03		228 15·6	· · 47·5	251 57·4	8·0	0 29·6	11·9	60·8
04		243 15·8	48·4	266 24·4	8·0	0 17·7	12·0	60·8
05		258 15·9	49·4	280 51·4	8·0	N 0 05·7	11·9	60·8
06		273 16·1	S 6 50·3	295 18·4	8·0	S 0 06·2	12·0	60·8
07	S	288 16·3	51·3	309 45·4	7·9	0 18·2	12·0	60·9
08		303 16·4	52·2	324 12·3	7·9	0 30·2	12·0	60·9
09	A	318 16·6	· · 53·2	338 39·2	7·8	0 42·2	12·0	60·9
10	T	333 16·7	54·1	353 06·0	7·9	0 54·2	12·0	60·9
11	U	348 16·9	55·1	7 32·9	7·8	1 06·2	12·0	61·0
12	R	3 17·1	S 6 56·0	21 59·7	7·7	S 1 18·2	12·0	61·0
13	D	18 17·2	56·9	36 26·4	7·7	1 30·2	12·0	61·0
14	A	33 17·4	57·9	50 53·1	7·7	1 42·2	12·0	61·0
15	Y	48 17·5	· · 58·8	65 19·8	7·7	1 54·2	12·0	61·0
16		63 17·7	6 59·8	79 46·5	7·6	2 06·2	12·0	61·1
17		78 17·9	7 00·7	94 13·1	7·6	2 18·2	12·0	61·1
18		93 18·0	S 7 01·7	108 39·7	7·6	S 2 30·2	12·0	61·1
19		108 18·2	02·6	123 06·3	7·5	2 42·2	11·9	61·1
20		123 18·3	03·5	137 32·8	7·5	2 54·1	12·0	61·1
21		138 18·5	· · 04·5	151 59·3	7·5	3 06·1	11·9	61·2
22		153 18·7	05·4	166 25·8	7·4	3 18·0	12·0	61·2
23		168 18·8	06·4	180 52·2	7·4	3 30·0	11·9	61·2
12 00		183 19·0	S 7 07·3	195 18·6	7·3	S 3 41·9	11·9	61·2
01		198 19·1	08·3	209 44·9	7·3	3 53·8	11·8	61·2
02		213 19·3	09·2	224 11·2	7·3	4 05·6	11·9	61·2
03		228 19·4	· · 10·2	238 37·5	7·3	4 17·5	11·8	61·2
04		243 19·6	11·1	253 03·8	7·2	4 29·3	11·8	61·3
05		258 19·8	12·0	267 30·0	7·1	4 41·1	11·8	61·3
06		273 19·9	S 7 13·0	281 56·1	7·2	S 4 52·9	11·7	61·3
07		288 20·1	13·9	296 22·3	7·1	5 04·6	11·7	61·3
08		303 20·2	14·9	310 48·4	7·0	5 16·3	11·7	61·3
09	S	318 20·4	· · 15·8	325 14·4	7·1	5 28·0	11·7	61·3
10	U	333 20·5	16·7	339 40·5	7·0	5 39·7	11·6	61·3
11	N	348 20·7	17·7	354 06·5	6·9	5 51·3	11·6	61·3
12	D	3 20·9	S 7 18·6	8 32·4	6·9	S 6 02·9	11·5	61·3
13	A	18 21·0	19·6	22 58·3	6·9	6 14·4	11·6	61·4
14	Y	33 21·2	20·5	37 24·2	6·9	6 26·0	11·4	61·4
15		48 21·3	· · 21·4	51 50·1	6·8	6 37·4	11·4	61·4
16		63 21·5	22·4	66 15·9	6·7	6 48·8	11·4	61·4
17		78 21·6	23·3	80 41·6	6·8	7 00·2	11·4	61·4
18		93 21·8	S 7 24·3					
19		108 21·9	25·2					
20		123 22·1	26·1					
21		138 22·2	· · 27·1					
22		153 22·4	28·0					
23		168 22·6	29·0					
		S.D. 16·0	d 0·9	S.D. 16·4		16·6		16·7

A Total Eclipse of the Sun occurs to-day. See page 5.

Lat.	Twilight Naut.	Twilight Civil	Sunrise	Moonrise 10	Moonrise 11	Moonrise 12	Moonrise 13
°	h m	h m	h m	h m	h m	h m	h m
N 72	04 35	05 54	07 03	02 14	04 14	06 17	08 24
N 70	04 41	05 52	06 54	02 22	04 14	06 09	08 07
68	04 46	05 51	06 47	02 28	04 14	06 03	07 54
66	04 50	05 50	06 41	02 33	04 14	05 58	07 43
64	04 54	05 49	06 36	02 38	04 14	05 54	07 34
62	04 56	05 48	06 32	02 41	04 15	05 50	07 27
60	04 59	05 47	06 28	02 44	04 15	05 47	07 20
N 58	05 00	05 46	06 25	02 47	04 15	05 44	07 14
56	05 02	05 45	06 22	02 50	04 15	05 42	07 09
54	05 03	05 44	06 19	02 52	04 15	05 39	07 05
52	05 04	05 43	06 17	02 54	04 15	05 37	07 01
50	05 05	05 42	06 15	02 56	04 15	05 35	06 57
45	05 06	05 40	06 10	03 00	04 15	05 31	06 49
N 40	05 07	05 38	06 06	03 04	04 15	05 28	06 42
35	05 07	05 37	06 02	03 07	04 15	05 25	06 36
30	05 07	05 35	05 59	03 09	04 15	05 23	06 31
20	05 06	05 31	05 53	03 14	04 15	05 18	06 23
N 10	05 03	05 27	05 48	03 18	04 15	05 15	06 15
0	04 59	05 23	05 44	03 21	04 16	05 11	06 08
S 10	04 53	05 18	05 39	03 25	04 16	05 08	06 01
20	04 45	05 11	05 33	03 29	04 16	05 04	05 54
30	04 35	05 03	05 27	03 34	04 16	05 00	05 46
35	04 28	04 58	05 24	03 36	04 16	04 58	05 41
40	04 19	04 52	05 19	03 39	04 17	04 55	05 35
45	04 09	04 44	05 15	03 43	04 17	04 52	05 29
S 50	03 55	04 36	05 09	03 47	04 17	04 48	05 22
52	03 49	04 31	05 06	03 49	04 17	04 47	05 18
54	03 42	04 26	05 03	03 51	04 18	04 45	05 15
56	03 33	04 21	05 00	03 53	04 18	04 43	05 11
58	03 24	04 15	04 56	03 55	04 18	04 41	05 06
S 60	03 12	04 08	04 52	03 58	04 18	04 39	05 01

Lat.	Sunset	Twilight Civil	Twilight Naut.	Moonset 10	Moonset 11	Moonset 12	Moonset 13
°	h m	h m	h m	h m	h m	h m	h m
N 72	16 30	17 38	18 56	16 19	16 13	16 06	15 59
N 70	16 38	17 40	18 50	16 16	16 16	16 16	16 17
68	16 45	17 41	18 46	16 13	16 18	16 24	16 32
66	16 51	17 43	18 42	16 10	16 20	16 31	16 44
64	16 56	17 44	18 39	16 09	16 22	16 37	16 54
62	17 01	17 45	18 36	16 07	16 24	16 42	17 03
60	17 04	17 46	18 34	16 05	16 25	16 46	17 11
N 58	17 08	17 47	18 33	16 04	16 26	16 50	17 17
56	17 11	17 48	18 31	16 03	16 27	16 54	17 23
54	17 14	17 49	18 30	16 02	16 29	16 57	17 28
52	17 16	17 50	18 29	16 01	16 30	17 00	17 33
50	17 18	17 51	18 28	16 00	16 30	17 02	17 37
45	17 23	17 53	18 26	15 58	16 32	17 08	17 47
N 40	17 27	17 55	18 26	15 56	16 34	17 13	17 55
35	17 31	17 56	18 26	15 55	16 35	17 17	18 02
30	17 34	17 58	18 26	15 53	16 36	17 21	18 08
20	17 40	18 02	18 28	15 51	16 38	17 27	18 18
N 10	17 45	18 06	18 31	15 49	16 40	17 33	18 27
0	17 50	18 11	18 35	15 47	16 42	17 38	18 36
S 10	17 55	18 16	18 41	15 45	16 43	17 43	18 44
20	18 01	18 23	18 49	15 43	16 45	17 49	18 53
30	18 07	18 31	19 00	15 40	16 47	17 55	19 04
35	18 11	18 37	19 07	15 39	16 48	17 59	19 10
40	18 15	18 43	19 15	15 37	16 49	18 03	19 17
45	18 20	18 50	19 26	15 35	16 51	18 08	19 25
S 50	18 26	18 59	19 40	15 33	16 53	18 14	19 35
52	18 29	19 04	19 47	15 32	16 53	18 16	19 39
54	18 32	19 09	19 54	15 31	16 54	18 19	19 44
56	18 35	19 14	20 03	15 29	16 55	18 23	19 50
58	18 39	19 20	20 13	15 28	16 56	18 26	19 56
S 60	18 43	19 28	20 24	15 26	16 57	18 30	20 03

Day	SUN Eqn. of Time 00h	SUN Eqn. of Time 12h	SUN Mer. Pass.	MOON Mer. Pass. Upper	MOON Mer. Pass. Lower	Age	Phase
	m s	m s	h m	h m	h m	d	
10	12 44	12 52	11 47	09 34	22 01	27	
11	13 00	13 08	11 47	10 29	22 56	28	●
12	13 16	13 23	11 47	11 24	23 53	29	

DECEMBER 21, 22, 23 (SUN., MON., TUES.)

G.M.T.		ARIES	VENUS −3·4		MARS −0·9		JUPITER −1·3		SATURN +0·6		STARS		
d	h	G.H.A. ° ′	G.H.A. ° ′	Dec. ° ′	G.H.A. ° ′	Dec. ° ′	G.H.A. ° ′	Dec. ° ′	G.H.A. ° ′	Dec. ° ′	Name	S.H.A. ° ′	Dec. ° ′
21	00	89 03·5	170 02·9	S24 01·8	45 27·3	N18 25·3	219 20·9	S17 20·2	181 01·9	S22 27·9	Acamar	315 49·2	S 40 28·3
	01	104 05·9	185 01·9	01·6	60 29·7	25·4	234 22·8	20·3	196 04·1	27·9	Achernar	335 57·1	S 57 27·0
	02	119 08·4	200 00·9	01·5	75 32·2	25·5	249 24·8	20·5	211 06·2	27·9	Acrux	173 55·1	S 62 52·0
	03	134 10·9	215 00·0	· 01·4	90 34·6	·· 25·5	264 26·8	·· 20·6	226 08·4	· 28·0	Adhara	255 44·3	S 28 55·0
	04	149 13·3	229 59·0	01·3	105 37·1	25·6	279 28·7	20·7	241 10·5	28·0	Aldebaran	291 36·1	N 16 25·6
	05	164 15·8	244 58·0	01·2	120 39·5	25·6	294 30·7	20·8	256 12·6	28·0			
	06	179 18·2	259 57·1	S24 01·1	135 42·0	N18 25·7	309 32·7	S17 21·0	271 14·8	S22 28·0	Alioth	166 56·9	N 56 10·7
	07	194 20·7	274 56·1	00·9	150 44·4	25·7	324 34·6	21·1	286 16·9	28·0	Alkaid	153 31·5	N 49 30·9
	08	209 23·2	289 55·1	00·8	165 46·9	25·8	339 36·6	21·2	301 19·1	28·0	Al Na'ir	28 35·4	S 47 09·7
S	09	224 25·6	304 54·1	·· 00·7	180 49·3	· 25·8	354 38·6	·· 21·3	316 21·2	·· 28·0	Alnilam	276 27·6	S 1 13·7
U	10	239 28·1	319 53·2	00·6	195 51·7	25·9	9 40·5	21·5	331 23·4	28·0	Alphard	218 36·1	S 8 28·9
N	11	254 30·6	334 52·2	00·5	210 54·2	26·0	24 42·5	21·6	346 25·5	28·0			
D	12	269 33·0	349 51·2	S24 00·3	225 56·6	N18 26·0	39 44·5	S17 21·7	1 27·7	S22 28·0	Alphecca	126 46·0	N 26 51·1
A	13	284 35·5	4 50·3	00·2	240 59·1	26·1	54 46·4	21·8	16 29·8	28·0	Alpheratz	358 25·9	N 28 52·0
Y	14	299 38·0	19 49·3	24 00·1	256 01·5	26·1	69 48·4	22·0	31 32·0	28·0	Altair	62 48·5	N 8 45·7
	15	314 40·4	34 48·3	23 59·9	271 03·9	·· 26·2	84 50·4	·· 22·1	46 34·1	·· 28·0	Ankaa	353 56·1	S 42 31·9
	16	329 42·9	49 47·3	59·8	286 06·4	26·2	99 52·3	22·2	61 36·2	28·0	Antares	113 16·9	S 26 20·4
	17	344 45·4	64 46·4	59·7	301 08·8	26·3	114 54·3	22·3	76 38·4	28·0			
	18	359 47·8	79 45·4	S23 59·5	316 11·2	N18 26·4	129 56·3	S17 22·5	91 40·5	S22 28·0	Arcturus	146 33·3	N 19 23·7
	19	14 50·3	94 44·4	59·4	331 13·7	26·4	144 58·2	22·6	106 42·7	28·1	Atria	108 56·1	S 68 57·1
	20	29 52·7	109 43·5	59·3	346 16·1	26·5	160 00·2	22·7	121 44·8	28·1	Avior	234 34·2	S 59 22·6
	21	44 55·2	124 42·5	· 59·1	1 18·5	· 26·5	175 02·2	·· 22·8	136 47·0	· 28·1	Bellatrix	279 15·6	N 6 18·7
	22	59 57·7	139 41·5	59·0	16 20·9	26·6	190 04·1	23·0	151 49·1	28·1	Betelgeuse	271 45·3	N 7 23·9
	23	75 00·1	154 40·5	58·9	31 23·4	26·7	205 06·1	23·1	166 51·3	28·1			
22	00	90 02·6	169 39·6	S23 58·7	46 25·8	N18 26·7	220 08·1	S17 23·2	181 53·4	S22 28·1	Canopus	264 13·8	S 52 40·5
	01	105 05·1	184 38·6	58·6	61 28·2	26·8	235 10·0	23·3	196 55·5	28·1	Capella	281 34·6	N 45 57·4
	02	120 07·5	199 37·6	58·4	76 30·6	26·8	250 12·0	23·5	211 57·7	28·1	Deneb	49 59·7	N 45 08·3
	03	135 10·0	214 36·7	58·3	91 33·0	·· 26·9	265 14·0	· 23·6	226 59·8	28·1	Denebola	183 15·4	N 14 48·0
	04	150 12·5	229 35·7	58·1	106 35·5	27·0	280 15·9	23·7	242 02·0	28·1	Diphda	349 36·9	S 18 12·8
	05	165 14·9	244 34·7	58·0	121 37·9	27·0	295 17·9	23·8	257 04·1	28·1			
	06	180 17·4	259 33·7	S23 57·8	136 40·3	N18 27·1	310 19·9	S17 23·9	272 06·3	S22 28·1	Dubhe	194 41·8	N 61 58·0
	07	195 19·8	274 32·8	57·7	151 42·7	27·2	325 21·8	24·1	287 08·4	28·1	Elnath	279 04·1	N 28 34·4
	08	210 22·3	289 31·8	57·5	166 45·1	27·2	340 23·8	24·2	302 10·6	28·1	Eltanin	91 05·7	N 51 29·7
M	09	225 24·8	304 30·8	· 57·4	181 47·5	· 27·3	355 25·8	·· 24·3	317 12·7	28·1	Enif	34 27·5	N 9 41·3
O	10	240 27·2	319 29·9	57·2	196 49·9	27·4	10 27·7	24·4	332 14·9	28·1	Fomalhaut	16 09·3	S 29 50·5
N	11	255 29·7	334 28·9	57·1	211 52·3	27·4	25 29·7	24·6	347 17·0	28·2			
D	12	270 32·2	349 27·9	S23 56·9	226 54·7	N18 27·5	40 31·7	S17 24·7	2 19·1	S22 28·2	Gacrux	172 46·6	S 56 52·7
A	13	285 34·6	4 27·0	56·8	241 57·1	27·5	55 33·7	24·8	17 21·3	28·2	Gienah	176 34·5	S 17 18·8
Y	14	300 37·1	19 26·0	56·6	256 59·5	27·6	70 35·6	24·9	32 23·4	28·2	Hadar	149 46·4	S 60 10·3
	15	315 39·6	34 25·0	·· 56·5	272 01·9	· 27·7	85 37·6	·· 25·1	47 25·6	·· 28·2	Hamal	328 46·8	N 23 16·2
	16	330 42·0	49 24·1	56·3	287 04·3	27·7	100 39·6	25·2	62 27·7	28·2	Kaus Aust.	84 38·6	S 34 24·2
	17	345 44·5	64 23·1	56·1	302 06·7	27·8	115 41·5	25·3	77 29·9	28·2			
	18	0 47·0	79 22·1	S23 56·0	317 09·1	N18 27·9	130 43·5	S17 25·4	92 32·0	S22 28·2	Kochab	137 19·2	N 74 19·2
	19	15 49·4	94 21·2	55·8	332 11·5	27·9	145 45·5	25·6	107 34·2	28·2	Markab	14 19·2	N 14 59·2
	20	30 51·9	109 20·2	55·6	347 13·9	28·0	160 47·4	25·7	122 36·3	28·2	Menkar	314 57·7	N 3 55·7
	21	45 54·3	124 19·2	·· 55·5	2 16·3	28·1	175 49·4	25·8	137 38·4	·· 28·2	Menkent	148 56·1	S 36 09·9
	22	60 56·8	139 18·3	55·3	17 18·7	28·1	190 51·4	25·9	152 40·6	28·2	Miaplacidus	221 47·6	S 69 32·8
	23	75 59·3	154 17·3	55·1	32 21·1	28·2	205 53·4	26·0	167 42·7	28·2			
23	00	91 01·7	169 16·3	S23 55·0	47 23·5	N18 28·3	220 55·3	S17 26·2	182 44·9	S22 28·2	Mirfak	309 38·6	N 49 43·1
	01	106 04·2	184 15·4	54·8	62 25·9	28·3	235 57·3	26·3	197 47·0	28·2	Nunki	76 49·5	S 26 20·8
	02	121 06·7	199 14·4	54·6	77 28·3	28·4	250 59·3	26·4	212 49·2	28·2	Peacock	54 24·3	S 56 52·1
	03	136 09·1	214 13·4	·· 54·4	92 30·7	· 28·5	266 01·2	·· 26·5	227 51·3	·· 28·2	Pollux	244 17·5	N 28 07·4
	04	151 11·6	229 12·5	54·3	107 33·0	28·6	281 03·2	26·7	242 53·5	28·3	Procyon	245 42·3	N 5 19·7
	05	166 14·1	244 11·5	54·1	122 35·4	28·6	296 05·2	26·8	257 55·6	28·3			
	06	181 16·5	259 10·5	S23 53·9	137 37·8	N18 28·7	311 07·2	S17 26·9	272 57·7	S22 28·3	Rasalhague	96 44·9	N 12 35·5
	07	196 19·0	274 09·6	53·7	152 40·2	28·8	326 09·1	27·0	287 59·9	28·3	Regulus	208 26·9	N 12 09·9
T	08	211 21·5	289 08·6	53·5	167 42·6	28·8	341 11·1	27·1	303 02·0	28·3	Rigel	281 51·1	S 8 15·0
U	09	226 23·9	304 07·6	· 53·4	182 44·9	· 28·9	356 13·1	· 27·3	318 04·2	·· 28·3	Rigil Kent.	140 48·1	S 60 39·7
E	10	241 26·4	319 06·7	53·2	197 47·3	29·0	11 15·0	27·4	333 06·3	28·3	Sabik	102 59·9	S 15 40·4
S	11	256 28·8	334 05·7	53·0	212 49·7	29·0	26 17·0	27·5	348 08·5	28·3			
D	12	271 31·3	349 04·7	S23 52·8	227 52·1	N18 29·1	41 19·0	S17 27·6	3 10·6	S22 28·3	Schedar	350 27·1	N 56 19·1
A	13	286 33·8	4 03·8	52·6	242 54·4	29·2	56 21·0	27·8	18 12·8	28·3	Shaula	97 18·0	S 37 04·4
Y	14	301 36·2	19 02·8	52·4	257 56·8	29·3	71 22·9	27·9	33 14·9	28·3	Sirius	259 09·5	S 16 39·7
	15	316 38·7	34 01·9	·· 52·2	272 59·2	·· 29·3	86 24·9	·· 28·0	48 17·0	·· 28·3	Spica	159 14·6	S 10 56·8
	16	331 41·2	49 00·9	52·0	288 01·5	29·4	101 26·9	28·1	63 19·2	28·3	Suhail	223 22·2	S 43 15·9
	17	346 43·6	63 59·9	51·8	303 03·9	29·5	116 28·8	28·2	78 21·3	28·3			
	18	1 46·1	78 59·0	S23 51·6	318 06·3	N18 29·6	131 30·8	S17 28·4	93 23·5	S22 28·3	Vega	81 07·1	N 38 44·9
	19	16 48·6	93 58·0	51·4	333 08·6	29·6	146 32·8	28·5	108 25·6	28·3	Zuben'ubi	137 51·1	S 15 52·2
	20	31 51·0	108 57·0	51·3	348 11·0	29·7	161 34·8	28·6	123 27·8	28·3			
	21	46 53·5	123 56·1	·· 51·1	3 13·4	· 29·8	176 36·7	·· 28·7	138 29·9	·· 28·4			
	22	61 55·9	138 55·1	50·9	18 15·7	29·8	191 38·7	28·8	153 32·1	28·4			
	23	76 58·4	153 54·1	50·7	33 18·1	29·9	206 40·7	29·0	168 34·2	28·4			
Mer. Pass.		17 56·9 (h m)	v −1·0	d 0·2	v 2·4	d 0·1	v 2·0	d 0·1	v 2·1	d 0·0			

	S.H.A. ° ′	Mer. Pass. h m
Venus	79 37·0	12 42
Mars	316 23·2	20 51
Jupiter	130 05·5	9 18
Saturn	91 50·8	11 51

DECEMBER 21, 22, 23 (SUN., MON., TUES.)

	G.M.T.	SUN G.H.A.	SUN Dec.	MOON G.H.A.	v	MOON Dec.	d	H.P.
	d h	° ′	° ′	° ′	′	° ′	′	′
	21 00	180 34·0	S23 26·1	59 40·8	14·4	N10 57·0	7·8	54·1
	01	195 33·7	26·1	74 14·2	14·4	11 04·8	7·8	54·1
	02	210 33·3	26·1	88 47·6	14·3	11 12·6	7·6	54·1
	03	225 33·0	· · 26·2	103 20·9	14·3	11 20·2	7·7	54·2
	04	240 32·7	26·2	117 54·2	14·3	11 27·9	7·6	54·2
	05	255 32·4	26·2	132 27·5	14·2	11 35·5	7·5	54·2
	06	270 32·1	S23 26·2	147 00·7	14·3	N11 43·0	7·5	54·2
	07	285 31·8	26·3	161 34·0	14·2	11 50·5	7·5	54·2
	08	300 31·5	26·3	176 07·2	14·1	11 58·0	7·4	54·2
S	09	315 31·2	· · 26·3	190 40·3	14·1	12 05·4	7·4	54·2
U	10	330 30·9	26·3	205 13·4	14·1	12 12·8	7·3	54·2
N	11	345 30·5	26·3	219 46·5	14·1	12 20·1	7·2	54·2
D	12	0 30·2	S23 26·4	234 19·6	14·0	N12 27·3	7·2	54·2
A	13	15 29·9	26·4	248 52·6	14·0	12 34·5	7·2	54·2
Y	14	30 29·6	26·4	263 25·6	13·9	12 41·7	7·1	54·2
	15	45 29·3	· · 26·4	277 58·5	13·9	12 48·8	7·0	54·2
	16	60 29·0	26·4	292 31·4	13·9	12 55·8	7·0	54·2
	17	75 28·7	26·4	307 04·3	13·9	13 02·8	7·0	54·2
	18	90 28·4	S23 26·4	321 37·2	13·8	N13 09·8	6·9	54·2
	19	105 28·1	26·5	336 10·0	13·7	13 16·7	6·8	54·2
	20	120 27·8	26·5	350 42·7	13·8	13 23·5	6·8	54·2
	21	135 27·4	· · 26·5	5 15·5	13·7	13 30·3	6·7	54·2
	22	150 27·1	26·5	19 48·2	13·6	13 37·0	6·7	54·2
	23	165 26·8	26·5	34 20·8	13·6	13 43·7	6·6	54·2
	22 00	180 26·5	S23 26·5	48 53·4	13·6	N13 50·3	6·5	54·2
	01	195 26·2	26·5	63 26·0	13·6	13 56·8	6·5	54·2
	02	210 25·9	26·5	77 58·6	13·5	14 03·3	6·4	54·3
	03	225 25·6	· · 26·5	92 31·1	13·4	14 09·7	6·4	54·3
	04	240 25·3	26·5	107 03·5	13·5	14 16·1	6·3	54·3
	05	255 25·0	26·5	121 36·0	13·3	14 22·4	6·3	54·3
	06	270 24·6	S23 26·5	136 08·3	13·4	N14 28·7	6·2	54·3
	07	285 24·3	26·5	150 40·7	13·3	14 34·9	6·1	54·3
	08	300 24·0	26·5	165 13·0	13·3	14 41·0	6·1	54·3
M	09	315 23·7	· · 26·5	179 45·3	13·2	14 47·1	6·0	54·3
O	10	330 23·4	26·5	194 17·5	13·2	14 53·1	5·9	54·3
N	11	345 23·1	26·5	208 49·7	13·1	14 59·0	5·9	54·3
D	12	0 22·8	S23 26·5	223 21·8	13·1	N15 04·9	5·8	54·3
A	13	15 22·5	26·5	237 53·9	13·1	15 10·7	5·8	54·3
Y	14	30 22·2	26·5	252 26·0	13·0	15 16·5	5·6	54·4
	15	45 21·8	· · 26·5	266 58·0	13·0	15 22·1	5·7	54·4
	16	60 21·5	26·5	281 30·0	12·9	15 27·8	5·5	54·4
	17	75 21·2	26·5	296 01·9	12·9	15 33·3	5·5	54·4
	18	90 20·9	S23 26·5	310 33·8	12·9	N15 38·8	5·4	54·4
	19	105 20·6	26·5	325 05·7	12·8	15 44·2	5·4	54·4
	20	120 20·3	26·5	339 37·5	12·8	15 49·6	5·2	54·4
	21	135 20·0	· · 26·5	354 09·3	12·7	15 54·8	5·2	54·4
	22	150 19·7	26·5	8 41·0	12·7	16 00·0	5·2	54·4
	23	165 19·4	26·4	23 12·7	12·6	16 05·2	5·1	54·5
	23 00	180 19·1	S23 26·4	37 44·3	12·6	N16 10·3	4·9	54·5
	01	195 18·7	26·4	52 15·9	12·6	16 15·2	5·0	54·5
	02	210 18·4	26·4	66 47·5	12·5	16 20·2	4·8	54·5
	03	225 18·1	· · 26·4	81 19·0	12·5	16 25·0	4·8	54·5
	04	240 17·8	26·4	95 50·5	12·4	16 29·8	4·7	54·5
	05	255 17·5	26·4	110 21·9	12·4	16 34·5	4·7	54·5
	06	270 17·2	S23 26·3	124 53·3	12·4	N16 39·2	4·5	54·5
	07	285 16·9	26·3	139 24·7	12·3	16 43·7	4·5	54·5
T	08	300 16·6	26·3	153 56·0	12·2	16 48·2	4·4	54·6
U	09	315 16·3	· · 26·3	168 27·2	12·3	16 52·6	4·3	54·6
E	10	330 15·9	26·3	182 58·5	12·1	16 56·9	4·3	54·6
S	11	345 15·6	26·2	197 29·6	12·2	17 01·2	4·2	54·6
D	12	0 15·3	S23 26·2	212 00·8	12·1	N17 05·4	4·1	54·6
A	13	15 15·0	26·2	226 31·9	12·1	17 09·5	4·0	54·6
Y	14	30 14·7	26·2	241 03·0	12·0	17 13·5	4·0	54·6
	15	45 14·4	· · 26·2	255 34·0	12·0	17 17·5	3·8	54·7
	16	60 14·1	26·1	270 05·0	11·9	17 21·3	3·8	54·7
	17	75 13·8	26·1	284 35·9	11·9	17 25·1	3·7	54·7
	18	90 13·5	S23 26·1	299 06·8	11·9	N17 28·8	3·6	54·7
	19	105 13·2	26·0	313 37·7	11·8	17 32·4	3·6	54·7
	20	120 12·8	26·0	328 08·5	11·7	17 36·0	3·5	54·7
	21	135 12·5	· · 26·0	342 39·2	11·8	17 39·5	3·3	54·7
	22	150 12·2	26·0	357 10·0	11·7	17 42·8	3·3	54·8
	23	165 11·9	25·9	11 40·7	11·6	17 46·1	3·3	54·8
		S.D. 16·3	*d* 0·0	S.D. 14·8		14·8		14·9

Lat.	Twilight Naut.	Twilight Civil	Sunrise	Moonrise 21	Moonrise 22	Moonrise 23	Moonrise 24
°	h m	h m	h m	h m	h m	h m	h m
N 72	08 26	10 57	■	11 33	11 25	11 12	▭
N 70	08 06	09 54	■	11 55	11 57	12 04	12 19
68	07 50	09 19	■	12 11	12 21	12 36	13 02
66	07 36	08 54	10 35	12 25	12 39	13 00	13 30
64	07 25	08 34	09 52	12 36	12 54	13 18	13 52
62	07 16	08 18	09 24	12 46	13 07	13 34	14 09
60	07 07	08 04	09 02	12 54	13 18	13 47	14 24
N 58	07 00	07 52	08 45	13 02	13 27	13 58	14 36
56	06 53	07 42	08 30	13 08	13 35	14 07	14 47
54	06 47	07 33	08 17	13 14	13 42	14 16	14 56
52	06 41	07 25	08 06	13 19	13 49	14 24	15 05
50	06 36	07 18	07 56	13 24	13 55	14 31	15 12
45	06 24	07 02	07 35	13 34	14 08	14 45	15 28
N 40	06 14	06 48	07 18	13 43	14 18	14 58	15 42
35	06 04	06 36	07 04	13 50	14 27	15 08	15 53
30	05 56	06 26	06 52	13 57	14 35	15 17	16 03
20	05 39	06 07	06 31	14 08	14 49	15 33	16 20
N 10	05 23	05 49	06 12	14 18	15 01	15 47	16 35
0	05 06	05 32	05 55	14 28	15 13	16 00	16 49
S 10	04 47	05 14	05 37	14 37	15 24	16 13	17 03
20	04 24	04 53	05 18	14 47	15 37	16 27	17 18
30	03 55	04 28	04 56	14 59	15 51	16 43	17 35
35	03 37	04 13	04 43	15 05	15 59	16 53	17 45
40	03 13	03 55	04 28	15 13	16 09	17 03	17 57
45	02 43	03 32	04 10	15 22	16 20	17 16	18 10
S 50	01 57	03 03	03 47	15 33	16 33	17 31	18 27
52	01 29	02 47	03 36	15 38	16 39	17 39	18 35
54	00 41	02 29	03 24	15 44	16 46	17 47	18 43
56	////	02 07	03 10	15 50	16 54	17 55	18 53
58	////	01 37	02 53	15 57	17 03	18 06	19 04
S 60	////	00 47	02 32	16 05	17 13	18 18	19 17

Lat.	Sunset	Twilight Civil	Twilight Naut.	Moonset 21	Moonset 22	Moonset 23	Moonset 24
°	h m	h m	h m	h m	h m	h m	h m
N 72	■	12 59	15 30	04 44	06 26	08 16	▭
N 70	■	14 02	15 51	04 24	05 54	07 24	08 49
68	■	14 37	16 07	04 08	05 31	06 53	08 08
66	13 21	15 03	16 20	03 55	05 14	06 30	07 39
64	14 04	15 22	16 31	03 45	04 59	06 11	07 18
62	14 33	15 39	16 41	03 36	04 47	05 56	07 00
60	14 55	15 52	16 49	03 28	04 37	05 44	06 46
N 58	15 12	16 04	16 56	03 22	04 28	05 33	06 34
56	15 27	16 14	17 03	03 16	04 21	05 24	06 23
54	15 40	16 24	17 09	03 10	04 14	05 15	06 14
52	15 51	16 32	17 16	03 06	04 07	05 08	06 06
50	16 01	16 39	17 21	03 01	04 02	05 01	05 58
45	16 21	16 55	17 33	02 52	03 50	04 47	05 42
N 40	16 38	17 09	17 43	02 44	03 40	04 35	05 29
35	16 52	17 21	17 52	02 38	03 31	04 25	05 18
30	17 05	17 31	18 01	02 32	03 24	04 16	05 08
20	17 26	17 50	18 18	02 22	03 11	04 01	04 52
N 10	17 45	18 08	18 34	02 13	03 00	03 48	04 37
0	18 02	18 25	18 51	02 05	02 49	03 35	04 23
S 10	18 20	18 43	19 10	01 56	02 39	03 23	04 10
20	18 39	19 03	19 33	01 48	02 28	03 10	03 55
30	19 01	19 28	20 02	01 38	02 15	02 55	03 39
35	19 14	19 44	20 20	01 32	02 07	02 46	03 29
40	19 29	20 02	20 44	01 25	01 59	02 36	03 18
45	19 47	20 25	21 14	01 18	01 49	02 25	03 05
S 50	20 10	20 54	22 00	01 09	01 38	02 11	02 49
52	20 21	21 10	22 28	01 04	01 32	02 04	02 42
54	20 33	21 27	23 15	01 00	01 26	01 57	02 34
56	20 47	21 50	////	00 55	01 19	01 49	02 24
58	21 04	22 20	////	00 49	01 12	01 40	02 14
S 60	21 25	23 10	////	00 43	01 04	01 29	02 02

Day	SUN Eqn. of Time 00h	SUN Eqn. of Time 12h	SUN Mer. Pass.	MOON Mer. Pass. Upper	MOON Mer. Pass. Lower	Age	Phase
	m s	m s	h m	h m	h m	d	
21	02 16	02 02	11 58	20 38	08 16	11	○
22	01 47	01 32	11 58	21 24	09 01	12	
23	01 17	01 02	11 59	22 12	09 48	13	

DECEMBER 24, 25, 26 (WED., THURS., FRI.)

G.M.T. d h	ARIES G.H.A. ° ′	VENUS −3·4 G.H.A. ° ′	VENUS Dec. ° ′	MARS −0·8 G.H.A. ′	MARS Dec. ° ′	JUPITER −1·3 G.H.A. ° ′	JUPITER Dec. ° ′	SATURN +0·6 G.H.A. ° ′	SATURN Dec. ° ′	STARS Name	S.H.A. ° ′	Dec. ° ′
24 00	92 00·9	163 53·2	S23 50·4	48 20·4	N18 30·0	221 42·7	S17 29·1	183 36·4	S22 28·4	Acamar	315 49·2	S 40 28·3
01	107 03·3	183·52·2	50·2	63 22·8	30·1	236 44·6	29·2	198 38·5	28·4	Achernar	335 57·1	S 57 27·0
02	122 05·8	198 51·3	50·0	78 25·1	30·1	251 46·6	29·3	213 40·6	28·4	Acrux	173 55·1	S 62 52·0
03	137 08·3	213 50·3	·· 49·8	93 27·5	·· 30·2	266 48·6	·· 29·5	228 42·8	·· 28·4	Adhara	255 44·3	S 28 55·0
04	152 10·7	228 49·3	49·6	108 29·9	30·3	281 50·5	29·6	243 44·9	28·4	Aldebaran	291 36·1	N 16 25·6
05	167 13·2	243 48·4	49·4	123 32·2	30·4	296 52·5	29·7	258 47·1	28·4			
06	182 15·7	258 47·4	S23 49·2	138 34·5	N18 30·5	311 54·5	S17 29·8	273 49·2	S22 28·4	Alioth	166 56·8	N 56 10·6
W 07	197 18·1	273 46·4	49·0	153 36·9	30·5	326 56·5	29·9	288 51·4	28·4	Alkaid	153 31·4	N 49 30·8
E 08	212 20·6	288 45·5	48·8	168 39·2	30·6	341 58·4	30·1	303 53·5	28·4	Al Na'ir	28 35·5	S 47 09·7
D 09	227 23·1	303 44·5	·· 48·6	183 41·6	·· 30·7	357 00·4	·· 30·2	318 55·7	·· 28·4	Alnilam	276 27·6	S 1 13·8
N 10	242 25·5	318 43·6	48·4	198 43·9	30·8	12 02·4	30·3	333 57·8	28·4	Alphard	218 36·1	S 8 28·9
E 11	257 28·0	333 42·6	48·1	213 46·3	30·8	27 04·4	30·4	349 00·0	28·4			
S 12	272 30·4	348 41·6	S23 47·9	228 48·6	N18 30·9	42 06·3	S17 30·5	4 02·1	S22 28·4	Alphecca	126 46·0	N 26 51·1
13	287 32·9	3 40·7	47·7	243 51·0	31·0	57 08·3	30·7	19 04·2	28·4	Alpheratz	358 25·9	N 28 52·0
D 14	302 35·4	18 39·7	47·5	258 53·3	31·1	72 10·3	30·8	34 06·4	28·5	Altair	62 48·5	N 8 45·7
A 15	317 37·8	33 38·8	·· 47·3	273 55·6	·· 31·2	87 12·3	·· 30·9	49 08·5	·· 28·5	Ankaa	353 56·1	S 42 31·9
Y 16	332 40·3	48 37·8	47·0	288 58·0	31·2	102 14·2	31·0	64 10·7	28·5	Antares	113 16·9	S 26 20·4
17	347 42·8	63 36·9	46·8	304 00·3	31·3	117 16·2	31·2	79 12·8	28·5			
18	2 45·2	78 35·9	S23 46·6	319 02·6	N18 31·4	132 18·2	S17 31·3	94 15·0	S22 28·5	Arcturus	146 33·3	N 19 23·7
19	17 47·7	93 34·9	46·4	334 05·0	31·5	147 20·2	31·4	109 17·1	28·5	Atria	108 56·0	S 68 57·1
20	32 50·2	108 34·0	46·1	349 07·3	31·6	162 22·1	31·5	124 19·3	28·5	Avior	234 34·2	S 59 22·6
21	47 52·6	123 33·0	·· 45·9	4 09·6	·· 31·7	177 24·1	·· 31·6	139 21·4	·· 28·5	Bellatrix	279 15·6	N 6 18·7
22	62 55·1	138 32·1	45·7	19 11·9	31·7	192 26·1	31·8	154 23·5	28·5	Betelgeuse	271 45·3	N 7 23·9
23	77 57·6	153 31·1	45·4	34 14·3	31·8	207 28·1	31·9	169 25·7	28·5			
25 00	93 00·0	168 30·1	S23 45·2	49 16·6	N18 31·9	222 30·0	S17 32·0	184 27·8	S22 28·5	Canopus	264 13·8	S 52 40·5
01	108 02·5	183 29·2	45·0	64 18·9	32·0	237 32·0	32·1	199 30·0	28·5	Capella	281 34·5	N 45 57·4
02	123 04·9	198 28·2	44·7	79 21·2	32·1	252 34·0	32·2	214 32·1	28·5	Deneb	49 59·7	N 45 08·3
03	138 07·4	213 27·3	·· 44·5	94 23·6	·· 32·1	267 36·0	·· 32·4	229 34·3	·· 28·5	Denebola	183 15·4	N 14 47·9
04	153 09·9	228 26·3	44·3	109 25·9	32·2	282 37·9	32·5	244 36·4	28·5	Diphda	349 37·0	S 18 12·8
05	168 12·3	243 25·4	44·0	124 28·2	32·3	297 39·9	32·6	259 38·6	28·5			
06	183 14·8	258 24·4	S23 43·8	139 30·5	N18 32·4	312 41·9	S17 32·7	274 40·7	S22 28·5	Dubhe	194 41·8	N 61 58·0
07	198 17·3	273 23·5	43·5	154 32·8	32·5	327 43·9	32·8	289 42·9	28·5	Elnath	279 04·1	N 28 34·4
T 08	213 19·7	288 22·5	43·3	169 35·1	32·6	342 45·8	33·0	304 45·0	28·6	Eltanin	91 05·7	N 51 29·7
H 09	228 22·2	303 21·5	·· 43·1	184 37·5	·· 32·7	357 47·8	·· 33·1	319 47·1	·· 28·6	Enif	34 27·6	N 9 41·3
U 10	243 24·7	318 20·6	42·8	199 39·8	32·7	12 49·8	33·2	334 49·3	28·6	Fomalhaut	16 09·3	S 29 50·5
R 11	258 27·1	333 19·6	42·6	214 42·1	32·8	27 51·8	33·3	349 51·4	28·6			
S 12	273 29·6	348 18·7	S23 42·3	229 44·4	N18 32·9	42 53·7	S17 33·4	4 53·6	S22 28·6	Gacrux	172 46·6	S 56 52·8
D 13	288 32·1	3 17·7	42·1	244 46·7	33·0	57 55·7	33·6	19 55·7	28·6	Gienah	176 34·5	S 17 18·8
A 14	303 34·5	18 16·8	41·8	259 49·0	33·1	72 57·7	33·7	34 57·9	28·6	Hadar	149 46·3	S 60 10·3
Y 15	318 37·0	33 15·8	·· 41·6	274 51·3	·· 33·2	87 59·7	·· 33·8	50 00·0	·· 28·6	Hamal	328 46·8	N 23 16·2
16	333 39·4	48 14·9	41·3	289 53·6	33·3	103 01·7	33·9	65 02·2	28·6	Kaus Aust.	84 38·6	S 34 24·2
17	348 41·9	63 13·9	41·0	304 55·9	33·3	118 03·6	34·0	80 04·3	28·6			
18	3 44·4	78 13·0	S23 40·8	319 58·2	N18 33·4	133 05·6	S17 34·1	95 06·5	S22 28·6	Kochab	137 19·2	N 74 19·2
19	18 46·8	93 12·0	40·5	335 00·5	33·5	148 07·6	34·3	110 08·6	28·6	Markab	14 19·3	N 14 59·2
20	33 49·3	108 11·1	40·3	350 02·8	33·6	163 09·6	34·4	125 10·8	28·6	Menkar	314 57·7	N 3 55·7
21	48 51·8	123 10·1	·· 40·0	5 05·1	·· 33·7	178 11·5	·· 34·5	140 12·9	·· 28·6	Menkent	148 56·1	S 36 09·9
22	63 54·2	138 09·1	39·8	20 07·4	33·8	193 13·5	34·6	155 15·0	28·6	Miaplacidus	221 47·6	S 69 32·8
23	78 56·7	153 08·2	39·5	35 09·7	33·9	208 15·5	34·7	170 17·2	28·6			
26 00	93 59·2	168 07·2	S23 39·2	50 12·0	N18 34·0	223 17·5	S17 34·9	185 19·3	S22 28·6	Mirfak	309 38·6	N 49 43·1
01	109 01·6	183 06·3	39·0	65 14·3	34·1	238 19·5	35·0	200 21·5	28·6	Nunki	76 49·5	S 26 20·8
02	124 04·1	198 05·3	38·7	80 16·6	34·1	253 21·4	35·1	215 23·6	28·6	Peacock	54 24·3	S 56 52·1
03	139 06·5	213 04·4	·· 38·4	95 18·9	·· 34·2	268 23·4	·· 35·2	230 25·8	·· 28·6	Pollux	244 17·5	N 28 07·4
04	154 09·0	228 03·4	38·2	110 21·2	34·3	283 25·4	35·3	245 27·9	28·7	Procyor	245 42·3	N 5 19·7
05	169 11·5	243 02·5	37·9	125 23·4	34·4	298 27·4	35·5	260 30·1	28·7			
06	184 13·9	258 01·5	S23 37·6	140 25·7	N18 34·5	313 29·4	S17 35·6	275 32·2	S22 28·7	Rasalhague	96 44·8	N 12 35·4
07	199 16·4	273 00·6	37·3	155 28·0	34·6	328 31·3	35·7	290 34·4	28·7	Regulus	208 26·9	N 12 09·9
08	214 18·9	287 59·6	37·1	170 30·3	34·7	343 33·3	35·8	305 36·5	28·7	Rigel	281 51·1	S 8 15·0
F 09	229 21·3	302 58·7	·· 36·8	185 32·6	·· 34·8	358 35·3	·· 35·9	320 38·6	·· 28·7	Rigil Kent.	140 48·1	S 60 39·7
R 10	244 23·8	317 57·7	36·5	200 34·9	34·9	13 37·3	36·0	335 40·8	28·7	Sabik	102 59·9	S 15 40·4
I 11	259 26·3	332 56·8	36·2	215 37·1	35·0	28 39·3	36·2	350 42·9	28·7			
D 12	274 28·7	347 55·8	S23 36·0	230 39·4	N18 35·1	43 41·2	S17 36·3	5 45·1	S22 28·7	Schedar	350 27·1	N 56 19·1
A 13	289 31·2	2 54·9	35·7	245 41·7	35·2	58 43·2	36·4	20 47·2	28·7	Shaula	97 18·0	S 37 04·4
Y 14	304 33·7	17 54·0	35·4	260 44·0	35·2	73 45·2	36·5	35 49·4	28·7	Sirius	259 09·5	S 16 39·7
15	319 36·1	32 53·0	·· 35·1	275 46·2	·· 35·3	88 47·2	·· 36·6	50 51·5	·· 28·7	Spica	159 14·6	S 10 56·8
16	334 38·6	47 52·1	34·8	290 48·5	35·4	103 49·1	36·8	65 53·7	28·7	Suhail	223 22·2	S 43 15·9
17	349 41·0	62 51·1	34·6	305 50·8	35·5	118 51·1	36·9	80 55·8	28·7			
18	4 43·5	77 50·2	S23 34·3	320 53·1	N18 35·6	133 53·1	S17 37·0	95 58·0	S22 28·7	Vega	81 07·1	N 38 44·8
19	19 46·0	92 49·2	34·0	335 55·3	35·7	148 55·1	37·1	111 00·1	28·7	Zuben'ubi	137 51·0	S 15 52·2
20	34 48·4	107 48·3	33·7	350 57·6	35·8	163 57·1	37·2	126 02·2	28·7			
21	49 50·9	122 47·3	·· 33·4	5 59·9	·· 35·9	178 59·1	·· 37·3	141 04·4	·· 28·7			
22	64 53·4	137 46·4	33·1	21 02·1	36·0	194 01·0	37·5	156 06·5	28·7			
23	79 55·8	152 45·4	32·8	36 04·4	36·1	209 03·0	37·6	171 08·7	28·7			
Mer. Pass. (h m)	17 45·1	v −1·0	d 0·2	v 2·3	d 0·1	v 2·0	d 0·1	v 2·1	d 0·0			

	S.H.A. ° ′	Mer. Pass. h m
Venus	75 30·1	12 47
Mars	316 16·6	20 40
Jupiter	129 30·0	9 09
Saturn	91 27·8	11 41

DECEMBER 24, 25, 26 (WED., THURS., FRI.)

	G.M.T. d h	SUN G.H.A. ° ′	SUN Dec. ° ′	MOON G.H.A. ° ′	MOON v ′	MOON Dec. ° ′	MOON d ′	MOON H.P. ′
	24 00	180 11·6	S 23 25·9	26 11·3	11·7	N17 49·4	3·1	54·8
	01	195 11·3	25·9	40 42·0	11·5	17 52·5	3·0	54·8
	02	210 11·0	25·8	55 12·5	11·6	17 55·5	3·0	54·8
	03	225 10·7	·· 25·8	69 43·1	11·5	17 58·5	2·9	54·8
	04	240 10·4	25·8	84 13·6	11·5	18 01·4	2·8	54·9
	05	255 10·0	25·7	98 44·1	11·4	18 04·2	2·7	54·9
	06	270 09·7	S 23 25·7	113 14·5	11·4	N18 06·9	2·6	54·9
W	07	285 09·4	25·7	127 44·9	11·3	18 09·5	2·6	54·9
E	08	300 09·1	25·6	142 15·2	11·4	18 12·1	2·4	54·9
D	09	315 08·8	·· 25·6	156 45·6	11·2	18 14·5	2·4	54·9
N	10	330 08·5	25·5	171 15·8	11·3	18 16·9	2·2	55·0
E	11	345 08·2	25·5	185 46·1	11·2	18 19·1	2·2	55·0
S	12	0 07·9	S 23 25·5	200 16·3	11·2	N18 21·3	2·1	55·0
D	13	15 07·6	25·4	214 46·5	11·1	18 23·4	2·0	55·0
A	14	30 07·3	25·4	229 16·6	11·1	18 25·4	2·0	55·0
Y	15	45 06·9	·· 25·3	243 46·7	11·1	18 27·4	1·8	55·0
	16	60 06·6	25·3	258 16·8	11·1	18 29·2	1·7	55·1
	17	75 06·3	25·2	272 46·9	11·0	18 30·9	1·7	55·1
	18	90 06·0	S 23 25·2	287 16·9	10·9	N18 32·6	1·5	55·1
	19	105 05·7	25·1	301 46·8	11·0	18 34·1	1·5	55·1
	20	120 05·4	25·1	316 16·8	10·9	18 35·6	1·4	55·1
	21	135 05·1	·· 25·0	330 46·7	10·9	18 37·0	1·3	55·1
	22	150 04·8	25·0	345 16·6	10·8	18 38·3	1·1	55·2
	23	165 04·5	24·9	359 46·4	10·9	18 39·4	1·1	55·2
	25 00	180 04·2	S 23 24·9	14 16·3	10·8	N18 40·5	1·1	55·2
	01	195 03·8	24·8	28 46·1	10·7	18 41·6	0·9	55·2
	02	210 03·5	24·8	43 15·8	10·8	18 42·5	0·8	55·2
	03	225 03·2	·· 24·7	57 45·6	10·7	18 43·3	0·7	55·2
	04	240 02·9	24·7	72 15·3	10·7	18 44·0	0·6	55·3
	05	255 02·6	24·6	86 45·0	10·6	18 44·6	0·6	55·3
	06	270 02·3	S 23 24·6	101 14·6	10·6	N18 45·2	0·4	55·3
	07	285 02·0	24·5	115 44·2	10·6	18 45·6	0·4	55·3
T	08	300 01·7	24·4	130 13·8	10·6	18 46·0	0·2	55·3
H	09	315 01·4	·· 24·4	144 43·4	10·6	18 46·2	0·2	55·4
U	10	330 01·1	24·3	159 13·0	10·5	18 46·4	0·0	55·4
R	11	345 00·8	24·3	173 42·5	10·5	18 46·4	0·0	55·4
S	12	0 00·4	S 23 24·2	188 12·0	10·5	N18 46·4	0·1	55·4
D	13	15 00·1	24·1	202 41·5	10·5	18 46·3	0·3	55·4
A	14	29 59·8	24·1	217 11·0	10·4	18 46·0	0·3	55·5
Y	15	44 59·5	·· 24·0	231 40·4	10·4	18 45·7	0·4	55·5
	16	59 59·2	24·0	246 09·8	10·4	18 45·3	0·5	55·5
	17	74 58·9	23·9	260 39·2	10·4	18 44·8	0·6	55·5
	18	89 58·6	S 23 23·8	275 08·6	10·4	N18 44·2	0·7	55·5
	19	104 58·3	23·8	289 38·0	10·3	18 43·5	0·8	55·5
	20	119 58·0	23·7	304 07·3	10·3	18 42·7	0·9	55·6
	21	134 57·7	·· 23·6	318 36·6	10·3	18 41·8	1·0	55·6
	22	149 57·3	23·6	333 05·9	10·3	18 40·8	1·1	55·6
	23	164 57·0	23·5	347 35·2	10·3	18 39·7	1·2	55·6
	26 00	179 56·7	S 23 23·4	2 04·5	10·3	N18 38·5	1·3	55·6
	01	194 56·4	23·3	16 33·8	10·2	18 37·2	1·4	55·7
	02	209 56·1	23·3	31 03·0	10·2	18 35·8	1·5	55·7
	03	224 55·8	·· 23·2	45 32·2	10·3	18 34·3	1·6	55·7
	04	239 55·5	23·1	60 01·5	10·2	18 32·7	1·7	55·7
	05	254 55·2	23·0	74 30·7	10·2	18 31·0	1·8	55·7
	06	269 54·9	S 23 23·0	88 59·9	10·1	N18 29·2	1·8	55·8
	07	284 54·6	22·9	103 29·0	10·2	18 27·4	2·0	55·8
	08	299 54·3	22·8	117 58·2	10·2	18 25·4	2·1	55·8
F	09	314 53·9	· 22·7	132 27·4	10·1	18 23·3	2·2	55·8
R	10	329 53·6	22·7	146 56·5	10·2	18 21·1	2·2	55·8
I	11	344 53·3	22·6	161 25·7	10·1	18 18·9	2·4	55·9
D	12	359 53·0	S 23 22·5	175 54·8	10·1	N18 16·5	2·5	55·9
A	13	14 52·7	22·4	190 23·9	10·1	18 14·0	2·5	55·9
Y	14	29 52·4	22·3	204 53·0	10·1	18 11·5	2·7	55·9
	15	44 52·1	·· 22·3	219 22·1	10·1	18 08·8	2·8	55·9
	16	59 51·8	22·2	233 51·2	10·1	18 06·0	2·8	56·0
	17	74 51·5	22·1	248 20·3	10·1	18 03·2	3·0	56·0
	18	89 51·2	S 23 22·0	262 49·4	10·1	N18 00·2	3·0	56·0
	19	104 50·9	21·9	277 18·5	10·1	17 57·2	3·2	56·0
	20	119 50·6	21·8	291 47·6	10·1	17 54·0	3·2	56·0
	21	134 50·2	·· 21·7	306 16·7	10·1	17 50·8	3·3	56·1
	22	149 49·9	21·6	320 45·8	10·0	17 47·5	3·5	56·1
	23	164 49·6	21·6	335 14·8	10·1	17 44·0	3·5	56·1
		S.D. 16·3	d 0·1	S.D.		15·0	15·1	15·2

Lat. °	Twilight Naut. h m	Twilight Civil h m	Sunrise h m	Moonrise 24 h m	Moonrise 25 h m	Moonrise 26 h m	Moonrise 27 h m
N 72	08 27	10 57	■	□	□	□	14 48
N 70	08 07	09 55	■	12 19	12 57	14 06	15 37
68	07 51	09 20	■	13 02	13 44	14 47	16 07
66	07 38	08 55	10 35	13 30	14 14	15 15	16 30
64	07 26	08 35	09 53	13 52	14 37	15 37	16 48
62	07 17	08 19	09 24	14 09	14 56	15 54	17 03
60	07 08	08 05	09 03	14 24	15 11	16 08	17 15
N 58	07 01	07 54	08 46	14 36	15 23	16 20	17 26
56	06 54	07 43	08 31	14 47	15 35	16 31	17 35
54	06 48	07 35	08 19	14 56	15 44	16 40	17 43
52	06 43	07 26	08 07	15 05	15 53	16 48	17 51
50	06 37	07 19	07 57	15 12	16 01	16 56	17 57
45	06 26	07 03	07 37	15 28	16 17	17 12	18 11
N 40	06 15	06 49	07 20	15 42	16 31	17 25	18 23
35	06 06	06 38	07 06	15 53	16 42	17 36	18 33
30	05 57	06 27	06 53	16 03	16 52	17 46	18 41
20	05 41	06 08	06 32	16 20	17 10	18 02	18 56
N 10	05 24	05 51	06 13	16 35	17 25	18 17	19 09
0	05 07	05 34	05 56	16 49	17 39	18 30	19 22
S 10	04 48	05 16	05 39	17 03	17 53	18 44	19 34
20	04 26	04 55	05 20	17 18	18 09	18 58	19 47
30	03 57	04 30	04 58	17 35	18 26	19 15	20 02
35	03 38	04 15	04 45	17 45	18 36	19 25	20 10
40	03 15	03 57	04 29	17 57	18 48	19 36	20 20
45	02 44	03 34	04 11	18 10	19 01	19 49	20 31
S 50	01 59	03 04	03 49	18 27	19 18	20 04	20 45
52	01 31	02 49	03 38	18 35	19 26	20 12	20 52
54	00 44	02 31	03 26	18 43	19 35	20 20	20 59
56	////	02 09	03 12	18 53	19 45	20 29	21 07
58	////	01 39	02 55	19 04	19 56	20 40	21 16
S 60	////	00 49	02 34	19 17	20 09	20 52	21 26

Lat. °	Sunset h m	Twilight Civil h m	Twilight Naut. h m	Moonset 24 h m	Moonset 25 h m	Moonset 26 h m	Moonset 27 h m
N 72	■	13 02	15 33	□	□	□	11 37
N 70	■	14 04	15 53	08 49	09 56	10 32	10 48
68	■	14 39	16 09	08 08	09 08	09 51	10 17
66	13 24	15 05	16 22	07 39	08 38	09 22	09 53
64	14 06	15 24	16 33	07 18	08 15	09 01	09 35
62	14 35	15 41	16 42	07 00	07 57	08 43	09 20
60	14 57	15 54	16 51	06 46	07 42	08 29	09 07
N 58	15 14	16 06	16 58	06 34	07 29	08 16	08 56
56	15 29	16 16	17 05	06 23	07 18	08 06	08 47
54	15 42	16 25	17 11	06 14	07 08	07 56	08 38
52	15 52	16 34	17 17	06 06	06 59	07 48	08 31
50	16 02	16 41	17 22	05 58	06 52	07 40	08 24
45	16 23	16 57	17 34	05 42	06 35	07 24	08 09
N 40	16 40	17 11	17 45	05 29	06 21	07 11	07 57
35	16 54	17 22	17 54	05 18	06 10	07 00	07 46
30	17 07	17 33	18 03	05 08	06 00	06 50	07 37
20	17 28	17 52	18 19	04 52	05 42	06 33	07 22
N 10	17 46	18 09	18 35	04 37	05 27	06 18	07 08
0	18 04	18 26	18 52	04 23	05 13	06 04	06 55
S 10	18 21	18 44	19 11	04 10	04 59	05 50	06 42
20	18 40	19 05	19 34	03 55	04 44	05 35	06 28
30	19 02	19 30	20 03	03 39	04 26	05 17	06 12
35	19 15	19 45	20 22	03 29	04 16	05 07	06 03
40	19 30	20 03	20 45	03 18	04 04	04 56	05 52
45	19 48	20 26	21 15	03 05	03 51	04 42	05 39
S 50	20 11	20 55	22 01	02 49	03 34	04 26	05 24
52	20 22	21 11	22 28	02 42	03 26	04 18	05 17
54	20 34	21 28	23 14	02 34	03 17	04 09	05 09
56	20 48	21 51	////	02 24	03 08	04 00	05 00
58	21 05	22 20	////	02 14	02 56	03 49	04 50
S 60	21 25	23 09	////	02 02	02 44	03 36	04 38

Day	SUN Eqn. of Time 00h m s	SUN Eqn. of Time 12h m s	SUN Mer. Pass. h m	MOON Mer. Pass. Upper h m	MOON Mer. Pass. Lower h m	MOON Age d	MOON Phase
24	00 47	00 32	11 59	23 01	10 36	14	○
25	00 17	00 02	12 00	23 51	11 26	15	
26	00 12	00 27	12 00	24 42	12 17	16	

DECEMBER 27, 28, 29 (SAT., SUN., MON.)

G.M.T.	ARIES	VENUS −3·4		MARS −0·7		JUPITER −1·3		SATURN +0·7	
d h	G.H.A. ° ′	G.H.A. ° ′	Dec. ° ′	G.H.A. ° ′	Dec. ° ′	G.H.A. ° ′	Dec. ° ′	G.H.A. ° ′	Dec. ° ′
27 00	94 58·3	167 44·5	S23 32·5	51 06·6	N18 36·2	224 05·0	S17 37·7	186 10·8	S22 28·8
01	110 00·8	182 43·5	32·2	66 08·9	36·3	239 07·0	37·8	201 13·0	28·8
02	125 03·2	197 42·6	31·9	81 11·2	36·4	254 09·0	37·9	216 15·1	28·8
03	140 05·7	212 41·6	31·6	96 13·4	36·5	269 10·9	38·0	231 17·3	28·8
04	155 08·2	227 40·7	31·3	111 15·7	36·6	284 12·9	38·2	246 19·4	28·8
05	170 10·6	242 39·8	31·0	126 17·9	36·7	299 14·9	38·3	261 21·6	28·8
06	185 13·1	257 38·8	S23 30·7	141 20·2	N18 36·8	314 16·9	S17 38·4	276 23·7	S22 28·8
07	200 15·5	272 37·9	30·4	156 22·4	36·9	329 18·9	38·5	291 25·9	28·8
S 08	215 18·0	287 36·9	30·1	171 24·7	37·0	344 20·8	38·6	306 28·0	28·8
A 09	230 20·5	302 36·0	29·8	186 27·0	37·1	359 22·8	38·8	321 30·1	28·8
T 10	245 22·9	317 35·0	29·5	201 29·2	37·2	14 24·8	38·9	336 32·3	28·8
U 11	260 25·4	332 34·1	29·2	216 31·4	37·3	29 26·8	39·0	351 34·4	28·8
R 12	275 27·9	347 33·2	S23 28·9	231 33·7	N18 37·4	44 28·8	S17 39·1	6 36·6	S22 28·8
D 13	290 30·3	2 32·2	28·6	246 35·9	37·5	59 30·8	39·2	21 38·7	28·8
A 14	305 32·8	17 31·3	28·3	261 38·2	37·6	74 32·7	39·3	36 40·9	28·8
Y 15	320 35·3	32 30·3	28·0	276 40·4	37·7	89 34·7	39·5	51 43·0	28·8
16	335 37·7	47 29·4	27·7	291 42·7	37·8	104 36·7	39·6	66 45·2	28·8
17	350 40·2	62 28·4	27·3	306 44·9	37·9	119 38·7	39·7	81 47·3	28·8
18	5 42·7	77 27·5	S23 27·0	321 47·2	N18 38·0	134 40·7	S17 39·8	96 49·5	S22 28·8
19	20 45·1	92 26·6	26·7	336 49·4	38·1	149 42·7	39·9	111 51·6	28·8
20	35 47·6	107 25·6	26·4	351 51·6	38·2	164 44·6	40·0	126 53·7	28·8
21	50 50·0	122 24·7	26·1	6 53·9	38·3	179 46·6	40·1	141 55·9	28·9
22	65 52·5	137 23·7	25·7	21 56·1	38·4	194 48·6	40·3	156 58·0	28·9
23	80 55·0	152 22·8	25·4	36 58·3	38·5	209 50·6	40·4	172 00·2	28·9
28 00	95 57·4	167 21·9	S23 25·1	52 00·6	N18 38·6	224 52·6	S17 40·5	187 02·3	S22 28·9
01	110 59·9	182 20·9	24·8	67 02·8	38·7	239 54·6	40·6	202 04·5	28·9
02	126 02·4	197 20·0	24·4	82 05·0	38·8	254 56·5	40·7	217 06·6	28·9
03	141 04·8	212 19·1	24·1	97 07·2	38·9	269 58·5	40·8	232 08·8	28·9
04	156 07·3	227 18·1	23·8	112 09·5	39·0	285 00·5	41·0	247 10·9	28·9
05	171 09·8	242 17·2	23·5	127 11·7	39·1	300 02·5	41·1	262 13·1	28·9
06	186 12·2	257 16·2	S23 23·1	142 13·9	N18 39·2	315 04·5	S17 41·2	277 15·2	S22 28·9
07	201 14·7	272 15·3	22·8	157 16·1	39·3	330 06·5	41·3	292 17·4	28·9
08	216 17·2	287 14·4	22·5	172 18·4	39·4	345 08·5	41·4	307 19·5	28·9
S 09	231 19·6	302 13·4	22·1	187 20·6	39·5	0 10·4	41·5	322 21·6	28·9
U 10	246 22·1	317 12·5	21·8	202 22·8	39·6	15 12·4	41·7	337 23·8	28·9
N 11	261 24·5	332 11·6	21·5	217 25·0	39·7	30 14·4	41·8	352 25·9	28·9
D 12	276 27·0	347 10·6	S23 21·1	232 27·2	N18 39·8	45 16·4	S17 41·9	7 28·1	S22 28·9
A 13	291 29·5	2 09·7	20·8	247 29·5	40·0	60 18·4	42·0	22 30·2	28·9
Y 14	306 31·9	17 08·8	20·4	262 31·7	40·1	75 20·4	42·1	37 32·4	28·9
15	321 34·4	32 07·8	20·1	277 33·9	40·2	90 22·4	42·2	52 34·5	28·9
16	336 36·9	47 06·9	19·7	292 36·1	40·3	105 24·3	42·4	67 36·7	28·9
17	351 39·3	62 06·0	19·4	307 38·3	40·4	120 26·3	42·5	82 38·8	28·9
18	6 41·8	77 05·0	S23 19·1	322 40·5	N18 40·5	135 28·3	S17 42·6	97 41·0	S22 28·9
19	21 44·3	92 04·1	18·7	337 42·7	40·6	150 30·3	42·7	112 43·1	28·9
20	36 46·7	107 03·2	18·4	352 44·9	40·7	165 32·3	42·8	127 45·3	29·0
21	51 49·2	122 02·2	18·0	7 47·1	40·8	180 34·3	42·9	142 47·4	29·0
22	66 51·6	137 01·3	17·7	22 49·3	40·9	195 36·3	43·0	157 49·5	29·0
23	81 54·1	152 00·4	17·3	37 51·5	41·0	210 38·2	43·2	172 51·7	29·0
29 00	96 56·6	166 59·4	S23 16·9	52 53·7	N18 41·1	225 40·2	S17 43·3	187 53·8	S22 29·0
01	111 59·0	181 58·5	16·6	67 55·9	41·2	240 42·2	43·4	202 56·0	29·0
02	127 01·5	196 57·6	16·2	82 58·1	41·4	255 44·2	43·5	217 58·1	29·0
03	142 04·0	211 56·6	15·9	98 00·3	41·5	270 46·2	43·6	233 00·3	29·0
04	157 06·4	226 55·7	15·5	113 02·5	41·6	285 48·2	43·7	248 02·4	29·0
05	172 08·9	241 54·8	15·2	128 04·7	41·7	300 50·2	43·8	263 04·6	29·0
06	187 11·4	256 53·9	S23 14·8	143 06·9	N18 41·8	315 52·2	S17 44·0	278 06·7	S22 29·0
07	202 13·8	271 52·9	14·4	158 09·1	41·9	330 54·1	44·1	293 08·9	29·0
08	217 16·3	286 52·0	14·1	173 11·3	42·0	345 56·1	44·2	308 11·0	29·0
M 09	232 18·8	301 51·1	13·7	188 13·5	42·1	0 58·1	44·3	323 13·2	29·0
O 10	247 21·2	316 50·1	13·3	203 15·7	42·2	16 00·1	44·4	338 15·3	29·0
N 11	262 23·7	331 49·2	13·0	218 17·9	42·4	31 02·1	44·5	353 17·5	29·0
D 12	277 26·1	346 48·3	S23 12·6	233 20·1	N18 42·5	46 04·1	S17 44·6	8 19·6	S22 29·0
A 13	292 28·6	1 47·4	12·2	248 22·3	42·6	61 06·1	44·8	23 21·7	29·0
Y 14	307 31·1	16 46·4	11·9	263 24·4	42·7	76 08·1	44·9	38 23·9	29·0
15	322 33·5	31 45·5	11·5	278 26·6	42·8	91 10·0	45·0	53 26·0	29·0
16	337 36·0	46 44·6	11·1	293 28·8	42·9	106 12·0	45·1	68 28·2	29·0
17	352 38·5	61 43·7	10·7	308 31·0	43·0	121 14·0	45·2	83 30·3	29·0
18	7 40·9	76 42·7	S23 10·4	323 33·2	N18 43·2	136 16·0	S17 45·3	98 32·5	S22 29·0
19	22 43·4	91 41·8	10·0	338 35·3	43·3	151 18·0	45·4	113 34·6	29·0
20	37 45·9	106 40·9	09·6	353 37·5	43·4	166 20·0	45·6	128 36·8	29·0
21	52 48·3	121 40·0	09·2	8 39·7	43·5	181 22·0	45·7	143 38·9	29·1
22	67 50·8	136 39·0	08·8	23 41·9	43·6	196 24·0	45·8	158 41·1	29·1
23	82 53·3	151 38·1	08·5	38 44·0	43·7	211 26·0	45·9	173 43·2	29·1
Mer. Pass.	h m 17 33·3	v −0·9	d 0·3	v 2·2	d 0·1	v 2·0	d 0·1	v 2·1	d 0·0

STARS

Name	S.H.A. ° ′	Dec. ° ′
Acamar	315 49·2	S 40 28·4
Achernar	335 57·1	S 57 27·0
Acrux	173 55·0	S 62 52·1
Adhara	255 44·3	S 28 55·1
Aldebaran	291 36·1	N 16 25·6
Alioth	166 56·8	N 56 10·6
Alkaid	153 31·4	N 49 30·8
Al Na'ir	28 35·5	S 47 09·7
Alnilam	276 27·6	S 1 13·8
Alphard	218 36·1	S 8 28·9
Alphecca	126 46·0	N 26 51·1
Alpheratz	358 25·9	N 28 52·0
Altair	62 48·5	N 8 45·7
Ankaa	353 56·2	S 42 31·9
Antares	113 16·9	S 26 20·4
Arcturus	146 33·3	N 19 23·7
Atria	108 56·0	S 68 57·1
Avior	234 34·1	S 59 22·6
Bellatrix	279 15·6	N 6 18·7
Betelgeuse	271 45·3	N 7 23·9
Canopus	264 13·8	S 52 40·5
Capella	281 34·5	N 45 57·4
Deneb	49 59·8	N 45 08·3
Denebola	183 15·4	N 14 47·9
Diphda	349 37·0	S 18 12·8
Dubhe	194 41·7	N 61 58·0
Elnath	279 04·1	N 28 34·4
Eltanin	91 05·7	N 51 29·7
Enif	34 27·6	N 9 41·3
Fomalhaut	16 09·3	S 29 50·5
Gacrux	172 46·5	S 56 52·8
Gienah	176 34·5	S 17 18·8
Hadar	149 46·3	S 60 10·3
Hamal	328 46·9	N 23 16·2
Kaus Aust.	84 38·6	S 34 24·2
Kochab	137 19·1	N 74 19·2
Markab	14 19·3	N 14 59·2
Menkar	314 57·7	N 3 55·7
Menkent	148 56·1	S 36 09·9
Miaplacidus	221 47·6	S 69 32·8
Mirfak	309 38·6	N 49 43·1
Nunki	76 49·5	S 26 20·8
Peacock	54 24·3	S 56 52·1
Pollux	244 17·4	N 28 07·4
Procyon	245 42·3	N 5 19·7
Rasalhague	96 44·8	N 12 35·4
Regulus	208 26·9	N 12 09·9
Rigel	281 51·1	S 8 15·0
Rigil Kent.	140 48·0	S 60 39·7
Sabik	102 59·9	S 15 40·4
Schedar	350 27·2	N 56 19·1
Shaula	97 18·0	S 37 04·4
Sirius	259 09·5	S 16 39·7
Spica	159 14·5	S 10 56·8
Suhail	223 22·2	S 43 16·0
Vega	81 07·1	N 38 44·8
Zuben'ubi	137 51·0	S 15 52·2

	S.H.A. ° ′	Mer. Pass. h m
Venus	71 24·5	12 51
Mars	316 03·2	20 29
Jupiter	128 55·2	8 59
Saturn	91 04·9	11 30

DECEMBER 27, 28, 29 (SAT., SUN., MON.)

G.M.T.		SUN G.H.A.	SUN Dec.	MOON G.H.A.	MOON v	MOON Dec.	MOON d	MOON H.P.
d h		° ′	° ′	° ′	′	° ′	′	′
27 00		179 49·3	S 23 21·5	349 43·9	10·1	N17 40·5	3·6	56·1
01		194 49·0	21·4	4 13·0	10·1	17 36·9	3·7	56·1
02		209 48·7	21·3	18 42·1	10·0	17 33·2	3·9	56·2
03		224 48·4	· · 21·2	33 11·1	10·1	17 29·3	3·9	56·2
04		239 48·1	21·1	47 40·2	10·1	17 25·4	4·0	56·2
05		254 47·8	21·0	62 09·3	10·1	17 21·4	4·1	56·2
06		269 47·5	S 23 20·9	76 38·4	10·0	N17 17·3	4·2	56·2
07		284 47·2	20·8	91 07·4	10·1	17 13·1	4·3	56·3
08	S	299 46·9	20·7	105 36·5	10·1	17 08·8	4·3	56·3
09	A	314 46·6	· · 20·6	120 05·6	10·1	17 04·5	4·5	56·3
10	T	329 46·2	20·5	134 34·7	10·1	17 00·0	4·6	56·3
11	U	344 45·9	20·4	149 03·8	10·1	16 55·4	4·6	56·3
12	R	359 45·6	S 23 20·3	163 32·9	10·1	N16 50·8	4·8	56·4
13	D	14 45·3	20·2	178 02·0	10·1	16 46·0	4·9	56·4
14	A	29 45·0	20·1	192 31·1	10·1	16 41·1	4·9	56·4
15	Y	44 44·7	· · 20·0	207 00·2	10·1	16 36·2	5·0	56·4
16		59 44·4	19·9	221 29·3	10·2	16 31·2	5·1	56·4
17		74 44·1	19·8	235 58·5	10·1	16 26·1	5·3	56·5
18		89 43·8	S 23 19·7	250 27·6	10·1	N16 20·8	5·3	56·5
19		104 43·5	19·6	264 56·7	10·2	16 15·5	5·4	56·5
20		119 43·2	19·5	279 25·9	10·2	16 10·1	5·4	56·5
21		134 42·9	· · 19·4	293 55·1	10·1	16 04·7	5·6	56·5
22		149 42·6	19·3	308 24·2	10·2	15 59·1	5·7	56·6
23		164 42·3	19·2	322 53·4	10·2	15 53·4	5·7	56·6
28 00		179 41·9	S 23 19·1	337 22·6	10·2	N15 47·7	5·9	56·6
01		194 41·6	18·9	351 51·8	10·2	15 41·8	5·9	56·6
02		209 41·3	18·8	6 21·0	10·2	15 35·9	6·0	56·6
03		224 41·0	· · 18·7	20 50·2	10·2	15 29·9	6·1	56·7
04		239 40·7	18·6	35 19·4	10·3	15 23·8	6·2	56·7
05		254 40·4	18·5	49 48·7	10·2	15 17·6	6·2	56·7
06		269 40·1	S 23 18·4	64 17·9	10·3	N15 11·4	6·4	56·7
07		284 39·8	18·3	78 47·2	10·3	15 05·0	6·4	56·7
08		299 39·5	18·1	93 16·5	10·3	14 58·6	6·6	56·8
09	S	314 39·2	· · 18·0	107 45·8	10·3	14 52·0	6·6	56·8
10	U	329 38·9	17·9	122 15·1	10·3	14 45·4	6·6	56·8
11	N	344 38·6	17·8	136 44·4	10·3	14 38·8	6·8	56·8
12	D	359 38·3	S 23 17·7	151 13·7	10·3	N14 32·0	6·9	56·8
13	A	14 38·0	17·6	165 43·0	10·4	14 25·1	6·9	56·9
14	Y	29 37·7	17·4	180 12·4	10·3	14 18·2	7·0	56·9
15		44 37·4	· · 17·3	194 41·7	10·4	14 11·2	7·1	56·9
16		59 37·1	17·2	209 11·1	10·4	14 04·1	7·2	56·9
17		74 36·7	17·1	223 40·5	10·4	13 56·9	7·2	56·9
18		89 36·4	S 23 16·9	238 09·9	10·4	N13 49·7	7·3	57·0
19		104 36·1	16·8	252 39·3	10·5	13 42·4	7·4	57·0
20		119 35·8	16·7	267 08·8	10·4	13 35·0	7·5	57·0
21		134 35·5	· · 16·6	281 38·2	10·5	13 27·5	7·6	57·0
22		149 35·2	16·4	296 07·7	10·4	13 19·9	7·6	57·0
23		164 34·9	16·3	310 37·1	10·5	13 12·3	7·7	57·1
29 00		179 34·6	S 23 16·2	325 06·6	10·5	N13 04·6	7·8	57·1
01		194 34·3	16·0	339 36·1	10·6	12 56·8	7·8	57·1
02		209 34·0	15·9	354 05·7	10·5	12 49·0	8·0	57·1
03		224 33·7	· · 15·8	8 35·2	10·5	12 41·0	8·0	57·1
04		239 33·4	15·6	23 04·7	10·6	12 33·0	8·0	57·2
05		254 33·1	15·5	37 34·3	10·6	12 25·0	8·2	57·2
06		269 32·8	S 23 15·4	52 03·9	10·6	N12 16·8	8·2	57·2
07		284 32·5	15·2	66 33·5	10·6	12 08·6	8·3	57·2
08		299 32·2	15·1	81 03·1	10·6	12 00·3	8·3	57·2
09	M	314 31·9	· · 15·0	95 32·7	10·6	11 52·0	8·4	57·3
10	O	329 31·6	14·8	110 02·3	10·6	11 43·6	8·5	57·3
11	N	344 31·3	14·7	124 31·9	10·7	11 35·1	8·6	57·3
12	D	359 31·0	S 23 14·5	139 01·6	10·7	N11 26·5	8·6	57·3
13	A	14 30·7	14·4	153 31·3	10·6	11 17·9	8·7	57·3
14	Y	29 30·4	14·3	168 00·9	10·7	11 09·2	8·7	57·4
15		44 30·0	· · 14·1	182 30·6	10·8	11 00·5	8·8	57·4
16		59 29·7	14·0	197 00·4	10·7	10 51·7	8·9	57·4
17		74 29·4	13·8	211 30·1	10·7	10 42·8	8·9	57·4
18		89 29·1	S 23 13·7	225 59·8	10·8	N10 33·9	9·0	57·4
19		104 28·8	13·6	240 29·6	10·7	10 24·9	9·1	57·5
20		119 28·5	13·4	254 59·3	10·8	10 15·8	9·1	57·5
21		134 28·2	· · 13·3	269 29·1	10·8	10 06·7	9·1	57·5
22		149 27·9	13·1	283 58·9	10·8	9 57·6	9·3	57·5
23		164 27·6	13·0	298 28·7	10·8	9 48·3	9·3	57·5
		S.D. 16·3	d 0·1	S.D. 15·4		15·5		15·6

Lat.	Twilight Naut.	Twilight Civil	Sunrise	Moonrise 27	Moonrise 28	Moonrise 29	Moonrise 30
°	h m	h m	h m	h m	h m	h m	h m
N 72	08 26	10 53	■	14 48	16 47	18 41	20 32
N 70	08 07	09 54	■	15 37	17 16	18 58	20 41
68	07 51	09 20	■	16 07	17 38	19 12	20 49
66	07 38	08 55	10 33	16 30	17 54	19 23	20 55
64	07 27	08 35	09 52	16 48	18 08	19 33	21 00
62	07 17	08 19	09 24	17 03	18 19	19 41	21 04
60	07 09	08 06	09 03	17 15	18 29	19 47	21 08
N 58	07 02	07 54	08 46	17 26	18 37	19 54	21 12
56	06 55	07 44	08 32	17 35	18 45	19 59	21 15
54	06 49	07 35	08 19	17 43	18 51	20 04	21 18
52	06 44	07 27	08 08	17 51	18 57	20 08	21 20
50	06 38	07 20	07 58	17 57	19 03	20 12	21 22
45	06 27	07 04	07 38	18 11	19 14	20 20	21 27
N 40	06 16	06 50	07 21	18 23	19 24	20 27	21 31
35	06 07	06 39	07 07	18 33	19 32	20 33	21 35
30	05 58	06 28	06 54	18 41	19 39	20 38	21 38
20	05 42	06 09	06 33	18 56	19 52	20 47	21 43
N 10	05 26	05 52	06 15	19 09	20 02	20 55	21 48
0	05 09	05 35	05 58	19 22	20 12	21 03	21 53
S 10	04 50	05 17	05 40	19 34	20 23	21 10	21 57
20	04 28	04 57	05 21	19 47	20 33	21 18	22 02
30	03 59	04 32	04 59	20 02	20 46	21 27	22 07
35	03 40	04 17	04 46	20 10	20 53	21 32	22 10
40	03 17	03 59	04 31	20 20	21 01	21 38	22 14
45	02 47	03 36	04 13	20 31	21 10	21 45	22 18
S 50	02 01	03 07	03 51	20 45	21 21	21 53	22 22
52	01 34	02 52	03 40	20 52	21 26	21 57	22 25
54	00 49	02 34	03 28	20 59	21 32	22 01	22 27
56	////	02 12	03 14	21 07	21 38	22 06	22 30
58	////	01 42	02 57	21 16	21 46	22 11	22 33
S 60	////	00 54	02 37	21 26	21 53	22 16	22 36

Lat.	Sunset	Twilight Civil	Twilight Naut.	Moonset 27	Moonset 28	Moonset 29	Moonset 30
°	h m	h m	h m	h m	h m	h m	h m
N 72	■	13 09	15 36	11 37	11 24	11 16	11 09
N 70	■	14 09	15 56	10 48	10 54	10 57	10 58
68	■	14 43	16 11	10 17	10 32	10 42	10 49
66	13 29	15 08	16 24	09 53	10 15	10 30	10 41
64	14 10	15 27	16 35	09 35	10 00	10 20	10 35
62	14 38	15 43	16 45	09 20	09 48	10 11	10 29
60	14 59	15 57	16 53	09 07	09 38	10 03	10 24
N 58	15 17	16 08	17 01	08 56	09 29	09 56	10 20
56	15 31	16 19	17 07	08 47	09 21	09 50	10 16
54	15 44	16 28	17 13	08 38	09 14	09 45	10 12
52	15 55	16 36	17 19	08 31	09 08	09 40	10 09
50	16 05	16 43	17 25	08 24	09 02	09 36	10 06
45	16 25	16 59	17 36	08 09	08 50	09 26	10 00
N 40	16 42	17 13	17 46	07 57	08 39	09 18	09 55
35	16 56	17 24	17 56	07 46	08 30	09 11	09 50
30	17 09	17 35	18 05	07 37	08 23	09 05	09 46
20	17 29	17 53	18 21	07 22	08 09	08 55	09 39
N 10	17 48	18 11	18 37	07 08	07 57	08 45	09 33
0	18 05	18 28	18 54	06 55	07 46	08 37	09 27
S 10	18 23	18 46	19 13	06 42	07 35	08 28	09 21
20	18 41	19 06	19 35	06 28	07 23	08 18	09 14
30	19 03	19 31	20 04	06 12	07 09	08 07	09 07
35	19 16	19 46	20 23	06 03	07 01	08 01	09 02
40	19 31	20 04	20 46	05 52	06 52	07 54	08 58
45	19 49	20 26	21 16	05 39	06 41	07 45	08 52
S 50	20 12	20 56	22 01	05 24	06 28	07 35	08 45
52	20 22	21 11	22 28	05 17	06 22	07 30	08 42
54	20 34	21 29	23 12	05 09	06 15	07 25	08 38
56	20 49	21 50	////	05 00	06 07	07 19	08 34
58	21 05	22 20	////	04 50	05 59	07 13	08 30
S 60	21 25	23 07	////	04 38	05 49	07 05	08 25

Day	SUN Eqn. of Time 00h	SUN Eqn. of Time 12h	SUN Mer. Pass.	MOON Mer. Pass. Upper	MOON Mer. Pass. Lower	MOON Age	MOON Phase
	m s	m s	h m	h m	h m	d	
27	00 42	00 57	12 01	00 42	13 08	17	
28	01 12	01 26	12 01	01 34	13 59	18	○
29	01 41	01 56	12 02	02 24	14 50	19	

POLARIS (POLE STAR) TABLES,

FOR DETERMINING LATITUDE FROM SEXTANT ALTITUDE AND FOR AZIMUTH

L.H.A. ARIES	0°–9°	10°–19°	20°–29°	30°–39°	40°–49°	50°–59°	60°–69°	70°–79°	80°–89°	90°–99°	100°–109°	110°–119°
	a_0	a_0	a_0	a_0	a_0	a_0	a_0	a_0	a_0	a_0	a_0	a_0
°	° ′	° ′	° ′	° ′	° ′	° ′	° ′	° ′	° ′	° ′	° ′	° ′
0	0 10·1	0 06·1	0 03·8	0 03·1	0 04·2	0 06·9	0 11·3	0 17·1	0 24·2	0 32·4	0 41·3	0 50·8
1	09·7	05·8	03·6	03·1	04·4	07·3	11·8	17·8	25·0	33·2	42·2	51·7
2	09·2	05·5	03·5	03·2	04·6	07·7	12·3	18·4	25·8	34·1	43·2	52·7
3	08·8	05·3	03·4	03·2	04·8	08·1	12·9	19·1	26·6	35·0	44·1	53·7
4	08·3	05·0	03·3	03·3	05·1	08·5	13·4	19·8	27·4	35·9	45·0	54·6
5	0 07·9	0 04·7	0 03·2	0 03·4	0 05·3	0 08·9	0 14·0	0 20·5	0 28·2	0 36·7	0 46·0	0 55·6
6	07·5	04·5	03·2	03·5	05·6	09·4	14·6	21·2	29·0	37·6	46·9	56·6
7	07·2	04·3	03·1	03·7	05·9	09·8	15·2	22·0	29·8	38·5	47·9	57·6
8	06·8	04·1	03·1	03·8	06·2	10·3	15·8	22·7	30·7	39·5	48·9	58·5
9	06·5	03·9	03·1	04·0	06·6	10·8	16·5	23·5	31·5	40·4	49·8	0 59·5
10	0 06·1	0 03·8	0 03·1	0 04·2	0 06·9	0 11·3	0 17·1	0 24·2	0 32·4	0 41·3	0 50·8	1 00·5
Lat.	a_1	a_1	a_1	a_1	a_1	a_1	a_1	a_1	a_1	a_1	a_1	a_1
°	′	′	′	′	′	′	′	′	′	′	′	′
0	0·5	0·6	0·6	0·6	0·6	0·5	0·4	0·3	0·2	0·2	0·1	0·1
10	·5	·6	·6	·6	·6	·5	·4	·4	·3	·2	·2	·1
20	·5	·6	·6	·6	·6	·5	·5	·4	·3	·3	·3	·2
30	·6	·6	·6	·6	·6	·5	·5	·5	·4	·4	·3	·3
40	0·6	0·6	0·6	0·6	0·6	0·6	0·5	0·5	0·5	0·5	0·5	0·4
45	·6	·6	·6	·6	·6	·6	·6	·6	·5	·5	·5	·5
50	·6	·6	·6	·6	·6	·6	·6	·6	·6	·6	·6	·6
55	·6	·6	·6	·6	·6	·6	·6	·7	·7	·7	·7	·7
60	·6	·6	·6	·6	·6	·6	·7	·7	·8	·8	·8	·8
62	0·7	0·6	0·6	0·6	0·6	0·7	0·7	0·8	0·8	0·9	0·9	0·9
64	·7	·6	·6	·6	·6	·7	·7	·8	·9	0·9	1·0	1·0
66	·7	·6	·6	·6	·6	·7	·8	·9	0·9	1·0	1·0	1·1
68	0·7	0·6	0·6	0·6	0·7	0·7	0·8	0·9	1·0	1·1	1·1	1·2
Month	a_2	a_2	a_2	a_2	a_2	a_2	a_2	a_2	a_2	a_2	a_2	a_2
	′	′	′	′	′	′	′	′	′	′	′	′
Jan.	0·7	0·7	0·7	0·7	0·7	0·7	0·7	0·7	0·7	0·7	0·7	0·7
Feb.	·6	·7	·7	·7	·8	·8	·8	·8	·8	·8	·8	·8
Mar.	·5	·5	·6	·6	·7	·8	·8	·8	·9	·9	·9	0·9
Apr.	0·3	0·4	0·4	0·5	0·6	0·6	0·7	0·8	0·8	0·9	0·9	1·0
May	·2	·3	·3	·4	·4	·5	·6	·6	·7	·8	·8	0·9
June	·2	·2	·2	·3	·3	·4	·4	·5	·5	·6	·7	·8
July	0·2	0·2	0·2	0·2	0·2	0·3	0·3	0·4	0·4	0·5	0·5	0·6
Aug.	·4	·3	·3	·3	·2	·2	·3	·3	·3	·4	·4	·4
Sept.	·5	·5	·4	·4	·3	·3	·3	·3	·3	·3	·3	·3
Oct.	0·7	0·6	0·6	0·5	0·5	0·4	0·4	0·3	0·3	0·3	0·3	0·3
Nov.	0·9	·8	·8	·7	·7	·6	·5	·5	·4	·4	·3	·3
Dec.	1·0	0·9	0·9	0·9	0·8	0·8	0·7	0·6	0·6	0·5	0·4	0·4
Lat.	AZIMUTH											
°	°	°	°	°	°	°	°	°	°	°	°	°
0	0·4	0·2	0·1	359·9	359·7	359·6	359·5	359·3	359·2	359·2	359·1	359·1
20	0·4	0·2	0·1	359·9	359·7	359·6	359·4	359·3	359·2	359·1	359·1	359·0
40	0·5	0·3	0·1	359·9	359·7	359·5	359·3	359·1	359·0	358·9	358·8	358·8
50	0·6	0·4	0·1	359·8	359·6	359·4	359·1	359·0	358·8	358·7	358·6	358·6
55	0·7	0·4	0·1	359·8	359·6	359·3	359·0	358·8	358·6	358·5	358·4	358·4
60	0·8	0·5	0·1	359·8	359·5	359·2	358·9	358·6	358·4	358·3	358·2	358·1
65	0·9	0·6	0·2	359·8	359·4	359·1	358·7	358·4	358·1	358·0	357·9	357·8

Latitude = corrected sextant altitude $- 1° + a_0 + a_1 + a_2$

The table is entered with L.H.A. Aries to determine the column to be used; each column refers to a range of 10°. a_0 is taken, with mental interpolation, from the upper table with the units of L.H.A. Aries in degrees as argument; a_1, a_2 are taken, without interpolation, from the second and third tables with arguments latitude and month respectively. a_0, a_1, a_2 are always positive. The final table gives the azimuth of *Polaris*.

POLARIS (POLE STAR) TABLES,

FOR DETERMINING LATITUDE FROM SEXTANT ALTITUDE AND FOR AZIMUTH

L.H.A. ARIES	120°–129°	130°–139°	140°–149°	150°–159°	160°–169°	170°–179°	180°–189°	190°–199°	200°–209°	210°–219°	220°–229°	230°–239°
	a_0	a_0	a_0	a_0	a_0	a_0	a_0	a_0	a_0	a_0	a_0	a_0
°	° ′	° ′	° ′	° ′	° ′	° ′	° ′	° ′	° ′	° ′	° ′	° ′
0	1 00·5	1 10·1	1 19·4	1 28·0	1 35·8	1 42·4	1 47·7	1 51·6	1 53·9	1 54·5	1 53·5	1 50·8
1	01·5	11·1	20·3	28·9	36·5	43·0	48·2	51·9	54·0	54·5	53·3	50·5
2	02·4	12·0	21·2	29·7	37·2	43·6	48·6	52·2	54·1	54·4	53·1	50·1
3	03·4	13·0	22·1	30·5	37·9	44·1	49·0	52·4	54·2	54·4	52·9	49·7
4	04·4	13·9	22·9	31·3	38·6	44·7	49·4	52·7	54·3	54·3	52·6	49·3
5	1 05·3	1 14·8	1 23·8	1 32·0	1 39·2	1 45·2	1 49·8	1 52·9	1 54·4	1 54·2	1 52·4	1 48·9
6	06·3	15·7	24·7	32·8	39·9	45·8	50·2	53·1	54·4	54·1	52·1	48·5
7	07·3	16·7	25·5	33·6	40·5	46·3	50·6	53·3	54·5	54·0	51·8	48·0
8	08·2	17·6	26·4	34·3	41·2	46·8	50·9	53·5	54·5	53·8	51·5	47·6
9	09·2	18·5	27·2	35·1	41·8	47·3	51·3	53·7	54·5	53·7	51·2	47·1
10	1 10·1	1 19·4	1 28·0	1 35·8	1 42·4	1 47·7	1 51·6	1 53·9	1 54·5	1 53·5	1 50·8	1 46·6
Lat.	a_1	a_1	a_1	a_1	a_1	a_1	a_1	a_1	a_1	a_1	a_1	a_1
°	′	′	′	′	′	′	′	′	′	′	′	′
0	0·1	0·1	0·2	0·3	0·3	0·4	0·5	0·6	0·6	0·6	0·6	0·5
10	·1	·2	·2	·3	·4	·5	·5	·6	·6	·6	·6	·5
20	·2	·3	·3	·4	·4	·5	·5	·6	·6	·6	·6	·5
30	·3	·3	·4	·4	·5	·5	·6	·6	·6	·6	·6	·5
40	0·4	0·5	0·5	0·5	0·5	0·5	0·6	0·6	0·6	0·6	0·6	0·6
45	·5	·5	·5	·5	·6	·6	·6	·6	·6	·6	·6	·6
50	·6	·6	·6	·6	·6	·6	·6	·6	·6	·6	·6	·6
55	·7	·7	·7	·7	·7	·6	·6	·6	·6	·6	·6	·6
60	·8	·8	·8	·8	·7	·7	·6	·6	·6	·6	·6	·6
62	0·9	0·9	0·8	0·8	0·8	0·7	0·7	0·6	0·6	0·6	0·6	0·7
64	1·0	1·0	0·9	·9	·8	·7	·7	·6	·6	·6	·6	·7
66	1·1	1·0	1·0	0·9	·8	·8	·7	·6	·6	·6	·6	·7
68	1·2	1·1	1·1	1·0	0·9	0·8	0·7	0·6	0·6	0·6	0·7	0·7
Month	a_2	a_2	a_2	a_2	a_2	a_2	a_2	a_2	a_2	a_2	a_2	a_2
	′	′	′	′	′	′	′	′	′	′	′	′
Jan.	0·6	0·6	0·6	0·6	0·6	0·5	0·5	0·5	0·5	0·5	0·5	0·5
Feb.	·8	·8	·7	·7	·7	·6	·6	·5	·5	·5	·4	·4
Mar.	0·9	0·9	0·9	0·8	0·8	·8	·7	·7	·6	·6	·5	·4
Apr.	1·0	1·0	1·0	1·0	1·0	0·9	0·9	0·8	0·8	0·7	0·6	0·6
May	0·9	1·0	1·0	1·0	1·0	1·0	1·0	0·9	0·9	·8	·8	·7
June	·8	0·9	0·9	1·0	1·0	1·0	1·0	1·0	1·0	0·9	0·9	·8
July	0·7	0·7	0·8	0·9	0·9	0·9	1·0	1·0	1·0	1·0	1·0	0·9
Aug.	·5	·6	·6	·7	·7	·8	0·8	0·9	0·9	0·9	1·0	1·0
Sept.	·4	·4	·5	·5	·6	·6	·7	·7	·8	·8	0·9	0·9
Oct.	0·3	0·3	0·3	0·4	0·4	0·4	0·5	0·6	0·6	0·7	0·7	0·8
Nov.	·2	·2	·2	·2	·3	·3	·3	·4	·4	·5	·5	·6
Dec.	0·3	0·3	0·2	0·2	0·2	0·2	0·2	0·3	0·3	0·3	0·4	0·4
Lat.	AZIMUTH											
°	°	°	°	°	°	°	°	°	°	°	°	°
0	359·1	359·1	359·2	359·3	359·4	359·5	359·6	359·8	359·9	0·1	0·3	0·4
20	359·0	359·1	359·1	359·2	359·3	359·5	359·6	359·8	359·9	0·1	0·3	0·4
40	358·8	358·8	358·9	359·0	359·2	359·3	359·5	359·7	359·9	0·1	0·3	0·5
50	358·6	358·6	358·7	358·8	359·0	359·2	359·4	359·7	359·9	0·2	0·4	0·6
55	358·4	358·5	358·6	358·7	358·9	359·1	359·4	359·6	359·9	0·2	0·4	0·7
60	358·2	358·2	358·3	358·5	358·7	359·0	359·3	359·6	359·9	0·2	0·5	0·8
65	357·8	357·9	358·1	358·3	358·5	358·8	359·1	359·5	359·9	0·2	0·6	0·9

ILLUSTRATION

On 1958 January 10 at G.M.T. 22ʰ 17ᵐ 50ˢ in longitude W. 27° 34′ the corrected sextant altitude of *Polaris* was 49° 31′·6.

From the daily pages	° ′
G.H.A. Aries (22ʰ)	79 54·9
Increment (17ᵐ 50ˢ)	4 28·2
Longitude (west)	−27 34
L.H.A. Aries	56 49

Corr. Sext. Alt.	49°31′·6
a_0 (argument 56° 49′)	0 09·7
a_1 (lat. 50° approx.)	0·6
a_2 (January)	0·7
Sum −1° = Lat. =	48 42·6

POLARIS (POLE STAR) TABLES,

FOR DETERMINING LATITUDE FROM SEXTANT ALTITUDE AND FOR AZIMUTH

L.H.A. ARIES	240°–249°	250°–259°	260°–269°	270°–279°	280°–289°	290°–299°	300°–309°	310°–319°	320°–329°	330°–339°	340°–349°	350°–359°
	a_0	a_0	a_0	a_0	a_0	a_0	a_0	a_0	a_0	a_0	a_0	a_0
°	° ′	° ′	° ′	° ′	° ′	° ′	° ′	° ′	° ′	° ′	° ′	° ′
0	1 46·6	1 41·0	1 34·0	1 26·1	1 17·3	1 07·9	0 58·2	0 48·5	0 39·1	0 30·4	0 22·4	0 15·6
1	46·1	40·3	33·3	25·2	16·3	06·9	57·2	47·6	38·2	29·5	21·7	15·0
2	45·6	39·7	32·5	24·4	15·4	06·0	56·3	46·6	37·3	28·7	21·0	14·4
3	45·1	39·0	31·7	23·5	14·5	05·0	55·3	45·7	36·4	27·9	20·2	13·8
4	44·5	38·3	31·0	22·6	13·6	04·0	54·3	44·7	35·5	27·1	19·5	13·2
5	1 43·9	1 37·6	1 30·2	1 21·7	1 12·6	1 03·1	0 53·3	0 43·8	0 34·7	0 26·3	0 18·9	0 12·7
6	43·4	36·9	29·4	20·9	11·7	02·1	52·4	42·8	33·8	25·5	18·2	12·1
7	42·8	36·2	28·6	20·0	10·7	01·1	51·4	41·9	32·9	24·7	17·5	11·6
8	42·2	35·5	27·7	19·1	09·8	1 00·1	50·4	41·0	32·1	23·9	16·9	11·1
9	41·6	34·8	26·9	18·2	08·8	0 59·2	49·5	40·1	31·2	23·2	16·2	10·6
10	1 41·0	1 34·0	1 26·1	1 17·3	1 07·9	0 58·2	0 48·5	0 39·1	0 30·4	0 22·4	0 15·6	0 10·1
Lat.	a_1	a_1	a_1	a_1	a_1	a_1	a_1	a_1	a_1	a_1	a_1	a_1
°	′	′	′	′	′	′	′	′	′	′	′	′
0	0·4	0·3	0·2	0·2	0·1	0·1	0·1	0·1	0·2	0·3	0·3	0·4
10	·4	·4	·3	·2	·2	·1	·1	·2	·2	·3	·4	·5
20	·5	·4	·3	·3	·3	·2	·2	·3	·3	·4	·4	·5
30	·5	·5	·4	·4	·3	·3	·3	·3	·4	·4	·5	·5
40	0·5	0·5	0·5	0·5	0·5	0·4	0·4	0·5	0·5	0·5	0·5	0·5
45	·6	·6	·5	·5	·5	·5	·5	·5	·5	·5	·6	·6
50	·6	·6	·6	·6	·6	·6	·6	·6	·6	·6	·6	·6
55	·6	·7	·7	·7	·7	·7	·7	·7	·7	·7	·7	·6
60	·7	·7	·8	·8	·8	·8	·8	·8	·8	·8	·7	·7
62	0·7	0·8	0·8	0·9	0·9	0·9	0·9	0·9	0·8	0·8	0·8	0·7
64	·7	·8	·9	0·9	1·0	1·0	1·0	1·0	0·9	·9	·8	·7
66	·8	·9	0·9	1·0	1·0	1·1	1·1	1·0	1·0	0·9	·8	·8
68	0·8	0·9	1·0	1·1	1·1	1·2	1·2	1·1	1·1	1·0	0·9	0·8
Month	a_2	a_2	a_2	a_2	a_2	a_2	a_2	a_2	a_2	a_2	a_2	a_2
	′	′	′	′	′	′	′	′	′	′	′	′
Jan.	0·5	0·5	0·5	0·5	0·5	0·5	0·6	0·6	0·6	0·6	0·6	0·7
Feb.	·4	·4	·4	·4	·4	·4	·4	·4	·5	·5	·5	·6
Mar.	·4	·4	·3	·3	·3	·3	·3	·3	·3	·4	·4	·4
Apr.	0·5	0·4	0·4	0·3	0·3	0·2	0·2	0·2	0·2	0·2	0·2	0·3
May	·6	·6	·5	·4	·4	·3	·3	·2	·2	·2	·2	·2
June	·8	·7	·7	·6	·5	·4	·4	·3	·3	·2	·2	·2
July	0·9	0·8	0·8	0·7	0·7	0·6	0·5	0·5	0·4	0·3	0·3	0·3
Aug.	·9	·9	·9	·8	·8	·8	·7	·6	·6	·5	·5	·4
Sept.	·9	·9	·9	·9	·9	·9	·8	·8	·7	·7	·6	·6
Oct.	0·8	0·9	0·9	0·9	0·9	0·9	0·9	0·9	0·9	0·8	0·8	0·8
Nov.	·7	·7	·8	·8	·9	·9	1·0	1·0	1·0	1·0	0·9	0·9
Dec.	0·5	0·6	0·6	0·7	0·8	0·8	0·9	0·9	1·0	1·0	1·0	1·0
Lat.	AZIMUTH											
°	°	°	°	°	°	°	°	°	°	°	°	°
0	0·5	0·7	0·8	0·8	0·9	0·9	0·9	0·9	0·8	0·7	0·6	0·5
20	0·6	0·7	0·8	0·9	0·9	1·0	1·0	0·9	0·9	0·8	0·7	0·5
40	0·7	0·9	1·0	1·1	1·2	1·2	1·2	1·2	1·1	1·0	0·8	0·7
50	0·8	1·0	1·2	1·3	1·4	1·4	1·4	1·4	1·3	1·2	1·0	0·8
55	0·9	1·1	1·3	1·5	1·6	1·6	1·6	1·6	1·5	1·3	1·1	0·9
60	1·1	1·3	1·5	1·7	1·8	1·8	1·8	1·8	1·7	1·5	1·3	1·0
65	1·3	1·5	1·8	1·9	2·1	2·2	2·2	2·1	2·0	1·8	1·6	1·3

Latitude = corrected sextant altitude $- 1° + a_0 + a_1 + a_2$

The table is entered with L.H.A. Aries to determine the column to be used; each column refers to a range of 10°. a_0 is taken, with mental interpolation, from the upper table with the units of L.H.A. Aries in degrees as argument; a_1, a_2 are taken, without interpolation, from the second and third tables with arguments latitude and month respectively. a_0, a_1, a_2 are always positive. The final table gives the azimuth of *Polaris*.

INCREMENTS AND CORRECTIONS

0ᵐ INCREMENTS AND CORRECTIONS 1ᵐ

0ᵐ s	SUN PLANETS ° ′	ARIES ° ′	MOON ° ′	v or d ′	Corrⁿ ′	v or d ′	Corrⁿ ′	v or d ′	Corrⁿ ′
00	0 00·0	0 00·0	0 00·0	0·0	0·0	6·0	0·1	12·0	0·1
01	0 00·3	0 00·3	0 00·2	0·1	0·0	6·1	0·1	12·1	0·1
02	0 00·5	0 00·5	0 00·5	0·2	0·0	6·2	0·1	12·2	0·1
03	0 00·8	0 00·8	0 00·7	0·3	0·0	6·3	0·1	12·3	0·1
04	0 01·0	0 01·0	0 01·0	0·4	0·0	6·4	0·1	12·4	0·1
05	0 01·3	0 01·3	0 01·2	0·5	0·0	6·5	0·1	12·5	0·1
06	0 01·5	0 01·5	0 01·4	0·6	0·0	6·6	0·1	12·6	0·1
07	0 01·8	0 01·8	0 01·7	0·7	0·0	6·7	0·1	12·7	0·1
08	0 02·0	0 02·0	0 01·9	0·8	0·0	6·8	0·1	12·8	0·1
09	0 02·3	0 02·3	0 02·1	0·9	0·0	6·9	0·1	12·9	0·1
10	0 02·5	0 02·5	0 02·4	1·0	0·0	7·0	0·1	13·0	0·1
11	0 02·8	0 02·8	0 02·6	1·1	0·0	7·1	0·1	13·1	0·1
12	0 03·0	0 03·0	0 02·9	1·2	0·0	7·2	0·1	13·2	0·1
13	0 03·3	0 03·3	0 03·1	1·3	0·0	7·3	0·1	13·3	0·1
14	0 03·5	0 03·5	0 03·3	1·4	0·0	7·4	0·1	13·4	0·1
15	0 03·8	0 03·8	0 03·6	1·5	0·0	7·5	0·1	13·5	0·1
16	0 04·0	0 04·0	0 03·8	1·6	0·0	7·6	0·1	13·6	0·1
17	0 04·3	0 04·3	0 04·1	1·7	0·0	7·7	0·1	13·7	0·1
18	0 04·5	0 04·5	0 04·3	1·8	0·0	7·8	0·1	13·8	0·1
19	0 04·8	0 04·8	0 04·5	1·9	0·0	7·9	0·1	13·9	0·1
20	0 05·0	0 05·0	0 04·8	2·0	0·0	8·0	0·1	14·0	0·1
21	0 05·3	0 05·3	0 05·0	2·1	0·0	8·1	0·1	14·1	0·1
22	0 05·5	0 05·5	0 05·2	2·2	0·0	8·2	0·1	14·2	0·1
23	0 05·8	0 05·8	0 05·5	2·3	0·0	8·3	0·1	14·3	0·1
24	0 06·0	0 06·0	0 05·7	2·4	0·0	8·4	0·1	14·4	0·1
25	0 06·3	0 06·3	0 06·0	2·5	0·0	8·5	0·1	14·5	0·1
26	0 06·5	0 06·5	0 06·2	2·6	0·0	8·6	0·1	14·6	0·1
27	0 06·8	0 06·8	0 06·4	2·7	0·0	8·7	0·1	14·7	0·1
28	0 07·0	0 07·0	0 06·7	2·8	0·0	8·8	0·1	14·8	0·1
29	0 07·3	0 07·3	0 06·9	2·9	0·0	8·9	0·1	14·9	0·1
30	0 07·5	0 07·5	0 07·2	3·0	0·0	9·0	0·1	15·0	0·1
31	0 07·8	0 07·8	0 07·4	3·1	0·0	9·1	0·1	15·1	0·1
32	0 08·0	0 08·0	0 07·6	3·2	0·0	9·2	0·1	15·2	0·1
33	0 08·3	0 08·3	0 07·9	3·3	0·0	9·3	0·1	15·3	0·1
34	0 08·5	0 08·5	0 08·1	3·4	0·0	9·4	0·1	15·4	0·1
35	0 08·8	0 08·8	0 08·4	3·5	0·0	9·5	0·1	15·5	0·1
36	0 09·0	0 09·0	0 08·6	3·6	0·0	9·6	0·1	15·6	0·1
37	0 09·3	0 09·3	0 08·8	3·7	0·0	9·7	0·1	15·7	0·1
38	0 09·5	0 09·5	0 09·1	3·8	0·0	9·8	0·1	15·8	0·1
39	0 09·8	0 09·8	0 09·3	3·9	0·0	9·9	0·1	15·9	0·1
40	0 10·0	0 10·0	0 09·5	4·0	0·0	10·0	0·1	16·0	0·1
41	0 10·3	0 10·3	0 09·8	4·1	0·0	10·1	0·1	16·1	0·1
42	0 10·5	0 10·5	0 10·0	4·2	0·0	10·2	0·1	16·2	0·1
43	0 10·8	0 10·8	0 10·3	4·3	0·0	10·3	0·1	16·3	0·1
44	0 11·0	0 11·0	0 10·5	4·4	0·0	10·4	0·1	16·4	0·1
45	0 11·3	0 11·3	0 10·7	4·5	0·0	10·5	0·1	16·5	0·1
46	0 11·5	0 11·5	0 11·0	4·6	0·0	10·6	0·1	16·6	0·1
47	0 11·8	0 11·8	0 11·2	4·7	0·0	10·7	0·1	16·7	0·1
48	0 12·0	0 12·0	0 11·5	4·8	0·0	10·8	0·1	16·8	0·1
49	0 12·3	0 12·3	0 11·7	4·9	0·0	10·9	0·1	16·9	0·1
50	0 12·5	0 12·5	0 11·9	5·0	0·0	11·0	0·1	17·0	0·1
51	0 12·8	0 12·8	0 12·2	5·1	0·0	11·1	0·1	17·1	0·1
52	0 13·0	0 13·0	0 12·4	5·2	0·0	11·2	0·1	17·2	0·1
53	0 13·3	0 13·3	0 12·6	5·3	0·0	11·3	0·1	17·3	0·1
54	0 13·5	0 13·5	0 12·9	5·4	0·0	11·4	0·1	17·4	0·1
55	0 13·8	0 13·8	0 13·1	5·5	0·0	11·5	0·1	17·5	0·1
56	0 14·0	0 14·0	0 13·4	5·6	0·0	11·6	0·1	17·6	0·1
57	0 14·3	0 14·3	0 13·6	5·7	0·0	11·7	0·1	17·7	0·1
58	0 14·5	0 14·5	0 13·8	5·8	0·0	11·8	0·1	17·8	0·1
59	0 14·8	0 14·8	0 14·1	5·9	0·0	11·9	0·1	17·9	0·1
60	0 15·0	0 15·0	0 14·3	6·0	0·1	12·0	0·1	18·0	0·2

1ᵐ s	SUN PLANETS ° ′	ARIES ° ′	MOON ° ′	v or d ′	Corrⁿ ′	v or d ′	Corrⁿ ′	v or d ′	Corrⁿ ′
00	0 15·0	0 15·0	0 14·3	0·0	0·0	6·0	0·2	12·0	0·3
01	0 15·3	0 15·3	0 14·6	0·1	0·0	6·1	0·2	12·1	0·3
02	0 15·5	0 15·5	0 14·8	0·2	0·0	6·2	0·2	12·2	0·3
03	0 15·8	0 15·8	0 15·0	0·3	0·0	6·3	0·2	12·3	0·3
04	0 16·0	0 16·0	0 15·3	0·4	0·0	6·4	0·2	12·4	0·3
05	0 16·3	0 16·3	0 15·5	0·5	0·0	6·5	0·2	12·5	0·3
06	0 16·5	0 16·5	0 15·7	0·6	0·0	6·6	0·2	12·6	0·3
07	0 16·8	0 16·8	0 16·0	0·7	0·0	6·7	0·2	12·7	0·3
08	0 17·0	0 17·0	0 16·2	0·8	0·0	6·8	0·2	12·8	0·3
09	0 17·3	0 17·3	0 16·5	0·9	0·0	6·9	0·2	12·9	0·3
10	0 17·5	0 17·5	0 16·7	1·0	0·0	7·0	0·2	13·0	0·3
11	0 17·8	0 17·8	0 16·9	1·1	0·0	7·1	0·2	13·1	0·3
12	0 18·0	0 18·0	0 17·2	1·2	0·0	7·2	0·2	13·2	0·3
13	0 18·3	0 18·3	0 17·4	1·3	0·0	7·3	0·2	13·3	0·3
14	0 18·5	0 18·6	0 17·7	1·4	0·0	7·4	0·2	13·4	0·3
15	0 18·8	0 18·8	0 17·9	1·5	0·0	7·5	0·2	13·5	0·3
16	0 19·0	0 19·1	0 18·1	1·6	0·0	7·6	0·2	13·6	0·3
17	0 19·3	0 19·3	0 18·4	1·7	0·0	7·7	0·2	13·7	0·3
18	0 19·5	0 19·6	0 18·6	1·8	0·0	7·8	0·2	13·8	0·3
19	0 19·8	0 19·8	0 18·9	1·9	0·0	7·9	0·2	13·9	0·3
20	0 20·0	0 20·1	0 19·1	2·0	0·1	8·0	0·2	14·0	0·4
21	0 20·3	0 20·3	0 19·3	2·1	0·1	8·1	0·2	14·1	0·4
22	0 20·5	0 20·6	0 19·6	2·2	0·1	8·2	0·2	14·2	0·4
23	0 20·8	0 20·8	0 19·8	2·3	0·1	8·3	0·2	14·3	0·4
24	0 21·0	0 21·1	0 20·0	2·4	0·1	8·4	0·2	14·4	0·4
25	0 21·3	0 21·3	0 20·3	2·5	0·1	8·5	0·2	14·5	0·4
26	0 21·5	0 21·6	0 20·5	2·6	0·1	8·6	0·2	14·6	0·4
27	0 21·8	0 21·8	0 20·8	2·7	0·1	8·7	0·2	14·7	0·4
28	0 22·0	0 22·1	0 21·0	2·8	0·1	8·8	0·2	14·8	0·4
29	0 22·3	0 22·3	0 21·2	2·9	0·1	8·9	0·2	14·9	0·4
30	0 22·5	0 22·6	0 21·5	3·0	0·1	9·0	0·2	15·0	0·4
31	0 22·8	0 22·8	0 21·7	3·1	0·1	9·1	0·2	15·1	0·4
32	0 23·0	0 23·1	0 22·0	3·2	0·1	9·2	0·2	15·2	0·4
33	0 23·3	0 23·3	0 22·2	3·3	0·1	9·3	0·2	15·3	0·4
34	0 23·5	0 23·6	0 22·4	3·4	0·1	9·4	0·2	15·4	0·4
35	0 23·8	0 23·8	0 22·7	3·5	0·1	9·5	0·2	15·5	0·4
36	0 24·0	0 24·1	0 22·9	3·6	0·1	9·6	0·2	15·6	0·4
37	0 24·3	0 24·3	0 23·1	3·7	0·1	9·7	0·2	15·7	0·4
38	0 24·5	0 24·6	0 23·4	3·8	0·1	9·8	0·2	15·8	0·4
39	0 24·8	0 24·8	0 23·6	3·9	0·1	9·9	0·2	15·9	0·4
40	0 25·0	0 25·1	0 23·9	4·0	0·1	10·0	0·3	16·0	0·4
41	0 25·3	0 25·3	0 24·1	4·1	0·1	10·1	0·3	16·1	0·4
42	0 25·5	0 25·6	0 24·3	4·2	0·1	10·2	0·3	16·2	0·4
43	0 25·8	0 25·8	0 24·6	4·3	0·1	10·3	0·3	16·3	0·4
44	0 26·0	0 26·1	0 24·8	4·4	0·1	10·4	0·3	16·4	0·4
45	0 26·3	0 26·3	0 25·1	4·5	0·1	10·5	0·3	16·5	0·4
46	0 26·5	0 26·6	0 25·3	4·6	0·1	10·6	0·3	16·6	0·4
47	0 26·8	0 26·8	0 25·5	4·7	0·1	10·7	0·3	16·7	0·4
48	0 27·0	0 27·1	0 25·8	4·8	0·1	10·8	0·3	16·8	0·4
49	0 27·3	0 27·3	0 26·0	4·9	0·1	10·9	0·3	16·9	0·4
50	0 27·5	0 27·6	0 26·2	5·0	0·1	11·0	0·3	17·0	0·4
51	0 27·8	0 27·8	0 26·5	5·1	0·1	11·1	0·3	17·1	0·4
52	0 28·0	0 28·1	0 26·7	5·2	0·1	11·2	0·3	17·2	0·4
53	0 28·3	0 28·3	0 27·0	5·3	0·1	11·3	0·3	17·3	0·4
54	0 28·5	0 28·6	0 27·2	5·4	0·1	11·4	0·3	17·4	0·4
55	0 28·8	0 28·8	0 27·4	5·5	0·1	11·5	0·3	17·5	0·4
56	0 29·0	0 29·1	0 27·7	5·6	0·1	11·6	0·3	17·6	0·4
57	0 29·3	0 29·3	0 27·9	5·7	0·1	11·7	0·3	17·7	0·4
58	0 29·5	0 29·6	0 28·2	5·8	0·1	11·8	0·3	17·8	0·4
59	0 29·8	0 29·8	0 28·4	5·9	0·1	11·9	0·3	17·9	0·4
60	0 30·0	0 30·1	0 28·6	6·0	0·2	12·0	0·3	18·0	0·5

2^m INCREMENTS AND CORRECTIONS 3^m

2^m	SUN PLANETS	ARIES	MOON	v or d	Corr^n	v or d	Corr^n	v or d	Corr^n
s	° ′	° ′	° ′	′	′	′	′	′	′
00	0 30·0	0 30·1	0 28·6	0·0	0·0	6·0	0·3	12·0	0·5
01	0 30·3	0 30·3	0 28·9	0·1	0·0	6·1	0·3	12·1	0·5
02	0 30·5	0 30·6	0 29·1	0·2	0·0	6·2	0·3	12·2	0·5
03	0 30·8	0 30·8	0 29·3	0·3	0·0	6·3	0·3	12·3	0·5
04	0 31·0	0 31·1	0 29·6	0·4	0·0	6·4	0·3	12·4	0·5
05	0 31·3	0 31·3	0 29·8	0·5	0·0	6·5	0·3	12·5	0·5
06	0 31·5	0 31·6	0 30·1	0·6	0·0	6·6	0·3	12·6	0·5
07	0 31·8	0 31·8	0 30·3	0·7	0·0	6·7	0·3	12·7	0·5
08	0 32·0	0 32·1	0 30·5	0·8	0·0	6·8	0·3	12·8	0·5
09	0 32·3	0 32·3	0 30·8	0·9	0·0	6·9	0·3	12·9	0·5
10	0 32·5	0 32·6	0 31·0	1·0	0·0	7·0	0·3	13·0	0·5
11	0 32·8	0 32·8	0 31·3	1·1	0·0	7·1	0·3	13·1	0·5
12	0 33·0	0 33·1	0 31·5	1·2	0·1	7·2	0·3	13·2	0·6
13	0 33·3	0 33·3	0 31·7	1·3	0·1	7·3	0·3	13·3	0·6
14	0 33·5	0 33·6	0 32·0	1·4	0·1	7·4	0·3	13·4	0·6
15	0 33·8	0 33·8	0 32·2	1·5	0·1	7·5	0·3	13·5	0·6
16	0 34·0	0 34·1	0 32·5	1·6	0·1	7·6	0·3	13·6	0·6
17	0 34·3	0 34·3	0 32·7	1·7	0·1	7·7	0·3	13·7	0·6
18	0 34·5	0 34·6	0 32·9	1·8	0·1	7·8	0·3	13·8	0·6
19	0 34·8	0 34·8	0 33·2	1·9	0·1	7·9	0·3	13·9	0·6
20	0 35·0	0 35·1	0 33·4	2·0	0·1	8·0	0·3	14·0	0·6
21	0 35·3	0 35·3	0 33·6	2·1	0·1	8·1	0·3	14·1	0·6
22	0 35·5	0 35·6	0 33·9	2·2	0·1	8·2	0·3	14·2	0·6
23	0 35·8	0 35·8	0 34·1	2·3	0·1	8·3	0·3	14·3	0·6
24	0 36·0	0 36·1	0 34·4	2·4	0·1	8·4	0·4	14·4	0·6
25	0 36·3	0 36·3	0 34·6	2·5	0·1	8·5	0·4	14·5	0·6
26	0 36·5	0 36·6	0 34·8	2·6	0·1	8·6	0·4	14·6	0·6
27	0 36·8	0 36·9	0 35·1	2·7	0·1	8·7	0·4	14·7	0·6
28	0 37·0	0 37·1	0 35·3	2·8	0·1	8·8	0·4	14·8	0·6
29	0 37·3	0 37·4	0 35·6	2·9	0·1	8·9	0·4	14·9	0·6
30	0 37·5	0 37·6	0 35·8	3·0	0·1	9·0	0·4	15·0	0·6
31	0 37·8	0 37·9	0 36·0	3·1	0·1	9·1	0·4	15·1	0·6
32	0 38·0	0 38·1	0 36·3	3·2	0·1	9·2	0·4	15·2	0·6
33	0 38·3	0 38·4	0 36·5	3·3	0·1	9·3	0·4	15·3	0·6
34	0 38·5	0 38·6	0 36·7	3·4	0·1	9·4	0·4	15·4	0·6
35	0 38·8	0 38·9	0 37·0	3·5	0·1	9·5	0·4	15·5	0·6
36	0 39·0	0 39·1	0 37·2	3·6	0·2	9·6	0·4	15·6	0·7
37	0 39·3	0 39·4	0 37·5	3·7	0·2	9·7	0·4	15·7	0·7
38	0 39·5	0 39·6	0 37·7	3·8	0·2	9·8	0·4	15·8	0·7
39	0 39·8	0 39·9	0 37·9	3·9	0·2	9·9	0·4	15·9	0·7
40	0 40·0	0 40·1	0 38·2	4·0	0·2	10·0	0·4	16·0	0·7
41	0 40·3	0 40·4	0 38·4	4·1	0·2	10·1	0·4	16·1	0·7
42	0 40·5	0 40·6	0 38·7	4·2	0·2	10·2	0·4	16·2	0·7
43	0 40·8	0 40·9	0 38·9	4·3	0·2	10·3	0·4	16·3	0·7
44	0 41·0	0 41·1	0 39·1	4·4	0·2	10·4	0·4	16·4	0·7
45	0 41·3	0 41·4	0 39·4	4·5	0·2	10·5	0·4	16·5	0·7
46	0 41·5	0 41·6	0 39·6	4·6	0·2	10·6	0·4	16·6	0·7
47	0 41·8	0 41·9	0 39·8	4·7	0·2	10·7	0·4	16·7	0·7
48	0 42·0	0 42·1	0 40·1	4·8	0·2	10·8	0·5	16·8	0·7
49	0 42·3	0 42·4	0 40·3	4·9	0·2	10·9	0·5	16·9	0·7
50	0 42·5	0 42·6	0 40·6	5·0	0·2	11·0	0·5	17·0	0·7
51	0 42·8	0 42·9	0 40·8	5·1	0·2	11·1	0·5	17·1	0·7
52	0 43·0	0 43·1	0 41·0	5·2	0·2	11·2	0·5	17·2	0·7
53	0 43·3	0 43·4	0 41·3	5·3	0·2	11·3	0·5	17·3	0·7
54	0 43·5	0 43·6	0 41·5	5·4	0·2	11·4	0·5	17·4	0·7
55	0 43·8	0 43·9	0 41·8	5·5	0·2	11·5	0·5	17·5	0·7
56	0 44·0	0 44·1	0 42·0	5·6	0·2	11·6	0·5	17·6	0·7
57	0 44·3	0 44·4	0 42·2	5·7	0·2	11·7	0·5	17·7	0·7
58	0 44·5	0 44·6	0 42·5	5·8	0·2	11·8	0·5	17·8	0·7
59	0 44·8	0 44·9	0 42·7	5·9	0·2	11·9	0·5	17·9	0·7
60	0 45·0	0 45·1	0 43·0	6·0	0·3	12·0	0·5	18·0	0·8

3^m	SUN PLANETS	ARIES	MOON	v or d	Corr^n	v or d	Corr^n	v or d	Corr^n
s	° ′	° ′	° ′	′	′	′	′	′	′
00	0 45·0	0 45·1	0 43·0	0·0	0·0	6·0	0·4	12·0	0·7
01	0 45·3	0 45·4	0 43·2	0·1	0·0	6·1	0·4	12·1	0·7
02	0 45·5	0 45·6	0 43·4	0·2	0·0	6·2	0·4	12·2	0·7
03	0 45·8	0 45·9	0 43·7	0·3	0·0	6·3	0·4	12·3	0·7
04	0 46·0	0 46·1	0 43·9	0·4	0·0	6·4	0·4	12·4	0·7
05	0 46·3	0 46·4	0 44·1	0·5	0·0	6·5	0·4	12·5	0·7
06	0 46·5	0 46·6	0 44·4	0·6	0·0	6·6	0·4	12·6	0·7
07	0 46·8	0 46·9	0 44·6	0·7	0·0	6·7	0·4	12·7	0·7
08	0 47·0	0 47·1	0 44·9	0·8	0·0	6·8	0·4	12·8	0·7
09	0 47·3	0 47·4	0 45·1	0·9	0·1	6·9	0·4	12·9	0·8
10	0 47·5	0 47·6	0 45·3	1·0	0·1	7·0	0·4	13·0	0·8
11	0 47·8	0 47·9	0 45·6	1·1	0·1	7·1	0·4	13·1	0·8
12	0 48·0	0 48·1	0 45·8	1·2	0·1	7·2	0·4	13·2	0·8
13	0 48·3	0 48·4	0 46·1	1·3	0·1	7·3	0·4	13·3	0·8
14	0 48·5	0 48·6	0 46·3	1·4	0·1	7·4	0·4	13·4	0·8
15	0 48·8	0 48·9	0 46·5	1·5	0·1	7·5	0·4	13·5	0·8
16	0 49·0	0 49·1	0 46·8	1·6	0·1	7·6	0·4	13·6	0·8
17	0 49·3	0 49·4	0 47·0	1·7	0·1	7·7	0·4	13·7	0·8
18	0 49·5	0 49·6	0 47·2	1·8	0·1	7·8	0·5	13·8	0·8
19	0 49·8	0 49·9	0 47·5	1·9	0·1	7·9	0·5	13·9	0·8
20	0 50·0	0 50·1	0 47·7	2·0	0·1	8·0	0·5	14·0	0·8
21	0 50·3	0 50·4	0 48·0	2·1	0·1	8·1	0·5	14·1	0·8
22	0 50·5	0 50·6	0 48·2	2·2	0·1	8·2	0·5	14·2	0·8
23	0 50·8	0 50·9	0 48·4	2·3	0·1	8·3	0·5	14·3	0·8
24	0 51·0	0 51·1	0 48·7	2·4	0·1	8·4	0·5	14·4	0·8
25	0 51·3	0 51·4	0 48·9	2·5	0·1	8·5	0·5	14·5	0·8
26	0 51·5	0 51·6	0 49·2	2·6	0·2	8·6	0·5	14·6	0·9
27	0 51·8	0 51·9	0 49·4	2·7	0·2	8·7	0·5	14·7	0·9
28	0 52·0	0 52·1	0 49·6	2·8	0·2	8·8	0·5	14·8	0·9
29	0 52·3	0 52·4	0 49·9	2·9	0·2	8·9	0·5	14·9	0·9
30	0 52·5	0 52·6	0 50·1	3·0	0·2	9·0	0·5	15·0	0·9
31	0 52·8	0 52·9	0 50·3	3·1	0·2	9·1	0·5	15·1	0·9
32	0 53·0	0 53·1	0 50·6	3·2	0·2	9·2	0·5	15·2	0·9
33	0 53·3	0 53·4	0 50·8	3·3	0·2	9·3	0·5	15·3	0·9
34	0 53·5	0 53·6	0 51·1	3·4	0·2	9·4	0·5	15·4	0·9
35	0 53·8	0 53·9	0 51·3	3·5	0·2	9·5	0·6	15·5	0·9
36	0 54·0	0 54·1	0 51·5	3·6	0·2	9·6	0·6	15·6	0·9
37	0 54·3	0 54·4	0 51·8	3·7	0·2	9·7	0·6	15·7	0·9
38	0 54·5	0 54·6	0 52·0	3·8	0·2	9·8	0·6	15·8	0·9
39	0 54·8	0 54·9	0 52·3	3·9	0·2	9·9	0·6	15·9	0·9
40	0 55·0	0 55·2	0 52·5	4·0	0·2	10·0	0·6	16·0	0·9
41	0 55·3	0 55·4	0 52·7	4·1	0·2	10·1	0·6	16·1	0·9
42	0 55·5	0 55·7	0 53·0	4·2	0·2	10·2	0·6	16·2	0·9
43	0 55·8	0 55·9	0 53·2	4·3	0·3	10·3	0·6	16·3	1·0
44	0 56·0	0 56·2	0 53·4	4·4	0·3	10·4	0·6	16·4	1·0
45	0 56·3	0 56·4	0 53·7	4·5	0·3	10·5	0·6	16·5	1·0
46	0 56·5	0 56·7	0 53·9	4·6	0·3	10·6	0·6	16·6	1·0
47	0 56·8	0 56·9	0 54·2	4·7	0·3	10·7	0·6	16·7	1·0
48	0 57·0	0 57·2	0 54·4	4·8	0·3	10·8	0·6	16·8	1·0
49	0 57·3	0 57·4	0 54·6	4·9	0·3	10·9	0·6	16·9	1·0
50	0 57·5	0 57·7	0 54·9	5·0	0·3	11·0	0·6	17·0	1·0
51	0 57·8	0 57·9	0 55·1	5·1	0·3	11·1	0·6	17·1	1·0
52	0 58·0	0 58·2	0 55·4	5·2	0·3	11·2	0·7	17·2	1·0
53	0 58·3	0 58·4	0 55·6	5·3	0·3	11·3	0·7	17·3	1·0
54	0 58·5	0 58·7	0 55·8	5·4	0·3	11·4	0·7	17·4	1·0
55	0 58·8	0 58·9	0 56·1	5·5	0·3	11·5	0·7	17·5	1·0
56	0 59·0	0 59·2	0 56·3	5·6	0·3	11·6	0·7	17·6	1·0
57	0 59·3	0 59·4	0 56·6	5·7	0·3	11·7	0·7	17·7	1·0
58	0 59·5	0 59·7	0 56·8	5·8	0·3	11·8	0·7	17·8	1·0
59	0 59·8	0 59·9	0 57·0	5·9	0·3	11·9	0·7	17·9	1·0
60	1 00·0	1 00·2	0 57·3	6·0	0·4	12·0	0·7	18·0	1·1

4m INCREMENTS AND CORRECTIONS 5m

m 4	SUN PLANETS	ARIES	MOON	v or d	Corrⁿ	v or d	Corrⁿ	v or d	Corrⁿ
s	° ′	° ′	° ′	′	′	′	′	′	′
00	1 00·0	1 00·2	0 57·3	0·0	0·0	6·0	0·5	12·0	0·9
01	1 00·3	1 00·4	0 57·5	0·1	0·0	6·1	0·5	12·1	0·9
02	1 00·5	1 00·7	0 57·7	0·2	0·0	6·2	0·5	12·2	0·9
03	1 00·8	1 00·9	0 58·0	0·3	0·0	6·3	0·5	12·3	0·9
04	1 01·0	1 01·2	0 58·2	0·4	0·0	6·4	0·5	12·4	0·9
05	1 01·3	1 01·4	0 58·5	0·5	0·0	6·5	0·5	12·5	0·9
06	1 01·5	1 01·7	0 58·7	0·6	0·0	6·6	0·5	12·6	0·9
07	1 01·8	1 01·9	0 58·9	0·7	0·1	6·7	0·5	12·7	1·0
08	1 02·0	1 02·2	0 59·2	0·8	0·1	6·8	0·5	12·8	1·0
09	1 02·3	1 02·4	0 59·4	0·9	0·1	6·9	0·5	12·9	1·0
10	1 02·5	1 02·7	0 59·7	1·0	0·1	7·0	0·5	13·0	1·0
11	1 02·8	1 02·9	0 59·9	1·1	0·1	7·1	0·5	13·1	1·0
12	1 03·0	1 03·2	1 00·1	1·2	0·1	7·2	0·5	13·2	1·0
13	1 03·3	1 03·4	1 00·4	1·3	0·1	7·3	0·5	13·3	1·0
14	1 03·5	1 03·7	1 00·6	1·4	0·1	7·4	0·6	13·4	1·0
15	1 03·8	1 03·9	1 00·8	1·5	0·1	7·5	0·6	13·5	1·0
16	1 04·0	1 04·2	1 01·1	1·6	0·1	7·6	0·6	13·6	1·0
17	1 04·3	1 04·4	1 01·3	1·7	0·1	7·7	0·6	13·7	1·0
18	1 04·5	1 04·7	1 01·6	1·8	0·1	7·8	0·6	13·8	1·0
19	1 04·8	1 04·9	1 01·8	1·9	0·1	7·9	0·6	13·9	1·0
20	1 05·0	1 05·2	1 02·0	2·0	0·2	8·0	0·6	14·0	1·1
21	1 05·3	1 05·4	1 02·3	2·1	0·2	8·1	0·6	14·1	1·1
22	1 05·5	1 05·7	1 02·5	2·2	0·2	8·2	0·6	14·2	1·1
23	1 05·8	1 05·9	1 02·8	2·3	0·2	8·3	0·6	14·3	1·1
24	1 06·0	1 06·2	1 03·0	2·4	0·2	8·4	0·6	14·4	1·1
25	1 06·3	1 06·4	1 03·2	2·5	0·2	8·5	0·6	14·5	1·1
26	1 06·5	1 06·7	1 03·5	2·6	0·2	8·6	0·6	14·6	1·1
27	1 06·8	1 06·9	1 03·7	2·7	0·2	8·7	0·7	14·7	1·1
28	1 07·0	1 07·2	1 03·9	2·8	0·2	8·8	0·7	14·8	1·1
29	1 07·3	1 07·4	1 04·2	2·9	0·2	8·9	0·7	14·9	1·1
30	1 07·5	1 07·7	1 04·4	3·0	0·2	9·0	0·7	15·0	1·1
31	1 07·8	1 07·9	1 04·7	3·1	0·2	9·1	0·7	15·1	1·1
32	1 08·0	1 08·2	1 04·9	3·2	0·2	9·2	0·7	15·2	1·1
33	1 08·3	1 08·4	1 05·1	3·3	0·2	9·3	0·7	15·3	1·1
34	1 08·5	1 08·7	1 05·4	3·4	0·3	9·4	0·7	15·4	1·2
35	1 08·8	1 08·9	1 05·6	3·5	0·3	9·5	0·7	15·5	1·2
36	1 09·0	1 09·2	1 05·9	3·6	0·3	9·6	0·7	15·6	1·2
37	1 09·3	1 09·4	1 06·1	3·7	0·3	9·7	0·7	15·7	1·2
38	1 09·5	1 09·7	1 06·3	3·8	0·3	9·8	0·7	15·8	1·2
39	1 09·8	1 09·9	1 06·6	3·9	0·3	9·9	0·7	15·9	1·2
40	1 10·0	1 10·2	1 06·8	4·0	0·3	10·0	0·8	16·0	1·2
41	1 10·3	1 10·4	1 07·0	4·1	0·3	10·1	0·8	16·1	1·2
42	1 10·5	1 10·7	1 07·3	4·2	0·3	10·2	0·8	16·2	1·2
43	1 10·8	1 10·9	1 07·5	4·3	0·3	10·3	0·8	16·3	1·2
44	1 11·0	1 11·2	1 07·8	4·4	0·3	10·4	0·8	16·4	1·2
45	1 11·3	1 11·4	1 08·0	4·5	0·3	10·5	0·8	16·5	1·2
46	1 11·5	1 11·7	1 08·2	4·6	0·3	10·6	0·8	16·6	1·2
47	1 11·8	1 11·9	1 08·5	4·7	0·4	10·7	0·8	16·7	1·3
48	1 12·0	1 12·2	1 08·7	4·8	0·4	10·8	0·8	16·8	1·3
49	1 12·3	1 12·4	1 09·0	4·9	0·4	10·9	0·8	16·9	1·3
50	1 12·5	1 12·7	1 09·2	5·0	0·4	11·0	0·8	17·0	1·3
51	1 12·8	1 12·9	1 09·4	5·1	0·4	11·1	0·8	17·1	1·3
52	1 13·0	1 13·2	1 09·7	5·2	0·4	11·2	0·8	17·2	1·3
53	1 13·3	1 13·5	1 09·9	5·3	0·4	11·3	0·8	17·3	1·3
54	1 13·5	1 13·7	1 10·2	5·4	0·4	11·4	0·9	17·4	1·3
55	1 13·8	1 14·0	1 10·4	5·5	0·4	11·5	0·9	17·5	1·3
56	1 14·0	1 14·2	1 10·6	5·6	0·4	11·6	0·9	17·6	1·3
57	1 14·3	1 14·5	1 10·9	5·7	0·4	11·7	0·9	17·7	1·3
58	1 14·5	1 14·7	1 11·1	5·8	0·4	11·8	0·9	17·8	1·3
59	1 14·8	1 15·0	1 11·3	5·9	0·4	11·9	0·9	17·9	1·3
60	1 15·0	1 15·2	1 11·6	6·0	0·5	12·0	0·9	18·0	1·4

m 5	SUN PLANETS	ARIES	MOON	v or d	Corrⁿ	v or d	Corrⁿ	v or d	Corrⁿ
s	° ′	° ′	° ′	′	′	′	′	′	′
00	1 15·0	1 15·2	1 11·6	0·0	0·0	6·0	0·6	12·0	1·1
01	1 15·3	1 15·5	1 11·8	0·1	0·0	6·1	0·6	12·1	1·1
02	1 15·5	1 15·7	1 12·1	0·2	0·0	6·2	0·6	12·2	1·1
03	1 15·8	1 16·0	1 12·3	0·3	0·0	6·3	0·6	12·3	1·1
04	1 16·0	1 16·2	1 12·5	0·4	0·0	6·4	0·6	12·4	1·1
05	1 16·3	1 16·5	1 12·8	0·5	0·0	6·5	0·6	12·5	1·1
06	1 16·5	1 16·7	1 13·0	0·6	0·1	6·6	0·6	12·6	1·2
07	1 16·8	1 17·0	1 13·3	0·7	0·1	6·7	0·6	12·7	1·2
08	1 17·0	1 17·2	1 13·5	0·8	0·1	6·8	0·6	12·8	1·2
09	1 17·3	1 17·5	1 13·7	0·9	0·1	6·9	0·6	12·9	1·2
10	1 17·5	1 17·7	1 14·0	1·0	0·1	7·0	0·6	13·0	1·2
11	1 17·8	1 18·0	1 14·2	1·1	0·1	7·1	0·7	13·1	1·2
12	1 18·0	1 18·2	1 14·4	1·2	0·1	7·2	0·7	13·2	1·2
13	1 18·3	1 18·5	1 14·7	1·3	0·1	7·3	0·7	13·3	1·2
14	1 18·5	1 18·7	1 14·9	1·4	0·1	7·4	0·7	13·4	1·2
15	1 18·8	1 19·0	1 15·2	1·5	0·1	7·5	0·7	13·5	1·2
16	1 19·0	1 19·2	1 15·4	1·6	0·1	7·6	0·7	13·6	1·2
17	1 19·3	1 19·5	1 15·6	1·7	0·2	7·7	0·7	13·7	1·3
18	1 19·5	1 19·7	1 15·9	1·8	0·2	7·8	0·7	13·8	1·3
19	1 19·8	1 20·0	1 16·1	1·9	0·2	7·9	0·7	13·9	1·3
20	1 20·0	1 20·2	1 16·4	2·0	0·2	8·0	0·7	14·0	1·3
21	1 20·3	1 20·5	1 16·6	2·1	0·2	8·1	0·7	14·1	1·3
22	1 20·5	1 20·7	1 16·8	2·2	0·2	8·2	0·8	14·2	1·3
23	1 20·8	1 21·0	1 17·1	2·3	0·2	8·3	0·8	14·3	1·3
24	1 21·0	1 21·2	1 17·3	2·4	0·2	8·4	0·8	14·4	1·3
25	1 21·3	1 21·5	1 17·5	2·5	0·2	8·5	0·8	14·5	1·3
26	1 21·5	1 21·7	1 17·8	2·6	0·2	8·6	0·8	14·6	1·3
27	1 21·8	1 22·0	1 18·0	2·7	0·2	8·7	0·8	14·7	1·3
28	1 22·0	1 22·2	1 18·3	2·8	0·3	8·8	0·8	14·8	1·4
29	1 22·3	1 22·5	1 18·5	2·9	0·3	8·9	0·8	14·9	1·4
30	1 22·5	1 22·7	1 18·7	3·0	0·3	9·0	0·8	15·0	1·4
31	1 22·8	1 23·0	1 19·0	3·1	0·3	9·1	0·8	15·1	1·4
32	1 23·0	1 23·2	1 19·2	3·2	0·3	9·2	0·8	15·2	1·4
33	1 23·3	1 23·5	1 19·5	3·3	0·3	9·3	0·9	15·3	1·4
34	1 23·5	1 23·7	1 19·7	3·4	0·3	9·4	0·9	15·4	1·4
35	1 23·8	1 24·0	1 19·9	3·5	0·3	9·5	0·9	15·5	1·4
36	1 24·0	1 24·2	1 20·2	3·6	0·3	9·6	0·9	15·6	1·4
37	1 24·3	1 24·5	1 20·4	3·7	0·3	9·7	0·9	15·7	1·4
38	1 24·5	1 24·7	1 20·7	3·8	0·3	9·8	0·9	15·8	1·4
39	1 24·8	1 25·0	1 20·9	3·9	0·4	9·9	0·9	15·9	1·5
40	1 25·0	1 25·2	1 21·1	4·0	0·4	10·0	0·9	16·0	1·5
41	1 25·3	1 25·5	1 21·4	4·1	0·4	10·1	0·9	16·1	1·5
42	1 25·5	1 25·7	1 21·6	4·2	0·4	10·2	0·9	16·2	1·5
43	1 25·8	1 26·0	1 21·8	4·3	0·4	10·3	0·9	16·3	1·5
44	1 26·0	1 26·2	1 22·1	4·4	0·4	10·4	1·0	16·4	1·5
45	1 26·3	1 26·5	1 22·3	4·5	0·4	10·5	1·0	16·5	1·5
46	1 26·5	1 26·7	1 22·6	4·6	0·4	10·6	1·0	16·6	1·5
47	1 26·8	1 27·0	1 22·8	4·7	0·4	10·7	1·0	16·7	1·5
48	1 27·0	1 27·2	1 23·0	4·8	0·4	10·8	1·0	16·8	1·5
49	1 27·3	1 27·5	1 23·3	4·9	0·4	10·9	1·0	16·9	1·5
50	1 27·5	1 27·7	1 23·5	5·0	0·5	11·0	1·0	17·0	1·6
51	1 27·8	1 28·0	1 23·8	5·1	0·5	11·1	1·0	17·1	1·6
52	1 28·0	1 28·2	1 24·0	5·2	0·5	11·2	1·0	17·2	1·6
53	1 28·3	1 28·5	1 24·2	5·3	0·5	11·3	1·0	17·3	1·6
54	1 28·5	1 28·7	1 24·5	5·4	0·5	11·4	1·0	17·4	1·6
55	1 28·8	1 29·0	1 24·7	5·5	0·5	11·5	1·1	17·5	1·6
56	1 29·0	1 29·2	1 24·9	5·6	0·5	11·6	1·1	17·6	1·6
57	1 29·3	1 29·5	1 25·2	5·7	0·5	11·7	1·1	17·7	1·6
58	1 29·5	1 29·7	1 25·4	5·8	0·5	11·8	1·1	17·8	1·6
59	1 29·8	1 30·0	1 25·7	5·9	0·5	11·9	1·1	17·9	1·6
60	1 30·0	1 30·2	1 25·9	6·0	0·6	12·0	1·1	18·0	1·7

6m INCREMENTS AND CORRECTIONS 7m

6m	SUN PLANETS	ARIES	MOON	v or d	Corrn	v or d	Corrn	v or d	Corrn
s	° ′	° ′	° ′	′	′	′	′	′	′
00	1 30·0	1 30·2	1 25·9	0·0	0·0	6·0	0·7	12·0	1·3
01	1 30·3	1 30·5	1 26·1	0·1	0·0	6·1	0·7	12·1	1·3
02	1 30·5	1 30·7	1 26·4	0·2	0·0	6·2	0·7	12·2	1·3
03	1 30·8	1 31·0	1 26·6	0·3	0·0	6·3	0·7	12·3	1·3
04	1 31·0	1 31·2	1 26·9	0·4	0·0	6·4	0·7	12·4	1·3
05	1 31·3	1 31·5	1 27·1	0·5	0·1	6·5	0·7	12·5	1·4
06	1 31·5	1 31·8	1 27·3	0·6	0·1	6·6	0·7	12·6	1·4
07	1 31·8	1 32·0	1 27·6	0·7	0·1	6·7	0·7	12·7	1·4
08	1 32·0	1 32·3	1 27·8	0·8	0·1	6·8	0·7	12·8	1·4
09	1 32·3	1 32·5	1 28·0	0·9	0·1	6·9	0·7	12·9	1·4
10	1 32·5	1 32·8	1 28·3	1·0	0·1	7·0	0·8	13·0	1·4
11	1 32·8	1 33·0	1 28·5	1·1	0·1	7·1	0·8	13·1	1·4
12	1 33·0	1 33·3	1 28·8	1·2	0·1	7·2	0·8	13·2	1·4
13	1 33·3	1 33·5	1 29·0	1·3	0·1	7·3	0·8	13·3	1·4
14	1 33·5	1 33·8	1 29·2	1·4	0·2	7·4	0·8	13·4	1·5
15	1 33·8	1 34·0	1 29·5	1·5	0·2	7·5	0·8	13·5	1·5
16	1 34·0	1 34·3	1 29·7	1·6	0·2	7·6	0·8	13·6	1·5
17	1 34·3	1 34·5	1 30·0	1·7	0·2	7·7	0·8	13·7	1·5
18	1 34·5	1 34·8	1 30·2	1·8	0·2	7·8	0·8	13·8	1·5
19	1 34·8	1 35·0	1 30·4	1·9	0·2	7·9	0·9	13·9	1·5
20	1 35·0	1 35·3	1 30·7	2·0	0·2	8·0	0·9	14·0	1·5
21	1 35·3	1 35·5	1 30·9	2·1	0·2	8·1	0·9	14·1	1·5
22	1 35·5	1 35·8	1 31·1	2·2	0·2	8·2	0·9	14·2	1·5
23	1 35·8	1 36·0	1 31·4	2·3	0·2	8·3	0·9	14·3	1·5
24	1 36·0	1 36·3	1 31·6	2·4	0·3	8·4	0·9	14·4	1·6
25	1 36·3	1 36·5	1 31·9	2·5	0·3	8·5	0·9	14·5	1·6
26	1 36·5	1 36·8	1 32·1	2·6	0·3	8·6	0·9	14·6	1·6
27	1 36·8	1 37·0	1 32·3	2·7	0·3	8·7	0·9	14·7	1·6
28	1 37·0	1 37·3	1 32·6	2·8	0·3	8·8	1·0	14·8	1·6
29	1 37·3	1 37·5	1 32·8	2·9	0·3	8·9	1·0	14·9	1·6
30	1 37·5	1 37·8	1 33·1	3·0	0·3	9·0	1·0	15·0	1·6
31	1 37·8	1 38·0	1 33·3	3·1	0·3	9·1	1·0	15·1	1·6
32	1 38·0	1 38·3	1 33·5	3·2	0·3	9·2	1·0	15·2	1·6
33	1 38·3	1 38·5	1 33·8	3·3	0·4	9·3	1·0	15·3	1·7
34	1 38·5	1 38·8	1 34·0	3·4	0·4	9·4	1·0	15·4	1·7
35	1 38·8	1 39·0	1 34·3	3·5	0·4	9·5	1·0	15·5	1·7
36	1 39·0	1 39·3	1 34·5	3·6	0·4	9·6	1·0	15·6	1·7
37	1 39·3	1 39·5	1 34·7	3·7	0·4	9·7	1·1	15·7	1·7
38	1 39·5	1 39·8	1 35·0	3·8	0·4	9·8	1·1	15·8	1·7
39	1 39·8	1 40·0	1 35·2	3·9	0·4	9·9	1·1	15·9	1·7
40	1 40·0	1 40·3	1 35·4	4·0	0·4	10·0	1·1	16·0	1·7
41	1 40·3	1 40·5	1 35·7	4·1	0·4	10·1	1·1	16·1	1·7
42	1 40·5	1 40·8	1 35·9	4·2	0·5	10·2	1·1	16·2	1·8
43	1 40·8	1 41·0	1 36·2	4·3	0·5	10·3	1·1	16·3	1·8
44	1 41·0	1 41·3	1 36·4	4·4	0·5	10·4	1·1	16·4	1·8
45	1 41·3	1 41·5	1 36·6	4·5	0·5	10·5	1·1	16·5	1·8
46	1 41·5	1 41·8	1 36·9	4·6	0·5	10·6	1·1	16·6	1·8
47	1 41·8	1 42·0	1 37·1	4·7	0·5	10·7	1·2	16·7	1·8
48	1 42·0	1 42·3	1 37·4	4·8	0·5	10·8	1·2	16·8	1·8
49	1 42·3	1 42·5	1 37·6	4·9	0·5	10·9	1·2	16·9	1·8
50	1 42·5	1 42·8	1 37·8	5·0	0·5	11·0	1·2	17·0	1·8
51	1 42·8	1 43·0	1 38·1	5·1	0·6	11·1	1·2	17·1	1·9
52	1 43·0	1 43·3	1 38·3	5·2	0·6	11·2	1·2	17·2	1·9
53	1 43·3	1 43·5	1 38·5	5·3	0·6	11·3	1·2	17·3	1·9
54	1 43·5	1 43·8	1 38·8	5·4	0·6	11·4	1·2	17·4	1·9
55	1 43·8	1 44·0	1 39·0	5·5	0·6	11·5	1·2	17·5	1·9
56	1 44·0	1 44·3	1 39·3	5·6	0·6	11·6	1·3	17·6	1·9
57	1 44·3	1 44·5	1 39·5	5·7	0·6	11·7	1·3	17·7	1·9
58	1 44·5	1 44·8	1 39·7	5·8	0·6	11·8	1·3	17·8	1·9
59	1 44·8	1 45·0	1 40·0	5·9	0·6	11·9	1·3	17·9	1·9
60	1 45·0	1 45·3	1 40·2	6·0	0·7	12·0	1·3	18·0	2·0

7m	SUN PLANETS	ARIES	MOON	v or d	Corrn	v or d	Corrn	v or d	Corrn
s	° ′	° ′	° ′	′	′	′	′	′	′
00	1 45·0	1 45·3	1 40·2	0·0	0·0	6·0	0·8	12·0	1·5
01	1 45·3	1 45·5	1 40·5	0·1	0·0	6·1	0·8	12·1	1·5
02	1 45·5	1 45·8	1 40·7	0·2	0·0	6·2	0·8	12·2	1·5
03	1 45·8	1 46·0	1 40·9	0·3	0·0	6·3	0·8	12·3	1·5
04	1 46·0	1 46·3	1 41·2	0·4	0·1	6·4	0·8	12·4	1·6
05	1 46·3	1 46·5	1 41·4	0·5	0·1	6·5	0·8	12·5	1·6
06	1 46·5	1 46·8	1 41·6	0·6	0·1	6·6	0·8	12·6	1·6
07	1 46·8	1 47·0	1 41·9	0·7	0·1	6·7	0·8	12·7	1·6
08	1 47·0	1 47·3	1 42·1	0·8	0·1	6·8	0·9	12·8	1·6
09	1 47·3	1 47·5	1 42·4	0·9	0·1	6·9	0·9	12·9	1·6
10	1 47·5	1 47·8	1 42·6	1·0	0·1	7·0	0·9	13·0	1·6
11	1 47·8	1 48·0	1 42·8	1·1	0·1	7·1	0·9	13·1	1·6
12	1 48·0	1 48·3	1 43·1	1·2	0·2	7·2	0·9	13·2	1·7
13	1 48·3	1 48·5	1 43·3	1·3	0·2	7·3	0·9	13·3	1·7
14	1 48·5	1 48·8	1 43·6	1·4	0·2	7·4	0·9	13·4	1·7
15	1 48·8	1 49·0	1 43·8	1·5	0·2	7·5	0·9	13·5	1·7
16	1 49·0	1 49·3	1 44·0	1·6	0·2	7·6	1·0	13·6	1·7
17	1 49·3	1 49·5	1 44·3	1·7	0·2	7·7	1·0	13·7	1·7
18	1 49·5	1 49·8	1 44·5	1·8	0·2	7·8	1·0	13·8	1·7
19	1 49·8	1 50·1	1 44·8	1·9	0·2	7·9	1·0	13·9	1·7
20	1 50·0	1 50·3	1 45·0	2·0	0·3	8·0	1·0	14·0	1·8
21	1 50·3	1 50·6	1 45·2	2·1	0·3	8·1	1·0	14·1	1·8
22	1 50·5	1 50·8	1 45·5	2·2	0·3	8·2	1·0	14·2	1·8
23	1 50·8	1 51·1	1 45·7	2·3	0·3	8·3	1·0	14·3	1·8
24	1 51·0	1 51·3	1 45·9	2·4	0·3	8·4	1·1	14·4	1·8
25	1 51·3	1 51·6	1 46·2	2·5	0·3	8·5	1·1	14·5	1·8
26	1 51·5	1 51·8	1 46·4	2·6	0·3	8·6	1·1	14·6	1·8
27	1 51·8	1 52·1	1 46·7	2·7	0·3	8·7	1·1	14·7	1·8
28	1 52·0	1 52·3	1 46·9	2·8	0·4	8·8	1·1	14·8	1·9
29	1 52·3	1 52·6	1 47·1	2·9	0·4	8·9	1·1	14·9	1·9
30	1 52·5	1 52·8	1 47·4	3·0	0·4	9·0	1·1	15·0	1·9
31	1 52·8	1 53·1	1 47·6	3·1	0·4	9·1	1·1	15·1	1·9
32	1 53·0	1 53·3	1 47·9	3·2	0·4	9·2	1·2	15·2	1·9
33	1 53·3	1 53·6	1 48·1	3·3	0·4	9·3	1·2	15·3	1·9
34	1 53·5	1 53·8	1 48·3	3·4	0·4	9·4	1·2	15·4	1·9
35	1 53·8	1 54·1	1 48·6	3·5	0·4	9·5	1·2	15·5	1·9
36	1 54·0	1 54·3	1 48·8	3·6	0·5	9·6	1·2	15·6	2·0
37	1 54·3	1 54·6	1 49·0	3·7	0·5	9·7	1·2	15·7	2·0
38	1 54·5	1 54·8	1 49·3	3·8	0·5	9·8	1·2	15·8	2·0
39	1 54·8	1 55·1	1 49·5	3·9	0·5	9·9	1·2	15·9	2·0
40	1 55·0	1 55·3	1 49·8	4·0	0·5	10·0	1·3	16·0	2·0
41	1 55·3	1 55·6	1 50·0	4·1	0·5	10·1	1·3	16·1	2·0
42	1 55·5	1 55·8	1 50·2	4·2	0·5	10·2	1·3	16·2	2·0
43	1 55·8	1 56·1	1 50·5	4·3	0·5	10·3	1·3	16·3	2·0
44	1 56·0	1 56·3	1 50·7	4·4	0·6	10·4	1·3	16·4	2·1
45	1 56·3	1 56·6	1 51·0	4·5	0·6	10·5	1·3	16·5	2·1
46	1 56·5	1 56·8	1 51·2	4·6	0·6	10·6	1·3	16·6	2·1
47	1 56·8	1 57·1	1 51·4	4·7	0·6	10·7	1·3	16·7	2·1
48	1 57·0	1 57·3	1 51·7	4·8	0·6	10·8	1·4	16·8	2·1
49	1 57·3	1 57·6	1 51·9	4·9	0·6	10·9	1·4	16·9	2·1
50	1 57·5	1 57·8	1 52·1	5·0	0·6	11·0	1·4	17·0	2·1
51	1 57·8	1 58·1	1 52·4	5·1	0·6	11·1	1·4	17·1	2·1
52	1 58·0	1 58·3	1 52·6	5·2	0·7	11·2	1·4	17·2	2·2
53	1 58·3	1 58·6	1 52·9	5·3	0·7	11·3	1·4	17·3	2·2
54	1 58·5	1 58·8	1 53·1	5·4	0·7	11·4	1·4	17·4	2·2
55	1 58·8	1 59·1	1 53·3	5·5	0·7	11·5	1·4	17·5	2·2
56	1 59·0	1 59·3	1 53·6	5·6	0·7	11·6	1·5	17·6	2·2
57	1 59·3	1 59·6	1 53·8	5·7	0·7	11·7	1·5	17·7	2·2
58	1 59·5	1 59·8	1 54·1	5·8	0·7	11·8	1·5	17·8	2·2
59	1 59·8	2 00·1	1 54·3	5·9	0·7	11·9	1·5	17·9	2·2
60	2 00·0	2 00·3	1 54·5	6·0	0·8	12·0	1·5	18·0	2·3

8ᵐ INCREMENTS AND CORRECTIONS 9ᵐ

8ᵐ	SUN PLANETS	ARIES	MOON	v or d	Corrⁿ	v or d	Corrⁿ	v or d	Corrⁿ
s	° ′	° ′	° ′	′	′	′	′	′	′
00	2 00·0	2 00·3	1 54·5	0·0	0·0	6·0	0·9	12·0	1·7
01	2 00·3	2 00·6	1 54·8	0·1	0·0	6·1	0·9	12·1	1·7
02	2 00·5	2 00·8	1 55·0	0·2	0·0	6·2	0·9	12·2	1·7
03	2 00·8	2 01·1	1 55·2	0·3	0·0	6·3	0·9	12·3	1·7
04	2 01·0	2 01·3	1 55·5	0·4	0·1	6·4	0·9	12·4	1·8
05	2 01·3	2 01·6	1 55·7	0·5	0·1	6·5	0·9	12·5	1·8
06	2 01·5	2 01·8	1 56·0	0·6	0·1	6·6	0·9	12·6	1·8
07	2 01·8	2 02·1	1 56·2	0·7	0·1	6·7	0·9	12·7	1·8
08	2 02·0	2 02·3	1 56·4	0·8	0·1	6·8	1·0	12·8	1·8
09	2 02·3	2 02·6	1 56·7	0·9	0·1	6·9	1·0	12·9	1·8
10	2 02·5	2 02·8	1 56·9	1·0	0·1	7·0	1·0	13·0	1·8
11	2 02·8	2 03·1	1 57·2	1·1	0·2	7·1	1·0	13·1	1·9
12	2 03·0	2 03·3	1 57·4	1·2	0·2	7·2	1·0	13·2	1·9
13	2 03·3	2 03·6	1 57·6	1·3	0·2	7·3	1·0	13·3	1·9
14	2 03·5	2 03·8	1 57·9	1·4	0·2	7·4	1·0	13·4	1·9
15	2 03·8	2 04·1	1 58·1	1·5	0·2	7·5	1·1	13·5	1·9
16	2 04·0	2 04·3	1 58·4	1·6	0·2	7·6	1·1	13·6	1·9
17	2 04·3	2 04·6	1 58·6	1·7	0·2	7·7	1·1	13·7	1·9
18	2 04·5	2 04·8	1 58·8	1·8	0·3	7·8	1·1	13·8	2·0
19	2 04·8	2 05·1	1 59·1	1·9	0·3	7·9	1·1	13·9	2·0
20	2 05·0	2 05·3	1 59·3	2·0	0·3	8·0	1·1	14·0	2·0
21	2 05·3	2 05·6	1 59·5	2·1	0·3	8·1	1·1	14·1	2·0
22	2 05·5	2 05·8	1 59·8	2·2	0·3	8·2	1·2	14·2	2·0
23	2 05·8	2 06·1	2 00·0	2·3	0·3	8·3	1·2	14·3	2·0
24	2 06·0	2 06·3	2 00·3	2·4	0·3	8·4	1·2	14·4	2·0
25	2 06·3	2 06·6	2 00·5	2·5	0·4	8·5	1·2	14·5	2·1
26	2 06·5	2 06·8	2 00·7	2·6	0·4	8·6	1·2	14·6	2·1
27	2 06·8	2 07·1	2 01·0	2·7	0·4	8·7	1·2	14·7	2·1
28	2 07·0	2 07·3	2 01·2	2·8	0·4	8·8	1·2	14·8	2·1
29	2 07·3	2 07·6	2 01·5	2·9	0·4	8·9	1·3	14·9	2·1
30	2 07·5	2 07·8	2 01·7	3·0	0·4	9·0	1·3	15·0	2·1
31	2 07·8	2 08·1	2 01·9	3·1	0·4	9·1	1·3	15·1	2·1
32	2 08·0	2 08·4	2 02·2	3·2	0·5	9·2	1·3	15·2	2·2
33	2 08·3	2 08·6	2 02·4	3·3	0·5	9·3	1·3	15·3	2·2
34	2 08·5	2 08·9	2 02·6	3·4	0·5	9·4	1·3	15·4	2·2
35	2 08·8	2 09·1	2 02·9	3·5	0·5	9·5	1·3	15·5	2·2
36	2 09·0	2 09·4	2 03·1	3·6	0·5	9·6	1·4	15·6	2·2
37	2 09·3	2 09·6	2 03·4	3·7	0·5	9·7	1·4	15·7	2·2
38	2 09·5	2 09·9	2 03·6	3·8	0·5	9·8	1·4	15·8	2·2
39	2 09·8	2 10·1	2 03·8	3·9	0·6	9·9	1·4	15·9	2·3
40	2 10·0	2 10·4	2 04·1	4·0	0·6	10·0	1·4	16·0	2·3
41	2 10·3	2 10·6	2 04·3	4·1	0·6	10·1	1·4	16·1	2·3
42	2 10·5	2 10·9	2 04·6	4·2	0·6	10·2	1·4	16·2	2·3
43	2 10·8	2 11·1	2 04·8	4·3	0·6	10·3	1·5	16·3	2·3
44	2 11·0	2 11·4	2 05·0	4·4	0·6	10·4	1·5	16·4	2·3
45	2 11·3	2 11·6	2 05·3	4·5	0·6	10·5	1·5	16·5	2·3
46	2 11·5	2 11·9	2 05·5	4·6	0·7	10·6	1·5	16·6	2·4
47	2 11·8	2 12·1	2 05·7	4·7	0·7	10·7	1·5	16·7	2·4
48	2 12·0	2 12·4	2 06·0	4·8	0·7	10·8	1·5	16·8	2·4
49	2 12·3	2 12·6	2 06·2	4·9	0·7	10·9	1·5	16·9	2·4
50	2 12·5	2 12·9	2 06·5	5·0	0·7	11·0	1·6	17·0	2·4
51	2 12·8	2 13·1	2 06·7	5·1	0·7	11·1	1·6	17·1	2·4
52	2 13·0	2 13·4	2 06·9	5·2	0·7	11·2	1·6	17·2	2·4
53	2 13·3	2 13·6	2 07·2	5·3	0·8	11·3	1·6	17·3	2·5
54	2 13·5	2 13·9	2 07·4	5·4	0·8	11·4	1·6	17·4	2·5
55	2 13·8	2 14·1	2 07·7	5·5	0·8	11·5	1·6	17·5	2·5
56	2 14·0	2 14·4	2 07·9	5·6	0·8	11·6	1·6	17·6	2·5
57	2 14·3	2 14·6	2 08·1	5·7	0·8	11·7	1·7	17·7	2·5
58	2 14·5	2 14·9	2 08·4	5·8	0·8	11·8	1·7	17·8	2·5
59	2 14·8	2 15·1	2 08·6	5·9	0·8	11·9	1·7	17·9	2·5
60	2 15·0	2 15·4	2 08·9	6·0	0·9	12·0	1·7	18·0	2·6

9ᵐ	SUN PLANETS	ARIES	MOON	v or d	Corrⁿ	v or d	Corrⁿ	v or d	Corrⁿ
s	° ′	° ′	° ′	′	′	′	′	′	′
00	2 15·0	2 15·4	2 08·9	0·0	0·0	6·0	1·0	12·0	1·9
01	2 15·3	2 15·6	2 09·1	0·1	0·0	6·1	1·0	12·1	1·9
02	2 15·5	2 15·9	2 09·3	0·2	0·0	6·2	1·0	12·2	1·9
03	2 15·8	2 16·1	2 09·6	0·3	0·0	6·3	1·0	12·3	1·9
04	2 16·0	2 16·4	2 09·8	0·4	0·1	6·4	1·0	12·4	2·0
05	2 16·3	2 16·6	2 10·0	0·5	0·1	6·5	1·0	12·5	2·0
06	2 16·5	2 16·9	2 10·3	0·6	0·1	6·6	1·0	12·6	2·0
07	2 16·8	2 17·1	2 10·5	0·7	0·1	6·7	1·1	12·7	2·0
08	2 17·0	2 17·4	2 10·8	0·8	0·1	6·8	1·1	12·8	2·0
09	2 17·3	2 17·6	2 11·0	0·9	0·1	6·9	1·1	12·9	2·0
10	2 17·5	2 17·9	2 11·2	1·0	0·2	7·0	1·1	13·0	2·1
11	2 17·8	2 18·1	2 11·5	1·1	0·2	7·1	1·1	13·1	2·1
12	2 18·0	2 18·4	2 11·7	1·2	0·2	7·2	1·1	13·2	2·1
13	2 18·3	2 18·6	2 12·0	1·3	0·2	7·3	1·2	13·3	2·1
14	2 18·5	2 18·9	2 12·2	1·4	0·2	7·4	1·2	13·4	2·1
15	2 18·8	2 19·1	2 12·4	1·5	0·2	7·5	1·2	13·5	2·1
16	2 19·0	2 19·4	2 12·7	1·6	0·3	7·6	1·2	13·6	2·2
17	2 19·3	2 19·6	2 12·9	1·7	0·3	7·7	1·2	13·7	2·2
18	2 19·5	2 19·9	2 13·1	1·8	0·3	7·8	1·2	13·8	2·2
19	2 19·8	2 20·1	2 13·4	1·9	0·3	7·9	1·3	13·9	2·2
20	2 20·0	2 20·4	2 13·6	2·0	0·3	8·0	1·3	14·0	2·2
21	2 20·3	2 20·6	2 13·9	2·1	0·3	8·1	1·3	14·1	2·2
22	2 20·5	2 20·9	2 14·1	2·2	0·3	8·2	1·3	14·2	2·2
23	2 20·8	2 21·1	2 14·3	2·3	0·4	8·3	1·3	14·3	2·3
24	2 21·0	2 21·4	2 14·6	2·4	0·4	8·4	1·3	14·4	2·3
25	2 21·3	2 21·6	2 14·8	2·5	0·4	8·5	1·3	14·5	2·3
26	2 21·5	2 21·9	2 15·1	2·6	0·4	8·6	1·4	14·6	2·3
27	2 21·8	2 22·1	2 15·3	2·7	0·4	8·7	1·4	14·7	2·3
28	2 22·0	2 22·4	2 15·5	2·8	0·4	8·8	1·4	14·8	2·3
29	2 22·3	2 22·6	2 15·8	2·9	0·5	8·9	1·4	14·9	2·4
30	2 22·5	2 22·9	2 16·0	3·0	0·5	9·0	1·4	15·0	2·4
31	2 22·8	2 23·1	2 16·2	3·1	0·5	9·1	1·4	15·1	2·4
32	2 23·0	2 23·4	2 16·5	3·2	0·5	9·2	1·5	15·2	2·4
33	2 23·3	2 23·6	2 16·7	3·3	0·5	9·3	1·5	15·3	2·4
34	2 23·5	2 23·9	2 17·0	3·4	0·5	9·4	1·5	15·4	2·4
35	2 23·8	2 24·1	2 17·2	3·5	0·6	9·5	1·5	15·5	2·5
36	2 24·0	2 24·4	2 17·4	3·6	0·6	9·6	1·5	15·6	2·5
37	2 24·3	2 24·6	2 17·7	3·7	0·6	9·7	1·5	15·7	2·5
38	2 24·5	2 24·9	2 17·9	3·8	0·6	9·8	1·6	15·8	2·5
39	2 24·8	2 25·1	2 18·2	3·9	0·6	9·9	1·6	15·9	2·5
40	2 25·0	2 25·4	2 18·4	4·0	0·6	10·0	1·6	16·0	2·5
41	2 25·3	2 25·6	2 18·6	4·1	0·6	10·1	1·6	16·1	2·5
42	2 25·5	2 25·9	2 18·9	4·2	0·7	10·2	1·6	16·2	2·6
43	2 25·8	2 26·1	2 19·1	4·3	0·7	10·3	1·6	16·3	2·6
44	2 26·0	2 26·4	2 19·3	4·4	0·7	10·4	1·6	16·4	2·6
45	2 26·3	2 26·7	2 19·6	4·5	0·7	10·5	1·7	16·5	2·6
46	2 26·5	2 26·9	2 19·8	4·6	0·7	10·6	1·7	16·6	2·6
47	2 26·8	2 27·2	2 20·1	4·7	0·7	10·7	1·7	16·7	2·6
48	2 27·0	2 27·4	2 20·3	4·8	0·8	10·8	1·7	16·8	2·7
49	2 27·3	2 27·7	2 20·5	4·9	0·8	10·9	1·7	16·9	2·7
50	2 27·5	2 27·9	2 20·8	5·0	0·8	11·0	1·7	17·0	2·7
51	2 27·8	2 28·2	2 21·0	5·1	0·8	11·1	1·8	17·1	2·7
52	2 28·0	2 28·4	2 21·3	5·2	0·8	11·2	1·8	17·2	2·7
53	2 28·3	2 28·7	2 21·5	5·3	0·8	11·3	1·8	17·3	2·7
54	2 28·5	2 28·9	2 21·7	5·4	0·9	11·4	1·8	17·4	2·8
55	2 28·8	2 29·2	2 22·0	5·5	0·9	11·5	1·8	17·5	2·8
56	2 29·0	2 29·4	2 22·2	5·6	0·9	11·6	1·8	17·6	2·8
57	2 29·3	2 29·7	2 22·5	5·7	0·9	11·7	1·9	17·7	2·8
58	2 29·5	2 29·9	2 22·7	5·8	0·9	11·8	1·9	17·8	2·8
59	2 29·8	2 30·2	2 22·9	5·9	0·9	11·9	1·9	17·9	2·8
60	2 30·0	2 30·4	2 23·2	6·0	1·0	12·0	1·9	18·0	2·9

10m INCREMENTS AND CORRECTIONS **11m**

10m s	SUN PLANETS ° ′	ARIES ° ′	MOON ° ′	v or d ′	Corr ′	v or d ′	Corr ′	v or d ′	Corr ′
00	2 30.0	2 30.4	2 23.2	0.0	0.0	6.0	1.1	12.0	2.1
01	2 30.3	2 30.7	2 23.4	0.1	0.0	6.1	1.1	12.1	2.1
02	2 30.5	2 30.9	2 23.6	0.2	0.0	6.2	1.1	12.2	2.1
03	2 30.8	2 31.2	2 23.9	0.3	0.1	6.3	1.1	12.3	2.2
04	2 31.0	2 31.4	2 24.1	0.4	0.1	6.4	1.1	12.4	2.2
05	2 31.3	2 31.7	2 24.4	0.5	0.1	6.5	1.1	12.5	2.2
06	2 31.5	2 31.9	2 24.6	0.6	0.1	6.6	1.2	12.6	2.2
07	2 31.8	2 32.2	2 24.8	0.7	0.1	6.7	1.2	12.7	2.2
08	2 32.0	2 32.4	2 25.1	0.8	0.1	6.8	1.2	12.8	2.2
09	2 32.3	2 32.7	2 25.3	0.9	0.2	6.9	1.2	12.9	2.3
10	2 32.5	2 32.9	2 25.6	1.0	0.2	7.0	1.2	13.0	2.3
11	2 32.8	2 33.2	2 25.8	1.1	0.2	7.1	1.2	13.1	2.3
12	2 33.0	2 33.4	2 26.0	1.2	0.2	7.2	1.3	13.2	2.3
13	2 33.3	2 33.7	2 26.3	1.3	0.2	7.3	1.3	13.3	2.3
14	2 33.5	2 33.9	2 26.5	1.4	0.2	7.4	1.3	13.4	2.3
15	2 33.8	2 34.2	2 26.7	1.5	0.3	7.5	1.3	13.5	2.4
16	2 34.0	2 34.4	2 27.0	1.6	0.3	7.6	1.3	13.6	2.4
17	2 34.3	2 34.7	2 27.2	1.7	0.3	7.7	1.3	13.7	2.4
18	2 34.5	2 34.9	2 27.5	1.8	0.3	7.8	1.4	13.8	2.4
19	2 34.8	2 35.2	2 27.7	1.9	0.3	7.9	1.4	13.9	2.4
20	2 35.0	2 35.4	2 27.9	2.0	0.4	8.0	1.4	14.0	2.5
21	2 35.3	2 35.7	2 28.2	2.1	0.4	8.1	1.4	14.1	2.5
22	2 35.5	2 35.9	2 28.4	2.2	0.4	8.2	1.4	14.2	2.5
23	2 35.8	2 36.2	2 28.7	2.3	0.4	8.3	1.5	14.3	2.5
24	2 36.0	2 36.4	2 28.9	2.4	0.4	8.4	1.5	14.4	2.5
25	2 36.3	2 36.7	2 29.1	2.5	0.4	8.5	1.5	14.5	2.5
26	2 36.5	2 36.9	2 29.4	2.6	0.5	8.6	1.5	14.6	2.6
27	2 36.8	2 37.2	2 29.6	2.7	0.5	8.7	1.5	14.7	2.6
28	2 37.0	2 37.4	2 29.8	2.8	0.5	8.8	1.5	14.8	2.6
29	2 37.3	2 37.7	2 30.1	2.9	0.5	8.9	1.6	14.9	2.6
30	2 37.5	2 37.9	2 30.3	3.0	0.5	9.0	1.6	15.0	2.6
31	2 37.8	2 38.2	2 30.6	3.1	0.5	9.1	1.6	15.1	2.6
32	2 38.0	2 38.4	2 30.8	3.2	0.6	9.2	1.6	15.2	2.7
33	2 38.3	2 38.7	2 31.0	3.3	0.6	9.3	1.6	15.3	2.7
34	2 38.5	2 38.9	2 31.3	3.4	0.6	9.4	1.6	15.4	2.7
35	2 38.8	2 39.2	2 31.5	3.5	0.6	9.5	1.7	15.5	2.7
36	2 39.0	2 39.4	2 31.8	3.6	0.6	9.6	1.7	15.6	2.7
37	2 39.3	2 39.7	2 32.0	3.7	0.6	9.7	1.7	15.7	2.7
38	2 39.5	2 39.9	2 32.2	3.8	0.7	9.8	1.7	15.8	2.8
39	2 39.8	2 40.2	2 32.5	3.9	0.7	9.9	1.7	15.9	2.8
40	2 40.0	2 40.4	2 32.7	4.0	0.7	10.0	1.8	16.0	2.8
41	2 40.3	2 40.7	2 32.9	4.1	0.7	10.1	1.8	16.1	2.8
42	2 40.5	2 40.9	2 33.2	4.2	0.7	10.2	1.8	16.2	2.8
43	2 40.8	2 41.2	2 33.4	4.3	0.8	10.3	1.8	16.3	2.9
44	2 41.0	2 41.4	2 33.7	4.4	0.8	10.4	1.8	16.4	2.9
45	2 41.3	2 41.7	2 33.9	4.5	0.8	10.5	1.8	16.5	2.9
46	2 41.5	2 41.9	2 34.1	4.6	0.8	10.6	1.9	16.6	2.9
47	2 41.8	2 42.2	2 34.4	4.7	0.8	10.7	1.9	16.7	2.9
48	2 42.0	2 42.4	2 34.6	4.8	0.8	10.8	1.9	16.8	2.9
49	2 42.3	2 42.7	2 34.9	4.9	0.9	10.9	1.9	16.9	3.0
50	2 42.5	2 42.9	2 35.1	5.0	0.9	11.0	1.9	17.0	3.0
51	2 42.8	2 43.2	2 35.3	5.1	0.9	11.1	1.9	17.1	3.0
52	2 43.0	2 43.4	2 35.6	5.2	0.9	11.2	2.0	17.2	3.0
53	2 43.3	2 43.7	2 35.8	5.3	0.9	11.3	2.0	17.3	3.0
54	2 43.5	2 43.9	2 36.1	5.4	0.9	11.4	2.0	17.4	3.0
55	2 43.8	2 44.2	2 36.3	5.5	1.0	11.5	2.0	17.5	3.1
56	2 44.0	2 44.4	2 36.5	5.6	1.0	11.6	2.0	17.6	3.1
57	2 44.3	2 44.7	2 36.8	5.7	1.0	11.7	2.0	17.7	3.1
58	2 44.5	2 45.0	2 37.0	5.8	1.0	11.8	2.1	17.8	3.1
59	2 44.8	2 45.2	2 37.2	5.9	1.0	11.9	2.1	17.9	3.1
60	2 45.0	2 45.5	2 37.5	6.0	1.1	12.0	2.1	18.0	3.2

11m s	SUN PLANETS ° ′	ARIES ° ′	MOON ° ′	v or d ′	Corr ′	v or d ′	Corr ′	v or d ′	Corr ′
00	2 45.0	2 45.5	2 37.5	0.0	0.0	6.0	1.2	12.0	2.3
01	2 45.3	2 45.7	2 37.7	0.1	0.0	6.1	1.2	12.1	2.3
02	2 45.5	2 46.0	2 38.0	0.2	0.0	6.2	1.2	12.2	2.3
03	2 45.8	2 46.2	2 38.2	0.3	0.1	6.3	1.2	12.3	2.4
04	2 46.0	2 46.5	2 38.4	0.4	0.1	6.4	1.2	12.4	2.4
05	2 46.3	2 46.7	2 38.7	0.5	0.1	6.5	1.2	12.5	2.4
06	2 46.5	2 47.0	2 38.9	0.6	0.1	6.6	1.3	12.6	2.4
07	2 46.8	2 47.2	2 39.2	0.7	0.1	6.7	1.3	12.7	2.4
08	2 47.0	2 47.5	2 39.4	0.8	0.2	6.8	1.3	12.8	2.5
09	2 47.3	2 47.7	2 39.6	0.9	0.2	6.9	1.3	12.9	2.5
10	2 47.5	2 48.0	2 39.9	1.0	0.2	7.0	1.3	13.0	2.5
11	2 47.8	2 48.2	2 40.1	1.1	0.2	7.1	1.4	13.1	2.5
12	2 48.0	2 48.5	2 40.3	1.2	0.2	7.2	1.4	13.2	2.5
13	2 48.3	2 48.7	2 40.6	1.3	0.2	7.3	1.4	13.3	2.5
14	2 48.5	2 49.0	2 40.8	1.4	0.3	7.4	1.4	13.4	2.6
15	2 48.8	2 49.2	2 41.1	1.5	0.3	7.5	1.4	13.5	2.6
16	2 49.0	2 49.5	2 41.3	1.6	0.3	7.6	1.5	13.6	2.6
17	2 49.3	2 49.7	2 41.5	1.7	0.3	7.7	1.5	13.7	2.6
18	2 49.5	2 50.0	2 41.8	1.8	0.3	7.8	1.5	13.8	2.6
19	2 49.8	2 50.2	2 42.0	1.9	0.4	7.9	1.5	13.9	2.7
20	2 50.0	2 50.5	2 42.3	2.0	0.4	8.0	1.5	14.0	2.7
21	2 50.3	2 50.7	2 42.5	2.1	0.4	8.1	1.6	14.1	2.7
22	2 50.5	2 51.0	2 42.7	2.2	0.4	8.2	1.6	14.2	2.7
23	2 50.8	2 51.2	2 43.0	2.3	0.4	8.3	1.6	14.3	2.7
24	2 51.0	2 51.5	2 43.2	2.4	0.5	8.4	1.6	14.4	2.8
25	2 51.3	2 51.7	2 43.4	2.5	0.5	8.5	1.6	14.5	2.8
26	2 51.5	2 52.0	2 43.7	2.6	0.5	8.6	1.6	14.6	2.8
27	2 51.8	2 52.2	2 43.9	2.7	0.5	8.7	1.7	14.7	2.8
28	2 52.0	2 52.5	2 44.2	2.8	0.5	8.8	1.7	14.8	2.8
29	2 52.3	2 52.7	2 44.4	2.9	0.6	8.9	1.7	14.9	2.9
30	2 52.5	2 53.0	2 44.6	3.0	0.6	9.0	1.7	15.0	2.9
31	2 52.8	2 53.2	2 44.9	3.1	0.6	9.1	1.7	15.1	2.9
32	2 53.0	2 53.5	2 45.1	3.2	0.6	9.2	1.8	15.2	2.9
33	2 53.3	2 53.7	2 45.4	3.3	0.6	9.3	1.8	15.3	2.9
34	2 53.5	2 54.0	2 45.6	3.4	0.7	9.4	1.8	15.4	3.0
35	2 53.8	2 54.2	2 45.8	3.5	0.7	9.5	1.8	15.5	3.0
36	2 54.0	2 54.5	2 46.1	3.6	0.7	9.6	1.8	15.6	3.0
37	2 54.3	2 54.7	2 46.3	3.7	0.7	9.7	1.9	15.7	3.0
38	2 54.5	2 55.0	2 46.6	3.8	0.7	9.8	1.9	15.8	3.0
39	2 54.8	2 55.2	2 46.8	3.9	0.7	9.9	1.9	15.9	3.0
40	2 55.0	2 55.5	2 47.0	4.0	0.8	10.0	1.9	16.0	3.1
41	2 55.3	2 55.7	2 47.3	4.1	0.8	10.1	1.9	16.1	3.1
42	2 55.5	2 56.0	2 47.5	4.2	0.8	10.2	2.0	16.2	3.1
43	2 55.8	2 56.2	2 47.7	4.3	0.8	10.3	2.0	16.3	3.1
44	2 56.0	2 56.5	2 48.0	4.4	0.8	10.4	2.0	16.4	3.1
45	2 56.3	2 56.7	2 48.2	4.5	0.9	10.5	2.0	16.5	3.2
46	2 56.5	2 57.0	2 48.5	4.6	0.9	10.6	2.0	16.6	3.2
47	2 56.8	2 57.2	2 48.7	4.7	0.9	10.7	2.1	16.7	3.2
48	2 57.0	2 57.5	2 48.9	4.8	0.9	10.8	2.1	16.8	3.2
49	2 57.3	2 57.7	2 49.2	4.9	0.9	10.9	2.1	16.9	3.2
50	2 57.5	2 58.0	2 49.4	5.0	1.0	11.0	2.1	17.0	3.3
51	2 57.8	2 58.2	2 49.7	5.1	1.0	11.1	2.1	17.1	3.3
52	2 58.0	2 58.5	2 49.9	5.2	1.0	11.2	2.1	17.2	3.3
53	2 58.3	2 58.7	2 50.1	5.3	1.0	11.3	2.2	17.3	3.3
54	2 58.5	2 59.0	2 50.4	5.4	1.0	11.4	2.2	17.4	3.3
55	2 58.8	2 59.2	2 50.6	5.5	1.1	11.5	2.2	17.5	3.4
56	2 59.0	2 59.5	2 50.8	5.6	1.1	11.6	2.2	17.6	3.4
57	2 59.3	2 59.7	2 51.1	5.7	1.1	11.7	2.2	17.7	3.4
58	2 59.5	3 00.0	2 51.3	5.8	1.1	11.8	2.3	17.8	3.4
59	2 59.8	3 00.2	2 51.6	5.9	1.1	11.9	2.3	17.9	3.4
60	3 00.0	3 00.5	2 51.8	6.0	1.2	12.0	2.3	18.0	3.5

12^m INCREMENTS AND CORRECTIONS 13^m

12^m	SUN PLANETS	ARIES	MOON	v or d	Corrⁿ	v or d	Corrⁿ	v or d	Corrⁿ
s	° ′	° ′	° ′	′	′	′	′	′	′
00	3 00·0	3 00·5	2 51·8	0·0	0·0	6·0	1·3	12·0	2·5
01	3 00·3	3 00·7	2 52·0	0·1	0·0	6·1	1·3	12·1	2·5
02	3 00·5	3 01·0	2 52·3	0·2	0·0	6·2	1·3	12·2	2·5
03	3 00·8	3 01·2	2 52·5	0·3	0·1	6·3	1·3	12·3	2·6
04	3 01·0	3 01·5	2 52·8	0·4	0·1	6·4	1·3	12·4	2·6
05	3 01·3	3 01·7	2 53·0	0·5	0·1	6·5	1·4	12·5	2·6
06	3 01·5	3 02·0	2 53·2	0·6	0·1	6·6	1·4	12·6	2·6
07	3 01·8	3 02·2	2 53·5	0·7	0·1	6·7	1·4	12·7	2·6
08	3 02·0	3 02·5	2 53·7	0·8	0·2	6·8	1·4	12·8	2·7
09	3 02·3	3 02·7	2 53·9	0·9	0·2	6·9	1·4	12·9	2·7
10	3 02·5	3 03·0	2 54·2	1·0	0·2	7·0	1·5	13·0	2·7
11	3 02·8	3 03·3	2 54·4	1·1	0·2	7·1	1·5	13·1	2·7
12	3 03·0	3 03·5	2 54·7	1·2	0·3	7·2	1·5	13·2	2·8
13	3 03·3	3 03·8	2 54·9	1·3	0·3	7·3	1·5	13·3	2·8
14	3 03·5	3 04·0	2 55·1	1·4	0·3	7·4	1·5	13·4	2·8
15	3 03·8	3 04·3	2 55·4	1·5	0·3	7·5	1·6	13·5	2·8
16	3 04·0	3 04·5	2 55·6	1·6	0·3	7·6	1·6	13·6	2·8
17	3 04·3	3 04·8	2 55·9	1·7	0·4	7·7	1·6	13·7	2·9
18	3 04·5	3 05·0	2 56·1	1·8	0·4	7·8	1·6	13·8	2·9
19	3 04·8	3 05·3	2 56·3	1·9	0·4	7·9	1·6	13·9	2·9
20	3 05·0	3 05·5	2 56·6	2·0	0·4	8·0	1·7	14·0	2·9
21	3 05·3	3 05·8	2 56·8	2·1	0·4	8·1	1·7	14·1	2·9
22	3 05·5	3 06·0	2 57·0	2·2	0·5	8·2	1·7	14·2	3·0
23	3 05·8	3 06·3	2 57·3	2·3	0·5	8·3	1·7	14·3	3·0
24	3 06·0	3 06·5	2 57·5	2·4	0·5	8·4	1·8	14·4	3·0
25	3 06·3	3 06·8	2 57·8	2·5	0·5	8·5	1·8	14·5	3·0
26	3 06·5	3 07·0	2 58·0	2·6	0·5	8·6	1·8	14·6	3·0
27	3 06·8	3 07·3	2 58·2	2·7	0·6	8·7	1·8	14·7	3·1
28	3 07·0	3 07·5	2 58·5	2·8	0·6	8·8	1·8	14·8	3·1
29	3 07·3	3 07·8	2 58·7	2·9	0·6	8·9	1·9	14·9	3·1
30	3 07·5	3 08·0	2 59·0	3·0	0·6	9·0	1·9	15·0	3·1
31	3 07·8	3 08·3	2 59·2	3·1	0·6	9·1	1·9	15·1	3·1
32	3 08·0	3 08·5	2 59·4	3·2	0·7	9·2	1·9	15·2	3·2
33	3 08·3	3 08·8	2 59·7	3·3	0·7	9·3	1·9	15·3	3·2
34	3 08·5	3 09·0	2 59·9	3·4	0·7	9·4	2·0	15·4	3·2
35	3 08·8	3 09·3	3 00·2	3·5	0·7	9·5	2·0	15·5	3·2
36	3 09·0	3 09·5	3 00·4	3·6	0·8	9·6	2·0	15·6	3·3
37	3 09·3	3 09·8	3 00·6	3·7	0·8	9·7	2·0	15·7	3·3
38	3 09·5	3 10·0	3 00·9	3·8	0·8	9·8	2·0	15·8	3·3
39	3 09·8	3 10·3	3 01·1	3·9	0·8	9·9	2·1	15·9	3·3
40	3 10·0	3 10·5	3 01·3	4·0	0·8	10·0	2·1	16·0	3·3
41	3 10·3	3 10·8	3 01·6	4·1	0·9	10·1	2·1	16·1	3·4
42	3 10·5	3 11·0	3 01·8	4·2	0·9	10·2	2·1	16·2	3·4
43	3 10·8	3 11·3	3 02·1	4·3	0·9	10·3	2·1	16·3	3·4
44	3 11·0	3 11·5	3 02·3	4·4	0·9	10·4	2·2	16·4	3·4
45	3 11·3	3 11·8	3 02·5	4·5	0·9	10·5	2·2	16·5	3·4
46	3 11·5	3 12·0	3 02·8	4·6	1·0	10·6	2·2	16·6	3·5
47	3 11·8	3 12·3	3 03·0	4·7	1·0	10·7	2·2	16·7	3·5
48	3 12·0	3 12·5	3 03·3	4·8	1·0	10·8	2·3	16·8	3·5
49	3 12·3	3 12·8	3 03·5	4·9	1·0	10·9	2·3	16·9	3·5
50	3 12·5	3 13·0	3 03·7	5·0	1·0	11·0	2·3	17·0	3·5
51	3 12·8	3 13·3	3 04·0	5·1	1·1	11·1	2·3	17·1	3·6
52	3 13·0	3 13·5	3 04·2	5·2	1·1	11·2	2·3	17·2	3·6
53	3 13·3	3 13·8	3 04·4	5·3	1·1	11·3	2·4	17·3	3·6
54	3 13·5	3 14·0	3 04·7	5·4	1·1	11·4	2·4	17·4	3·6
55	3 13·8	3 14·3	3 04·9	5·5	1·1	11·5	2·4	17·5	3·6
56	3 14·0	3 14·5	3 05·2	5·6	1·2	11·6	2·4	17·6	3·7
57	3 14·3	3 14·8	3 05·4	5·7	1·2	11·7	2·4	17·7	3·7
58	3 14·5	3 15·0	3 05·6	5·8	1·2	11·8	2·5	17·8	3·7
59	3 14·8	3 15·3	3 05·9	5·9	1·2	11·9	2·5	17·9	3·7
60	3 15·0	3 15·5	3 06·1	6·0	1·3	12·0	2·5	18·0	3·8

13^m	SUN PLANETS	ARIES	MOON	v or d	Corrⁿ	v or d	Corrⁿ	v or d	Corrⁿ
s	° ′	° ′	° ′	′	′	′	′	′	′
00	3 15·0	3 15·5	3 06·1	0·0	0·0	6·0	1·4	12·0	2·7
01	3 15·3	3 15·8	3 06·4	0·1	0·0	6·1	1·4	12·1	2·7
02	3 15·5	3 16·0	3 06·6	0·2	0·0	6·2	1·4	12·2	2·7
03	3 15·8	3 16·3	3 06·8	0·3	0·1	6·3	1·4	12·3	2·8
04	3 16·0	3 16·5	3 07·1	0·4	0·1	6·4	1·4	12·4	2·8
05	3 16·3	3 16·8	3 07·3	0·5	0·1	6·5	1·5	12·5	2·8
06	3 16·5	3 17·0	3 07·5	0·6	0·1	6·6	1·5	12·6	2·8
07	3 16·8	3 17·3	3 07·8	0·7	0·2	6·7	1·5	12·7	2·9
08	3 17·0	3 17·5	3 08·0	0·8	0·2	6·8	1·5	12·8	2·9
09	3 17·3	3 17·8	3 08·3	0·9	0·2	6·9	1·6	12·9	2·9
10	3 17·5	3 18·0	3 08·5	1·0	0·2	7·0	1·6	13·0	2·9
11	3 17·8	3 18·3	3 08·7	1·1	0·2	7·1	1·6	13·1	2·9
12	3 18·0	3 18·5	3 09·0	1·2	0·3	7·2	1·6	13·2	3·0
13	3 18·3	3 18·8	3 09·2	1·3	0·3	7·3	1·6	13·3	3·0
14	3 18·5	3 19·0	3 09·5	1·4	0·3	7·4	1·7	13·4	3·0
15	3 18·8	3 19·3	3 09·7	1·5	0·3	7·5	1·7	13·5	3·0
16	3 19·0	3 19·5	3 09·9	1·6	0·4	7·6	1·7	13·6	3·1
17	3 19·3	3 19·8	3 10·2	1·7	0·4	7·7	1·7	13·7	3·1
18	3 19·5	3 20·0	3 10·4	1·8	0·4	7·8	1·8	13·8	3·1
19	3 19·8	3 20·3	3 10·7	1·9	0·4	7·9	1·8	13·9	3·1
20	3 20·0	3 20·5	3 10·9	2·0	0·5	8·0	1·8	14·0	3·2
21	3 20·3	3 20·8	3 11·1	2·1	0·5	8·1	1·8	14·1	3·2
22	3 20·5	3 21·0	3 11·4	2·2	0·5	8·2	1·8	14·2	3·2
23	3 20·8	3 21·3	3 11·6	2·3	0·5	8·3	1·9	14·3	3·2
24	3 21·0	3 21·6	3 11·8	2·4	0·5	8·4	1·9	14·4	3·2
25	3 21·3	3 21·8	3 12·1	2·5	0·6	8·5	1·9	14·5	3·3
26	3 21·5	3 22·1	3 12·3	2·6	0·6	8·6	1·9	14·6	3·3
27	3 21·8	3 22·3	3 12·6	2·7	0·6	8·7	2·0	14·7	3·3
28	3 22·0	3 22·6	3 12·8	2·8	0·6	8·8	2·0	14·8	3·3
29	3 22·3	3 22·8	3 13·0	2·9	0·7	8·9	2·0	14·9	3·4
30	3 22·5	3 23·1	3 13·3	3·0	0·7	9·0	2·0	15·0	3·4
31	3 22·8	3 23·3	3 13·5	3·1	0·7	9·1	2·0	15·1	3·4
32	3 23·0	3 23·6	3 13·8	3·2	0·7	9·2	2·1	15·2	3·4
33	3 23·3	3 23·8	3 14·0	3·3	0·7	9·3	2·1	15·3	3·4
34	3 23·5	3 24·1	3 14·2	3·4	0·8	9·4	2·1	15·4	3·5
35	3 23·8	3 24·3	3 14·5	3·5	0·8	9·5	2·1	15·5	3·5
36	3 24·0	3 24·6	3 14·7	3·6	0·8	9·6	2·2	15·6	3·5
37	3 24·3	3 24·8	3 14·9	3·7	0·8	9·7	2·2	15·7	3·5
38	3 24·5	3 25·1	3 15·2	3·8	0·9	9·8	2·2	15·8	3·6
39	3 24·8	3 25·3	3 15·4	3·9	0·9	9·9	2·2	15·9	3·6
40	3 25·0	3 25·6	3 15·7	4·0	0·9	10·0	2·3	16·0	3·6
41	3 25·3	3 25·8	3 15·9	4·1	0·9	10·1	2·3	16·1	3·6
42	3 25·5	3 26·1	3 16·1	4·2	0·9	10·2	2·3	16·2	3·6
43	3 25·8	3 26·3	3 16·4	4·3	1·0	10·3	2·3	16·3	3·7
44	3 26·0	3 26·6	3 16·6	4·4	1·0	10·4	2·3	16·4	3·7
45	3 26·3	3 26·8	3 16·9	4·5	1·0	10·5	2·4	16·5	3·7
46	3 26·5	3 27·1	3 17·1	4·6	1·0	10·6	2·4	16·6	3·7
47	3 26·8	3 27·3	3 17·3	4·7	1·1	10·7	2·4	16·7	3·8
48	3 27·0	3 27·6	3 17·6	4·8	1·1	10·8	2·4	16·8	3·8
49	3 27·3	3 27·8	3 17·8	4·9	1·1	10·9	2·5	16·9	3·8
50	3 27·5	3 28·1	3 18·0	5·0	1·1	11·0	2·5	17·0	3·8
51	3 27·8	3 28·3	3 18·3	5·1	1·1	11·1	2·5	17·1	3·8
52	3 28·0	3 28·6	3 18·5	5·2	1·2	11·2	2·5	17·2	3·9
53	3 28·3	3 28·8	3 18·8	5·3	1·2	11·3	2·5	17·3	3·9
54	3 28·5	3 29·1	3 19·0	5·4	1·2	11·4	2·6	17·4	3·9
55	3 28·8	3 29·3	3 19·2	5·5	1·2	11·5	2·6	17·5	3·9
56	3 29·0	3 29·6	3 19·5	5·6	1·3	11·6	2·6	17·6	4·0
57	3 29·3	3 29·8	3 19·7	5·7	1·3	11·7	2·6	17·7	4·0
58	3 29·5	3 30·1	3 20·0	5·8	1·3	11·8	2·7	17·8	4·0
59	3 29·8	3 30·3	3 20·2	5·9	1·3	11·9	2·7	17·9	4·0
60	3 30·0	3 30·6	3 20·4	6·0	1·4	12·0	2·7	18·0	4·1

14ᵐ INCREMENTS AND CORRECTIONS 15ᵐ

14ᵐ	SUN PLANETS	ARIES	MOON	v or d	Corrⁿ	v or d	Corrⁿ	v or d	Corrⁿ
s	° ′	° ′	° ′	′	′	′	′	′	′
00	3 30·0	3 30·6	3 20·4	0·0	0·0	6·0	1·5	12·0	2·9
01	3 30·3	3 30·8	3 20·7	0·1	0·0	6·1	1·5	12·1	2·9
02	3 30·5	3 31·1	3 20·9	0·2	0·0	6·2	1·5	12·2	2·9
03	3 30·8	3 31·3	3 21·1	0·3	0·1	6·3	1·5	12·3	3·0
04	3 31·0	3 31·6	3 21·4	0·4	0·1	6·4	1·5	12·4	3·0
05	3 31·3	3 31·8	3 21·6	0·5	0·1	6·5	1·6	12·5	3·0
06	3 31·5	3 32·1	3 21·9	0·6	0·1	6·6	1·6	12·6	3·0
07	3 31·8	3 32·3	3 22·1	0·7	0·2	6·7	1·6	12·7	3·1
08	3 32·0	3 32·6	3 22·3	0·8	0·2	6·8	1·6	12·8	3·1
09	3 32·3	3 32·8	3 22·6	0·9	0·2	6·9	1·7	12·9	3·1
10	3 32·5	3 33·1	3 22·8	1·0	0·2	7·0	1·7	13·0	3·1
11	3 32·8	3 33·3	3 23·1	1·1	0·3	7·1	1·7	13·1	3·2
12	3 33·0	3 33·6	3 23·3	1·2	0·3	7·2	1·7	13·2	3·2
13	3 33·3	3 33·8	3 23·5	1·3	0·3	7·3	1·8	13·3	3·2
14	3 33·5	3 34·1	3 23·8	1·4	0·3	7·4	1·8	13·4	3·2
15	3 33·8	3 34·3	3 24·0	1·5	0·4	7·5	1·8	13·5	3·3
16	3 34·0	3 34·6	3 24·3	1·6	0·4	7·6	1·8	13·6	3·3
17	3 34·3	3 34·8	3 24·5	1·7	0·4	7·7	1·9	13·7	3·3
18	3 34·5	3 35·1	3 24·7	1·8	0·4	7·8	1·9	13·8	3·3
19	3 34·8	3 35·3	3 25·0	1·9	0·5	7·9	1·9	13·9	3·4
20	3 35·0	3 35·6	3 25·2	2·0	0·5	8·0	1·9	14·0	3·4
21	3 35·3	3 35·8	3 25·4	2·1	0·5	8·1	2·0	14·1	3·4
22	3 35·5	3 36·1	3 25·7	2·2	0·5	8·2	2·0	14·2	3·4
23	3 35·8	3 36·3	3 25·9	2·3	0·6	8·3	2·0	14·3	3·5
24	3 36·0	3 36·6	3 26·2	2·4	0·6	8·4	2·0	14·4	3·5
25	3 36·3	3 36·8	3 26·4	2·5	0·6	8·5	2·1	14·5	3·5
26	3 36·5	3 37·1	3 26·6	2·6	0·6	8·6	2·1	14·6	3·5
27	3 36·8	3 37·3	3 26·9	2·7	0·7	8·7	2·1	14·7	3·6
28	3 37·0	3 37·6	3 27·1	2·8	0·7	8·8	2·1	14·8	3·6
29	3 37·3	3 37·8	3 27·4	2·9	0·7	8·9	2·2	14·9	3·6
30	3 37·5	3 38·1	3 27·6	3·0	0·7	9·0	2·2	15·0	3·6
31	3 37·8	3 38·3	3 27·8	3·1	0·7	9·1	2·2	15·1	3·6
32	3 38·0	3 38·6	3 28·1	3·2	0·8	9·2	2·2	15·2	3·7
33	3 38·3	3 38·8	3 28·3	3·3	0·8	9·3	2·2	15·3	3·7
34	3 38·5	3 39·1	3 28·5	3·4	0·8	9·4	2·3	15·4	3·7
35	3 38·8	3 39·3	3 28·8	3·5	0·8	9·5	2·3	15·5	3·7
36	3 39·0	3 39·6	3 29·0	3·6	0·9	9·6	2·3	15·6	3·8
37	3 39·3	3 39·9	3 29·3	3·7	0·9	9·7	2·3	15·7	3·8
38	3 39·5	3 40·1	3 29·5	3·8	0·9	9·8	2·4	15·8	3·8
39	3 39·8	3 40·4	3 29·7	3·9	0·9	9·9	2·4	15·9	3·8
40	3 40·0	3 40·6	3 30·0	4·0	1·0	10·0	2·4	16·0	3·9
41	3 40·3	3 40·9	3 30·2	4·1	1·0	10·1	2·4	16·1	3·9
42	3 40·5	3 41·1	3 30·5	4·2	1·0	10·2	2·5	16·2	3·9
43	3 40·8	3 41·4	3 30·7	4·3	1·0	10·3	2·5	16·3	3·9
44	3 41·0	3 41·6	3 30·9	4·4	1·1	10·4	2·5	16·4	4·0
45	3 41·3	3 41·9	3 31·2	4·5	1·1	10·5	2·5	16·5	4·0
46	3 41·5	3 42·1	3 31·4	4·6	1·1	10·6	2·6	16·6	4·0
47	3 41·8	3 42·4	3 31·6	4·7	1·1	10·7	2·6	16·7	4·0
48	3 42·0	3 42·6	3 31·9	4·8	1·2	10·8	2·6	16·8	4·1
49	3 42·3	3 42·9	3 32·1	4·9	1·2	10·9	2·6	16·9	4·1
50	3 42·5	3 43·1	3 32·4	5·0	1·2	11·0	2·7	17·0	4·1
51	3 42·8	3 43·4	3 32·6	5·1	1·2	11·1	2·7	17·1	4·1
52	3 43·0	3 43·6	3 32·8	5·2	1·3	11·2	2·7	17·2	4·2
53	3 43·3	3 43·9	3 33·1	5·3	1·3	11·3	2·7	17·3	4·2
54	3 43·5	3 44·1	3 33·3	5·4	1·3	11·4	2·8	17·4	4·2
55	3 43·8	3 44·4	3 33·6	5·5	1·3	11·5	2·8	17·5	4·2
56	3 44·0	3 44·6	3 33·8	5·6	1·4	11·6	2·8	17·6	4·3
57	3 44·3	3 44·9	3 34·0	5·7	1·4	11·7	2·8	17·7	4·3
58	3 44·5	3 45·1	3 34·3	5·8	1·4	11·8	2·9	17·8	4·3
59	3 44·8	3 45·4	3 34·5	5·9	1·4	11·9	2·9	17·9	4·3
60	3 45·0	3 45·6	3 34·8	6·0	1·5	12·0	2·9	18·0	4·4

15ᵐ	SUN PLANETS	ARIES	MOON	v or d	Corrⁿ	v or d	Corrⁿ	v or d	Corrⁿ
s	° ′	° ′	° ′	′	′	′	′	′	′
00	3 45·0	3 45·6	3 34·8	0·0	0·0	6·0	1·6	12·0	3·1
01	3 45·3	3 45·9	3 35·0	0·1	0·0	6·1	1·6	12·1	3·1
02	3 45·5	3 46·1	3 35·2	0·2	0·1	6·2	1·6	12·2	3·2
03	3 45·8	3 46·4	3 35·5	0·3	0·1	6·3	1·6	12·3	3·2
04	3 46·0	3 46·6	3 35·7	0·4	0·1	6·4	1·7	12·4	3·2
05	3 46·3	3 46·9	3 35·9	0·5	0·1	6·5	1·7	12·5	3·2
06	3 46·5	3 47·1	3 36·2	0·6	0·2	6·6	1·7	12·6	3·3
07	3 46·8	3 47·4	3 36·4	0·7	0·2	6·7	1·7	12·7	3·3
08	3 47·0	3 47·6	3 36·7	0·8	0·2	6·8	1·8	12·8	3·3
09	3 47·3	3 47·9	3 36·9	0·9	0·2	6·9	1·8	12·9	3·3
10	3 47·5	3 48·1	3 37·1	1·0	0·3	7·0	1·8	13·0	3·4
11	3 47·8	3 48·4	3 37·4	1·1	0·3	7·1	1·8	13·1	3·4
12	3 48·0	3 48·6	3 37·6	1·2	0·3	7·2	1·9	13·2	3·4
13	3 48·3	3 48·9	3 37·9	1·3	0·3	7·3	1·9	13·3	3·4
14	3 48·5	3 49·1	3 38·1	1·4	0·4	7·4	1·9	13·4	3·5
15	3 48·8	3 49·4	3 38·3	1·5	0·4	7·5	1·9	13·5	3·5
16	3 49·0	3 49·6	3 38·6	1·6	0·4	7·6	2·0	13·6	3·5
17	3 49·3	3 49·9	3 38·8	1·7	0·4	7·7	2·0	13·7	3·5
18	3 49·5	3 50·1	3 39·0	1·8	0·5	7·8	2·0	13·8	3·6
19	3 49·8	3 50·4	3 39·3	1·9	0·5	7·9	2·0	13·9	3·6
20	3 50·0	3 50·6	3 39·5	2·0	0·5	8·0	2·1	14·0	3·6
21	3 50·3	3 50·9	3 39·8	2·1	0·5	8·1	2·1	14·1	3·6
22	3 50·5	3 51·1	3 40·0	2·2	0·6	8·2	2·1	14·2	3·7
23	3 50·8	3 51·4	3 40·2	2·3	0·6	8·3	2·1	14·3	3·7
24	3 51·0	3 51·6	3 40·5	2·4	0·6	8·4	2·2	14·4	3·7
25	3 51·3	3 51·9	3 40·7	2·5	0·6	8·5	2·2	14·5	3·7
26	3 51·5	3 52·1	3 41·0	2·6	0·7	8·6	2·2	14·6	3·8
27	3 51·8	3 52·4	3 41·2	2·7	0·7	8·7	2·2	14·7	3·8
28	3 52·0	3 52·6	3 41·4	2·8	0·7	8·8	2·3	14·8	3·8
29	3 52·3	3 52·9	3 41·7	2·9	0·7	8·9	2·3	14·9	3·8
30	3 52·5	3 53·1	3 41·9	3·0	0·8	9·0	2·3	15·0	3·9
31	3 52·8	3 53·4	3 42·1	3·1	0·8	9·1	2·4	15·1	3·9
32	3 53·0	3 53·6	3 42·4	3·2	0·8	9·2	2·4	15·2	3·9
33	3 53·3	3 53·9	3 42·6	3·3	0·9	9·3	2·4	15·3	4·0
34	3 53·5	3 54·1	3 42·9	3·4	0·9	9·4	2·4	15·4	4·0
35	3 53·8	3 54·4	3 43·1	3·5	0·9	9·5	2·5	15·5	4·0
36	3 54·0	3 54·6	3 43·3	3·6	0·9	9·6	2·5	15·6	4·0
37	3 54·3	3 54·9	3 43·6	3·7	1·0	9·7	2·5	15·7	4·1
38	3 54·5	3 55·1	3 43·8	3·8	1·0	9·8	2·5	15·8	4·1
39	3 54·8	3 55·4	3 44·1	3·9	1·0	9·9	2·6	15·9	4·1
40	3 55·0	3 55·6	3 44·3	4·0	1·0	10·0	2·6	16·0	4·1
41	3 55·3	3 55·9	3 44·5	4·1	1·1	10·1	2·6	16·1	4·2
42	3 55·5	3 56·1	3 44·8	4·2	1·1	10·2	2·6	16·2	4·2
43	3 55·8	3 56·4	3 45·0	4·3	1·1	10·3	2·7	16·3	4·2
44	3 56·0	3 56·6	3 45·2	4·4	1·1	10·4	2·7	16·4	4·2
45	3 56·3	3 56·9	3 45·5	4·5	1·2	10·5	2·7	16·5	4·3
46	3 56·5	3 57·1	3 45·7	4·6	1·2	10·6	2·7	16·6	4·3
47	3 56·8	3 57·4	3 46·0	4·7	1·2	10·7	2·8	16·7	4·3
48	3 57·0	3 57·6	3 46·2	4·8	1·2	10·8	2·8	16·8	4·3
49	3 57·3	3 57·9	3 46·4	4·9	1·3	10·9	2·8	16·9	4·4
50	3 57·5	3 58·2	3 46·7	5·0	1·3	11·0	2·8	17·0	4·4
51	3 57·8	3 58·4	3 46·9	5·1	1·3	11·1	2·9	17·1	4·4
52	3 58·0	3 58·7	3 47·2	5·2	1·3	11·2	2·9	17·2	4·4
53	3 58·3	3 58·9	3 47·4	5·3	1·4	11·3	2·9	17·3	4·5
54	3 58·5	3 59·2	3 47·6	5·4	1·4	11·4	2·9	17·4	4·5
55	3 58·8	3 59·4	3 47·9	5·5	1·4	11·5	3·0	17·5	4·5
56	3 59·0	3 59·7	3 48·1	5·6	1·4	11·6	3·0	17·6	4·5
57	3 59·3	3 59·9	3 48·4	5·7	1·5	11·7	3·0	17·7	4·6
58	3 59·5	4 00·2	3 48·6	5·8	1·5	11·8	3·0	17·8	4·6
59	3 59·8	4 00·4	3 48·8	5·9	1·5	11·9	3·1	17·9	4·6
60	4 00·0	4 00·7	3 49·1	6·0	1·6	12·0	3·1	18·0	4·7

16m INCREMENTS AND CORRECTIONS **17m**

16m	SUN PLANETS	ARIES	MOON	v or d	Corrn	v or d	Corrn	v or d	Corrn
s	° ′	° ′	° ′	′	′	′	′	′	′
00	4 00·0	4 00·7	3 49·1	0·0	0·0	6·0	1·7	12·0	3·3
01	4 00·3	4 00·9	3 49·3	0·1	0·0	6·1	1·7	12·1	3·3
02	4 00·5	4 01·2	3 49·5	0·2	0·1	6·2	1·7	12·2	3·4
03	4 00·8	4 01·4	3 49·8	0·3	0·1	6·3	1·7	12·3	3·4
04	4 01·0	4 01·7	3 50·0	0·4	0·1	6·4	1·8	12·4	3·4
05	4 01·3	4 01·9	3 50·3	0·5	0·1	6·5	1·8	12·5	3·4
06	4 01·5	4 02·2	3 50·5	0·6	0·2	6·6	1·8	12·6	3·5
07	4 01·8	4 02·4	3 50·7	0·7	0·2	6·7	1·8	12·7	3·5
08	4 02·0	4 02·7	3 51·0	0·8	0·2	6·8	1·9	12·8	3·5
09	4 02·3	4 02·9	3 51·2	0·9	0·2	6·9	1·9	12·9	3·5
10	4 02·5	4 03·2	3 51·5	1·0	0·3	7·0	1·9	13·0	3·6
11	4 02·8	4 03·4	3 51·7	1·1	0·3	7·1	2·0	13·1	3·6
12	4 03·0	4 03·7	3 51·9	1·2	0·3	7·2	2·0	13·2	3·6
13	4 03·3	4 03·9	3 52·2	1·3	0·4	7·3	2·0	13·3	3·7
14	4 03·5	4 04·2	3 52·4	1·4	0·4	7·4	2·0	13·4	3·7
15	4 03·8	4 04·4	3 52·6	1·5	0·4	7·5	2·1	13·5	3·7
16	4 04·0	4 04·7	3 52·9	1·6	0·4	7·6	2·1	13·6	3·7
17	4 04·3	4 04·9	3 53·1	1·7	0·5	7·7	2·1	13·7	3·8
18	4 04·5	4 05·2	3 53·4	1·8	0·5	7·8	2·1	13·8	3·8
19	4 04·8	4 05·4	3 53·6	1·9	0·5	7·9	2·2	13·9	3·8
20	4 05·0	4 05·7	3 53·8	2·0	0·6	8·0	2·2	14·0	3·9
21	4 05·3	4 05·9	3 54·1	2·1	0·6	8·1	2·2	14·1	3·9
22	4 05·5	4 06·2	3 54·3	2·2	0·6	8·2	2·3	14·2	3·9
23	4 05·8	4 06·4	3 54·6	2·3	0·6	8·3	2·3	14·3	3·9
24	4 06·0	4 06·7	3 54·8	2·4	0·7	8·4	2·3	14·4	4·0
25	4 06·3	4 06·9	3 55·0	2·5	0·7	8·5	2·3	14·5	4·0
26	4 06·5	4 07·2	3 55·3	2·6	0·7	8·6	2·4	14·6	4·0
27	4 06·8	4 07·4	3 55·5	2·7	0·7	8·7	2·4	14·7	4·0
28	4 07·0	4 07·7	3 55·7	2·8	0·8	8·8	2·4	14·8	4·1
29	4 07·3	4 07·9	3 56·0	2·9	0·8	8·9	2·4	14·9	4·1
30	4 07·5	4 08·2	3 56·2	3·0	0·8	9·0	2·5	15·0	4·1
31	4 07·8	4 08·4	3 56·5	3·1	0·9	9·1	2·5	15·1	4·2
32	4 08·0	4 08·7	3 56·7	3·2	0·9	9·2	2·5	15·2	4·2
33	4 08·3	4 08·9	3 56·9	3·3	0·9	9·3	2·6	15·3	4·2
34	4 08·5	4 09·2	3 57·2	3·4	0·9	9·4	2·6	15·4	4·2
35	4 08·8	4 09·4	3 57·4	3·5	1·0	9·5	2·6	15·5	4·3
36	4 09·0	4 09·7	3 57·7	3·6	1·0	9·6	2·6	15·6	4·3
37	4 09·3	4 09·9	3 57·9	3·7	1·0	9·7	2·7	15·7	4·3
38	4 09·5	4 10·2	3 58·1	3·8	1·0	9·8	2·7	15·8	4·3
39	4 09·8	4 10·4	3 58·4	3·9	1·1	9·9	2·7	15·9	4·4
40	4 10·0	4 10·7	3 58·6	4·0	1·1	10·0	2·8	16·0	4·4
41	4 10·3	4 10·9	3 58·8	4·1	1·1	10·1	2·8	16·1	4·4
42	4 10·5	4 11·2	3 59·1	4·2	1·2	10·2	2·8	16·2	4·5
43	4 10·8	4 11·4	3 59·3	4·3	1·2	10·3	2·8	16·3	4·5
44	4 11·0	4 11·7	3 59·6	4·4	1·2	10·4	2·9	16·4	4·5
45	4 11·3	4 11·9	3 59·8	4·5	1·2	10·5	2·9	16·5	4·5
46	4 11·5	4 12·2	4 00·0	4·6	1·3	10·6	2·9	16·6	4·6
47	4 11·8	4 12·4	4 00·3	4·7	1·3	10·7	2·9	16·7	4·6
48	4 12·0	4 12·7	4 00·5	4·8	1·3	10·8	3·0	16·8	4·6
49	4 12·3	4 12·9	4 00·8	4·9	1·3	10·9	3·0	16·9	4·6
50	4 12·5	4 13·2	4 01·0	5·0	1·4	11·0	3·0	17·0	4·7
51	4 12·8	4 13·4	4 01·2	5·1	1·4	11·1	3·1	17·1	4·7
52	4 13·0	4 13·7	4 01·5	5·2	1·4	11·2	3·1	17·2	4·7
53	4 13·3	4 13·9	4 01·7	5·3	1·5	11·3	3·1	17·3	4·8
54	4 13·5	4 14·2	4 02·0	5·4	1·5	11·4	3·1	17·4	4·8
55	4 13·8	4 14·4	4 02·2	5·5	1·5	11·5	3·2	17·5	4·8
56	4 14·0	4 14·7	4 02·4	5·6	1·5	11·6	3·2	17·6	4·8
57	4 14·3	4 14·9	4 02·7	5·7	1·6	11·7	3·2	17·7	4·9
58	4 14·5	4 15·2	4 02·9	5·8	1·6	11·8	3·2	17·8	4·9
59	4 14·8	4 15·4	4 03·1	5·9	1·6	11·9	3·3	17·9	4·9
60	4 15·0	4 15·7	4 03·4	6·0	1·7	12·0	3·3	18·0	5·0

17m	SUN PLANETS	ARIES	MOON	v or d	Corrn	v or d	Corrn	v or d	Corrn
s	° ′	° ′	° ′	′	′	′	′	′	′
00	4 15·0	4 15·7	4 03·4	0·0	0·0	6·0	1·8	12·0	3·5
01	4 15·3	4 15·9	4 03·6	0·1	0·0	6·1	1·8	12·1	3·5
02	4 15·5	4 16·2	4 03·9	0·2	0·1	6·2	1·8	12·2	3·6
03	4 15·8	4 16·5	4 04·1	0·3	0·1	6·3	1·8	12·3	3·6
04	4 16·0	4 16·7	4 04·3	0·4	0·1	6·4	1·9	12·4	3·6
05	4 16·3	4 17·0	4 04·6	0·5	0·1	6·5	1·9	12·5	3·6
06	4 16·5	4 17·2	4 04·8	0·6	0·2	6·6	1·9	12·6	3·7
07	4 16·8	4 17·5	4 05·1	0·7	0·2	6·7	2·0	12·7	3·7
08	4 17·0	4 17·7	4 05·3	0·8	0·2	6·8	2·0	12·8	3·7
09	4 17·3	4 18·0	4 05·5	0·9	0·3	6·9	2·0	12·9	3·8
10	4 17·5	4 18·2	4 05·8	1·0	0·3	7·0	2·0	13·0	3·8
11	4 17·8	4 18·5	4 06·0	1·1	0·3	7·1	2·1	13·1	3·8
12	4 18·0	4 18·7	4 06·2	1·2	0·4	7·2	2·1	13·2	3·9
13	4 18·3	4 19·0	4 06·5	1·3	0·4	7·3	2·1	13·3	3·9
14	4 18·5	4 19·2	4 06·7	1·4	0·4	7·4	2·2	13·4	3·9
15	4 18·8	4 19·5	4 07·0	1·5	0·4	7·5	2·2	13·5	3·9
16	4 19·0	4 19·7	4 07·2	1·6	0·5	7·6	2·2	13·6	4·0
17	4 19·3	4 20·0	4 07·4	1·7	0·5	7·7	2·2	13·7	4·0
18	4 19·5	4 20·2	4 07·7	1·8	0·5	7·8	2·3	13·8	4·0
19	4 19·8	4 20·5	4 07·9	1·9	0·6	7·9	2·3	13·9	4·1
20	4 20·0	4 20·7	4 08·2	2·0	0·6	8·0	2·3	14·0	4·1
21	4 20·3	4 21·0	4 08·4	2·1	0·6	8·1	2·4	14·1	4·1
22	4 20·5	4 21·2	4 08·6	2·2	0·6	8·2	2·4	14·2	4·1
23	4 20·8	4 21·5	4 08·9	2·3	0·7	8·3	2·4	14·3	4·2
24	4 21·0	4 21·7	4 09·1	2·4	0·7	8·4	2·5	14·4	4·2
25	4 21·3	4 22·0	4 09·3	2·5	0·7	8·5	2·5	14·5	4·2
26	4 21·5	4 22·2	4 09·6	2·6	0·8	8·6	2·5	14·6	4·3
27	4 21·8	4 22·5	4 09·8	2·7	0·8	8·7	2·5	14·7	4·3
28	4 22·0	4 22·7	4 10·1	2·8	0·8	8·8	2·6	14·8	4·3
29	4 22·3	4 23·0	4 10·3	2·9	0·8	8·9	2·6	14·9	4·3
30	4 22·5	4 23·2	4 10·5	3·0	0·9	9·0	2·6	15·0	4·4
31	4 22·8	4 23·5	4 10·8	3·1	0·9	9·1	2·7	15·1	4·4
32	4 23·0	4 23·7	4 11·0	3·2	0·9	9·2	2·7	15·2	4·4
33	4 23·3	4 24·0	4 11·3	3·3	1·0	9·3	2·7	15·3	4·5
34	4 23·5	4 24·2	4 11·5	3·4	1·0	9·4	2·7	15·4	4·5
35	4 23·8	4 24·5	4 11·7	3·5	1·0	9·5	2·8	15·5	4·5
36	4 24·0	4 24·7	4 12·0	3·6	1·1	9·6	2·8	15·6	4·6
37	4 24·3	4 25·0	4 12·2	3·7	1·1	9·7	2·8	15·7	4·6
38	4 24·5	4 25·2	4 12·5	3·8	1·1	9·8	2·9	15·8	4·6
39	4 24·8	4 25·5	4 12·7	3·9	1·1	9·9	2·9	15·9	4·6
40	4 25·0	4 25·7	4 12·9	4·0	1·2	10·0	2·9	16·0	4·7
41	4 25·3	4 26·0	4 13·2	4·1	1·2	10·1	2·9	16·1	4·7
42	4 25·5	4 26·2	4 13·4	4·2	1·2	10·2	3·0	16·2	4·7
43	4 25·8	4 26·5	4 13·6	4·3	1·3	10·3	3·0	16·3	4·8
44	4 26·0	4 26·7	4 13·9	4·4	1·3	10·4	3·0	16·4	4·8
45	4 26·3	4 27·0	4 14·1	4·5	1·3	10·5	3·1	16·5	4·8
46	4 26·5	4 27·2	4 14·4	4·6	1·3	10·6	3·1	16·6	4·8
47	4 26·8	4 27·5	4 14·6	4·7	1·4	10·7	3·1	16·7	4·9
48	4 27·0	4 27·7	4 14·8	4·8	1·4	10·8	3·2	16·8	4·9
49	4 27·3	4 28·0	4 15·1	4·9	1·4	10·9	3·2	16·9	4·9
50	4 27·5	4 28·2	4 15·3	5·0	1·5	11·0	3·2	17·0	5·0
51	4 27·8	4 28·5	4 15·6	5·1	1·5	11·1	3·2	17·1	5·0
52	4 28·0	4 28·7	4 15·8	5·2	1·5	11·2	3·3	17·2	5·0
53	4 28·3	4 29·0	4 16·0	5·3	1·5	11·3	3·3	17·3	5·0
54	4 28·5	4 29·2	4 16·3	5·4	1·6	11·4	3·3	17·4	5·1
55	4 28·8	4 29·5	4 16·5	5·5	1·6	11·5	3·4	17·5	5·1
56	4 29·0	4 29·7	4 16·7	5·6	1·6	11·6	3·4	17·6	5·1
57	4 29·3	4 30·0	4 17·0	5·7	1·7	11·7	3·4	17·7	5·2
58	4 29·5	4 30·2	4 17·2	5·8	1·7	11·8	3·4	17·8	5·2
59	4 29·8	4 30·5	4 17·5	5·9	1·7	11·9	3·5	17·9	5·2
60	4 30·0	4 30·7	4 17·7	6·0	1·8	12·0	3·5	18·0	5·3

INCREMENTS AND CORRECTIONS

18m

18m	SUN PLANETS	ARIES	MOON	v or d	Corrn	v or d	Corrn	v or d	Corrn
s	° ′	° ′	° ′	′	′	′	′	′	′
00	4 30·0	4 30·7	4 17·7	0·0	0·0	6·0	1·9	12·0	3·7
01	4 30·3	4 31·0	4 17·9	0·1	0·0	6·1	1·9	12·1	3·7
02	4 30·5	4 31·2	4 18·2	0·2	0·1	6·2	1·9	12·2	3·8
03	4 30·8	4 31·5	4 18·4	0·3	0·1	6·3	1·9	12·3	3·8
04	4 31·0	4 31·7	4 18·7	0·4	0·1	6·4	2·0	12·4	3·8
05	4 31·3	4 32·0	4 18·9	0·5	0·2	6·5	2·0	12·5	3·9
06	4 31·5	4 32·2	4 19·1	0·6	0·2	6·6	2·0	12·6	3·9
07	4 31·8	4 32·5	4 19·4	0·7	0·2	6·7	2·1	12·7	3·9
08	4 32·0	4 32·7	4 19·6	0·8	0·2	6·8	2·1	12·8	3·9
09	4 32·3	4 33·0	4 19·8	0·9	0·3	6·9	2·1	12·9	4·0
10	4 32·5	4 33·2	4 20·1	1·0	0·3	7·0	2·2	13·0	4·0
11	4 32·8	4 33·5	4 20·3	1·1	0·3	7·1	2·2	13·1	4·0
12	4 33·0	4 33·7	4 20·6	1·2	0·4	7·2	2·2	13·2	4·1
13	4 33·3	4 34·0	4 20·8	1·3	0·4	7·3	2·3	13·3	4·1
14	4 33·5	4 34·2	4 21·0	1·4	0·4	7·4	2·3	13·4	4·1
15	4 33·8	4 34·5	4 21·3	1·5	0·5	7·5	2·3	13·5	4·2
16	4 34·0	4 34·8	4 21·5	1·6	0·5	7·6	2·3	13·6	4·2
17	4 34·3	4 35·0	4 21·8	1·7	0·5	7·7	2·4	13·7	4·2
18	4 34·5	4 35·3	4 22·0	1·8	0·6	7·8	2·4	13·8	4·3
19	4 34·8	4 35·5	4 22·2	1·9	0·6	7·9	2·4	13·9	4·3
20	4 35·0	4 35·8	4 22·5	2·0	0·6	8·0	2·5	14·0	4·3
21	4 35·3	4 36·0	4 22·7	2·1	0·6	8·1	2·5	14·1	4·3
22	4 35·5	4 36·3	4 22·9	2·2	0·7	8·2	2·5	14·2	4·4
23	4 35·8	4 36·5	4 23·2	2·3	0·7	8·3	2·6	14·3	4·4
24	4 36·0	4 36·8	4 23·4	2·4	0·7	8·4	2·6	14·4	4·4
25	4 36·3	4 37·0	4 23·7	2·5	0·8	8·5	2·6	14·5	4·5
26	4 36·5	4 37·3	4 23·9	2·6	0·8	8·6	2·7	14·6	4·5
27	4 36·8	4 37·5	4 24·1	2·7	0·8	8·7	2·7	14·7	4·5
28	4 37·0	4 37·8	4 24·4	2·8	0·9	8·8	2·7	14·8	4·6
29	4 37·3	4 38·0	4 24·6	2·9	0·9	8·9	2·7	14·9	4·6
30	4 37·5	4 38·3	4 24·9	3·0	0·9	9·0	2·8	15·0	4·6
31	4 37·8	4 38·5	4 25·1	3·1	1·0	9·1	2·8	15·1	4·7
32	4 38·0	4 38·8	4 25·3	3·2	1·0	9·2	2·8	15·2	4·7
33	4 38·3	4 39·0	4 25·6	3·3	1·0	9·3	2·9	15·3	4·7
34	4 38·5	4 39·3	4 25·8	3·4	1·0	9·4	2·9	15·4	4·7
35	4 38·8	4 39·5	4 26·1	3·5	1·1	9·5	2·9	15·5	4·8
36	4 39·0	4 39·8	4 26·3	3·6	1·1	9·6	3·0	15·6	4·8
37	4 39·3	4 40·0	4 26·5	3·7	1·1	9·7	3·0	15·7	4·8
38	4 39·5	4 40·3	4 26·8	3·8	1·2	9·8	3·0	15·8	4·9
39	4 39·8	4 40·5	4 27·0	3·9	1·2	9·9	3·1	15·9	4·9
40	4 40·0	4 40·8	4 27·2	4·0	1·2	10·0	3·1	16·0	4·9
41	4 40·3	4 41·0	4 27·5	4·1	1·3	10·1	3·1	16·1	5·0
42	4 40·5	4 41·3	4 27·7	4·2	1·3	10·2	3·1	16·2	5·0
43	4 40·8	4 41·5	4 28·0	4·3	1·3	10·3	3·2	16·3	5·0
44	4 41·0	4 41·8	4 28·2	4·4	1·4	10·4	3·2	16·4	5·1
45	4 41·3	4 42·0	4 28·4	4·5	1·4	10·5	3·2	16·5	5·1
46	4 41·5	4 42·3	4 28·7	4·6	1·4	10·6	3·3	16·6	5·1
47	4 41·8	4 42·5	4 28·9	4·7	1·4	10·7	3·3	16·7	5·1
48	4 42·0	4 42·8	4 29·2	4·8	1·5	10·8	3·3	16·8	5·2
49	4 42·3	4 43·0	4 29·4	4·9	1·5	10·9	3·4	16·9	5·2
50	4 42·5	4 43·3	4 29·6	5·0	1·5	11·0	3·4	17·0	5·2
51	4 42·8	4 43·5	4 29·9	5·1	1·6	11·1	3·4	17·1	5·3
52	4 43·0	4 43·8	4 30·1	5·2	1·6	11·2	3·5	17·2	5·3
53	4 43·3	4 44·0	4 30·3	5·3	1·6	11·3	3·5	17·3	5·3
54	4 43·5	4 44·3	4 30·6	5·4	1·7	11·4	3·5	17·4	5·4
55	4 43·8	4 44·5	4 30·8	5·5	1·7	11·5	3·5	17·5	5·4
56	4 44·0	4 44·8	4 31·1	5·6	1·7	11·6	3·6	17·6	5·4
57	4 44·3	4 45·0	4 31·3	5·7	1·8	11·7	3·6	17·7	5·5
58	4 44·5	4 45·3	4 31·5	5·8	1·8	11·8	3·6	17·8	5·5
59	4 44·8	4 45·5	4 31·8	5·9	1·8	11·9	3·7	17·9	5·5
60	4 45·0	4 45·8	4 32·0	6·0	1·9	12·0	3·7	18·0	5·6

19m

19m	SUN PLANETS	ARIES	MOON	v or d	Corrn	v or d	Corrn	v or d	Corrn
s	° ′	° ′	° ′	′	′	′	′	′	′
00	4 45·0	4 45·8	4 32·0	0·0	0·0	6·0	2·0	12·0	3·9
01	4 45·3	4 46·0	4 32·3	0·1	0·0	6·1	2·0	12·1	3·9
02	4 45·5	4 46·3	4 32·5	0·2	0·1	6·2	2·0	12·2	4·0
03	4 45·8	4 46·5	4 32·7	0·3	0·1	6·3	2·0	12·3	4·0
04	4 46·0	4 46·8	4 33·0	0·4	0·1	6·4	2·1	12·4	4·0
05	4 46·3	4 47·0	4 33·2	0·5	0·2	6·5	2·1	12·5	4·1
06	4 46·5	4 47·3	4 33·4	0·6	0·2	6·6	2·1	12·6	4·1
07	4 46·8	4 47·5	4 33·7	0·7	0·2	6·7	2·2	12·7	4·1
08	4 47·0	4 47·8	4 33·9	0·8	0·3	6·8	2·2	12·8	4·2
09	4 47·3	4 48·0	4 34·2	0·9	0·3	6·9	2·2	12·9	4·2
10	4 47·5	4 48·3	4 34·4	1·0	0·3	7·0	2·3	13·0	4·2
11	4 47·8	4 48·5	4 34·6	1·1	0·4	7·1	2·3	13·1	4·3
12	4 48·0	4 48·8	4 34·9	1·2	0·4	7·2	2·3	13·2	4·3
13	4 48·3	4 49·0	4 35·1	1·3	0·4	7·3	2·4	13·3	4·3
14	4 48·5	4 49·3	4 35·4	1·4	0·5	7·4	2·4	13·4	4·4
15	4 48·8	4 49·5	4 35·6	1·5	0·5	7·5	2·4	13·5	4·4
16	4 49·0	4 49·8	4 35·8	1·6	0·5	7·6	2·5	13·6	4·4
17	4 49·3	4 50·0	4 36·1	1·7	0·6	7·7	2·5	13·7	4·5
18	4 49·5	4 50·3	4 36·3	1·8	0·6	7·8	2·5	13·8	4·5
19	4 49·8	4 50·5	4 36·6	1·9	0·6	7·9	2·6	13·9	4·5
20	4 50·0	4 50·8	4 36·8	2·0	0·7	8·0	2·6	14·0	4·6
21	4 50·3	4 51·0	4 37·0	2·1	0·7	8·1	2·6	14·1	4·6
22	4 50·5	4 51·3	4 37·3	2·2	0·7	8·2	2·7	14·2	4·6
23	4 50·8	4 51·5	4 37·5	2·3	0·7	8·3	2·7	14·3	4·6
24	4 51·0	4 51·8	4 37·7	2·4	0·8	8·4	2·7	14·4	4·7
25	4 51·3	4 52·0	4 38·0	2·5	0·8	8·5	2·8	14·5	4·7
26	4 51·5	4 52·3	4 38·2	2·6	0·8	8·6	2·8	14·6	4·7
27	4 51·8	4 52·5	4 38·5	2·7	0·9	8·7	2·8	14·7	4·8
28	4 52·0	4 52·8	4 38·7	2·8	0·9	8·8	2·9	14·8	4·8
29	4 52·3	4 53·1	4 38·9	2·9	0·9	8·9	2·9	14·9	4·8
30	4 52·5	4 53·3	4 39·2	3·0	1·0	9·0	2·9	15·0	4·9
31	4 52·8	4 53·6	4 39·4	3·1	1·0	9·1	3·0	15·1	4·9
32	4 53·0	4 53·8	4 39·7	3·2	1·0	9·2	3·0	15·2	4·9
33	4 53·3	4 54·1	4 39·9	3·3	1·1	9·3	3·0	15·3	5·0
34	4 53·5	4 54·3	4 40·1	3·4	1·1	9·4	3·1	15·4	5·0
35	4 53·8	4 54·6	4 40·4	3·5	1·1	9·5	3·1	15·5	5·0
36	4 54·0	4 54·8	4 40·6	3·6	1·2	9·6	3·1	15·6	5·1
37	4 54·3	4 55·1	4 40·8	3·7	1·2	9·7	3·2	15·7	5·1
38	4 54·5	4 55·3	4 41·1	3·8	1·2	9·8	3·2	15·8	5·1
39	4 54·8	4 55·6	4 41·3	3·9	1·3	9·9	3·2	15·9	5·2
40	4 55·0	4 55·8	4 41·6	4·0	1·3	10·0	3·3	16·0	5·2
41	4 55·3	4 56·1	4 41·8	4·1	1·3	10·1	3·3	16·1	5·2
42	4 55·5	4 56·3	4 42·0	4·2	1·4	10·2	3·3	16·2	5·3
43	4 55·8	4 56·6	4 42·3	4·3	1·4	10·3	3·3	16·3	5·3
44	4 56·0	4 56·8	4 42·5	4·4	1·4	10·4	3·4	16·4	5·3
45	4 56·3	4 57·1	4 42·8	4·5	1·5	10·5	3·4	16·5	5·4
46	4 56·5	4 57·3	4 43·0	4·6	1·5	10·6	3·4	16·6	5·4
47	4 56·8	4 57·6	4 43·2	4·7	1·5	10·7	3·5	16·7	5·4
48	4 57·0	4 57·8	4 43·5	4·8	1·6	10·8	3·5	16·8	5·5
49	4 57·3	4 58·1	4 43·7	4·9	1·6	10·9	3·5	16·9	5·5
50	4 57·5	4 58·3	4 43·9	5·0	1·6	11·0	3·6	17·0	5·5
51	4 57·8	4 58·6	4 44·2	5·1	1·7	11·1	3·6	17·1	5·6
52	4 58·0	4 58·8	4 44·4	5·2	1·7	11·2	3·6	17·2	5·6
53	4 58·3	4 59·1	4 44·7	5·3	1·7	11·3	3·7	17·3	5·6
54	4 58·5	4 59·3	4 44·9	5·4	1·8	11·4	3·7	17·4	5·7
55	4 58·8	4 59·6	4 45·1	5·5	1·8	11·5	3·7	17·5	5·7
56	4 59·0	4 59·8	4 45·4	5·6	1·8	11·6	3·8	17·6	5·7
57	4 59·3	5 00·1	4 45·6	5·7	1·9	11·7	3·8	17·7	5·8
58	4 59·5	5 00·3	4 45·9	5·8	1·9	11·8	3·8	17·8	5·8
59	4 59·8	5 00·6	4 46·1	5·9	1·9	11·9	3·9	17·9	5·8
60	5 00·0	5 00·8	4 46·3	6·0	2·0	12·0	3·9	18·0	5·9

20m INCREMENTS AND CORRECTIONS 21m

20m	SUN PLANETS	ARIES	MOON	v or d	Corrn	v or d	Corrn	v or d	Corrn
s	° ′	° ′	° ′	′	′	′	′	′	′
00	5 00·0	5 00·8	4 46·3	0·0	0·0	6·0	2·1	12·0	4·1
01	5 00·3	5 01·1	4 46·6	0·1	0·0	6·1	2·1	12·1	4·1
02	5 00·5	5 01·3	4 46·8	0·2	0·1	6·2	2·1	12·2	4·2
03	5 00·8	5 01·6	4 47·0	0·3	0·1	6·3	2·2	12·3	4·2
04	5 01·0	5 01·8	4 47·3	0·4	0·1	6·4	2·2	12·4	4·2
05	5 01·3	5 02·1	4 47·5	0·5	0·2	6·5	2·2	12·5	4·3
06	5 01·5	5 02·3	4 47·8	0·6	0·2	6·6	2·3	12·6	4·3
07	5 01·8	5 02·6	4 48·0	0·7	0·2	6·7	2·3	12·7	4·3
08	5 02·0	5 02·8	4 48·2	0·8	0·3	6·8	2·3	12·8	4·4
09	5 02·3	5 03·1	4 48·5	0·9	0·3	6·9	2·4	12·9	4·4
10	5 02·5	5 03·3	4 48·7	1·0	0·3	7·0	2·4	13·0	4·4
11	5 02·8	5 03·6	4 49·0	1·1	0·4	7·1	2·4	13·1	4·5
12	5 03·0	5 03·8	4 49·2	1·2	0·4	7·2	2·5	13·2	4·5
13	5 03·3	5 04·1	4 49·4	1·3	0·4	7·3	2·5	13·3	4·5
14	5 03·5	5 04·3	4 49·7	1·4	0·5	7·4	2·5	13·4	4·6
15	5 03·8	5 04·6	4 49·9	1·5	0·5	7·5	2·6	13·5	4·6
16	5 04·0	5 04·8	4 50·2	1·6	0·5	7·6	2·6	13·6	4·6
17	5 04·3	5 05·1	4 50·4	1·7	0·6	7·7	2·6	13·7	4·7
18	5 04·5	5 05·3	4 50·6	1·8	0·6	7·8	2·7	13·8	4·7
19	5 04·8	5 05·6	4 50·9	1·9	0·6	7·9	2·7	13·9	4·7
20	5 05·0	5 05·8	4 51·1	2·0	0·7	8·0	2·7	14·0	4·8
21	5 05·3	5 06·1	4 51·3	2·1	0·7	8·1	2·8	14·1	4·8
22	5 05·5	5 06·3	4 51·6	2·2	0·8	8·2	2·8	14·2	4·9
23	5 05·8	5 06·6	4 51·8	2·3	0·8	8·3	2·8	14·3	4·9
24	5 06·0	5 06·8	4 52·1	2·4	0·8	8·4	2·9	14·4	4·9
25	5 06·3	5 07·1	4 52·3	2·5	0·9	8·5	2·9	14·5	5·0
26	5 06·5	5 07·3	4 52·5	2·6	0·9	8·6	2·9	14·6	5·0
27	5 06·8	5 07·6	4 52·8	2·7	0·9	8·7	3·0	14·7	5·0
28	5 07·0	5 07·8	4 53·0	2·8	1·0	8·8	3·0	14·8	5·1
29	5 07·3	5 08·1	4 53·3	2·9	1·0	8·9	3·0	14·9	5·1
30	5 07·5	5 08·3	4 53·5	3·0	1·0	9·0	3·1	15·0	5·1
31	5 07·8	5 08·6	4 53·7	3·1	1·1	9·1	3·1	15·1	5·2
32	5 08·0	5 08·8	4 54·0	3·2	1·1	9·2	3·1	15·2	5·2
33	5 08·3	5 09·1	4 54·2	3·3	1·1	9·3	3·2	15·3	5·2
34	5 08·5	5 09·3	4 54·4	3·4	1·2	9·4	3·2	15·4	5·3
35	5 08·8	5 09·6	4 54·7	3·5	1·2	9·5	3·2	15·5	5·3
36	5 09·0	5 09·8	4 54·9	3·6	1·2	9·6	3·3	15·6	5·3
37	5 09·3	5 10·1	4 55·2	3·7	1·3	9·7	3·3	15·7	5·4
38	5 09·5	5 10·3	4 55·4	3·8	1·3	9·8	3·3	15·8	5·4
39	5 09·8	5 10·6	4 55·6	3·9	1·3	9·9	3·4	15·9	5·4
40	5 10·0	5 10·8	4 55·9	4·0	1·4	10·0	3·4	16·0	5·5
41	5 10·3	5 11·1	4 56·1	4·1	1·4	10·1	3·5	16·1	5·5
42	5 10·5	5 11·4	4 56·4	4·2	1·4	10·2	3·5	16·2	5·5
43	5 10·8	5 11·6	4 56·6	4·3	1·5	10·3	3·5	16·3	5·6
44	5 11·0	5 11·9	4 56·8	4·4	1·5	10·4	3·6	16·4	5·6
45	5 11·3	5 12·1	4 57·1	4·5	1·5	10·5	3·6	16·5	5·6
46	5 11·5	5 12·4	4 57·3	4·6	1·6	10·6	3·6	16·6	5·7
47	5 11·8	5 12·6	4 57·5	4·7	1·6	10·7	3·7	16·7	5·7
48	5 12·0	5 12·9	4 57·8	4·8	1·6	10·8	3·7	16·8	5·7
49	5 12·3	5 13·1	4 58·0	4·9	1·7	10·9	3·7	16·9	5·8
50	5 12·5	5 13·4	4 58·3	5·0	1·7	11·0	3·8	17·0	5·8
51	5 12·8	5 13·6	4 58·5	5·1	1·7	11·1	3·8	17·1	5·8
52	5 13·0	5 13·9	4 58·7	5·2	1·8	11·2	3·8	17·2	5·9
53	5 13·3	5 14·1	4 59·0	5·3	1·8	11·3	3·9	17·3	5·9
54	5 13·5	5 14·4	4 59·2	5·4	1·8	11·4	3·9	17·4	5·9
55	5 13·8	5 14·6	4 59·5	5·5	1·9	11·5	3·9	17·5	6·0
56	5 14·0	5 14·9	4 59·7	5·6	1·9	11·6	4·0	17·6	6·0
57	5 14·3	5 15·1	4 59·9	5·7	1·9	11·7	4·0	17·7	6·0
58	5 14·5	5 15·4	5 00·2	5·8	2·0	11·8	4·0	17·8	6·1
59	5 14·8	5 15·6	5 00·4	5·9	2·0	11·9	4·1	17·9	6·1
60	5 15·0	5 15·9	5 00·7	6·0	2·1	12·0	4·1	18·0	6·2

21m	SUN PLANETS	ARIES	MOON	v or d	Corrn	v or d	Corrn	v or d	Corrn
s	° ′	° ′	° ′	′	′	′	′	′	′
00	5 15·0	5 15·9	5 00·7	0·0	0·0	6·0	2·2	12·0	4·3
01	5 15·3	5 16·1	5 00·9	0·1	0·0	6·1	2·2	12·1	4·3
02	5 15·5	5 16·4	5 01·1	0·2	0·1	6·2	2·2	12·2	4·4
03	5 15·8	5 16·6	5 01·4	0·3	0·1	6·3	2·3	12·3	4·4
04	5 16·0	5 16·9	5 01·6	0·4	0·1	6·4	2·3	12·4	4·4
05	5 16·3	5 17·1	5 01·8	0·5	0·2	6·5	2·3	12·5	4·5
06	5 16·5	5 17·4	5 02·1	0·6	0·2	6·6	2·4	12·6	4·5
07	5 16·8	5 17·6	5 02·3	0·7	0·3	6·7	2·4	12·7	4·6
08	5 17·0	5 17·9	5 02·6	0·8	0·3	6·8	2·4	12·8	4·6
09	5 17·3	5 18·1	5 02·8	0·9	0·3	6·9	2·5	12·9	4·6
10	5 17·5	5 18·4	5 03·0	1·0	0·4	7·0	2·5	13·0	4·7
11	5 17·8	5 18·6	5 03·3	1·1	0·4	7·1	2·5	13·1	4·7
12	5 18·0	5 18·9	5 03·5	1·2	0·4	7·2	2·6	13·2	4·7
13	5 18·3	5 19·1	5 03·8	1·3	0·5	7·3	2·6	13·3	4·8
14	5 18·5	5 19·4	5 04·0	1·4	0·5	7·4	2·7	13·4	4·8
15	5 18·8	5 19·6	5 04·2	1·5	0·5	7·5	2·7	13·5	4·8
16	5 19·0	5 19·9	5 04·5	1·6	0·6	7·6	2·7	13·6	4·9
17	5 19·3	5 20·1	5 04·7	1·7	0·6	7·7	2·8	13·7	4·9
18	5 19·5	5 20·4	5 04·9	1·8	0·6	7·8	2·8	13·8	4·9
19	5 19·8	5 20·6	5 05·2	1·9	0·7	7·9	2·8	13·9	5·0
20	5 20·0	5 20·9	5 05·4	2·0	0·7	8·0	2·9	14·0	5·0
21	5 20·3	5 21·1	5 05·7	2·1	0·8	8·1	2·9	14·1	5·1
22	5 20·5	5 21·4	5 05·9	2·2	0·8	8·2	2·9	14·2	5·1
23	5 20·8	5 21·6	5 06·1	2·3	0·8	8·3	3·0	14·3	5·1
24	5 21·0	5 21·9	5 06·4	2·4	0·9	8·4	3·0	14·4	5·2
25	5 21·3	5 22·1	5 06·6	2·5	0·9	8·5	3·0	14·5	5·2
26	5 21·5	5 22·4	5 06·9	2·6	0·9	8·6	3·1	14·6	5·2
27	5 21·8	5 22·6	5 07·1	2·7	1·0	8·7	3·1	14·7	5·3
28	5 22·0	5 22·9	5 07·3	2·8	1·0	8·8	3·2	14·8	5·3
29	5 22·3	5 23·1	5 07·6	2·9	1·0	8·9	3·2	14·9	5·3
30	5 22·5	5 23·4	5 07·8	3·0	1·1	9·0	3·2	15·0	5·4
31	5 22·8	5 23·6	5 08·0	3·1	1·1	9·1	3·3	15·1	5·4
32	5 23·0	5 23·9	5 08·3	3·2	1·1	9·2	3·3	15·2	5·4
33	5 23·3	5 24·1	5 08·5	3·3	1·2	9·3	3·3	15·3	5·5
34	5 23·5	5 24·4	5 08·8	3·4	1·2	9·4	3·4	15·4	5·5
35	5 23·8	5 24·6	5 09·0	3·5	1·3	9·5	3·4	15·5	5·6
36	5 24·0	5 24·9	5 09·2	3·6	1·3	9·6	3·4	15·6	5·6
37	5 24·3	5 25·1	5 09·5	3·7	1·3	9·7	3·5	15·7	5·6
38	5 24·5	5 25·4	5 09·7	3·8	1·4	9·8	3·5	15·8	5·7
39	5 24·8	5 25·6	5 10·0	3·9	1·4	9·9	3·5	15·9	5·7
40	5 25·0	5 25·9	5 10·2	4·0	1·4	10·0	3·6	16·0	5·7
41	5 25·3	5 26·1	5 10·4	4·1	1·5	10·1	3·6	16·1	5·8
42	5 25·5	5 26·4	5 10·7	4·2	1·5	10·2	3·7	16·2	5·8
43	5 25·8	5 26·6	5 10·9	4·3	1·5	10·3	3·7	16·3	5·8
44	5 26·0	5 26·9	5 11·1	4·4	1·6	10·4	3·7	16·4	5·9
45	5 26·3	5 27·1	5 11·4	4·5	1·6	10·5	3·8	16·5	5·9
46	5 26·5	5 27·4	5 11·6	4·6	1·6	10·6	3·8	16·6	5·9
47	5 26·8	5 27·6	5 11·9	4·7	1·7	10·7	3·8	16·7	6·0
48	5 27·0	5 27·9	5 12·1	4·8	1·7	10·8	3·9	16·8	6·0
49	5 27·3	5 28·1	5 12·3	4·9	1·8	10·9	3·9	16·9	6·1
50	5 27·5	5 28·4	5 12·6	5·0	1·8	11·0	3·9	17·0	6·1
51	5 27·8	5 28·6	5 12·8	5·1	1·8	11·1	4·0	17·1	6·1
52	5 28·0	5 28·9	5 13·1	5·2	1·9	11·2	4·0	17·2	6·2
53	5 28·3	5 29·1	5 13·3	5·3	1·9	11·3	4·0	17·3	6·2
54	5 28·5	5 29·4	5 13·5	5·4	1·9	11·4	4·1	17·4	6·2
55	5 28·8	5 29·7	5 13·8	5·5	2·0	11·5	4·1	17·5	6·3
56	5 29·0	5 29·9	5 14·0	5·6	2·0	11·6	4·2	17·6	6·3
57	5 29·3	5 30·2	5 14·3	5·7	2·0	11·7	4·2	17·7	6·3
58	5 29·5	5 30·4	5 14·5	5·8	2·1	11·8	4·2	17·8	6·4
59	5 29·8	5 30·7	5 14·7	5·9	2·1	11·9	4·3	17·9	6·4
60	5 30·0	5 30·9	5 15·0	6·0	2·2	12·0	4·3	18·0	6·5

22m INCREMENTS AND CORRECTIONS 23m

22m	SUN PLANETS	ARIES	MOON	v or d	Corrn	v or d	Corrn	v or d	Corrn
s	° ′	° ′	° ′	′	′	′	′	′	′
00	5 30·0	5 30·9	5 15·0	0·0	0·0	6·0	2·3	12·0	4·5
01	5 30·3	5 31·2	5 15·2	0·1	0·0	6·1	2·3	12·1	4·5
02	5 30·5	5 31·4	5 15·4	0·2	0·1	6·2	2·3	12·2	4·6
03	5 30·8	5 31·7	5 15·7	0·3	0·1	6·3	2·4	12·3	4·6
04	5 31·0	5 31·9	5 15·9	0·4	0·2	6·4	2·4	12·4	4·7
05	5 31·3	5 32·2	5 16·2	0·5	0·2	6·5	2·4	12·5	4·7
06	5 31·5	5 32·4	5 16·4	0·6	0·2	6·6	2·5	12·6	4·7
07	5 31·8	5 32·7	5 16·6	0·7	0·3	6·7	2·5	12·7	4·8
08	5 32·0	5 32·9	5 16·9	0·8	0·3	6·8	2·6	12·8	4·8
09	5 32·3	5 33·2	5 17·1	0·9	0·3	6·9	2·6	12·9	4·8
10	5 32·5	5 33·4	5 17·4	1·0	0·4	7·0	2·6	13·0	4·9
11	5 32·8	5 33·7	5 17·6	1·1	0·4	7·1	2·7	13·1	4·9
12	5 33·0	5 33·9	5 17·8	1·2	0·5	7·2	2·7	13·2	5·0
13	5 33·3	5 34·2	5 18·1	1·3	0·5	7·3	2·7	13·3	5·0
14	5 33·5	5 34·4	5 18·3	1·4	0·5	7·4	2·8	13·4	5·0
15	5 33·8	5 34·7	5 18·5	1·5	0·6	7·5	2·8	13·5	5·1
16	5 34·0	5 34·9	5 18·8	1·6	0·6	7·6	2·9	13·6	5·1
17	5 34·3	5 35·2	5 19·0	1·7	0·6	7·7	2·9	13·7	5·1
18	5 34·5	5 35·4	5 19·3	1·8	0·7	7·8	2·9	13·8	5·2
19	5 34·8	5 35·7	5 19·5	1·9	0·7	7·9	3·0	13·9	5·2
20	5 35·0	5 35·9	5 19·7	2·0	0·8	8·0	3·0	14·0	5·3
21	5 35·3	5 36·2	5 20·0	2·1	0·8	8·1	3·0	14·1	5·3
22	5 35·5	5 36·4	5 20·2	2·2	0·8	8·2	3·1	14·2	5·3
23	5 35·8	5 36·7	5 20·5	2·3	0·9	8·3	3·1	14·3	5·4
24	5 36·0	5 36·9	5 20·7	2·4	0·9	8·4	3·2	14·4	5·4
25	5 36·3	5 37·2	5 20·9	2·5	0·9	8·5	3·2	14·5	5·4
26	5 36·5	5 37·4	5 21·2	2·6	1·0	8·6	3·2	14·6	5·5
27	5 36·8	5 37·7	5 21·4	2·7	1·0	8·7	3·3	14·7	5·5
28	5 37·0	5 37·9	5 21·6	2·8	1·1	8·8	3·3	14·8	5·6
29	5 37·3	5 38·2	5 21·9	2·9	1·1	8·9	3·3	14·9	5·6
30	5 37·5	5 38·4	5 22·1	3·0	1·1	9·0	3·4	15·0	5·6
31	5 37·8	5 38·7	5 22·4	3·1	1·2	9·1	3·4	15·1	5·7
32	5 38·0	5 38·9	5 22·6	3·2	1·2	9·2	3·5	15·2	5·7
33	5 38·3	5 39·2	5 22·8	3·3	1·2	9·3	3·5	15·3	5·7
34	5 38·5	5 39·4	5 23·1	3·4	1·3	9·4	3·5	15·4	5·8
35	5 38·8	5 39·7	5 23·3	3·5	1·3	9·5	3·6	15·5	5·8
36	5 39·0	5 39·9	5 23·6	3·6	1·4	9·6	3·6	15·6	5·9
37	5 39·3	5 40·2	5 23·8	3·7	1·4	9·7	3·6	15·7	5·9
38	5 39·5	5 40·4	5 24·0	3·8	1·4	9·8	3·7	15·8	5·9
39	5 39·8	5 40·7	5 24·3	3·9	1·5	9·9	3·7	15·9	6·0
40	5 40·0	5 40·9	5 24·5	4·0	1·5	10·0	3·8	16·0	6·0
41	5 40·3	5 41·2	5 24·7	4·1	1·5	10·1	3·8	16·1	6·0
42	5 40·5	5 41·4	5 25·0	4·2	1·6	10·2	3·8	16·2	6·1
43	5 40·8	5 41·7	5 25·2	4·3	1·6	10·3	3·9	16·3	6·1
44	5 41·0	5 41·9	5 25·5	4·4	1·7	10·4	3·9	16·4	6·2
45	5 41·3	5 42·2	5 25·7	4·5	1·7	10·5	3·9	16·5	6·2
46	5 41·5	5 42·4	5 25·9	4·6	1·7	10·6	4·0	16·6	6·2
47	5 41·8	5 42·7	5 26·2	4·7	1·8	10·7	4·0	16·7	6·3
48	5 42·0	5 42·9	5 26·4	4·8	1·8	10·8	4·1	16·8	6·3
49	5 42·3	5 43·2	5 26·7	4·9	1·8	10·9	4·1	16·9	6·3
50	5 42·5	5 43·4	5 26·9	5·0	1·9	11·0	4·1	17·0	6·4
51	5 42·8	5 43·7	5 27·1	5·1	1·9	11·1	4·2	17·1	6·4
52	5 43·0	5 43·9	5 27·4	5·2	2·0	11·2	4·2	17·2	6·5
53	5 43·3	5 44·2	5 27·6	5·3	2·0	11·3	4·2	17·3	6·5
54	5 43·5	5 44·4	5 27·9	5·4	2·0	11·4	4·3	17·4	6·5
55	5 43·8	5 44·7	5 28·1	5·5	2·1	11·5	4·3	17·5	6·6
56	5 44·0	5 44·9	5 28·3	5·6	2·1	11·6	4·4	17·6	6·6
57	5 44·3	5 45·2	5 28·6	5·7	2·1	11·7	4·4	17·7	6·6
58	5 44·5	5 45·4	5 28·8	5·8	2·2	11·8	4·4	17·8	6·7
59	5 44·8	5 45·7	5 29·0	5·9	2·2	11·9	4·5	17·9	6·7
60	5 45·0	5 45·9	5 29·3	6·0	2·3	12·0	4·5	18·0	6·8

23m	SUN PLANETS	ARIES	MOON	v or d	Corrn	v or d	Corrn	v or d	Corrn
s	° ′	° ′	° ′	′	′	′	′	′	′
00	5 45·0	5 45·9	5 29·3	0·0	0·0	6·0	2·4	12·0	4·7
01	5 45·3	5 46·2	5 29·5	0·1	0·0	6·1	2·4	12·1	4·7
02	5 45·5	5 46·4	5 29·8	0·2	0·1	6·2	2·4	12·2	4·8
03	5 45·8	5 46·7	5 30·0	0·3	0·1	6·3	2·5	12·3	4·8
04	5 46·0	5 46·9	5 30·2	0·4	0·2	6·4	2·5	12·4	4·9
05	5 46·3	5 47·2	5 30·5	0·5	0·2	6·5	2·5	12·5	4·9
06	5 46·5	5 47·4	5 30·7	0·6	0·2	6·6	2·6	12·6	4·9
07	5 46·8	5 47·7	5 31·0	0·7	0·3	6·7	2·6	12·7	5·0
08	5 47·0	5 48·0	5 31·2	0·8	0·3	6·8	2·7	12·8	5·0
09	5 47·3	5 48·2	5 31·4	0·9	0·4	6·9	2·7	12·9	5·1
10	5 47·5	5 48·5	5 31·7	1·0	0·4	7·0	2·7	13·0	5·1
11	5 47·8	5 48·7	5 31·9	1·1	0·4	7·1	2·8	13·1	5·1
12	5 48·0	5 49·0	5 32·1	1·2	0·5	7·2	2·8	13·2	5·2
13	5 48·3	5 49·2	5 32·4	1·3	0·5	7·3	2·9	13·3	5·2
14	5 48·5	5 49·5	5 32·6	1·4	0·5	7·4	2·9	13·4	5·2
15	5 48·8	5 49·7	5 32·9	1·5	0·6	7·5	2·9	13·5	5·3
16	5 49·0	5 50·0	5 33·1	1·6	0·6	7·6	3·0	13·6	5·3
17	5 49·3	5 50·2	5 33·3	1·7	0·7	7·7	3·0	13·7	5·4
18	5 49·5	5 50·5	5 33·6	1·8	0·7	7·8	3·1	13·8	5·4
19	5 49·8	5 50·7	5 33·8	1·9	0·7	7·9	3·1	13·9	5·4
20	5 50·0	5 51·0	5 34·1	2·0	0·8	8·0	3·1	14·0	5·5
21	5 50·3	5 51·2	5 34·3	2·1	0·8	8·1	3·2	14·1	5·5
22	5 50·5	5 51·5	5 34·5	2·2	0·9	8·2	3·2	14·2	5·6
23	5 50·8	5 51·7	5 34·8	2·3	0·9	8·3	3·3	14·3	5·6
24	5 51·0	5 52·0	5 35·0	2·4	0·9	8·4	3·3	14·4	5·6
25	5 51·3	5 52·2	5 35·2	2·5	1·0	8·5	3·3	14·5	5·7
26	5 51·5	5 52·5	5 35·5	2·6	1·0	8·6	3·4	14·6	5·7
27	5 51·8	5 52·7	5 35·7	2·7	1·1	8·7	3·4	14·7	5·8
28	5 52·0	5 53·0	5 36·0	2·8	1·1	8·8	3·4	14·8	5·8
29	5 52·3	5 53·2	5 36·2	2·9	1·1	8·9	3·5	14·9	5·8
30	5 52·5	5 53·5	5 36·4	3·0	1·2	9·0	3·5	15·0	5·9
31	5 52·8	5 53·7	5 36·7	3·1	1·2	9·1	3·6	15·1	5·9
32	5 53·0	5 54·0	5 36·9	3·2	1·3	9·2	3·6	15·2	6·0
33	5 53·3	5 54·2	5 37·2	3·3	1·3	9·3	3·6	15·3	6·0
34	5 53·5	5 54·5	5 37·4	3·4	1·3	9·4	3·7	15·4	6·0
35	5 53·8	5 54·7	5 37·6	3·5	1·4	9·5	3·7	15·5	6·1
36	5 54·0	5 55·0	5 37·9	3·6	1·4	9·6	3·8	15·6	6·1
37	5 54·3	5 55·2	5 38·1	3·7	1·4	9·7	3·8	15·7	6·1
38	5 54·5	5 55·5	5 38·4	3·8	1·5	9·8	3·8	15·8	6·2
39	5 54·8	5 55·7	5 38·6	3·9	1·5	9·9	3·9	15·9	6·2
40	5 55·0	5 56·0	5 38·8	4·0	1·6	10·0	3·9	16·0	6·3
41	5 55·3	5 56·2	5 39·1	4·1	1·6	10·1	4·0	16·1	6·3
42	5 55·5	5 56·5	5 39·3	4·2	1·6	10·2	4·0	16·2	6·3
43	5 55·8	5 56·7	5 39·5	4·3	1·7	10·3	4·0	16·3	6·4
44	5 56·0	5 57·0	5 39·8	4·4	1·7	10·4	4·1	16·4	6·4
45	5 56·3	5 57·2	5 40·0	4·5	1·8	10·5	4·1	16·5	6·5
46	5 56·5	5 57·5	5 40·3	4·6	1·8	10·6	4·2	16·6	6·5
47	5 56·8	5 57·7	5 40·5	4·7	1·8	10·7	4·2	16·7	6·5
48	5 57·0	5 58·0	5 40·7	4·8	1·9	10·8	4·2	16·8	6·6
49	5 57·3	5 58·2	5 41·0	4·9	1·9	10·9	4·3	16·9	6·6
50	5 57·5	5 58·5	5 41·2	5·0	2·0	11·0	4·3	17·0	6·7
51	5 57·8	5 58·7	5 41·5	5·1	2·0	11·1	4·3	17·1	6·7
52	5 58·0	5 59·0	5 41·7	5·2	2·0	11·2	4·4	17·2	6·7
53	5 58·3	5 59·2	5 41·9	5·3	2·1	11·3	4·4	17·3	6·8
54	5 58·5	5 59·5	5 42·2	5·4	2·1	11·4	4·5	17·4	6·8
55	5 58·8	5 59·7	5 42·4	5·5	2·2	11·5	4·5	17·5	6·9
56	5 59·0	6 00·0	5 42·6	5·6	2·2	11·6	4·5	17·6	6·9
57	5 59·3	6 00·2	5 42·9	5·7	2·2	11·7	4·6	17·7	6·9
58	5 59·5	6 00·5	5 43·1	5·8	2·3	11·8	4·6	17·8	7·0
59	5 59·8	6 00·7	5 43·4	5·9	2·3	11·9	4·7	17·9	7·0
60	6 00·0	6 01·0	5 43·6	6·0	2·4	12·0	4·7	18·0	7·1

24m INCREMENTS AND CORRECTIONS 25m

24m	SUN PLANETS	ARIES	MOON	v or d	Corrⁿ	v or d	Corrⁿ	v or d	Corrⁿ
s	° ′	° ′	° ′	′	′	′	′	′	′
00	6 00·0	6 01·0	5 43·6	0·0	0·0	6·0	2·5	12·0	4·9
01	6 00·3	6 01·2	5 43·8	0·1	0·0	6·1	2·5	12·1	4·9
02	6 00·5	6 01·5	5 44·1	0·2	0·1	6·2	2·5	12·2	5·0
03	6 00·8	6 01·7	5 44·3	0·3	0·1	6·3	2·6	12·3	5·0
04	6 01·0	6 02·0	5 44·6	0·4	0·2	6·4	2·6	12·4	5·1
05	6 01·3	6 02·2	5 44·8	0·5	0·2	6·5	2·7	12·5	5·1
06	6 01·5	6 02·5	5 45·0	0·6	0·2	6·6	2·7	12·6	5·1
07	6 01·8	6 02·7	5 45·3	0·7	0·3	6·7	2·7	12·7	5·2
08	6 02·0	6 03·0	5 45·5	0·8	0·3	6·8	2·8	12·8	5·2
09	6 02·3	6 03·2	5 45·7	0·9	0·4	6·9	2·8	12·9	5·3
10	6 02·5	6 03·5	5 46·0	1·0	0·4	7·0	2·9	13·0	5·3
11	6 02·8	6 03·7	5 46·2	1·1	0·4	7·1	2·9	13·1	5·3
12	6 03·0	6 04·0	5 46·5	1·2	0·5	7·2	2·9	13·2	5·4
13	6 03·3	6 04·2	5 46·7	1·3	0·5	7·3	3·0	13·3	5·4
14	6 03·5	6 04·5	5 46·9	1·4	0·6	7·4	3·0	13·4	5·5
15	6 03·8	6 04·7	5 47·2	1·5	0·6	7·5	3·1	13·5	5·5
16	6 04·0	6 05·0	5 47·4	1·6	0·7	7·6	3·1	13·6	5·6
17	6 04·3	6 05·2	5 47·7	1·7	0·7	7·7	3·1	13·7	5·6
18	6 04·5	6 05·5	5 47·9	1·8	0·7	7·8	3·2	13·8	5·6
19	6 04·8	6 05·7	5 48·1	1·9	0·8	7·9	3·2	13·9	5·7
20	6 05·0	6 06·0	5 48·4	2·0	0·8	8·0	3·3	14·0	5·7
21	6 05·3	6 06·3	5 48·6	2·1	0·9	8·1	3·3	14·1	5·8
22	6 05·5	6 06·5	5 48·8	2·2	0·9	8·2	3·3	14·2	5·8
23	6 05·8	6 06·8	5 49·1	2·3	0·9	8·3	3·4	14·3	5·8
24	6 06·0	6 07·0	5 49·3	2·4	1·0	8·4	3·4	14·4	5·9
25	6 06·3	6 07·3	5 49·6	2·5	1·0	8·5	3·5	14·5	5·9
26	6 06·5	6 07·5	5 49·8	2·6	1·1	8·6	3·5	14·6	6·0
27	6 06·8	6 07·8	5 50·0	2·7	1·1	8·7	3·6	14·7	6·0
28	6 07·0	6 08·0	5 50·3	2·8	1·1	8·8	3·6	14·8	6·0
29	6 07·3	6 08·3	5 50·5	2·9	1·2	8·9	3·6	14·9	6·1
30	6 07·5	6 08·5	5 50·8	3·0	1·2	9·0	3·7	15·0	6·1
31	6 07·8	6 08·8	5 51·0	3·1	1·3	9·1	3·7	15·1	6·2
32	6 08·0	6 09·0	5 51·2	3·2	1·3	9·2	3·8	15·2	6·2
33	6 08·3	6 09·3	5 51·5	3·3	1·3	9·3	3·8	15·3	6·2
34	6 08·5	6 09·5	5 51·7	3·4	1·4	9·4	3·8	15·4	6·3
35	6 08·8	6 09·8	5 52·0	3·5	1·4	9·5	3·9	15·5	6·3
36	6 09·0	6 10·0	5 52·2	3·6	1·5	9·6	3·9	15·6	6·4
37	6 09·3	6 10·3	5 52·4	3·7	1·5	9·7	4·0	15·7	6·4
38	6 09·5	6 10·5	5 52·7	3·8	1·6	9·8	4·0	15·8	6·5
39	6 09·8	6 10·8	5 52·9	3·9	1·6	9·9	4·0	15·9	6·5
40	6 10·0	6 11·0	5 53·1	4·0	1·6	10·0	4·1	16·0	6·5
41	6 10·3	6 11·3	5 53·4	4·1	1·7	10·1	4·1	16·1	6·6
42	6 10·5	6 11·5	5 53·6	4·2	1·7	10·2	4·2	16·2	6·6
43	6 10·8	6 11·8	5 53·9	4·3	1·8	10·3	4·2	16·3	6·7
44	6 11·0	6 12·0	5 54·1	4·4	1·8	10·4	4·2	16·4	6·7
45	6 11·3	6 12·3	5 54·3	4·5	1·8	10·5	4·3	16·5	6·7
46	6 11·5	6 12·5	5 54·6	4·6	1·9	10·6	4·3	16·6	6·8
47	6 11·8	6 12·8	5 54·8	4·7	1·9	10·7	4·4	16·7	6·8
48	6 12·0	6 13·0	5 55·1	4·8	2·0	10·8	4·4	16·8	6·9
49	6 12·3	6 13·3	5 55·3	4·9	2·0	10·9	4·5	16·9	6·9
50	6 12·5	6 13·5	5 55·5	5·0	2·0	11·0	4·5	17·0	6·9
51	6 12·8	6 13·8	5 55·8	5·1	2·1	11·1	4·5	17·1	7·0
52	6 13·0	6 14·0	5 56·0	5·2	2·1	11·2	4·6	17·2	7·0
53	6 13·3	6 14·3	5 56·2	5·3	2·2	11·3	4·6	17·3	7·1
54	6 13·5	6 14·5	5 56·5	5·4	2·2	11·4	4·7	17·4	7·1
55	6 13·8	6 14·8	5 56·7	5·5	2·2	11·5	4·7	17·5	7·1
56	6 14·0	6 15·0	5 57·0	5·6	2·3	11·6	4·7	17·6	7·2
57	6 14·3	6 15·3	5 57·2	5·7	2·3	11·7	4·8	17·7	7·2
58	6 14·5	6 15·5	5 57·4	5·8	2·4	11·8	4·8	17·8	7·3
59	6 14·8	6 15·8	5 57·7	5·9	2·4	11·9	4·9	17·9	7·3
60	6 15·0	6 16·0	5 57·9	6·0	2·5	12·0	4·9	18·0	7·4

25m	SUN PLANETS	ARIES	MOON	v or d	Corrⁿ	v or d	Corrⁿ	v or d	Corrⁿ
s	° ′	° ′	° ′	′	′	′	′	′	′
00	6 15·0	6 16·0	5 57·9	0·0	0·0	6·0	2·6	12·0	5·1
01	6 15·3	6 16·3	5 58·2	0·1	0·0	6·1	2·6	12·1	5·1
02	6 15·5	6 16·5	5 58·4	0·2	0·1	6·2	2·6	12·2	5·2
03	6 15·8	6 16·8	5 58·6	0·3	0·1	6·3	2·7	12·3	5·2
04	6 16·0	6 17·0	5 58·9	0·4	0·2	6·4	2·7	12·4	5·3
05	6 16·3	6 17·3	5 59·1	0·5	0·2	6·5	2·8	12·5	5·3
06	6 16·5	6 17·5	5 59·3	0·6	0·3	6·6	2·8	12·6	5·4
07	6 16·8	6 17·8	5 59·6	0·7	0·3	6·7	2·8	12·7	5·4
08	6 17·0	6 18·0	5 59·8	0·8	0·3	6·8	2·9	12·8	5·4
09	6 17·3	6 18·3	6 00·1	0·9	0·4	6·9	2·9	12·9	5·5
10	6 17·5	6 18·5	6 00·3	1·0	0·4	7·0	3·0	13·0	5·5
11	6 17·8	6 18·8	6 00·5	1·1	0·5	7·1	3·0	13·1	5·6
12	6 18·0	6 19·0	6 00·8	1·2	0·5	7·2	3·1	13·2	5·6
13	6 18·3	6 19·3	6 01·0	1·3	0·6	7·3	3·1	13·3	5·7
14	6 18·5	6 19·5	6 01·3	1·4	0·6	7·4	3·1	13·4	5·7
15	6 18·8	6 19·8	6 01·5	1·5	0·6	7·5	3·2	13·5	5·7
16	6 19·0	6 20·0	6 01·7	1·6	0·7	7·6	3·2	13·6	5·8
17	6 19·3	6 20·3	6 02·0	1·7	0·7	7·7	3·3	13·7	5·8
18	6 19·5	6 20·5	6 02·2	1·8	0·8	7·8	3·3	13·8	5·9
19	6 19·8	6 20·8	6 02·5	1·9	0·8	7·9	3·4	13·9	5·9
20	6 20·0	6 21·0	6 02·7	2·0	0·9	8·0	3·4	14·0	6·0
21	6 20·3	6 21·3	6 02·9	2·1	0·9	8·1	3·4	14·1	6·0
22	6 20·5	6 21·5	6 03·2	2·2	0·9	8·2	3·5	14·2	6·0
23	6 20·8	6 21·8	6 03·4	2·3	1·0	8·3	3·5	14·3	6·1
24	6 21·0	6 22·0	6 03·6	2·4	1·0	8·4	3·6	14·4	6·1
25	6 21·3	6 22·3	6 03·9	2·5	1·1	8·5	3·6	14·5	6·2
26	6 21·5	6 22·5	6 04·1	2·6	1·1	8·6	3·7	14·6	6·2
27	6 21·8	6 22·8	6 04·4	2·7	1·1	8·7	3·7	14·7	6·2
28	6 22·0	6 23·0	6 04·6	2·8	1·2	8·8	3·7	14·8	6·3
29	6 22·3	6 23·3	6 04·8	2·9	1·2	8·9	3·8	14·9	6·3
30	6 22·5	6 23·5	6 05·1	3·0	1·3	9·0	3·8	15·0	6·4
31	6 22·8	6 23·8	6 05·3	3·1	1·3	9·1	3·9	15·1	6·4
32	6 23·0	6 24·0	6 05·6	3·2	1·4	9·2	3·9	15·2	6·5
33	6 23·3	6 24·3	6 05·8	3·3	1·4	9·3	4·0	15·3	6·5
34	6 23·5	6 24·5	6 06·0	3·4	1·4	9·4	4·0	15·4	6·5
35	6 23·8	6 24·8	6 06·3	3·5	1·5	9·5	4·0	15·5	6·6
36	6 24·0	6 25·1	6 06·5	3·6	1·5	9·6	4·1	15·6	6·6
37	6 24·3	6 25·3	6 06·7	3·7	1·6	9·7	4·1	15·7	6·7
38	6 24·5	6 25·6	6 07·0	3·8	1·6	9·8	4·2	15·8	6·7
39	6 24·8	6 25·8	6 07·2	3·9	1·7	9·9	4·2	15·9	6·8
40	6 25·0	6 26·1	6 07·5	4·0	1·7	10·0	4·3	16·0	6·8
41	6 25·3	6 26·3	6 07·7	4·1	1·7	10·1	4·3	16·1	6·8
42	6 25·5	6 26·6	6 07·9	4·2	1·8	10·2	4·3	16·2	6·9
43	6 25·8	6 26·8	6 08·2	4·3	1·8	10·3	4·4	16·3	6·9
44	6 26·0	6 27·1	6 08·4	4·4	1·9	10·4	4·4	16·4	7·0
45	6 26·3	6 27·3	6 08·7	4·5	1·9	10·5	4·5	16·5	7·0
46	6 26·5	6 27·6	6 08·9	4·6	2·0	10·6	4·5	16·6	7·1
47	6 26·8	6 27·8	6 09·1	4·7	2·0	10·7	4·5	16·7	7·1
48	6 27·0	6 28·1	6 09·4	4·8	2·0	10·8	4·6	16·8	7·1
49	6 27·3	6 28·3	6 09·6	4·9	2·1	10·9	4·6	16·9	7·2
50	6 27·5	6 28·6	6 09·8	5·0	2·1	11·0	4·7	17·0	7·2
51	6 27·8	6 28·8	6 10·1	5·1	2·2	11·1	4·7	17·1	7·3
52	6 28·0	6 29·1	6 10·3	5·2	2·2	11·2	4·8	17·2	7·3
53	6 28·3	6 29·3	6 10·6	5·3	2·3	11·3	4·8	17·3	7·4
54	6 28·5	6 29·6	6 10·8	5·4	2·3	11·4	4·8	17·4	7·4
55	6 28·8	6 29·8	6 11·0	5·5	2·3	11·5	4·9	17·5	7·4
56	6 29·0	6 30·1	6 11·3	5·6	2·4	11·6	4·9	17·6	7·5
57	6 29·3	6 30·3	6 11·5	5·7	2·4	11·7	5·0	17·7	7·5
58	6 29·5	6 30·6	6 11·8	5·8	2·5	11·8	5·0	17·8	7·6
59	6 29·8	6 30·8	6 12·0	5·9	2·5	11·9	5·1	17·9	7·6
60	6 30·0	6 31·1	6 12·2	6·0	2·6	12·0	5·1	18·0	7·7

26ᵐ INCREMENTS AND CORRECTIONS 27ᵐ

26ᵐ	SUN PLANETS	ARIES	MOON	v or d	Corrⁿ	v or d	Corrⁿ	v or d	Corrⁿ
s	° ′	° ′	° ′	′	′	′	′	′	′
00	6 30·0	6 31·1	6 12·2	0·0	0·0	6·0	2·7	12·0	5·3
01	6 30·3	6 31·3	6 12·5	0·1	0·0	6·1	2·7	12·1	5·3
02	6 30·5	6 31·6	6 12·7	0·2	0·1	6·2	2·7	12·2	5·4
03	6 30·8	6 31·8	6 12·9	0·3	0·1	6·3	2·8	12·3	5·4
04	6 31·0	6 32·1	6 13·2	0·4	0·2	6·4	2·8	12·4	5·5
05	6 31·3	6 32·3	6 13·4	0·5	0·2	6·5	2·9	12·5	5·5
06	6 31·5	6 32·6	6 13·7	0·6	0·3	6·6	2·9	12·6	5·6
07	6 31·8	6 32·8	6 13·9	0·7	0·3	6·7	3·0	12·7	5·6
08	6 32·0	6 33·1	6 14·1	0·8	0·4	6·8	3·0	12·8	5·7
09	6 32·3	6 33·3	6 14·4	0·9	0·4	6·9	3·0	12·9	5·7
10	6 32·5	6 33·6	6 14·6	1·0	0·4	7·0	3·1	13·0	5·7
11	6 32·8	6 33·8	6 14·9	1·1	0·5	7·1	3·1	13·1	5·8
12	6 33·0	6 34·1	6 15·1	1·2	0·5	7·2	3·2	13·2	5·8
13	6 33·3	6 34·3	6 15·3	1·3	0·6	7·3	3·2	13·3	5·9
14	6 33·5	6 34·6	6 15·6	1·4	0·6	7·4	3·3	13·4	5·9
15	6 33·8	6 34·8	6 15·8	1·5	0·7	7·5	3·3	13·5	6·0
16	6 34·0	6 35·1	6 16·1	1·6	0·7	7·6	3·4	13·6	6·0
17	6 34·3	6 35·3	6 16·3	1·7	0·8	7·7	3·4	13·7	6·1
18	6 34·5	6 35·6	6 16·5	1·8	0·8	7·8	3·4	13·8	6·1
19	6 34·8	6 35·8	6 16·8	1·9	0·8	7·9	3·5	13·9	6·1
20	6 35·0	6 36·1	6 17·0	2·0	0·9	8·0	3·5	14·0	6·2
21	6 35·3	6 36·3	6 17·2	2·1	0·9	8·1	3·6	14·1	6·2
22	6 35·5	6 36·6	6 17·5	2·2	1·0	8·2	3·6	14·2	6·3
23	6 35·8	6 36·8	6 17·7	2·3	1·0	8·3	3·7	14·3	6·3
24	6 36·0	6 37·1	6 18·0	2·4	1·1	8·4	3·7	14·4	6·4
25	6 36·3	6 37·3	6 18·2	2·5	1·1	8·5	3·8	14·5	6·4
26	6 36·5	6 37·6	6 18·4	2·6	1·1	8·6	3·8	14·6	6·4
27	6 36·8	6 37·8	6 18·7	2·7	1·2	8·7	3·8	14·7	6·5
28	6 37·0	6 38·1	6 18·9	2·8	1·2	8·8	3·9	14·8	6·5
29	6 37·3	6 38·3	6 19·2	2·9	1·3	8·9	3·9	14·9	6·6
30	6 37·5	6 38·6	6 19·4	3·0	1·3	9·0	4·0	15·0	6·6
31	6 37·8	6 38·8	6 19·6	3·1	1·4	9·1	4·0	15·1	6·7
32	6 38·0	6 39·1	6 19·9	3·2	1·4	9·2	4·1	15·2	6·7
33	6 38·3	6 39·3	6 20·1	3·3	1·5	9·3	4·1	15·3	6·8
34	6 38·5	6 39·6	6 20·3	3·4	1·5	9·4	4·2	15·4	6·8
35	6 38·8	6 39·8	6 20·6	3·5	1·5	9·5	4·2	15·5	6·8
36	6 39·0	6 40·1	6 20·8	3·6	1·6	9·6	4·2	15·6	6·9
37	6 39·3	6 40·3	6 21·1	3·7	1·6	9·7	4·3	15·7	6·9
38	6 39·5	6 40·6	6 21·3	3·8	1·7	9·8	4·3	15·8	7·0
39	6 39·8	6 40·8	6 21·5	3·9	1·7	9·9	4·4	15·9	7·0
40	6 40·0	6 41·1	6 21·8	4·0	1·8	10·0	4·4	16·0	7·1
41	6 40·3	6 41·3	6 22·0	4·1	1·8	10·1	4·5	16·1	7·1
42	6 40·5	6 41·6	6 22·3	4·2	1·9	10·2	4·5	16·2	7·2
43	6 40·8	6 41·8	6 22·5	4·3	1·9	10·3	4·5	16·3	7·2
44	6 41·0	6 42·1	6 22·7	4·4	1·9	10·4	4·6	16·4	7·2
45	6 41·3	6 42·3	6 23·0	4·5	2·0	10·5	4·6	16·5	7·3
46	6 41·5	6 42·6	6 23·2	4·6	2·0	10·6	4·7	16·6	7·3
47	6 41·8	6 42·8	6 23·4	4·7	2·1	10·7	4·7	16·7	7·4
48	6 42·0	6 43·1	6 23·7	4·8	2·1	10·8	4·8	16·8	7·4
49	6 42·3	6 43·4	6 23·9	4·9	2·2	10·9	4·8	16·9	7·5
50	6 42·5	6 43·6	6 24·2	5·0	2·2	11·0	4·9	17·0	7·5
51	6 42·8	6 43·9	6 24·4	5·1	2·3	11·1	4·9	17·1	7·6
52	6 43·0	6 44·1	6 24·6	5·2	2·3	11·2	4·9	17·2	7·6
53	6 43·3	6 44·4	6 24·9	5·3	2·3	11·3	5·0	17·3	7·6
54	6 43·5	6 44·6	6 25·1	5·4	2·4	11·4	5·0	17·4	7·7
55	6 43·8	6 44·9	6 25·4	5·5	2·4	11·5	5·1	17·5	7·7
56	6 44·0	6 45·1	6 25·6	5·6	2·5	11·6	5·1	17·6	7·8
57	6 44·3	6 45·4	6 25·8	5·7	2·5	11·7	5·2	17·7	7·8
58	6 44·5	6 45·6	6 26·1	5·8	2·6	11·8	5·2	17·8	7·9
59	6 44·8	6 45·9	6 26·3	5·9	2·6	11·9	5·3	17·9	7·9
60	6 45·0	6 46·1	6 26·6	6·0	2·7	12·0	5·3	18·0	8·0

27ᵐ	SUN PLANETS	ARIES	MOON	v or d	Corrⁿ	v or d	Corrⁿ	v or d	Corrⁿ
s	° ′	° ′	° ′	′	′	′	′	′	′
00	6 45·0	6 46·1	6 26·6	0·0	0·0	6·0	2·8	12·0	5·5
01	6 45·3	6 46·4	6 26·8	0·1	0·0	6·1	2·8	12·1	5·5
02	6 45·5	6 46·6	6 27·0	0·2	0·1	6·2	2·8	12·2	5·6
03	6 45·8	6 46·9	6 27·3	0·3	0·1	6·3	2·9	12·3	5·6
04	6 46·0	6 47·1	6 27·5	0·4	0·2	6·4	2·9	12·4	5·7
05	6 46·3	6 47·4	6 27·7	0·5	0·2	6·5	3·0	12·5	5·7
06	6 46·5	6 47·6	6 28·0	0·6	0·3	6·6	3·0	12·6	5·8
07	6 46·8	6 47·9	6 28·2	0·7	0·3	6·7	3·1	12·7	5·8
08	6 47·0	6 48·1	6 28·5	0·8	0·4	6·8	3·1	12·8	5·9
09	6 47·3	6 48·4	6 28·7	0·9	0·4	6·9	3·2	12·9	5·9
10	6 47·5	6 48·6	6 28·9	1·0	0·5	7·0	3·2	13·0	6·0
11	6 47·8	6 48·9	6 29·2	1·1	0·5	7·1	3·3	13·1	6·0
12	6 48·0	6 49·1	6 29·4	1·2	0·6	7·2	3·3	13·2	6·1
13	6 48·3	6 49·4	6 29·7	1·3	0·6	7·3	3·3	13·3	6·1
14	6 48·5	6 49·6	6 29·9	1·4	0·6	7·4	3·4	13·4	6·1
15	6 48·8	6 49·9	6 30·1	1·5	0·7	7·5	3·4	13·5	6·2
16	6 49·0	6 50·1	6 30·4	1·6	0·7	7·6	3·5	13·6	6·2
17	6 49·3	6 50·4	6 30·6	1·7	0·8	7·7	3·5	13·7	6·3
18	6 49·5	6 50·6	6 30·8	1·8	0·8	7·8	3·6	13·8	6·3
19	6 49·8	6 50·9	6 31·1	1·9	0·9	7·9	3·6	13·9	6·4
20	6 50·0	6 51·1	6 31·3	2·0	0·9	8·0	3·7	14·0	6·4
21	6 50·3	6 51·4	6 31·6	2·1	1·0	8·1	3·7	14·1	6·5
22	6 50·5	6 51·6	6 31·8	2·2	1·0	8·2	3·8	14·2	6·5
23	6 50·8	6 51·9	6 32·0	2·3	1·1	8·3	3·8	14·3	6·6
24	6 51·0	6 52·1	6 32·3	2·4	1·1	8·4	3·9	14·4	6·6
25	6 51·3	6 52·4	6 32·5	2·5	1·1	8·5	3·9	14·5	6·6
26	6 51·5	6 52·6	6 32·8	2·6	1·2	8·6	3·9	14·6	6·7
27	6 51·8	6 52·9	6 33·0	2·7	1·2	8·7	4·0	14·7	6·7
28	6 52·0	6 53·1	6 33·2	2·8	1·3	8·8	4·0	14·8	6·8
29	6 52·3	6 53·4	6 33·5	2·9	1·3	8·9	4·1	14·9	6·8
30	6 52·5	6 53·6	6 33·7	3·0	1·4	9·0	4·1	15·0	6·9
31	6 52·8	6 53·9	6 33·9	3·1	1·4	9·1	4·2	15·1	6·9
32	6 53·0	6 54·1	6 34·2	3·2	1·5	9·2	4·2	15·2	7·0
33	6 53·3	6 54·4	6 34·4	3·3	1·5	9·3	4·3	15·3	7·0
34	6 53·5	6 54·6	6 34·7	3·4	1·6	9·4	4·3	15·4	7·1
35	6 53·8	6 54·9	6 34·9	3·5	1·6	9·5	4·4	15·5	7·1
36	6 54·0	6 55·1	6 35·1	3·6	1·7	9·6	4·4	15·6	7·2
37	6 54·3	6 55·4	6 35·4	3·7	1·7	9·7	4·4	15·7	7·2
38	6 54·5	6 55·6	6 35·6	3·8	1·7	9·8	4·5	15·8	7·2
39	6 54·8	6 55·9	6 35·9	3·9	1·8	9·9	4·5	15·9	7·3
40	6 55·0	6 56·1	6 36·1	4·0	1·8	10·0	4·6	16·0	7·3
41	6 55·3	6 56·4	6 36·3	4·1	1·9	10·1	4·6	16·1	7·4
42	6 55·5	6 56·6	6 36·6	4·2	1·9	10·2	4·7	16·2	7·4
43	6 55·8	6 56·9	6 36·8	4·3	2·0	10·3	4·7	16·3	7·5
44	6 56·0	6 57·1	6 37·0	4·4	2·0	10·4	4·8	16·4	7·5
45	6 56·3	6 57·4	6 37·3	4·5	2·1	10·5	4·8	16·5	7·6
46	6 56·5	6 57·6	6 37·5	4·6	2·1	10·6	4·9	16·6	7·6
47	6 56·8	6 57·9	6 37·8	4·7	2·2	10·7	4·9	16·7	7·7
48	6 57·0	6 58·1	6 38·0	4·8	2·2	10·8	5·0	16·8	7·7
49	6 57·3	6 58·4	6 38·2	4·9	2·2	10·9	5·0	16·9	7·7
50	6 57·5	6 58·6	6 38·5	5·0	2·3	11·0	5·0	17·0	7·8
51	6 57·8	6 58·9	6 38·7	5·1	2·3	11·1	5·1	17·1	7·8
52	6 58·0	6 59·1	6 39·0	5·2	2·4	11·2	5·1	17·2	7·9
53	6 58·3	6 59·4	6 39·2	5·3	2·4	11·3	5·2	17·3	7·9
54	6 58·5	6 59·6	6 39·4	5·4	2·5	11·4	5·2	17·4	8·0
55	6 58·8	6 59·9	6 39·7	5·5	2·5	11·5	5·3	17·5	8·0
56	6 59·0	7 00·1	6 39·9	5·6	2·6	11·6	5·3	17·6	8·1
57	6 59·3	7 00·4	6 40·2	5·7	2·6	11·7	5·4	17·7	8·1
58	6 59·5	7 00·6	6 40·4	5·8	2·7	11·8	5·4	17·8	8·2
59	6 59·8	7 00·9	6 40·6	5·9	2·7	11·9	5·5	17·9	8·2
60	7 00·0	7 01·1	6 40·9	6·0	2·8	12·0	5·5	18·0	8·3

INCREMENTS AND CORRECTIONS

28m

28m s	SUN PLANETS ° ′	ARIES ° ′	MOON ° ′	v or d ′	Corrn ′	v or d ′	Corrn ′	v or d ′	Corrn ′
00	7 00·0	7 01·1	6 40·9	0·0	0·0	6·0	2·9	12·0	5·7
01	7 00·3	7 01·4	6 41·1	0·1	0·0	6·1	2·9	12·1	5·7
02	7 00·5	7 01·7	6 41·3	0·2	0·1	6·2	2·9	12·2	5·8
03	7 00·8	7 01·9	6 41·6	0·3	0·1	6·3	3·0	12·3	5·8
04	7 01·0	7 02·2	6 41·8	0·4	0·2	6·4	3·0	12·4	5·9
05	7 01·3	7 02·4	6 42·1	0·5	0·2	6·5	3·1	12·5	5·9
06	7 01·5	7 02·7	6 42·3	0·6	0·3	6·6	3·1	12·6	6·0
07	7 01·8	7 02·9	6 42·5	0·7	0·3	6·7	3·2	12·7	6·0
08	7 02·0	7 03·2	6 42·8	0·8	0·4	6·8	3·2	12·8	6·1
09	7 02·3	7 03·4	6 43·0	0·9	0·4	6·9	3·3	12·9	6·1
10	7 02·5	7 03·7	6 43·3	1·0	0·5	7·0	3·3	13·0	6·2
11	7 02·8	7 03·9	6 43·5	1·1	0·5	7·1	3·4	13·1	6·2
12	7 03·0	7 04·2	6 43·7	1·2	0·6	7·2	3·4	13·2	6·3
13	7 03·3	7 04·4	6 44·0	1·3	0·6	7·3	3·5	13·3	6·3
14	7 03·5	7 04·7	6 44·2	1·4	0·7	7·4	3·5	13·4	6·4
15	7 03·8	7 04·9	6 44·4	1·5	0·7	7·5	3·6	13·5	6·4
16	7 04·0	7 05·2	6 44·7	1·6	0·8	7·6	3·6	13·6	6·5
17	7 04·3	7 05·4	6 44·9	1·7	0·8	7·7	3·7	13·7	6·5
18	7 04·5	7 05·7	6 45·2	1·8	0·9	7·8	3·7	13·8	6·6
19	7 04·8	7 05·9	6 45·4	1·9	0·9	7·9	3·8	13·9	6·6
20	7 05·0	7 06·2	6 45·6	2·0	1·0	8·0	3·8	14·0	6·7
21	7 05·3	7 06·4	6 45·9	2·1	1·0	8·1	3·8	14·1	6·7
22	7 05·5	7 06·7	6 46·1	2·2	1·0	8·2	3·9	14·2	6·7
23	7 05·8	7 06·9	6 46·4	2·3	1·1	8·3	3·9	14·3	6·8
24	7 06·0	7 07·2	6 46·6	2·4	1·1	8·4	4·0	14·4	6·8
25	7 06·3	7 07·4	6 46·8	2·5	1·2	8·5	4·0	14·5	6·9
26	7 06·5	7 07·7	6 47·1	2·6	1·2	8·6	4·1	14·6	6·9
27	7 06·8	7 07·9	6 47·3	2·7	1·3	8·7	4·1	14·7	7·0
28	7 07·0	7 08·2	6 47·5	2·8	1·3	8·8	4·2	14·8	7·0
29	7 07·3	7 08·4	6 47·8	2·9	1·4	8·9	4·2	14·9	7·1
30	7 07·5	7 08·7	6 48·0	3·0	1·4	9·0	4·3	15·0	7·1
31	7 07·8	7 08·9	6 48·3	3·1	1·5	9·1	4·3	15·1	7·2
32	7 08·0	7 09·2	6 48·5	3·2	1·5	9·2	4·4	15·2	7·2
33	7 08·3	7 09·4	6 48·7	3·3	1·6	9·3	4·4	15·3	7·3
34	7 08·5	7 09·7	6 49·0	3·4	1·6	9·4	4·5	15·4	7·3
35	7 08·8	7 09·9	6 49·2	3·5	1·7	9·5	4·5	15·5	7·4
36	7 09·0	7 10·2	6 49·5	3·6	1·7	9·6	4·6	15·6	7·4
37	7 09·3	7 10·4	6 49·7	3·7	1·8	9·7	4·6	15·7	7·5
38	7 09·5	7 10·7	6 49·9	3·8	1·8	9·8	4·7	15·8	7·5
39	7 09·8	7 10·9	6 50·2	3·9	1·9	9·9	4·7	15·9	7·6
40	7 10·0	7 11·2	6 50·4	4·0	1·9	10·0	4·8	16·0	7·6
41	7 10·3	7 11·4	6 50·6	4·1	1·9	10·1	4·8	16·1	7·6
42	7 10·5	7 11·7	6 50·9	4·2	2·0	10·2	4·8	16·2	7·7
43	7 10·8	7 11·9	6 51·1	4·3	2·0	10·3	4·9	16·3	7·7
44	7 11·0	7 12·2	6 51·4	4·4	2·1	10·4	4·9	16·4	7·8
45	7 11·3	7 12·4	6 51·6	4·5	2·1	10·5	5·0	16·5	7·8
46	7 11·5	7 12·7	6 51·8	4·6	2·2	10·6	5·0	16·6	7·9
47	7 11·8	7 12·9	6 52·1	4·7	2·2	10·7	5·1	16·7	7·9
48	7 12·0	7 13·2	6 52·3	4·8	2·3	10·8	5·1	16·8	8·0
49	7 12·3	7 13·4	6 52·6	4·9	2·3	10·9	5·2	16·9	8·0
50	7 12·5	7 13·7	6 52·8	5·0	2·4	11·0	5·2	17·0	8·1
51	7 12·8	7 13·9	6 53·0	5·1	2·4	11·1	5·3	17·1	8·1
52	7 13·0	7 14·2	6 53·3	5·2	2·5	11·2	5·3	17·2	8·2
53	7 13·3	7 14·4	6 53·5	5·3	2·5	11·3	5·4	17·3	8·2
54	7 13·5	7 14·7	6 53·8	5·4	2·6	11·4	5·4	17·4	8·3
55	7 13·8	7 14·9	6 54·0	5·5	2·6	11·5	5·5	17·5	8·3
56	7 14·0	7 15·2	6 54·2	5·6	2·7	11·6	5·5	17·6	8·4
57	7 14·3	7 15·4	6 54·5	5·7	2·7	11·7	5·6	17·7	8·4
58	7 14·5	7 15·7	6 54·7	5·8	2·8	11·8	5·6	17·8	8·5
59	7 14·8	7 15·9	6 54·9	5·9	2·8	11·9	5·7	17·9	8·5
60	7 15·0	7 16·2	6 55·2	6·0	2·9	12·0	5·7	18·0	8·6

29m

29m s	SUN PLANETS ° ′	ARIES ° ′	MOON ° ′	v or d ′	Corrn ′	v or d ′	Corrn ′	v or d ′	Corrn ′
00	7 15·0	7 16·2	6 55·2	0·0	0·0	6·0	3·0	12·0	5·9
01	7 15·3	7 16·4	6 55·4	0·1	0·0	6·1	3·0	12·1	5·9
02	7 15·5	7 16·7	6 55·7	0·2	0·1	6·2	3·0	12·2	6·0
03	7 15·8	7 16·9	6 55·9	0·3	0·1	6·3	3·1	12·3	6·0
04	7 16·0	7 17·2	6 56·1	0·4	0·2	6·4	3·1	12·4	6·1
05	7 16·3	7 17·4	6 56·4	0·5	0·2	6·5	3·2	12·5	6·1
06	7 16·5	7 17·7	6 56·6	0·6	0·3	6·6	3·2	12·6	6·2
07	7 16·8	7 17·9	6 56·9	0·7	0·3	6·7	3·3	12·7	6·2
08	7 17·0	7 18·2	6 57·1	0·8	0·4	6·8	3·3	12·8	6·3
09	7 17·3	7 18·4	6 57·3	0·9	0·4	6·9	3·4	12·9	6·3
10	7 17·5	7 18·7	6 57·6	1·0	0·5	7·0	3·4	13·0	6·4
11	7 17·8	7 18·9	6 57·8	1·1	0·5	7·1	3·5	13·1	6·4
12	7 18·0	7 19·2	6 58·0	1·2	0·6	7·2	3·5	13·2	6·5
13	7 18·3	7 19·4	6 58·3	1·3	0·6	7·3	3·6	13·3	6·5
14	7 18·5	7 19·7	6 58·5	1·4	0·7	7·4	3·6	13·4	6·6
15	7 18·8	7 20·0	6 58·8	1·5	0·7	7·5	3·7	13·5	6·6
16	7 19·0	7 20·2	6 59·0	1·6	0·8	7·6	3·7	13·6	6·7
17	7 19·3	7 20·5	6 59·2	1·7	0·8	7·7	3·8	13·7	6·7
18	7 19·5	7 20·7	6 59·5	1·8	0·9	7·8	3·8	13·8	6·8
19	7 19·8	7 21·0	6 59·7	1·9	0·9	7·9	3·9	13·9	6·8
20	7 20·0	7 21·2	7 00·0	2·0	1·0	8·0	3·9	14·0	6·9
21	7 20·3	7 21·5	7 00·2	2·1	1·0	8·1	4·0	14·1	6·9
22	7 20·5	7 21·7	7 00·4	2·2	1·1	8·2	4·0	14·2	7·0
23	7 20·8	7 22·0	7 00·7	2·3	1·1	8·3	4·1	14·3	7·0
24	7 21·0	7 22·2	7 00·9	2·4	1·2	8·4	4·1	14·4	7·1
25	7 21·3	7 22·5	7 01·1	2·5	1·2	8·5	4·2	14·5	7·1
26	7 21·5	7 22·7	7 01·4	2·6	1·3	8·6	4·2	14·6	7·2
27	7 21·8	7 23·0	7 01·6	2·7	1·3	8·7	4·3	14·7	7·2
28	7 22·0	7 23·2	7 01·9	2·8	1·4	8·8	4·3	14·8	7·3
29	7 22·3	7 23·5	7 02·1	2·9	1·4	8·9	4·4	14·9	7·3
30	7 22·5	7 23·7	7 02·3	3·0	1·5	9·0	4·4	15·0	7·4
31	7 22·8	7 24·0	7 02·6	3·1	1·5	9·1	4·5	15·1	7·4
32	7 23·0	7 24·2	7 02·8	3·2	1·6	9·2	4·5	15·2	7·5
33	7 23·3	7 24·5	7 03·1	3·3	1·6	9·3	4·6	15·3	7·5
34	7 23·5	7 24·7	7 03·3	3·4	1·7	9·4	4·6	15·4	7·6
35	7 23·8	7 25·0	7 03·5	3·5	1·7	9·5	4·7	15·5	7·6
36	7 24·0	7 25·2	7 03·8	3·6	1·8	9·6	4·7	15·6	7·7
37	7 24·3	7 25·5	7 04·0	3·7	1·8	9·7	4·8	15·7	7·7
38	7 24·5	7 25·7	7 04·3	3·8	1·9	9·8	4·8	15·8	7·8
39	7 24·8	7 26·0	7 04·5	3·9	1·9	9·9	4·9	15·9	7·8
40	7 25·0	7 26·2	7 04·7	4·0	2·0	10·0	4·9	16·0	7·9
41	7 25·3	7 26·5	7 05·0	4·1	2·0	10·1	5·0	16·1	7·9
42	7 25·5	7 26·7	7 05·2	4·2	2·1	10·2	5·0	16·2	8·0
43	7 25·8	7 27·0	7 05·4	4·3	2·1	10·3	5·1	16·3	8·0
44	7 26·0	7 27·2	7 05·7	4·4	2·2	10·4	5·1	16·4	8·1
45	7 26·3	7 27·5	7 05·9	4·5	2·2	10·5	5·2	16·5	8·1
46	7 26·5	7 27·7	7 06·2	4·6	2·3	10·6	5·2	16·6	8·2
47	7 26·8	7 28·0	7 06·4	4·7	2·3	10·7	5·3	16·7	8·2
48	7 27·0	7 28·2	7 06·6	4·8	2·4	10·8	5·3	16·8	8·3
49	7 27·3	7 28·5	7 06·9	4·9	2·4	10·9	5·4	16·9	8·3
50	7 27·5	7 28·7	7 07·1	5·0	2·5	11·0	5·4	17·0	8·4
51	7 27·8	7 29·0	7 07·4	5·1	2·5	11·1	5·5	17·1	8·4
52	7 28·0	7 29·2	7 07·6	5·2	2·6	11·2	5·5	17·2	8·5
53	7 28·3	7 29·5	7 07·8	5·3	2·6	11·3	5·6	17·3	8·5
54	7 28·5	7 29·7	7 08·1	5·4	2·7	11·4	5·6	17·4	8·6
55	7 28·8	7 30·0	7 08·3	5·5	2·7	11·5	5·7	17·5	8·6
56	7 29·0	7 30·2	7 08·5	5·6	2·8	11·6	5·7	17·6	8·7
57	7 29·3	7 30·5	7 08·8	5·7	2·8	11·7	5·8	17·7	8·7
58	7 29·5	7 30·7	7 09·0	5·8	2·9	11·8	5·8	17·8	8·8
59	7 29·8	7 31·0	7 09·3	5·9	2·9	11·9	5·9	17·9	8·8
60	7 30·0	7 31·2	7 09·5	6·0	3·0	12·0	5·9	18·0	8·9

30m INCREMENTS AND CORRECTIONS

30m	SUN PLANETS	ARIES	MOON	v or d	Corr	v or d	Corr	v or d	Corr
s	° ′	° ′	° ′	′	′	′	′	′	′
00	7 30·0	7 31·2	7 09·5	0·0	0·0	6·0	3·1	12·0	6·1
01	7 30·3	7 31·5	7 09·7	0·1	0·1	6·1	3·1	12·1	6·2
02	7 30·5	7 31·7	7 10·0	0·2	0·1	6·2	3·2	12·2	6·2
03	7 30·8	7 32·0	7 10·2	0·3	0·2	6·3	3·2	12·3	6·3
04	7 31·0	7 32·2	7 10·5	0·4	0·2	6·4	3·3	12·4	6·3
05	7 31·3	7 32·5	7 10·7	0·5	0·3	6·5	3·3	12·5	6·4
06	7 31·5	7 32·7	7 10·9	0·6	0·3	6·6	3·4	12·6	6·4
07	7 31·8	7 33·0	7 11·2	0·7	0·4	6·7	3·4	12·7	6·5
08	7 32·0	7 33·2	7 11·4	0·8	0·4	6·8	3·5	12·8	6·5
09	7 32·3	7 33·5	7 11·6	0·9	0·5	6·9	3·5	12·9	6·6
10	7 32·5	7 33·7	7 11·9	1·0	0·5	7·0	3·6	13·0	6·6
11	7 32·8	7 34·0	7 12·1	1·1	0·6	7·1	3·6	13·1	6·7
12	7 33·0	7 34·2	7 12·4	1·2	0·6	7·2	3·7	13·2	6·7
13	7 33·3	7 34·5	7 12·6	1·3	0·7	7·3	3·7	13·3	6·8
14	7 33·5	7 34·7	7 12·8	1·4	0·7	7·4	3·8	13·4	6·8
15	7 33·8	7 35·0	7 13·1	1·5	0·8	7·5	3·8	13·5	6·9
16	7 34·0	7 35·2	7 13·3	1·6	0·8	7·6	3·9	13·6	6·9
17	7 34·3	7 35·5	7 13·6	1·7	0·9	7·7	3·9	13·7	7·0
18	7 34·5	7 35·7	7 13·8	1·8	0·9	7·8	4·0	13·8	7·0
19	7 34·8	7 36·0	7 14·0	1·9	1·0	7·9	4·0	13·9	7·1
20	7 35·0	7 36·2	7 14·3	2·0	1·0	8·0	4·1	14·0	7·1
21	7 35·3	7 36·5	7 14·5	2·1	1·1	8·1	4·1	14·1	7·2
22	7 35·5	7 36·7	7 14·7	2·2	1·1	8·2	4·2	14·2	7·2
23	7 35·8	7 37·0	7 15·0	2·3	1·2	8·3	4·2	14·3	7·3
24	7 36·0	7 37·2	7 15·2	2·4	1·2	8·4	4·3	14·4	7·3
25	7 36·3	7 37·5	7 15·5	2·5	1·3	8·5	4·3	14·5	7·4
26	7 36·5	7 37·7	7 15·7	2·6	1·3	8·6	4·4	14·6	7·4
27	7 36·8	7 38·0	7 15·9	2·7	1·4	8·7	4·4	14·7	7·5
28	7 37·0	7 38·3	7 16·2	2·8	1·4	8·8	4·5	14·8	7·5
29	7 37·3	7 38·5	7 16·4	2·9	1·5	8·9	4·5	14·9	7·6
30	7 37·5	7 38·8	7 16·7	3·0	1·5	9·0	4·6	15·0	7·6
31	7 37·8	7 39·0	7 16·9	3·1	1·6	9·1	4·6	15·1	7·7
32	7 38·0	7 39·3	7 17·1	3·2	1·6	9·2	4·7	15·2	7·7
33	7 38·3	7 39·5	7 17·4	3·3	1·7	9·3	4·7	15·3	7·8
34	7 38·5	7 39·8	7 17·6	3·4	1·7	9·4	4·8	15·4	7·8
35	7 38·8	7 40·0	7 17·9	3·5	1·8	9·5	4·8	15·5	7·9
36	7 39·0	7 40·3	7 18·1	3·6	1·8	9·6	4·9	15·6	7·9
37	7 39·3	7 40·5	7 18·3	3·7	1·9	9·7	4·9	15·7	8·0
38	7 39·5	7 40·8	7 18·6	3·8	1·9	9·8	5·0	15·8	8·0
39	7 39·8	7 41·0	7 18·8	3·9	2·0	9·9	5·0	15·9	8·1
40	7 40·0	7 41·3	7 19·0	4·0	2·0	10·0	5·1	16·0	8·1
41	7 40·3	7 41·5	7 19·3	4·1	2·1	10·1	5·1	16·1	8·2
42	7 40·5	7 41·8	7 19·5	4·2	2·1	10·2	5·2	16·2	8·2
43	7 40·8	7 42·0	7 19·8	4·3	2·2	10·3	5·2	16·3	8·3
44	7 41·0	7 42·3	7 20·0	4·4	2·2	10·4	5·3	16·4	8·3
45	7 41·3	7 42·5	7 20·2	4·5	2·3	10·5	5·3	16·5	8·4
46	7 41·5	7 42·8	7 20·5	4·6	2·3	10·6	5·4	16·6	8·4
47	7 41·8	7 43·0	7 20·7	4·7	2·4	10·7	5·4	16·7	8·5
48	7 42·0	7 43·3	7 21·0	4·8	2·4	10·8	5·5	16·8	8·5
49	7 42·3	7 43·5	7 21·2	4·9	2·5	10·9	5·5	16·9	8·6
50	7 42·5	7 43·8	7 21·4	5·0	2·5	11·0	5·6	17·0	8·6
51	7 42·8	7 44·0	7 21·7	5·1	2·6	11·1	5·6	17·1	8·7
52	7 43·0	7 44·3	7 21·9	5·2	2·6	11·2	5·7	17·2	8·7
53	7 43·3	7 44·5	7 22·1	5·3	2·7	11·3	5·7	17·3	8·8
54	7 43·5	7 44·8	7 22·4	5·4	2·7	11·4	5·8	17·4	8·8
55	7 43·8	7 45·0	7 22·6	5·5	2·8	11·5	5·8	17·5	8·9
56	7 44·0	7 45·3	7 22·9	5·6	2·8	11·6	5·9	17·6	8·9
57	7 44·3	7 45·5	7 23·1	5·7	2·9	11·7	5·9	17·7	9·0
58	7 44·5	7 45·8	7 23·3	5·8	2·9	11·8	6·0	17·8	9·0
59	7 44·8	7 46·0	7 23·6	5·9	3·0	11·9	6·0	17·9	9·1
60	7 45·0	7 46·3	7 23·8	6·0	3·1	12·0	6·1	18·0	9·2

31m

31m	SUN PLANETS	ARIES	MOON	v or d	Corr	v or d	Corr	v or d	Corr
s	° ′	° ′	° ′	′	′	′	′	′	′
00	7 45·0	7 46·3	7 23·8	0·0	0·0	6·0	3·2	12·0	6·3
01	7 45·3	7 46·5	7 24·1	0·1	0·1	6·1	3·2	12·1	6·4
02	7 45·5	7 46·8	7 24·3	0·2	0·1	6·2	3·3	12·2	6·4
03	7 45·8	7 47·0	7 24·5	0·3	0·2	6·3	3·3	12·3	6·5
04	7 46·0	7 47·3	7 24·8	0·4	0·2	6·4	3·4	12·4	6·5
05	7 46·3	7 47·5	7 25·0	0·5	0·3	6·5	3·4	12·5	6·6
06	7 46·5	7 47·8	7 25·2	0·6	0·3	6·6	3·5	12·6	6·6
07	7 46·8	7 48·0	7 25·5	0·7	0·4	6·7	3·5	12·7	6·7
08	7 47·0	7 48·3	7 25·7	0·8	0·4	6·8	3·6	12·8	6·7
09	7 47·3	7 48·5	7 26·0	0·9	0·5	6·9	3·6	12·9	6·8
10	7 47·5	7 48·8	7 26·2	1·0	0·5	7·0	3·7	13·0	6·8
11	7 47·8	7 49·0	7 26·4	1·1	0·6	7·1	3·7	13·1	6·9
12	7 48·0	7 49·3	7 26·7	1·2	0·6	7·2	3·8	13·2	6·9
13	7 48·3	7 49·5	7 26·9	1·3	0·7	7·3	3·8	13·3	7·0
14	7 48·5	7 49·8	7 27·2	1·4	0·7	7·4	3·9	13·4	7·0
15	7 48·8	7 50·0	7 27·4	1·5	0·8	7·5	3·9	13·5	7·1
16	7 49·0	7 50·3	7 27·6	1·6	0·8	7·6	4·0	13·6	7·1
17	7 49·3	7 50·5	7 27·9	1·7	0·9	7·7	4·0	13·7	7·2
18	7 49·5	7 50·8	7 28·1	1·8	0·9	7·8	4·1	13·8	7·2
19	7 49·8	7 51·0	7 28·4	1·9	1·0	7·9	4·1	13·9	7·3
20	7 50·0	7 51·3	7 28·6	2·0	1·1	8·0	4·2	14·0	7·4
21	7 50·3	7 51·5	7 28·8	2·1	1·1	8·1	4·3	14·1	7·4
22	7 50·5	7 51·8	7 29·1	2·2	1·2	8·2	4·3	14·2	7·5
23	7 50·8	7 52·0	7 29·3	2·3	1·2	8·3	4·4	14·3	7·5
24	7 51·0	7 52·3	7 29·5	2·4	1·3	8·4	4·4	14·4	7·6
25	7 51·3	7 52·5	7 29·8	2·5	1·3	8·5	4·5	14·5	7·6
26	7 51·5	7 52·8	7 30·0	2·6	1·4	8·6	4·5	14·6	7·7
27	7 51·8	7 53·0	7 30·3	2·7	1·4	8·7	4·6	14·7	7·7
28	7 52·0	7 53·3	7 30·5	2·8	1·5	8·8	4·6	14·8	7·8
29	7 52·3	7 53·5	7 30·7	2·9	1·5	8·9	4·7	14·9	7·8
30	7 52·5	7 53·8	7 31·0	3·0	1·6	9·0	4·7	15·0	7·9
31	7 52·8	7 54·0	7 31·2	3·1	1·6	9·1	4·8	15·1	7·9
32	7 53·0	7 54·3	7 31·5	3·2	1·7	9·2	4·8	15·2	8·0
33	7 53·3	7 54·5	7 31·7	3·3	1·7	9·3	4·9	15·3	8·0
34	7 53·5	7 54·8	7 31·9	3·4	1·8	9·4	4·9	15·4	8·1
35	7 53·8	7 55·0	7 32·2	3·5	1·8	9·5	5·0	15·5	8·1
36	7 54·0	7 55·3	7 32·4	3·6	1·9	9·6	5·0	15·6	8·2
37	7 54·3	7 55·5	7 32·6	3·7	1·9	9·7	5·1	15·7	8·2
38	7 54·5	7 55·8	7 32·9	3·8	2·0	9·8	5·1	15·8	8·3
39	7 54·8	7 56·0	7 33·1	3·9	2·0	9·9	5·2	15·9	8·3
40	7 55·0	7 56·3	7 33·4	4·0	2·1	10·0	5·3	16·0	8·4
41	7 55·3	7 56·6	7 33·6	4·1	2·2	10·1	5·3	16·1	8·5
42	7 55·5	7 56·8	7 33·8	4·2	2·2	10·2	5·4	16·2	8·5
43	7 55·8	7 57·1	7 34·1	4·3	2·3	10·3	5·4	16·3	8·6
44	7 56·0	7 57·3	7 34·3	4·4	2·3	10·4	5·5	16·4	8·6
45	7 56·3	7 57·6	7 34·6	4·5	2·4	10·5	5·5	16·5	8·7
46	7 56·5	7 57·8	7 34·8	4·6	2·4	10·6	5·6	16·6	8·7
47	7 56·8	7 58·1	7 35·0	4·7	2·5	10·7	5·6	16·7	8·8
48	7 57·0	7 58·3	7 35·3	4·8	2·5	10·8	5·7	16·8	8·8
49	7 57·3	7 58·6	7 35·5	4·9	2·6	10·9	5·7	16·9	8·9
50	7 57·5	7 58·8	7 35·7	5·0	2·6	11·0	5·8	17·0	8·9
51	7 57·8	7 59·1	7 36·0	5·1	2·7	11·1	5·8	17·1	9·0
52	7 58·0	7 59·3	7 36·2	5·2	2·7	11·2	5·9	17·2	9·0
53	7 58·3	7 59·6	7 36·5	5·3	2·8	11·3	5·9	17·3	9·1
54	7 58·5	7 59·8	7 36·7	5·4	2·8	11·4	6·0	17·4	9·1
55	7 58·8	8 00·1	7 36·9	5·5	2·9	11·5	6·0	17·5	9·2
56	7 59·0	8 00·3	7 37·2	5·6	2·9	11·6	6·1	17·6	9·2
57	7 59·3	8 00·6	7 37·4	5·7	3·0	11·7	6·1	17·7	9·3
58	7 59·5	8 00·8	7 37·7	5·8	3·0	11·8	6·2	17·8	9·3
59	7 59·8	8 01·1	7 37·9	5·9	3·1	11·9	6·2	17·9	9·4
60	8 00·0	8 01·3	7 38·1	6·0	3·2	12·0	6·3	18·0	9·5

32m INCREMENTS AND CORRECTIONS 33m

32m	SUN PLANETS	ARIES	MOON	v or d	Corrn	v or d	Corrn	v or d	Corrn
s	° ′	° ′	° ′	′	′	′	′	′	′
00	8 00·0	8 01·3	7 38·1	0·0	0·0	6·0	3·3	12·0	6·5
01	8 00·3	8 01·6	7 38·4	0·1	0·1	6·1	3·3	12·1	6·6
02	8 00·5	8 01·8	7 38·6	0·2	0·1	6·2	3·4	12·2	6·6
03	8 00·8	8 02·1	7 38·8	0·3	0·2	6·3	3·4	12·3	6·7
04	8 01·0	8 02·3	7 39·1	0·4	0·2	6·4	3·5	12·4	6·7
05	8 01·3	8 02·6	7 39·3	0·5	0·3	6·5	3·5	12·5	6·8
06	8 01·5	8 02·8	7 39·6	0·6	0·3	6·6	3·6	12·6	6·8
07	8 01·8	8 03·1	7 39·8	0·7	0·4	6·7	3·6	12·7	6·9
08	8 02·0	8 03·3	7 40·0	0·8	0·4	6·8	3·7	12·8	6·9
09	8 02·3	8 03·6	7 40·3	0·9	0·5	6·9	3·7	12·9	7·0
10	8 02·5	8 03·8	7 40·5	1·0	0·5	7·0	3·8	13·0	7·0
11	8 02·8	8 04·1	7 40·8	1·1	0·6	7·1	3·8	13·1	7·1
12	8 03·0	8 04·3	7 41·0	1·2	0·7	7·2	3·9	13·2	7·2
13	8 03·3	8 04·6	7 41·2	1·3	0·7	7·3	4·0	13·3	7·2
14	8 03·5	8 04·8	7 41·5	1·4	0·8	7·4	4·0	13·4	7·3
15	8 03·8	8 05·1	7 41·7	1·5	0·8	7·5	4·1	13·5	7·3
16	8 04·0	8 05·3	7 42·0	1·6	0·9	7·6	4·1	13·6	7·4
17	8 04·3	8 05·6	7 42·2	1·7	0·9	7·7	4·2	13·7	7·4
18	8 04·5	8 05·8	7 42·4	1·8	1·0	7·8	4·2	13·8	7·5
19	8 04·8	8 06·1	7 42·7	1·9	1·0	7·9	4·3	13·9	7·5
20	8 05·0	8 06·3	7 42·9	2·0	1·1	8·0	4·3	14·0	7·6
21	8 05·3	8 06·6	7 43·1	2·1	1·1	8·1	4·4	14·1	7·6
22	8 05·5	8 06·8	7 43·4	2·2	1·2	8·2	4·4	14·2	7·7
23	8 05·8	8 07·1	7 43·6	2·3	1·2	8·3	4·5	14·3	7·7
24	8 06·0	8 07·3	7 43·9	2·4	1·3	8·4	4·6	14·4	7·8
25	8 06·3	8 07·6	7 44·1	2·5	1·4	8·5	4·6	14·5	7·9
26	8 06·5	8 07·8	7 44·3	2·6	1·4	8·6	4·7	14·6	7·9
27	8 06·8	8 08·1	7 44·6	2·7	1·5	8·7	4·7	14·7	8·0
28	8 07·0	8 08·3	7 44·8	2·8	1·5	8·8	4·8	14·8	8·0
29	8 07·3	8 08·6	7 45·1	2·9	1·6	8·9	4·8	14·9	8·1
30	8 07·5	8 08·8	7 45·3	3·0	1·6	9·0	4·9	15·0	8·1
31	8 07·8	8 09·1	7 45·5	3·1	1·7	9·1	4·9	15·1	8·2
32	8 08·0	8 09·3	7 45·8	3·2	1·7	9·2	5·0	15·2	8·2
33	8 08·3	8 09·6	7 46·0	3·3	1·8	9·3	5·0	15·3	8·3
34	8 08·5	8 09·8	7 46·2	3·4	1·8	9·4	5·1	15·4	8·3
35	8 08·8	8 10·1	7 46·5	3·5	1·9	9·5	5·1	15·5	8·4
36	8 09·0	8 10·3	7 46·7	3·6	2·0	9·6	5·2	15·6	8·5
37	8 09·3	8 10·6	7 47·0	3·7	2·0	9·7	5·3	15·7	8·5
38	8 09·5	8 10·8	7 47·2	3·8	2·1	9·8	5·3	15·8	8·6
39	8 09·8	8 11·1	7 47·4	3·9	2·1	9·9	5·4	15·9	8·6
40	8 10·0	8 11·3	7 47·7	4·0	2·2	10·0	5·4	16·0	8·7
41	8 10·3	8 11·6	7 47·9	4·1	2·2	10·1	5·5	16·1	8·7
42	8 10·5	8 11·8	7 48·2	4·2	2·3	10·2	5·5	16·2	8·8
43	8 10·8	8 12·1	7 48·4	4·3	2·3	10·3	5·6	16·3	8·8
44	8 11·0	8 12·3	7 48·6	4·4	2·4	10·4	5·6	16·4	8·9
45	8 11·3	8 12·6	7 48·9	4·5	2·4	10·5	5·7	16·5	8·9
46	8 11·5	8 12·8	7 49·1	4·6	2·5	10·6	5·7	16·6	9·0
47	8 11·8	8 13·1	7 49·3	4·7	2·5	10·7	5·8	16·7	9·0
48	8 12·0	8 13·3	7 49·6	4·8	2·6	10·8	5·9	16·8	9·1
49	8 12·3	8 13·6	7 49·8	4·9	2·7	10·9	5·9	16·9	9·2
50	8 12·5	8 13·8	7 50·1	5·0	2·7	11·0	6·0	17·0	9·2
51	8 12·8	8 14·1	7 50·3	5·1	2·8	11·1	6·0	17·1	9·3
52	8 13·0	8 14·3	7 50·5	5·2	2·8	11·2	6·1	17·2	9·3
53	8 13·3	8 14·6	7 50·8	5·3	2·9	11·3	6·1	17·3	9·4
54	8 13·5	8 14·9	7 51·0	5·4	2·9	11·4	6·2	17·4	9·4
55	8 13·8	8 15·1	7 51·3	5·5	3·0	11·5	6·2	17·5	9·5
56	8 14·0	8 15·4	7 51·5	5·6	3·0	11·6	6·3	17·6	9·5
57	8 14·3	8 15·6	7 51·7	5·7	3·1	11·7	6·3	17·7	9·6
58	8 14·5	8 15·9	7 52·0	5·8	3·1	11·8	6·4	17·8	9·6
59	8 14·8	8 16·1	7 52·2	5·9	3·2	11·9	6·4	17·9	9·7
60	8 15·0	8 16·4	7 52·5	6·0	3·3	12·0	6·5	18·0	9·8

33m	SUN PLANETS	ARIES	MOON	v or d	Corrn	v or d	Corrn	v or d	Corrn
s	° ′	° ′	° ′	′	′	′	′	′	′
00	8 15·0	8 16·4	7 52·5	0·0	0·0	6·0	3·4	12·0	6·7
01	8 15·3	8 16·6	7 52·7	0·1	0·1	6·1	3·4	12·1	6·8
02	8 15·5	8 16·9	7 52·9	0·2	0·1	6·2	3·5	12·2	6·8
03	8 15·8	8 17·1	7 53·2	0·3	0·2	6·3	3·5	12·3	6·9
04	8 16·0	8 17·4	7 53·4	0·4	0·2	6·4	3·6	12·4	6·9
05	8 16·3	8 17·6	7 53·6	0·5	0·3	6·5	3·6	12·5	7·0
06	8 16·5	8 17·9	7 53·9	0·6	0·3	6·6	3·7	12·6	7·0
07	8 16·8	8 18·1	7 54·1	0·7	0·4	6·7	3·7	12·7	7·1
08	8 17·0	8 18·4	7 54·4	0·8	0·4	6·8	3·8	12·8	7·1
09	8 17·3	8 18·6	7 54·6	0·9	0·5	6·9	3·9	12·9	7·2
10	8 17·5	8 18·9	7 54·8	1·0	0·6	7·0	3·9	13·0	7·3
11	8 17·8	8 19·1	7 55·1	1·1	0·6	7·1	4·0	13·1	7·3
12	8 18·0	8 19·4	7 55·3	1·2	0·7	7·2	4·0	13·2	7·4
13	8 18·3	8 19·6	7 55·6	1·3	0·7	7·3	4·1	13·3	7·4
14	8 18·5	8 19·9	7 55·8	1·4	0·8	7·4	4·1	13·4	7·5
15	8 18·8	8 20·1	7 56·0	1·5	0·8	7·5	4·2	13·5	7·5
16	8 19·0	8 20·4	7 56·3	1·6	0·9	7·6	4·2	13·6	7·6
17	8 19·3	8 20·6	7 56·5	1·7	0·9	7·7	4·3	13·7	7·6
18	8 19·5	8 20·9	7 56·7	1·8	1·0	7·8	4·4	13·8	7·7
19	8 19·8	8 21·1	7 57·0	1·9	1·1	7·9	4·4	13·9	7·8
20	8 20·0	8 21·4	7 57·2	2·0	1·1	8·0	4·5	14·0	7·8
21	8 20·3	8 21·6	7 57·5	2·1	1·2	8·1	4·5	14·1	7·9
22	8 20·5	8 21·9	7 57·7	2·2	1·2	8·2	4·6	14·2	7·9
23	8 20·8	8 22·1	7 57·9	2·3	1·3	8·3	4·6	14·3	8·0
24	8 21·0	8 22·4	7 58·2	2·4	1·3	8·4	4·7	14·4	8·0
25	8 21·3	8 22·6	7 58·4	2·5	1·4	8·5	4·7	14·5	8·1
26	8 21·5	8 22·9	7 58·7	2·6	1·5	8·6	4·8	14·6	8·2
27	8 21·8	8 23·1	7 58·9	2·7	1·5	8·7	4·9	14·7	8·2
28	8 22·0	8 23·4	7 59·1	2·8	1·6	8·8	4·9	14·8	8·3
29	8 22·3	8 23·6	7 59·4	2·9	1·6	8·9	5·0	14·9	8·3
30	8 22·5	8 23·9	7 59·6	3·0	1·7	9·0	5·0	15·0	8·4
31	8 22·8	8 24·1	7 59·8	3·1	1·7	9·1	5·1	15·1	8·4
32	8 23·0	8 24·4	8 00·1	3·2	1·8	9·2	5·1	15·2	8·5
33	8 23·3	8 24·6	8 00·3	3·3	1·8	9·3	5·2	15·3	8·5
34	8 23·5	8 24·9	8 00·6	3·4	1·9	9·4	5·2	15·4	8·6
35	8 23·8	8 25·1	8 00·8	3·5	2·0	9·5	5·3	15·5	8·7
36	8 24·0	8 25·4	8 01·0	3·6	2·0	9·6	5·4	15·6	8·7
37	8 24·3	8 25·6	8 01·3	3·7	2·1	9·7	5·4	15·7	8·8
38	8 24·5	8 25·9	8 01·5	3·8	2·1	9·8	5·5	15·8	8·8
39	8 24·8	8 26·1	8 01·8	3·9	2·2	9·9	5·5	15·9	8·9
40	8 25·0	8 26·4	8 02·0	4·0	2·2	10·0	5·6	16·0	8·9
41	8 25·3	8 26·6	8 02·2	4·1	2·3	10·1	5·6	16·1	9·0
42	8 25·5	8 26·9	8 02·5	4·2	2·3	10·2	5·7	16·2	9·0
43	8 25·8	8 27·1	8 02·7	4·3	2·4	10·3	5·8	16·3	9·1
44	8 26·0	8 27·4	8 02·9	4·4	2·5	10·4	5·8	16·4	9·2
45	8 26·3	8 27·6	8 03·2	4·5	2·5	10·5	5·9	16·5	9·2
46	8 26·5	8 27·9	8 03·4	4·6	2·6	10·6	5·9	16·6	9·3
47	8 26·8	8 28·1	8 03·7	4·7	2·6	10·7	6·0	16·7	9·3
48	8 27·0	8 28·4	8 03·9	4·8	2·7	10·8	6·0	16·8	9·4
49	8 27·3	8 28·6	8 04·1	4·9	2·7	10·9	6·1	16·9	9·4
50	8 27·5	8 28·9	8 04·4	5·0	2·8	11·0	6·1	17·0	9·5
51	8 27·8	8 29·1	8 04·6	5·1	2·8	11·1	6·2	17·1	9·5
52	8 28·0	8 29·4	8 04·9	5·2	2·9	11·2	6·3	17·2	9·6
53	8 28·3	8 29·6	8 05·1	5·3	3·0	11·3	6·3	17·3	9·7
54	8 28·5	8 29·9	8 05·3	5·4	3·0	11·4	6·4	17·4	9·7
55	8 28·8	8 30·1	8 05·6	5·5	3·1	11·5	6·4	17·5	9·8
56	8 29·0	8 30·4	8 05·8	5·6	3·1	11·6	6·5	17·6	9·8
57	8 29·3	8 30·6	8 06·1	5·7	3·2	11·7	6·5	17·7	9·9
58	8 29·5	8 30·9	8 06·3	5·8	3·2	11·8	6·6	17·8	9·9
59	8 29·8	8 31·1	8 06·5	5·9	3·3	11·9	6·6	17·9	10·0
60	8 30·0	8 31·4	8 06·8	6·0	3·4	12·0	6·7	18·0	10·1

34ᵐ INCREMENTS AND CORRECTIONS 35ᵐ

34ᵐ	SUN PLANETS	ARIES	MOON	v or d	Corrⁿ	v or d	Corrⁿ	v or d	Corrⁿ
s	° ′	° ′	° ′	′	′	′	′	′	′
00	8 30·0	8 31·4	8 06·8	0·0	0·0	6·0	3·5	12·0	6·9
01	8 30·3	8 31·6	8 07·0	0·1	0·1	6·1	3·5	12·1	7·0
02	8 30·5	8 31·9	8 07·2	0·2	0·1	6·2	3·6	12·2	7·0
03	8 30·8	8 32·1	8 07·5	0·3	0·2	6·3	3·6	12·3	7·1
04	8 31·0	8 32·4	8 07·7	0·4	0·2	6·4	3·7	12·4	7·1
05	8 31·3	8 32·6	8 08·0	0·5	0·3	6·5	3·7	12·5	7·2
06	8 31·5	8 32·9	8 08·2	0·6	0·3	6·6	3·8	12·6	7·2
07	8 31·8	8 33·2	8 08·4	0·7	0·4	6·7	3·9	12·7	7·3
08	8 32·0	8 33·4	8 08·7	0·8	0·5	6·8	3·9	12·8	7·4
09	8 32·3	8 33·7	8 08·9	0·9	0·5	6·9	4·0	12·9	7·4
10	8 32·5	8 33·9	8 09·2	1·0	0·6	7·0	4·0	13·0	7·5
11	8 32·8	8 34·2	8 09·4	1·1	0·6	7·1	4·1	13·1	7·5
12	8 33·0	8 34·4	8 09·6	1·2	0·7	7·2	4·1	13·2	7·6
13	8 33·3	8 34·7	8 09·9	1·3	0·7	7·3	4·2	13·3	7·6
14	8 33·5	8 34·9	8 10·1	1·4	0·8	7·4	4·3	13·4	7·7
15	8 33·8	8 35·2	8 10·3	1·5	0·9	7·5	4·3	13·5	7·8
16	8 34·0	8 35·4	8 10·6	1·6	0·9	7·6	4·4	13·6	7·8
17	8 34·3	8 35·7	8 10·8	1·7	1·0	7·7	4·4	13·7	7·9
18	8 34·5	8 35·9	8 11·1	1·8	1·0	7·8	4·5	13·8	7·9
19	8 34·8	8 36·2	8 11·3	1·9	1·1	7·9	4·5	13·9	8·0
20	8 35·0	8 36·4	8 11·5	2·0	1·2	8·0	4·6	14·0	8·1
21	8 35·3	8 36·7	8 11·8	2·1	1·2	8·1	4·7	14·1	8·1
22	8 35·5	8 36·9	8 12·0	2·2	1·3	8·2	4·7	14·2	8·2
23	8 35·8	8 37·2	8 12·3	2·3	1·3	8·3	4·8	14·3	8·2
24	8 36·0	8 37·4	8 12·5	2·4	1·4	8·4	4·8	14·4	8·3
25	8 36·3	8 37·7	8 12·7	2·5	1·4	8·5	4·9	14·5	8·3
26	8 36·5	8 37·9	8 13·0	2·6	1·5	8·6	4·9	14·6	8·4
27	8 36·8	8 38·2	8 13·2	2·7	1·6	8·7	5·0	14·7	8·5
28	8 37·0	8 38·4	8 13·4	2·8	1·6	8·8	5·1	14·8	8·5
29	8 37·3	8 38·7	8 13·7	2·9	1·7	8·9	5·1	14·9	8·6
30	8 37·5	8 38·9	8 13·9	3·0	1·7	9·0	5·2	15·0	8·6
31	8 37·8	8 39·2	8 14·2	3·1	1·8	9·1	5·2	15·1	8·7
32	8 38·0	8 39·4	8 14·4	3·2	1·8	9·2	5·3	15·2	8·7
33	8 38·3	8 39·7	8 14·6	3·3	1·9	9·3	5·3	15·3	8·8
34	8 38·5	8 39·9	8 14·9	3·4	2·0	9·4	5·4	15·4	8·9
35	8 38·8	8 40·2	8 15·1	3·5	2·0	9·5	5·5	15·5	8·9
36	8 39·0	8 40·4	8 15·4	3·6	2·1	9·6	5·5	15·6	9·0
37	8 39·3	8 40·7	8 15·6	3·7	2·1	9·7	5·6	15·7	9·0
38	8 39·5	8 40·9	8 15·8	3·8	2·2	9·8	5·6	15·8	9·1
39	8 39·8	8 41·2	8 16·1	3·9	2·2	9·9	5·7	15·9	9·1
40	8 40·0	8 41·4	8 16·3	4·0	2·3	10·0	5·8	16·0	9·2
41	8 40·3	8 41·7	8 16·5	4·1	2·4	10·1	5·8	16·1	9·3
42	8 40·5	8 41·9	8 16·8	4·2	2·4	10·2	5·9	16·2	9·3
43	8 40·8	8 42·2	8 17·0	4·3	2·5	10·3	5·9	16·3	9·4
44	8 41·0	8 42·4	8 17·3	4·4	2·5	10·4	6·0	16·4	9·4
45	8 41·3	8 42·7	8 17·5	4·5	2·6	10·5	6·0	16·5	9·5
46	8 41·5	8 42·9	8 17·7	4·6	2·6	10·6	6·1	16·6	9·5
47	8 41·8	8 43·2	8 18·0	4·7	2·7	10·7	6·2	16·7	9·6
48	8 42·0	8 43·4	8 18·2	4·8	2·8	10·8	6·2	16·8	9·7
49	8 42·3	8 43·7	8 18·5	4·9	2·8	10·9	6·3	16·9	9·7
50	8 42·5	8 43·9	8 18·7	5·0	2·9	11·0	6·3	17·0	9·8
51	8 42·8	8 44·2	8 18·9	5·1	2·9	11·1	6·4	17·1	9·8
52	8 43·0	8 44·4	8 19·2	5·2	3·0	11·2	6·4	17·2	9·9
53	8 43·3	8 44·7	8 19·4	5·3	3·0	11·3	6·5	17·3	9·9
54	8 43·5	8 44·9	8 19·7	5·4	3·1	11·4	6·6	17·4	10·0
55	8 43·8	8 45·2	8 19·9	5·5	3·2	11·5	6·6	17·5	10·1
56	8 44·0	8 45·4	8 20·1	5·6	3·2	11·6	6·7	17·6	10·1
57	8 44·3	8 45·7	8 20·4	5·7	3·3	11·7	6·7	17·7	10·2
58	8 44·5	8 45·9	8 20·6	5·8	3·3	11·8	6·8	17·8	10·2
59	8 44·8	8 46·2	8 20·8	5·9	3·4	11·9	6·8	17·9	10·3
60	8 45·0	8 46·4	8 21·1	6·0	3·5	12·0	6·9	18·0	10·4

35ᵐ	SUN PLANETS	ARIES	MOON	v or d	Corrⁿ	v or d	Corrⁿ	v or d	Corrⁿ
s	° ′	° ′	° ′	′	′	′	′	′	′
00	8 45·0	8 46·4	8 21·1	0·0	0·0	6·0	3·6	12·0	7·1
01	8 45·3	8 46·7	8 21·3	0·1	0·1	6·1	3·6	12·1	7·2
02	8 45·5	8 46·9	8 21·6	0·2	0·1	6·2	3·7	12·2	7·2
03	8 45·8	8 47·2	8 21·8	0·3	0·2	6·3	3·7	12·3	7·3
04	8 46·0	8 47·4	8 22·0	0·4	0·2	6·4	3·8	12·4	7·3
05	8 46·3	8 47·7	8 22·3	0·5	0·3	6·5	3·8	12·5	7·4
06	8 46·5	8 47·9	8 22·5	0·6	0·4	6·6	3·9	12·6	7·5
07	8 46·8	8 48·2	8 22·8	0·7	0·4	6·7	4·0	12·7	7·5
08	8 47·0	8 48·4	8 23·0	0·8	0·5	6·8	4·0	12·8	7·6
09	8 47·3	8 48·7	8 23·2	0·9	0·5	6·9	4·1	12·9	7·6
10	8 47·5	8 48·9	8 23·5	1·0	0·6	7·0	4·1	13·0	7·7
11	8 47·8	8 49·2	8 23·7	1·1	0·7	7·1	4·2	13·1	7·8
12	8 48·0	8 49·4	8 23·9	1·2	0·7	7·2	4·3	13·2	7·8
13	8 48·3	8 49·7	8 24·2	1·3	0·8	7·3	4·3	13·3	7·9
14	8 48·5	8 49·9	8 24·4	1·4	0·8	7·4	4·4	13·4	7·9
15	8 48·8	8 50·2	8 24·7	1·5	0·9	7·5	4·4	13·5	8·0
16	8 49·0	8 50·4	8 24·9	1·6	0·9	7·6	4·5	13·6	8·0
17	8 49·3	8 50·7	8 25·1	1·7	1·0	7·7	4·6	13·7	8·1
18	8 49·5	8 50·9	8 25·4	1·8	1·1	7·8	4·6	13·8	8·2
19	8 49·8	8 51·2	8 25·6	1·9	1·1	7·9	4·7	13·9	8·2
20	8 50·0	8 51·5	8 25·9	2·0	1·2	8·0	4·7	14·0	8·3
21	8 50·3	8 51·7	8 26·1	2·1	1·2	8·1	4·8	14·1	8·3
22	8 50·5	8 52·0	8 26·3	2·2	1·3	8·2	4·9	14·2	8·4
23	8 50·8	8 52·2	8 26·6	2·3	1·4	8·3	4·9	14·3	8·5
24	8 51·0	8 52·5	8 26·8	2·4	1·4	8·4	5·0	14·4	8·5
25	8 51·3	8 52·7	8 27·0	2·5	1·5	8·5	5·0	14·5	8·6
26	8 51·5	8 53·0	8 27·3	2·6	1·5	8·6	5·1	14·6	8·6
27	8 51·8	8 53·2	8 27·5	2·7	1·6	8·7	5·1	14·7	8·7
28	8 52·0	8 53·5	8 27·8	2·8	1·7	8·8	5·2	14·8	8·8
29	8 52·3	8 53·7	8 28·0	2·9	1·7	8·9	5·3	14·9	8·8
30	8 52·5	8 54·0	8 28·2	3·0	1·8	9·0	5·3	15·0	8·9
31	8 52·8	8 54·2	8 28·5	3·1	1·8	9·1	5·4	15·1	8·9
32	8 53·0	8 54·5	8 28·7	3·2	1·9	9·2	5·4	15·2	9·0
33	8 53·3	8 54·7	8 29·0	3·3	2·0	9·3	5·5	15·3	9·1
34	8 53·5	8 55·0	8 29·2	3·4	2·0	9·4	5·6	15·4	9·1
35	8 53·8	8 55·2	8 29·4	3·5	2·1	9·5	5·6	15·5	9·2
36	8 54·0	8 55·5	8 29·7	3·6	2·1	9·6	5·7	15·6	9·2
37	8 54·3	8 55·7	8 29·9	3·7	2·2	9·7	5·7	15·7	9·3
38	8 54·5	8 56·0	8 30·2	3·8	2·2	9·8	5·8	15·8	9·3
39	8 54·8	8 56·2	8 30·4	3·9	2·3	9·9	5·9	15·9	9·4
40	8 55·0	8 56·5	8 30·6	4·0	2·4	10·0	5·9	16·0	9·5
41	8 55·3	8 56·7	8 30·9	4·1	2·4	10·1	6·0	16·1	9·5
42	8 55·5	8 57·0	8 31·1	4·2	2·5	10·2	6·0	16·2	9·6
43	8 55·8	8 57·2	8 31·3	4·3	2·5	10·3	6·1	16·3	9·6
44	8 56·0	8 57·5	8 31·6	4·4	2·6	10·4	6·2	16·4	9·7
45	8 56·3	8 57·7	8 31·8	4·5	2·7	10·5	6·2	16·5	9·8
46	8 56·5	8 58·0	8 32·1	4·6	2·7	10·6	6·3	16·6	9·8
47	8 56·8	8 58·2	8 32·3	4·7	2·8	10·7	6·3	16·7	9·9
48	8 57·0	8 58·5	8 32·5	4·8	2·8	10·8	6·4	16·8	9·9
49	8 57·3	8 58·7	8 32·8	4·9	2·9	10·9	6·4	16·9	10·0
50	8 57·5	8 59·0	8 33·0	5·0	3·0	11·0	6·5	17·0	10·1
51	8 57·8	8 59·2	8 33·3	5·1	3·0	11·1	6·6	17·1	10·1
52	8 58·0	8 59·5	8 33·5	5·2	3·1	11·2	6·6	17·2	10·2
53	8 58·3	8 59·7	8 33·7	5·3	3·1	11·3	6·7	17·3	10·2
54	8 58·5	9 00·0	8 34·0	5·4	3·2	11·4	6·7	17·4	10·3
55	8 58·8	9 00·2	8 34·2	5·5	3·3	11·5	6·8	17·5	10·4
56	8 59·0	9 00·5	8 34·4	5·6	3·3	11·6	6·9	17·6	10·4
57	8 59·3	9 00·7	8 34·7	5·7	3·4	11·7	6·9	17·7	10·5
58	8 59·5	9 01·0	8 34·9	5·8	3·4	11·8	7·0	17·8	10·5
59	8 59·8	9 01·2	8 35·2	5·9	3·5	11·9	7·0	17·9	10·6
60	9 00·0	9 01·5	8 35·4	6·0	3·6	12·0	7·1	18·0	10·7

36m INCREMENTS AND CORRECTIONS 37m

36m	SUN PLANETS	ARIES	MOON	v or d	Corrn	v or d	Corrn	v or d	Corrn
s	° ′	° ′	° ′	′	′	′	′	′	′
00	9 00·0	9 01·5	8 35·4	0·0	0·0	6·0	3·7	12·0	7·3
01	9 00·3	9 01·7	8 35·6	0·1	0·1	6·1	3·7	12·1	7·4
02	9 00·5	9 02·0	8 35·9	0·2	0·1	6·2	3·8	12·2	7·4
03	9 00·8	9 02·2	8 36·1	0·3	0·2	6·3	3·8	12·3	7·5
04	9 01·0	9 02·5	8 36·4	0·4	0·2	6·4	3·9	12·4	7·5
05	9 01·3	9 02·7	8 36·6	0·5	0·3	6·5	4·0	12·5	7·6
06	9 01·5	9 03·0	8 36·8	0·6	0·4	6·6	4·0	12·6	7·7
07	9 01·8	9 03·2	8 37·1	0·7	0·4	6·7	4·1	12·7	7·7
08	9 02·0	9 03·5	8 37·3	0·8	0·5	6·8	4·1	12·8	7·8
09	9 02·3	9 03·7	8 37·5	0·9	0·5	6·9	4·2	12·9	7·8
10	9 02·5	9 04·0	8 37·8	1·0	0·6	7·0	4·3	13·0	7·9
11	9 02·8	9 04·2	8 38·0	1·1	0·7	7·1	4·3	13·1	8·0
12	9 03·0	9 04·5	8 38·3	1·2	0·7	7·2	4·4	13·2	8·0
13	9 03·3	9 04·7	8 38·5	1·3	0·8	7·3	4·4	13·3	8·1
14	9 03·5	9 05·0	8 38·7	1·4	0·9	7·4	4·5	13·4	8·2
15	9 03·8	9 05·2	8 39·0	1·5	0·9	7·5	4·6	13·5	8·2
16	9 04·0	9 05·5	8 39·2	1·6	1·0	7·6	4·6	13·6	8·3
17	9 04·3	9 05·7	8 39·5	1·7	1·0	7·7	4·7	13·7	8·3
18	9 04·5	9 06·0	8 39·7	1·8	1·1	7·8	4·7	13·8	8·4
19	9 04·8	9 06·2	8 39·9	1·9	1·2	7·9	4·8	13·9	8·5
20	9 05·0	9 06·5	8 40·2	2·0	1·2	8·0	4·9	14·0	8·5
21	9 05·3	9 06·7	8 40·4	2·1	1·3	8·1	4·9	14·1	8·6
22	9 05·5	9 07·0	8 40·6	2·2	1·3	8·2	5·0	14·2	8·6
23	9 05·8	9 07·2	8 40·9	2·3	1·4	8·3	5·0	14·3	8·7
24	9 06·0	9 07·5	8 41·1	2·4	1·5	8·4	5·1	14·4	8·8
25	9 06·3	9 07·7	8 41·4	2·5	1·5	8·5	5·2	14·5	8·8
26	9 06·5	9 08·0	8 41·6	2·6	1·6	8·6	5·2	14·6	8·9
27	9 06·8	9 08·2	8 41·8	2·7	1·6	8·7	5·3	14·7	8·9
28	9 07·0	9 08·5	8 42·1	2·8	1·7	8·8	5·4	14·8	9·0
29	9 07·3	9 08·7	8 42·3	2·9	1·8	8·9	5·4	14·9	9·1
30	9 07·5	9 09·0	8 42·6	3·0	1·8	9·0	5·5	15·0	9·1
31	9 07·8	9 09·2	8 42·8	3·1	1·9	9·1	5·5	15·1	9·2
32	9 08·0	9 09·5	8 43·0	3·2	1·9	9·2	5·6	15·2	9·2
33	9 08·3	9 09·8	8 43·3	3·3	2·0	9·3	5·7	15·3	9·3
34	9 08·5	9 10·0	8 43·5	3·4	2·1	9·4	5·7	15·4	9·4
35	9 08·8	9 10·3	8 43·8	3·5	2·1	9·5	5·8	15·5	9·4
36	9 09·0	9 10·5	8 44·0	3·6	2·2	9·6	5·8	15·6	9·5
37	9 09·3	9 10·8	8 44·2	3·7	2·3	9·7	5·9	15·7	9·6
38	9 09·5	9 11·0	8 44·5	3·8	2·3	9·8	6·0	15·8	9·6
39	9 09·8	9 11·3	8 44·7	3·9	2·4	9·9	6·0	15·9	9·7
40	9 10·0	9 11·5	8 44·9	4·0	2·4	10·0	6·1	16·0	9·7
41	9 10·3	9 11·8	8 45·2	4·1	2·5	10·1	6·1	16·1	9·8
42	9 10·5	9 12·0	8 45·4	4·2	2·6	10·2	6·2	16·2	9·9
43	9 10·8	9 12·3	8 45·7	4·3	2·6	10·3	6·3	16·3	9·9
44	9 11·0	9 12·5	8 45·9	4·4	2·7	10·4	6·3	16·4	10·0
45	9 11·3	9 12·8	8 46·1	4·5	2·7	10·5	6·4	16·5	10·0
46	9 11·5	9 13·0	8 46·4	4·6	2·8	10·6	6·4	16·6	10·1
47	9 11·8	9 13·3	8 46·6	4·7	2·9	10·7	6·5	16·7	10·2
48	9 12·0	9 13·5	8 46·9	4·8	2·9	10·8	6·6	16·8	10·2
49	9 12·3	9 13·8	8 47·1	4·9	3·0	10·9	6·6	16·9	10·3
50	9 12·5	9 14·0	8 47·3	5·0	3·0	11·0	6·7	17·0	10·3
51	9 12·8	9 14·3	8 47·6	5·1	3·1	11·1	6·8	17·1	10·4
52	9 13·0	9 14·5	8 47·8	5·2	3·2	11·2	6·8	17·2	10·5
53	9 13·3	9 14·8	8 48·0	5·3	3·2	11·3	6·9	17·3	10·5
54	9 13·5	9 15·0	8 48·3	5·4	3·3	11·4	6·9	17·4	10·6
55	9 13·8	9 15·3	8 48·5	5·5	3·3	11·5	7·0	17·5	10·6
56	9 14·0	9 15·5	8 48·8	5·6	3·4	11·6	7·1	17·6	10·7
57	9 14·3	9 15·8	8 49·0	5·7	3·5	11·7	7·1	17·7	10·8
58	9 14·5	9 16·0	8 49·2	5·8	3·5	11·8	7·2	17·8	10·8
59	9 14·8	9 16·3	8 49·5	5·9	3·6	11·9	7·2	17·9	10·9
60	9 15·0	9 16·5	8 49·7	6·0	3·7	12·0	7·3	18·0	11·0

37m	SUN PLANETS	ARIES	MOON	v or d	Corrn	v or d	Corrn	v or d	Corrn
s	° ′	° ′	° ′	′	′	′	′	′	′
00	9 15·0	9 16·5	8 49·7	0·0	0·0	6·0	3·8	12·0	7·5
01	9 15·3	9 16·8	8 50·0	0·1	0·1	6·1	3·8	12·1	7·6
02	9 15·5	9 17·0	8 50·2	0·2	0·1	6·2	3·9	12·2	7·6
03	9 15·8	9 17·3	8 50·4	0·3	0·2	6·3	3·9	12·3	7·7
04	9 16·0	9 17·5	8 50·7	0·4	0·3	6·4	4·0	12·4	7·8
05	9 16·3	9 17·8	8 50·9	0·5	0·3	6·5	4·1	12·5	7·8
06	9 16·5	9 18·0	8 51·1	0·6	0·4	6·6	4·1	12·6	7·9
07	9 16·8	9 18·3	8 51·4	0·7	0·4	6·7	4·2	12·7	7·9
08	9 17·0	9 18·5	8 51·6	0·8	0·5	6·8	4·3	12·8	8·0
09	9 17·3	9 18·8	8 51·9	0·9	0·6	6·9	4·3	12·9	8·1
10	9 17·5	9 19·0	8 52·1	1·0	0·6	7·0	4·4	13·0	8·1
11	9 17·8	9 19·3	8 52·3	1·1	0·7	7·1	4·4	13·1	8·2
12	9 18·0	9 19·5	8 52·6	1·2	0·8	7·2	4·5	13·2	8·3
13	9 18·3	9 19·8	8 52·8	1·3	0·8	7·3	4·6	13·3	8·3
14	9 18·5	9 20·0	8 53·1	1·4	0·9	7·4	4·6	13·4	8·4
15	9 18·8	9 20·3	8 53·3	1·5	0·9	7·5	4·7	13·5	8·4
16	9 19·0	9 20·5	8 53·5	1·6	1·0	7·6	4·8	13·6	8·5
17	9 19·3	9 20·8	8 53·8	1·7	1·1	7·7	4·8	13·7	8·6
18	9 19·5	9 21·0	8 54·0	1·8	1·1	7·8	4·9	13·8	8·6
19	9 19·8	9 21·3	8 54·3	1·9	1·2	7·9	4·9	13·9	8·7
20	9 20·0	9 21·5	8 54·5	2·0	1·3	8·0	5·0	14·0	8·8
21	9 20·3	9 21·8	8 54·7	2·1	1·3	8·1	5·1	14·1	8·8
22	9 20·5	9 22·0	8 55·0	2·2	1·4	8·2	5·1	14·2	8·9
23	9 20·8	9 22·3	8 55·2	2·3	1·4	8·3	5·2	14·3	8·9
24	9 21·0	9 22·5	8 55·4	2·4	1·5	8·4	5·3	14·4	9·0
25	9 21·3	9 22·8	8 55·7	2·5	1·6	8·5	5·3	14·5	9·1
26	9 21·5	9 23·0	8 55·9	2·6	1·6	8·6	5·4	14·6	9·1
27	9 21·8	9 23·3	8 56·2	2·7	1·7	8·7	5·4	14·7	9·2
28	9 22·0	9 23·5	8 56·4	2·8	1·8	8·8	5·5	14·8	9·3
29	9 22·3	9 23·8	8 56·6	2·9	1·8	8·9	5·6	14·9	9·3
30	9 22·5	9 24·0	8 56·9	3·0	1·9	9·0	5·6	15·0	9·4
31	9 22·8	9 24·3	8 57·1	3·1	1·9	9·1	5·7	15·1	9·4
32	9 23·0	9 24·5	8 57·4	3·2	2·0	9·2	5·8	15·2	9·5
33	9 23·3	9 24·8	8 57·6	3·3	2·1	9·3	5·8	15·3	9·6
34	9 23·5	9 25·0	8 57·8	3·4	2·1	9·4	5·9	15·4	9·6
35	9 23·8	9 25·3	8 58·1	3·5	2·2	9·5	5·9	15·5	9·7
36	9 24·0	9 25·5	8 58·3	3·6	2·3	9·6	6·0	15·6	9·8
37	9 24·3	9 25·8	8 58·5	3·7	2·3	9·7	6·1	15·7	9·8
38	9 24·5	9 26·0	8 58·8	3·8	2·4	9·8	6·1	15·8	9·9
39	9 24·8	9 26·3	8 59·0	3·9	2·4	9·9	6·2	15·9	9·9
40	9 25·0	9 26·5	8 59·3	4·0	2·5	10·0	6·3	16·0	10·0
41	9 25·3	9 26·8	8 59·5	4·1	2·6	10·1	6·3	16·1	10·1
42	9 25·5	9 27·0	8 59·7	4·2	2·6	10·2	6·4	16·2	10·1
43	9 25·8	9 27·3	9 00·0	4·3	2·7	10·3	6·4	16·3	10·2
44	9 26·0	9 27·5	9 00·2	4·4	2·8	10·4	6·5	16·4	10·3
45	9 26·3	9 27·8	9 00·5	4·5	2·8	10·5	6·6	16·5	10·3
46	9 26·5	9 28·1	9 00·7	4·6	2·9	10·6	6·6	16·6	10·4
47	9 26·8	9 28·3	9 00·9	4·7	2·9	10·7	6·7	16·7	10·4
48	9 27·0	9 28·6	9 01·2	4·8	3·0	10·8	6·8	16·8	10·5
49	9 27·3	9 28·8	9 01·4	4·9	3·1	10·9	6·8	16·9	10·6
50	9 27·5	9 29·1	9 01·6	5·0	3·1	11·0	6·9	17·0	10·6
51	9 27·8	9 29·3	9 01·9	5·1	3·2	11·1	6·9	17·1	10·7
52	9 28·0	9 29·6	9 02·1	5·2	3·3	11·2	7·0	17·2	10·8
53	9 28·3	9 29·8	9 02·4	5·3	3·3	11·3	7·1	17·3	10·8
54	9 28·5	9 30·1	9 02·6	5·4	3·4	11·4	7·1	17·4	10·9
55	9 28·8	9 30·3	9 02·8	5·5	3·4	11·5	7·2	17·5	10·9
56	9 29·0	9 30·6	9 03·1	5·6	3·5	11·6	7·3	17·6	11·0
57	9 29·3	9 30·8	9 03·3	5·7	3·6	11·7	7·3	17·7	11·1
58	9 29·5	9 31·1	9 03·6	5·8	3·6	11·8	7·4	17·8	11·1
59	9 29·8	9 31·3	9 03·8	5·9	3·7	11·9	7·4	17·9	11·2
60	9 30·0	9 31·6	9 04·0	6·0	3·8	12·0	7·5	18·0	11·3

38ᵐ INCREMENTS AND CORRECTIONS 39ᵐ

38ᵐ	SUN PLANETS	ARIES	MOON	v or d	Corrⁿ	v or d	Corrⁿ	v or d	Corrⁿ
s	° ′	° ′	° ′	′	′	′	′	′	′
00	9 30·0	9 31·6	9 04·0	0·0	0·0	6·0	3·9	12·0	7·7
01	9 30·3	9 31·8	9 04·3	0·1	0·1	6·1	3·9	12·1	7·8
02	9 30·5	9 32·1	9 04·5	0·2	0·1	6·2	4·0	12·2	7·8
03	9 30·8	9 32·3	9 04·7	0·3	0·2	6·3	4·0	12·3	7·9
04	9 31·0	9 32·6	9 05·0	0·4	0·3	6·4	4·1	12·4	8·0
05	9 31·3	9 32·8	9 05·2	0·5	0·3	6·5	4·2	12·5	8·0
06	9 31·5	9 33·1	9 05·5	0·6	0·4	6·6	4·2	12·6	8·1
07	9 31·8	9 33·3	9 05·7	0·7	0·4	6·7	4·3	12·7	8·1
08	9 32·0	9 33·6	9 05·9	0·8	0·5	6·8	4·4	12·8	8·2
09	9 32·3	9 33·8	9 06·2	0·9	0·6	6·9	4·4	12·9	8·3
10	9 32·5	9 34·1	9 06·4	1·0	0·6	7·0	4·5	13·0	8·3
11	9 32·8	9 34·3	9 06·7	1·1	0·7	7·1	4·6	13·1	8·4
12	9 33·0	9 34·6	9 06·9	1·2	0·8	7·2	4·6	13·2	8·5
13	9 33·3	9 34·8	9 07·1	1·3	0·8	7·3	4·7	13·3	8·5
14	9 33·5	9 35·1	9 07·4	1·4	0·9	7·4	4·7	13·4	8·6
15	9 33·8	9 35·3	9 07·6	1·5	1·0	7·5	4·8	13·5	8·7
16	9 34·0	9 35·6	9 07·9	1·6	1·0	7·6	4·9	13·6	8·7
17	9 34·3	9 35·8	9 08·1	1·7	1·1	7·7	4·9	13·7	8·8
18	9 34·5	9 36·1	9 08·3	1·8	1·2	7·8	5·0	13·8	8·9
19	9 34·8	9 36·3	9 08·6	1·9	1·2	7·9	5·1	13·9	8·9
20	9 35·0	9 36·6	9 08·8	2·0	1·3	8·0	5·1	14·0	9·0
21	9 35·3	9 36·8	9 09·0	2·1	1·3	8·1	5·2	14·1	9·0
22	9 35·5	9 37·1	9 09·3	2·2	1·4	8·2	5·3	14·2	9·1
23	9 35·8	9 37·3	9 09·5	2·3	1·5	8·3	5·3	14·3	9·2
24	9 36·0	9 37·6	9 09·8	2·4	1·5	8·4	5·4	14·4	9·2
25	9 36·3	9 37·8	9 10·0	2·5	1·6	8·5	5·5	14·5	9·3
26	9 36·5	9 38·1	9 10·2	2·6	1·7	8·6	5·5	14·6	9·4
27	9 36·8	9 38·3	9 10·5	2·7	1·7	8·7	5·6	14·7	9·4
28	9 37·0	9 38·6	9 10·7	2·8	1·8	8·8	5·6	14·8	9·5
29	9 37·3	9 38·8	9 11·0	2·9	1·9	8·9	5·7	14·9	9·6
30	9 37·5	9 39·1	9 11·2	3·0	1·9	9·0	5·8	15·0	9·6
31	9 37·8	9 39·3	9 11·4	3·1	2·0	9·1	5·8	15·1	9·7
32	9 38·0	9 39·6	9 11·7	3·2	2·1	9·2	5·9	15·2	9·8
33	9 38·3	9 39·8	9 11·9	3·3	2·1	9·3	6·0	15·3	9·8
34	9 38·5	9 40·1	9 12·1	3·4	2·2	9·4	6·0	15·4	9·9
35	9 38·8	9 40·3	9 12·4	3·5	2·2	9·5	6·1	15·5	9·9
36	9 39·0	9 40·6	9 12·6	3·6	2·3	9·6	6·2	15·6	10·0
37	9 39·3	9 40·8	9 12·9	3·7	2·4	9·7	6·2	15·7	10·1
38	9 39·5	9 41·1	9 13·1	3·8	2·4	9·8	6·3	15·8	10·1
39	9 39·8	9 41·3	9 13·3	3·9	2·5	9·9	6·4	15·9	10·2
40	9 40·0	9 41·6	9 13·6	4·0	2·6	10·0	6·4	16·0	10·3
41	9 40·3	9 41·8	9 13·8	4·1	2·6	10·1	6·5	16·1	10·3
42	9 40·5	9 42·1	9 14·1	4·2	2·7	10·2	6·5	16·2	10·4
43	9 40·8	9 42·3	9 14·3	4·3	2·8	10·3	6·6	16·3	10·5
44	9 41·0	9 42·6	9 14·5	4·4	2·8	10·4	6·7	16·4	10·5
45	9 41·3	9 42·8	9 14·8	4·5	2·9	10·5	6·7	16·5	10·6
46	9 41·5	9 43·1	9 15·0	4·6	3·0	10·6	6·8	16·6	10·7
47	9 41·8	9 43·3	9 15·2	4·7	3·0	10·7	6·9	16·7	10·7
48	9 42·0	9 43·6	9 15·5	4·8	3·1	10·8	6·9	16·8	10·8
49	9 42·3	9 43·8	9 15·7	4·9	3·1	10·9	7·0	16·9	10·8
50	9 42·5	9 44·1	9 16·0	5·0	3·2	11·0	7·1	17·0	10·9
51	9 42·8	9 44·3	9 16·2	5·1	3·3	11·1	7·1	17·1	11·0
52	9 43·0	9 44·6	9 16·4	5·2	3·3	11·2	7·2	17·2	11·0
53	9 43·3	9 44·8	9 16·7	5·3	3·4	11·3	7·3	17·3	11·1
54	9 43·5	9 45·1	9 16·9	5·4	3·5	11·4	7·3	17·4	11·2
55	9 43·8	9 45·3	9 17·2	5·5	3·5	11·5	7·4	17·5	11·2
56	9 44·0	9 45·6	9 17·4	5·6	3·6	11·6	7·4	17·6	11·3
57	9 44·3	9 45·8	9 17·6	5·7	3·7	11·7	7·5	17·7	11·4
58	9 44·5	9 46·1	9 17·9	5·8	3·7	11·8	7·6	17·8	11·4
59	9 44·8	9 46·4	9 18·1	5·9	3·8	11·9	7·6	17·9	11·5
60	9 45·0	9 46·6	9 18·4	6·0	3·9	12·0	7·7	18·0	11·6

39ᵐ	SUN PLANETS	ARIES	MOON	v or d	Corrⁿ	v or d	Corrⁿ	v or d	Corrⁿ
s	° ′	° ′	° ′	′	′	′	′	′	′
00	9 45·0	9 46·6	9 18·4	0·0	0·0	6·0	4·0	12·0	7·9
01	9 45·3	9 46·9	9 18·6	0·1	0·1	6·1	4·0	12·1	8·0
02	9 45·5	9 47·1	9 18·8	0·2	0·1	6·2	4·1	12·2	8·0
03	9 45·8	9 47·4	9 19·1	0·3	0·2	6·3	4·1	12·3	8·1
04	9 46·0	9 47·6	9 19·3	0·4	0·3	6·4	4·2	12·4	8·2
05	9 46·3	9 47·9	9 19·5	0·5	0·3	6·5	4·3	12·5	8·2
06	9 46·5	9 48·1	9 19·8	0·6	0·4	6·6	4·3	12·6	8·3
07	9 46·8	9 48·4	9 20·0	0·7	0·5	6·7	4·4	12·7	8·4
08	9 47·0	9 48·6	9 20·3	0·8	0·5	6·8	4·5	12·8	8·4
09	9 47·3	9 48·9	9 20·5	0·9	0·6	6·9	4·5	12·9	8·5
10	9 47·5	9 49·1	9 20·7	1·0	0·7	7·0	4·6	13·0	8·6
11	9 47·8	9 49·4	9 21·0	1·1	0·7	7·1	4·7	13·1	8·6
12	9 48·0	9 49·6	9 21·2	1·2	0·8	7·2	4·7	13·2	8·7
13	9 48·3	9 49·9	9 21·5	1·3	0·9	7·3	4·8	13·3	8·8
14	9 48·5	9 50·1	9 21·7	1·4	0·9	7·4	4·9	13·4	8·8
15	9 48·8	9 50·4	9 21·9	1·5	1·0	7·5	4·9	13·5	8·9
16	9 49·0	9 50·6	9 22·2	1·6	1·1	7·6	5·0	13·6	9·0
17	9 49·3	9 50·9	9 22·4	1·7	1·1	7·7	5·1	13·7	9·0
18	9 49·5	9 51·1	9 22·6	1·8	1·2	7·8	5·1	13·8	9·1
19	9 49·8	9 51·4	9 22·9	1·9	1·3	7·9	5·2	13·9	9·2
20	9 50·0	9 51·6	9 23·1	2·0	1·3	8·0	5·3	14·0	9·2
21	9 50·3	9 51·9	9 23·4	2·1	1·4	8·1	5·3	14·1	9·3
22	9 50·5	9 52·1	9 23·6	2·2	1·4	8·2	5·4	14·2	9·3
23	9 50·8	9 52·4	9 23·8	2·3	1·5	8·3	5·5	14·3	9·4
24	9 51·0	9 52·6	9 24·1	2·4	1·6	8·4	5·5	14·4	9·5
25	9 51·3	9 52·9	9 24·3	2·5	1·6	8·5	5·6	14·5	9·5
26	9 51·5	9 53·1	9 24·6	2·6	1·7	8·6	5·7	14·6	9·6
27	9 51·8	9 53·4	9 24·8	2·7	1·8	8·7	5·7	14·7	9·7
28	9 52·0	9 53·6	9 25·0	2·8	1·8	8·8	5·8	14·8	9·7
29	9 52·3	9 53·9	9 25·3	2·9	1·9	8·9	5·9	14·9	9·8
30	9 52·5	9 54·1	9 25·5	3·0	2·0	9·0	5·9	15·0	9·9
31	9 52·8	9 54·4	9 25·7	3·1	2·0	9·1	6·0	15·1	9·9
32	9 53·0	9 54·6	9 26·0	3·2	2·1	9·2	6·1	15·2	10·0
33	9 53·3	9 54·9	9 26·2	3·3	2·2	9·3	6·1	15·3	10·1
34	9 53·5	9 55·1	9 26·5	3·4	2·2	9·4	6·2	15·4	10·1
35	9 53·8	9 55·4	9 26·7	3·5	2·3	9·5	6·3	15·5	10·2
36	9 54·0	9 55·6	9 26·9	3·6	2·4	9·6	6·3	15·6	10·3
37	9 54·3	9 55·9	9 27·2	3·7	2·4	9·7	6·4	15·7	10·3
38	9 54·5	9 56·1	9 27·4	3·8	2·5	9·8	6·5	15·8	10·4
39	9 54·8	9 56·4	9 27·7	3·9	2·6	9·9	6·5	15·9	10·5
40	9 55·0	9 56·6	9 27·9	4·0	2·6	10·0	6·6	16·0	10·5
41	9 55·3	9 56·9	9 28·1	4·1	2·7	10·1	6·6	16·1	10·6
42	9 55·5	9 57·1	9 28·4	4·2	2·8	10·2	6·7	16·2	10·7
43	9 55·8	9 57·4	9 28·6	4·3	2·8	10·3	6·8	16·3	10·7
44	9 56·0	9 57·6	9 28·8	4·4	2·9	10·4	6·8	16·4	10·8
45	9 56·3	9 57·9	9 29·1	4·5	3·0	10·5	6·9	16·5	10·9
46	9 56·5	9 58·1	9 29·3	4·6	3·0	10·6	7·0	16·6	10·9
47	9 56·8	9 58·4	9 29·6	4·7	3·1	10·7	7·0	16·7	11·0
48	9 57·0	9 58·6	9 29·8	4·8	3·2	10·8	7·1	16·8	11·1
49	9 57·3	9 58·9	9 30·0	4·9	3·2	10·9	7·2	16·9	11·1
50	9 57·5	9 59·1	9 30·3	5·0	3·3	11·0	7·2	17·0	11·2
51	9 57·8	9 59·4	9 30·5	5·1	3·4	11·1	7·3	17·1	11·3
52	9 58·0	9 59·6	9 30·8	5·2	3·4	11·2	7·4	17·2	11·3
53	9 58·3	9 59·9	9 31·0	5·3	3·5	11·3	7·4	17·3	11·4
54	9 58·5	10 00·1	9 31·2	5·4	3·6	11·4	7·5	17·4	11·5
55	9 58·8	10 00·4	9 31·5	5·5	3·6	11·5	7·6	17·5	11·5
56	9 59·0	10 00·6	9 31·7	5·6	3·7	11·6	7·6	17·6	11·6
57	9 59·3	10 00·9	9 32·0	5·7	3·8	11·7	7·7	17·7	11·7
58	9 59·5	10 01·1	9 32·2	5·8	3·8	11·8	7·8	17·8	11·7
59	9 59·8	10 01·4	9 32·4	5·9	3·9	11·9	7·8	17·9	11·8
60	10 00·0	10 01·6	9 32·7	6·0	4·0	12·0	7·9	18·0	11·9

40^m INCREMENTS AND CORRECTIONS 41^m

40^m	SUN PLANETS	ARIES	MOON	v or d	Corrⁿ	v or d	Corrⁿ	v or d	Corrⁿ
s	° ′	° ′	° ′	′	′	′	′	′	′
00	10 00.0	10 01.6	9 32.7	0.0	0.0	6.0	4.1	12.0	8.1
01	10 00.3	10 01.9	9 32.9	0.1	0.1	6.1	4.1	12.1	8.2
02	10 00.5	10 02.1	9 33.1	0.2	0.1	6.2	4.2	12.2	8.2
03	10 00.8	10 02.4	9 33.4	0.3	0.2	6.3	4.3	12.3	8.3
04	10 01.0	10 02.6	9 33.6	0.4	0.3	6.4	4.3	12.4	8.4
05	10 01.3	10 02.9	9 33.9	0.5	0.3	6.5	4.4	12.5	8.4
06	10 01.5	10 03.1	9 34.1	0.6	0.4	6.6	4.5	12.6	8.5
07	10 01.8	10 03.4	9 34.3	0.7	0.5	6.7	4.5	12.7	8.6
08	10 02.0	10 03.6	9 34.6	0.8	0.5	6.8	4.6	12.8	8.6
09	10 02.3	10 03.9	9 34.8	0.9	0.6	6.9	4.7	12.9	8.7
10	10 02.5	10 04.1	9 35.1	1.0	0.7	7.0	4.7	13.0	8.8
11	10 02.8	10 04.4	9 35.3	1.1	0.7	7.1	4.8	13.1	8.8
12	10 03.0	10 04.7	9 35.5	1.2	0.8	7.2	4.9	13.2	8.9
13	10 03.3	10 04.9	9 35.8	1.3	0.9	7.3	4.9	13.3	9.0
14	10 03.5	10 05.2	9 36.0	1.4	0.9	7.4	5.0	13.4	9.0
15	10 03.8	10 05.4	9 36.2	1.5	1.0	7.5	5.1	13.5	9.1
16	10 04.0	10 05.7	9 36.5	1.6	1.1	7.6	5.1	13.6	9.2
17	10 04.3	10 05.9	9 36.7	1.7	1.1	7.7	5.2	13.7	9.2
18	10 04.5	10 06.2	9 37.0	1.8	1.2	7.8	5.3	13.8	9.3
19	10 04.8	10 06.4	9 37.2	1.9	1.3	7.9	5.3	13.9	9.4
20	10 05.0	10 06.7	9 37.4	2.0	1.4	8.0	5.4	14.0	9.5
21	10 05.3	10 06.9	9 37.7	2.1	1.4	8.1	5.5	14.1	9.5
22	10 05.5	10 07.2	9 37.9	2.2	1.5	8.2	5.5	14.2	9.6
23	10 05.8	10 07.4	9 38.2	2.3	1.6	8.3	5.6	14.3	9.7
24	10 06.0	10 07.7	9 38.4	2.4	1.6	8.4	5.7	14.4	9.7
25	10 06.3	10 07.9	9 38.6	2.5	1.7	8.5	5.7	14.5	9.8
26	10 06.5	10 08.2	9 38.9	2.6	1.8	8.6	5.8	14.6	9.9
27	10 06.8	10 08.4	9 39.1	2.7	1.8	8.7	5.9	14.7	9.9
28	10 07.0	10 08.7	9 39.3	2.8	1.9	8.8	5.9	14.8	10.0
29	10 07.3	10 08.9	9 39.6	2.9	2.0	8.9	6.0	14.9	10.1
30	10 07.5	10 09.2	9 39.8	3.0	2.0	9.0	6.1	15.0	10.1
31	10 07.8	10 09.4	9 40.1	3.1	2.1	9.1	6.1	15.1	10.2
32	10 08.0	10 09.7	9 40.3	3.2	2.2	9.2	6.2	15.2	10.3
33	10 08.3	10 09.9	9 40.5	3.3	2.2	9.3	6.3	15.3	10.3
34	10 08.5	10 10.2	9 40.8	3.4	2.3	9.4	6.3	15.4	10.4
35	10 08.8	10 10.4	9 41.0	3.5	2.4	9.5	6.4	15.5	10.5
36	10 09.0	10 10.7	9 41.3	3.6	2.4	9.6	6.5	15.6	10.5
37	10 09.3	10 10.9	9 41.5	3.7	2.5	9.7	6.5	15.7	10.6
38	10 09.5	10 11.2	9 41.7	3.8	2.6	9.8	6.6	15.8	10.7
39	10 09.8	10 11.4	9 42.0	3.9	2.6	9.9	6.7	15.9	10.7
40	10 10.0	10 11.7	9 42.2	4.0	2.7	10.0	6.8	16.0	10.8
41	10 10.3	10 11.9	9 42.4	4.1	2.8	10.1	6.8	16.1	10.9
42	10 10.5	10 12.2	9 42.7	4.2	2.8	10.2	6.9	16.2	10.9
43	10 10.8	10 12.4	9 42.9	4.3	2.9	10.3	7.0	16.3	11.0
44	10 11.0	10 12.7	9 43.2	4.4	3.0	10.4	7.0	16.4	11.1
45	10 11.3	10 12.9	9 43.4	4.5	3.0	10.5	7.1	16.5	11.1
46	10 11.5	10 13.2	9 43.6	4.6	3.1	10.6	7.2	16.6	11.2
47	10 11.8	10 13.4	9 43.9	4.7	3.2	10.7	7.2	16.7	11.3
48	10 12.0	10 13.7	9 44.1	4.8	3.2	10.8	7.3	16.8	11.3
49	10 12.3	10 13.9	9 44.4	4.9	3.3	10.9	7.4	16.9	11.4
50	10 12.5	10 14.2	9 44.6	5.0	3.4	11.0	7.4	17.0	11.5
51	10 12.8	10 14.4	9 44.8	5.1	3.4	11.1	7.5	17.1	11.5
52	10 13.0	10 14.7	9 45.1	5.2	3.5	11.2	7.6	17.2	11.6
53	10 13.3	10 14.9	9 45.3	5.3	3.6	11.3	7.6	17.3	11.7
54	10 13.5	10 15.2	9 45.6	5.4	3.6	11.4	7.7	17.4	11.7
55	10 13.8	10 15.4	9 45.8	5.5	3.7	11.5	7.8	17.5	11.8
56	10 14.0	10 15.7	9 46.0	5.6	3.8	11.6	7.8	17.6	11.9
57	10 14.3	10 15.9	9 46.3	5.7	3.8	11.7	7.9	17.7	11.9
58	10 14.5	10 16.2	9 46.5	5.8	3.9	11.8	8.0	17.8	12.0
59	10 14.8	10 16.4	9 46.7	5.9	4.0	11.9	8.0	17.9	12.1
60	10 15.0	10 16.7	9 47.0	6.0	4.1	12.0	8.1	18.0	12.2

41^m	SUN PLANETS	ARIES	MOON	v or d	Corrⁿ	v or d	Corrⁿ	v or d	Corrⁿ
s	° ′	° ′	° ′	′	′	′	′	′	′
00	10 15.0	10 16.7	9 47.0	0.0	0.0	6.0	4.2	12.0	8.3
01	10 15.3	10 16.9	9 47.2	0.1	0.1	6.1	4.2	12.1	8.4
02	10 15.5	10 17.2	9 47.5	0.2	0.1	6.2	4.3	12.2	8.4
03	10 15.8	10 17.4	9 47.7	0.3	0.2	6.3	4.4	12.3	8.5
04	10 16.0	10 17.7	9 47.9	0.4	0.3	6.4	4.4	12.4	8.6
05	10 16.3	10 17.9	9 48.2	0.5	0.3	6.5	4.5	12.5	8.6
06	10 16.5	10 18.2	9 48.4	0.6	0.4	6.6	4.6	12.6	8.7
07	10 16.8	10 18.4	9 48.7	0.7	0.5	6.7	4.6	12.7	8.8
08	10 17.0	10 18.7	9 48.9	0.8	0.6	6.8	4.7	12.8	8.9
09	10 17.3	10 18.9	9 49.1	0.9	0.6	6.9	4.8	12.9	8.9
10	10 17.5	10 19.2	9 49.4	1.0	0.7	7.0	4.8	13.0	9.0
11	10 17.8	10 19.4	9 49.6	1.1	0.8	7.1	4.9	13.1	9.1
12	10 18.0	10 19.7	9 49.8	1.2	0.8	7.2	5.0	13.2	9.1
13	10 18.3	10 19.9	9 50.1	1.3	0.9	7.3	5.0	13.3	9.2
14	10 18.5	10 20.2	9 50.3	1.4	1.0	7.4	5.1	13.4	9.3
15	10 18.8	10 20.4	9 50.6	1.5	1.0	7.5	5.2	13.5	9.3
16	10 19.0	10 20.7	9 50.8	1.6	1.1	7.6	5.3	13.6	9.4
17	10 19.3	10 20.9	9 51.0	1.7	1.2	7.7	5.3	13.7	9.5
18	10 19.5	10 21.2	9 51.3	1.8	1.2	7.8	5.4	13.8	9.5
19	10 19.8	10 21.4	9 51.5	1.9	1.3	7.9	5.5	13.9	9.6
20	10 20.0	10 21.7	9 51.8	2.0	1.4	8.0	5.5	14.0	9.7
21	10 20.3	10 21.9	9 52.0	2.1	1.5	8.1	5.6	14.1	9.8
22	10 20.5	10 22.2	9 52.2	2.2	1.5	8.2	5.7	14.2	9.8
23	10 20.8	10 22.4	9 52.5	2.3	1.6	8.3	5.7	14.3	9.9
24	10 21.0	10 22.7	9 52.7	2.4	1.7	8.4	5.8	14.4	10.0
25	10 21.3	10 23.0	9 52.9	2.5	1.7	8.5	5.9	14.5	10.0
26	10 21.5	10 23.2	9 53.2	2.6	1.8	8.6	5.9	14.6	10.1
27	10 21.8	10 23.5	9 53.4	2.7	1.9	8.7	6.0	14.7	10.2
28	10 22.0	10 23.7	9 53.7	2.8	1.9	8.8	6.1	14.8	10.2
29	10 22.3	10 24.0	9 53.9	2.9	2.0	8.9	6.2	14.9	10.3
30	10 22.5	10 24.2	9 54.1	3.0	2.1	9.0	6.2	15.0	10.4
31	10 22.8	10 24.5	9 54.4	3.1	2.1	9.1	6.3	15.1	10.4
32	10 23.0	10 24.7	9 54.6	3.2	2.2	9.2	6.4	15.2	10.5
33	10 23.3	10 25.0	9 54.9	3.3	2.3	9.3	6.4	15.3	10.6
34	10 23.5	10 25.2	9 55.1	3.4	2.4	9.4	6.5	15.4	10.7
35	10 23.8	10 25.5	9 55.3	3.5	2.4	9.5	6.6	15.5	10.7
36	10 24.0	10 25.7	9 55.6	3.6	2.5	9.6	6.6	15.6	10.8
37	10 24.3	10 26.0	9 55.8	3.7	2.6	9.7	6.7	15.7	10.9
38	10 24.5	10 26.2	9 56.1	3.8	2.6	9.8	6.8	15.8	10.9
39	10 24.8	10 26.5	9 56.3	3.9	2.7	9.9	6.8	15.9	11.0
40	10 25.0	10 26.7	9 56.5	4.0	2.8	10.0	6.9	16.0	11.1
41	10 25.3	10 27.0	9 56.8	4.1	2.8	10.1	7.0	16.1	11.1
42	10 25.5	10 27.2	9 57.0	4.2	2.9	10.2	7.1	16.2	11.2
43	10 25.8	10 27.5	9 57.2	4.3	3.0	10.3	7.1	16.3	11.3
44	10 26.0	10 27.7	9 57.5	4.4	3.0	10.4	7.2	16.4	11.3
45	10 26.3	10 28.0	9 57.7	4.5	3.1	10.5	7.3	16.5	11.4
46	10 26.5	10 28.2	9 58.0	4.6	3.2	10.6	7.3	16.6	11.5
47	10 26.8	10 28.5	9 58.2	4.7	3.3	10.7	7.4	16.7	11.6
48	10 27.0	10 28.7	9 58.4	4.8	3.3	10.8	7.5	16.8	11.6
49	10 27.3	10 29.0	9 58.7	4.9	3.4	10.9	7.5	16.9	11.7
50	10 27.5	10 29.2	9 58.9	5.0	3.5	11.0	7.6	17.0	11.8
51	10 27.8	10 29.5	9 59.2	5.1	3.5	11.1	7.7	17.1	11.8
52	10 28.0	10 29.7	9 59.4	5.2	3.6	11.2	7.7	17.2	11.9
53	10 28.3	10 30.0	9 59.6	5.3	3.7	11.3	7.8	17.3	12.0
54	10 28.5	10 30.2	9 59.9	5.4	3.7	11.4	7.9	17.4	12.0
55	10 28.8	10 30.5	10 00.1	5.5	3.8	11.5	8.0	17.5	12.1
56	10 29.0	10 30.7	10 00.3	5.6	3.9	11.6	8.0	17.6	12.2
57	10 29.3	10 31.0	10 00.6	5.7	3.9	11.7	8.1	17.7	12.2
58	10 29.5	10 31.2	10 00.8	5.8	4.0	11.8	8.2	17.8	12.3
59	10 29.8	10 31.5	10 01.1	5.9	4.1	11.9	8.2	17.9	12.4
60	10 30.0	10 31.7	10 01.3	6.0	4.2	12.0	8.3	18.0	12.5

42m INCREMENTS AND CORRECTIONS **43m**

42m s	SUN PLANETS ° ′	ARIES ° ′	MOON ° ′	v or d ′	Corrn ′	v or d ′	Corrn ′	v or d ′	Corrn ′
00	10 30·0	10 31·7	10 01·3	0·0	0·0	6·0	4·3	12·0	8·5
01	10 30·3	10 32·0	10 01·5	0·1	0·1	6·1	4·3	12·1	8·6
02	10 30·5	10 32·2	10 01·8	0·2	0·1	6·2	4·4	12·2	8·6
03	10 30·8	10 32·5	10 02·0	0·3	0·2	6·3	4·5	12·3	8·7
04	10 31·0	10 32·7	10 02·3	0·4	0·3	6·4	4·5	12·4	8·8
05	10 31·3	10 33·0	10 02·5	0·5	0·4	6·5	4·6	12·5	8·9
06	10 31·5	10 33·2	10 02·7	0·6	0·4	6·6	4·7	12·6	8·9
07	10 31·8	10 33·5	10 03·0	0·7	0·5	6·7	4·7	12·7	9·0
08	10 32·0	10 33·7	10 03·2	0·8	0·6	6·8	4·8	12·8	9·1
09	10 32·3	10 34·0	10 03·4	0·9	0·6	6·9	4·9	12·9	9·1
10	10 32·5	10 34·2	10 03·7	1·0	0·7	7·0	5·0	13·0	9·2
11	10 32·8	10 34·5	10 03·9	1·1	0·8	7·1	5·0	13·1	9·3
12	10 33·0	10 34·7	10 04·2	1·2	0·9	7·2	5·1	13·2	9·4
13	10 33·3	10 35·0	10 04·4	1·3	0·9	7·3	5·2	13·3	9·4
14	10 33·5	10 35·2	10 04·6	1·4	1·0	7·4	5·2	13·4	9·5
15	10 33·8	10 35·5	10 04·9	1·5	1·1	7·5	5·3	13·5	9·6
16	10 34·0	10 35·7	10 05·1	1·6	1·1	7·6	5·4	13·6	9·6
17	10 34·3	10 36·0	10 05·4	1·7	1·2	7·7	5·5	13·7	9·7
18	10 34·5	10 36·2	10 05·6	1·8	1·3	7·8	5·5	13·8	9·8
19	10 34·8	10 36·5	10 05·8	1·9	1·3	7·9	5·6	13·9	9·8
20	10 35·0	10 36·7	10 06·1	2·0	1·4	8·0	5·7	14·0	9·9
21	10 35·3	10 37·0	10 06·3	2·1	1·5	8·1	5·7	14·1	10·0
22	10 35·5	10 37·2	10 06·5	2·2	1·6	8·2	5·8	14·2	10·1
23	10 35·8	10 37·5	10 06·8	2·3	1·6	8·3	5·9	14·3	10·1
24	10 36·0	10 37·7	10 07·0	2·4	1·7	8·4	6·0	14·4	10·2
25	10 36·3	10 38·0	10 07·3	2·5	1·8	8·5	6·0	14·5	10·3
26	10 36·5	10 38·2	10 07·5	2·6	1·8	8·6	6·1	14·6	10·3
27	10 36·8	10 38·5	10 07·7	2·7	1·9	8·7	6·2	14·7	10·4
28	10 37·0	10 38·7	10 08·0	2·8	2·0	8·8	6·2	14·8	10·5
29	10 37·3	10 39·0	10 08·2	2·9	2·1	8·9	6·3	14·9	10·6
30	10 37·5	10 39·2	10 08·5	3·0	2·1	9·0	6·4	15·0	10·6
31	10 37·8	10 39·5	10 08·7	3·1	2·2	9·1	6·4	15·1	10·7
32	10 38·0	10 39·7	10 08·9	3·2	2·3	9·2	6·5	15·2	10·8
33	10 38·3	10 40·0	10 09·2	3·3	2·3	9·3	6·6	15·3	10·8
34	10 38·5	10 40·2	10 09·4	3·4	2·4	9·4	6·7	15·4	10·9
35	10 38·8	10 40·5	10 09·7	3·5	2·5	9·5	6·7	15·5	11·0
36	10 39·0	10 40·7	10 09·9	3·6	2·6	9·6	6·8	15·6	11·1
37	10 39·3	10 41·0	10 10·1	3·7	2·6	9·7	6·9	15·7	11·1
38	10 39·5	10 41·3	10 10·4	3·8	2·7	9·8	6·9	15·8	11·2
39	10 39·8	10 41·5	10 10·6	3·9	2·8	9·9	7·0	15·9	11·3
40	10 40·0	10 41·8	10 10·8	4·0	2·8	10·0	7·1	16·0	11·3
41	10 40·3	10 42·0	10 11·1	4·1	2·9	10·1	7·2	16·1	11·4
42	10 40·5	10 42·3	10 11·3	4·2	3·0	10·2	7·2	16·2	11·5
43	10 40·8	10 42·5	10 11·6	4·3	3·0	10·3	7·3	16·3	11·5
44	10 41·0	10 42·8	10 11·8	4·4	3·1	10·4	7·4	16·4	11·6
45	10 41·3	10 43·0	10 12·0	4·5	3·2	10·5	7·4	16·5	11·7
46	10 41·5	10 43·3	10 12·3	4·6	3·3	10·6	7·5	16·6	11·8
47	10 41·8	10 43·5	10 12·5	4·7	3·3	10·7	7·6	16·7	11·8
48	10 42·0	10 43·8	10 12·8	4·8	3·4	10·8	7·7	16·8	11·9
49	10 42·3	10 44·0	10 13·0	4·9	3·5	10·9	7·7	16·9	12·0
50	10 42·5	10 44·3	10 13·2	5·0	3·5	11·0	7·8	17·0	12·0
51	10 42·8	10 44·5	10 13·5	5·1	3·6	11·1	7·9	17·1	12·1
52	10 43·0	10 44·8	10 13·7	5·2	3·7	11·2	7·9	17·2	12·2
53	10 43·3	10 45·0	10 13·9	5·3	3·8	11·3	8·0	17·3	12·3
54	10 43·5	10 45·3	10 14·2	5·4	3·8	11·4	8·1	17·4	12·3
55	10 43·8	10 45·5	10 14·4	5·5	3·9	11·5	8·1	17·5	12·4
56	10 44·0	10 45·8	10 14·7	5·6	4·0	11·6	8·2	17·6	12·5
57	10 44·3	10 46·0	10 14·9	5·7	4·0	11·7	8·3	17·7	12·5
58	10 44·5	10 46·3	10 15·1	5·8	4·1	11·8	8·4	17·8	12·6
59	10 44·8	10 46·5	10 15·4	5·9	4·2	11·9	8·4	17·9	12·7
60	10 45·0	10 46·8	10 15·6	6·0	4·3	12·0	8·5	18·0	12·8

43m s	SUN PLANETS ° ′	ARIES ° ′	MOON ° ′	v or d ′	Corrn ′	v or d ′	Corrn ′	v or d ′	Corrn ′
00	10 45·0	10 46·8	10 15·6	0·0	0·0	6·0	4·4	12·0	8·7
01	10 45·3	10 47·0	10 15·9	0·1	0·1	6·1	4·4	12·1	8·8
02	10 45·5	10 47·3	10 16·1	0·2	0·1	6·2	4·5	12·2	8·8
03	10 45·8	10 47·5	10 16·3	0·3	0·2	6·3	4·6	12·3	8·9
04	10 46·0	10 47·8	10 16·6	0·4	0·3	6·4	4·6	12·4	9·0
05	10 46·3	10 48·0	10 16·8	0·5	0·4	6·5	4·7	12·5	9·1
06	10 46·5	10 48·3	10 17·0	0·6	0·4	6·6	4·8	12·6	9·1
07	10 46·8	10 48·5	10 17·3	0·7	0·5	6·7	4·9	12·7	9·2
08	10 47·0	10 48·8	10 17·5	0·8	0·6	6·8	4·9	12·8	9·3
09	10 47·3	10 49·0	10 17·8	0·9	0·7	6·9	5·0	12·9	9·4
10	10 47·5	10 49·3	10 18·0	1·0	0·7	7·0	5·1	13·0	9·4
11	10 47·8	10 49·5	10 18·2	1·1	0·8	7·1	5·1	13·1	9·5
12	10 48·0	10 49·8	10 18·5	1·2	0·9	7·2	5·2	13·2	9·6
13	10 48·3	10 50·0	10 18·7	1·3	0·9	7·3	5·3	13·3	9·6
14	10 48·5	10 50·3	10 19·0	1·4	1·0	7·4	5·4	13·4	9·7
15	10 48·8	10 50·5	10 19·2	1·5	1·1	7·5	5·4	13·5	9·8
16	10 49·0	10 50·8	10 19·4	1·6	1·2	7·6	5·5	13·6	9·9
17	10 49·3	10 51·0	10 19·7	1·7	1·2	7·7	5·6	13·7	9·9
18	10 49·5	10 51·3	10 19·9	1·8	1·3	7·8	5·7	13·8	10·0
19	10 49·8	10 51·5	10 20·2	1·9	1·4	7·9	5·7	13·9	10·1
20	10 50·0	10 51·8	10 20·4	2·0	1·5	8·0	5·8	14·0	10·2
21	10 50·3	10 52·0	10 20·6	2·1	1·5	8·1	5·9	14·1	10·2
22	10 50·5	10 52·3	10 20·9	2·2	1·6	8·2	5·9	14·2	10·3
23	10 50·8	10 52·5	10 21·1	2·3	1·7	8·3	6·0	14·3	10·4
24	10 51·0	10 52·8	10 21·3	2·4	1·7	8·4	6·1	14·4	10·4
25	10 51·3	10 53·0	10 21·6	2·5	1·8	8·5	6·2	14·5	10·5
26	10 51·5	10 53·3	10 21·8	2·6	1·9	8·6	6·2	14·6	10·6
27	10 51·8	10 53·5	10 22·1	2·7	2·0	8·7	6·3	14·7	10·7
28	10 52·0	10 53·8	10 22·3	2·8	2·0	8·8	6·4	14·8	10·7
29	10 52·3	10 54·0	10 22·5	2·9	2·1	8·9	6·5	14·9	10·8
30	10 52·5	10 54·3	10 22·8	3·0	2·2	9·0	6·5	15·0	10·9
31	10 52·8	10 54·5	10 23·0	3·1	2·2	9·1	6·6	15·1	10·9
32	10 53·0	10 54·8	10 23·3	3·2	2·3	9·2	6·7	15·2	11·0
33	10 53·3	10 55·0	10 23·5	3·3	2·4	9·3	6·7	15·3	11·1
34	10 53·5	10 55·3	10 23·7	3·4	2·5	9·4	6·8	15·4	11·2
35	10 53·8	10 55·5	10 24·0	3·5	2·5	9·5	6·9	15·5	11·2
36	10 54·0	10 55·8	10 24·2	3·6	2·6	9·6	7·0	15·6	11·3
37	10 54·3	10 56·0	10 24·4	3·7	2·7	9·7	7·0	15·7	11·4
38	10 54·5	10 56·3	10 24·7	3·8	2·8	9·8	7·1	15·8	11·5
39	10 54·8	10 56·5	10 24·9	3·9	2·8	9·9	7·2	15·9	11·5
40	10 55·0	10 56·8	10 25·2	4·0	2·9	10·0	7·3	16·0	11·6
41	10 55·3	10 57·0	10 25·4	4·1	3·0	10·1	7·3	16·1	11·7
42	10 55·5	10 57·3	10 25·6	4·2	3·0	10·2	7·4	16·2	11·7
43	10 55·8	10 57·5	10 25·9	4·3	3·1	10·3	7·5	16·3	11·8
44	10 56·0	10 57·8	10 26·1	4·4	3·2	10·4	7·5	16·4	11·9
45	10 56·3	10 58·0	10 26·4	4·5	3·3	10·5	7·6	16·5	12·0
46	10 56·5	10 58·3	10 26·6	4·6	3·3	10·6	7·7	16·6	12·0
47	10 56·8	10 58·5	10 26·8	4·7	3·4	10·7	7·8	16·7	12·1
48	10 57·0	10 58·8	10 27·1	4·8	3·5	10·8	7·8	16·8	12·2
49	10 57·3	10 59·0	10 27·3	4·9	3·6	10·9	7·9	16·9	12·3
50	10 57·5	10 59·3	10 27·5	5·0	3·6	11·0	8·0	17·0	12·3
51	10 57·8	10 59·6	10 27·8	5·1	3·7	11·1	8·0	17·1	12·4
52	10 58·0	10 59·8	10 28·0	5·2	3·8	11·2	8·1	17·2	12·5
53	10 58·3	11 00·1	10 28·3	5·3	3·8	11·3	8·2	17·3	12·5
54	10 58·5	11 00·3	10 28·5	5·4	3·9	11·4	8·3	17·4	12·6
55	10 58·8	11 00·6	10 28·7	5·5	4·0	11·5	8·3	17·5	12·7
56	10 59·0	11 00·8	10 29·0	5·6	4·1	11·6	8·4	17·6	12·8
57	10 59·3	11 01·1	10 29·2	5·7	4·1	11·7	8·5	17·7	12·8
58	10 59·5	11 01·3	10 29·5	5·8	4·2	11·8	8·6	17·8	12·9
59	10 59·8	11 01·6	10 29·7	5·9	4·3	11·9	8·6	17·9	13·0
60	11 00·0	11 01·8	10 29·9	6·0	4·4	12·0	8·7	18·0	13·1

44m INCREMENTS AND CORRECTIONS 45m

44m	SUN PLANETS	ARIES	MOON	v or d	Corrⁿ	v or d	Corrⁿ	v or d	Corrⁿ
s	° ′	° ′	° ′	′	′	′	′	′	′
00	11 00·0	11 01·8	10 29·9	0·0	0·0	6·0	4·5	12·0	8·9
01	11 00·3	11 02·1	10 30·2	0·1	0·1	6·1	4·5	12·1	9·0
02	11 00·5	11 02·3	10 30·4	0·2	0·1	6·2	4·6	12·2	9·0
03	11 00·8	11 02·6	10 30·6	0·3	0·2	6·3	4·7	12·3	9·1
04	11 01·0	11 02·8	10 30·9	0·4	0·3	6·4	4·7	12·4	9·2
05	11 01·3	11 03·1	10 31·1	0·5	0·4	6·5	4·8	12·5	9·3
06	11 01·5	11 03·3	10 31·4	0·6	0·4	6·6	4·9	12·6	9·3
07	11 01·8	11 03·6	10 31·6	0·7	0·5	6·7	5·0	12·7	9·4
08	11 02·0	11 03·8	10 31·8	0·8	0·6	6·8	5·0	12·8	9·5
09	11 02·3	11 04·1	10 32·1	0·9	0·7	6·9	5·1	12·9	9·6
10	11 02·5	11 04·3	10 32·3	1·0	0·7	7·0	5·2	13·0	9·6
11	11 02·8	11 04·6	10 32·6	1·1	0·8	7·1	5·3	13·1	9·7
12	11 03·0	11 04·8	10 32·8	1·2	0·9	7·2	5·3	13·2	9·8
13	11 03·3	11 05·1	10 33·0	1·3	1·0	7·3	5·4	13·3	9·9
14	11 03·5	11 05·3	10 33·3	1·4	1·0	7·4	5·5	13·4	9·9
15	11 03·8	11 05·6	10 33·5	1·5	1·1	7·5	5·6	13·5	10·0
16	11 04·0	11 05·8	10 33·8	1·6	1·2	7·6	5·6	13·6	10·1
17	11 04·3	11 06·1	10 34·0	1·7	1·3	7·7	5·7	13·7	10·2
18	11 04·5	11 06·3	10 34·2	1·8	1·3	7·8	5·8	13·8	10·2
19	11 04·8	11 06·6	10 34·5	1·9	1·4	7·9	5·9	13·9	10·3
20	11 05·0	11 06·8	10 34·7	2·0	1·5	8·0	5·9	14·0	10·4
21	11 05·3	11 07·1	10 34·9	2·1	1·6	8·1	6·0	14·1	10·5
22	11 05·5	11 07·3	10 35·2	2·2	1·6	8·2	6·1	14·2	10·5
23	11 05·8	11 07·6	10 35·4	2·3	1·7	8·3	6·2	14·3	10·6
24	11 06·0	11 07·8	10 35·7	2·4	1·8	8·4	6·2	14·4	10·7
25	11 06·3	11 08·1	10 35·9	2·5	1·9	8·5	6·3	14·5	10·8
26	11 06·5	11 08·3	10 36·1	2·6	1·9	8·6	6·4	14·6	10·8
27	11 06·8	11 08·6	10 36·4	2·7	2·0	8·7	6·5	14·7	10·9
28	11 07·0	11 08·8	10 36·6	2·8	2·1	8·8	6·5	14·8	11·0
29	11 07·3	11 09·1	10 36·9	2·9	2·2	8·9	6·6	14·9	11·1
30	11 07·5	11 09·3	10 37·1	3·0	2·2	9·0	6·7	15·0	11·1
31	11 07·8	11 09·6	10 37·3	3·1	2·3	9·1	6·7	15·1	11·2
32	11 08·0	11 09·8	10 37·6	3·2	2·4	9·2	6·8	15·2	11·3
33	11 08·3	11 10·1	10 37·8	3·3	2·4	9·3	6·9	15·3	11·3
34	11 08·5	11 10·3	10 38·0	3·4	2·5	9·4	7·0	15·4	11·4
35	11 08·8	11 10·6	10 38·3	3·5	2·6	9·5	7·0	15·5	11·5
36	11 09·0	11 10·8	10 38·5	3·6	2·7	9·6	7·1	15·6	11·6
37	11 09·3	11 11·1	10 38·8	3·7	2·7	9·7	7·2	15·7	11·6
38	11 09·5	11 11·3	10 39·0	3·8	2·8	9·8	7·3	15·8	11·7
39	11 09·8	11 11·6	10 39·2	3·9	2·9	9·9	7·3	15·9	11·8
40	11 10·0	11 11·8	10 39·5	4·0	3·0	10·0	7·4	16·0	11·9
41	11 10·3	11 12·1	10 39·7	4·1	3·0	10·1	7·5	16·1	11·9
42	11 10·5	11 12·3	10 40·0	4·2	3·1	10·2	7·6	16·2	12·0
43	11 10·8	11 12·6	10 40·2	4·3	3·2	10·3	7·6	16·3	12·1
44	11 11·0	11 12·8	10 40·4	4·4	3·3	10·4	7·7	16·4	12·2
45	11 11·3	11 13·1	10 40·7	4·5	3·3	10·5	7·8	16·5	12·2
46	11 11·5	11 13·3	10 40·9	4·6	3·4	10·6	7·9	16·6	12·3
47	11 11·8	11 13·6	10 41·1	4·7	3·5	10·7	7·9	16·7	12·4
48	11 12·0	11 13·8	10 41·4	4·8	3·6	10·8	8·0	16·8	12·5
49	11 12·3	11 14·1	10 41·6	4·9	3·6	10·9	8·1	16·9	12·5
50	11 12·5	11 14·3	10 41·9	5·0	3·7	11·0	8·2	17·0	12·6
51	11 12·8	11 14·6	10 42·1	5·1	3·8	11·1	8·2	17·1	12·7
52	11 13·0	11 14·8	10 42·3	5·2	3·9	11·2	8·3	17·2	12·8
53	11 13·3	11 15·1	10 42·6	5·3	3·9	11·3	8·4	17·3	12·8
54	11 13·5	11 15·3	10 42·8	5·4	4·0	11·4	8·5	17·4	12·9
55	11 13·8	11 15·6	10 43·1	5·5	4·1	11·5	8·5	17·5	13·0
56	11 14·0	11 15·8	10 43·3	5·6	4·2	11·6	8·6	17·6	13·1
57	11 14·3	11 16·1	10 43·5	5·7	4·2	11·7	8·7	17·7	13·1
58	11 14·5	11 16·3	10 43·8	5·8	4·3	11·8	8·8	17·8	13·2
59	11 14·8	11 16·6	10 44·0	5·9	4·4	11·9	8·8	17·9	13·3
60	11 15·0	11 16·8	10 44·3	6·0	4·5	12·0	8·9	18·0	13·4

45m	SUN PLANETS	ARIES	MOON	v or d	Corrⁿ	v or d	Corrⁿ	v or d	Corrⁿ
s	° ′	° ′	° ′	′	′	′	′	′	′
00	11 15·0	11 16·8	10 44·3	0·0	0·0	6·0	4·6	12·0	9·1
01	11 15·3	11 17·1	10 44·5	0·1	0·1	6·1	4·6	12·1	9·2
02	11 15·5	11 17·3	10 44·7	0·2	0·2	6·2	4·7	12·2	9·3
03	11 15·8	11 17·6	10 45·0	0·3	0·2	6·3	4·8	12·3	9·3
04	11 16·0	11 17·9	10 45·2	0·4	0·3	6·4	4·9	12·4	9·4
05	11 16·3	11 18·1	10 45·4	0·5	0·4	6·5	4·9	12·5	9·5
06	11 16·5	11 18·4	10 45·7	0·6	0·5	6·6	5·0	12·6	9·6
07	11 16·8	11 18·6	10 45·9	0·7	0·5	6·7	5·1	12·7	9·6
08	11 17·0	11 18·9	10 46·2	0·8	0·6	6·8	5·2	12·8	9·7
09	11 17·3	11 19·1	10 46·4	0·9	0·7	6·9	5·2	12·9	9·8
10	11 17·5	11 19·4	10 46·6	1·0	0·8	7·0	5·3	13·0	9·9
11	11 17·8	11 19·6	10 46·9	1·1	0·8	7·1	5·4	13·1	9·9
12	11 18·0	11 19·9	10 47·1	1·2	0·9	7·2	5·5	13·2	10·0
13	11 18·3	11 20·1	10 47·4	1·3	1·0	7·3	5·5	13·3	10·1
14	11 18·5	11 20·4	10 47·6	1·4	1·1	7·4	5·6	13·4	10·2
15	11 18·8	11 20·6	10 47·8	1·5	1·1	7·5	5·7	13·5	10·2
16	11 19·0	11 20·9	10 48·1	1·6	1·2	7·6	5·8	13·6	10·3
17	11 19·3	11 21·1	10 48·3	1·7	1·3	7·7	5·8	13·7	10·4
18	11 19·5	11 21·4	10 48·5	1·8	1·4	7·8	5·9	13·8	10·5
19	11 19·8	11 21·6	10 48·8	1·9	1·4	7·9	6·0	13·9	10·5
20	11 20·0	11 21·9	10 49·0	2·0	1·5	8·0	6·1	14·0	10·6
21	11 20·3	11 22·1	10 49·3	2·1	1·6	8·1	6·1	14·1	10·7
22	11 20·5	11 22·4	10 49·5	2·2	1·7	8·2	6·2	14·2	10·8
23	11 20·8	11 22·6	10 49·7	2·3	1·7	8·3	6·3	14·3	10·8
24	11 21·0	11 22·9	10 50·0	2·4	1·8	8·4	6·4	14·4	10·9
25	11 21·3	11 23·1	10 50·2	2·5	1·9	8·5	6·4	14·5	11·0
26	11 21·5	11 23·4	10 50·5	2·6	2·0	8·6	6·5	14·6	11·1
27	11 21·8	11 23·6	10 50·7	2·7	2·0	8·7	6·6	14·7	11·1
28	11 22·0	11 23·9	10 50·9	2·8	2·1	8·8	6·7	14·8	11·2
29	11 22·3	11 24·1	10 51·2	2·9	2·2	8·9	6·7	14·9	11·3
30	11 22·5	11 24·4	10 51·4	3·0	2·3	9·0	6·8	15·0	11·4
31	11 22·8	11 24·6	10 51·6	3·1	2·4	9·1	6·9	15·1	11·5
32	11 23·0	11 24·9	10 51·9	3·2	2·4	9·2	7·0	15·2	11·5
33	11 23·3	11 25·1	10 52·1	3·3	2·5	9·3	7·1	15·3	11·6
34	11 23·5	11 25·4	10 52·4	3·4	2·6	9·4	7·1	15·4	11·7
35	11 23·8	11 25·6	10 52·6	3·5	2·7	9·5	7·2	15·5	11·8
36	11 24·0	11 25·9	10 52·8	3·6	2·7	9·6	7·3	15·6	11·8
37	11 24·3	11 26·1	10 53·1	3·7	2·8	9·7	7·4	15·7	11·9
38	11 24·5	11 26·4	10 53·3	3·8	2·9	9·8	7·4	15·8	12·0
39	11 24·8	11 26·6	10 53·6	3·9	3·0	9·9	7·5	15·9	12·1
40	11 25·0	11 26·9	10 53·8	4·0	3·0	10·0	7·6	16·0	12·1
41	11 25·3	11 27·1	10 54·0	4·1	3·1	10·1	7·7	16·1	12·2
42	11 25·5	11 27·4	10 54·3	4·2	3·2	10·2	7·7	16·2	12·3
43	11 25·8	11 27·6	10 54·5	4·3	3·3	10·3	7·8	16·3	12·4
44	11 26·0	11 27·9	10 54·7	4·4	3·3	10·4	7·9	16·4	12·4
45	11 26·3	11 28·1	10 55·0	4·5	3·4	10·5	8·0	16·5	12·5
46	11 26·5	11 28·4	10 55·2	4·6	3·5	10·6	8·0	16·6	12·6
47	11 26·8	11 28·6	10 55·5	4·7	3·6	10·7	8·1	16·7	12·7
48	11 27·0	11 28·9	10 55·7	4·8	3·6	10·8	8·2	16·8	12·7
49	11 27·3	11 29·1	10 55·9	4·9	3·7	10·9	8·3	16·9	12·8
50	11 27·5	11 29·4	10 56·2	5·0	3·8	11·0	8·3	17·0	12·9
51	11 27·8	11 29·6	10 56·4	5·1	3·9	11·1	8·4	17·1	13·0
52	11 28·0	11 29·9	10 56·7	5·2	3·9	11·2	8·5	17·2	13·0
53	11 28·3	11 30·1	10 56·9	5·3	4·0	11·3	8·6	17·3	13·1
54	11 28·5	11 30·4	10 57·1	5·4	4·1	11·4	8·6	17·4	13·2
55	11 28·8	11 30·6	10 57·4	5·5	4·2	11·5	8·7	17·5	13·3
56	11 29·0	11 30·9	10 57·6	5·6	4·2	11·6	8·8	17·6	13·3
57	11 29·3	11 31·1	10 57·9	5·7	4·3	11·7	8·9	17·7	13·4
58	11 29·5	11 31·4	10 58·1	5·8	4·4	11·8	8·9	17·8	13·5
59	11 29·8	11 31·6	10 58·3	5·9	4·5	11·9	9·0	17·9	13·6
60	11 30·0	11 31·9	10 58·6	6·0	4·6	12·0	9·1	18·0	13·7

46ᵐ INCREMENTS AND CORRECTIONS 47ᵐ

46ᵐ s	SUN PLANETS ° ′	ARIES ° ′	MOON ° ′	v or d ′	Corrⁿ ′	v or d ′	Corrⁿ ′	v or d ′	Corrⁿ ′
00	11 30·0	11 31·9	10 58·6	0·0	0·0	6·0	4·7	12·0	9·3
01	11 30·3	11 32·1	10 58·8	0·1	0·1	6·1	4·7	12·1	9·4
02	11 30·5	11 32·4	10 59·0	0·2	0·2	6·2	4·8	12·2	9·5
03	11 30·8	11 32·6	10 59·3	0·3	0·2	6·3	4·9	12·3	9·5
04	11 31·0	11 32·9	10 59·5	0·4	0·3	6·4	5·0	12·4	9·6
05	11 31·3	11 33·1	10 59·8	0·5	0·4	6·5	5·0	12·5	9·7
06	11 31·5	11 33·4	11 00·0	0·6	0·5	6·6	5·1	12·6	9·8
07	11 31·8	11 33·6	11 00·2	0·7	0·5	6·7	5·2	12·7	9·8
08	11 32·0	11 33·9	11 00·5	0·8	0·6	6·8	5·3	12·8	9·9
09	11 32·3	11 34·1	11 00·7	0·9	0·7	6·9	5·3	12·9	10·0
10	11 32·5	11 34·4	11 01·0	1·0	0·8	7·0	5·4	13·0	10·1
11	11 32·8	11 34·6	11 01·2	1·1	0·9	7·1	5·5	13·1	10·2
12	11 33·0	11 34·9	11 01·4	1·2	0·9	7·2	5·6	13·2	10·2
13	11 33·3	11 35·1	11 01·7	1·3	1·0	7·3	5·7	13·3	10·3
14	11 33·5	11 35·4	11 01·9	1·4	1·1	7·4	5·7	13·4	10·4
15	11 33·8	11 35·6	11 02·1	1·5	1·2	7·5	5·8	13·5	10·5
16	11 34·0	11 35·9	11 02·4	1·6	1·2	7·6	5·9	13·6	10·5
17	11 34·3	11 36·2	11 02·6	1·7	1·3	7·7	6·0	13·7	10·6
18	11 34·5	11 36·4	11 02·9	1·8	1·4	7·8	6·0	13·8	10·7
19	11 34·8	11 36·7	11 03·1	1·9	1·5	7·9	6·1	13·9	10·8
20	11 35·0	11 36·9	11 03·3	2·0	1·6	8·0	6·2	14·0	10·9
21	11 35·3	11 37·2	11 03·6	2·1	1·6	8·1	6·3	14·1	10·9
22	11 35·5	11 37·4	11 03·8	2·2	1·7	8·2	6·4	14·2	11·0
23	11 35·8	11 37·7	11 04·1	2·3	1·8	8·3	6·4	14·3	11·1
24	11 36·0	11 37·9	11 04·3	2·4	1·9	8·4	6·5	14·4	11·2
25	11 36·3	11 38·2	11 04·5	2·5	1·9	8·5	6·6	14·5	11·2
26	11 36·5	11 38·4	11 04·8	2·6	2·0	8·6	6·7	14·6	11·3
27	11 36·8	11 38·7	11 05·0	2·7	2·1	8·7	6·7	14·7	11·4
28	11 37·0	11 38·9	11 05·2	2·8	2·2	8·8	6·8	14·8	11·5
29	11 37·3	11 39·2	11 05·5	2·9	2·2	8·9	6·9	14·9	11·5
30	11 37·5	11 39·4	11 05·7	3·0	2·3	9·0	7·0	15·0	11·6
31	11 37·8	11 39·7	11 06·0	3·1	2·4	9·1	7·1	15·1	11·7
32	11 38·0	11 39·9	11 06·2	3·2	2·5	9·2	7·1	15·2	11·8
33	11 38·3	11 40·2	11 06·4	3·3	2·6	9·3	7·2	15·3	11·9
34	11 38·5	11 40·4	11 06·7	3·4	2·6	9·4	7·3	15·4	11·9
35	11 38·8	11 40·7	11 06·9	3·5	2·7	9·5	7·4	15·5	12·0
36	11 39·0	11 40·9	11 07·2	3·6	2·8	9·6	7·4	15·6	12·1
37	11 39·3	11 41·2	11 07·4	3·7	2·9	9·7	7·5	15·7	12·2
38	11 39·5	11 41·4	11 07·6	3·8	2·9	9·8	7·6	15·8	12·2
39	11 39·8	11 41·7	11 07·9	3·9	3·0	9·9	7·7	15·9	12·3
40	11 40·0	11 41·9	11 08·1	4·0	3·1	10·0	7·8	16·0	12·4
41	11 40·3	11 42·2	11 08·3	4·1	3·2	10·1	7·8	16·1	12·5
42	11 40·5	11 42·4	11 08·6	4·2	3·3	10·2	7·9	16·2	12·6
43	11 40·8	11 42·7	11 08·8	4·3	3·3	10·3	8·0	16·3	12·6
44	11 41·0	11 42·9	11 09·1	4·4	3·4	10·4	8·1	16·4	12·7
45	11 41·3	11 43·2	11 09·3	4·5	3·5	10·5	8·1	16·5	12·8
46	11 41·5	11 43·4	11 09·5	4·6	3·6	10·6	8·2	16·6	12·9
47	11 41·8	11 43·7	11 09·8	4·7	3·6	10·7	8·3	16·7	12·9
48	11 42·0	11 43·9	11 10·0	4·8	3·7	10·8	8·4	16·8	13·0
49	11 42·3	11 44·2	11 10·3	4·9	3·8	10·9	8·4	16·9	13·1
50	11 42·5	11 44·4	11 10·5	5·0	3·9	11·0	8·5	17·0	13·2
51	11 42·8	11 44·7	11 10·7	5·1	4·0	11·1	8·6	17·1	13·3
52	11 43·0	11 44·9	11 11·0	5·2	4·0	11·2	8·7	17·2	13·3
53	11 43·3	11 45·2	11 11·2	5·3	4·1	11·3	8·8	17·3	13·4
54	11 43·5	11 45·4	11 11·5	5·4	4·2	11·4	8·8	17·4	13·5
55	11 43·8	11 45·7	11 11·7	5·5	4·3	11·5	8·9	17·5	13·6
56	11 44·0	11 45·9	11 11·9	5·6	4·3	11·6	9·0	17·6	13·6
57	11 44·3	11 46·2	11 12·2	5·7	4·4	11·7	9·1	17·7	13·7
58	11 44·5	11 46·4	11 12·4	5·8	4·5	11·8	9·1	17·8	13·8
59	11 44·8	11 46·7	11 12·6	5·9	4·6	11·9	9·2	17·9	13·9
60	11 45·0	11 46·9	11 12·9	6·0	4·7	12·0	9·3	18·0	14·0

47ᵐ s	SUN PLANETS ° ′	ARIES ° ′	MOON ° ′	v or d ′	Corrⁿ ′	v or d ′	Corrⁿ ′	v or d ′	Corrⁿ ′
00	11 45·0	11 46·9	11 12·9	0·0	0·0	6·0	4·8	12·0	9·5
01	11 45·3	11 47·2	11 13·1	0·1	0·1	6·1	4·8	12·1	9·6
02	11 45·5	11 47·4	11 13·4	0·2	0·2	6·2	4·9	12·2	9·7
03	11 45·8	11 47·7	11 13·6	0·3	0·2	6·3	5·0	12·3	9·7
04	11 46·0	11 47·9	11 13·8	0·4	0·3	6·4	5·1	12·4	9·8
05	11 46·3	11 48·2	11 14·1	0·5	0·4	6·5	5·1	12·5	9·9
06	11 46·5	11 48·4	11 14·3	0·6	0·5	6·6	5·2	12·6	10·0
07	11 46·8	11 48·7	11 14·6	0·7	0·6	6·7	5·3	12·7	10·1
08	11 47·0	11 48·9	11 14·8	0·8	0·6	6·8	5·4	12·8	10·1
09	11 47·3	11 49·2	11 15·0	0·9	0·7	6·9	5·5	12·9	10·2
10	11 47·5	11 49·4	11 15·3	1·0	0·8	7·0	5·5	13·0	10·3
11	11 47·8	11 49·7	11 15·5	1·1	0·9	7·1	5·6	13·1	10·4
12	11 48·0	11 49·9	11 15·7	1·2	1·0	7·2	5·7	13·2	10·5
13	11 48·3	11 50·2	11 16·0	1·3	1·0	7·3	5·8	13·3	10·5
14	11 48·5	11 50·4	11 16·2	1·4	1·1	7·4	5·9	13·4	10·6
15	11 48·8	11 50·7	11 16·5	1·5	1·2	7·5	5·9	13·5	10·7
16	11 49·0	11 50·9	11 16·7	1·6	1·3	7·6	6·0	13·6	10·8
17	11 49·3	11 51·2	11 16·9	1·7	1·3	7·7	6·1	13·7	10·8
18	11 49·5	11 51·4	11 17·2	1·8	1·4	7·8	6·2	13·8	10·9
19	11 49·8	11 51·7	11 17·4	1·9	1·5	7·9	6·3	13·9	11·0
20	11 50·0	11 51·9	11 17·7	2·0	1·6	8·0	6·3	14·0	11·1
21	11 50·3	11 52·2	11 17·9	2·1	1·7	8·1	6·4	14·1	11·2
22	11 50·5	11 52·4	11 18·1	2·2	1·7	8·2	6·5	14·2	11·2
23	11 50·8	11 52·7	11 18·4	2·3	1·8	8·3	6·6	14·3	11·3
24	11 51·0	11 52·9	11 18·6	2·4	1·9	8·4	6·7	14·4	11·4
25	11 51·3	11 53·2	11 18·8	2·5	2·0	8·5	6·7	14·5	11·5
26	11 51·5	11 53·4	11 19·1	2·6	2·1	8·6	6·8	14·6	11·6
27	11 51·8	11 53·7	11 19·3	2·7	2·1	8·7	6·9	14·7	11·6
28	11 52·0	11 53·9	11 19·6	2·8	2·2	8·8	7·0	14·8	11·7
29	11 52·3	11 54·2	11 19·8	2·9	2·3	8·9	7·0	14·9	11·8
30	11 52·5	11 54·5	11 20·0	3·0	2·4	9·0	7·1	15·0	11·9
31	11 52·8	11 54·7	11 20·3	3·1	2·5	9·1	7·2	15·1	12·0
32	11 53·0	11 55·0	11 20·5	3·2	2·5	9·2	7·3	15·2	12·0
33	11 53·3	11 55·2	11 20·8	3·3	2·6	9·3	7·4	15·3	12·1
34	11 53·5	11 55·5	11 21·0	3·4	2·7	9·4	7·4	15·4	12·2
35	11 53·8	11 55·7	11 21·2	3·5	2·8	9·5	7·5	15·5	12·3
36	11 54·0	11 56·0	11 21·5	3·6	2·9	9·6	7·6	15·6	12·4
37	11 54·3	11 56·2	11 21·7	3·7	2·9	9·7	7·7	15·7	12·4
38	11 54·5	11 56·5	11 22·0	3·8	3·0	9·8	7·8	15·8	12·5
39	11 54·8	11 56·7	11 22·2	3·9	3·1	9·9	7·8	15·9	12·6
40	11 55·0	11 57·0	11 22·4	4·0	3·2	10·0	7·9	16·0	12·7
41	11 55·3	11 57·2	11 22·7	4·1	3·2	10·1	8·0	16·1	12·7
42	11 55·5	11 57·5	11 22·9	4·2	3·3	10·2	8·1	16·2	12·8
43	11 55·8	11 57·7	11 23·1	4·3	3·4	10·3	8·2	16·3	12·9
44	11 56·0	11 58·0	11 23·4	4·4	3·5	10·4	8·2	16·4	13·0
45	11 56·3	11 58·2	11 23·6	4·5	3·6	10·5	8·3	16·5	13·1
46	11 56·5	11 58·5	11 23·9	4·6	3·6	10·6	8·4	16·6	13·1
47	11 56·8	11 58·7	11 24·1	4·7	3·7	10·7	8·5	16·7	13·2
48	11 57·0	11 59·0	11 24·3	4·8	3·8	10·8	8·6	16·8	13·3
49	11 57·3	11 59·2	11 24·6	4·9	3·9	10·9	8·6	16·9	13·4
50	11 57·5	11 59·5	11 24·8	5·0	4·0	11·0	8·7	17·0	13·5
51	11 57·8	11 59·7	11 25·1	5·1	4·0	11·1	8·8	17·1	13·5
52	11 58·0	12 00·0	11 25·3	5·2	4·1	11·2	8·9	17·2	13·6
53	11 58·3	12 00·2	11 25·5	5·3	4·2	11·3	8·9	17·3	13·7
54	11 58·5	12 00·5	11 25·8	5·4	4·3	11·4	9·0	17·4	13·8
55	11 58·8	12 00·7	11 26·0	5·5	4·4	11·5	9·1	17·5	13·9
56	11 59·0	12 01·0	11 26·2	5·6	4·4	11·6	9·2	17·6	13·9
57	11 59·3	12 01·2	11 26·5	5·7	4·5	11·7	9·3	17·7	14·0
58	11 59·5	12 01·5	11 26·7	5·8	4·6	11·8	9·3	17·8	14·1
59	11 59·8	12 01·7	11 27·0	5·9	4·7	11·9	9·4	17·9	14·2
60	12 00·0	12 02·0	11 27·2	6·0	4·8	12·0	9·5	18·0	14·3

Chart 5011

Edition 1 – 1991

SYMBOLS AND ABBREVIATIONS USED ON ADMIRALTY CHARTS

Reproduced from British Admiralty Chart No. 5011 with the permission of the Controller of H.M. Stationery Office and of the Hydrographer of the Navy.
Note:—As these pages have been printed in black some colours have not reproduced successfully, students are advised to purchase a copy of 5011 if required.

IA Chart Number, Title, Marginal Notes

Schematic Layout of an Admiralty INT chart (reduced in size)

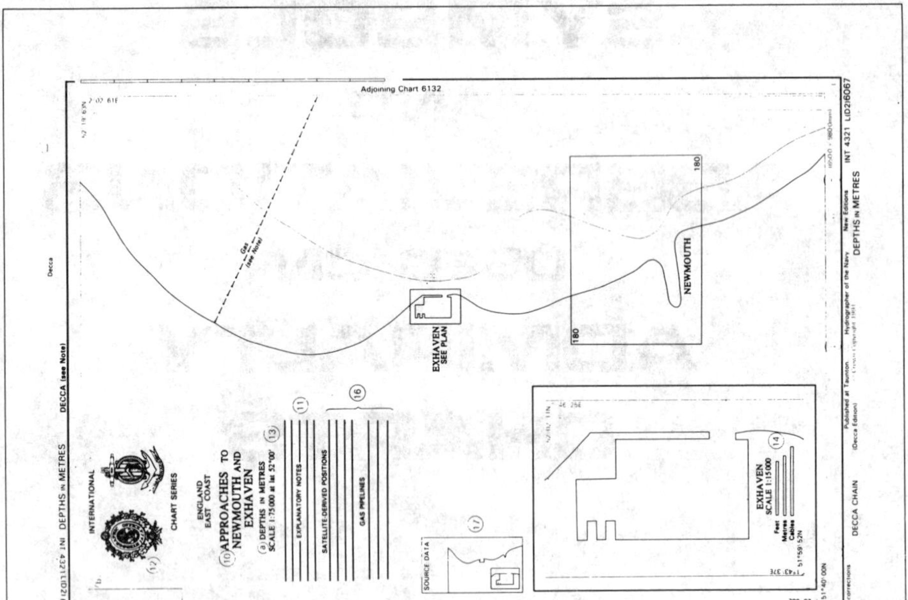

IB Positions, Distances, Directions, Compass

Geographical Positions

1	Lat	*Latitude*			*E13*
2	Long	*Longitude*			*E14*
3		*International Meridian (Greenwich)*			*E14a*
4	°	*Degree(s)*			130 *E20*
5	′	*Minute(s) of arc*			130 *E21*
6	″	*Second(s) of arc*			130 *E22*
7	PA	*Position approximate*	† (PA)	† (P.A.)	417 424 1 *Q41*
8	PD	*Position doubtful*	† (PD)	† (P.D.)	417 424 2 *Q42*
9	N	*North*			*U1*
10	E	*East*			*U2*
11	S	*South*			*U3*
12	W	*West*			*U4*
13	NE	*Northeast*			*U5*
14	SE	*Southeast*			*U6*
15	NW	*Northwest*			*U8*
16	SW	*Southwest*			*U7*

Control Points

20	▲	*Triangulation point*			304 1 *D1*
21	⊕	*Observation spot*	† + Obs Spot	† + *Obsn. Spot*	304 2 *D4*
22	○ •	*Fixed point*			305 1 340 5 *D2*
23		*Benchmark*	† BM	† B.M.	304 3 *D5*
24		*Boundary mark*			306 *D14*
a		*Viewpoint*	∘ See View		390 2 *D6*

Symbolised Positions (Examples)

30	# (18) Wk	*Symbols in plan: position is centre of primary symbol*			305 1
31	P	*Symbols in profile: position is at bottom of symbol*			305 1
32	∘ Mast ○ MAST ★	*Point symbols (accurate positions)*			305 1 340 5
33	∘ Mast PA	*Approximate position*	∘ Mast PA		305 1

Chart Number, Title, Marginal Notes IA

Magnetic Features → IB *Tidal Data* → IH *Decca, Loran-C, Omega* → IS

(1)	*Chart number in the Admiralty series. Where it is necessary to distinguish Admiralty charts from others, it is usual to add the prefix 'BA' (British Admiralty) to the number.*	251
(2)	*Identification of a latticed chart (if any). Charts for which Decca, Loran-C or Omega lattice-overprinted versions are available bear the letter 'L' in the lower right-hand corner inside the outer border. The lattice prefix number is shown against the national number.*	603
(3)	*Chart number in the International (INT) Chart series.*	251 1
(4)	*Publication note (imprint) showing the date of publication as a New Chart.*	252 1 252 4
(5)	*Copyright note. All Admiralty charts are subject to Crown Copyright restrictions.*	253
(6)	*Dates of (a) New Editions and (b) Large Corrections (abandoned as a revision category in 1972).*	252 2
(7)	*Small corrections: (a) the year dates and numbers of Notices to Mariners and (b) the dates (usually bracketed) of minor corrections included in reprints but not formally promulgated (abandoned as a method of correction in 1986).*	252 3
(8)	*Dimensions of the inner neat-lines of the chart border. In the case of charts on Transverse Mercator and Gnomonic projections, dimensions may be quoted for all borders of the chart which differ. Fathoms charts show the dimensions in inches eg (38.40 × 25.40).*	222 3 222 4
(9)	*Corner co-ordinates*	232
(10)	*Chart title. This should be quoted, in addition to the chart number, when ordering a chart.*	241 3
(11)	*Explanatory notes on chart content; to be read before using the chart.*	242
(12)	*Seals. Where an Admiralty chart is in the International Chart series, the seal of the International Hydrographic Organization (IHO) is shown in addition to the national seal. Modified facsimile reproductions of the charts of other nations show the seals of the original producer (left), the publisher (centre), and the IHO (right).*	241 1 241 2
(13)	*Scale of chart; on Mercator projection, at a stated latitude.*	211 241 4
(14)	*Linear scales on large-scale plan.*	221
(15)	*Linear border scales (metres). On smaller-scale charts, the latitude border should be used to measure Sea miles and Cables.*	221 1
(16)	*Cautionary notes (if any) on charted detail, to be read before using the chart.*	242
(17)	*Source Diagram (if any). If a Source Diagram is not shown, details of the sources used in the compilation of the chart are given in the explanatory notes (see 11). The Source Diagram or notes should be studied carefully before using the chart in order to assess the reliability of the data used.*	170 - 241 9
(18)	*Reference to a larger-scale chart or plan.*	254
(19)	*Reference to an adjoining chart of similar scale.*	254
(20)	*Instruction to refer to related Admiralty Publications.*	243
(a)	*Reference to the units used for depth measurement. The legend, 'DEPTHS IN FATHOMS', is shown on certain more recent fathoms charts where confusion might otherwise arise.*	241 5 255 2
(b)	*Conversion scales. To allow approximate conversions between metric and fathoms and feet units. On older charts, conversion tables are given instead.*	280

70	Compass Roses, True and Magnetic. 4°30'W 1991 (9'E) on magnetic north arrow means Magnetic Variation 4°30'W in 1991, annual change 9'E (i.e. magnetic variation decreasing 9' annually). Magnetic Variation is expressed to the nearest 5' and relates to 1 January of the year stated. Annual change E or W is given to the nearest minute. This style is obsolescent. True Compass Rose Magnetic North indicated by arrow. The arrow indicating Magnetic North is omitted on charts comprising separate plans and on charts showing isogonals.		260-262.2 272.3 U
71	Isogonals (lines of equal magnetic variation)	Magnetic Variation Curves are for 1990. The Magnetic Variation is shown in degrees, followed by the letter E or W, as appropriate, at certain positions on the curves. The annual change is expressed in minutes with the letter E or W and is given in brackets, immediately following the variation.	272.1
82.1	±15°	Local Magnetic Anomaly. Within the enclosed area the magnetic variation may deviate from the normal by the value shown.	274 U26
82.2	Local Magnetic Anomaly (see Note)	Where the area affected cannot be easily defined, a legend only is shown at the position.	

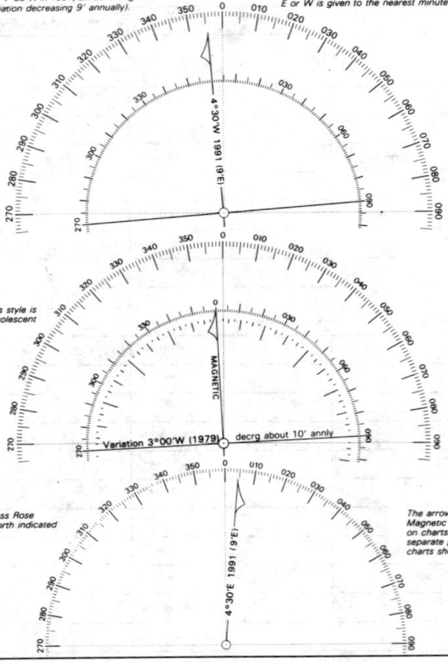

Units

No.	Symbol	Description	Abbreviation	Ref.
40	km	Kilometre(s)		E5
41	m	Metre(s)		130 E4
42	dm	Decimetre(s)		130 E4a
43	cm	Centimetre(s)		E4c
44	mm	Millimetre(s)		130 E4c
45	M	International Nautical Mile(s) (1852m) Sea mile(s)	n mile(s) M	130 E11
46		Cable		130 E10
47	ft	Foot/feet		E7
48		Fathom(s)	fm., fms.	E9
49	h	Hour		130 E1
50	# m min	Minute(s) of time		130 E2
51	s # sec	Second(s) of time	† sec	130 E3
52	kn	Knot(s)		130 E12
53	t	Tonne(s) or Ton(s)		E12a
54	# cd	Candela		E12b

Magnetic Compass

No.		Description	Abbreviation	Ref.
60		Variation	Var	U24
61		Magnetic	Mag	U23
62		Bearing		263 U21
63		true		U22
64		decreasing	decrg	
65		increasing	incrg	
66		Annual change		U25
67		Deviation		U28
68.1		Note of magnetic variation, in position	Magnetic Variation: 4°30'W 1988 (10'E)	U
68.2		Note of magnetic variation, out of position		

13	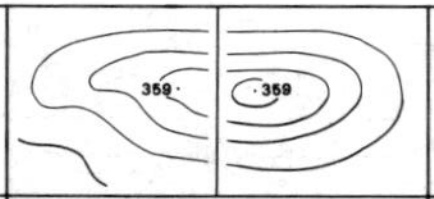	*Form lines with spot height*	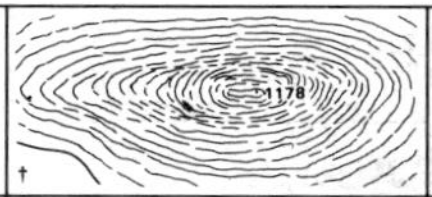	351 2 351 7 352 2 *C2a* *D3*
14	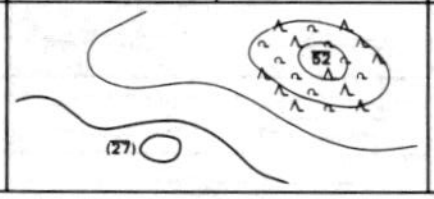	*Approximate height of top of trees (above height datum)*	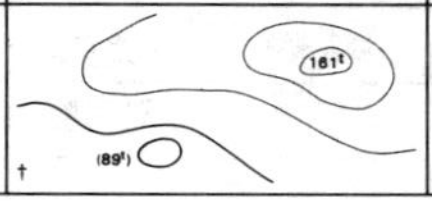	352 4 *C11*

Water Features, Lava				
20	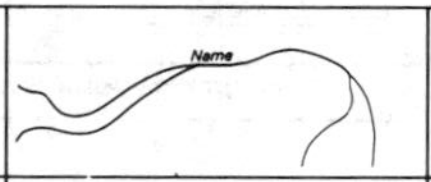	*River, Stream*		353 1 353 2 353 4 *C13*
21	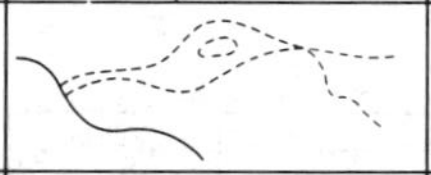	*Intermittent river*		353 3 *C14*
22	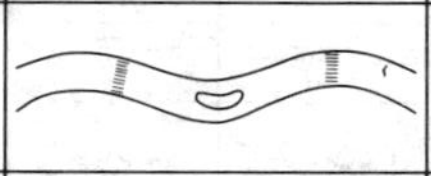	*Rapids, Waterfalls*		353 5 *C19* *C20*
23	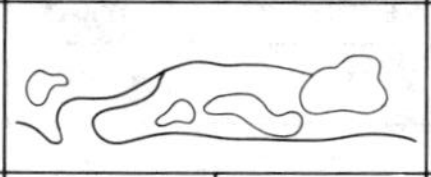	*Lakes*		353 6 *C15*
24	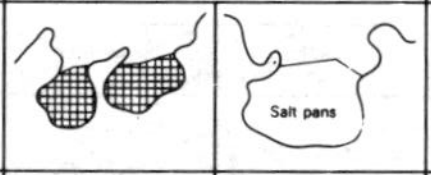	*Salt pans*		353 7 *C4*
25	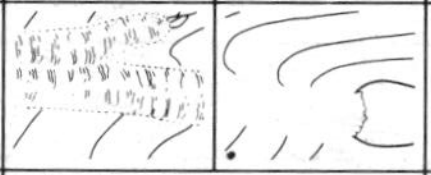	*Glacier*	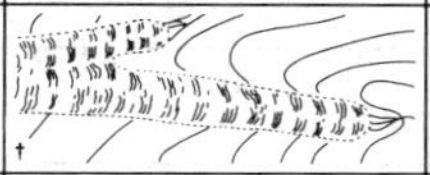	353 8 *C3*
26	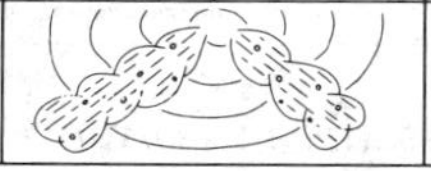	*Lava flow*		355 *C12*

Foreshore → II, IJ				Coastline
1	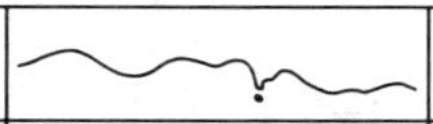	*Coastline, surveyed*		310 1 310 2 *A8*
2	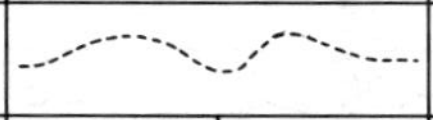	*Coastline, unsurveyed*		311 *A1*
3		*Steep coast, Cliffs*		312 1 *A2* *A3*
4	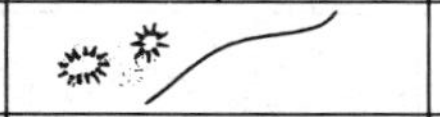	*Coastal hillocks*		312 1
5	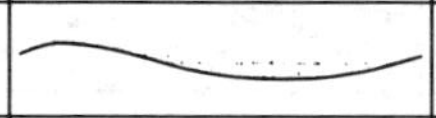	*Flat coast*		312 2 *A2a*
6	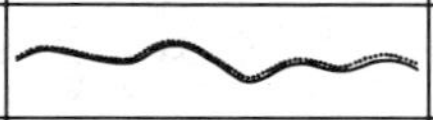	*Sandy shore*	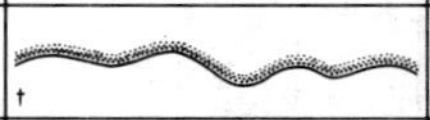	312 2 *A6*
7	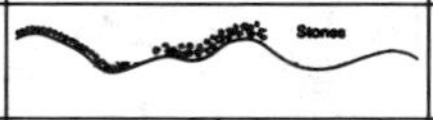	*Stony shore, Shingly shore*		312 2 *A5*
8		*Sandhills, Dunes*	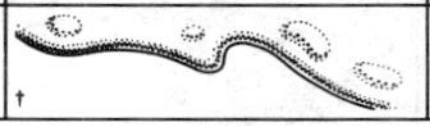	312 3 *A4*

Plane of Reference for Heights → IH				Relief
10	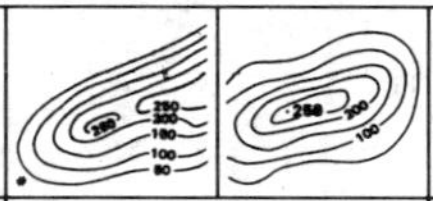	*Contour lines with spot height*		351 3 351 4 351 5 351 6 352 2 *C1* *D3*
11	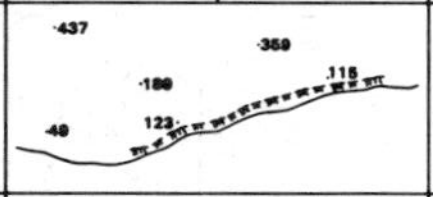	*Spot heights*		352 1 352 2 *D3*
12	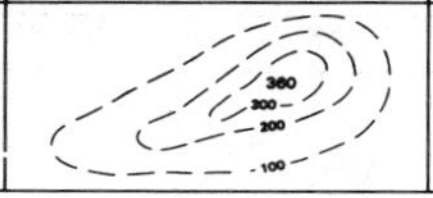	*Approximate contour lines with approximate height*		351 3 351 4 351 5 351 6 352 3 *C1a*

ID Cultural Features

Natural Features IC

Settlements, Buildings

Height of objects → IE   Landmarks → IE

1			Urban area		370.3 370.4 I1
2			Settlement with scattered buildings		370.5 I2
3	Name	Name	Settlement (on medium and small-scale charts)	Name	370.7 I1 I2 I3
4	Name	Name HOTEL	Inland village		370.6 I3
5			Building	Bldg	370.5 I3a I66
6	Name Hotel	Name Hotel	Important building in built-up area		370.3
7	NAME	NAME	Street name, Road name		371 I26
8	Ru	Ru	Ruin, Ruined landmark	(ru)	378 378.2 I40

Roads, Railways, Airfields

10			Motorway		365.1 H10
11			Road (hard surfaced)		365.2 H1
12			Track, Path (loose or unsurfaced)		365.3 H2
13			Railway, with station	† Rly † Ry † Sta † Stn	328.4 362.1 362.2 H3 H3b
14			Cutting	† †	363.2 H3e
15			Embankment	† †	364.1 H3d
16			Tunnel		363.1 H3c
17		Airfield Airport	Airport, Airfield Large-scale charts Small-scale charts		366.1 366.2 I23 I24
a			Tramway		H3a

Vegetation

30		Wooded	Woods in general		[illegible]
31			Example of prominent trees (in groups or isolated)		[illegible]
31.1			Deciduous tree, unknown or unspecified tree		[illegible]
31.2			Evergreen (except conifer)		[illegible]
31.3			Conifer		[illegible]
31.4			Palm		[illegible]
31.5			Nipa palm		[illegible]
31.6			Casuarina		[illegible]
31.7			Filao		[illegible]
31.8			Eucalypt		[illegible]
32			Mangrove	†	312.4 A7
33		Marsh Marsh	Marsh, Swamp, Saltmarsh	Saltings Saltings	312.2 C1

General

Plane of Reference for Heights → IH Lighthouses → IP Beacons → IQ

1	Factory Hotel	Examples of landmarks		340.1 340.2 340.5 L63
2	FACTORY HOTEL WATER TOWER	Examples of conspicuous landmarks. A legend in capital letters indicates that a feature is conspicuous	† conspic	340.1 340.2 340.3 340.5 L63
3.1		Pictorial symbols (in true position)		340.7 373.1 390 456.5 457.3
3.2		Sketches, Views (out of position)		
4	(30)	Height of top of a structure above High Water		302.3
5	(30)	Height of top of a structure above ground level		303

Landmarks

10.1		Ch	Church	† Cath	373.1 373.2 I8 I8a
10.2	Tr	Tr	Church tower		373.2
10.3	Sp	Sp	Church spire		373.2 I8b
10.4	Cup	Cup	Church cupola		373.2 I36
11			Chapel	Ch	I11
12			Cross, Calvary		I17
13			Temple	†	373.3 I10
14			Pagoda	Pag	373.3 I14
15			Shinto shrine, Josshouse		373.3 I15a
16			Buddhist temple	†	373.3 I15
17			Mosque, Minaret	†	373.4 I12 I12a
18	Marabout		Marabout	Tomb †	373.5 I13
19			Cemetery (all religious denominations)	† Cemy	373.6 I17a I[illegible]

Other Cultural Features

20	20	(8·9)	Vertical clearance above High Water	(17)† (H17m)† (Headway 55ft)†	380.1 380.2 H18a
21	23		Horizontal clearance		380.3 H18b
22	20		Fixed bridge with vertical clearance	(20)†	381.1 H14 H18a
23.1	20		Opening bridge (in general) with vertical clearance	† (20)	381.3 H18a
23.2	7·8 Swing Bridge		Swing bridge with vertical clearance		H16
23.3	4·2 Lifting Bridge (open 12)		Lifting bridge with vertical clearance (closed and open)		H16a
23.4	12 Bascule Bridge		Bascule bridge with vertical clearance		H[illegible]
23.5	Pontoon Bridge		Pontoon bridge	†	H17
23.6	5·5 Draw Bridge		Draw bridge with vertical clearance		H15
24	20 Transporter Bridge		Transporter bridge with vertical clearance between HW and lowest part of structure		381.7 H18 H18a
25	20		Overhead transporter, Aerial cableway with vertical clearance	† Transporter (7)	382.1 H4 H18a
26	Pyl 28 Pyl		Power transmission line with pylons and safe overhead clearance (see NOTE)	† Power (H30m) Power Overhead (H 98ft)	382.1 H5 H5a
27	20		Overhead cable, Telephone line, Telegraph line with vertical clearance	† H20m Overhead (H 64ft)	382 382.2 H6 H18a
28	20 Overhead pipe		Overhead pipe with vertical clearance		383 H7 H18a
29			Pipeline on land	† Pipeline	377 H6a

NOTE: **Safe overhead clearance.** *The safe overhead clearance above HW, as defined by the responsible authority, is given in magenta where known; otherwise the physical vertical clearance is shown in black as in ID20*

Artificial Features

No.	Symbol		Description	Symbol (other)		Ref.
1			*Dyke, Levee*	†	†	313.1 G6a
2.1			*Seawall (on large-scale charts)*			313.2
2.2			*Seawall (on small-scale charts)*			
3	Causeway		*Causeway*			313.3 H3l
4.1			*Breakwater (in general)*			322.1 G6
4.2			*Breakwater (loose boulders, tetrapods, etc)*			
4.3			*Breakwater (slope of concrete or masonry)*			
5	Training Wall (covers); Training Wall; Training Wall (covers); Training Bank (covers)		*Training wall*			322.2 H23
6.1			*Groyne (always dry)*			313.4 324 G11
6.2			*Groyne (intertidal)*			
6.3			*Groyne (always underwater)*			

Harbour Installations

Depths ⟶ II   *Anchorages, Limits* ⟶ IN   *Beacons and other fixed marks* ⟶ IQ   *Marina* ⟶ IU

No.	Symbol		Description	Symbol (other)	Ref.
10			*Fishing harbour*		320.1 G30
12			*Mole (with berthing facility)*		321.3 G7
13			*Quay, Wharf*	Whf	321.1 G18 G19

Landmarks

No.	Symbol		Description	Symbol (other)		Ref.
20		Tr	*Tower*			[illegible]
21			*Water tower, Water tank on a tower*	○ Water Tr		[illegible]
22		Chy	*Chimney*			[illegible]
23			*Flare stack (on land)*			[illegible]
24		Mon	*Monument (including column, pillar, obelisk, statue)*	† Mont	† Col	[illegible]
25.1			*Windmill*			[illegible]
25.2		Ru	*Windmill (without sails)*	(ru)		[illegible]
26			*Windmotor*			[illegible]
27		FS	*Flagstaff, Flagpole*			[illegible]
28			*Radio mast, Television mast*	○ Radio mast ○ TV mast		[illegible] M9 M9a
29			*Radio tower, Television tower*	○ Radio Tr ○ TV Tr		[illegible] M9 M9a
30.1	○ Radar Mast		*Radar mast*			487.1
30.2	○ Radar Tr		*Radar tower*			
30.3	○ Radar Sc		*Radar scanner*			
30.4	○ Radome		*Radar dome*			
31			*Dish aerial*	† ○ Dish aerial		[illegible]
32		Tanks	*Tanks*	†		[illegible]
33	○ Silo	○ Silo	*Silo*			[illegible]
34.1	Fort (disused)		*Fortified structure (on large-scale charts)*			[illegible]
34.2			*Castle, Fort, Blockhouse (on smaller-scale charts)*	† Ft	Cas	[illegible]
34.3			*Battery, Small fort (on smaller-scale charts)*	† Batt	Baty	[illegible]
35.1			*Quarry (on large-scale charts)*	†		[illegible]
35.2			*Quarry (on smaller-scale charts)*			[illegible]
36			*Mine*			[illegible]

33.1	Ru	Ruin		378.1
33.2	Pier (Ru)	Ruined pier, partly submerged at high water	Pier (ru)	
34	Hulk	Hulk		G45
a		Bollard	Bol	G22

Canals, Barrages

Clearances → ID *Signal Stations* → IT

40	km 32	km 36	Canal, with distance mark. Distance shown in black indicates a physical structure e.g. a notice board	†	361.3 361.6 307 H13 E5
41.1	Lock		Lock (on large-scale charts)		326.6 361.6 G40 H13
41.2			Lock (on smaller-scale charts)	†	
42			Caisson		326.5
43	Flood Barrage		Flood barrage		326.7
44	Dam		Dam → Direction of flow		364.2 H21

Transhipment Facilities

Roads → ID *Railways* → ID *Tanks* → IE

50	RoRo		Roll-on, Roll-off Ferry Terminal		321.5
51	3 3	2 3	Transit shed, Warehouse (with designation)		328.1 I34a I38
52	#		Timber yard		328.2 G43
53.1	(3t)		Crane (with lifting capacity) Travelling crane (on railway)		328.3 G24
53.2	(50t)		Container crane (with lifting capacity)		
53.3	SHEERLEGS		Sheerlegs (conspicuous)		

Public Buildings

60		Harbour Master's office	Hr Mr	325.1 G28
61		Custom office		325.2 G29
62.1		Health office, Quarantine building		325.3 G26 G44 I32
62.2	Hospital	Hospital	Hosp † Hospl	
63		Post office	PO	372.1 I29

14		Pier	Pier, Jetty		321.2 321.4 G8 G9
15	Promenade Pier		Promenade pier		321.2
16		Pontoon	Pontoon		326.9
17	Lndg	Lndg	Landing for boats	† Ldg	324.2 G16
18		Steps	Steps, Landing stairs		G25a
19	4	B	Designation of berth	† 4	323.1 G20 G20b
20	Dn	Dns	Dolphin		327.1 G21
21			Deviation dolphin		327.2
22			Minor post or pile		327.3 H9
23		Slip	Slipway, Patent slip, Ramp		324.1 G39 G39a
24		Gridiron	Gridiron, Scrubbing grid		326.8 G38
25		Dry Dock	Dry dock, Graving dock	†	326.1 G36
26		Floating Dock	Floating dock	† † †	326.2 G37
27			Non-tidal basin, Wet dock		326.3 G41
28			Tidal basin, Tidal harbour		326.4 G34
29.1			Floating oil barrier	Floating Barrier	
29.2			Oil retention barrier (high pressure pipe)		
30	Dock under construction (1991)		Works on land, with year date		[illegible] G4[illegible] G5[illegible]
31	Being reclaimed (1991)		Works at sea, Area under reclamation, with year date		[illegible] G5[illegible]
32	Under construction (1991) Works in progress (1991)		Works under construction, with year date	const † constrn † constn	[illegible] [illegible] G4[illegible] G5[illegible]

Buildings

60		Structure			
61		House	Ho		I5
62		Hut			I68
63		Multi-storey building			
64		Castle	Cas		I4
65		Pyramid			I74
66		Column	Col		
67		Mast			H9a
68		Lattice tower			
69		Mooring mast			I25
70		Floodlight			
71		Town Hall			I31
72		Office		†Off	J22
73		Observatory		†Obsy	J21
74		Institute			I61
75		Cathedral		†Cath	I8a
76		Monastery, Convent		†Mony	I16
77		Lookout station, Watch tower			J4
78		Navigation school			
79		Naval college			
80		Factory			I47
81		Brick kiln, Brick works			I49
82		Cement works			
83		Water mill			I43
84		Greenhouse			
85		Warehouse, Storehouse			I34a
86		Cold store, Refrigerated storage house			
87		Refinery			
88		Power station			
89		Electric works			
90		Gas works			
91		Water works			
92		Sewage works			
93		Machine house, Pump house			
94		Well			I51
95		Telegraph office		†Tel	I28
96		Hotel			
97		Sailors' home			
98		Spa hotel			
a		School	Sch		I65

Road, Rail and Air Traffic

110		Street, Road	St	I26
111		Avenue	Ave	I26a
112		Tramway		H3a
113		Viaduct		H8
114		Suspension bridge		H14d
115		Footbridge		H17a
116		Runway		
117		Landing lights		
118		Helicopter landing site		

Coast

1		Island	I.		B18
2		Islet	I.	†It	B19
3		Cay			
4		Peninsula	Pen.	†Penla	B21
5		Archipelago	Arch.	†Archo	B20
6		Atoll			
7		Cape	C.		B22
8		Head, Headland	Hd.		B24
9		Point	Pt.		B25
10		Spit			
11		Rock	Rk.		B35
12		Saltmarsh, Saltings			
13		Lagoon	Lag.	†Lagn	C16

Natural Inland Features

20		Promontory	Prom.	†Promy	B23
21		Range			B27
22		Ridge			
23		Mountain, Mount	Mt.		B26
24		Summit			B28
25		Peak	Pk.		B29
26		Volcano	Vol.		B30
27		Hill			B31
28		Boulder			B32
29		Table-land			B34
30		Plateau			
31		Valley			B27a
32		Ravine, Cut			
33		Gorge			
34		Vegetation			
35		Grassland			C6a
36		Paddyfield			C7
37		Bushes			C8
38		Deciduous woodland			C9
39		Coniferous woodland			C10

Settlements

50		City, Town			I1
51		Village			I3
52		Fishing village			
53		Farm		†Fm	I7
54		Saint	S.	†St	F11

Terms Relating to Tidal Levels

1	CD	*Chart Datum* *Datum for sounding reduction*		405 *T5*
2	LAT	*Lowest Astronomical Tide*		405.3
3	HAT	*Highest Astronomical Tide*		
4	MLW	*Mean Low Water*		*T8c*
5	MHW	*Mean High Water*		*T7a*
6	MSL	*Mean Sea Level*		*T4*
7		*Land survey datum*		
8	MLWS	*Mean Low Water Springs*		*T9*
9	MHWS	*Mean High Water Springs*		*T8*
10	MLWN	*Mean Low Water Neaps*		*T9a*
11	MHWN	*Mean High Water Neaps*		*T8a*
12	MLLW	*Mean Lower Low Water*		*T9b*
13	MHHW	*Mean Higher High Water*		*T8b*
14	MHLW	*Mean Higher Low Water*		
15	MLHW	*Mean Lower High Water*		
16	Sp	*Spring Tide*	† Spr.	*T6*
17	Np	*Neap Tide*		*T7*
a		*High Water*	HW	*T1*
b		*Low Water*	LW	*T2*
c		*Mean Tide Level*	MTL	*T3*
d		*Ordnance Datum*	OD	

Ports, Harbours

130		*Tidal barrier*		
131		*Boat lift, Ship lift*		
132		*Minor canal*		
133		*Sluice*		*H13*
134		*Basin*		
135		*Reservoir*		
136		*Reclamation area*		
137		*Port*	**P.**	*G5*
138		*Harbour*	**Hr.**	*G3*
139		*Haven*	**Hn.**	*G4*
140		*Inner harbour*		
141		*Outer harbour*		
142		*Deep water harbour*		
143		*Free port*		
144		*Customs harbour*		
145		*Naval port*		
146		*Industrial harbour*		
147		*Commercial port, Trade port*		
148		*Building harbour*		
149		*Oil harbour*		
150		*Ore harbour*		
151		*Grain harbour*		
152		*Container harbour*		
153		*Timber harbour*		
154		*Coal harbour*		
155		*Ferry harbour*		
156		*Police*		
b		*Dock*	**Dk**	*G35*
c		*Wharf*	**Whf**	*G18*

Harbour Installations

170		*Terminal*		
171		*Building slip*		
172		*Building yard*		
173		*Buoy yard, Buoy dump*		
174		*Bunker station*		
175		*Reception facilities for oily wastes*		
176		*Tanker cleaning facilities*		
177		*Cooling water intake/outfall*		
178		*Floating barrier, Boom*		
179		*Piling*		*H9*
180		*Row of piles*		
181		*Bollard*	**Bol**	*G22*
182		*Conveyor*		
183		*Storage tanker*		
184		*Lighter Aboard Ship*	LASH	
185		*Liquefied Natural Gas*	LNG	
186		*Liquefied Petroleum Gas*	LPG	
187		*Very Large Crude Carrier*	VLCC	

Tidal Streams and Currents

Breakers → IK   Tide Gauge → IT

40		Flood tide stream (with rate)	† The number of black dots on the tidal stream arrows indicates the number of hours after High or Low Water at which the streams are running	407.4 408.2 T19
41		Ebb tide stream (with rate)	†	407.4 408.2 T20
42		Current in restricted waters	†	408.2 T18
43	(see Note)	Ocean current. Details of current strength and seasonal variations may be shown		408.3 T18
44		Overfalls, tide rips, races	†	423.1 O18
45		Eddies		423.3 O19
46		Position of tabulated tidal stream data with designation	†	407.2 T34
e		Wave recorder	Wave recorder	
f		Current meter	Current meter	

Tidal Levels and Charted Data

Vertical clearance → ID   Tide Gauge → IT

20 *Planes of reference are not exactly as shown below for all charts. They are usually defined in notes under chart titles* — 302.2, 380.1, 405

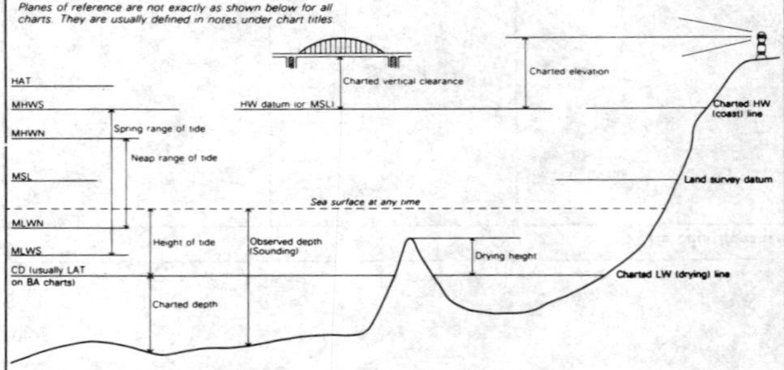

Tide Tables

30 *Tabular statement of semi-diurnal or diurnal tides* — 406.2, 406.3, 406.4, 406.5

Tidal Levels referred to Datum of Soundings

Place	Lat. N/S	Long. E/W	Heights in metres/feet above datum				Datum and Remarks
			MHWS	MHWN	MLWN	MLWS	
			MHHW	MLHW	MHLW	MLLW	

Offshore position for which tidal levels are tabulated a

31 *Tidal stream table* — 407.2, 407.3

Tidal streams referred to....

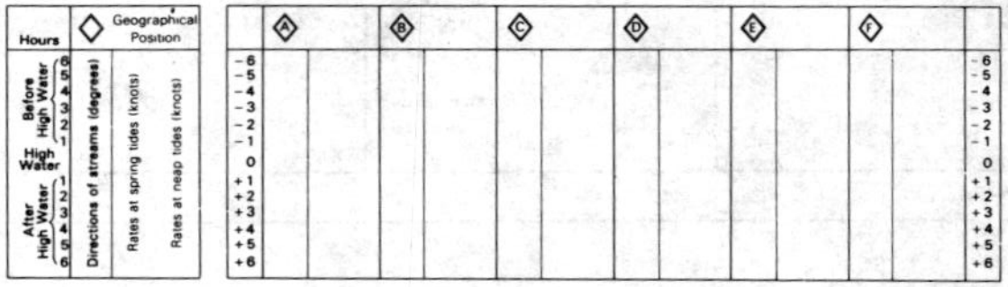

General

1	ED	Existence doubtful	† (ED)	417 424.3 O43
2	(40) SD	Sounding of doubtful depth		417 424.4 Q1
3.1	Rep	Reported, but not confirmed	† Repd	417 424.5 O35
3.2	Rep (1973)	Reported, with year of report, but not confirmed	† Repd (1973)	
4	(184) (212)	Reported, but not confirmed, sounding or danger (on small-scale INT charts only)		Small-scale INT charts 403E
a		Unexamined	unexam † unexamd	O46

Soundings and Drying Heights

Plane of Reference for Depths ⟶IH Plane of Reference for Heights ⟶IH

10	12 9_2 * 9.7	Sounding in true position		403.1 410/412 412.1 Q10a
11	* +(12) 3349	Sounding out of position		412 412.1 412.2 Q3
12	(14_7)	Least depth in narrow channel		412 412.1 412.2 Q4
13	330	No bottom found at depth shown		412.3 Q2
14	12 9_1	Soundings taken from old or smaller-scale sources shown in upright, hairline figures		412.4 417.3 Q10 Q12
15		Drying heights and contours above chart datum		413 413.1 413.2 Q8
b		Half-tide channel (in intertidal area)	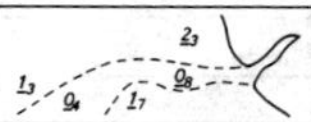	

Depths in Fairways and Areas

Plane of Reference for Depths ⟶IH

20		Limit of dredged area (major and minor)		414.3 Q6
21	7·0m 3·5m	Dredged channel or area with depth of dredging in metres and decimetres		414 Q5
22	7·0m (1991) Dredged to 3·5m (1991)	Dredged channel or area with depth of dredging and year of the latest control survey		414.1 Q5
23	7·0m Maintained depth 3·5m	Dredged channel or area with depth regularly maintained		414.2 Q5
24	10_2 10_8 9_6 (1990) 9_8 11	Depth at chart datum, to which an area has been swept by wire drag. The latest date of sweeping may be shown in parentheses	† 6_4 (1990)	415 415.1 Q7 Q9 Q9a
25	Unsurveyed (see Note) Inadequately surveyed (see Note) Unsurveyed (see Note) Inadequately surveyed (see Note) *	Unsurveyed or inadequately surveyed area; area with inadequate depth information		410 417 417.6 417.7 Q22 A14

Depth Contours

30	2, 0, 2, 5, 10, 20, 30, 50, 100, 200, 300, 400, 500, 1000, 2000, 3000, 4000, 5000 etc	Drying contour Low Water (LW) Line, Chart Datum (CD) Blue tint, in one or more shades, and tint ribbons, are shown to different limits according to the scale and purpose of the chart and the nature of the bathymetry. On some charts, the standard set of contours is augmented by additional contours in order to delimit particular bathymetric features or for the benefit of particular categories of shipping. However, in some instances where the provision of additional contours would be helpful, the survey data available does not permit it. * Occasionally, contours and figures are printed in blue	On charts showing depths in fathoms/feet, the following contours are used: 1 †, 2 †, 3 †, 4 †, 5 †, 6, 10, 20, 30, 40, 50, 60, 100 †, 200, 500, 1000 †, 2000, 3000, 5000 1, 3, 6, 10, 20, 30, 40, 50, 60, 100, 200, 500, 1000, 2000, 3000, 5000 On some recently-corrected charts, contours may be shown by continuous lines.	404.2 410 411 R
31	10, 50	Approximate depth contours		411.2 417.5 R

o		Manganese	† Mn man	S22
p		Glauconite	† Gc	
q		Oysters	† Oy oys	S24
r		Mussels	† Ms mus	S25
s		Sponge	† Sp	S26
t		Algae	† Al	
u		Foraminifera	† Fr for	S32
v		Globigerina	† Gl	S33
w		Diatoms	† Di	S34
x		Radiolaria	† Rd rad	S35
y		Pteropods	† Pt	S36
z		Polyzoa	† Po pol	S37

Intertidal Areas				
20	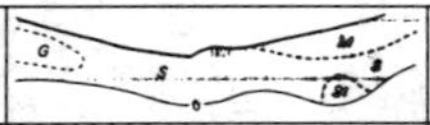	Area of sand and mud with patches of stones or gravel	† 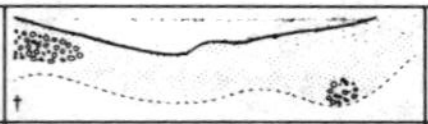	426.1 A11c A11e
21	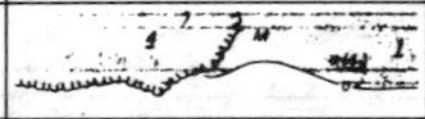	Rocky area	† 	426.2 A11d
22	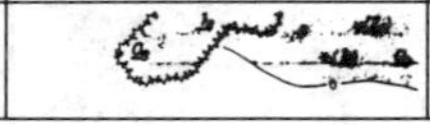	Coral reef	† 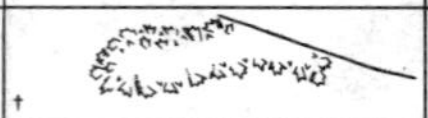	426.3 A11g

Qualifying Terms				
30	f	Fine } only used in relation to sand		425 427 S39
31	m	Medium		
32	c	Coarse		S40
33	bk	Broken	† brk	S47
34	sy	Sticky	† stk	S46
35	so	Soft	† sft	S41
36	sf	Stiff	† stf	S43
37	v	Volcanic	† vol	S16
38	ca	Calcareous	† cal	S12a
39	h	Hard		425.7 S42

Rocks → IK				Types of Seabed
1	S	Sand	† s	425 427 S2
2	M	Mud	† m	S3
3	Cy	Clay	† cl	S6
4	Si	Silt		S4
5	St	Stones	† st	S10
6	G	Gravel	† g	S7
7	P	Pebbles	† peb	S9
8	Cb	Cobbles		S11a
9	R	Rock	† r	S11
10	Co	Coral	† crl	S14
11	Sh	Shells	† sh	S23
12.1	S/M	Two layers e.g. Sand over Mud		425.8
12.2	fS M Sh	Mixed bottom: where the seabed comprises a mixture of materials, the main constituent is given first, e.g. fine Sand with Mud and Shells		425.9
13.1	Wd	Weed (including Kelp)	† wd	425.5 S27 S28
13.2		Kelp		428.2 O20
14		Sandwaves		428.1 O17a
15		Spring in seabed		428.3 S76
a		Ground	† Gd grd	S1
b		Ooze	† Oz	S4
c		Marl	† Ml	S5
d		Shingle	† Sn shin	S8
e		Boulders	† Bo	S11a
f		Chalk	† Ck chk	S12
g		Quartz	† Qz qrtz	S13
h		Madrepore	† Md mad	S15
i		Basalt	† Ba	
j		Lava	† Lv	S17
k		Pumice	† Pm pum	S18
l		Tufa	† T	S19
m		Scoriæ	† Sc	S20
n		Cinders	† Cn cin	S21

IK Rocks, Wrecks, Obstructions

General				
1		Danger line: A danger line draws attention to a danger which would not stand out clearly enough if represented solely by its symbol (e.g. isolated rock) or delimits an area containing numerous dangers, through which it is unsafe to navigate		420.1 O38
2		Depth cleared by wire drag sweep		415 422.3 422.9 O6a
a		Dries	†Dr †dr	O32
b		Covers	†cov	O33
c		Uncovers	†uncov	O34

Rocks	Plane of Reference for Heights →IH		Plane of Reference for Depths →IH	
10	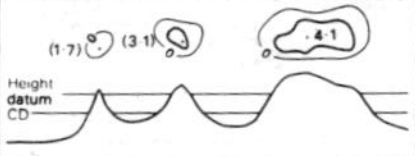	Rock which does not cover, height above High Water		421.1 O1
11	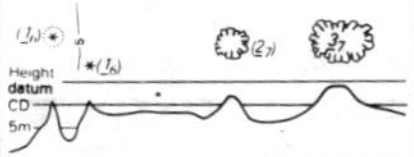	Rock which covers and uncovers, height above Chart Datum		421.2 O2
12	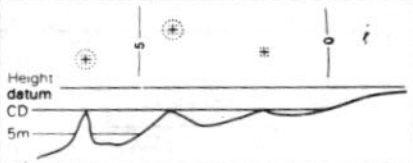	Rock awash at the level of Chart Datum		421.3 O3
13	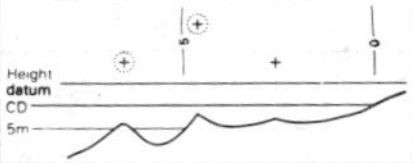	Underwater rock over which the depth is unknown, but which is considered dangerous to surface navigation		421.4 O4
14		Dangerous underwater rock of known depth:		421.4 O5
14.1	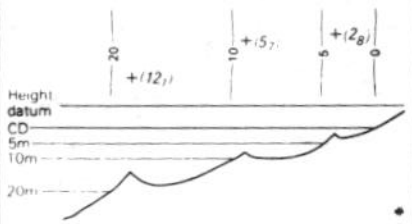	inside the corresponding depth area	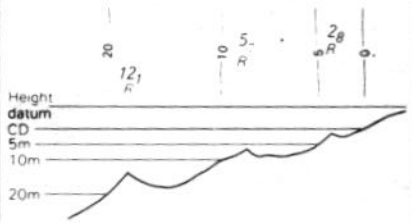	
14.2	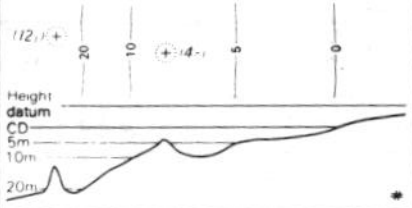	outside the corresponding depth area	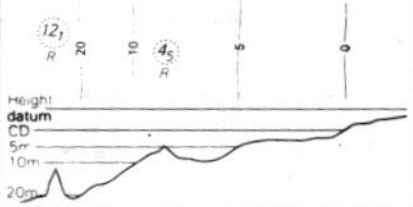	
15	30 R	Underwater rock not dangerous to surface navigation		421.4 O6

Nature of the Seabed IJ

aa		Small	† sm	S44
ab		Large	† l	S45
ac		Glacial	† ga glac	S54
ad		Speckled	† sk spk	S50
ae		White	† w	S56
af		Black	† bl blk	S57
ag		Blue	† b	S59
ah		Green	† gn	S60
ai		Yellow	† y	S61
aj		Red	† rd	S63
ak		Brown	† br	S64
al		Chocolate	† ch choc	S65
am		Grey	† gy	S66
an		Light	† lt	S67
ao		Dark	† d	S68

Obstructions — *Plane of Reference for Depths* → IH — *Kelp, Sea-Weed* → IJ — *Wells* → IL

No.	Symbol	Description	Other symbols	Ref.
40	Obstn · Obstn	Obstruction or danger to navigation the exact nature of which is not specified or has not been determined, depth unknown		422.9 O27
41	4_6 Obstn · 16_8 Obstn	Obstruction, depth known		422.9
42	4_6 Obstn · 16_8 Obstn	Obstruction which has been swept by wire to the depth shown		422.9 O6a
43.1	Obstn	Stumps of posts or piles, wholly submerged		327.5 O30 O30a
43.2		Submerged pile, stake, snag, well or stump (with exact position)		
44.1		Fishing stakes		447.1 G14
44.2		Fish trap, fish weir, tunny nets		447.2 G14a G15
45	Fish traps · Tunny nets	Fish trap area, tunny nets area	(U.S. waters only)	447.3 G14a G15
46.1		Fish haven		447.5
46.2	(2_1) · 2_4	Fish haven, depth known		
47	Shellfish Beds	Shellfish beds		447.4 G15a
48.1		Marine farm (on large-scale charts)	Fish farm · Fish cages	447.6
48.2		Marine farm (on small-scale charts)		

No.	Symbol	Description	Other symbols / remarks	Ref.
16	Co · Co · 5_8	Coral reef which is always covered		421.5 O10
17	18 5_8 Br 19	Breakers		423.2 O25
d		Discoloured water	Discol · †Discold	424.6 O9

Plane of Reference for Depths → IH — *Historic Wreck* → IN — **Wrecks**

No.	Symbol	Description	Other symbols / remarks	Ref.
20	Mast (1·2) Wk	On large-scale charts, wreck which does not cover, height above height datum		422.1 O11
21	Mast (1_2) Wk	On large-scale charts, wreck which covers and uncovers, height above Chart Datum	† Wk · † Wk	
22	5_7 Wk	On large-scale charts, submerged wreck, depth known	† 5_7 Wk	422.1 O15
23	Wk	On large-scale charts, submerged wreck, depth unknown	† Wk	422.1
24		Wreck showing any part of hull or superstructure at the level of Chart Datum		422.2 O11 O13a
25	Mast (1·2) · Funnel · Mast (1_2) · Masts	Wreck of which the mast(s) only are visible at Chart Datum		422.2 O12
26	4_6 Wk · 25 Wk	Wreck over which the depth has been obtained by sounding but not by wire sweep		422.4 O15
27	4_6 Wk · 25 Wk	Wreck which has been swept by wire to the depth shown		422.3 O15a
28		Wreck, depth unknown, which is considered dangerous to surface navigation	On modern Admiralty charts, this symbol is used when the depth over the wreck is thought to be 28 metres (15 fathoms) or less. The limiting depth at which a wreck is categorised "dangerous" was changed from 8 to 10 fathoms in 1960, to 11 fathoms in 1963 and 15 fathoms/28 metres in 1968.	422.5 O14
29		Wreck, depth unknown, which is not considered dangerous to surface navigation	On modern Admiralty charts, this symbol is used when the depth over the wreck is thought to be more than 28 metres (15 fathoms) [see IK 28 above for previous depth limits] OR when the depth over the wreck is thought to be 28 metres or less, but the wreck is not considered dangerous to surface vessels capable of navigating in the vicinity	422.6 O16
30	20 Wk	Wreck over which the exact depth is unknown, but which is considered to have a safe clearance at the depth shown		422.7
31	# · Foul	Remains of a wreck, or other foul area, no longer dangerous to surface navigation, but to be avoided by vessels anchoring, trawling, etc	† Foul · † Foul · 22 Foul (Where depth known)	422.8 O17 O29a

22		Site of cleared platform		
23	Pipe	Above-water wellhead (lit and unlit). The drying height or height above height datum is charted if known		445.7
c		Suspended well which has been swept by wire to the depth shown	15 Well	
d		Suspended well over which the exact depth is unknown, but which is considered to have a safe clearance at the depth shown	15 Well	
e		Single Well Oil Production System. The depth shown is the least depth over the wellhead. For substantial periods of time a loading tanker is positioned over the wellhead	93 SWOPS	

Submarine Cables

30.1		Submarine cable		443.1 P7
30.2		Submarine cable area	Cable Area	443.2 439.3 P7a
31.1		Submarine power cable	Power Power	443.3 P7
31.2		Submarine power cable area	Power Cable Area	443.2 439.3
32		Disused submarine cable		443.7 P7b

Submarine Pipelines

40.1	Oil Gas Chem Water	Supply pipeline: unspecified, oil, gas, chemicals, water	Pipeline	444 444.1 P8 H8a
40.2	Oil Gas Chem Water	Supply pipeline area: unspecified, oil, gas, chemicals, water	Pipeline Area; Pipeline Area	444.2 439.3 P8a
41.1	Water Sewer Outfall Intake	Discharge pipe: unspecified, water, sewer, outfall, intake	Sewer Outfall	444 444.4 P8 H11
41.2	Water Sewer Outfall Intake	Discharge pipe area: unspecified, water, sewer, outfall, intake	Pipeline Area; Pipeline Area	444.2 439.3 P8a
42	Buried 1·6m	Buried pipeline/pipe (with nominal depth to which buried)		444.5
43	3_2 Obstn	Diffuser	3_2 Diffuser	
44		Disused pipeline/pipe		444.7

Areas, Limits →IN			Oilfields and gasfields	
1	EKOFISK OILFIELD	Name of oilfield or gasfield		445.3
2	Z-44	Platform with designation/name		445.3 I76 O27a
3		Limit of safety zone around offshore installation		439.2 445.2
4		Limit of development area		
a		Limit of oilfield or gasfield		

Mooring Buoys →IQ			Platforms and Moorings	
10		Production platform, Platform, Oil derrick		445.2 I76 O27a
11	Fla	Flare stack (at sea)		445.6
12	SPM	Single Point Mooring, including Single Anchor Leg Mooring (SALM), Articulated Loading Column (ALC)		445.2 445.4
13		Observation/research platform (with name)	Name	
14		Disused platform	(disused)	
15		Artificial Island	Name	
16		Tanker mooring buoy of superbuoy size, including Catenary Anchor Leg Mooring (CALM), Single Buoy Mooring (SBM)		445.4 L66
17		Moored storage tanker		
b		Anchoring system for floating production platform		

Plane of Reference for Depths →IH	Obstructions →IK		Underwater Installations	
20	15 Prod Well; Prod Well	Production well, with depth where known	Well	445.5
21.1	Well	Suspended well (wellhead and pipes projecting from the seabed) over which the depth is unknown		445.1
21.2	15 Well	Suspended well over which the depth is known		445.1
21.3		Suspended well with height of wellhead above the seabed		

Examples of Routeing Measures

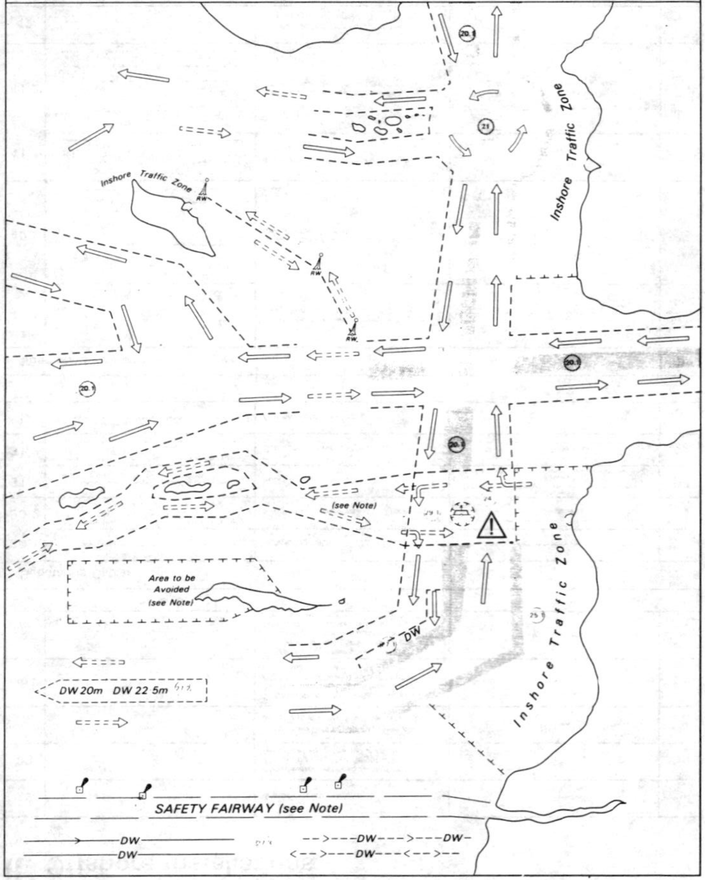

Tracks Marked by Lights → IP	Leading Beacons → IQ			Tracks
1	2 Bns ≠ 270°30′; 2 Bns ≠ 270°30′	Leading line (the firm line is the track to be followed) * The bearing may be shown in degrees and tenths of a degree	Bn Bn Bns in line 270°30′; Leading Bns 270°30′	433 1 433 2 433 3 P1
2	2 Bns ≠ 270°30′; Island open of Headland 270°30′	Transit (other than leading line) Clearing line * The bearing may be shown in degrees and tenths of a degree	Bns in line 270°30′	433 4 433 5 P2
3	270°-090°	Recommended track based on a system of fixed marks		434 1 434 2 P5
4	270°-090°	Recommended track not based on a system of fixed marks		434 1 434 2 P21
5.1		One-way track combined with routeing element		432 3
5.2	SEE NOTE; SEE NOTE	Two-way track combined with routeing element (including a regulation described in a note)		
6	7·3m; 7·3m	Recommended track with maximum authorised draught		432 4 434 3 434 4 P5 P21

Routeing Measures

Basic Symbols

10		Established direction of traffic flow		435 1 P6a
11		Recommended direction of traffic flow		435 5 P6a
12		Separation line		435 1 436 3 P6a
13		Separation zone		435 1 436 3 P6a
14		Limit of restricted area (eg Inshore Traffic Zone, Area to be Avoided)		435 1 436 3 439 2 P25
15		Limit of routeing measure		435 1 436 3 P9
16	Precautionary Area	Precautionary area		435 2

General	*Dredged and Swept Areas* → II	*Submarine Cables, Submarine Pipelines* → IL	*Tracks Routes* → IM	
1.1		*Maritime limit in general, usually implying physical obstructions*		439.1 *P9*
1.2		*Maritime limit in general, usually implying no physical obstructions*		
2.1		*Limit of restricted area*		439.2 439.3 441.6 *G46* *P25*
2.2	Entry Prohibited	*Limit of restricted area into which entry is prohibited*		

Anchorages, Anchorage Areas				
10		*Recommended anchorage (no defined limits)*		431.1 *G1*
11.1	A N53 14	*Anchor berths*		431.2 *G20a*
11.2	N53 14	*Anchor berths with swinging circle shown*		
12.1		*Anchorage area in general*		431.3 *G12a* *G12b* *P12*
12.2	No 1	*Numbered anchorage area*	1 1	
12.3	Oaze	*Named anchorage area*		
12.4	DW	*Deep water anchorage area anchorage area for deep-draught vessels*		
12.5	Tanker	*Tanker anchorage area*		
12.6	24 h	*Anchorage area for periods up to 24 hours*		
12.7		*Explosives anchorage area*		
12.8		*Quarantine anchorage area*		
12.9	Reserved (see Note)	*Reserved anchorage area*		
13		*Seaplane landing area*	Seaplane Landing Area (see Note)	449.6 *G48*
14		*Anchorage for seaplanes*		449.6 *G47*
a		*Anchorage area for small craft*	Small Craft Anchorage	*P12*

Examples of Routeing Measures (see diagram on page 36)

20.1	*Traffic separation scheme, traffic separated by separation zone*	435.1
20.2	*Traffic separation scheme, traffic separated by natural obstructions*	435.1
20.3	*Traffic separation scheme, with outer separation zone, separating traffic using scheme from traffic not using it*	435.1
21	*Traffic separation scheme, roundabout*	435.1
22	*Traffic separation scheme with 'crossing gates'*	435.1
23	*Traffic separation schemes crossing, without designated precautionary area*	435.1
24	*Precautionary area*	435.2
25.1	*Inshore traffic zone, with defined end-limits*	435.1
25.2	*Inshore traffic zone, without defined end-limits*	435.1
26.1	*Recommended direction of traffic flow, between traffic separation schemes*	435.5
26.2	*Recommended direction of traffic flow, for ships not needing a deep water route*	435.5
27.1	*Deep water route, as part of one-way traffic lane*	435.3
27.2	*Two-way deep water route, with minimum depth stated*	435.3
27.3	*Deep water route, centreline shown as recommended one-way or two-way track*	435.3
28.1	*Recommended route (often marked by centreline buoys)*	435.4
28.2	*Two-way route with one-way sections*	435.6
29.1	*Area to be avoided, around navigational aid*	435.7
29.2	*Area to be avoided, because of danger of stranding*	435.7
a	*Safety fairway*	432.2

Radar Surveillance System				
30	Radar Surveillance Station	*Radar traffic surveillance station*		487 487.3
31	Ra Cuxhaven	*Radar range*		487.1
32.1	Ra	*Radar reference line*		487.2 *P6*
32.2	Ra 270°–090°	*Radar reference line coinciding with a leading line*		

Radio Reporting				
40	7 B	*Radio calling-in point, way point, or reporting point (with designation, if any) showing direction(s) of vessel movement*	7 B	488 *G46a*
b		*Radio reporting line (with designation, if any) showing direction(s) of vessel movement*		

Ferries				
50		*Ferry*	Ferry Ferry	438.1 *H19*
51	Cable Ferry	*Cable Ferry*		438.2

International Boundaries and National Limits

No.	Symbol	Description	Other symbol	Ref.
40	DENMARK GERMANY	International boundary on land	†	440.1 P16
41	UNITED KINGDOM NORWAY	International maritime boundary	† Continental Shelf Boundary	440.3 P16
42		Territorial Sea straight baseline		440.4
43		Limit of Territorial Sea	*	440.5 P14
44		Limit of Contiguous Zone		440.6
45		National fishery limits		440.7 P10
46		Limit of Continental Shelf		440.8
47	EEZ	Limit of Exclusive Economic Zone	*	440.9
48		Customs limit		440.2 P15
49	Harbour Limit	Harbour limit	† Harbour Limit	430.1
c		National fishery limit and limit of Territorial Sea		

Various Limits

No.	Symbol		Description	Other symbol	Ref.
60.1		*	Limit of fast ice, Ice front		449.1 P18
60.2		*	Limit of sea ice (pack ice)-seasonal	†	
61	Log Pond		Log pond	† Booming Ground † Timber	449.2 H24
62.1	Spoil Ground		Spoil ground		446.1 446.2 G13 P11
62.2	Spoil Ground (disused)		Spoil ground (disused)		
63	Dredging Area		Dredging area		446.4 446.5
64	Cargo Transhipment Area		Cargo transhipment area		449.4
65	Incineration Area		Incineration area	† Area for burning refuse material	449.3

Restricted Areas

No.	Symbol	Description	Other symbol	Ref.
20		Anchoring prohibited	† Anchoring Prohibited †	431.4 439.3 439.4 G12
21		Fishing prohibited		439.3 439.4
22		Limit of nature reserve, bird sanctuary, game preserve, seal sanctuary	* Marine Nature Reserve (see Note) (shown on certain charts of the British Isles)	
23.1	Explosives Dumping Ground	Explosives dumping ground	† Explosives Dumping Ground	442.1 442.2 442.3 442.4 P11
23.2	Explosives Dumping Ground (disused)	Explosives dumping ground (disused)	† Explosives Dumping Ground (disused)	
24	Dumping Ground for Chemicals	Dumping ground for chemical waste		442.1 442.2 442.3 P11
25	Degaussing Range	Degaussing range	† D.G. Range D.G. Range	448.1 448.2
26	Historic Wk	Historic wreck and restricted area		449.5
b		Explosives dump used to temporarily deposit explosives which are recovered at a later date	Explosives Dump (temp)	

Military Practice Areas

No.	Symbol	Description	Other symbol	Ref.
30		Firing practice area	Limits not usually charted	441.1 441.2 441.3 P13a
31	Entry Prohibited	Military restricted area into which entry is prohibited		441.6 P13a P25 G46
32		Mine-laying practice area		441.4 P13a
33	SUBMARINE EXERCISE AREA (see Note)	Submarine transit lane and exercise area		441.5 P13a
34	Minefield (see Note)	Minefield	Mine Danger Area (see Note)	441.8

No.		Term	Abbreviation		Ref.
58		*Fan*			
59		*Aprön*			
60		*Fracture zone*			
61		*Scarp, Escarpment*			
62		*Sill*			
63		*Cap*			
64		*Saddle*			
65		*Levee*			
66		*Province*			
67		*Tideway, Tidal gully*			
68		*Sidearm*			

Other Terms

No.		Term	Abbreviation		Ref.
80		*projected*	proj	† projd	F15
81		*lighted*			F19
82		*buoyed*			
83		*marked*			
84		*ancient*		† Anct	F9
85		*distant*	dist		F16
86		*lesser*			
87		*closed*			
88		*partly*			
89		*approximate*	approx		F34
90		*submerged*	subm	† submd	F33
91		*shoaled*			
92		*experimental*	exper	† experl	F24
93		*destroyed*	dest	† destd	F14
a		*about*	abt		F17
b		*discontinued*	discont	† discontd	F25
c		*prohibited*	prohib	† prohibd	F26
d		*prominent*	prom	† promt	F31

Hydrographic Terms

No.		Term	Abbreviation		Ref.
1		*Ocean*			
2		*Sea*			
3		*Gulf*	*G.*		*B1*
4		*Bay*	*B.*		*B2*
5		*Fjord*	*Fj.*	† *Fd*	*B3*
6		*Lake, Loch, Lough*	*L.*		*B4*
7		*Creek*	*Cr.*		*B5*
8		*Lagoon*	*Lag.*	† *Lagn*	*C16*
9		*Cove*			*B5a*
10		*Inlet*			*B6*
11		*Strait*	*Str.*		*B7*
12		*Sound*	*Sd.*		*B8*
13		*Passage*	*Pass.*		*B9*
14		*Channel*	*Chan.*		*B10*
15		*Narrows*			*B10a*
16		*Entrance*	*Ent.*	† *Entce*	*B11*
17		*Estuary*	*Est.*	† *Esty.*	*B12*
18		*Delta*			*B12a*
19		*Mouth*	*Mth.*		*B13*
20		*Roads, Roadstead*	*Rds.*		*B14*
21		*Anchorage*	*Anch.*	† *Anchge.*	*B15*
22		*Approach*	*Appr.*	† *Apprs*	
23		*Bank*	*Bk.*		*O21*
25		*Shoal*	*Sh.*		*O22*
26		*Reef*	*Rf.*		*O23*
27		*Sunken rock*			*O26*
28		*Ledge*	*Le.*		*O24*
29		*Pinnacle*			
30		*Ridge*			*O23a*
31		*Rise*			
32		*Mountain*			
33	SMt	*Seamount*			429.1 O8a
34		*Seamount chain*			
35		*Peak*			
36		*Knoll*			
37		*Abyssal hill*			
38		*Tablemount*			
39		*Plateau*			
40		*Terrace*			
41		*Spur*			
42		*Continental shelf*			
43		*Shelf-edge*			
44		*Slope*			
45		*Continental slope*			
46		*Continental rise*			
47		*Continental borderland*			
48		*Basin*			
49		*Abyssal plain*			
50		*Hole*			
51		*Trench*			
52		*Trough*			
53		*Valley*			
54		*Median valley*			
55		*Canyon*			
56		*Seachannel*			
57		*Moat, Sea moat*			

This page is also published separately as chartlet 5045

Light Characters

Light Characters on Light Buoys → IQ — 471.2 K21-30e

	Abbreviation: International	Abbreviation: National	Class of Light	Illustration / Period shown
10.1	F		Fixed	
10.2	*Occulting (total duration of light longer than total duration of darkness)*			
	Oc	† Occ	Single-occulting	
	Oc(2) *Example*	† Gp Occ(2) *Example*	Group-occulting	
	Oc(2+3) *Example*	† Gp Occ(2+3) *Example*	Composite group-occulting	
10.3	*Isophase (duration of light and darkness equal)*			
	Iso		Isophase	
10.4	*Flashing (total duration of light shorter than total duration of darkness)*			
	Fl		Single-flashing	
	Fl(3) *Example*	† Gp Fl(3) *Example*	Group-flashing	
	Fl(2+1) *Example*	† Gp Fl(2+1) *Example*	Composite group-flashing	
10.5	L Fl		Long-flashing (flash 2s or longer)	
10.6	*Quick (repetition rate of 50 to 79—usually either 50 or 60—flashes per minute)*			
	Q	† Qk Fl	Continuous quick	
	Q(3) *Example*	† Qk Fl(3) *Example*	Group quick	
	IQ	† Int Qk Fl	Interrupted quick	
10.7	*Very quick (repetition rate of 80 to 159—usually either 100 or 120—flashes per minute)*			
	VQ	† V Qk Fl	Continuous very quick	
	VQ(3) *Example*	† V Qk Fl(3) *Example*	Group very quick	
	IVQ	† Int V Qk Fl	Interrupted very quick	
10.8	*Ultra quick (repetition rate of 160 or more—usually 240 to 300—flashes per minute)*			
	UQ		Continuous ultra quick	
	IUQ		Interrupted ultra quick	
10.9	Mo(K) *Example*		Morse Code	
10.10	FFl		Fixed and flashing	
10.11	Al.WR *Example*	† Alt.WR *Example*	Alternating	R W R W R W

Light Structures, Major Floating Lights

Beacons → IQ

	Symbols	Description	Other symbols	Ref.
1	Lt — Lt Ho	*Major light, minor light‡, light, lighthouse*		470.5 *K1-3*
2		*Lighted offshore platform*		445.2 *I76*
3	BnTr	*Lighted beacon tower‡*	† Bn Tower — † Bn Tr	456.4 457.1 457.2 *L54*
4	Bn	*Lighted beacon‡ On smaller-scale charts, where navigation within recognition range of the daymark is unlikely, lighted beacons are charted solely as lights*		457.1 457.2 *K5*
5	Bn	*Buoyant beacon, resilient beacon‡*		459.1 459.2
6	Lt V	*Light-vessel*	†	474.1/3 474.5/6 474.7 *K6*
7		*Light-float‡*	† B R B	474.1 462.8 *L12*
8		*LANBY Large Automatic Navigational Buoy (superbuoy as navigational aid)*		462.9 474.2 474.4 474.5 474.6 *L67*

‡ *Minor lights, fixed and floating, usually conform to IALA Maritime Buoyage System characteristics.*

Lights marking Fairways

Note: Quoted bearings are always from seaward

Leading Lights and Lights in line

20.1	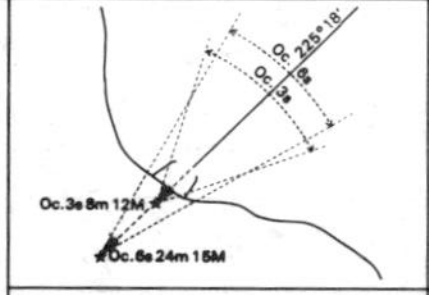	*Leading lights with leading line (the firm line is the track to be followed) and arcs of visibility.* *∗ The bearing may be shown in degrees and tenths of a degree*	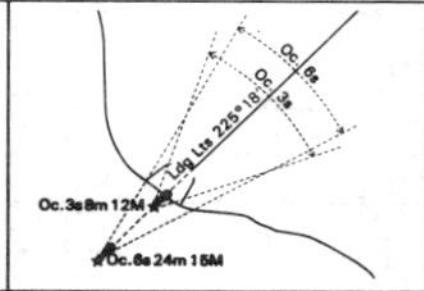	433 433.1 433.2 433.3 475.1 475.6 *K11*
20.2		*Leading lights* *≠ means "in line"* *† Lights in line in this sense is obsolescent*	† 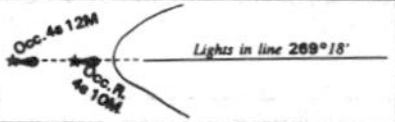	433.2 433.3 475.6 *K11*
20.3		*Leading lights on small-scale charts*		433.1 475.6 *K11*
21	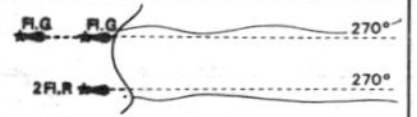	*Lights in line, marking the sides of a channel*	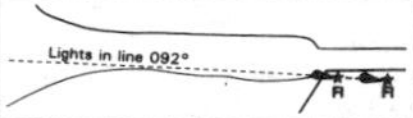	433.4 475.6
22	Rear Lt or Upper Lt	*Rear or upper light*	† *Upr.*	470.7 *K76* *K78*
23	Front Lt or Lower Lt	*Front or lower light*	† Lr	470.7 *F23a* *K77* *K79*

Direction Lights

30.1		*Direction light with narrow sector and course to be followed, flanked by darkness or unintensified light* *∗ The bearing may be shown in degrees and tenths of a degree*	† Dir Lt	475 475.1 475.7 *K13*
30.2	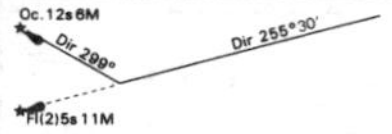	*Direction light with course to be followed, uncharted sector is flanked by darkness or unintensified light*	† Dir Lt	475 475.7 *K13*
30.3		*Direction light with narrow fairway sector flanked by light sectors of different character*		471.3 471.9 475 475.1 475.5 475.7 *K13*
31		*Moiré effect light (day and night), variable arrow mark*		475.8

Colours of Lights

11.1	W	*White (may be omitted)*		450.2 450.3 470.4 470.6 471.4 475.1 *K61* *K63-K67a*
11.2	R	*Red*		
11.3	G	*Green*		
11.4	Bu	*Blue*	† Bl	
11.5	Vi	*Violet*		
11.6	Y	*Yellow*		
11.7	Y Or	*Orange*	† Or	
11.8	Y Am	*Amber*		
		*Colours of lights shown on:* *standard charts* *on multicoloured charts* *on multicoloured charts at sector lights*		

Period

12	**90s** *Example*	*Period in seconds*	† 90 sec	471.5 *K41*

Elevation

Plane of Reference for Heights ⟶ IH   *Tidal Levels* ⟶ IH

13	**12m** *Example*	*Elevation of light given in metres*	† *On fathoms charts, the elevation of a light is given in feet e.g.* **40ft**	471.6

Range

Note: Charted ranges are nominal ranges given in sea miles

14	**15M** *Example*	*Light with single range*		471.7 471.9 475.5 *K44*
	15/10M *Example*	*Light with two different ranges*	† 15, 10M	
	15-7M *Example*	*Light with three or more ranges*	† 15, 10, 7M	

Disposition

15	(hor)	*horizontally disposed*	† *(horl.)*	471.8 *K81*
	(vert)	*vertically disposed*	† *(vertl.)*	471.8 *K80*

Example of a full Light Description — 471.9

16

Example *of a light description on a* **metric** *chart using international abbreviations:* **★ Fl(3)WRG.15s 13m 7-5M**

- **Fl(3)** — **Class** *or* **character** *of light: in this example a group-flashing light, regularly repeating a group of three flashes.*
- **WRG.** — **Colours** *of light: white, red and green, exhibiting the different colours in defined sectors.*
- **15s** — **Period** *of light in seconds, ie, the time taken to exhibit one full sequence of 3 flashes and eclipses: 15 seconds.*
- **13m** — **Elevation** *of focal plane above MHWS or MHHW or, where there is no tide, above MSL: 13 metres.*
- **7-5M** — **Luminous range** *in sea miles: the distance at which a light of a particular intensity can be seen in 'clear' visibility, taking no account of earth curvature. In those countries (eg United Kingdom) where the term 'clear' is defined as a meteorological visibility of 10 sea miles, the range may be termed* **"nominal"**. *In this example the ranges of the colours are: white 7 miles, green 5 miles, red between 7 and 5 miles.*

Example *of a light description on a* **fathoms** *chart using international abbreviations:* **★ Al.Fl.WR.30s 110ft 23/22M**

- **Al.Fl.** — **Class** *or* **character** *of light: in this example exhibiting single flashes of differing colours alternately.*
- **WR.** — **Colours** *of light shown alternately: white and red all-round (ie, not a sector light).*
- **30s** — **Period** *of light in seconds, ie, the time taken to exhibit the sequence of two flashes and two eclipses: 30 seconds*
- **110ft** — **Elevation** *of focal plane above MHWS or MHHW or, where there is no tide, above MSL: 110 feet.*
- **23/22M** — **Range** *in sea miles. Until 1971 the lesser of* **geographical** *range (based on a height of eye of 15 feet) and* **luminous** *range was charted. Now, when the charts are corrected, luminous (or nominal) range is given. In this example the luminous ranges of the colours are: white 23 miles, red 22 miles. The geographical range can be found from the table in the Admiralty List of Lights (for the elevation of 110 feet, it would be 16 miles).*

Lights with limited Times of Exhibition

50	F.R(occas)	Lights exhibited only when specially needed (for fishing vessels, ferries) and some private lights	† (fishg.) † (Priv.) † (occasl.)	473.2 K15 K17 K70
51	Fl.10s 40m 27M (F.37m 11M by day)	Daytime light (charted only where the character shown by day differs from that shown at night)		473.4
52	Q.WRG.5m 10-3M Fl.5s(in fog)	Fog light (exhibited only in fog, or character changes in fog)		473.5
53	Fl.5s (U)	Unwatched (unmanned) light with no standby or emergency arrangements	(U)	473.1
54	(temp)	Temporary	† (temp) † (tempy.)	K73
55	(exting)	Extinguished	† (exting) † (exting d.)	K74
b		Tidal light(s)	(tidal)	K16

Special Lights

Flare Stack (at Sea) → IL *Flare Stack (on Land)* → IE *Signal Stations* → IT

60	Aero Al.Fl.WG.7·5s 11M	Aero light (may be unreliable)		476.1 K4
61.1	Aero F.R.353m 11M RADIO MAST (353)	Air obstruction light of high intensity		476.2
61.2	(89) (R Lts)	Air obstruction lights of low intensity	(Red Lt.)	476.2
62	Fog Det Lt	Fog detector light		477 K68a
63	(Illuminated)	Floodlit, floodlighting of a structure	(illum) † (lit)	478.2
64	F F.R Iso	Strip light		478.5
65	F.R (priv)	Private light other than one exhibited occasionally	○ Y Lt ○ R Lt † (Priv)	473.2 K17

Sector Lights

40	Fl.WRG.4s 21m 18-12M; G W R	Sector light on standard charts		475 475.1 475.2 475.5 K12
41.1	Oc.WRG 10-6M; Fl.G Name Fl.W.3s Fl.R; G W R; Oc.R Oc.W Oc.G Name	Sector lights on standard charts, the white sector limits marking the sides of the fairway		475 475.1 475.5 K12
41.2	Oc.WRG 10-6M; Fl.G Name Fl.W.3s Fl.R; G W R; Oc.R Name Oc.W Oc.G	Sector lights on multicoloured charts, the white sector limits marking the sides of the fairway		475 475.1 475.5 470.4 K12
42	Fl(3)10s 62m 25M F.R.55m 12M; F.R	Main light visible all-round with red subsidiary light seen over danger		471.8 475.4
43	Fl.5s 41m 30M; Obscd	All-round light with obscured sector		475.3 K68
44	Iso.WRG; G W R; G	Light with arc of visibility deliberately restricted		475.3 K68
45	Q.14m 5M; Faint	Light with faint sector		475.3 K75
46	Oc.R.8s; R.5M R.9M R.5M Oc.R.8s 5M; R R. Intens R	Light with intensified sector		475.5
a		Light with unintensified sector	Oc.R.8s; R.5M R.2M / Oc.R.8s 5M; R R. Unintens	

Buoys

Features Common to Buoys and Beacons → IQ 1–11

Shapes

No.	Symbol	Description	Non-IALA System	Ref.
20		Conical buoy, nun buoy, ogival buoy	Non-IALA System etc.	462.2 L6
21		Can buoy, cylindrical buoy	Non-IALA System etc.	462.3 L5
22		Spherical buoy	Non-IALA System etc.	462.4 L7
23		Pillar buoy	Non-IALA System	462.5 L8a
24		Spar buoy, spindle buoy	Non-IALA System	462.6 L8 L51
25		Barrel buoy, tun buoy		462.7 L10
26		Superbuoy. Superbuoys are very large buoys, e.g. a LANBY (IP8) is a navigational aid mounted on a circular hull of about 12m diameter. Oil or gas installation buoys (IL16) and ODAS buoys (IQ58), of similar size, are shown by variations of the superbuoy symbol		462.9 L67

Light Floats

No.	Symbol	Description	Non-IALA System	Ref.
30	Fl.G.3s Name G	Light float as part of IALA System		462.8 L12
31	Fl.10s 12m 26M	Light float not part of IALA System	† B R B	462.8 L12

Mooring Buoys

Oil or Gas Installation Buoy → IL Visitors' (Small Craft) Mooring → IU

No.	Symbol	Description	Other	Ref.
40		Mooring buoy	†	431.5 L22
41	Fl.Y.2·5s	Lighted mooring buoy		431.5 466.1 466.2 466.3 466.4 L22
42	* ① ②	Trot, mooring buoys with ground tackle and berth numbers	① ②	323.1 431.6
43		Mooring buoy with telegraphic or telephonic communications		431.5 L22b L22c
44	Small Craft Moorings	Numerous moorings		431.7 L22a

Buoys and Beacons

IALA Maritime Buoyage System, which includes Beacons → IQ 130

General

No.	Symbol	Description	Other	Ref.
1		Position of buoy or beacon		455.3 460.1 462.1 L1

Colours of Buoys and Beacon Topmarks

No.	Symbol	Description	Non-IALA System	Ref.
2	G B G G G	Single colour: green (G) and black (B)	Non-IALA System B G	450 450.1 450.2 450.3 464 464.1 464.2 464.3 L31 L48b
3	R R Y Y Or R	Single colour other than green and black: red (R), yellow (Y), orange (Or)		
a		Single colour other than green and black (non-IALA system: white (W), red (R), yellow (Y), grey (Gy), blue (Bu), orange (Or))	Non-IALA System W R Y Gy Bu Or	
4	BY GRG BRB	Multiple colours in horizontal bands; the colour sequence is from top to bottom	Non-IALA System BW RW BR BW	
5	RW RW RW	Multiple colours in vertical or diagonal stripes; the darker colour is given first. In these examples, red(R), white(W) & black(B)	Non-IALA System RW BR BW BW	
6		Retroreflecting material may be fitted to some unlit marks. Charts do not usually show it. Black bands will appear dark blue under a spotlight	Refl	464 L64
b		Wreck buoy (not used in the IALA System)	G G G G	L20
c		Chequered	BR BW RW BW	L33

Lighted Marks

Marks with Fog Signals → IR

No.	Symbol	Description	Non-IALA System	Ref.
7	Fl.G G; Fl.R R	Lighted marks on standard charts	Non-IALA System	457.1 466 466.1 K5 L2
8	* Fl.R R; Iso RW; Fl.G G	Lighted marks on multicoloured charts		

Topmarks and Radar Reflectors

For Application of Topmarks within the IALA System → IQ 130 Radar reflector → IS
For other Topmarks (Special Purpose Buoys and Beacons) → IQ

No.	Symbol	Description	Non-IALA System	Ref.
9		IALA System buoy topmarks (beacon topmarks shown upright)	Non-IALA System etc.	463 463.1 L57
10	Name 2 R	Beacon with topmark, colour, radar reflector and designation	† Ra. Refl "2"	450 455.2 455.7 M13
11	Name 3 G	Buoy with topmark, colour, radar reflector and designation. Radar reflectors are not generally charted on IALA System buoys	† Ra. Refl "No. 3"	460.3 460.6 465.1 465.2 L9 M13

Beacons

Lighted Beacons → IP   Features Common to Beacons and Buoys → IQ 1–11

General

80		Bn	Beacon in general, characteristics unknown or chart scale too small to show		455.5 *L52* *L53*
81	BW	Bn BW	Beacon with colour, no distinctive topmark		455.4 456 456.3 *L53*
82	R BY BRB		Beacon with colour and topmark	*Non-IALA System* W B R BW etc.	455.4 456 463 463.1 *L41–L48b*
83	BRB	BRB	Beacon on submerged rock (topmark as appropriate)		455.6

Minor Impermanent Marks usually in Drying Areas (Lateral Mark for Minor Channel)

Minor Pile → IF

	PORT HAND	STARBOARD HAND			
90			Stake, pole		456.1 *L59*
91			Perch, stake		456.1 *L59*
92			Withy		456.1 *L59*

Minor Marks, usually on Land

Landmarks → IE

100		Cairn	Cairn	456.2 *L61*
101	Mk	Coloured or white mark (the colour may be indicated)		456.2 *L62*
102.1	RW	Coloured topmark (colour known or unknown) with function of a beacon	R G	456.3 *L53*
102.2	RW RW	Painted boards with function of leading beacons		456.3 *L53*

Beacon Towers

110	R G R G BY BRB	Beacon towers without and with topmarks and colours	Bn Tower Bn Tr etc.	456.4 *L54*
111		Lattice beacon		456.4

Special Purpose Buoys

The symbols shown below are examples: shapes of buoys may differ; lateral or cardinal buoys may be used in some situations; the use of the cross topmark is optional.

50	Y DZ	Firing danger area (Danger Zone) buoy		441.2 *L24a*
51	Y Target	Target		*L65*
52	Y Marker Ship	Marker Ship		*L65*
53	Y Barge	Barge		
54	Y	Degaussing Range buoy	Y DG	448.3
55	Y	Cable buoy	*Non-IALA System* Cable	443.6 *L21*
56	Y	Spoil ground buoy		446.3 *L27a*
57	Y	Buoy marking outfall		444.4
58	ODAS	Data collection buoy (Ocean Data-Acquisition System) of superbuoy size		462.9 *L67*
59	Y	Buoy marking wave recorder or current meter		
60		Seaplane anchorage buoy		*L25a*
61		Buoy marking traffic separation scheme		
62	Y	Buoy marking recreation zone		

Seasonal Buoys

70	Y (priv)	Buoy privately maintained		*F30* *L29*
71	Y (Apr–Oct)	Seasonal buoy. (The example shows a yellow spherical buoy in use during the months April to October inclusive).		460.5 *F30* *F39* *L29* *L30*

This page is also published separately as chartlet 5044

130 IALA Maritime Buoyage System

IALA International Association of Lighthouse Authorities

Where in force, the IALA System applies to all fixed and floating marks except lighthouses, sector lights, leading lights and leading marks, light vessels and LANBYs. The standard buoy shapes are cylindrical (can), conical, spherical, pillar, and spar, but variations may occur, for example: light floats. In the illustrations below, only the standard buoy shapes are used. In the case of fixed beacons (lit or unlit) only the shape of the topmark is of navigational significance.

130.1 Lateral marks

are generally for well-defined channels. There are two international Buoyage Regions-A and B-where Lateral marks differ.

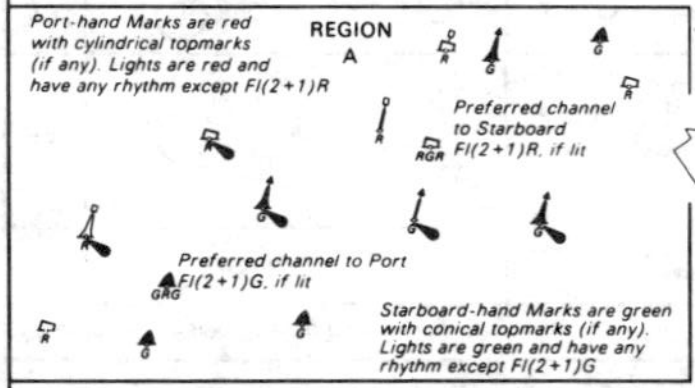

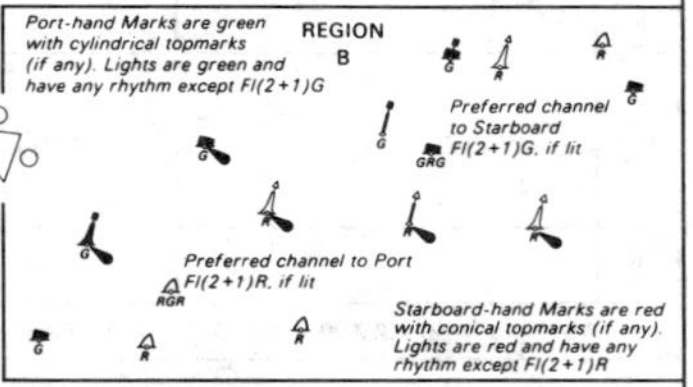

A preferred channel buoy may also be a pillar or a spar. All preferred channel marks have three horizontal bands of colour. Where for exceptional reasons an Authority considers that a green colour for buoys is not satisfactory, black may be used.

130.2

Symbol showing direction of buoyage where not obvious

Symbol showing direction of buoyage where not obvious, on multicoloured charts (red and green circles coloured as appropriate).

130.3 Cardinal Marks

indicating navigable water to the named side of the marks. Cardinal marks have the same meaning in Regions A and B

UNLIT MARKS

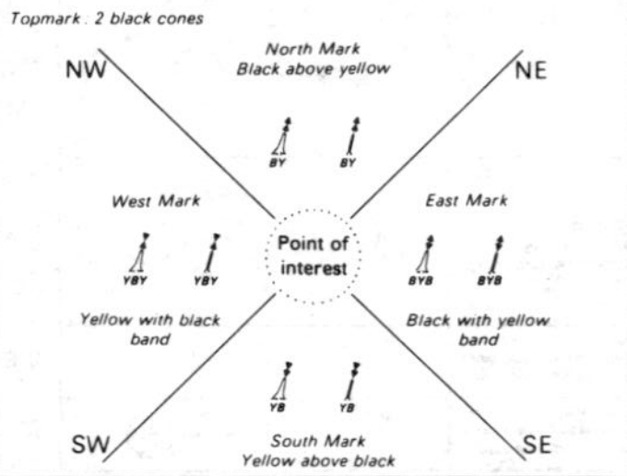

LIGHTED MARKS

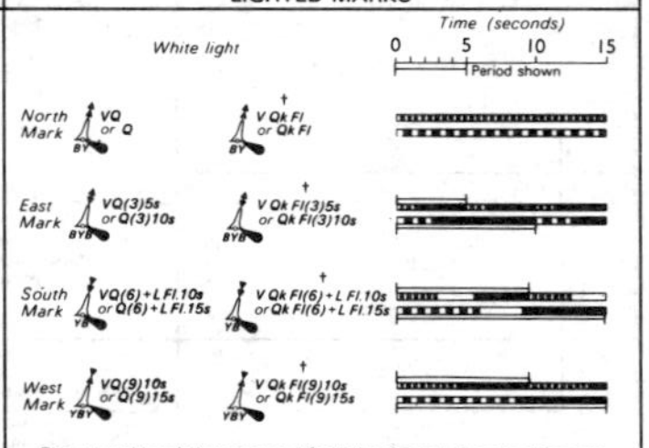

The same abbreviations are used for lights on spar buoys and beacons. The periods, 5s, 10s and 15s, may not always be charted.

130.4

Isolated Danger Marks, stationed over dangers with navigable water around them.

Body: black with red horizontal band(s)
Topmark: 2 black spheres

Fl(2) † Gp Fl(2) — *White light*

130.5

Safe Water Marks, such as mid-channel and landfall marks.

Body: red and white vertical stripes
Topmark (if any): red sphere

Iso, or Oc, or L Fl.10s, or Mo(A) † Occ — *White light*

130.6

Special Marks, not primarily to assist navigation but to indicate special features.

Body (shape optional): yellow
Topmark (if any): yellow X

Fl.Y etc — *Yellow light*

Leading Lines, Clearing Lines → IM

Special Purpose Beacons

Note: Topmarks and colours are shown where scale permits

120		*Leading beacons*		458 *P1*
121		*Beacons marking a clearing line or transit*		458 *P2*
122	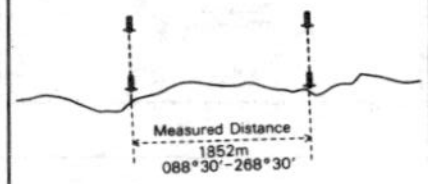	*Beacons marking measured distance with quoted bearings. The track is shown as a firm line if it is to be followed precisely*		458 *P24*
123		*Cable landing beacon*		443.5 458 *L58*
124	 	*Refuge beacon*		456.4
125		*Firing practice area beacons*		
126		*Notice board*	NB	456.2

IS Radar, Radio, Electronic Position-Fixing Systems

Radar	Radar Structures Forming Landmarks → IE		Radar Surveillance Systems → IM	
1	Ra	*Coast radar station providing range and bearing from station on request*		485.1 *M11*
2	Ramark	*Ramark, radar beacon transmitting continuously*		486.1 *M14a*
3.1	Racon(Z)	*Radar transponder beacon, with morse identification, responding within the 3cm (X) band*		486.2 486.3 *M12*
3.2	Racon(Z)(10cm)	*Radar transponder beacon, with morse identification, responding within the 10cm (S) band*		486.3
3.3	Racon(Z)(3 & 10cm)	*Radar transponder beacon, with morse identification, responding within the 3cm (X) and the 10cm (S) band*		
3.4	F Racon	*Radar transponder beacon, responding on a fixed frequency outside the marine band*		486.4
3.5	Racon Racon Racons / 270°	*Radar transponder beacons with bearing line*		486.2 433.3
3.6	Racon Racon Racon	*Floating marks with radar transponder beacons*		486.2 *M12*
4		*Radar reflector (not usually charted on IALA System buoys)*	† *Ra Refl.*	460.3 465 *M13*
5		*Radar-conspicuous feature*	† Ra conspic	485.2 *M14*

Radio	Radio Structures Forming Landmarks → IE		Radio Reporting (Calling-in or Way) Points → IM	
10	Name RC	*Non-directional marine or aeromarine radiobeacon*		481.1 480.1 *M4*
11	RD RD 270°	*Directional radiobeacon with bearing line*	† Dir.Ro.Bn. Dir.Ro.Bn.269°30′	481.2 433.6 *M5*
12	RW	*Rotating pattern radiobeacon*		481.1 *M6*
13	Consol	*Consol beacon*		481.3 *M19*
14	RG	*Radio direction-finding station*	† Ro.D.F	483 *M7*
15	R	*Coast radio station providing QTG service*	† Ro.	484 *M10a*
16	Aero RC	*Aeronautical radiobeacon*	† Aero Ro.Bn.	482 *M16*

Fog Signals IR

Fog Detector Light → IP *Fog Light* → IP				General
1		*Position of fog signal. Type of fog signal not stated*	†Fog Sig	451 451.2 452.8 453 *N1*

Types of Fog Signals, with Abbreviations

10	Explos	*Explosive*	†Gun	452.1 *N3* *N10*
11	Dia	*Diaphone*		452.2 *N9*
12	Siren	*Siren*		452.3 *N11*
13	Horn	*Horn (nautophone, reed, tyfon)*	†Nauto †E.F.Horn †Tyfon †Reed	452.4 *NH* *N12 13a* *N16*
14	Bell	*Bell*		452.5 *N14* *L3*
15	Whis	*Whistle*		452.6 *N15* *L4*
16	Gong	*Gong*		452.7 *N17* *L3a*

Examples of Fog Signal Descriptions

20	‡ Fl.3s 70m 29M Siren Mo(N)60s	*Siren at a lighthouse, giving a long blast followed by a short one (N), repeated every 60 seconds*		452.3 453.3 *N1* *N11*
21	‡ Bell	*Wave-actuated bell buoy* *The provision of a legend indicating number of emissions, and sometimes the period, distinguishes automatic bell or whistle buoys from those actuated by waves*		425.5 453 454.1 *L3* *N14*
22	‡ Q(6)+LFl.15s Horn(1)15s Whis	*Light buoy, with horn giving a single blast every 15 seconds, in conjunction with a wave-actuated whistle*	Reserve fog signals are fitted to certain buoys. Only those actuated by waves are charted	452.4 453.1 454.3 *L4* *N1* *N13*

‡ *The Fog Signal symbol may be omitted when a description of the signal is given*

Pilotage

1.1	⦶	Pilot boarding place; position of pilot cruising vessel	† Pilots † Pilots	491.1 491.2 491.6 J8
1.2	⦶ Name	Pilot boarding place; position of pilot cruising vessel, with name (e.g. District, Port)		
1.3	⦶ Note	Pilot boarding place; position of pilot cruising vessel, with note (e.g. Tanker, Disembarkation)		
1.4	⦶ H	Pilots transferred by helicopter		491.2
2	▬ Pilot look-out	Pilot office with Pilot look-out; Pilot look-out station		491.3
3	▬ Pilots	Pilot office		491.4 J8
4	Port Name (Pilots)	Port with pilotage service (location not shown)		491.2 491.5 J8

Coastguard, Rescue

10	▬CG ●CG ⚑CG	Coastguard station	▬CGFS	492 492.1 492.2 J3
11	▬CG✚ ●CG✚ ⚑CG✚	Coastguard station with Rescue station	▬CGFS ✚	493.3
12	✚	Rescue station; Lifeboat station; Rocket station	† LB	493 493.1 J5,6,7
13	▲✚ ✚	Lifeboat lying at a mooring		493.2
14	Ref	Refuge for shipwrecked mariners		456.4

Electronic Position-Fixing Systems

Decca

20	AB AD	Identification of Lattice Patterns		614.1
21	———	Line of Position (LOP)		605.1
22	———	LOP representing Zone Limit or, on larger-scale charts, other intermediate LOPs		615.1
23	— — — —	Half-lane LOP		615.2 615.3
24		LOP from adjoining Chain (on Interchain Fixing Charts)		611.4 611.5
25	(6) A12	Lane value, with Chain designator (Interchain Fixing Charts only) and Zone designator		616.1 616.2 616.3

Note: A Decca Chain Coverage Diagram is given when patterns from more than one Chain appear on a chart. LOPs are normally theoretical ones; if Fixed Error is included, an explanatory note is given.

Loran-C

30	7970-V 7970-W 7970-Y 7970 X	Identification of Loran-C Rates		624.1
31	———	Line of Position (LOP)		605.1
32	———	LOP representing time difference value of an integral hundred or thousand µs (microseconds)		625.1
33	— — —	LOP beyond reliable groundwave service area		625.4
34	- - - - - - - - - -	LOP from adjoining Chain		625.3
35	- - - - - - - -	LOP from adjoining Chain beyond reliable groundwave service area		625.4
36	7970-X 33000	LOP labelled with Rate and full time difference (µs) value		626.2 626.3
37	050	LOP labelled with final three digits only		626.2

Note: A Loran-C Chain Diagram may be given if Rates from more than one Chain appear on a chart. An explanatory note is given if LOPs include propagation delays.

Omega

40	AG AH DG	Charted station pairs		633
41	———	Line of Position (LOP)		605.1 634.1
42	897 DG-900	Lane values		635.1 635.2

Note: A cautionary note draws attention to the need to consult Propagation Prediction Correction (PPC) Tables. An explanatory note draws attention to the unreliability of LOPs.

Satellite Navigation Systems

50	WGS WGS 72 WGS 84	World Geodetic System, 1972 or 1984		

Note: A note may be shown to indicate the shifts of latitude and longitude, in hundredths or tenths of a minute, which should be made to satellite-derived positions (which are referred to WGS) to relate them to the chart.

Small Craft Facilities	*Transport Features, Bridges* →ID *Public Buildings, Cranes* →IF	*Pilots, Coastguard, Rescue, Signal Stations* →IT		
1.1		*Yacht harbour, Marina*		320.2 *G33*
1.2		*Yacht berths without facilities*		
2		*Visitors' berth*		
3		*Visitors' mooring*		
4		*Yacht club; Sailing club*		
5		*Public slipway*		
6		*Boat hoist*		
7		*Public landing; Steps; Ladder*		
8		*Sailmaker*		
9		*Boatyard*		
10		*Public house, Inn*		
11		*Restaurant*		
12		*Chandler*		
13		*Provisions*		
14		*Bank; Bureau de change*		
15		*Physician; Doctor*		
16		*Pharmacy; Chemist*		
17		*Water tap*		
18		*Fuel station (Petrol, Diesel)*		
19		*Electricity*		

Stations					
20	○SS	*Signal station in general*	†Sig Sta	†Sig Stn	490.3 *J9*
21	○SS (INT)	*Signal station showing International Port Traffic Signals*			495.5
22	○SS (Traffic)	*Traffic signal station*			495.1
23	○SS (Port Control)	*Port control signal station*			495.1
24	○SS (Lock)	*Lock signal station*			495.2
25.1	○SS (Bridge)	*Bridge passage signal station*			495.3
25.2	F. Traffic Sig	*Bridge lights including traffic signals*			495.4
26	○SS	*Distress signal station*			497
27	○SS	*Telegraph station*			497.1 *I28*
28	○SS (Storm)	*Storm signal station*	†Storm Sig	†*Stm. Sig. Stn.*	494.1 *J11*
29	○SS (Weather)	*Weather signal station, Wind signal station*			494.1 *J12*
30	○SS (Ice)	*Ice signal station*			494.1 *J15*
31	○SS (Time)	*Time signal station*			494.2 *J16*
32.1		*Tide scale or gauge*			496.1 *T21*
32.2	○Tide gauge	*Automatically recording tide gauge*			
33	○SS (Tide)	*Tide signal station*			496.2 *J13*
34	○SS (Stream)	*Tidal stream signal station*			496.3 *J14*
35	○SS (Danger)	*Danger signal station*			490.1
36	○SS (Firing)	*Firing practice signal station*			490.1

IV Abbreviations of Principal Foreign Terms

Glossaries of foreign terms will be found in the volumes of Admiralty Sailing Directions.

On metric Admiralty charts, foreign terms are generally given in full wherever space and information permits. Where abbreviations are used on metric charts they accord with the following list, apart from those on charts published before 1980 where full stops are omitted. Obsolescent forms of abbreviations may also be found on these charts.

CURRENT FORM	OBSOLESCENT FORM(S)	TERM	ENGLISH MEANING
ALBANIAN			
	K	Kodër, Kodra	*Hill*
ARABIC			
	Djeb, Dj	Djebel	*Mountain, Hill*
Geb.	G	Gebel	*Mountain, Hill*
J.	Jab, J^{l}	Jabal, Jibāl, Jebel	*Mountain(s), Hill(s)*
Jaz.	Jazt	Jazīrat, Jazā'ir Jazīreh	*Island(s), Peninsula*
Jeb.	J, J^{l}	Jebel	*Mountain, Hill*
Jez.	Jez^{t}	Jezirat	*Island, Peninsula*
Kh.	K	Khawr, Khōr	*Inlet, Channel*
	Si, S^{i}	Sidi	*Tomb*
W.		Wād, Wādī, Wed	*Valley, River, River bed*
CHINESE			
Chg.	Ch^{g}	Chiang	*River, Shoal, Harbour, Inlet, Channel, Sound*
DANISH			
B.		Bugt	*Bay, Bight*
Bk.	B^{k}	Banke	*Bank*
Fj.	F^{d}	Fjord	*Inlet*
Gr.	Grd, Gr^{d}, G^{d}	Grund	*Shoal*
H.	Hm, H^{m}, Hne, H^{ne}	Holm, Holmene	*Islet(s)*
Hd.	H^{d}	Hoved	*Headland*
Hn.	H^{n}	Havn, Havnen	*Harbour*
Ll.		Lille	*Little*
N.		Nord, Nordre	*North, Northern*
Ø.		Øst, Østre, Østlig	*East, Eastern*
	Øy, Øne, Ø^{ne}, Ö^{ne}	Øyane, Øyene, Öyane Öyene	*Islands*
Pt.	P^{t}	Pynt	*Point*
S.		Syd, Sydlig	*South, Southern*
Sd.	S^{d}	Sund, Sundet	*Sound*
Sk.	Skr, Sk^{r}	Skær, Skjær	*Rock above water*
St.		Stor	*Great*
V.		Vest, Vestre	*West*
DUTCH			
B.	B^{i}	Baai	*Bay*
Bg.	B^{g}	Berg	*Mountain*
Bk.	B^{k}	Bank	*Bank*
Eil.	Eiln, Eil^{n}	Eiland, Eilanden	*Island(s)*
G.		Golf	*Gulf*
	Gt, Grt, G^{t}, G^{rt}	Groot, Groote	*Great*
H.		Hoek	*Cape, Hook*
Pt.	P^{t}	Punt	*Point*
R.		Rivier	*River*
Rf.	R^{f}	Rif	*Reef*
Str.	Stn, St^{r}, St^{n}	Straat, Straten	*Strait(s)*
FINNISH			
K.		Kari, Kallio, Kivi, Kivet	*Rock, Reef*
	Lu.	Luoto, Luodet	*Rock(s)*
Ma.		Matalikko	*Shoal*
	P	Pieni, Pikku	*Small*
Sa.	S^{a}	Saari, Saaret	*Island(s)*
Tr.	T^{r}	Torni	*Tower*
FRENCH			
B.	B^{e}	Baie	*Bay*
Bas.	B	Basse	*Shoal*
Bc.	B^{c}	Banc	*Bank*
	Bssn, Bn, B^{n}	Bassin	*Basin*
C.		Cap	*Cape*
Ch.	Chap, Chap^{e}	Chapelle	*Chapel*
Chât.	Chât^{u}, Ch^{au}	Château	*Castle*
Chen.	Ch^{al}	Chenal	*Channel*
F.	Fl	Fleuve	*Large river*
Ft.	F^{t}	Fort	*Fort*
G.		Golfe	*Gulf*
	Gd, G^{d}, Gde, G^{de}	Grand, Grande	*Great*
H.F.	Ht fd, H^{t}f^{d}, H^{t} fond	Haut-fond	*Shoal*
Î.	I, I^{t}	Île, Îles, Îlot	*Island(s), Islet*
L.		Lac	*Lake*
	Mn, M^{in}	Moulin	*Mill*
Mou.	Mge, M^{age}	Mouillage	*Anchorage*
Mt.	M^{t}	Mont	*Mount, Mountain*
	N.D.	Notre Dame	*Our Lady*
P.		Port	*Port*
	Pet, P^{it}, P^{ite}, P^{t}	Petit, Petite	*Small*
Pit.	Pn, P^{on}	Piton	*Peak*
Pl.		Plage	*Beach*
Plat.	Pla, Plat^{u}	Plateau	*Tableland, Sunken flat*
Pte.	P^{te}	Pointe	*Point*
Qu.	Q	Quai	*Quay*
R.	Rau, Riv, R^{au}	Rivière, Ruisseau	*River, Stream*
	Rav, R^{ne}	Ravine	*Ravine*
Roc.	Re, R^{e}, Rer, R^{er}	Roche, Rocher	*Rock*
S.	St, S^{t}, Ste, S^{te}	Saint, Sainte	*Saint, Holy*
Som.		Sommet	*Summit*
Tr.	T^{r}	Tour	*Tower*
	Vi, V^{x}	Vieux, Vieil, Vielle	*Old*
GAELIC			
Bo.		Bogha	*Below water rock*
Eil.	E, En, E^{n}	Eilean, Eileanan	*Island(s), Islet(s)*
Ru.	R^{u}	Rubha	*Point*
Sg.	Sgr, Sg^{r}	Sgeir	*Rock*
GERMAN			
B.		Bucht	*Bay*
Bg.	B^{g}	Berg	*Mountain*
Gr.	Grd, Grd, G^{d}	Grund	*Shoal*
Hn.	H^{n}	Hafen	*Harbour*
K.		Kap	*Cape*
Rf.	R^{f}	Riff	*Reef*
	Schl	Schloss	*Castle*
GREEK			
Ák., Ak.		Ákra, Akrotírion	*Cape*
Ang.	Ali, A^{li}	Angáli	*Bight, Open bay*
Angir.	Ang, Ang^{n}	Angirovólion	*Anchorage*
Áy., Ay.		Áyios, Ayía	*Saint, Holy*
Íf.	If, I^{f}	Ífalos, Ífaloi	*Reef(s)*
Kól.	Kol, Ks, K^{os}	Kólpos	*Gulf*
Lim.	Ln, L^{n}	Limín, Liménas	*Harbour*
N.	Noi, N^{oi}	Nísos, Nísoi	*Island(s)*
N.	Nis, N^{is}, Nes, N^{es}	Nisídha, Nisídhes	*Islet(s)*
Ó.	O	Órmos	*Bay*
Or.		Ormískos	*Creek*
Ór.	Or	Óros, Óroi	*Mountain*
Pot.		Potamós	*River*
	Prof	Profítis	*Prophet*
Sk.		Skópelos, Skópeloi	*Reef(s), Drying rock(s)*
Vrak.	Vis, V^{is}, Ves, V^{es}	Vrakhonisídha, Vrakhonisídhes	*Rocky islet(s)*
Vrák.	Vrak, Vos, V^{os}, Voi, V^{oi}	Vrákhos, Vrákhoi	*Rock(s)*
ICELANDIC			
Fj.	Fjr, F^{dr}	Fjördhur	*Fjord*
Gr.		Grunn	*Shoal*
INDONESIAN, MALAY and SINGAPOREAN			
A.		Air, Ajer, Ayer	*Stream*
B.	Bu, B^{u}	Batu	*Rock*
Bat.	Btg, B^{tg}	Batang	*River*
	Bdr, B^{dr}	Bandar, Bendar	*Port*
	Br, B^{r}	Besar	*Great*
Buk.	Bt, B^{t}	Bukit	*Hill*

Small Craft Facilities IU

20		*Bottle gas*		
21		*Showers*		
22		*Laundrette*		
23		*Public toilets*		
24		*Post box*		
25		*Public telephone*		*170*
26		*Refuse bin*		
27		*Public car park*		
28		*Parking for boats and trailers*		
29		*Caravan site*		
30		*Camping site*		
31		*Water police*		

32

Marina Facilities *may be tabulated on harbour charts and large-scale coastal charts.*
● indicates that the facility is available at the marina itself. Laundrettes etc, located outside the marina are not included. The facilities may not be available outside normal working hours. All marinas have water, toilets and rubbish disposal.

MARINA FACILITIES

	Diesel	Petrol	Bottle gas	Electricity	Showers	Laundrette	Launching slip	Boat hoist	Provisions	VHF Radio	Telephone numbers
Bank's Quay	●	●	●	●	●	●	●	●			(02013) 4299
Beach Harbour Yacht Club	●	●	●	●	●		●	●	●	●	(0202) 707321
Grain Wharf Yacht Centre	●	●	●	●	●	●	●	●	●		(09295) 2650
Quay East					●		●	●			(0202) 685335

Corrections
Information on facilities for small craft will be updated as charts are revised by New Edition The Hydrographic Office would be pleased to receive reports of alterations or additions to small craft facilities.

CURRENT FORM	OBSOLESCENT FORM(S)	TERM	ENGLISH MEANING
PORTUGUESE (continued)			
Estr.		Estreito	*Strait*
Estu.	Est, Est^{o}	Estuario	*Estuary*
Fte.	F^{te}	Forte	*Fort*
Fte.	Ftza, F^{tza}	Fortaleza	*Fortress*
Fund.		Fundeadouro	*Anchorage*
G.		Golfo	*Gulf*
	Gde, G^{de}	Grande	*Great*
I.		Ilhéu, Ilhéus, Ilhota	*Islet(s)*
I.		Ilha, Ilhas	*Island(s)*
L.		Lago	*Lake*
L.		Lagoa	*Small lake, Marsh*
La.	Le, L^{e}	Laje	*Flat-topped rock*
Lag.	La, L^{a}	Laguna	*Lagoon*
Mol.	Me, M^{e}	Molhe	*Mole*
Mor.	Mo, M^{o}	Morro	*Headland, Hill*
Mte.	M^{te}	Monte, Montanha	*Mount, Mountain*
NS.	Na.Sa, N^{a}S^{a}	Nosso Senhor, Nossa Senhora	*Our Lord, Our Lady*
P.	Pto, P^{to}	Porto	*Port*
Pal.	Pals, Pal^{s}	Palheiros	*Fishing village*
Par.	Pel, P^{el}	Parcel	*Shoal, Reef*
Pas.		Passagem, Passo	*Passage, Pass*
	Pco, P^{co}, P^{o}	Pico	*Peak*
Pda.	P^{da}	Pedra	*Rock*
	Peq	Pequeno, Pequena	*Small*
Pr.	Pa, P^{a}	Praia	*Beach*
Pta.	P^{ta}	Ponta	*Point*
Queb.		Quebrada, Quebrado	*Cut, Ravine*
R.		Rio	*River*
Rch.		Riacho, Ribeira, Ribeirão	*Creek, Stream, River*
Rf.		Recife	*Reef*
Roc.	Ra, R^{a}	Rocha, Rochedo	*Rock*
S.	Sto, S^{to}, Sta, Sta	São, Santo, Santa	*Saint, Holy*
Sr.	Sa, S^{a}	Serra, Cordilheira	*Mountain range*
	Va, V^{a}	Vila	*Town, Village, Villa*
ROMANIAN			
Br.		Braţ, Braţul, Braţu	*Branch, Arm (of the sea)*
C.		Cap, Capul, Capu	*Cape*
I.		Insulă, Insula	*Island*
L.		Lac, Lacul, Lacu	*Lake*
O.		Ostrov, Ostrovul, Ostrovu	*Island*
S.	Sf	Sfînt, Sfîntu, Sfîntul, Sfînta	*Saint, Holy*
RUSSIAN			
B.		Bukhta	*Bay, Inlet*
Bk.	Bka, B^{ka}, Bki, B^{ki}	Banka, Banki	*Bank(s)*
	Bol	Bol'shoy, Bol'shaya Bol'shoye	*Great, Large*
Gb.	G, Ga, G^{a}	Guba	*Gulf, Bay, Inlet*
G		Gora	*Mountain, Hill*
Gav.	G	Gavan'	*Harbour, Basin*
Kam.		Kamen'	*Rock*
M.		Mys	*Cape, Headland*
	Mal	Malyy, Malaya, Maloye	*Little*
O.	O^{va}	Ostrov, Ostrova	*Island(s)*
Oz.		Ozero	*Lake*
Pol.	Pol^{ov}, P^{ov}	Poluostrov	*Peninsula*
Pr.	Prv, Pr^{v}	Proliv	*Channel, Strait*
R.		Reka	*River*
Zal.		Zaliv	*Gulf, Bay*
SINGAPOREAN (see INDONESIAN)			
SPANISH			
A.	Arro, Arr^{o}	Arroyo	*Stream*
Arch.	Arch^{o}	Archipiélago	*Archipelago*
Arr.	Arr^{e}, Arr^{fe}	Arrecife	*Reef*
B.	B^{a}	Bahía	*Bay*
Ba.	Bo, B^{o}	Bajo	*Shoal*
Bco.	B^{co}	Banco	*Bank*
Br.	Bzo, B^{zo}	Brazo	*Arm (of the sea)*
C.		Cabo	*Cape*
Cal.	Cta	Caleta	*Cove*

CURRENT FORM	OBSOLESCENT FORM(S)	TERM	ENGLISH MEANING
Can.		Canal	*Channel*
Cer.	Co, C^{o}	Cerro	*Hill*
Cre.		Cumbre, Cima	*Summit*
	Cy	Cayo	*Cay, Key*
Ens.	Ens^{a}	Ensenada	*Bay, Creek*
Est.	Est^{o}	Estero	*Creek, Inlet*
Estr.		Estrecho	*Strait*
Estu.	Est, Est^{o}	Estuario	*Estuary*
Fond.	Fond^{o}	Fondeadero	*Anchorage*
Fte.	F^{te}	Fuerte	*Fort*
G.		Golfo	*Gulf*
	Gde, G^{de}	Grande	*Great*
I.	I^{a}	Isla, Islas	*Island(s)*
I.	I^{te}	Islote, Isleta	*Islet*
L.		Lago	*Lake*
Lag.	La, L^{a}	Laguna	*Lagoon*
Mor.	Mo, M^{o}	Morro	*Headland, Hill*
Mte.	M^{te}	Monte	*Mount, Mountain*
Mu.	Me, M^{e}, Mlle	Muelle	*Mole*
	Na.Sa, N^{a}S^{a}	Nuestra Señora	*Our Lady*
P.	Pto, P^{to}	Puerto	*Port*
Pco.	P^{co}, P^{o}	Pico	*Peak*
Pda.	P^{da}	Piedra	*Rock*
Pen.	Pen^{la}	Península	*Peninsula*
	Peq	Pequeño, Pequeña	*Small*
Pl.	Pa, P^{a}	Playa	*Beach*
Prom.	Prom^{to}	Promontorio	*Promontory*
Pta.	P^{ta}	Punta	*Point*
Queb.		Quebrada	*Cut, Ravine*
R.		Río	*River*
Rga.		Restinga	*Shoal, Sandbank*
Roc.	Ra, R^{a}	Roca	*Rock*
S.	Sn, S^{n}, Sto, S^{to}, *Sta, Sta*	San, Santo, Santa	*Saint, Holy*
Sr.	Sa, S^{a}	Sierra	*Mountain range*
Surg.	Surgo, Surg^{o}	Surgidero	*Anchorage, Roadstead*
Tr.	T^{re}	Torre	*Tower*
	Va, V^{a}	Villa	*Villa, Small town*
SWEDISH			
	B.	Bukt	*Bay, Bight*
Bg.	Bgt, B^{g}	Berg, Berget	*Mountain*
	Bk, B^{k}	Bank	*Bank*
Fj.	F^{d}	Fjärd, Fjord	*Fjord*
	Gla, G^{la}	Gamla	*Old*
Gr.	Grn, Grd, Grd, G^{d}	Grund	*Shoal*
H.	Hm, H^{m}	Holme, Holmarna	*Islet*
	Hd, H^{d}	Huvud	*Headland*
	Hn, H^{n}	Hamn, Hamnen	*Harbour*
I.		Inre	*Inner*
L.		Lilla, Liten	*Little, Small*
N.		Nord, Norra	*North, Northern*
Ö.		Öst, Östra	*East, Eastern*
S.		Syd, Södra	*South, Southern*
Sk.	Sk^{r}	Skär, Skäret, Skären	*Rock above water*
St.		Stor	*Great, Large*
V.		Väst, Västra	*West, Western*
Y.	Y^{t}	Yttre	*Outer*
THAI			
Kh.		Khao	*Hill, Mountain*
L.	Lm, L^{m}	Laem	*Cape, Point*
M.N.		Mae Nam	*River*
TURKISH			
Ad.		Ada, Adası	*Island*
Adc.	Ad	Adacık	*Islet*
Boğ.		Boğaz, Boğazı	*Strait*
Br.	Bn, Bu	Burun, Burnu	*Point, Cape*
Ç.	Ça	Çay, Çayı	*Stream, River*
	Da	Dağ, Dağı	*Mountain*
D.	De	Dere, Deresi	*Valley, Stream*
G.		Göl, Gölü	*Lake*
Isk		Iskele, Iskelesi	*Jetty*
Kf.		Körfez, Körfezi	*Gulf*
Ky.		Kaya, Kayası	*Rock*

CURRENT FORM	OBSOLESCENT FORM(S)	TERM	ENGLISH MEANING
INDONESIAN, MALAY and SINGAPOREAN (continued)			
G.	Gg, G^{g}	Gosong, Gosung, Gusong, Gusung	*Shoal, Reef, Islet*
Gun.	Gg, G^{g}	Gunong, Gunung	*Mountain*
K.	Ki, K^{i}	Kali	*River*
K.	Kr	Kroeng, Krueng	*River*
Kam.	Kg, K^{g}	Kampong, Kampung	*Village*
Kar.	Kg, K^{g}	Karang	*Coral reef, Reef*
Kep.	Kpn, K^{pn}	Kepulauan	*Archipelago*
Kl.	K^{l}	Kachil, Kechil, Ketjil, Kecil	*Small*
Ku.	Kla, K^{la}	Kuala	*River mouth*
Lab.	Labn, Lab^{n}	Labuan, Labuhan	*Anchorage, Harbour*
Mu.	Ma, M^{a}	Muara	*River mouth*
P.	Pu, P^{u}, P^{o}	Pulau, Pulu, Pulo	*Island*
Peg.		Pegunungan	*Mountain range*
Pel.	Pln, P^{ln}	Pelabuan, Pelabuhan	*Roadstead, Anchorage*
P.-P.	P.P.	Pulau-pulau	*Group of islands*
	Prt, P^{rt}	Parit	*Stream, Canal, Ditch*
S.	Si, S^{i}	Sungai, Sungei	*River*
Sel.	Slt, S^{lt}	Selat	*Strait*
T.	Tg, T^{g}	Tandjong, Tandjung, Tanjong, Tanjung Tanjing	*Cape*
Tel.	Tal, Tk, T^{k}	Taluk, Telok, Teluk	*Bay*
U.	Ug, U^{g}	Udjung, Ujung	*Cape*
W.		Wai	*River*
ITALIAN			
Anc.		Ancoraggio	*Anchorage*
B.		Baia	*Bay*
Banch.	Bna, B^{na}	Banchina	*Quay*
Bco.	B^{co}	Banco	*Bank*
C.		Capo	*Cape*
Cal.		Calata	*Wharf*
Can.		Canale	*Channel*
Cas.		Castello	*Castle*
F.		Fiume	*River*
Fte.	F^{te}	Forte	*Fort*
G.		Golfo	*Gulf*
	Gde, G^{de}	Grande	*Great*
I.	I^{a}, I^{e}	Isola, Isole	*Island(s)*
I.	I^{to}, I^{ti}	Isolotto, Isolotti	*Islet(s)*
L.		Lago	*Lake*
Lag.	La, L^{a}	Laguna	*Lagoon*
	Mda, Mad, Mad^{a}, Mad^{na}	Madonna	*Our Lady*
Mte.	M^{te}	Monte	*Mount, Mountain*
P.	Pto, P^{to}	Porto	*Port*
P.	Portlo, Port^{lo}	Porticciolo	*Small port*
Pco.	P^{co}	Picco	*Peak*
Pog.	Pgio, P^{gio}	Poggio	*Mound, Small hill*
Pta.	P^{ta}	Punta	*Point, Summit*
	Pte, P^{te}	Ponte	*Bridge*
	Pzo, P^{zo}	Pizzo	*Peak*
S.	Sto, S^{to}, Sta, Sta	San, Santo, Santa	*Saint, Holy*
S.	SS, S.S.	Santi	*Saints*
Scog.	Sco, Sci, Sc, Sc^{i}	Scoglio, Scogli	*Rock(s), Reef(s)*
Scog.	Sc, Scra	Scogliera	*Ridge of rocks, Breakwater*
Sec.	Se	Secca, Secche	*Shoal(s)*
	T, T^{te}	Torrente	*Intermittent stream*
Tr.	Tre, T^{re}	Torre	*Tower*
	Va, V^{la}	Villa	*Villa*
JAPANESE			
B.	B^{a}	Bana	*Cape, Point*
By.	Bi, B^{i}	Byōchi	*Anchorage*
	D^{e}	Dake	*Mountain, Hill*
G.	G^{a}	Gawa	*River*
H.	Ha, H^{a}	Hana	*Cape, Point*
Hak.	Hi, H^{i}	Hakuchi	*Roadstead*
J.	J^{a}	Jima	*Island*
K.	Ka, K^{a}	Kawa	*River*
	Kaik, Ko, K^{o}	Kaikyō	*Strait*
M.	Mki, M^{ki}, M^{i}	Misaki	*Cape*
	Ma, M^{a}	Mura	*Village*
	Mi, M^{i}	Machi	*Town*

CURRENT FORM	OBSOLESCENT FORM(S)	TERM	ENGLISH MEANING
S.	Si, S^{i}	Saki	*Cape, Point*
Sh.	Sa, S^{a}	Shima	*Island*
	Sn, S^{n}	San	*Mountain*
	So, S^{o}	Seto	*Strait*
Su.	Sdo, S^{do}	Suidō	*Channel*
	Te, T^{e}	Take	*Hill, Mountain*
	Ya, Y^{a}	Yama	*Mountain*
Z.	Z^{i}	Zaki	*Cape, Point*
	Z^{n}	Zan	*Mountain*
MALAY (see INDONESIAN)			
NORWEGIAN			
B.	B, B^{kt}	Bukt, Bukta	*Bay, Bight*
Bg.	B^{g}	Berg, Bierg, Bjerg	*Mountain, Hill*
Fj.	F^{d}	Fjord, Fjorden	*Fjord*
Fjel.	Fj	Fjell, Fjellet, Fjeld, Fjeldet	*Mountain*
Fl.	Flne, Fl^{ne}	Flu, Flua, Fluen, Fluane, Fluene	*Below water rock(s)*
Gr.	Grne, Gr^{ne}	Grunn, Grunnen, Grunnane	*Shoal(s)*
H.	Hm, H^{m}, Hne, H^{ne}	Holm, Holmen, Holmane	*Islet(s)*
Hn.	H^{n}	Hamn, Havn	*Harbour*
I.	In^{r}	Indre, Inre, Inste	*Inner*
L.		Lille, Liten, Litla, Litle	*Little*
Lag.	La, L^{a}	Laguna	*Lagoon*
N.		Nord, Nordre	*North, Northern*
Ø.	Ö	Øst, Østre, Öst, Östre	*East, Eastern*
Od.	O	Odde, Odden,	*Point*
Øy.	Ø, Ö, O	Øy, Øya, Öy, Öya	*Island*
Øy.	Øne, Ø^{ne}, Öne, Ö^{ne}	Øyane, Øyene, Öyane Öyene	*Islands*
Pt.	P^{t}	Pynt, Pynten	*Point*
S.		Syd, Søre, Søndre	*South, Southern*
Sd.	S^{d}	Sund, Sundet	*Sound*
Sk.	Skr, Sk^{r}	Skjær, Skjer, Skjeret	*Rock above water*
Sk.	Skne, Sk^{ne}	Skjerane, Skjærane	*Rocks above water*
St.		Stor, Stora, Store	*Great*
Tar.	Tn, T^{n}	Taren	*Below water rock*
V.		Vest, Vestre	*West*
Våg.	Vg, V^{g}	Våg, Vågen	*Bay, Cove*
	Vd, V^{d}	Vand	*Lake*
Vik.	Vk, V^{k}	Vik, Vika, Viken	*Bay, Inlet*
	Vn, V^{n}	Vann, Vatn	*Lake*
Y.	Y^{t}	Ytre, Ytter, Yttre	*Outer*
PERSIAN			
B.		Bandar	*Harbour*
Jab.		Jabal	*Mountain, Hill*
Jaz.	Jazh, Jaz^{h}	Jazireh	*Island, Peninsula*
Kh.	K	Khowr	*Inlet, Channel*
R.		Rūd	*River*
POLISH			
Jez.		Jezioro	*Lake*
Kan.		Kanal	*Channel*
Miel.		Mielizna	*Shoal*
R.		Rzeka	*River*
W.	Wys, Wa, W^{a}	Wyspa	*Island*
Zat.		Zatoka	*Gulf, Bay*
PORTUGUESE			
Anc.		Ancoradouro	*Anchorage*
Arq.	Arqu^{o}	Arquipélago	*Archipelago*
B.		Baía	*Bay*
Ba.	Bxo, B^{xo}, Bxa, B^{xa}	Baixo, Baixa, Baixia, Baixio	*Shoal*
Bco.	B^{co}	Banco	*Bank*
C.		Cabo	*Cape*
Can.		Canal	*Channel*
Ens.	Ens^{a}	Enseada	*Bay, Creek*
Est.	Est^{o}	Esteiro	*Creek, Inlet*

IV Abbreviations of Principal English Terms

CURRENT FORM	OBSOLESCENT FORM(S)	TERM	REFERENCES
abt	ab^{t}	About	IO a
Aero		Aeronautical	IP 60, 61
	Al	Algae	IJ t
Al.	Alt	Alternating light	IP 10.11
ALC		Articulated Loading Column	IL 12
ALL		Admiralty List of Lights and Fog Signals	—
ALRS		Admiralty List of Radio Signals	—
Am		Amber	IP 11.8
Anch.	Anch^{e}	Anchorage	IO 21
	Anct, Anc^{t}	Ancient	IO 84
ANM		Annual Summary of Admiralty Notices to Mariners	—
Annly	Ann^{ly}	Annually	—
Appr.	Apprs, Appr^{s}	Approaches	IO 22
approx	Approx	Approximate	IO 89
Arch.	Archo, Arch^{o}	Archipelago	IG 5
ASD		Admiralty Sailing Directions	—
	Astr. Astrl, Astr^{l}	Astronomical	—
ATT		Admiralty Tide Tables	—
Aus		Australia	—
Ave	Av^{e}	Avenue	IG 111
B.		Bay	IO 4
B	bl, blk	Black	IJ af, IQ 2
	Ba	Basalt	IJ i
	Batt, Baty, Bat^{y}	Battery	IE 34.3
Bk.	B^{k}	Bank	IO 23
bk	brk	Broken	IJ 33
Bldg	B^{ldg}	Building	ID 5
	BM, B.M.	Bench Mark	IB 23
Bn, Bns		Beacon, beacons	IM 1-2, IP 4-5, IQ 80-81
Bn Tr	Bn Tower	Beacon Tower	IP 3, IQ 110
	Bo	Boulders	IJ e
Bol	Boll.	Bollard	IF a, IG 181
Br		Breakers	IK 17
	br	Brown	IJ ak
Bu	Bl, Bl., b	Blue	IJ ag, IP 11.4, IQ a
C.		Cape	IG 7
c		Coarse	IJ 32
ca	cal	Calcareous	IJ 38
CALM		Catenary Anchor Leg Mooring	IL 16
Cas	Cas.	Castle	IE 34.2, IG 64
	Cath, Cath.	Cathedral	IE 10.1, IG 75.
Cb		Cobbles	IJ 8
cd		Candela	IB 54
CD		Chart Datum	IH 1
	Cemy, Cem^{y}	Cemetery	IE 19
CG	C.G.	Coastguard station	IT 10-11
Ch	Ch.	Church, chapel	IE 10.1, IE 11
	ch, choc	Chocolate	IJ al
Chan.		Channel	IO 14
Chem		Chemical	IL 40
	chk, Ck	Chalk	IJ f
Chy	Ch^{y}	Chimney	IE 22
	cin, Cn	Cinders	IJ n
cm	cm.	Centimetre(s)	IB 43
Co	crl	Coral	IJ 10, IK 16
	Col	Column, pillar, obelisk	IE 24, IG 66
	conspic·	Conspicuous	IE 2
const	constn, constr^{n}	Construction	IF 32
cov	cov.	Covers	IK b
Cr.		Creek	IO 7
Cup	Cup.	Cupola	IE 10.4
Cy	cl	Clay	IJ 3
	(D.)	Doubtful	—
	d	Dark	IJ ao
decrg	decr^{g}	Decreasing	IB 64
dest	destd, Dest^{d}	Destroyed	IO 93
Det		(see Fog Det Lt)	—
DG, DG Range	D. G. Range	Degaussing Range	IN 25,IQ 54
	Di, di	Diatoms	IJ w
Dia		Diaphone	IR 11
Dir	Dir^{n}	Direction	—
Dir	Dir Lt	Directional light	IP 30-31
Discol	Discol^{d}	Discoloured water	IK d
discont	discontd, discont^{d}	Discontinued	IO b
dist	Dist	Distant	IO 85
Dk	D^{k}	Dock	IG b
dm	dm.	Decimetre(s)	IB 42
Dn, Dns	D^{n}	Dolphin(s)	IF 20
dr	dr., Dr.	Dries	IK a
DW		Deep-water, Deep-draught	IM 27, IN 12.4
DZ		Danger Zone	IQ 50
E	E.	East	IB 10
ED	(ED), (E.D.)	Existence doubtful	II 1
EEZ		Exclusive Economic Zone	IN 47
	E.F. Horn	Electric fog horn	IR 13
Ent.	Entce, Ent^{ce}	Entrance	IO 16
	Equin^{l}	Equinoctial	—
Est.	Est^{y}	Estuary	IO 17
	Estab^{t}	Establishment	—
	ev.	Every	—
exper	experl, Exper^{l}	Experimental	IO 92
explos	explos.	Explosive	IR 10
(exting)	(exting^{d})	Extinguished	IP 55
f		Fine	IJ 30
F		Fixed light	IP 10.1
FAD		Fish Aggregating Device	—
F Racon		Fixed frequency radar transponder beacon	IS 3.4
FFL		Fixed and flashing light	IP 10.10
Fj.	Fd, F^{d}	Fjord	IO 5
	(fish^{g})	Fishing light	IP 50
Fl	fl.	Flashing light	IP 10.4
	Fl., fl.	Flood	—
Fla		Flare stack (at sea)	IL 11
	Fm, F^{m}	Farm	IG 53
fm, fms	f^{m}, f^{ms}	Fathom, fathoms	IB 48
Fog Det Lt		Fog detector light	IP 62
	Fog Sig.	Fog signal station	IR 1
	Fog W/T	Radio fog signal	—
	Fr, for	Foraminifera	IJ u
FS	F.S.	Flagstaff, Flagpole	IE 27
	Ft, F^{t}	Fort	IE 34.2
ft	f^{t}	Foot, feet	IB 47, IP 13
G	g	Gravel	IJ 6
G	gn	Green	IJ ah, IP 11.3, IQ 2
G.		Gulf	IO 3
	ga, glac	Glacial	IJ ac
	Gc	Glauconite	IJ p
	Gd, grd	Ground	IJ a
	Gl, gl	Globigerina	IJ v
	Govt Ho, Gov^{t} Ho	Government House	—
	GpFl, Gp.Fl.	Group-flashing light	IP 10.4
	GpOcc, Gp.Occ.	Group-occulting light	IP 10.2
grt		Gross Registered Tons	—
	Gt, Grt, G^{t}, Gr^{t}	Great	—
	G.T.S.	Great Trigonometrical Survey Station (India)	—
	Gy, gy	Grey	IJ an, IQ a
h		Hard	IJ 39
	H, H.	Headway	ID 20, ID 26-27
H		Helicopter transfer (Pilots)	IT 1.4
h	h., H.	Hour	IB 49
HAT		Highest Astronomical Tide	IH 3
Hd.	H^{d}	Headland	IG 8
Hn.	H^{n}	Haven	IG 139
Ho		House	IG 61
(hor)	(hor^{l})	Horizontally disposed	IP 15
Hosp	Hospl, Hosp^{l}	Hospital	IF 62.2
Hr.	H^{r}	Harbour	IG 138
	Hr, H^{r}	Higher	—
Hr Mr		Harbour Master	IF 60
	Ht, H^{t}	Height	—
HW	H.W.	High Water	IH a
	H.W.F. & C.	High Water Full and Change	—
	H.W.O.S.	High Water Ordinary Springs	—

Abbreviations of Principal Foreign Terms IV

CURRENT FORM	OBSOLESCENT FORM(S)	TERM	ENGLISH MEANING
TURKISH (continued)			
Kyl.		Kayalık, Kayalığı	*Rock*
Lim.	Li	Liman, Limanı	*Harbour*
N.		Nehir, Nehri, Irmak, Irmaği	*River*
T.	Te, T^{e}	Tepe, Tepesi	*Hill, Peak*
Yad.		Yarımada, Yarımadası	*Peninsula*
YUGOSLAV			
Br.		Brdo, Brda	*Mountain(s)*
Gr.		Greben, Grebeni	*Rock, Reef, Cliff, Ridge*
Hr.		Hrid, Hridi	*Rock*
L.		Luka	*Harbour, Port*
M.		Mali, Mala, Malo, Malen	*Small*
O.		Otočić, Otočići	*Islet(s)*
O.		Otok, Otoci	*Island(s)*
Pl.		Plićina	*Shoal*
Pr.		Prolaz	*Passage*
S.	Sv	Sveti, Sveta, Sveto	*Saint, Holy*
Sk.		Školj, Školjic	*Island, Reef*
U.		Uvala, Uvalica	*Inlet*
V.		Veli, Vela, Velo, Velik, Veliki, Velika, Veliko	*Great*
Z.	Zal	Zaliv, Zaljev, Zaton	*Gulf, Bay*

CURRENT FORM	OBSOLESCENT FORM(S)	TERM	REFERENCES
Ramark		Radar Beacon	IS 2
RC		Non-directional Radiobeacon	IS 10
RD	Dir.Ro.Bn.	Directional Radiobeacon	IS 11
Rds.	R^{ds}	Roads, Roadstead	IO 20
Ref		Refuge	IQ 124, IT 14
Refl	Refl.	Retroreflecting material	IQ 6
	Rem^{ble}	Remarkable	—
Rep	Repd, Rep^{d}	Reported	II 3
Rf.	R^{f}	Reef	IO 26
RG	R^{o} D.F.	Radio Direction-Finding Station	IS 14
Rk.	R^{k}	Rock	IG 11
(R Lts)	(Red Lts)	Air Obstruction Lights (low intensity)	IP 61.2
	Rly, Ry, R^{y}	Railway	ID 13
	R^{o} B^{n}	Radiobeacon in general	IS 10
RoRo	Ro-Ro	Roll-on Roll-off ferry terminal	IF 50
	R.S.	Rocket station	—
Ru, (ru)	Ru.	Ruins	ID 8, IE 25.2, IF 33
RW		Rotating Pattern Radiobeacon	IS 12
S.	St, S^{t}	Saint	IG 54
S	s	Sand	IJ 1
S	S.	South	IB 11
s	sec, sec.	Second(s) of time	IB 51, IP 12
SALM		Single Anchor Leg Mooring	IL 12
SBM		Single Buoy Mooring	IL 16
SC	S.C.	Sailing Club	IU 4
	Sc, sc	Scoriæ	IJ m
Sc	Sc.	Scanner	IE 30.3
Sch	Sch.	School	IG a
SD	S.D.	Sailing Directions	—
SD		Sounding of doubtful depth	II 2
Sd.	S^{d}	Sound	IO 12
SE	S.E.	South-east	IB 14
	Sem, Sem.	Semaphore	—
sf	stf	Stiff	IJ 36
Sh	sh	Shells	IJ 11
Sh.		Shoal	IO 25
Si		Silt	IJ 4
Sig	Sig.	Signal	IR 1, IT 25.2
	sk, spk	Speckled	IJ ad
	sm	Small	IJ aa
SMt	SM^{t}	Seamount	IO 33
	Sn, shin	Shingle	IJ d
so	sft	Soft	IJ 35
Sp	Sp.	Spire	IE 10.3
	Sp, sp	Sponge	IJ s
Sp	Sp., Spr.	Spring Tides	IH 16
SPM		Single Point Mooring	IL 12
SS	Sig Sta, Sig Stn	Signal Station	IT 20-36
St	st	Stones	IJ 5
St	St.	Street	IG 110
Sta	Sta., Stn, St^{n}	Station	ID 13
	Stm.Sig.Stn.	Storm Signal Station	IT 28
Str.		Strait	IO 11
subm	submd, Subm^{d}	Submerged	IO 90
SW	S.W.	South-west	IB 16
SWOPS		Single Well Oil Production System	IL e
sy	stk	Sticky	IJ 34
	T, t	Tufa	IJ l
(T)		Temporary (NM)	—
t		Ton, tonne	IB 53, IF 53
	t	Elevation of top of trees	IC 14
Tel	Tel.	Telephone, Telegraph	IG 95
(temp)	(tempy), (temp^{y})	Temporary	IN b, IP 54
Tr	T^{r}	Tower	IE 10.2, IE 20
TSS		Traffic Separation Scheme	—
TV Tr	T.V. T^{r}	Television Tower	IE 28-29
	(U)	Unwatched, unmanned (light)	IP 53
ULCC		Ultra Large Crude Carrier	—
uncov	uncov.	Uncovers	IK c
unexam	unexamd, unexam^{d}	Unexamined	II a

CURRENT FORM	OBSOLESCENT FORM(S)	TERM	REFERENCES
Unintens		Unintensified	IP a
	Up^{r}	Upper	IP 22
UQ		Ultra quick-flashing light	IP 10.8
V	vol	Volcanic	IJ 37
	Va, V^{a}	Villa	—
Var	Var^{n}	Variation	IB 60
	var	Varying	—
Vel	Vel.	Velocity	—
(vert)	(vert^{l})	Vertically disposed	IP 15
Vi		Violet	IP 11.5
	vis.	Visible	—
VLCC		Very Large Crude Carrier	IG 187
Vol.		Volcano	IG 26
VQ	VQkFl, V.Qk.Fl	Very quick-flashing light	IP 10.7
W	W.	West	IB 12
W	w	White	IJ ae, IP 11.1, IQ a
Water Tr	Water T^{r}	Water tower	IE 21
Wd	wd	Weed	IJ 13.1
WGS		World Geodetic System	IS 50
Whf	Wh^{f}	Wharf	IF 13, IG c
Whis	Whis.	Whistle	IR 15
Wk	W^{k}	Wreck	IK 20-30
	W/T	Radio (Wireless/Telegraphy)	—
Y	y	Yellow, amber, orange	IJ ai, IP 11, IQ 3
YC	Y.C.	Yacht Club	IU 4
	y^{d}, y^{ds}	Yard(s)	—

CURRENT FORM	OBSOLESCENT FORM(S)	TERM	REFERENCES
I.	It	Island, islet	IG 1-2
IALA		International Association of Lighthouse Authorities	IQ 130
IHO		International Hydrographic Organization	—
(illum)	Illum., (lit)	Illuminated	IP 63
IMO		International Maritime Organization	—
	in., ins.	Inch, inches	—
incrg	incr^{g}	Increasing	IB 65
INT		International	IA 3, IT 21
Intens	(intens)	Intensified	IP 46
IQ	IntQkFl, Int.Qk.Fl.	Interrupted quick-flashing light	IP 10.6
	(irreg.)	Irregular	—
	ISLW, I.S.L.W.	Indian Spring Low Water	—
Iso		Isophase light	IP 10.3
	It	Islet	IG 2
ITZ		Inshore Traffic Zone	—
IUQ		Interrupted ultra quick-flashing light	IP 10.8
IVQ	IntVQkFl, Int.V.Qk.Fl	Interrupted very quick-flashing light	IP 10.7
km	km.	Kilometre(s)	IB 40, IF 40
kn	kn.	Knot(s)	IB 52, IH 40-41
L.		Lake, Loch, Lough	IO 6
	l	Large	IJ ab
Lag.	Lagn, Lag^{n}	Lagoon	IG 13, IO 8
LANBY		Large Automatic Navigational Buoy	IP 8
LASH		Lighter Aboard Ship	IG 184
LAT		Lowest Astronomical Tide	IH 2
Lat	Lat.	Latitude	IB 1
	LB, L.B.	Lifeboat station	IT 12
Ldg	L^{dg}	Leading	IP 20.3
Le.	L^{e}	Ledge	IO 28
LFl		Long-flashing light	IP 10.5
	Lit, Lit.	Little	—
	(lit)	Floodlit	IP 63
LL	L.L.	List of Lights	—
Lndg.	L^{dg}	Landing place	IF 17
LNG		Liquefied Natural Gas	IG 185
Long	Long.	Longitude	IB 2
LPG		Liquefied Petroleum Gas	IG 186
	Lr, L^{r}	Lower	IP 23
	L.S.S.	Lifesaving station	—
Lt	L^{t}, lt	Light	IJ an, IP 1
Lts		Lights	IP 20.1, 61.2
Lt Ho	L^{t} Ho	Lighthouse	IP 1
Lt V	L^{t} V	Light-vessel	IP 6
	Lv, lv	Lava	IJ j
LW	L.W.	Low Water	IH b
	L.W.F. & C.	Low Water Full and Change	—
	L.W.O.S.	Low Water Ordinary Springs	—
M	m	Mud	IJ 2
M	M.	Sea Mile(s)	IB 45, IP 14
m-		Medium	IJ 31
m	m.	Metre(s)	IB 41, IP 13
	mad, Md	Madrepore	IJ h
Mag	Mag.	Magnetic	IB 61
	Magz, Mag^{z}	Magazine	—
	man, Mn	Manganese	IJ o
MHHW	M.H.H.W.	Mean Higher High Water	IH 13
MHLW	M.H.L.W.	Mean Higher Low Water	IH 14
MHW		Mean High Water	IH 5
MHWN	M.H.W.N.	Mean High Water Neaps	IH 11
MHWS	M.H.W.S.	Mean High Water Springs	IH 9
	Mid, Mid.	Middle	—
min	min., m.	Minute(s) of time	IB 50
Mk		Mark	IQ 101
	Ml, ml	Marl	IJ c
MLHW	M.L.H.W.	Mean Lower High Water	IH 15
MLLW	M.L.L.W.	Mean Lower Low Water	IH 12
MLW		Mean Low Water	IH 4
MLWN	M.L.W.N.	Mean Low Water Neaps	IH 10
MLWS	M.L.W.S.	Mean Low Water Springs	IH 8

CURRENT FORM	OBSOLESCENT FORM(S)	TERM	REFERENCES
mm	mm.	Millimetre(s)	IB 44
Mo		Morse	IP 10.9, IR 20
Mon	Mont, Mon^{t}	Monument	IE 24
	Mony, Mon^{y}	Monastery	IG 76
	Ms, mus	Mussels	IJ r
MSL	M.S.L.	Mean Sea Level	IH 6
Mt.	M^{t}	Mountain, mount	IG 23
Mth.	Mth	Mouth	IO 19
MTL	M.T.L.	Mean Tide Level	IH c
N	N.	North	IB 9
	Nauto	Nautophone	IR 13
NB	N.B.	Notice Board	IQ 126
NE	N.E.	North-east	IB 13
NM	N.M.	Notice(s) to Mariners	—
n mile		International Nautical Mile	IB 45
No	N^{o}	Number	IN 12.2
Np	Np.	Neap Tides	IH 17
NW	N.W.	North-west	IB 15
NZ		New Zealand	—
	Obs Spot, Obstn Spot, Obs^{n} Spot	Observation Spot	IB 21
Obscd	Obsc^{d}	Obscured	IP 43
Obstn	Obst^{n}	Obstruction, Diffuser	IK 40-43, IL 43
	Obsy, Obs^{y}	Observatory	IG 73
Oc	Occ, Occ.	Occulting light	IP 10.2
(occas)	(occas^{l})	Occasional	IP 50
OD	O.D.	Ordnance Datum	IH d
ODAS		Ocean Data-Acquisition System	IQ 58
	Off, Off.	Office	IG 72
Or	Or.	Orange	IP 11.7, IQ 3
	ord.	Ordinary	—
	Oy, oys	Oysters	IJ q
	Oz, oz	Ooze	IJ b
P	peb	Pebbles	IJ 7
P.		Port	IG 137
(P)		Preliminary (NM)	—
PA	(PA), (P.A.)	Position approximate	IB 7
Pag	Pag.	Pagoda	IE 14
Pass.		Passage	IO 13
PD	(PD), (P.D.)	Position doubtful	IB 8
Pen.	Penla, Pen^{la}	Peninsula	IG 4
Pk.	P^{k}	Peak	IG 25
	Pm, pum	Pumice	IJ k
PO	P.O.	Post Office	IF 63
	Po, pol	Polyzoa	IJ z
pos	posn, pos^{n}	Position	—
(priv)	priv., (Priv.)	Private	IP 50, IP 65, IQ 70
Prod Well		Production Well	IL 20
prohib	Prohib^{d}	Prohibited	IO c
proj	projd, Proj^{d}	Projected	IO 80
prom	promt, Prom^{t}	Prominent	IO d
Prom.	Promy, Prom^{y}	Promontory	IG 20
	(prov), (prov^{l})	Provisional	—
Pt.	P^{t}	Point	IG 9
	Pt, pt	Pteropods	IJ y
Pyl		Pylon	ID 26
Q	QkFl, Qk.Fl.	Quick-flashing light	IP 10.6
	Q^{r}	Quarter	—
	Qz, qrtz	Quartz	IJ g
R	rd	Red	IJ aj, IP 11.2, IQ 3
R.		River	—
R	r	Rock	IJ 9, IK 15
R	R^{o}	Coast Radio Station providing QTG service	IS 15
Ra		Radar Range, Radar Reference Line, Coast Radar Station	IM 31-32, IS 1
	Ra (conspic), Ra. (conspic)	Radar conspicuous object	IS 5
	Ra. Refl.	Radar Reflector	IQ 10-11, IS 4
Racon		Radar Transponder Beacon	IS 3
	rad, Rd	Radiolaria	IJ x
Radome		Radar dome	IE 30.4

IP Lights

_t	Light	IP 1
F	Fixed	IP 10.1
Oc	Occulting	IP 10.2
Iso	Isophase	IP 10.3
Fl	Flashing	IP 10.4
LFl	Long-flashing	IP 10.5
Q	Quick	IP 10.6
IQ	Interrupted quick	IP 10.6
VQ	Very quick	IP 10.7
IVQ	Interrupted very quick	IP 10.7
UQ	Ultra quick	IP 10.8
IUQ	Interrupted ultra quick	IP 10.8
Mo	Morse Code	IP 10.9
Al	Alternating	IP 10.11
W	White	IP 11.1
R	Red	IP 11.2
G	Green	IP 11.3
Bu	Blue	IP 11.4
Vi	Violet	IP 11.5
Y	Yellow/Orange/Amber	IP 11.6-11.8
Or	Orange	IP 11.7
Am	Amber	IP 11.8
Ldg	Leading light	IP 20
Dir	Direction light	IP 30
occas	Occasional	IP 50
Aero	Aeronautical	IP 60,61.1
R Lts	Air obstruction lights	IP 61.2
Fog Det Lt	Fog detector light	IP 62

IQ Buoys, Beacons

B	Black	IQ 2
Bn	Beacon	IQ 80
Mk	Mark	IQ 101
IALA	International Association of Lighthouse Authorities	IQ 130

IR Fog Signals

Explos	Explosive	IR 10
Dia	Diaphone	IR 11
Whis	Whistle	IR 15

IS Radar, Radio, Electronic Position-Fixing Systems

Ra	Coast radar station	IS 1
Ramark	Radar beacon	IS 2
Racon	Radar transponder beacon	IS 3
F Racon	Radar transponder beacon, responding on a fixed frequency outside the marine band	IS 3.4
RC	Non-directional marine radiobeacon	IS 10
RD	Rotating pattern radiobeacon	IS 12
RG	Radio direction-finding station	IS 14
R	Coast radio station providing QTG service	IS 15
Aero RC	Aeronautical radiobeacon	IS 16
WGS	World Geodetic System	IS 50

IT Services

H	Helicopter	IT 1.4
SS	Signal station	IT 20
INT	International	IT 21

IB Positions, Distances, Directions, Compass

PA	Position approximate	IB 7
PD	Position doubtful	IB 8
N	North	IB 9
E	East	IB 10
S	South	IB 11
W	West	IB 12
NE	Northeast	IB 13
SE	Southeast	IB 14
NW	Northwest	IB 15
SW	Southwest	IB 16
km	Kilometre(s)	IB 40
m	Metre(s)	IB 41
dm	Decimetre(s)	IB 42
cm	Centimetre(s)	IB 43
mm	Millimetre(s)	IB 44
M	International Nautical Mile(s), Sea mile(s)	IB 45
ft	Foot/feet	IB 47
h	Hour	IB 49
m, min	Minute(s) of time	IB 50
s, sec	Second(s) of time	IB 51
kn	Knot(s)	IB 52
t	Ton(s) tonne(s)	IB 53
cd	Candela	IB 54

ID Cultural Features

Ru	Ruin	ID 8

IF Ports

Lndg	Landing for boats	IF 17
RoRo	Roll-on, Roll-off Ferry Terminal	IF 50

II Depths

ED	Existence doubtful	II 1
SD	Sounding of doubtful depth	II 2

IK Rocks, Wrecks, Obstructions

Br	Breakers	IK 17
Wk	Wreck	IK 20-30
Obstn	Obstruction	IK 40-43

IL Offshore Installations, Submarine Cables, Submarine Pipelines

Fla	Flare stack (at sea)	IL 11
Prod. Well	Submerged Production Well	IL 20

IM Tracks, Routes

Ra	Radar	IM 31
DW	Deep-Water	IM 27

IN Areas, Limits

No	Number	IN 12.2
DW	Deep Water	IN 12.4

IO Hydrographic Terms

SMt	Seamount	IO 33

CONTENTS

Selection of Symbols: